José Rolim et al. (Eds.)

Parallel and Distributed Processing

11th IPPS/SPDP'99 Workshops
Held in Conjunction with the
13th International Parallel Processing Symposium
and 10th Symposium on
Parallel and Distributed Processing
San Juan, Puerto Rico, USA, April 12-16, 1999
Proceedings

 Springer

Series Editors

Gerhard Goos, Karlsruhe University, Germany
Juris Hartmanis, Cornell University, NY, USA
Jan van Leeuwen, Utrecht University, The Netherlands

Managing Volume Editor

José Rolim
Université de Genève , Centre Universitaire d'Informatique
24, rue General Dufour, CH-1211 Genève 4, Switzerland
E-mail: Jose.Rolim@cui.unige.ch

Cataloging-in-Publication data applied for

Die Deutsche Bibliothek - CIP-Einheitsaufnahme

Parallel and distributed processing : 11 IPPS/SPDP '99 workshops held in conjunction with the 13th International Parallel Processing Symposium and 10th Symposium on Parallel and Distributetd Processing, San Juan, Puerto Rico, USA, April 12 - 16, 1999 ; proceedings / José Rolim et al. (ed.). - Berlin ; Heidelberg ; New York ; Barcelona ; Hong Kong ; London ; Milan ; Paris ; Singapore ; Tokyo : Springer, 1999
(Lecture notes in computer science ; Vol. 1586)
ISBN 3-540-65831-9

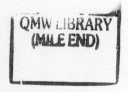

CR Subject Classification (1998): C.1-4, B.1-7, D.1-4, F.1-2, G.1-2, E.1, H.2
ISSN 0302-9743
ISBN 3-540-65831-9 Springer-Verlag Berlin Heidelberg New York

This work is subject to copyright. All rights are reserved, whether the whole or part of the material is concerned, specifically the rights of translation, reprinting, re-use of illustrations, recitation, broadcasting, reproduction on microfilms or in any other way, and storage in data banks. Duplication of this publication or parts thereof is permitted only under the provisions of the German Copyright Law of September 9, 1965, in its current version, and permission for use must always be obtained from Springer-Verlag. Violations are liable for prosecution under the German Copyright Law.

© Springer-Verlag Berlin Heidelberg 1999
Printed in Germany

Typesetting: Camera-ready by author
SPIN: 10703197   06/3142 – 5 4 3 2 1 0   Printed on acid-free paper

Volume Editors:

- José D.P. Rolim
- Frank Mueller
- Albert Y. Zomaya
- Fikret Ercal
- Stephan Olariu
- Binoy Ravindran
- Jan Gustafsson
- Hiroaki Takada
- Ron Olsson
- Laxmikant V. Kale
- Pete Beckman
- Matthew Haines
- Hossam ElGindy
- Denis Caromel
- Serge Chaumette
- Geoffrey Fox
- Yi Pan
- Keqin Li
- Tao Yang
- G. Chiola
- G. Conte
- L.V. Mancini
- Domenique Méry
- Beverly Sanders
- Devesh Bhatt
- Viktor Prasanna

Foreword

This volume contains proceedings from eleven workshops held in conjunction with the 13th International Parallel Symposium and the 10th Symposium on Parallel and Distributed Processing, *1999 IPPS/SPDP*, on 12-16 April 1999 in San Juan, Puerto Rico.

The workshops provide a forum for bringing together researchers, practitioners and designers from various backgrounds to discuss the state of the art in parallelism. They focus on different aspects of parallelism, from run-time systems to formal methods, from optics to irregular problems, from biology to PC networks, from embedded systems to programming environments. The Workshops on the following topics are represented in this volume:

- High-Level Parallel Programming Models and Supportive Environments
- Biologically Inspired Solutions to Parallel Processing Problems
- Parallel and Distributed Real-Time Systems
- Run-Time Systems for Parallel Programming
- Reconfigurable Architectures
- Java for Parallel and Distributed Computing
- Optics and Computer Science
- Solving Irregularly Structured Problems in Parallel
- Personal Computer Based Workstation Networks
- Formal Methods for Parallel Programming
- Embedded HPC Systems and Applications

All papers published in the workshops proceedings were selected by the program committee on the basis of referee reports. Each paper was reviewed by independent referees who judged the papers for originality, quality, and consistency with the themes of the workshops.

We would like to thank the General Co-Chair Charles Weems and the General Vice-Chair John Antonio for their support and encouragement, the Steering Committee Chairs, George Westrom and Victor Prasanna, for their guidance and vision, the Program Commitee and its chair, Mikhail Atallah, for its technical leadership in organizing the conference and the Finance Chair, Bill Pitts, for making this publication possible. Special thanks are due to Sally Jelinek, for her assistance with meeting publicity, to Susamma Barua for making local arrangements and to Prashanth Bhat for his tireless efforts in interfacing with the organizers.

We gratefully acknowledge sponsorship from the IEEE Computer Society and its Technical Committee of Parallel Processing and the cooperation of the ACM SIGARCH. Finally, we would like to thank Danuta Sosnowska and Germaine Gusthiot for their help in the preparation of this volume.

February 1999   José D. P. Rolim, Workshops Chair and General Co-Chair

Contents

Workshop on High-Level Parallel Programming Models and Supportive Environments   1
Frank Mueller

Efficient Program Partitioning Based on Compiler Controlled Communication   4
Ram Subramanian and Santosh Pande

SCI-VM: A Flexible Base for Transparent Shared Memory Programming Models on Clusters of PCs   19
Martin Schulz

Flexible Collective Operations for Distributed Object Groups   34
Joerg Nolte

SCALA: A Framework for Performance Evaluation of Scalable Computing   49
Xian-He Sun, Mario Pantano, Thomas Fahringer and Zhaohua Zhan

Recursive Individually Distributed Objects   63
Z. George Mou

The MuSE System: A Flexible Combination of On-Stack Execution and Work-Stealing   79
Markus Leberecht

Pangaea: An Automatic Distribution Front-End for Java   93
André Spiegel

Concurrent Language Support for Interoperable Applications   100
Eugene F. Fodor and Ronald A. Olsson

On the Distributed Implementation of Aggregate Data Structures by Program Transformation   108
Gabriele Keller and Manuel M. T. Chakravarty

A Transformational Framework for Skeletal Programs: Overview and Case Study   123
Sergei Gorlatch and Susanna Pelagatti

Implementing a Non-strict Functional Programming Language on a Threaded Architecture   138
Shigeru Kusakabe, Kentaro Inenaga, Makoto Amamiya, Xinan Tang, Andres Marquez and Guang R. Gao

Workshop on Biologically Inspired Solutions to Parallel   153
Processing Problems
Albert Y. Zomaya, Fikret Ercal, Stephan Olariu

The Biological Basis of the Immune System as a Model for Intelligent Agents   156
Roger L. King, Aric B. Lambert, Samuel H. Russ, Donna S. Reese

A Formal Definition of the Phenomenon of Collective Intelligence
and Its IQ Measure   165
Tadeusz Szuba

Implementation of Data Flow Logical Operations via Self-Assembly of DNA   174
Piotr Wąsiewicz, Piotr Borsuk, Jan J. Mulawka, Piotr Węgleński

A Parallel Hybrid Evolutionary Metaheuristic for the Period Vehicle   183
Routing Problem
Dalessandro Soares Vianna, Luiz S. Ochi, Lúcia M.A. Drummond

Distributed Scheduling with Decomposed Optimization Criterion:
Genetic Programming Approach   192
Franciszek Seredynski, Jacek Koronacki, Cezary Z. Janikow

A Parallel Genetic Algorithm for Task Mapping on
Parallel Machines   201
S. Mounir Alaoui, O. Frieder, T. El-Ghazawi

Evolution-Based Scheduling of Fault-Tolerant Programs on
Multiple Processors   210
*Piotr Jedrzejowicz, Ireneusz Czarnowski, Henryk Szreder,
Aleksander Skakowski*

A Genetic-Based Fault-Tolerant Routing Strategy for
Multiprocessor Networks   220
Peter K. K. Loh and Venson Shaw

Regularity Considerations in Instance-Based Locality Optimization   230
Claudia Leopold

Parallel Ant Colonies for Combinatorial Optimization Problems   239
El-ghazali Talbi, Olivier Roux, Cyril Fonlupt, Denis Robillard

An Analysis of Synchronous and Asynchronous Parallel Distributed
Genetic Algorithms with Structured and Panmictic Islands   248
Enrique Alba, Jos M. Troya

GA-based Parallel Image Registration on Parallel Clusters 257
Prachya Chalermwat, Tarek El-Ghazawi, Jacqueline LeMoigne

Implementation of a Parallel Genetic Algorithm on a Cluster of 266
Workstations: The Traveling Salesman Problem, A Case Study
Giuseppe Sena, Germinal Isem, Dalila Megherbi

Structural Biology Metaphors Applied to the Design of a
Distributed Object System 275
Ladislau Bölöni, Ruibing Hao, Kyungkoo Jun, Dan C. Marinescu

Workshop on Parallel and Distributed Real-Time Systems 284
Binoy Ravindran, Jan Gustafsson, Hiroaki Takada

Building an Adaptive Multimedia System Using the Utility Mode 289
Lei Chen, Shahadat Khan, Kin F. Li, Eric G. Manning

Evaluation of Real-Time Fiber Communications for
Parallel Collective Operations 299
P. Rajagopal and A. W. Apon

The Case for Prediction-Based Best-Effort Real-Time Systems 309
Peter A. Dinda, Loukas F. Kallivokas, Bruce Lowekamp,
David R. O'Hallaron*

Dynamic Real-Time Channel Establishment in Multiple
Access Bus Networks 319
Anita Mittal, G. Manimaran, C. Siva Ram Murthy

A Similarity-Based Protocol for Concurrency Control in
Mobile Distributed Real-Time Database Systems 329
Kam-yiu Lam, Tei-Wei Kuo, Gary C.K. Law, Wai-Hung Tsang

From Task Scheduling in Single Processor Environments
to Message Scheduling in a PROFIBUS Fieldbus Network 339
Eduardo Tovar, Francisco Vasques

An Adaptive Distributed Airborne Tracking System 353
*Raymond Clark, E. Douglas Jensen, Arkady Kanevsky, John Maurer,
Paul Wallace, Thomas Wheeler, Yun Zhang, Douglas Wells, Tom Lawrence,
Pat Hurley*

Non-preemptive Scheduling of Real-Time Threads
on Multi-Level-Context Architectures 363
Jan Jonsson, Henrik Lönn, Kang G. Shin

QoS Control and Adaptation in Distributed Multimedia Systems 375
Farid Nat-Abdesselam, Nazim Agoulmine

Dependability Evaluation of Fault Tolerant Distributed
Industrial Control Systems 384
J.C. Campelo, P. Yuste, F. Rodríguez, P.J. Gil, J.J. Serrano

An Approach for Measuring IP Security Performance in
a Distributed Environment 389
Brett L. Chappell, David T. Marlow, Philip M. Irey IV, Karen O'Donoghue

An Environment for Generating Applications Involving Remote
Manipulation of Parallel Machines 395
Luciano G. Fagundes, Rodrigo F. Mello, Clio E. Móron

Real-Time Image Processing on a Focal Plane SIMD Array 400
*Antonio Gentile, Jos L. Cruz-Rivera, D. Scott Wills, Leugim Bustelo,
Jos J. Figueroa, Javier E. Fonseca-Camacho, Wilfredo E. Lugo-Beauchamp,
Ricardo Olivieri, Marlyn Quiñones-Cerpa, Alexis H. Rivera-Ríos, Iomar
Vargas-Gonzáles, Michelle Viera-Vera*

Metrics for the Evaluation of Multicast Communications 406
Philip M. Irey IV, David T. Marlow

Distributing Periodic Workload Uniformly Across Time to
Achieve Better Service Quality 413
Jaeyong Koh, Kihan Kim and Heonshik Shin

A Dynamic Fault-Tolerant Mesh Architecture 418
Jyh-Ming Huang, Ted C. Yang

Evaluation of a Hybrid Real-Time Bus Scheduling Mechanism for CAN 425
Mohammad Ali Livani, Jörg Kaiser

System Support for Migratory Continuous Media Applications in
Distributed Real-Time Environments 430
Tatsuo Nakajima, Mamadou Tadiou Kone, Hiroyuki Aizu

Dynamic Application Structuring on Heterogeneous, Distributed Systems 442
Saurav Chatterjee

Improving Support for Multimedia System Experimentation
and Deployment 454
Douglas Niehaus

Workshop on Run-Time Systems for Parallel Programming   466
Ron Olsson, Laxmikant V. Kalé, Pete Beckman, Matthew Haines

Efficient Communications in Multithreaded Runtime Systems   468
Luc Bougé, Jean-François Méhaut, Raymond Namyst

Application Performance of a Linux Cluster Using Converse   483
Laxmikant Kalë, Robert Brunner, James Phillips, Krishnan Varadarajan

An Efficient and Transparent Thread Migration Scheme in the PM2
Runtime System   496
Gabriel Antoniu, Luc Bougë, Raymond Namyst

Communication-Intensive Parallel Applications and Non-Dedicated   511
Clusters of Workstations
Kritchalach Thitikamol, Peter Keleher

A Framework for Adaptive Storage Input/Output on Computational Grids 519
Huseyin Simitci, Daniel A. Reed, Ryan Fox, Mario Medina, James Oly, Nancy Tran, Guoyi Wang

ARMCI: A Portable Remote Memory Copy Library for Distributed
Array Libraries and Compiler Run-Time Systems   533
Jarek Nieplocha, Bryan Carpenter

Multicast-Based Runtime System for Highly Efficient Causally
Consistent Software-Only DSM   547
Thomas Seidmann

Adaptive DSM-Runtime Behavior via Speculative Data Distribution   553
Frank Mueller

Reconfigurable Architectures Workshop   568
Hossam Elgindy

DEFACTO: A Design Environment for Adaptive Computing Technology   570
Kiran Bondalapati, Pedro Diniz, Phillip Duncan, John Granacki, Mary Hall, Rajeev Jain, Heidi Ziegler

A Web-Based Multiuser Operating System for Reconfigurable Computing   579
Oliver Diessel, David Kearney, Grant Wigley

Interconnect Synthesis for Reconfigurable Multi-FPGA Architectures   588
Vinoo Srinivasan, Shankar Radhakrishnan, Ranga Vemuri, Jeff Walrath

Hardwired-Clusters Partial-Crossbar: A Hierarchical Routing
Architecture for Multi-FPGA Systems 597
Mohammed A.S. Khalid, Jonathan Rose

Integrated Block-Processing and Design-Space Exploration
in Temporal Partitionning for RTR Architectures 606
Meenakshi Kaul, Ranga Vemuri

Improved Scaling Simulation of the General Reconfigurable Mesh 616
José Alberto Fernández-Zepeda, Ramachandran Vaidyanathan, Jerry L. Trahan

Bit Summation on the Reconfigurable Mesh 625
Martin Middendorf

Scalable Hardware-Algorithms for Binary Prefix Sums 634
R. Lin, K. Nakano, S. Olariu, M.C. Pinotti, J.L. Schwing, A.Y. Zomaya

Configuration Sequencing with Self Configurable Binary Multipliers 643
Mathew Wojko, Hossam ElGindy

Domain Specific Mapping for Solving Graph Problems
on Reconfigurable Devices 652
Andreas Dandalis, Alessandro Mei, Victor K. Prasanna

MorphoSys: A Reconfigurable Processor Targeted to High Performance 661
Image Application
Guangming Lu, Ming-hau Lee, Hertej Singh, Nader Bagherzadeh, Fadi J. Kurdahi, Eliseu M. Filho

An Efficient Implementation Method of Fractal Image Compression on 670
Dynamically Reconfigurable Architecture
Hidehisa Nagano, Akihiro Matsuura, Akira Nagoya

Plastic Cell Architecture: A Dynamically Reconfigurable
Hardware-Based Computer 679
*Hiroshi Nakada, Kiyoshi Oguri, Norbert Imlig,
Minoru Inamori, Ryusuke Konishi, Hideyuki Ita, Kouichi Nagami,
Tsunemichi Shiozawa*

Leonardo and Discipulus Simplex: An Autonomous, Evolvable
Six-Legged Walking Robot 688
Gilles Ritter, Jean-Michel Puiatti, Eduardo Sanchez

Reusable Internal Hardware Templates 697
Ka-an Agun, Morris Chang

An On-Line Arithmetic-Based Reconfigurable Neuroprocessor 700
Jean-Luc Beuchat, Eduardo Sanchez

The Re-Configurable Delay-Intensive FLYSIG Architecture 703
Wolfram Hardt, Achim Rettberg, Bernd Kleinjohann

Digital Signal Processing with General Purpose Microprocessors, DSP and Reconfigurable Logic 706
Steffen Köhler, Sergej Sawitzki, Achim Gratz, Rainer G.Spallek

Solving Satisfiability Problems on FPGAs Using Experimental Unit Propagation Heuristic 709
Takayuki Suyama, Makoto Yokoo, Akira Nagoya

FPGA Implementation of Modular Exponentiation 712
Alexander Tiountchik, Elena Trichina

Workshop on Java for Parallel and Distributed Computing 716
Denis Caromel, Serge Chaumette, Geoffrey Fox

More Efficient Object Serialization 718
Michael Philippsen, Bernhard Haumacher

A Customizable Implementation of RMI for High Performance Computing 733
Fabian Breg, Dennis Gannon

mpiJava: An Object-Oriented Java Interface to MPI 748
Mark Baker, Bryan Carpenter, Geoffrey Fox, Sung Hoon Koo, Sang Lim

An Adaptive, Fault-Tolerant Implementation of BSP for Java-Based Volunteer Computing Systems 763
Luis F.G. Sarmenta

High Performance Computing for the Masses 781
Mark Clement, Quinn Snell, Glenn Judd

Process Networks as a High-Level Notation for Metacomputing 797
Darren Webb, Andrew Wendelborn, Kevin Maciunas

Developing Parallel Applications Using the JavaPorts Environment 813
Demetris G. Galatopoullos and Elias S. Manolakos

Workshop on Optics and Computer Science 829
Yi Pan, Keqin Li

Permutation Routing in All-Optical Product Networks 831
Weifa Liang, Xiaojun Shen

NWCache: Optimizing Disk Accesses via an
Optical Network/Write Cache Hybrid 845
Enrique V. Carrera, Ricardo Bianchini

NetCache: A Network/Cache Hybrid for Multiprocessors 859
Enrique V. Carrera, Ricardo Bianchini

A Multi-Wavelength Optical Content-Addressable Parallel Processor 873
(MW-OCAPP) for High-Speed Parallel Relational Database Processing:
Architectural Concepts and Preliminary Experimental System
Peng Yin Choo, Abram Detofsky, Ahmed Louri

Optimal Scheduling Algorithms in WDM Optical Passive Star Networks 887
Hongjin Yeh, Kyubum Wee, Manpyo Hong

OTIS-Based Multi-Hop Multi-OPS Lightwave Networks 897
David Coudert, Afonso Ferreira, Xavier Muñoz

Solving Graph Theory Problems Using Reconfigurable
Pipelined Optical Buses 911
Keqin Li, Yi Pan, Mounir Hamdi

High Speed, High Capacity Bused Interconnects Using
Optical Slab Waveguides 924
Martin Feldman, Ramachandran Vaidyanathan, Ahmed El-Amawy

A New Architecture for Multihop Optical Networks 938
A. Jaekel, S. Bandyopadhyay, A. Sengupta

Pipelined Versus Non-pipelined Traffic Scheduling in
Unidirectional WDM Rings 950
Xijun Zhang, Chunming Qiao

Workshop on Solving Irregularly Structured Problems in Parallel 964
Tao Yang

Self-Avoiding Walks over Adaptive Unstructured Grids 968
Gerd Heber, Rupak Biswas, Guang R. Gao

A Graph Based Method for Generating the Fiedler Vector of
Irregular Problems 978
Michael Holzrichter, Suely Oliveira

Hybridizing Nested Dissection and Halo Approximate Minimum
Degree for Efficient Sparse Matrix Ordering   986
Franois Pellegrini, Jean Roman, Patrick Amestoy

ParaPART: Parallel Mesh Partitioning Tool for Distributed Systems   996
Jian Chen, Valerie E. Taylor

Sparse Computations with PEI   1006
Frédérique Voisin, Guy-René Perrin

Optimizing Irregular HPF Applications Using Halos   1015
Siegfried Benkner

From EARTH to HTMT: An Evolution of a Multitheaded
Architecture Model   1025
Guang R. Gao

Irregular Parallel Algorithms in Java   1026
Brian Blount, Siddhartha Chatterjee, Michael Philippsen

A Simple Framework to Calculate the Reaching Definition of Array
References and Its Use in Subscript Array Analysis   1036
Yuan Lin, David Padua

Dynamic Process Composition and Communication Patterns in
Irregularly Structured Applications   1046
C.T.H. Everaars, B. Koren, F. Arbab

Scalable Parallelization of Harmonic Balance Simulation   1055
David L. Rhodes, Apostolos Gerasoulis

Towards an Effective Task Clustering Heuristic for LogP Machines   1065
Cristina Boeres, Aline Nascimento, Vinod E.F. Rebello

A Range Minima Parallel Algorithm for Coarse Grained Multicomputers   1075
H. Mongelli, W. Song

Deterministic Branch-and-Bound on Distributed Memory Machines   1085
Kieran T. Herley, Andrea Pietracaprina, Geppino Pucci

**Workshop on Personal Computer Based Networks of
Workstations**   1095
G. Chiola, G. Conte, L.V. Mancini

Performance Results for a Reliable Low-Latency Cluster
Communication Protocol   1097
Stephen R. Donaldson, Jonathan M.D. Hill, David B. Skillicorn

Coscheduling through Synchronized Scheduling Servers -
A Prototype and Experiments 1115
Holger Karl

High-Performance Knowledge Extraction from Data on
PC-Based Networks of Workstations 1130
Cosimo Anglano, Attilio Giordana, Giuseppe Lo Bello

Addressing Communication Latency Issues on Clusters for
Fine Grained Asynchronous Applications - A Case Study 1145
*Umesh Kumar V. Rajasekaran, Malolan Chetlur, Girindra D. Sharma,
Radharamanan Radhakrishnan, Philip A. Wilsey*

Low Cost Databases for NOW 1163
*Gianni Conte, Michele Mazzeo, Agostino Poggi, Pietro Rossi,
Michele Vignali*

Implementation and Evaluation of MPI on an SMP Cluster 1178
*Toshiyuki Takahashi, Francis O'Carroll, Hiroshi Tezuka, Atsushi Hori,
Shinji Sumimoto, Hiroshi Harada, Yutaka Ishikawa, Peter H. Beckman*

**Workshop on Formal Methods for Parallel Programming:
Theory and Applications** 1193
Dominique Méry, Beverly Sanders

From a Specification to an Equivalence Proof in Object-Oriented
Parallelism 1197
Isabelle Attali, Denis Caromel, Sylvain Lippi

Examples of Program Composition Illustrating the Use of
Universal Properties 1215
Michel Charpentier, K. Mani Chandy

A Formal Framework for Specifying and Verifying
Time Warp Optimizations 1228
*Victoria Chernyakhovsky, Peter Frey, Radharamanan Radhakrishnan,
Philip A. Wilsey, Perry Alexander, Harold W. Carter*

Verifying End-to-End Protocols Using Induction with CSP/FDR 1243
S.J. Creese, Joy Reed

Mechanical Verification of a Garbage Collector 1258
Klaus Havelund

A Structured Approach to Parallel Programming: Methodology and Models 1284
Berna L. Massingill

BSP in CSP: Easy as ABC 1299
Andrew C. Simpson, Jonathan M.D. Hill, Stephen R. Donaldson

Workshop on Embedded HPC Systems and Applications 1314
Devesh Bhatt, Viktor Prasanna

A Distributed System Reference Architecture for Adaptive QoS and 1316
Resource Management
Lonnie R. Welch, Michael W. Masters, Leslie A. Madden, David T. Marlow, Philip M. Irey IV, Paul V. Werme, Behrooz A. Shirazi

Transparent Real-Time Monitoring in MPI 1327
Samuel H. Russ, Jean-Baptiste Rashid, Tangirala Shailendra, Krishna Kumar, Marion Harmon

DynBench: A Dynamic Benchmark Suite for
Distributed Real-Time Systems 1335
Behrooz Shirazi, Lonnie Welch, Binoy Ravindran, Charles Cavanaugh, Bharath Yanamula, Russ Brucks, Eui-nam Huh

Reflections on the Creation of a Real-Time Parallel Benchmark Suite 1350
Brian Van Voorst, Rakash Jha, Subbu Ponnuswamy, Luiz Pires, Chirag Nanavati, David Castanon

Tailor-Made Operating Systems for Embedded Parallel Applications 1361
Antônio Augusto Fröhlich, Wolfgang Schröder-Preikschat

Fiber-Optic Interconnection Networks for Signal Processing Applications 1374
Magnus Jonsson

Reconfigurable Parallel Sorting and Load Balancing: HeteroSort 1386
Emmett Davis, Bonnie Holte Bennett, Bill Wren, Linda Davis

Addressing Real-Time Requirements of Automatic Vehicle
Guidance with MMX Technology 1407
Massimo Bertozzi, Alberto Broggi, Alessandra Fascioli, Stefano Tommesani

Condition-Based Maitenance: Algorithms and Applications
for Embedded High Performance Computing 1418
Bonnie Holte Bennett, George Hadden

Author Index 1439

Fourth International Workshop on High-Level Parallel Programming Models and Supportive Environments (HIPS'99)

http://www.informatik.hu-berlin.de/~mueller/hips99

Chair: Frank Mueller

Humboldt-Universität zu Berlin, Institut für Informatik, 10099 Berlin (Germany)
e-mail: mueller@informatik.hu-berlin.de   *phone: (+49) (30) 2093-3011, fax:-3011*

Preface

HIPS'99 is the 4th workshop on High-Level Parallel Programming Models and Supportive Environments held in conjunction with IPPS/SPDP 1999 and focusing on high-level programming of networks of workstations and of massively-parallel machines. Its goal is to bring together researchers working in the areas of applications, language design, compilers, system architecture, and programming tools to discuss new developments in programming such systems.

Today, a number of high-performance networks, such as Myrinet, SCI and Gigabit-Ethernet, have become available and provide the means to make cluster computing an attractive alternative to specialized parallel computers, both in terms of cost and efficiency. These network architectures are generally used through network protocols in a message-passing style.

Parallel programming using the message-passing style has reached a certain level of maturity due to the advent of the de facto standards Parallel Virtual Machine (PVM) and Message Passing Interface (MPI). However, in terms of convenience and productivity, this parallel programming style is often considered to correspond to assembler-level programming of sequential computers.

One of the keys for a (commercial) breakthrough of parallel processing, therefore, are high-level programming models that allow to produce truly efficient code. Along this way, languages and packages have been established, which are more convenient than explicit message passing and allow higher productivity in software development; examples are High Performance Fortran (HPF), thread packages for shared memory-based programming, and Distributed Shared Memory (DSM) environments.

Yet, current implementations of high-level programming models often suffer from low performance of the generated code, from the lack of corresponding high-level development tools, e.g. for performance analysis, and from restricted applicability, e.g. to the data parallel programming style. This situation requires strong research efforts in the design of parallel programming models and languages that are both at a high conceptual level and implemented efficiently, in the development of supportive tools, and in the integration of languages and tools into convenient programming environments.

Hardware and operating system support for high-level programming, e.g. distributed shared memory and monitoring interfaces, are further areas of interest.

This workshop provides a forum for researchers and commercial developers to meet and discuss the various hardware and software issues involved in the design and use of high-level programming models and supportive environments.

HIPS'99 features an invited talk by Jayadev Misra (University of Texas at Austin, USA) titled "A Discipline of Multiprogramming" as well as nine long and two short papers. The papers were selected out of 23 submissions. Each paper was reviewed by three members of the program committee. The recommendations of the referees determined an initial set of papers selected for acceptance. The members of the program committee were then given the option to resolve differences in opinion for the papers they refereed based on the written reviews. After the resolution process, the 11 papers included in the workshop proceedings were selected for presentation and publication. The papers cover the topics of:

- Concepts and languages for high-level parallel programming,
- Concurrent object-oriented programming,
- Automatic parallelization and optimization,
- High-level programming environments,
- Performance analysis, debugging techniques and tools,
- Distributed shared memory,
- Implementation techniques for high-level programming models,
- Operating system support for runtime systems,
- Integrating high-speed communication media into programming environments and
- Architectural support for high-level programming models.

Steering Committee

- Hermann Hellwagner, Universität Klagenfurt, Austria
- Hans Michael Gerndt, Forschungszentrum Jülich, Germany

Program Committee

- Arndt Bode, Technische Universität München, Germany
- Helmar Burkhart, Universität Basel, Switzerland
- John Carter, University of Utah, USA
- Karsten Decker, CSCS/SCSC, Switzerland
- Hans Michael Gerndt, Forschungszentrum Jülich, Germany
- Hermann Hellwagner, Universität Klagenfurt, Austria
- Francois Irigoin, Ecole des Mines de Paris, France
- Vijay Karamcheti, New York University, USA
- Peter Keleher, University of Maryland, USA
- Ulrich Kremer, Rutgers University, USA
- James Larus, University of Wisconsin-Madison, USA
- Ronald Perrott, Queen's University of Belfast, United Kingdom
- Domenico Talia, ISI-CNR, Italy

- Thierry Priol, INRIA, France
- Hans Zima, University of Vienna, Austria

Acknowledgments

The workshop chair would like to acknowledge the following people:

- the invited speaker, Jayadev Misra, for his voluntary contribution to the workshop;
- Hermann Hellwagner and Michael Gerndt for their assistance throughout the planning of the workshop;
- Richard Gerber for making the START software available to assist the submission and review process;
- Jose Rolim and Danuta Sosnowska for their organizational support;
- and last but not least the eager members of the program committee and the anonymous external reviewers.

The Chair,

January 1999 Frank Mueller

Efficient Program Partitioning Based on Compiler Controlled Communication[1]

Ram Subramanian and Santosh Pande

Dept. of ECECS, P.O. Box 210030, Univ. of Cincinnati, Cincinnati, OH45221
E-mail: {rsubrama, santosh}@ececs.uc.edu, correspondence: santosh@ececs.uc.edu

Abstract. In this paper, we present an efficient framework for intra-procedural performance based program partitioning for sequential loop nests. Due to the limitations of static dependence analysis especially in the inter-procedural sense, many loop nests are identified as sequential but available task parallelism amongst them could be potentially exploited. Since this available parallelism is quite limited, performance based program analysis and partitioning which carefully analyzes the interaction between the loop nests and the underlying architectural characteristics must be undertaken to effectively use this parallelism.

We propose a compiler driven approach that configures underlying architecture to support a given communication mechanism. We then devise an iterative program partitioning algorithm that generates efficient program partitioning by analyzing interaction between effective cost of communication and the corresponding partitions. We model this problem as one of partitioning a directed acyclic task graph (DAG) in which each node is identified with a sequential loop nest and the edges denote the precedences and communication between the nodes corresponding to data transfer between loop nests. We introduce the concept of **behavioral edges** between edges and nodes in the task graph for capturing the interactions between computation and communication through parametric functions. We present an efficient iterative partitioning algorithm using the behavioral edge augmented PDG to incrementally compute and improve the schedule. A significant performance improvement (factor of 10 in many cases) is demonstrated by using our framework on some applications which exhibit this type of parallelism.

1 Introduction and Related Work

Parallelizing compilers play a key role in translating sequential or parallel programs written in high level language into object code for highly efficient concurrent execution on a multi-processor system. Due to the limitations of static dependence analysis especially in the inter-procedural sense, many loop nests are identified as *sequential*. However, available task parallelism could be potentially exploited. Since this type of available parallelism is quite limited, performance

---

[1] This work is supported by the National Science Foundation grant no. CCR-9696129 and DARPA contract ARMY DABT63-97-C-0029

based program analysis and partitioning which carefully analyzes the interaction between the loop nests and the underlying architectural characteristics (that affect costs of communication and computation) must be undertaken to effectively use this parallelism. In this work, we show that the compiler, if given control over the underlying communication mechanism, can significantly improve runtime performance through careful analysis of program partitions.

When the object code executes on the underlying hardware and software subsystems, a complex interaction occurs amongst them at different levels. These interactions could be from a much finer level such as the multi-function pipelines within a CPU, to a more coarser level such as at the interconnection network, cache and/or at the memory hierarchy level. Unfortunately, parallelizing compilers have very little or no knowledge of the actual run time behavior of the synthesized code. More precisely, most optimizing parallel compilers do not have a control over the underlying architecture characteristics such as communication mechanism to match their sophisticated program analysis and restructuring ability. Due to this reason, the actual performance of the code may be different (often worse) than the one anticipated at compile time [11, 12]. The above problem of exploiting task parallelism amongst sequential loop nests is especially susceptible to this limitation since the amount of available parallelism is small. However, it can be effectively exploited if the compiler is given control over key architecture features. The compiler could then undertake program partitioning driven by performance model of the architecture. This is the motivation behind our work in this paper.

Most of the reported work in the parallelizing compilers literature focuses on analyzing the program characteristics such as the dependences, loop structures, memory reference patterns etc. to optimize the generated parallel code. This is evident from the theory of loop parallelization [1] and other techniques for detecting, transforming and mapping both the loop and task parallelism [3, 5, 8, 6, 13].

Current approaches to performance enhancement involve the use of a separate performance evaluation tool such as Paradyn [15], to gather performance statistics. Another important effort in this area is Pablo [16] which uses a more traditional approach of generating and analyzing the trace of program execution to identify performance bottlenecks. Wisconsin Wind Tunnel (WWT) project has contributed an important discrete event simulation technique that results in accurate and fast performance estimation through simulation [20]. However, the drawback of these tools is that they demand that a user be responsible for guiding the estimator tool to elicit high performance.

Performance estimation of a parallel program has been the focus of research by parallelizing compiler researchers for a while. Sarkar [7] first developed a framework based on operation counts and profile data to estimate the program execution costs for Single Assignment Languages. Another important effort from Rice University involved development of *Parascope Editor* [17] in which the user annotates the program and is given a feed-back based on performance estimates to help the development process. However, the use of compilation techniques for estimation was a complex problem due to the complex run time behaviors of the architectures. The communication overhead is especially tricky to model and the complexity of modeling its software component has been shown [18].

Some studies have indicated the potential of compiler analyses for performance prediction. The P3T system based on Vienna Fortran [14] was the first attempt in using compiler techniques for performance estimation for such systems. However, the approach focussed more on the effect of the compiler optimizations and transformations on the performance rather than on relating it to the behavior of the underlying architectural components. For example, P3T based on Vienna Fortran Compiler (VFC) estimates the effect of loop unrolling, tiling, strip-mining and a whole variety of loop transformations to choose their best ordering on a given architecture. Our work can be contrasted with theirs in that we study the relationship between architecture configuration and partitioning to effectively map a limited amount of task parallelism available across sequential loop nests. Our approach allows the interactions between costs and partitions to be analyzed to refine partitions. The input to our scheme can be any scheduling algorithm such as STDS [2], Chretienne's [9], DSC (Dominant Sequence Clustering) [10], and so on.

The organization of this paper is as follows: section 2 presents a motivating example, section 3 describes our solution methodology, section 4 explains our algorithm, section 5 includes results and discussion, followed by conclusions in section 6.

2 Motivating Example

We now introduce the motivation behind our work through an example. Figure 1(a) shows a code segment consisting of loop nests. The loop nests execute sequentially due to the nature of data dependencies inside the loops. Figure 1(b) gives the task graph (DAG) such that each node $v_i \in V$ represents a task and the directed edge $e(v_i, v_j) \in E$ represents the precedence constraint from v_i to v_j, that corresponds to the actual data dependence edge. The node computation costs and the edge communication costs are shown next to each node and edge respectively. The computation cost of node v_i is denoted by $t(v_i)$ and the communication cost of edge $e(v_i, v_j)$ is denoted by $c(v_i, v_j)$. The node and edge costs are found using operation counts and the sizes of the messages and assuming a certain cost model. Typical cost models for message passing use a fixed start-up cost and a component proportional to size of message. We first show how to find optimal partitioning assuming an ideal architecture in which cost of each message is independent of any other concurrent messages. We then show how a compiler could choose a communication protocol and improve the performance.

In order to minimize the completion time, it is desirable to schedule the successor of a node on the same processor to eliminate the communication cost on the edge. We, however, still continue to pay communication costs on the other edges. The completion time found by using this strategy is given by

$$t(v_i) + \min_{j: e(v_i, v_j) \in E} \max_{k: k \neq j, e(v_i, v_k) \in E} (t(v_j), t(v_k) + c(v_i, v_k)) \qquad (1)$$

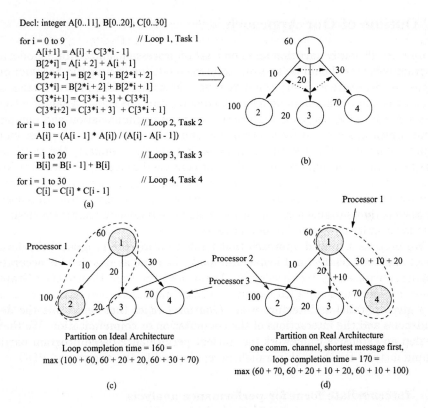

Fig. 1. An Example Graph

Using the ideal architecture (in which the cost of each message is independent of any other concurrent messages), the optimal partition found using the above strategy is given in Figure 1(c). However, on a real machine, the costs of concurrent messages could be heavily influenced by limitations of number of channels, communication protocol used, and so on. The behavior is shown by dotted lines in Figure 1(b). For example, if there is only one communication channel available per processor such as in a cluster of workstations topology, it will be time-multiplexed between different concurrent messages which will have an influence on costs of communication. In such cases, to increase throughput, the compiler could choose a *Shortest Message First* protocol. Under this condition, the optimal partition changes to the one shown in Figure 1(c). The actual completion time of the partition shown in Figure 1(c) using this protocol is 180 as against 160 on ideal machine; whereas the completion time of partition shown in Figure 1(d) found using actual message-passing protocol used (shortest message first) is 170, which is optimal. This shows that the choice of communication protocol also allows the compiler to refine its partition to get better performance.

3 Outline of Our Approach

In order to efficiently partition tasks on a set of processors, an accurate model of program behavior should be known. Accurate modeling of program behavior on a given architecture is a very complex task. The complexities involve latencies, cache behaviors, unpredictable network conflicts and hot spots. A factor which affects partitioning decisions are relative costs of communication and computation which are decided by loop bounds and data sizes. In many cases, loop bounds could be estimated through symbolic tests [19]; similarly the maximal data footprint sizes could be determined at compile time using Fourier-Motzkin elimination and bounds on static array declarations. As illustrated in the motivating example, these costs are themselves dependent on partitioning decisions. We assume that suitable analysis outlined as above is done to expose the desired costs for a series of sequential loop nests.

We propose a unified approach that tends to bring the program behavioral model and the architectural behavioral model as close as possible for accurate program partitioning. We develop an Augmented Program Dependence Graph that can represent a detailed execution behavioral description of the program on a given parallel architecture using *behavioral edges* that represents the dependencies and the interactions of the computation or communication. We then develop an extensive framework for efficient performance-based program partitioning method for the task parallelism represented in the augmented PDG.

3.1 Intermediate form for performance analysis

Sequential loops in the program are identified as tasks to be executed on processors. A dependence relationship between various tasks (sequential loops) is captured by the task graph, or the Program Dependence Graph (PDG). The PDG is a tuple, $G(V,E,C(V), C(E))$, where, V is the set of nodes, E is the set of edges, $C(V)$ is the set of node computation costs and $C(E)$ is the set of edge communication/synchronization costs. These costs are primarily based on the operation counts and the number of data items to be communicated and hence are simplistic and error prone. We develop a more accurate model of each architectural component such as buffer set-up delays, network set-up delays and transmission delays and based on this model, add several parametric annotations to the basic PDG to reflect the complex inter-dependent behaviors. In order to capture the dependences of these parameters on each other, we introduce an extra set of edges called **behavioral edges** that denote the direction and degree of the interaction. The degree of interaction is captured by a partial function of all the involved variables. Behavioral edges are shown in Figure 2 by dotted lines.

Behavioral edges could be of four types which denote the interaction of computation on both computation(nodes) and communication(edges), and interaction of communication on both computation and communication. Thus, a behavioral edge could have a source or sink on a PDG computation node or on a PDG precedence edge. For example, if communication on a PDG precedence edge is affected by communication on another PDG precedence edge, we add a behavioral edge between the two. This edge bears a partial function involving the number of available channels, buffers, network delays, sizes of the messages and the routing parameters that reflects the interactions of these two communications. If the interaction is symmetric, the edge added is bidirectional with the same function, otherwise, it is unidirectional and there are two such edges and two partial functions. The partial functions map the original cost on an edge or a node to an appropriately modified cost using the underlying parameters and the other interacting costs. For example, a partial function $f(a,b,c)$ on a behavioral edge going from precedence edge $e1$ to $e2$, designates the modification in the cost of edge $e2$ due to the cost of edge $e1$ and the underlying parameters a, b and c. The augmented PDG is thus, a tuple $G(V, E, C(V), C(E), B(f))$, where $B(f)$ denotes a set of behavioral edges with the respective partial functions. These behavioral edges are used only for augmenting the PDG to perform cost-based program partitioning. They are not part of the precedence edges of the PDG. Thus, through this elaborate model of augmented PDG, we capture the interactions of computation and communication. Analyzing these interactions, a compiler is able to control a key architectural feature such as communication mechanism and perform program partitioning accurately. In the next subsection, we develop an example to illustrate the Augmented PDG.

3.2 Augmented PDG Example

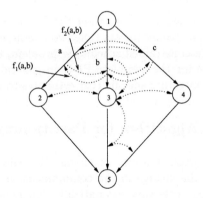

Fig. 2. An Example Augmented PDG

The solid edges in Figure 2 represent the precedence constraints on the computation and the dotted ones represent the behavioral edges between two nodes or two precedence edges or between a node and an edge. In this example, the communication costs on edges (1,2), (1,3) and (1,4) are inter-dependent since the communication may occur concurrently. This inter-dependence is designated by six directed behavioral edges between these edges. We need to have two directed behavioral edges instead of a bidirectional behavioral edge between the precedence edges since the interaction is not symmetric. Here, the partial functions f_1 and f_2 are shown representing two of these behavioral edges. If edges (1,2), (1,3) and (1,4) have costs a, b and c respectively, then the partial functions are functions of these variables. For example, in case the underlying architecture is configured to select the underlying message passing mechanism as "shortest message first", then

$$f_1(a,b) = \begin{cases} a+b & \text{if } b < a \\ a & \text{otherwise} \end{cases} \text{ and } f_2(a,b) = \begin{cases} a+b & \text{if } a < b \\ b & \text{otherwise} \end{cases}$$

The incoming communication on node 3 can perturb its ongoing computation and thus, there is an edge between edge (1,3) and node 3. Similarly the cache performance of computation at node 2 will be affected if merged with 3 and the same is true with the merger of nodes 3 and 4. This is designated by the respective behavioral edges between these nodes. However, no such performance degradation possibly occurs for nodes 2 and 4 and hence there is no edge. Similarly, maybe the outgoing communication on edge (3,5) can be improved if the data is directly written in the communication buffer and thus, there is a behavioral edge between the node 3 and edge (3,5). There is also a behavioral edge between edge (3,5) and (4,5) to denote the interaction of communication between them.

3.3 Program Partitioning

The key goal of this phase is to partition the PDG using the results of the previous analysis so that partitions are tuned to communication latencies.

Being a precedence constrained scheduling problem, the problem of finding the optimum schedule for the complete PDG in general is NP-hard [21]. Our goal here is to use architecture specific cost estimates to refine the partition.

4 An Iterative Algorithm for Partitioning

One of the most important reasons for changes in schedule times of different nodes at run-time, is the change in the communication costs of the edges. As mentioned before, edge costs vary dramatically depending on the communication mechanisms of the architecture. Concurrent communication may be affected by *Shortest Message First, First Come First Served (FCFS)* protocols etc. A compiler may choose a particular communication scheme to configure the architecture depending upon the communication pattern. For example, in case of concurrent communication being known at compile time, a shortest first ordering may be useful to minimize the average message passing delays. On the other hand, if at compile time, it is not known whether concurrent communication will occur, then an FCFS scheme may be relied upon to increase throughput. The choice of communication mechanism potentially affects the edge costs and hence affects the task schedules. The schedule chosen is also closely related to the hardware support. An example of this can be seen in the IBM SP2 scalable cluster. It is a cluster of processors connected by a single high performance switch (HPS). This means that if a processor sends two messages to two processors concurrently, the messages will not be routed the same way, but instead, will be routed sequentially, depending on the underlying communication layer. This changes the communication costs on the edges depending on the message size. On the other hand, on a CRAY T3E, the 3D torus interconnect makes it possible for each processor to communicate to 6 neighboring processors concurrently. Once these communications are fixed, the compiler attempts to refine the schedule. Here, the costs of the communication edges will not change as long as there is no contention. Hence, our algorithm tries to involve these communication cost changes to iteratively determine a schedule tuned to the costs.

In our algorithm, pairs of concurrent communication edges are identified and behavioral edges are associated with each pair. The effect of the underlying communication model (shortest message first or FCFS) is used to determine the communication costs of these edges. For example, assume that the compiler chooses shortest message first communication mechanism. Let an edge e_1 in the PDG be connected to edges $e_2, e_3, \ldots, e_i$ by means of behavioral edges. If edge cost $c(e_1)$ is the least among the above edges, e_1 would be communicated first and the communication costs of the rest of the edges will increase by $c(e_1)$. The increase in edge costs would, in turn, worsen the schedule. A recomputation of the schedule has to be done under this circumstance, and the new schedule could in turn affect the behavioral edges. A two way dependence exists between the schedule and the edge costs. Thus, this algorithm iteratively refines the dependence between behavioral edges and schedule, by computing a schedule using behavioral edges and uses this schedule in turn to compute the effect of cost changes due to behavioral edges. In this way the algorithm arrives at a stable schedule. A given scheduling algorithm thus, can be used as an input to our iterative algorithm, which our algorithm invokes to compute the schedule. Other inputs to our iterative refinement algorithm are: the task graph G, message passing model M chosen by the compiler analyzing the communication pattern, and the convergence limit Ψ [2]. Our algorithm demonstrates an effective method for finding operational behavioral edges, changing costs appropriately and iteratively minimize the schedule for an input task graph.

Algorithm:

```
Iterative_Partition()
Input: A ≡ scheduling algorithm, G ≡ task graph - G(V,E,C(V),C(E))
       M ≡ message passing model (SMF, FCFS), Ψ ≡ limit
Output: schedule S
Begin
    C1(E) ← C(E);           {Initial edge costs are denoted as C1(E)}
```

---

[2] The convergence of this algorithm is slow in some cases; thus, we impose Ψ as a limit imposed on the differences between successive schedule lengths.

```
Decl: integer A[0..11], B[0..11], C[0..11]

for i = 0 to 9           // Task 1
    A[i+1] = A[i] + C[i]
    B[i+1] = A[i+1] + A[i+2]
    C[i+1] = B[i+1] + B[i+2]

for i = 1 to 10          // Task 2
    A[i] = A[i-1] * A[i]

for i = 1 to 10          // Task 3
    B[i] = B[i-1] / 3 * B[i]

for i = 1 to 10          // Task 4
    C[i] = C[i-1] / C[i]

for i = 1 to 10          // Task 5
    A[i] = B[3*i/4] / A[i]

for i = 1 to 10          // Task 6
    B[i] = C[i+1] / B[i] * C[i]

for i = 1 to 10          // Task 7
    A[i] = B[i] / A[i/2]
```
(a)

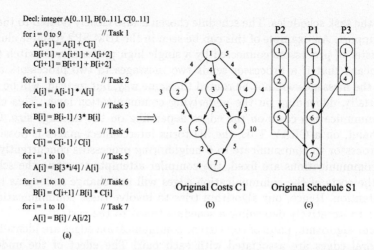

Original Costs C1   Original Schedule S1
(b)   (c)

Fig. 3. (a) An example code segment, (b) its task graph with costs C1 and (c) its schedule S1

$S1 = \text{call\_A} \; (\; G(V,E,C(V),C1(E)) \;);$ {call A, with cost C1 for sche. S1}
call iter_schedule($G(V,E,C(V),C1(E))$, S1); {iteratively refine schedule}
End Iterative_Partition

```
iter_schedule()
```
Input: S1 ≡ schedule, G ≡ task graph - $G(V,E,C(V),C1(E))$
Output: S2 ≡ Final schedule, L2 ≡ Final schedule length
Denotation: *An edge from node i to node j is denoted by (i,j)*
Begin
    {First find behavioral edge pairs from schedule S1. beh_edge_pairs[i] is a list of edges connected to edge i by a behavioral edge. List ended with -1 }
    repeat
      for (all edges (i,j) in G) do
        for (all other edges (u,v) in G) do
        { A directed edge (i,j) has behavioral edges to all directed edges (u,v) such that node $u \neq j$ and u is not the successor of node j}
        if $((u \neq j) \wedge (u \neq SUCC(j)))$ then
          if $((ect(i) \leq ect(u) < last(j)) \vee (ect(u) \leq ect(i) < last(v)))$ then
            add edge (u,v) to beh_edge_pairs$[(i,j)]$
          endif
        endif
        endfor
      endfor
    { $SUCC(j)$ - successor node of node j; ect - earliest completion time; }
    { last - latest allowable start time of node }
    {modify costs of edges connected by beh. edges}

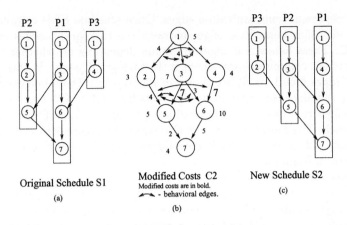

Fig. 4. Illustration of Algorithm

```
    for (each row r of beh_edge_pairs) do
      If (M = SMF) then
         sort edges of row r in ascending order of costs
      endif
      for (each edge (u, v) in beh_edge_pairs[(i, j)]) do
         { modify costs if u and v not assigned to the same processor }
         If ( Processor(u) ≠ Processor(v) ) then
            c((u, v)) = c((u, v)) + c((i, j));
         endif
      endfor
    endfor
    C2(E) ← C1(E); {Denote modified cost of edges as C2}
    S2 = call_A ( G(V, E, C(V), C2(E)) ); {call A, using C2 to get sche. S2}
    L1 = schedule_length(S2, C2);
    L2 = schedule_length(S1, C2); {Using C2 and S1 find schedule length L2}
  until ((convergence) ∨ (L2 - L1 > Ψ))
End iter_schedule
```

4.1 An Example

We now illustrate the working of the above algorithm through an example.

An example code segment consisting of sequential loops, its PDG and its associated schedule is shown in Figure 3(a), Figure 3(b) and Figure 3(c) respectively. It can easily be seen that each loop in the code segment is sequential and cannot be parallelized. When the graph in Figure 3(b) is input to the algorithm with costs C1 and assuming that the compiler has chosen First Come First Served to increase throughput, the schedule obtained is S1, using a scheduling algorithm A. In the figures shown, P1, P2 and P3 are 3 processors on which the corresponding nodes are scheduled. The arrows between the nodes in different processors represent communication edges. Once schedule S1 is computed, the algorithm identifies behavioral edges between the task graph edges using the schedule S1. These behavioral edges are found depending on whether there exists a concurrent communication amongst the task graph edges. In this case, the behavioral edges found for S1 are shown in Figure 4(b). These behavioral edges are used by the algorithm to modify edge costs from C1 to C2. For the example DAG, the edges that have the costs modified are edges $(3,5)$ and $(4,6)$, with new costs of 7 each. The modified costs are also shown in Figure 4(b) in bold. Using the new costs C2, a schedule S2 is determined by the scheduling algorithm A, as illustrated in Figure 4(c).

The schedule S1 has schedule length 26 and uses 3 processors. Suppose the costs of the edges change to costs C2 due to interaction of communications. The schedule length obtained when S1 operates with C2 costs is 30. Our algorithm uses the modified costs C2 and gives a schedule length of 29, which is 1 less than the schedule S1 operating with costs C2. When the iteration progresses to completion, the costs are further refined and schedule lengths are minimized. The final schedule length obtained is 26.

The complexity of finding the pairs of behavioral edges is $O(e^2)$. If there are e edges in the graph, we have a worst case of $(e-1) + (e-2) + \ldots + 1 = e(e-1)/2$ behavioral edges. Since the algorithm examines every pair of edges, its complexity is $O(e^2)$. Modifying the costs incurs a worst case complexity of $O(e^2)$. The step of finding out concurrent communication edges using the earliest and latest, start and completion times, reduces the number of actual behavioral edges. This improves the average case of the algorithm to a large extent. Assuming the algorithm converges in k iterations, the overall complexity of the algorithm is $O(ke^2)$.

5 Results and Discussion

In this work, we used the STDS algorithm developed by Darbha et al. [2] as an input to our iterative algorithm for performance evaluation. We have chosen STDS algorithm due to its relative insensitivity to cost changes as demonstrated in [11]. Our motivation here is that the use of such a stable algorithm would demonstrate the essential merit of our iterative based approach, by factoring out the effect of sensitivity of the schedule to the algorithm.

Our algorithm was implemented on a SPARCstation 5. We first tested the algorithm with the BTRIX test program taken from NAS benchmarks [22] as input in the form of a task graph. The BTRIX test program consists of a series of sequential loop nests with one potentially parallel loop nest. Figure 5 shows the task graph obtained from the program. Due to limitations of inter-procedural analysis in the presence of procedure calls, loops corresponding to tasks 2 and 3 are marked as sequential loop nests. The loop corresponding to task 6 can be parallelized. Hence, we assume that the processor allocated to task 6 is responsible for scatter and gather of data to perform task 6. The computation time for task 6 is taken as the effective parallel time if all the steps are executed in

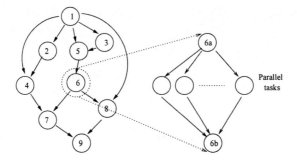

Fig. 5. Task Graph obtained from NAS benchmark code given in Appendix A

BTRIX Test Program					
Loop	L2	NP2	L1	NP1	Factor
30	38225	3	41225	3	1.08
50	66725	3	72825	3	1.09
80	111350	3	123600	3	1.11
100	142350	3	159700	3	1.12
130	190725	3	217225	3	1.14

Table 1. NAS kernel benchmark program: BTRIX - with varying loop counts

parallel. This is clearly illustrated in the graph in Figure 5. Task 6 has been expanded as synchronization nodes **6a** and **6b**, and intermediate parallel nodes, which can be partitioned using any partitioning approach. Results were obtained for the FCFS model by varying the loop counts in BTRIX test program These results are tabulated in Table 1. In all result tables, the following notations and derivations have been made: S1 is the schedule obtained from the STDS algorithm and C1 is the original costs of the task graph. When the costs change to C2 in one iteration of our algorithm, **L1** is the schedule length obtained when **S1** operates on **C2**. The rational for this is that in the STDS algorithm, without our iterative tuning, the tasks are scheduled in the order given by S1, but the actual underlying costs are C2. S2 is the schedule obtained by our algorithm, with costs C2 and the schedule length is L2. NP1 is the number of processors required to schedule the tasks for schedule length L1 and NP2 for schedule length L2. The *Factor* column shows the factor of improvement of the new schedule S2 over S1 using the new costs C2, as discussed before.

We tested our algorithm with different types of task graphs with different node and edge sizes. Examples of the graphs that were given as input are illustrated in Figure 6. These graphs occur in various numerical analysis computations [4]. The node sizes were varied from 50 to 400 nodes and the average values for the schedule lengths have been tabulated in Table 2 and Table 3.

The NAS benchmark test program shows improvement in the schedule lengths for all loop counts. It can also be seen that the improvements increase as loop

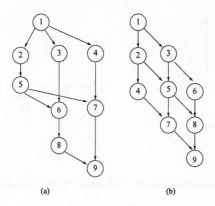

Fig. 6. (a) Cholesky Decomposition, (b) a Laplace Equation solver task graphs

Laplace Solver Graphs						Cholesky Decomposition Graphs					
Node	Edges	L2	NP2	L1	NP1	Factor	Node	Edges	L2	NP2	L1	NP1	Factor
49	84	237	12	1214	12	4.12	54	89	624	24	10099	24	16.18
100	180	402	27	3000	27	6.46	119	209	2149	69	36473	69	16.97
144	264	539	38	4455	38	7.26	152	271	3144	94	64466	94	20.50
225	420	768	55	6907	55	8.00	230	419	5970	156	166490	156	27.88
324	612	1050	78	10969	78	9.44	324	599	10124	234	358428	234	35.40
400	760	1267	100	15952	100	11.59	405	755	14265	303	585928	303	41.07

Table 2. FCFS model: Laplace Equation solver and Cholesky Decomposition graphs

counts increase. Typically for such numerical computing programs, loop counts are very high, and hence, the algorithm generates a better schedule. Table 2 shows results for the FCFS model of communication when the various graphs are input. Table 3 shows results for the SMF model of communication. Table 2 and Table 3 show results for Laplace Equation solver graph and Cholesky Decomposition graph inputs. It can be seen from the tables that the larger the size of the input, the better is the performance improvement due to our algorithm. This could be attributed to the fact that a larger graph size would increase the possibility of concurrent communication and this effect is captured more efficiently by our algorithm leading to a performance enhancement. The drawback, of course, is that more analysis is needed during each iteration. The results also show that the SMF model of communication results in lesser schedule lengths than the FCFS model. This can be easily explained by the fact that the increase in communication costs due to concurrent communication is minimum, if the messages are sent in shortest message first manner.

When the costs of the edges are modified, the schedule also changes and vice versa. Hence, a convergence is not always guaranteed but rather the schedules may oscillate between two values. The reason is that the edge costs are modified using the previous schedule and behavioral edges, and the new schedule is in

Laplace solver graphs							Cholesky Decomposition Graphs						
Node	Edges	L2	NP2	L1	NP1	Factor	Node	Edges	L2	NP2	L1	NP1	Factor
49	84	237	12	1214	12	4.12	54	89	193	24	1255	24	5.50
100	180	402	27	3000	27	6.46	119	209	443	65	5647	64	11.74
144	264	539	38	4455	38	7.26	152	271	571	88	9193	88	15.10
225	420	768	55	6907	55	8.00	230	419	875	146	13316	146	14.22
324	612	1050	78	10969	78	9.44	324	599	1243	216	37485	216	29.16
400	760	1267	100	15952	100	11.59	405	755	1561	279	53560	279	33.31

Table 3. SMF model: Laplace equation solver and Cholesky Decomposition graphs

turn dependent on the modified edge costs. In such a situation, the algorithm chooses the better of two schedules and terminates.

As can be seen, in all cases, the algorithm adapts schedules and costs to each other using behavioral edges and minimizes schedule length. The improvement in many schedules is of the order of 10.

6 Conclusion

We have developed a framework based on behavioral edges and have also developed an iterative algorithm to successively refine the schedule to minimize schedule lengths. The compiler also analyzes the communication behavior of the program and appropriately chooses the communication mechanism. Using this mechanism and the changed costs, the schedule is further refined by the algorithm. Using a performance study undertaken, it can be seen that an improvement of over 10 times is possible for many graphs encountered in numerical computations. Thus, if the compiler is given control over choosing the underlying communication mechanism, it can effectively analyze the partitioning decisions and significantly improve the performance.

References

1. Banerjee, Utpal, *Loop Parallelization*, Kluwer Academic Publishers, 1994 (Loop Transformations for Restructuring Compilers Series).
2. Darbha S. and Agrawal D. P., "Optimal Scheduling Algorithm for Distributed-Memory Machines", *IEEE Transactions on Parallel and Distributed Systems*, Vol. 9, No. 1, January 1998, pp. 87–95.
3. High Performance Fortran Forum. High Performance Fortran Language Specification, Version 1.0, Technical Report, CRPC-TR92225, Center for Research on Parallel Computation, Rice University, Houston, TX, 1992 (revised January 1993).
4. Kwok Y-K and Ahmad I., "Dynamic Critical-Path Scheduling: An Effective Technique for Allocating Task Graphs to Multiprocessors", *IEEE Transactions on Parallel and Distributed Systems*, May 1996, Vol. 7, No. 5, pp. 506–521.
5. Bau D., Kodukula I., Kotlyar V., Pingali K. and Stodghill P., "Solving Alignment Using Elementary Linear Algebra", *Proceedings of 7th International Workshop on Languages and Compilers for Parallel Computing*, LNCS 892, 1994, pp. 46–60.

6. Gerasoulis A. and Yang T., "On Granularity and Clustering of Directed Acyclic Task Graphs", *IEEE Transactions on Parallel and Distributed Systems*, Vol. 4, Number 6, June 1993, pp. 686-701.
7. Sarkar V., *Partitioning and Scheduling Parallel Programs for Multiprocessors*, MIT Press, Cambridge, Mass. 1989.
8. Subhlok Jaspal and Vondran Gary, "Optimal Mapping of Sequences of Data Parallel Tasks", *Proceedings of Principles and Practice of Parallel Programming (PPoPP) '95*, pp. 134–143.
9. Chretienne P., 'Tree Scheduling with Communication Delays', *Discrete Applied Mathematics*, vol. 49, no. 1-3, p 129-141, 1994.
10. Yang, T. and Gerasoulis, A., 'DSC: scheduling parallel tasks on an unbounded number of processors', *IEEE Transactions on Parallel and Distributed Systems*, vol. 5, no. 9, 951-967, 1994.
11. Darbha S. and Pande S. S., 'A Robust Compile Time Method for Scheduling Task Parallelism on Distributed Memory Systems', *Proceedings of the 1996 ACM/IEEE Conference on Parallel Architectures and Compilation Techniques (PACT '96)*, pp. 156–162.
12. Pande S. S. and Psarris K., 'Program Repartitioning on Varying Communication Cost Parallel Architectures', *Journal of Parallel and Distributed Computing 33*, March 1996, pp. 205–213.
13. Pande S. S., Agrawal D. P., and Mauney J., 'A Scalable Scheduling Method for Functional Parallelism on Distributed Memory Multiprocessors', *IEEE Transactions on Parallel and Distributed Systems*, Vol. 6, No. 4, April 1995, pp. 388–399.
14. Fahringer, T., 'Estimating and Optimizing Performance of Parallel Programs', *IEEE Computer: Special Issue on Parallel and Distributed Processing Tools*, Vol. 28, No. 11, November 1995, pp. 47–56.
15. Miller Barton P., Callaghan M., Cargille J., et. al. The Paradyn Parallel Performance Measurement Tool', *IEEE Computer: Special Issue on Parallel and Distributed Processing Tools*, Vol. 28, No. 11, November 1995, pp. 37–46.
16. Reed D. A., et. al.,'Scalable Performance Analysis: The Pablo Performance Analysis Environment', *Proceedings of Scalable Parallel Libraries Conference*, IEEE CS Press, 1993, pp. 104–113.
17. Balasundaram V., Fox G., Kennedy K. and Kremer U., 'A Static Performance Estimator to Guide Data Partitioning Decisions', *Proceedings of 3rd ACM SIGPLAN Symposium on Principles and Practice of Parallel Programming*, 1991, pp. 213–223.
18. Karamcheti V. and Chien A., 'Software Overhead in Messaging Layers: Where Does the Time Go?', *Proceedings of the 6th ACM International Conference on Architectural Support for Programming Languages and Systems (ASPLOS VI)*, pp. 51–60.
19. Blume W. and Eigenmann R., 'Symbolic Range Propagation', *Proceedings of the 9th International Parallel Processing Symposium*, April 1995.
20. Reinhardt, S., Hill M. D., Larus J. R., Lebeck A. et. al., 'The Wisconsin Wind Tunnel: Virtual Prototyping of Parallel Computers', *Proceedings of the 1993 ACM Sigmetrics Conference on Measurement and Modeling of Computer Systems*, pp. 48–60, May 1993.
21. Garey, M.R. and Johnson, D.S., 'Computers and Intractability: A guide to the theory of NP-Completeness', *Freeman and Company*, 1979.
22. NAS Parallel Benchmarks, http://science.nas.nasa.gov/Software/NPB/

SCI-VM: A Flexible Base for Transparent Shared Memory Programming Models on Clusters of PCs

Martin Schulz
schulzm@in.tum.de

Lehrstuhl für Rechnertechnik und Rechnerorganisation, LRR–TUM
Institut für Informatik, Technische Universität München

Abstract. Clusters of PCs are traditionally programmed using the message passing paradigm as this is directly supported by their loosely coupled architecture. Shared memory programming is mostly neglected although it is commonly seen as the easier and more intuitive way of parallel programming. Based on the user-level remote memory capabilities of the Scalable Coherent Interface, this paper presents the concept of the SCI Virtual Memory which allows a cluster-wide virtual memory abstraction. This SCI Virtual Memory offers a flexible basis for a large variety of shared memory programming models which will be demonstrated in this paper based on an SPMD model.

1 Introduction

Due to their excellent price–performance, clusters built out of commodity–off–the–shelf PCs and connected with new low–latency interconnection networks are becoming increasingly commonplace and are even starting to replace traditional massively parallel systems. According to their loosely coupled architecture, they are traditionally programmed using the message passing paradigm. This trend was further supported by the wide availability of high–level message passing libraries like PVM and MPI and intensive research in low–latency, user–level messaging [17, 12].

Next to the message passing paradigm, which relies on explicit data distribution and communication, a second parallel programming paradigm exists, the shared memory paradigm. This paradigm, which offers a global virtual address space for sharing data between processes or threads, is preferably utilized in tightly coupled machines like SMPs that offer special hardware support for a globally shared virtual address space. It is generally seen as the easier programming model, especially for programmers that are not used to parallel programming, but it comes with the price of higher implementation complexity either in the form of special hardware support or in the form of complex software layers. This prohibits their widespread use on cluster architectures as the necessary hardware support is generally missing.

In order to overcome this gap and provide a bridge between tightly and loosely coupled machines, we utilize the Scalable Coherent Interface (SCI) [3,

13] as the interconnection technology. This network fabric, which is standardized in the IEEE standard 1596–1992 [14], utilizes a state of the art split transaction protocol and currently offers bandwidths of up to 400 MB/s. The base topology for SCI networks are small ringlets of up to 8 nodes which hierarchically can be organized to configurations of up to 64K nodes.

The core feature of SCI based networks is the ability to perform remote memory operations through direct hardware distributed shared memory (DSM) support. Figure 1 gives a general overview of how this capability can be applied. The basis is formed by the SCI physical address space which allows addressing of any physical memory location on any connected node through a 64 bit identifier (16 bit to specify the node, 48 bit to specify the physical address). From this global address space, each node can import pieces of remote memory into a designated address window within the PCI address space using special address translation tables on the SCI adapter cards. After mapping the PCI address space into the virtual memory of a process, the remote memory can be directly accessed using standard user–level read and write operations. The SCI hardware forwards these operations transparently to the remote node and, in case of a read operation, returns the result. Due to the pure hardware implementation avoiding any software overhead, extremely low latencies of about 2.3 μs (one way) can be achieved.

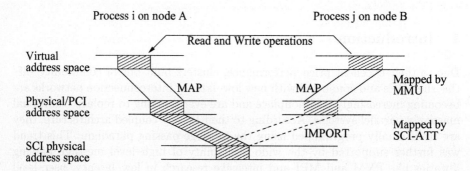

Fig. 1. Mapping remote memory via SCI from node A into the virtual address space of process j on node B.

Unfortunately, this hardware DSM mechanism can not directly be utilized to construct a true shared memory programming model as it is only based on sharing physical memory. It misses an integration into the virtual memory management of the underlying operating system. Extra software mechanisms have to be applied to overcome this limitation and to build a fully transparent global virtual address space. The SCI Virtual Memory layer, presented in this paper, implements these concepts on a Windows NT platform forming the basis for any kind of true shared memory programming on clusters with hardware DSM support.

On top of this virtual memory, it is then possible to implement a large variety of shared memory programming models with different properties customized for the needs of the programmers. This can range from distributed thread libraries over SPMD–style programming libraries, as presented later in this paper, to models with explicit data and thread placement. This flexibility forms a thorough basis for the application of the shared memory paradigm in cluster environments and therefore opens this promising architecture to whole new group of applications and users.

The reminder of this paper is organized as follows: Section 2 gives a brief overview of some related work. Section 3 introduces the concept and design of the SCI Virtual Memory a software layer that provides a global virtual address space which forms the basis for the cluster wide shared memory. Section 4 introduces one possible programming model supported by the SCI-VM. This paper is then rounded up by a brief outlook into the future in Section 6 and by some concluding remarks in Section 7.

2 Related work

Work on shared memory models for clusters of PCs is mostly done with the help of pure software DSM systems. A well–known representative of this group is the TreadMarks [1] implementation. Here, shared pages are replicated across the cluster and synchronized with the help of a complex multiple-writers protocol. To minimize the cost of maintaining the memory consistency, a relaxed consistency model is applied, the Lazy Release Consistency [7]. However, unlike in the SCI-VM approach, where the sharing of the memory is transparently embedded into a global abstraction of a process, TreadMarks requires the programmer to explicitly specify the shared segment. The same is also valid for any other currently existing software DSM package, although they vary with respect to their APIs, consistency models, and usage focus. Examples for those kind of systems that have been developed for the same platform as the SCI Virtual Memory, Windows NT, are Millipede [6], Brazos [15], and the SVMlib [11].

Besides these pure software DSM systems that just deal with providing a global memory abstraction on a cluster of workstations, there are also projects that provide an execution model in the form of a thread library on top of the constructed global memory. One example for this are the DSM-Threads [10], which are based on a POSIX 1003.1c conforming API. This thread package, based on the FSU-Pthreads [9], allows the distribution of threads across a cluster interconnected by conventional interconnection networks. However, it also does not provide complete transparency. It therefore forces the user to modify and partly rewrite the application's source code.

Only a few projects utilize similar techniques than the SCI Virtual Memory and try to deploy SCI's hardware DSM capabilities directly. In the SciOS project [8], a global memory abstraction is created using SCI's DSM capabilities to implement fast swapping to remote memory. The University of California at Santa Barbara is also working on providing transparent global address space with the help of SCI [5]. Their approach, however, does not target a pure programming API, but rather to a hybrid approach of shared memory and message passing in the context of a runtime system for Split-C [4].

3 Global virtual memory for SCI based clusters

The main prerequisite for shared memory programming on clusters is a fully transparent, global virtual address space. This section describes a software layer that provides such a cluster-wide memory abstraction, the SCI Virtual Memory or SCI-VM. This forms the basis for any kind of shared memory programming model on top of an SCI based PC cluster.

3.1 Design goals for a transparent global memory abstraction

The design and the development of the SCI-VM was governed by several design goals and guidelines. They mostly originated from the initial project goal of creating an SMP-like environment for clusters as well as from the idea of providing a flexible basis for a large variety of shared memory programming models.

In order to fulfill these requirements, one of the main goals was to enable the SCI-VM to provide a fully transparent global address space hiding the distribution of the memory resources from the programmer. This includes the distribution and global availability of any static variables that are provided within the original process image. As a final result, the SCI-VM has to provide every process on each node with exactly the same view onto the complete memory system as only this allows the illusion of a single virtual address space that spawns the whole cluster.

Another important goal for the SCI-VM is the implementation of a relaxed consistency model together with the appropriate routines to control it. The SCI-VM should not fix a certain model, though, but rather provide mechanisms that allow higher software layers to adapt the model to their needs and requirements.

The implementation of the SCI-VM should be directly based on the HW-DSM provided by SCI rather than deploying a traditional SW-DSM system with all its problems like false sharing and complex differential page update protocols. Only this enables the SCI-VM to benefit from the special feature and the full performance of the interconnection fabric. Additionally, the implementation of synchronization primitives should directly utilize atomic transactions provided by SCI to ensure greatest possible efficiency.

3.2 The concept of the SCI-VM

As already mentioned in Section 1, SCI's hardware DSM support cannot directly be used to create a global virtual abstraction as it is required according to the goals stated above. For this endeavor, the remote memory capabilities have to be merged with well–known mechanisms from traditional DSM systems. The memory is distributed on the granularity of pages and these distributed pages are then combined into a global virtual address space. In contrast to pure software mechanisms though, no page has to be migrated or replicated. All remote pages are simply mapped using SCI's HW-DSM mechanisms and then accessed directly.

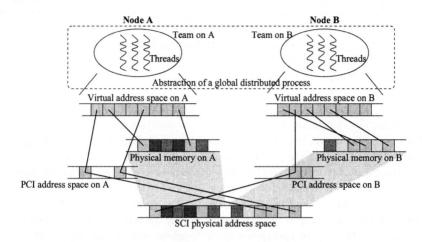

Fig. 2. Principle design of the SCI Virtual Memory

This concept is further illustrated in Figure 2 for a two–node system. In order to build a global virtual memory abstraction, a global process abstraction has to also be built with team processes as placeholders for the global process on each node. These processes are running on top of the global address space which is created by mapping the appropriate pages from either the local physical memory in the traditional way or from remote memory using SCI HW-DSM.

The mapping of the individual pages in done in a two–step process. First, the page has to be located in the SCI physical address space from where it can be mapped in the PCI address space using the address translation tables (ATT) of the SCI adapter cards. From there, the page can be mapped with the help of the processor's page tables into the virtual address space. Problematic is the different mapping granularity in these two steps; while the latter mapping can be done at page granularity, the SCI mappings can only be done at the basis of 512 KByte or 128 pages segment. To overcome this difference, the SCI-VM layer has to manage the mappings of several pages from one single SCI segment. The mappings of the SCI segments themselves will be managed with an on-demand, dynamic scheme very similar to paging mechanisms in operating systems.

3.3 Technical challenges of the SCI-VM design on Windows NT

The design discussed above presents several interesting implementation challenges. The most severe is related to the integration of the concepts presented above into the underlying operating system, in this case Windows NT and its virtual memory manager (VMM). Windows NT does not provide the capabilities to map individual page frames into the virtual address space of a process nor to intercept exceptions like the page fault before the VMM treats it. Currently both issues can only be solved by bypassing the OS and directly manipulating the hardware (the page table and the interrupt vector respectively). Experiments have shown that this approach, applied carefully, does not impair the stability of the operating system. However, we are looking for ways of cleanly integrating our concepts into the Virtual Memory Management of Windows NT to improve the robustness of the SCI-VM implementation.

Also, the integration of the SCI-VM concept into the currently existing hard- and software infrastructure imposes problems as it is mainly intended to provide support for sharing large, pinned segments of contiguous memory. To solve this problem, the hardware here also has to be addressed directly through the physical abstraction layer interface provided within the SCI device drivers.

A special problem is caused by the requirement of sharing the global variables within a program to complete the transparency for the user. These variables are part of the process image that is loaded at process creation time and are therefore treated by the operating system in a special way. In order to transparently share them, it is necessary to determine their location within the image file and to distinguish them from global variables used within the standard C++ runtime system as those must not be shared among the cluster nodes. To accomplish this, the concept of sections in Windows NT object files can be used. All global variables are located by the C++ code generator in sections called ".bss" or ".data". In all object files of the application, these sections have to be renamed with the help of a special patch program using a new, by default unused section name. After linking these modified object files with the unmodified runtime system, this renamed section in the executable containing all global variables can easily be identified (Figure 3.1).

In order to share them across all nodes, a global memory is allocated at program start in the standard SCI-VM manner and the initial information from the image file is copied into the global memory segment. From there it can be mapped into the virtual location of the global variables covering the information from the image file and replacing it with its globally shared counterpart (see also Figure 3.2).

3.4 Caching and consistency

An additional problem is imposed by the fact that, although the SCI standard [14] defines a complex and efficient cache coherency protocol, it is not possible to cache remotely mapped memory in standard PC based systems. This is caused by the inability of PCI based cards to perform bus snooping on the system bus. However, caching is necessary to overcome the problem of the large latencies involved when reading transparently from remote memory (around $6\mu s$ per access). The only solution here is to apply a relaxed consistency model that allows to enable caching while coping with the possible cache inconsistency.

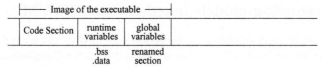

Figure 3.1: Address space of modified executable image

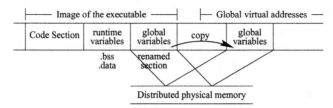

Figure 3.2: Address space after distributing the global variables

Fig. 3. Distributing an application's global variables.

In order to allow for the greatest possible flexibility, the SCI-VM layer does not enforce a specific cache consistency protocol, but merely provides mechanisms that allows applications or higher level programming models to construct application specific consistency protocols. For this purpose, the SCI-VM offers the functionality to flush the current memory state of the local node to the SCI network and to synchronize the memory of individual nodes with the global state by invalidating all local buffers including the caches. This functionality is implemented by routines that allow to control both the SCI adapter card with its internal buffers and the CPU's memory model by providing routines to flush the local caches and write buffers.

3.5 SCI-VM implementation

The main part of the SCI-VM is currently implemented in user level. This part is responsible for the setup of the SCI-VM, the distributed memory allocation, and the appropriate mappings from SCI to PCI and from PCI to the virtual address space. It is supported by two kernel mode drivers: the SCI device driver and a special VMM driver. The latter one was especially developed for the SCI-VM and helps in manipulating the virtual memory management. It provides the appropriate routines to map individual pages through maintaining the page tables, to intercept the page fault exception, and to configure the cacheable area to enable caching of the PCI address window used to map remote memory. This driver also includes cleanup routines that remove all modifications to the process's virtual memory configuration at process termination. This is necessary to maintain the stability of Windows NT.

4 Programming models based on the SCI-VM

The cluster-wide virtual memory abstraction provided by the SCI-VM offers the basis for the implementation of various shared memory programming models. Each of these models can differ in the manner of how threads of activities are created and controlled, in the memory consistency model, and in the degree of control the programmer has regarding the locality properties of underlying virtual memory.

4.1 An SPMD model on top of the SCI-VM

This section introduces in detail one example of these programming models that are based on the functionality of the SCI-VM. This specific model realizes a very simple concept of parallelism by allowing one thread of activity based on the same image file per node. Due to this execution mode and due to the synchronous manner of allocating global memory resources, which will be discussed below, this programming model can be classified as a Single Program Multiple Data (SPMD) model. Within the actual parallel execution, however, this synchronism is not fully enforced; it is possible to implement independent threads of control on each node as long as the restrictions mentioned above are observed.

4.2 Allocating shared memory

The SPMD programming model offers a central routine to allocate shared memory. It reserves a virtual address segment that is valid on all nodes within the cluster and performs the appropriate mappings of interleaved local and remote memory into this address range. The memory resources are thereby distributed in a round robin fashion at page granularity, i.e. at a granularity of 4 KBytes (on x86 architectures). This results in an even distribution of the memory resources across the cluster while trying to avoid locality hot spots.

This allocation routine has to be called by all nodes within the cluster simultaneously. It is therefore the main constraint that causes the SPMD properties of the programming model. This is currently necessary to request memory resources from all nodes during the allocation process. In the future, this constraint will be eliminated by a more complex SCI-VM implementation which allows to preempt the execution on remote nodes to request memory.

The allocation itself is a three step process. In the first step, the contribution of the local node to the new global segment is determined and an appropriate amount of local physical memory is allocated. In the second step, each node enters the physical locations of all local pages together with its node ID into a global list of pages. After the completion of this list, a virtual address segment is allocated at the same location on each node and all pages in the global list are mapped into this newly created address segment. Local pages are mapped directly from physical memory, whereas remote pages are first mapped into the PCI address range and from there into the virtual address space. Upon completion of the page mappings on all nodes, the allocation process is completed and the new virtual address is returned to the caller.

4.3 Synchronization

Shared memory programming models need mechanisms to coordinate accesses to shared data and to synchronize the execution of the individual threads. For this purpose, the SPMD SCI-VM programming model provides the following three mechanisms: global locks without any association to data structures, cluster-wide barriers, and atomic counters that can be used for the implementation of further custom synchronization mechanisms.

All of these mechanisms are implemented based on atomic fetch and increment transactions provided directly in hardware by the SCI adapter cards in the form of designated SCI remote memory segments. Read accesses to these segments automatically trigger an atomic fetch and increment. This transaction increments the value at the memory location specified through the read transaction atomically and returns the original value of the memory location as the result of the read operation.

While the atomic counters can directly be implemented using the atomic SCI transactions, locks and barriers need additional concepts. Locks are implemented using two separate counters, a ticket counter and an access counter. Each thread trying to acquire a lock requests a ticket by incrementing the ticket counter. The lock is granted to the thread if and only if the value of the ticket counter equals the value of the access counter. On release of the lock, the thread increments the access counter and thereby grants the next waiting thread access to the lock.

The implementation of barriers utilizes one global atomic counters that is incremented by each thread on entrance into the barrier. When the counter reaches a value that is divisible by the number of threads in the cluster, all threads have reached a barrier and are now allowed to continue. To distinguish multiple generations of one barrier an additional local, non-shared counter is maintained. This counter helps keeping track of the number of times the barrier has been executed.

Currently, all waiting operations required by locks and barriers are implemented using simple spin waits. This is not a major problem as there is currently only one thread per node allowed. In the future, however, this will be replaced using blocked and spin-blocked waits with a tight integration into the thread scheduling policies to allow more flexibility.

4.4 Coherency model

The SPMD programming model utilizes a relaxed consistency model to enable the caching of remote memory. Most of the synchronization mechanisms are implicitly hidden within the synchronization operations to ease the use of the programming models by hiding the relaxed consistency model from the programmer.

Barrier operations perform a full synchronization of all memories within the cluster. This is achieved by performing a flush of all local buffers (on the processor, in the memory system, and on the network adapter) to the network together with a complete invalidation of all caches and local buffers.

Locks provide a semantics similar to release consistency; after acquiring a lock, the local memory state is invalidated, and before the unlock or release operation, all local buffers are flushed to network to force the propagation of the local memory state to remote nodes. This semantics guarantees that accesses to shared data structures, which are guarded by locks to implement mutual exclusion, are always performed correctly in a consistent manner.

In addition to this implicit synchronization, the SPMD programming model also provides a routine for explicit synchronization. This routine performs a full flush and invalidation of the local memory on the node it has been called. However, this routine should only be necessary in special cases when working with unguarded data structures as all other cases are already covered by the implicit synchronization mechanisms.

4.5 Implementation

The SPMD programming model described in this section is implemented as a Win32 DLL which has to be linked to all images on all nodes. The startup of the SCI-VM and the SPMD data structures is automatically done from the main routine of that DLL without requiring an interaction with the application. This routine initializes all internal mappings and synchronizes the individual processes on all nodes.

Also, the termination of the application is controlled by the main routine of the DLL which is executed after the termination of the application. At this point, the DLL removes all mappings created during the runtime of the application and frees all memory resources that have been allocated.

5 Experiments and results

In order to evaluate the concept presented above, we implemented a few numerical mini kernels and ported an existing shared memory based volume rendering application using the described SPMD programming model. All codes utilized the shared memory in a fully transparent fashion without any locality optimizations. Also, the relaxed consistency model was applied only through the implicit mechanisms described above without requiring any code modifications.

5.1 Experimental setup

All of the following experiments were conducted on our SCI cluster in a two node configuration. This reduced version of our cluster was chosen due to current limitations in the implementation. Future version will overcome this and are expected to scale to distinctly larger cluster configurations. Each compute node in the cluster is a Pentium II (233 MHz) based PC with a 512 KByte L2 cache. The motherboards of these systems are based on the Intel 440FX chipsets. We deployed the PCI–SCI adapter cards, Revision D from Dolphin ICS [2]. These cards are equipped with the Link Controller LC-2 which allows to operate the SCI network at a raw bandwidth of up to 400 MB/s.

The base operating system for all experiments was a standard Windows NT 4.0 (Build 1381, SP 3). The driver software for the SCI adapter cards was also supplied by Dolphin ICS through the SCI Demokit in version 2.06.

5.2 Results: Numerical mini kernels

The first experiments where conducted using three numerical mini kernels: linear sum, standard dense matrix multiplication, and one iteration of the successive over-relaxation (SOR) method. All of these kernels are applied to one contiguous virtual memory segment in which the respective data structures are mapped.

For each of these codes, three different experiments were performed: a parallel version on top of the global memory provided by the SCI-VM, a sequential version on top of local memory to get a baseline comparison for the speed–up values, and a sequential version on top of global memory to get information about the overhead connected to using the global memory abstraction. Based on the results from these experiments, three key values were computed: speed–up as the ratio of parallel to sequential execution on local memory, fake speed–up as the ratio of parallel to sequential execution on global memory, and overhead as the ratio of sequential execution on local to global memory. The results for two different sizes of the virtual segment are presented in Table 1.

	Memory size	Local seq.	Speed–up	Fake s.u.	Overhead
Linear sum	256 KByte	4128 μs	0.92	1.50	63.01 %
	1024 KByte	16549 μs	1.06	1.63	53.56 %
Matrix multiplication	256 KByte	4912 ms	1.97	1.98	0.48 %
	1024 KByte	42731 ms	1.89	1.99	5.2 %
SOR iteration	256 KByte	296 ms	1.53	1.71	11.1 %
	1024 KByte	1225 ms	1.58	1.73	5.3 %

Table 1. Performance results for the numerical mini kernels.

The linear sum performs poorly; applied to small problem sizes, no speed–up, but rather a slight slow–down can be observed. Only when applied to larger problem sizes a small speed–up is possible. This can be explained through the fact that this algorithm traverses the global memory range only once and therefore does not utilize any temporal locality through the caches. Only the implicit prefetching of cache lines and the prefetching abilities that are implemented within the SCI hardware help the performance. The result is a rather high overhead and consequently a low speed-up. In addition, the linear sum is an extremely short benchmark which causes the parallel version of the algorithm to infer significant overhead which can be seen in the rather low fake speed–up value. This can only be improved by applying the algorithm to larger data sizes.

The matrix multiplication algorithm, however, behaves almost optimally by achieving a nearly perfect speed–up. This can be achieved by the high exploitation of both spatial and temporal cache locality which is documented in the extremely low–overhead number. When applying the scheme to larger data sizes, however, this overhead increases as the working set no longer fits into the L2 cache. Due to this, an increased number of cache misses has to be satisfied through the SCI network which causes additional overhead. This also leads to a slight decrease of the overall speed–up.

Also, the third numerical mini kernel, the SOR iteration, shows a very good performance. This is again the result of an efficient cache utilization for temporal and spatial locality. In contrast to the matrix multiplication, however, it is not necessary to keep the whole virtual address segment within the cache, but rather only a small environment around the current position within the matrix. Due to this, the algorithm does not suffer in performance when applied to data sizes larger than the L2 cache. It even benefits from the reduced overhead due to the larger data size resulting in a higher speed–up.

5.3 Results: Volume rendering

While the experiments using the numerical mini kernels provides detailed information about the raw performance and the problems of the SCI-VM and the SPMD model implementation, it also necessary to perform experiments using large complex applications to evaluate the full impact of the SCI-VM. For this purpose we utilized a volume rendering code from the SPLASH-II suite [18]. It was ported to the SPMD programming model by providing an adapted version of the ANL macros. The actual volume rendering code itself was not modified.

To evaluate this application the same experiments were conducted as described above for the numerical kernels. The results are summarized in Table 2. The values correspond to rendering times of single images. Pre– and postprocessing of the data was not included in the measurements. The data set used for all experiments is the standard test case provided together with the SPLASH distribution and has a raw size of roughly 7 MB. The work sharing granularity is set to blocks of 50 by 50 pixels to achieve the optimal tradeoff between management overhead and load balancing.

Sequential execution (local memory)	3522 ms
Sequential execution (global memory)	4136 ms
Parallel execution (global memory)	2381 ms
Speed–up	1.49
Fake speed–up	1.74
Overhead	17.43 %

Table 2. Performance results for the volume rendering application.

Also this complex application which handles large data sets exhibits a good performance on top of the transparent virtual address space. It is possible to achieve a speed–up of about 1.5 due to a low overhead caused by utilizing the transparent memory of only about 18 %. Most of the overhead prohibiting a larger speed–up is caused by the management and locking overhead of a central work queue which is a common bottleneck in any shared memory environment. This experiment shows that the concepts presented in this paper can be directly and efficiently applied to a large existing shared memory code without a major porting effort. Future experiments will extend this to further applications from different domains and will evaluate the SCI-VM in a large variety of scenarios.

6 Future work

The numbers above show that the SCI-VM concept together with the SPMD programming model allows to run parallel shared memory codes on clusters of PCs in a fully transparent fashion with good performance. However, the performance is not optimal, a price that has to be paid for the transparency. Further research will therefore investigate on adding optional mechanisms to the programming model that allow the programmer to incrementally optimize the data distribution. This enables a gradual transition of codes from a fully transparent model, which is easier for the programmer, to a more explicit model with more locality control for the programmer. This will be beneficial for porting existing codes to clusters as well as for implementing new applications using the shared memory paradigm because implementation and locality optimization can be treated as two orthogonal phases.

A second focal point of future research will lie on the development and implementation of other programming models for the SCI-VM that overcome the limitations of the current SPMD–like model. Of special interest will be the development of a distributed thread package on the basis of the POSIX thread API [16], which features full transparency in an SMP–like manner, and its counterpart, a completely explicit programming model which forces the programmer to manually place data and threads on specific nodes. On the basis of these two most extreme models, the tradeoffs between total transparency with its ease of use and full control with the possibility of ideal optimizations will be evaluated.

7 Conclusions

Clusters of commodity PCs have traditionally been exploited using applications built according to the message passing paradigm. Shared memory programming models, which are generally seen as easier and closer to sequential programming, are only available through software DSM systems. Currently, however, these systems lack performance and/or transparency. With the help of the SCI Virtual Memory presented in this work, it is now possible to create a fully transparent global virtual address space across multiple nodes and operating system instances. This is achieved by merging hardware DSM mechanisms for physical memory with page based techniques from traditional software DSM systems to provide the participating processes on the cluster nodes with an equal view onto the distributed memory resources.

On top of the global virtual address space provided by the SCI-VM, it is then possible to implement various programming models. One example of such a model, an SPMD-style programming environment, has also been presented here. It allows the programmer to easily develop parallel application in a synchronous manner which are then executed in top of a global address space. This programming model also provides appropriate synchronization mechanisms for shared memory programming in the form of global locks, barriers, and atomic counters. In order to allow the greatest possible efficiency, a relaxed memory consistency model is applied. The coherency is maintained using synchronization mechanisms which are, invisible for the programmer, implicitly implemented within the synchronization mechanisms in release consistency style.

This paper also presented the results of several experiments using this SPMD programming model on top of the SCI-VM to implement three numerical mini kernels. This evaluation shows, in most cases, a good performance with significant speed-ups. In some cases it is even possible to achieve a nearly optimal speed-up. Further experiments prove that also large applications, like the presented volume rendering code, can efficiently take advantage of the shared memory programming model. Without any locality optimization and with a minimal porting effort a speed-up of about 1.5 is achieved on a two node system.

In summary, the SCI-VM is the key component to open the cluster architecture to the shared memory programming paradigm which has been traditionally the domain of tightly coupled parallel systems like SMPs. Together with appropriate programming models on top, it will ease the programmability of clusters and allow to utilize the large number of existing shared memory codes directly in cluster environments.

Acknowledgments

This work is supported by the European Commission in the Fourth Framework Programme under ESPRIT HPCN Project EP23174 (Standard Software Infrastructure for SCI based parallel systems – SISCI). The SCI device driver is kindly provided in source by Dolphin ICS and the driver software to augment the Windows NT memory management has been implemented by Detlef Fliegl (LRR–TUM, TU–München).

References

1. C. Amza, A. Cox, S. Dwarkadas, P. Keleher, H. Lu, R. Rajamony, W. Yu, and W. Zwaenepoel. TreadMarks: Shared Memory Computing on Networks of Workstations. *IEEE Computer*, February 1995.
2. Dolphin Interconnect Solutions, AS. *PCI-SCI Cluster Adapter Specification*, May 1996. Version 1.2.

3. D. B. Gustavson and Q. Li. Local-Area MultiProcessor: the Scalable Coherent Interface. In S. F. Lundstrom, editor, *Defining the Global Information Infrastructure: Infrastructure, Systems, and Services*, pages 131–160. SPIE Press, 1994.
4. M. Ibel, K. Schauser, C. Scheiman, and M. Weis. Implementing Active Messages and Split-C for SCI Clusters and Some Architectural Implications. In *Sixth International Workshop on SCI-based Low-cost/High-performance Computing*, September 1996.
5. M. Ibel, K. Schauser, C. Scheiman, and M. Weis. High-Performance Cluster Computing Using SCI. In *Hot Interconnects V*, August 1997.
6. A. Itzkovitz, A. Schuster, and L. Shalev. Millipede: a User-Level NT-Based Distributed Shared Memory System with Thread Migration and Dynamic Run-Time Optimization of Memory References. In *Proceedings of the 1st USENIX Windows NT Workshop*, August 1997.
7. P. Keleher. *Lazy Release Consistency for Distributed Shared Memory*. PhD thesis, Rice University, January, 1995.
8. P. Koch, E. Cecchet, and X. de Pina. Global Management of Coherent Shared Memory on an SCI Cluster. In *Proceedings of SCI-Europe '98, a conference stream of EMMSEC '98*, pages 51–57, September 1998.
9. F. Müller. A Library Implementation of POSIX Threads under UNIX. In *Proceedings of USENIX*, pages 29–42, January 1993.
10. F. Müller. Distributed Shared–Memory Threads: DSM–Threads, Description of Work in Progress. In *Proceedings of the Workshop on Run–Time Systems for Parallel Programming*, pages 31–40, April 1997.
11. S. Paas, M. Dormanns, T. Bemmerl, K. Scholtyssik, and S. Lankes. Computing on a Cluster of PCs: Project Overview and Early Experiences. In W. Rehm, editor, *Tagungsband zum 1. Workshop Cluster Computing*, number CSR-97-05 in Chemnitzer Informatik–Berichte, pages 217–229, November 1997.
12. S. Pakin, V. Karamcheti, and A. Chien. Fast Messages (FM): Efficient, Portable Communication for Workstation Clusters and Massively-Parallel Processors. *IEEE Concurrency*, 1997.
13. S. J. Ryan, S. Gjessing, and M. Liaaen. Cluster Communication using a PCI to SCI Interface. In *IASTED 8th International Conference on Parallel and Distributed Computing and Systems*, Chicago, Illinois, October 1996.
14. IEEE Computer Society. *IEEE Std 1596-1992: IEEE Standard for Scalable Coherent Interface*. The Institute of Electrical and Electronics Engineers, Inc., 345 East 47th Street, New York, NY 10017, USA, August 1993.
15. E. Speight and J. Bennett. Brazos: A Third Generation DSM System. In *Proceedings of the 1st USENIX Windows NT Workshop*, August 1997.
16. Technical Committee on Operating Systems and Application Environments of the IEEE. *Portable Operating Systems Interface (POSIX) — Part 1: System Application Interface (API)*, chapter including 1003.1c: Amendment 2: Threads Extension [C Language]. IEEE, 1995 edition, 1996. ANSI/IEEE Std. 1003.1.
17. T. von Eicken, D. Culler, S. Goldstein, and K. Schauser. Active Messages: a Mechanism for Integrated Communication and Computation. In *Proc. of the 19th Int'l Symposium on Computer Architecture*, Gold Coast, Australia, May 1992.
18. S. Woo, M. Ohara, E. Torrie, J. Singh, and A. Gupta. The SPLASH-2 Programs: Characterization and Methodological Considerations. In *Proceedings of the 22nd International Symposium on Computer Architecture*, pages 24–36, June 1995.

Flexible Collective Operations for Distributed Object Groups

Jörg Nolte

GMD FIRST
Rudower Chaussee 5
D-12489 Berlin, Germany
jon@first.gmd.de

Abstract. Collective operations on multiple distributed objects are a powerful means to coordinate parallel computations. In this paper we present an inheritance based approach to implement parallel collective operations on distributed object groups. Object groups are described as reusable application-specific classes that coordinate both operation propagation to group members as well as the global collection (reduction) of the computed results. Thus collective operations can be controlled by applications using language-level inheritance mechanisms. Existing group classes as well as global coordination patterns can therefore effectively be reused.

1 Introduction

Collective operations are a powerful means to implement coordinated operations on distributed data-sets as well as synchronized reductions of multiple computed results. In this paper we present an inheritance based approach to implement parallel operations on distributed data-sets as multicasts based on point-to-point communication. We want to be able to control, extend and adapt global operations on user-level by means of standard inheritance mechanisms. Multicast groups are therefore described as reusable application-specific topology classes that coordinate both the spreading of multicast messages and the collection (reduction) of the computed results. Thus global operations are controllable through applications and existing communication topologies can potentially be reused for global operations.

Several common multicast patterns are supported and a generator tool generates suitable client and server-stubs for group communications from annotated C++-classes. Our system was initially designed as an extension to the remote object invocation (ROI) [9] system of the PEACE[12] operating system family mainly targeting tasks like parallel signal propagations, name server queries and page-(in)validations in our VOTE[3] VSM system. Consequently our work has a system biased scope like early work on fragmented objects in SOS[13] with language-level support (FOG[5]) for group operations. Nevertheless, medium to coarse grained data-parallel computations can also be expressed with modest effort and also executed efficiently.

We neither rely on complex compiler technology for data-parallel languages nor do we imply applications to be structured as SPMD programs. Therefore we are able to support a parallel server paradigm that allows multiple clients to access available parallel services running either on parallel computers or homogeneous workstation clusters. This scenario is specifically useful to increase multi-user performance when developers or scientists work on related topics and use parallel computers as domain-specific parallel servers.

In the following sections we discuss the design of our multicast framework. Furthermore, we describe our language-level support and how it is compiled. Finally we present performance results and compare our work with other approaches.

2 Collective Operation Patterns

Collective operations typically follow three basic patterns with possibly multiple variations (fig. 1).

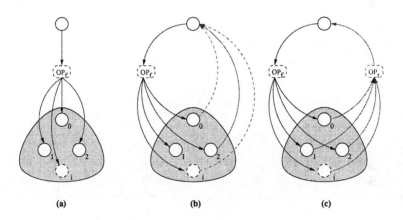

Fig. 1. Collective Operation Patterns

Pattern (a) is applicable to initiate a parallel computation or to propagate values to all members of the group. Pattern (b) is a typical search pattern. A request is issued to the group and only some specific objects deliver a result. Pattern (c) is a global reduction pattern. All group members deliver a result that is reduced according to a reduction function. Such kind of operations are useful to compute global maxima, minima as well as global sums and the like.

It should be noted that pattern (a) and (b) can in principle be supported by most remote object invocation systems. (a) can easily be introduced by means of *repeater* objects that act as representatives of object groups and spread a request message to all group members. (b) can be implemented by passing a reference to a remote result object to an operation according to pattern (a). (c) can be implemented in a similar way. However, (c) will become very inefficient when object groups become larger because a single result object has to process the result messages of all group members.

Pattern (a) and (b) can also be problematic if it must be known when the operation is terminated or the number of expected results is a priori unknown. Considering these potential problems we extended the PEACE ROI system with flexible collective operations based on multicast mechanisms.

2.1 Multicast Groups

In our ROI system multicast groups are implemented as distributed linked object sets. These sets can be addressed as a whole by issuing a multicast operation to a *group leader* representing the group. For example, if the group is organized as a tree (fig. 2), the root acts as the group leader and multicast operations issued to the root are executed on each object within the group. In principle, a multicast group can have any user defined topology such as n-dimensional grids, lists or trees.

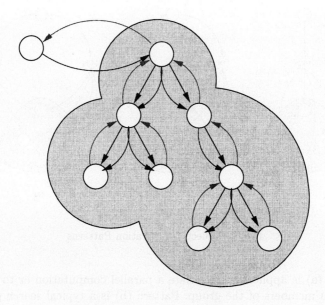

Fig. 2. A Multicast Topology

These topologies are described through reusable C++ topology classes. Members of a multicast group are usually derived from both a topology class and an arbitrary number of application classes. For instance, to model a flock of sheep a `Sheep` class describing an individual sheep would simply be combined by means of multiple inheritance with a topology class `Group` to a `FlockMember` class thus describing the relations between an individual sheep and a (structured) group of sheep (fig. 3).

The members of the group do not need to be of the same exact type but need to be derived from the same member class. Thus full polymorphism is retained and each group member can potentially redefine all those multicast methods that have been declared as virtual methods.

3 Language-level Support

Figure 3 gives an impression how collective operations are supported on language-level. The `FlockMember` class inherits a `Sheep` class as well as the topology class `Group` to indicate that `FlockMember` is a group member class being able to process collective operations on groups of `Sheep`. Such classes can specify both group operations that are executed as collective operations on the whole group as well as (remote) methods that refer to the specific members only.

```
class FlockMember: public Sheep,
                   public Group</*!dual!*/FlockMember>
{
    public:
        // simple methods
        int rank() { return number; }

        // pattern (a) multicast without reduction
        void sayHello(String who) /*! multi !*/
            { cout << "Hello " << who << endl; }

        // pattern (b)(c) multicast with manual reduction
        void count(/*!dual!*/Counter result) /*! multi !*/
            { result.increment(); }

        // pattern (c) multicast with automatic reduction
        int countMembers() /*! multi:+ !*/
            { return 1; }
        ...
    private:
        int number;
}/*!dual!*/;
```

Fig. 3. A Member Class

Thus the `rank()` method of the `FlockMember` class refers to one member object only and returns a group related identifier of the object within the group. In contrast the `sayHello()`, `count()` and `countMembers()` methods are group related. While `sayHello()` causes all members to print a message to the screen, `count()` and `countMembers()` both calculate the number of objects in the group (fig. 3) in parallel.

Multicast methods can have arbitrary input parameters including proxy objects that allow the passing of global references to other objects. Proxy objects are identified by a /*!dual!*/ annotation. Thus the count() method passes a /*!dual!*/Counter proxy to all group members who increment the counter object remotely. This is a simple example of a manual global reduction. In contrast the countMembers() method automatically reduces the multiple results in parallel. The syntactical notation /*! multi:+ !*/ tells the stub generator that countMembers() is a multicast method and that its results should be reduced using the standard +-operator.

Multicasts are propagated in parallel according to the strategy defined by the topology class from which the member class is derived. The same holds for the reduction of the results. Thus the countMembers() method first is propagated in parallel to all members of the group according to the group's propagation strategy defined by the Group topology class. Then the multiple results are reduced in parallel according to the specified reduction method in the same order in which the multicast has been propagated. Provided this order is mathematically correct (refer also to section 4.2), any *binary* user defined reduction method that is compatible to the type of the result can be applied. The reduction method is applied iteratively when necessary. This makes the generated code independent from n-ary topologies producing n > 2 subresults.

Instead of specifying the reduction method statically we could also have used a more dynamic scheme for reduction. Passing pointers to reduction methods to each multicast is even more flexible. This would have worked well for SPMD-programs but it is otherwise cumbersome to implement because the same method does not have the same address on each node. Furthermore, since also virtual methods of the group member class or any virtual method of the base classes can be specified for reduction, we think we found a scheme that is still very flexible and can effectively be implemented in any distributed or parallel system. This scheme can also be applied easily to implement parallel servers that are used by multiple clients.

3.1 Applying Object Groups

Whole object groups can be conveniently created through a GroupOf template class as depicted in figure 4. The declaration of flock as an instance of class

```
// create a flock of sheep with NumSheep members
GroupOf</*!dual!*/FlockMember> flock (NumSheep,"Sheep");

/*!dual!*/Counter counter;               // a counter Proxy

flock.sayHello("jon");                   // pattern (a)
flock.count(counter);                    // pattern (b)(c)
int sum = flock.countMembers();          // pattern (c)
```

Fig. 4. Applying Object Groups

GroupOf</*!dual!*/FlockMember> creates NumSheep instances of FlockMember at the parallel "Sheep" service and joins them to a distributed object group that now can be collectively addressed. Collective operations are issued to the group by simply invoking a method on the flock object representing the group. Thus the common client-server paradigm is extended to parallel services too. In principle group objects can also be passed as parameters to other (collective) methods thus granting access to existing object groups.

```
template<class T> class GroupOf: public T {
public:
    GroupOf(int nodes, const char* name):
        T(DomainAtNode(name, 0))
    {
        for (register int i = 1; i < nodes; i++) {
            join(new T (DomainAtNode(name,i)));
        }
    }
};
```

Fig. 5. The GroupOf Template

In figure 5 a simplified implementation for the GroupOf class is given. The template parameter T denotes the type of the group members. Each instance of T is created at the location denoted by DomainAtNode() that resolves the service identified by "name" at node i. Services are implemented by active server objects that are capable to handle object instantiations as well as invocations on existing objects. The code for the servers is automatically generated. Many clients can create objects at the same server and address these objects through proxy objects.

3.2 Topology Classes

If an application or distributed system service is already structured according to a specific topology, the existing communication infrastructure can easily be reused for global operations too. More specialized topology classes can be optimized for specific physical network topologies such as meshes or hierarchical crossbar networks. Tree topologies typically provide the best performance results in networks that offer roughly the same communication speed between any two nodes.

The methods of a topology class are used by automatically generated server stubs to control the order of the spreading of multicast messages and the number of results to expect. Topology classes (fig. 6) only need to define two methods used to control the spreading of multicast messages and the collection of the results.

The members()-method returns the number of group members or subgroup members in case of recursively defined topologies like lists or trees as shown in

```
class Topology {
    public:
        virtual int members()=0;
        GlobalReference* iterate ()=0;
};
```

Fig. 6. Interface of a Topology Class

(fig. 7). The `iterate()`-method is used to determine the next group member to pass a multicast message to. This iterator iterates all group members and returns NIL when all members are iterated (fig. 7). Subsequent calls to the iterator start the iteration from scratch.

```
template<class T> class Group: public Topology {
public:
    void join(/*!in!*/ T* obj)
    {
        if (used == slots) {
            children[balance++%slots]->join(obj);
        } else {
            children[used++] = new T(*obj);
        }
    }

    int members() { return used; }
    GlobalReference* iterate()
    {
        T* result;
        if (iterated < members()) {
            result = children[iterated++];
        } else {
            result = NIL;
            iterated = 0;
        }
        return referenceOf(result);
    }
    ...
private:
    int used, iterated, slots, balance, ...;
    T** children;
}/*!dual!*/;
```

Fig. 7. A Group Class

Although only the `iterate()` and `members()` methods are needed by the automatically generated stubs, topology classes typically need to provide additional methods to add and remove members dynamically. The `GroupOf` class (fig. 5) e.g. requires all topology classes to provide at least a `join()` method as shown in figure 7 to build topologies dynamically. The `/*!in!*/` annotation of `obj` indicates that the argument `obj` needs to be passed by value. The `obj` parameter is typically a pointer to a proxy to construct a distributed tree topology. When `obj` is locally inserted a copy of the proxy `obj` is made and locally stored. Otherwise `obj` is passed on to existing (remote) children applying `join` recursively.

Furthermore, user-level naming schemes that allow to address individual members of a multicast group by their group specific identifier instead of their system-level `GlobalReference` identifier often need to be supported. This is especially true for regular array topologies used for data parallel numerical computations that require indexing operations within n-dimensional index spaces. Such user-level naming can directly be supported, when the `GroupOf` class is slightly changed such that the remote references of the group members (the proxies) are stored in a table that maps user-level names to system-level references.

4 Multicast Processing

Multicast processing is entirely bound to the server's stub. Client stubs are not even aware that a specific method is executed as a multicast operation. In the most general case a server stub processes a multicast message as follows:

1. initialize a (local) barrier for the number of expected results.
2. allocate buffer space for the results.
3. spread the multicast message concurrently to other group members in the order defined by the `iterate()` method of the topology class
4. execute the desired method on the local object instance
5. wait for all results to arrive
6. reduce the results (including local result) according to the reduction method in the order previously defined by the `iterate()` method
7. reply reduced result to client

In case of simple multicast operations without results, steps 1-2 as well as 5-7 are omitted. The following subsections discuss the various handling schemes in detail and present several stub code examples.

4.1 Asynchronous and Triggered Multicasts

Simple multicasts without return values are handled in two ways. Either the multicast message is passed asynchronously to all (sub) group members or the multicast message is passed synchronously and each member sends an acknowledge before it starts to spread the message further and execute the desired method. We call the latter *triggered* multicasts and indicate such methods with a `/*! multi, trigger !*/` annotation. Triggered multicasts are non-buffered. When multiple casts are issued to a group a client only succeeds as far as the a previous cast has been completed. Thus multiple casts are pipelined into the group. Therefore a single source for multicast messages cannot overflow the network with numerous asynchronously sent messages.

```
void FlockMember::decode (ObjectRequest* msg ...)
{
    switch(msg->code) {
    ...
        case 1:
        {
            register FlockMember_sayHello_in1 *in;
            register GlobalReference *next;
            in = (FlockMember_sayHello_in1*) msg;
            while (next = iterate()) {
                next->pass(in->code, in, sizeof(*in));
            }
            this->sayHello(in->who);
        }
        break;
    ...
    }
}
```

Fig. 8. Server Stub for Asynchronous Multicasts

In contrast asynchronous multicasts are buffered and cause exceptions when the buffer space is exhausted. In figure 8 the server stub for an asynchronous multicast is depicted. The `decode()` method is executed when the basic ROI system propagates a message to an object. The stub generator generates a statically typed message format for each method and dependent on the method's numerical code the generic `ObjectRequest` message is casted to the specific format such as `FlockMember_sayHello_in1` in this example (fig. 8). Then the members of the group are iterated applying the `iterate()` method of the `Group` class. The message is propagated by a `pass()` method executed in the `while` loop. After propagation the requested multicast method is executed locally.

In case of triggered multicasts an acknowledge is sent to the invoker before the loop is executed and instead of the asynchronous `pass()` method a synchronous `invoke()` method is applied within the loop.

4.2 Global Reductions

Global reductions are implemented in two ways. Since proxy objects of other objects can be passed as arguments to multicasts each member can access these objects by reference and invoke methods on the remote object that will later hold the reduced result. This scheme can be extended with implicit lazy synchronization mechanisms if the passed object is a kind of future[6]. The benefit here is that the client can go on with computations while the multicast is processed. The disadvantage of this scheme is that the reduction of the result is serialized and can cost too much when either big multicast groups are involved (see section 5) or reduction methods are very complex.

Multicasts with parallel global reductions are more efficient in this case. Figure 9 shows a slightly simplified server stub for the `countMembers()` method with automatic parallel reduction. This is an example for a multicast that applies all

```
void FlockMember::decode (ObjectRequest* msg, DomainReference* invoker)
{ ...
        case 3:
        {
            register GlobalReference *next;
            register FlockMember_countMembers_in3 *in;
            register FlockMember_countMembers_out3 *out;
            FlockMember_countMembers_out3 reply_msg;
            in = (FlockMember_countMembers_in3*) msg;
            out = (FlockMember_countMembers_out3*) &reply_msg;
/* 1. */    Barrier results(members());
/* 2. */    FlockMember_countMembers_out3* replies =
                    new FlockMember_countMembers_out3 [members()];

/* 3. */    { for (register int i=0; next = iterate(); i++)
                next->delegate(results, in->code,
                in, sizeof(*in),
                replies[i], sizeof(replies[i])); }

/* 4. */    out->result = this->countMembers();

/* 5. */    results.wait();
/* 6. */    { for (register int i=0;  i<members(); i++)
                out->result = out->result+replies[i].result; }

/* 7. */    invoker->done(out, sizeof(*out));
            delete replies;
        }
        break;
... }
```

Fig. 9. Multicast with Reduction

steps described in section 4. After preparation of input and result messages a local barrier instance is initialized for the number of expected results (step 1 in fig. 9). Then buffers for all expected results from the (sub)group members are allocated (step 2). In the next step the message is casted to all (sub)group members by means of a `delegate()` method (step 3). This method sends a request and collects the result. When the result is available the passed barrier object is signaled. Our current implementation is straightforward and creates for each `delegate()` call a thread[1] that invokes the method synchronously. Thus multiple threads that invoke multiple methods concurrently are created implicitly in the spreading loop. After spreading the multicast message the `countMembers()` method is invoked on the local member (step 4.). Meanwhile the other group members handle the multicast concurrently.

When the computation of the local result is finished the flow of control is blocked by the `wait()` method until all results are available. We first wait for all results to arrive because we want to retain strictly the order in which results are processed. This is a strong requirement of numerical simulations. If we would compute the results in the order of arrival the result of a multicast would be

---

[1] Actually threads are managed efficiently in a pool and will be reused multiple times.

non-deterministic and two runs of the same program would not yield the same result in case of floating point operations executed in non-deterministic order. The order of reduction is the same as the order of message propagation and thus strictly defined by the user-controllable `iterate()` method of the topology class.

When all results are available the (partial) reduction of the results is computed in a `for` loop (step 6 in fig. 9) applying the specified reduction operation

(the standard +-operator in this case). Finally the local result is replied (step 7) either to the client if the respective object was the group leader or to a parent group member that recursively continues the reduction until the group leader is reached.

5 Performance

All performance measurements have been performed on a 16 node Manna computer. The Manna is a scalable architecture based on a 50MHz i860 processor and is interconnected via a hierarchical crossbar network with 50MBytes/s links. Up to 16 nodes can communicate directly over a single crossbar switch. Therefore the communication latency between any two nodes is the same in the single cluster we used.

Our measurements where performed using the `FlockMember` class (fig. 3) introduced in section 3. The `Group` class topology is based on balanced n-ary trees. We have chosen a balanced binary tree for our measurements. This topology is not optimal as several researchers[4, 11] already have pointed out. Since we only want to show that our basic runtime mechanisms are scalable and efficient, we can neglect this fact. The objects of the group have been mapped to the computing nodes using a straightforward modulo 16 wrap around strategy.

We measured the five different types of multicasts that have been described in the previous sections:

1. an empty *triggered* multicast (`ping()`)
2. an empty *asynchronous* multicast (`pingAsync()`)
3. a *triggered* multicast with sequential reduction (`count()`)
4. an *asynchronous* multicast with sequential reduction (`countAsync()`)
5. a multicast with parallel global reduction (`countMembers()`)

All multicast messages fit into a single packet of 64 Bytes. Bigger messages are automatically transmitted using the bulk data transfer primitives supplied by the PEACE OS family. When messages do not fit into a single packet the startup-time for a single remote invocation is roughly doubled and additionally the transfer costs of the passed data need to be added on the basis of 40MBytes/sec (end-to-end on the Manna network).

Each multicast method has been issued 1000 times to achieve higher accuracy. Note that with two exceptions there have not been any pipeline effects caused by the fact that the same multicast was issued in a loop. The reason is that after each multicast with a result we always synchronized on the availability of the

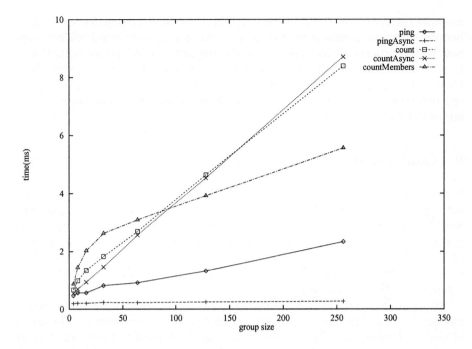

Fig. 10. Multicast performance for a balanced binary tree

result before we continued the loop. The asynchronous `pingAsync()` multicast possibly shows such effects, because we naturally could not synchronize exactly on the end of the asynchronous multicast. Therefore we issued only one multicast with one global reduction at the end to make sure, that all operations really had been performed. Consequently the measurement reveals some of the additional effects achievable with multiply overlapped multicasts issued to the same group of objects. The same holds for the triggered `ping()` multicast, which is implicitly flow-controlled but allows modest pipelining.

Figure 10 shows the results in detail. The `pingAsync()` method is extremely fast in the vicinity of 250 microseconds due to the pipelining reasons explained above. Notably the triggered `ping()` method that requires an additional acknowledge for each propagation step is still quite fast in relation to the amount of messages needed.

When looking at the methods that include reduction operations, the `countMembers()` method with parallel global reduction outperforms the other operations when object groups become larger. Initially the higher synchronization overhead for the parallel reduction is clearly the bottleneck for smaller groups. The `count()` and `countAsync()` methods with simple manual reduction via a proxy object passed as an argument in the multicast message only have minimal overhead. Therefore these methods are very efficient when applied to relatively small object groups. As expected, the sequential reduction of the multiple results becomes the bottleneck when the object groups become larger because all group members send their local result to a single result object, that now has to process all messages. Here the implicit parallel global reduction is significantly faster and the graph shows a logarithmic shape as expected for a binary propagation tree. In contrast the curves for the methods with sequential reduction quickly convert to a linear shape revealing the bottleneck of the sequential reduction.

6 Conclusion

Multicast communication is neither new nor did we invent the notion of multicast topologies as such. Nevertheless we have shown that it is possible to implement flexible and efficient multicast communication applying relatively simple stub generator techniques in combination with the standard inheritance mechanisms of C++. The generator is designed to be used in conjunction with off the shelf C++ compilers and basically provides proxy based language-level support for computations in a global object space.

We found an easy way to provide effective control over group communication mechanisms through a simple iterator method that controls both the spreading of multicast messages as well as the reduction of the results. Therefore existing topologies can be reused for globally synchronized operations with modest effort. Furthermore, very specialized multicast patterns with minimal message exchanges can be implemented to solve e.g. problems like global page-invalidations in VSM systems.

Our collective operation patterns have much in common with those provided by the Ocore[8] programming language. Multicast groups have a similar functionality as *Communities* in Ocore or *Collections* in pC++[1]. The major difference is here, that e.g. communities have regular array topologies and message propagation cannot be controlled directly on application level using language-level constructs. Furthermore Ocore is based on a SPMD model like many other approaches targeting data-parallel programming such as pC++ or PROMOTER[10]. In PROMOTER topologies can be defined as constraints over n-dimensional index spaces. These topologies are both used to describe distributed variables (such as matrices) as well as coordinated communications between those variables. The PROMOTER compiler extracts application patterns from topology definitions to optimize both alignment and mapping. Data parallel expressions and statements simultaneously involve parallel computations, (implicit) communications as well as reductions. Typical numerical problems can therefore be expressed with a minimum of programming effort on a very high level of abstraction.

We target more general parallel and distributed systems and like to support parallel servers that can be used by multiple applications. This scenario is specifically useful to increase multi-user performance when developers or scientists work on related topics and use parallel computers as domain-specific parallel servers for common parallel calculations. Therefore we do not initially support complex data-parallel expressions based on index spaces. Nevertheless, topology classes allowing index based addressing can also be implemented with modest effort. Furthermore, many aspects of data-parallel programming are also covered by our approach as far as medium to coarse grained parallelism is concerned. Fine grained parallel computing like in ICC++[2] is possible in principle, but we do not expect satisfactory performance results since we don't apply aggressive compiler optimizations like in ICC++.

In the future we are considering to incorporate meta-object technology similar to MPC++[7] or Europa C++[14] in our design. Especially frameworks like Europa C++ could offer a reasonable basis for general purpose parallel and distributed systems since it does not rely on complex compiler technology but is still open to multiple computing paradigms.

References

1. Francois Bordin, Peter Beckman, Dennis Gannon, Srinivas Narayana, and Shelby X. Yang. Distributed pC++: Basic Ideas for an Object Parallel Language. *Scientific Programming*, 2(3), Fall 1993.
2. A. Chien, U.S. Reddy, J.Plevyak, and J. Dolby. ICC++ – A C++ Dialect for High Performance Parallel Computing. In *Proceedings of the 2nd JSSST International Symposium on Object Technologies for Advanced Software, ISOTAS'96*, Kanazawa, Japan, March 1996. Springer.
3. J. Cordsen. Basing Virtually Shared Memory on a Family of Consistency Models. In *Proceedings of the IPPS Workshop on Support for Large-Scale Shared Memory Architectures*, pages 58–72, Cancun, Mexico, April 26th, 1994.
4. J. Cordsen, H. W. Pohl, and W. Schröder-Preikschat. Performance Considerations in Software Multicasts. In *Proceedings of the 11th ACM International Conference on Supercomputing (ICS '97)*, pages 213–220. ACM Inc., July 1997.
5. Yvon Gourhant and Marc Shapiro. FOG/C++: a Fragmented–Object Generator. In *C++ Conference*, pages 63–74, San Francisco, CA (USA), April 1990. Usenix.
6. R. H. Jr. Halstead. Multilisp: A Language for Concurrent Symbolic Computation. *ACM Transactions on Programming Languages and Systems*, 7(4), 1985.
7. Yutaka Ishikawa, Atsushi Hori, Mitsuhisa Sato, Motohiko Matsuda, Jörg Nolte, Hiroshi Tezuka, Hiroki Konaka, Munenori Maeda, and Kazuto Kubota. Design and Implementation of Metalevel Architecture in C++ – MPC++ Approach –. In *Reflection '96*, 1996.
8. H. Konaka, M. Maeda, Y. Ishikawa, T. Tomokiyo, and A. Hori. Community in Massively Parallel Object-based Language Ocore. In *Proc. Intl. EUROSIM Conf. Massively Parallel Processing Applications and Development*, pages 305–312. Elsevier Science B.V., 1994.
9. Henry M. Levy and Ewan D. Tempero. Modules, Objects, and Distributed Programming: Issues in RPC and Remote Object Invocation. *Software—Practice and Experience*, 21(1):77–90, January 1991.
10. Besch M., Bi H., Enskonatus P., Heber G., and Wilhelmi M. High-Level Data Parallel Programming in PROMOTER. In *Proc. Second International Workschop on High-level Parallel Programming Models and Supportive Environments HIPS'97*, Geneva, Switzerland, April 1997. IEEE-CS Press.

11. J.-Y. L. Park, H.-A. Choi, N. Nupairoj, and L.M. Ni. Construction of Optimal Multicast Trees Based on the Parameterized Communication Model. Technical Report MSU-CPS-ACS-109, Michigan State University, 1996.
12. W. Schröder-Preikschat. *The Logical Design of Parallel Operating Systems*. Prentice Hall International, 1994. ISBN 0-13-183369-3.
13. Marc Shapiro, Yvon Gourhant, Sabine Habert, Laurence Mosseri, Michel Ruffin, and Céline Valot. SOS: An object–oriented operating system – assessment and perspectives. *Computing Systems*, 2(4):287–338, December 1989.
14. The Europa WG. EUROPA Parallel C++ Specification. Technical report, http://www.dcs.kcl.ac.uk/EUROPA, 1997.

SCALA: A Framework for Performance Evaluation of Scalable Computing

Xian-He Sun[1], Mario Pantano[2], Thomas Fahringer[3], and Zhaohua Zhan[1]

[1] Department of Computer Science, Louisiana State University, LA 70803-4020
{sun,zzhan}@csc.lsu.edu
[2] Department of Computer Science, University of Illinois, Urbana, IL 61801
pantano@cs.uiuc.edu
[3] Institute for Software Technology and Parallel Systems
University of Vienna, Liechtensteinstr. 22 1090, Vienna, Austria
tf@par.univie.ac.at

Abstract. Conventional performance environments are based on profiling and event instrumentation. It becomes problematic as parallel systems scale to hundreds of nodes and beyond. A framework of developing an integrated performance modeling and prediction system, SCALability Analyzer (SCALA), is presented in this study. In contrast to existing performance tools, the program performance model generated by SCALA is based on scalability analysis. SCALA assumes the availability of modern compiler technology, adopts statistical methodologies, and has the support of browser interface. These technologies, together with a new approach of scalability analysis, enable SCALA to provide the user with a higher and more intuitive level of performance analysis. A prototype SCALA system has been implemented. Initial experimental results show that SCALA is unique in its ability of revealing the scaling properties of a computing system.

1 Introduction

Although rapid advances in highly scalable multiprocessing systems are bringing petaflops performance within grasp, software infrastructure for massive parallelism simply has not kept pace. Modern compiler technology has been developed to reduce the burden of parallel programming through automatic program restructuring. Among others, notable examples include the Vienna HPF+ compiler [2] and the Fortran D compilation system [6]. There are many ways to parallelize an application, however, and the relative performance of different parallelizations vary with problem size and system ensemble size. Parallelization of programs is far from being fully automated optimization. Predicting performance variation and connecting dynamic behavior to original source code are two major challenges facing researchers in the field [1]. Emerging architectures such as petaflop computers and next generation networks demand sophisticated performance environments to provide the user with useful information on the scaling behavior and to identify possible improvements.

In this study we introduce the framework and initial implementation of the SCALability Analyzer (SCALA) system, a performance system designed for scalable computers. The goal of SCALA is to provide an integrated, concrete and robust performance system for performance modeling and prediction of real value to the high performance computing community. Distinguished from other existing performance tools, the program performance model generated by SCALA is based on scalability analysis [15, 19]. SCALA is a system which correlates static, dynamic and symbolic information to reveal the scaling properties of parallel programs and machines and for predicting their performance in a scalable computing environment. SCALA serves three purposes: predict performance, support performance debugging and program restructuring, and estimate the influence of hardware variations. Symbolic and scalability analysis are integrated into advanced compiler systems to explore knowledge from the compiler and to guide the program restructuring process for optimal performance. A graphical user interface is developed to visualize the performance variations of different program implementations with different hardware resources. Users can choose the performance metrics and parameters they are interested in.

SCALA supports range comparison [15], a new concept which allows the performance of different code-machine combinations to be compared over a range of system and problem sizes. SCALA is an experimental system which is designed to explore the plausibility and credibility of new techniques, and to use them collectively to bring performance analysis environments to their most advanced level. In this paper, we present the design and initial implementation of SCALA. Section 2 presents the structure and primary components of SCALA. Section 3 and 4 gives a short description of scalability analysis and data analysis respectively. Implementation results are reported in Section 5. Finally, Section 6 gives the summary and conclusion.

2 The Design of SCALA

The design of SCALA is based on the integration of symbolic cost models and dynamic metrics collected at run-time. The resulting code-machine model is then analyzed to predict performance for different software (i.e. program parameters) and architecture characteristics variations. To accomplice this process, SCALA combines performance prediction techniques, data analysis and scalability analysis with modern compiler techniques. An important feature of SCALA is its capability to relate performance information back to the source code in order to guide the user and the compiler in the selection of transformation and optimization strategies. Although SCALA is an integrated system designed to predict the scaling behavior of parallel applications, it supports standard performance information via data analysis and visualization techniques and can be used without compiler support.

The general structure of SCALA comprises several modules which combined together provide a robust environment for advanced performance analysis. SCALA has three input sources of informations (compiler, measurement system, and user) and two outputs (compiler and user). The compiler provides symbolic models of the program and information on the code regions measured. Dynamic performance information collected are provided by the measurement system in terms of tracefiles. The user can also be involved in the process supplying specific information on software parameters and on the characteristics of the input data. The three input sources can be used collectively for the best result or can be used separately based on physical constraints and target applications. On the output side, performance indices can be given directly to the compiler annotating syntax tree and call graph of the application so that the compiler can automatically select the most appropriate transformations to be applied to each code region. Detailed information on the predicted behavioral characteristics of the application will be given to the user by employing visualization techniques. Figure 1 depicts the design structure of SCALA which is composed of the following modules:

- *Data Management:* This module implements data filtering and reduction techniques for extrapolating detailed information on each code region measured. The input of this module is a tracefile in a specific format while its output is given to the analysis modules and to the interface for visualization.
- *Statistical Analysis:* The statistical component of SCALA determines code and/or machine effect, finds the correlation between different program phases, identifies the scaling behavior of "critical-segments", and provides statistical data for the user interface.
- *Symbolic Analysis:* The symbolic module gathers cost models for alternative code variants computed at compile-time. Furthermore, by employing dynamic data, symbolic analysis evaluates the resulting expressions for specific code-machine combinationes.
- *Scalability Analysis:* The scalability analysis module is the most important part of SCALA. The development of this module is based on highly sophisticated analytical results on scalability analysis as described in Section 3. The scalability module implements novel algorithms for predicting performance in terms of execution time and scalability of a code-machine combination.
- *Model Generator and Database:* The performance model generator automatically determines the model of system execution. The database stores previously measured information which will be used to find appropriate values of symbolic constants and statistic data for performance prediction.
- *Interface:* The measured and predicted results can be visualized via an user-friendly graphical interface. The appearance of the interface can be justified by the user and the visualization can be "zoomed" in and out for a specific computing phase and component. The results also can be propagated through the interface to the compiler for automatic optimization purposes.

While each module of SCALA has its specific design goal, they are closely interconnected for a cooperative analysis process that provides the user advanced performance information for a thorough investigation of the application itself as well as of the interaction with the underlying parallel machine. The other important point in this design is the complete integration of SCALA in a restructuring

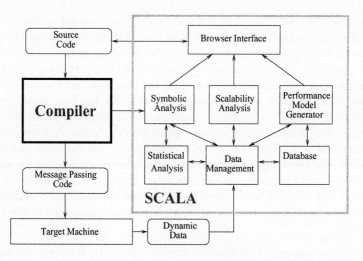

Fig. 1. Structure of SCALA.

compiler. This integration allows the tool to directly gather detailed information on the source code being transformed. As a consequence, performance predictions can be linked back to the user source code providing suggestions for code optimizations. For the sake of brevity, only a brief discussion of the scalability analysis and data analysis modules is given in the following sections. The implementation of these modules, as well as others, can be found in Section 5 where the implementation issues are addressed.

3 Scalability Analysis

Several scalability metrics and prediction methodologies have been proposed [7, 9, 13, 8]. The SCALA's approach is based on *isospeed* scalability [19]: A code-machine combination (CMC) is scalable if the achieved average unit speed can remain constant when the number of processors scales up, provided the problem size can be increased with the system size.

The performance index considered in isospeed scalability, therefore, is *speed* which is defined as work divided by time. *Average unit speed* (or, average speed, in short) is the achieved speed of the given computing system divided by p, the number of processors. The speed represents a quantity that ideally would increase linearly with the system size. Average speed is a measure of efficiency of the the underlying computing system.

For a large class of CMCs, the average speed can be maintained by increasing the problem size. The necessary problem size increase varies with code-machine combinations. This variation provides a quantitative measurement of scalability. Let W be the amount of work of a code when p processors are employed in a machine, and let W' be the amount of work of the code when $p' > p$ processors are employed to maintain the average speed, then the scalability from system size p to system size p' of the code-machine combination is:

$$\psi(p, p') = \frac{p' \cdot W}{p \cdot W'} \qquad (1)$$

Where the work W' is determined by the isospeed constraint.

In addition to measuring and computing scalability, the prediction of scalability and the relation between scalability and execution time have been well studied. Theoretical and experimental results show that scalability combined with initial execution time can provide good performance prediction, in terms of execution times.

New concepts of crossing-point analysis and range comparison are introduced. Crossing-point analysis finds fast/slow performance crossing points of parallel programs and machines. In contrast with execution time which is measured for a particular pair of problem and system size, range comparison compares performance over a wide range of ensemble and problem size via scalability and crossing-point analysis. Only two most relevant theoretical results are given here. More results can be found in [14–16].

Result 1: *If a code-machine combination is faster at the initial state and has a better scalability than that of other code-machine combinations, then it will remain superior over the scalable range.*

Range comparison becomes more challenging when the initial faster CMC has a smaller scalability. When the system ensemble size scales up, an originally faster code with smaller scalability can become slower than a code that has a better scalability. Finding the fast/slow crossing point is critical for achieving optimal performance. Definition 1 gives an informal definition of crossing point based on the isospeed scalability. The fomral definition can be found in [15, 16].

Definition 1. *(scaled crossing point) Let code-machine combination 1 have execution time t, scalability $\Phi(p, p')$, and scaled problem size W'. Let code-machine combination 2 have execution time T, scalability $\Psi(p, p')$, and scaled problem size W^*. If $t_p(W) = \alpha T_p(W)$ at the initial ensemble size p and problem size W for some $\alpha >$, then p' is a crossing point of code-machine combinations 1 and 2 if and only if*

$$\frac{\Phi(p, p')}{\Psi(p, p')} > \alpha. \qquad (2)$$

In fact, as given by [15, 16], when $\Phi(p, p') > \alpha \Psi(p, p')$ we have $t_{p'}(W') < T_{p'}(W^*)$. Notice that since $\alpha > 1$ combination 2 has a smaller execution time at the initial state $t_p(W) > T_p(W)$. This fast/slow changing in execution time gives the meaning of crossing point.

Result 2: *Assume code-machine combination 1 has a larger execution time than code-machine combination 2 at the initial state, then the scaled ensemble size p' is not a scaled crossing point if and only if combination 1 has a larger execution time than that of combination 2 for solving any scaled problem $W^\dagger$*

such that $W^\dagger$ is between W' and W^* at p', where W' and W^* is the scaled problem size of combination 1 and combination 2 respectively.

Result 2 gives the necessary condition for range comparison of scalable computing: p' is not a crossing point of p if and only if the fast/slow relation of the codes does not change for any scaled problem size within the scalable range of the two compared code-machine combinations. Based on this theoretical finding, with the comparison of scalability, we can predict the relative performance of codes over a range of problem sizes and machine sizes. This special property of scalability comparison is practically valuable. Restructuring compilation, or programming optimization in general, in a large sense is to compare different programming options and choses the best option available. Result 2 provides a foundation for the database component of SCALA. Figure 2 gives the range comparison algorithm in terms of finding the smallest scaled crossing point via scalability comparison. While being not listed here, an alternative range comparison algorithm that finds the smallest scaled crossing point via execution time comparison can be found in [15]. In general, there could have more than one scaled crossing point over the consideration range for a given pair of CMCs. These two algorithms can be used iteratively to find successive scaled crossing points.

Assumption of the Algorithm: Assume code-machine combinations 1 and 2 have execution time $t(p, W)$ and $T(p, W)$ respectively, and $t(p, W) = \alpha T(p, W)$ at the initial state, where $\alpha > 1$.

Objective of the Algorithm: Find the superior range of combination 2 starting at the ensemble size p

Range Comparison
Begin
  $p' = p$;
  Repeat
    $p' = p' + 1$;
    Compute the scalability of combination 1 $\Phi(p, p')$;
    Compute the scalability of combination 2 $\Psi(p, p')$;
  Until($\Phi(p, p') > \alpha\Psi(p, p')$ **or** $p' =$ the limit of ensemble size)
  If $\Phi(p, p') > \alpha\Psi(p, p')$ **then**
    p' is the smallest scaled crossing point;
    Combination 2 is superior at any ensemble size $p^\dagger$, $p \leq p^\dagger < p'$;
  Else
    Combination 2 is superior at any ensemble size $p^\dagger$, $p \leq p^\dagger \leq p'$
  End{If}
End{Range Comparison }

Fig. 2. Range Comparison Via Crossing Point Analysis

4 Data Management and Statistical Analysis

The dynamic properties of an application can be investigated by measuring the run time of the code and the interactions with the underlying parallel system. Tracing the occurring run time events can lead to large amount of performance information which needs to be filtered and reduced to provide the users with a compact and easy-to-interpret set of indices describing the characteristics of the application. The level of detail of this set depends mainly on the scope of the performance analysis. As an example, if the communication behavior of the application is under investigation, a performance analysis tool should be able to extrapolate from measured data only the most significant information regarding the communication activities such as total time spent transmitting a message, amount of data transmitted, and communication protocols used in each transmission. For modeling purpose, a performance analysis system must analyze the data and combine dynamic indices with symbolic models.

SCALA is an advanced tool that combines both static performance information and measurement/analysis to investigate the scaling behavior of an application varying architecture characteristics and application parameters. The data collected by the measurement system need to be analyzed and reduced so that they can be sufficiently detailed for evaluating the static model generated at compile time. This evaluation is the base for applying scalability analysis techniques and predicting the scaling behavior of the code-machine combination. It is important to note that this prediction need not to be quantitatively exact, only qualitatively with a magnitude of errors as high as 10-20 percent.

The *data management* module of SCALA implements several data-reduction techniques that can be applied in isolation or in combination for data management. Filtering is the simplest form of data reduction to eliminate data that do not meet specified performance analysis criteria retaining only the pertinent characteristics of performance data. Among the range of statistical techniques that can be applied to a data set, mean, standard deviation, percentiles, minimum, and maximum are the most common and provide a preliminary insight on the application performance, while reducing the amount of data. Other statistical methods such as the analysis of distribution and variability provide more detailed information on the dynamic behavior of parallel program. For example, the analysis of distribution gives information on how the communication and computation are distributed across processors and the user or compiler can apply transformations to improve the performance. Moreover, the coefficients of variation of communication time and computation time are good metrics to express the "goodness" of work and communication distributions across processors. In most cases, the data measured need to be scaled in a common interval so that further statistical techniques can be successfully applied. Timing indices also can be scaled in a more significant metric for the analysis such as from seconds to microseconds or nanoseconds.

All these statistics can be applied to the entire data set as well as to a subset of data such as a reduced number of processors or sliding windows of the most recent data. However, to manage large amount of data and large number of performance indices, advanced statistical techniques need to be applied. Clustering is one of the most common techniques used to reduce the amount of performance data [3, 12] and consequently reduce the number of performance indices needed to identify sources of performance losses. Statistical clustering is a multidimensional technique useful to automatically identify equivalence classes. For example, clustering classifies processors such that all processors in one cluster have similar behavior.

Unlike the data management module, the statistical analysis module is designed to handle more sophisticate statistical tasks. For instance, some computing phases are more sensitive than others. The performance of some program segments or run-time parameters may change dramatically at some hardware/software thresholds. In addition to conventional statistical methods such as synthetic perturbation screening and curve fitting [10], new methods are also introduced to SCALA. Noticeably, based on factorial analysis, we have proposed a methodology for examining latency hiding techniques of hierarchical memory systems [17]. This methodology consists of four levels of evaluation. For a set of codes and a set of machines, we first determine the effect of code, machine, and code-machine interaction on performance respectively. If a main effect exists, then, in the second level of evaluation, the code and/or machine is classified based on certain criteria to determine the cause of the effect. The first two levels of evaluation are based on average performances over the ranges of problem size of current interest. The last two levels of evaluation determine the performance variation when problem sizes scale up. Level three evaluation is the scalability evaluation of the underlying memory system for a given code. Level four evaluation conducts detailed examination on component contributions toward the performance variation. The combination of the four levels of evaluation makes the proposed methodology adaptive and more appropriate for advanced memory systems than existing methodologies.

Complicated with architectural techniques and speed and capacity of different storage devices, how to measure and compare the performance variation of modern memory systems remains a research issue. The statistical method is introduced to provide a relative comparison for memory systems. It can be extended to other contexts where the scalability results given in Section 3 are not feasible.

5 Prototype Implementation

The implementation of SCALA involves several steps, which include the development of each component module, integration of these components, testing, modification and enhancement of each component as well as of the integrated system. As a collective effort, we have conducted the implementation of each component concurrently and have successfully tested a prototype of the SCALA system.

5.1 Integrated Range Comparison for Restructuring Compilation

Restructuring a program can be seen as an iterative process in which a parallel program is transformed at each iteration. The performance of the current parallel program is analyzed and predicted at each iteration. Then, based on the performance result, the next restructuring transformation is selected for improving the performance of the current parallel program. Integrating performance analysis with a restructuring system is critical to support automatic performance tuning in the iterative restructuring process. As a collaborative effort between Louisiana State University and University of Vienna, a first prototype of SCALA has been developed and tested under the *VFCS* as an integrated performance tool for revealing scaling behavior of simple and structured codes [18]. In this framework, the static performance data have been provided by P^3T [5], a performance estimator that assists users in performance tuning of programs at compile time. The integration of P^3T into VFCS enables the user to exploit considerable knowledge about the compiler's program structural analysis information and provides performance estimates for code restructuring. The SCALA system has been integrated to predict the scaling behavior of the program by using static information provided by P^3T and the dynamic performance information collected at run-time. In particular, the Data Management Module of SCALA, filters the raw performance data obtained, executing the code on a parallel machine and provides selected metrics which, combined with the static models of P^3T allows an accurate scalability prediction of the algorithm-machine combination. Naturally, the interactions between SCALA and the estimator P^3T is an interactive process where at each iteration the scalability analysis model is refined to converge at the most precise prediction. For this purpose we have designed and implemented within SCALA an iterative algorithm to automatically predict the scalability of code-machine combinations and compare the performance over a range of algorithm implementation (range comparison). As a result, the automatic range comparison is computed within a data-parallel compilation system. Figure 3 shows the predicted crossing point for solving a scientific application with two different data-distributions in VFCS environment on an Intel iPSC/860 machine. We can see that the column distribution becomes superior when the number of processors is greater than or equal to 8. The superiority of 2-D block-block distribution ends when the system size equals 4. The prediction has been confirmed by experimental testing.

In order to verify Result 2 we measured both codes with $n = 33$ and $n = 50$, respectively. In accordance with Result 2 before $p = 8$ there is no performance crossing point, and $p = 8$ may correspond to a crossing point for a given problem size in the scalable range. The results are shown in Figure 3.b. Since the results obtained with this implementation demonstrate the feasibility and high potential of range comparison in performance evaluation, we extended SCALA for its integration in the newly developed Vienna High Performance Fortran Compiler (VFC) [2]. VFC is a source-to-source compilation system from HPF codes to explicit message passing codes. The parallelization strategies of VFC are very gen-

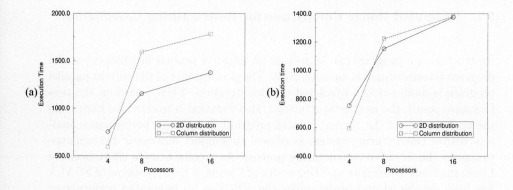

Fig. 3. Scaled crossing point (a) and crossing point (b) for the Jacobi with n=20

eral and applicable to programs with dynamically allocated or distributed arrays. VFC combines compile-time parallelization strategies with run-time analysis.

As a first step for investigating the performance of parallelized codes, we embedded in the VFC the SCALA Instrumentation System (SIS). SIS [4] is a tool that allows the automatic instrumentation, via command line options, of various code regions such as subroutines, loops, independent loops and any arbitrary sequence of executable statements. SIS has been extensively tested in connection with the Medea tool [3] for investigating the performance of applications with irregular data accesses. SCALA is being integrated within new compilation framework. The data management module accepts tracefiles obtained executing the code instrumented with SIS and computes a set of statistical metrics for each code region measured. Here, P^3T is substituted by a new tool that will provide SCALA with symbolic expressions of the code regions. The value of the parameters used in the symbolic expression, however, may vary when system or problem size increase. For this purpose in SCALA we compute and save run-time information in SCALA database (see Figure 1) for specific code-machine combinations. For example, on Cray T3E and QSW (Quadrics Supercomputers World) scalable computing system, the communication cost is given as $T_{comm} = \rho + \beta \cdot D$. D is the length of a single message, ρ, β are values describing the startup time and the time required for the transmission of a single byte. These machine dependent values are either supplied by the manufactures or can be experimentally measured. In order to reproduce the interference between the messages we compute ρ and β for each specific communication pattern used in the application. Table 1 presents the communication values computed for the communication patterns 1-D and 2-D Torus used in the Jacobi implementation. As shown in Table 2, SCALA accurately predicts the scalability of the algorithm and therefore range comparison. More detailed information can be found in [11].

Table 1. Communication Models (in μs)

	Cray T3E				QWS CS-2			
	1-d		2-d		1-d		2-d	
	short	long	short	long	short	long	short	long
ρ	35.6	71.2	40.3	77.5	220.3	297.2	150.4	192.9
β	$2.0 \cdot 10^{-2}$	$12.5 \cdot 10^{-3}$	$2.1 \cdot 10^{-2}$	$14.1 \cdot 10^{-3}$	$5.5 \cdot 10^{-2}$	$3.5 \cdot 10^{-2}$	$6.7 \cdot 10^{-2}$	$4.4 \cdot 10^{-2}$

Table 2. Bidirectional 2-d torus on CRAY T3E: Predicted (P) and measured (M) scalability

	Jacobi 2-d											
$\psi(p,p')$	$p'=8, n=1519$			$p'=16, N=2186$			$p'=32, N=3124$			$p'=64, N=4372$		
	P	M	%	P	M	%	P	M	%	P	M	%
p=4	0.908	0.899	1.0	0.877	0.871	0.6	0.859	0.856	0.3	0.877	0.871	0.6
p=8	1.000	1.000	-	0.966	0.969	0.3	0.946	0.952	0.6	0.966	0.969	0.3
p=16	-	-	-	1.000	1.000	-	0.979	0.982	0.3	1.000	1.000	-
p=32	-	-	-	-	-	-	1.000	1.000	-	1.021	1.018	0.2

5.2 Browser Interface

A Java 3D visualization environment is developed for the SCALA interface. This visualization tool is designed based on a client-server model. The server side mainly provides data services. At startup, the server accesses certain data files, creates data objects, and waits for the client to call. Currently the server supports two data file formats: Self-Defining Data Format (SDDF) [12] and a simple text format used by SCALA. The driving force behind the client/server approach is to increase accessibility, as most users may not have SDDF installed on their machines and may like to use SCALA over the net. Moreover, current distribution of SDDF only supports a few of computing platforms. Partitioning our tool into separate objects based on their services makes it possible to deploy the data server in a central site and the client objects anywhere on the Internet. The client is implemented in pure Java and the data servant is implemented as a CORBA compliant object so that it can be accessed by clients coded in other programming languages.

For effective visualization, a good environment should allow users to interact with graphical objects and reconfigure the interface. The basic configuration supported by the Java interface includes changing scale and color, and selecting specific performance metrics and views. At the run-time, the user can rotate, translate, and zoom the graphical objects. Automatic rotation similar to the animation is also an enabled feature. The graphical objects are built using Java 2D/3D API, which are parts of the JavaMedia suite of APIs [20].

We have implemented a custom canvas which serves as a drawing area for graphical objects. The paint() method in canvas class is extended to the full capability of drawing a complete 2D performance graph. The classes for 2D performance graphics are generic in the sense that the resulting graph depends only on the performance data. Besides 2D performance graphical objects, Java 3D is used to build three-dimension graphical objects. Java 3D uses the scene-graph based programming model in which individual application graphical elements are constructed as separate objects and connected together into a tree-like structure. It gives us high-level constructs for creating and manipulating 3D geometry and for constructing the structures used in rendering that geometry. As in developing 2D graphical classes, we analyze the 3D performance data and build several generic 3D graphical classes. A terrain grid class based on IndexedQuadArray can be used to describe the performance surface. The rotation, translation, and zooming by mouse are supported.

Figure 4 is the kiviat view which shows the variation of *cpi* (sequential computing capacity), communication latence, and parallel speed when problem size increases. Figure 5 shows the execution time as a function of work and number of processors on a given machine. These views and more are currently supported by the JAVA 3D visualization tool. While this Java visualization environment is developed for SCALA, its components are loosely coupled with other SCALA components and can be easily integrated into other performance tools. It separates data objects from GUI objects and is portable and reconfigurable.

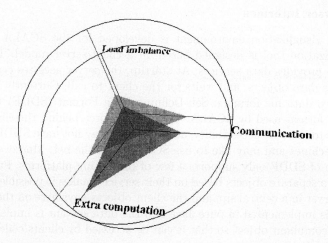

Fig. 4. *cpi*, Communication Latency, and parallel speed as a Function of Problem Size

6 Conclusion

The purpose of the SCALA project is to improve the state of the art in performance analysis of parallel codes by extending the current methodologies and to

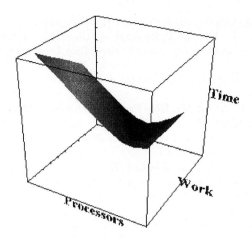

Fig. 5. Execution Time as a Function of Work and Number of Processors

develop a prototype system of real value for the high performance computing community. SCALA combines symbolic and static analysis of the parallelized code with dynamic performance data. The static and symbolic analysis are provided by the restructuring compiler, while dynamic data are collected by executing the parallel code on a small number of processors of the target machine. A performance model is then generated for predicting the scaling behavior of the code with varying input data size and machine characteristics. This approach to performance prediction and analysis provides the user with detailed information regarding the behavior of the parallelized code and the influences of the underlying architecture.

References

1. ADVE, V. S., CRUMMEY, J. M., ANDERSON, M., KENNEDY, K., WANG, J.-C., AND REED, D. A. Integrating compilation and performance analysis for data parallel programs. In *Proc. of Workshop on Debugging and Performance Tuning for Parallel Computing Systems* (Jan. 1996).
2. BENKNER, S., SANJARI, K., SIPKOVA, V., AND VELKOV, B. Parallelizing irregular applications with vienna HPF+ compiler VFC. In *HPCN Europe* (April 1998), Lecture Notes in Computer Science, Springer-Verlag.
3. CALZAROSSA, M., MASSARI, L., MERLO, A., PANTANO, M., AND TESSERA, D. Medea: A tool for workload characterization of parallel systems. *IEEE Parallel & Distributed Technology Winter* (1995), 72–80.
4. CALZAROSSA, M., MASSARI, L., MERLO, A., PANTANO, M., AND TESSERA, D. Integration of a compilation system and a performance tool: The HPF+ approach. In *LNCS - HPCN98* (Amsterdam, NL, 1998).
5. FAHRINGER, T. *Automatic Performance Prediction of Parallel Programs*. Kluwer Academic Publishers, Boston, USA, ISBN 0-7923-9708-8, March 1996.

6. FOX, G., HIRANANDANI, S., KENNEDY, K., KOELBEL, C., KREMER, U., TSENG, C., AND WU, M. Fortran D language specification. Technical Report,COMP TR90079, Department of Computer Science, Rice University, Mar. 1991.
7. GUSTAFSON, J., MONTRY, G., AND BENNER, R. Development of parallel methods for a 1024-processor hypercube. *SIAM J. of Sci. and Stat. Computing 9*, 4 (July 1988), 609–638.
8. HWANG, K., AND XU, Z. *Scalable Parallel Computing*. McGraw-Hill WCB, 1998.
9. KUMAR, V., GRAMA, A., GUPTA, A., AND KARYPIS, G. *Introduction to Parallel Computing, Design and Analysis of Algorithms*. The Benjamin/Cummings Publishing Company, Inc., 1994.
10. LYON, G., SNELICK, R., AND KACKER, R. Synthetic-perturbation tuning of mimd programs. *Journal of Supercomputing 8*, 1 (1994), 5–8.
11. NOELLE, M., PANTANO, M., AND SUN, X.-H. Communication overhead: Prediction and its influence on scalability. In *Proc. the International Conference on Parallel and Distributed Processing Techniques and Applications* (July 1998).
12. REED, D., AYDT, R., MADHYASTHA, T., NOE, R., SHIELDS, K., AND SCHWARTZ, B. An overview of the Pablo performance analysis environment. In *Technical Report*. UIUCCS, Nov. 1992.
13. SAHNI, S., AND THANVANTRI, V. Performance metrics: Keeping the focus on runtime. *IEEE Parallel & Distributed Technology* (Spring 1996), 43–56.
14. SUN, X.-H. The relation of scalability and execution time. In *Proc. of the International Parallel Processing Symposium'96* (April 1996).
15. SUN, X.-H. Scalability versus execution time in scalable systems. TR-97-003 (Revised May 1998), Louisiana State Uiversity, Department of Computer Science, 1997.
16. SUN, X.-H. Performance range comparison via crossing point analysis. In *Lecture Notes in Computer Science, No 1388*, J. Rolim, Ed. Springer, March 1998. Parallel and Distributed Processing.
17. SUN, X.-H., HE, D., CAMERON, K., AND LUO, Y. A factorial performance evaluation for hierarchical memory systems. In *Proc. of the IEEE Int'l Parallel Processing Symposium* (Apr. 1999).
18. SUN, X.-H., PANTANO, M., AND FAHRINGER, T. Integrated range comparison for data-parallel compilation systems. *IEEE Transactions on Parallel and Distributed Systems* (accepted to appear, 1999).
19. SUN, X.-H., AND ROVER, D. Scalability of parallel algorithm-machine combinations. *IEEE Transactions on Parallel and Distributed Systems* (June 1994), 599–613.
20. SUN MICROSYSTEMS INC. Java 3D API specification. http://java.sun.com/products/java-media/3D, 1998.

Recursive Individually Distributed Object

Z. George Mou

Applied Physics Laboratory
Johns Hopkins University, USA
george.mou@jhuapl.edu

Abstract. Distributed Objects (DO) as defined by OMG's CORBA architecture provide a model for object-oriented parallel distributed computing. The parallelism in this model however is limited in that the distribution refers to the mappings of different objects to different hosts, and not to the distribution of any individual object. We propose in this paper an alternative model called Individually Distributed Object (IDO) which allows a single large object to be distributed over a network, thus providing a high level interface for the exploitation of parallelism inside the computation of each object which was left out of the distributed objects model. Moreover, we propose a set of functionally orthogonal operations for the objects which allow the objects to be recursively divided, combined, and communicate over recursively divided address space. Programming by divide-and-conquer is therefore effectively supported under this framework. The Recursive Individually Distributed Object (RIDO) has been adopted as the primary parallel programming model in the Brokered Objects for Ragged-network Gigaflops (BORG) project at the Applied Physics Laboratory of Johns Hopkins University, and applied to large-scale real-world problems.

1 Introduction

A methodology to decompose a computational task into a sequence of concurrent subtasks is the essence of any parallel programming model. Function-level parallelism takes advantage of the partial ordering of the dependencies whereas data-level parallelism [13] exploits the parallelism inside a single operation over a large data set. The OMG's *distributed objects* should be viewed as the result of function-level decomposition. Each object is responsible for a set of operations over a given collection of data, and the messages between objects realize the dependencies between functions. It should be noted that in the distributed-objects (DO) model of the OMG's Common Object Request Broker Architecture

(CORBA) [12, 26] or Microsoft's DCOM [25], each object is assigned to one and only one host. Therefore, a collection of distributed objects is distributed, but any object in the collection is not.

The *individually distributed object* (IDO) model we propose is not at all the same as one of the distributed objects in the OMG's model. A single data field of an IDO can be distributed over multiple host processors, and each operation over the distributed data field(s) is implemented by the collective operations of the hosts. The IDO implementation preserves the object semantics to give the programmers the illusion of working with a single object. The distributed nature of the object therefore is hidden from the programmers. IDO's are particularly well suited for computations with operations over large data sets as often found in scientific computing.

A data structure is *recursive* if it can be recursively divided and combined. The *list* in the language Lisp and other functional programming languages is an example of a recursive data structure. Recursive data structures can undoubtedly contribute to an elegant high-level programming style, particularly for programming by divide-and-conquer. The importance of divide-and-conquer as a programming paradigm has been discussed in many textbooks on algorithms [1, 6] or numerical analysis [9]. The significance and potential of divide-and-conquer in parallel processing have also been emphasized over the years by Smith [27], Preparata [24], Mou [23, 21, 20, 22], Axford [2], Dongarra [8], Gorlatch and Christian [11], and many others. Unfortunately, balanced recursive data structures (i.e., those with balanced division trees) are unsupported in most, if not all, existing major programming languages. Research efforts to address the problem have been made over the years, including David Wise's block arrays [30], Mou's recursive array [21], and Misra's Powerlist [19]. A *recursive individually distributed object* (RIDO) proposed here is an IDO that supports divide, combine, and communication operations.

In response to the demand for high-level and high-performance computing, the Applied Physics Laboratory of Johns Hopkins University has launched a research project under the name of the Brokered Objects for Ragged-network Gigaflops (BORG). The RIDO model is adopted as the foundation of the BORG programming system. The BORG system architecture (Figure 5) is described in Section 5.

In Section 2, we introduce the notion of distributed function and individually distributed object, and show why and how functions and objects can be distributed. In Section 3, we focus on recursive array objects, and their functionally orthogonal operations. Higher order programming frameworks that work with RIDO's are discussed in Section 4 with some examples. A systematic method for the implementation of the RIDO's and the BORG architecture are described in Section 5. An application to the problem of radar scattering over ocean surface is presented as a test case in Section 6.

2 Distributed Function and Individually Distributed Object

Let $f : X \to Y$ be a function from X to Y, and H a partition of the domain X and Y, such that $Hx = [x_1, \ldots, x_m]$ and $Hy = [y_1, \ldots, y_m]$ for any $x \in X$ and $y \in Y$. We say $F : X^m \to Y^m$ is a *distribution* of f if and only if $f = H^{-1}FH$ (Figure 2).

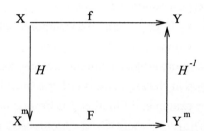

Fig. 1. The relationship between a function f and its distribution F with respect to a partition of H.

For example, let f be the function that maps an array of integers to their squares, then its distribution is simply the construct $(f, \ldots f)$ which applies to a tuple of arrays of the same length $(x_1, \ldots, x_m)$, and returns $(fx_1, \ldots, fx_m)$. As another example, let us consider the inner product ip of two vectors. Its distributed form is $IP = \sum_{i=1}^{m}(ip, ip, \ldots, ip)$. The second example shows that inter-block[1] dependencies or communications is generally a component of the distribution of a function except for trivial cases.

Formally, a *communication* $C : S \to S \times S$ defines a directed graph of the form (S, A), where an arc $a = (x, y) \in A$ if and only if $C(x) = (x, y)$ for $x, y \in S$.

---

[1] Blocks here refer to the blocks generated by a partition.

Note that an arc (x, y) implies that the element x receives a value (message) from element y. Given a partition H over S, a communication C is said to be *intra-* (*inter-*)*block* if none (all) of the arcs in C crosses the boundaries between the blocks. A communication which is neither intra- nor inter block is *mixed*. Obviously, a mixed communication can always be decomposed into inter- and intra-block communications.

A function $L : X \to Y$ is *local* with respect to a partition H over X, if there there exists a tuple of functions $(L_1, \ldots, L_k)$ such that

$$L\ x = (L_1, \ldots, L_k)\ H\ x$$
$$= (L_1, \ldots, L_k)\ (x_1, \ldots, x_k)$$
$$= (L_1\ x_1, \ldots, L_k\ x_k)$$

A function is *strongly local* if it is local with respect to the maximum partition by which each element of X is a block. Note that a strong local function is necessarily local with respect to any partition.

It is easy to see that any function can be decomposed into a sequence of communications and local functions. Since any mixed communications can in turn be decomposed into inter and intra-block communications, we can re-group the sequence so that each group in the sequence is either a inter-block communication or a local operation. For example, if function f is decomposed into the sequence of $f = \cdots L_i C_i L_{(i-1)} C_{(i-1)} \cdots$ where C_j is decomposed into C_{j1} followed by C_{j2} which are respectively its inter- and intra-communication components. The above sequence can be rewritten as

$$f = \cdots L_i C_{i2} C_{i1} L_{(i-1)} C_{(i-1)2} C_{(i-1)1} \cdots$$
$$= \cdots (L_i C_{i2}) C_{i1} (L_{(i-1)} C_{(i-1)2}) C_{(i-1)1} \cdots$$

We conclude that any function f over a given set has a distribution F that is a sequence of inter-block communications and local operations.

An object is no more than a collection of functions over a collection of data sets. Given that its data sets are all partitioned, we say an object is individually distributed if all its functions are distributed with respect to the partitions of its data set. The above discussion therefore shows that an object can always be individually distributed. Observe that the components of the local operations can be done in parallel, and data movement implied by the arcs in the inter-block communication can be performed in parallel by definition.

3 Array Recursive Individually Distributed Object

There are three types of *functionally orthogonal operations* that can be performed over data: manipulations of the structure, data exchanges or communications between the nodes, and mappings of the nodes to new values. There is no reason to think any type of the operations is more important than others and none can replace another since they are mutually *orthogonal* in functionality. However, there is apparently an operation orientation in the design of most programming languages. As a result, communication functions are treated as second class citizens buried in other operations. Structural operations (e.g., array division in merge-sort) are often absent, and can only be "simulated". Functional programming languages provide structural operations (e.g., car, cdr, and cons in Lisp) for its main data structure of lists. However, the division is fixed and unbalanced, and the communication operation is still implicit.

According to many textbooks on algorithms [1,6,9], divide-and-conquer is the single most important design paradigm for both sequential and parallel algorithms. The ubiquity of divide-and-conquer indeed cannot be overstated. Moreover, many algorithms and paradigms not always considered as divide-and-conquer can be naturally modeled by divide-and-conquer, including greedy algorithms, dynamic programming, recursive doubling [28], and cyclic reduction [14, 16]. All of the above share a need for balanced recursive data structures, which are supported in few, if any, programming languages.

We will focus our discussion on the *array recursive individually distributed object* (ARIDO), which is an individually distributed array that can recursively divide and combine itself, communicate over a recursively divided address space, and perform local operations in parallel to all the entries.

In the following we list the operations of an ARIDO array, which are supported in the BORG Recursive Individually Distributed Object Library (BRIDOL). We use A[s:k:e] to denote the sub-array of A that contains entries of index s, s+k, s+2k, and so on, but excluding any entries with index greater than e.

dlr: divides an array of even length into its left and right sub-arrays:

$$dlr\ A = (A_l, A_r),\ \text{where}\ A_l = A[0:1:n/2-1],\ A_r = A[n/2:1:n-1]$$

deo: divides an array of even length into its even and odd sub-arrays:

$$deo\ A = (A_e, A_o),\ \text{where}\ A_e = A[0:2:n-1],\ A_o = A[1:2:n-1]$$

dht: head-tail division of an array:

$$dht\ A = (A_h, A_t),\ \text{where}\ A_h = A[0:1:1],\ A_t = A[1,\ 1,\ n-1]$$

clr: Inverse of dlr[2].
ceo: Inverse of deo.
cht: Inverse of dht.

Higher dimensional arrays can divide along one or more of its dimensions. For example, [dlr, dlr] divides a matrix into four sub-blocks as expected.

A communication can be either *intra-array* or *inter-array*. In the case of inter-array communication, we assume the two arrays have the same shape.

corr: *Correspondent* inter-array communication which sends the value of every entry to the entry of the same index in the other array, i.e.,

$$corr\ (A, B) = (A', B'),\ \text{where}\ A'(i) = (A(i),\ B(i)),\ B'(i) = (B(i),\ A(i))$$

mirr: *Mirror-image* inter-array communication which sends the value of every entry to the entry with "mirror-image" index in the other array, i.e.,

$$mirr\ (A, B) = (A', B'),\ \text{where}\ A'(i) = (A(i),\ B(n-1-i)),$$
$$B'(i) = (B(i),\ A(n-1-i))$$

br: *Broadcast* inter-array communication which sends the value of last entry to all entries in the other array, i.e.,

$$br\ (A, B) = (A', B'),\ \text{where}\ A'(i) = (A(i),\ B(n-1)),\ B'(i) = (B(i),\ A(n-1))$$

Similar to br, the broadcast inter-array communication *brr* sends values only from left to right, *brl* sends values only from right to left, and (*br k*) sends the *k*-th last value instead of the very last value.

pred: *Predecessor* intra-array communication, i.e.,

$$pred\ (A) = A'\ \text{where}\ A'(i) = (A(i),\ A(i-1)),\ (i \neq 0)$$

succ: *Successor* intra-array communication, which is symmetric to predecessor.

---

[2] The inverse of the division in fact is not unique [23]. We here refer to the minimum inverse, i.e., the one with the smallest domain. The same applies to the inverses of other divide functions.

By overloading, an intra-array communication can turn into an inter-array communication when supplied with two, instead of one, array as arguments, and vice versa. For example, mirr(A) can be used to reverse a one-dimensional array[3].

For an array of type T, an unary (binary) operator $\oplus$ for T can be converted to a *local operator* for array of type T before (after) the communication by prefixing the operator with a "!". For example

$$! + .corr\ ([1,2],[3,4]) = ([4,6],[4,6])$$
$$!second.mirr\ ([1,2,3,4]) = !second\ ([(1,4),(2,3),(3,2),(4,1)]) = [4,3,2,1]$$

where "." denotes *function composition*, and second(a,b) = b.

4 The Divide-and-Conquer Frameworks

In object-oriented terminology, *frameworks* are equal to components plus patterns [15]. A framework is therefore an abstract pattern that can be instantiated by binding to given components. This is not an idea inherently tied to object technology, and is referred to as *high order constructs* [3, 23] in functional programming. Divide-and-conquer is a powerful framework which can be instantiated to a wide range of algorithms.

The BORG library currently provides the following DC constructs:

Pre-DC: Premorphism [21, 20, 22] divide-and-conquer, defined as

   f = PreDC (d, c, g, p, f_b)
      where
      f A = if p A then return f_b A
            else c.(map f).g.d A

where g is the *pre-function* applied before the recursion and is a composition of communication followed by a local operation, d and c respectively the divide and combine operators, p the base predicate, f_b a base function for f, and (map f)$(x_1, x_2) = (fx_1, fx_2)$.

Post-DC: Postmorphism divide-and-conquer, defined as

---
[3] More precisely, the reversed array after the communication only exists as the second element of the pairs generated by the communication.

f = PostDC (d, c, h, p, f_b)
    where
        f A = if P A then return f_b A,
                else c.h. (map f). d A

where h is the *post-function* applied after the recursion and is a composition of communication followed by a local operation.

RD: Recursive doubling which is divide-and-conquer with even-odd division, and intra-array communication, defined as

f = RD ($\oplus$)
    where
        f A = if (SizeOne? A) return A
                else ceo.(map f).deo.!$\oplus$.pred.A

For example, the function scan $\oplus$ (associative binary operator) defined by

$$\text{scan} \oplus A = A', \text{ where } A(i) = \oplus_{j=0}^{i} A(j) \text{ for } i = 0 \text{ to } |A| - 1$$

can be programmed as a postmorphism or a recurisve doubling:

$$\text{scan} \oplus = \text{PostDC (dlr, clr, } \oplus.\text{brr, SizeOne?, id)} = \text{RD } (\oplus)$$

where *id* is the identity function. The PostDC algorithm[4] is illustrated in Figure 2.

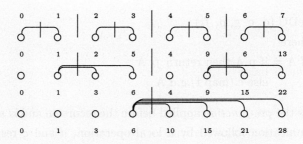

Fig. 2. Scan with broadcast communication

---

[4] This algorithm is simple but not most efficient, and can be improved by replacing its *broadcast* communication by *correspondent* communication [21]. In terms of total number of operations, a cyclic reduction algorithm is more efficient [17].

A divide-and-conquer with other divide-and-conquer as its components is a *higher order divide-and-conquer*. Figure 3 gives the *monotonic sort* as an example of second order divide-and-conquer, which resembles but differs from Batcher's bitonic sort [4].

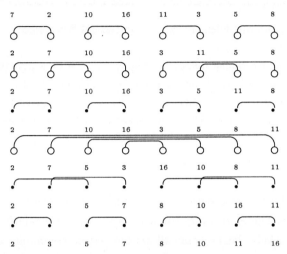

Fig. 3. Monotonic sort is a second order divide-and-conquer algorithm where the top level is PostDC with mirror-image communication (circles), and the nested level is PreDC with correspondent communication (solid dots).

5 Implementation

An ARIDO array is distributed over multiple processors. By default, the most significant $\log p$ bits of an index are used to map the entry with the index to a host processor.

Observe that the left-right (even-odd) division recursively partitions an array from left to right (right to left) by the remaining most (least) significant bit. The implementation of balanced division (left-right and even-odd) uses a "pointer" of type integer, whose only non-zero bit points to the current dividing bit. Each divide operation thus reduces to a shift of the pointer by one-bit. Left-right division shifts to right, even odd to left (Figure 4). Note that an implementation with this approach involves no copying or sending of sub-arrays as often suggested in literature (e.g., [18]), and takes $O(1)$ time. A combine operation is simply to shift towards the opposite direction from that of the corresponding division[5].

---

[5] If you noticed that the odd-even division is the same as left-right combine, you are right. In fact, an isomorphism can be formalized which allows DC algorithms with

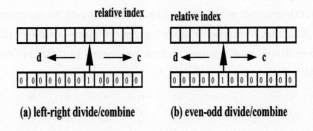

Fig. 4. Division and combine operations implemented with a pointer. The bottom integer in binary contains exactly one non-zero bit, which points to the current dividing bit. For the left-right (even-odd) division, the relative index is given by the bits to the right (left) of the dividing bits. This example shows an array of size $8k$ divided at the eighth level.

To support high-level programming, the indexing of the recursively divided arrays is relative. Therefore, every sub-array starts with the relative index of zero at all recursion levels. This scheme allows the system communication operators to be specified with very simple functions. For example, the *correspondent* communication is specified by the identity function[6]. The implementation however needs to translate the relative indices to global indices. It turned out that the relative index is given precisely by the remaining bits of the index at each level of the division (Figure 4). To translate a given relative index to its global index, all we need to do is to concatenate the remaining bits with the masked bits. If the communication is inter-array, the dividing bit needs to be complemented. It follows that the translation time between relative and global indices is $O(1)$, the actual time to carry out the communication depends on the communication pattern and the underlying communication network platform.

A clear distinction should be drawn between the communications over recursive distributed arrays and the inter-processor communications over the platforms. The former is defined over the logic domain, and the latter the physical domain. The programmers should only be concerned with the former, and only the system implementors are concerned with the latter. Also, observe that

---

the two types of divisions to be transformed into each other under certain conditions. This topic will be left for future discussion elsewhere.

[6] The seemingly complicated butterfly communication used in FFT and many other applications is therefore no more than the recursive applications of communications generated by the identity function.

communications over recursive arrays may and may not involve inter-processor communications. Let n be the size of a recursive array, p the number of processors, then inter-processor communications only occur for $\log(p)$ times in a first-order DC algorithm. In the remaining $\log(n/p)$ times, communications over the recursive arrays are mapped to memory accesses. Whether a given recursive array communication maps to inter-processor communication can be easily determined by a comparison of the dividing pointer and the bits used to index the processors. If the pointer points to a bit used to index the processors, the communication will be translated into an inter-processor communication.

Unlike Beowulf systems [5], BORG's implementation of inter-processor communication is based on the Common Object Request Broker Architecture (CORBA). The benefits of this approach in comparison with MPI or PVM include:

- An open environment that is neutral to different programming languages, operating systems, and hardware platforms.
- Host location transparency.
- Portability.
- Capability of dynamically adding computational resources available to the system through the Internet.
- Higher level system programming.

The BORG hardware platform currently consists of 17 Pentium boxes, of which 12 has single 300MHz Pentium II processor, and the rest dual 233Mhz Pentium II processors. Each machine is equipped with a Fast Ethernet card with 100 Mb/s bandwidth. All the nodes are connected by a high bandwidth Ethernet switch with an aggregate bandwidth of 3 Gb/s. The inter-processor communications are implemented with CORBA ORB as illustrated in Figure 5. BORG allows a mixture of Unix and NT workstations to work together. We have made a distinction between the individually distributed object proposed here and the distributed objects defined in literature. Our CORBA implementation indeed means that an individually distributed object is implemented, not surprisingly, with a set of distributed objects. Our preliminary experience also showed that a communication transaction using CORBA takes time in the order of millisecond, which we consider acceptable relative to the performance of MPI or PVM.

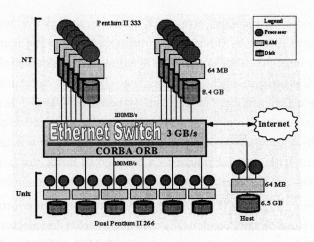

Fig. 5. The Architecture of the Brokered Objects for Ragged-network Gigaflops (BORG).

6 Applications

A number of real world applications are being developed with the recursive individually distributed object model on BORG system. In this section, we will focus on one – the ocean surface radar scattering problem [7]. A multi-grid method for the problem iteratively solves the following equation on each grid level:

$$ZX^{n+1} = ZX^n + \alpha(C - ZX^n)$$

where n is the iteration number, α a free parameter, Z a preconditioner which can be approximated by a banded matrix. To solve for X^{n-1}, the inverse of Z is computed, and a matrix-vector multiplication (ZX^n) needs to be performed.

We use the N-merge algorithm illustrated in Figure 6 to solve the banded linear system (with a bandwidth of hundreds). The division used is left-right along the rows, and the communication used is broadcast. This algorithm is a generalization of the algorithm presented in [10], which is similar but different from Wang's algorithm in [29].

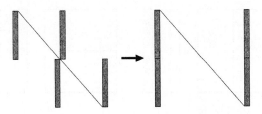

Fig. 6. N-Merge algorithm for banded linear systems merges two N-shaped non-zero bands of height m to one N-shape of height 2m, and thus eliminates all the non-diagonal elements in $\log(n)$ parallel steps for a system of size n.

The matrix-vector multiplication can be solved by DC in two ways. The first divides the matrix along the column dimension, the other the row dimension (Figure 7). Although the total operation counts of the two algorithms are the same, the first is more memory efficient than the second. The performance of the two algorithms can be quite different depending on the layout method of two dimensional arrays of the underlying programming language.

Let the sample space size be n, the total number of operations for the above two key procedures is in the order of $O(n^2)$. Since this is to be performed for many iterations with complex double precision floating point numbers, the total number of operation for a moderate value of $n = 10^4$ is in the order of Teraflops (10^{12} floating point operations per second) for just one (of the hundreds of) realizations in the Monte Carlo simulation.

The problem was solved on Sun Ultra workstations. Since the memory is not large enough to hold the entire preconditioner matrix, the preconditioner was recalculated at each realization. The total time taken is in the order of 36 hours for one realization. The BORG version of the algorithm is currently being implemented, and the initial performance data suggests that the same computation can be completed in a couple of hours with 16 Pentium II processors.

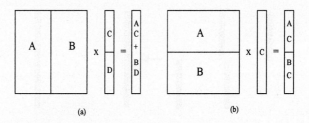

Fig. 7. Two DC algorithms for matrix vector multiplication. In (a), the recursive application results in column vector and scalar operations, and the final result is obtained by row reduction over all rows. On the other hand, (b) leads to many inner products between the matrix rows and the input vector.

7 Conclusion

The recursive individually distributed object model can effectively exploit the parallelism in the computation internal to an object, and support high level programming by divide-and-conquer. The implementation of the model under the BORG project at the Applied Physics Laboratory is built on top of the Java language and the CORBA architecture. The resultant system is neutral to programming languages, operating systems, and hardware platforms. A high performance scientific computing library with individually distributed object model is being developed, and the preliminary results have shown that the model can be effectively applied to large-scale real problems with acceptable parallel performance.

References

1. A. V. Aho, J. E. Hopcroft, and J. D. Ullman. *The Design and Analysis of Computer Algorithms*. Addision-Wesley, 1974.
2. Tom Axford. *The Divide-and-Conquer Paradigm as a Basis for Parallel Language Design*, pages 26–65. John Wiley and Sons, Inc., 1992. edited by Kronsjo, Lydia and Shumsheruddin, Dean.
3. J. Backus. Can programming be liberated from the von Neumann style? A functional style and its algebra of programs. *Communications of the ACM*, 21(8):613–641, August 1978.
4. K. E. Batcher. Sorting networks and their applications. In *Proceedings of the AFIPS Spring Joint Computer Conference*, volume 32, pages 307–314, 1968.
5. CESDIS. Beowulf project at cesdis. *http://www.beowulf.org*.

6. T. H. Corman, C. E. Leiserson, and R. L. Rivest. *Introduction to Algorithms*. MIT Press, 1990.
7. D.J.Donohue, H-C. Ku, and D.R.Thomson. Application of iterative moment-method solution to ocean surface radar scattering. *IEEE Transactions on Antennas and Propagation*, 46(1):121–132, January 1998.
8. Jack Dongarra. Linear library liberaries for high-performance computer: A personal perspective. *IEEE Parallel & Distributed Technology*, (Premiere Issue):17–24, February 1993.
9. W. H. Press et al. *Numerical Recipes – The Art of Scientific Computing*. Cambridge University Press, 1986.
10. Y. Wang et al. The 3dp: A processor architecture for three dimensional applications. *Computer, IEEE*, 25(1):25–38, January 1992.
11. Sergei Gorlatch and Christian Lengauer. Parallelization of divide-and-conquer in the bird-meertens formalizm. Technical Report MIP-9315, Universitat Passau, Dec. 1993.
12. Object Management Group, editor. *CORBA Specifications*. www.omg.org.
13. W. D. Hillis and G. L. Steele Jr. Data parallel algorithms. *Communications of the ACM*, 29(12):1170–1183, December 1986.
14. R. W. Hockney. The potential calculation and some applications. *Methods in Computational Physics*, 9:135 – 211, 1970.
15. Ralph E. Johnson. Frameworks = components + patterns. *Communications of the ACM*, 40(10):39–42, October 1997.
16. S. Lennart Johnsson. Cyclic reduction on a binary tree. *Computer Physics Communication 37 (1985)*, pages 195–203, 1985.
17. R. E. Ladner and M. J. Fischer. Parallel prefix computation. *Journal of the ACM*, 27(4):831–838, 1980.
18. V. Lo and et. al. Mapping divide-and-conquer algorithms to parallel architectures. Technical Report CIS-TR-89-19, Dept. of Computer and Info. Science, University of Oregon, January 1990.
19. Jayadev Misra. Powerlist: A structure for parallel recursion. *ACM Transactions on Programming Languages and Systems*, 16(6), November 1994.
20. Z. G. Mou. Divacon: A parallel language for scientific computing based on divide-and-conquer. In *Proceedings of the Third Symposium on the Frontiers of Massively Parallel Computation*, pages 451–461. IEEE Computer Society Press, October 1990.
21. Z. G. Mou. *A Formal Model for Divide-and-Conquer and Its Parallel Realization*. PhD thesis, Yale University, May 1990.
22. Z. G. Mou. The elements, structure, and taxonomy of divide-and-conquer. In *Theory and Proactice of Higher-Order Parallel Programming, Dagstul Seminar Report 169*, pages 13–14, February 1997.

23. Z. G. Mou and Paul Hudak. An algebraic model for divide-and-conquer algorithms and its parallelism. *The Journal of Supercomputing*, 2(3):257–278, November 1988.
24. F. P. Preparata and J. Vuillemin. The cube-connected cycles: A versatile network for parallel computation. *Communications of the ACM*, 8(5):300–309, May 1981.
25. Roger Sessions. *Com and Dcom: Microsofts's Vision of Distributed Objects*. John Willey and Sons, 1987.
26. Jon Siegel. Omg overview: Corba and the oma in enterprise computing. *Communications of the ACM*, 41(10):37–43, October 1998.
27. D. R. Smith. Applications of a strategy for designing divide-and-conquer algorithms. *Science of Computer Programming*, 8:213–229, 1987.
28. Harold S. Stone. Parallel tridiagonal equation solvers. *ACM Transactions on Mathematical Software*, 1(4), 12 1975.
29. H. H. Wang. A parallel method for tridiagonal equations. *ACM Transactions on Mathematical Software*, 7(2):170–183, June 1981.
30. David S. Wise. Matrix algebra and applicative programming. In *Functional Programming Languages and Computer Architecture, Lecture Notes in Computer Science*, pages 134–153. Springer, 1987.

The MuSE System: A Flexible Combination of On-Stack Execution and Work-Stealing

Markus Leberecht

Institut für Informatik
Lehrstuhl für Rechnertechnik und Rechnerorganisation (LRR-TUM)
Technische Universität München, Germany
Markus.Leberecht@in.tum.de

Abstract. Executing subordinate activities by pushing return addresses on the stack is the most efficient working mode for sequential programs. It is supported by all current processors, yet in most cases is inappropriate for parallel execution of indepented threads of control. This paper describes an approach of dynamically switching between efficient on-stack execution of sequential threads and off-stack spawning of parallel activities. The presented method allows to incorporate work-stealing into the scheduler, letting the system profit from its near-to-optimal load-balancing properties.

Keywords: Multithreading, Work-Stealing, Scalable Coherent Interface, PC Cluster

1 Introduction

For sequential execution, the most efficient method of calling a subroutine is to push the return address onto the program stack and to perform a branch operation. On return to the superordinate function, the previously stored continuation, the return address, can simply be retrieved by fetching it from the location pointed to by the stack pointer. A following stack-pointer correction finishes the necessary actions. Using the stack for storing continuations only works as the program order for sequential programs can be mapped onto a sequential traversal of the tree of function invocations, the *call tree*. A parallel program execution, on the other hand, has to be mapped onto a parallel traversal of the call tree that is now often named *spawn tree*. A simple stack here does not suffice to hold the continuations of more than one nested execution as can be seen from Fig. 1.

As interconnects for networks or clusters of PCs and workstations exhibit shrinking roundtrip latencies well below the 20 μs mark for small messages and bandwidths up to hundreds of MBytes/s, fine-grained parallelism has become a definite possibility even on these types of architectures. In order to utilize most of the offered communication performance, runtime system services have to be

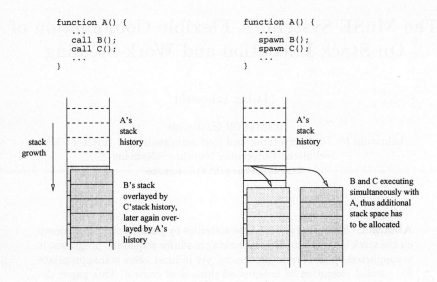

Fig. 1. Additional stack space is needed for parallel execution.

comparably efficient yet flexible. They e. g. need to adapt to the load situation and availability of nodes in a frequently changing cluster environment.

Spawning as a common runtime mechanism enables the migration of activities to underloaded nodes in a parallel or distributed system. Techniques of load balancing built on top of this feature are particularly important in the context of clusters of PCs with varying interactive background load.

The following text first presents the SMiLE cluster, a network of PCs connected by a contemporary low-latency network. Some communication performance figures are given in order to motivate the need for low-overhead runtime mechanisms, in particular with respect to work migration. The following section then describes the Multithreaded Scheduling Environment, a scheduler prototype with the ability to dynamically switch between on-stack and on-heap allocation of contexts. This property is utilized to implement a distributed work-stealing algorithm that achieves a significant improvement over those on runtime system using purely heap-based execution. A number of synthetic benchmarks are used to assess the performance of the proposed system. It can be concluded that speed-up still scales well in the given small test environment. A comparison with already existing systems and an identification of further improvements finish the text.

2 The MuSE System

2.1 The SMiLE Cluster of PCs

The name SMiLE is an acronym for *Shared Memory in a LAN-like Environment* and represents the basic outline of the project. PCs are clustered with the

Scalable Coherent Interface (SCI) interconnect. Being an open standard [11], SCI specifies the hardware and protocols allowing to connect up to 64 K nodes (processors, workstations, PCs, bus bridges, switches) in a high-speed network. A 64-Bit global address space across SCI nodes is defined as well as a set of read-, write-, and synchronization transactions enabling SCI-based systems to provide hardware-supported DSM with low-latency remote memory accesses. SCI nodes are interconnected via point-to-point links in ring-like arrangements or are attached to switches.

The lack of commercially available SCI interface hardware for PCs during the initiation period of the SMiLE project led to the development of our custom PCI-SCI hardware. Its primary goal is to serve as the basis of the SMiLE cluster by bridging the PCs I/O bus with the SCI virtual bus. Additionally, monitoring functions can be easily integrated into this extensible system. A good overview of the SMiLE project is given in [6]. The adapter is described in great detail in [1].

The current SMiLE configuration features four Pentium PCs running at 133 MHz with 32 MBytes of RAM and 2 GBytes of external storage coupled by four SMiLE PCI-SCI adapters in a ring configuration. The operating system is Linux 2.0.33.

Active Messages 2.0 [8] have been implemented on SMiLE as a communication layer particularly suited to be a building block for distributed runtime system communication. Its 0-byte message latency was measured to be $14\,\mu s$ while a maximum throughput of 25 MBytes/s can be reached with messages larger than 1 KByte.

2.2 The Multithreaded Scheduling Environment

The *Multithreaded Scheduling Environment (MuSE)* is a distributed runtime system specifically targeting the SMiLE platform. Programs running on MuSE are structured in a particular way in order to enable its own load balancing strategy.

Two basic methods exist in general for this purpose: while in *work-sharing* the load-generating nodes decide about the placement of new activities, *work-stealing* lets underloaded nodes request for new work. The *Cilk* project [2] highlighted that the technique of work-stealing is well suited for load balancing on distributed systems such as clusters of workstations. Its spawn-tree based scheduling ensures provably optimal on-line schedules with respect to time and space requirements. However, in order to permit the application of this technique, Cilk requires every subthread to be executed with a heap-allocated context. For program parts that are not actually executed in parallel due to runtime decisions or an excessive number of concurrent activities for the given number of processors, this means a potential source of inefficiency as unnecessary decomposition cost is paid.

MuSE in turn provides on-stack execution by default, trying to profit from its high sequential efficiency. In order to permit a Cilk-like work-stealing flexibility nonetheless, a different policy of spawning on-heap activities is required. The following structure of the runtime system and its active entities provides the basis for this.

Organization of Active Entities. MuSE programs are compiled from the dataflow language SISAL. In intermediate steps, appropriate dataflow graphs are generated which are subsequently converted into MuSE-compliant C code. The front end of the *Optimizing SISAL Compiler (OSC)* [3] combined with a custom-made code generator and the SMiLE system's gcc are utilized to do this.

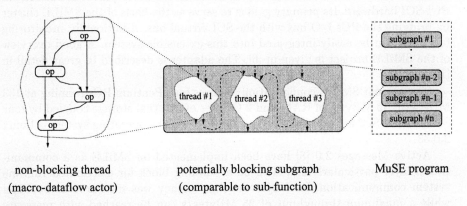

Fig. 2. MuSE program structure: non-blocking threads are combined into potentially blocking subgraphs a collection of which forms the MuSE program.

MuSE Graphs. A MuSE *graph* is the C code representation of IF1/2 dataflow graphs and follows their semantics. IF1/2 are intermediate dataflow representations put out by OSC and are described in [10]. The graph is represented by a regular C function with an appropriate declaration of parameters as well as return values. Input data is brought into the graphs via regular C function parameters, as is return data. Access to these within the graph is performed via a context reference, in this case pointing to the appropriate stack location. Thus, sequential execution of MuSE graphs is able to utilize the simple and efficient on-stack parameter passing of conventional sequential languages.

For parallel execution, i. e. on-heap allocation of graph contexts, the same mechanism is used. By allocating heap storage and changing the context reference to heap storage, spawning is prepared whenever necessary but transparently to the MuSE application code. The rationale behind the decision whether to either dynamically spawn or call a subgraph is explained below.

As spawning on a deeply nested level blocks at least one continuation within the caller until the callee has finished, heap context allocation will be performed for calling graphs on demand, resulting in lazily popping contexts off the stack.

In order to allow for blocking and to exploit intra-graph concurrency as well, graphs are subdivided into non-blocking activities called *threads*. Blocking by default has to occur at thread borders at which execution will later also resume. A graphical representation of graphs and their threads within a MuSE program is shown in Fig. 2.

MuSE Threads. A MuSE *thread* is a non-blocking sequence of C statements that is executed strictly, i. e. once all input data is present, threads can be run to completion without interruption. Threads are declared within a MuSE graph with the `THREAD_START(thread_number)` statement and finished with the following `THREAD_END` macro. These macros implement a simple guarded execution of the C statements forming the thread. The guard corresponds to a numerical synchronization counter responsible for implementing the dataflow firing rule.

MuSE threads are arranged by a simple as-soon-as-possible schedule within the graph as their underlying dataflow code is guaranteed by the compiler to be non-cyclic. Thus, whenever no blocking occurs, a MuSE graph executes like a regular C function.

Contexts. As already pointed out, in the case of blocking, MuSE graphs need to save their local data for later invocations. By default, graph parameters are located on the stack as are graph-local variables. Thus, this data has to be removed from the stack and put into a heap-allocated structure, the graph *context*. A context has to reflect the data also present on the stack during sequential execution and thus has to contain at least a pointer to the graph's C function, its input values as well as a container for the return value, some local variables, the threads' synchronization counters, and potentially the graph's continuation.

By choosing this form of implementing dataflow execution, MuSE effectively forms a two-level scheduling hierarchy. While the initial decision about which graph to execute is performed by the MuSE scheduler, the scheduling of threads within a graph is statically compiled in through the appropriate thread order. However, for nested sequential execution, the scheduler is only invoked once at the root of the call tree.

An advantage of this organization is the possibility to optimistically use calls instead of spawns to execute any MuSE graph. This means that only in those cases in which a spawn is occuring, the penalty for saving the context on the heap, memory allocation and copy operations, is paid.

MuSE Stack Handling. MuSE treats execution by default as being sequential first, thus attempting to use on-stack execution as long as it is possible. Only when the actual need arises, heap storage is requested and execution can proceed concurrently:

1. In the non-blocking case (a) of Fig. 3, execution can completely be stack-based.
2. In (b), some subgraph C blocks. The blocking can occur several levels deeper on the stack than where the sequential starting procedure or graph A is

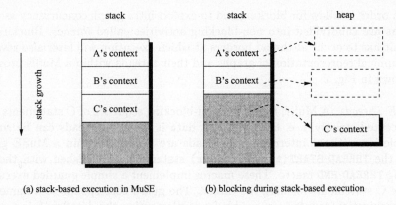

Fig. 3. Two cases during stack-based execution in MuSE: completion without blocking (a), or blocking at an arbitrary nesting level.

located. As C blocks, its data is transferred to a heap-allocated context. C returns to B, telling it through its return value that it has blocked. For B this means that at least one of its threads, C's continuation, will not be executed. This results in subsequent blocking and heap allocation operations until A is reached and all intermediate levels have blocked. During this, however, all remaining work that is executable in the spawn subtree below A has already been performed.

The described situation points out MuSE's advantage: being stack-based by default, this execution model offers a basically unlimited stack space to each new subgraph call. Even in the case of a deeply nested blocking graph, all remaining enabled work can proceed sequentially. Only those graphs that really ever block need context space from the heap.

Load Balancing and Parallelism Generation in MuSE. On the thread level of execution, the dataflow firing rule provides the basis for an effective self scheduling.

MuSE graphs are executed strictly, too. As a newly spawned context contains all input data and the graph's continuation, a context can basically be executed on any node. This property is used by the MuSE scheduler to distribute work across all participating nodes of a MuSE system.

MuSE Work-Stealing. While MuSE threads correspond roughly to Cilk threads, their grouping together into procedures or graphs has consequences regarding work-stealing. In Cilk, the grouping of threads into procedures does not exist for the work-stealing algorithm. The method is purely based on closures, i. e. migrating input data and continuations of ready threads.

In MuSE, threads do not have separate closures. Their parameters and synchronization counters are combined within the graph context. MuSE thus bases work-stealing on contexts instead of single thread closures. It is hoped that allocating a context once and performing a single migration per stolen graph, the slightly larger communication and memory management overhead can be amortized.

MuSE Queue Design. In Cilk, threads run to completion without being interrupted. The graphs in MuSE, however, can block. This means that a single ready queue does not suffice in MuSE. A graph that is currently not being executed can basically be in one of the three states READY, BLOCKED, and UNBLOCKED and is either placed in the *ready queue*, a *working queue*, or in a *notworking pool*.

The ready queue has to service potential thief nodes as well as the local scheduler in the same way as in Cilk. Therefore, a structure comparable to the LIFO/FIFO token queue of the ADAM machine [4] was chosen, its functionality, however, is equivalent to Cilk's *ready dequeue*.

Communication Communication across nodes in MuSE occurs in mainly three places:

- while passing return values,
- during compound data accesses, and
- during work-stealing of ready contexts.

Simple SCI transactions do not completely cover the functionality required to implement all three cases mentioned above. Thus, they have to be complemented by actions taking place on the local and the remote node. For instance, passing of dataflow values requires updating of synchronization counters, while work-stealing with its migration of complete contexts requires queue-handling capabilities connected to these communication operations.

In order to accomplish these tasks with the least overhead, MuSE bases all communication on the SMiLE Active Messages already mentioned in Sec 2.1. This is being done in accordance with other, similar projects. Cilk as well as the Concert system [9] all rely on comparable messaging layers. Active Messages have distinct advantages in this case over other communication techniques:

1. The definition of request/response pairs of handler functions enables the emulation of most communication paradigms through Active Messages. For MuSE, AMs allow to define a dataflow naming scheme of slots for dataflow tokens within contexts, and still operate on these in a memory-oriented way.
2. The handler concept of AMs allows to define sender-initiated actions at the receiver without the receiver side being actively involved apart from a regular polling operation.
3. Due to the zero-copy implementation on the SMiLE cluster, Active Messages actually offer the least overhead possible in extending SCI's memory oriented transactions to a complete messaging layer, as can be seen from its performance figures.

Polling Versus Interrupts. Technical implementations of SCI-generated remote interrupts unfortunetely exhibit high notification latencies, as was documented in [7] for the Solaris operating system and Dolphin's SBus-2 SCI adapter. The reason for this lies in context switch costs and signal-delivery times within the operating system. Thus, in order to avoid these, MuSE has to deal with repeatedly polling for incoming messages. In its current implementation, MuSE therefore checks the AM layer once during every scheduler invocation.

3 Experimental Evaluation

3.1 Basic Runtime System Performance

Table 1 summarizes the performance of several basic runtime services of MuSE on the SMiLE cluster that are explained below in more detail:

t_{AM_Poll} is the time required for the Active Messages poll when no incoming messages are pending. This effort is spent during every scheduler invocation.

t_{get_graph} represents the time typically spent for finding the a graph in the queues to be executed next.

$t_{exec,best}$ is the duration of an invocation of a graph by the scheduler whenever the context data is cached.

$t_{exec,worst}$ in contrast denotes the same time spent with all context data being non-cached.

t_{call} is the amount of time being spent for the on-stack calling mechanism.

t_{spawn} on the other hand represents the amount of time required for spawning a graph, i. e. allocating new context memory from the heap, initializing it, and placing it into the appropriate queue.

t_{return_data} summarizes the time of the operations necessary for passing return values on the same node whenever this cannot be done on stack.

t_{queue_mgmt} finally represents such actions as removing contexts from one queue and placing into another.

The most important result of Table 1 is the fact that the on-stack call overhead is smaller than the spawn overhead by more than an order of magnitude.

Unfortunately, the uncached execution of an empty graph points to a potential source of inefficiency of the current MuSE implementation: taking more than $20\,\mu s$ doing no actual application processing may be too high for significantly fine-grained parallel programs.

3.2 Load Balancing and Parallelism Generation

Load balancing: knary(k, m, i) is a synthetic test program that dynamically builds a spawn tree. Three parameters define its behaviour, of which the spawn degree k describes the number of subgraphs being spawned in each graph, the tree height m specifies the maximum recursion level to which spawning occurs,

Function	Symbol	Time [μs]
empty poll operation	t_{AM_Poll}	0.9
obtain graph reference	t_{get_graph}	0.25
empty graph execution (cached)	$t_{exec,best}$	0.92
graph call overhead	t_{call}	0.26
graph spawn overhead	t_{spawn}	3.9
empty graph execution (non-cached)	$t_{exec,worst}$	21.9
local return-data passing	t_{return_data}	5.1
queue management	t_{queue_mgmt}	2.95

Table 1. Performance of various runtime system services.

and i is proportional to the number of idle operations performed in each graph before the spawn operation gets executed.

The first experiment consists of running knary on MuSE in a pure *on-heap* execution mode. The work-stealing mechanism performs load balancing among the participating nodes. This mode of operation is essentially equivalent to Cilk's. The case with $k = 4$ and $m = 10$ showed the most representative behaviour and is thus used in the remainder of this section.

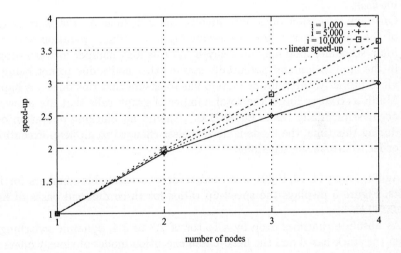

Fig. 4. Speed-up ratios of knary(4, 10, 1000), knary(4, 10, 5000), and knary(4, 10, 10000) with sequential execution times of 20.6 s, 43.0 s, and 73.1 s respectively.

Fig. 4 displays the gathered speed-up curves related to the respective single-node execution time. The result is as expected: work-stealing with per-default on-heap allocation of contexts achieves good speed-up ratios. MuSE thus conserves the positive Cilk-like properties. The drop compared to a linear speed-up can be explained with the significant costs of work-stealing. The work-stealing latency is dominated by the time required for migrating a graph context from the victim to the steal node. Sending this 364 byte-sized chunk of memory is performed via a medium-length Active Message and takes approximately $t_{migrate} \simeq 120\,\mu s$ using the SMiLE AMs, representing a throughput of about 3.3 Mbytes/s for this message size. Only a few microseconds have to be accounted for the ready queue management on both sides and for the sending of the return data token. The gathered data supports this reasoning. As soon as the average execution time of a graph grows beyond the typical work-stealing latency — a value of $i = 5000$ represents an approximate graph runtime of $123\,\mu s$ — a work-steal becomes increasingly beneficial.

Load balancing with dynamic on-stack execution. It is expected that resorting to stack-based execution for mostly sequential parts of a program will help to speed-up a given application as the overhead for spawning in relation to the actual computation time shrinks. Switching these modes is based upon the following guidelines:

– MuSE uses by default the on-stack execution. Blocking is handled as described above and changes the execution mode for the blocked subgraphs to *on-heap*.
– On-stack execution does only involve the scheduler on the topmost level of the current spawn tree. Consequently, the polling operation within the main scheduler loop may be delayed. As this may hamper forward progress in the application and potentially starve other nodes due to not being serviced properly, the runtime system has to ensure that this does not happen. MuSE's strategy limits the total number of graph calls that are allowed to occur sequentially (MAX_SEQUENTIAL). Should no blocking operation occur during this time, the mode is nevertheless changed to on-heap execution in order to invoke the scheduler.

Again, knary(4, 10, 1000) was used to derive reasonable numbers for both limits. Figure 5 displays the speed-up ratios for three different cases of MAX_SEQUENTIAL.

As absolute runtimes drop by a factor of 2.8 to 3.4, dynamic switching between the stack-based and the heap-based execution modes obviously offers fundamental improvements over the purely off-stack execution of the previous experiments.

In the single-node case MuSE takes care of executing the application purely on-stack, requiring only the initial graph plus its first level of four calls to be executed with a spawned context. Its overhead is easily amortized over the 349531 total subgraph calls in this program and thus represents the truly fastest single-node implementation. In contrast to this, the single-node runs of the previous experiments had to pay the unnecessary spawn overhead.

Increasing MAX_SEQUENTIAL from 10 to higher values also improves speed-up ratios, as can be seen from Fig. 5. MAX_SEQUENTIAL = 100 was chosen as a

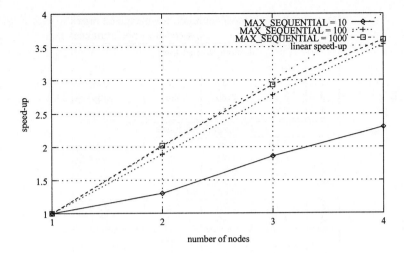

Fig. 5. Speed-up of knary(4, 10, 1000) for different values of MAX_SEQUENTIAL. Sequential runtime was $t = 7.3\,s$.

compromise value since it realizes good speed-up while at the same time keeping the polling latency low. For instance, an average graph runtime of $2\,\mu s$ would yield a maximum of $200\,\mu s$ between two polls invoked by the scheduler. Higher values for MAX_SEQUENTIAL can still improve the speed-up somewhat, yet also increase the polling latency.

Load balancing under heterogeneous loads. In order to assess MuSE's capability of dealing with unevenly distributed background load while retaining its ability of switching execution modes dynamically, artificial load was systematically placed on each SMiLE node in the form of infinetely looping separate Linux processes. Assuming a normalized processing throughput of $Y_i = \frac{1}{p}$ for each node with p equally active processes on the node, the maximum ideal speed-up for n nodes would be

$$S_{max} = \sum_{i=1}^{n} Y_i. \qquad (1)$$

Thus, for a single background load on one node and two otherwise unloaded nodes, MuSE execution can exploit a cumulated throughput of $\sum Y = \frac{1}{2}+1+1 = 2.5$ and therefore hope for a maximum speed-up of $S_{max} = 2.5$.

Again, knary(4, 10, 1000) served as the test case. Table 2 summarizes the experiments with up to 4 background loads.

It is clearly visible that MuSE is able to balance the load even when the processing performance of the nodes is uneven. However, when the number of background processes increases, less and less of the offered throughput can be utilized. This can be attributed to the uncoupled Linux schedulers which are

nodes	wall-clock time [s]	speed-up	max. ideal speed-up	utilized max. performance
1 background load				
1	14.05	0.52	0.5	> 100 %
2	5.33	1.37	1.5	91.3 %
3	3.25	2.25	2.5	90.0 %
4	2.47	2.96	3.5	84.4 %
2 background loads				
2	7.70	0.95	1	95.0 %
3	4.26	1.71	2	85.7 %
4	3.42	2.13	3	71.2 %
3 background loads				
3	5.71	1.28	1.5	85.3 %
4	4.05	1.80	2.5	72.1 %
4 background loads				
4	4.84	1.51	2	75.4 %

Table 2. Absolute runtimes and speed-up values for knary(4, 10, 1000) and 1...4 background loads.

bound to impede communication among the nodes more than necessary: while node A sends a request to node B, the latter is no longer being executed, forcing node A to wait for the next time slice of B.

Running more realistic SISAL applications unfortunately requires further tuning of the current prototypical and mostly C macro-based runtime system mechanisms. In particular, the prevailing implementation of MuSE's compound data handling suffers from significant inefficiencies that prevent actual application speed-up. This work therefore only covers synthetic benchmarks that nevertheless represent typical model cases.

4 Related Work and Conclusion

Load balancing actions usually introduce overhead due to two tasks: *decomposition cost* for setting up independent activities and *load migration cost* for transporting an activity to a different node. This distinction helps to differentiate the results of the seemingly similar execution models of MuSE and Cilk.

While in Cilk on-heap execution is the only and default execution mode, MuSE's default paradigm lets execution run on the stack and only switches back to on-heap execution when demanded. Or, to rephrase this differently, it means that decomposition cost for parallel execution in Cilk is implicitly hidden in the execution model and proportional to the number of Cilk threads being spawned — even for the sequential case in which this would not be necessary — while MuSE's costs for generating parallelism are made explicit by mode changes due to work-steal or fairness demands. This also means that Cilk's good speed-up values are achieved partly due to a non-optimal sequential version while MuSE exhibits the fairer comparison: decomposition into concurrent active entities, the MuSE graphs with their contexts, only occurs when it is really necessary. In Cilk, only load migration cost has to be offset by the performance improvement while MuSE has to offset both costs yet still performs well.

Runtime systems suitable for fine-grained parallelism have to be efficient when supporting sequential as well as parallel programs. They thus have to overcome the difficulty of trading off efficient on-stack execution modes against more flexible but less lightweigt on-heap execution modes. Several methods for this have been presented so far:

- The *Lazy Threads* project [5] attempts to amortize heap allocation overhead by utilizing *stacklets*, independent and fixed-size chunks of heap space serving as stack for intermediate sequential execution. On stacklet overrun or a spawn operation, new stacklets are allocated, thus requiring static knowledge about when to spawn and when to call.

  In Lazy Threads, the decision about spawning and calling is non-reversible and has to be made at compile-time. MuSE in contrast offers a truly transparent model with the possibility to change execution modes at any time during the application's run.

- The *StackThreads* approach [12] defaults to sequential on-stack execution. In case of a spawn operation, a *reply box* in the calling thread serves as the point of synchronization for the spawned thread and its caller, blocking the caller until the spawned thread has completed.

  MuSE advances beyond StackThreads by not necessarily blocking a higher level activity as soon as a procedure has been spawned. Intra-graph parallelism through threads ensures that all ready execution can proceed on the stack while in StackThreads a single blocking operation always blocks an entire call tree.

- In the *Concert* system [9], differently compiled versions of the same code supporting calling as well as spawning are utilized dynamically.

  In contrast to Concert, MuSE's method requires only a single object code, avoiding the storage overhead of multiple versions.

MuSE thus offers distinct advantages over these described runtime systems. Unfortunately, its prototype implementation suffers from a number of inefficiencies. C macros that are used to implement certain runtime system functions and to enhance manual readability are not flexible enough for the presented purpose. They do not perform sufficiently well to allow for even more fine-grained parallelism and realistic dataflow applications. Although this is no actual surprise, they behaved interestingly well as a testbed for the presented mechanisms. Machine-level implementations of the runtime mechanisms, however, seem more appropriate as well as using more compiler knowledge to resort to less general and more optimized functionality whenever possible.

References

1. G. Acher, H. Hellwagner, W. Karl, and M. Leberecht. A PCI-SCI Bridge for Building a PC Cluster with Distributed Shared Memory. In *Proceedings of the 6th International Workshop on SCI-Based High-Performance Low-Cost Computing*, Santa Clara, CA, September 1996.
2. R. D. Blumofe and C. E. Leiserson. Scheduling Multithreaded Computations by Work Stealing. In *Proceedings of the 35th Annual Symposium on Foundations of Computer Science (FOCS '94)*, pages 356–368, Santa Fe, NM, USA, Nov 1994.
3. D. C. Cann. The Optimizing SISAL Compiler: Version 12.0. Technical Report UCRL-MA-110080, Lawrence Livermore National Laboratory, April 1992.
4. P. Färber. *Execution Architecture of the Multithreaded ADAM Prototype*. PhD thesis, Eidgenössische Technische Hochschule, Zurich, Switzerland, 1996.
5. S. C. Goldstein, K. E. Schauser, and D. E. Culler. Lazy Threads: Implementing a Fast Parallel Call. *Journal of Parallel and Distributed Computing*, 37(1):5–20, 25 August 1996.
6. H. Hellwagner, W. Karl, and M. Leberecht. Enabling a PC Cluster for High-Performance Computing. *SPEEDUP Journal*, June 1997.
7. M. Ibel, K. E. Schauser, C. J. Scheiman, and M. Weis. High-Performance Cluster Computing Using Scalable Coherent Interface. In *Proceedings of the 7th Workshop on Low-Cost/High-Performance Computing (SCIzzL-7)*, Santa Clara, USA, March 1997. SCIzzL.
8. A. M. Mainwaring and D. E. Culler. *Active Messages: Organization and Applications Programming Interface*. Computer Science Division, University of California at Berkeley, 1995. http://now.cs.berkeley.edu/Papers/Papers/am-spec.ps.
9. J. Plevyak, V. Karamcheti, X. Zhang, and A. Chien. A Hybrid Execution Model for Fine-Grained Languages on Distributed Memory Multicomputers. In *Proceedings of the 1995 ACM/IEEE Supercomputing Conference*, San Diego, CA, December 1995. ACM/IEEE.
10. S. Skedzielewski and J. Glauert. IF1 - An Intermediate Form for Applicative Languages. Technical Report TR M-170, Lawrence Livermore National Laboratory, July 1985.
11. IEEE Computer Society. *IEEE Standard for Scalable Coherent Interface (SCI)*. The Institute of Electrical and Electronics Engineers, Inc., 345 East 47th Street, New York, NY 10017, USA, August 1993.
12. K. Taura, S. Matsuoka, and A. Yonezawa. StackThreads: An Abstract Machine for Scheduling Fine-Grain Threads on Stock CPUs. In T. Ito and A. Yonezawa, editors, *Proceedings of the International Workshop on the Theory and Practice of Parallel Programming*, volume 907 of *Lecture Notes of Computer Science*, pages 121–136, Sendai, Japan, November 1994. Springer Verlag.

Pangaea:
An Automatic Distribution Front-End for Java

André Spiegel

Freie Universität Berlin, Germany
spiegel@inf.fu-berlin.de

Abstract. Pangaea is a system that can distribute centralized Java programs, based on static source code analysis and using arbitrary distribution middleware as a back-end. As Pangaea handles the entire distribution aspect transparently and automatically, it helps to reduce the complexity of parallel programming, and also shows how static analysis can be used to optimize distribution policies in ways that would be difficult for programmers to find, and impossible to detect dynamically by the run-time system.

1 Introduction

Java is receiving increasing attention as a language for scientific computing, partly because the performance of some implementations starts to rival that of traditional, compiled languages, and most notably because of its clean object-oriented programming model with support for concurrency [4]. However, as soon as programs need to be *distributed* across loosely coupled machines for parallel execution, the clarity of the model is often diminished by a wealth of additional, purely distribution-related complexity.

Pangaea[1] is a system that can distribute Java programs automatically. It allows the programmer to write parallel programs in a centralized fashion, ignoring all distribution-related issues such as where to place object instances, when and how to employ object migration etc. With Pangaea handling these issues automatically, the clarity of the programming model is maintained for the distributed case, thus reducing the complexities of parallel programming in a way very similar to the Distributed Shared Memory (DSM) approach. Pangaea, however, is able to use arbitrary distribution middleware, such as Java/RMI [9] or CORBA [6], to accomplish this abstraction.

Pangaea uses static analysis to determine distribution policies for a program, showing how distribution can be handled not only transparently, but also automatically. However, Pangaea does *not* attempt to transform sequential programs into concurrent ones (sometimes also called "parallelizing" programs). We leave it to the programmer to express concurrency using standard language features; it is the distribution aspect that Pangaea focuses on.

Pangaea is work-in-progress, currently being designed and implemented as a PhD project. This paper gives a general overview of the Pangaea architecture, and outlines some of the problems and solutions we have identified so far.

---

[1] The name *Pangaea* refers to the ancient continent in which the landmass of the earth was centralized until about 200 million years ago, when it broke up and drifted apart to create the distributed world we know today.

2 Motivation

We assume the object-oriented programming model of languages such as Java or C++, with coarse-grained concurrency being expressed by means of threads and the necessary synchronization primitives. To achieve parallel execution on serveral loosely coupled processors, the threads and object instances of a program must be *distributed* appropriately across these processors.

To distribute a program in this way actually involves three distinct steps. First, it means to *decide* about distribution policies: where to place object instances, and what dynamic strategies for object placement, like migration or caching, should be employed. Second, it means to *accomplish* these decisions by tailoring the program towards a given distribution platform, e.g. modifying the program code, creating interface definition files or configuration files. Third, the platform must *implement* the desired distribution, e.g. by generating proxies, compiling the system for distributed execution, and execute it under the control of a distributed run-time system.

It is the first and the second of these steps that Pangaea handles automatically, which pays off in a number of respects. First, it frees the programmer from the routine task of distributing programs, which may seem simple at first sight (and therefore deserves to be automated), while a closer look reveals that many optimization opportunities exist that are likely to be missed even by experienced programmers, and tedious to work out in detail.

Second, while there is no doubt that distribution platforms should have run-time systems that can monitor object interaction patterns and re-adjust distribution policies *dynamically*, it turns out that static program analysis can strongly assist the run-time system's work by providing information that can only be derived from the source code. Examples for this include:

- identifying *immutable objects* (or places where objects are used in a read-only fashion), because such objects can be freely replicated in the entire system, passed to remote callees by value (serialization) and needn't be monitored by the run-time system at all,
- finding the *dynamic scope* of object references, discovering for example that certain objects are only used privately, inside other objects or subsystems, and therefore needn't be remotely invokable, nor be considered in the run-time system's placement decisions,
- recognizing opportunities for *synchronous object migration* (pass-by-move, pass-by-visit), which is preferable to the run-time system's dynamic adjustments because such adjustments can only be made *after* sub-optimal object placement has already manifested itself for a certain amount of time.

3 Related Work

There are a number of projects that aim to facilitate parallel programming in Java by providing different, more structured *programming models*. An example is the *Do!* project [5], where special kinds of distributed *collections* (of tasks or of data) are used to express concurrency and distribution. Do! handles distribution in a highly transparent fashion; it does not, however, use static analysis to detect optimizations (at least not as extensively as we propose). While it may be true that under a different programming model, some distribution choices are more naturally expressed or even implied in the source code, thus maybe eliminating the need for some kinds of static analysis, most of our techniques would still be beneficial in such environments as well.

We know of two more similar projects in which static analysis has actually been used to automate distribution decisions, one based on the Orca language [1], the other using the JavaParty platform [8]. The *Orca* project successfully demonstrated that a static analyzer can guide the run-time system to make better placement and replication choices than the RTS could have made on its own; the performance achieved is very close to that of manually distributed programs. While Orca is an object-*based* language that has been kept simple precisely to facilitate static analysis, Pangaea makes a similar attempt in an unrestricted, object-oriented language (Java).

We consider the *JavaParty* project as a first step in this direction. Its authors, however, acknowledge many remaining problems of their analysis algorithm, some of which we hope to solve in Pangaea. What also distinguishes our work from both projects is that Pangaea covers additional distribution concepts besides mere placement and replication decisions (see previous section). Finally, Pangaea does not depend on any particular distribution platform, it rather provides an abstraction mechanism that harnesses the capabilities of many existing platforms, and is specifically designed to be extensible to future distribution technology.

4 Pangaea

The architecture of the Pangaea system is shown in Fig. 1. We will first give an overview of the system in a top-down fashion, then have a closer look at the individual components, or stages, in bottom-up order.

4.1 Overview

Pangaea's input is a centralized Java program, plus requirements regarding the desired distribution, e.g. the number of separate machines to use. The *Analyzer* basically derives an *object graph* from the source code (which approximates the program's run-time structure and communication patterns), then it decides about distribution policies by analyzing this graph. The algorithm is parameterized with an abstract view of the capabilities of the given distribution platform, as provided by the corresponding *Back-End Adapter*. Such capabilities, which the platform may or may not possess, include the ability to invoke objects remotely, to transfer object state to another machine, or to perform object migration or caching. If the platform does provide a certain feature, the Analyzer tries to make use of it for the given program, if the feature is absent, the Analyzer tries to make do without it.

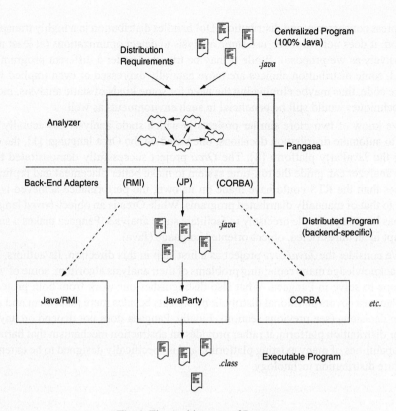

Fig. 1. The Architecture of Pangaea

After the Analyzer has completed its work, the Back-End Adapter re-generates the program's source code, changing or adding whatever is necessary to implement the decisions indicated by the Analyzer. The result is a distributed program, possibly augmented with configuration files or interface definition files, specifically tailored for the given back-end platform. The distribution platform is then responsible for transforming the generated code into an executable program (involving steps like stub-generation and compilation), and to execute it under the control of its run-time system.

4.2 Distribution Back-Ends

As Pangaea will be able to work with arbitrary distribution platforms, one goal of our project is to develop a general understanding and classification of platforms according to the distribution concepts they provide, showing both differences and similarities, and how to abstract from these to handle distribution transparently.

Fig. 2 shows an initial attempt at such a classification. We consider the ability to *invoke* objects remotely as the essential common denominator of object-oriented distribution platforms. The ability to *create* objects remotely is present on some platforms, e.g. the JavaParty system, while it is absent on others such as CORBA or RMI. The

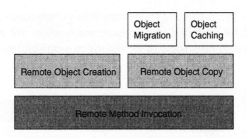

Fig. 2. A Taxonomy of Distribution Platform Capabilities

ability to *copy* objects remotely is present on RMI-based platforms through Java's serialization mechanism; it is currently being added to CORBA as well, in the proposed "Objects-by-Value" extension [7]. To pass objects by value is an optimization technique for distributed programs in its own right; we also consider it as the basis for more advanced mobility concepts like object *migration* (as in JavaParty [8]) and object *caching* (e.g. in the Javanaise system [2]).

Pangaea's Back-End Adapters provide a set of boolean flags that indicate whether the corresponding platform does or does not provide each of the above features; this information is used by the Analyzer to fine-tune its distribution decisions. After the analysis has completed, the Analyzer passes annotated versions of the program's abstract syntax trees to the Back-End Adapter; the annotations indicate, for example, whether instances of a given class need to be remotely invokable, or should be considered in dynamic placement decisions. The Back-End Adapter re-generates the program's source code based on this information, applying whatever transformations are necessary to accomplish the Analyzer's decisions, while preserving the program's original semantics. The Back-End Adapter may in fact also *simulate* some distribution concepts through code generation, if the corresponding platform does not already offer them.

4.3 Distribution Analysis

The problem of placing objects onto a set of distributed processors is a variant of the well-studied graph partitioning problem, for which, although NP-hard in general, good approximation techniques exist [3]. Essentially, Pangaea's Analyzer will try to co-locate thread objects and the data objects used by these threads to minimize communication cost.

In addition to these macroscopic placement decisions, Pangaea will also perform various other analyses to find out whether objects are immutable, or to choose appropriate parameter passing mechanisms, etc.

4.4 Constructing the Object Graph

The task to derive an object graph from a program's source code is probably the most challenging one for the Analyzer, and the one on which the rest of the analysis most critically depends. Fortunately, it is also the one for which we have the most concrete and promising results yet. A thorough presentation of our approach is beyond the scope of this paper; it will be described in a separate publication. We merely give a brief example here, which illustrates some of the algorithm's capabilities.

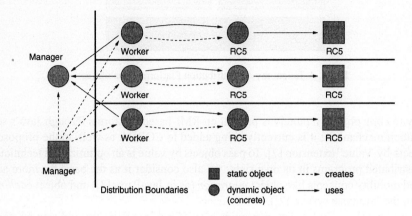

Fig. 3. An Object Graph

Fig. 3 shows the object graph of a simple, parallel RSA cracking program, designed in a manager/worker fashion. The workers (which are Java thread objects) retrieve encryption keys to try from the manager object, and report any success back to the manager, which then terminates the entire program. Each worker internally uses an RC5 object that encapsulates the actual encryption algorithm.

The example illustrates the following properties of our approach:

- We consider a Java class to define, optionally, a *static type* (comprising the static members of the class), and a *dynamic type* (the non-static members). Of a static type, precisely one instance exists at run-time, while for a dynamic type, the number of instances may or may not be known beforehand. If it is not known, a *summary node* is inserted into the object graph, otherwise *concrete* object nodes are used.
- In the example graph, static objects are shown as grey boxes. Examining object creation statements in the program, the Analyzer is also able to tell that three concrete worker objects and corresponding RC5 objects will exist at run-time (shown as grey circles); there are no summary nodes needed in this example.
- The dynamic RC5 objects each use the static part of their class, which, according to what was said above, should only exist in a single instance. However, Pangaea realizes that the static part of the RC5 class happens to be *immutable* (it contains only constants), and can therefore freely be replicated for each worker.
- Using data flow analysis, Pangaea detects that the worker objects use their corresponding RC5 objects *privately*. Consequently, there are no usage edges from a worker to the other two RC5 objects, which allows Pangaea to find the right distribution boundaries shown in the figure.

An important result of the analysis is that only the dynamic manager object is ever invoked remotely, and must therefore have stubs and skeletons generated. All other objects can be ordinary Java objects, and this is true in particular for the dynamic RC5 objects, which are used heavily by each worker. It turns out that on RMI-based distribution platforms, the impact of this seemingly modest optimization is tremendous. This is because of the (admittedly odd) fact that a local invocation of an RMI "remote" object still takes on the order of a few milliseconds. To our own surprise, we found that in the given example, to optimize the stubs away speeds up the program by no less than a factor of *four*.

5 Summary

We hope to make three distinct research contributions with the Pangaea system. First, we will show how to abstract from the features and technicalities of various distribution platforms, and how to use them in a transparent fashion, reducing the complexities of parallel programming. Second, Pangaea will show that even the task of *deciding* about distribution policies for a program can be automated by means of static analysis. Third, it will be shown that the role of static analysis cannot be taken over by dynamic approaches entirely, because many optimizations are impossible to accomplish by a run-time system alone.

Pangaea is work-in-progress. We have so far implemented a complete analysis frontend for the Java language; based on this, the algorithm to construct an object graph from the source code has been completed and successfully tested on a number of real-world programs. We are now moving towards the actual distribution analysis, and hope to create one or two Back-End Adapters for different distribution platforms to demonstrate the feasibility of our approach.

References

1. Bal, H.E. et al.: Performance Evaluation of the Orca Shared-Object System. ACM Trans. CS, Vol. 16, No. 1 (1998) 1–40
2. Hagimont, D., Louvegnies, D.: Javanaise: distributed shared objects for Internet cooperative applications. Middleware '98, The Lake District, England (1998) 339–354
3. Hendrickson, B., Leland, R.: A Multilevel Algorithm for Partitioning Graphs. Tech. Rep. SAND93-1301, Sandia National Laboratories (1993)
4. Jacob, M. et al.: Large-Scale Parallel Geophysical Algorithms in Java: A Feasibility Study. Concurrency: Practice & Experience, Vol. 10, No. 11–13 (1998) 1143–1153
5. Launay, P., Pazat, J.-L.: A Framework for Parallel Programming in Java. Tech. Rep. 1154, IRISA, Rennes (1997)
6. Object Management Group: The Common Object Request Broker: Architecture and Speicification. Revision 2.0 (1995)
7. Object Management Group: Objects-by-Value: Joint Revised Submission. OMG document orbos/98-01-01 (1998)
8. Philippsen, M., Haumacher, B.: Locality optimization in JavaParty by means of static type analysis. First UK Workshop on Java for High Performance Network Computing at EuroPar '98, Southhampton (1998)
9. Sun Microsystems: RMI — Remote Method Invocation. http://java.sun.com

Concurrent Language Support for Interoperable Applications [*]

Eugene F. Fodor and Ronald A. Olsson

Department of Computer Science
University of California, Davis, USA, 95616
{fodor, olsson}@cs.ucdavis.edu
http://avalon.cs.ucdavis.edu

Abstract. Current language mechanisms for concurrency are largely isolated to the domain of single programs. Furthermore, increasing interest in concurrent programming encourages further research into interoperability in multi-lingual, multi-program settings. This paper introduces our work on Synchronizing Interoperable Resources (SIR), a language approach to begin achieving such interoperability with respect to concurrent language mechanisms.

1 Introduction

Developers of distributed software systems need simpler, more efficient ways of providing communication between different, remote programs in multi-lingual environments. Figure 1 shows an example of this multi-lingual interaction problem in a federation of databases and a collection of client programs.

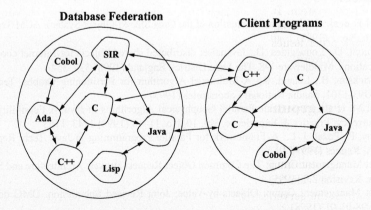

Fig. 1. Interoperability in a Database Federation

---

[*] This work is partially supported by the United States National Security Agency University Research Program and by ZWorld, Inc. and the State of California under the UC MICRO Program.

Many systems contain mechanisms for performing distributed communication. They typically provide some kind of Remote Procedure Call (RPC) [1]. For example, DCOM[2], Java RMI[3], ILU[4], and CORBA[5] are forms of object request brokers that allow remote methods to be invoked through remote objects. Direct support for interoperability of concurrency control mechanisms under these systems is limited. For example, CORBA's Concurrency Control Service [6] restricts concurrent control to a transaction based two-phase locking protocol commonly used in database systems. Converse[7] provides multi-lingual interoperability for concurrent languages via run-time support within the scope of a single, parallel application.

Unfortunately, these (and other) existing mechanisms that address inter-program communication are complicated, so system designers continue to seek novel, simpler methods to express such communication. This paper presents an alternative approach to this problem. Synchronizing Interoperable Resources (SIR) extends the communication mechanisms in the SR concurrent programming language [8]. SR provides concurrent communication mechanisms like RPC, semaphores, asynchronous message passing, rendezvous, and dynamic process creation. This variety of mechanisms allows the programmer to more intuitively design and implement the desired application. Although SIR does not provide the full functionality of, say, CORBA, it does provide more flexible inter-program communication.

SIR is intended to benefit distributed system researchers and developers. SIR's expressive communication allows the programmer to work at a high conceptual level, without getting bogged down in the many details of other approaches. The SIR extensions to the SR programming model and implementation are minor. SIR also facilitates rapid prototype development.

The rest of this paper presents an overview of the design and implementation of SIR, including the requirements and interactions of the SIR compiler and runtime support system. It presents some preliminary performance results, which are promising. This paper also discusses trade-offs with other approaches and some of the other issues that have arisen so far in this work, which will be the basis of future work.

2 SR Background

The expressive power of SR's communication mechanisms helps the programmer create complex communication patterns with relative ease [8]. For example, a file server can service "open" invocations only when file descriptors are available. Furthermore, the server might want to schedule invocation servicing by some priority. The SR code fragment in Figure 2 accomplishes this easily through SR's **in** statement, which provides rendezvous-style servicing. Both synchronization and scheduling expressions (**st** and **by**, respectively) constrain the execution of each arm of the **in** statement. The priority (`pri`) could be mapped to an array of host names, thus allowing preference for specific hosts. In SR, the types of an operation's parameters and its return type are collectively called the operation's *signature*.

```
process fileserve()
  do true ->
    in rmt_open(fname, pri) st fdcnt > 0 by pri ->
      rfs_service(fname, OPEN)
      fdcnt--
    [] rmt_close(fname, pri) by pri ->
      rfs_service(fname, CLOSE)
      fdcnt++
    ...
    ni
  od
end
```

Fig. 2. SR file server

3 Design

SIR aims to provide interoperability by using the full concurrent communication mechanisms of SR. Therefore, only minimal changes were to be made to these mechanisms. For instance, SIR retains SR's mechanisms for invoking and servicing operations so that processes in SIR programs can communicate with one another via rendezvous, asynchronous message passing, or RPC expressed using the same syntax and semantics as in SR. Unlike in SR, however, SIR invocations may cross program boundaries.

At a high level, our model consists of communication between three programs: the remote service requester (RSR), the remote service host (RSH), and the registry program (RP). The RSR is the client, the RSH is the server, and the RP negotiates the connection between the other two. More details appear in Section 5.

To enable inter-program communication among SR programs, we added registration (making operations available to other programs) and bind (connecting to another program's registered operations) mechanisms. Specifically, we added the following built-in functions:

register(regname, oper) Registers the operation **oper** with the RP on the local host under the name **regname**.
unregister(regname) Removes the registry entry from the RP.
bind(regname, host, opcap) Contacts the RP on the specified host. Assigns the remote operation to the capability (i.e., pointer to operation) **opcap**.
unbind(opcap) Disassociates **opcap** from the remote operation.

This design implies that type checking must be performed at run time because the parameters **opcap** and **oper** of the functions **bind** and **register**, respectively, may have any signature and are bound at run time. Our initial design has two limitations: operations may only be registered on the local machine, and the registration name must be unique for each operation. These problems can be solved, respectively, by adding another parameter to the register function and by using path names as Java RMI does. The former limitation also occurs in approaches like Java RMI; security issues motivate this limitation[3].

4 Example

SIR communication naturally extends traditional SR mechanisms and allows for rapid prototyping. Consider, for example, building a server for some online database. The clients and server both create remote queues using SR's send and receive mechanisms to establish a complex dialog. Each client first registers its queue to the RP and then binds the known server queue ("kq") to its capability (**capqueue**). In turn, the server binds to the unique name transferred by the client and replies on the corresponding capability (**remqueue**) after performing some data processing. The SIR code fragments in Figures 3 and 4 show this interaction. Note that the program could actually be written more directly using RPC; however, this more complicated version better illustrates the expressive power of SIR.

```
#Each client requests a unique name
#from its local RP.
uniqname(uid)
register(uid, myqueue)
do not done->
   bind("kq", "host", capqueue)
   #myhost() returns local host name.
   send capqueue(uid, myhost(), data)
   #Receive processed data.
   receive myqueue(data)
od
unbind(capqueue)
unregister(uid)
```

```
register ("kq", queue)
do not done ->
   receive queue(name, host, data)
   #Invoke datahndlr as
   #separate thread.
   send datahndlr(name, host, data)
od
unregister("kq")

proc datahndlr(name, host, data)
   var remqueue:cap (datatype)
   bind(name, host, remqueue)
   #Process data (not shown).
   #Send data after processing.
   send remqueue(data)
   unbind(remqueue)
end
```

Fig. 3. Code for multiple, different clients

Fig. 4. Code for single server

5 Implementation

We have built a working version of SIR's compiler and runtime system. The implementation, which extends the standard SR compiler and run-time system (RTS), required additional code to map identifiers for remote programs into existing RTS data structures and to support the new built-in functions. Implementing these changes required approximately 340 lines of new code in the RTS and 263 additional lines of code in the compiler. Also, a special program called the registry program (discussed below) was added to the RTS. The implementation of the registry program totals about 669 lines of code. The implementation took approximately 1 programmer month.

For register and bind, the compiler-generated code encodes the signatures of capabilities and operations into RTS calls to provide dynamic type checking. When an operation is registered, this encoding is sent as part of the registry entry to the RP. When an operation is bound, an RSR initiates the RTS comparison of the encoded signature. Once bound, an operation can be invoked without requiring further type checking. The encoding is a simple text representation of the signature.

The bind function provides a direct connection between the RSR and RSH. The RP establishes this connection by forwarding the capability of the RSH's registered operation to the RSR. The RSR then sends back an acknowledgment to the RP to complete the request and thus frees it to establish other connections. If the RSR times-out or receives an error from the RP, the call to bind returns false. Upon invocation, an invocation block is generated at the client side and is sent over a connection to the RSH. Such connections between clients and servers are created only upon invocation and are cached for later use. Unbind simply disconnects and cleans up any data structures as necessary. Figure 5 shows this interaction at a high level.

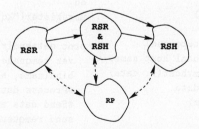

Fig. 5. Sample interactions among RSR (Remote Service Requester), RSH (Remote Service Host), and RP (Registry Program). Note that a program can act both as an RSR and an RSH. The solid lines indicate invocations; the dashed lines imply the low-level communication with the RTS for register and bind.

6 Performance

We have obtained preliminary SIR performance results. We compared SIR with two freely available implementations of the CORBA 2.x specification: Mico[9] and OmniORB2[10]. The former serves primarily as an educational 2.2 compliant ORB; the latter serves as a high-performance 2.0 compliant ORB. The following performance comparisons with CORBA are meant to only serve as a rough guide of SIR performance.

The tests were run on an isolated 10 Mbps Ethernet LAN with 3 machines. The machines used in the experiment were a Dual SMP 233 MHz Pentium, a 120 MHz Pentium laptop, and a 120 MHz Pentium PC. The CORBA programs were written in C++. The test application is a simple time service application that uses RPC. Each test client runs for 1 minute counting the number of remote invocations of the time service. Table 1 shows the average number of invocations per second for five test runs under three different configurations. The last column shows the total lines of code required. As shown, the SIR code requires significantly less effort due to the details that are required to setup communication in CORBA.

These performance results are intended to give a general idea of SIR's performance versus the performance of other approaches. However, direct comparisons are not necessarily appropriate (without further work) because CORBA ORB's and SIR are different in important aspects. For example, ORB's support the notion of remote objects, but SIR does not. On the other hand, SIR supports more communications paradigms. CORBA also provides a much richer set of naming and other services, while SIR is relatively simple.

Table 1. Invocations per second of OmniORB, Mico and SIR. Configurations 1, 2, and 3 consist of a single client and single server running on one machine, a single client and single server running on different machines, and 3 clients and a single server running on 3 machines, respectively.

	Config 1	Config 2	Config 3			Lines of code
			120MHz	120MHz	233MHz SMP	
Mico	560	366	198	203	400	88
SIR	1531	595	443	438	891	33
OmniORB2	2110	928	711	725	856	84

7 Issues

The design and implementation of SIR has raised some interesting issues. In traditional SR, capabilities can be casually copied and passed within a single program. For example, in the following program fragment, process A sends process B a capability to foo. Process B assigns this capability to y and then invokes foo via y.

```
op x(cap ())                       process B
                                     var y:cap ()
process A                            receive x(y)
  op foo()                           y()
                                   end
  send f(foo)
  in foo() ->
    write("bar")
  ni
end
```

Interoperating SIR programs do not provide this sort of functionality *if* limited to passing capabilities via register and bind. Under this limitation, register and bind are required for inter-program communication, since they explicitly pass the context information needed for remote invocation. However, SIR escapes this problem by introducing the notion of implicit binding. Implicit binding allows a program to pass a capability to another program as done in traditional SR without requiring a call to bind. Explicit binding allows the same communication mechanisms to be used regardless of whether they are local or remote—a property known as location transparency [5]. Implicit binding enhances location transparency by reducing the need for explicit binds.

Implicit binding complicates other aspects of distributed computation. For instance, termination detection mechanisms present in traditional SR become more complicated in a multi-program environment. In standard SR, the RTS detects program quiescence and decides to terminate the program by keeping track of invocation requests through a special monitoring process called **srx**. However, performing accounting on invocation requests between SIR programs complicates matters since an additional mechanism is required to monitor pending invocations between programs. Furthermore, the semantics for quiescence change. A program with registered operations that would be quiescent in traditional SR may have another program bind and use those registered operations. Alternatively, a program might be waiting on an operation that it has implicitly passed to another program. Even if all explicitly bound operations (obviously, these are easier to track) are resolved, we should not allow the program to terminate until all external processes using it are blocked. Since the conditions for termination may be only partially satisfied, a new, more global view is required to detect such conditions. A possible solution to this problem would be to create a monitor process to help manage termination.

Finally, we are investigating ways to map concurrent communication mechanisms between programming languages. We are considering several approaches. One approach would use SIR as a definition language to explicitly define mappings. Another would simulate mechanisms through generated stub code and run-time system calls.

8 Conclusion

We have presented SIR, a flexible development system for interoperable, concurrent applications. Since SIR naturally extends the SR language mechanisms, it helps make communication between programs more intuitive.

Currently, we have only investigated SIR to SIR communication, but intend to extend this work to allow other languages to utilize SIR operations. SIR also currently does not have remote resource (i.e., object) support, and we hope to address this issue in future work as well. Also, we will look further at SIR performance issues against other interoperable packages. Specifically, we are looking at the performance of binds, registers, and invocations as well as doing qualitative comparisons of communication mechanisms. We are also investigating better methods for partial, distributed termination, implicit binding, and exception handling. We are also extending SIR to include resources (objects) to provide a better basis for comparison with other object-oriented approaches.

We are also developing some practical applications with SIR. For example, we wrote a distributed, HTTP server in SR, and are currently augmenting it with SIR mechanisms. Additionally, we have also written a simple SIR distributed file server.

Acknowledgements

We thank Greg Benson for early technical discussions on this work. We thank the anonymous referees for helpful comments on the paper.

References

1. Andrew Birrell and Bruce Nelson. Implementing remote procedure calls. *ACM Trans. Computer Systems*, 2(1):39–59, February 1984.
2. Nat Brown and Charlie Kindel. *Distributed Component Object Model Protocol (DCOM/1.0)*, January 1998. Microsoft.
3. Sun Microsystems. Java remote method invocation specification. July 1998.
4. Bill Janssen, Mike Spreitzer, Dan Larner, and Chris Jacobi. *ILU 2.0alpha12 Reference Manual*, November 1997. XEROX PARC.
5. Object Management Group. *The Common Object Request Broker: Architecture and Specification.*, Feb 1998.
6. Object Management Group. *CORBAservices: Common Object Services Specification.*, July 1998.
7. L. V. Kale, Milind Bhandarkar, Robert Brunner, and Joshua Yelon. Multiparadigm, Multilingual Interoperability: Experience with Converse. In *Proceedings of 2nd Workshop on Runtime Systems for Parallel Programming (RTSPP) Orlando, Florida - USA*, Lecture Notes in Computer Science, March 1998.
8. Gregory Andrews and Ronald Olsson. *The SR Programming Language: Concurrency in Practice*. Benjamin/Cummings, Redwood City, CA, 1993.
9. Kay Römer, Arno Puder, and Frank Pilhofer. Mico 2.2.1. http://diamant-atm.vsb.cs.uni-frankfurt.de/~mico/, October 1998.
10. Olivetti & Oricale Reasearch Laboratory. Omniorb2. http://www.orl.co.uk/omniORB/, October 1998.

On the Distributed Implementation of Aggregate Data Structures by Program Transformation

Gabriele Keller[1] and Manuel M. T. Chakravarty[2]

[1] Fachbereich Informatik, Technische Universität Berlin, Germany
keller@cs.tu-berlin.de
www.cs.tu-berlin.de/~keller/
[2] Inst. of Inform. Sciences and Electronics, University of Tsukuba, Japan
chak@is.tsukuba.ac.jp
www.score.is.tsukuba.ac.jp/~chak/

Abstract. A critical component of many data-parallel programming languages are operations that manipulate *aggregate* data structures as a whole—this includes Fortran 90, Nesl, and languages based on BMF. These operations are commonly implemented by a library whose routines operate on a distributed representation of the aggregate structure; the compiler merely generates the control code invoking the library routines and all machine-dependent code is encapsulated in the library. While this approach is convenient, we argue that by breaking the abstraction enforced by the library and by presenting some of internals in the form of a new intermediate language to the compiler back-end, we can optimize on all levels of the memory hierarchy and achieve more flexible data distribution. The new intermediate language allows us to present these optimisations elegantly as program transformations. We report on first results obtained by our approach in the implementation of nested data parallelism on distributed-memory machines.

1 Introduction

A *collection-oriented* programming style is characterized by the use of operations on *aggregate* data structures (such as, arrays, vectors, lists, or sets) that manipulate the structure as a whole [21]. In parallel collection-oriented languages, a basic set of such operations is provided, where each operation has a straight-forward parallel semantics, and more complex parallel computations are realized by composition of the basic operations. Typically, the basic operations include elementwise arithmetic and logic operations, reductions, prescans, re-ordering operations, and so forth. We find collection-oriented operations in languages, such as, Fortran 90/95 [18], Nesl [1], and Gold-FISh [15] as well as parallel languages based on the Bird-Meertens Formalism (BMF) and the principle of algorithmic skeletons [9, 6]. Usually, higher-order operations on the aggregate structure are provided, e.g., apply-to-each in Nesl; FORALL in Fortran 90/95[1]; and map, filter, and so on in BMF. Collection-oriented parallelism supports clear code, the use of program transformation for both program development and compilation, and simplifies the definition of cost models [23, 1].

---

[1] Fortran's FORALL loops are index-based, which complicates the implementation in comparison to, e.g., Nesl's apply-to-each, but still, many of the basic principles are the same [8].

The aggregate data structures of collection-oriented languages are often implemented by a library whose routines are in functionality close to the basic operations of the source language. Consequently, the library is both high-level and complex, but carefully optimized for the targeted parallel hardware; furthermore, it encapsulates most machine dependencies of the implementation. We call this the *library approach*—for Nesl's implementation, see [4]; Fortran 90/95's collection-oriented operations are usually directly realized by a library, and even, FORALL loops can benefit from a collection-oriented implementation [8]. The library approach, while simplifying the compiler, blocks optimizations by the rigid interface imposed by the library—most importantly, it inhibits the use of standard optimizations for processor-local code and hinders the efficient use of the memory hierarchy, i.e., optimizations for the processor cache and for minimizing communication operations.

We propose to continue compilation, where the library approach stops and delegates an efficient implementation to the library. As a result, we can optimize on all levels of the memory hierarchy (registers, caches, local memory, and distributed memory)—after breaking the abstraction enforced by the library, optimizations across multiple operations become possible and data distribution gets more flexible. Our main technical contribution is an *intermediate language* that distinguishes local and distributed data structures and separates local from global operations. We present the internals of the library routines to the back-end of the compiler in the form of this intermediate language, which allows us to implement optimizations by program transformation—thus, the optimizations are easy to comprehend, the compiler can be well structured, optimizations are easily re-ordered and can be proved correct individually. Although we provide some technical detail, we do not have enough room for discussing all important points; the proposed technique has been applied to the implementation of nested data parallelism in the first author's PhD thesis [17], where the missing detail can be found.

Our main contributions are the following three:

- We introduce the notion of *distributed types*, to statically distinguish between local and distributed data structures.
- We outline an intermediate language based on distributed types that allows to statically separate purely local computations, where program fusion can be applied to improve register and cache utilization, from global operations, where program transformations can be used to minimize communication.
- We present experimental evidence for the efficiency of our method.

The paper is structured as follows: Section 2 details the benefits of an intermediate language that explicitly distinguishes between local and global data structures and computations. Section 3 introduces the concept of distributed types and outlines a declarative intermediate language based on these types. Section 4 briefly outlines the use of the intermediate language. Section 5 presents results obtained by applying our method to the implementation of nested data parallelism. Finally, Section 6 discusses related work and summaries.

2 Global versus Local Computations

Before outlining our proposal, let us discuss the disadvantages of the library approach.

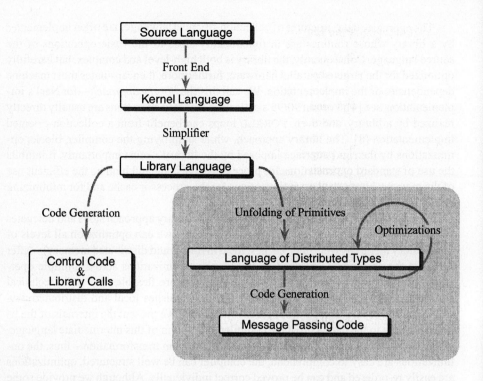

Fig. 1. Compiler structure (the grey area marks our new components)

2.1 The Problems of High-level Libraries

The structure of a compiler based on the library approach is displayed in Figure 1 when the grey area is omitted. The source language is analyzed and desugared, yielding a kernel language, which forms the input to the back-end (optimizations on the kernel language are left out in the figure). The back-end simplifies the kernel language and generates the control code that implements the source program with calls to the library implementing the parallel aggregate structure.

As an example, consider the implementation of Nesl [1]. In the case of Nesl, the simplifier implements the *flattening transformation* [5, 20, 16], which transforms all nested into flat data parallelism. In CMU's implementation of the language [4], the library language is called *VCODE* [2]—it implements a variety of operations on simple and segmented vectors, which are always uniformly distributed over the available processing elements. In fact, CMU's implementation does not generate an executable, but instead emits VCODE and interprets it at runtime by an interpreter linked to their *C Vector Library (CVL)* [3]. Unfortunately, this approach, while working fine for vector computers, is not satisfying for shared-memory [7] and distributed-memory machines [14].

The problems with this approach are mainly for three reasons: (1) Processor caches are badly utilized, (2) communication operations cannot be merged, and (3) data distribution is too rigid. In short, the program cannot be optimized for the memory hierarchy.

2.2 Zooming Into the Library

The library routines provide a *macroscopic view* of the program, where many details of the parallel implementation of these routines are hidden from the compiler. We propose to break the macroscopic view in favour of a *microscopic view*, where each of the library routines is unfolded into some *local computations* and *global communication* operations. The grey area of Figure 1 outlines an improved compiler structure including this unfolding and an optimization phase, where the microscopic view allows new optimizations realized as program transformations that minimize access to data elements that are located in expensive regions of the memory hierarchy. In the following, we outline the most important optimizations. Please note that the optimizations themselves are not new—our contribution is to *enable* the automatic use of these optimizations for the parallel implementation of aggregate structures in collection-oriented languages.

Optimizing local computations. The microscopic view allows to jointly optimize adjacent local computations that were part of different library routines; in the extreme, purely local routines, such as, elementwise operations on aggregate structures, can be fully merged. Such optimizations are crucial to exploit machine registers and caches.

For example, the following Nesl [1] function summing all squares from 1 to n proceeds in three steps: First, a list of numbers from 1 to n is generated; second, the square for each number is computed; and third, all squares are summed up.

```
sumSq1 (n)  : Int -> Int =
  let nums = [1:n+1];              — numbers from 1 to n
      sqs  = {i*i : i in nums}     — square each i from nums
  in
  sum (sqs);                       — perform a reduction on sqs
```

Each of the three steps corresponds to one VCODE operation, i.e., one library routine, in CMU's Nesl implementation. Even on a single processor machine, this program performs poorly when the value of n is sufficiently large, i.e., when the intermediate results do not fit into the cache anymore. Much more efficient is the following explicitly recursive version that can keep all intermediate results in registers.

```
sumSq2 (n)  : Int -> Int = sumSq2' (1, n);

sumSq2' (x, n)  : (Int, Int) -> Int =
  if (x > n) then 0
  else x*x + sumSq2' (x + 1, n);
```

In sequential programming, *fusion* (or deforestation) techniques are applied to derive sumSq2 automatically from sumSq1 [25, 11, 24, 19]. Unfortunately, these techniques are not directly applicable to parallel programming, because they replace the aggregate-based computations fully by sequential recursive traversals—sumSq2 completely lost the parallel interpretation of sumSq1. However, as soon as we have clearly separated local from global computations, we can apply fusion and restrict it to the local computations, so that the global, parallel structure of the program is left intact.

Optimizing global communication. The separation of local and global operations, which becomes possible in the microscopic view, allows to reduce communication overhead in two important ways: (1) We can merge communication operations and (2) we can trade locality of reference for a reduction of communication needed for load balancing. A simple example for the first point is the Nesl expression for concatenating three vectors: (xs ++ ys) ++ zs. It specifies pure communication, which re-distributes xs, ys, and zs such that the result vector is adequately distributed. Hence, the intermediate result xs ++ ys should ideally not be computed at all, but xs and ys should immediately be distributed correctly for the final result. Unfortunately, the rigid interface of a high-level library hinders such an optimization.

Regarding the second point, i.e., trading locality of reference for reduced load balancing costs, it is well known that optimal load balance does not necessarily lead to minimal runtime. Often, optimal load balance requires so much communication that an imbalanced computation is faster. Again, the compiler would often be able to optimize this tradeoff, but it is hindered in generating optimal code in the library approach, as the library encapsulates machine specific factors and library routines require fixed data distributions for their arguments.

3 Distributed Types Identify Local Values

A macroscopic parallel primitive, i.e., an operation on an aggregate structure in the library approach, is in the microscopic view implemented as a combination of purely local and global subcomputations. The latter induce communication to either combine the local sub-results into a global result or to propagate intermediate results to other processors, which need this information to execute the next local operation. For example, consider the summation of all values of an aggregate structure on a distributed-memory machine. In a first step, each processor adds up all the values in its local memory; in the second step, the processors communicate to sum all local results, yielding the global result, which is usually distributed to all processors. To model local and global operations in the microscopic view, we need to distinguish between local and global values. We define a *global value* as a value whose layout is known by all processors; thus, each processor is able to access the components of a global value without explicitly co-operating with other processors.[2] In contrast, a *local value* is represented by an instance on every processor, where the various instances are completely independent and accessible only by the processor on which they reside. We use a type system including *distributed types* to statically model this difference.

3.1 Distributed Types

In the source language, we distinguish between scalar values of basic type and compound values of aggregate type, where the latter serve to express parallelism. Both kinds

---

[2] We assume a global address space, where a processor can get non-local data from another processor as long as it knows the exact layout of the whole structure across all processors. Such a global address space is supported in hardware by machines, such as the Cray T3E, and is generally available in MPI-2 by one-sided communication [13].

of values are global in the sense defined before; each processor either has a private copy of the value (this is usually the case for scalar values) or it is aware of the global layout and can, thus, access any component of the value if necessary. Such values are often not suitable for expressing the results computed by local operations, as for example, the local sums computed in the first step of the previous summation example. The result of the local summation is actually an ordered collection of scalar values, with as many elements as processors. The ordering, which is essential for some computations, corresponds to the ordering of the processors. We introduce *distributed types*, denoted by double angle brackets $\langle\langle \cdot \rangle\rangle$ to represent such local values; to be precise, such a value is an ordered collection of fixed, but unknown arity. The arity reflects the number of processors, which is unknown at compile-time, but fixed throughout one program execution. For any type α, $\langle\langle \alpha \rangle\rangle$ represents such a collection of processor-local values of type α.

The usefulness of distributed types for local results of basic type can easily be seen in the sum example. However, it is probably not as obvious for values of aggregate type, as those values are usually not replicated across processors, but distributed over the processors. Still, the distinction between global and local values is also important for aggregate data types. For example, consider a filter operation that removes all elements that do not satisfy a given predicate from a collection. After each processor computed the filter operation on its local share of the aggregate structure, the overall size of the result is not yet known; thus, the intermediate result of the local operation is not a global aggregate structure according to our definition, but a collection of local structures.

3.2 Distributed Computations

Together with the type constructor $\langle\langle \cdot \rangle\rangle$, we introduce a higher-order operation $\langle\langle \cdot \rangle\rangle$. For a function f, we denote by $\langle\langle f \rangle\rangle$ the operation that applies f in parallel to each local component of a value of distributed type. Thus, for

$$f : \alpha_1 \times \cdots \times \alpha_n \to \beta_1 \times \cdots \times \beta_m \tag{1}$$

the operation $\langle\langle f \rangle\rangle$ has type

$$\langle\langle f \rangle\rangle : \langle\langle \alpha_1 \rangle\rangle \times \cdots \times \langle\langle \alpha_n \rangle\rangle \to \langle\langle \beta_1 \rangle\rangle \times \cdots \times \langle\langle \beta_m \rangle\rangle \tag{2}$$

In other words, $\langle\langle f \rangle\rangle (a_1, \ldots, a_n)$, with the a_i of distributed type $\langle\langle \alpha_i \rangle\rangle$, applies f on each processor P to the n-tuple formed from the components of a_1 to a_n that are local to P and yields an m-tuple of local results on P. The jth components of the result tuples of all processors form, then, a local value of distributed type $\langle\langle \beta_j \rangle\rangle$. This formalism distinguishes local computations (inside $\langle\langle \cdot \rangle\rangle$) from global computations and allows to calculate with them, while still maintaining a high level of abstraction.

3.3 A Language of Distributed Types

Before continuing with the implementation of aggregate structures using distributed types, let us become more concrete by fixing a syntax for an intermediate language that

$$
\begin{align*}
\mathsf{F} \to{}& \mathsf{V}\ (\ \mathsf{V}_1\ ,\ \ldots\ ,\ \mathsf{V}_n\ )\ :\ \mathsf{T}\ =\ \mathsf{E} && \text{(function definition)}\\
\mathsf{T} \to{}& \langle\!\langle\ \mathsf{T}\ \rangle\!\rangle && \text{(distributed type)}\\
\mid{}& [\ \mathsf{T}\ ] && \text{(aggregate type)}\\
\mid{}& \mathsf{T}_1 \times \cdots \times \mathsf{T}_n && \text{(tuple type)}\\
\mid{}& \mathsf{T}_1\ \to\ \mathsf{T}_2 && \text{(function type)}\\
\mid{}& \mathit{Int}\ \mid\ \mathit{Float}\ \mid\ \mathit{Char} && \text{(scalar types)}\\
\mathsf{E} \to{}& \mathsf{let}\ \mathsf{V}\ =\ \mathsf{E}\ \mathsf{in}\ \mathsf{E} && \text{(local definition)}\\
\mid{}& \mathsf{if}\ \mathsf{E}_1\ \mathsf{then}\ \mathsf{E}_2\ \mathsf{else}\ \mathsf{E}_3 && \text{(conditional)}\\
\mid{}& \mathsf{E}_1\ \mathsf{E}_2 && \text{(application)}\\
\mid{}& \langle\!\langle\ \mathsf{E}\ \rangle\!\rangle && \text{(replication)}\\
\mid{}& [\ \mathsf{E}_1\ ,\ \ldots\ ,\ \mathsf{E}_n\ ] && \text{(aggregate constructor)}\\
\mid{}& (\ \mathsf{E}_1\ ,\ \ldots\ ,\ \mathsf{E}_n\ ) && \text{(tuple)}\\
\mid{}& \mathsf{V} && \text{(variable)}
\end{align*}
$$

Fig. 2. Syntax of $\mathcal{L}_{\mathrm{DT}}$

supports distributed types. The language $\mathcal{L}_{\mathrm{DT}}$, whose syntax is displayed in Figure 2, allows a functional representation of the compiled program. A functional representation is well suited for optimization by program transformation and does not restrict the applicability of our approach, as parallel computations are usually required to be side-effect free, even in imperative languages (see, for example, the *pure functions* of Fortran 95 and HPF [10]).

A program is a collection of typed function definitions produced by **F**. Types **T** include a set of scalar types as well as more complex types formed by the tuple and function constructors. Furthermore, $[\alpha]$ denotes an aggregate type whose elements are of type α and $\langle\!\langle \alpha \rangle\!\rangle$ is a distributed type as discussed previously.

Expressions **E** include the standard constructs for local definitions, conditionals, and function application as well as tuple and aggregate construction. Moreover, we have $\langle\!\langle E \rangle\!\rangle$ for some expression E, where we require that E is of functional type and the function represented by E is made into a parallel operation as discussed in the previous subsection. Furthermore, we use $E_1 \circ E_2$ to denote function composition and $E_1 \times E_2$ to denote the application of a pair of functions to a pair of arguments, i.e., $(f \times g)(x, y) \equiv (f(x), g(y))$. We omit a formal definition of $\mathcal{L}_{\mathrm{DT}}$'s type system, as it is standard apart from the addition of distributed types, which essentially means to cast Equations 1 and 2 into a typing rule. Replication distributes over $\circ$ and $\times$, i.e., $\langle\!\langle f \rangle\!\rangle \circ \langle\!\langle g \rangle\!\rangle = \langle\!\langle f \circ g \rangle\!\rangle$ and $\langle\!\langle f \rangle\!\rangle \times \langle\!\langle g \rangle\!\rangle = \langle\!\langle f \times g \rangle\!\rangle$.

3.4 Operations on Distributed Types

To re-phrase the primitives of the library approach in the microscopic view, we need operations that convert values of global type into values of distributed type and vice versa. In general, we call a function of type $\alpha \to \langle\!\langle \alpha \rangle\!\rangle$ a *split operation* and a function of type $\langle\!\langle \alpha \rangle\!\rangle \to \alpha$ a *join operation*. Intuitively, a split operation decomposes a global value into its local components for the individual processors. Conversely, a join operation combines the components of a local value into one global value. Note that especially split operations do not necessarily perform work; sometimes they merely shift the perspective from a monolithic view of a data structure to one where the distributed components of the same structure can be manipulated independently.

The operation $split\_scalar_A : A \to \langle\langle A \rangle\rangle$ takes a value of basic type A (i.e., *Bool*, *Int*, or *Float*) and yields a local value whose components are identical to the global value. It is an example of a function, which typically does not imply any work, as it is common to replicate scalar data over all processors, to minimize communication; but, the shift in perspective is essential, as $\langle\langle A \rangle\rangle$ allows to alter the local copies of the value independently, whereas A implies the consistent change of all local copies.

On values of type integer, we also have $split\_size : Int \to \langle\langle Int \rangle\rangle$. The argument, say n, is assumed to be the size of an aggregate structure; $split\_size$ yields a collection of local integer values, whose sum is equal to n. Each component of the collection is the size of the processor-local chunk of an aggregate structure of size n when it is distributed according to a fixed load balancing strategy.

$\mathcal{L}_{DT}$ requires a join operation for each reduction on the aggregate structure:

$join\_+_A \quad : \langle\langle A \rangle\rangle \to A \qquad$ — for integers and floats
$join\_max_A : \langle\langle A \rangle\rangle \to A \qquad$ — for integers and floats
$join\_and \quad : \langle\langle Bool \rangle\rangle \to Bool \quad$ — and so on ...

They are used to combine the local results of the parallel application of the reduction to yield a global result—in the summation example from the beginning of this section, we would need $join\_+_{Int}$ if we sum integer values. Furthermore, we can use $join\_+_{Int} \circ \langle\langle \# \rangle\rangle$ to compute the number of elements in a structure of type $\langle\langle [\alpha] \rangle\rangle$, where $\#$ gives the size of a structure $[\alpha]$.

Furthermore, for aggregate structures, the operation $split\_agg : [\alpha] \to \langle\langle [\alpha] \rangle\rangle$ provides access to the local components of the distributed structure and, conversely, the operation $join\_agg : \langle\langle [\alpha] \rangle\rangle \to [\alpha]$ combines the local components of a structure into a global structure. The operation $split\_agg$ induces no communication or other costs; it merely provides an explicit view on the distribution of a global structure. In contrast, $join\_agg$ re-arranges the elements, if applied to a structure of distributed type, whose components are not properly distributed according to the fixed load balancing strategy. Thus, the combined application $split\_agg \circ join\_agg$, while computationally being the identity, ensures that the resulting structure is distributed according to the load-balancing strategy of the concrete implementation. To illustrate the use of split and join operations for the definition of more complex routines, we formalize the summation that we already discussed in the beginning of this section:

$$sum\ (xs) : [Int] \to Int = (join\_+_{Int} \circ \langle\langle reduce\_+_{Int} \rangle\rangle \circ split\_agg)\ xs$$

After obtaining the local view of the aggregate structure xs with $split\_agg$, summation proceeds in two steps: First, the purely local reduction $\langle\langle reduce\_+_{Int} \rangle\rangle$, which yields a local integer result on each processing element, i.e., a value of type $\langle\langle Int \rangle\rangle$; and second, the global reduction of the local results with $join\_+_{Int}$.

Apart from join and split operations, which compute global values from distributed values, and vice versa, we need operations that *propagate* information over the components of a distributed value; they receive a value of distributed type as input parameter and also return a distributed type. Although the parallel application of local operations

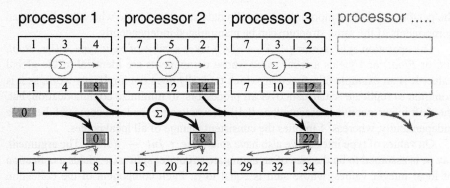

Fig. 3. Local and global operations in *prescan_+*

has the same type, namely $\langle\!\langle[\alpha]\rangle\!\rangle \to \langle\!\langle[\beta]\rangle\!\rangle$, propagation is fundamentally different, since, in the case of the parallel application of local operations, there is no communication between the processing elements. Similar to the join operations, where $\mathcal{L}_{DT}$ offers an operation for every form of reduction, there is a propagate operation in $\mathcal{L}_{DT}$ for each prescan operation that we need on the aggregate structure. A propagate operation prescans over the components of the distributed type according to the implicit order of the components. Some examples of this type of operation are the following:

$propagate\_+_A \quad : \langle\!\langle A\rangle\!\rangle \to \langle\!\langle A\rangle\!\rangle \quad$ — for integers and floats
$propagate\_max_A : \langle\!\langle A\rangle\!\rangle \to \langle\!\langle A\rangle\!\rangle \quad$ — for integers and floats
$propagate\_and \quad : \langle\!\langle Bool\rangle\!\rangle \to \langle\!\langle Bool\rangle\!\rangle \quad$ — and so on ...

Formally, the semantics of the split, join, and propagate operations can be characterized by a set of axioms [17].

4 Optimizing with Distributed Types

With $\mathcal{L}_{DT}$ we can realize the three compiler phases marked by the grey area in Figure 1: unfolding of the library primitives, optimizations, and code generation. We do not have the space to discuss them in detail, but like to outline some important points of the first two phases, i.e., unfolding and optimizations, in the next two subsections. Afterwards, we discuss a small example to illustrate the technique. More details can be found in [17].

4.1 Unfolding Library Routines

The whole point of $\mathcal{L}_{DT}$ is to allow defining the internal behaviour of the routines that are primitive in the library approach (the unfolding step in Figure 1). We already defined the summation function *sum* in $\mathcal{L}_{DT}$; now, we discuss the related, but more complex pre-scan operation. We do not detail the local operations, because they depend on the implemented aggregate type and source language—in [17], the details for the case of nested parallel languages are given.

Figure 3 sketches the steps of a parallel *prescan_+* routine. The dotted lines mark processor boundaries and indicate that each processor has a local copy of a part of the aggregate structure. Local computations are marked by grey, global computations by black arrows. The $prescan_+$ algorithm proceeds in three steps:

1. Each processor computes a local pre-scan on its share of the aggregate structure. The local sum of each processor is marked with a grey box.
2. A global pre-scan (the black arrow) over all local sums provides each processor with the sum of all elements residing on preceding processors.
3. Each processor adds its result of the previous operation to each element of the intermediate vector computed in the first step.

Overall, the $prescan_+$ operation consists of two purely local computations (Step 1 and 3) and one global operation (Step 2), entailing communication.

Next, we express this computation in terms of distributed types, *split*, *join*, *propagate*, and local operations. Before applying the first local operation, we have to use $split\_agg$ to obtain the local view on the input structure xs. Then, we apply the local pre-scan operation, $prescan_+\_local$ to each component of the local value. As can be seen in Figure 3, $prescan_+\_local$ has to return two values on each processor: the new local structure and the local sum—that is, $prescan_+\_local : [Int] \rightarrow ([Int] \times Int)$. Overall, Step 1 can be expressed in $\mathcal{L}_{DT}$ by the composition $\langle\!\langle prescan_+\_local \rangle\!\rangle \circ split\_agg$.

Step 2 essentially is the previously discussed $propagate_+ : \langle\!\langle Int \rangle\!\rangle \rightarrow \langle\!\langle Int \rangle\!\rangle$. It realizes the black arrow of Figure 3, propagating the local results from left to right. In Step 3, we add, on each processor, the local result of the previous step to each element of the local fragment of the intermediate aggregate structure that we obtained in Step 1. Therefore, we define an auxiliary function $+\_offset$ that gets the results of the previous two steps as arguments.

$$+\_offset\ (zs,\ off) : [Int] \times Int \rightarrow [Int] = elem_+\ (zs,\ dist\ (off,\ \#zs))$$

The expression $dist\ (off,\ \#zs)$ replicates the value off as often as the size of zs requires. Elementwise addition $elem_+$ computes the local contributions to the overall result, which in turn is obtained from the local results with $join\_agg$.

Overall, we have the following definition for $prescan_+$, where local and global computations are explicit—local computations are enclosed in $\langle\!\langle \cdot \rangle\!\rangle$:

$prescan_+\ (xs) : [Int] \rightarrow [Int] =$
  $(join\_agg \circ \langle\!\langle +\_offset \rangle\!\rangle \circ (id \times propagate_+)$
    $\circ \langle\!\langle prescan_+\_local \rangle\!\rangle \circ split\_agg)\ xs$

4.2 Optimizations on $\mathcal{L}_{DT}$

The unfolding of the library routines in $\mathcal{L}_{DT}$ enables new kinds of optimizations, which we can categorize as follows: (1) optimizations of local computations, (2) optimizations of global operations, and (3) enabling optimizations. We outlined the first two kinds already informally in Section 2 and refer to [17] for a formal definition of applicable fusion techniques, optimizations minimizing communication, and example derivations. In the previous subsection, we avoided fixing a notation for specifying local computations, as there is a wide design space, which is largely independent from the main points of this paper. One possibility is to provide a set of primitives functions for manipulating local fragments of the aggregate structure; another possibility is the use of loop constructs that define iterators over the local fragments of aggregate structures (this the approach taken in [17]). In any case, a set of equational transformation rules should be provided that implements fusion (see Section 2) and possibly also other optimizations.

The third form of optimizations, *enabling optimizations*, are independent of the notation for local computations; they use essential properties of $\mathcal{L}_{DT}$ to re-order local and global computations such that subcomputations are nicely clustered for the first two forms of optimizations. Space limitations again force us to refer to [17] for the complete set of these optimization rules; however, we want to state one crucial point: As mentioned, $split\_agg \circ join\_agg$ does not change the contents of the value of type $\langle\!\langle[\alpha]\rangle\!\rangle$ to which it is applied, but it ensures that the aggregate structure is well balanced. After unfolding the library primitives, all intermediate results are re-balanced in this way, which does not necessarily minimize the runtime (see Section 2). Thus, the optimizer removes many of these re-distributions, which saves communication and moves local computations, which were originally separated by re-balancing, next to each other. Thus, new local optimizations becomes possible, after applying the enabling optimization $\langle\!\langle f \rangle\!\rangle \circ \langle\!\langle g \rangle\!\rangle = \langle\!\langle f \circ g \rangle\!\rangle$.

4.3 An Example

We use sumSq1 from Section 2.2 for an example transformation. Let us start with the definition of the function as it appears in the library language, i.e., after the font-end:

$sumSq1\,(n)\,:\,Int\,\to\,Int\,=\,(sum\,\circ\,map\,(sq)\,\circ\,range)\,(1,\,n)$

where we have $sq\,(x)\,:\,Int\,\to\,Int\,=\,x\,*\,x$. We already provided the definition for sum in Section 3.4 and define map and $range$ as follows:

$map\,(f)\,:\,(a \to b) \to [a] \to [b]\,=\,join\_agg \circ \langle\!\langle seq\_map\,(f) \rangle\!\rangle \circ split\_agg$
    where
        $seq\_map\,(f)\,[x_1,\,\ldots,\,x_n]\,=\,[f\,(x_1),\,\ldots,\,f\,(x_n)]$
$range\,:\,Int \times Int \to [Int]\,=\,join\_agg \circ \langle\!\langle seq\_range \rangle\!\rangle \circ split\_range$
    where
        $seq\_range\,(n,\,m)\,=\,[n,\,n+1,\,\ldots,\,m]$

For the local functions seq_map and seq_range, we only provide a specification; the exact definition depends on how we choose to represent the processor-local, sequential code. However, the function $split\_range$ deserves a little more attention; it computes for each processing element the local range of values and may be defined as

$split\_range\,(from,\,to)\,:\,Int \times Int \to \langle\!\langle Int \rangle\!\rangle \times \langle\!\langle Int \rangle\!\rangle\,=$
    let
        $lens\,=\,split\_size\,(to - from + 1)$
        $froms\,=\,\langle\!\langle + \rangle\!\rangle\,(split\_scalar_{Int}\,(from),\,propagate\_+\,(lens))$
    in
        $(froms,\,\langle\!\langle \lambda f, l. f + l - 1 \rangle\!\rangle\,(froms,\,lens))$

Nevertheless, it makes sense to regard this function as a primitive, as it can, in dependence on the distribution policy for the concrete aggregate structure, usually be implemented without any communication.

When the compiler unfolds the library primitives, it can transform the function compositions within the body of $sumSq1$ as follows:

$(join\_ +_{Int} \circ \langle\langle reduce\_ +_{Int} \rangle\rangle) \circ split\_agg) \circ (join\_agg \circ \langle\langle seq\_map \, (sq) \rangle\rangle)$
$\circ split\_agg) \circ (join\_agg \circ \langle\langle seq\_range \rangle\rangle) \circ split\_range)$
$= \{\text{eliminate } split\_agg \circ join\_agg \quad \text{(there is no load imbalance anyway)}\}$
$join\_ +_{Int} \circ \langle\langle reduce\_ +_{Int} \rangle\rangle \circ \langle\langle seq\_map \, (sq) \rangle\rangle \circ \langle\langle seq\_range \rangle\rangle \circ split\_range$
$= \{\text{distribution law for replication} \quad \text{(Section 3.3)}\}$
$join\_ +_{Int} \circ \langle\langle reduce\_ +_{Int} \circ seq\_map \, (sq) \circ seq\_range \rangle\rangle \circ split\_range$

Now, the local computation $reduce\_ +_{Int} \circ seq\_map \, (sq) \circ seq\_range$ is a slight generalization of the original $sumSq1$ function (it has an explicit start value instead of starting with 1). As it is purely local, we can fuse it into essentially the same code as sumSq2' (the auxiliary function in the fused version of Section 2.2), i.e., we get

$$sumSq1\,(n) \,:\, Int \to Int = (join\_ +_{Int} \circ \langle\langle sumSq2' \rangle\rangle \circ split\_range)\,(1, n)$$

We achieved our goal of applying fusion to the local code in a controlled manner, which leaves the parallel semantics of our code intact.

5 Benchmarks

We summarize results collected by applying our method to the implementation of Nesl-like [1] nested data parallelism [17]. However, we do not directly compare our code and that of CMU's implementation of Nesl [4], because the implementation of CMU's vector library CVL [3] is already an order of magnitude slower and scales worse than our vector primitives on the Cray T3E, which we used for the experiments. Our code is hand-generated using the compilation and transformation rules of [17]—we are currently implementing a full compiler.

Local optimizations. In Section 2, we mentioned fusion as an important technique for optimizing local computations. For sufficiently large vectors, the optimization achieves a speedup linear to the numbers of vector traversals that become obsolete as main memory access is reduced. Comparing sumSq1 and sumSq2 from Section 2, we find a speedup of about a factor four for large vectors—see Figure 4 (left part). Fusion is particularly important for compiling nested parallelism, as the flattening transformation [5, 20, 16] leads to extremely fine-grained code.

Global optimizations. Concerning global optimizations, we measured the savings from reducing load balancing, i.e., a $split\_agg \circ join\_agg$ combination (see previous section). In Figure 4 (right part), we applied a filter operation removing all elements from a vector except those on a single processor, thus, creating an extreme load imbalance. After filtering, we applied one scalar operation elementwise. The figure shows that, if this operation (due to fusion) finds its input in a register, re-distribution is up to 16 processors significantly slower than operating on the unbalanced vector. In the case of a more complex operation, the runtime becomes nearly identical for 16 processors. As the flattening transformation introduces many filter operations it is essential to avoid re-distribution where possible.

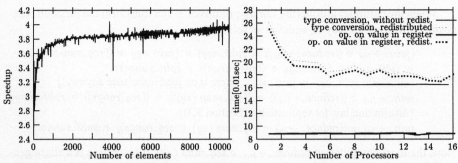

The left figure displays the speedup of *sumSq2* over *sumSq1* (by loop fusion). The right figure contains the runtime of filtering plus one successive operation; with and without re-distribution.

Fig. 4. Effects of local and global optimizations

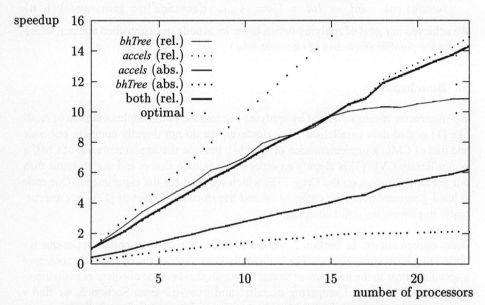

Fig. 5. Relative and absolute speedup of n-body algorithm for 20,000 particle

A complete application. We implemented an n-body code, the 2-dimensional Barnes-Hut algorithm, to get a feeling for the overall quality of our code. The absolute and relative speedup for 20,000 particles are shown in Figure 5. The runtime is factored into the two main routines, which build the Barnes-Hut tree (*bhTree*) and compute the accelerations of the particles (*accels*), respectively. The routine *bhTree* needs a lot of communication, and thus, scales worse, but as it contributes only to a small fraction of the overall runtime, the overall algorithm shows a speedup nearly identical to *accels*.

The absolute speedup is against a purely sequential C implementation of the algorithm. We expect to be able to improve significantly on the absolute speedup, which is about a factor of 6 for 24 processors, because we did not exploit all possible optimizations. To see the deficiency of the library approach, consider that in the two-dimensional Barnes-Hut algorithm, we achieve a speedup of about a factor of 10 by fusion.

6 Related Work and Conclusions

With regard to implementing nested parallelism, Chatterjee et al's work, which covers shared-memory machines [7], is probably closest to ours; however, their work is less general and not easily adapted for distributed memory. In the BMF community, *p-bounded lists* are used to model block distributions of lists [22]; this technique is related to distributed types, but the latter are more general as they (a) can be applied to any type and (b) do not make any assumption about the distribution, especially, $split\_agg \circ join\_agg$ is in general not the identity. Gorlatch's *distributed homomorphisms* appear to be well suited to systematically derive the function definitions, which we need for the library unfoldings—like, e.g., the $prescan\_+$ definition [12]. To the best of our knowledge, we are the first to propose a *generic, transformation-based* alternative to the library approach. Our method enables a whole class of optimizations by exposing the internals of the library routines to the compiler's optimizer; the transformation-based approach systematically distinguishes between local and global operations using a type system and makes it easy to formalize the optimizations. Most other work on implementing aggregate structures either focuses on the library approach or describes language-specific optimizations. We applied our method to the implementation of nested data parallelism and achieved satisfying results in first experiments.

Acknowledgements. We thank the Nepal project members, Martin Simons and Wolf Pfannenstiel, for providing the context of our work and feedback on a first version of the paper. We are also grateful to Sergei Gorlatch and the anonymous referees for their comments on an earlier version. The first author likes to thank Tetsuo Ida for his hospitality during her stay in Japan. Her work was funded by a PhD scholarship of the Deutsche Forschungsgemeinschaft (DFG).

References

1. Guy E. Blelloch. Programming parallel algorithms. *Communications of the ACM*, 39(3):85–97, 1996.
2. Guy E. Blelloch and Siddhartha Chatterjee. VCODE: A data-parallel intermediate language. In *Proceedings Frontiers of Massively Parallel Computation*, pages 471–480, October 1990.
3. Guy E. Blelloch, Siddhartha Chatterjee, Jonathan C. Hardwick, Margaret Reid-Miller, Jay Sipelstein, and Marco Zagha. CVL: A C vector library. Technical Report CMU-CS-93-114, Carnegie Mellon University, 1993.
4. Guy E. Blelloch, Siddhartha Chatterjee, Jonathan C. Hardwick, Jay Sipelstein, and Marco Zagha. Implementation of a portable nested data-parallel language. In *4th ACM SIGPLAN Symposium on Principles and Practice of Parallel Programming*, 1993.
5. Guy E. Blelloch and Gary W. Sabot. Compiling collection-oriented languages onto massively parallel computers. *Journal of Parallel and Distributed Computing*, 8:119–134, 1990.
6. George Horatiu Botorog and Herbert Kuchen. Using algorithmic skeletons with dynamic data structures. In *IRREGULAR 1996*, volume 1117 of *Lecture Notes in Computer Science*, pages 263–276. Springer Verlag, 1996.
7. S. Chatterjee. Compiling nested data-parallel programs for shared-memory multiprocessors. *ACM Transactions on Programming Languages and Systems*, 15(3):400–462, july 1993.

8. S. Chatterjee, Jan F. Prins, and M. Simons. Expressing irregular computations in modern Fortran dialects. In *Fourth Workshop on Languages, Compilers, and Run-Time Systems for Scalable Computers*, Lecture Notes in Computer Science. Springer Verlag, 1998.
9. J. Darlington, A. J. Field, P. G. Harrison, P. H. J. Kelly, D. W. N. Sharp, Q. Wu, and R. L. While. Parallel programming using skeleton functions. In A. Bode, M. Reeve, and G. Wolf, editors, *PARLE '93: Parallel Architectures and Languages Europe*, number 694 in Lecture Notes in Computer Science, pages 146–160, Berlin, Germany, 1993. Springer-Verlag.
10. High Performance Fortran Forum. High Performance Fortran language specification. Technical report, Rice University, 1993. Version 1.0.
11. Andrew J. Gill, John Launchbury, and Simon L. Peyton Jones. A short cut to deforestation. In Arvind, editor, *Functional Programming and Computer Architecture*, pages 223–232. ACM, 1993.
12. S. Gorlatch. Systematic efficient parallelization of scan and other list homomorphisms. In L. Bouge, P. Fraigniaud, A. Mignotte, and Y. Robert, editors, *Euro-Par'96, Parallel Processing*, number 1124 in Lecture Notes in Computer Science, pages 401–408. Springer-Verlag, 1996.
13. William Gropp, Steven Huss-Lederman, Andrew Lumsdaine, Ewing Lusk, Bill Nitzberg, William Saphir, and Marc Snir. *MPI: The Complete Reference*, volume 2—The MPI-2 Extensions. The MIT Press, second edition, 1998.
14. Jonathan C. Hardwick. An efficient implementation of nested data parallelism for irregular divide-and-conquer algorithms. In *First International Workshop on High-Level Programming Models and Supportive Environments*, 1996.
15. C. Barry Jay. Costing parallel programs as a function of shapes. Invited submission to Science of Computer Programming, September 1998.
16. G. Keller and M. Simons. A calculational approach to flattening nested data parallelism in functional languages. In J. Jaffar, editor, *The 1996 Asian Computing Science Conference*, Lecture Notes in Computer Science. Springer Verlag, 1996.
17. Gabriele Keller. *Transformation-based Implementation of Nested Data Parallelism for Distributed Memory Machines*. PhD thesis, Technische Universität Berlin, Fachbereich Informatik, 1998. To appear.
18. Michael Medcalf and John Reid. *Fortran 90 Explained*. Oxford Science Publications, 1990.
19. Y. Onue, Zhenjiang Hu, Hideya Iwasaki, and Masato Takeichi. A calculational fusion system HYLO. In R. Bird and L. Meertens, editors, *Proceedings IFIP TC 2 WG 2.1 Working Conf. on Algorithmic Languages and Calculi, Le Bischenberg, France, 17–22 Feb 1997*, pages 76–106. Chapman & Hall, London, 1997.
20. Jan Prins and Daniel Palmer. Transforming high-level data-parallel programs into vector operations. In *Proceedings of the Fourth ACM SIGPLAN Symposium on Principles and Practice of Parallel Programming*, pages 119–128, San Diego, CA., May 19-22, 1993. ACM.
21. Jay M. Sipelstein and Guy E. Blelloch. Collection-oriented languages. *Proceedings of the IEEE*, 79(4):504–523, April 1991.
22. D. Skillicorn and W. Cai. A cost calculus for parallel functional programming. *Journal of Parallel and Distributed Computing*, 28:65–83, 1995.
23. D.B. Skillicorn. *Foundations of Parallel Programming*. Cambridge Series in Parallel Computation 6. Cambridge University Press, 1994.
24. Akihiko Takano and Erik Meijer. Shortcut deforestation in calculational form. In *Conf. Record 7th ACM SIGPLAN/SIGARCH Intl. Conf. on Functional Programming Languages and Computer Architecture*, pages 306–316. ACM Press, New York, 1995.
25. Philip Wadler. Deforestation: Transforming programs to eliminate trees. *Theoretical Computer Science*, 73:231–248, 1990.

A Transformational Framework for Skeletal Programs: Overview and Case Study

Sergei Gorlatch[1] and Susanna Pelagatti[2]

[1] Universität Passau, D-94030 Passau, Germany
[2] Universitá di Pisa, Corso Italia 40, I-56125 Pisa, Italy

Abstract. A structured approach to parallel programming allows to construct applications by composing *skeletons*, i.e., recurring patterns of task- and data-parallelism. First academic and commercial experience with skeleton-based systems has demonstrated both the benefits of the approach and the lack of a special methodology for algorithm design and performance prediction. In the paper, we take a first step toward such a methodology, by developing a general transformational framework named FAN, and integrating it with an existing skeleton-based programming system, P3L. The framework includes a new functional abstract notation for expressing parallel algorithms, a set of semantics-preserving transformation rules, and analytical estimates of the rules' impact on the program performance. The use of FAN is demonstrated on a case study: we design a parallel algorithm for the maximum segment sum problem, translate the algorithm in P3L, and experiment with the target C+MPI code on a Fujitsu AP1000 parallel machine.

1 Introduction

Current difficulties in the low-level parallel programming, using e.g., MPI [17], can be addressed by developing higher-level programming models together with convenient programming environments.

One of popular higher-level approaches is based on so-called *skeletons*, which can be viewed as recurring algorithmic and communication patterns, expressed in a rigorous way [19]. Representatives of the skeleton-based systems are the P3L system at the University of Pisa [3], its commercial analog SKIECL at QSW Ltd., and SCL at Imperial College [10]. These systems provide the user with a fixed number of higher-order skeletons, which can be customized for a particular application and used to construct a parallel program. A skeletal program is then (semi)automatically translated into some target language, e.g., C+MPI, using prepackaged parallel implementations of particular skeletons. The abstraction from communication and other details makes skeletal programs considerably better structured and less error-prone than their low-level counterparts.

The experience with skeletal programming, both in academia and industry, has brought also a new challenge. Since abstract constructs are usually very "coarse-grained", a slight change in one of skeletons, and especially in the way they are composed with each other, can lead to dramatic changes in the target program performance. Analyze the possible changes, predicting their influence on performance and optimizing the high-level program appropriately, has proven to be a non-trivial task for the user. In the long term, the challenge of designing efficient high-level programs should be addressed by providing a *methodology of skeleton-based programming* which would include methods and tools for choosing suitable skeletons, composing them in a program, estimating its expected performance, and making changes in design if necessary.

In this paper, we take a first step towards such a methodology, by presenting a general transformational framework for designing parallel algorithms on a high level of abstraction. The aim of the framework is to support the user in producing a high-quality skeletal program. We integrate this framework with the current P3L implementation, and assess the whole system in a case study.

The contributions and structure of the paper are as follows:

- Section 2 gives an overview of the framework, and puts it into the context of the P3L programming system, which we use as a testbed in our work.
- In Section 3, we present a new notation for parallel algorithms, called **FAN** (Functional Abstract Notation), which facilitates semantics-preserving transformations of algorithms in the design process.
- Section 4 describes a case study: designing a parallel algorithm for the maximum segment sum problem (a programming pearl [4]). We start with an intuitively clear **FAN**-formulation of the algorithm, and proceed, by applying **FAN**-transformations, towards an efficient but intricate parallel version, which hardly could be written by the user from scratch.
- In Section 5, we describe our cost model, which helps to decide under which conditions a particular transformation rule improves target performance.
- Section 6 revisits our case study: from the **FAN**-algorithm for the maximum segment sum, a P3L skeletal program is generated. This program is then automatically translated by the P3L compiler into a C+MPI code. We report some experimental result with thus obtained target program on a Fujitsu AP1000 parallel machine.

In conclusion, we compare to related work and outline our future efforts.

2 System Structure Overview

This section takes the P3L system [3] as a representative of skeleton-based systems, and outlines how it is augmented by the **FAN** transformational framework. The overall structure of the resulting **FAN**-P3L system is presented in Figure 1.

The figure shows how the user, depicted on the left side, communicates with the high-level programming system, which is partitioned using horizontal bold, solid lines into three parts, top-down: the transformational framework (presented in this paper), the (current) P3L system, and the target machines. Solid arrows show the connections between the parts of the system, dashed arrows depict the user interaction with the system, with bold, dashed arrows for the new interactions added by the transformational framework.

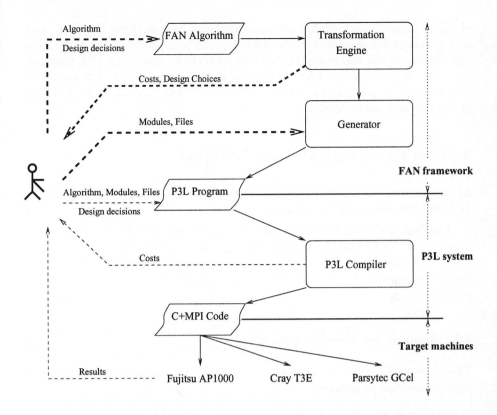

Fig. 1. Skeleton-based system P3L, augmented by the FAN framework

In the current P3L system, the user starts the development by writing a complete skeletal P3L program (in the middle of Figure 1). The user must provide the skeleton-based algorithm itself and also supply all necessary sequential modules, input and output files. This makes the corresponding arrow in the figure look quite overloaded. The program is optimized and translated by the P3L compiler, which provides the user with preliminary cost estimates for the program. On the base of these estimates, the user has to make design decisions which may require changing the algorithm, rewriting modules and reorganizing the files. If the user is satisfied with costs, the C+MPI code produced by the compiler can be run on an available target machine (some current platforms targeted by the P3L compiler are shown in the figure). The translation/estimation/rewriting process, which may have to be repeated several times, represents the challenge mentioned in the introduction and addressed in this paper.

The transformational **FAN** framework, shown in the upper part of Figure 1, offers the user an additional support in designing a skeletal program. The design process now starts by writing a functional version of the algorithm, without providing concrete modules and files. The algorithm is analyzed by the transformation engine which attempts to apply transformations from its depository of rules, thereby suggesting a choice of design alternatives for the user, with cost estimates for the alternatives. After possibly several iterations with different versions of the algorithm, the user may decide to generate a P3L program, for which the descriptors of files and modules becomes necessary. The generated P3L program is given to the current P3L system, which proceeds as described in the previous paragraphs.

3 FAN: Functional Abstract Notation

In this section, we briefly introduce an abstract notation, **FAN**, for specifying and transforming parallel programs.

The data space in a **FAN** program is viewed globally and there is no notion of processor at this abstract level. **FAN** allows the user to concentrate on the parallel structure of the program, abstracting from the syntactic details of a particular skeleton environment and from the application-specific sequential parts of the program.

3.1 Support for Transformations

The current P3L system implements several basic transformations and optimizations of skeleton programs:

- Data redistributions: if collecting data in one place becomes redundant due to a subsequent data scattering, the compiler eliminates the unnecessary data movements. Similarly, no rearrangement is done when data are produced and consumed by two consecutive skeletons requiring the same data distribution.
- Skeleton compositions: several important sequences of skeletons are optimized in order to redistribute the minimal amount of data among two consecutive local computations.
- Computation fusion: local pieces of computations, corresponding to different skeletons in a program, are composed within the sequential processor programs, which are subsequently optimized by native compilers, e.g., C.

The motivation of **FAN** is to allow much more powerful transformations, and to make them controllable by the program designer. Programming with skeletons deserves scrutiny and requires a method. The **FAN** framework supports such a method, based on a set of transformation rules, augmented with a cost calculus.

The **FAN** framework is based on the paradigm of functional programming. In particular, all algorithmic skeletons are captured as higher-order functions on data structures. Hence, program transformations can be proved as semantic equalities in the corresponding functional formalism. Functional framework offers effective support in prototyping and analysis, and does not limit the choice of the final implementation.

3.2 Notation Overview and Example

A FAN program contains a *header*, specifying the program name, a list of variables (arrays, tuples and scalars) which are taken as input and produced as output, and a *body*, defining a sequence of skeleton applications.

The declaration of a program header reads as follows:

progname (in *inlist* , out *outlist*)

where *inlist* describes the input variables on which the program works and *outlist* describes the variables computed by the program. Both input and output lists contain variables along with their types such as

$$v_1 : T_1, \ldots v_n : T_n$$

where $v_i : T_i$ means that variable v_i has type T_i. Data types allowed in variable declarations are scalars (scalar), vectors (arr[n]) and multidimensional arrays (arr[n][m]). There are also tuples which are arrays of a fixed dimension. A program body is a list of skeleton calls expressed in functional notation using single assignment variables. FAN skeletons are second order functions defined on arrays and scalars.

As an example, let us consider the following algorithm which computes the inner product of two vectors a and b of length n:

alg-inprod (in a, b : arr[n], out c : scalar)
   $t = $ map (*) (arrange (a,b));
   $c = $ reduce (+) t;

The input of program alg-inprod are two arrays of length n, and the output is a scalar, c. The body describes how the parallel computation takes place. First, the elements of a and b are paired using the arrange skeleton. In the example, (a,b) arranges a and b in a vector s of length n, in which each element $s[i]$ is a pair $(a[i], b[i])$. Then, the map skeleton multiplies all the pairs in s and stores the results in array t, so that $t[i] = a[i] * b[i]$. Finally, elements of t are summed up using the reduce skeleton.

3.3 Data- and Task-parallel Skeletons

There are two kinds of skeletons in FAN: data- and task-parallel skeletons. Task-parallel skeletons, including farm and pipeline, exploit parallelism among independent parts of an application. Independent parts typically exploit data parallelism internally and are more coarse grain [10,19]. Data-parallel skeletons express parallel structures in which the same operation has to be applied to a number of related data. Such skeletons in FAN include data arrangements expressed by arrange, and computations expressed by map, reduce, scanL, and scanR. In the rest of the paper we will concentrate on data parallel skeletons.

Data-parallel skeleton arrange is used to express data manipulations (replication, alignment etc.) which are needed to put data in place for subsequent data parallel computations. Different kinds of data arrangement are specified by providing particular predefined functions as parameters of **arrange** skeleton.

Some legal arrangements which will be used in the rest of the paper are shown in Table 3.3, where a and b are vectors of length n, c is a vector of pairs of length n and s is a scalar. (a,b) builds a vector of pairs as described before and **proj [1]** returns the first element of a particular pair:

arrangement	description	possible result
c = (a,b);	creates tuples (pairs)	$c[i] = (a[i], b[i])$, c :arr[n][2]
s = proj [1] c;	projects tuples/arrays	$s = (a[1], b[1])$, s :arr[2]

Table 1. Some legal arrangement patterns. a and b are vectors of length n.

The simplest computational skeleton, **map**, applies a function, f, to all the elements of an array, arranging results in an array with the same shape, for instance, if $a = [a[1], a[2], \ldots, a[n]]$ is a vector of length n then

map $f\ [a[1], a[2], \ldots, a[n]]\ =\ [f\ a[1], f\ a[2], \ldots, f\ a[n]]$

Finally, skeletons **reduce**, **scanL** and **scanR** provide the usual data-parallel reduce and scan operations on vectors:

reduce $(\oplus)\ a\ =\ a[0] \oplus \ldots \oplus a[n-1]$
scanL $(\oplus)\ a\ =\ [a[0],\ a[0] \oplus a[1],\ \ldots,\ a[0] \oplus \ldots \oplus a[n-1]]$
scanR $(\oplus)\ a\ =\ [a[0] \oplus \ldots \oplus a[n-1],\ \ldots,\ a[n-2] \oplus a[n-1],\ a[n-1]]$

The base operator, $\oplus$, plays an important role in the reduction and scan skeletons: if the operator is associative, the corresponding skeleton can be parallelized, with parallel time logarithmic in the size of the argument array.

4 Case Study: Maximum Segment Sum

In this section, we demonstrate the use of the **FAN** notation and the transformation rules in the program design process on a particular case study. We present only three transformations which are used in the example; a rich set of transformations, with details about their performance, is developed in a more general context in paper [16] being presented at IPPS/SPDP'99.

We consider the famous *Maximum Segment Sum* (MSS) problem – a *programming pearl* [4], studied by many authors [5, 8, 20, 22, 23]. Given a one-dimensional array of integers, function *mss* finds a contiguous array segment, whose members have the largest sum among all such segments, and returns this sum. For example:

$$mss\ [2, -4, 2, -1, 6, -3]\ =\ 7$$

where the result is contributed by the segment $[2, -1, 6]$.

4.1 Intuitive algorithm

The first version of the program for computing mss could be quite simple:

    alg-mss1 (in x : arr[n], out r : scalar);
      s = scanL $(op1)$ x;
      r = reduce (max) s;

where operator op1 is defined as follows:

$$a_1 \text{ op1 } a_2 = max\left((a_1 + a_2), a_2\right) \qquad (1)$$

Algorithm alg-mss1 takes array x as input, and proceeds intuitively as follows:

- The first stage, scanL $(op1)$, produces array s of type arr[n], whose i-th element is the maximum sum of segments of x ending in position i.
- The second stage, reduce (max), computes the maximum element of array s, thus yielding the resulting value, r, so that $r = mss\ x$.

Our first algorithm expressed by alg-mss1 has several good features. First, it is intuitively clear, and its correctness with respect to the specification of the mss problem can be proved. Second, it consists of only two skeletons, scan and reduction, both being well parallelizable. However, a closer look reveals also a serious drawback: the scan skeleton in alg-mss1 cannot be parallelized since operator $op1$ defined by (1) is not associative.

4.2 Parallelizable algorithm

In order to enable parallelization, we must construct an associative operator for the scan skeleton. An ubiquitous technique used for that is the introduction of auxiliary variables. In our case, we define the new, associative operator on pairs, so that the original operator is modeled by the first element of a pair:

$$(a_1, b_1)\, op2\, (a_2, b_2) = \left(\ max\left((a_1 + b_2), a_2\right),\ b_1 + b_2\ \right) \qquad (2)$$

Using this operator, we can rewrite the initial program as follows:

    alg-mss2 (in x : arr[n], out r : scalar)
      y = scanL $(op2)$ (arrange (x,x));
      s = arrange (proj [1]) y;
      r = reduce (max) s;

Note that the first two statements of alg-mss2 are semantically equivalent to the first statement of alg-mss1.

The asymptotic parallel complexity of algorithm alg-mss2 is logarithmic in the size of the input array. This follows from the logarithmic complexity of both the scan and the reduction stage with associative operators, and from the constant parallel complexity of map. Thus, theoretically the algorithm is optimal.

Algorithm alg-mss2 can be programmed directly in any parallel language which provides primitives for scans and reduction, i.e. P3L, MPI, etc. However, its performance in practice may suffer from the fact that it exploits two collective operations, each of which involves a considerable amount of interprocessor communication, with start-up time etc. Note that the algorithm presented in textbook [1] may have an even higher cost than alg-mss2, since it exploits three collective operations, two scans and one reduce.

Thus, the user's problem is not only to produce an algorithm but also to understand whether the obtained algorithm is really usable in practice, and, even more important, how it can be transformed and how the possibly different versions can be compared.

4.3 Program transformations

The goal of our transformations is to try and reduce the number of collective operations in alg-mss2. A rich set of transformation rules for collective operations has been developed recently [15, 16]. The rule we could try for alg-mss2 is the scan-reduce fusion:

Rule SR-ARA

b = scanL $Op1$ a
c = reduce $Op2$ b
b = reduce $(New\ (Op1, Op2))$ (arrange (a,a))
c = arrange (proj [1]) b
If $Op1$ distributes forward over $Op2$
$(a_1, b_1)\ New(Op1, Op2)\ (a_2, b_2) = (a_1\ Op2\ (b_1\ Op1\ a_2), b_1\ Op2\ b_2)$

The rule is named SR-ARA, which reflects the transformation expressed by it: "**S**can;**R**educe → **A**rrange;**R**educe;**A**rrange". We present this and other rules in a format that consists of four boxes, top-down: (1) a program fragment before transformation (the "left-hand side" of the rule), (2) the fragment after transformation (the "right-hand side"), (3) optional: the If-condition expressing when the rule is applicable, (4) optional: the definition(s) of new function(s) used by the rule. The transformation rules use parameter operators $Op1$, $Op2$, etc. Note their difference to the concrete operators $op1$, $op2$, which we have used in the programs alg-mss1, etc.: those are written with a small "o". In the SR-ARA rule, the base operation of the reduction on the right-hand side is constructed by applying (higher-order) function New to the parameter operators of the left-hand side; New is defined in the bottom box of the rule.

Trying to apply the SR-ARA rule to our algorithm alg-mss2, the transformation engine (see Figure 1) immediately finds a mismatch: there is an extra arrange statement between scan and reduction in the algorithm.

To make the SR-ARA rule applicable, we can use the following transformation:

Rule AR-RA

$c =$ reduce Op (arrange (proj [1]) b)
$c =$ arrange(proj [1]) (reduce $Op.pair\ b$)
$(a_1, b_1)\ Op.pair\ (a_2, b_2) = (a_1\ Op\ a_2,\ b_1\ Op\ b_2)$

In this rule, it is assumed that the input, b, is an array of tuples, and that the projection arrangement refers to a correct position in the tuple. We provide here the instance of the rule when the tuples are pairs, and the projection yields the first position. The rule is universally applicable since it has no If-part.

Note that the AR-RA transformation is applicable in both directions, with completely different consequences: from right to left, the transformation eliminates redundant computations, thereby reducing the cost of the overall composition; from left to right, it restructures the program by pushing the projection after the reduction, at the price of redundant computations. In our case study, we have the latter situation for **alg-mss2**: we need to push the projection, which would enable us to apply the scan-reduce fusion afterwards.

Since the left-to-right application of AR-RA makes sense only when enabling the SR-ARA transformation afterwards, it is useful to combine both rules – the one pushing the projection and the scan-reduce fusion – into one bigger rule:

Rule SAR-ARA

$b =$ scanL $Op1\ a$
$c =$ reduce $Op2$ (arrange (proj [1]) b)
$b =$ reduce $(New(Op1, Op2.pair))$ (arrange (a,a))
$c =$ arrange (proj [1]) b
If $Op1$ distributes forward over $Op2.pair$
$(a_1, b_1)\ New(Op1, Op2)\ (a_2, b_2) = (a_1\ Op2\ (b_1\ Op1\ a_2), b_1\ Op2\ b_2)$
$(a_1, b_1)\ Op.pair\ (a_2, b_2) = (a_1\ Op\ a_2,\ b_1\ Op\ b_2)$

4.4 Target algorithm

Now we are ready to apply the SAR-ARA rule to **alg-mss2**, with the following instantiations of the parameter operators: $Op1 = op2$ and $Op2 = max$. The applicability condition is the forward distributivity of $Op1$ over $Op2.pair$, which in case of **alg-mss2** are the following operations:

$$(a_1, b_1)\ Op1\ (a_2, b_2) = (\ max\ ((a_1 + b_2), a_2),\ b_1 + b_2\ ) \quad (3)$$
$$(a_1, b_1)\ Op2.pair\ (a_2, b_2) = (max(a_1, a_2), max(b_1, b_2)) \quad (4)$$

The distributivity required by the rule can be checked straightforwardly, under the obviously correct assumption that operator max is commutative.

After applying the SAR-ARA rule to program alg-mss2, we obtain the following *target program for the maximum segment sum problem*:

alg-mss3 (in x : arr[n], out r : scalar);
  $s =$ reduce $(New(op2, max.pair))$ (arrange (arrange (x,x), arrange (x,x))) ;
  $r =$ arrange (proj [1]) (arrange proj [1] x);

Target program alg-mss3 contains only one computational skeleton, reduction, with an associative base operator. Such a reduction is easily parallelizable. Thus, we arrive at a better solution than both the intuitive and the first parallelizable version. Note that data arrangements and the reduction are quite intricate: $max.pair$ is constructed according to (4) for $Op2 = max$, and $op2$ is defined by (2). It is rather unlikely that the programmer could write such a program without support from the system.

5 Execution Model and Cost Estimates

In this section, we sketch the execution model and execution costs of FAN programs in the current implementation. For the sake of simplicity, we consider only data-parallel skeletons on one-dimensional arrays. We assume all input arrays to be distributed block-wise on processors in advance, while input scalars are replicated. Skeletons in the body are executed one after the other on all available processors, and the total cost of a program is the sum of costs of all skeletons in the body. The costs of skeletons are given in Table 2 and explained below:

FAN Operation	Time estimate
map f	mt_f
arrange proj [i]	0
arrange (x,x)	$2mt_{copy}$
reduce $(\oplus)$	$mt_\oplus + \log p(t_s + t_w + t_\oplus)$
scanR $(\oplus)$	$2mt_\oplus + \log p(t_s + t_w + 2t_\oplus)$
scanL $(\oplus)$	$2mt_\oplus + \log p(t_s + t_w + 2t_\oplus)$

Table 2. Costs for basic operations and arrangement functions.

We assume to work on vectors of length n, on p processors so that each processor has $m = n/p$ contiguous elements to compute. map f applies function f in parallel to all the array elements residing on each processor. If t_f is the cost of applying f to an array element than the cost of map f is mt_f.

The execution and cost of arrange depend on its particular parameters. For instance, arrange proj [i] does not imply costs as we simply select the needed value, ignoring the rest. However, projection may sometimes incur costs, e.g., if the user wants to save memory on the program. On the other hand, arrange (a,b) requires copying, which can be done in parallel on all processors, leading to a cost of $2mt_{copy}$.

Skeleton **reduce** $\oplus$ is executed in two steps. The first step reduces data locally to each processor according to the operator $\oplus$; this costs $mt_\oplus$. Then a global reduce is computed in $\log p$ phases using a butterfly-like communication pattern [14]. In each phase, two elements of the vector are exchanged pairwise between processes. Assuming that $T_{sr} = t_s + mt_w$ is the cost for exchanging a two m sized message between two processes (t_s is the start-up time and t_w is the per-element transfer time), the cost estimate if the global phase is $\log p(t_s + t_w + t_\oplus)$. Adding up the two phases, we obtain the cost of **reduce** ($\oplus$), as $mt_\oplus + \log p(t_s + t_w + t_\oplus)$.

The execution model for **scanR** differs from reduction in that we need three steps. The first step computes **scanR** locally in parallel on the blocks of data resident on processors (cost $mt_\oplus$). The second step executes a global **scanR** using the last element of a local vector, on a butterfly-like schema similar to the one used for **reduce**, except that **reduce** requires one operation per element received whereas **scanR** requires two. This phase costs $\log p(t_s + t_w + 2t_\oplus)$. Finally, each processor combines the received value to local values which costs $mt_\oplus$. Thus, the total cost is: $2mt_\oplus + \log p(t_s + t_w + 2t_\oplus)$. **scanL** works in a similar way and has the same cost.

Costs for transformation rules are given in Tab. 3. To make things simpler, costs for copying and making local operations are assumed to be all equal, and t_s and t_w are normalized to this cost.

Rule	Time before	Time after rule application
SR-ARA	$3m + \log p(2(t_s + t_w) + 3))$	$2m + \log p(t_s + 2t_w + 2)$
AR-RA	$m + \log p(t_s + t_w + 1)$	$2m + \log p(t_s + 2(t_w + 1))$
SAR-ARA	$3m + \log p(2(t_s + t_w) + 3)$	$5m + \log p(t_s + t_w + 1)$

Table 3. Performance estimates for optimization rules

The cost of the left-hand side of the SR-ARA rule is obtained by adding the costs of **reduce** and **scanL** (vectors a, b have length n). On the right-hand side, initial pairing doubles the size of elements in vector b with respect to vector a. Thus, **reduce** works on elements of length 2 (and thus the cost of sending/computing is $2(t_w+1)$). The cost of rule SAR-ARA are derived assuming that before the application a and b are vectors of pairs of length n, and after the application a is a vector of pairs of length n and b is a vector of quadruples of length n. Thus **scanL** and **reduce** before application work on elements on length 2, while after application, the composed **reduce** works on elements of size 4.

We can use the costs derived for rules to estimate the conditions under which a particular rule is worth applying. Rule SR-ARA always improves costs, whereas rule AR-RA, applied alone, worsens costs (as expected). The combined rule SAR-ARA improves costs if

$$(t_s + t_w + 2) \log p > 2m$$

For our case study, this result indicates that program alg-mss3 is better than alg-mss2 for larger numbers of processors, and for systems with comparatively slow communication, i.e. big values for t_w and t_s.

6 Experimental Results

Our target program alg-mss3 has been manually rewritten in the P3L language as shown in Fig 2:

```
#define N 100000
seq op4 in(int x[4], int y[4]) out(int z[4])
 ${  z[0]=max(max(x[0],y[0]),(x[2]+y[1]));
     z[1]=max(x[1],(x[3]+y[1]));
     z[2]=max(y[2],(x[2]+y[3]));
     z[3]=x[3]+y[3];   }$
end seq

seq w_make_quad in(int x) out(int y[4])
 ${ int i;
    for(i=0;i<4;i++)
      y[i]=x;  }$
end seq

reduce red_four in(int x[N][4]) out(int y[4])
    op4 in(x[*][]) out(y)
end reduce

map make_quadruple in(int a[N]) out(int a4[N][4])
    w_make_quad in(a[*i]) out(a4[*i][])
end map

comp mss in(int a[N]) out(int b[4])
  make_quadruple in(a) out(int a4[N][4])
  red_four in(a4) out(b)
end comp
```

Fig. 2. The testbed P3L version of the alg-mss3 program.

Here the sequential skeleton (introduced by seq) encapsulated a fragment of C code defining an operator or a function which is to be applied in a skeleton instance. The sequential op4 defines the $New(...)$ operation in terms of C assignments and the sequential w_make_quad creates a 4-tuple replicating an integer value. The map skeleton in general implements a FAN arrange followed by a map. In this case, the map make_quad then implements the arrange(...) and creates in parallel the array of tuples. The P3L reduce skeleton implements the FAN reduce using the binary operator op4.

The P3L program was automatically translated into a C+MPI code, and run on a Fujitsu AP1000 located at the Imperial College London. The results are summarized by the plots in Fig. 3. They show both the predicted and the measured performance, for two problem sizes: 10^5 and 10^6.

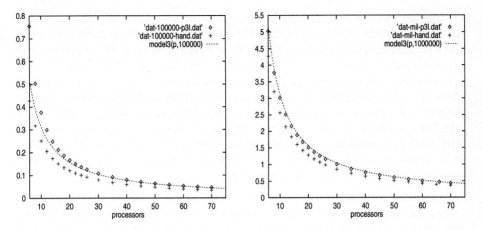

Fig. 3. Run times (sec) of alg-mss3 on the problem size 10^5 (left) and 10^6 (right).

In our experiments, the data are already distributed on processors. We compare the C+MPI code, xxxp3l.dat, generated by the p3l compiler [7], and a corresponding hand-coded MPI version, xxxhand.dat. Curve model3 shows the costs predicted using our performance estimate of Section 5. The plots demonstrate that the p3l code is a bit slower than the hand-coded one, but still reaches good efficiency. The code analysis indicates that the main sources of inefficiency are function calls inserted to compute quadruples and copying in the communication buffers. The predictability of performance is strikingly good, which we explain by a very simple structure of the target program, which consists of just one reduction. In addition, computations are very regular: the cost of the reduction operator New (op2, max.pair) does not depend on the input data.

7 Conclusion

In this paper, we argue that the implementations of high-level languages should be extended by special frameworks to support the process of designing efficient high-level programs. We presented the FAN transformational framework, showed how it can be integrated with the P3L programming system, and demonstrated its use on a case study with machine experiments. The integrated system consisting of FAN and P3L provides the user with a rapid prototyping tool by automatically producing an executable C+MPI code for the algorithm under design, together with expected performance estimates.

Because of the lack of space we could only sketch some aspects of the framework which deserve special consideration: the rich collection of transformations available (more details in [16]), details of cost estimates and benchmarking on different platforms, complete syntax and exact semantics of FAN notation, task-parallel skeletons like pipe, farm, etc.

Related work which inspired our research includes both higher-order functional programming [20] and high-level imperative programming environments. Data parallel languages such as NESL [6] and HPF [18] supports implicit parallelization over bulk data structures. Coordination languages like PCN [12] take a much more explicit approach than ours: tasks are composed by connecting pairs of communication ports, using primitive composition operators. Further examples are the use of programmer-oriented abstractions on top of MPI in the PEMPI system at Basel University [11], and a high-level extension of C++, PROMOTER, developed at GMD FIRST Berlin [13]. For an excellent survey of a variety of high-level parallel models, see [21]. Some ongoing projects were presented at a recent Dagstuhl seminar [9].

The main novelty of our **FAN** framework is the intensive use of program transformations in the early stages of design process, supported by corresponding cost models and programming tools. The framework is language-independent, and can be integrated easily with the existing high-level parallel programming environments, as our experience with P3L demonstrates.

Our current work and future plans address the following aspects:

- to develop design strategies for applying sequences of transformations, and to incorporate these strategies into the **FAN** transformation engine;
- providing that there is a choice of parallel implementations for particular **FAN** operations, to exploit this in the transformation process, and to pass the resulting algorithms together with implementation hints to the P3L compiler;
- to lift some of the arrangement optimizations, currently done by the P3L compiler, to the **FAN** level;
- to incorporate transformations for task-parallel skeletons of P3L [2] into our framework.

Thanks to Silvia Ciarpaglini for helping with the computer experiments. We also thank IFPC London for the access to the Fujitsu AP1000 used in the experiments. The anonymous referees helped to improve the presentation.

This work has been supported by a travel grant from the German-Italian academic exchange programme VIGONI.

References

1. S. G. Akl. *Parallel Computation: models and methods.* Prentice-Hall, 1997.
2. M. Aldinucci, M. Coppola, and M. Danelutto. Rewriting skeleton programs: How to evaluate the data-parallel stream-parallel tradeoff. In S. Gorlatch, editor, *Proceedings of First Int. Workshop on Constructive Methods for Parallel Programming (CMPP'98)*, pages 44–58. Techreport 9805, University of Passau, May 1998.
3. B. Bacci, M. Danelutto, S. Orlando, S. Pelagatti, and M. Vanneschi. P^{3}L: A structured high level programming language and its structured support. *Concurrency: Practice and Experience*, 7(3):225–255, 1995.
4. J. Bentley. Programming pearls. *Comm. ACM*, 27:865–871, 1984.

5. R. Bird. Lectures on constructive functional programming. In M. Broy, editor, *Constructive Methods in Computing Science*, NATO ASI Series F: Computer and Systems Sciences. Vol. 55, pages 151–216. Springer Verlag, 1988.
6. G. Blelloch. NESL: a nested data-parallel language (Version 2.6). Technical Report CMU-CS-93-129, School of Computer Science, Carnegie Mellon University, 1993.
7. S. Ciarpaglini, M. Danelutto, L. Folchi, C. Manconi, and S. Pelagatti. Anacleto: a template-based P^3L compiler. In *Proceedings of the Seventh Parallel Computing Workshop (PCW '97)*, Australian National University, Canberra, 1997.
8. M. Cole. Parallel programming with list homomorphisms. *Par. Proc. Letters*, 5(2):191–204, 1994.
9. M. Cole, S. Gorlatch, C. Lengauer, and D. Skillicorn, editors. *Theory and Practice of Higher-Order Parallel Programming*. Dagstuhl-Seminar Report 169, Schlo Dagstuhl. 1997.
10. J. Darlington, Y. Guo, H. W. To, and Y. Jing. Skeletons for structured parallel composition. In *Proc. of the 15th ACM SIGPLAN Symposium on Principles and Practice of Parallel Programming*, 1995.
11. N. Fang and H. Burkhart. Structured parallel programming using mpi. In *Proceedings of HPCN'96*, pages 840–847. Springer-Verlag, 1996.
12. I. Foster and S. Taylor. A compiler approach to scalable concurrent-program design. *ACM TOPLAS*, 16(3):577–604, 1994.
13. W. K. Giloi, M. Kessler, and A. Schramm. A high-level, massively parallel programming environment and its realization. In *Proceedings of the Real World Computing Symposium (RWC'97)*. Tokyo, 1997.
14. S. Gorlatch. Systematic efficient parallelization of scan and other list homomorphisms. In L. Bougé, P. Fraigniaud, A. Mignotte, and Y. Robert, editors, *Parallel Processing. Euro-Par'96, Vol. II*, Lecture Notes in Computer Science 1124, pages 401–408. Springer-Verlag, 1996.
15. S. Gorlatch and C. Lengauer. (De)Composition rules for parallel scan and reduction. In *Proc. 3rd Int. Working Conf. on Massively Parallel Programming Models (MPPM'97)*, pages 23–32. IEEE Computer Society Press, 1998. Available at http://brahms.fmi.uni-passau.de/cl/papers/GorLe97c.html.
16. S. Gorlatch, C. Wedler, and C. Lengauer. Optimization rules for programming with collective operations. Report MIP-9813, Universität Passau, October 1998. http://brahms.fmi.uni-passau.de/cl/papers/Gor.Wed.Len:report.oct98.html.
17. W. Gropp, E. Lusk, and A. Skijellum. *Using MPI: Portable Parallel Programming with the Message Passing*. MIT Press, 1994.
18. High Performance Fortran Forum. High Performance Fortran language specification. Version 2.0, Department of Computer Science, Rice University, Jan. 1997.
19. S. Pelagatti. *Structured development of parallel programs*. Taylor&Francis, 1998.
20. D. Skillicorn. *Foundations of Parallel Programming*. Cambridge University Press, 1994.
21. D. Skillicorn and D. Talia. Models and languages for parallel computation. *ACM Computing Surveys*, 1998.
22. D. Smith. Applications of a strategy for designing divide-and-conquer algorithms. *Science of Computer Programming*, 8(3):213–229, 1987.
23. D. Swierstra and O. de Moor. Virtual data structures. In B. Möller, H. Partsch, and S. Schuman, editors, *Formal Program Development*, Lecture Notes in Computer Science 755, pages 355–371. Springer-Verlag.

Implementing a Non-strict Functional Programming Language on a Threaded Architecture

Shigeru Kusakabe†, Kentaro Inenaga†, Makoto Amamiya†,
Xinan Tang‡, Andres Marquez‡, and Guang R. Gao‡

†Dept. of Intelligent Systems, Kyushu University,
‡EE & CE Dept. University of Delaware
E-mail: kusakabe@is.kyushu-u.ac.jp

Abstract. The combination of a language with fine-grain implicit parallelism and a dataflow evaluation scheme is suitable for high-level programming on massively parallel architectures. We are developing a compiler of V, a non-strict functional programming language, for EARTH(Efficient Architecture for Running THreads). Our compiler generates codes in Threaded-C, which is a lower-level programming language for EARTH. We have developed translation rules, and integrated them into the compiler. Since overhead caused by fine-grain processing may degrade performance for programs with little parallelism, we have adopted a thread merging rule. The preliminary performance results are encouraging. Although further improvement is required for non-strict data-structures, some codes generated from V programs by our compiler achieved comparable performance with the performance of hand-written Threaded-C codes.

1 Introduction

Many fine-grain multithreaded architectures have been proposed as promising multiprocessor architectures because of their ability to tolerate communication and synchronization latencies inherent in massively parallel machines[1][3][9][12][14][18]. By providing a lot of threads and supporting fast switches among threads, multithreaded architectures can hide communication and synchronization latencies. Control of execution is switched to a new thread when a long-latency operation is encountered. Many explicit parallel languages have been proposed for fine-grain multithreaded programming[7][8][17][19]. Overlapping computation and communication is programmer's task when using this kind of explicit languages. Writing explicit parallel programs on parallel machines is still a skilled job.

We are developing a high-level parallel programming language, called "V," which would minimize the difficulties in writing massively parallel programs[15]. In order to provide a high-level abstraction, the language is a non-strict functional programming language with implicit parallelism. There is no anti-dependence in V programs and it is easy to extract parallelism at various levels including fine-grain parallelism from V programs. The underlying computation model is an optimized dataflow computation model, Datarol[4]. The combination of a functional language with implicit parallelism and a fine-grain multithread evaluation scheme is suitable for massively parallel computation. The languages abstracts the timing problems in writing massively parallel programs, while fine-grain multithread evaluation supports efficient execution of a large number of fine-grain processes for implicit parallel functional programs in a highly concurrent way.

This approach can exploit irregular and dynamic parallelism, thus support efficient execution of "non-optimal" codes. Since our language does not support explicit descriptions for parallel execution and data-mapping, it is necessary for the compiler to automatically extract parallel codes of optimal grain size, partition data-structures and map them to each processing node. However, this is not an easy task, and the compiler may generate non-optimal codes of non-uniform grain size and ill-mapped data-structures. The multithreaded architectures with the ability to tolerate computation and synchronization latencies alleviate the problems of non-optimal code.

In this paper, we discuss implementation issues of V on a multithreaded architecture EARTH (Efficient Architecture for Running THreads). In order to show the feasibility of our language, we have implemented our language on commercially available machines such as Sequent Symmetry and Fujitsu AP1000[16]. However, since our basic execution model is a multithreaded execution model extended from Datarol model, multithreaded architectures which have special support for fine-grain parallel processing are preferable for our language. EARTH realizes efficient multithreading with off-the-shell microprocessor-based processing nodes[12]. We are implementing V on EARTH, while there are two explicit parallel languages for EARTH, EARTH-C[11] as a higher-level language and Threaded-C[21] as a lower-level language. The goal of this work is to show adequate compiler support makes our implicit parallel language as efficient as the explicit parallel languages on the multithreaded architecture. Our final aim is to realize a high-level programming environment on a high performance architecture.

In compilation, we use a virtual machine code, DVMC (Datarol Virtual Machine Code), as an intermediate code, and Threaded-C as a target code. Threaded-C is an explicit multi-threaded language, targeted for the EARTH model. We have developed translation rules, and integrated them into the compiler. Since overhead caused by fine-grain processing may degrade performance for programs with little parallelism, we have adopted a thread merging rule. The preliminary performance results on EARTH-MANNA[12][10] are encouraging. Although further improvement is required for non-strict data-structures, some codes generated from V programs by our compiler achieved comparable performance with the performance of hand-written Threaded-C codes.

This paper is organized as follows: section 2 and section 3 introduce our language V and EARTH respectively. Section 4 explains our compilation method. Section 5 discusses implementation issues of fine-grain data-structures on EARTH. Section 6 makes concluding remarks.

2 Language V

2.1 Language feature

Originally, V is a functional programming language Valid[2], developed at NTT ECL for colored token dataflow architectures as a part of the Dataflow Project. The fundamental design policies are:

- Recursive functional programming was adopted as the basic structure.
- A static binding rule was applied to variables.
- A static type system was adopted.
- Call-by-value was adopted as the fundamental computation rule.

Valid provided facilities for highly parallel list processing, and demonstrated the feasibility of functional programming and dataflow computation.

A program consists of a set of user-defined functions, and their invocations. A function instance is invoked using a function definition. The following is a form of function definition:

```
function ⟨function_definition_name⟩ (⟨formal_parameters⟩)
return (⟨types_of_return_values⟩)
= ⟨body_expression⟩ ;
```

Notation ⟨function_definition_name⟩ is an identifier for the function. Notation ⟨formal_parameters⟩ specifies the names and types of formal parameters, and ⟨types_of_return_values⟩ the types of return values. As an example, we consider the following program which calculates the summation from low to high:

```
function summ(low,high:integer) return(integer)
= if low=high then low
    else {let mid:integer=(low+high)/2 in summ(low,mid)+summ(mid+1,high)};
```

This function has two integer inputs low and high, and one integer return value. Within the function body, two recursive calls may be invoked.

2.2 Intermediate code

There is no anti-dependence in V programs and it is easy to extract parallelism at various levels including fine-grain parallelism from V programs. Parallelism is two-fold: function applications and subexpressions. Subexpressions executable in parallel are compiled to independent threads.

As an intermediate code, our compiler uses DVMC, which is a dataflow virtual machine code based on the Datarol model. The idea of Datarol is to remove redundant dataflow by introducing registers and a by-reference data access mechanism. A Datarol program (DVMC) is a multithread control-flow program, which reflects the dataflow inherent in the source program. While Datarol reflects the underlying data-flow structure in the given program, it eliminates the redundant data-flow operations such as switch and gate controls, and also eliminates the operand matching overhead. In Datarol, a function (procedure) consists of multiple threads, and an instance frame (virtual register file) is allocated in function invocation. The threads within the function share the context on the frame. While a data-driven mechanism is used to activate threads, result data and operand data are not passed as explicit data tokens, but managed on the instance frame.

Lenient evaluation (non-strict and eager dataflow evaluation) can maximally exploit fine-grain parallelism and realize flexible execution order. Argument expressions for a function invocation can be evaluated in parallel, and the callee computation can proceed if either of argument value is prepared. the "call $f\ r$" creates an instance of the function f, and stores the pointer to the instance in r. The "rins" releases the current instance. The "link $r\ v\ slot$" sends the value of v as the $slot$-th parameter data to the callee instance specified by r. The "rlink $r\ cont\ slot$" sends the continuation, $cont$ for the returned value and the continuation threads in the callee instance, as the $slot$-th parameter data to the callee instance specified by r. The "receive $slot\ v$" receives the data transferred from the caller instance through the $slot$-th slot, and stores the data into v. The " return $rp\ v$" triggers the continuation threads specified by the rlink operation after setting the value of v to the content of rp.

Fig.1 shows DVMC for the summation program, generated by a straightforward compilation method. In DVMC, execution control among threads proceeds along continuation points. In the figure, a solid box represents a thread, an arrow a continuation arc, and a wavy arrow an implicit dependence.

In the figure, three RECEIVEs at the top are nodes to receive input parameters, two integers low and high, and one continuation point to return the result value. Within the function body, two shaded parts correspond to two recursive calls. A LINK is a node to send parameter value, and an RLINK to send a continuation point to receive a result value.

3 EARTH

EARTH architecture is a multiprocessor architecture designed for the efficient parallel execution of both numerical and non-numerical programs. The basic EARTH design begins with a conventional processor, and adds the minimal external hardware necessary for efficient support of multithreaded programs.

In the EARTH model, a multiprocessor consists of multiple EARTH nodes and an interconnection network. The EARTH node architecture is derived from the McGill Data Flow Architecture, which was based on the idea that synchronization of operations and the execution of the operations themselves could be performed more efficiently in separate units rather than in the same processor. Each EARTH node consists of an Execution Unit (EU) and a Synchronization Unit (SU), linked together by buffers. The SU and EU share a local memory, which is part of a distributed shared memory architecture in which the aggregate of the local memories of all the nodes represents a global memory address space.

The EU processes instructions in an active thread, where an active thread is initiated for execution when the EU fetches its thread id from the ready queue.

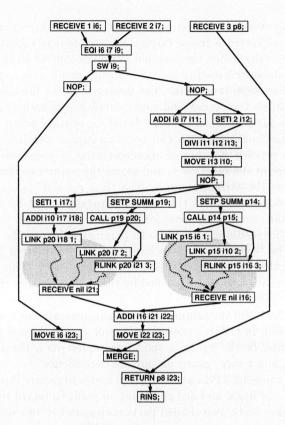

Fig. 1. Fine-grain abstract machine code of summ.

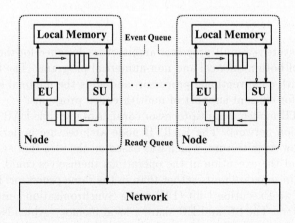

Fig. 2. EARTH architecture

The EU executes a thread to completion before moving to another thread. It interacts with the SU and the network by placing messages in the event queue. The SU fetches these messages, plus messages coming from remote processors via the network. The SU responds to remote synchronization commands and requests for data, and also determines which threads are to be run and adds their thread ids to the ready queue.

The EARTH programming model is implemented as an extension to the C language. The resulting C dialect, EARTH Threaded-C, is an explicitly parallel language that allows the programmer to directly specify the partitioning into threads and the EARTH operations[21]. The higher-level C dialect for EARTH is EARTH-C[11], which has simple extensions to express control parallelism, data locality, and collective communication. The EARTH-C compiler generates the Threaded-C code as output.

For preliminary performance evaluation, we use an implementation of EARTH on MANNA (EARTH-MANNA). The MANNA (Massively parallel Architecture for Non-numerical and Numerical Applications) architecture was developed at GMD FIRST in Berlin, Germany. Each node of a MANNA multiprocessor consists of two Intel i860 XP RISC processors clocked at 50 MHz, 32 MB of dynamic RAM and a bidirectional network interface. The link interface is capable of transferring 50 MB/s in each direction simultaneously, for a total bandwidth of 100 MB/s per node. The network is based on 16 x 16 crossbar chips which support the full 50 MB/s link speed.

4 Compilation

In this section, we explain thread level compilation rules, and discuss effectiveness of static thread scheduling.

4.1 Abstract machine

Fig.3 shows the schematic view of our abstract runtime model. We employ a two-level scheduling: instance frame level scheduling, and thread level scheduling within an instance frame. Instance frames hold the local variables and synchronization variables. Threads within a function share the context on an instance frame for the function. A thread is a schedulable unit, and each thread has a synchronization counter variable whose initial value is determined at compile time. When a thread is triggered, the counter is decremented. If it reaches zero, the thread becomes ready and enqueued into the thread queue. Operations with unpredictably long latency are realized as a split-phase transaction, in which one thread initiates the operation, while the operation that uses the returned value is in another thread and triggered by the returned value.

The thread queue is provided in order to dynamically schedule intra-instance threads. If the execution of the current thread terminates, the next ready thread is dequeued from the thread queue for the next execution. If the thread queue is exhausted before the instance terminates, the instance becomes a "suspended instance" and is stored into the idle pool before the execution switches to another frame. If the instance terminates, the runtime system releases the instance frame. The runtime system manages runnable instance frames by using the instance queue. The runtime system picks up one of the runnable frames and passes the execution control to the corresponding code to activate the instance frame. When a frame in the idle pool (a suspended instance frame) receives data from

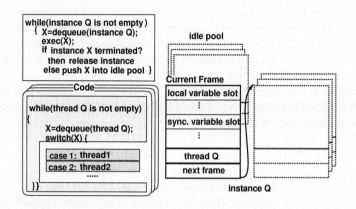

Fig. 3. Overview of abstract machine.

another active frame, the corresponding thread in the suspended frame is triggered. If the thread becomes ready and is enqueued into the thread queue, the instance frame also becomes ready and is moved to the instance queue.

4.2 Compiling to Threaded-C

The compiler must generate an explicit threaded code from an implicit parallel V program. In compilation, we use a virtual machine code DVMC as an intermediate code, and Threaded-C as a target code.

We explain the compilation rules for DVMC to generate Threaded-C code. As shown in the previous sections, both Datarol virtual machine and EARTH architecture support threaded execution. However, the two-level scheduling in DVMC, instance frame level scheduling and thread level scheduling, is merged and mapped to the ready queue in EARTH.

Table 1 shows basic translation rules. Basically, since a function instance in DVMC is a parallel entity, a function in DVMC is translated to a THREADED function in Threaded-C. By using a TOKEN instruction for a THREADED function, the function instance is forked in Threaded-C. Both DVMC functions and THREADED functions use heap frame as their activation record. There are two types of function calls in DVMC: strict call and non-strict call. As a usual convention, a strict call does not start until all the arguments become available. In con-

DVMC	Threaded-C
function	$\Rightarrow$ THREADED function
thread	$\Rightarrow$ THREAD
basic instruction	$\Rightarrow$ basic instruction
strict call	$\Rightarrow$ TOKEN instruction
non-strict call	$\Rightarrow$ TOKEN (with I-structure)

Table 1. Translation rules

trast, a non-strict call starts its execution before the arguments become available. This execution style overlaps computations among caller and callee side, maximally exploits fine-grain parallelism, and realizes flexible execution order, since computation can be triggered by a subset of the arguments. Although function call in Threaded-C is basically strict, by using I-structure library[5], we can realize non-strict call in Threaded-C.

In Threaded-C, a thread is enclosed by THREAD_id and END_THREAD(). In DVMC, a thread is started with a thread label and ended with ! mark. Each instruction has a continuation tag, showing which threads are waiting for this instruction to finish. In contrast to Threaded-C, the number of the synchronization signals that a thread waits for is not explicitly coded in the instruction sequence in DVMC. It relies on the compiler to scan the sequence to compute synchronization information for each thread. On contrary, Threaded-C needs the programmer to explicitly specify *sync* signals that a thread waits for by the INIT_SYNC primitive. In Threaded-C, each thread is associated with a *sync* slot, which is a quadruple: *(slot-number, sync-count, reset-count, thread-number)*. The *reset-count* is necessary when *sync-count* becomes zero and the thread needs to restart, for example a thread within a loop. The *sync* slots need to be initialized in advance. Thus, the first few instructions in any Threaded-C function are usually INIT_SYNC primitives, assigning *sync* slots to threads and setting *sync* counts accordingly. (DVMC and Threaded-C versions of Fibonacci program `fib` are shown in Fig.7 in appendix)

4.3 Static scheduling

Although EARTH supports thread level execution, excessively fine-grained threads may incur heavy overhead. Compile-time scheduling is effective to reduce run-time cost of scheduling fine-grain activities. In scheduling a non-strict program, cares must be taken for achieving high performance while keeping non-strict semantics of the program.

Our scheduling method pays attention not to violate non-strict semantics. Although another scheduling is possible for software implemented DVMC system[13], in which blocking threads can be realized by software support, the compiler for EARTH, in which a thread is no-preemptive, adopt a compile-time scheduling similar to separation constraint partitioning[20]. The basic rule for the thread scheduling is that two nodes must reside in different threads if there exists any indirect dependency between them. Indirect dependences, such as a certain indirect dependence due to access to asynchronous data and a potential indirect dependence due to function call, may require dynamic scheduling. In our scheduling, a merging policy is introduced to make more functions fit conventional invocation style.

Fine-grain codes such as shown in Fig.1 will incur heavy overhead. Our static scheduler can translate such a fine-grain code into a coarse grain code as shown in Fig.4. In the figure, shaded parts are two recursive calls. As a result of static thread merge, operations concerning each function call are merged into a single thread, and the function is invoked as a strict function. This code can be translated to a Threaded-C program, and can be executed efficiently on EARTH.

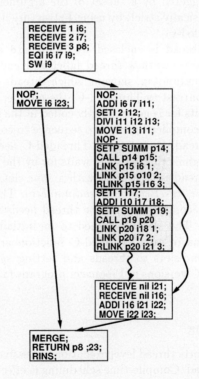

Fig. 4. Code after static thread merging.

Following is a result of preliminary performance evaluation on EARTH-MANNA using Fibonacci program. The compiled version is a Threaded-C program generated from V program by the compiler. The hand-coded version is a Threaded-C program written by a programmer using the same algorithm as the V source program. As shown in the table, the automatically generated compiled version runs as fast as the hand-coded version on EARTH-MANNA.

	n		
	16	24	32
compiled[ms]	2	59	1984
hand-coded[ms]	2	56	1970

Table 2. Elapsed time of fib(n) on EARTH-MANNA

5 Fine-grain parallel data-structures

Non-strict data-structures alleviate the difficulty of programming for complex data-structures[6]. Due to non-strictness, programmers can concentrate on the essential dependencies in the problems without writing explicit synchronizations between producers and consumers of data-structure elements. In addition to this advantage of abstracting timing problems, non-strict data-structures have another advantage to communication overhead, which is one of the problems in parallel processing. An eager evaluation scheme of non-strict data-structures increases potential parallelism during program execution. This evaluation scheme is effective for hiding communication overhead, by switching multiple contexts while waiting for the result of communication results.

Partitioning and distribution of large data-structures have a large impact on the parallel processing performance. Although the owner computes rule is known to be effective for SPMD style programming to some extent, such a monolithic rule is not suitable for MIMD style programming, whose computation and communication pattern may complex and unpredictable. The combination of non-strict data-structures and the eager evaluation can eliminate descriptions of explicit partitioning and distribution of large data-structures, as well as explicit synchronization in MIMD style programming. The multithreaded execution with the ability to tolerate computation and synchronization latencies alleviate the problems of non-optimal code such as non-uniform grain size and ill-mapped data-structures. However, non-strict data-structures require frequent dynamic scheduling at a fine-grain level during program execution. Fine-grain dynamic scheduling causes heavy overhead, which offsets the gain of latency hiding. In this section, we discuss on the implementation issues of non-strict data-structures.

5.1 Implementation model

An array is generated by a bulk operator, mkarray, in V. In creating a non-strict array, the allocation of the array block and the computations to fill the elements are separated. In a dataflow computation scheme, a mkarray immediately returns an array block (pointer) and calls the filling functions in parallel. Fig.5 outlines the `mkarray` computation model on a distributed memory machine. In the figure, the `mkarray` node is an instance representing an array, a `mksubarray` node is an instance representing a subarray block on a node, and `pf`s are instances of filling functions for array elements. On distributed-memory machines, an array is often distributed over the processor nodes, and filling functions are scattered among the processor nodes. We map the filling functions using the owner computes rule as a basic rule. Although the `pf`s may activated in parallel, they may be scheduled according to data dependencies.

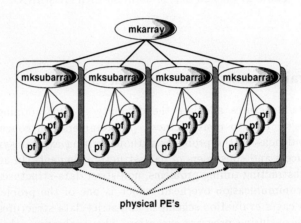

Fig. 5. Schematic view of a `mkarray` computation model on a distributed memory machine.

Array descriptors support array distribution and realize global address access for distributed arrays. Array descriptors consist of a body, a decomposition table, and a mapping table, in order to manage the top-level information, the decomposition scheme, and the mapping scheme, respectively. The body has a total size, the number of dimension, the number of processors, and various flags. The decomposition table has the lower and the upper bounds of each dimension, and some parameters to specify the decomposition scheme: the number of logical processors, the block size, and the base point. The mapping table has correspondences between the logical and physical processors for the sub-arrays, the top of the local address for the block.

5.2 Discussion

According to this model, we implemented arrays in V. We also measured preliminary performance on EARTH-MANNA using matrix multiplication program `matmul`.

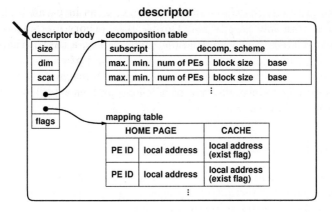

Fig. 6. Array descriptor

matrix size	10	20	30	40	50
elapsed time[sec]	$3.14*10^{-2}$	$4.10*10^{-1}$	3.27	10.0	52.1

Table 3. Elapsed time of `matmul` on EARTH-MANNA

This result is very slow. Although EARTH-MANNA has the external hardware necessary for efficient support of multithreaded execution, naive implementation of non-strict data-structures incurs heavy overhead. Arrays in V are non-strict, and support element-by-element synchronization. Ideally, fine-grain computation and communication are pipelined and overlapped in a dataflow computation scheme, without exposing the synchronization and communication overhead. However, the overheads of fine-grain processing degrade the performance.

Although our language is a fine-grain non-strict dataflow language and our implementation scheme employs a multithread execution model, access to a data-structure element is performed by means of pointers to heap areas. Thus, many optimization techniques proposed for conventional languages are also applicable to our implementation of fine-grain parallel data-structures. In order to reduce the overhead caused by the frequent fine-grain data access, we are considering to incorporate following optimization techniques:

- caching mechanism for fine-grain data accesses, and
- grouping mechanism for fine-grain data accesses.

The second technique is to transform non-strict access to data-structure into scheduled strict access, by data dependency analysis between producers and consumers at compilation phase. In this technique, as many as possible non-strict accesses could be transformed to strict accesses, while the parallelism would fall victim. So that, the compiler has the trade off between non-strictness with high parallelism and strictness with low parallelism. During the optimization process, following issues are considered: Fine-grain parallel data-structures of our language allow element-by-element access. This feature is one of the key points in order to write non-strict programs as well as parallel programs. Optimizations involving indiscreet grouping may reduce effective parallelism, and may lead to a deadlock at runtime in the worst case.

6 Concluding remarks

We discussed implementation issues of V, a non-strict functional programming language, on EARTH. Functional languages are attractive for writing parallel programs due to their expressive power and clean semantics. The goal is to implement V efficiently on EARTH, realizing a high-level programming language on a high performance architecture. The combination of a language with fine-grain parallelism and a dataflow evaluation scheme is suitable for high-level programming for massively parallel computation. In compiling V to EARTH, we use a virtual machine code DVMC as an intermediate code, and Threaded-C as target code. Threaded-C is an explicit multi-threaded language, targeted for the EARTH model. We executed sample programs on EARTH-MANNA. The preliminary performance results are encouraging. Although further improvement is required for non-strict data-structures, the codes generated from V programs by our compiler achieved comparable performance with the performance of hand-written Threaded-C codes.

References

1. Robert Alverson, David Callahan, Daniel Cummings, Brian Koblenz, Allan Porterfield, and Burton Smith. The Tera computer system. In *Proceedings of the 1990 ACM International Conference on Supercomputing*, pages 1–6, 1990.
2. M. Amamiya, R. Hasegawa, and S. Ono. Valid: A High-Level Functional Programming Language for Data Flow Machine. *Review of Electrical Communication Laboratories*, 32(5):793–802, 1984.
3. M. Amamiya, T. Kawano, H. Tomiyasu, and S. Kusakabe. A Practical Processor Design For Multithreading. In *Proc. of the Sixth Symposium on Frontiers of Masssively Parallel Computing*, pages 23–32, Annapolis, October 1996.
4. M. Amamiya and R. Taniguchi. Datarol:A Massively Parallel Architecture for Functional Languages. In *the second IEEE symposium on Parallel and Distributed Processing*, pages 726–735, December 1990.
5. Jose Nelson Amaral and Gunag R. Gao. Implementation of I-Structures as a Library of Functions in Portable Threaded-C. Technical report, University of Delaware, 1998.
6. Arvind, R. S. Nikhil, and K. K. Pingali. I-structures: data structures for parallel computing. *ACM Trans. Prog. Lang. and Sys.*, 11(4):598–632, October 1989.
7. Robert D. Blumofe, Christopher F. Joerg, Bradley C. Kuszmaul, Charles E. Leiserson, Keith H. Randall, and Yuli Zhou. Cilk: An Efficient Multithreaded Runtime System. In *Proceedings of the 5th Symposium on Principles and Practice of Parallel Programming*, pages 207–216, July 1995.

8. David E. Culler, Andrea Dusseau, Seth Copen Goldstein, Arvind Krishnamurthy, Steven Lumetta, Thorsten von Eicken, and Katherine Yelick. Parallel programming in Split-C. In IEEE, editor, *Proceedings, Supercomputing '93: Portland, Oregon, November 15–19, 1993*, pages 262–273, 1109 Spring Street, Suite 300, Silver Spring, MD 20910, USA, 1993. IEEE Computer Society Press.
9. Jack B. Dennis and Guang R. Gao. *Multithreaded computer architecture: A summary of the state of the art*, chapter Multithreaded Architectures: Principles, Projects, and Issues, pages 1–74. Kluwer academic, 1994.
10. W. K. Giloi. Towards the Next Generation Parallel Computers: MANNA and META. In *Proceedings of ZEUS '95*, June 1995.
11. Laurie J. Hendren, Xinan Tang, Yingchun Zhu, Guang R. Gao, Xun Xue, Haiying Cai, and Pierre Ouellet. Compiling C for the EARTH Multithreaded Architecture. In *Proceedings of the 1996 Conference on Parallel Architectures and Compilation Techniques (PACT '96)*, pages 12–23, Boston, Massachusetts, October 20–23, 1996. IEEE Computer Society Press.
12. Herbert H. J. Hum, Olivier Maquelin, Kevin B. Theobald, Xinmin Tian, Guang R. Gao, and Laurie J. Hendren. A Study of the EARTH-MANNA Multithreaded System. *International Journal of Parallel Programming*, 24(4):319–348, August 1996.
13. K. Inenaga, S. Kusakabe, T. Morimoto, and M. Amamiya. Hybrid Approach for Non-strict Dataflow Program on Commodity Machine. In *International Symposium on High Performance Computiong (ISHPC)*, pages 243–254, November 1997.
14. Yuetsu Kodama, Hirohumi Sakane, Mitsuhisa Sato, Hayato Yamana, Shuichi Sakai, and Yoshinori Yamaguchi. The EM-X Parallel Computer: Architecture and Basic Performance. In *Proceedings of the 22th Annual International Symposium on Computer Architecture*, 1995.
15. S. Kusakabe and M. Amamiya. *Dataflow-based Language V*, chapter 3.3, pages 98–111. Ohmsha Ltd., 1995.
16. S. Kusakabe, T. Nagai, Y. Yamashita, R. Taniguchi, and M. Amamiya. A Dataflow Language with Object-based Extension and its Implementation on a Commercially Available Parallel Machine. In *Proc. of Int'l Conf. on Supercomputing'95*, pages 308–317, Barcelona, Spain, July 1995.
17. R. S. Nikhil. Cid: A Parallel, "Shared-Memory" C for Distributed-Memory Machines. *Lecture Notes in Computer Science*, 892:376–, 1995.
18. R. S. Nikhil, G. M. Papadopoulos, and Arvind. *T: A multithreaded massively parallel architecture. In *Proceedings of the 19th Annual International Symposium on Computer Architecture*, pages 156–167, May 1992. Also as CSG-memo-325-1, Massachusetts Institute of Technology, Laboratory for Computer Science.
19. Mitsuhisa Sato, Yuetsu Kodama, Shuichi Sakai, and Yoshinori Yamaguchi. EM-C: Programming with Explicit Parallelism and Locality for the EM-4 Multiprocessor. In Michel Cosnard, Guang R. Gao, and Gabriel M. Silberman, editors, *Proceedings of the IFIP WG 10.3 Working Conference on Parallel Architectures and Compilation Techniques, PACT '94*, pages 3–14, Montréal, Québec, August 24–26, 1994. North-Holland Publishing Co.
20. K. E. Schauser, D. E. Culler, and S. C. Goldstein. Separation Constraint Partitioning — A New Algorithm for Partitioning Non-strict Programs into Sequential Threads. In *Proc. Principles of Programming Languages*, January 1995.
21. Xinan Tang, Oliver Maquelin, Gunag R. Gao, and Prasad Kakulavarapu. An Overview of the Portable Threaded-C Language. Technical report, McGill University, 1997.

Appendix

| | ```
THREADED fib(SPTR done, long int i5,
               long int *GLOBAL p6)
{
  SLOT SYNC_SLOTS [2];
  long int i7,i8,i9,i10,i13,i14,
           i15,i18,i19,i20,i21;

  INIT_SYNC(0, 2, 2, 2);
  INIT_SYNC(1, 1, 1, 3);
  i7 = 2;
  i8 = i5 < i7;
  if (i8)
  {
    i20 = 1;
    i21 = i20;
    SYNC(1);
  }
  else {
    i9 = 1;
    i10 = i5 - i9;
    TOKEN(fib, SLOT_ADR(0), i10,
                TO_GLOBAL(&i13));
    i14 = 2;
    i15 = i5 - i14;
    TOKEN(fib, SLOT_ADR(0), i15,
                TO_GLOBAL(&i18));
    END_THREAD();
THREAD_2:
    i19 = i13 + i18;
    i21 = i19;
  }
  END_THREAD();
THREAD_3:
  DATA_RSYNC_L(i21, p6, done);
  END_FUNCTION();
}
``` |

```
1:  RECEIVE 1 i5;
    RECEIVE 2 p6;
    SETI 1 i7;
    LTI i5 i7 i8;
    SWN i8 15;
    __SW_B 15 T;
    NOP ;
    SETI 1 i20;
    MOVE i20 i21 ->(3);
    __SW_E 15 T;
    __SW_B 15 E;
    NOP;
    SETI 1 i9;
    SUBI i5 i9 i10;
    SETN FIB n11;
    CALLS n11 f12 (i10) (i13)
                      -> (2);
    SETI 2 i14;
    SUBI i5 i14 i15;
    SETN FIB n16;
    CALLS n16 f17 (i15) (i18)
                      -> (2);
!   __SW_E 15 E;
2:  RECEIVE NIL i13;
    RECEIVE NIL i18;
    ADDI i13 i18 i19;
!   MOVE i19 i21 -> (3);
3:  MERGE 2;
    RETURN p6 i21;
!   RINS;
```

| (a) DMVC | (b) Threaded-C |

Fig. 7. DVMC and Threaded-C code of function fib
```

Second Workshop on Bio-Inspired Solutions to Parallel Processing Problems (BioSP3)

Workshop Chairs
Albert Y. Zomaya, The University of Western Australia
Fikret Ercal, University of Missouri-Rolla
Stephan Olariu, Old Dominion University

Steering Committee
Peter Fleming, University of Sheffield, United Kingdom
Frank Hsu, Fordham University
Oscar Ibarra, University of California, Santa Barbara
Viktor Prasanna, University of Southern California
Sartaj Sahni, University of Florida - Gainsville
Hartmut Schmeck, University of Karlsruhe, Germany
H.J. Siegel, Purdue University
Les Valiant, Harvard University

Program Committee
Ishfaq Ahmad, Hong Kong University of Science and Technology
David Andrews, University of Arkansas
Prithviraj Banerjee, Northwestern University
Juergen Branke, University of Karlsruhe, Germany
Jehoshua (Shuki) Bruck, California Institute of Technology
Jens Clausen, Technical University of Denmark, Denmark
Sajal Das, University of North Texas
Hossam El-Gindy, University of Newcastle, Australia
Hesham El-Rewini, University of Nebraska at Omaha
Afonso Ferreira, CNRS, LIP-ENS, France
Ophir Frieder, Illinois Institute of Technology
Eileen Kraemer, Washington University in St. Louis
Mohan Kumar, Curtin University of Technology, Australia
Richard Lipton, Princeton University
John H. Reif, Duke University
P. Sadayappan, Ohio State University
Peter M. A. Sloot, University of Amsterdam, The Netherlands
Assaf Schuster, Technion, Israel
Franciszek Seredynski, Polish Academy of Sciences, Poland
Ivan Stojmenovic, Ottawa University, Canada
El-ghazali Talbi, Laboratoire d'Informatique Fondamentale de Lille, France
Weiping Zhu, University of Queensland, Australia

Bio-Inspired Solutions to Parallel Processing Problems

Preface

This section contains papers presented at the Second Workshop on Bio-Inspired Solutions to Parallel Processing Problems (BioSP3) which is held in conjunction with the Second Merged IPPS/SPDP'99 Conference, San Juan, Puerto Rico, April 12-16, 1999. The workshop aimed to provide an opportunity for researchers to explore the connection between biologically-based techniques and the development of solutions to problems that arise in parallel processing. Techniques based on bio-inspired paradigms can provide efficient solutions to a wide variety of problems in parallel processing. A vast literature exists on bio-inspired approaches to solving an impressive array of problems and, more recently, a number of studies have reported on the success of such techniques for solving difficult problems in all key areas of parallel processing. Rather remarkably, most biologically-based techniques are inherently parallel. Thus, solutions based on such methods can be conveniently implemented on parallel architectures.

In the first group of three papers, the authors introduce computational models that are inspired by biological-based paradigms. King et al. discuss the biological basis for an artificial immune system and draws inferences from its operation that can be exploited in intelligent systems. Szuba presents a molecular quasi-random model of computation known as the Random Prolog Processor. The model is used to study collective intelligence and behavior of social structures. In an interesting paper, Wasiewicz et al. show how simple logic circuits can be implemented via self-assembly of DNA. This approach has promise to be used in the realization of complex molecular circuits.

In the next three papers, the authors investigate solutions to different optimization problems. Vianna et al. present a hybrid evolutionary metaheuristic to solve the popular vehicle routing problem. They use parallel genetic algorithms and local search heuristics to generate good suboptimal solutions. Seredynski et al. propose a multi-agent based optimization technique for task scheduling. Their approach is based on local decision making units (agents) which take part in an iterated game to find directions of migration in the system graph. Alaoui et al. tackle a similar problem using a slightly different approach; Genetic Local Neighborhood Search (GLNS). They demonstrate through experimental results that their approach exhibit good performance and it is easily parallelizable. Third group of papers also deal with the solution of optimization problems. Paper by Jedrzejowicz et al. propose evolution-based algorithms for the scheduling of fault-tolerant programs under various constraints. In the next paper, Peter Loh and Venson Shaw investigate the adaptation of a combined genetic-heuristic algorithm as a fault-tolerant routing strategy. With this hybrid approach to routing, they are able to achieve a significant reduction in the number of redundant nodes. The third paper in this group is by C. Leopold which introduces a semi-automatic program restructuring method in order to reduce the number of cache misses. The proposed method uses local search algorithms and uncovers many sources for locality improvement.

The last group of four papers mostly deal with applications. Alba and Troja demonstrate the importance of the synchronization in the migration step of parallel distributed GAs with different reproduction strategies. Chalermwat et al. describe a coarse-grained parallel image registration algorithm based on GAs. Their technique is based on first finding a set of good solutions using low-resolution images and then registering the full-resolution images using these results. The authors report scalable and accurate results in image registration. Sena et al. use GAs and a master-slave paradigm to solve the well known Traveling Salesman Problem in parallel. They analyze the effect of different parameters on performance such as population size, mutation rate/interval, number of slaves, etc. The results presented show the potential value of the proposed techniques in tackling other NP-Complete problems of similar nature. In the last paper of the workshop, Boloni et al. first present the basic ideas of a distributed object middleware for network computing. And then, they try to answer the question whether the biologically-inspired computational techniques used so far can be expanded to cover the way we build complex systems out of components, and whether we can emulate the genetic mechanisms and the immune system to achieve such a goal.

In closing, we would like to thank the authors, the program committee members and the referees for their valuable contributions in making this workshop a success. With their dedication and support, we hope to continue the Workshop on Biologically-Inspired Solutions to Parallel Processing Problems in an annual basis.

Albert Y. Zomaya
Fikret Ercal
Stephan Olariu

The Biological Basis of the Immune System as a Model for Intelligent Agents

Roger L. King[1], Aric B. Lambert[1], Samuel H. Russ[1], and Donna S. Reese[1]

[1] MSU/NSF Engineering Research Center for Computational Field Simulation,
Box 9627, Mississippi State, MS 39762-9627, U.S.A.
Rking@erc.mstate.edu

Abstract. This paper describes the human immune system and its functionalities from a computational viewpoint. The objective of this paper is to provide the biological basis for an artificial immune system. This paper will also serve to illustrate how a biological system can be studied and how inferences can be drawn from its operation that can be exploited in intelligent agents. Functionalities of the biological immune system (e.g., content addressable memory, adaptation, etc.) are identified for use in intelligent agents. Specifically, in this paper, an intelligent agent will be described for task allocation in a heterogeneous computing environment. This research is not intended to develop an explicit model of the human immune system, but to exploit some of its functionalities in designing agent-based parallel and distributed control systems.

1 Introduction

Biological systems are serving as inspirations for a variety of computationally-based learning systems (e.g., artificial neural networks and genetic algorithms). This research uses as its biological inspiration the human immune system. From a computing standpoint, the immune system can be viewed as a parallel, distributed system that has the capability to control a complex system over time [1]. The primary objectives of this research are to gain an understanding of the recognition process, learning, and memory mechanisms of the immune system and how to make use of these functionalities in the design of intelligent agents [2, 3].

The immune system is comprised of several different layers of defense -- physical, physiological, innate (non-adaptive), and adaptive. The adaptive defense mechanism is sometimes referred to as *acquired*. Any action of the immune system against a pathogen is known as an *immune response*. The most basic defense mechanism is the skin. It serves as a physical barrier to many pathogens. For pathogens that elude the skin barrier, there are physiological barriers. These are immune responses that provide environments non-conducive to pathogen survival (e.g., an area of low pH). These static (physical and physiological) immune responses are of minor interest in this research.

The important difference between innate and adaptive responses is that an adaptive response is highly specific for a particular pathogen. This occurs because the innate system agent will not alter itself on repeated exposure to a given pathogen; however, the adaptive system agent improves its response to the pathogen with each encounter of the same pathogen. Also, this adapted cell will remain within the body and will be able to attack the pathogen if it re-enters the system in the future. Therefore, two key functionalities to exploit of the adaptive immune response system are its abilities for *adaptation* and *memory*. The following sections explore in more detail how adaptation and memory occur as a result of the immune response process.

2 Immune Response

Innate and adaptive immune responses are produced primarily by leukocytes. There are several different types of leukocytes, but the most important for our consideration are phagocytes and lymphocytes. The phagocytes are the first lines of defense for the innate immune system. These cells use primitive, non-specific recognition systems that allow them to bind to a variety of organisms, engulf them, and then internalize and destroy them. An interesting strategic maneuver on the part of the immune system is that it positions phagocytes at sites where they are more likely to encounter the organisms that they are most suitable to control. In other words, the immune system does a judicious job in controlling its limited resources. Lymphocytes initiate all adaptive immune system responses since they specifically recognize individual pathogens. The two main categories of lymphocytes are B lymphocytes (B-cells) and T lymphocytes (T-cells). The B-cells develop in the bone marrow and the T-cells develop in the thymus.

Immune response is a two step process - *recognition* followed by *resolution*. It is also proposed that immune system based intelligent agents for task allocation utilize a two step process - *recognition* of a specific hardware and/or software instance followed by an *allocation response* (i.e. a resolution) that will better utilize the computing environment's resources to perform on-going and planned tasks. Two types of intelligent agents are proposed to accomplish the recognition process - H-cells specialized for hardware and S-cells specialized for software. Details of their functionalities and use are described later.

Pathogen recognition is the first step in the adaptive system's response to an organism. There are several mechanisms through which this can occur - the complement system, cytokines, and antibodies. Most applications of the immune system to computing system research have concentrated on the antibody recognition system [4-9]. Although this will play an important role in this research, it is believed that the cytokine system offers other desirable functionalities that should be exploited.

Cytokines are a class of soluble mediators released by T-helper cells (a specific type of T lymphocyte) to communicate with other cells in the innate and adaptive immune systems. For example, these T-helper cells work with the innate system's phagocytes or the adaptive system's B-cells. In the innate system case, phagocytes take up antigens and present them to the T-helper cells to aid in antigen recognition. In return, if the T-cell recognizes it as an antigen, it will release cytokines to activate the phagocyte to destroy the internalized pathogen. T-cells also work with B-cells to help them divide, differentiate, and make antibody. The principal to recognize here is that although the agents are distributed, many of them communicate with each other to facilitate optimal control. The functionality exhibited by the cytokine system which we would like to exploit for our agents are the *division of responsibility among distributed agents for command, control, and communication*.

3 Recognition, Adaptation, and Antibodies

B-cells manufacture antibodies. However, each B-cell is monoclonal (i.e., it only manufactures one type of antibody) and so it is limited to recognizing antigens that are similar in structure to itself. Recognition of an antigen by an antibody takes place at a special region on the antibody molecule – the paratope. The paratope is the

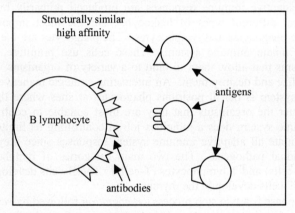

Fig. 1. Antibody – antigen recognition and binding.

genetically encoded surface receptor on an antibody molecule. The area on the antigen where a paratope can attach itself is known as the epitope. When a particular antigen is recognized or detected by a B-cell, it combats the antigen by reproducing in numbers (i.e., cloning) and by releasing its soluble antibodies to bind to the target antigen. It should be noted that an antibody produced by a B-cell is specific for an epitope and not the entire antigen molecule. In fact, several different antibodies from different B-cells may bind to a single antigen.

Antigen recognition or detection is accomplished via a process known as binding. The binding of an antibody to an antigen is a result of multiple non-covalent bonds. The analytical representation often used for this antibody-antigen bond is a complementary binary string (a visual representation would be a lock and key) [2]. For example, if a paratope on a B-cell was represented as 01100110, then the epitope on an antigen that would exactly fit this lock would be 10011001 and the binding energy would be at its maximum. Antibody affinity is a measure of the strength of the bond between the antibody's paratope and a single epitope. Therefore, it can be surmised that structurally similar epitopes will also bind to this antibody, however, with lesser energies. This gives rise to the concept of an affinity threshold. If the threshold is exceeded, recognition has occurred and the B-cell is cloned. For an intelligent agent, if real values are used it may be necessary to develop a fuzzy metric relationship. However, if binary representations are used for resource loading, software preferences, etc., then Hamming distance or an XOR approach may be a good metric for threshold affinity. This is illustrated in Figure 2. In this example, if the threshold affinity was 10, then antibody 2 would have exceeded the threshold and cloned itself.

```
Antigen      0 1 0 0 1 0 1 1 1 0 1 1
Antibody 1   1 1 1 0 1 0 1 1 0 1 1 1
Antibody 2   1 0 1 1 0 1 0 0 0 0 0 0

XOR 1        1 0 1 0 0 0 0 0 1 1 0 0
XOR 2        1 1 1 1 1 1 1 1 1 0 1 1

∑ antigen ⊗ antibody 1 = 4
∑ antigen ⊗ antibody 1 = 11
```

Fig. 2. Illustration of binding energy calculation.

It was just observed that when a B-cell exceeds its affinity threshold it clones itself. An important aspect of this cloning process is that it does not produce exact clones. The offspring produced by the cloning process are mutated. This variation from child to parent is important in adapting B-cells to better fight antigens. The rationale behind this inexact cloning can be explained as follows. Because the parent probably had a less than maximum affinity when it was stimulated, an exact copy of the B-cell would not improve the response of the immune system. However, if the parent's offspring are mutated, then there is a chance that the children will have a higher affinity for the antigen, and then, subsequent generations can improve their response. Therefore, the immune system exhibits the functionality of *evolutionary adaptation* (i.e., learning ala Darwin). Learning through adaptation is a functionality of intelligent agents for task allocation.

4 Immunological Memory

Memory is another important functionality existing within the human immune system. The above description of evolving B-cells to better fight a particular antigen is known as *the immune system's primary response*. This initial encounter with a new antigen involves a finite amount of time for the immune system to adapt its B-cells to best combat the intruder. And, when fighting a pathogen, time is of the essence. To improve response time, the immune system has developed a way to maintain a memory of this encounter so it will be able to respond faster in the future (i.e., have antibodies already available). Then, when the body is re-infected with the same antigen, the secondary response of the immune system, based on *immunological memory*, is very specific and rapid. This ability of the immune system to remember instances of previously encountered activities and the learned response is another functionality of immune systems to be exploited in intelligent agents. In fact, immunological memory is the functionality of the human immune system that vaccinations exploit. With this analogy in mind, it very quickly follows that *a priori* information about a heterogeneous computing environment can be initially known via vaccinations with appropriate H-cells and S-cells. The inoculation could provide H-cells containing information about the system's initial hardware resources and the S-cells could contain explicitly declared information about the code (e.g., what tasks are read-write intensive or how much processor time will the task need).

Immunological memory may best be described as a *content addressable memory* (CAM). The CAM functions fundamentally different from the classical location addressing memory of most computer architectures (i.e., von Neumann architecture). In the immune system an antigen may be addressed for recognition purposes simultaneously by many B-cells (i.e., in parallel). When the affinity threshold of a specific B-cell is exceeded by either an exact or approximate match, then a specific response is elicited from memory (clone, release antibodies, destroy). The key functionality of a CAM is that the content of the stimulus is important in eliciting a response specific to that stimulus. In-other-words, an instance of a previously encountered hardware configuration and/or software requirement will be recognized by the system and the learned response (task allocation) associated with this instance will be specifically and rapidly dispatched from memory. Therefore, the secondary response of the intelligent agents (H-cells and S-cells) should improve over their primary response due to the retained memory of their previous encounter.

Another important point about immunological memory is that it is *self-organizing*. It must have this capability because the immune system is limited in the number of cells it may have available to it at any one time. Since the number of cells is limited, the immune system replaces a percentage of its lymphocytes on a continuing basis. This replacement or self-organizing takes place during the learning phase of the cell. It was explained previously that immune system learning was evolutionary in the sense of Darwin (i.e., survival of the fittest). Therefore, if cells are not periodically matched with a pathogen, then they will die out to be replaced by a new cell that may better match existing or previously encountered pathogens in the body. Since the number of instances of hardware configuration and software architecture can be infinite, it will be important for the intelligent agents to have the capability of purging poorly performing H-cells and S-cells from the system. The measure of quality of performance used by the self-organizing functionality of the resource allocator may be based on a performance index, a time since last used, or some other set of metrics.

One of the primary functions of the body's immune system is to perform classification of organisms as either self or non-self (i.e., your own versus foreign). This capability is distributed among a variety of agents (adaptive and non-adaptive) whose function it is to continually monitor the body's environment for these undesirable intruders. If an organism is classified as self, then no actions are taken. However, if an entity is classified as non-self, the immune system's agents work autonomously and/or in concert to eliminate the organism.

5 H-cells and S-cells as Intelligent Agents for Task Allocation

H-cells and S-cells are envisioned as being designed as complementary intelligent agents. The S-cells have the role of defining the long term schedule of resources for execution of a program (i.e., a road map) and the H-cells provide near real-time control of the schedule (i.e., adjustments for the bumps and curves in the road). Where H-cells interrogate and monitor hardware resources during execution of a program for information, the S-cells interrogate the parallel program code for information prior to execution. This information is then used by the S-cells in planning how to schedule the code on the known distributed system hardware resources. The H-cells and S-cells are also distributed throughout the heterogeneous computing environment to focus on their assigned tasks, thus decentralizing the control process.

Initially, the S-cells may determine whether the code is data domain intensive or functional domain intensive. This determination can be made through either programmer supplied information or as a learned response to the program as it is executed. For example, in a data domain intensive code, the tasks to execute on all the hardware resources is functionally the same. However, the data may be different. In this case, knowledge about the nuances of the data (file size, precision requirements, etc.) will be important in assigning resources. In the case where the functionality of tasks differs (i.e., functional domain intensive), it is important to have knowledge about the resource requirements of tasks (e.g., memory requirements, execution time, etc.). A simulation is a good example of a functionally intensive domain. Tasks will vary among the resources with different constraints for communication, execution, etc. The S-cell has the responsibility of determining critical paths for the simulation and allocating its distributed resource base in a judicious manner.

In the case of the H-cells, they are initially inoculated with basic knowledge about system hardware resources and behavioral properties. They will then have the tasks of learning about new hardware resources connected into the environment, monitoring resource usage, and managing system resources and processes. Inputs to the H-cells may include memory availability, CPU usage, disk traffic, and other performance related information. For the prototype S-cells it is envisioned that the software programmer will embed information about the code into pre-defined header slots. For example, information about critical paths may be specified, along with which tasks are computationally or communicationally intensive. Then, when a program is launched, the S-cells continue to monitor the activity of resources that they have planned to utilize along the execution path. If one of these resources becomes unavailable, it will be the responsibility of the H-cells to find an appropriate available resource in the short term. However, the S-cells will then reassess the road map and determine what resources it had planned to use to complete the tasks that may have to be remapped. After making any necessary adjustments to the road map, the new routes will be passed to the master unit for any appropriate action.

At present, research is on-going to develop intelligent agents to perform task allocation for a heterogeneous computing environment [10]. Implementation of many of the functions of the human immune system described in this paper are being demonstrated in an H-cell implementation based on a hierarchical ART architecture [11]. Data for inoculation of the H-cells and for defining the content addressable memory are being collected at both the system and task levels. At the system level, information is both static and dynamic. The static information includes: machine type, processor speed, operating system, internal memory available (both physical and virtual), page information, ticks per second, and number of processors. At the task level, all of the information is dynamic in nature. Task information includes: memory usage and a breakdown of a task's CPU usage into time for computing and communicating while executing an MPI task.

6 Implementation in the HECTOR Environment

An intelligent control agent (ICA) based on the H-cell is being implemented to replace the slave agent used in the HECTOR system that helps globally manage the heterogeneous system [10]. The artificial immune based scheduling system is designed around the decentralized decision-making of the ICAs. The ICAs perform all the decision-making functions of the HECTOR system. The ICAs are distributed among the nodes running parallel tasks, monitoring run-time performance and performing task control functions such as starting, migrating, restarting, and notifying tasks. System information such as memory, CPU, network, and disk usage are monitored to determent resource utilization of the node.

The ICAs run on all platforms that are candidates for executing parallel jobs. The primary focus of the ICA is performance optimization of the parallel tasks running on its node. Once every five seconds, the ICA takes performance measurements from the node it is residing on. A five-second interval was selected, because it provides a reasonable compromise between frequent updates and excessive system usage. Maintaining awareness of system performance is important for two reasons. First, it is the means by which the ICAs can detect external load and determine the appropriate action. Second, it permits the agent to identify the machines that are most appropriate to run tasks. Because of the need of constant performance information a new capability has been added to the ICA. This phase, known as the monitoring phase, gathers performance information, and structures it as a string of floating values for input into the intelligent controller portion of the agent.

The system calls used to gather the performance information were used from the original structure that was established for the slave agents used in HECTOR. On an unloaded single-processor Sun 110 MHz Sparc 5 workstation, the ICA uses 2.14 Mbytes of memory. Out of that, 617 Kbytes of it is resident. The measurements increased as the number of tasks increased, and the wall-clock indicated an additional 8.3 ms due to the extra tasks.

7 Conclusions

This paper has described the biological basis for intelligent agents based upon the human immune system. The primary functionalities of the human immune system to be captured in intelligent agents for an artificial immune system are:

- two step process of recognition followed by resolution (similar to content addressable memory)
- distributed responsibility among agents
- communication among agents
- evolutionary adaptation
- immunological memory
- self-organization
- judicious use of resources

Also, it has been proposed that two different types of agents be developed – H-cells for hardware management and S-cells for software. An implementation of this proposed methodology is presently under development.

References

1. Farmer, J. Doyne, Norman H. Packard, and Alan S. Perlson (1986). The Immune System, Adaption, and Machine Learning, *Physica* 22D, pp. 187-204.
2. Hunt, John E. and Denise E. Cooke (1996). Learning Using an Artificial Immune System, *Journal of Network and Computer Applications*, 19, pp 189-212.
3. Roitt, Ivan, Jonathan Brostoff, and David Male (1996). *Immunology* 4th Ed., Mosby.
4. D'haeseleer, P., Stephanie Forrest, and Paul Helman (1996). An Immunological Approach to Change Detection: Algorithms, Analysis, and Implications, *Proceedings of the 1996 IEEE Symposium on Security and Privacy,* pp. 110-119.
5. Ishiguro, A., Yuji Watanabe, Toshiyuki Kondo, and Yoshiki Uchikawa (1996). Decentralized Consensus-Making Mechanisms Based on Immune System, *Proceedings of the 1996 IEEE International Conference on Evolutionary Computation*, pp. 82-87.
6. Ishida, Y. and N. Adachi (1996). An Immune Algorithm for Multiagent: Application to Adaptive Noise Neutralization, *Proceedings of the 1996 IEEE/RSJ International Conference on Intelligent Robots and Systems*, pp. 1739-1746.
7. Kephart, Jeffrey O. (1994). A Biologically Inspired Immune System for Computers, *Artificial Life IV: Proceedings of the 4th International Workshop on the Synthesis and Simulation of Living Systems*, pp. 130-139.
8. Kumar, K. K. and James Neidhoffer (1997). Immunized Neurocontrol, *Expert Systems with Applications,* Vol. 13, No. 3, pp. 201-214.
9. Xanthakis, S., S. Karapoulios, R. Pajot, and A. Rozz (1995). Immune System and Fault Tolerant Computing, *Artificial Evolution: European Conference AE 95, Lecture Notes in Computer Science*, Vol.1063, Springer-Verlag, pp. 181-197.
10. Russ, S. R., Jonathan Robinson, Brian K. Flachs, and Bjorn Heckel. The Hector Parallel Run-Time Environment, accepted for publication in *IEEE Transactions on Parallel and Distributed Systems.*

11. Bartfai, Guszti (1994). Hierarchical Clustering with ART Neural Networks, *Proceedings of the IEEE International Conference on Neural Networks*, Vol. 2, IEEE Press, pp. 940-944.

Appendix: Glossary

antibody - a molecule produced by a B cell in response to an antigen that has the particular property of combining specifically with the antigen which induced its formation

antigen - any molecule that can be specifically recognized by the adaptive elements of the immune system (either B or T cells) (originally - antibody generator)

cytokines - the general term for a large group of molecules involved in signaling between cells during immune responses

epitope - molecular shapes on an antigen recognized by antibodies and T cell receptors (i.e., the portion of an antigen that binds with the antibody paratope)

paratope - the specialize portion on the antibody molecule used for identifying other molecules

pathogen - an organism which causes disease

A Formal Definition of the Phenomenon of Collective Intelligence and Its IQ Measure[1]

Tadeusz SZUBA

Department of Mathematics and Computer Science, Kuwait University,
E-mail: szuba@mcs.sci.kuniv.edu.kw

Abstract. This paper formalizes the concept of Collective Intelligence *(C-I)*. Application of the Random PROLOG Processor (*RPP*) has allowed us to model the phenomenon of *C-I* in social structures, and to define a *C-I* measure (*IQS*). This approach works for various beings: bacterial ÷ insect colonies to human social structures. It gives formal justification to well-known patterns of behavior in social structures. Some other social phenomenon can be explained as optimization toward higher IQS. The definition of *C-I* is based on the assumption that it is a specific property of a social structure, initialized when individuals organize, acquiring the ability to solve more complex problems than individuals can. This property amplifies if the social structure improves its synergy. The definition covers both cases when *C-I* results in physical synergy or in logical cooperative problem-solving.

1. Introduction

Individual Intelligence (I-I) and *Artificial Intelligence (A-I)* have obtained research results, but for *Collective Intelligence (C-I)*, the bibliography is poor. Only a few books can be found, e.g. [8], [12], but there is no formal approach. There are also books describing *C-I* biological behavior, e.g. of ant colonies [7], and papers on abstract ant colonies are available, e.g. [5]. The intelligence of humans is widely analyzed in psychology (psychometrics provides different IQ tests for *I-I*, e.g. [4]); sociology, and neuropsychology where there are several models of human intelligence, e.g. [3]. From the point of view of social structures, there is a library of experiments and observations of behavior and cooperation there, (group dynamics, e.g. [11]). The situation in the research on *C-I* might be explained by a widespread fear that *C-I* must be something much more complex than *I-I*. In addition, some top scientists support the conclusion of A. Newell, who objected to *C-I* [10], arguing: *"A social system, whether a small group or a large formal organization, ceases to act, even approximately, as a single rational agent. This failure is guaranteed by the very small communication bandwidth between humans compared with the large amount of knowledge available in each human's head. Modeling groups as if they had a group mind is too far from the truth to be useful scientific approximation very often."* This is a logical result of a priori assumed models of computations. The power and rapid growth of computers impress us; thus, we assume the determinism of computations, and that the information must have an

---

[1] This research is supported by Grant No. SM 174 funded by Kuwait University.

address, or at least must be rationally (in a deterministic way) moved around a computer system. However, alternative models of computations have been proposed, e.g. [1], and applied to *C-I* [13], [14]. Experiments with modeling of chaotic collective inferences in a social structure with use of such models demonstrate [14] that the group (under some restrictions) can complete an inference with an appropriate conclusion, even facing internal inconsistencies, i.e. it can work as a *group mind*.

2. The Basic Concepts of Modeling Collective Intelligence

It is a paradox that the evaluation of *C-I* seems easier than the evaluation of the *I-I* of a single being, which can only be evaluated on the external results of behavior during the problem-solving process. Neuropsychological processes for problem-solving are still far from being observable. Consequently, it is necessary to create abstract models of brain activity [3] based on neuropsychological hypotheses or computer-oriented models like *A-I*. In contrast, many more elements of collectively intelligent activity can be observed and evaluated in a social structure. We can observe displacements/actions of beings, as well as exchange of information between them (e.g. language or the pheromone communication system). *I-I* and behavior is scaled down as a factor – to accidental, local processes. <u>*C-I* can be evaluated through chaotic models of computations like the RPP, and statistical evaluation of the behavior of beings in structured environments.</u> *C-I* evokes some basic observations: • In a social structure, it is difficult to differentiate thinking and non-thinking beings. • The individuals usually cooperate in chaotic, often non-continuous ways. Beings move randomly because real life forces them to do so. Inference processes are random, starting when there is a need, when higher level needs are temporarily not disturbing a given being, or when there is a chance to make inference(s). Most inferences are not finished at all. This suggests using quasi-Brownian [15] movements for modeling social structure behavior [14]. • Individual inferences are also accidental and chaotic. • Resources for inference are distributed in space, time, and among beings. • The performance of a social structure is highly dependent on its organization. • Facts and rules can create inconsistent systems. Multiple copies are allowed. • Probability must be used as an IQ measure for a social structure.

3. Formal Definition of Collective Intelligence

The entry assumption is that *C-I* itself is a property of a group of beings and is expressed/observable and measurable. It is <u>not assumed</u> that beings are cooperating or are conscious or not; nothing is assumed about the communication; we don't even assume that these beings are alive. To better understand the above issues, let's look at some examples. Suppose that we observe a group of ants which have discovered a heavy prey that must be transported, and we also observe a group of humans who gather to transport heavy cargo. Ant intelligence is very low, and a simple perception/communication system is used – however, it is clear that ants display *C-I*. On the other hand, humans, after a lot of time, thought, and discussion will also move the cargo; this is also *C-I*. Because of such situations, the definition of *C-I* must be abstracted from possible methods of thinking and communication between individuals. The definition must be based on the results of group behavior. Let's look into another case. In medieval cities there were streets with shoemaker shops only. They gravitated there because the benefits gained overwhelmed the disadvantages, e.g. when some customers decided to buy shoes from a neighbor. Some shoemakers were sometimes in fact, even enemies. In this example, *C-I* emerges without any doubt; this is obvious just looking at the amount and quality of shoes produced on such a street. Thus we cannot assume willful cooperation for *C-I*, or the definition of cooperation would have to be very vague. Bacteria and viruses cooperate through exchange of (genetic) information; they use *C-I* against antibiotics, but the question is whether they are alive. Thus, the assumption about the existence of live agents must also be dropped. The definition we give now is based on these assumptions, and will formally cover any type of being, any communication system, and any form of synergy, virtual or real.

Let there be given a set *S* of individuals $indiv_1, ..., indiv_n$ existing in any environment *Env*. No specific nature is assumed for the individuals nor for their environment. It is necessary only to assume the existence of a method to distinguish $indiv_i$ $i = 1,...,n$ from the *Env*. Let there be also given a testing period $t_{start} - t_{end}$ to judge/evaluate the property of *C-I* of *S{...}* in *Env*. Let there now be given any universe *U* of possible problems $Probl_i$ proper for the environment *Env*, and be given the complexity evaluation for every problem $Probl_i$ denoted by $f_o^{Probl_i}(n)$. *C-I* deals with both *formal* and *physical* problems, thus we should write the following:

$$f_o^{Probl_i}(n) \stackrel{def}{=} \begin{cases} \text{if } Probl_i \text{ is a computational problem, then apply a standard} \\ \text{complexity definition with n defining the size of the problem;} \\ \text{if } Probl_i \text{ is a physical problem, then use as a complexity} \\ \text{measure any proper physical units, e.g. weight, size, time,} \\ \text{etc to express n} \end{cases} \quad (1)$$

Let's also denote in the formula the ability to solve the problems of our set of individuals *S* over *U* when working/thinking without any mutual interaction (absolutely alone, far from each other, without exchange of information):

$$Abl_U^{all\ indiv} \stackrel{def}{=} \bigcup_{Probl_i \in U} \max_S \left(\max_n f_o^{Probl_i}(n) \right) \quad (2)$$

This set defines the maximum possibilities of *S*, if individuals are asked e.g. one by one to display their abilities through all the problems. Observe that if any problem is beyond the abilities of any individual from *S*, this problem is not included in the set.

DEFINITION 1. COLLECTIVE INTELLIGENCE AS A PROPERTY

Now assume that individuals coexist together, and interact some way. We say that *C-I* emerges because of cooperation or coexistence in S, iff at least one problem $Probl'$ can be pointed to, such that it can be solved by a lone individual but supported by the group, or by some individuals working together, such that

$$f_O^{Probl_i}(n') \overset{significantly}{>} f_O^{Probl_i}(n) \in Abl_U^{all\ indiv} \tag{3}$$

or

$$\exists\, Probl'\ such\ that\ \left(\forall n\ Probl' \notin Abl_U^{all\ indiv}\right) \wedge \left(Probl' \in U\right)$$

The basic concept of definition (3) is that the property of *C-I* emerges for a set of individuals *S* in an environment *U* iff there emerges a quite new problem $\in U$ which can be solved from that point, or similar but even more complex problems can be solved. Even a small modification in the structure of a social group, or in its communication system, or even in, e.g. the education of some individuals can result in *C-I* emergence, or increase. An example is when the shoemakers move their shops from remote villages into the City. This can be, e.g. the result of a king's order, or creation of a "free trade zone". The important thing is that the distance between them has been reduced so much that it triggers new communication channels of some nature (e.g. spying). Defining *C-I* seems simpler but measuring it is quite a different problem. The difficulty with measuring *C-I* lies in the necessity of using a specific model of computations, which is not based on the DTM Turing Machine.

4. The Random PROLOG Model of Computations

Upon observing the social structures of humans, it is striking that the inferences are parallel, chaotic, and at the same time different algorithms of inference are applied. The same agent can be the processor, the message carrier, and the message, from the point of view of different parallel-running processes. The inferring process can be observed as global, but the role of specific elements cannot be easily identified and interpreted. Thus, a mathematical logic-based representation of a social structure is necessary for analysis, separate from interpretation. We have to rely on only approximations of the inferences we observe, and we must be able to easily improve an approximation; thus declarative, loosely coupled inference systems are necessary. PROLOG fulfills the requirements after some modifications. PROLOG is declarative, i.e. clauses can be real-time added/retracted; independent variants of facts and procedures are allowed. This is basic for chaotic systems. Symbols and interpretation of clauses are independent because PROLOG is based on 1st order predicate calculus. Thus, clauses can describe roughly observable aspects of behavior (e.g. of ants) without understanding them very well. In PROLOG, it is easy to adopt a molecular model of computations [1]. The RPP model of computations includes these properties: • *Information_molecules* infer randomly and are independent which implies full parallelism. • The RPP allows us to implement more inference patterns than standard PROLOG, i.e. concurrent forward, backward, and rule-rule inferences, e.g. where an inference is backward only from the formula of the goal. • With simple stabilizing methods (e.g. self-elimination, self-destruction of inconsistent clauses) the RPP is resistant to inconsistent sets of clauses. • Different parallel threads of inferences can be run at the same time in the RPP. • Parallelism of multiple inference diagrams highly compensates for the low efficiency of the random inference process.

4.1. Clauses and Computational PROLOG Space in the RPP

The 1st level Computational Space (*CS*) with inside quasi-random traveling Clause Molecules (*CMs*) of facts, rules, and goals c_i is denoted as the multiset $CS^1 = \{c_1,\ldots,c_n\}$. Thus, clauses of facts, rules, and goals are themselves 0-level *CS*. For a given *CS*, we define a membrane similar to that of the Chemical Abstract Machine (*CHAM*) [2] denoted by $|\cdot|$ which encloses inherent facts, rules, and goals. It is obvious that $CS^1 = \{c_1,\ldots,c_n\} \equiv \{|c_1,\ldots,c_n|\}$. For a certain kind of membrane $|\cdot|$ its type p_i is given, which will be denoted $|\cdot|_{p_i}$ to define which *CMs* can pass through it. Such an act is considered Input/Output for the given *CS* with a given $|\cdot|$ It is also allowable in the RPP to define degenerated membranes marked with $\cdot|$ or $|\cdot$ i.e. a collision-free (with membrane) path can be found going from exterior to interior of an area enclosed by such a membrane, for all types of *CMs*. The simplest possible application of degenerated membranes in the *CS* simulating a given social structure is to make, e.g. streets or other boundaries. If the *CS* contains clauses as well as other *CSs*, then it is considered a higher order one, depending on the level of internal *CS*. Such internal *CS* will be also labeled with $\hat{v}_j$ e.g.

$$CS^2 = \{|c_1,\ldots CS^1_{\hat{v}_j},\ldots c_n|\} \quad\text{iff}\quad CS^1_{\hat{v}_j} \equiv \{|b_1,\ldots,b_n|\} \tag{4}$$

where $b_i \; i=1\ldots m$ and $c_j \; j=1\ldots n$ are clauses

Every c_i can be labeled with $\hat{v}_j$ to denote characteristics of its individual quasi-random displacements. The general practice will be that higher level *CSs* will take fixed positions, i.e. will create structures, and lower level *CSs* will perform displacements. For a given *CS* there is a defined position function *pos*:

$pos: O_i \rightarrow \langle position\ description \rangle \cup undefined$ where $O_i \in CS$

If there are any two internal *CS* objects O_i, O_j in the given *CS*, then there is a defined distance function $D(pos(O_i), pos(O_j)) \rightarrow \Re$ and a rendezvous distance d. We say that during the computational process, at any time t or time period Δt, two objects O_i, O_j come to rendezvous iff $D(pos(O_i), pos(O_j)) \leq d$. The rendezvous act will be denoted by the rendezvous relation ®, e.g. O_i ® O_j which is reflexive and symmetric, but not transitive. The computational PROLOG process for the given *CS* is defined as the sequence of frames F labeled by t or Δt, interpreted as the time (given in standard time units or simulation cycles) with a well-defined *start* and *end*, e.g. $F_{t_0},\ldots,F_{t_e}$. For every frame its multiset $F_j \equiv (|c_1,\ldots,c_m|)$ is explicitly given, with all related specifications: *pos(.)*, membrane types p, and movement specifications v if available. The RPP is initialized for the start of inference process when the set of clauses, facts, rules, and goals (defined by the programmer) is injected into this *CS*. When modeling the *C-I* of certain closed social structures, interpretations in the structure will be given for all CS^m_n, i.e. "this *CS* is a message"; "this is a single human"; "this is a village, a city", etc. The importance of properly defining $\hat{v}_j$ for every CS^i_j should be emphasized. As has been mentioned, the higher level CS^i_j will take a fixed position to model substructures like villages or cities. If we model a single human as CS^1_j, then $\hat{v}_j$ will reflect displacement of the being. Characteristics of the given $\hat{v}_j$ can be purely Brownian or can be quasi-random, e.g. in lattice, but it is profitable to subject it to the present form of CS^i_j. Proper characteristics of $\hat{v}_j$ provide the following essential tools; • The goal clause, when it reaches the final form, can migrate toward the defined *Output* location (membrane of the main *CS* or even a specific *CS*). Thus, the appearance of a solution of a problem in the *CS* can be observable. • Temporarily, the density of some *CMs* can be increased in the given area of *CS*. After the given CS^i_j reaches the necessary form, it migrates to specific area(s) to increase the speed of selected inferences.

The right side of the rule and goal clause in the RPP is any set *S(...)* of unit clauses, contrary to standard PROLOG, where there is an ordered list of unit clauses. Such a set can even be considered a local, invariable *CS* of rules, containing unit clauses. This approach is compatible with the philosophy of the RPP and allows us to define traveling clusters of facts enclosed by membranes. It is allowed to introduce the *configuration*, to order, e.g. spatially, the unit clauses in the set *S(...)*.

4.2. The Inference Model in the RPP

The pattern of inference in Random PROLOG generalized for any CS has the form:
DEFINITION 2. GENERALIZED INFERENCE IN CS^N

Assuming that $CS = \{\ldots CS^i_j \ldots CS^k_l \ldots\}$, we give these definitions:

$CS^i_j \circledR CS^k_l$ and $U(CS^i_j, CS^k_l)$ and C(one or more CS^m_n of conclusions) $\vdash$ one or more CS^m_n of conclusions, $R(CS^i_j \text{ or } CS^k_l)$ **where:**

$CS^i_j \circledR CS^k_l$ denotes rendezvous relation; $U(CS^i_j, CS^k_l)$ denotes unification of the necessary type successfully applied; C(*one or more CS^m_n of conclusions*) denotes that CS^m_n are satisfiable. Notice that the reaction $\rightarrow$ in (*CHAM* [2]) is equivalent to $\vdash$ inference here. $R(CS^i_j \text{ or } CS^k_l)$ denotes that any parent *CMs* are retracted if necessary. The standard PROLOG inferences are simple cases of the above definition. Later, when discussing N-element inference, we will only be interested in "constructive" inferences, i.e. when a full chain of inferences exists. Thus the above diagram will be abbreviated as $CS^i_j; CS^k_l \xrightarrow{RPP} \sum_n CS^m_n$ without mentioning the retracted *CMs* given by $R(CS^i_j \text{ or } CS^k_l)$. In general, successful rendezvous can result in the "birth" of one or more *CMs*. All of them must fulfill a *C(...)* condition;

otherwise, they are aborted. Because our RPP is designed to evaluate the *C-I* of closed social structures, simplifying assumptions based on real life can be made. It is difficult to find cases of direct rendezvous and inference between two CS_i^m and CS_j^n if $m, n \geq 1$ without an intermediary involved CS_k^0 $k = 1, 2, \ldots$ (messages, pheromones, observation of behavior, e.g. the bee's dance, etc.). Even in *GA* [9], the crossover of genes can be considered the inference of the two genomes CS_i^0 and CS_j^0. Only if we consider CS^n on the level of nations, where exchange (migration) of humans takes place, can such a case be considered an approximation to such high level rendezvous and inferences. This is, however, just approximation, because finally, this exchange is implemented at the level of personal contact of humans, which are just rendezvous and inferences of two CS_i^0 and CS_j^0 with the help of CS_k^0 $k = 1, 2, \ldots$. Thus, rendezvous and direct inference between two CS_j^i if $i \geq 1$ are rare. We will only make use of a single CS_{main}^n for $n > 1$ as the main *CS*. Single beings (humans, ants) can be represented as $CS_{individual}^1$. Such beings perform internal brain-inferences, independently of higher, cooperative inferences inside CS_{main} and exchange of messages of the type CS^0. It will be allowable to have internal CS^k inside the main *CS*, as static ones (taking fixed positions) to define sub-structures such as streets, companies, villages, cities, etc. For simplicity, we will try to approximate beings as CS^0; otherwise, even statistical analysis would be too complicated. It is also important to assume that the results of inference are not allowed to infer recursively. They must immediately disperse after inference; however, after a certain time, inferences between them are allowed again (this is called *refraction* [6]).

5. Formal Definition of the Collective Intelligence Measure

The two basic definitions for *C-I* and its measure IQS have the form:

<u>DEFINITION 3: N-ELEMENT INFERENCE IN CS^N</u>

There is a given *CS* at any level $CS^n = \{CS_1^{a_1}, \ldots CS_m^{a_m}\}$, and an allowed Set of Inferences *SI* of the form $\{set\ of\ premises\ CS\} \xrightarrow{I_j} \{set\ of\ conclusions\ CS\}$, and one or more CS_{goal} of a goal. We say that $\{I_{a_0}, \ldots, I_{a_{N-1}}\} \subseteq SI$ is an N-element inference in CS^n, if for all $I \in \{I_{a_0}, \ldots, I_{a_{N-1}}\}$ the premises $\in$ *the present state of CS^n* at the moment of firing this inference, all $\{I_{a_0}, \ldots, I_{a_{N-1}}\}$ can be connected into one tree by common conclusions and premises, and $CS_{goal} \in \{set\ of\ conclusions\ for\ I_{a_{N-1}}\}$.

<u>DEFINITION 4: *COLLECTIVE INTELLIGENCE* QUOTIENT (IQS)</u>

IQS is measured by the probability P that after time t, the conclusion CM_{goal} will be reached from the starting *state of CS^t*, as a result of the assumed N-element inference. This is denoted $IQS = P(t, N)$.

For evaluating *C-I* the last two definitions fulfill these requirements: • N-element inference must be allowed to be interpreted as any problem-solving process in a social structure or inside a single being, where N inferences are necessary to get a result; or any production process, where N-technologies/elements have to be found and unified into one final technology or product. Therefore in the same uniform way we model inferring processes or production processes within a social structure. This is very important because some inference processes can be observed only through resultant production processes or specific logical behavior (of ants, bees, bacteria). • Simulating N-element inference in the RPP allows us to model the distribution of inference resources between individuals, dissipation in space or time, or movements (or temporary concentration) in the *CS*. This reflects well the dissipated, moving, or concentrated resources in a social structure of any type. • Cases can be simulated where some elements of the inference chain are temporarily not available, but at a certain time t, another inference running in the background or in parallel will produce the missing components. This is well known in human social structures, e.g. when a given research or technological discovery is blocked until missing theorems or sub-technology is discovered. • Humans infer in all directions: forward, e.g. improvements of existing technology; backward, e.g. searching how to manufacture a given product going back from the known formula; and also through generalization, e.g. two or more technologies can be combined into one more general and powerful technology or algorithm. N-element inference simulated in the RPP reflects all these cases clearly.

<u>EXAMPLE: HUMAN SOCIAL STRUCTURE FACING A SIMPLE PROBLEM</u>

Let us consider a human social structure suffering from cold weather. Assume that a) there is one person with matches moving somewhere inside the social structure; b) another has a matchbox; c) a third person has an idea of how to use the matches and matchbox to make a fire, if both components are available; d) a fourth person is conscious that heat from fire is a solution for the cold weather problem in this social structure; e) a fifth person is conscious that fire is necessary. Other humans are logically void. Nobody has total knowledge of all the necessary elements and their present position; everybody can infer only locally. For this social structure the CS^t (for the RPP) can be defined as set of CS^0: CS^1 = {matches. matchbox. matches $\wedge$ matchbox $\rightarrow$ fire. fire $\rightarrow$ heat. ?heat.} Now it is necessary to make assignments: • The displacement abilities v for above defined CS^0. We can use pure or quasi-Brownian movements, e.g. beings run chaotically around on foot looking for the necessary components. Public transportation can change the characteristics of displacement. Virtual travel by telephone or Internet search can also be defined. Traveling may be on a day/week/year scale. • The internal structure of CS^t which affect rendezvous probability should be defined (internal membranes reflecting boundaries, e.g. streets). • Artificial restrictions can be imposed on unification, e.g. different languages, casts, racial conflicts, etc. This system is a 4-element inference. The IQS can be computed here for this social structure on the basis of simulations as a curve showing how probability changes in the given time period, or as the general analytical function $P(4, t)$, if we can manage to derive it.

The proposed theory of *C-I* allows deriving formal definitions of some important properties of Social Structures obvious in real life: • LEVELS OF *C-I*; • SPECIALIZATION OF SOCIAL STRUCTURES; • EQUIVALENCE OF SOCIAL STRUCTURES OVER DOMAIN D.

6. Conclusion

The formal approach to the *C-I* presented here targets the future *C-I* Engineering of humans, and similar applications: • The Collective Intelligence of bacterial colonies against new drugs; • The Collective Intelligence of social animals and insects that create a threat, e.g. for farming. We should remember that the science of *C-I* is very sensitive, i.e. any definition or assumption can immediately be verified through numerous results of an observational or experimental nature, e.g. [7], [11], etc. Thus, every element from the formal approach has interpretation. So far this formalism has done very well when applied to human social structures [19], [14]. It has allowed us to confirm theoretically some behavior and properties of the social structures discussed in [11], as well as to discuss some new ones which until now had seemed "beyond" theoretical possibilities for evaluation, e.g. "why we build cities'.

Literature

1. Adleman L.: Molecular computations of solutions to combinatorial problems. *Science*. 11. 1994.
2. Berry G., Boudol G.: The chemical abstract machine. Theoretical Computer Science. 1992.
3. Das J. P. et al: Assessment of cognitive processes. Allyn and Bacon. 1994.
4. Goldstein A. P., Krasner L.: Handbook of Psychological Assessment. Pergamon. 1990.
5. Dorigo M., Gambardella L.M.: Ant Colonies... BioSystems, 1997.
6. Giarratano J., Riley G.: *Expert Systems*. PWS Pub. II ed. 1994.
7. Hölldobler E., Wilson O.: The ants. Harvard University. 1990.
8. Levy P.: Collective Intelligence. Plenum Trade 1997.
9. Michalewicz Z.: GA + data structures = evolution programs. Springer 1992.
10. Newell A.: Unified theories of cognition. Harvard University Press. 1990.
11. Shaw M. E.: Group Dynamics. IV ed. McGraw-Hill. 1971.
12. Smith J. B.: Collective Intelligence in Computer-Based Collaboration. Lawrence 1994.
13. Szuba T.: Evaluation measures for the collective intelligence of closed social structures. ISAS'97 Gaithersburg/Washington D.C., USA, 22-25 Sept., 1997
14. Szuba T.: A Molecular Quasi-random Model of Computations Applied to Evaluate Collective Intelligence. Future Generation Computing Journal. Volume/issue: 14/5-6. 1998.
15. Whitney C. A.: Random processes in physical systems. Wiley & Sons. 1990.

Implementation of Data Flow Logical Operations via Self-Assembly of DNA

Piotr Wąsiewicz[1], Piotr Borsuk[2], Jan J. Mulawka[1], Piotr Węgleński[2]

[1] Institute of Electronics Systems, Warsaw University of Technology
Nowowiejska 15/19, Warsaw, Poland
{jml,pwas}@ipe.pw.edu.pl
[2] University of Warsaw, Pawińskiego 5A, Warsaw, Poland
wegle@ibb.waw.pl

Abstract: Self-assembly of DNA is considered a fundamental operation in realization of molecular logic circuits. We propose a new approach to implementation of data flow logical operations based on manipulating DNA strands. In our method the logic gates, input, and output signals are represented by DNA molecules. Each logical operation is carried out as soon as the operands are ready. This technique employs standard operations of genetic engineering including radioactive labeling. To check practical utility of the method a series of genetic engineering experiments have been performed. The obtained results confirm interesting properties of the DNA-based molecular data flow logic gates. This technique may be utilized in massively parallel computers.

1 Introduction

Digital logic circuits create the base of computer architecture. Digital systems consist sometimes of single logic gates and switches [1]. In constructing computer hardware different types of logic gates e.g. OR, AND, XOR, NAND are utilized. Since computers are very important in our life and more powerful machines are still needed it is worth considering various implementations of their architectures. In conventional von Neumann machines, a program counter is used to perform in sequence the execution of instructions. These machines are based on a control-flow mechanism by which the order of program execution is explicitly stated in the user program.

Data flow machines do not fall into any of the classifications described above. In a data flow machine the flow of the program is not determined sequentially, but rather each operation is carried out as soon as the operands are ready. This data-driven mechanism allows the execution of any instruction depending on data availability. This permits of a high degree of parallelism at the instruction level and can be used in recent superscalar microprocessor architectures. However, in the field of parallel computing, dataflow techniques are in the research stage and industry has not yet adopted these techniques[3]. One of approaches suitable for massively parallel operations seems to be DNA based molecular computing [4].

The primary objective of this contribution is to show striking adequacy of molecular computing for dataflow techniques. Logical operations considered here are based on self-assembly of DNA strands. By self-assembly operation we understand putting together fragments of DNA during the process of hybridization [5]. Since processing units as well as input and output signals are represented by hybridizing DNA strands, data-driven mechanisms of performance are assured. We present a new approach to implementation of logical operations suitable for dataflow architectures. Our method is an extension of works reported by Ogihara, Ray and Amos [6, 7, 8].

2 Computing Properties of DNA Molecules

A number of papers have appeared on general concept of molecular computers [9, 10, 11, 12]. In such approach computations are performed on molecular level and different chemical compounds can be utilized. Deoxyribonucleic acid (DNA) is usually used in molecular computing because this compound serves as a carrier of information in living matter and is easy to manipulate in genetic engineering laboratory.

The DNA molecule is composed of definite sequence of nucleotides. There are four different nucleotides in such molecule depending on purine and pyrimidine bases present in the given nucleotides, which we denote by the letter: A, T, C, G (adenine, thymine, cytosine, guanine). Thus the DNA molecule can be considered as a polynucleotide or a strand or a string of letters. Since the strands may be very long, they can carry an immense number of bits. Therefore particular molecules may serve as information carriers.

Another interesting property of DNA is that particular bases join together forming pairs: A with T and C with G. This process known as hybridization or annealing causes self-assembly of DNA fragments. It occurs when two DNA strands are composed of matching (complementary) nucleotide sequences. Due to favored intermolecular interactions particular molecules can recognize each other. As a result a kind of key-lock decoding of information is possible. The self-assembly property may be utilized to implement the memory of an associate type. According to Baum [13] this memory may be of greater capacity than that of human brain. The other important feature of self-assembling is that particular molecules may be considered as processing units performing some computing.

The following techniques of the genetic engineering will be utilized in our approach: synthesis of DNA strands, labeling, hybridization, analysis by DNA electrophoresis, digestion of double strands by restriction enzymes, synthesis of DNA strands by Polymerase Chain Reaction (PCR) [14, 15].

3 Dataflow Parallel Processing Units

In a dataflow machines [16], data availability rather than a program counter is used to drive the execution of instructions. It means that every instruction that has all of its operands available can be executed in parallel. Such an evaluation of instructions permits the instructions to be unordered in a data-driven program [3].

Instead of being stored in a shared memory, data are directly hold inside instructions. Computational results are passed directly between instructions via data tokens. A single result is forwarded via separate tokens for each instruction that needs it. Results are not stored elsewhere and are not available to instructions that do not receive tokens. The asynchronous nature of parallel execution with the data-driven scheme requires no shared memory and no program counter. However, it implies the need for special handshaking or token-matching operations. If a pure dataflow machine could be cost-effectively implemented with a low instruction execution overhead, massive fine-grain parallelism at the instruction level could be exploited.

The computation is triggered by the demand for an operation's result. Consider the evaluation of a nested arithmetic expression $z=(x+y)*(w-5)$. The data-driven computation chooses a bottom-up approach, starting from the innermost operations $x+y$ and $w-5$, then proceeding to the outermost operation *. Such operations are carried out immediately after all their operands become available.

In a data flow processor the stored program is represented as a directed graph as shown for example in Fig. 1. The graph is reduced by evaluation of branches or subgraphs. Different parts of a graph or subgraph can be reduced or evaluated in parallel upon demand. Nodes labeled with particular operations calculate a result whenever the inputs are valid, and pass that result on to other nodes as soon as it is valid which is known as firing a node. To illustrate this operation some process of computation is described in Fig. 2. As is seen a node representing a performed process has two incoming branches, which represent inputs. The flow of data is shown in these pictures by introducing data tokens denoted by block dots. In Fig. 2a there is a moment when data on a single input has appeared and the processor is waiting for a second data token while in Fig. 2b the processor is ready to execute its function and finally in Fig. 2c there is a situation after firing the node. The dataflow graphs usually are more complicated and there are different types of operations applied. Note that there are no variables in a data flow program, and there can be no notion of entities such as a program counter or global memory.

Generally, molecular logic gates may be utilized in constructing computer architecture based on dataflow graphs. Such graphs consisting of DNA strands can exchange tokens as DNA single strings representing gate inputs and outputs. Some sectors of strands can transport values of variables or special constants. In such solution a number of gates-nodes (processing units) and a number of data and control tokens may be theoretically illimitable. In the next point we demonstrate how to realize data flow logical processing units using DNA molecules.

4 A New Concept of Data Flow Logic Gates

In classical computer the logic gate is an electronic circuit that has one or more inputs and one output. In such circuit the electrical condition of the output at any time is dependent on those of the inputs at that time [1]. An alternative approach to implement logic operations may be performed on molecular level by self-assembly of DNA strands. In such method DNA molecules as is depicted in Figs 3a respectively represent the logic gates, input and output signals.

Assume that strands representing gates consist of three sectors belonging to the following groups: $\underline{X}=\{\underline{x}_1, \underline{x}_2, \underline{x}_3,...,\underline{x}_N\}$, $\underline{Z}=\{\underline{z}_1, \underline{z}_2, \underline{z}_3,...,\underline{z}_N\}$, $\underline{Y}=\{\underline{y}_1, \underline{y}_2, \underline{y}_3,...,\underline{y}_N\}$. Representing inputs strands are composed of two sectors from groups: $X=\{x_1, x_2, x_3,...,x_N\}$, $Y=\{y_1, y_2, y_3,...,y_N\}$. Representing outputs strands consist of three sectors belonging to groups: $X=\{x_1, x_2, x_3,...,x_N\}$, $Z=\{z_1, z_2, z_3,...,z_N\}$, $Y=\{y_1, y_2, y_3,...,y_N\}$. The composition of DNA strands are depicted in Fig. 3b. To enable hybridization process, input and output signals sectors from the groups X, Z, Y are complementary to adequate sectors in logic gates strands: $\underline{X}$, $\underline{Z}$, $\underline{Y}$. This means that two DNA strands can be connected together with the two sequences in the following way: x_i with $\underline{x}_i$ or z_i with $\underline{z}_i$ or y_i with $\underline{y}_i$. There must be one sequence: x_i or z_i or y_i in the first string and another complementary sequence: $\underline{x}_i$ or $\underline{z}_i$ or $\underline{y}_i$ in the second string. The mentioned two sequences hybridize each other forming a double-strand fragment.

Thus the sectors x_1 and x_2 of input I_1 and I_2 strands pair with matching sections ($\underline{x}_1$ and $\underline{x}_2$) in the AND strand. Output signals are composed of input strands and other DNA fragments. These fragments ($\underline{z}_i$) are reserved to accomplish special tasks. For example they can transport additional information about logical gates, nodes in a computer program etc. The complete molecules after annealing are shown in Fig. 3c and Fig. 4c. Upper strand is labeled. Labeling of this strand can be achieved by introducing radioactive, biotinylated or fluorescent nucleotide. In our experiments (see section 6) we have employed the technique of filling the gap (A*) left between the sectors (oligonucleotides) z_1 and x_2 by addition of ^{32}PdATP using the Klenov polymerase. All strands are ligated with DNA ligase and then the output strands are disconnected from gate strands as is shown in Fig. 3d and then analyzed using standard method of the DNA electrophoresis in acrylamide gel. Single stranded radioactive oligonucleotides representing output of the logic gate are detected by autoradiography. Fragments of proper length could be isolated from mentioned gel and used as signal and data tokens in the next layer of molecular gates. As follows from Fig. 3 both AND and OR gates have similar construction, but when OR gates have two or more inputs, then two or more different molecules each with one input are created.

5 DNA-Based Implementation of Simple Logic Network

A Boolean function is composed of binary variables, operation symbols, brackets and a symbol of equivalence relation. The Boolean function describes usually more complicated logic networks composed of logic gates. Some number of connected logic gates with sets of inputs and outputs is called a combinational network. However, in a sequential network there are not only logic gates, but also flip-flops with memory [1].

We have considered to implement simple combinational network by operations on DNA strands. An example of a simple combinational network and its implementation in DNA as well as its performance is shown in Fig. 4. The logic gates are DNA oligonucleotides shown in Fig. 4b. Two gates AND$_1$, AND$_2$ are self-assembling molecules from Fig. 4c, d. In Fig. 4e these two molecules anneal to DNA strand of the

AND$_3$ gate. The regions named A* are for adding labeled radioactive adenine by the Klenov polymerase to a 3' end of oligonucleotides as described in Fig. 3. It should be noted that $E^{1,2}$ restriction sites are ready for digestion if there is a double DNA strand there. After hybridization and ligation a process of digestion a double strand molecule by restriction enzymes is performed. The enzymes cut the double strand in areas E^1 and E^2 yielding three fragments from which output strands could be potentially rescued after strand separation. It is necessary to remove enzymes and redundant strings before further computation. The redundant strings are removed in a process of electrophoresis.

Neighbouring logic gates layers could be connected with each other to form the combinational network. A number of layers may be theoretically illimitable if the oligonucleotides are recycled during computing after an execution of some number of layers. For construction of our molecular gate layers one can envisage use of solid support. These gates can be for example attached to DNA chips [17]. E.g. if in Fig. 4f the AND$_3$ gate strand was attached to the DNA chip and the AND$_1$ and AND$_2$ gate strands – to the magnetic beads with biotin, then output strings could be easily separated.

6 Results of Experiments

To confirm oligonucleotide self-assembling as a technique useful in data flow logical operations several experiments were performed.

Four oligonucleotides were designed for self-assembling. The lower strand is composed of one longest oligonucleotide named AND (30 bp) – 5' AGTAACTCCCTCACCATCGCTCGCCAACTC 3' and the upper strand of three other: I1 – 5' GGTGTCGATCATTGAG 3', SR – 5' GGAGTGGTAGCG 3' and I2 – 5'CGGTTGAGCTTGGAAG 3'. In our experiments I1 and/or I2 were hybridized to AND in presence of SR oligonucleotide. Typically 5 – 10 pM of each oligonucleotide were used and hybridization was performed in ligase buffer for 5 – 30 min. in 40^{0}C. Under these conditions formation of the SR/AND hybrid was preferred. In next step 5 µCi of ^{32}PdATP (3000 Ci/mM, Amersham-USB), 100 pM dGTP and 10u of Klenov polymerase (Amersham-USB) were added, and samples were incubated for 30 min at 37°C. Afterwards the temperature was lowered to 25°C (to hybridize I1 and/or I2 to AND) and samples were incubated for 30 min. 5u of ligase (Promega) were added. Ligation were performed at 25°C for 2 h, and at 14°C for 2 – 24 h. Ligation reactions were terminated by addition of the equal volume of the stop mixture (formamide with 0,05% bromophenol blue and 0,05% xylene cyanol). Samples were denatured by 5 min. incubation in 80^{0}C and loaded onto poliacrylamide (8 or 20%) denaturing (8M urea) gel. Electrophoresis was performed under standard conditions. Radioactive single stranded DNA were detected in electrophoretograms by autoradiography (Fig. 5 and 6). Typically only proper single stranded DNA fragments were obtained (Fig. 5). Oligonucleotide self-assembling was not disturbed by addition of the random oligonucleotide (100 pM per reaction) (Fig. 6 A1, B1 and C1).

7 Conclusions

We have explored the possibility of implementation of data flow logical operations by DNA manipulations. We have demonstrated that self-assembly of DNA can be utilized to provide the flow of data. In our approach standard genetic operations are employed. An appropriate DNA reactions have been performed and their results confirmed our assumptions. We think that our methodology has several advantages when compared with that proposed by Ogihara and Amos [6,7,8] and can be further improved by the use of DNA chips. This would permit for easy separation of output and input strands and would provide an important step towards massively parallel molecular computer.

References

[1] Biswas, N.N.: Logic Design Theory, Prentice-Hall International Editions, USA, 1993
[2] Hill, F. J., Peterson, G.P.: Switching Theory and Logical Design. Wiley New York (1974)
[3] Landry E., Kishida Y.: A Survey of Dataflow Architectures. Internet.
[4] Mulawka, J. J., Wąsiewicz, P.: Molecular Computing, (in Polish) Informatyka, 4, April (1998) 36-39
[5] Mulawka, J.J., Borsuk, P., Węgleński, P.: Implementation of the Inference Engine Based on Molecular Computing Technique. Proc. IEEE Int. Conf. on Evolutionary Computation (ICEC'98), Anchorage USA (1998) 493-496
[6] Ogihara M. and Ray, A.: Simulating Boolean Circuits On a DNA Computer. Technical Report TR 631, University of Rochester, Computer Science Department, August (1996)
[7] Ogihara M. and Ray, A.: The Minimum DNA Computation Model and Its Computational Power. Technical Report TR 672, University of Rochester, Singapur (1998)
[8] Martyn Amos and Paul E. Dunne. DNA Simulation of Boolean Circuits. Technical Report CTAG-97009, Department of Computer Science, University of Liverpool UK (1997)
[9] Adleman, L.M.: Molecular Computation of Solutions to Combinatorial Problems. Science, vol. 266. (1994) 1021-1024
[10] Adleman, L.M.: On Constructing a Molecular Computer, Internet
[11] Kurtz, S., Mahaney, S., Royer, J.S., Simon, J.: Biological Computer, Internet
[12] Dassen, R.: A Bibliography of Molecular Computation and Splicing Systems, at the site WWW http://liinwww.ira.uka.de/bibliography/Misc/dna.html
[13] E.B. Baum, Building an Associative Memory Vastly Larger than the Brain, Science, vol. 268, 583-585, 1995.
[14] Sambrook, J., Fritsch, E.F., Maniatis, T.: Molecular Cloning. A Laboratory Manual. Second Edition, Cold Spring Harbor Laboratory Press (1989)
[15] Amos, M.: DNA Computation. PhD thesis, Department of Computer Science, University of Warwick UK, September (1997)
[16] Papadopoulos G.M.: Implementation of a General Purpose Dataflow Mulitprocessor. PhD thesis, MIT Laboratory for Computer Science, 545 Technology Square, Cambridge, MA 02139, August 1988.
[17] Wodicka, L., et al.: Genome-wide Expression Monitoring in Saccharomyces cerevisiae. Nature Biotechnology 15 Dec. (1997) 1359-1367

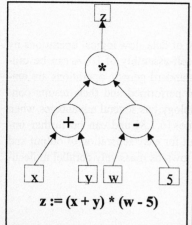

Fig. 1. A dataflow graph of a simple program

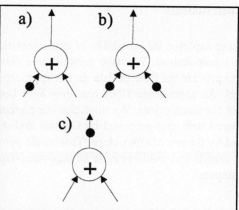

Fig. 2. Successive phases of firing a node: a) waiting for a second data token, b) ready to execute an operation, c) after firing

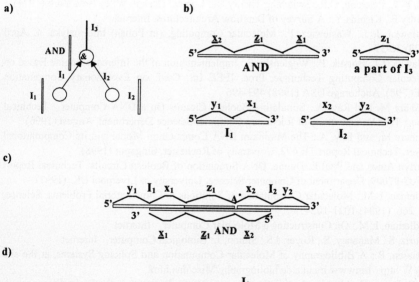

Fig. 3. A data flow logic AND gate: a) a scheme of the gate, b) oligonucleotides representing the AND gate, input signals I_1, I_2 and a part of an output I_3, c) hybridized oligos, d) disconnecting of the AND gate and an output I_3; A^* - regions of adding labeled radioactive Adenine by Klenov polymerase to a 3' end of z_1, so there must be several free Tymine nucleotides in the AND gate strand in the area between $\underline{z_1}$, I_2.

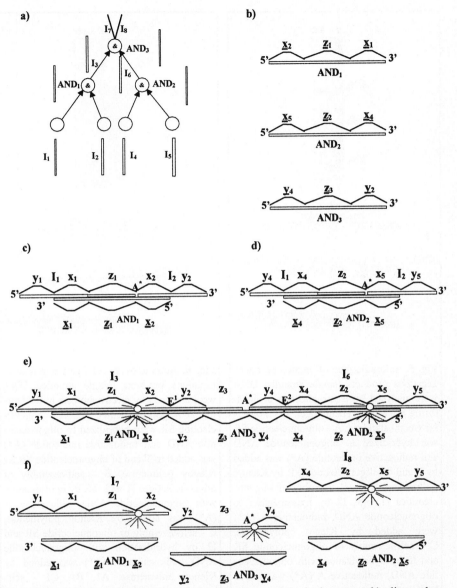

Fig. 4. A data flow combinational network: a) a scheme of the logic system, b) oligonucleotides representing logic gates, c) and d) hybridized oligonucleotides of gates AND_1, AND_2 with oligos of inputs signals and parts of output signals, e) hybridized an oligonucleotide of the AND_3 gate with hybridized earlier oligos of gates AND_1, AND_2, f) digestion by restriction enzyme of the restriction sites E^1, E^2, disconnecting of the AND gates and outputs I_7 and I_8; A^* regions of adding labeled radioactive Adenine by Klenov polymerase to a 3' end of oligonucleotides as described in Fig.1 and Fig.2. ; $E^{1,2}$ restriction sites ready for digestion if there is a double DNA strand there.

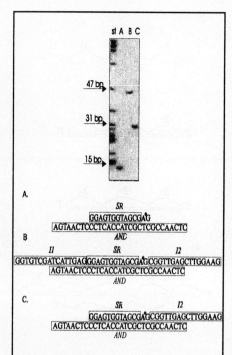

Fig. 5. Autoradiogram of ligation products. Ligations products (single stranded DNA were analysed by electrophoresis in denaturing 8% polyacrylamide gel. st - molecular weight standard, A - oligonucleotide SR was hybridized to oligonucleotide AND and radioactive nucleotide (A*) was added to 3' end of oligonucleotide SR by Klenov polimerase, B - self-assembly of oligonucleotides I1, SR, I2 by hybridization to oligonucleotide AND, radioactive A (A*) were added by Klenov polymerase, C - self-assembly of oligonucleotide SR and I2 by hybridization with oligonucleotide AND, radioactive A (A*) was added by Klenov polymerase.

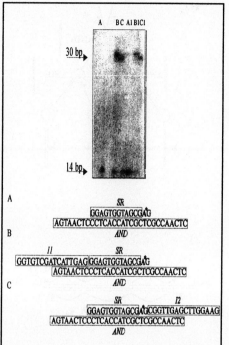

Fig. 6. Autoradiogram of ligation products. Ligations products (single stranded DNA were analysed by electrophoresis in denaturing 20% polyacrylamide gel. A - oligonucleotide SR were hybridized to oligonucleotide AND and radioactive nucleotide (A*) was added to 3' end of oligonucleotide SR by Klenov polimerase, B - self-assembly of oligonucleotides I1 and SR by hybridization with oligonucleotide AND, radioactive A (A*) was added by Klenov polymerase, C - self-assembly of oligonucleotide SR and I2 by hybridization with oligonucleotide AND, radioactive A (A*) was added by Klenov polymerase. A1, B1, C1 - self-assembly and labeling reactions were performed as in A, B and C additionally to each reaction random oligonucleotide were added.

A Parallel Hybrid Evolutionary Metaheuristic for the Period Vehicle Routing Problem

Dalessandro Soares Vianna, Luiz S. Ochi
and Lúcia M. A. Drummond

Universidade Federal Fluminense - Departamento de Ciência da Computação
Travessa Capitão Zeferino 56/808
Niterói - Rio de Janeiro - Brazil - 24220-230
e-mail: dada@pgcc.uff.br, {satoru,lucia}@dcc.uff.br

Abstract. This paper presents a Parallel Hybrid Evolutionary Metaheuristic for the Period Vehicle Routing Problem (PVRP). The PRVP generalizes the classical Vehicle Routing Problem by extending the planning period from a single day to M days. The algorithm proposed is based on concepts used in Parallel Genetic Algorithms and Local Search Heuristics. The algorithm employs the island model in which the migration frequency must not be very high. The results of computational experiments carried out on problems taken from the literature indicate that the proposed approach outperforms existing heuristics in most cases.

1 Introduction

The classical Vehicle Routing Problem (VRP) is defined as follows: vehicles with a fixed capacity Q must deliver order quantities q_i ($i = 1, \ldots, n$) of goods to n customers from a single depot ($i = 0$). Knowing the distance d_{ij} between customers i and j ($i, j = 0, \ldots, n$), the objective of the problem is to minimize the total distance travelled by the vehicles in such a way that only one vehicle handles the deliveries for a given customer and the total quantity of goods that a single vehicle delivers do not be larger than Q.

In classical VRPs, typically the planning period is a single day. The period Vehicle Routing Problem (PVRP) generalizes the classical VRP by extending the planning period to M days. Over the M-day period, each customer must be visited at least once during the considered period. The classical PVRP consists of a homogeneous vehicle fleet (vehicles with same capacities) which must visit a group of customers from a depot where the vehicles must start and return to at the end of their journeys. Each vehicle has a fixed capacity that cannot be exceeded and each customer has a known daily demand that must be completely satisfied in only one visit by exactly one vehicle. The planning period is M days. If M = 1, then PVRP becomes an instance of the classical VRP. Each customer in PVRP must be visited k times, where $1 \leq k \leq$ M. In the classical model of PVRP, the daily demand of a customer is always fixed. The PVRP can be seen as a problem of generating a group of routes for each day so that the constraints involved are satisfied and the global costs are minimized.

PVRP can also be seen as a multi-level combinatorial optimization problem. In the first level, the objective is to generate a group of feasible alternatives (combinations) for each customer. For example, if the planning period has $t = 3$ days $\{d_1, d_2, d_3\}$ then the possible combinations are: $0 \to 000$; $1 \to 001$; $2 \to 010$; $3 \to 011$; $4 \to 100$; $5 \to 101$; $6 \to 110$ and $7 \to 111$. If a customer requests two visits, then this customer has the following visiting alternatives: $\{d_1, d_2\}$, $\{d_1, d_3\}$, and $\{d_2, d_3\}$ (or the options: 3, 5 and 6 of Table 1). In the second level, we must select one of the alternatives for each customer, so that the daily constraints are satisfied. Thus we must select the customers to be visited in each day. In the third level, we solve the vehicle routing problem for each day. In this paper, our algorithm is applied to the basic model of PVRP with an additional constraint: the number of vehicles is limited, although this limit is not necessarily the same every day. The technique proposed here can be applied to various models of PVRP.

Customer	Diary Demand	# of Visits	# of Combinations	Possible Combinations
1	30	1	3	1, 2 e 4
2	20	2	3	3, 5 e 6
3	20	2	3	3, 5 e 6
4	30	1	3	1, 2 e 4
5	10	3	1	7

Table 1. A PVRP with $t=3$ days

In the last years, Evolutionary Metaheuristics and in particular Genetic Algorithms (GA) have been used successfully in solution of NP-Complete and NP-HARD problems of high dimensions. Hard problems require large search spaces resulting in high computational costs. In this context, Evolutionary Metaheuristics including GA may require a large amount of time to find good feasible solutions, encouranging the use of parallel techniques [10]. Although the main goal of a Parallel Evolutionary Metaheuristic is the reduction of the execution time necessary to find an acceptable solution, sometimes it can also be used to improve the results obtained by sequential versions.

Most of Evolutionary Metaheuristics, such as GA, Scatter Search, Ant Systems and Neural Nets, are easy to parallelize because of their intrinsic parallelism. There are differents ways to parallelize GA. The generally used classification divides parallel GAs in three categories: Island or stepping stone, Fine Grain and Panmitic Models [10].

In this paper, we propose a parallel hybrid evolutionary metaheuristic based on parallel GA, scatter search and local search methods. The algorithm is based on the Island Model and it was implemented on a cluster of workstations with 4 RISC/6000 processors.

The remainder of this paper is organized as follows. Section 2 presents the related literature about PVRP. Section 3 presents the algorithm proposed. Experimental results are shown in Section 4. Finally, Section 5 concludes the paper.

2 Related Literature

Although the PVRP has been used in many applications, it has not been extensively studied in related literature. Golden, Chao and Wasil [7] developed a heuristic based on the concept of "record-to-record" proposed by Dueck [5]. To the best of our knowledge, there are not many papers using Metaheuristics to solve PVRP. Cordeau, Gendreau and Laporte [4] presented a Tabu Search Metaheuristic to solve this problem. Computational results taken from the literature indicate that their algorithm outperforms all previous conventional heuristics.

GA have been used with great success to solve several problems with high degrees of complexity in Combinatorial Optimization, including models of daily VRP ([1], [8], [9], [10]). More recently, Rocha, Ochi and Glover[11] proposed a Hybrid Genetic Algorithm (HGA) for the PVRP. Their algorithm is based on concepts of GA and Local Search heuristics. Computational experiments using tests available in literature showed the superiority of this method when it is compared with other existing metaheuristics.

3 A Parallel Hybrid Evolutionary Metaheuristic

This paper presents a Parallel Hybrid Evolutionary Metaheuristic (PAR-HEM) for the PVRP. The parallel algorithm is based on the Island Model and it was implemented on a cluster of workstations with 4 RISC/6000 processors. In order to reduce communication overhead, which is usual in this kind of model [9] and ensure the occupation of all processors during the execution of the algorithm, a new criteria of migration and termination were employed. In Island Model, the population of chromosomes of Genetic Algorithms is partitioned into several subpopulations, which evolve in parallel and periodically migrate their individuals(chromosomes) among themselves. Because of the high cost of communication in this model, migration frequency must not be very high. Thus migration is only executed when subpopulation renewal is necessary. The criterion of termination of processes is based on a global condition, which involves every process that composes the PAR-HEM, in order to prevent a process from being idle while the others are still executing.

In our algorithm, each processor executes a pair of tasks m_i and q_i. Each task m_i executes the following steps, based on the sequential algorithm proposed by Rocha, Ochi and Glover [11], described in next subsections: generation of a feasible alternative for each customer; selection of an alternative for each customer; representation of solutions through chromosomes; generation of an initial population of chromosomes; evaluation and reproduction; and diversification.

In order to execute the diversification step, m_i sends a migration request to q_i if population renewal is required. The task q_i is responsible for migration and termination of the algorithm above. See [9] for more details.

3.1 Representation of Chromosomes

The representation of a feasible solution in a chromosome structure may be much more complex for the PVRP than it is for the VRP. In addition to the problem of finding an optimal assignment of customers to vehicles and defining the best route for each vehicle, there is also the problem of distributing the number of visits required by each customer in the planning horizon, satisfying the daily constraints. Each chromosome in our GA, is represented by two vectors. The first vector is associated with the n customers of PVRP, as shown in Table 2. The ith position (i^{th} gene) of this vector is a non-negative integer k, such that $0 \leq k \leq 2t - 1$, where t is the number of days in the current planning horizon. The number k represents a feasible alternative (combination) of visits to the associated customer. The binary representation of k represents the days when the associated customer receives a visit. The second vector (associated to the first) corresponds to the accumulated demand of each day in the planning horizon. For example, considering 3 days, we could have the following served daily demand {60, 30, 50}, which means that in the first day 60 goods were delivered, in the second day 30 goods and finally in the last day 50 goods were delivered.

L = (1	2	3	4	5) reference vector
P = (4	5	3	1	7) chromosome
⇕	⇕	⇕	⇕	⇕
100	101	011	001	111 } combination in binary form

Table 2. Representation of a chromosome

3.2 Initial Population

The initial population of our GA is generated as follows. Initially, a set of feasible alternative visits for each customer is generated. The customers are ordered according to their number of alternatives. For example, in a period of three days ($t = 3$), a customer which requires three visits has only one feasible alternative, while another which requires only one visit has three feasible alternatives. In order to generate an initial chromosome, we randomly select a customer from the set of customers which present the smallest number of feasible alternatives. Then one of the alternatives of this customer is chosen. The integer number associated to this alternative is placed in the corresponding gene. Then successively, an alternative is selected from each of the remaining customers with the smallest number of alternatives. For each alternative selected, we update the data in the second vector (vector of served daily demand). If the introduction of an alternative of a new customer violates the global capacity of the fleet in any day, this alternative is discarded and another alternative for this customer is sought. When the customer does not have other alternatives, the chromosome is abandoned. The chromosome is complete when every gene has been filled.

In the first allocation, the customers selected hardly violate the maximum capacity constraint of the daily fleet, because few customers have been allocated. On the other hand, according to the addition of each new customer, the accumulated demand for each day increases and so does the risk of exceeding the maximum capacity allowed for any particular day. Therefore, starting the selection of customers with the one with greatest number of alternatives, may reduce the risk of discarding this chromosome.

3.3 Evaluation and Reproduction

The fitness level of a chromosome in our GA corresponds to the objective function value for the PVRP. The cost of the objective function is obtained by solving a set of VRPs, one for each day. Although this technique is expensive, it is the simplest form of evaluating a chromosome in most optimization problems. In order to minimize this effort, we implemented a fast and efficient heuristic, based on saving concepts [3] to solve a VRP for each day.

The reproduction mechanism in our GA uses classical crossover and mutation operators.

In the crossover operator, given two parents from the population, we choose randomly a position (gene) associated with a customer. Initially, we subtract from the second vector (the accumulated demand vector associated with this chromosome) the respective demand values corresponding to the selected alternatives. Next, it is tried to change the alternatives of this customer selected in each parent (if these alternatives are distinct). If this change does not produce a new feasible solution, considering the maximum daily capacity constraints of the fleet, another attempt with a new gene is made. This procedure is applied p times, where p is an input parameter of the problem.

The second operator used, is a Mutation operator, which executes feasible changes in p points selected randomly in the chromosome.

3.4 Diversification and Termination

In order to diversify a population with a small tax of renewal, our algorithm uses the migration operator.

The strategy adopted in this algorithm, consists of associating to each task m_i, which executes the PAR-HEM, a task q_i. A task q_i, for $1 \leq i \leq w$, where w is the number of processors, communicate only by message-passing.

A task q_i is activated by the associated task m_i in the following cases:

- When the renewal tax is less than 5 per cent (tax of replacement of parents by offspring's), triggering the migration of individuals.
- When the task m_i is capable of finishing, initiating the termination of the algorithm.

In the first case, q_i executes a broadcast of requests so that all tasks q_j ($\forall j : j \neq i$) send their local best solutions (LBS) generated so far. When this occurs, a new population is generated based on the k-LBS. If a chromosome received is better than one of the k-LBS, then the chromosome received replaces the worst of the k-LBS. Usually the number k is less than the size of a population. Thus, in order to mantain the same number of chromosomes in the new population, the following procedure is executed. Copies of the k-best solutions are generated using the roulette criterion, where the most suitable individuals give a bigger number of copies. For each copy, a "window" is opened in a random position and new feasible alternatives for each customer are selected swapping genes in these windows. Built a new population with characteristics from the LBS, the genetic operator is restarted.

After y migrations (y is an entry parameter), m_i becomes capable of finishing its execution and consequently it activates q_i, which keeps an array of n elements representing the state of the system. Each position i of the array of state may assume values true or false, indicating that the process m_i is capable of finishing or not. When q_i is activated, it updates the positon i in the array of state with true and initiates a broadcast so that the other arrays represent the current state of the system. This strategy of termination is based on a simplification of the algorithm for detection of stable conjunctive predicates presented in [6]. A task m_i will continue its execution, even when capable of finishing, until all processors are able to finish. This prevents a processor from being idle while other processors remain executing and still allows the improvement of the optimal solution while the application does not finish. When all elements of the array of state are true, termination is detected by q_i and m_i is informed to finish its execution. A task m_i informs to q_i the optimal solution so far whenever its champion chromosome is improved.

The strategy described above allows that the process, which is responsible for the PAR-HEM execution does not remain occupied with the communication required by migration of individuals, and termination of the PAR-HEM, simplifying its design and implementation.

4 Computational Results

Our numerical experiments were run on a cluster of workstations with 4 RISC/6000 processors using the Programming Language C and MPI (Message-Passing Interface) for parallelism [12]. The performance of PAR-HEM was evaluated as follows. The algorithm was applied to several instances of the problems presented by Christofides and Beasley [2] and Golden, Chao and Wasil [7]. These test problems have dimensions ranging from 50 to 417 customers and planning period varying from 2 to 10 days as shown in Table 3.

We compared PAR-HEM with the following heuristics: CGL - Metaheuristic proposed by Cordeau, Gendreau e Laporte [4]; CGW - Heuristic proposed by Golden, Chao, e Wasil [7] and HGA - Metaheuristic proposed by Rocha, Ochi and Glover [11].

Problem	Customers	Period	Vehicles
2	50	5	3
3	50	5	1
4	75	2	5
5	75	5	6
6	75	10	1
7	100	2	4
8	100	5	5
9	100	8	1
10	100	5	4
11	126	5	4
12	163	5	3
13	417	7	9
14	20	4	2
15	38	4	2
16	56	4	2
17	40	4	4
18	76	4	4
19	112	4	4
20	184	4	4

Table 3. Problems

Problem	HEM	PAR-HEM
2	1355.37	1310.25
3	525.96	555.07
4	826.69	884.55
5	2102.83	2103.84
6	799.04	786.70
7	862.26	884.72
8	2158.83	2129.73
9	844.72	839.76
10	1717.60	1705.82
11	790.92	779.73
12	1227.37	1227.00
13	4558.57	4626.11
14	865.96	904.32
15	1792.14	1792.14
16	2749.68	2780.30
17	1631.53	1623.47
18	3180.04	3161.84
19	4809.64	4792.15
20	8570.94	8378.77

Table 4. Average cost: (HEM X PAR-HEM)

As showed in [11], their algorihtm (HGA) presents the best results in terms of quality of solution among the existing heuristics in literature. Thus, we compared PAR-HEM with an adapted version of HGA, which is called here as Hybrid Evolutionary Metaheuristic (HEM).

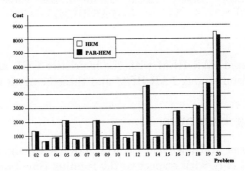

Fig. 1. Average cost: (HEM X PAR-HEM)

Figure 1 and Table 4 show a comparison in terms of average costs between the algorithms HEM and PAR-HEM. Each problem of the Table 3 was executed 3 times. Data corresponding to problem 01 were not avalibale. Figure 1 and

Table 4 show that PAR-HEM achieved the best results for 11 problems while HEM achieved the best results in 7 problems. Problem 15 presented identical results. Figure 2 and Table 5 show the average time in seconds required by the algorithms HEM and PAR-HEM. Figure 3 shows the speedup obtained by PAR-HEM. Sometimes the speedup exceeds the number of processors. In these cases the PAR-HEM converges faster than HEM. This may happen because the populations generated are different.

Problem	HEM	PAR-HEM
2	110.67	36.67
3	27.00	5.33
4	418.33	105.67
5	997.67	166.33
6	67.00	16.33
7	1365.67	362.67
8	1391.67	606.67
9	906.00	203.33
10	2470.67	950.33
11	2826.33	1415.33
12	10474.33	5844.33
13	370.33	169.00
14	14.00	4.00
15	67.00	20.33
16	269.67	87.00
17	108.67	34.33
18	541.00	215.67
19	3475.00	1205.50
20	13873.00	5444.00

Table 5. Average time: (HEM X PAR-HEM)

Problem	CGL	CGW	PAR-HEM
2	1330.09	1337.20	**1291.10**
3	524.61	**524.60**	533.91
4	**837.93**	860.90	871.71
5	**2061.36**	2089.00	2089.29
6	840.30	881.10	**770.82**
7	**829.37**	832.00	844.74
8	**2054.90**	2075.10	2112.96
9	**829.45**	829.90	836.74
10	**1629.96**	1633.20	1660.90
11	817.56	791.30	**775.89**
12	1239.58	1237.40	**1215.37**
13	**3602.76**	3629.80	4604.74
14	954.81	954.80	**864.06**
15	1862.63	1862.60	**1792.14**
16	2875.24	2875.20	**2749.68**
17	**1597.75**	1614.40	1613.67
18	3159.22	3217.70	**3143.23**
19	4902.64	4846.50	**4792.15**
20	8367.40	8367.40	**8299.71**

Table 6. Best Cost: (CGL X CGW X PAR-HEM)

We also compared PAR-HEM with the heuristics CGL and CGW. Table 6 presents a comparison among PAR-HEM and these heuristics in terms of cost.

5 Concluding Remarks

Our results presented in Tables and Figures above, show significant advantages for PAR-HEM, not only with respect to the running time but also with respect to the search quality.

The partial results exhibited by PAR-HEM qualify it as a promising way to obtain good aproximate solutions of PVRP since the solutions shown in this paper were obtained by initial tests, without an accurate definition of the best parameters for PAR-HEM. On the other hand, the results presented by sequential heuristics were obtained by exhaustive tests with several parameters.

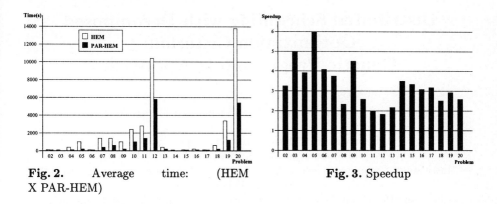

Fig. 2. Average time: (HEM X PAR-HEM)

Fig. 3. Speedup

References

1. Back, T., Fogel, D.B., Michalewicz, Z.: Handbook of Evolutionary Computation, University Oxford Press, NY, (1996).
2. Christofides, N., Beasley, J.E.: The Period Routing Problem, Networks **14** (1984) 237 - 256.
3. Clarke, G., Wright, J.: Scheduling of Vehicles From a Central Depot to a Number of Delivery Points. Operations Research **12-4** (1964) 568-581.
4. Cordeau, J.F., Gendreau, M., Laporte, G.: A Tabu Search Heuristic for Periodic and Multi-Depot Vehicle Routing Problems. Technical Report 95-75, Center for Research on Transportation, Montréal (1995).
5. Dueck, G.: New Optimization Heuristics: The Great Deluge Algorithm and the Record-to-Record Travel, Scientific Center Technical Report, IBM Germany, Heidelberg Scientific Center (1990).
6. Drummond, L.M.A., Barbosa, V.C.: Distributed Breakpoint Detection in Message-passing programs, J. of Parallel and Distributed Computing **39** (1996) 153-167.
7. Golden, B.L., Chao, I.M., Wasil, E.: An Improved Heuristic for the Period Vehicle Routing Problem. Networks (1995), 25-44.
8. Ochi, L.S., Figueiredo, R.M.V., Drummond, L.M.A.: Design and Implementation of a Parallel Genetic Algorithm for the Travelling Purchaser Problem, ACM symposium on Applied Computing (1997) 257-263.
9. Ochi, L.S., Vianna, D.S., Drummond, L.M.A.: A Parallel Genetic Algorithm for the Vehicle Routing Problem with Heterogeneous Fleet, Proc. of the BioSP3, Lecture Notes in Computer Science, Springer-Verlag **1388** (1998) 187-195.
10. Ribeiro, J.L.: An Object-oriented Programming Environment for Parallel Genetic Algorithms, Ph.D. Thesis, Dep. of Comp. Science, Univ. College London (1995).
11. Rocha, M.L., Ochi, L.S., Glover, F.: A Hybrid Genetic Algorithm for the Periodic Vehicle Routing Problem (1998) - (To be published).
12. Snir, M., Otto, S.W., Huss-Lederman, S., Walker, D.W., Dongarra, J.: MPI: The Complete Reference, The MIT Press (1996).

Distributed Scheduling with Decomposed Optimization Criterion: Genetic Programming Approach

Franciszek Seredyński[1,2], Jacek Koronacki[2] and Cezary Z. Janikow[1]

[1] Department of Mathematics and Computer Science
University of Missouri at St. Louis, St. Louis, MO 63121, USA
[2] Institute of Computer Science, Polish Academy of Sciences, Ordona 21, 01-237 Warsaw, Poland
e-mail: sered@arch.umsl.edu, korona@ipipan.waw.pl, janikow@radom.umsl.edu

Abstract. A new approach to develop parallel and distributed scheduling algorithms for multiprocessor systems is proposed. Its main innovation lies in evolving a decomposition of the global optimization criteria. For this purpose a program graph is interpreted as a multi-agent system. A game-theoretic model of interaction between agents is applied. Competetive coevolutionary genetic algorithm, termed loosely coupled genetic algorithm, is used to implement the multi-agent system. To make the algoritm truly distributed, decomposition of the global optimization criterion into local criteria is proposed. This decomposition is evolved with genetic programming. Results of succesive experimental study of the proposed algorithm are presented.

1 Introduction

In this paper, we propose a new approach to develop parallel and distributed algorithms for the problem of multiprocessor scheduling [1, 4]. Parallel and distributed scheduling algorithms represent a new direction in the area of multiprocessor scheduling [2, 8, 9]. Here we propose an approach based on a multi-agent system paradigm, which offers both parallelism and distributed control of executed algorithms. To apply a multi-agent system methodology we will interpret the graph of a parallel program as a multi-agent system, and we will investigate the global behavior of such a system in terms of minimization of the total execution time of the program graph on a given multiprocessor system. To implement such a multi-agent system, we will use competetive coevolutionary genetic algorithm termed *loosely coupled genetic algorithm* (LCGA) [8], which provides evolutionary algorithm for a local decision making, and a game-theoretic model of interaction between agents.

The specific open question in the area of multi-agent systems is the problem of incorporating a global goal of the multi-agent system into local interests of individual agents. We propose to apply an evolutionary technique, known as genetic programming (GP) [7], to decompose a global optimization criterion of the scheduling problem into local optimization criteria of agents.

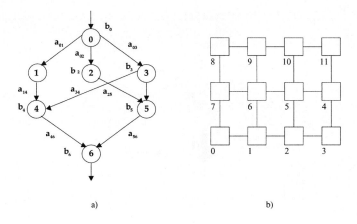

Fig. 1. Examples of a system graph (a), and a precedence task graph (b)

The paper is organized as follows. Section 2 discusses accepted models of parallel programs and systems in the context of scheduling problems, and proposes a multi-agent interpretation of the scheduling problem. Section 3 presents a parallel and distributed algorithm interpreted as a multi-agent system and implemented with LCGA. Section 4 describes an approach to decompose the global scheduling optimization criterion into local optimization criteria – based on GP. Results of experimental study of the proposed scheduler are presented in Section 5.

2 Multi-agent Approach to Multiprocessor Scheduling

A multiprocessor system is represented by an undirected unweighted graph $G_s = (V_s, E_s)$ called a *system graph*. V_s is the set of N_s nodes representing processors and E_s is the set of edges representing bidirectional channels between processors.

A parallel program is represented by a weighted directed acyclic graph $G_p = <V_p, E_p>$, called a *precedence task graph* or a *program graph*. V_p is the set of N_p nodes of the graph, representing elementary tasks. The weight b_k of the node k describes the processing time needed to execute task k on any processor of the homogeneous system. E_p is the set of edges of the precedence task graph describing the communication patterns between the tasks. The weight a_{kl}, associated with the edge (k, l), defines the communication time between the ordered pair of tasks k and l when they are located in neighboring processors. Figure 1 shows examples of the program graph and the system graph representing a homogeneous multiprocessor system with a grid topology.

The purpose of *scheduling* is to distribute the tasks among the processors in such a way that the precedence constraints are preserved, and the *response time* T (the total execution time) is minimized. The response time T depends on

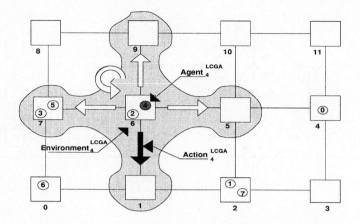

Fig. 2. Multi-agent interpretation of the scheduling problem.

the *allocation* of tasks in the multiprocessor topology and on *scheduling policy* applied in individual processors:

$$T = f(allocation, scheduling\_policy). \qquad (1)$$

We assume that the scheduling policy is fixed for a given run and the same for all processors.

We propose an approach to multiprocessor scheduling based on a multi-agent interpretation of the parallel program. An interaction between agents is based on applying a game-theoretic model termed a game with limited interaction [8]. The agents migrate in the topology of the parallel system environment, searching for an optimal allocation of program tasks into the processors. To evolve strategies for for the migration of agents in the system graph, we use competetive coevolutionary genetic algorithm LCGA (see Section 3).

We assume that a collection of agents is assigned to tasks of the precedence task graph in a such way that one agent is assigned to one task. An agent A_k, associated with task k, can perform a number of migration actions S_k on the system graph (see Figure 2). These local actions change the task allocation, thus influence the global optimization criterion (1).

The scheduling algorithm can be described in the following way:

- Agents are assigned to tasks of the program graph and located on the system graph (e.g., randomly).
- Agents are considered as players taking part in a game with limited interaction. Each agent-player has $r_k + 1$ possible actions interpreted as migrations to r_k nearest neighbors (plus 1 for stying in place). Figure 2 shows the precedence task graph from Figure 1 allocated in the system graph of the same Figure 1.
- Each player has a global criterion (1) describing the total execution time of the scheduling problem.

- The objective of players in the game is to alocate their corresponding tasks to processors in such a way that the global execution time is minimized (1).
- After a predefined number of single games, players are expected to reach an equilibrium state corresponding to a Nash point - a solution of this noncooperative game, providing the maximal payoff of the team of players in the game.

3 Distributed Scheduling with LCGA

The LCGA [8] implementing the parallel and distributed algorithm interpreted as a multi-agent system can be described in the following way:

#1: For each agent-player A_k, associated with task k and initially allocated in node n of the multiprocessor system, create an initial subpopulation of size N with its actions. S_k^n defines the chromosome phenotype, as each chromosome is randonly initialized with an action from S_k^n.

#2: Play a single game
- in a discrete moment of time, each player randomly selects one action from the set of actions represented in its subpopulation and not used until now
- calculate the output of the game: each player evaluates its payoff in the game; the payoff corresponds to the value of the criterion (1)

#3: Repeat step #2 until N games are played.

#4: Use GA operators on individual subpopulations
- after playing N individual games, each player knows the value of its payoff received for each action from its subpopulation,
- these payoff values become fitness values for GA; genetic operators [6] of selection (S), crossover (Cr) and mutation (M) are applied locally to subpopulations of actions; the new subpopulations of actions will be used by players in games played in the next game horizon.

#5: Return to step #2 until the termination condition is satisfied.

Potential possibilities of this multi-agent based scheduler are not fully explored because of the nature of the global criterion (1). To make the algorithm fully distributed, decomposition of the global criterion (1) into local interests of agents is proposed in the next section. This decomposition will be accomplished with use of genetic programming.

4 Decomposing Global Function with Genetic Programming

4.1 Genetic Programming

Genetic programming [7] was proposed as an evolutionary method for evolving solutions represented as computer programs. Its main innovation lies in variable-length, tree-like representation. For a particular problem, sets of *functions* and *terminals* are determined. They define labels for the nodes of the trees – terminals can labels leaves, and functions can label nodes according to their arities. Initial structures (programs), with randomly generated labels, are created. These undergo silumated evolution, with the same Darwinian selection, and with tree-based mutation and crossover. The evolutionary process is continued until either a solution is found or the maximum number of generations is reached.

4.2 Genetic Programming Environment

To design a fully distributed scheduling algorithm we need to decompose the global scheduling criterion (1) into local criteria of individual agents of the multi-agent based scheduler. This means that we need to express (1) as

$$T \approx g(h_0(X_{env_0}, sp), h_1(X_{env_1}, sp), ..., h_k(X_{env_k}, sp), ..., h_{N-1}(X_{env_{N-1}}, sp)), \quad (2)$$

where X_{env_k} is the set of local variables of agent A_k, defined in its local environment env_k, sp is the homogeneous scheduling policy, $h()$ is the local scheduling criterion, and $g()$ is some composition of functions $h()$, approximating the criterion (1).

To simplify a computational complexity of the problem we will assume that

$$h_0() = h_1() =, ... = h_k() = ... = h_{N-1}() = h(). \quad (3)$$

We will assign agent A_k^{GP} to task k, which will execute in some local environment env_k.

4.3 Terminals and Functions

The set of *terminals* is composed of

- **StartTime** - beginning time of execution of the visited task.
- **MrtTime** - a Message Ready Time, i.e. time of receiving last data from some predecessor, by a task actually visited by A_k^{GP}.
- **ExecTime** - a computational time b_l (see Section 2) of a task actually visited by agent A_k^{GP}.
- **PredStartTime** - the sum of time-starts of execution of predecessors of a task actually visited by an agent A_k^{GP}.
- **SuccStartTime** - the sum of the time-starts of successors of a task actually visited by an agent A_k^{GP}.

The set of *functions* contains the following function:

- **add, sub** - two argument function returning a sum, a difference of argument, respectively.
- **min, max** - two argument function returning smaller, greater value from the set of arguments, respectively.

- **forward** - one argument function which causes moving an agent from a current position at the actually visited node (task), corresponding to some incomiming or outcoming edge, to a node adjacent by this edge.
- **right, left** - one argument function which causes turning right, left respectively, from an actual position corresponding to some incoming or outcoming edge of the currently visited node, to the next edge of the precedence task graph.
- **ifMRTNodeAhead** - two argument function which returns the value of the first argument if the agent currently visiting a given node is in a position corresponding to some edge pointing a Message Ready Time node (MRT Node), i.e. the node which is last one from which the visited task receives data, and returns a value of the second argument, if an agent has another position.
- **ifMaxDynLevelNodeAhead** - two argument function which returns the value of the first argument if the agent currently visiting a given node is in a position pointing to a node with the locally greatest value of a dynamic level, and returns a value of the second argument if an agent has another position.
- **ifMaxDynCoLevelNodeAhead** - two argument function which returns the value of the first argument if the agent currently visiting a given node is in a position pointing to a node with the locally greatest value of a dynamic co-level, and returns a value of the second argument if an agent has another position.

5 Experiments

In this section we report some initial results of an experimental study of the proposed scheduling system. The system is composed of two modules: the LCGA scheduler working with either a global criterion (1) or local functions $h()$, and the GP system evolving local functions. The system was written in C++, with use of a library [5] to implement the modul of GP.

Because of large computational requirements of the GP module, the conducted experiments were limited to program graphs with the number of tasks not exceeding 40, population size of GP not exceeding 50, and the number of generation of GP not exciding 30. The size of each subpopulation of the LCGA was 80, the crossover probability was $p_c = 0.6$, and the mutation probability $p_m = 0.001$.

A precedence task graph [9] referred as *gauss*18, corresponding to a parallel version of the gaussian elimination algorithm was used in the experiment aiming to decompose a global criterion (1) into local criteria. It was assumed that this program is executed in a multiprocessor system of the MIMD architecture consisting of 8 processors arranged into *cube* topology.

Figure 3a presents an evolutionary GP process of decomposition a global function into local functions. Each individual of GP, representing a local scheduling function, is used by the LCGA module. The total execution time T (a global

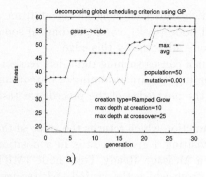 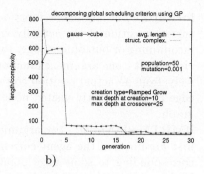

Fig. 3. Changing fitness function during GP evolutionary process (a), changing complexity of discovered local functions during evolutionary processing (b)

scheduling criterion) is evaluated for the final allocation found by the LCGA module. $T^* = const - T$ serves as the fitness function of a given individual from the population of GP. Figure 3a shows the best (max) and the average (avg) values of the fitness function in each generation of GP.

Figure 3b shows how the complexity of evaluated local functions are changing during the evolutionary process. An initial randomly created population of programs describing local scheduling functions, expressed in a LISP-like notation, contains programs with the average length ($avg.lenght$) equal to 504.75, and with structural complexity ($struct.complex.$) (the number of terminals and functions) of the best program in a population equal to 797. After 30 generation of GP the average length and the complexity drop to values 20.2 and 7, respectively. The best program in the last generation corresponds to the following local function $h()$:

((left(left(+(right SuccStartTime)(forward SuccStartTime))))).

The function can be interpreted in the following way:

1. Turn the agent right, to the next edge
2. Compute the sum of the time-start (**SuccStartTime**) of successors from the environment of the task visited by the agent.
3. Move the agent to a task, according to the current position of the agent.
4. Compute the sum of the time start (**SuccStartTime**) of succesors from the environment of the task curently being visited
5. Calculate the sum of the values from the points 2 and 4, and use it as the value of h().
6. Turn left two times; these operations do not influence the value of h().

How good is the found local function $h()$? To find it out, we conducted a number of experiments with the LCGA scheduling module using both global and local scheduling optimization criteria.

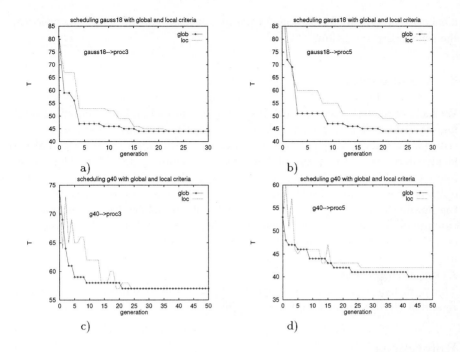

Fig. 4. Performance of scheduling algorithms with global and local optimization criteria

Figure 4a illustrates the work of the LCGA module using both optimization methods to schedule *gauss*18 in 3-processor system. Both methods find the optimal solution equal to 44, after relatively small number of generations. The speed of convergence of the algorithm with local optimization criteria is slightly slower than that of the algorithm with the global optimization criterion. Increasing the number of processors to 5 (fully connected system) results in achieving a suboptimal solution by the algorithm with local criteria (see Figure 4b).

The purpose of the next experiment was to see if the evolved local criterion is suitable for scheduling programs different that *gauss*18. For this purpose, we used the precedednce task graph *g*40 [9] with 40 tasks. Behavior of both algorithms (see Figure 4c,d) for this new problem is similar as in the previous experiment. The algorithm with local criteria finds optimal (3 processors) and suboptimal (5 processors) solutions.

Results of experiments conducted with other program graphs available in the literature show similar behavior of the scheduling algorithm. It finds either optimal or suboptimal solution, with slightly more generations than the algorithm with the global criterion. To evaluate each individual from the population of the LCGA working with the global criterion, calculation global value of T is required. Time of calculation of T increases with groving number of tasks. The algorithm with local optimization is based on locally performed evaluation of local criteria, which can be done in parallel at least partiall, and may speed up the scheduling algorithm.

6 Conclusions

We have presented a new approach to develop parallel and distributed scheduling algorithms. The approach is based on decomposing the global optimization criterion into local criteria, with use of genetic programming. Both global and local criteria are used by the LCGA - based scheduling algorithm. The preliminary results indicate that the algorithm with the local criteria is able to find optimal or near optimal solutions in a number of generations comparable with the number required by the algorithm with the global criterion. The main advantage of the proposed approach is parallelization and distributed control of evaluated optimization criterion, what may speed up the scheduling algorithm implemented on a parallel machine.

Acknowledgement

The work has been partially supported by the State Committee for Scientific Research (KBN) under Grant 8 T11A 009 13.

References

1. I. Ahmad (Ed.). Special Issue on Resource Management in Parallel and Distributed Systems with Dynamic Scheduling: Dynamic Scheduling, *Concurrency: Practice and Experience*, 7(7), 1995
2. I. Ahmad and Y. Kwok, A Parallel Approach for Multiprocessing Scheduling. *9th Int. Parallel Processing Symposium*, Santa Barbara, CA, April 25-28, 1995
3. V. C. Barbosa, *An Introduction to Distributed Algorithms*, The MIT Press, 1996
4. J. Błażewicz, M. Dror and J. Węglarz, Mathematical Programming Formulations for Machine Scheduling: A Survey, *European Journal of Operational Research* 51, pp. 283-300, 1991
5. A. P. Fraser, *Genetic Programming in C++. A manual in progress for gpc++, a public domain genetic programming system*, Technical Report 040, University of Salford, Cybernetics Research Institute, 1994
6. D. E. Goldberg, *Genetic Algorithms in Search, Optimization and Machine Learning*, Addison-Wesley, Reading, MA, 1989
7. J. R. Koza, Genetic Programming, MIT Press, 1992
8. F. Seredynski, Competitive Coevolutionary Multi-Agent Systems: The Application to Mapping and Scheduling Problems, *Journal of Parallel and Distributed Computing*, **47**, 1997, 39-57.
9. F. Seredynski, Discovery with Genetic Algorithm Scheduling Strategies for Cellular Automata, in *Parallel Problem Solving from Nature-PPSNV*, A. E. Eiben, T. Back, M. Schoenauer and H.-P. Schwefel (Eds.), Lecture Notes in Computer Science 1498, Springer, 1998, 643-652.

A Parallel Genetic Algorithm for Task Mapping on Parallel Machines

S. Mounir Alaoui[1], O. Frieder[2], and T. El-Ghazawi[3]

[1] Florida Institute of Technology, Melbourne, Florida, USA
salim@ee.fit.edu,
[2] Illinois Institute of Technology, Chicago, Illinois, USA
ophir@csam.iit.edu,
[3] George Mason University, Washington D.C, USA
tarek@gmu.edu

Abstract. In parallel processing systems, a fundamental consideration is the maximization of system performance through task mapping. A good allocation strategy may improve resource utilization and increase significantly the throughput of the system. We demonstrate how to map the tasks among the processors to meet performance criteria, such as minimizing execution time or communication delays. We review the Local Neighborhhod Search (LNS) strategy for the mapping problem.We base our approach on LNS since it was shown that this method outperforms a large number of heuristic-based algorithms. We call our mapping algorithm, that is based on LNS, Genetic Local Neighborhood Search (GLNS), and its parallel version, Parallel Genetic Local Neighborhood Search (P-GLNS). We implemented and compared all three of these mapping strategies. The experimental results demonstrate that 1) GLNS algorithm has better performance than LNS and, 2) The P-GLNS algorithm achieves near linear speedup.

1 Introduction

Parallel processor systems offer a promising and powerful alternative for high performance computing. Inefficient mapping or scheduling of parallel programs on architectures, however, limit our success. A parallel program is a collection of separate cooperating and communicating modules called tasks or processes. Tasks can execute in sequence or at the same time on two or more processors. Task mapping distributes the load of the system among its processors so that the overall processing objective according to given criteria is maximized. An efficient allocation strategy avoids the situation where some processors are idle while others have multiple jobs queued up.
A study in 1994 [8] demonstrated the power of the *Local Neighborhood Search* (LNS) algorithm for task allocation over a wide range of methods. We develop two new load balancing algorithms based on LNS. The first, called *Genetic Local Neighborhood Search* (GLNS), is a genetic algorithm based on LNS. The second is a coarse grain parallel implementation of GLNS, we call

it **P-GLNS**. In the reminder of this paper, we present our algorithm GLNS and a parallelization of this algorithm called P-GLNS. Our experimental results demonstrate that GLNS and P-GLNS achieve better mappings than LNS and P-GLNS yields near linear speedup.

2 Background

A popular classification scheme for task scheduling algorithms was introduced by Casavant [2]. This classification scheme, at the highest level, distinguishes between local and global scheduling. Local scheduling schedules the execution of processes within one processor . Global scheduling relates to parallel and distributed machines and determines on which processor to execute a job.

The next level in the hierarchy differentiates between dynamic and static scheduling. In other words, it defines the time where the scheduling decisions are made. In dynamic scheduling, the decisions are made during the execution of the program. The scheduler dynamically balances the workload each time there is unbalance. In this case, the process of load balancing can be made by a single processor (nondistributed), or it can be distributed physically among processors (distributed).

In the case of distributed dynamic global scheduling, we distinguish between cooperative and noncooperative scheduling. In cooperative scheduling, a processor balances the load cooperatively with other processors. In a noncooperative scheduling, a processor balances the load based only on the information it has locally. Each processor is independent of the actions of other processors. The major disadvantages of dynamic distributed load balancing algorithms is the runtime overhead for load information distribution, making placement decisions, and transferring a job to a target host continually during program execution.

In static scheduling, the decisions are made at the compilation stage, before the program execution. The resource needs of processes and the information regarding the state of the system are known. An "optimal" assignment can be made based on selected criteria [16], [20]. This problem, however, is generally NP-complete [15]. Sub-optimal solutions can be divided into two categories. Algorithms in the first category (approximate) of the "suboptimal" class consist of searching in a sub-part of the solution space, and stopping when a "good" solution is obtained. The second category is made up of heuristic methods, which assume a priori knowledge concerning processes, communication, and system characteristics.

Another classification scheme is Sender- or Receiver-initiated [5]. A load balancing algorithm is said to be sender-initiated if the load balancing process is initiated by the overloaded nodes; whereas for receiver-initiated, the underloaded nodes initiate the algorithm. A combination of both sender and receiver initiated yields symmetrically-initiated algorithms. The load balancing can be initiated either by the overloaded node if the load exceeds a pre-determined threshold, or by the underloaded node if the load index drops below a given threshold.

In 1994, Benmohammed-Mahieddine, et al [1] developed a variant of symmetrically initiated algorithm: the PSI (Periodic Symmetrically-Initiated) that outperformed its predecessor.

We focus on global, static and sub-optimal heuristic scheduling algorithms. Talbi and Bessiere developed a parallel genetic algorithm for load balancing[6]. Their implementation is based on the assumption that all the communication latencies between processes are known. They also compared their results with those found by the simulated annealing method. They demonstrated that genetic approach gives better results. A paper by Watts and Taylor [11] exposes a serious deficiency in current load balancing strategies, motivating further work in this area.

3 A Genetic Local Neighborhood Search

As described in Golberg [4], in general terms, a genetic algorithm consists of four parts.

1) Generate an initial population
2) Select pair of individuals based on the fitness function.
3) Produce next generation from the selected pairs by applying pre-selected genetic operators.
4) If the termination condition is satisfied stop, else go to step 2.

The termination condition can be either:
1) No improvement in the solution after a certain number of generations.
2) The solution converges to a pre-determined threshold.

3.1 Initial Population

The representation of an individual (mapping) is similar to the one in [6]. This representation satisfies two constraints: 1) We must be able to represent all the possible mapping solutions, and 2) We must be able to apply genetic operators on those individuals. A mapping is represented by a vector where number p in position q means that process q has been placed on processor p. Based on the three cost functions of LNS, we define our general cost function (GCF). GCF is the fitness function in our genetic approach. The goal is to minimize this general cost function. GCF is defined as a weighted sum of the three LNS cost functions.

$$GCF = w1 \times H + w2 \times L + w3 \times D$$

The weights w1, w2, and w3 depend on the mapping problem. In our experiment, these weights are chosen equal to those used in the LNS algorithm [19] to be able to compare the two algorithms.

3.2 Individual representation

3.3 selection scheme

Two common Genetic Algorithms selection schemes were tested:
rank selection and roulette wheel selection [4]. Rank selection selects the fittest individuals to take part in the reproductive steps. This method opts for more exploitation at the expense of exploration. The second scheme is the roulette wheel selection. In this selection the search is more randomized but we still give to the fittest individuals more probability to be selected. These two different schemes are used in Genetic Algorithms depending on the problem. In our genetic algorithm we chose rank selection. We found experimentaly that this selection converges faster than the roulette wheel selection for the mapping problem[19].

3.4 crossover

The best individuals resulting from the selection step are paired up. The two elements of each give birth to two offsprings. The crossover operator combines two elements of a pair to produce two new individuals. This operator picks randomly a crossover point number (between 0 and the size of an individual). It combines the pair, depending on this number, as shown in Figure 3. This new pair constitutes a pair replacing the parent pair in the new population. The crossover operator combine all the pairs to form the new population.

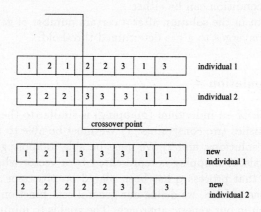

Fig. 1. Crossover

3.5 mutation

A mutation is to applied to a random transformation on the individuals after a certain number of generations. This mutation potentially explores new solutions and introduces random behavior into the search. The use of mutation avoids potentially limiting the search space due to local minimum values.

Mutation occurs after a certain number of generations. Its objective is to change randomly one bit or gene. In addition to traditional mutation, we also evaluate our adaptive mutation. That is, in adaptive mutation we mutade only if the fitness of the best individual is not increased after N generations. The value of N is determined experimentally. We ran our genetic algorithm 50 times starting with different initial populations each time. Six different mutation rates were taken into consideration: 0.1, 0.3, 0.5, 0.7, 0.9 and the adaptive method, we found experimentally that our adaptive mutation yields a better general cost function value.

3.6 Experimental Results

We experiment with five different data sets (DATA1, DATA2, DATA3, DATA4, DATA5). Each data set corresponds to a different communication graph. LNS and GLNS algorithms are runned on these five data sets. We plot the results after 100,200...800 generations for each algorithm (Table 2).

Generations	DATA1 LNS	GA	DATA2 LNS	GA	DATA3 LNS	GA	DATA4 LNS	GA	DATA5 LNS	GA
0	7.28	7.25	10.93	10.93	6.88	6.72	6.44	6.44	7.28	7.03
100	6.87	6.88	10	8.90	4.71	3.11	5.60	3.14	6.87	6.88
200	6.87	6.88	9.83	8.82	4.65	2.87	4.82	3.08	6.87	6.88
300	6.66	6.32	9.83	8.73	4.20	2.60	4.28	3.08	6.66	6.72
400	6.11	5.89	9.44	8.55	3.88	2.60	4.04	3.02	6.11	6.02
500	5.95	5.69	9.23	8.55	3.88	2.43	3.88	3.02	5.95	5.69
600	5.22	4.89	9.36	8.48	3.27	2.40	3.64	2.97	5.95	5.64
700	5.22	4.75	8.85	8.48	3.20	2.40	3.64	2.97	5.22	5.18
800	5.22	4.75	8.85	8.33	3.20	2.33	3.64	2.86	5.22	5.18

Table 1. performance of GLNS VS LNS

Table 2 summarizes the performance of both LNS and GLNS. We run the two algorithms on five different data sets. The weights used for the genetic algorithm are the same than for the LNS algorithm: W1=1, W2=0.1, W3=0.01 The GA is the GCF value of the best member of the population at termination.

For all the data sets the Genetic Algorithm converge to a better solution than LNS. For the data set DATA1 for example the GA converges to a value of 4.75 after 800 generations. The LNS algorithm converges to 5.22 which is much greater than 4.75. The worst case is with the last data set. The GA converge to a value of 5.22 after 800 generations not much better than the LNS value of 5.18, but before 400 generations LNS found better results than GA. The Genetic Algorithm outperforms the LNS algorithm which is known to be better than a wide range of methods including simulated annealing.

4 Parallel Genetic Local Neighborhood Search

4.1 Parallelization strategies

Evolution is a highly parallel process. Each individual is selected according to its fitness to survive and reproduce. Genetic algorithms are an abstraction of the evolutionary process and are indeed very easy to parallelize. Nowadays, the coarse grain parallelization is the most applied model for parallel genetic algorithms [13], [12] [7] [18]. This model divides the population into few sub-populations and an individual can migrate from one sub-population to another one. It is characterized by three parameters, namely, the *sub-population size*, the *migration rate* wich defines how many individuals migrate, and the *migration interval* that represents the interval between two migrations. The main problem is to define the migration rate and the migration interval correctly [7]. We use this model in our computation. We have implemented both GLNS and P-GLNS algorithms. P-GLNS is implemented on a network of workstations using Parallel Virtual Machine (PVM) language.

For this parallelization we define three parameters. The first one is the sub-populations sizes. The second is the migration rate which define how many individuals migrate. The third is the migration interval that represents the interval between two migrations.

4.2 strategy

We use a master-slave model. The processors that are executing the genetic algorithm on a subpopulation send their results to a master after a fixed number of generations. This master will select the best individuals and broadcast them to slaves. Those slaves will work on those new individuals combined with the previous ones. This method is quite similar to the one used in [14].

4.3 migration rate

The migration rate corresponds to the number of individuals that migrate each time there is a migration. The migration rate is taken as half of the population size. In other words at each migration the processors send to the master half of their population. The individuals sent are the best of the corresponding subpopulation. A larger migration rate will increase the communication cost between processes. It will be shown that a migration rate of 1/2 is the best for the problem parameters considered.

4.4 migration interval

The migration interval represents the interval between two migrations. This migration interval is determined experimentally. After a certain number of generations the processors send their best individuals to the master. It will be shown that a migration interval of 110 is the best for the problem parameters considered.

5 Experimental Results

5.1 Experiments

The experiments were conducted on a network of workstations. For our parallel implementation we used the Parallel Virtual Machine message passing library (PVM). We run P-GLNS to define the best migration rate and the best migration interval for the problem parameters used. To compute the speedup we runned P-GLNS and GLNS on three different data sets.

We found experimentally that the minimum cost is achieved with a migration rate of 0.5. The more individuals we send, the bigger is the migration rate. This results in more interaction between the nodes. This explains the decreasing cost between rates 0 and 0.5. However, when the number of individuals sent becomes larger, the communication cost becomes higher. This explains the increasing cost between 0.5 and 0.8 migration rates. For the problem parameters used, 0.5 is the perfect trade off point between communication cost and interaction between nodes. The position of such trade off point should be determined experimentally for each case. In the same way, the best migration interval found is 110.

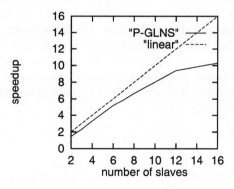

Fig. 2. speedup

For the parallel implementation three different mapping problems are taken into consideration. For each one we perform 100 runs and compute the sequential and the parallel time to find an acceptable solution (fitness greater than some threshold). The speedup is the ratio between these two time measurements. Our algorithm was run on a 10 Base-T ethernet network. We can see in (Figure 5) that for 2,4 and 8 slaves we have a near linear speedup with our experiments on a 10 Base-T network. Assuming, we are using a 100 Base-T network the speedup will be much better when larger number of slaves are used. Thus, near linear speedup can be obtained in any parallel processing environement with a reasonable interprocessor communication network. Each slave works independently of other slaves. They search during a certain number of generations in their search space until they reached a sub-solution. The best sub-solution will be the start of new searchs. In the sequential genetic algorithm the best individuals mate together. There is no possibility with a ranking selection that two locally-optimal solutions mate together. On the other hand, in a parallel implementation each individual has its own sub-optimal solutions, and they are mixed without other sub-populations interferences. This fact makes the parallel search powerfull and efficient. For a large number of stations the communication cost become higher, and this explain the decreasing speedup with 12 and 16 processors in a 10 Base-T network.

6 Conclusion

Based on the Local Neighborhood Search algorithm we developed a coarse graine parallel genetic algorithm called Parallel Genetic Local Neighborhood Search (P-GLNS). The fitness function of P-GLNS is a weighted sum of the three LNS cost functions[19]. Our algorithm balances communication requests and takes into consideration the case where the communication latencies are unknown. Both GLNS and P-GLNS were implemented for different data sets. Simulation results show that GLNS outperforms the LNS algorithm wich is known to be better than a wide range of methods. The parallelization of G-LNS, P-GLNS achieves a near linear speedup and thus is scalable. We have shown that by increasing the migration rate and migration interval, the performance of the parallel algorithm improves. However, increasing these parameters above certain operation points can result in larger overhead that may decrease the benefit from using parallel genetic algorithms. A trade off can be easily obtained experimentally as demonstrated in this work.

References

1. Benmohammed-Mahieddine, K., P. M. Dew and M. Kara. 1994. " A Periodic Symmetrically-initiated Load Balancing Algorithm for Distributed Systems.
2. Casavant, T. L., and J.G. Kuhl. 1988. " A taxonomy of scheduling in General-purpose Distributed Computing Systems" IEEE Trans. Software Eng., vol. 14, No. 2,
3. Cybenko, G. 1989 "Load balancing for distributed memory multiprocessors" Journal of Parallel and distributed computing. 7: 279-301, 1989.
4. D.E.Goldberg. Genetic Algorithms in search Optimization and Machine learning. Addison Wesley, New York, 1989.

5. Eager, D. L., E. D. Lazowska, and J. Zahorjan. 1986. " A comparison of receiver initiated and sender initiated adaptive load sharing" Performance evaluation, Vol. 6, 1986, pp. 53-68.
6. E-G.Talbi, P.Bessiere : Parallel genetic algorithm for the graph partitioning problem. ACM international Conference on SuperComputing, Cologne, Germany, June 1991
7. Erik Cantu-Paz. A summary of Research on Parallel Genetic Algorithms. Illinois Genetic Algorithms Laboratory http://www-illigal.ge.uiuc.edu/ cantu-paz/publications.html, 01/20/98.
8. F.Berman and Bernd Stramm. Mapping Function-Parallel Programs with the Prep-P Automatic Mapping Preprocessor. University of California, San Diego. January 7,1994.
9. Grosso, P. (1985). Computer simulations of genetic adaptation: Parallel Subcomponent Interaction in a Multilocus Model. PhD thesis, University of Michigan.
10. Horton, G. 1993. "A multi-level diffusion method for dynamic load balancing." Par. Comp. 19, pp. 209-218 (1993). In proc 9th International.
11. J. Watts and S. Taylor. A Practical Approach to Dynamic Load Balancing. Submitted to IEEE Transactions on Parallel and Distributed Systems. 1998.
12. Levine, D. (1994). A parallel genetic algorithm for the set partitionning problem. PhD thesis, Illinois Institute of technology.
13. Li T. and Mashford J. (1990) A parallel genetic algorithm for quadratic assignement. In R.A Ammar, editor, Proceedings of the ISMM International Conference. Parallel and Distributed Computing and Systems, pages 391-394, New York, NY: Acta Press, Anaheim, CA.
14. Marin, F., Trelles-Salazar, O., and Sandoval, F. (1994). Genetic algorithms on lan-message passing architectures using pvm: Application to the routing problem. In Y.Davidor, H-P. Shwefel, and R.Manner, editors, Parallel Problem Solving from Nature -PPSN III, volume 866 of Lecture Notes in Computer Science, pages 534-543, berlin, Germany: Springer-Verlag.
15. M. R. Garey and D.S. Johnson. Computers and Intractability: A guide to the theory of NP-completeness. W.H. Freeman and Company, New York 1979.
16. Ni, L. M. and K. Hwang. 1981. "Optimal load balancing strategies for a multiple processors system" in Proc. Int. Conf. Parallel Proc., 1981, pp. 352-357
17. P.J.M van Laarhoven and E.H.L Aarts. Simulated Annealing: Theory and Applications. D.Reidel Publishing Company, Boston, 1987.
18. S. Mounir Alaoui, A. Bellaachia, A.Bensaid, O. Frieder "A Parallel Genetic Load Balancing Strategy" Cluster Computing Conference - CCC '97 March 9-11, 1997 Emory University, Atlanta, Georgia.
19. S. Mounir Alaoui, Master Thesis, Alakhawayn University, Rabat Morocco. June 1997
20. Shen, C. and W. Tsai. 1985. "A graph matching approach to optimal tasks assignment in distributed computing systems using a minimax criterion" IEEE Trans.

Evolution-Based Scheduling of Fault-Tolerant Programs on Multiple Processors

Piotr Jedrzejowicz[1], Ireneusz Czarnowski[1], Henryk Szreder[1], and Aleksander Skakowski[1]

[1]Chair of Computer Science, Gdynia Maritime Academy, ul.Morska 83
81-225 Gdynia, Poland
{pj,irek,hsz,askakow}@wsm.gdynia.pl

Abstract. The paper introduces a family of scheduling problems called fault-tolerant programs scheduling (FTPS). Since FTPS problems are, in general, computationally difficult, a challenge is to find effective scheduling procedures. Three evolution-based algorithms solving three basic kinds of FTPS problems have been proposed. The problems involve scheduling multiple variant tasks on multiple identical processors under time constraints. To validate the algorithms computational experiment has been carried. Experiment results show that evolution based algorithms produce satisfactory to good solutions in reasonable time.

1 Introduction

Complex computer systems are often facing the conflicting requirements. One of the frequently encountered conflicts in the fault-tolerant computing involves high dependability and safety standards required versus system performance and cost.

A unit of software is fault-tolerant (f-t) if it can continue delivering the required service after dormant imperfections called software faults have become active by producing errors. To make simplex software units fault-tolerant, the corresponding solution is to add one, two or more program variants to form a set of $N \geq 2$ units. The redundant units are intended to compensate for, or mask a failed software unit. The evolution of techniques for building f-t software out of simplex units has taken several directions. Basic models of f-t software are N-version programming (NVP) [1], recovery block (RB) [15], [12], and N-version self-checking programming (NSCP) [18], [13]. A hybrid solution integrating NVP and recovery block is known as the consensus recovery block [16]. In all the above listed techniques, the required fault tolerance is achieved by increasing the number of independently developed program variants, which in turn leads to a higher dependability at a cost of the resources used. Self-configuring optimal programming [5] is an attempt to combine some techniques used in RB and NVP in order to exploit a trade-off between software dependability and efficiency. In fact, the approach does attempt to dynamically optimize a number of program variants run but it offers local optimization capabilities rather then global ones.

Relation between use of resources and reliability is not straightforward, and often difficult to identify [16], [2]. Yet, besides the already mentioned self - configuring optimal programming, some direct optimization models of software with redundancy have been proposed [3]. Rationale for optimizing a single f-t program structure is clearly limited. Hence, it is suggested to manage complexity and costs issues of the fault-tolerant programs not at a single program level, and thus, to introduce a family of scheduling problems called fault-tolerant programs scheduling (FTPS). The paper focuses on multiple-variant tasks. Concept of the multiple-variant task (or program) serves here as a general model of the fault-tolerant program structure. Each task is characterized by: ready time, due date or deadline, number of variants and variant processing time. FTPS problem is how to distribute (or schedule) the f-t components among processing elements deciding also how to choose their structure or execution mode to achieve some overall performance goals. FTPS problems differ from the traditional scheduling problems. Their decision space is extended to include additional variables describing f-t components structure or execution mode (i.e. number of variants and adjudication algorithm used).

Considering important application areas for the f-t programs (real-time systems, highly dependable systems, safety systems), some useful optimality criteria for FTPS problems could include (among others):

- total number of the executed program variants scheduled without violating deadlines V ($V = \sum V_j$, where V_j is the number of variants of task $j, j = 1, 2, \ldots, n$, used to construct a schedule);
- sum of penalties for violating time and dependability constraints - PF ($PF = \sum PF_j$, where PF_j is the penalty associated with scheduling task j).

It should be noted that scheduling fault-tolerant, multiple variant programs requires a different approach than scheduling programs to achieve system fault tolerance in presence of hardware faults (the latter has been investigated in, for example, [14],[7]).

2 FTPS Problem Formulation

To identify and classify scheduling problems Graham notation [8] is used. In this notation, three fields represent, respectively: processor environment, task with resource characteristics, and optimality criterion. The paper considers three basic FTPS problems. The first one, denoted as $P|m-v, r_j|V$, is characterized by multiple, identical processors (P) and a set of multiple-variant ($m-v$) tasks. Each task is a set of independent, non-preemptable variants with ready times -(r_j) and deadlines differing per task. Variants have arbitrary processing times. Optimization criterion is the number of the program variants scheduled without violating deadlines. Decision variables include assignment of tasks to processors and number of scheduled variants for each task. Tasks can not be delayed. There might be a constraint imposed on minimal number of variants of each task to be scheduled.

As it was pointed out in [4], there exist scheduling problems where tasks have to be processed on more than one processor at a time. During an execution of these multiple-processor tasks communication among processors working on the same task is implicitly hidden in a "black box" denoting an assignment of this task to a subset of processors during some time interval. To represent multiprocessor problems a $size_j$ parameter denoting processor requirement is used. Thus, the second of the considered problems denoted as $P|r_j, size_j|V$ is characterized by multiple, identical processors and a set of multiple-processor tasks. Each task is a set of independent, non-preemptable variants with ready times and deadlines differing per task which have to be processed in parallel. Tasks have arbitrary processing times, which may differ per size chosen. Optimization criterion remains the number of the program variants scheduled. Decision variables include assignment of tasks to processors and size of each task. Tasks can not be delayed. There might be a constraint imposed on minimal size of each task to be scheduled.

The third FTPS problem denoted as $P|b-t-b, r_j|PF$ is characterized by multiple, identical processors and a set of two-variant tasks. Each task is a set of two independent, non-preemptable variants with ready times and deadlines differing per task, which have to be processed in one of the three possible modes. In the first mode only the first variant is executed. The second mode involves execution of only the second variant. Third mode requires parallel, that is back-to-back $(b-t-b)$ processing of both variants. Tasks have arbitrary processing times, which may differ per execution mode chosen. Optimization criterion is the expected sum of penalties associated with task delays and failure probabilities. Decision variables include assignment of tasks to processors and execution mode for each task. Tasks can be delayed. $P|b-t-b, r_j|PF$ assumes that the following additional information with respect to each task is available: version reliabilities, penalty for delay, penalty for the safe failure and penalty for the unsafe failure.

The problem $P|m-v, r_j|V$ can be considered as a general model of scheduling N-version programs under hard real-time constraints. The problem $P|r_j, size_j|V$ can be considered as a model of scheduling N-version programs or recovery block schemes, or consensus recovery block schemes, or mixture of these under hard real-time constraints. Finally, $P|b-t-b, r_j|PF$ can be considered as a model of scheduling fail-safe tasks under soft real-time constraints (see for example [17]).

3 Evolution-based Approach to Solving FTPS Problems

In general, scheduling of independent, non-preemptable tasks with arbitrary arrival and processing times is NP-hard [4]. FTPS problems are not computationally easier. It is easily shown by polynomial transformation from the result of [6] that FTPS problems of maximizing number of variants executed even on one processor $(1|r_j|V)$ is NP-hard. Similar conclusion can be inferred with respect to complexity of the $P|r_j, size_j|V$ problem. $P|b-t-b, r_j|PF$ problems with penalty function aggregating dependability/reliability properties with schedule length, maximum lateness, number of tardy tasks, etc., remain NP-hard. Since the discussed FTPS problems are, in general, NP-hard it is not likely that polynomial-time algorithms solving them can be found. The following subsections include a description of the three evolution-based algorithms developed to deal with the $P|m-v,r_j|V$, $P|r_j,size_j|V$, and $P|b-t-b,r_j|PF$ problems, respectively.

3.1 Algorithm EVOLUTION $P|m-v,r_j|V$

Algorithm EVOLUTION $P|m-v,r_j|V$ is based on the following assumptions:

- an individual within a population is represented by vector $x_1,\ldots,x_n$ where $x_j, (1 \leq x_j \leq n_j)$ is the number of variants of the task j that will be executed;
- initial population is composed from random individuals;
- each individual can be transformed into a solution by applying $APPROX$ $P|m-v,r_j|V$, which is a specially designed scheduling heuristic;
- each solution produced by the $APPROX$ $P|m-v,r_j|V$ can be directly evaluated in terms of its fitness, that is number of variants scheduled;
- new population is formed by applying three evolution operators: crossover, offspring random generation, and transfer of some more "fit" individuals. Some members of the new population undergo further transformation by means of mutation.

$EVOLUTION$ $P|m-v,r_j|V$ involves executing steps shown in the following pseudo-code:

> **set** population size PS, generate randomly an initial population I_0;
> **set** evolution parameters x,y,z $(0 < x,y,z < 1, x+y+z = 1)$;
> **set** $i := 0$; (iteration counter);
> **While** no stoping criterion is met **do**
>   set i:=i+1;
>   calculate, using $APPROX$ $P|r_j|V$, fitness factor for each individual in I_{i-1};
>   form new population I_i by: selecting randomly x/PS individuals from I_{i-1} (probability of selection depends on fitness of an individual); producing y/PS individuals by applying crossover operator to x/PS previously selected individuals; randomly generating z/PS individuals; applying mutation operators to a random number of just created individuals;
> **End While**

$APPROX$ $P|m-v,r_j|V$ heuristics used within $EVOLUTION$ $P|m-v,r_j|V$ algorithm is carried in three steps:

Step 1. Set of variants that belong to an individual is ordered using earliest ready time as the first, and earliest deadline as the second criterion.

Step 2. List-schedule variants, allocating them one-by-one to one of the processors, minimizing at each stage processors idle-time, and not violating time constraints. If there are more than one feasible allocation with identical, minimal, idle time, choose a solution where scheduled variant finishes **later**. Repeat list scheduling with the opposite rule for breaking times (i.e. choosing a solution where scheduled variant finishes **earlier**).

Step 3. If variants can not be allocated to processors without violating deadlines than fitness of the solution is set to 0. Otherwise evaluation function generates fitness factor equal to max $\{ \sum x_i$ (**later**), $\sum x_i$ (**earlier**) $\}$.

To summarize main features of the *EVOLUTION* $P|m-v,r_j|V$, a sheme proposed by [9] is used: *individuals* - part of solution; *evolution process* - steady state evolution with variable fixed part; *neighborhood* - unstructured; *information source* - two parents; *unfeasibility* - penalized; *intensification* - none; *diversification* - noise procedure (mutation).

3.2 Algorithm EVOLUTION $P|r_j,size_j|V$

Algorithm EVOLUTION $P|r_j,size_j|V$ is based on the following assumptions:

- an individual is represented by n-element vector $y_{\sigma(1)},\ldots,y_{\sigma(n)}$ where σ is a permutation, y_j $(1 \leq y_j \leq max\{size_j\})$ is a size of the task j to be executed;
- the place of an element within $y_{\sigma(1)},\ldots,y_{\sigma(n)}$ tells in which order the algorithm *APPROX* $P|r_j,size_j|V$ will allocate it to processors;
- an initial population is composed in part from individuals with random size and order of tasks, and in part from individuals with random size and order determined by non-decreasing ready time as the first criterion and non-decreasing deadlines as the second;
- each individual can be transformed into a solution by applying *APPROX* $P|r_j,size_j|V$, which is a specially designed algorithm for scheduling multiple-variant and multiple-processor tasks;
- each solution produced by the *APPROX* $P|r_j,size_j|V$ can be directly evaluated in terms of its fitness, that is total of task sizes scheduled without violating deadlines;
- new population is formed as in *EVOLUTION* $P|m-v,r_j|V$

EVOLUTION $P|r_j,size_j|V$ involves executing steps shown in the following pseudo-code:

set population size PS, generate an initial population I_0 (part of the individuals with random size and order, and part with random size and fixed order);
set $i := 0$; (iteration counter);
While no stoping criterion is met **do**
    set i:=i+1;
    calculate, using *APPROX* $P|r_j,size_j|V$, fitness factor for each individual in I_{i-1};
    form new population I_i by: selecting randomly a fixed number of individuals from I_{i-1} (probability of selection depends on fitness of an individual); producing one part of individuals by applying crossover operator to previously selected individuals from I_{i-1}; generating one part of individuals with random size, and order; generating another part with

random size, and fixed order; applying mutation operators to a random number of just created individuals;
EndWhile

$APPROX\ P|r_j, size_j|V$ algorithm used within $EVOLUTION\ P|r_j, size_j|V$ is carried in three steps:
Step 1. Set loop over tasks.
Step 2. Within the loop, allocate current task to multiple processors minimizing beginning time of its processing. Continue with tasks until all have been allocated.
Step 3. If the resulting schedule has task delays, fitness factor of the individual from which the schedule is generated is set to 0. Otherwise, fitness factor is set to $\sum size_i$.

Main features of the $EVOLUTION\ P|r_j, size_j|V$ include: *individuals* - part of solution; *evolution process* - steady state evolution, population of constant size; *neighborhood* - unstructured; *information source* - two parents; *unfeasibility* - penalized for a solution, repaired for an individual; *intensification* - none; *diversification* - noise procedure (mutation).

3.3 Algorithm $EVOLUTION\ P|b-t-b, r_j|PF$

Algorithm $EVOLUTION\ P|b-t-b, r_j|PF$ is based on the following assumptions:

- an individual within a population is the solution represented by matrix **S** of four rows and n columns where element $s_{1,j}$ - represents number of task, $s_{2,j}$ - its execution mode, $s_{3,j}$ - number of processor on which task is to be executed, and $s_{4,j}$ - number of another processor allocated to execute a task in fail-safe mode, otherwise $s_{4,j} = 0, j = 1, 2, \ldots, n$;
- an initial population is composed, in part, from random individuals, and in part, from individuals with values in the first row, that is numbers of task, ordered by their non-decreasing ready time as the first criterion and non-decreasing deadlines as the second. Values of the remaining rows are random;
- each solution can be directly evaluated in terms of its fitness using $EVALUATION P|b-t-b, r_j|PF$ algorithm;
- new population is formed by applying the some principles as in $EVOLUTION\ P|m-v, r_j|V$.

$EVOLUTION\ P|b-t-b, r_j|PF$ involves executing steps shown in the following pseudo-code:

set population size PS, generate an initial population I_0 (part of the individuals with random order of tasks, and part with the fixed one);
set $i := 0$; (iteration counter);
While no stopping criterion is met **do**
    set $i := i + 1$;
    calculate, using $EVALUATION\ P|r_j|PF$ algorithm, fitness factor for each individual in I_{i-1};

form new population I_i by: selecting randomly fixed number of individuals from I_{i-1} (probability of selection depends on fitness of an individual); producing part individuals by applying crossover operator to previously selected individuals from I_{i-1}; generating another part with random size and order; generating further part with random size and fixed order; applying mutation operators to a random number of just created individuals;
End While

EVALUATION $P|b - t - b, r_j|PF$ algorithm is carried in three steps:
Step 1. Set loop over tasks.
Step 2. Within the loop, allocate current task to a processor (or processors) specified in the second row of matrix representing an individual, assuming processing begins as early as possible. Calculate penalty for delay and expected penalties for failing safe and failing unsafe. Update total incurred penalties. Continue with tasks until all have been allocated.
Step 3. Set fitness factor of the individual from which the schedule is generated to total incurred penalties.

Main features of the *EVOLUTION* $P|b - t - b, r_j|PF$ include: *individuals* - feasible solutions of the problem; *evolution process* - steady state evolution, population of constant size; *neighborhood* - unstructured; *information source* - two parents; *unfeasibility* - repaired; *intensification* - none; *diversification* - noise procedure (mutation).

4 Computational Experiment Results

To verify and evaluate the presented evolution-based algorithms several computational experiments have been carried. The first experiment included 50 randomly generated scheduling problems belonging to the $P|m-v, r_j|V$ class. Problems involved scheduling 10 - 29 multiple (3 5) variant tasks on 2 - 5 processors. Problems have been solved using *EVOLUTION* $P|m - v, r_j|V$ algorithm and simple *PJ*-heuristics proposed in [10]. Solutions have been compared with an upper bound on optimal solution, which can be calculated in polynomial time. In Table 1 relative distances from upper bound on optimal solutions are shown.

To evaluate the performance of *EVOLUTION* $P|m-v, r_j|V$ against optimal algorithm a subset of 20 problems has been selected from the previously used 50 cases. For these problems exact solutions have been obtained using a VISUAL C++ application, based on CPLEX callable library. Mean, smallest and largest relative errors for the *EVOLUTION* $P|m - v, r_j|V$ algorithm and *PJ*-heuristics are, respectively, 1.34% and 4.05% (mre), 0.00% and 4.76% (sre) and 0.00% and 14.63% (lre).

Next experiment included 50 randomly generated scheduling problems belonging to the $P|r_j, size_j|V$ class. Problems involved scheduling 10 - 15 tasks, each task with maximum size of 4, on 3 - 10 processors. All problems have been solved using *EVOLUTION* $P|r_j, size_j|V$ algorithm, and APPROX heuristics proposed in [11]. Solutions have been compared with the optimal ones obtained

Table 1. Mean, smallest and largest relative distances (mrd, srd, lrd) from an upper bound on optimum solution for the *EVOLUTION* $P|m-v,r_j|V$ algorithm and *PJ-*heuristics

No of proc.	mrd EVOL.	mrd PJ-HEUR.	srd EVOL.	lrd EVOL.	srd PJ-HEUR.	lrd PJ-HEUR.
2	13.80%	20.17%	5.00%	22.45%	7.50%	38.78%
3	15.68%	18.17%	10.53%	24.14%	14.29%	27.59%
4	18.39%	21.50%	10.81%	25.53%	15.52%	31.91%
5	18.52%	19.88%	11.43%	22.86%	13.16%	26.47%
Overall	16.60%	19.93%				

using branch-and-bound algorithm. In Table 2 the respective mean relative errors are shown.

Table 2. Mean, smallest and largest relative errors (mre, sre, lre) from an optimum solution for the *EVOLUTION* $P|r_j, size_j|V$ algorithm, and APPROX heuristics.

No of proc.	mre EVOL.	mre APPROX	sre EVOL.	lre EVOL.	sre APPROX	lre APPROX
3	0.00%	4.55%	0.00%	0.00%	0.00%	4.55%
4	0.00%	4.84%	0.00%	0.00%	0.00%	9.68%
5	1.04%	1.84%	0.00%	4.17%	0.00%	5.56%
6	2.07%	2.32%	0.00%	5.00%	0.00%	8.00%
7	2.77%	7.66%	0.00%	8.33%	0.00%	24.14%
8	1.13%	4.36%	0.00%	3.33%	0.00%	14.81%
9	0.61%	5.01%	0.00%	3.03%	0.00%	12.12%
10	0.00%	8.26%	0.00%	0.00%	2.44%	15.63%
Overall	0.95%	4.85%				

The last experiment included 50 randomly generated scheduling problems belonging to the $P|b-t-b,r_j|PF$ class. Problems involved scheduling 10 tasks on 2 - 4 processors. All problems have been solved using *EVALUATION* $P|b-t-b,r_j|PF$ algorithm, and *GREEDY* heuristics. In *GREEDY* heuristics tasks are ordered in accordance with non-decreasing ready-times as the first criterion and non-decreasing due dates as the second one. Then, all possible combinations of execution modes are considered. Solutions have been compared with the optimal ones obtained using branch-and-bound algorithm. In Table 3 the respective relative errors are shown.

Table 3. Mean, smallest and largest relative errors (mre, sre, lre) from an optimum solution for the $EVALUATION$ $P|r_j|PF$ algorithm, and GREEDY heuristics.

No of proc.	mre EVOL.	mre GREEDY	sre EVOL.	lre EVOL.	sre GREEDY	lre GREEDY
2	24.02%	11.92%	0.00%	46.09%	0.00%	36.84%
3	28.30%	11.16%	14.59%	53.59%	0.00%	39.46%
4	31.92%	12.54%	7.51%	69.69%	0.00%	34.46%
Overall	28.08%	11.87%				

5 Conclusions

In this paper fault tolerant programs scheduling problems are considered. Since, as it was shown, FTPS problems are, in general, computationally difficult, a challenge is to find an effective scheduling procedure. Three evolution-based algorithms solving, respectively, $P|m-v,r_j|V$, $P|r_j,size_j|V$ and $P|b-t-b,r_j|PF$ problems have been proposed. The approach seems promising and the algorithms produce excellent to satisfactory solutions. In case of $P|m-v,r_j|V$ experiment results proved that the evolution-based approach generates considerably better results then a good polynomial-time heuristic algorithm. It also produces results not too far from an upper bound on optimal solution. In case of $P|r_j,size_j|V$ evolution based algorithm produced, on average, solutions not worse more than 1% from optimum. For $P|b-t-b,r_j|PF$ problems evolution based algorithm generates only satisfactory solutions (average error margin is about 28% and this performance is worse than greedy heuristics). For all three evolution-based algorithms there is still room for improvements which can be expected from introducing some better solution-improvement strategies at a self-adaptation phase. Another promising direction is searching for effective hybrid evolutionary-heuristics algorithms.

References

1. Avizienis A., L.Chen: On the implementation of the N-version programming for software fault tolerance during execution, *Proc. IEEE COMPSAC 77*, 1977, pp. 149 - 155.
2. Belli F., P.Jedrzejowicz: Fault Tolerant Programs and Their Reliability, *IEEE Trans. on Reliability*, 39(2), 1990, pp. 184 - 192.
3. Belli F., P.Jedrzejowicz: An Approach to the Reliability Optimization of Software with Redundancy, *IEEE Trans. on Software Engineering*, 17(3), 1991, pp. 310 - 312.
4. Blazewicz J., K.H.Ecker, E.Pesch, G.Schmidt, J.Weglarz: Scheduling Computer and Manufacturing Processes, *Springer*, Berlin, 1996.
5. Bondavalli A., F.Di Giandomenico, J.Xu: Cost-effective and flexible scheme for software fault tolerance, *Computer System Science & Engineering*, 4, 1993, pp. 234 - 244.

6. Garey M.R., D.S.Johnson: Computers and Intractability: A Guide to the Theory of NP-Completeness, *W.H.Freeman*, New York, 1979.
7. Ghosh S., R.Melham, D.Mosse: Fault-Tolerance through Scheduling of Aperiodic Tasks in Hard Real-Time Multiprocessor Systems, *IEEE Trans. on Parallel and Distributed Systems*, 8(13),1997, pp. 272-284.
8. Graham R.L., E.L.Lawler, J.K.Lenstra, A.H.G.Rinnooy Kan: Optimization and approximation in deterministic sequencing and scheduling: a survey, *Annals Discrete Math.*, 5, 1979, pp. 287 - 326.
9. Hertz A., D.Kobler: A Framework for the Description of Population Based Methods, *16th European Conference on Operational Research - Tutorials and Research Reviews*, Brussels, 1998, pp. 1 - 23
10. Jedrzejowicz P.: Scheduling multiple-variant programs under hard real-time constraints, *Proc. International Workshop Project Management and Scheduling*, Istambul, 1998, pp.
11. Jedrzejowicz P.: Maximizing number of program variants run under hard time constarints, *Proc. International Symposium Software Reliability*, Paderborn, 1998, pp.
12. Kim K.H.: Distributed execution of recovery blocks: an approach to uniform treatment of hardware and software faults, *Proc. 4th International Conference on Distributed Computing Systems*, IEEE Computer Society Press, 1984, pp. 526 - 532.
13. Laprie J.C., J.Arlat, C.Beounes, K.Kanoun: Definition and Analysis of Hardware-and-Software Fault-Tolerant Architectures, *IEEE Computer*, 23(7), 1990, pp. 39-51.
14. Liestman A.L., R.H.Campbell: A Fault-tolerant Scheduling Problem, *IEEE Trans. on Software Engineering*, SE-12(11), 1988, pp. 1089-1095.
15. Melliar-Smith P.M., B.Randell: Software reliability: the role of programmed exception handling, *SIGPLAN Notices* 12(3), 1977, pp. 95 - 100.
16. Scott R.K., J.W.Gault, D.F. Mc Allister: Fault tolerant software reliability modelling, *IEEE Trans. on Software Engineering*, 13(5), 1987, pp. 582 - 592.
17. Vouk, M.A.: Back-to-back testing, *Information and software technology*, vol.32, no 1, 1990, pp. 34-35.
18. Yau S.S., R.C.Cheung: Design of Self-Checking Software, *Proc. Int. Conf. on Reliable Software*, 1975, pp. 450 - 457.

Acknowledgement: This research has been supported by the KBN grant NR 301/T11/97/12

A Genetic-Based Fault-Tolerant Routing Strategy for Multiprocessor Networks

Peter K. K. Loh and Venson Shaw

Nanyang Technological University
School of Applied Science
Nanyang Avenue, Singapore 639798
Tel: 65-7991242, Fax: 65-7926559
E-mail: askkloh@ntu.edu.sg

Abstract. We have investigated the adaptation of AI-based search techniques as topology-independent fault-tolerant routing strategies on multiprocessor networks [9]. The results showed that these search techniques are suitable for adaptation, as fault-tolerant routing strategies with the exception that the routes obtained were non-minimal. In this research, we investigate the adaptation of a genetic-heuristic algorithm combination as a fault-tolerant routing strategy. Our results show that such a hybrid strategy results in a viable fault-tolerant routing strategy, which produces minimal or near-minimal routes with a corresponding significant reduction in the number of redundant node traversals. Under certain fault conditions, this new hybrid routing strategy outperforms the purely heuristic ones.

1. Introduction

Much research has been centred on the use of genetic algorithms [2,4,15,16]. Since a genetic algorithm is essentially an optimization technique, many have applied it to applications requiring improvements in their end-results e.g. producing efficient VLSI layouts, enhancing factory automation, network design and reducing transportation problems. Few have applied genetic algorithms to communications to manage routing [11-13]. In [13], a Markov model for an unreliable link is used as part of three heuristic routing table optimisation algorithms. Routing tables are obtained such that the overall network grade-of-service (NGOS) is minimised under different call control rules. These algorithms are then used to demonstrate the improvements in trunk network fault-tolerance (assessed in terms of NGOS) that can be achieved through routing table updating in the event of link (trunk group) failure. In [12], the three algorithms are then used to demonstrate the improvements in trunk network fault-tolerance that can be achieved through network augmentation i.e. the addition of extra links to a network, either in parallel with existing links, or between nodes that previously had no direct connection. In [11], an alternative approach to routing table optimisation is taken, which makes use of a genetic algorithm. The genetic algorithm is applied indirectly as an optimization search technique to achieve dynamic routing control rather than adapted for use directly as a routing strategy. Here, the network model is assumed to be constant with no mechanism incorporated to tolerate faulty network components. Instead, based on traffic conditions at a given time, the routing control is invoked if the maximum *call loss-rate* (fraction of traffic that a link is unable to carry) exceeds a specified threshold. In this paper, we describe a new hybrid fault-tolerant routing strategy formed by the combination of genetic and heuristic algorithms. Our results show that our new routing strategy produces minimal or near-minimal routes, which reduce significantly the redundant traversals of nodes [3]. In addition, under certain fault patterns, the hybrid strategy outperforms the purely heuristic ones.

2. The Network Model

We can represent, without loss of generality, a multiprocessor network of arbitrary topology as an undirected graph G defined as $G = (N, L)$ where N is the set of nodes and L the set of bidirectional links. If n_i and l_j denote a specific node and a specific link respectively, we have $n_i \in N$ ($i = 0, 1, ..., |N| - 1$) and $l_j \in L$ ($i = 0, 1, ..., |L| - 1$) where $L = \{ (n_i \rightarrow n_j), (n_j \rightarrow n_i) \mid \forall (n_i, n_j) \in N \}$. Some assumptions made with our network model are network nodes are homogeneous, communication links have identical communication cost, each message fits into a node buffer and each message is transmitted as a complete unit. We adopt the *fail-stop* mode where a network component is either operational or not (faulty). A faulty node is deemed equivalent to having all its links faulty. Routing information in message header is shown in Figure 1.

Control Flag	Destination Node	Pending Nodes	Traversed Nodes
1 bit	6 bits	20 bytes	20 bytes

Figure 1: Format of Message Header

Control Flag- indicates whether through routing or backtracking should be performed.
Destination Node - This field indicates the identity of the destination node.
Pending Nodes - Stack containing nodes that form the route in visitation order.
Traversed Nodes - This is a stack containing visited nodes in the route.

For the software implementation of the current prototype, the length of each field is indicated in Figure 1. However, it is apparent that with this implementation, the header size does not scale well with increase in network size. In an actual implementation, for reduced inter-processor latency, the header can be encoded-decoded and/or compressed-decompressed with additional routing hardware [10].

Modelling Network Routing

A search Θ performed on the problem space S, comprising a set of objectives O and a set of search paths P, may be defined from a given starting objective o_s with a specified target objective o_t as $\Theta(o_s, o_t) : S \times S \rightarrow S^*$, where $S = (O, P)$ and $S^* = (O^*,$

P^*). Since a search path may be traversed both ways and a physical network link supports bi-directional communications, our problem space is a logical representation of a multiprocessor network. An objective in our problem space corresponds to a node while a specific search path corresponds to a physical route in the multiprocessor network. We further define $(o_s, o_t) \in O$ and $(o_s, o_t) \in O^* \Rightarrow$ successful search and $(o_s, o_t) \in O$ and $(o_s \in O^*, o_t \notin O^*) \Rightarrow$ unsuccessful search. In a multiprocessor network, a route $\Re$ from a source node n_s to a destination node n_d may be defined as $\Re(n_s, n_d) : G \times G \to G^*$, where $G = (N, L)$ and $G^* = (N^*, L^*)$. We can also define: $(n_s, n_d) \in N$ and $(n_s, n_d) \in N^* \Rightarrow$ complete route and $(n_s, n_d) \in N$ and $(n_s \in N^*, n_d \notin N^*) \Rightarrow$ incomplete route. Conceptually, a search is therefore logically equivalent to a routing attempt. A successful search corresponds to a complete route, which includes the source and destination nodes. On the other hand, an unsuccessful search corresponds to an incomplete route, which does not reach the destination node. The definition, however, does not specify the search technique and hence the routing strategy. The definition also does not constrain $\Re(n_s, n_d)$ to be minimal or non-minimal. Finally, fault tolerance is not explicitly specified.

3. Design of Hybrid Routing Strategy

Since a genetic-based strategy is used, we consider first the generation of the initial population of routes. The initial population should contain some *valid* routes (which lead to the destination). However, in a large-scale network, the large number of possible routes between each node pair may exceed memory constraints specified by routing support. To keep the route tables small, the number of initial candidate routes are limited. This limit is defined as the *population size*. The initial population is generated randomly for a fault-free network of N nodes using the following algorithm:

```
for source node 1 to N      // N is the network size //
   for destination node 1 to N
      if current node = destination node, quit
         else
            for route 1 to R    // population of routes //
               current node = start node
               save source node to route table
               counter = 0     // node can't be found //
               do
                  select new node randomly
                  check whether new node is in route table
                  if new node connected, save in route table
                  if new node = destination node, quit
               while (no new node found and counter <> MAX)
            endfor
   endfor
endfor
```

3.1 The Chromosome

The chromosome in the routing strategy represents the route and each gene represents a node in the route. Formally, for any route $\Re(n_i, n_j)$, we have an associated chromosome $\varsigma(\mathsf{F})$, where F is the set of genes. Let such a $\Re(n_i, n_j)$ be defined on $G = (N, L)$. Then, for any $n_k \in \Re(n_i, n_j)$, we have an associated gene $g_k \in \mathsf{F}$. The length of the chromosome therefore plays an important role in the routing strategy as it affects the communications performance directly. Since the number of nodes to reach the destination varies on different routes, a specific length limit cannot be defined. To maintain unique identities for each gene (node), we can set $\|\mathsf{F}\| \leq N$, the network size. If the route involves any backtracking, the path will be longer and the same node(s) will be traversed. Despite backtracking, the distance it travels to the same node(s) will be calculated. The total distance traversed is thus used to compute the fitness value as a selection criterion. The fitness function is discussed in the next sub-section.

3.2 Fitness function

The GA uses the fitness function to perform route selection. In this application, the fitness value is determined by computing the cost of distance travelled to reach the destination. Fitness value for individual route is determined by the following equation:

$$F = \frac{(D_{max} - D_{route})}{D_{max}} + \Delta,$$

where F is the fitness value
  D_{max} is the maximum distance between source and destination nodes in route,
  D_{route} is the distance travelled by the specified route, and
  Δ is a predefined small positive value to ensure non-zero fitness scores
Thus, the higher the fitness value, the more optimum the route.

3.3 Selection Scheme

We investigate three different selection schemes [17] used to selected the routes:
- *Roulette Wheel Selection*
- *Tournament Selection*
- *Stochastic Remainder*

Each selection scheme employs a different approach. Roulette Wheel Selection is based on the random selection of generating a possible value between 0 and 1, multiplied by the sum of fitness values. The result is called the *roulette value*. The route is then determined by adding the fitness value of each individual route until the sum is equal to or greater than the roulette value. In Tournament Selection, fitness values are compared and the highest selected. Finally, Stochastic Remainder uses probabilities to select the individual route. One of the objectives in our performance analysis is to identify which selection schemes exhibit the best performance in determining optimum route.

3.4 Crossover Operation

Since chromosomes represent solutions to routes, a valid gene sequence of the chromosome (valid route) cannot contain two similar nodes. Different crossover types for permutation, order-based problems are covered in [4,5,17]. Crossover techniques like PMX and Order Crossover will not only destroy the route but also result in invalid source and destination nodes. A secondary constraint is selected routes used to perform crossover operation might contain different route lengths (in terms of number of nodes required to reach the destination). To solve these problems, we use a *one-point random crossover* in our implementation. Figure 2 illustrates this. With this technique, only selected nodes from the crossover point are switched. This means that source and destination nodes can remain intact. If selected route contains only two nodes (source and destination are immediate neighbours), no crossover is performed and specified route is included in next generation if it passes the reroute routine.

				Crossover Point ↓			
Route m:	1	2	3	4	5		
Route n:	1	2	7	8	9	10	5
Route m':	1	2	3	4	9	10	5
Route n':	1	2	7	8	5		

Figure 2: One-Point Random Crossover

3.5 Mutation Operation

Population at this stage may still contain invalid routes. This can be solved by using mutation and reroute operations. Mutation helps maintain diversity of population pool. E.g., in route m' of Figure 3, swapping the 5th and 7th nodes will result in a shorter route to the destination. At this stage, although the complete route still needs to be verified for correctness, the source and destination nodes are intact. In the implementation stage, two types of mutation are employed: *swap mutation* and *shift mutation* [1].

3.6 Reroute Operation

```
REROUTE
   select 2 routes from old population
   perform crossover and mutation
   start the repair function
   check validity of generated offspring (new route) for:
     • node duplication
     • route validity
     • destination validity
   if valid new route, save in new population
   else perform backtracking or discard route
```

The hybrid routing strategy is as follows:

```
initialise population
select two parents
perform one-point crossover and swap- or shift-mutation
REROUTE // verify correctness of route //
BACKTRACKING
remove invalid nodes
start from valid node
repeat
        select randomly new node connected to valid node
        if new node selected = destination node
           save route to new population
           quit
until (new node = destination node) or counter = MAX
```

If the counter reaches MAX, the destination is not found. In this case, the new population will contain partial and possibly full routes. The above step is repeated for a number of generations until new population contains optimal solution. In this case, the routes are valid and reach the destination. The backtracking operation enables the message packet to "retract" when it encounters a *deadend* (a node that is almost completely disconnected save for a single link).

4. Performance Analysis

With population generation and fitness evaluation defined and specified, we next identify the combination of selection and exploration techniques exhibit the best performance. Roulette Wheel (RW), Tournament Selection (TS), and Stochastic Remainder (SR) are first analysed, in combination with both *random* reroute (RR) and *hill climbing* reroute (HCR) strategies [6]. A 25-node mesh network is modelled with varying fault percentage. Ten tests were conducted, each over 20 generations with variations in crossover and mutation probabilities. Generation number is determined

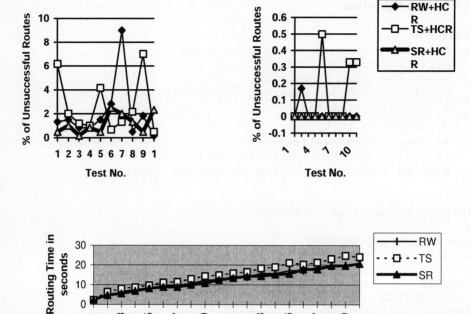

Figure 3 : Performance Comparison of Genetic Selection and Exploration Combinations

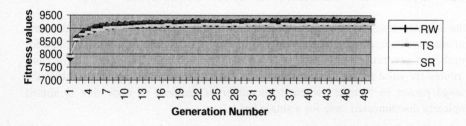

Figure 4: Variation of Fitness Values with Genetic Selection Scheme

from observation that fitness values for all selection schemes reach optimal at around 10 generations (Figure 4). More comprehensive testing with different network models and fault percentages [14] showed the same trend. An unsuccessful reroute means that the destination cannot be reached despite existence of valid path. Different crossover and mutation rates influence the number of successful routes. Inherent randomness in RR reduces its suitability for rerouting [6,9] and it has been shown that number of unsuccessful routes increase significantly as generation number increases [14]. With only shift mutation (tests 3, 5 and 9), number of unsuccessful routes may be reduced. Further analysis shows when swap mutation is used with probability rates above 0.5 and 0.2, number of unsuccessful routes increases and decreases respectively [14]. Using crossover rate of 0.2 and mutation rate of 0.1 shows an increase in unsuccessful routes. This indicates that swap mutation must be used with low probability and a lower bound on crossover and mutation probabilities exists. Omitting crossover necessitates raising mutation rates especially for SR as evinced by the routing times.

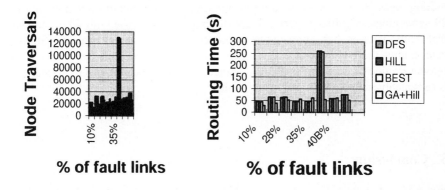

Figure 5: Fault-Tolerant Performance Comparison of Heuristic vs GA-based strategies

SR with HCR clearly outperforms other combinations in having the lowest average percentage of unsuccessful reroutes as well as routing time. With a minimal percentage of unsuccessful routes, SR contributes significantly to the reduction of premature convergence as well as maintaining diversity in the route population. With random one-point crossover, the result does not perform to expectation as it increases the number of unsuccessful routes appreciably (see tests 6 to 8 and 10 for 20 generations). Of 10 tests performed, test 3 provides the best solution using only swap mutation of 0.1. Omitting crossover and having a high shift mutation rate also works favourably as mentioned previously. Mutations rate for both swap and shift mutations should be low (below 0.5) for better results. Higher mutation rates result in more node traversals. However, applying shift mutation with zero crossovers also work favourably. Notably, tests for roulette wheel selection show that applying shift mutation with no crossover, result obtained compared favourably. Tests for TS used tournament size range 2 to 7. Analysis shows that lower rates of mutation perform much better (with exception of omitting crossover which necessitates raising mutation rates). In RW, no comparison is done between between individual routes. It is expected that the individual route selected might not be optimum. TS should provide a better performance as it does comparisons before selection. Overall results incline favourably towards SR selection, with least time required when using shift mutation alone. In the following tests, we compare performance of the GA-based hybrid strategy with Depth-First Search (DFS), Hill-Climbing (HILL) and Best-First Search (BEST) [8]. A larger mesh model of 49 nodes is used. For 40% fault links, two different tests are run, each with a different fault pattern. A *uniform traffic* model is utilised to filter off variations in message traffic over the network. Under the uniform traffic model, a node sends messages to every other node in the mesh with equal probability. Results in Figure 5 shows that, in several cases, the GA-based hybrid strategy outperforms the heuristic strategies in terms of reduced node traversals and routing times. Penalty is the need for progressive optimisation of route population over 20 generations. Significant performance difference in test case with 40% faults shows that fault patterns has appreciable influence on the routing strategy. Success in determining the routes, especially when deadends are encountered, depend on the reroute routine to accomplish fault-tolerance. Trends indicate that GA-based strategy is reasonably scalable with increasing faults.

5. Conclusion

This research has shown that it is possible to adapt and apply a GA-based search technique as a fault-tolerant routing strategy in a multiprocessor network. To reduce overall communications latency, however, dedicated hardware support is required.

References

1. Buckles,B.P., Petry,F.E., Kuester,R.L., "Schema survival rates and heuristic search in GAs", *2nd Intl Conf Tools Art. Intelligence*, '90, pp322-327.
2. Chen,M. and Zalzala, A.M.S., "Safety considerations in the optimization of paths for mobile robots using genetic algorithms", *Proc. 1st Intl Conf. on Genetic Algorithms in Engring Systems: Innovations and Applications*, 1995, pp 299-306.
3. Dally, J. and Seitz,C.L., Deadlock-Free Message Routing in Multiprocessor Interconnection Networks, *IEEE Tran. Comput.*, VC36, N5, May 87, pp 547-553.
4. Goldberg, D.E., "Genetic and evolutionary algorithms come of age", Communications of the ACM, Vol. 37, No. 3, March '94, pp 113-119.
5. Jog, P., Suh, J. Y., and Van Gucht, D., "The Effects of Population Size, Heuristic Crossover and Local Improvement on a GA for Travelling Salesman Problem", *Proc. 1st Intl Conf. Genetic Algorithms and Applications*, July85, pp 110-115.
6. Korf, R.E., *Encyclopedia of Artificial Intelligence*, John Wiley, V2, 1987.
7. Lin,S.., Punch, W.F.,Goodman, E.D.,"Coarse-grain parallel genetic algorithms: categorization and new approach", *Proc. Symp Par and Dist Proc*, '94, pp 28-37.

8. Loh, P.K.K., "Heuristic fault-tolerant routing strategies for a multiprocessor network, *Microprocessors and Microsystems*, V19, N10, Dec. 1995, pp 591-597.
9. Loh, P.K.K., "Artificial intelligence search techniques as fault-tolerant routing strategies", *Parallel Computing*, Vol 22, No, 8, October 1996, pp 1127-1147.
10. Loh, P.K.K. and Wahab, A.,"A Fault-Tolerant Communications Switch Prototype", *Microelectronics and Reliability*, V37, N8, July 1997, pp 1193-1196.
11. Sinclair, M.C., "Application of a Genetic Algorithm to Trunk Network Routing Table Optimisation", *IEE Proc. 10th UK Teletraffic Symp*, April 93, pp 2/1-2/6.
12. Sinclair, M.C., "Trunk Network Fault-Tolerance through Network Augmentation", *Proc. 4th Bangor Symp on Comms*, Bangor, May92, pp 252-256.
13. Sinclair, M.C., "TrunkNetwork Fault-Tolerance through Routing Table Updating in the Event of Trunk Failure", *Proc. 9th UK Teletraffic Symp*, Apr92, pp 1-9.
14. Tay, K.P., "An Indestructible Worm", P114-97, NTech. University, 1997.
15. De Jong, K.A. Spears, W.M., and Gordon, D.F., "Using genetic algorithms for concept learning", *Machine Learning*, V13, No. 2-3, Nov-Dec 1993, pp 161-188.
16. Srinivas,M. & Patnaik,L.M.,"GAs: a survey",*Computer*,V27N6, Jun'94, pp17-26.
17. Srinivas, M. & Patnaik, L.M., "Adaptive probabilities of crossover and mutation in genetic algorithms", *Tran Sys, Man Cybernetics*, V24 N4, Apr94, pp 656-667.

Regularity Considerations in Instance–Based Locality Optimization

Claudia Leopold

Fakultät für Mathematik und Informatik
Friedrich-Schiller-Universität Jena
07740 Jena, Germany
claudia@informatik.uni-jena.de

Abstract. Instance–based locality optimization [6] is a semi–automatic program restructuring method that reduces the number of cache misses. The method imitates the human approach of considering several small program instances, optimizing the instances, and generalizing the structure of the solutions to the program under consideration. A program instance appears as a sequence of statement instances that are reordered for better locality. In [6], a local search algorithm was introduced that reorders the statement instances automatically.
This paper supplements [6] by introducing a second objective into the optimization algorithm for program instances: regularity. A sequence of statement instances is called regular if it can be written compactly using loops. We quantify the intuitive notion of regularity in an objective function, and incorporate this function into the local search algorithm. The functionality of the compound algorithm is demonstrated with examples, that also show trade–offs between locality and regularity.

1 Introduction

In memory hierarchies, (data) locality is a gradual property of programs that reflects the level of concentration of the accesses to the same memory block, during program execution. Locality optimization increases the degree of locality, and hence reduces the number of cache misses (or memory accesses in general). A major technique for locality optimization is program restructuring, i.e., reordering the statements of a given program.

In Ref. [6], a semi–automatic locality optimization method was introduced that is based on the optimization of program *instances* (PI). A PI is obtained from a program by fixing the input size and possibly more parameters. After unrolling loops and resolving recursions, the PI can be written as a sequence of statement instances (SIs) each of which is carried out once. An SI can be an elementary statement instance or a complex operation. In both cases, variables are instantiated, i.e., replaced by constant values.

The instance–based method follows a human approach to program optimization in which one first selects some small program instances, then optimizes the instances, and finally generalizes the structure of the optimized instances to the program. Correspondingly, the instance–based method consists of three phases: an instantiation, an optimization and a generalization phase.

In the instantiation phase, the user fixes a sufficient number of parameters. The user must also choose the granularity of the SIs, i.e., decide if elementary or complex operations (such as function calls) should be considered as SIs. The parameters, particularly the input size, must be fixed at *small* values so that the SI sequence is of manageable length. When the parameters are known, the PI is automatically transformed into the required representation, and a data dependence graph is extracted.

Next, in the optimization phase, the SIs are reordered[1] such that the PI is improved with respect to locality. During the reordering, data dependencies are strictly obeyed. The reordering is controlled by a local search algorithm that is outlined in Sect. 2.1. The output of the optimization phase is a locally optimal SI ordering for the PI.

Finally, in the generalization phase, the user inspects the SI orderings produced for one or several PIs of the program. He or she has to recognize the structural differences that make the suggested orderings superior to the initial orderings, and generalizes them to the program. The user writes down the modified program and makes sure that data dependencies are not violated in the general case.

To facilitate the generalization phase, the output SI sequences of the optimization phase should preferably be structured, i.e., the indices attached to the names of the SIs should follow some regular patterns, as much as possible. This brings up a second objective in the optimization phase: regularity. Although [6] briefly mentions the implementation of a regularity–improving postprocessing phase, that implementation was preliminary and could only achieve few improvements.

This paper concentrates on the issue of regularity optimization, and supplements [6] in this respect. The paper makes two contributions: First, it derives a quite involved objective function that covers the intuitive notion of regularity, for a class of programs. Second, it incorporates the function into the local search algorithm of [6], such that the new algorithm optimizes PIs with respect to both locality and regularity. We report on experiments with nested loop programs and show trade–offs between locality and regularity. Extending the method of [6], our output is represented in a structured form, as a sequence of nested for-loops with constant bounds.

The paper is organized as follows. First, Sect. 2 recalls the formal statement of the optimization problem from Ref. [6], and states the regularity optimization problem. Section 3 defines the new objective function for regularity. Next, Sect. 4 explains how the function is integrated into the local search algorithm. In particular, the neighbourhood of the local search algorithm is extended by various loop transformations, and the two objective functions are combined. Sect. 5 demonstrates the functionality of the algorithm by examples. The paper finishes with an overview of related work in Sect. 6.

---

[1] [6] additionally reassigns data to memory blocks, but this is out of our scope.

2 The Problem

2.1 The Local Search Algorithm of Ref. [6]

According to Ref. [6], the input of the optimization phase consists of a set S of SIs, a data dependence graph *dag*, a set D of data instances (DIs), a function $SD : S \rightarrow D^*$, and memory hierarchy parameters C, L. The *dag* (a directed acyclic graph) contains precendence constraints between the SIs. A DI is a concrete location in memory, e.g. $A[5]$ is a DI while A or $A[i]$ are not. The function SD assigns to each SI s the sequence of DIs that are accessed by s. The parameters C and L are the capacity and line size of a hypothetical cache. They are chosen in relation to the size of the PI.

The output of the optimization algorithm is an SI sequence where each $s \in S$ appears once. It is called schedule. Ref. [6] optimizes the PIs with respect to an objective function OFL_{loc} that quantifies the intuitive notion of locality. The function is quite involved, and omitted here.

The optimization problem stated above is solved with a local search algorithm. Although we restrict ourselves to local search, an interesting consequence of this work is the provision of a framework that makes locality optimization accessible for various optimization techniques (including bio-inspired approaches such as genetic algorithms).

Local search is a well-known general heuristic solution approach to combinatorial optimization problems [1]. For each schedule *Sched* that agrees with the *dag*, a neighbourhood *NB(Sched)* is defined. In Ref. [6], *NB(Sched)* consists of all *dag*-compatible schedules that are obtained from *Sched* by moving a group of consecutive SIs as a whole from one position in *Sched* to another. Starting with a random schedule, the neighbours of the current schedule *Sched* are considered in some fixed order. If a neighbour with a higher objective function value is encountered, *Sched* is replaced by that neighbour immediately. The search stops when no improving neighbour exists.

2.2 The Regularity Optimization Problem

For regularity optimization, we need additional information about the SIs. In a structured representation, two or several SIs can be represented by a single loop statement only if they carry out the same operation. The operation may be modified by parameters that explicitly appear in the name of the statement. These parameters may but need not coincide with the indices of data that are accessed by the SIs.

Hence, our input additionally comprises functions $op \colon S \rightarrow \mathbb{N}$ and $V \colon S \rightarrow \mathbb{Z}^*$. Here, $op(s)$ characterizes the operation carried out by s, and $V(s) = (v_0(s), v_1(s), \ldots v_{r-1}(s))$ are the parameters that may modify the operation. If $op(s_1) = op(s_2)$, then s_1 and s_2 have the same number of parameters (written $r(s_1) = r(s_2)$).

The regularity optimization problem consists in reordering the SIs such that the final schedule can be written in a structured and intuitively simple form.

Currently, we restrict the set of permitted control structures to for-loops with constant bounds. We consider array accesses with index expressions that are affine functions of the loop variables. The final PI is represented by a sequence of possibly nested for-loops, using a PASCAL–like notation and synthetic loop variables.

3 An Objective Function for Regularity

The objective function is based on our intuitive notion of regularity, according to which the degree of regularity is higher,

- the more compact the schedule can be written with possibly consecutive and nested for–loops,
- the more common the step sizes in the loops are, and
- the simpler the index expressions are.

Thus, to measure the degree of regularity, we must identify loops (i.e., for–loops) in the schedule under consideration (denoted by *Sched*).

In *Sched*, a (non–nested) loop appears as a SI subsequence of the form

$$G = (s_{0,0}, s_{0,1}, \ldots, s_{0,w-1}, s_{1,0}, s_{1,1}, \ldots, s_{1,w-1}, \ldots, s_{l-1,0}, s_{l-1,1}, \ldots s_{l-1,w-1})$$

with $s_{i,j} \in S$, $l, w \in \mathbb{N}$, $l \geq 2$ and $w \geq 1$. $s_{i,j}$ and $s_{i,j+1}$, as well as $s_{i,w-1}$ and $s_{i+1,0}$ are consecutive in *Sched*. G can alternatively be written as

$$G = A_0 \circ A_1 \circ \cdots \circ A_{l-1} \quad \text{with} \quad A_i = (s_{i,0}, s_{i,1}, \ldots s_{i,w-1})$$

The A_i's are called aggregates. Let $V(s_{i,j}) = (v_{i,j,0}, v_{i,j,1}, \ldots, v_{i,j,r(s_{i,j})-1})$. G is denoted as *group* if it fulfills the following compatibility conditions:

1. $op(s_{i_1,j}) = op(s_{i_2,j})$, for all i_1, i_2, and
2. $v_{i,j,k} - v_{i-1,j,k} = v_{1,j,k} - v_{0,j,k} \underset{D}{=} D^G_{j,k}$ (or $D_{j,k}$, for short).

Each group can be written as a loop, where A_i corresponds to one iteration of the loop, l is the number of iterations, and w is the number of statements in the loop body. The correspondence between loops and groups is not one–to–one, since inner loops of the program text stand for several groups. G corresponds to the following loop:

FOR $i := 0$ TO $l - 1$ DO
  $op(s_{0,0})((f_{0,0}(i), f_{0,1}(i), \ldots f_{0,r(s_{0,0})-1}(i)))$
  $op(s_{0,1})((f_{1,0}(i), f_{1,1}(i), \ldots f_{1,r(s_{0,1})-1}(i)))$
  $\vdots$
  $op(s_{0,w-1})((f_{w-1,0}(i), f_{w-1,1}(i), \ldots f_{w-1,r(s_{0,w-1})-1}(i)))$.

where $f_{j,k}(i) = v_{0,j,k} + i \cdot D_{j,k}$. We always represent loops in the above normalized form, where the step size is one and the loop starts from zero. Normalization converts the requirement of common step sizes into the requirement of simple index expressions (manifested by common $D_{j,k}$ values).

We will see below that the groups within *Sched* can be easily identified. Unfortunately, groups may conflict, i.e., it may be impossible to realize them simultaneously in a structured formulation of the program. Fig. 1 shows three typical cases of conflicts that may occur. The conflict in Fig. 1c) arises since a proper nesting of the groups $s_0 \ldots s_{11}$ and $s_0 \ldots s_5$ is not possible, as $s_0 \ldots s_5$ and $s_6 \ldots s_{11}$ are structured differently.

a)
```
A[0]++;   (s0)
A[1]++;   (s1)
A[2]++;   (s2)
A[1]++;   (s3)
A[0]++;   (s4)
```

b)
```
A[0]++;   (s0)
A[1]++;   (s1)
A[2]++;   (s2)
A[3]++;   (s3)
A[4]++;   (s4)
A[2]++;   (s5)
A[3]++;   (s6)
A[4]++;   (s7)
A[2]++;   (s8)
A[3]++;   (s9)
A[4]++;   (s10)
```

c)
```
C[0,1]+=A[0,0]*B[0,1];   (s0)
C[0,1]+=A[0,1]*B[1,1];   (s1)
C[0,1]+=A[0,2]*B[2,1];   (s2)
C[0,2]+=A[0,0]*B[0,2];   (s3)
C[0,2]+=A[0,1]*B[1,2];   (s4)
C[0,2]+=A[0,2]*B[2,2];   (s5)
C[0,0]+=A[0,0]*B[0,0];   (s6)
C[0,0]+=A[0,1]*B[1,0];   (s7)
C[0,0]+=A[0,2]*B[2,0];   (s8)
C[1,0]+=A[1,0]*B[0,0];   (s9)
C[1,0]+=A[1,1]*B[1,0];   (s10)
C[1,0]+=A[1,2]*B[2,0];   (s11)
```

Fig. 1. Examples of conflicts

If there are no conflicts, groups can be nested. A nesting is permitted if all aggregates of the outer group have the same internal organization. Let $G = A_0 \circ \cdots \circ A_{l-1}$ be the outer group, and let $A_i = g_{i,0} \circ \cdots \circ g_{i,m_i}$ where $g_{i,j}$ stands for either a subgroup or a single SI. Then, A_{i_1} and A_{i_2} have the same internal organization iff

- $m_{i_1} = m_{i_2}$,
- $g_{i_1,d}$ and $g_{i_2,d}$ are either both groups or both SIs.
- If they are groups, $g_{i_1,d}$ agrees with $g_{i_2,d}$ in l, w, and $D_{j,k}^{g_{i_1},d} = D_{j,k}^{g_{i_2},d}$.
- If they are groups, the aggregates of $g_{i_1,d}$ and $g_{i_2,d}$ have the same internal organization.
- If they are SIs, $op(g_{i_1,d}) = op(g_{i_2,d})$.

The definition of the objective function for regularity (*OFR*) refers to a maximal set of conflict–free groups that can be identified in *Sched*. We define

$$OFR = \max_{\mathcal{G}} \sum_{G \in \mathcal{G}} Eval(G)$$

where $\mathcal{G}$ stands for any set of mutually conflict–free groups, and *Eval* is defined below.

Let G be a group characterized by l, w, and by the sequence $(D_{j,k})$ of essential $D_{j,k}$ values. A value $D_{j,k}$ is essential if the corresponding SI $s_{0,j}$ is either not contained in a subgroup of G, or if $s_{0,j}$ recursively belongs to the first aggregates of all subgroups it is contained in. We define

$$Eval(G) = Dval^{1+2/l^2} \cdot l^{1+p_0} \ .$$

Here, p_0 is a parameter (default 0.5), and $Dval$ evaluates the $(D_{j,k})$ sequence according to our intuitive feeling of how a loop should preferably look like. We use the function

$$Dval(D_{j,k}) = W_{j,k} \cdot \begin{cases} 1 & D_{j,k} = 0 \\ 0.7 & D_{j,k} \neq 0 \text{ and} \\ & D_{j,k} = D_{j',k'} \text{ for some } (j',k') <_{\text{lex}} (j,k) \\ 1/(D_{j,k}+1) & D_{j,k} > 0 \\ 1/(|D_{j,k}|+2) & D_{j,k} < 0 \end{cases}$$

where

$$W_{j,k} = 0.9 + 0.1 \cdot \left(\sum_{d=0}^{j-1} r(s_{0,d}) + k + 1 \right)^{-1}$$

and where

$$Dval = \frac{1}{2} \cdot \left(\operatorname*{avg}_{j,k} Dval(D_{j,k}) + \operatorname*{max}_{j,k} Dval(D_{j,k}) \right) \ .$$

Both *Eval* and *Dval* were designed empirically. We started with some initial functions that roughly corresponded to our intuition. Then, we repeatedly considered two groups. If the relation of the values assigned by the current functions did not agree with that which we had intuitively expected to see, the functions were adjusted appropriately. Additionally, the functions were fine-tuned during their use in the local search algorithm.

For a given SI sequence and width w, it is easy to decide if the sequence forms a group of width w, by a simple test of the compatibility conditions. If we systematically consider all SI subsequences and widths, we find all groups contained in *Sched*. The selection of a conflict-free set of groups $\mathcal{G}$ with maximum $\sum_{G \in \mathcal{G}} Eval(G)$ value is, however, computationally expensive. Hence, we have implemented a greedy algorithm that finds a conflict-free subset $\mathcal{G}$ with a high value of $\sum_{G \in \mathcal{G}} Eval(G)$, but not necessarily with a maximum value. The algorithm is omitted for brevity.

4 Incorporating Regularity Optimization in the Local Search Algorithm

Since it is difficult to repair regularity deficiencies after the locality optimization process has finished, we combine locality and regularity optimization. Our algorithm consists of two phases. In the first phase, emphasis is given to locality optimization, but regularity is considered as a secondary goal. The second phase optimizes the output schedule of the first phase with respect to regularity.

More detailed, the algorithm for the first phase differs from the algorithm of Ref. [6] in that the current schedule is no longer replaced by the first locality-improving neighbour encountered. Instead, several locality-improving neighbours are recorded. Whenever some number of candidates was found, the one that performs best is selected with respect to the combined objective function

$$p_1 \cdot OFL_{\text{loc}} + (1 - p_1) \cdot OFR \quad \text{(for some parameter } p_1 \in [0 \ldots 1]\text{)}.$$

In phase 2, the user can influence the trade-off between locality and regularity via a parameter $p_2 \geq 1$. The current schedule *Sched* is replaced by a neighbour *Sched'* if $OFR(\textit{Sched'}) > OFR(\textit{Sched})$, and if additionally $OFL_{\text{cc}}(\textit{Sched'}) \leq p_2 \cdot OFL_{\text{cc}}(\textit{Sched}_0)$. Here, $\textit{Sched}_0$ is the schedule after phase 1, and OFL_{cc} measures the number of simulated cache misses.

In experiments, the algorithm described so far was typically stuck with a local optimum. We analyzed the particular local optima, and could overcome the problem by extending the neighbourhood of the local search algorithm. In particular, the following transformations were added:

- *Move with Reversal:* This transformation corresponds to the neighbourhood of Ref. [6], except that the order of the moved SI subsequence is additionally reversed.
- *Loop Reversal, Permutation and Distribution*: These transformations are well-known compiler transformations [7].
- *Loop Extension:* This transformation simultaneously adds one or more new aggregates to all groups represented by a loop. It is based on the observation that the old aggregates $A_0 \ldots A_{l-1}$ uniquely imply how possible new aggregates (e.g. A_{-1} or A_l) must look like. In particular, they imply the *op* and V values of SIs that may form the new aggregate. If all the desired SIs are available somewhere in *Sched* (outside the loop under consideration), they are moved to their new positions, forming an extended loop.

The new neighbourhood consists of all SI sequences that are produced by the above transformations, and that additionally agree with the DAG.

After the second phase, the schedule is output in a structured representation. This representation is a by-product of the calculation of *OFR*, where the index expressions of the program statements are composed of the $v_{i,j,k}$ values of the first SI represented by the statement, together with the corresponding essential $D_{j,k}$ values of all surrounding loops.

As regularity optimization aims at supporting the generalization phase, regularity optimization should highlight the locality-relevant structure. Many optimized PIs have a tiled structure, where most reuse of cached data occurs within tiles. *OFR* is modified to prefer groups that are located within a tile, by scaling down $Eval(G)$ for groups G that cross a tile boundary. The experiments in Sect. 5 are based on a straightforward tile recognition scheme.

5 Examples

Figures 2–4 give the outputs of our algorithm for some example PIs. They show that the algorithm is able to produce structured outputs with a high degree of locality. They also show that the algorithm can make several reasonable suggestions that represent different trade–offs between locality and regularity. Currently, such trade–offs can not be found with other automatic locality optimization methods. Compiler transformations find one of the solutions, e.g. b) for Fig. 3, and c) for Fig. 4.

```
a) FOR i2:=0 TO 1 DO
       FOR i1:=0 TO 3 DO
           FOR i0:=0 TO 1 DO
               A[i1,i0+2*i2]=B[i0+2*i2,i1]
```

$OFL_{cc} = 16$
(vs. 24 for input PI)

Fig. 2. optimized PI for 4×4 matrix transpose with $C = 8$ and $L = 2$.

```
a) FOR i1:=0 TO 1 DO
       FOR i0:=0 TO 1 DO
           C[0,i0]+=A[0,i1]*B[i1,i0]
   FOR i1:=0 TO 1 DO
       FOR i0:=0 TO 1 DO
           C[1,i0]+=A[1,1-i1]*B[1-i1,i0]

   OFL_cc: 7
```

```
b) FOR i2:=0 TO 1 DO
       FOR i1:=0 TO 1 DO
           FOR i0:=0 TO 1 DO
               C[i2,i0]+=A[i2,i1]*B[i1,i0]

   OFL_cc: 8
```

Fig. 3. optimized PI for 2×2 matrix multiplication with $C = 6$ and $L = 2$. a) $p_2 = 1.0$, b) $p_2 = 1.2$.

6 Related Work

The instance–based approach to locality optimization differs significantly from other approaches to locality optimization [4,7,8], as discussed in [6]. Since the instance–based method is new, the regularity optimization problem posed in this paper has not been investigated before. Some rather loosely related work exists in the areas of decompilers, software metrics, and genetic programming.

Decompilers [2] identify structures in a statement sequence. The structure already exists and the task is to make it explicit. The design of an objective function is not an issue.

Software metrics, particularly metrics for structural complexity [3], resemble our objective function in that they assess the structure of programs. However, software metrics do not consider the assignment of a meaningful value to an unstructured program, and are less sensitive to minor differences between programs.

a) ```
FOR i1:=0 TO 2 DO
  FOR i0:=0 TO 2 DO
    C[0,i0]+=A[0,i1]*B[i1,i0]
FOR i1:=0 TO 2 DO
  FOR i0:=0 TO 2 DO
    C[1,i0]+=A[1,2-i1]*B[2-i1,i0]
FOR i1:=0 TO 2 DO
  FOR i0:=0 TO 2 DO
    C[2,i0]+=A[2,i1]*B[i1,i0]
```

$OFL_{cc}$: 13  (vs. 33 for input PI)

b) ```
FOR i2:=0 TO 1 DO
  FOR i1:=0 TO 2 DO
    FOR i0:=0 TO 2 DO
      C[2*i2,i0]+=A[2*i2,i1]*B[i1,i0]
FOR i1:=0 TO 2 DO
  FOR i0:=0 TO 2 DO
    C[1,i0]+=A[1,2-i1]*B[2-i1,i0]
```

OFL_{cc}: 14

c) ```
FOR i2:=0 TO 2 DO
  FOR i1:=0 TO 2 DO
    FOR i0:=0 TO 2 DO
      C[i2,i0]+=A[i2,i1]*B[i1,i0]
```

$OFL_{cc}$: 15

**Fig. 4.** optimized PI for $3 \times 3$ matrix multiplication with $C = 9$ and $L = 3$. a) $p_2 = 1.0$, b) $p_2 = 1.1$, c) $p_2 = 1.2$

Genetic Programming also uses fitness measures that assess structural complexity. In Ref. [5], p. 91, structural complexity is defined as the number of elementary operations that appear in a structured representation of the program. This function is too rough, for our purposes, since we must compare PIs that are very similar to each other. The ability to react sensitively to minor differences is particularly important in local search, where it helps to escape from local optima. Furthermore, $OFR$ is more detailed in that it reflects desirable properties such as the intuitive preference of upwards–counting over downwards–counting loops. In genetic programming, programs are in a structured form by construction, i.e., the recovery of the structure is not an issue.

# References

1. Aarts, E.H.L. and Lenstra, J.K.: Local Search in Combinatorial Optimization. Wiley, 1997
2. Cifuentes, C.: Structuring Decompiled Graphs. Int. Conf. on Compiler Construction, 1996, LNCS 1060, 91–105
3. Fenton, N.E.: Software Metrics – A Rigorous Approach. Chapman & Hall, 1991, Chapter 10
4. LaMarca, A. and Ladner, R.E.: The Influence of Caches on the Performance of Sorting. Proc. ACM SIGPLAN Symp. on Discrete Algorithms, 1997, 370–379
5. Koza, J.R.: Genetic Programming II . MIT Press, 1994
6. Leopold, C.: Arranging Statements and Data of Program Instances for Locality. Future Generation Computer Systems (to appear)
7. Muchnick, S.S.: Advanced Compiler Design and Implementation. Morgan Kaufmann, 1997
8. Wolf, M.E. and Lam, M.S.: A Data Locality Optimizing Algorithm. Proc. ACM SIGPLAN'91 Conf. on Programming Language Design and Implementation, 1991, 30–44
```

Parallel Ant Colonies for Combinatorial Optimization Problems

El-ghazali Talbi *, Olivier Roux, Cyril Fonlupt, Denis Robillard **

Abstract. Ant Colonies (AC) optimization take inspiration from the behavior of real ant colonies to solve optimization problems. This paper presents a parallel model for ant colonies to solve the quadratic assignment problem (QAP). Parallelism demonstrates that cooperation between communicating agents improve the obtained results in solving the QAP. It demonstrates also that high-performance computing is feasible to solve large optimization problems.

1 Introduction

Many interesting combinatorial optimization problems are NP-hard, and then they cannot be solved exactly. Consequently, heuristics must be used to solve real-world problems within a reasonable amount of time. There has been a recent interest in the field of the Ant Colony Optimization (ACO). The basic idea is to imitate the cooperative behavior of ant colonies in order to solve combinatorial optimization problems within a reasonable amount of time. Ant Colonies (AC) is a general purpose heuristic (meta-heuristic) that has been proposed by Dorigo [1]. AC has achieved widespread success in solving different optimization problems (traveling salesman [2], quadratic assignment [3], vehicle routing [4], job-shop scheduling [5], telecommunication routing [6], etc.).

Our aim is to develop parallel models for ACs to solve large combinatorial optimization problems. The parallel AC algorithm has been combined with a local search method based on tabu search. The testbed optimization problem we used is the quadratic assignment problem (QAP), one of the hardest among the NP-hard combinatorial optimization problems.

2 Ant colonies for combinatorial optimization

The ants based algorithms have been introduced with Marco Dorigo's PhD [1]. They are based on the principle that using very simple communication mechanisms, an ant is able to find the shortest path between two points. During the trip a chemical trail (pheromone) is left on the ground. The role of this trail is to guide the other ants towards the target point. For one ant, the path is chosen according to the quantity of pheromone. Furthermore, this chemical substance has a decreasing action over time, and the quantity left by one ant depends on the amount of food found and the

---

* LIFL URA-369 CNRS / Université de Lille 1, Bât.M3 59655 Villeneuve d'Ascq Cedex FRANCE. E-mail: talbi@lifl.fr
** LIL, University of Littoral, Calais, France. E-mail: fonlupt@lifl.fr

number of ants using this trail. As indicated in Fig.1, when facing an obstacle, there is an equal probability for every ant to choose the left or right path. As the left trail is shorter than the right one, the left will end up with higher level of pheromone. The more the ants will take the left path, the higher the pheromone trail is. This fact will be increased by the evaporation stage.

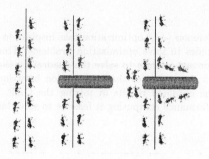

Fig. 1. Ants facing an obstacle

This principle of communicating ants has been used as a framework for solving combinatorial optimization problems. Figure 2 presents the generic ant algorithm. The first step consists mainly in the initialization of the pheromone trail. In the iteration step, each ant constructs a complete solution to the problem according to a probabilistic state transition rule. The state transition rule depends mainly on the state of the pheromone. Once all ants generate a solution, a global pheromone updating rule is applied in two phases: an evaporation phase where a fraction of the pheromone evaporates, and a reinforcement phase where each ant deposits an amount of pheromone which is proportional to the fitness of its solution. This process is iterated until a stopping criteria.

```
Step 1 : Initialization
  - Initialize the pheromone trail
Step 2 : Iteration
  - For each Ant Repeat
    - Solution construction using the pheromone trail
    - Update the pheromone trail
  - Until stopping criteria
```

Fig. 2. A generic ant algorithm.

3 Ant colonies for the quadratic assignement problem

The AC algorithm has been used to solve the quadratic assignment problem (QAP). The QAP represents an important class of NP-hard combinatorial optimization problems with many applications in different domains (facility location, data analysis, task scheduling, image synthesis, etc.).

3.1 The quadratic assignment problem

The QAP can be defined as follows. Given a set of n objects $O = \{O_1, O_2, ..., O_n\}$, a set of n locations $L = \{L_1, L_2, ..., L_n\}$, a flow matrix C, where each element c_{ij} denotes a flow cost between the objects O_i and O_j, a distance matrix D, where each element d_{kl} denotes a distance between location L_k and L_l, find an object-location bijective mapping $M : O \longrightarrow L$, which minimizes the objective function f:

$$f = \sum_{i=1}^{n} \sum_{j=1}^{n} c_{ij} \cdot d_{M(i)M(j)}$$

3.2 Application to the QAP

Our method is based on an hybridization of the ant system with a local search method, each ant being associated with an integer permutation. Modifications based on the pheromone trail are then applied to each permutation. The solutions (ants) found so far are then optimized using a local search method; update of the pheromone trail simulates the evaporation and takes into account the solutions produced in the search strategy. In some way, the pheromone matrix can be seen as shared memory holding the assignments of the best found solutions. The different steps of the ANTabu algorithm are detailed below (fig.3):

Generate m (number of ants) permutations π^k of size n
Initialization of the pheromone matrix F
FOR $i = 1$ to I^{max} ($I^{max}=n/2$)
  FOR all permutations π^k ($1 \leq k \leq m$)
    Solution construction :
    $\hat{\pi}^k = n/3$ transformations of π^k based on the pheromone matrix
    Local search :
    $\tilde{\pi}^k = $ tabu search $(\hat{\pi}^k)$
    IF $\tilde{\pi}^k < \pi^*$
      THEN the best solution found $\pi^* = \tilde{\pi}^k$
  Update of the pheromone matrix
  IF $n/2$ iterations applied without amelioration of π^*
    THEN **Diversification**

Fig. 3. ANTabu: Ant colonies algorithm for the QAP.

Initialization: We have used a representation which is based on a permutation of n integers:

$$s = (l_1, l_2, ..., l_n)$$

where l_i denotes the location of the object O_i. The initial solution for each ant is initialized randomly. Three phases are necessary to initialize the matrix of pheromones. First, we apply a local search optimization of the m initial solutions. Then, we identify the best solution of the population π^*. Finally, the matrix of pheromones F is initialized as follows:

$$\tau_{ij}^0 = 1/100 \cdot f(\pi^*), i,j \in [1..n]$$

Solution construction: The current solution of each ant is transformed function of the pheromone matrix. We use a pair exchange move as a local transformation in which two objects of a permutation are swapped. $n/3$ swapping exchanges (n is the problem size) are applied as follows: the first element r is selected randomly, and the second one s is selected with a probability 0.9 such as $\tau_{r\pi_s}^k + \tau_{s\pi_r}^k$ is maximal ($\tau_{r\pi_s}^k$ is the pheromone for the position r containing the element s where π is a solution). In the other cases, the element is selected with a probability proportional to the associated pheromone:

$$\frac{\tau_{r\pi_s}^k + \tau_{s\pi_r}^k}{\sum_{r \neq s}(\tau_{r\pi_s}^k + \tau_{s\pi_r}^k)}$$

Local search: We have designed a local search procedure based on the tabu search (TS) method [7]. To apply TS to the QAP, we must define the short-term memory to avoid cycling. The long-term memory for the intensification/diversification phase has not been used, because it was handled by the ant system. The tabu list contains pairs (i,j) of objects that cannot be exchanged (recency-based restriction). The efficiency of the algorithm depends on the choice of the size of the tabu list. Our experiments indicate that choosing a size which varies between $\frac{n}{2}$ and $\frac{3n}{2}$ gives very good results. Each TS task is initialized with a random tabu list size in the interval $\frac{n}{2}$ to $\frac{3n}{2}$. The aspiration function allows a tabu move if it generates a solution better than the best found solution.

Update of the pheromone matrix: First, we update the pheromone matrix to simulate the evaporation process, which consists in reducing the matrix values F with the following formulae:

$$\tau_{ij}^{k+1} = (1-\alpha)\tau_{ij}^k, i,j \in [1..n].$$

where $0 < \alpha < 1$ ($\alpha = 0.1$ in our experiments). If α is close to 0 the influence of the pheromone will be efficient for a long time. However, if α is close to 1 his action will be short-lived. In a second phase, the pheromone is reinforced function of the solution found. Instead of only taking into account the best found solution for updating the pheromone matrix as it is done in the HAS-QAP algorithm [3], we have devised a new strategy where each ant adds a contribution inversely proportional to the fitness of its solution. This contribution is weakened by dividing the difference between the solution and the worst one with the best one. So far, the update formula is:

$$\tau_{i\pi(i)}^{k+1} = (1-\alpha)\tau_{i\pi(i)}^k + \frac{\alpha}{f(\pi)}\frac{f(\pi^-)-f(\pi)}{f(\pi^*)}, i \in [1..n]$$

where $\tau_{i\pi(i)}^{k+1}$ is the pheromone trail at the (k+1)th iteration associated to the element $(i, \pi(i))$, π^- is the worst solution found so far, π^* the best one, and π the current solution.

Diversification: In the case where the best solution found has not been improved in $n/2$ iterations, the diversification task is started. This diversification scheme will force the ant system to start from totally new solutions with new structures. The diversification in the HAS-QAP system consisted in re-initializing the pheromone matrix and randomly generating new solutions [3]. In our case, we have created a long-term memory called *frequency matrix*. This matrix will be used to hold the frequency of all previous assignments, and will be used when a diversification phase will be triggered. During the diversification phase, the least chosen affectations will be used to generate new solutions.

4 Parallel ant colonies

To solve efficiently large optimization problems, a parallel model of ant colonies has been developed. The programming style used is a synchronous master/workers paradigm. The master implements a central memory through which passes all communication, and that captures the global knowledge acquired during the search. The worker implements the search process. The parallel algorithm works as follows (fig.4): The pheromone matrix and the best found solution will be managed by the master. At each iteration, the master broadcasts the pheromone matrix to all the workers. Each worker handles an ant process. He receives the pheromone matrix, constructs a complete solution, applies a tabu search for this solution and send the solution found to the master. When the master receives all the solutions, he update the pheromone matrix and the best solution found, and then the process is iterated.

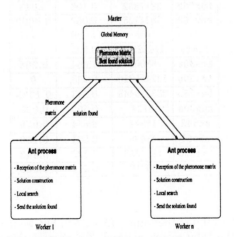

Fig. 4. Synchronous master/workers model for parallel ant colonies

Our ant system has been implemented for a parallel/distributed architecture using the programming environment C/PVM (Parallel Virtual Machine). Each ant is managed by one machine. In all the experiments, we work with only 10 ants (10 machines, network of SGIs Indy workstations).

5 Experimental Results

We first compare our metaheuristic ANTabu with the HAS-QAP method, which is also based on ant colonies. Then, we compare it with parallel independent tabu search to assess the cooperation paradigm of ant colonies. Finally, the performances of our approach is compared with other methods based on hill-climbing and genetic algorithms. For those comparisons, we studied instances from three classes of problems of the QAP-library [8]: random uniform cost and distance matrices, random cost matrix and a grid for the distance matrix, and real problems.

5.1 Comparison with HAS-QAP

The parameters for the ANTabu algorithm are set identical to HAS-QAP ones ($\alpha = 0.1$). The Tabu search method is restricted to $5n$ iterations in order to evaluate the performance of the ant system. We have selected about 20 problems from the QAPlib [8] either from the irregular class problems (bur26d, chr25a, els19, tai20b, tai35b) or from the regular class problems (nug30, sko42, sko64, tai25a and wil50). We have allowed 10 iterations for HAS-QAP and ANTabu. In Table 1, we compare the results of ANTabu and HAS-QAP. Results for HAS-QAP are directly taken from [3].

Table 1. Solution quality of HAS-QAP and ANTabu. Best results are in boldface

Problem	Best known	HAS-QAP	ANTabu
bur26b	3817852	0.106	**0.018**
bur26d	3821225	0.002	**0.0002**
chr25a	3796	15.69	**0.047**
els19	17212548	0.923	**0**
kra30a	88900	1.664	**0.208**
tai20b	122455319	0.243	**0**
tai35b	283315445	0.343	**0.1333**
nug30a	6124	0.565	**0.029**
sko42	15812	0.654	**0.076**
sko64	48498	0.504	**0.156**
tai25a	1167256	2.527	**0.843**
wil50	48816	0.211	**0.066**

It is clear that ANTabu outperforms HAS-QAP, but we may wonder if the gain is not only due to the tabu search method. We address this problem in the next section.

5.2 Impact of the cooperation between agents

In order to evaluate the gain brought by the ants cooperation system, we have compared the results with those of the PATS algorithm [9]. The PATS is a parallel adaptive tabu search, and it consists in a set of independent tabu algorithms running in a distributed fashion on a network of heterogeneous workstations (including Intel PCs, Sun and Alpha workstations).

Results are shown in Tab. 2. Best found for PATS and ANTabu is the best solution out of 10 runs. The difference relative to the QAPlib optimum is given as a percentage gap. Running times are in minutes, and correspond to the mean execution time over 10 runs on a network of 126 machines for PATS, and on a network of 10 machines for ANTabu. Thus, the ANTabu algorithm was at a great disadvantage in this regard.

Table 2. Comparaison between PATS and ANTabu algorithms.
Best results are in boldface.

	tai100a	sko100a	sko100b	sko100c	sko100d	sko100e	wil100	esc128	tai256c
Best known	21125314	152002	153890	147862	149576	149150	273038	64	44759294
PATS									
Best found	21193246	152036	153914	147862	149610	149170	273074	64	44810866
Gap	0.322	0.022	0.016	0	0.022	0.013	0.013	0	0.115
Time (mn)	117	142	155	132	152	124	389	230	593
ANTabu									
Best found	21184062	152002	153890	147862	149578	149150	273054	64	44797100
Gap	0.278	0	0	0	0.001	0	0.006	0	0.085
Time (mn)	139	137	139	137	201	139	478	258	741

Results show that the ANTabu system finds better results using less computing resources and less search agents. Notice that the best known solutions have been found in three out of four sko100 instances.

5.3 Comparison with other algorithms

The performances of the ANTabu algorithm have been compared with other metaheuristics on 28 problems:

- ant colonies: FANT [10], HAS-QAP [3].
- genetic algorithms: GDH [10], GTSH [11].
- hill-climbing heuristics: VNS-QAP and RVNS-QAP [12].

The algorithms are run 10 times to obtain an average performance estimate. The table 3 shows the average ecart-type value of the solutions obtained for instances of sizes between 19 and 80. The results of the other algorithms are obtained from [10].

Our algorithm ANTabu always succeed in finding better solutions than other algorithms. This result shows the efficiency and the robustness of the algorithm. The best known solutions are always found for 9 instances. According to the results, we observe that the algorithm is well fitted to a large number of instances.

Table 3. Comparaison of algorithms with long execution times
(execution equivalent to 100 calls of the local search procedure).

Problem	Best known value	ANTabu	FANT	HAS-QAP	GDH	GTSH	VNS-QAP	RVNS-QAP
bur26a	5426670	**0.0169**	0.0449	0.0275	0.0581	0.0903	0.055	0.198
bur26b	3817852	**0.0343**	0.0717	0.1065	0.0571	0.1401	0.053	0.346
bur26c	5426795	**0.0000**	0.0000	0.0090	0.0021	0.0112	0.001	0.347
bur26d	3821225	**0.0001**	0.0022	0.0017	0.0039	0.0059	0.003	0.223
bur26e	5386879	**0.0000**	0.0068	0.0039	0.0059	0.0066	0.005	0.121
bur26f	3782044	**0.0000**	0.0007	0.0000	0.0034	0.0097	0.001	0.357
bur26g	10117172	**0.0000**	0.0025	0.0002	0.0027	0.0094	0.003	0.153
bur26h	7098658	**0.0000**	0.0003	0.0007	0.0021	0.0063	0.001	0.305
els19	17212548	**0.0000**	0.5148	0.9225	1.6507	2.2383	0.421	14.056
kra30a	88900	**0.0000**	2.4623	1.6637	2.9269	2.0070	1.660	5.139
kra30b	91420	**0.0328**	0.6607	0.5043	1.1748	0.6082	0.677	4.137
nug20	2570	**0.0000**	0.6226	0.1556	0.4825	0.1774	0.638	2.903
nug30	6124	**0.0065**	0.6172	0.5650	1.1920	0.3971	0.686	2.972
sko42	15812	**0.0089**	0.8158	0.6539	1.2168	0.6964	0.971	3.045
sko49	23386	**0.0462**	0.7740	0.6611	1.4530	0.4936	0.547	2.518
sko56	34458	**0.0070**	1.0941	0.7290	1.4272	0.5869	0.923	2.215
sko64	48498	**0.0132**	0.9213	0.5035	1.2739	0.6678	0.804	2.310
sko72	66256	**0.1138**	0.8826	0.7015	1.4224	0.7130	0.762	2.166
sko81	90998	**0.0769**	0.9950	0.4925	1.3218	0.7125	0.561	1.666
sko90	115534	**0.1207**	0.9241	0.5912	1.2518	0.7606	0.842	1.773
tai20b	122455319	**0.0000**	0.1810	0.2426	0.1807	1.0859	0.226	1.252
tai25b	344355646	**0.0000**	0.1470	0.1326	0.3454	1.9563	0.196	7.792
tai30b	637117113	**0.0343**	0.2550	0.2603	0.3762	3.2438	0.510	4.197
tai35b	283315445	**0.0869**	0.3286	0.3429	0.7067	1.5810	0.248	5.795
tai40b	637250948	0.6997	1.3401	**0.2795**	0.9439	3.9029	1.559	7.421
tai50b	458821517	**0.2839**	0.5412	0.2906	1.1281	1.6545	0.642	4.063
tai60b	608215054	**0.1621**	0.6699	0.3133	1.2756	2.6585	0.880	6.309
tai80b	818415043	**0.4841**	1.4379	1.1078	2.3015	2.7702	1.569	4.909
Mean		**0.080**	0.583	0.402	0.864	0.936	0.552	3.168

6 Conclusion and future work

In this paper we have proposed a powerful and robust algorithm for the QAP, based on ants colonies. Compared with previous ants systems for the QAP (HAS-QAP algorithm), we have refined the ants cooperation mechanism, both in the pheromone matrix update phase and in the exploitation/diversification phase by using a frequency matrix. The search process of each ant has also been reinforced with a local search procedure based on tabu search.

Results show a noticeable increase in performance compared to HAS-QAP and also to parallel tabu search, thus demonstrating the complementary gains brought by the combined use of a powerful local search and ants-like cooperation. This last comparison, with the parallel tabu search algorithm, pleads for a more widely spread use of cooperation in parallel heuristics.

Future works include an application of the metaheuristic to other optimization problems (set covering, graph coloring, multi-objective optimization), and an implementation under the execution support MARS (Multi-user Adaptive Resource Scheduler) to allow efficient fault-tolerant runs on larger heterogeneous networks of workstations [13].

References

1. M. Dorigo. *Optimization, learning and natural algorithms*. PhD thesis, Politecnico di Milano, Italy, 1992.
2. M. Dorigo, V. Maniezzo, and A. Colorni. The ant system: Optimization by a colony of cooperating agents. *IEEE Transactions on Systems, Mans, and Cybernetics*, 1(26):29–41, 1996.
3. L. Gambardella, E. Taillard, and M. Dorigo. Ant colonies for the qap. Technical Report 97-4, IDSIA, Lugano, Switzerland, 1997.
4. B. Bullnheimer, R. F. Hartl, and C. Strauss. Applying the ant system to the vehicle routing problem. In *2nd Metaheuristics Int. Conf. MIC'97*, Sophia-Antipolis, France, July 1997.
5. A. Colorni, M. Dorigo, V. Maniezzo, and M. Trubian. Ant system for job-shop scheduling. *JORBEL - Belgian Journal of Operations Research, Statistics, and Computer Science*, 34(1):39–53, 1994.
6. R. Schoonderwoerd, O. Holland, J. Bruten, and L. Rothkrantz. Ant-based load balancing in telecommunications networks. *Adaptive behavior*, 5(2):169–207, 1997.
7. F. Glover. Tabu search - part I. *ORSA Journal of Computing*, 1(3):190–206, 1989.
8. R. E. Burkard, S.Karisch, and F. Rendl. Qaplib: A quadratic assignment problem library. *European Journal of Operational Research*, 55:115–119, 1991.
9. E. G. Talbi, Z. Hafidi, and J-M. Geib. Parallel adaptive tabu search for large optimization problems. In *2nd Metaheuristics International Conference MIC'97*, pages 137–142, Sophia-Antipolis, France, July 1997.
10. E. D. Taillard and L. Gambardella. Adaptive memories for the quadratic assignment problem. Technical Report 87-97, IDSIA, Lugano, Switzerland, 1997.
11. C. Fleurent and J. A. Ferland. Genetic hybrids for the quadratic assignment problem. *DIMACS Series in discrete Mathematics and Theoretical Computer Science*, 16:173–188, 1994.
12. P. Hansen and N. Mladenovic. An introduction to variable neighborhood search. In *2nd Metaheuristic Int. Conf.*, Sophia-Antipolis, France, 1997.
13. E-G. Talbi, J-M. Geib, Z. Hafidi, and D. Kebbal. A fault-tolerant parallel heuristic for assignment problems. In *BioSP3 Workshop on Biologically Inspired Solutions to Parallel Processing Systems, in IEEE IPPS/SPDP'98 (Int. Parallel Processing Symposium / Symposium on Parallel and Distributed Processing*.

An Analysis of Synchronous and Asynchronous Parallel Distributed Genetic Algorithms with Structured and Panmictic Islands

Enrique Alba, José Mª Troya

Dpto. de Lenguajes y Ciencias de la Computación, Univ. de Málaga
Campus de Teatinos (2.2.A.6), 29071- MÁLAGA, España
{eat,troya}@lcc.uma.es

Abstract. In a parallel genetic algorithm (PGA) several communicating nodal GAs evolve in parallel to solve the same problem. PGAs have been traditionally used to extend the power of serial GAs since they often can be tailored to provide a larger efficiency on complex search tasks. This has led to a considerable number of different models and implementations that preclude direct comparisons and knowledge exchange. To fill this gap we begin by providing a common framework for studying PGAs. This allows us to analyze the importance of the synchronism in the migration step of parallel distributed GAs. We will show how this implementation issue affects the evaluation effort as well as the search time and the speedup. In addition, we consider popular evolution schemes of panmictic (steady-state) and structured-population (cellular) GAs for the islands. The evaluated PGAs demonstrate linear and even super-linear speedup when run in a cluster of workstations. They also show important numerical benefits when compared with their sequential counterparts. In addition, we always report lower search times for the asynchronous versions.

1 Introduction

Genetic algorithms (GAs) are stochastic search methods that have been successfully applied in many search, optimization, and machine learning problems [2]. Unlike most other optimization techniques, GAs maintain a population of encoded tentative solutions that are *competitively* manipulated by applying some *variation operators* to find a global optimum.

A sequential GA proceeds in an iterative manner by generating new populations of strings from the old ones. Every string is the encoded (binary, real, ...) version of a tentative solution. An evaluation function associates a fitness measure to every string indicating its suitability to the problem. The canonical GA applies stochastic operators such as selection, crossover, and mutation on an initially random population in order to compute a whole generation of new strings.

For non-trivial problems this process might require high computational resources, and thus, a variety of algorithmic issues are being studied to design efficient GAs. With this goal in mind, numerous advances are continuously being achieved by designing new operators, hybrid algorithms, and more. We extend one of such improvements consisting in using parallel models of GAs (PGAs).

Several arguments justify our work. First of all, PGAs are naturally prone to parallelism since the operations on the strings can be easily undertaken in parallel. The evidences of a higher efficiency [3], larger diversity maintenance, additional availability of memory/cpu, and multi-solution capabilities, reinforce the importance of the research advances with PGAs [9].

Using a PGA often leads to superior numerical performance and not only to a faster algorithm [5]. However, the truly interesting observation is that the use of a structured population, either in the form of a set of islands [3] or a diffusion grid [11], is the responsible of such numerical benefits. As a consequence, many authors do not use a parallel machine at all to run structured-population models, and still get better results than with serial GAs [5].

In traditional PGA works it is assumed that the model maps directly onto the parallel hardware, thus making no distinction between the *model* and its *implementation*. However, once a structured-population model has been defined it can be implemented in any monoprocessor or parallel machine. This separate vision of model vs. implementation raises several questions. Firstly, any GA can be run in parallel, although a linear speedup is not always possible. In this sense, our contribution is to extend the existing work on distributed GAs (we will use dGA for short) from generational island evolution to steady-state [10], and cellular GAs [11].

Secondly, this suggests the necessity of using a difficult and heterogeneous test suite. In this sense, we use a test suite composed of separable, non-separable, multimodal, deceptive, and epistatic problems (main difficulties of real applications).

Thirdly, the experiments should be replicable to help future extensions of our work. This led us to using a readily available parallel hardware such as a cluster of workstations and to describe the references, parameters, and techniques.

Finally, some questions are open in relation to the physically parallel execution of the models. In our case we study a decisive implementation issue: the synchronism in the communication step [6]. Parameters such as the communication frequency used in a PGA are also analyzed in this paper.

The high number of non-standard or machine-dependent PGAs has led to efficient algorithms in many domains. However, these unstructured approaches often hide their canonical behavior, thus making it difficult to forecast further behavior and to make knowledge exchange. This is why we adopt a unified study of the studied models.

The paper is organized as follows. In Section 2 we present sequential and parallel GAs as especial cases of a more general meta-heuristic. In Section 3 we describe the benchmark used. Section 4 presents a study on the search time of sync/async versions for different basic techniques and problems. Section 5 contains an analysis from the point of view of the speedup, and, finally, Section 6 offers some concluding remarks.

2 A Common Framework for Parallel Genetic Algorithms

In this section we briefly want to show how coarse (cgPGA) and fine grain (fgPGA) PGAs are subclasses of the same kind of parallel GA consisting in a set of communicating sub-algorithms. We propose a change in the nomenclature to call them *distributed* and *cellular* GAs (dGA and cGA), since the grain is usually intended to refer to their computation/communication ratio, while the actual differences can be also found in the way in which they both structure their populations (see Figure 1).

While a distributed GA has large sub-populations (>>1) a cGA has typically one single string in every sub-algorithm. For a dGA the sub-algorithms are loosely connected, while for a cGA they are tightly connected. In addition, in a dGA there exist only a few sub-algorithms, while in a cGA there is a large number of them.

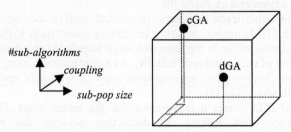

Figure 1. The Structured-Population Genetic Algorithm Cube

This approach to PGAs is *directly* supported in only a few real implementations of PGA software. However, it is very important since we can study the full spectrum of parallel implementations by only considering different coding, operators, and communication details. In fact, this is a natural vision of PGAs that is inspired in the initial work of Holland, in which a *granulated adaptive system* is a set of grains with shared structures. Every grain has its own reproductive plan and internal/external representations for its structures.

The outline of a general PGA is presented in the Algorithm 1. It begins by randomly creating a population $P(t=0)$ of μ structures -strings-, each one encoding the p problem variables, usually as a vector over $I\!B=\{0,1\}$ ($I=I\!B^{p\cdot l_x}$) or $I\!R$ ($I=I\!R^p$). An evaluation function Φ is needed to associate a quality real value to every structure.

The stopping criterion ι of the reproductive loop is to fulfill some condition. The solution is identified as the best structure ever found.

The *selection* s_{Θ_s} operator makes copies of every structure from $P(t)$ to $P'(t)$ attending to the fitness values of the structures (some parameters Θ_s might be required). The typical variation operators in GAs are *crossover* ($\otimes$), and *mutation* (m). They both are stochastic techniques whose behavior is governed by a set of parameters, e.g. an application probability: $\Theta_c=\{p_c\}$ -high- and $\Theta_m=\{p_m\}$ -low-.

Finally, each iteration ends by selecting the μ new individuals for the new population (*replacement policy r*). For this purpose the new pool $P'''(t)$ plus a set Q are considered. The best structure usually survives deterministically (*elitism*). Many panmictic variants exist, but two of them are especially popular [10]: the *generational GA* -genGA- ($\lambda=\mu$, $Q=\emptyset$), and the *steady-state GA* -ssGA- ($\lambda=1$, $Q=P(t)$).

In a parallel GA there exist many elementary GAs (grains) working on separate sub-populations $P^i(t)$. Each sub-algorithm includes an additional phase of periodic *communication* with a set of neighboring sub-algorithms located on some topology.

This communication usually consists in exchanging a set of individuals or population statistics. All the sub-algorithms are though to perform the same reproductive plan. Otherwise the PGA is heterogeneous [2].

In particular, our steady-state panmictic algorithm (Figure 2a) generates one single individual in every iteration. It is inserted back in the population only if it is better (larger fitness than) the worst existing individual.

ALGORITHM 1: PARALLEL GENETIC ALGORITHM

$t := 0$;
initialize: $\quad P(0) := \{\vec{a}_1(0), ..., \vec{a}_\mu(0)\} \in I^\mu$;
evaluate: $\quad P(0) : \{\Phi(\vec{a}_1(0)), ..., \Phi(\vec{a}_\mu(0))\}$;

while not $\iota(P(t))$ **do**   *// Reproductive Loop*
   select: $\quad\quad\quad P'(t) := s_{\Theta_s}(P(t))$;
   recombine: $\quad P''(t) := \otimes_{\Theta_c}(P'(t))$;
   mutate: $\quad\quad P'''(t) := m_{\Theta_m}(P''(t))$;
   evaluate: $\quad P'''(t) : \{\Phi(\vec{a}_1'''(0)), ..., \Phi(\vec{a}_\lambda'''(0))\}$;
   replace: $\quad P(t+1) := r_{\Theta_r}(P'''(t) \cup Q)$;
   <communication step>
   $t := t+1$;

end while

In all the cGAs we use (Figure 2b) a NEWS neighborhood is defined (North-East-West-South in a toroidal grid [11]) in which overlapping demes of 5 strings (4+1) execute the same reproductive plan. In every deme the new string computed after selection, crossover, and mutation replaces the current one only if it is better.

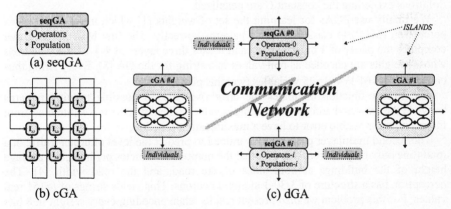

Figure 2. Two basic non-distributed models (a) (b), and a possible -merged!- distribution (c)

In all this paper we deal with MIMD implementations (Figure 2c) of homogeneous dGAs in which migrants are selected *randomly*, and the target island replaces its *worst* string with the incoming one only if it is better. The sub-algorithms are disposed in a unidirectional ring (easy implementation and other advantages).

In all the cases we will study the frequency of migrations in multiples of the global population size: $1\cdot\mu, 2\cdot\mu, 4\cdot\mu, 8\cdot\mu, ...$ The graphs in this paper show these multiples: 1, 2, 4, 8,... A value of 0 means an idle (*partitioned*) distributed evolution. Hence, the mig. frequency depends on the population size, and has not "general recipe" values.

3   The Test Suite

All the problems we test maximize a fitness function. We use the *generalized sphere problem* to help us in understanding the PGA models by defining a baseline for comparisons. The *subset sum problem* is NP-Complete and difficult for the algorithms. Finally, *training a Neural Network* needs an efficient exploration of the search space, final local tuning of the parameters in the strings, and shows high epistasis (weight correlation), and multimodality.

The first problem is an instance of the well-known sphere function (Equation 1). We use 16 variables encoded in 32 bits each one (l=512 bits), and call it SPH16-32.

$$f(\vec{x}) = \sum_{i=1}^{n} x_i^2 \qquad x_i \in [-5.12, 5.12] \tag{1}$$

The second optimization task we address is the subset sum problem [7]. It consists in finding a subset of values $V \subseteq W$ from a set of integers $W=\{\omega_1, \omega_2, ..., \omega_n\}$, such that the sum of this subset approaches, without exceeding, a constant C.

We use instances of n=128 integers generated as explained in [7] to increase the instance difficulty, and call it the SSS128 problem.

In order to formulate it as a maximization problem we first compute the sum $P(\vec{x}) = \sum_{i=1}^{n} x_i \cdot \omega_i$ for a tentative solution $\vec{x}$, and then use the fitness function:

$$f(\vec{x}) = a \cdot (P(\vec{x})) + (1-a) \cdot \lfloor (C - 0,1 \cdot P(\vec{x})) \rfloor_0 \tag{2}$$

where a=1 when $\vec{x}$ is admissible, i.e., when $C - P(\vec{x}) \geq 0$, and a=0 otherwise. Solutions exceeding the constant C are penalized.

We finally use PGAs for learning the set of weights [1] which make a multilayer perceptron (MLP) to classify a set of patterns correctly. The first MLP we consider computes the parity of 4 binary inputs and it uses three layers of 4-4-1 binary neurons whose weights are encoded as real values in a string for the GA [8]. Every string thus contains l=4·4+4·1+4+1=25 variables (weights plus biases).

The fitness function computes the error over the 16 possible training patterns between the expected and the actual output values. Then, this error is subtracted from the maximum possible error to have a maximization function.

The second multilayer perceptron is trained to predict the level of urban traffic in a road (one only output) from three inputs: the number of vehicles per hour, the average height of the buildings at both sides of the road, and the road width [4]. The perceptron has a structure of 3-30-1 sigmoid neurons. This yields strings of l=151 real values. For this problem we *also* present results when encoding every weight in 8 bits (strings of l=1208 bit length) to experiment with very long strings. Every function evaluation needs to compute the error along a set of 41 patterns.

All the experiments compare sync/async versions of parallel dGAs (labeled with **s** and **a**, respectively, in the graphs). The experiments use equivalent parameterizations for all the algorithms solving the same problem (same global population, operators and probabilities). In addition, standard operators are used: proportional selection, double point crossover, and bit flip mutation for SPH16-32 and SSS128, and proportional plus random parent selection (to enhance diversity and hyper-plane recombination), uniform crossover, and real additive mutation for the FP-genes.

4 Real Time Analysis of the Impact of the Synchronism

Synchronous and asynchronous parallel dGAs perform both the same algorithm. However, sync islands wait for every incoming string they must accept, while async ones do not. In general, no differences should appear in the search, provided that all the machines are of the same type (our case). On the contrary, the execution time may be modified due to the continuous waits induced in a synchronous model.

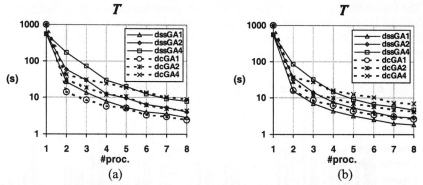

Figure 3. Time in solving SPH16-32 with synchronous (a) and asynchronous versions (b)

In Figure 3 we show the expected execution time (50 independent runs) to solve SPH16-32. The asynchronous models (3b) consistently lessen the search time with respect to their equivalent synchronous algorithms (3a) for any of the tested migration frequencies (1, 2, 4), and island evolution modes (steady-state or cellular).

The effects of the synchronization on the subset sum (Figures 4a and 4b) clearly demonstrate the drawbacks of a tight coupling for non-trivial problems. The sync and async models (100 independent runs) show similar numeric performances for a variety of techniques and processors, thus confirming their similarities.

Sync and async versions needed a similar effort except when migration frequency is 1. Values of 16 and 32 provided a better numeric efficiency and 100% efficacy in all the experiments. Also, it is detected a better resistance of dcGA to bad migration frequencies. For example, for 2 processors and migration frequency 1, dssGA had a 57% of success while dcGA had 100% with a smaller number of evaluations.

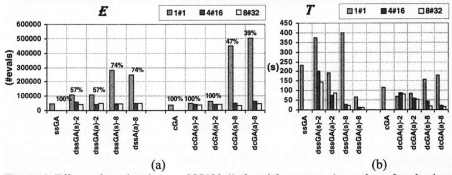

Figure 4. Effects of synchronism on SSS128 (1, 2 and 8 processors): number of evaluations (a), and time to get a solution (b)

In almost all the cases the time to get a solution is consistently reduced by the parallel distributed models. As an exception, we can see a scenario for SSS128: excessive interaction among the parallel islands is slower than the sequential version.

The same holds for the experiments with the "traffic" problem (Figure 5). Only a parallel version (8 processors) is showed since no sequential version was able to solve it. For this network the migration interval has a minor impact on the effort (Figure 5a). The main difficulty is in making the final tuning of the real genes, and in overcoming the lost of functionality in the resulting crossed children strings [1].

The search time of dssGA (Figure 5b) is very good due to its high exploitation. For dcGA, the loosely connected and the partitioned variants are the better ones. This is true since the cGA islands are able of a timely sustained exploitation without important diversity losses, and since they have a relatively faster reproductive loop.

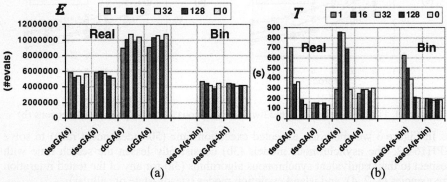

Figure 5. Effects of synchronism on the "traffic" MLP: number of evaluations (a), and time to get a solution (b). We plot dssGA and dcGA with real encoding, and dssGA with a binary code

Notice that, for the "traffic" problem, the search time is also influenced by the encoding (only dssGA is showed): a binary representation is slightly slower than a real encoding due to the longer strings the algorithm has to manage. However, the number of evaluations is somewhat smaller since binary strings induce a pseudo-Boolean search in a discrete space.

It can be seen that the number of needed evaluations for SSS128 and "traffic" (Figures 4a and 5a) is almost not affected by the synchronization (also for SPH16-32). On the contrary, the search time is considerably lessen by all the async algorithms.

5 Speedup

Before discussing the results on the speedup of the analyzed algorithms, we need to make some considerations. As many authors have established [6], sequential and parallel GAs *must* be compared by running them until a solution of the same quality has been found, *and not* until the same number of steps has been completed. In deterministic algorithms the speedup (Equation 3) is upper bounded by the number of processors n_{proc}, but for a PGA it might not, since it reduces *both* the number of necessary steps *and* the expected execution time $T_{n_{proc}}$ in relation to the sequential one T_1. This means that super-linear speedups are possible [9].

$$S(n_{nproc}) = \frac{T_1}{T_{n_{proc}}} \tag{3}$$

In Figure 6 we show the speedup of dssGA and dcGA in solving SPH16-32 (50 independent runs). Two conclusions can be drawn due to the fact that the problem is GA-easy. On the one hand, the speedup is clearly super-linear. On the other hand, highly coupled islands are better (like using a single population, but faster). This does not hold for other problems, as shown in the previous section (see also [3]). Besides that, the asynchronous versions show definitely better speedup values than the synchronous ones for dssGA (Figure 6a), and slightly better for dcGA (Figure 6b) for equivalent migration frequencies (a1>s1, a2>s2, ...).

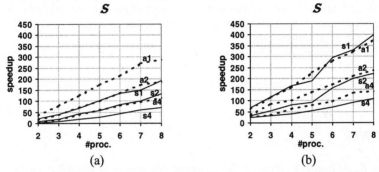

Figure 6. Speedup in solving SPH16-32 with dssGA (a) and dcGA (b)

The speedup for the subset sum problem (Figure 7a) confirms the benefits of a sparse migration (100 independent runs): the speedup is better for larger isolation times (16 and 32). Super-linear speedups for dssGA, and almost linear speedup for dcGAs with migration frequency of 32 have been obtained. More reasonable speedup values are observed due to the difficulty of this problem. As expected, a tight coupling sometimes prevent of having linear speedups.

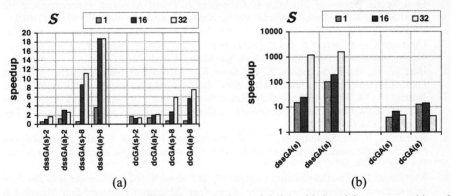

Figure 7. Speedup in solving SSS128 with dssGA and dcGA with 2 and 8 processors (a), and speedup in solving the "parity4" training problem with 8 processors (b)

We finally plot in Figure 7b the results in training the "parity4" neural network. Again, the asynchronous models get a larger speedup (and lower absolute execution times). Also, a large isolated evolution is the best global choice.

6 Concluding Remarks

In this paper we have stated the advantages of using parallel distributed GAs and of performing a common analysis of different models. Parallel distributed GAs almost always outperformed sequential GAs. We have distributed both, a sequential steady-state GA, and a cellular GA with the goal of extending the traditional studies.

In all the experiments asynchronous algorithms outperformed their equivalent synchronous counterparts in real time. This confirms other existing results with different PGAs and problems, clearly stating the advantages of the asynchronous communications [6].

Our results take into account the kind of performance analysis proposed also in [6]. After that, we are able to confirm that these algorithms are sometimes capable of showing super-linear speedup, as shown in some other precedent works [3] [9] that did not follow these indications.

As a future work it would be interesting to study (1) the influence of the migration policy, (2) a heterogeneous PGA search, and (3) the importance of the selected island model for a given fitness landscape (generational, steady-state or cellular).

References

1. Alba E., Aldana J. F., Troya J. M.: "Genetic Algorithms as Heuristics for Optimizing ANN Design". In Albrecht R. F., Reeves C. R., Steele N. C. (eds.): *Artificial Neural Nets and Genetic Algorithms*. Springer-Verlag (1993) 683-690
2. Bäck T., Fogel D., Michalewicz Z. (eds.): *Handbook of Evolutionary Computation*. Oxford University Press (1997)
3. Belding T. C.: "The Distributed Genetic Algorithm Revisited". In Eshelman L. J. (ed.): *Proceedings of the Sixth International Conference on Genetic Algorithms*. Morgan Kaufmann, San Francisco, CA (1995) 114-121
4. Cammarata G., Cavalieri S., Fichera A., Marletta L.: "Noise Prediction in Urban Traffic by a Neural Approach". In Mira J., Cabestany J., Prieto A. (eds.): *Proceedings of the International Workshop on Artificial Neural Networks*, Springer-Verlag (1993) 611-619
5. Gordon V. S., Whitley D.: "Serial and Parallel Genetic Algorithms as Function Optimizers". In Forrest S. (ed.): *Proceedings of the Fifth International Conference on Genetic Algorithms*. Morgan Kaufmann, San Mateo, CA (1993) 177-183
6. Hart W. E., Baden S., Belew R. K., Kohn S.: "Analysis of the Numerical Effects of Parallelism on a Parallel Genetic Algorithm". In IEEE (ed.): CD-ROM IPPS97 (1997)
7. Jelasity M.: "A Wave Analysis of the Subset Sum Problem". In Bäck T. (ed.): *Proceedings of the Seventh International Conference on Genetic Algorithms*. Morgan Kaufmann, San Francisco, CA (1997) 89-96
8. Romaniuk, S. G.: "Evolutionary Growth Perceptrons". In Forrest S. (ed.): *Proceedings of the Fifth International Conference on Genetic Algorithms*. Morgan Kaufmann, San Mateo, CA (1993) 334-341
9. Shonkwiler R. "Parallel Genetic Algorithms". In Forrest S. (ed.): *Proceedings of the Fifth International Conference on Genetic Algorithms*. Morgan Kaufmann, San Mateo, CA (1993) 199-205
10. Syswerda G.: "A Study of Reproduction in Generational and Steady-State Genetic Algorithms". In Rawlins G. (ed.): *Foundations of GAs*, Morgan Kaufmann (1991) 94-101
11. Whitley D.: "Cellular Genetic Algorithms". In Forrest S. (ed.): *Proceedings of the Fifth International Conference on GAs*. Morgan Kaufmann, San Mateo, CA (1993) 658

GA-based Parallel Image Registration on Parallel Clusters

Prachya Chalermwat[1], Tarek El-Ghazawi[1], and Jacqueline LeMoigne[2]

1 Institute for Computational Sciences and Informatics
George Mason University
prachya@science.gmu.edu, tarek@gmu.edu

2 Code 935
NASA Goddard Space Flight Center
lemoigne@cesdis.gsfc.nasa.gov

Abstract. Genetic Algorithms (GAs) have been known to be robust for search and optimization problems. Image registration can take advantage of the robustness of GAs in finding best transformation between two images, of the same location with slightly different orientation, produced by moving space-borne remote sensing instruments. In this paper, we have developed sequential and coarse-grained parallel image registration algorithms using GA as an optimization mechanism. In its first phase the algorithm finds a small set of good solutions using low-resolution versions of the images. Based on the results from the first phase, the algorithm uses full resolution image data to refine the final registration results in the second phase. Experimental results are presented and we found that our algorithms yield very accurate registration results and the parallel algorithm scales quite well on the Beowulf parallel cluster.

1 Introduction

Digital image registration is very important in many applications, such as medical imagery, robotics, visual inspection, and remotely sensed data processing. The NASA's Earth Observing System (EOS) project will be producing enormous Earth global change data, reaching hundreds of Gigabytes per day, that are collected from different spacecrafts and different perspectives using many sensors with diverse resolutions and characteristics. The analysis of such data requires integration, therefore, accurate registration of these data. Image registration is defined as the process that determines the most accurate relative orientation between two or more images, acquired at the same or different times by different or identical sensors. Registration can also provide the orientation between a remotely sensed image of the Earth and a map.

Given a reference and an input images, the image registration process determines the amount of rotation and the amount of translation (in both the x-axis and y-axis), the input image has with respect to the reference image, figure 1. Due to the tremendous amount of data generated from the EOS remote sensing satellites and the amount of images to be registered, the time required to register input images to existing reference images is quite large. This prompted the need for parallel processing to conquer the lengthy computation time[1][2][3]. In [3] we discussed the computational savings in image registration resulted from using the Wavelet-based technique which exploits the multiresolution property of Wavelet.

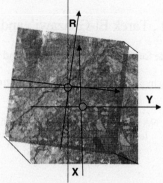

Fig. 1. Image Registration is to find transformations (rotation R, X and Y translation) between input and reference images

Genetic Algorithms (GAs) have been long known to be very robust for search and optimization problems. It is based on the principles of natural biological evolution that operates on a population of potential solutions applying the principle of survival of the fittest to produce better and better approximations to a solution [4][5]. In this paper, two-phase GA-based image registration algorithms (sequential and parallel) are proposed. In the first phase near optimal transformations of the registration are found using low-resolution version of the images (32x32). In the second phase we employ full resolution images to refine the results from the first phase and obtain the best solution. The transformation, consisting of rotation, translations in X-axis and Y-axis (R, X, Y), is encoded into a bit string of chromosome and is decoded for an evaluation of the fitness values. We will show that our GA-based sequential registration is very computationally efficient and robust for the subset of image test suite. Additionally, we will demonstrate that our GA-based parallel image registration scales and provides fast registration. Our implementation is based on SuGal library[1] that allows us to design our image registration evaluation function. We will discuss this evaluation function in details in section 2. In this paper, the used Beowulf cluster (at NASA GSFC) has 50 200MHz Pentium Pro, and runs Linux kernel 2.0.35. Interprocessor communication is provided by a combination of GigabitEthernet and switched FastEthernet.

This paper is organized as follows. Section 2 describes GA-based sequential image registration. Section 3 discusses parallel approach to GA-based image registration followed by some experimental results in section 4. Conclusions and future research are listed in section 5.

---

[1] A library supports a large number of variants of Genetic Algorithms, and has extensive features to support customization and extension. Sugal can be freely downloaded at http://www.trajan-software.demon.co.uk/sugal.htm

2 GA-based Image Registration

2.1 Genetic Algorithms (GAs): Concept and Definitions

Genetic algorithm (GA) was formally introduced by John Holland and his colleague [4]. It is based on the natural concept that diversity helps to ensure a population's survival under changing environmental conditions. GAs are simple and robust methods for optimization and search and have intrinsic parallelism. GAs are iterative procedures that maintain a population of candidate solutions encoded in form of chromosome string. The initial population can be selected heuristically or randomly. For each generation, each candidate is evaluated and is assigned the fitness value that is generally a function of the decoded bits contained in each candidate's chromosome. These candidates will be selected for the reproduction in the next generation based on their fitness values. The selected candidates are combined using the genetic recombination operation "crossover". The crossover operator exchanges portions of bit string hopefully to produce better candidates with higher fitness for the next generation. The "mutation" is then applied to perturb the string of chromosome as to guarantee that the probability of searching a particular subspace of the problem space is never zero [5]. It also prevents the algorithm from becoming trapped on local optima [8][9]. Then, the whole population is evaluated again in the next generation and the process continues until it reaches the termination criteria. The termination criteria may be triggered by finding an acceptable approximate solution, reaching a specific number of generations, or until the solution converges. Table 1 shows some basic GAs terminology used in this paper.

Table 1. Some important GAs definitions

Chromosome	Individual encoded strings of bits, integer, etc. that are uniquely mapped onto the decision domain
Population	Number of individuals to be evaluated for each generation
Fitness value	Performance of each individual derived from the objective function
Crossover	An operator applied to pairs of individuals to exchange their characteristics
Selection	A process of determining which individual will be chosen for reproduction and the number of offspring that an individual will produce
Mutation	An operator used to alter bits in the chromosomes to possibly avoid the local optima convergence

2.2 GA-based Image Registration

GA-based image registrations for medical images have been reported in [10][11][12]. The first two targeted finding transformation vector between two medical images, while in [12] Turton applied the GAs to real-time computer vision. The transformation vector used is called "bilinear geometric transformation" which is basically a combination of four values indicating the distances between the four corners of the two images [11]. The work by Ozkan is based on a 5-processor transputer architecture applied to images of 100x100 pixels and using 80-bit chromosome. Turton's work is based on VLSI-level hardware architecture focusing on vision systems that require fast and robust processing. The effort, however, did not address MIMD parallel GAs or multiresolution image registration. These studies have also provided no quantitative insight into the accuracy of the registration results.

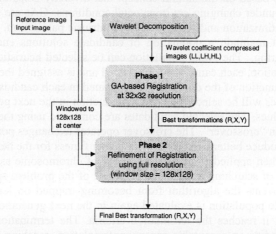

Fig. 2. Flow diagram of GA-based image registration using wavelets

In this paper, a two-phase approach is proposed. In the first phase, called GA32, we apply GAs technique to the 32x32-pixel compressed version of the image, there the image is compressed using the wavelet technique [13]. As shown in figure 2, using the results from GA32, we refine the registration using full resolution images in the second phase. At full resolution of 512x512, registration is computationally very expensive. We found, however, that 128x128 of full resolution image, the registration requires much less computations but still maintains the accuracy of the registration [14]. While the rotation is not sensitive to the image size the translation in both x-axis and y-axis must be adjusted to the corresponding original resolution. In this second phase, called FULL phase, we only use a 128x128 window at the center of the original reference and input images for registration. In this work, compression using the wavelet decomposition is done off-line. Our work in [13] has shown promising scalability of wavelet decomposition on various parallel platforms and we believe that incorporating parallel wavelet decomposition to the current implementation of GA-based parallel image registration in the future will not degrade the overall performance.

2.3 Accuracy of GA-based Image Registration

The accuracy of our 2-phase GA-based image registration is shown in table 2. The table shows that the GA-based image registration at 32x32 pixels is very accurate despite the fact that it has been compressed by a factor of 256. The final results from the two-phase algorithm is very accurate and robust, and have found the solution in thematic mapper and digitized photo images with an average error of 0.3 degrees, 1.7 pixels, and 0.4 pixels for rotation, translation x, translation y respectively. Note also that in the second phase, it is sufficient to use a 128x128 subimage to refine the results. This subimage is selected from the center of the full image, where image sensor distortion is minimal.

Table 2. Accuracy using full resolution to refine the registration (with 128x128 centered window) and FP FACTOR = 200

	ORIGINAL			32x32						512x5152					
				REGISTERED			ERROR			REGISTERED			ERROR		
	R	X	Y	R	X	Y	R-ERR	X-ERR	Y-ERR	R	X	Y	R-ERR	X-ERR	Y-ERR
tm.ref	0	0	0	-0.2	3.2	3.2	0.2	-3.2	-3.2	0	0	0	0	0	0
tm.r4	4	0	0	5.17	-1.12	-0.64	-1.17	1.12	0.64	4	0	0	0	0	0
tm.r4tx50	4	50	0	3.43	46.08	1.92	0.57	3.92	-1.92	4	50	0	0	0	0
tm.r4tx5ty2	4	5	2	4.2	3.52	1.92	-0.2	1.48	0.08	4	5	2	0	0	0
tm.tx50	0	50	0	0.58	43.84	-4.48	-0.58	6.16	4.48	0	50	0	0	0	0
tm.tx5ty2	0	5	2	0.1	5.28	2.56	-0.1	-0.28	-0.56	0	5	2	0	0	0
	R	X	Y	R	X	Y	R-ERR	X-ERR	Y-ERR	R	X	Y	R-ERR	X-ERR	Y-ERR
girl.ref	0	0	0	0.33	-2.4	3.2	-0.33	2.4	-3.2	0	0	0	0	0	0
girl.r5	5	0	0	5.21	0.8	1.92	-0.21	-0.8	-1.92	5	0	0	0	0	0
girl.r5tx20ty60	5	20	60	4.65	22.88	59.2	0.35	-2.88	0.8	5	19	56	0	1	4
girl.r5tx6ty4	5	6	4	5.8	2.4	4.8	-0.8	3.6	-0.8	5	6	4	0	0	0
girl.tx20ty60	0	20	60	1.1	12.8	49.6	-1.1	7.2	10.4	0	19	57	0	1	3
Average							-0.31	1.70	0.44				0	0.182	0.636

3 GA-based Parallel Image Registration

3.1 Mapping Image Registration to GAs

3.1.1 Chromosome Encoding

Using a bit encoding scheme for chromosome string, our transformation R, X, and Y is encoded as shown in figure 3. A 12-bit field is used to represent possible relative rotation of the input image to the reference image. Likewise, 10 bits are used to express translation in x-axis and 10 more for the y-axis. All representations are signed magnitude, using one bit for the sign and the rest of the bits to represent the magnitude of the rotation or translation. Thus, relative rotation has the range of ±2048 degrees, while relative translation in the x (or y) direction has the range of ±512 pixels. As shown in table 3, to increase the precision of the search into sub-degrees using floating-point numbers, we divide these numbers by FP_FACTOR. For example if FP_FACTOR is equal to 400 then we have the search scope within ± 5.2, ±1.28, and ± 1.28 for R, X, and Y respectively. Note the translation at 32x32 pixels is compressed by a factor of 16. This means the translation X and Y will be equivalent to $1.28 \times 2^4 = \pm 20.5$ pixels at full resolution (512x512).

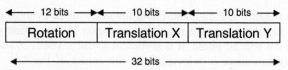

Fig. 3. Encoding of transformation R, X, and Y into 32-bit chromosome string

3.1.2 Objective Function and Other GA Operations

Our objective function is to maximize the correlation between the reference and input images. This means the fitness function is simply the correlation function of the input (transformed) image and the reference image. This works quite well regardless of the simplicity of the GAs operation. The correlation function between two images A and B is shown in equation (1).

$$C(A,B) = \frac{\sum (A_i - \bar{A}) \times (B_i - \bar{B})}{\sqrt{\sum (A_i - \bar{A})^2} \times \sqrt{\sum (B_i - \bar{B})^2}} \quad (1)$$

For selection of individuals, we select highly fit individuals with higher correlation (fitness) value. Thus, the objective function is to maximize the correlation values between the reference image and the input image. The crossover used is a single-point crossover and the termination condition is to stop GA after the solution converges or reaches a pre-specified number of generations.

3.2 Parallel Genetic Algorithms

Chipperfield and Fleming [5] classified parallel GAs into three categories: Global GAs, Migration GAs, and Diffusion GAs.

Global GAs treat the entire populations as a single breeding unit. Near linear speed-up may be obtained using global GAs when the objective function is significantly more computationally expensive than the GA itself (our image registration fits this). This is even true on networks of workstations that demand low communication overhead.

Migration GAs divide the population into a number of sub-populations, each of which is treated as a separate breeding unit under the control of a conventional GA. To encourage the proliferation of good genetic material throughout the whole population, individuals migrate between the sub-populations from time to time. It is also considered coarse-grained. In this category, the population is distributed into sub-population and the local breeding is used (sometimes called *island model*)[5]. The highly fit individuals migrate to other subgroups within specific paths to be mated with other individuals in other subgroups to produce better-fit offspring. The Migration GAs are suitable for MIMD architectures.

Diffusion GAs treat each individual as a separate breeding unit, the individuals it might mate with being selected from within a small local neighborhood. The use of local selection and reproduction rules leads to a continuous diffusion of individuals over the population. The Diffusion GAs are considered fine-grained GAs. It is different from the Migration GAs in that it migrates only between local neighborhood.

Local selection results in performance superior to global and migration GAs. It is suitable for SIMD machines but can also be implemented efficiently on MIMD systems.

We found, from the timing profile of our sequential GA-based image registration, that the time to compute the objective function dominates the overall execution time (97.8%). The fraction of other GA-related functions is very small (2.2%). Therefore, our parallel GA-based image registration can take advantage of the simplicity of the global GAs approach while providing good scalability. In our implementation, a master node keeps track of all genetic-related processes except for the evaluation process of the chromosomes. The parallelism resides in the evaluation of the fitness of the chromosomes can be exploited.

4   Experimental Results

We used subset of images from test suite of NASA and NOAA images [3][1]. The selected images include LandSat/Thematic Mapper (TM) images and digitized photo images (GIRL). Our implementation is based on SUGAL (V-2.1)—SUnderland Genetic Algorithm package. It has been extended for sequential and parallel image registration using MPI. We used two pool sizes (32 and 100) as to observe the effects of increasing computation to the performance of GA-based image registration. We ran our code on a Beowulf parallel computer (ecgtheow.gsfc.nasa.gov), a 50-node Pentium Pro 200 MHz cluster with a fat-tree network topology[2] [6][7].

Figure 4 shows experimental results of GA-based parallel image registration with final refinement on a Beowulf parallel computer. The legend GA32 represents the first phase of our parallel algorithm that uses wavelet compressed images of size 32x32 pixels. The legend FULL denotes the second phase that uses the windowed image (128x128) of the full resolution image to refine the registration results obtained from the GA32. The legend ALL represents the measurements of the whole algorithm (GA32+FULL). This figure shows scalability, communication overhead, and computation time for pool size of 32 and 100, on 32 and 50 nodes respectively.

Note that as the number of nodes increases, the communication overhead of the GA32 phase increases, figure 4 c and d. However, the communication overhead of the FULL phase only slightly increases with the number of processors. This is due to the fact that the FULL phase has much fewer communication transactions after the GA32 has narrowed the solution search space dramatically. Figure 4 f shows the pool size effect on the ratio between the computation of GA32 and the FULL phases. In figure

---

[2] See http://beowulf.gsfc.nasa.gov for more details.

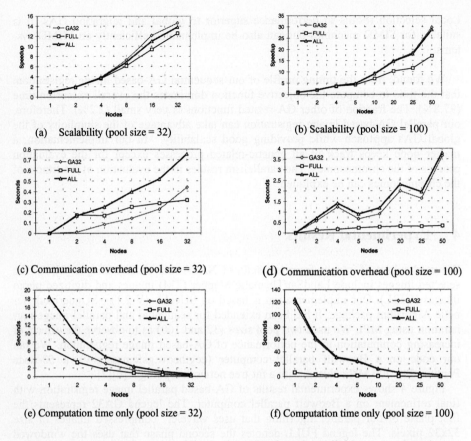

Fig. 4. Experimental results of the 2-phase GA-based parallel image registration on Beowulf parallel computer: Scalability (a and b), Communication overhead (c and d), Computation only (e and f).

4 f, the computation time of the FULL phase is almost identical to the computation in figure 4 e for the pool size of 32. This is because the FULL case does not use GA's and is independent to the GA operations and population size. Figure 4 c and b show also (for the same reason) that communication increases for the GA phase only by increasing the pool size.

5 Conclusions and Future Works

Despite the low resolution of the compressed wavelet images, the GA-based image has a high degree of accuracy. Additionally, using the full-resolution registration on windowed images at the center in the second phase, the algorithm leads to perfect registration in case of TM images and minor errors with the digitized image. Our parallel approach scales with the number of processors almost with linear speedup. High efficiency has been obtained for both pool size = 32 and pool size = 100. For pool size = 32, the algorithm maintain an average of 90% efficiency up to 8 nodes. At 16 nodes, efficiency is about 70% before dropping to 43% at 32 nodes, due to increasing communication overhead.

The current algorithm expects the pool size to be divisible by number of processors used to evenly distribute the load among all processors. Future work will incorporate dynamic load balancing to eliminate the dependency between the pool size and number of processors and will test against a larger set of images.

Acknowledgment The authors would like to thank Donald Becker, Erik Hendrik, and Phil Merky at CESDIS GSFC for their help with the Ecgtheow Beowulf cluster at CESDIS, NASA Goddard Space Flight Center.

References

1. J. Le Moigne. "Towards a Parallel Registration of Multiple Resolution Remote Sensing Data", Proceedings of IGARSS'95, Firenze, Italy, July 10-14, 1995.
2. M. Corvi and G. Nicchiotti, "Multiresolution Image Registration," in Proceedings 1995 IEEE International Conference on Image Processing, Washington, D.C., Oct. 23-26, 1995.
3. T. El-Ghazawi, P. Chalermwat, and J. LeMoigne, "Wavelet-based image Registration on parallel computers," in SC'97: High Performance Networking and Computing: Proceedings of the 1997 ACM/IEEE SC97 Conference: November 1521, 1997.
4. J. H. Holland, "Adaptation in Natural and Artificial System," University of Michigan Press, Ann Arbor, 1975.
5. A. Chipperfield and P. Fleming, "Parallel Genetic Algorithms," in *Parallel & Distributed Computing Handbook* by A. Y. H. Zomaya, McGraw-Hill, 1996, pp. 1118-1143.
6. Mike Berry and Tarek El-Ghazawi. "Parallel Input/Output Characteristics of NASA Science Applications". Proceedings of the International Parallel Processing Symposium (IPPS'96), IEEE Computer Society Press. Honolulu, April 1996.
7. D. Ridge, D. Becker, P. Merkey, T. Sterling, "Beowulf: Harnessing the Power of Parallelism in a Pile-of-PCs," Proceedings, IEEE Aerospace, 1997.
8. P. Husbands, "Genetic Algorithms in Optimisation and Adaptation," in *Advances in Parallel Algorithms* Kronsjo and Shumsheruddin ed., 1990, pp. 227-276.
9. D. E. Goldburg, Genetic Algorithms in Search: optimization and machine learning, Reading, Mass. Addison-Wesley, 1989.
10. J. M. Fitzpatrick, J. J. Grefenstette, and D. Van-Gucht. "Image registration by genetic search," Proceedings of Southeastcon 84, pp. 460-464, 1984.
11. M. Ozkan, J. M. Fitzpatrick, and K. Kawamura, "Image Registration for a Transputer-Based Distributed System," in proceedings of the 2nd International Conference on Industrial & Engineering Applications of AI & Expert Systems (IEA/AIE-89), June 6-9, 1989, pp. 908-915.
12. B. Turton, T. Arslan, and D. Horrocks, "A hardware architecture for a parallel genetic algorithm for image registration," in Proceedings of IEE Colloquium on Genetic Algorithms in Image Processing and Vision, pp. 111-116, Oct. 1994.
13. T. El-Ghazawi and J. L. Moigne, "Wavelet decomposition on high-performance computing systems," in Proceedings of the 1996 International Conference on Parallel Processing, vol. II, pp. 19-23, May 1996.
14. P. Chalermwat, T. El-Ghazawi, and J. LeMoigne, "Image registration by parts," in Image Registration Workshop (IRW97), pp. 299-306, NASA Goddard Space Flight Center, MD, Nov. 1997.

Implementation of a Parallel Genetic Algorithm on a Cluster of Workstations: The Travelling Salesman Problem, a Case Study

Giuseppe Sena[1]*, Germinal Isern[1]**, and Dalila Megherbi[2]

[1] Northeastern University, College of Computer Science,
Boston, MA 02115, U.S.A.
{tonysena,isern}@ccs.neu.edu
[2] University of Denver, Division of Engineering,
Denver, CO 80208, U.S.A.
dmegherb@du.edu

Abstract. A parallel version of a Genetic Algorithm is presented and implemented on a cluster of workstations. Even though our algorithm is general enough to be applied to a wide variety of problems, we used it to obtain optimal/suboptimal solutions to the well known Traveling Salesman Problem. The proposed algorithm is implemented using the Parallel Virtual Machine library over a network of workstations, and it is based on a master-slave paradigm and a distributed-memory approach. Tests were performed with clusters of 1, 2, 4, 8, and 16 workstations, using several real problems and population sizes. Results are presented to show how the performance of the algorithm is affected by variations on the number of slaves, population size, mutation rate, and mutation interval. The results presented show the utility, efficiency and potential value of the proposed algorithm to tackle similar NP-Complete problems.

1 Introduction

In this paper a parallel/distributed version of a *Genetic Algorithm* (GA) is presented and implemented over a cluster of workstations to obtain quasi-optimal solutions to the widely known *Traveling Salesman Problem* (TSP).

Genetic Algorithms are computational models inspired by the idea of evolution [15], and they were introduced by John Holland and his students [7, 3]. GAs encode solutions to a specific problem using a chromosome-like data structure and apply recombination operators to produce new individuals. GAs are often used as function optimizers, and are also considered as global search methods that do not use gradient information. Therefore, they might be applied to problems in which the function to be optimized is non-differentiable or with multiple local optima.

---
* ON Technology Corp., Cambridge, MA 02142, U.S.A., tsena@on.com.
** On leave from Universidad Central de Venezuela, Caracas, Venezuela.

A GA is inherently parallel [9], and at every iteration individuals are independently selected for crossover and mutation following some probability distribution. The proposed parallel and distributed implementation introduced in this paper uniformly decomposes the population among the available processors (hosts), so that genetic operators can be applied in parallel to multiple individuals. A *master-slave paradigm* is used to implement the Parallel Distributed Genetic Algorithm (PDGA).

The proposed PDGA is applied to the TSP [10, 4, 8]. The problem at hand is an optimization problem which consists of finding a *Hamiltonian* cycle of minimum length. Individuals in the population are Hamiltonian cycles represented as a sequence of vertices, which makes the mutation and crossover operators simpler and efficient. New heuristics as well as decisional rules are introduced to support computationally efficient migration of individuals between sub-populations in order to improve the gene pool.

Our PDGA is implemented on a network of SUN workstations (Ultra Sparc 1) running the Parallel Virtual Machine (PVM) library, and tests are performed using 1, 2, 4, 8, and 16 slave workstations. Several real problems and population sizes are used in our experiments, which are obtained from a library of TSPs called *TSPLIB* from Rice University.

The experimental results obtained are very promising. In particular, we show that as we increase the size of the population the performance of the PDGA improves as we increase the number of slave tasks used. We additionally show that with small population sizes (i.e. $p = 128$) the communication overhead overcomes the advantage of using our PDGA.

The rest of this paper is organized as follows: in section 2 we give an overview of the GA methodology used, present the problem encoding and the fitness function used. In addition, we show the general design of our PDGA, the problem parameters and the GA operators used. Section 3 describes our test environment, and the test results and analysis are presented in section 4. The conclusions and future research is presented in section 5.

2 GA Methodology

Usually, only two components of a GA are problem dependent: the *problem encoding* and the *evaluation (fitness) function*. The following subsections present the encoding, mutation and crossover operators, and the fitness function used for the TSP. In addition, an overview of the proposed PDGA is presented.

Three different ways of exploiting parallelism in GAs can be defined [15]: (a) a parallel GA similar to the canonical GA [7], (b) an *Island Model* that uses distinct sub-populations, and (c) a fine-grain massively parallel implementation where each individual resides in a processor. In this paper the *island model* is used in order to exploit a more coarse-grain parallelism, which can be easily extended to a distributed system. The general idea behind the *island model* is to distribute the total population among the available processors, and to execute a classical GA with each sub-population. Hence every few generations (based on well defined criteria), sub-populations could swap few individuals (chosen at random among the best fit). This migration process allows sub-populations to share "genetic material".

2.1 Problem Encoding

Traditional GAs use binary bit-strings to encode individuals in the population. However, many researchers [2, 10, 8] use real-valued or non-binary encodings instead of binary encodings, and employ *recombination operators* or *specialized operations* that may be problem or domain specific. For the TSP, individuals in the population are Hamiltonian cycles. If we try to represent Hamiltonian cycles as bit-strings we will soon find out that the mutation and crossover operators can generate invalid individuals with a very high probability. Hence the need for repairing algorithms to convert invalid bit-strings into valid individuals. Therefore, we represent a cycle as sequence of vertices, so that each Hamiltonian cycle could be one of the possible $n!$ permutations of the vertices of $G = (V, E)$.

2.2 Fitness Function

Let us assume that $G = (V, E)$ is the graph representing the problem, with $V = \{v_0, v_1, \ldots, v_{n-1}\}$. Each individual in the population is a Hamiltonian cycle HC_k ($k = 1, 2, \ldots, n!$) in G. Let us define $V_n = \{p_1, p_2, \ldots, p_{n!}\}$ as the set of all possible permutations of the elements of V. The *fitness* function F is defined as:

$$F(HC_k) = \sum_{i=0}^{n-1} Cost(v_{p_k(i)}, v_{p_k((i+1) \bmod n)}) \quad \forall k \in [1, n!] \;. \tag{1}$$

where $HC_k = \langle v_{p_k(0)}, v_{p_k(1)}, \ldots, v_{p_k(n-1)} \rangle$ and $Cost(v_i, v_j)$ is the distance between each pair of cities v_i and v_j. Therefore, $F(HC_k)$ would be the tour-length of the Hamiltonian cycle HC_k.

2.3 PDGA: Distributed-Memory Approach

The basic idea of a GA is simply to generate a random population of individuals and apply mutation and crossover operators [4, 17] to a sub-population in each generation until an individual is found to satisfy some criteria for *fitness* or a maximum number of generations (iterations) is reached. It should be clear at this point, that there is an implicit parallelism in this basic GA and one could apply this basic GA to sub-populations in parallel in order to improve the performance of the algorithm.

Several *Parallel Genetic Algorithms* (PGA) have been proposed in the literature [14, 16, 12, 10, 6, 13]. In this paper we present the design and implementation of a *Parallel-Distributed Genetic Algorithm* (PDGA) based on a distributed-memory approach. Figure 1 shows the general architecture of the proposed PDGA, and the interaction between the MASTER and SLAVE tasks. The idea is to start-up a MASTER task whose main purpose is to control the

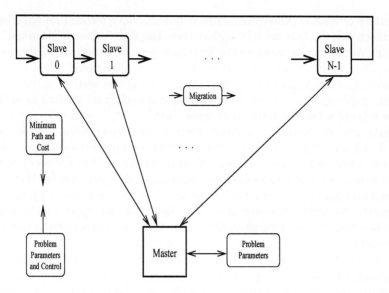

Fig. 1. PDGA: General Design

operations of the system by broadcasting the problem parameters and by synchronizing the operation of the SLAVE tasks. The master task spawns as many slave tasks as specified by the user and uniformly decomposes the population among those slaves. Every slave task receives the problem parameters from the master task, generates a random sub-population and starts executing a basic GA [9] within its own sub-population. During every generation and after each slave task has performed the crossover and mutation operations (and probably migration), each slave task sends back to the master task the minimum-cost Hamiltonian cycle found in that generation. Based on the results received, the master decides if it has reached the goal or if it is necessary to compute the next generation.

2.4 Problem Parameters

The input parameters required by the PDGA to solve the TSP are: a) **nslaves**: number of slave tasks, b) **n**: number of vertices (cities) in the graph, c) **p**: total number of Hamiltonian cycles in the population, d) **r**: crossover rate, e) **m**: mutation rate, f) **mut_int**: mutation interval, g) **mig_rate**: migration rate, h) **mig_int**: migration interval, and i) **min_threshold**: minimum tour-length accepted to stop the algorithm.

2.5 GA Operators

In this section we present the GA operators used for mutation and crossover, as well as the way we perform migration of sub-populations. The proposed PDGA is an example of what is called a *steady state GA* [2], which means that offsprings do not replace parents, instead they replace some less fit members of the population.

Many mutation operators for the TSP have been proposed in the literature [4, 10, 17, 8], and we selected one that is simple to implement when cycles are represented as a sequence of vertices: choose two of the vertices in the cycle at random and swap them. A mutation operation is performed as often as indicated by the mutation interval parameter (**mut_int**).

There are also several crossover operators that have been studied for the TSP [4, 10, 17, 8]. In order to apply a GA to permutation problems (like the TSP), it is necessary to use special operators to ensure that the recombination of two permutations (individuals) also yields another permutation [11]. Again, due to its simplicity, we chose here a crossover operator called *OX operator* [17], also known as *order crossover operator*. The crossover operation (mating) is performed every generation (iteration) with some percentage of the population (**r**) selected at random.

As we mentioned before, we use an *island model* to exploit differences among sub-populations. Each sub-population represents an island and we move the genetic material between islands in a specific way. The migration of a percentage of the population (**mig_rate**) is performed every **mig_int** generations between consecutive slave tasks as if they were connected in a logical ring topology. Each slave task S_i chooses the best individuals in their sub-population and sends them to its successor $S_{(i+1) \bmod nslaves}$ in the ring, and receives those individuals from its predecessor task $S_{(i-1) \bmod nslaves}$ and replaces the worst fit individuals with the ones just received. After several migrations one may observe that every slave task will have most of the best individuals found by the others. In addition this process will provide more diversity to the gene pool used for the search.

3 Test Environment (Experiments)

The proposed PDGA was implemented using PVM [5], which is a message passing library that allows programmers to write parallel/distributed applications using a NoW [1]. Figure 1 shows the message exchange that occurs between the MASTER and SLAVE tasks.

Tests were performed on a cluster of twenty (20) SUN SparcStations (Ultra Sparc 1) running *Solaris 2.6* and *PVM 3.4b*. For debugging and virtual machine configuration we used *XPVM 1.1*, an X Windows interface for the PVM console program. We tested our algorithm using 1, 2, 4, 8 and 16 slave tasks running in parallel, and different population sizes (p =128, 512, 1024, and 2048) changing only the number of slave tasks in order to observe and study the increase in performance (speedup). In addition, we studied the effect of variations in the mutation rate (**m**) and mutation interval (**mut_int**) on the total execution time and convergence rate of the proposed algorithm.

We used a library of TSP problems (TSPLIB) from Rice University that contains real data for graphs of different sizes. Vertices in the graphs represent cities in countries around the world, and the cost associated with every edge represents distances between each pair of cities. Test runs were performed with data from cities in Germany and Switzerland. We present the results obtained from data in Bavaria, Germany (**bavaria29.data**), represented by a complete undirected graph of $n = 29$ vertices, and with optimal tour-length of 2020 (problem parameters: $r = m = 0.5$, $mut\_int = 20$, $mig\_int = 25$, and $mig\_rate = 0.1$).

In order to determine the effectiveness of the proposed PDGA we selected TSP problems for which the minimum tour-length cost was known. In addition, we required the PDGA to stop if and only if the derived solution was within 1% off the optimal solution provided (for **bavaria29.data** $min\_threshold = 2040$).

4 Test Results and Analysis

4.1 Effect of Population Size and Number of Slave Tasks

In light of the results presented in figures 2 and 3, one can observe that as the size of the population increased the performance of the parallel version improved (proportionally) as the number of slave tasks increased. One may also notice that for small populations (i.e. $p = 128$) the performance decreased as the number of slaves increased. In addition, when the number of slaves increased the communication and synchronization overhead increased and the program spent more time in communication than performing useful computations.

We also notice from fig. 2 that one cannot keep increasing arbitrarily the number of slave tasks used. There is a maximum number of slaves after which the execution time increases instead of decreasing (i.e. 8 slaves). This result can be justified by and is in line with the fact that, depending on the population size and the number of slaves, if the number of slaves keeps increasing the communication and synchronization overhead will overcome the advantage of using the PDGA. The worst case happened when we used very small populations (i.e. $p = 128$). In these cases with a single slave (sequential program) we obtained better results (most of the time) than with 4, 8, or 16 slaves. This shows again that the proposed PDGA is better suited for large populations. It is also worth and important to mention that with small populations a parallel and distributed version of a GA is most likely to converge to a local minima due to a small gene pool. Figure 3 shows that the number of generations is virtually invariant to the increase in the number of slave tasks used.

We should emphasize that all tests were performed on a cluster of workstations connected through an Ethernet LAN (baseband network), which becomes a bottleneck when the number of workstations is increased. In order to obtain better performance as the number of workstations is increased, it would be necessary to replace the interconnection network used (Ethernet) by a high-performance network (like ATM, FastEthernet, etc.).

4.2 Effect of the Mutation Interval (mut_int)

Figure 4 presents the results of varying the mutation interval, and how it affects the convergence and total execution time of the PDGA. These tests were per-

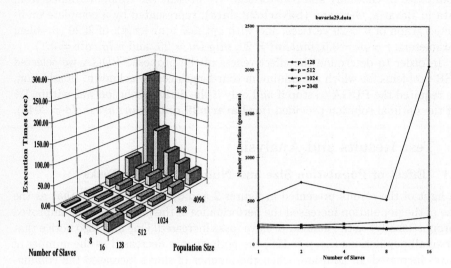

Fig. 2. (*Execution time*) as a function of (*number of slaves*) and (*population size*)

Fig. 3. (*Number of generations*) as a function of (*number of slaves*)

formed with 8 slaves, a population of $p = 1024$, and a mutation rate of $m = 10\%$. In the top of the graph we observe that the execution time decreases rapidly at first and then slowly reaches a minimum around $mut\_int = 30$. After that it begins to increase again. It is worthwhile to mention that for $mut\_int \leq 2$ the PDGA diverges. This behavior can be justified and explained by the fact that in this case mutations are performed almost on every generation, which does not allow the PDGA to get stabilized towards an optimal solution. The bottom of the graph shows a behavior similar to the behavior illustrated in the top of the graph, but with respect to the number of generations (iterations).

4.3 Effect of the Mutation Rate (m)

Figure 5 presents the results of varying the mutation rate, and how this affects the convergence and total execution time of the PDGA. These tests were performed with 8 slaves, a population of $p = 1024$, and a mutation interval, mut_int, equal to 20. In the top of the graph we observe that the execution time does not change drastically with variations in the mutation rate. However, the execution time slightly improves as we reduce the mutation rate below 30%, which seems logical because we are reducing the amount of work each slave must do to perform mutations. The bottom of the graph shows a behavior similar to the behavior illustrated in the top of the graph, but with respect to the number of generations.

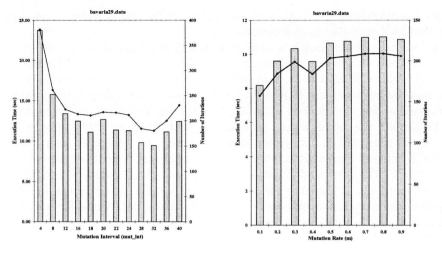

Fig. 4. Effect of the Mutation Interval (*mut_int*) on the PDGA

Fig. 5. Effect of the Mutation Rate (m) on the PDGA

5 Conclusions and Future Work

In this paper we introduced a parallel/distributed paradigm of a GA (PDGA) and we successfully applied it to the TSP. We used PVM to implement the proposed PDGA and we tested it on real data. We compared the proposed PDGA with a sequential GA and showed that said PDGA yielded very good results for large populations. Moreover, we also showed that, for a given population size, it is not recommended to increase the number of slave tasks arbitrarily. In particular, it has been shown in this paper that there is an optimum combination, in terms of population size and number of slave tasks, which leads to a better computational speed.

There are still many unanswered questions remaining. It is necessary to experiment with different encoding, crossover and mutation operators. We should also study in more detail how the mutation and migration rates affect the convergence speed of our PDGA. It would be interesting to test the PDGA with very large size problems (i.e. graphs with thousands of nodes), as well as other variations of the classical TSP. We are experimenting with several new heuristics as well as decisional rules to support computationally efficient migration to improve the gene pool. In particular, we are studying several migration schemes (i.e. random migration, alternating slaves, etc.). We are also analyzing other implementations of the basic PDGA using a different programming paradigm (i.e. MPMD) in order to reduce the intrinsic communication overhead and the need of synchronization exposed by the master-slave paradigm. We also must produce a deeper performance analysis considering granularity (communication vs. computation time and frequency) and scalability factors, in order to tune up our parallel/distributed implementation.

We will also be studying how to apply GAs and our experience with the TSP to other scheduling and networking problems (i.e. general routing, ATM routing and bandwidth allocation, load balancing, etc.). We are also developing a new implementation of our PDGA that will run over a network of heterogenous workstations directly attached to an ATM switch and with the latest version of PVM. Despite the lack of a deep performance analysis of the PDGA proposed, the results presented in this paper show the utility, versatility, efficiency and potential value of the proposed PDGA to solve and tackle NP-Complete problems. The PDGA presented is not only a good example of parallel/distributed programming, but it also shows how an inherent parallel algorithm can be efficiently implemented using well understood programming paradigms, flow control techniques, and load balancing approaches.

References

[1] T.E. Anderson, D.E. Culler, D.A. Patterson, and the NOW team. A Case of NOW (Networks of Workstations). *IEEE Micro*, 15(1):54–64, February 1995.
[2] L. Davis, editor. *Handbook of GA*. Van Nostrand Reinhold, NY, 1991.
[3] K. DeJong. *An Analysis of the Behavior of a Class of Genetic Adaptive Systems*. Phd's thesis, University of Michigan, Ann Arbor, MI, 1975.
[4] B.R. Fox and M.B. McMahon. Genetic Operators for Sequencing Problems. In Rawlins [12], pages 284–300.
[5] A. Geist, A. Beguellin, J. Dongarra, et al. *PVM: Parallel Virtual Machine. A User's Guide and Tutorial for Net. Parallel Comp.* The MIT Press, 1994.
[6] M. Gorges-Schleuter. Explicit Parallelism of Genetic Algorithms through Population Structures. *Parallel Problem Solving from Nature*, pages 150–159, 1991.
[7] J. Holland. *Adaptation in Natural and Artificial Systems*. U. Mich. Press, 1975.
[8] Z. Michalewicz. *Genetic Algorithms + Data Structures = Evolutionary Programs*. Springer-Verlag, Berlin, Germany, 1993.
[9] T.M. Mitchell. *Machine Learning*. Series in CS. McGraw-Hill, 1997.
[10] H. Mühlenbein. Evolution in Time and Space - The Parallel Genetic Algorithm. In Rawlins [12], pages 317–337.
[11] S. Rana, A.E. Howe, D. Whitley, and K. Mathias. Comparing Heuristic, Evolutionary and Local Search Approaches to Scheduling. In *Proc. of the 3rd Artificial Intelligence Planning Systems Conference*, 1996.
[12] G.J.E. Rawlins, editor. *Foundations of Genetic Algorithms*. Morgan-Kaufmann Publishers, Inc., San Mateo, California, 1991.
[13] T. Starkweather, D. Whitley, and K. Mathias. Optimization Using Distributed Genetic Algorithms. *Parallel Problem Solving from Nature*, 1991.
[14] R. Tanese. Distributed Genetic Algorithms. In *Proc. of the 3rd Int. Conf. on Genetic Algorithms*, pages 434–439. Morgan-Kaufmann Publishers, Inc., 1989.
[15] D. Whitley. A Genetic Algorithm Tutorial. *Stat. and Computing*, 4:65–85, 1994.
[16] D. Whitley and T. Starkweather. Genitor II: a Distributed Genetic Algorithm. *Journal Expt. Theory Artificial Intelligence*, 2:189–214, 1990.
[17] D. Whitley, T. Starkweather, and D. Shaner. The Traveling Salesman and Sequence Scheduling: Quality Solutions Using Genetic Edge Recombination. In Davis [2], chapter 22, pages 350–372.

Structural Biology Metaphors Applied to the Design of a Distributed Object System

Ladislau Bölöni, Ruibing Hao, Kyungkoo Jun, and Dan C. Marinescu

Computer Sciences Department
Purdue University
West Lafayette, IN 47907, USA
`boloni, hao, junkk, and dcm@cs.purdue.edu`

Abstract. In this paper we present the basic ideas of a distributed object middleware for network computing. We argue that when building complex systems out of components one can emulate the lock and key mechanisms used by proteins to recognize each other.

1 Overview

A number of biological analogies have found their way into computer science. Neural networks provide an alternative to von Neumann architecture [2–4], genetic algorithms [5] are used to solve optimization problems, mutation analysis was proposed for software engineering. The question we are concerned with is if these analogies can be further expanded to cover the way we build complex systems out of components and emulate genetic mechanisms and the immune system [10].

In this paper we present a design philosophy for network computing middleware, within the larger context of building complex systems out of ready-made components. Inherently, a complex computing system is heterogeneous and accommodating the heterogeneity of the hardware and the diversity of the software is a main concern of such a design. This heterogeneity is a constant source of pain for those intent on solving problems with computers. The alternative to heterogeneity and diversity is uniformity, but is this an exciting or a feasible alternative? Though we thoroughly enjoy the diversity of biological and social systems we complain when faced with the diversity of computer systems. Here we argue that hardware heterogeneity and software diversity can be accommodated provided that we transfer to computers themselves the tedious tasks of this endevour. Nature uses composition to build very complex forms of life and as we understand the principles and mechanisms that form the foundation of life we should try to emulate them to build more dependable and easy to use computing systems.

The cornerstones of the architecture we propose are metaobjects, entities that provide a description of the network objects, and software agents that perform operations on network objects. Examples of network objects are programs, data, hardware components including hosts and communication links, services, and so on. Metaobjects allows us to annotate the network objects.

The Bond project was triggered by a collaboration with structural biologists who provided the problems, the motivation to design a Virtual Structural Biology Laboratory, and need to learn some basic facts about the structure of biological macromolecules. The complex procedures needed for data acquisition, data analysis and model building for x-ray crystallography and electron microscopy are discussed elsewhere [9, 12–14]. Here we only note that processing of structural biology data involves large groups and facilities scattered around the world, complex programs that are modified frequently. Most computations are data intensive, they require the use of parallel and distributed systems.

2 Biological Systems

Nature uses composition to build extremely complex structures [1]. There are 20 amino-acids, the basic building blocks of life. The amino-acid sequence of a protein's peptide chain is called a *primary structure*. Different regions of the structure form local regular *secondary structure* such as alpha helices and beta strands. The *tertiary structure* is formed by packing such structural elements onto globular units called *domains*. The final protein may contain several polypeptide chains arranged in a *quaternary structure*. By formation of such tertiary and quaternary structures amino-acids far apart in the sequence are brought together in three dimensions to form a functional region, an *active site* [1]. The three dimensional structure of a protein determines its function, the disposition in space and the type of the atoms in a region of the protein provide a *lock* that can be recognized by other proteins that may bind to it, provided that they have the proper *key*. Living organisms *mutate*, the atomic structure of their cells changes and a *selection* mechanisms ensures the survival of those able to perform best their function.

Biological cells carry with them *genetic material, RNA and DNA* that describe the sequence of amino-acids in every protein. How this information is used to actually build the protein, the so-called *folding problem* is not elucidated yet. Moreover proteins with different sequences of amino-acids may fold to identical or very similar 3D atomic structures. They may have different properties, e.g. thermal stability, but they will perform the same functions because a fundamental principle is that the *structure determines the function of a biological specimen*. Another fundamental principle in genetics is *genetic economy*. *Symmetry* plays an important role in building complex cells. Spherical viruses have an icosahedral symmetry, they look like a soccer ball, are composed out of wedges with identical structure. The virus core contains the genetic material and has only one copy of the DNA or RNA sequence of an "unit cell", the wedge mentioned above.

3 Composition, Reflections, and Introspection

Let us now turn our attention to software composition. The idea of building a program out of ready made components has been around since the dawn of the computing age, backworldsmen have practiced it very successfully. Most scientific programs we are familiar with, use mathematical libraries, parallel programs use communication libraries, graphics programs rely on graphics libraries, and so on.

Modern programming languages like Java, take the composition process one step further. A software component, be it a package, or a function, carries with itself a number of properties that can be queried and/or set to specific values to customize the component according to the needs of an application which wishes to embed the component. The mechanism supporting these functions is called introspection. Properties can even be queried at execution time. *Reflection* mechanisms allow us to determine run time conditions, for example the source of an event generated during the computation. The reader may recognize the reference to the Java Beans but other *component architectures* exists, Active X based on Microsoft's COM and LiveConnect from Netscape to name a few.

Can these ideas be extended to other types of computational objects besides software components, for example to data, services, and hardware components? What can be achieved by creating *metaobjects* describing *network objects* like programs, data or hardware? We use here the term "object" rather loosely, but later on it will become clear that the architecture we envision is intimately tied to object-oriented concepts. We talk about network objects to acknowledge that we are concerned with an environment where programs and data are distributed on autonomous nodes interconnected by high speed networks.

The set of properties embedded into a metaobject provide a "lock and key" mechanism similar with the one used by proteins to recognize one another. In a *workspace* populated with metaobjects, one can envision mechanisms to link network objects together to form a *metaprogram*, a new metaobject, capable of carrying out a well defined computational task. Example: to link a data object to a software object we need to search the workspace for a metaobject describing a crystallographic FFT program able to compute the 3D Fourier transform of a symmetric object with a known symmetry, given a lattice of real numbers describing the "unit cell", the building block of the object. At each step, the selection process succeeds if the target metaobject possesses a set of "keys", corresponding to the desired properties needed for composition. This is a deliberate example, anyone familiar with FFTs recognizes that creating the metaobjects describing the elements discussed above is a non trivial task.

We expect a metaobject to contain *genetic information*, i.e. a description of all relevant properties of a network object . This information must be in a form suitable for machine processing. For example a metaobject associated with a software component should include a description of the functions and interfaces of that component using a descriptive language like IDL, Interface Description Language. IDL reveals only the interfaces of an object and does not specify how these properties are to be *folded* into an actual implementation. The genetic information associated with a data object should reveal its ancestry, the characteristics of the sensor that has generated the data, or provide links to the program that has produced the data and to its input data objects. The genetic information would allow an autonomous agent to generate an actual implementation of the code in case of a software component, or a human contemplating the results of a sequence of computations to trace back decisions made at some point in the past.

We also need to add new properties to a metaobject. For example when an agent discovers that a processor executes incorrectly a sequence of instructions this new property should be added to the metaobject describing that particular class of processors or processors with Ids within a certain range. A fundamental principle is that acquired traits take precedence over inherited ones.

Both hardware and software are created as a result of an *evolutionary* process. Occasionally, new programs, or microprocessors, are created from scratch, but often, new versions are upgrades of existing objects, that inherit many characteristics of the older versions. Inheritance has the potential to simplify the task of building metaobjects and agents capable to manipulate them intelligently.

Care must be taken to expose in the metaobject only *stable properties* of the network resources. For example the amount of main memory installed on a system is a stable property, though it may change, such changes are likely to be infrequent, while the load placed upon the system varies rapidly, it is a *transient property* and should not be exposed.

Once we recognize that creation of metaobjects is a difficult task we have to ask ourselves if metaobjects and the network resources need to be tightly or loosely coupled with each other. The tightly coupled approach would require that the network object and the metaobject be kept together to ensure consistency. There are fundamental flaws with the tightly coupled argument. First, this "ab-initio" approach would require re-creation of legacy components, software, hardware and data, to fit our scheme. Second, information would be unnecessarily duplicated and confidential information compromised.

By virtue of the arguments discussed above a metaobject should only have a *soft link* to the network object it describes. Different properties of the network object may be accessible on a need to know basis, e.g. the source code may be available only to those who have the need for it. The metaobject associated with a program may contain attributes describing the function of the program, its input and output. It should also provide references or pointers to the: source code, the executables for different platforms, the human readable documentation of the program, the implementation notes, an error log, and so on. Following the principle of genetic economy, components of the metaobject would be shared among all the users of the object.

The price to pay for the distributed object approach is that inconsistency between the metaobjects and the network objects are occasionally unavoidable. We argue that in a network rich environment catastrophic consequences of such inconsistencies are unlikely to occur. Moreover, whenever possible, discovery agents should update the metaobjects.

In summary, we propose to create metaobjects describing properties of network objects. These properties should reveal how objects can be composed together using a lock and key mechanism and support a selection mechanism to eliminate components that do not perform their functions well. Linking metaobjects together can only be done based upon a universally accepted *taxonomy of objects* and properties, and a given context. We acknowledge the fact that a considerable amount of work in the area of knowledge sharing still remains to be done, but we believe that a system built along the principles discussed above will allow a larger segment of the population to use computers to solve complex tasks.

4 An Architecture for Communicating Objects

This section introduces Bond, an agent-based, object-oriented middleware for network computing. The goals of the Bond system are to: (a) facilitate access to network resources, (b) support collaborative activities, and (c) accommodate software diversity and hardware heterogeneity.

Metaobjects and agents are important classes of Bond objects. Metaobjects provide synthetic information about network objects and agents use this information to access and manipulate network objects on behalf of users.

The first question we have to answer when designing an agent-based, distributed-object architecture is the level of integration of agents and objects. Should the agents must be tightly integrated into the architecture, or can they be added on to an existing distributed-object framework? To answer this question we examine two critical issues, communication and the complexity of the system.

Communication is a critical issue in a distributed system and though procedure-oriented and message-oriented paradigms are arguably dual to each other, most distributed-object systems favor remote method invocation and synchronous communication. The conventional wisdom is that remote method invocation is more efficient than message passing and that asynchronous communication is considerably more difficult to program and debug than synchronous one. Thus if we want to design a complex system, capable to evolve and to exhibit a good level of performance the only choice is synchronous remote method invocation.

On the other hand, *agent-oriented programming* favors the message-oriented approach and asynchronous communication. Since agents can exhibit a complex behavior, remote method invocation is too restrictive and inter-agent communication is based upon message passing, using agent-communication languages like KQML. An agent has the ability to make autonomous decisions and should be able to adapt its behavior to the response to a message or to the lack of a response. In a *wide-area distributed system*, components including other agents or objects, and communication links may fail, and the lack of a response should still lead to predictable behavior. The ability to respond to component failure is an important argument in favor of integrating software agents into network computing systems.

An approach often encountered in the literature, is to support remote method invocation for regular objects and message passing for agents. We believe that this dichotomy increase the software complexity and creates additional problems when regular objects and agents need to communicate with each other. Our solution is to design a pure message passing system with all objects communicating using KQML. An agent is just another object extended with the ability to make autonomous decisions, it can communicate with any other object and the communication fabric is common to the entire object population.

Important as it may be, the fact that every object "speaks" KQML, in other words understands the syntax of a KQML message and is able to parse it, does not guarantee that two objects can cooperate with each other. An object should be able to "understand" semantically a message and respond to it in a coherent manner. The granularity of objects can differ widely, ranging from simple objects like an address, a date, or a message, to complex objects like a scheduling agent or an authentication server. A server is an active object capable to provide a well-defined set of services, though it may be very complex a server does not have the autonomy an agent must exhibit. The ability of an object to "understand" a message is also very different.

To solve the problem of semantical understanding of a message we introduce the concept of *subprotocols*. A subprotocol is a "dialect" understood by a subset of objects. A subprotocol is a closed set of messages, an object understands semantically every single message in a subprotocol and it is able to respond with a message in the same subprotocol. This guarantees the ability of agents to cooperate. An agent may use a *discovery subprotocol* and determine a set of common subprotocols with another agent or object. Once a group of agents discover that they all understand a subprotocol any agent sending a message belonging to the common subprotocol is guaranteed to get a meaningful response. An object can determine with ease if it understands a message because every message is stamped with the subprotocol it belongs to. A subprotocol is inherited based upon object hierarchy. When a new object extending an existing one is created, in addition to the subprotocols inherited, new ones can be defined. The only subprotocol understood by all objects is the *property access subprotocol*, all agents understand the *agent control subprotocol*. The ability of an existing agent may be enhanced with *probes*, special objects capable to handle other aspects of a complex system, e.g. security, monitoring, accounting, and so on. We also have the ability to generate subprotocols dynamically once the communication patterns of a group of agents are known.

4.1 Component-based agents

A first distinctive feature of the Bond architecture is that agents are native components of the system. This guarantees that agents and objects can communicate with one another and the same communication fabric is used by the entire population of objects. We are thus forced to pay close attentions to the software engineering aspects of agent development, in particular to software reuse. We decided to provide a framework for assembly of agents out of components, some of them reusable. This is possible due to the agent model we overview now.

We view an agent as a *finite state machine*, with a *strategy* associated with every state, *a model of the world*, and an *agenda*. Upon entering a state the strategy or strategies associated with that state are activated and various actions are triggered. The model is the "memory" of the agent, it reflects the knowledge the agent has access to, as well as the state of the agent. Transitions from one state to another are triggered by internal conditions determined by the completion code of the strategy, e.g. success or failure, or by messages from other agents or objects.

The finite state machine description of an agent can be provided at multiple *granularity levels*, a course-grain description contains a few states with complex strategies, a fine-grain description consists of a large number of states with simple strategies. The strategies are the reusable elements in our software architecture and granularity of the finite state machine of an agent should be determined to maximize the number of ready-made strategies used for the agent. We have identified a number of common actions and we started building a strategy database. Examples of actions packed into strategies are: starting up one or more agents, writing into the model of another agent, starting up a legacy application, data staging and so on. Ideally, we would like to assemble an agent without the need to program, using ready-made strategies.

Another feature of our software agent model is the ability to assemble an agent dynamically from a *blueprint*, a text file describing the states, the transitions, and the model of the agent. Every Bond-enabled site has an *agent factory* capable to create an agent from its blueprint. The blueprint can be embedded into a message, or the URL of the blueprint can be provided to the agent factory. Once an agent was created, the agent control subprotocol can be used to control it from a remote site.

In addition to favoring reusability, the software agent model we propose has other useful features. First, it allows a smooth integration of increasingly complex behavior into agents. For example, consider a scheduling agent with a mapping state and a mapping strategy. Given a task and a set of target hosts capable to execute the task, the agent will map the task to one of the hosts subject to some optimization criteria. We may start with a simple strategy, select randomly one of the target hosts. Once we are convinced that the scheduling agent works well, we may replace the mapping strategy with one based upon an inference engine with access to a database of past performance. The scheduling agent will perform a more intelligent mapping with the new strategy, without altering other aspects of its behavior, provided that the model is not altered by the new strategy.

Second, the model supports agent mobility. A blueprint can be modified dynamically and an additional state can be inserted before a transition takes place. For example a *suspend state* can be added and the *suspend strategy* be concatenated with the strategy associated with any state. Upon entering the suspend state the agent can be migrated elsewhere. All we need is to send the blueprint and the model to the new site and make sure that the new site has access to the strategies associated with the states the agent may traverse in the future. The dynamic alteration of the finite state machine of an agent can be used to create a *snapshot* of a group of collaborating agents and help debug a complex system.

4.2 Metaobjects

Metaobjects are other critical components of our system, they glue together network objects and agents. Network objects or resources are: hardware platforms, communication links, sensors, programs, libraries, services, data files, and so on. The function of metaobjects is to provide the knowledge needed by the software agents to manipulate network objects.

The metaobjects form a distributed knowledge base and their defining attribute is that besides a name and a unique Id their structure is fluid, it changes with time to reflect our knowledge about a world populated with network objects. The ever-changing relationship amongst them is the other important trait of metaobjects. Metaobjects are suitable to represent knowledge while traditional objects in the object-oriented programming sense have well-defined inheritance and best represent entities with a well-known set of properties. There are no methods specific to a metaobject other than to set and reset their attributes. Agents are capable to manipulate metaobjects and invoke functions to determine if the metaobject provides the information needed by a specific strategy. The metaobjects provide the means to integrate object-oriented and agent-oriented programming.

To illustrate these ideas consider the scheduling agent example discussed earlier. The agent must be able to determine if a particular host is suitable for a computational task. We have the choice of creating a regular object with a number of attributes to reflect the most relevant properties of a host e.g. its IP address, the type and Id of the processor, the number of processors, the amount of main memory, and so on. We may ask ourselves what is a relevant property for a host, for example are the type of the graphics cards, or the number of geometric engines important? The answer is that in some cases the graphics hardware may not be present and in other cases we may not be interested in graphics when scheduling the task. Rather than defining a *host object* with a potentially very large set of attributes we define a metaobject with no other property but a name and add dynamically properties to it as needed.

Some of the traits of the network object are discovered at the time the metaobject was created others reflect changes in the network object itself discovered by some agent at a later point in time. For example a *discovery agent* will fill in the metaobject associated with a host, a number of ¡attribute, attribute-value¿ pairs, including the type and the Id of the processor. A selection strategy of the scheduling agent will match the attributes desired by the task with the ones offered by a specific host as reflected by the corresponding metaobject.

In this paper we describe an object-oriented, agent-based, distributed object system based upon active messages. The system exercises the metaphors discussed above. An alpha release of the Bond system is available for downloading at http://bond.cs.purdue.edu.

5 Acknowledgments

The work reported in this paper is partially supported by the National Science Foundation grants BIR-9301210 and MCB-9527131, by the California Institute of Technology, under the Scalable I/O Initiative, by a grant from Intel Corporation, and by the Computational Science Alliance and the NCSA at the University of Illinois.

References

1. C. Branden and J. Toose. *"Introduction to Protein Structure"* Garland Publishing 1991.
2. F. Rosenblatt, *"The perceptron: A perceiving and recognizing automaton"*, 85-460-1, Project PARA Cornell Aeronautical Laboratory Ithaca, NY, 1957.
3. J. Hopfield *Neural Networks and Physical Systems with Emergent "Collective Computational Abilities"* Proceedings of the National Academy of Sciences, 1982 vol.79, pp. 2554.
4. J. L. McClelland and D. E. Rumelhart, *"Parallel Distributed Processing: Explorations in the Microstructures of Cognition, Volume 2: Psychological and Biological Models"*, MIT Press Cambridge, Mass. 1986
5. J. R. Koza, *"Genetic Programming: On the Programming of Computers by Means of Natural Selection"*, MIT Press 1992.
6. L. Bölöni. *"Bond Objects – A White Paper."* Department of Computer Sciences, Purdue University CSD-TR #98-002.
7. L. Bölöni, Ruibing Hao, K. K. Jun, and D.C. Marinescu. *"Subprotocols: an object-oriented solution for semantic understanding of messages in a distributed object system"*, September 1998, (submitted).
8. K. Mani Chandy. *"Caltech Infospheres Project Overview: Information Infrastructures for Task Forces."*
9. M.C. Cornea Hasegan, Z. Zhang, R.E. Lynch, D. C. Marinescu, A. Hadfield, J.K. Muckelbauer, S. Munshi, L. Tong and M.G. Rossmann. *"Phase Refinement and Extension by Means of Non-crystallographic Symmetry Averaging using Parallel Computers."* Acta Cryst D51, pp 749-759, 1995
10. J.C. Fabre and T. Perennou. *"A Metaobject Architecture for Fault Tolerant Distributed Systems: The FRIENDS Approach"* IEEE Trans. on Computers, Vol 47, No 1, pp 78-95, 1998.
11. T. Finin, et al. *"Specification of the KQML Agent-Communication Language"* DARPA Knowledge Sharing Initiative draft, June 1993.
12. R. E. Lynch, D.C. Marinescu, H. Lin, and T. S. Baker. *"Parallel Algorithms for 3D Reconstruction of Asymmetric Objects from Electron Micrographs"* September 1998, (submitted).
13. I. Martin, D.C. Marinescu, R. E. Lynch, and T. S. Baker. *"Identification of Spherical Virus Particles in Digitized Images of Entire Micrographs"* Journal of Structural Biology, 120, pp 146-157, 1997.
14. I. Martin and D.C. Marinescu *"Concurrent Computations and Data Visualization for Structure Determination of Spherical Viruses"* IEEE Computational Science and Engineering, 1998 (in press).
15. R. J. Stroud. *"Transparency and Reflection in Distributed Systems"* ACM Operating Systems vol. 22, no. 2 pp. 99-103, 1993.

Proceedings of the

Seventh International Workshop on Parallel and Distributed Real-Time Systems

held in conjunction with
Second Merged Symposium IPPS/SPDP 1999
13th International Parallel Processing Symposium &
10th Symposium on Parallel and Distributed Processing

April 12–13, 1999
San Juan, Perto Rico

Sponsored by
U.S. Naval Surface Warfare Center

Message from the Program Chairs

The following section of this volume contains the proceedings of the Seventh International Workshop on Parallel and Distributed Real-Time Systems (WPDRTS). It features manuscripts that demonstrate refereed, original, unpublished research pertaining to real-time systems that are parallel and/or distributed, representing experimental and commercial systems, their scientific and commercial applications, and theoretical foundations. Topics addressed in this section include:

Fault Tolerance	Multimedia
Quality of Service	Applications
Scheduling	Databases
Communication and Networking	Tools

In addition to the regular paper presentation, the workshop also features three Keynote Speeches, "Distributed Real-Time Systems Architecture - Rules & Tools" by Douglass Locke, Lockheed Martin, "Towards Flexible Real-Time Processing by developing Imprecise Computation Technique" by Kenji Toda, Electrotechnical Laboratory, and "Composability in the Time-Triggered Architecture" by Hermann Kopetz, Technical University of Vienna; two Panel Sessions, "Fundamental Problems in Distributed Real-Time Systems" organized by Maarten Boasson, Hollands Signaalapparaten B.V. and "Distributed Real-Time Computer Systems in the Real World" organized by E. Douglas Jensen, MITRE Corporation; and a session comprised of Invited Papers.

We would like to thank all who have helped to make WPDRTS a success. In particular, The Program Committee members carefully reviewed the submitted papers. The General Chair, David Andrews, and the Steering Committee members provided valuable guidance and support. The efforts of the publicity chairs are also greatly appreciated. We would like to thank the U.S. Naval Surface Warfare Center who generously sponsored the workshop. Finally, we thank the IPPS organizers for providing an ideal environment in San Juan.

Binoy Ravindran	**Jan Gustafsson**	**Hiroaki Takada**
binoy@vt.edu	jan.gustafsson@mdh.se	hiro@ertl.ics.tut.ac.jp

Program Chairs
7th International Workshop on Parallel and Distributed Real-Time Systems

Program and Organizing Committees

General Chair
David Andrews, University of Arkansas, USA

Steering Committee
Dieter K. Hammer, Eindhoven University of Technology, The Netherlands
Bruce Lewis, US Army, USA
Mike Masters, NSWCDD, USA
Lalit M. Patnaik, Indian Institute of Science, India
Behrooz A. Shirazi, The University of Texas at Arlington, USA
John A. Stankovic, University of Virginia, USA
Chair: Lonnie R. Welch, The University of Texas at Arlington, USA

Program Chairs
Binoy Ravindran, Virginia Tech, USA
Jan Gustafsson, Mälardalens University, Sweden
Hiroaki Takada, Toyohashi University of Technology, Japan

Program Committee
Emile Aarts, Eindhoven University of Technology, The Netherlands
Bruno Achauer, University of Linz, Austria
Gul Agha, University of Illinois at Urbana-Champaign, USA
Mehmet Aksit, Twente University of Technology, The Netherlands
Sriramulu Balaji, Indian Space Research Organization, India
Azer Bestavros, Boston University, USA
Maarten Boasson, Hollands Signaalapparaten B.V., The Netherlands
Carl Bruggeman, The University of Texas at Arlington, USA
Kyung-Hee Choi, Ajou University, Korea
Jin-Young Choi, Korea University, Korea
Juan A. De La Puente, Universidad Politecnica de Madrid, Spain
Klaus Ecker, University of Clausthal, Germany
Martin Gergeleit, GMD, Germany
Wolfgang Halang, University of Hagen, Germany
Hans Hansson, Mälardalens University, Sweden
Kenji Ishida, Hiroshima City University, Japan
Farnam Jahanian, University of Michigan, USA
Kevin Jeffay, University of North Carolina, Chapel Hill, USA
Rakesh Jha, Honeywell Technology Center, USA
Edwin de Jong, Hollands Signaal Apparaten, The Netherlands

Jan Jonsson, Chalmers University of Technology, Sweden
Joerg Kaiser, University of Ulm, Germany
Yoshiaki Kakuda, Hiroshima City University, Japan
Carl Kesselman, California Institute of Technology, USA
Tei-Wei Kuo, National Chung Cheng University, Taiwan
Kwei-Jay Lin, University of California, Irvine, USA
Miroslaw Malek, Humboldt University, Germany
Rajib Mall, Indian Institute of Technology, India
Scott Midkiff, Virginia Tech, USA
Sang Lyul Min, Seoul National University, Korea
Ichiro Mizunuma, Mitsubishi Electric, Japan
Matt Mutka, Michigan State University, USA
Klara Nahrstedt, University of Illinois at Urbana-Champaign, USA
Tatsuo Nakajima, JAIST, Japan
Yukikazu Nakamoto, NEC, Japan
Edgar Nett, GMD, Germany
Joseph Kee-Yin Ng, Hong Kong Baptist University, Hong Kong
Brian Noble, University of Michigan, USA
David O'Hallaron, Carnegie Mellon University, USA
Peter Puschner, Vienna University of Technology, Austria
Ragunathan Rajkumar, Carnegie Mellon University, USA
Krithi Ramamritham, University of Massachusetts, USA
Onno van Roosmalen, Eindhoven University of Technology, The Netherlands
Diane Rover, Michigan State University, USA
Helmut Rzehak, University of the Federal Armed Forces, Germany
William H. Sanders, University of Illinois at Urbana-Champaign, USA
Karsten Schwan, Georgia Tech, USA
Sang Son, University of Virginia, USA
Neeraj Suri, Boston University, USA
Kazunori Takashio, University of Electro-Communications, Japan
Kenji Toda, Electrotechnical Laboratory, Japan
Wang Yi, Uppsala University, Sweden
Tomohiro Yoneda, Tokyo Institute of Technology, Japan
Farn Wang, Academia Sinica, Taiwan
Paul Werme, NSWCDD, USA
Wei Zhao, Texas A&M University, USA

Publicity Chairs

Marie-Agnès Peraldi-Frati, I.U.T Nice, France
Mitch Thornton, University of Arkansas, USA
Yoshinori Yamaguchi, Electrotechnical Laboratory, Japan

Publications Chair

John Huh, University of Texas at Arlington, USA

Reviewers

Emile Aarts
Bruno Achauer
Mehmet Aksit
James M. Baker, Jr.
Azer Bestavros
Maarten Boasson
Carl Bruggeman
Jin-Young Choi
Kyunghee Choi
Klaus Ecker
Andreas Ermedahl
Martin Gergeleit
P. C. N. van Gorp
Satrajit Gupta
Jan Gustafsson
Wolfgang A. Halang
Dieter Hammer
Hans Hansson
Pao-Ann Hsiung
Kenji Ishida
Mark Jones
Edwin de Jong
Jan Jonsson
Joerg Kaiser
Yoshiaki Kakuda
Tei-Wei Kuo
David C. Lee
Chenyang Lu
Miroslaw Malek
Rajib Mall
Scott F. Midkiff

Sang Lyul Min
Ichiro Mizunuma
Matt W. Mutka
Klara Nahrstedt
Tatsuo Nakajima
Yukikazu Nakamoto
Edgar Nett
Joseph Kee-Yin Ng
Brian Noble
Ernst Nordstrom
David O'Hallaron
Lalit Patnaik
Juan A. De La Puente
Peter Puschner
Krithi Ramamritham
Binoy Ravindran
Onno van Roosmalen
Diane T. Rover
Helmut Rzehak
Karsten Schwan
Behrooz Shirazi
Sang H. Son
Jack Stankovic
Hiroaki Takada
Kazunori Takashio
Christian Tschudin
Farn Wang
Wang Yi
Levent Yilmaz
Tomohiro Yoneda

Building an Adaptive Multimedia System Using the Utility Model

Lei Chen[1], Shahadat Khan[2], Kin F. Li[3], and Eric G. Manning[3]

[1] Siara Research Canada, 305-8988 Fraserton Court, Burnaby, B.C., Canada V5J 5H8
glchen@siara.com
[2] Infranet Solutions, Suite 409-100 Park Royal, West Vancouver, B.C. Canada V7T 1A2
skhan@infranetsolutions.com
[3] University of Victoria, Victoria B.C., Canada V8W 3P6
{kinli, emanning}@sirius.uvic.ca

Abstract. We present our experience of building a prototype system based on the Utility Model for adaptive multimedia. The Utility Model is proposed to capture the issues and dynamics in multi-session multimedia systems where the quality of service (QoS) of individual sessions is adapted to dynamic changes of available resources and of user preferences. We present the design and implementation of our prototype multimedia system, and report experimental results. We demonstrate that the Utility Model supports two types of adaptation: reactive adaptation for systems where only a subset of the applications follows the adaptation model, and proactive adaptation for systems where all the applications follow the adaptation model. Our results demonstrate that the Utility Model may be effectively used for dynamic quality adaptation in real-time multimedia systems.

1 Introduction

Multimedia applications require system support with quality of service (QoS) guarantees. For instance, in order to support a video stream at a refresh rate of 10 frame/sec, the stream handler has to be scheduled onto the CPU once every 0.1 second. In a traditional video transmission system, it is possible that the session is getting sufficient network bandwidth, but cannot utilize this bandwidth effectively because the system cannot allocate sufficient CPU cycles to decode and render the video stream. Thus, to enforce QoS guarantees effectively, resources in a multimedia system must be managed in an integrated way. Designing an adaptive multimedia system with integrated resource management requires a comprehensive understanding of the problems, issues and dynamics within such a system.

In a dynamic system environment, multimedia applications are potentially adaptive. For example, if there is enough network bandwidth available, one can have real-time video transmitted at 30 frame/sec; otherwise, one may reduce the bandwidth requirement to half by reducing the rate to 15 frame/sec. How do we design a multimedia system to exploit the potential for adaptation of multimedia applications? In this article, we present the design, implementation and experimental results for a prototype system based on the Utility Model for adaptive multimedia.

The rest of this article is organized as follows. Section 2 discusses some related work. Section 3 presents a brief overview of the Utility Model. Section 4 presents our prototype implementation. Section 5 presents our experiments and performance results. Finally, section 6 concludes the article.

2 Related Work

In recent years, there has been significant interest in the understanding of adaptive multimedia systems[1,3,7,8,9,11,12]. Schreier and Davis proposed the *Benefit Model* for adaptation of quality attributes of a single-user multimedia application[12]. The user expresses her media quality preferences by associating benefit functions with performance attributes such as audio quality, video frame rate and audio/video synchronization. The objective is to maximize the benefit by adjusting the amount of processing, communication and storage resources provided to the application.

Moser presented an Optimally Graceful QoS Degradation (OGQD) Model, where the realized QoS of a multimedia session is gracefully degraded to conform to changes in resource availability[9]. Suppose a multimedia service consists of several elementary services such as audio input, audio output, video input and video output. For each elementary service, the system has to find a suitable implementation and the operating QoS of that implementation, depending on available resources. The goal of the OGQD system is to provide multimedia service with minimum degradation from desired levels of QoS.

Since the Benefit Model and OGQD are both targeted to the adaptation problem of a *single* multimedia session, they do not capture issues peculiar to systems with multiple concurrent sessions. For example, they do not address the following:
1. how to specify and achieve the adaptation objective of a multi-session system,
2. how to incorporate the notion of relative importance of different sessions, and
3. how to deal with session admission.

In [4], we presented the Utility Model — a mathematical model to capture the issues of resource management in adaptive multimedia systems with multiple concurrent sessions, where the quality of individual sessions is dynamically adapted to the available resources and to the run-time user preferences. In this paper, we present our experience with a prototype multimedia system designed using the Utility Model.

In the context of operating system design to support continuous media, Kawachiya *et al.* of the Keio Multimedia Project have proposed a processor execution model, called Q-Thread, where a periodic continuous media task can specify the tolerable ranges for the period of scheduling and computation time in each period, and the system controls the operating quality, specified by period and computation time, of tasks dynamically[3]. One of the main challenges for such a system is as follows: At a given system state, how does the system find an appropriate operating quality (such as period and computation time) for each task? Our proposal of the Utility Model is an effort to address this issue.

McCanne proposed an adaptive multimedia transport system using the Receiver-driven Layered Multicast (RLM)[8]. In RLM, a multimedia source transmits a stream into multiple subscription groups, each with different components of the stream, and receivers adapt the quality of reception by joining and leaving subscription groups. We use the concepts of RLM for implementing a video-on-demand application in our prototype system.

3 The Utility Model

Figure 1 illustrates an adaptive multimedia system (AMS) where n sessions share m resources. When sufficient resources are available, the system should provide every session the highest desired quality for everyone. However, when available resources cannot support the highest desired quality, the system must compromise the operating quality of certain sessions. For example, when a higher-priority video-conference session needs better quality, compromising the quality of a lower-priority virtual environment session is in order. For a video session, this may mean reduced refresh rate or reduced resolution, and for a virtual environment, this may mean lower levels of detail of three dimensional objects.

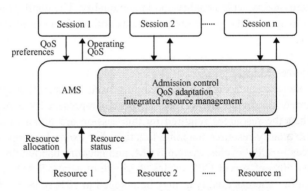

Figure 1: Adaptive multimedia system requirements.

Adaptation in a multimedia system may be triggered by many events, such as change of system load, change of network load or congestion/failure in the network, and dynamic change of users' requirements. An AMS has the following requirements:
- The users must be able to specify their quality requirements/preferences in a flexible way.
- The system must know the current status of system resources.
- The system must have the option to accept or to reject new session requests.
- The system must provide quality adaptation and integrated resource management in order to make best use of the available resources.

In order to capture the issues of resource management in adaptive multimedia systems with multiple concurrent sessions, where the quality of individual sessions is dynamically adapted to the available resources and to the run-time user preferences, we proposed a mathematical model, called the Utility Model[4]. It provides a unified and computationally feasible way to solve the admission problem for new multimedia sessions, and the dynamic quality adaptation and integrated resource allocation problems for existing sessions.

3.1 Concepts

The Utility Model is based on the concepts of quality profile, quality-resource mapping, session and system utility, and system resource constraints.

Quality Profile
The quality profile of a session is a sequence of acceptable operating qualities in increasing order of preference (from minimum acceptable quality to highest desired quality). For example, suppose a multimedia session consists of three media: audio, video and still image. Table 1 shows a possible set of user choices where the session may have one of three operating qualities: gold, silver or bronze. Suppose user i's quality profile is expressed as $\mathbf{P}_i=(q_3, q_2, q_1)$ where q_3 is the highest desired quality, and q_1 is the worst quality the user is ready to accept and pay for. Figure 2(a) shows the user profile where each operating quality is represented as a vector of three media qualities: audio quality, video quality and image quality. The quality profile provides a flexible way to specify the quality requirements and preferences of the user of a particular session.

Quality-resource Mapping
We assume the existence of a unique mapping from an operating quality to the resources required to provide that quality. Suppose q_i represent the operating quality of session i. Then the resources required by this session are given by $r(q_i)$ where $r(\cdot)$ represents the quality-resource mapping function. For example, suppose a system has three resources — CPU cycles, system RAM, and network bandwidth. Quality q_3 may be mapped to 50% of server CPU cycles, 1 Mbyte of server RAM, and 5 Mbps network bandwidth; whereas quality q_1 may be mapped to 10% of server CPU cycles, 10 Kbyte of server RAM, and 1 Mbps network bandwidth. (Quality-resource mappings are addressed in research projects at Keio University. In [10], Nishio *et al.* present a quality-resource table based on measurements of the RT-Mach system.)

Table 1: Flexible choice for the user of adaptive multimedia system.

	Bronze (q_1)	Silver (q_2)	Gold (q_3)
Audio	Mono	Stereo	Surround
Video	Low resolution, no color	High resolution, no color	High resolution, Color
Image	Low resolution	Medium resolution	High resolution

Session and System Utility
Any adaptive system must have an objective, which decides the direction of adaptation in any given situation. We assume that the objective of an AMS is to maximize *system*

utility, which is a weighted sum of the individual *session utilities*. The concept of utility is taken from economics, where it means the capacity of a commodity or a service to satisfy some human want, and the goal of an economic system is to allocate resources to maximize utility.

We assume the existence of an utility function which maps a session's operating quality q_i to session utility $u_i(q_i)$. For simplicity, we assume that system utility objective is to maximize the system utility function U given by:

$$U = \Sigma\, u_i(q_i). \qquad (1)$$

For example, in a multimedia service provider application, the session utility may be the revenue paid by individual sessions (such as $100 for gold, $60 for silver and $40 for bronze), and the system utility is the total revenue earned by the system.

System Resource Constraints
At any moment, the system state has to obey the following *system resource constraints*: For each resource, the sum of the quantities of the resource allocated to all the sessions cannot exceed the total available quantity of the resource. Suppose the available system resources are expressed as a vector R. The resource constraints are expressed as

$$\Sigma\, r(q_i) \le R. \qquad (2)$$

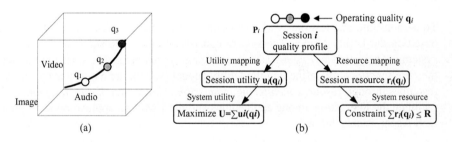

Figure 2: (a) Quality profile and (b) the main concepts of the Utility Model.

3.2 Using the Model

The main concepts of the Utility Model are illustrated in Figure 2(b):
- Each user specifies a quality profile, set of acceptable session operating qualities.
- A session's operating qualities are mapped uniquely to required resources.
- A session's operating qualities are mapped uniquely to session utilities.
- The system utility is the sum of all session utilities.
- The system is subject to the system resource constraints.

Now the main problem in an adaptive multimedia system is as follows: *Find* the operating quality q_i of each session i which *maximizes* the system utility U under the system *resource constraints*. In [4], we show that this problem can be mapped to a multiple-choice multi-dimension knapsack problem (MMKP). We also provide two solutions for the MMKP: an exact algorithm BBLP using branch-and bound searching, and a heuristic HEU for fast and approximate solution. The latter solution is suitable for time-critical applications such as real-time multimedia systems.

The Utility Model may be used in session admission, dynamic quality adaptation and integrated resource management in real-time multimedia systems.

- *Session Admission*: Suppose the system has n sessions, and a user requests another session. Provided that there is enough system resources available for the new session, the achievable system utility U' with $(n+1)$ sessions is compared to current system utility U with n sessions. If $U'>U$, then the new session should be admitted; otherwise the new session should be rejected.
- *Quality Adaptation*: Suppose a session is dropped or the amount of resource available is changed due to external/uncontrollable reasons. The system computes the solution of the Utility Model under the new constraint, and some of the existing sessions may have to change their operating qualities.
- *Integrated Resource Management*: Since the Utility Model considers allocation of all the system resources in an integrated way, it would avoid situations such as a session getting sufficient network bandwidth, but cannot utilize this bandwidth effectively because the system cannot allocate sufficient CPU cycles to decode and render the video stream.

4 Prototype Implementation

To demonstrate and test our work, we designed and implemented a prototype system based on the Utility Model. The system consists of an off-the-shelf Linux operating system, a *Utility Model Engine* (UME), and some applications. The UME[1] implements the core functionality defined by the Utility Model (see Figure 3): it manages system resources through *resource monitoring*, *admission control*, and *quality adaptation*. To facilitate the interaction among applications and the UME, an application level proxy, QoS Agent, is also defined.

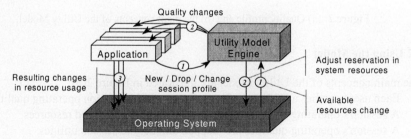

Figure 3: Components and interactions in the prototype system.

In our prototype system, applications are *admitted*, system resources are *monitored*, and applications' operating qualities are dynamically *adapted*. The admission

---

[1] For performance reasons, HUE is implemented in this prototype system, but BBLP is not.

control requires an application to submit a quality profile to the UME. The profile, maintained by a QoS Agent, encapsulates *quality-utility, quality-methods* and *method-resources* mappings. The quality-utility mapping assigns a dollar value to each of the three quality levels: *gold, silver,* and *bronze*; the quality-methods mapping associates quality levels with implementations that deliver them; and finally, method-resources mapping defines the resource requirements for each implementation. An application is admitted only if the system will yield a higher utility value as a result.

For implementation convenience, only three system resources are monitored: *CPU* (in percentages of cycles per second), *inbound bandwidth* and *outbound bandwidth* (in packets per second) of the network interface. Since we are constrained by the lack of reservation capabilities of the operating system, we allow 5% of the *raw* system resources to be used by system functions, while the remaining 95% is *manageable* by the UME. At times of admission control, the UME *reserves* the amount of resources required for applications' minimal admitted quality; during quality adaptation, the UME *allocates* the amount of resources for applications' operating quality.

The goal of the UME is to maximize the total system utility, while obeying user preferences and system resource constraints (i.e., not over-subscribing total available resource). Specifically, it needs to adapt the amount of manageable resource, based on the monitored resource usage, in order to avoid contention for resource use[2] The UME implements the following algorithm for adapting manageable resources and avoiding usage contention:

if (monitored$_t$ < allocated$_t$)   then manageable$_t$ = 95% * raw$_t$ (i.e. initial condition)
else if (monitored$_t$ < reserved$_t$) then manageable$_t$ = raw$_t$ - (monitored$_t$ - allocated$_{t-1}$)
else if (monitored$_t$ == raw$_t$)   then manageable$_t$ = raw$_t$ - (monitored$_t$ - allocated$_{t-1}$) - CR[3]
if (manageable$_t$ < reserved$_t$)[4] then manageable$_t$ = reserved$_t$
do quality adaptation and calculate new allocated$_t$

With the above mechanisms in place, our system is able to provide reactive adaptation in a best-effort mixed environment, and to enforce proactive adaptation in an entirely reservation-based system. The results of the following section will demonstrate this point.

5 Experiments and Results

We experimented with two kinds of systems: a best-effort mixed system and an entirely reservation-based system. In *a best-effort mixed system*, (e.g., the Internet) only a subset of the applications follows the adaptation model, whereas *in a reservation-based system*, all the applications follow the adaptation model. Applications not fol-

---

[2] Contention is possible since we are using a best-effort system where some applications are out of the Utility Model domain.

[3] CR (*Cushioning Region*), defined as 5% of the raw resource in our system, decides the aggressiveness of the UME in scaling back the manageable resource to avoid contention.

[4] Note that because the UME must guarantee the minimal qualities of admitted applications, the manageable resources are not allowed to reduce beyond the reserved resources.

lowing the Utility Model may cause resource contention. When this happens, we want the UME to scale back the quality of the applications following the model (see Figure 4a). In an entirely reservation-based system, the UME will adapt operating qualities of the applications in order to optimize the system utility (see Figure 4b).

Our experiments use two types of applications: a file transfer application and a video-on-demand application. For the case of a best-effort mixed system, the common ftp program is used together with a video-on-demand application (VoD[5]). A similar setup is used for an entirely reservation-based system, but with the ftp program replaced by a rate-controllable file transfer program (QFTP[6]). In the latter case, both applications conform to the Utility Model, and the receivers adjust their operating qualities based on adaptation signals received from the UME.

The topology of our experiments involved two server hosts (A and B), and one client host (C). The UME is implemented only at the client host, where VoD and FTP clients run. For simplicity, the inbound bandwidth is considered to be the only constrained resource in our experiments.

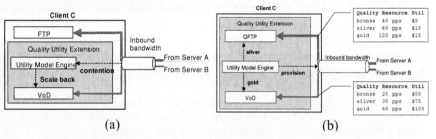

Figure 4: (a) Best-effort mixed system and (b) Reservation-based system.

5.1 Best-effort Mixed System

Figure 5 shows the adaptation events in this experiment. We note that the UME adapts to changes of system state by scaling the amount of manageable resource and dynamically adapting the resource allocation to the VoD application. We observe the following:

- Before event A, the UME monitored some background network activities. It adjusted the amount of manageable resource, so that resource would not be over-subscribed when admitting new sessions.
- At event A, the VoD application was admitted (not started!). The UME reserves the bandwidth required for its minimum quality (bronze) so that it can deliver the minimum quality as long as the session lives. However, since there is no contention, the UME allocates bandwidth for its desired quality (gold). The monitored resource, at event A, shows the resource consumption of VoD; the amount of manageable resource remained steady in the absence of resource contention.

---

[5] In VoD, a client receives a hierarchical layered video stream over some distinct IP multicast groups. Video quality varies as the client incrementally joins or leaves the groups.
[6] Unlike FTP, which uses the slow-start algorithm, a QFTP client controls the server's transmission rate, thus varying the transfer throughput quality.

- At event B, the FTP application is started. Our resource monitor shows that this application seizes large amounts of network bandwidth. To avoid contention, the UME scaled back the amount of manageable resource, and degraded the operating quality of VoD to bronze level by reducing its bandwidth allocation.
- At event C, the FTP terminated, and the UME scaled up the amount of manageable resource to 95%, and upgraded the operating quality of VoD.

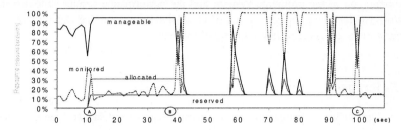

Figure 5: Scaling manageable resource and quality adaptation[7].

5.2 Reservation-Based System

Figure 6 shows the resource allocation and quality adaptation of the system during our experiment. Since both applications followed the adaptation model, the UME allocated the system resource in a systematic way; there is no need to scale back the manageable resource. We observe the following events:
- At event A, the UME admitted VoD, and allocated resource for gold quality.
- At event B, the UME admitted QFTP. Since the available resource is not enough to support gold quality for both applications, the UME has allocated resource for silver quality for QFTP. At this time, VoD gets gold quality because it has offered more for the gold quality.
- Now at event C, QFTP raised its offer for gold quality to $150. The UME adapts to this change by upgrading the operating quality of QFTP to gold, and by degrading the operating quality of VoD to silver.
- Finally at event D, application QFTP terminates, and frees the resource allocated to it. The UME adapts to this by upgrading the operating quality of VoD to gold.

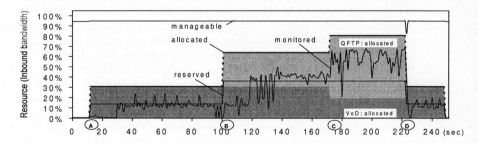

Figure 6: Optimal quality adaptation in a reservation-based system.

---

[7] The area above the manageable line represents unmanageable amount of resource.

6 Conclusions

We have presented the design and implementation of a prototype system using the concepts of the Utility Model for adaptive multimedia. We reported experimental results with two types of adaptation: reactive adaptation in a best-effort mixed system and proactive adaptation in an entirely reservation-based system. The design and implementation of the prototype system, and the performance results validated that the Utility Model can be effectively used for session admission and dynamic quality adaptation in practical real-time multimedia systems.

7 Reference

1. A. Campbell, G. Coulson, and D. Hutchison, "Supporting Adaptive Flows in Quality of Service Architecture", ACM Multimedia Systems Journal, 1996.
2. Lei Chen, "Utility Model Applied to Layered-coded Sources", M.Sc. Thesis, Department of Computer Science, University of Victoria, October 1998.
3. Kiyokuni Kawachiya and Hideyuki Tokuda, "QOS-Ticket: A New Resource-Management Mechanism for Dynamic QOS Control of Multimedia", In Proceedings of Multimedia Japan '96, April 1996.
4. Shahadat Khan, "Quality Adaptation in a Multisession Multimedia System: Model, Algorithms and Architecture", Ph.D. Dissertation, Department of ECE, University of Victoria, May 1998.
5. Shahadat Khan, Kin F. Li and Eric G. Manning, "Padma: An End-System Architecture for Distributed Adaptive Multimedia", IEEE Pacific Rim Conference on Communications, Computers, and Signal Processing, Victoria, August 1997.
6. Shahadat Khan, Kin F. Li and Eric G. Manning, "Optimal and Fast Approximate Solutions for the Multiconstraint Multiple-choice Knapsack Problem", Technical Report ECE-97-3, Department of ECE, University of Victoria, September 1997.
7. Shahadat Khan, Kin F. Li and Eric G. Manning, "The Utility Model for Adaptive Multimedia Systems", Presented at The International Conference on Multimedia Modeling, Singapore, November 1997.
8. Steve McCanne, Van Jacobson and Martin Vetterli, M., "Receiver-driven Layered Multicast", ACM SIGCOMM, Stanford, CA, August 1996.
9. Martin Moser, "Declarative Scheduling for Optimally Graceful QoS Degradation", In Proceedings of IEEE Multimedia Systems, Hiroshima, Japan, 1996.
10. Nobushiko Nishio and Hideyuki Tokuda, QOS Translation and Session Coordination Techniques for Multimedia Systems", In Proceedings of Sixth International Workshop on Network and Operating System Support for Digital Audio and Video, Jushi, Japan, 1996.
11. Klara Nahrstedt, "An Architecture for End-to-End Quality of Service Provision and its Experimental Validation", Ph.D. Thesis, Department of Computer and Information Science, University of Pennsylvania, August 1995.
12. Louis C. Schreier and Michael B. Davis, "System-level Resource Management for Network-based Multimedia Applications", In Fifth International Workshop on Network and Operating System Support for Digital Audio and Video, NOSSDAV 95, Durham, NH, April 1995.

Evaluation of Real-Time Fiber Communications for Parallel Collective Operations*

Parvathi Rajagopal[1] and Amy W. Apon[2]

[1] Vanderbilt University, Nashville, TN 37235 parvathi@dg-rtp.dg.com
[2] University of Arkansas, Fayetteville, AR 72701 aapon@comp.uark.edu

Abstract. Real-Time Fiber Communications (RTFC) is a gigabit speed network that has been designed for damage tolerant local area networks. In addition to its damage tolerant characteristics, it has several features that make it attractive as a possible interconnection technology for parallel applications in a cluster of workstations. These characteristics include support for broadcast and multicast messaging, memory cache in the network interface card, and support for very fine grain writes to the network cache. Broadcast data is captured in network cache of all workstations in the network providing a distributed shared memory capability. In this paper, RTFC is introduced. The performance of standard MPI collective communications using TCP protocols over RTFC are evaluated and compared experimentally with that of Fast Ethernet. It is found that the MPI message passing libraries over traditional TCP protocols over RTFC perform well with respect to Fast Ethernet. Also, a new approach that uses direct network cache movement of buffers for collective operations is evaluated. It is found that execution time for parallel collective communications may be improved via effective use of network cache.

1 Introduction

Real-Time Fiber Communications (RTFC) is a unique local and campus area network with damage tolerant characteristics necessary for high availability in mission critical applications. The RTFC standard [5] has been in development over a period of several years and first generation products are available through two independent vendors. RTFC arose from the requirements of military surface ship combat systems. Existing communication systems, while fast enough, did not have the necessary fault and damage tolerance characteristics. When constructed in a damage-tolerant configuration, RTFC will automatically reconfigure itself via a process known as rostering whenever damage to a node, link, or hub occurs.

In addition to its damage-tolerant features, RTFC is a ring that uses a connectionless data-link layer that has minimum overhead [9]. Unlike switched technologies such as Myrinet and ATM that may perform multicast and broadcast

---

* This work was supported in part by NSF Research Instrumentation Grant #9617384 and by equipment support from Belobox Systems, Inc., Irvine, California.

communications by sending unicast messages to all destinations, RTFC directly supports the multicast and broadcast communications that are important in parallel applications. Therefore, it seems appropriate to explore the advantages that RTFC may have in parallel collective communications (i.e., those that are variants of broadcast and multicast messaging).

The purpose of this paper is to introduce Real-Time Fiber Communications and to evaluate its applicability for parallel collective operations. The evaluation is based on measurements and experimentation on a cluster of workstations that is interconnected via RTFC. The paper is organized as follows.

- Section 2 introduces Real-Time Fiber Communications,
- Section 3 describes the experimental workstation and network platform,
- Section 4 describes the software platform for collective communications,
- Section 5 gives the measurement results, and
- Section 6 gives conclusions and a description of related and future work.

2 Overview of Real-Time Fiber Communications

The basic network topology of RTFC is a set of nodes that are connected via a hub or switch in a simple star configuration. The nodes form a logical ring, and each node has an assigned number in increasing order around the ring. One physical star forms a single non-redundant fault-tolerant segment. The maximum number of nodes in a segment is 254. The distance from a node to a hub or switch is normally up to 300 meters for 62.5 micro fiber-optic media. For applications that do not require high levels of fault tolerance, it is also possible to configure RTFC as a simple physical ring using only the nodes and the network interface cards in the nodes. At the physical layer RTFC is fully compliant with FC-0 and FC-1 of the Fibre Channel FC-PH specification.

The RTFC ring is a variant of a register insertion ring [1], and guarantees real-time, in-order, reliable delivery of data. During normal operation, packets are removed from the ring by the sending node (i.e., a source removal policy is used). While this limits the maximum throughput on the ring to the bandwidth of the links only, this policy has certain advantages in that sending nodes are always aware of traffic on the entire ring. Since each node can see all of the ring traffic, it is possible for nodes to monitor their individual contribution to ring traffic to ensure that the ring buffers do not fill, and that ring tour time is bounded. The unique purge node in the segment removes any orphan packets that traverse the ring more than one time.

Bridge nodes may be used to connect segments, and redundant bridge nodes may be used to connect segments to meet the damage-tolerance requirements of a multi-segmented network. For applications that require high levels of fault tolerance, it is possible to construct redundant copies of the ring. By adding channels to the network interface cards at each node and adding additional hubs or switches, up to four physically identical rings may be constructed with the same set of nodes. In this manner, RTFC can be configured to be dual, triple, or quad-redundant for fault tolerance. The communication path from one node to the next higher-numbered node in the ring goes through a single hub. While other hubs receive a copy of all packets and forward them, the receiving node only listens to one hub, usually the one with the lowest number if no damage has been sustained. If a node, hub, or communication link fails in an RTFC segment, then a process known as rostering determines the new logical ring. This ability to reconfigure the route between working nodes (i.e., the logical ring), in real-time, after sustaining wide-scale damage, is a distinguishing characteristic of RTFC.

RTFC supports traditional unicast, multicast, and broadcast messaging between nodes. In addition, RTFC supports the concept of network cache. Network cache is memory in the network interface card that is mapped to the user address space and is accessible to the user either via direct memory access (DMA) or via copy by the CPU. One use of network cache is to provide an efficient hardware implementation of distributed shared memory across the nodes in a segment. Each network interface card can support up to 256MB of network cache. Some portion of network cache is reserved for system use, depending on the version of system software running, but the remaining portion of network cache can be read or written by user processes running on any workstation in the network.

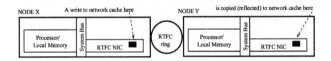

Fig. 1. Operation of the RTFC network cache

Broadcast data written to network cache of any node appears in the network cache of all nodes on that segment within one ring-tour, as illustrated in Figure 1. The ring-tour time is the time that it takes a micropacket to travel from the source around the ring one time. The total hardware latency to travel from the source around the ring one time can be calculated to be less than one millisecond on the largest ring at maximum distance [5].

The data link layer in RTFC is defined by the RTFC standard and replaces the FC-2 layer defined in Fibre Channel. Data at the data link layer is transferred primarily using two sizes of hardware-generated micropackets. For signals and small messages, a fixed-size, 12-byte (96 bit), micropacket is used [9].

The data rate of RTFC can be calculated. RTFC sends at the standard Fibre Channel rate of approximately 1 Gbps. Fibre Channel uses 8B/10B data encoding, giving a user data rate of 1 Gbps * 80 % = 800 Mbps. For each micropacket the percentage of data bits is (32 data bits)/(96 bits/micropacket + 32 bits/idle) = 25%. Therefore, the user data rate is approximately 200 Mbps. For longer messages, a high-payload micropacket longer than 12 bytes is supported in hardware that obtains up to 80% efficiency of the theoretical fiber optic bandwidth. All testing was performed with first generation hardware that only supports 96 bit micropackets.

3 Experimental Workstation and Network Platform

The workstations used in this study consists of three dual-processor VME bus-based workstations, connected by a 3Com SuperStack II Hub 100 TX Fast Ethernet hub for 100Mbps tests and a VMEbus RTFC gigabit FiberHub. Each workstation is identified by an Ethernet network IP address and host name, and similarly by an RTFC network IP address and host name.

A simple block diagram of the workstation hardware is shown in Figure 2. Each of the three workstations has a Motorola MVME3600 VME processor module, as outlined by the dashed line in the figure. The MVME3600 consists of a processor/memory module and a base module, assembled together to form a VMEbus board-level computer. The base module and processor/memory modules are both 6U VME boards. The combination occupies two slots in a VME rack. The processor/memory modules consists of two 200MHz PowerPC 604 processors and 64MB of memory. The processor modules have a 64-bit PCI connection to allow mating with the MVME3600 series of baseboards. The base I/O module consists of a digital 2114X 10/100 Ethernet/Fast Ethernet interface and video, I/O, keyboard, and mouse connectors. The MVME761 transition module (not shown in Figure 2) provides a connector access to 10/100 BaseT port. A P2 adapter (not shown in Figure 2) routes the Ethernet signals to the MVME761 transition module. Each workstation is connected to the RTFC network via a V01PAK1-01A FiberLAN SBC network interface, each with 1MB of SRAM network cache memory.

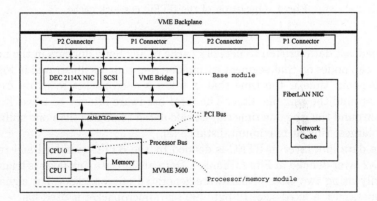

Fig. 2. Simple block diagram of the workstation architecture

System software on the workstations is the IBM AIX 4.1.5 operating system. The MPICH implementation of the MPI (Message Passing Interface) is chosen as the application platform [4]. MPICH network communication that is based on TCP/IP can be made to run over Fast Ethernet or over RTFC, depending on the host names that are specified in the proc group configuration file. The MPICH implementation of MPI provides unicast (point-to-point) communications, broadcast, and multicast (collective) communications. MPICH collective communications are implemented using point-to-point, which are implemented through message passing libraries built for the specific hardware. The performance of the collective operations in MPICH is enhanced by the use of asynchronous (non-blocking) sends and receives where possible, and by using a tree-structured communication model where possible.

In addition to TCP/IP, messaging is provided over RTFC via the Advanced MultiProcessor/Multicomputer Distributed Computing (AmpDC) environment [2], a product of Belobox Systems, Inc. However, an implementation of AmpDC middleware for RTFC was not yet available for AIX. At the time of this study, a partial implementation of some RTFC primitive components for AIX were available, including a function to write data directly to network cache memory from local workstation memory using DMA or CPU access, and a function to read data directly from network cache to local workstation memory. Also available were two functions that return a pointer to the first and last available long word location of user-managed network cache, respectively.

Standard benchmarking techniques were used throughout this study. All runs were executed for at least one hundred trials. The timings obtained were cleared of outliers by eliminating the bottom and top ten percent of readings. The network was isolated in that other traffic was eliminated and no other applications were running on the workstations at the time. Each call was executed a few times before the actual timed call was executed in order to eliminate time taken for setting up connections, loading of the cache, and other housekeeping functions. Measurements were made on both RTFC and Fast Ethernet.

4 Software Platform for Collective Communications

An experimental software platform was developed for evaluating the performance of MPI collective communications by using the network cache architecture [8]. Collective calls for broadcast, scatter, gather, allgather, alltoall were designed to work similarly to their corresponding MPI calls. These calls were implemented on the experimental platform by using the basic RTFC read and write calls.

In order to allow processes to not over-write each other's data, network cache was partitioned among the processes [6]. Each process writes to its area of network cache. Each process may read from one or more areas of network cache, depending on the specific collective operation. An initialize function finds the first and last location on network cache that is available to the user applications, and partitions the available memory equally among the processes. The assumption is made that at any given time there is only one parallel application (i.e., MPI program) executing, and that this application has access to all of network cache that is available to user programs. If there is a need to execute more than one parallel program at a time then an external scheduler should be used that will be responsible for the additional partitioning of network cache that would be required.

Collective calls are considered complete when the area of network cache used by the process can be reused. Since the experimental platform does not yet support interrupts at the receiving node via the RTFC primitives available at the time of these tests, there is a difficulty in how to signal a receiving node that a message has arrived. In order to demonstrate proof-of-concept, MPI messaging over Fast Ethernet is used in the prototype code to notify a set of nodes of the start of each collective operation and to synchronize nodes between collective operations.

In MPI, processes involved in a collective call can return as soon as their participation in the routine is completed. Completion means that the buffer (i.e., the area of network cache) used by the process is free to be used by other processes. The method for signaling completion of each of the calls varies depending on the operations being performed. In general, the sending processes write to network cache and inform all other processes where to find the data in network cache, the receiving processes read the data from the network cache, and an MPI synchronization call ensures that the buffer is safe before it is re-used.

For example, in case of the broadcast, the root writes the contents of the send buffer to its area in network cache and sends the starting address to all other processes in the group. The other processes receive the address, read the information from network cache and write it to their local memory. Once the processes write the message to local memory, the send a completed message to the root which exits upon receing the complete message from all receivers.

As another example, in case of allgather, each process sends the start address to all other processes. That is, the address is allgathered to other processes. Once a process reads the contents of network cache it sends a completed message to the sending process, which then exits upon receiving the completed message. Figure 3 illustrates the flow of execution for allgather. Figure 4 illustrates the partitioning of network cache and the use of local memory for allgather. Similar algorithms are implemented for scatter, gather, and alltoall messaging.

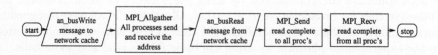

Fig. 3. Flowchart for the allgather operation

5 Measurement Results for Collective Communication

Performance measurements were conducted for the experimental code running over RTFC, and also for MPI over TCP/IP on both RTFC and Fast Ethernet. The time from the start of the collective call until the completion of the last process is the metric used in this study [7]. Measurements were performed for broadcast, alltoall, allgather, scatter and gather, for two and three processes.

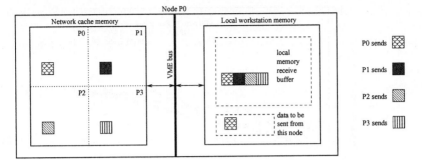

Fig. 4. Partitioning of network cache for the allgather operation (shown at node P0)

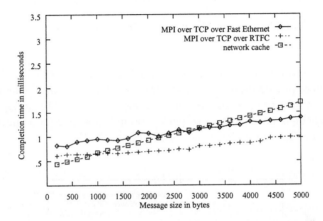

Fig. 5. Completion time of collective calls for broadcast, two nodes

Figure 5 gives the measurements for the broadcast operation between two nodes. With two processes the messaging reduces to sends and receives. Figure 5 shows that MPI performs well over TCP over RTFC in comparison to MPI over Fast Ethernet. The network cache implementation performs well for very small messages, but then poorly as the message size increases. This is expected, since with only two nodes involved in the messaging there is less advantage in using a broadcast network. There is some advantage in using network cache messaging in the case of small messages, but the prototype code for collective operations via network cache uses a separate step for distributing addresses and signalling that the buffer is free, even in the case of two processes. This extra step causes the network cache messaging to perform worse than the MPI over TCP over RTFC equivalent in the case of two nodes as the message size increases.

Figure 6 illustrates the measurements for the broadcast operation between three nodes. Again, MPI performs well over TCP on RTFC in comparison to MPI on Fast Ethernet. This graph shows noticeable changes in slope in the MPI results at packet boundaries of approximately 1500, 3000, 4500, etc. The same

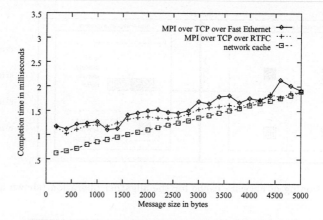

Fig. 6. Completion time of collective calls for broadcast, three nodes

buffer size is used internally for TCP/IP for both Fast Ethernet and RTFC. In the case of three processes, one per node, the network cache implementation does have a relative advantage over MPI messaging over TCP. Messaging via writes to network cache is not sensitive to packet boundaries and the completion time is smaller than the completion time for MPI over TCP.

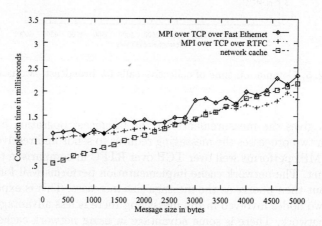

Fig. 7. Completion time of collective calls for gather, three nodes

Figure 7 illustrates the measurements for the gather operation between three nodes. As before, Figure 7 shows that MPI performs well over TCP over RTFC in comparison to MPI over Fast Ethernet. Also, as in the case of broadcast for three processes, one per node, the network cache implementation does have a relative advantage over MPI messaging over TCP. Figure 8 illustrates the measurements for the scatter operation between three nodes, with similar results. Graphs for the other collective operations also show similar results but are not included here due to space limitations.

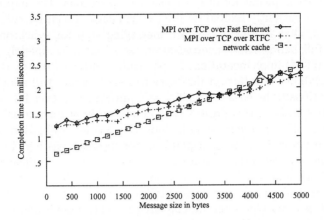

Fig. 8. Completion time of collective calls for scatter, three nodes

The relatively higher rise in the line for messaging via writes to network cache is noticeable in each of the graphs and is the subject of further study. This effect may be due to the combination of the beta nature of the current hardware and software platform and the experimental synchronization technique. In the current platform, a data transfer between host memory and network cache on the interface card must complete before control returns to the user application, even in the case of DMA access. As a result, in the current setup, notification of a sent message begins after the completion of the network cache write. In contrast, the TCP drivers on both platforms move the message in units of a framesize at a time, and allow for return to the user application while transfer is still being completed. This overlap gives greater efficiency for TCP over RTFC for larger message sizes as compared to direct network cache writes in the current setup. Effective use of DMA access to the network cache memory will be provided in the full implementation of the AmpDC middleware and final RTFC product and will help to eliminate this source of inefficiency.

6 Conclusions and Future Work

The intent of this work is to investigate the performance of Real-Time Fiber Communications and its applicability as an interconnection technology for parallel computing. MPI over TCP on RTFC performs well, and better than MPI over TCP on Fast Ethernet. As in the case of other gigabit-speed networks, the improvement in performance is limited by the use of high-overhead protocols such as TCP.

Improvements for parallel collective operations may be obtained using the network cache architecture. The broadcast, multicast, and simple hardware distributed shared memory mechanisms provide the potential for improvement of traditional MPI implementations of collective operations. The improvement in performance is currently limited by the experimental design for collective communication in that it uses a separate messaging step for synchronization. In contrast, a full handshake synchronization technique is completely eliminated with a full RTFC implementation. In particular, effective use of DMA access and the availability of interrupt messages to remote nodes will allow these collective operations to be easily and efficiently implemented over RTFC.

One of the goals of this project is to create an implementation of MPI for network cache. Future efforts also include the further evaluation of methods for using the network cache as a platform for message passing. Related work includes study of the implementation and measurement of efficient collective communications and multicast mechanisms [3]. The RTFC Team is exploring many of the features and capabilities of RTFC, including the use of the AmpDC environment as a parallel programming environment, bounds and limitations of the RTFC rostering algorithm, and new techniques for flow control.

Acknowledgement is given to Larry Wilbur, Dave Belo, Mike Jeffers, Glynn Smith, and Joan Clark of Belobox Systems, Inc. and to Joe Gwinn of Raytheon Company for their helpful suggestions and indispensable support on this project.

References

1. BAHAA-EL-DIN, W. H., AND LIU, M. T. Register insertion: A protocol for the next generation of ring local-area networks. *Computer Networks and ISDN Systems 24* (1992), 349–366.
2. Belobox Systems Inc., AmpDC Advanced Multiprocessor Distributed Computing Environment, 1997. http://www.belobox.com/ampdc.html.
3. DUNIGAN, T. H., AND A.HALL, K. PVM and IP multicast. Tech. Rep. ORNL Report ORNL/TM-13030, Oak Ridge National Laboratory, December 1996.
4. GROPP, W., LUSK, E., DOSS, N., AND SKJELLUM, A. A High-Performance, Portable Implementation of the MPI Message Passing Interface Standard. Tech. Rep. Preprint MCS-P567-0296, Argonne National Laboratory, March 1996.
5. GWINN, J. RTFC Principles of Operation, RTFC-2001. Tech. rep., Belobox Systems Inc., Irvine, California, January 2 1998.
6. MARTIN, R. P. HPAM: An active message layer for a network of HP workstations. In *Hot Interconnects II* (August 1994).
7. NUPAIROJ, N., AND NI, L. Benchmarking of multicast communication services. Tech. Rep. Technical Report MSU-CPS-ACS-103, Department of Computer Science, Michigan State University, April 1995.
8. RAJAGOPAL, P. Study of the network cache architecture as a platform for parallel applications. Master's thesis, Vanderbilt University, August 1998.
9. RTFC-TEAM. RTFC Micropacket Data Link Layer, RTFC-2002. Tech. rep., Belobox Systems Inc., Irvine, California, October 1997.

The Case for Prediction-Based Best-Effort Real-Time Systems

Peter A. Dinda   Bruce Lowekamp
Loukas F. Kallivokas   David R. O'Hallaron

Carnegie Mellon University
5000 Forbes Avenue, Pittsburgh, PA 15213, USA
(pdinda,lowekamp,loukas,droh)@cs.cmu.edu

Abstract. We propose a prediction-based best-effort real-time service to support distributed, interactive applications in shared, unreserved computing environments. These applications have timing requirements, but can continue to function when deadlines are missed. In addition, they expose two kinds of adaptability: tasks can be run on any host, and their resource demands can be adjusted based on user-perceived quality. After defining this class of applications, we describe a significant example, an earthquake visualization tool, and show how it could benefit from the service. Finally, we present evidence that the service is feasible in the form of two studies of algorithms for host load prediction and for predictive task mapping.

1 Introduction

There is an interesting class of interactive applications that could benefit from a real-time service, but which must run on conventional reservation-less networks and hosts where traditional forms of real-time service are difficult or impossible to implement. However, these applications do not require either deterministic or statistical guarantees about the number of tasks that will meet their deadlines. Further, they are adaptable in two ways: their components are distributed, so a task can effectively be run on any available host, and they can adjust the computations tasks perform and the communication between tasks, trading off user-perceived degradation and the chances of meeting deadlines.

We propose a best-effort real-time service that uses history-based prediction to choose how to exploit these two levels of adaptation in order to cause most tasks to meet their deadlines and for the application to present reasonable quality to the user. We believe that such a service is feasible, and, while providing no guarantees, would nonetheless simplify building responsive interactive applications and greatly improve user experience.

The paper has two main thrusts. The first is to define the class of resilient, adaptable, interactive applications we have described above and to show that it contains significant real applications. As an example of typical user demands and application adaptivity we analyze QuakeViz, a distributed visualization system for earthquake simulations being developed at CMU. We are also studying OpenMap, a framework for interactively presenting map-based information, developed at BBN [4]. The extended version of this paper [10] covers OpenMap in more detail.

Effort sponsored in part by the Advanced Research Projects Agency and Rome Laboratory, Air Force Materiel Command, USAF, under agreement number F30602-96-1-0287, in part by the National Science Foundation under Grant CMS-9318163, and in part by a grant from the Intel Corporation.

The second main thrust of the paper is to support the claim that history-based prediction is an effective way to implement a best-effort real-time service for this class of applications. We present two pieces of supporting evidence here. The first is a trace-based simulation study of simple algorithms for predicting on which host a task is most likely to meet its deadline based on a history of past running times. One of the algorithms provides near optimal performance in the environments we simulated. The second piece of supporting evidence is a study of linear time series models for predicting host load that shows that simple, practical models can provide very good load predictions. Because running time is strongly correlated with the load experienced during execution, the fact that load is predictable suggests that information that can be collected in a scalable way can be used to make decisions about where to execute tasks. Together, these two results suggest that prediction is a feasible approach to implementing a best-effort real-time service.

2 Application characteristics

The applications we are interested in supporting have the following four characteristics. First, they exhibit **interactivity** — computation takes the form of tasks that are initiated or guided by a human being who desires responsiveness and predictable behavior. Research has shown that people have difficulty using an interactive application which does not achieve timely, consistent, and predictable feedback [12, 16]. Our mechanism for specifying interactive performance is the task deadline. The second application characteristic is **resilience** in the face of missed deadlines — such failures do not make these applications unusable, but merely result in lowered quality. Resilience is the characteristic that suggests a best-effort real-time approach, instead of traditional "soft" (statistically guaranteed) [21, 6, 15] and "hard" (deterministically guaranteed) real-time approaches [20]. The third characteristic is **distributability** — we assume that the applications are implemented in a distributable manner, with data movement exposed for scheduling purposes. Finally, our applications are characterized by **adaptability** — they expose controls, called *application-specific quality parameters*, that can be adjusted to change the amount of computation and communication resources a task requires.

3 QuakeViz

The Quake project developed a toolchain capable of detailed simulation of large geographic areas during strong earthquakes [2]. Thorough assessment of the seismicity associated with a geographic region requires accurate, interactive visualization of the data produced by the simulation. Visualization of this data is complex because the full data for even a small region such as the San Fernando Valley requires approximately 6TB of data. Even selective output results in tens of gigabytes of data. Nevertheless, with proper management of the visualization data, it is possible to achieve reasonable quality in a resource-limited environment.

The visualization toolchain is shown in Figure 1(a). The raw simulation data is typically read from storage, rather than from the running application, because of the simulation's high cost, scheduling complexities, and the lack of any runtime tunable parameters in the simulation. The raw irregular mesh data is first interpolated onto a regular grid to facilitate processing and downsampled to a more manageable size. The resulting grid is then used to calculate isosurfaces corresponding to various response intensities. The displacement and density isosurfaces are combined with topology information to produce a 3D image, which is then drawn on the user's desktop display.

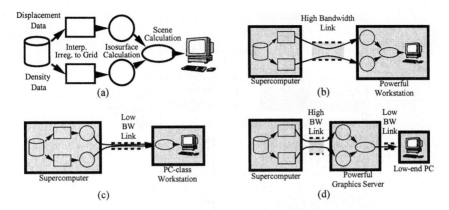

Fig. 1. QuakeViz Application. (a) Stages of the Quake visualization toolchain. Different placements of toolchain tasks corresponding to different resource situations: (b) A high performance network and workstation are available for the visualization. (c) A low-bandwidth network and PC with a 3D graphics accelerator are used. (d) A low-end PC is used as a framebuffer displaying the result of a completely remote visualization process.

Because the Quake visualization is done offline, the user's input consists of controlling the parameters of the visualization. These operations may include rotating the display, removing surface layers, zooming in and out, and adjusting the parameters for isosurface calculation.

In the visualization diagram shown in Figure 1(a), we generally expect the resources to retrieve and do the initial interpolation to be available on only a few high-performance machines, such as those commonly used at remote supercomputing facilities. Because the output device is determined by the user's location, the isosurface calculation and scene calculation are the two components which can be freely distributed to optimize performance.

3.1 Task location

The location of these two phases is governed by the combination of network bandwidth, and computing and graphics resources available to the user. In fact, there are reasons to divide the pipeline at any of the three remaining edges, depending on predicted processor and network capacity. Three examples are shown in Figure 1(b)-(d).

Figure 1(b) is the ideal case where resources allow the full regular mesh produced by the interpolation phase to be sent directly to the user's desktop. This situation is unlikely in all but the most powerful environments, but may offer the best opportunity for interactive use. In Figure 1(c), the isosurface calculation is performed in the supercomputer doing the interpolation. The scene calculation is done in the user's desktop machine. The isosurface calculation is also fairly expensive, whereas the scene calculation can be done quite effectively by a variety of commodity graphics cards currently available. A very limited case is shown in Figure 1(d), where the user's desktop is acting only as a framebuffer for the images. This setup may be useful if the size of the final pixmap is smaller than the size of the 3D representation, which depends on the complexity of the scene, or if the user's workstation does not have 3D hardware.

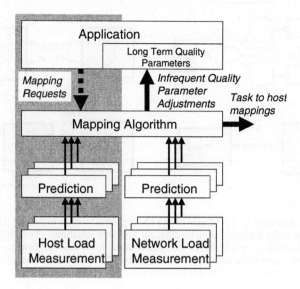

Fig. 2. Structure of best-effort real-time system. The shaded portion is discussed.

3.2 Quality parameters

Frame rate, resolution, and interactive response time are the primary quality parameters which can be adjusted to meet the user's requirements. Because it may take some time for adjustments to these parameters to be propagated through the pipeline, other mechanisms may be used to maintain responsiveness. For example, resolution can be lowered along any edge in the toolchain. Similarly, zooming in on the image can be accomplished along edges, while waiting for the greater detail to be propagated from the beginning of the pipeline. Finally, even if only a pixmap of the image is available on the user's workstation, that image can be used for temporary adjustments, providing the effects of zooming and rotation, while waiting for actual data to fill in the unavailable information.

4 Structure of a best-effort real-time system

Figure 2 illustrates the structure of our proposed best-effort real-time system. The boxes represent system components. Thin arrows represent the flow of measured and predicted information, thick arrows represent the control the system exerts over the application, and the dotted arrow represents the flow of application requests. We have shaded the parts of the design that we discuss in this paper.

The system accepts requests to run tasks from the application and then uses a mapping algorithm to assign these tasks to the hosts where they are most likely to meet their deadlines. A task mapping request includes a description of the data the task needs, its resource requirements (either supplied by the application programmer or predicted based on past executions of the task), and the required deadline. The mapping algorithm uses predictions of the load on each of the available hosts and on the network to determine the expected running time of the task on each of the hosts, and then runs the task on one of the hosts (the target host) where it is likely to meet its deadline with high confidence. If no such host exists, the system can either map the task suboptimally, hoping that the resource crunch is transient behavior which should soon disappear, or adjust the quality parameters of the application to attempt to reduce the task's computation and/or communication requirements or to add more slack to the deadline.

The choice of target host and the choice of application-specific quality parameters are adaptation mechanisms that are used at different time scales. A target host is chosen for every task, but quality parameters are only changed when many tasks have failed to meet their deadlines, or when it becomes clear that sufficient resources exist to improve quality while still insuring that most deadlines can be met by task mapping. Intuitively, the system tries to meet deadlines by using adaptation mechanisms that do not affect the user's experience, falling back on mechanisms that do only on failure. Of course, the system can also make these adjustments pre-emptively based on predictions.

The performance of the system largely reflects the quality of its measurements and predictions. The measurement and prediction stages run continuously, taking periodic measurements and fitting predictive models to series of these measurements. For example, each host runs a daemon that measures the load with some periodicity. As the daemon produces measurements, they are passed to a prediction stage which uses them to improve its predictive model. When servicing a mapping request, the mapping algorithm requests predictions from these models. The predictions are in the form of a series of estimated future values of the sequence of measurements, annotated with estimates of their quality and measures of the past performance of the predictive model. Given the predictions, the mapping algorithm computes the running time of the task as a confidence interval whose length is determined by the quality measures of the prediction. Higher quality predictions lead to shorter confidence intervals, which makes it easier for the mapping algorithm to choose between its various options.

5 Evidence for a history-based prediction approach

We present two pieces of evidence that suggest history-based prediction is a feasible approach to implementing a best-effort real-time service for distributed, interactive applications. The first is a trace-based simulation study of simple mapping algorithms that use past task running times to predict on which host a task is most likely to meet its deadline. The study shows that being able to map a task to any host in the system exposes significant opportunities to meet its deadline. Further, one of the algorithms provides near optimal performance in the environments we simulated. The second piece of evidence is a study of linear time series models for predicting host load. We found that simple, practical models can provide very good load predictions, and these good predictions lead to short confidence intervals on the expected running times of tasks, making it easier for a mapping algorithm to choose between the hosts. Network prediction is of considerable current interest, so we conclude with a short discussion of representative results from the literature.

5.1 Relationship of load and running time

The results in this section depend on the strong relationship that exists between host load and running time for CPU-bound tasks. We measured the host load as the Unix five second load average and sample it every second. We collected week-long traces of such measurements on 38 different machines in August 1997 and March 1998. We use these extensive traces to compute realistic running times for the simulation experiments of Section 5.2. In Section 5.3 we directly predict them with an eye to using the predictions to estimate running times. A detailed statistical analysis of the traces is available in [9]. Considering load as a continuous signal $z(t)$ the relationship between load and running

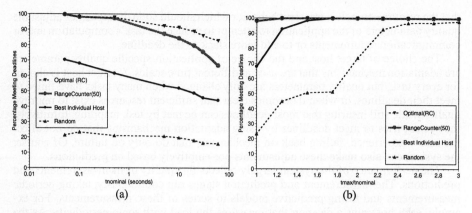

Fig. 3. Representative data from our study of simple prediction-based mapping algorithms: (a) the effect of varying nominal execution time, (b) the effect of increasing deadline slack.

time $t_{running}$ is $\int_0^{t_{running}} \frac{1}{1+z(t)} dt = t_{nominal}$ where $t_{nominal}$ is the running time of the task on a completely unloaded host. Notice that the integral simply computes the inverse of the average load during execution. Further details on how this relationship was derived and verified can be found in [11].

5.2 Simple prediction-based mapping algorithms

We developed and evaluated nine different mapping algorithms that use a past history of task running times on the different hosts to predict the host on which the current task's deadline is most likely to be met. These algorithms included several different window-average schemes, schemes based on confidence intervals computed using stationary IID stochastic models of running times, and simple neural networks. Due to space limits, we will focus on the performance of *RangeCounter(W)*, the most successful of the algorithms, and only describe the evaluation procedure at a high level. For more details on the algorithms and how we evaluated them, consult the extended version of this paper [10].

We evaluated the mapping algorithms with a trace-driven simulator that uses the load traces described in Section 5.1 to synthesize highly realistic execution times. The simulator repeatedly requested that a task with a nominal execution time $t_{nominal}$ be completed in t_{max} seconds and counted the number of successful mapping requests. Communication costs were ignored—the task is entirely compute-bound and requires no inputs or outputs to be communicated. In addition to simulating the mappings chosen by the algorithm being tested, the simulator also simultaneously simulated a random mapping, the best static mapping to an individual host, and the optimal mapping.

Figure 3(a), which is representative of our overall results, illustrates the effect of varying the nominal time, $t_{nominal}$ with a tight deadline of $t_{max} = t_{nominal}$. Notice that there is a substantial gap between the performance of the optimal mapping algorithm and the performance of random mappings. Further, it is clear that always mapping to the one host does not result in an impressive number of tasks meeting their deadlines. These differences suggest that a good mapping algorithm can make a large difference. We can also see that *RangeCounter(W)* performs quite well, even for fairly long tasks.

Even with relaxed deadlines, a good mapping algorithm can make a significant difference. Figure 3(b), which is representative of our results, illustrates the effect of relax-

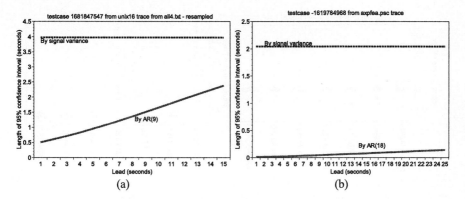

Fig. 4. Benefits of prediction: (a) server machine with 40 users and long term load average of 10, (b) interactive cluster machine with long term load average of 0.17.

ing the deadline t_{max} (normalized to $t_{nominal}$ on the x-axis) on the percentage of tasks that meet their deadlines. Notice that there is a large, persistent difference between random mapping and optimal mapping even with a great deal of slack. Further, using the best single host is suboptimal until the deadline is almost doubled. On the other hand, *RangeCounter(W)* presents nearly optimal performance even with the tightest possible deadline.

In a real system, the cost of communication will result in fewer deadlines being possible to be met and will tend to encourage the placement of tasks near the data they would consume, and thus possibly reduce the benefits of prediction. Still, this study shows that it is possible to meet many deadlines by using prediction to exploit the degree of freedom that being able to run a task on any host gives us.

5.3 Linear time series models for host load prediction

Implicitly predicting the current task's running time on a particular host based on previous task running times on that host, as we did in the previous section, limits scalability because each application must keep track of running times for every task on every node, and the predictions are all made on a single host.

Because running time is so strongly related to load, as we discussed in Section 5.1, another possibility is to have each host directly measure and predict its own load and then make these predictions available to other hosts. When mapping a task submitted on some host, the best-effort real-time system can use the load predictions and the resource requirements of the task to estimate its running time on each of the other hosts and then choose one where the task is likely to meet its deadline with high confidence.

Load predictions are unlikely to be perfect, so the estimates of running time are actually confidence intervals. Better load predictions lead to smaller confidence intervals, which makes it easier for the mapping algorithm to decide between the available hosts. We have found that very good load predictions that lead to acceptably small confidence intervals can be made using relatively simple linear time series models. Due to space limits, we do not discuss these models here, but interested readers will find Box, et al. [8] to be a worthy introduction. Figure 4 illustrates the benefits of such models on (a) a heavily loaded server and (b) a lightly loaded interactive cluster machine. In each of the graphs we plot the length of the confidence interval for the running time of

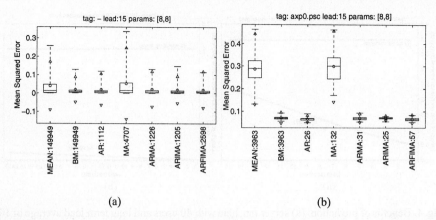

Fig. 5. Performance of 8 parameter host load prediction models for 15 second ahead predictions: (a) All traces, (b) Moderately loaded interactive host.

a one second task as a function of how far ahead the load predictions are made. Figure 4(a) compares the confidence intervals for a predictive AR(9) model and for the raw variance of the load. Notice that for short-term predictions, the AR(9) model provides confidence intervals that are almost an order of magnitude smaller than the raw signal. For example, when predicting 5 seconds ahead, the confidence interval for the AR(9) model is less than 1 second while the confidence interval for the raw load signal is about 4 seconds. Figure 4(b) shows that such benefits are also possible on a very lightly loaded machine.

The prediction process begins by fitting a *model* to a history of periodically sampled load measurements. As new measurements become available, they are fed to the fitted model in order to refine its predictions. At any point after the model is fitted, we can request predictions for an arbitrary number of samples into the future. The quality of predictions for k samples into the future is measured by the k-*step ahead mean squared error*, which is the average of the squares of the differences between the k-step ahead predictions and their corresponding actual measurements. After some number of new samples have been incorporated, a decision is made to refit the model and the process starts again. We refer to one iteration of this process as a *testcase*.

A good model provides *consistent predictability* of load, by which we mean it satisfies the following two requirements. First, for the average testcase, the model must have a considerably lower expected mean squared error than the expected raw variance of the load. The second requirement is that this expectation is also very likely, or that there is little variability from testcase to testcase. Intuitively, the first requirement says that the model provides good predictions on average, while the second says that most predictions are close to that average.

In a previous paper [11], we evaluated the performance of linear time series models for predicting our load traces using the criterion of consistent predictability discussed above. Although load exhibits statistical properties such as self-similarity (Beran [5] provides a good introduction to self-similarity and long-range dependence) and epochal behavior [9] that suggest that complex, expensive models such as ARFIMA [14] models might be necessary to ensure consistent predictability, we found that relatively simple models actually performed just about as well.

Figure 5(a) shows the performance of 8 parameter versions of the models we studied for 15 second ahead predictions, aggregated over all of our traces, while Figure 5(b) shows the performance on the trace of a single, moderately loaded interactive host. Each of the graphs in Figure 5 is a Box plot that shows the distribution of 15-step-ahead (predicting load 15 seconds into the future) mean squared error. The data for the figure comes from running a large number of randomized testcases. Each testcase fits a model to a random section of a trace and then tests the model on a consecutive ection of random length. In the figure, each category is a specific model and is annotated with the number of testcases used. For each model, the circle indicates the expected mean squared error, while the triangles indicated the 2.5th and 97.5th percentiles assuming a normal distribution. The center line of each box shows the median while the lower and upper limits of the box show the 25th and 75th percentiles and the lower and upper whiskers show the actual 2.5th and 97.5th percentiles.

Each of the predictive models has a significantly lower expected mean squared error than the expected raw variance of the load (measured by the MEAN model) and there is also far less variation in mean square error from testcase to testcase. These are the criteria for consistent predictability that we outlined earlier. Another important point is that there is little variation in the performance across the predictive models, other than that the MA model does not perform well. This is interesting because the ARMA, ARIMA, and especially the long-range dependence capturing ARFIMA models are vastly more expensive to fit and use than the simpler AR and BM models. An important conclusion of our study is that reasonably high order AR models are sufficient for predicting host load. We recommend AR(16) models or better.

5.4 Network prediction

Predicting network traffic levels is a challenging task due to the large numbers of machines which can create traffic over a shared network. Statistical models are providing a better understanding of how both wide-area [1, 18] and local-area [22] network traffic behave. Successes have been reported in using linear time series models to predict both long term [13] and short term [3] Internet traffic. These results and systems such as NWS [23] and Remos [17] are being developed to provide the predictions of network performance that are needed to provide best effort real-time service. For applications which are interested in predicting the behavior of a link on which they are already communicating on, passive monitoring may be appropriate [7, 19].

6 Conclusion

We have identified two applications (QuakeViz, which we analyzed here, and Open-Map, analyzed in [10]) which can benefit from a best-effort real-time service. Without such a service, people using such interactive applications would face a choice between acquiring other, possibly reserved or dedicated resources or running the application at degraded quality.

The success of best-effort real-time depends on the accuracy of the predictions of resource availability. We have shown that CPU load can be predicted with a high degree of accuracy using simple history-based time series models. When combined with currently available network information systems, these resources allow decisions to be made for locating tasks and selecting application parameters to provide a usable system.

References

1. BALAKRISHNAN, H., STEMM, M., SESHAN, S., AND KATZ, R. H. Analyzing stability in wide-area network performance. In *Proceedings of SIGMETRICS'97* (1997), ACM, pp. 2–12.

2. BAO, H., BIELAK, J., GHATTAS, O., KALLIVOKAS, L. F., O'HALLARON, D. R., SHEWCHUK, J. R., AND XU, J. Large-scale Simulation of Elastic Wave Propagation in Heterogeneous Media on Parallel Computers. *Computer Methods in Applied Mechanics and Engineering 152*, 1–2 (Jan. 1998), 85–102.
3. BASU, S., MUKHERJEE, A., AND KLIVANSKY, S. Time series models for internet traffic. Tech. Rep. GIT-CC-95-27, College of Computing, Georgia Institute of Technology, February 1995.
4. BBN CORPORATION. Distributed spatial technology laboratories: Openmap. (web page). http://javamap.bbn.com/.
5. BERAN, J. Statistical methods for data with long-range dependence. *Statistical Science 7*, 4 (1992), 404–427.
6. BESTAVROS, A., AND SPARTIOTIS, D. Probabilistic job scheduling for distributed real-time applications. In *Proceedings of the First IEEE Workshop on Real-Time Applications* (May 1993).
7. BOLLIGER, J., GROSS, T., AND HENGARTNER, U. Bandwidth modelling for network-aware applications. In *Proceedings of Infocomm '99* (1999). to appear.
8. BOX, G. E. P., JENKINS, G. M., AND REINSEL, G. *Time Series Analysis: Forecasting and Control*, 3rd ed. Prentice Hall, 1994.
9. DINDA, P. A. The statistical properties of host load. In *Proc. of 4th Workshop on Languages, Compilers, and Run-time Systems for Scalable Computers (LCR '98)* (Pittsburgh, PA, 1998), vol. 1511 of *Lecture Notes in Computer Science*, Springer-Verlag, pp. 319–334. Extended version available as CMU Technical Report CMU-CS-TR-98-143.
10. DINDA, P. A., LOWEKAMP, B., KALLIVOKAS, L. F., AND O'HALLARON, D. R. The case for prediction-based best-effort real-time systems. Tech. Rep. CMU-CS-TR-98-174, School of Computer science, Carnegie Mellon University, 1998.
11. DINDA, P. A., AND O'HALLARON, D. R. An evaluation of linear models for host load prediction. Tech. Rep. CMU-CS-TR-98-148, School of Computer Science, Carnegie Mellon University, November 1998.
12. EMBLEY, D. W., AND NAGY, G. Behavioral aspects of text editors. *ACM Computing Surveys 13*, 1 (January 1981), 33–70.
13. GROSCHWITZ, N. C., AND POLYZOS, G. C. A time series model of long-term NSFNET backbone traffic. In *Proceedings of the IEEE International Conference on Communications (ICC '94)* (May 1994), vol. 3, pp. 1400–4.
14. HOSKING, J. R. M. Fractional differencing. *Biometrika 68*, 1 (1981), 165–176.
15. JENSEN, E. D., LOCK, C. D., AND TOKUDA, H. A time-driven scheduling model for real-time operating systems. In *Proceedings of the Real-Time Systems Symposium* (February 1985), pp. 112–122.
16. KOMATSUBARA, A. Psychological upper and lower limits of system response time and user's preferance on skill level. In *Proceedings of the 7th International Conference on Human Computer Interaction (HCI International 97)* (August 1997), G. Salvendy, M. J. Smith, and R. J. Koubek, Eds., vol. 1, IEE, pp. 829–832.
17. LOWEKAMP, B., MILLER, N., SUTHERLAND, D., GROSS, T., STEENKISTE, P., AND SUBHLOK, J. A resource monitoring system for network-aware applications. In *Proceedings of the 7th IEEE International Symposium on High Performance Distributed Computing (HPDC)* (July 1998), IEEE, pp. 189–196.
18. PAXSON, V., AND FLOYD, S. Wide-area traffic: The failure of poisson modeling. *IEEE/ACM Transactions on Networking 3*, 3 (June 1995), 226–244.
19. SESHAN, S., STEMM, M., AND KATZ, R. H. SPAND: Shared passing network performance discovery. In *Proceedings of the USENIX Symposium on Internet Technologies and Systems* (December 1997), pp. 135–46.
20. STANKOVIC, J., AND RAMAMRITHAM, K. *Hard Real-Time Systems*. IEEE Computer Society Press, 1988.
21. WALDSPURGER, C. A., AND WEIHL, W. E. Lottery scheduling: Flexible proportional-share resource management. In *Proceedings of the First Symposium on Operating Systems Design and Implementation* (1994), Usenix.
22. WILLINGER, W., MURAD S, T., SHERMAN, R., AND WILSON, D. V. Self-similarity through high-variability: Statistical analysis of ethernet lan traffic at the source level. In *Proceedings of ACM SIGCOMM '95* (1995), pp. 100–113.
23. WOLSKI, R. Dynamically forecasting network performance to support dynamic scheduling using the network weather service. In *Proceedings of the 6th High-Performance Distributed Computing Conference (HPDC)* (August 1997).

Dynamic Real-Time Channel Establishment in Multiple Access Bus Networks

Anita Mittal[1], G. Manimaran[2], and C. Siva Ram Murthy[1]

[1] Dept.of Computer Science and Engineering,
Indian Institute of Technology, Madras 600 036, INDIA
{naman@hpc., murthy@}iitm.ernet.in
[2] Dept. of Electrical and Computer Engineering,
Iowa State Univ., Ames, IA 50011, U.S.A.
gmani@shasta.ee.iastate.edu

Abstract. Real-time communication with performance guarantees is becoming very important to many applications, like computer integrated manufacturing, multimedia, and many embedded systems. Though several real-time protocols have been proposed for the multiple access bus networks, there is no guarantee based protocol which addresses the problem of integrated scheduling of dynamically arriving periodic and aperiodic messages. In this paper, we propose two guarantee based protocols, EDF and BUS protocols, which address this problem. In the simulation studies, the performance metrics, success ratio (measure of schedulability) and channel utilization are used to study the performance of the two protocols. It is observed that the EDF protocol offers higher schedulability as compared to the BUS protocol for periodic messages, while the BUS protocol offers better channel utilization for aperiodic messages as compared to the EDF protocol.

1 Introduction

Predictable inter-task communication is of prime importance in real-time systems, because unpredictable delays in the delivery of the messages can affect the completion time of the tasks participating in message communication. Also in multimedia applications, timely delivery of messages is required to guarantee the bounds on delay and delay jitter in delivering video/audio frames consistent with human perception. In order to support real-time communication with performance guarantees, real-time channels[1] are established with the specified traffic characteristics and Quality of Service (QoS) requirements.

Distributed real-time applications are implemented using either real-time multi-hop networks or real-time multiple access networks. While schemes for channel establishment and run-time scheduling algorithms have been proposed for real-time multi-hop networks [1, 2], not much work has been done for real-time multiple access networks. Multiple access bus networks are being increasingly used, for real-time applications, because they are simple, economical, and

---
[1] A real-time channel is a unidirectional virtual circuit with specified QoS guarantees.

their propagation delays are small. In this type of network, scheduling is the responsibility of Medium Access Control (MAC) protocol, which arbitrates access to the shared medium and determines which message is to be transmitted at any given time.

Several schemes for real-time communication on multiple access networks have been proposed [3, 4], but they generally belong to the best effort category. In this paper, we propose two guarantee based protocols for dynamic real-time channel establishment with support for scheduling of aperiodic messages in multiple access bus networks.

2 Background

2.1 Desirable Properties of Real-time MAC Protocol

A real-time MAC protocol should possess the following desirable properties.

1. Support Real-time Channel Establishment: A real-time channel is established by reserving the resources as per the specified traffic characteristics of the periodic message, if the admission of this connection does not jeopardize the guarantees given to the already admitted channels. The protocol should have an associated admission test which is to be performed while admitting a periodic message stream.

2. Provide Bounded Response Time: In many real-time applications such as air-traffic control and multimedia, periodic message requests for channel establishment arrive dynamically. The protocol should provide bounded response time to a channel establishment request (i.e, time taken to decide acceptance/rejection of a request based on admission test should be bounded) so that some alternate recovery action can be initiated in case the request is rejected.

3. Integrated Scheduling of Periodic and Aperiodic Messages: Besides these periodic messages which require hard real-time guarantees, the protocol should also support scheduling of aperiodic messages which may have soft deadlines or no deadlines at all.

4. Distributed in Nature: Since the channel is shared by a number of nodes in the multiple access networks, the nodes need to cooperate and coordinate with one another (i) for transmitting channel establishment requests, (ii) for admission test and resource reservation, and (iii) for transmission of messages on the real-time channels so that all the messages meet their deadlines. These functions can be performed either in centralized or in distributed manner. Centralized scheme is prone to single point failure and also incurs substantial overhead for giving channel access rights to the nodes. Therefore distributed scheme is preferred wherein every node participates in decision making in a decentralized manner.

2.2 Prior Work

A guarantee based protocol for scheduling of aperiodic messages has been proposed in [5]. This protocol does not support scheduling of periodic messages. (ii)

A centralized guarantee based protocol for scheduling of periodic messages in multiple access bus networks has been proposed in [6]. It does not support integrated scheduling of periodic and aperiodic messages and also suffers from the disadvantages of a centralized protocol. (iii) A centralized scheduling algorithm for scheduling of periodic and aperiodic messages has also been proposed for the backplane bus of a workstation [7]. In the workstation, instant global knowledge of all the messages contending for the backplane bus is available through separate hard-wired control channels. Hence this scheduling algorithm cannot be used as such for multiple access networks. Thus we see that there is no protocol which addresses all the four requirements of a guarantee based real-time MAC protocol. This motivated us to propose guarantee based protocols for integrated scheduling of periodic and aperiodic messages in multiple access bus networks.

3 Proposed Protocols

3.1 Channel Model

- A population of N nodes share the channel.
- Channel access is slotted. Transmission can start only at the beginning of a slot. The clocks of all the nodes are synchronized.
- The slot time is equal to the maximum end-to-end round trip delay for a bit.
- Carrier sensing is instantaneous. Each node is capable of detecting an idle slot, a successful transmission, and a collision as in CSMA based protocols.
- All the nodes monitor the activity on the shared channel.

3.2 Proposed Protocols

We have proposed two protocols EDF protocol and BUS protocol. These protocols are distributed in nature with each node executing the same algorithm. The protocol alternates between two phases: reservation phase and transmission phase. The proposed protocols have the following components:

- Request Server (RS): This is a periodic server used for collecting periodic and aperiodic message requests and for giving bounded response time to channel establishment requests. During the activation of RS, in the reservation phase, message notifiers are transmitted. The notifier contains information such as, type of request - periodic/aperiodic, service time, period/deadline, and node number of the message source. One slot is required for transmission of a message notifier. Since all the nodes monitor the channel, and the channel is a broadcast medium, all the nodes get to know about the message traffic at the other nodes through these message notifiers. The preorder DCR protocol [8] is used for transmitting message notifiers. Preorder DCR protocol is an efficient protocol wherein when the load is light, say, only one message request is ready in the entire network, then only one slot of RS is used and the remaining slots can be used for transmission of messages. When the load is heavy, collision is resolved deterministically to bound the worst case channel access delay, giving every node a chance to broadcast at least one message notifier in one RS period.

- Aperiodic Server (AS): This is also a periodic server used for servicing aperiodic messages. The service could be either guaranteed service for which the guarantee based DCR protocol [5] can be used or could be a best effort service. In the best effort service category, the service policy can be either fairness oriented, where the AS bandwidth is fairly given to all the contending nodes, or aimed at maximizing throughput of aperiodic messages. The percentage of bandwidth reserved for aperiodic traffic is application dependent.
- Run-time Scheduling: The run-time scheduling algorithms, for the EDF and BUS protocols, are based on the Earliest Deadline First (EDF) algorithm [9] and backplane bus scheduling algorithm [7], respectively. These algorithms are used during the transmission phase of the protocols.
- Admission Tests: There is an associated admission test with each protocol for admission control of real-time channel establishment requests.

3.3 Protocols Description

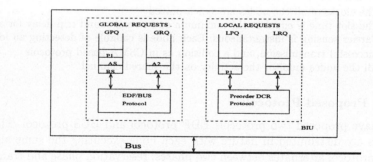

Fig. 1. A node in the multiple access network

Each node maintains two queues (local) for storing the local periodic (LPQ) and aperiodic message (LRQ) notifiers, as shown in Figure 1. A message notifier has to be broadcast before the transmission of the message. Each node also maintains two global queues, aperiodic queue (GRQ) for storing the aperiodic message notifiers which have been broadcast successfully on the bus and periodic queue (GPQ) for storing the notifiers of admitted real-time channels (periodic messages). The RS and AS notifiers, are put in the GPQ of each node. In the EDF/BUS protocol, channel access rights are given to the real-time channels for say, n slots, based on the EDF/BUS run-time scheduling algorithm. Each node executes the following protocol once the channel access is granted to a real-time channel.

Protocol(*RTchannel,n***)** /* *RTchannel* - the channel which has access rights for n slots */

1. If $RTchannel$ = RS,
   Repeat steps (a)-(c) until a slot goes idle or RS service time is exhausted.

(a) Transmit message notifiers from the local queues with priority to LPQ over LRQ as per the preorder DCR protocol.
(b) If notifier transmission is successful and the notifier is for an aperiodic message then insert the notifier in the GRQ.
(c) else if notifier transmission is successful and the notifier is for a periodic message then
   i. $result$ = Admit(notifier) /* Perform admission test using the EDF/BUS protocol admission test */
   ii. If $result = success$ then
      A. Insert the notifier in GPQ.
      B. Reserve resources for this real-time channel and update system utilization information.
   iii. else reject the request.
2. If $RTchannel$ = AS,
   (a) Grant channel access rights to messages in GRQ based on predetermined aperiodic message service policy/protocol.
3. If $RTchannel = P_i$, /* real-time channel i */
   (a) Grant channel access rights to the node which is the source of the real-time channel P_i.

When a node wishes to tear down the real-time channel, the node transmits channel termination request as the last message on this channel. All the nodes then update their GPQ and channel utilization information accordingly.

When none of the periodic messages are using the channel (no message is ready in the GPQ), then aperiodic messages (if any) are serviced. In the event of GRQ also being empty, (i.e, when the system load is light) the RS is activated, thus resulting in very less average response time to channel establishment requests compared to RS period. It can be noted that, when the system load is light, these protocols behave like a contention based protocol, since RS follows a CSMA/CD based DCR protocol. When the system load is heavy, the channel is granted to the admitted requests without any collisions based on the EDF/BUS policy. Thus the protocols adapt well to the system load.

3.4 EDF Protocol

The nodes follow the EDF algorithm of serving the earliest deadline first message in the global periodic queue. EDF [9] is a dynamic priority scheduling algorithm that always schedules the message with the earliest deadline, preempting any currently scheduled message if required. Preemption of messages takes place at packet boundaries.

Admission Test

A periodic message is admitted if it satisfies the following condition:
$$util_{curr} + util_{new} \leq 1$$
where $util_{curr}$ is the current utilization of the channel and $util_{new}$ is the channel utilization requirement of the new message, given by the expression:
$$util_{new} = \frac{\text{maximum message size/slot size}}{\text{period of the message/slot time}(\tau)}$$

3.5 BUS Protocol

The BUS protocol is based on backplane bus scheduling algorithm [7] and uses RS for request collection and AS for servicing aperiodic messages. In a multiple access bus network, the periodic traffic arrives at each node independently unlike in backplane bus where it is centralized. Moreover, there is no central Bus Arbiter (BA) which performs access arbitration based on channel reservation of the admitted requests. Also there are no control lines to convey bus request and bus grant information. In the BUS protocol, the nodes follow a periodic service policy similar to the one executed by the BA with S slots and cycle time (period) $S\tau$. The service times and periods of the RS and AS are converted to the number of slots required per service cycle to R_s and A_s, respectively.

Admission Test

A periodic message stream with message size M_i and period T_i is admitted if the following condition holds:
$$c_i + Q \leq S - \alpha$$
where c_i is the smallest integer that satisfies
$$M_i \leq c_i(\lceil T_i/(S*\tau)\rceil - 2)$$
Q is the number of time slots in a cycle reserved for periodic messages, and α denotes the number of slots dedicated for random (aperiodic) messages. α is a system parameter that should be changed only when a change in the fraction of allowable periodic traffic is desired.

The algorithm executed by each node is:

Algorithm BUS(R_s, A_s, S, Q) /* Initially n and q counters are set equal to S and Q, respectively */

1. If $n = S$, service RS for R_s slots or until a slot goes idle whichever is earlier.
2. If $n > q$ and aperiodic queue is not empty, service AS for $(n - q)$ slots or until aperiodic queue becomes empty whichever is earlier.
3. If $q > 0$ and $n > q$, service periodic requests and decrement q accordingly.
4. If $n > 0$ and $q = 0$, service AS if aperiodic queue is not empty otherwise service RS till a request arrives.
5. if $n = 0$, set n and q to S and Q, respectively.

3.6 Protocols Comparison

Schedulability of Periodic Messages: The EDF protocol offers higher schedulability than the BUS protocol. This is because the admission test of BUS protocol slightly over allocates bandwidth to periodic messages and as a result of this over allocation lesser number of real-time channels can be established. However, at run time this over allocated channel bandwidth is made available for servicing aperiodic messages.

Response Time to Aperiodic Messages: The BUS scheduling scheme gives fast response to aperiodic messages without jeopardizing the guarantees given to admitted periodic messages. Aperiodic messages (if any) are serviced when there are no periodic messages to be transmitted, by activating the AS for idle channel duration. In the BUS protocol, the idle slots are uniformly distributed since the service cycle time is much less than the deadline of aperiodic messages, thus resulting in better channel utilization for aperiodic messages. In the EDF protocol, besides lack of over allocated bandwidth being made available for aperiodic messages, occurrence of idle channel instances, when there are no periodic messages to be transmitted, is not uniform (the idle channel bandwidth is made available in bursts) and hence aperiodic messages cannot utilize the idle channel bandwidth efficiently.

Granularity of AS: In EDF protocol, though the idle channel time cannot be given uniformly, the AS time can be made available uniformly by changing the granularity of the AS, i.e., the period and service time would be divided by the same constant factor. For example, service time of 4 units and period 10 units can be transformed into a server with service time 2 units and period 5 units. Similarly, the granularity of RS can be reduced to improve response to message requests.

The differences mentioned above between the two protocols can be clearly observed in the simulation results.

4 Simulation Study

The simulation parameters are given in Table 1. The slot time is assumed to be 5 micro seconds and slot size 64 bytes. The bandwidth reserved for AS/RS can be generalized as $(C_s/m_{as/rs})/(P_s/m_{as/rs})$, where C_s and P_s gives the service time and period of the AS/RS, respectively and $m_{as/rs}$ parameter (m is an integer, $m \geq 1$) decides the granularity. As $m_{as/rs}$ increases, granularity decreases which means decrease in period and service time of the AS. When $m = 1$, granularity is maximum.

parameter	explanation	distribution	value/range
N	number of nodes		16
RS_BW	bandwidth reserved for RS		1%
AS_BW	bandwidth reserved for AS		10%
PerMesgSize	periodic message size (bytes)	uniform	4500-6400
Period	period of messages	uniform	4000-6000
AperMesgSize	aperiodic message size (bytes)	uniform	1000-5000
AperLaxity	laxity of aperiodic messages	uniform	20-100
Instances	number of instances of periodic messages	uniform	30-50
PLoad	periodic load	Poisson	0.1-3
ALoad	aperiodic load	Poisson	0.5-3

Table 1. Simulation parameters

4.1 Performance Metrics

1. Effective Channel Utilization, ECU, defined as
$$ECU = \frac{\text{total time units channel used for transmitting messages}}{\text{total time units simulated}}$$
This metric when applied to periodic, aperiodic messages and RS individually gives the metrics PCU, ACU, and RCU respectively. RCU captures the overhead for collecting requests and this should be small.
2. Success Ratio, SR, defined as
$$SR = \frac{\text{number of messages serviced}}{\text{total number of messages generated}}$$
This metric when applied to periodic and aperiodic messages individually gives the metrics PSR and ASR, respectively.

4.2 Simulation Results

Variation of system load: Figure 2 shows the channel utilization by periodic and aperiodic messages and RS for varying system load for both the BUS and EDF protocols.

Effect on RCU: As the load increases, RCU decreases. This is because the RS is active when the channel is idle and hence when load is low RCU is high. When system becomes overloaded, idle time decreases and hence RCU decreases, and approaches the reserved bandwidth utilization bound.

Effect on PCU: The PCU for the EDF protocol is slightly higher than that of BUS protocol. Reason is the same as mentioned earlier. It may be noted that at low loads PCU is same for both the protocols, since channel bandwidth exceeds the total bandwidth demand of all the message requests.

Effect on ACU: ACU increases with increasing load up to a point and then starts decreasing. Aperiodic messages use the bandwidth reserved for them and also use the channel when it is idle. When the periodic load increases, PCU increases and hence idle time decreases. Therefore aperiodic message utilization decreases. ACU for the BUS protocol is higher than that of the EDF protocol. Reason is the same as mentioned in Sect. 3.6.

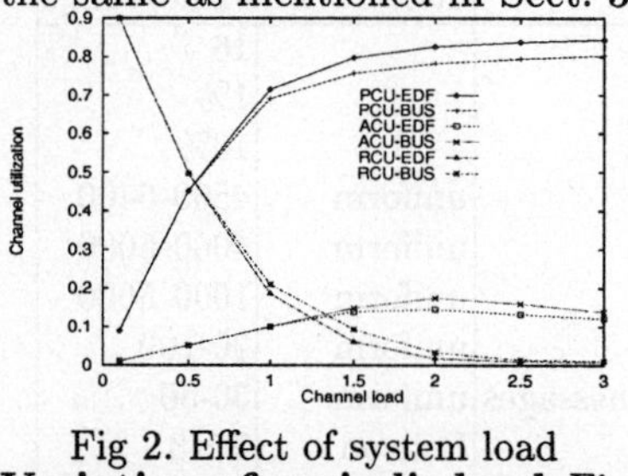

Fig 2. Effect of system load 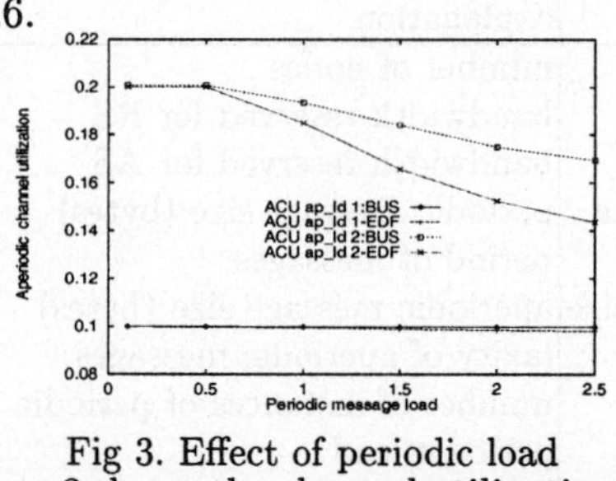   Fig 3. Effect of periodic load

Variation of periodic load: Figure 3 shows the channel utilization by aperiodic messages for varying periodic message load. ACU has been plotted for fixed aperiodic loads of 1 and 2. As the periodic load increases, ASR remains same as long as the aperiodic load does not exceed the reserved AS capacity. When aperiodic load is twice the AS capacity, then ACU decreases with increasing periodic load. When periodic load is negligible (0.1) all the aperiodic messages get serviced. This is because the entire channel bandwidth is available for RS and AS. As periodic load increases, the idle time decreases and hence ACU decreases.

Variation of aperiodic load: Figure 4 shows the PCU for varying aperiodic message load plotted for fixed periodic loads of 0.5, 1 and 2. We see that PCU is independent of aperiodic load. This is expected since the presence or absence of aperiodic requests does not affect the bandwidth reserved for periodic messages and hence insensitivity of metric PCU to aperiodic load. When the periodic load is 0.5 both the EDF and BUS protocols have the same PCU, but as the load increases, the EDF protocol performs better and the difference in PCU between these two protocols increases.

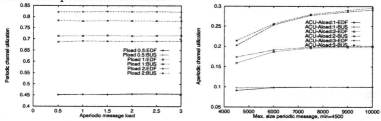

Fig 4. Effect of aperiodic load Fig 5. Effect of stochastic periodic message size

Effect of stochastic nature of periodic message size: Figure 5 shows the effect on ACU by variation in the range (difference between the maximum and minimum size of messages of a periodic message stream) of periodic request message size. It is to be noted that the admission test is based on the maximum message size of the periodic message request, and hence some of the reserved bandwidth is not used when instances of that periodic message stream have shorter message sizes. This bandwidth is made available for servicing aperiodic messages and hence the aperiodic channel utilization increases. The implication of this result is that the bandwidth reservation for aperiodic messages needs to be less, if the periodic message sizes are stochastic in nature.

Effect of server granularity: Here, we examine the effect on ASR, by varying the AS granularity for the EDF protocol.

(i)Effect of RS granularity: Figure 6 studies the influence on ASR by variation in AS granularity for RS granularity m_{rs} values equal to 1, 12, and 24. The curves are plotted for periodic and aperiodic message load of 1, aperiodic message size of 1000 bytes, and laxity value of 20. It can be observed that RS granularity significantly effects ASR. This is because as the granularity of RS decreases (m_{rs} increases), responsiveness to newly arrived requests increases. Since the period of RS becomes small, the number of new requests in that period decreases, decreasing the probability of collisions and as a result the efficiency of RS increases. ASR is maximum when the RS granularity is least i.e, $m_{rs} = 24$. The ASR increases with decrease in granularity, till $m_{as} = 12$ and then decreases for further decrease in granularity ($m_{as} = 24$). This is because the number of slots required for transmission of an aperiodic message equals the slots serviced per AS period when $m_{as} = 12$, and on further increase in m_{as} no aperiodic message is serviced in entirety in one period of AS and hence the decrease in ASR. It is to be observed that ASR is maximum for $m_{rs} = 24$ and $m_{as} = 3$.

Appropriate choice of these two parameters needs to be made depending on the characteristics of aperiodic messages.

(ii)**Effect of laxity**: Figure 7 shows the effect on ASR for variation in AS granularity for different laxities of aperiodic messages. It can be seen that the granularity effect is almost negligible for higher laxities. This is because with increase in laxity of aperiodic messages, the deadline of messages becomes larger than the highest granularity of AS with $m_{as} = 1$. The RS granularity effect can still be observed though it is moderate for higher laxities.

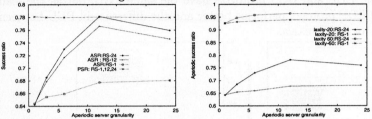

Fig 6. Effect of request server granularity   Fig 7. Effect of laxity

5 Conclusion

We proposed the EDF and BUS protocols, for integrated scheduling of dynamically arriving periodic and aperiodic messages in multiple access bus networks. The proposed protocols can be extended to exploit the variable QoS requirements of messages to improve schedulability and fault tolerance of the system.

References

1. D.Ferrari,and D.C.Verma,"A Scheme for real-time channel establishment in wide area networks",*IEEE J.on Selected Areas in Commun.*, vol.8, no.4, 368-379, 1990.
2. D.D.Kandlur, K.G.Shin, and D.Ferrari, "Real-time communication in multi-hop networks", *IEEE Trans. on Parallel and Distributed Systems*, vol.5, no.10, 1044-1055, 1994.
3. J.F.Kurose, M.Schwartz, and Y.Yemini, "Multiple- access protocols and time constrained communication", *ACM Computing Surveys*, vol.16, no.1,43-70, 1984.
4. N.Malcom and W.Zhao, "Hard real-time communication in mutiple-access networks", *Real-Time Systems*, no.8, 35-77, 1995.
5. S.Norden, G.Manimaran, and Siva Ram Murthy, "Dynamic Planning based protocols for real-time communication in multiple access networks", *Technical Report, Dept.of Computer Science and Engg., Indian Institute of Technology Madras*, 1997.
6. C.C.Chou, and K.G.Shin,"Statistical real-time channels on multi-access bus networks",*IEEE Trans. on Parallel and Distributed Systems*,vol.8,no.8,769-780, 1997.
7. S.H.Khayat, and B.D.Bovopoulos, "A simple and efficient bus management scheme that supports continuous streams", *ACM Trans. on Computer Systems*, vol.13, no.2,122-140,1995.
8. V.Upadhya, "Design and analysis of real-time LAN protocols", *M.S.Thesis, Dept. of Computer Science and Engg., Indian Institute of Technology, Madras*, 1996.
9. C.L.Liu, and J.W.Layland, "Scheduling algorithms for multiprogramming in a hard-real-time environment", *Journal of the ACM*, vol.20, no.1,45-61, 1973.

A Similarity-Based Protocol for Concurrency Control in Mobile Distributed Real-Time Database Systems[*]

Kam-yiu Lam[1], Tei-Wei Kuo[2], Gary C.K. Law[1] and Wai-Hung Tsang[1]

Department of Computer Science[1]
City University of Hong Kong
83 Tat Chee Avenue, Kowloon
HONG KONG
Email: cskylam@cityu.edu.hk

Department of Computer Science and
Information Engineering[2]
National Chung Cheng University
Chiayi, 621 Taiwan, ROC
email: ktw@cs.ccu.edu.tw

Abstract

Research in the concurrency control of real-time data access over mobile networks is receiving growing attention. With possibly lengthy transmission delay and frequent disconnection, traditional concurrency control mechanisms may become very costly and time-consuming in mobile distributed real-time database systems (MDRTDBS). Due to limited bandwidth and unpredictable behavior of mobile networks, the past research on concurrency control for distributed real-time database systems (DRTDBS) can not be directly applied to MDRTDBS. In this paper, we propose a distributed real-time locking protocol, called SDHP-2PL, based on the High Priority Two Phase Locking (HP-2PL) scheme and the concept of similarity for MDRTDBS. We consider the characteristics of mobile networks and adopt the concept of similarity to reduce the possibility of lock conflicts. A detailed model of a MDRTDBS has been developed, and a series of simulation experiments have been conducted to evaluate the capability of SDHP-2PL and the effectiveness of using similarity for MDRTDBS. The simulation results have confirmed our belief that the use of similarity as the correctness criterion for concurrency control in MDRTDBS can significantly improve the system performance.

1 Introduction

In recent years, the research in *mobile distributed real-time database systems (MDRTDBS)* is receiving growing attention due to a large number of potential mobile computing applications, such as real-time traffic monitoring systems and mobile internet stock trading systems [3]. Owing to the intrinsic limitations of mobile computing systems, such as limited bandwidth and frequent disconnection, the design of an efficient and cost-effective MDRTDBS requires techniques, which are quite different from those developed for *distributed real-time database systems* (DRTDBS) over wired networks [7,9,10]. How to meet the timing constraints of transactions over a mobile network and, at the same time, to maintain the external and temporal consistency of the databases involves various challenging research issues [10]. Up to now, to our knowledge, it is still lack of a detailed study on the design of MDRTDBS.

Concurrency control has been an active research topic for *real-time database systems (RTDBS)* [6,11]. In the last decade, researchers have proposed various real-time concurrency control protocols, e.g., [1,5], for single-site as well as for DRTDBS. Efficient techniques were proposed to bound the number of priority inversions [1]. In RTDBS, priority inversion is, in general, resolved by either transaction restarts or by priority inheritance [2]. For example, the High Priority Two Phase Locking (HP-2PL) protocol [1] restarts a lower-priority transaction if there exists a higher-priority transaction that wants to set a lock on a data object currently locked by the lower-priority transaction in a conflicting mode.

---

[*] Supported in part by a research grant from the National Science Council under Grants NSC88-2213-E-194-034 and Strategic Grant # 7000584 City University of Hong Kong.

Although many excellent concurrency control protocols have been proposed for RTDBS and DRTDBS, the proposed protocols, in general, can not be directly applied to MDRTDBS due to the unique characteristics of MDRTDBS [7,10]. In a MDRTDBS, whether a transaction can meet its deadline highly depends on the properties of the communication network. The unpredictable propagation delay of mobile networks imposes a serious time burden on the performance of MDRTDBS, and the delay can greatly affect the performance of any adopted concurrency control protocol. Compared with wired networks, a mobile network is often much slower, unreliable, and unpredictable. The mobility of clients in a mobile computing system also greatly affects the distribution of workload in the communication network. Disconnection between clients and stations is common [3,7]. The poor quality of service provided by a mobile network usually seriously increases the overheads in resolving data conflicts and affects the performance of existing concurrency control protocols which do not consider the characteristics of mobile networks.

This paper proposes a distributed real-time locking protocol, called SDHP-2PL which is based on the High Priority Two Phase Locking (HP-2PL) scheme [1] and the concept of *similarity* [4,5], for MDRTDBS. We consider the characteristics of mobile networks and adopt the concept of *similarity* to reduce the possibility of lock conflicts and to increase the system concurrency, which is vital to systems "virtually divided" by slow and unreliable mobile networks. In the SDHP-2PL, the priority inheritance mechanism is mixed with semantics-based concurrency control mechanisms to boost the performance of the proposed locking protocol for data access over a mobile network.

2 Model of a Mobile Distributed Real-Time Database System
2.1 System Architecture

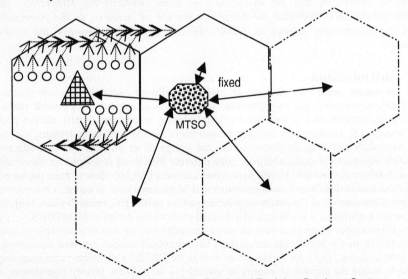

Figure 1: System Architecture of the Mobile Distributed Real-time Database System

A MDRTDBS consists of three major components: mobile clients (MC's), base stations, and a main terminal switching office (MTSO), as shown in Figure 1. The entire service area is divided into a number of connected cells. Within each cell, there is a base station that is augmented with a wireless interface to communicate with the MC's within its cell. The base stations in different cells are connected to the MTSO by a point-to-point wired network.

Attached to each base station is a database system containing a local database which may be accessed by transactions issued by MC's within its cell or from the other cells via the MTSO. The MTSO is responsible for active call information maintenance, handoff procedure, channel allocation, and message routing.

An MC may move around within a cell or cross the cell border to enter another cell. Periodically, an MC sends a location signal to its assigned base station. The strength of the location signal received by the base station depends on the distance between the MC and the base station. When an MC is crossing the cell border, the signal may become very weak. When the signal received by the base station is lower than a certain threshold level, the MTSO will be notified. The MTSO will perform a handoff procedure. It sends out requests to all the base stations to acquire their responses regarding the strength of the location signal received by them from the MC. The MTSO will then re-assign the MC to the base station which has received the strongest signal from the MC. Usually this is the base station which is responsible for the cell which the MC is entering.

When a transaction is generated from an MC, the MC will first attempt to issue a call request to the base station of its cell. A channel is granted to the MC after the completion of a setup procedure. The execution of the setup procedure incurs a fixed overhead as it involves communication amongst the MC, the base station, and the MTSO. Since the number of channels for communication between a base station and its MC's is limited, it is possible that the channel request is refused due to no free channel available. If the number of attempts issued by the MC exceeds a certain specified threshold number, the call and the transaction will be aborted. Due to slow and unreliable communication, and channel contention, the time required to establish a channel is highly unpredictable. Once a channel is established, the transaction will be sent out through the RF transmitter (which is connected to the antenna) from the MC to the base station of the cell where the MC currently resides. When an MC is crossing the cell border while it is communicating with its base station, a new channel will be created with its newly assigned base station after a setup procedure is done by the MTSO.

2.2   Database and Transaction Models

The database is distributed among the base stations. At each base station, there is a local database that consists of two types of data objects: *temporal* and *non-temporal* data objects. Temporal data objects are used to record the status of external objects in the real world. Each temporal data object is associated with a timestamp that denotes the age of the data object. A transaction is given a timestamp, when the transaction is initiated, if the transaction may update a temporal data object. If the transaction commits successfully before its deadline, the timestamp of the data object updated by the transaction will be set as the timestamp of the transaction. The validity of a temporal data object is indicated by an *absolute validity interval (avi)* [8]. A temporal data object satisfies the *avi* constraint if the age of the data object is up to date, i.e., the difference of the current time and the age is no more than *avi*. A *relative validity interval (rvi)* may also be given to a transaction which requires that the maximum age difference of data objects read by the transaction is no larger than *rvi* [8].

An MC may initiate a transaction sporadically. Transactions from the MC's are consisting of a sequence of read and write operations. Each transaction issued from an MC is associated with a deadline and a numerical value expressing its criticality. The priority of a transaction is derived from these values. The operations may access data objects residing at several base stations. When an operation of a transaction accesses data objects residing on a base station other than the MC's current one, the operation will be routed to the base station via the MTSO. When all the operations of a transaction have been processed, a commit protocol will be performed to ensure the failure atomicity of the transaction.

3 Similarity-Based Concurrency Control Protocol
3.1 The Basic Strategy
3.1.1 Data Similarity

Similarity is closely related to the important idea of imprecise computation in real-time systems and to the idea of partial computation for database. For many real-time applications, the value of a data object that models an entity in the real world cannot, in general, be updated continuously to perfectly track the dynamics of the real-world entities. It is also *unnecessary* for data values to be perfectly up-to-date or precise to be useful. In particular, data values of a data object that are slightly different are often interchangeable. The concept of *similarity* can be used to extend the conventional correctness criteria for concurrency control in database systems [4,5] and to balance data precision (based on *similarity*) and system workload. If two operations in a schedule are strongly similar, then they can always be used interchangeably in a schedule without violating the integrity and consistency of the database.

3.1.2 The Basic Strategy for Conflict Resolution

Specifically, we assume that the application semantics allows us to derive a similarity bound for each data object such that two write operations on a data object must be similar if their time-stamps differ by an amount not greater than the similarity bound. The basic strategy for conflict resolution in the similarity-based Distributed High Priority Two Phase Locking (SDHP-2PL) can be summarized as follows:

(i) Suppose that a transaction T_i issues a read-lock request on data object D_k, and D_k is already write-locked by a collection of transactions TH = $\{T_j, ..., T_y\}$. Let T_j be the most recently committed transaction which updates D_k. If the write operation (on D_k) of T_j and the write operations (on D_k) of *all* transactions in TH are *similar*, then the read-lock request on data object D_k should be granted because T_i will always read *similar* data values, regardless of which transaction in TH commits (or updates D_k first).

(ii) Suppose a transaction T_i issues a write-lock request on data object D_k, and D_k is already write-locked by a collection of transactions TW and read-locked by a collection of transactions TR. Let T_j be the most recently committed transaction which updates D_k. There are two cases to consider:

 (1) If TR is empty, and the conflicting write operation of T_i and the conflicting write operations of all transactions in TW are similar, then the write-lock of T_i should be granted. It is because the final database state will be similar regardless of which write operation is performed first.

 (2) If TR is not empty, and the conflicting write operation of T_i, the conflicting write operation of T_j (the last committed transaction on D_k), and the conflicting write operations of all transactions in TW are similar, then the write-lock of T_i should be granted. It is because the final database state will always be *similar* and the read operations will read *similar* data values regardless of which write operation is performed first.

Obviously, if the above *similarity* test of a lock request fails, the lock-requesting transaction should be blocked or the transactions which cause the blocking might be restarted. In the SDHP-2PL, we adopt the following policies:

Suppose that the *similarity* test fails for the lock request of transaction T_i on data object D_k, and TH is a set of transactions which currently read-lock or write-lock D_k. Let CT be a subset of TH which consists of transactions that lock D_k in a mode conflicting and non-similar to the lock request of T_i, and SCT be a subset of CT which consists of transactions with a priority lower than T_i. If the restart of all transactions in SCT will make T_i passing the *similarity* test, then the transactions in SCT are restarted, and the lock request of T_i is granted. Otherwise, T_i is blocked. Thus, blocking will be used if the priority of any of the lock-holding transactions is higher than the priority of the lock requesting transaction.

3.2 The Protocol

In SDHP-2PL, it is assumed that a distributed lock scheduler is deployed at each base station. The lock scheduler of a base station manages the lock requests for the data items residing at the base station. Let a transaction T_r request a lock in a specified mode, i.e., read or write, on a data object O_k residing at a base station. T_r should invoke the lock request procedure of SDHP-2PL, which is defined as follows, to receive the lock. The invocation may cause the restart of lower-priority transactions, the blocking of T_r, or the granting of the lock request.

```
Lock-Request (T_r, Mode, O_k)
Begin
    If the lock request of T_r is similar to existing locks on
    data object O_k
    Then
    /* Please see Section 3.1.2 for the similarity of locks */
        Return(Success)
    Else
        Let CT be the set of transactions that lock O_k in a mode
        conflicting and non-similar to the lock request of T_r
        For each transaction T_c in CT Do
            If   Priority(T_r) > Priority(T_c)
                Then
                    If T_c is not committing and restarting
                       Then
                          Restart T_c locally (or globally) if T_c is a
                          local (or global) transaction.
                          Remove T_c from CT
                       Else
                          Priority(T_c) = Priority(T_r) + C
                    EndIf
            EndIf
        EndFor
        If  CT is still not empty
            Then
                Block T_r until all transactions in CT release
                conflicting and non-similar locks.
        EndIf
        Return(Success)
    EndIf
End
```

If the lock request is *similar* to the existing locks on the same data object, then the lock request is granted. Otherwise, the set of transactions, which lock O_k in a mode conflicting and *non-similar* to the lock request of T_r, is considered for restart or priority inheritance. If a transaction T_c in CT has a lower priority than the priority of T_r, and the transaction is committing or restarting, then we should accelerate the execution of T_c to release the lock that blocks T_r. The acceleration is done by letting T_c inherit the priority of T_r, where C is a tunable constant to further boost the priority of transactions that blocks T_r such that T_c can complete as soon as possible.

If a transaction T_c in CT has a priority lower than T_r, and T_r is not committing or restarting, then we simply restart T_c. If T_c is a local (or global) transaction, we restart it locally (or globally). When a transaction is restarted, no lock-requesting transactions can get any lock on any data object locked by the restarted transaction until the undo operations of the restarted transaction are completed. The time required to complete the undo operations can be shortened by using of the idea of priority inheritance.

Note that the SDHP-2PL is a variation of HP-2PL, and HP-2PL is deadlock-free. Obviously, the priority inheritance mechanism in the lock request procedure of SDHP-2PL will not introduce any type of deadlocks to HP-2PL because the priority inheritance mechanism is only used to resolve lock conflicts between T_r and committing transactions. When there exist transactions which lock some data object O_k in a mode conflicting and non-similar to the lock request of T_r, the restarting policy of SDHP-2PL also will not introduce deadlocks to HP-2PL. Furthermore, the *similarity* consideration of locks only increases the probability of a lock request to be granted, and it surely will not introduce any deadlock to the system. Therefore, SDHP-2PL is deadlock-free.

4 Performance Studies
4.1 Simulation Model

We have built a detailed simulation program for the MDRTDBS model discussed in Section 2. The simulation program implements the essential functionality of a mobile personal communication system (PCS) and a DRTDBS. For instance, the call setup and tear down features are modeled explicitly. The mobile clients move around and may cross the cell border into another cell. Based on the received signal strength from each base station, they can self-locate themselves. The handoff procedure is handled by the MTSO. In the system level, we have implemented a DRTDBS.

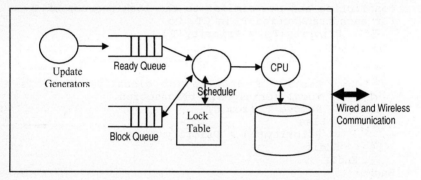

Figure 2: Model of the database system in a base station

The database system at each base station consists of a scheduler, a CPU with a ready queue, a local database, a lock table, and a block queue, as shown in Figure 2. Since a main-memory database system usually better supports real-time applications [1], it is assumed that the database resides in the main memory to eliminate the impact of disk scheduling on the system performance. In order to focus our simulation objectives and exclude unnecessary factors, it is assumed that all the temporal data objects have the same *avi and rvi*. In each local database system, an update generator creates updates periodically to refresh the validity of the temporal data objects. Updates are single operation transactions. They do not have deadlines. The priorities of the updates are assigned to be higher than the priorities of all the transactions in order to maintain the validity of the temporal data objects [8].

Transactions are generated from the mobile clients sporadically. It is assumed that all the transactions are firm real-time transactions. Each transaction is defined as a sequence of operations. Once a transaction is generated at a mobile client, it will be forwarded to that base station. A coordinator process will be created at the base station. It is responsible for the processing and committing of the transaction. The operations in a transaction will be processed one after one. If the required data item of an operation is located at another base station, the transaction will be forwarded to the base station. Occasionally, an operation may need to return to its originating mobile client for getting input data. We define the probability of an operation returning to the originating mobile client as the *return probability* (RR). Return probabilities can greatly affect the performance of the system as a higher return probability means a greater dependency on the mobile network.

Transactions wait in the ready queue for CPU allocation, and the CPU is scheduled according to the priorities of the transactions. The scheduler will check the deadline before a transaction is allocated the CPU. If the deadline is missed, the transaction will be aborted immediately. If any of the temporal data objects accessed by the transaction is invalid before the commit of the transaction, the transaction will also be aborted as it has read some out-dated data objects. After the completion of a transaction, the mobile client will generate another transaction after a think time.

4.2 Model Parameters and Performance Measures

Similar to many previous studies on RTDBS and DRTDBS, the deadline of a transaction, T, is defined according to the expected execution time of a transaction [1,9]. The following table lists the model parameters and their base values.

Parameters	Baseline Values
Number of MTSO	1
Number of Cells	7
Location Update Interval	0.2 second
Transmission Speed for Channel	10 kbps
Number of Channels for each Cell	10
Think Time	8 seconds
Transaction Size	7 to 14 operations, uniform distribution
Write operation probability (WOP)	1
Return Probability	0.6
Slack Factor	Between 8 and 12 (uniformly distributed)
Similarity bound (SB)	15 seconds
Number of Mobile Clients	84
Channel Connection time (T_{comm})	1 second
Call Update Interval	0.2 second
Number of Local Databases	7 (1 in each base station)
Database Size	100 data objects per local database
Fraction of Temporal Data Objects	10%
Temporal Data Object Update Interval	0.5 update per second
Absolute Threshold	12 seconds
Relative Threshold	8 seconds

The channel connection time is the time required to send a message from an MC to its base station. It is defined on the transmission speed of the channel and the message size. For a transmission speed of 10Kbps, the channel connection time is 1 second for a message of 1Kbytes. The location update interval is smaller than the channel connection time. Otherwise, a channel will not be able to be created after an MC has crossed the border into another cell.

The primary performance measure used is the miss rate (MR) which is defined as the number of deadline-missing transactions over the total number of transactions generated in the system. In addition to the miss rate, we also measure the restart probability which is defined as the total number of restarts over the number of committed transactions.

4.3 Simulation Results and Discussions

In the experiments, we test the performance of the SDHP-2PL under different similarity bounds to investigate the effectiveness of using the concept of similarity as the correctness criterion for concurrency control. When the similarity bound is set to zero, the protocol is similar to a distributed HP-2PL except that priority inheritance is used to resolve the conflict when the lock holder is committing.

Figure 3 gives the impact of different similarity bounds on MR at return probabilities of 0.6 and 0.8. As explained in section 4.1, the return probability defines the probability of a transaction returning to its originating mobile client after the completion of an operation. An increase in return probability increases the workload on the mobile network as a transaction will have a higher probability to go back to its originating mobile client. Thus, the miss rate is higher at a higher return probability. In Figure 3, we can see that although the improvement in MR is marginal if the similarity bound is small, a great improvement is achieved for large similarity bounds. The miss rate at similarity bound of 30 sec is only about 40% of that at the similarity bound of 5 sec. The improvement is due to smaller number of blocking and restart probability. When the similarity bound is large enough, e.g., 15 sec or greater, more transactions may be allowed to access the locks in conflicting modes in the SDHP-2PL if the conditions for the similarity test (defined in section 3.2) are satisfied. However, if the similarity bound is zero or small, more transaction restarts, and blocking will occur. The probability of deadline missing will be much higher. The restarted transactions increase the system workload, especially the communication channels between the mobile clients and the base stations. Thus, the channel blocking probability will be higher and the transactions will require a much longer time for completion. The small improvement in MR at small similarity bounds may come from the fact that even though a transaction may be allowed to set a lock (if it is similar to all the existing lock holders), other transactions may not be similar. Thus, blocking and restarts may still occur. Furthermore, the number of transaction restart will be greater and the blocking time will be longer due to a greater number of lock holders.

The smaller probability of transaction restart due to similarity can be observed in Figure 4. One interesting fact we can see from Figure 4 is that the restart probability at return probability of 0.6 is higher than that at return probability of 0.8 except at large similarity bounds, e.g., greater than 15 sec. The reason is that at a higher return probability, more transactions will be aborted before they set any locks due to deadline missing.

The impact of return probability on the performance of the system at a similarity bound of 15 sec is shown in Figure 5 and Figure 6. It can be seen that the use of *similarity* can consistently improve the system performance at different return probabilities. Greater improvement is obtained at smaller return probabilities. The reason is that at a larger return probability, the transactions will have a higher probability of deadline missing due to fewer channel contention (even though they do not have any data conflict). This can be observed in Figure 6 in which we can see that the restart probability is higher for similarity bound of 0 second than a similarity bound of 15 seconds.

Figure 7 depicts the miss rate at different write operation probabilities of transactions. It can be observed that the improvement will be greater for the write operation probabilities between 0.3 to 0.5. If the operations are mainly write operations, the conditions is more difficult to be satisfied as the conditions is more restrictive for write operations. (Please refer to section 3 for the details.) If most of the operations are read operations, the probability of lock conflict will be low. At a write operation probability of zero, there is no lock conflict as all the lock requests are read locks. Thus, the performance of the system at similarity bound of 15 seconds is the same as at similarity bound of 0 second. Increasing the write operation probability increases the lock conflict probability. Thus, the difference between similarity bound = 15 and similarity bound = 0 becomes greater. When there are a large number of write operations, the probability of satisfying the conditions in *similarity* check will be smaller. Thus, the improvement becomes smaller.

5   Conclusions

This paper proposes a real-time locking protocol, called SDHP-2PL, based on the High Priority Two Phase Locking (HP-2PL) scheme [1] and the concept of *similarity* for MDRTDBS. We consider the characteristics of mobile networks and adopt the concept of *similarity* to reduce the possibility of lock conflicts and to increase the concurrency degree of the system, which is vital to systems "virtually divided" by slow and unreliable mobile networks. The priority inheritance mechanism is mixed with semantics-based concurrency control mechanisms to boost the performance of the proposed locking protocol for data access over a mobile network. We propose a detailed model for MDRTDBS and conduct a series of simulation experiments to evaluate the capability of SDHP-2PL and the effectiveness of using *similarity* for MDRTDBS. Since bandwidth is one of the most scarce resources in a mobile environment, the study shows that the use of *similarity* can significantly improve the system performance by reducing the number of lock conflicts. We also observe that the effectiveness of *similarity* depends highly on the return probability of the transactions and the read operation probability. The use of *similarity* can significantly improve the system performance in terms of miss rate. This is especially true when the probability of write operations is higher and the return probability is not very high.

References

[1] R. Abbott and Hector Garcia-Molina, "Scheduling Real-Time Transactions: A Performance Evaluation," *ACM Trans. on Database Systems*, vol. 17, no. 3, pp. 513-560, Sep. 1992.
[2] J. Huang, J. Stankovic & K. Ramamritham, "Priority Inheritance in Soft Real-Time Databases", Journal of Real-Time Systems, vol. 4, no. 3, pp. 243-268, 1992.
[3] T. Imielinski and B. R. Badrinath, "Mobile Wireless Computing: Challenges in Data Management," *Communications of the ACM*, vol. 37, no. 10, Oct. 1994.
[4] T.W. Kuo and A.K. Mok, "Application Semantics and Concurrency Control of Real-Time Data-Intensive Applications," *Proceedings of IEEE 13th Real-Time Systems Symposium*, 1992.
[5] T.W. Kuo and A.K. Mok, "SSP: a Semantics-Based Protocol for Real-Time Data Access," *Proceedings of IEEE 14th Real-Time Systems Symposium*, December 1993.
[6] G. Ozsoyoglu and R. Snodgrass, "Temporal and Real-Time Databases: A Survey", *IEEE Transactions on Knowledge and Data Engineering*, vol. 7, no. 4, pp. 513-532, 1995.
[7] E. Pitoura and Bharat Bhargava, "Revising Transaction Concepts for Mobile Computing," *Proc. Workshop on Mobile Computing Systems and Applications*, pp. 164-168, USA, 1995.
[8] K. Ramamritham, "Real-time Databases", *International Journal of Distributed and Parallel Databases*, vol. 1, no. 2, 1993.
[9] O. Ulusoy, "Processing of Real-time Transactions in a Replicated Database Systems", *Journal of Distributed and Parallel Databases*, vol. 2, no. 4, pp. 405-436 1994.
[10] O. Ulusoy, "Real-Time Data Management for Mobile Computing", *Proceedings of International Workshop on Issues and Applications of Database Technology (IADT'98)*, Berlin, Germany, July 1998.
[11] P.S. Yu, K.L. Wu, K.J. Lin & S.H.Son, "On Real-Time Databases: Concurrency Control and Scheduling," *Proceedings of IEEE*, vol. 82, no. 1, pp. 140-57, 1994.

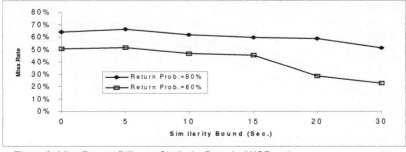

Figure 3. Miss Rate at Different Similarity Bounds (WOP = 1)

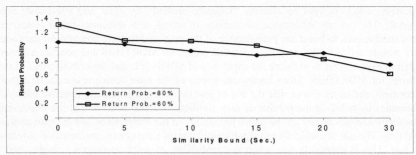

Figure 4. Restart Probability at Different Similarity Bounds (WOP =1)

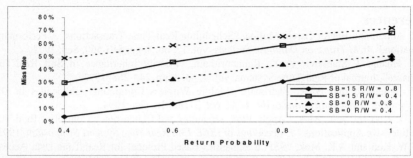

Figure 5. Miss Rate at Different Return Probability

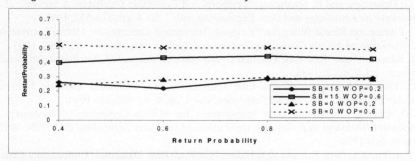

Figure 6. Restart Probability at Different Return Probability

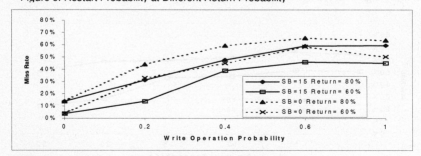

Figure 7. Miss Rate at Different Write Operation Probability

From Task Scheduling in Single Processor Environments to Message Scheduling in a PROFIBUS Fieldbus Network

Eduardo Tovar[1], Francisco Vasques[2]

[1] Department of Computer Engineering, Polytechnic Institute of Porto, Rua de São Tomé, 4200 Porto, Portugal
emt@dei.isep.ipp.pt
[2] Department of Mechanical Engineering, University of Porto, Rua dos Bragas, 4099 Porto Codex, Portugal
vasques@fe.up.pt

Abstract. In this paper we survey the most relevant results for the priority--based schedulability analysis of real-time tasks, both for the fixed and dynamic priority assignment schemes. We give emphasis to the worst-case response time analysis in non-preemptive contexts, which is fundamental for the communication schedulability analysis. We define an architecture to support priority-based scheduling of messages at the application process level of a specific fieldbus communication network, the PROFIBUS. The proposed architecture improves the worst-case messages' response time, overcoming the limitation of the first-come-first-served (FCFS) PROFIBUS queue implementations.

1 Introduction

Real-time computing systems are defined as those systems in which the correctness of the system depends not only on the logical result of computation, but also on the time at which the results are produced [1]. There are various examples of real-time computing systems, ranging from distributed computer control to robotics. In this paper we particularly address distributed computer-controlled systems (DCCS) applications.

A recent trend in DCCS is to interconnect distributed elements by means of a multi-point broadcast network, instead of using traditional point-to-point links. As the network bus is shared between a number of network nodes, there is an access contention, which must be solved by the Medium Access Control (MAC) protocol.

Usually, a DCCS application imposes real-time constraints. In essence, by real-time constraints we mean that traffic must be sent and received within a bounded interval, otherwise a timing fault is said to occur. This motivates the use of communication networks within which the MAC protocol is able to schedule messages streams according to its real-time requirements.

During the past decade a reasonable number of commercial solutions, usually called fieldbus networks, have been proposed to support DCCS applications. Some distinguished examples are FIP [2], PROFIBUS [3], CAN [4] or P-NET [5]. In parallel, several international standardisation efforts have been and are still being carried out. One of the most relevant resulted into the European Standard EN 50170 [6], which basically encompasses three well-proven fieldbus national standards: PROFIBUS, FIP and P-NET.

A potential leap towards the use of fieldbus networks to support DCCS applications lies in the evaluation of its temporal behaviour. Several studies have been performed, such as those on CAN [7,8], on FIP [9,10] and on P-NET [11,12].

Contrarily to other networks [15-16] which are based in the timed token protocol [17], in PROFIBUS it is not possible to define, in each master, its synchronous bandwidth allocation[1]. Thus, it is not possible to use analysis similar to those proposed in [18-19]. In [13] the authors take two different approaches to guarantee real-time traffic using PROFIBUS networks. One of the approaches considers a worst-case scenario with, at most, one high priority message cycle[2] transmission per token visit. Thus, if there are m high priority messages pending to access the bus, in the worst-case it will take m token visits to execute all those high priority message cycles. Considering the maximum token cycle (has derived in [14]), it is possible to evaluate the maximum queuing delay of any message request.

In PROFIBUS, pending requests are queued in a First-Come-First-Served (FCFS) queue. Hence, a message request will have a worst-case queuing delay that depends on the maximum priority inversion of messages. This motivated a study on adding local priority-based scheduling mechanisms to PROFIBUS masters. One possibility relies on the implementation of a priority-based queue at the application process level, and limits the FCFS communication stack queue to one pending request. There are several results available for the pre-run-time schedulability analysis of tasks in a single processor environment. In this paper we will show how those results can be adapted for message pre-run-time schedulability analysis in PROFIBUS communication networks.

The remaining of the paper is organised as follows. In the next section we will survey the most relevant results of tasks' pre-run-time schedulability analysis for both fixed and dynamic priority assignment schemes. We will give emphasis to the worst-case response time analysis in non-preemptive contexts. In section 3, PROFIBUS protocol is presented and the timing analysis proposed in [13] is detailed. Finally, in section 4, an architecture to support priority-based scheduling of messages at the PROFIBUS application process level is defined, and the single processor worst-case tasks' response time analysis is adapted to obtain the worst-case messages' response time.

---

[1] In PROFIBUS there is no synchronous bandwidth allocation (H_i). If a master receives a late token only one high priority message may be transmitted. Contrarily, in the timed token protocol the station can transmit real-time (high-priority) traffic during H_i, even if the token is late.

[2] In PROFIBUS a message cycle consists on a master's request frame and responder's immediate response frame.

2 Schedulability Analysis of Real-Time Tasks

One of the most used priority assignment schemes is to give tasks a priority level based on its period: the shorter the period, the higher the priority. This assignment is intuitively explained by the fact that more critical devices will provide inputs more frequently (asynchronous interrupts), or indeed will be polled more frequently. Thus, if they have shorter periods, the worst-case response time should also be shorter. This type of priority assignment is known as the rate monotonic (RM) priority assignment.

If some of the tasks are sporadic, it may not be reasonable to consider its relative deadline equal to the period. A different priority assignment can be to give the tasks a priority level based on its deadline: the shorter the relative deadline, the higher the priority. This type of priority assignment is known as the deadline monotonic (DM) priority assignment [20].

In both RM and DM priority assignments, priorities are fixed, in the sense that they do not vary along time. At run-time, tasks are dispatched highest-priority-first. A similar dispatching policy can be used if the task that is chosen to run, is the one with the earliest deadline. This corresponds to a priority-driven scheduling where priorities of the tasks vary along time, hence corresponding to a dynamic priority assignment. This type of dynamic priority assignment is known as earliest deadline first (EDF) priority assignment [21].

In all three cases, the dispatching will take place when either a new task is released or the execution of the running task ends. This however may not stand in a non pre--emptive context. With priority-based scheduling, a higher-priority task may be released during the execution of a lower priority one. In a pre-emptive system, the higher-priority task will pre-empt the lower-priority one. Contrarily, in non-preemptive systems, the lower-priority task will be allowed to complete its execution before the other executes.

For the remaining of this paper, we characterise a task set (or a message stream set) by its maximum execution time (transmission time), its relative deadline and its period, denoted, respectively, as C_i, D_i and T_i.

2.1 Analysis for the Fixed-Priority Assignment

For the RM priority assignment, Liu and Layland [21] derived an utilisation-based pre-run-time schedulability test, which, if satisfied, guarantees that all n tasks will meet their deadlines: $\sum_{i=1,..,n} C_i/T_i < n \times (2^{1/n} - 1)$[3]. This utilisation-based test is valid for periodic independent tasks with relative deadlines equal to the period and for pre-emptive systems. For the non pre-emptive case, a similar analysis can be adapted from results available in [22]. The response time tests (by opposition to the utilisation-based tests) are more advantageous for pre-run-time schedulability analysis, since individual results can be obtained for each task.

Joseph and Pandaya [23] proved that the worst-case response time r_i of a task τ_i is found when all tasks are synchronously released at their maximum rate (critical

---

[3] C is worst-case computation time of the task an T is the minimum time between task releases (period). If the task is sporadic, the minimum period between any two aperiodic releases is considered.

instant). r_i is computed by the following recursive equation (where $hp(i)$ denotes the set of tasks of higher priority than τ_i): $r_i^{m+1} = C_i + \sum_{j \in hp(i)} (\lceil r_i^m / T_j \rceil \times C_j)$. The recursion ends when $r_i^{m+1} = r_i^m = r_i$ and can be solved by successive iterations starting from $r_i^0 = C_i$. Indeed, it is easy to show that r_i^m is non-decreasing. Consequently, the series either converges or exceeds T_i (in the case of RM) or D_i (in the case of DM). If the series exceeds T_i (D_i), the task τ_i is not schedulable.

This result is valid for the preemptive context. Fewer results are known about fixed priority-based non-preemptive scheduling. In [24] Audsley et al. updated the analysis of Joseph and Pandaya [23] to include blocking factors introduced by periods of non-preemption. The worst-case response time is updated to:

$$r_i = w_i + C_i \tag{1}$$

where w_i is given by $w_i^{m+1} = B_i + \sum_{j \in hp(i)} (\lceil w_i^m / T_j \rceil \times C_j)$. B_i is the blocking factor of task τ_i, that is, an upper bound on the time a lower priority (thus resulting in priority inversion) task can execute and prevent the execution of task τ_i. A methodology for deriving B_i was introduced in [22] to solve the problem of non-independence of tasks, which can also be used for the schedulability analysis of non pre-emptive tasks:

$$B_i = \max_{\forall j \in lp(i)} \{C_j\} \tag{2}$$

where $lp(i)$ denotes the tasks with lower priority than i.

2.2 Dynamic-Priority Assignment

The EDF schedulability test was introduced in [21], for periodic tasks with relative deadlines equal to task periods. The main result from that work is the following utilisation-based test: $\sum_{i=1...n} C_i / T_i < 1$. This result is hardly useful for real-time systems with sporadic tasks. If we consider the more general case of $D_i \leq T_i$ [25], the inequality can be updated to: $\forall_{t \geq 0}, \sum_{i=1...n} (\lceil (t - D_i)/T_i \rceil^+ \times C_i) \leq t$, where $\lceil x \rceil^+ = 0$ if $x < 0$.

The proof for this inequality is intuitive. Assume that at time $t = 0$, there are no pending tasks. Then, a necessary condition to guarantee the tasks deadlines is that the amount of time, T, needed to transmit all tasks generated during $[0, t]$ with absolute deadlines $\leq t$ is not greater than t. Since the minimal inter-arrival time for a task τ_i is T_i, there are at most $\lceil (t - D_i)/T_i \rceil^+$ requests for that task during $[0, t]$ which have a deadline $\leq t$. Those requests will need, at most, $\lceil (t - D_i)/T_i \rceil^+ \times C_i$ time to be completed. Thus the maximum value for T is given by $\sum_{i=1...n} (\lceil (t - D_i)/T_i \rceil^+ \times C_i)$.

Note that if $D_i = T_i$, inequality $\forall_{t \geq 0}, \sum_{i=1...n} (\lceil (t - D_i)/T_i \rceil^+ \times C_i) \leq t$ is satisfied if the utilisation-based test, $\sum_{i=1...n} C_i / T_i < 1$, is satisfied, since in this case $\lceil (t - T_i)/T_i \rceil^+ \leq t/T_i$. The inequality $\forall_{t \geq 0}, \sum_{i=1...n} (\lceil (t - D_i)/T_i \rceil^+ \times C_i) \leq t$ has a practical problem, since it must be checked over an infinite length interval $[0, \infty)$. But, if we look to its left-hand interval, we can easily understand that its value only changes at $k \times T_i + D_i$ steps. Additionally, there exists a point t_{max}, such that $\sum_{i=1...n} (\lceil (t - D_i)/T_i \rceil^+ \times C_i) \leq t$ always hold for $\forall_{t \geq t_{max}}$ (under the condition that the total utilisation $\sum_{i=1...n} C_i / T_i < 1$). Hence, the following represents the feasibility test for EDF dispatched tasks in a single processor pre-emptive context:

$$\forall_{t\in S}, \sum_{i=1}^{n}\left\lceil\frac{t-D_i}{T_i}\right\rceil^{+} \times C_i \leq t, \text{ with } S = \left(\bigcup_{i=1}^{n}\{D_i + k\times T_i, k\in \aleph\}\right)\cap [0, t_{max}) \qquad (3)$$

The determination of t_{max} has been addressed in several works [26-29]. Note that when the utilisation approaches 1, t_{max} becomes very large. This is the main difficulty for such utilisation-based schedulability test.

For the non-preemptive case, a similar feasibility test was derived in [25,30]:

$$\forall_{t\geq D_{min}}, \sum_{i=1}^{n}\left\lceil\frac{t-D_i}{T_i}\right\rceil^{+} \times C_i + \max_{i=1,\ldots,n}\{C_i\}\leq t, \text{ with } D_{min} = \min_{i=1,\ldots,n}\{D_i\} \qquad (4)$$

In [31], George et al. argue about the pessimism of (4). The main argument is that in [30] Zheng and Shin consider that the cost of possible priority inversions, caused by task non pre-emptability, is always initiated by the longest task and, moreover, is effective during all the studied interval. To reduce the pessimism level of (4), they suggest the following update (the values for S are precisely defined in [31]):

$$\forall_{t\in S}, \sum_{i=1}^{n}\left\lceil\frac{t-D_i}{T_i}\right\rceil^{+} \times C_i + \max_{\substack{i=1,\ldots,n \\ D_i>t}}\{C_i-1\}\leq t, \text{ with } \max_{\substack{i=1,\ldots,n \\ D_i>t}}\{C_i-1\}=0 \text{ if } \nexists_i : D_i > t \qquad (5)$$

The worst-case response time analysis for preemptive EDF scheduling was first studied in [32]. The starting basis for such analysis was that the worst-case response time for a general task set is not necessarily obtained with a synchronous pattern arrival (that is, considering the critical instant as defined for the fixed priority case). In that work, Spuri showed that the worst-case response time of a task τ_i is found in the deadline busy period[4] of the processor.

This means that, in order to find the worst-case response time of τ_i, we need to examine several scenarios within which, while τ_i has an instance released at time a, all other tasks are synchronously released. If the start time of the *asap*[5] pattern of τ_j is denoted by S_j then $S_j = 0$, $\forall_{j \neq i}$. In general, τ_i may also have other instances released earlier than a. In particular, its start time is $S_i = a - \lfloor a/T_i\rfloor \times T_i$. Thus, given a value of a, the response time of the τ_i's instance released at time a is:

$$r_i(a) = \max\{C_i, L_i(a)-a\} \qquad (6)$$

where $L_i(a)$ is the length of the busy period. $L_i(a)$ can be evaluated by the following iterative computation (starting with $L_i^0=0$): $L_i^{m+1}(a) = W_i(a, L_i^m(a)) + (1+\lfloor a/T_i\rfloor)\times C_i$, with $W_i(a, t) = \sum_{j\neq i, \text{ and } D_j \leq a+D_i}(\min\{\lceil t/T_j\rceil, 1+\lfloor (a+D_i-D_j)/T_j\rfloor\}\times C_j)$.

---

[4] The longest deadline busy period (the processor is fully utilised) appears when all tasks but task i are synchronously released and at their maximum rate.
[5] as soon as possible

The worst-case response time for a task τ_i is then given:

$$r_i = \max_{a \geq 0}\{r_i(a)\} \tag{7}$$

The remaining problem is for which values of a r_i must be evaluated? If we look to the right-hand side of the $W_i(a, t)$ equation, we can easily understand that its value does only changes at $k \times T_j + D_j - D_i$ steps. Thus,

$$a \in \bigcup_{j=1}^{n}\{k \times T_j + D_j - D_i, k \in \aleph\} \cap [0, L_i] \tag{8}$$

where L_i is the maximum length of the deadline busy periods, as defined in [32].

The worst-case response time analysis for the non pre-emptive EDF scheduling was introduced in [31]. The main difference to the analysis made for the pre-emptive case is that a task instance with a later absolute deadline can possibly cause a priority inversion. Thus, instead of analysing the busy period preceding its completion time, we must analyse the busy period preceding the execution start time of the task's instance. Consequently, the response time of the τ_i's instance released at time a is:

$$r_i(a) = \max\{C_i, L_i(a) + C_i - a\} \tag{9}$$

where $L_i(a)$ is now the length of the busy period (preceding execution).

Thus, $r_i(a)$ can be evaluated by means of the following iterative computation (also starting with $L_i^0 = 0$): $L_i^{m+1}(a) = \max_{Dj > a+Di}\{C_j - 1\} + W_i^*(a, L_i^m(a)) + \lfloor a/T_i \rfloor \times C_i$, with $W_i^*(a, t) = \sum_{j \neq i, \text{ and } Dj \leq a+Di}(\min\{1 + \lfloor t/T_j \rfloor, 1 + \lfloor (a+D_i-D_j)/T_j \rfloor\} \times C_j)$.

Again, a should be evaluated in the following set of values:

$$a \in \bigcup_{j=1}^{n}\{k \times T_j + D_j - D_i, k \in \aleph\} \cap [0, L] \tag{10}$$

where, in this case, L is the length of the sychrounous busy period, which can be evaluated by (starting with $L^0 = \sum_{i=1...n} C_i$): $L_i^{m+1} = W(L^m)$, with $W(t) = \sum_{i=1...n} \lceil t/T_i \rceil \times C_i$, which is solved by reoccurrence.

3. PROFIBUS Timing Analysis

3.1 PROFIBUS Timed Token Protocol

The PROFIBUS MAC protocol is based on a token passing procedure, used by master stations to grant the bus access to each one of them, and a master-slave procedure used by master stations to communicate with slave stations. One of the PROFIBUS MAC main functions is the control of the token cycle time, which will now be briefly explained.

After receiving the token, the measurement of the token rotation time begins. This measurement expires at the next token arrival and results in the real token rotation time (T_{RR}). A target token rotation time (T_{TR}) must be defined in a PROFIBUS network. The value of this parameter is common to all masters, and is used as follows. When a station receives the token, the token holding time (T_{TH}) timer is given the value corresponding to the difference, if positive, between T_{TR} and T_{RR}. PROFIBUS defines two categories of messages: high priority and low priority. These two categories of messages use two independent outgoing queues. If at the arrival, the token is late, that is, the real token rotation time (T_{RR}) was greater than the target rotation time (T_{TR}), the master station may execute, at most, one high priority message cycle. Otherwise, the master station may execute high priority message cycles while $T_{TH} > 0$. T_{TH} is always tested at the beginning of the message cycle execution. This means that once a message cycle is started it is always completed, including any required retries, even if T_{TH} expires during the execution. We denote this occurrence as a T_{TH} overrun. The low priority message cycles are executed if there are no high priority messages pending, and while $T_{TH} > 0$ (also evaluated at the start of the message cycle execution, thus leading to a possible T_{TH} overrun).

Below is a description of the PROFIBUS token passing algorithm:

```
/* initialisation procedure */
At each station k, DO:
T_TH ← 0 ;
T_RR ← 0 ;
Start T_RR ;          /* count-up timer */

/* run-time procedure */
At each station k, at the Token arrival, DO:
T_TH ← T_TR - T_RR ;
T_RR ← 0 ;
Start T_RR ;  /* count-up timer */
IF T_TH > 0 THEN
         Start T_TH   /* count-down timer */
ENDIF;
IF waiting High priority messages THEN:
         Execute one High priority message cycle
ENDIF;
WHILE T_TH > 0 AND pending High priority message cycles DO
Execute High priority message cycles
ENDWHILE;
WHILE T_TH > 0 AND pending Low priority message cycles DO
         Execute Low priority message cycles
ENDWHILE;
Pass the token to station (k + 1) (modulo n);
```

As mentioned in section 1, in PROFIBUS, a message cycle consists on a master's action frame (request or send/request frame) and the responder's immediate acknowledgement or response frame. User data may be transmitted in the action frame or in the response frame. Note that in PROFIBUS a master station is allowed to send up to a limited number of retries, if the response does not come within a predefined time. Hence, the message cycle time length must also include the time needed to process the allowed retries.

3.2 Message Worst-Case Response Time

As a master station is able to transmit, at least, one high priority message per received token (no matter if there is enough token holding time left), a maximum queuing delay can be guaranteed for PROFIBUS messages. Defining T_{cycle} as the upper bound between two consecutive token arrivals to a particular master, the maximum queuing delay of a single message request (Q) is equal to T_{cycle}. Note that this only guarantees a maximum transmission delay for the first high priority message in the outgoing queue. If there are m pending messages in the outgoing queue it will take, in the worst-case, m token visits to execute all those high priority messages.

It is obvious that the queuing delay depends on the outgoing high priority queue implementation. PROFIBUS implements First-Come-First-Served (FCFS) outgoing queues. Consequently, if nh^k represents the number of high priority message streams[6] in a master k, then the maximum number of pending messages will be nh^k, corresponding to one message per each high priority message stream (Sh_i^k) (two messages from the same stream would mean that a deadline for that message stream was missed). Thus, an upper bound for the message queuing delay in a master k is: $Q_i^k = nh^k \times T_{cycle} - Ch_i^k$, where Ch_i^k denotes the maximum length (request, response, turnaround time and maximum allowable retries) of a message from stream Sh_i^k. The worst-case response time for a message cycle is given by:

$$R_i^k = Q_i^k + Ch_i^k = nh^k \times T_{cycle} \qquad (11)$$

Thus, a set of PROFIBUS high priority message streams is guaranteed if the following pre-run-time schedulability condition is verified:

$$Dh_i^k \geq R_i^k, \; \forall_{master\; k,\; stream\; Sh_i^k} \qquad (12)$$

that is, if relative deadlines of all high priority message streams are greater than or equal to their worst-case communication response time. If (12) is not satisfied, the high priority traffic is not schedulable. In order to solve inequality (12), we need to evaluate T_{cycle}.

3.3 Cycle Time Evaluation

As shown in [13,14], T_{cycle} can be expressed as a function of T_{TR}. Thus, having defined T_{cycle}, it is possible to set the T_{TR} parameter in order to satisfy the pre-run-time schedulability condition (12).

Considering that T_{cycle} is the upper bound of the real token rotation time (T_{RR}), we must reason about T_{RR} in order to evaluate the value of T_{cycle}. The T_{RR} value will be smaller than T_{TR} (that is, the token will always be in advance to its schedule), except if one or more masters in the logical ring cause the token to be late. The main cause for the token lateness (T_{del}) is the T_{TH} overrun. Such token lateness may be worsened if the following masters, having received a late token, still transmit, each one, one high priority message.

---

[6] A message stream corresponds to a temporal sequence of message cycles related, for instance, with the reading of a process sensor or the updating of a process actuator.

Considering the worst-case scenario in which a master k overruns its T_{TH} and the following masters until master $k-1$ (modulo n) receive and use a late token, T_{del} can be evaluated as follows:

$$T_{del} = \sum_{k=1}^{n}\left(C_M^k\right) \tag{13}$$

where C_M^k stands for the longest message cycle associated to each master k: $C_M^k = \max\{\max_{i=1...nh^k}\{Ch_i^k\}, Cl^k\}$, and Cl^k is the longest low priority message cycle in a master k. Thus, T_{cycle} can be evaluated as follows:

$$T_{cycle} = T_{TR} + T_{del} \tag{14}$$

To illustrate the T_{cycle} evaluation, assume the following scenario: after a token cycle without message transmissions, master k receives the token ($T_{TH}^{(k)} = T_{TR} - \tau$, since $T_{RR}^{(k)} = \tau$)[7]. This master can actually hold the token during $T_{TH}^{(k)}$ plus the duration of its longest message. In this situation all the following masters in the logical ring will receive a late token, thus only being able to process, at most, one high priority message cycle.

A more accurate definition for T_{cycle} can be found in [14], where factors such as the different message cycle lengths and the relative position of each master in the logical ring are taken into consideration for the definition of T_{cycle}.

3.4 Setting the PROFIBUS T_{TR} Parameter

We can now update the pre-run-time schedulability condition: $Dh_i^k \geq nh^k \times (T_{TR} + T_{del})$. Therefore, the T_{TR} parameter can be set as follows:

$$0 \leq T_{TR} \leq \frac{Dh_i^k}{nh^k} - T_{del}^k, \forall_{\text{master } k, \text{ stream } Sh_i^k} \tag{15}$$

4 Adding Priority-Based Message Scheduling

The first-come-first-served (FCFS) dispatching of the PROFIBUS outgoing queue may induce in a "priority inversion" with the length ns^k-1 as, in the worst-case, a message request with a more stringent deadline may be placed in the outgoing queue after the other ns^k-1 message requests of the same master station.

A priority-based queue would solve this problem: more stringent messages would have lower worst-case response times whereas less stringent messages would have higher worst-case response times (as compared to the FCFS dispatching policy).

Without changing the PROFIBUS FCFS implementation, a priority-based queue may be implemented at the application process level, provided that the communication stack outgoing queue (FCFS based) is limited to one pending request.

---

[7] The parameter τ includes the ring latency and other protocol and network overheads.

In PROFIBUS, this length control of the communication stack outgoing queue can be trivially achieved by the proper use of a local management service.

4.1 Message's Release Jitter

In the analysis we made in section 3, the periods of the messages were not relevant for the evaluation of the worst-case message's response time. Independently of the model of the tasks that generate the message requests, the worst-case queuing delay occurs when ns^k requests are placed at the outgoing queue just before master k passes the token to the subsequent station.

If, in each master, messages are dispatched using a priority-based scheduling policy (either DM or EDF can be considered), then, by using similar worst-case response time analysis as for the case of non pre-emptive[8] task scheduling in single processor environments, periods of the message request are now relevant. In fact, both equation (1) (for the case of pre-run-time schedulability analysis of non pre-emptive DM) and equation (9) (for the case of pre-run-time schedulability analysis of non pre-emptive EDF) very much depend on the periods of the tasks (now message requests).

This rises the problem of message release jitter, which in the context of communication networks has been addressed in several studies such as by Tindell and Clark [33] and by Spuri [34].

In the case of PROFIBUS we can assume that message requests are placed in the priority-ordered application process queue by an application task, and a task's instance will generate a message stream request. In this sense we can say that messages inherit from sending tasks both their period and priority level (if fixed priorities are used). It is implicit that tasks at the application process level are scheduled according to a priority-based policy, most probably in a preemptive context. It results from this inheritance approach that the message requests generated by tasks will have a minimum inter-arrival time smaller than the period of the tasks, which generate them.

In PROFIBUS, the sending task and the response's receiving task are in the same host processor. In fact, they can even be the same task. There is an initial part of the task responsible for placing the request in the priority-based AP queue (during this process, this task competes with the other tasks for the processor). Then the task auto-suspends itself until the response arrives. Finally, after receiving the response, it processes it the task until completion (again in this phase competing with other running tasks). If this is the case, the messages's release jitter is the worst-case response time of the first part of the task (which includes placing the request in the priority ordered queue).

We can think also about a model where sending and receiving tasks are separate tasks (in fact each pair of sending/receiving tasks do not correspond to independent tasks, since if one is runnable the other is not in the runnable queue). In this case, the release jitter for a message request corresponds to the worst-case response time of the sending task. That is, in a particular instance of the task, the message can be released close to the worst-case response time of the task; and in the subsequent release of the task, the message can be released as soon as the arrival of that new task's instance.

---

[8] The pre-emptive schedulability analysis is not useful for message scheduling in PROFIBUS networks, as a message cycle is non pre-emptable.

Independently of the task model adopted for the application level, we denote the release jitter of a message stream i in a master k as J_i^k. The evaluation of its value depends on the adopted task model and also on the scheduling policy of the tasks. Both task models demand some careful for analysing tasks' worst-case response times.

4.2 End-To-End Communication Delay

The inclusion of an application process task model motivates the definition of the end-to-end communication delay [35]. With this concept, we associate timing requirements to tasks, and messages inherit periods, priorities and release jitter from tasks. If we use EDF scheduling, messages will be placed in the priority-based queue according to the earliness of the absolute deadline of the message's generating task. Note that the AP priority-based queue should only be re-ordered when a new request is generated (this stands for both DM and EDF).

The end-to-end communication delay (E) can be generically defined as: $E=g+Q+C+d$. g represents the worst-case generation delay for the master application task to generate and queue a specific message request. This would also corresponds to the message release jitter, which will be used for computing Q. The term Q corresponds to the worst-case delay for that request to gain access to the communications device after being queued. As previously mentioned, using a priority-based dispatching, this parameter would depend on the release jitter of each message stream. The term C corresponds not only to the worst-case for transmitting the request, but also to the time needed to receive the response from the slave (including processing at the slave side, slave turnaround time, propagation delay and other network latencies). Finally, d represents the delivery delay, that is, the time needed to process the response before finally delivering it to the destination task, which in the case of PROFIBUS is in the same host processor as the sending task. Variation to this results from whether separated sending and receiving tasks are considered.

In the next sub-section we will simply focus on the Q (or $Q+C$) parameter, and redefine the worst-case message's response time. Basically we are going to update equation (11) for embodying a priority-based dispatching policy.

4.3 Priority-Based Message's Response Time

Basically, the Cs in equations (1) and (9) will be replaced by T_{cycle}, and for considering one "blocking" the requests can appear marginally after receiving the token and marginally before passing the token. Assume also that all message cycles are equal. Hence, considering DM, equation (11) can be updated to:

$$R_i^k = T_{cycle}^* + \sum_{\forall j \in hp(i)} \left(\left\lceil \frac{R_i^k + J_j^k}{T_j^k} \right\rceil \times T_{cycle} \right) \qquad (16)$$

where $T^*_{cycle}=T_{cycle}$ except for the case of the message with the lowest priority (in this case $T^*_{cycle}=0$).

Considering EDF, equation (11) can be updated to:

$$R_i^k(a) = \max\{T_{cycle}, L_i(a) + T_{cycle} - a\} \tag{17}$$

where (starting with $L_i^0(a)=0$):

$$L_i^{m+1}(a) = \max_{D_j^k > a + D_i^k}\{T_{cycle} - 1\} + W_i^*(a, L_i^m(a)) + \left\lfloor\frac{a}{T_i^k}\right\rfloor \times T_{cycle}, \text{ with}$$

$$W_i^*(a,t) = \sum_{\substack{j \neq i \\ D_j^k \leq a + D_i^k}} \left(\min\left\{1 + \left\lfloor\frac{t + J_j^k}{T_j^k}\right\rfloor, 1 + \left\lfloor\frac{a + D_i - D_j + J_j^k}{T_j^k}\right\rfloor\right\} \times T_{cycle}\right) \tag{18}$$

5 Conclusions

In this paper we have drawn a comprehensive study on how to use PROFIBUS to support real-time communications. We surveyed previous relevant work on how to evaluate the worst-case response time of tasks in a single processor environment. Particular relevance was devoted to the non pre-emptive case.

As in PROFIBUS the message requests are processed in a first-come-first-served basis, we reasoned about the possibility of supporting priority-based scheduling of message requests at the application process level of PROFIBUS.

Finally, we have derived the worst-case response time of PROFIBUS message requests for both the cases of message requests dispatched using a DM priority assignment and an EDF priority assignment.

The relevance of this work results from the obvious conclusion that the use of priority-based dispatching mechanism at the application process level allows the support of messages with more tight deadlines.

Acknowledgements

This work was partially supported by FLAD under the project SISTER 471/97 and by ISEP under the project REMETER.

References

1. Stankovic, J.: "Real-Time Computing Systems: the Next Generation", in STANKOVIC J., RAMAMRITHAM, K. (Eds.) "Tutorial: Hard Real-Time Systems" (IEEE, 1988), pp. 14-38, 1988.
2. Normes FIP NF C46-601 to NF C46-607, Union Technique de l'Electricité, AFNOR, 1990.

3. Profibus Standard DIN 19245 part I and II. Translated from German, Profibus Nutzerorganisation e.V., 1992.
4. SAE J1583, Controller Area Network (CAN), an In-Vehicle Serial Communication Protocol. SAE Handbook, Vol. II, 1992.
5. The P-NET Standard. International P-NET User Organisation ApS, 1994.
6. General Purpose Field Communication System, Vol. 1/3 (P-NET), Vol. 2/3 (Profibus), Vol. 3/3 (FIP), CENELEC, 1996.
7. Tindell, K., Hansson, H., Wellings, A.: "Analysing Real-Time Communications: Controller Area Network (CAN)". Proceedings of the IEEE Real Time Systems Symposium (RTSS'94), S.Juan, Puerto Rico, pp. 259-263, IEEE Press, 1994.
8. Tindell, K., Burns, A., Wellings, A.: "Calculating Controller Area Network (CAN) Message Response Times". Control Engineering Practice, Vol. 3, No. 8, pp. 1163-1169, Pergamon, 1995.
9. Raja, P., Ruiz, L., Decotignie, J.-D.: "On the Necessary Real-Time Conditions for the Producer-Distributer-Consumer Model". Proceedings of 1st IEEE Workshop on Factory Communication Systems (WFCS'95), Leysin, Switzerland, 1995.
10. Pedro, P., Burns, A.: "Worst Case Response Time Analysis of Hard Real-Time Sporadic Traffic in FIP Networks". Proceedings of 9th Euromicro Workshop on Real-time Systems, Toledo, Spain, pp. 5-12, 1997.
11. Tovar, E., Vasques, F.: "Pre-run-time Schedulability Analysis of P-NET Networks". Proceedings of 24th Annual Conference of the IEEE Industrial Electronics Society (IECON'98), Aachen, Germany, pp. 236-241, 1998.
12. Tovar, E., Vasques, F., Burns, A.: "Real-Time Communications in Multihop P-NET Networks". Submitted to Control Engineering Practice, 1998.
13. Tovar, E., Vasques, F.: "Real-Time Fieldbus Communications Using Profibus Networks". To appear in the IEEE Transactions on Industrial Electronics, 1998.
14. TOVAR, E., VASQUES, F.: "Cycle Time Properties of the Profibus Timed Token Protocol", submitted to IEE Proceedings - Software, 1998.
15. ISO, Information Processing Systems - Fibre Distributed Data Interface (FDDI) - Part 2: Token Ring Media Access Control (MAC), ISO International Standard 9314-2, 1989.
16. IEEE, IEEE Standard 802.4: token passing bus access method and physical layer specification, 1985.
17. Grow, R.: "A Timed Token Protocol for Local Area Networks". Proceedings of Electro'82, May 1982, Token Access Protocols, Paper 17/3.
18. Agrawal, G., Chen, B., Zhao, W., Davari, S.: "Guaranteeing Synchronous Message Deadlines with the Timed Token Protocol". Proceedings of the 12th IEEE International Conference on Distributed Computing Systems, June 1992.
19. Montuschi, P., Ciminiera, L., Valenzano, A.: "Time Characteristics of IEE802.4 Token Bus Protocol". IEE Proceedings, January 1992, 139 (1), pp. 81-87.
20. Burns, A.: "Scheduling Hard Real-Time Systems". Software Engineering Journal - Special Issue on Real-time Systems, pp. 116-128, May 1991.
21. Liu, C., Layland, J.: "Scheduling Algorithms for Multiprogramming in a Hard-Real-Time Environment". Journal of the Association for Computing Machinery (ACM), Vol. 20, NO. 1, pp. 46-61, January 1973.
22. Sha, L., Rajkumar, R., Lehoczky, J.: "Priority Inheritance Protocols: an Approach to Real-Time Synchronisation". IEEE Transactions on Computers, Vol. 39, NO. 9, pp. 1175-1185, September 1990.
23. Joseph, M., Pandya, P.: "Finding Response Times in a Real-Time System". The Computer Journal, Vol. 29, NO. 5, pp. 390-395, 1986.
24. Audsley, N., Burns, A., Richardson, M., Tindell, K, Wellings, A.: "Applying New Scheduling Theory to Static Priority Pre-emptive Scheduling". Software Engineering Journal, Vol. 8, NO. 5, pp. 285-292, September 1993.
25. Zheng, Q.: "Real-Time Fault-Tolerant Communication in Computer Networks". PhD Thesis, University of Michigan, 1993.

26. Baruah, S., Howell, R., Rosier, L.: "Algorithms and Complexity Concerning the Preemptive Scheduling of Periodic Real-time Tasks on One Processor". Real-Time Systems, 2, pp. 301-324, 1990.
27. Baruah, S., Mok, A., Rosier, L.: "Preemptively Scheduling Hard-Real-Time Sporadic Tasks on One Processor". Proceedings of the 11th Real-Time Systems Symposium (RTSS'90), pp. 182-190, 1990.
28. Ripoll, I., Crespo, A., Mok, A.: "Improvement in Feasibility Testing for Real-time Systems". Real-Time Systems, 11, pp. 19-39, 1996.
29. Spuri, M.: "Earliest deadline Scheduling in Real-time Systems". PhD Thesis, Scuola Superiore Santa Anna, Pisa, 1995.
30. Zheng, Q., Shin, K.: "On the Ability of Establishing Real-Time Channels in Point-to-Point Packet-Switched Networks". IEEE Transactions on Communications, Vol. 42, no. 2/3/4, pp. 1096-1105, 1994.
31. George, L., Rivierre, N., Spuri, M.: "Preemptive and Non-Preemptive Real-Time Uni-Processor Scheduling". Technical Report No. 2966, INRIA, September 1996.
32. Spuri, M.: "Analysis of Deadline Scheduled Real-Time Systems". Technical Report No. 2772, INRIA, January 1996.
33. Tindell, K., Clark, J.: "Holistic Schedulability Analysis for Distributed Hard Real-Time Systems", in Microprocessors and Microprogramming, No. 40, 1994.
34. Spuri, M.: "Holistic Analysis for Deadline Scheduled Real-Time Distributed Systems". INRIA, Technical Report no. 2873, April 1996.
35. Tindell, K., Burns, A., Wellings, A.: "Analysis of Hard Real-Time Communications". Real-Time Systems, 1995, 9, pp. 147-171.

An Adaptive, Distributed Airborne Tracking System
("Process the Right Tracks at the Right Time")

Raymond Clark[1], E. Douglas Jensen[1], Arkady Kanevsky[1], John Maurer[1], Paul Wallace[1], Thomas Wheeler[1], Yun Zhang[1], Douglas Wells[2], Tom Lawrence[3], and Pat Hurley[3]

[1] The MITRE Corporation, Bedford, MA, USA
rkc@mitre.org
[2] The Open Group Research Institute, Woburn, MA, USA
d.wells@opengroup.org
[3] Air Force Research Laboratory/IFGA, Rome, NY 13441, USA
hurleyp@rl.af.mil

Abstract. This paper describes a United States Air Force Advanced Technology Demonstration (ATD) that applied value-based scheduling to produce an adaptive, distributed tracking component appropriate for consideration by the Airborne Warning and Control System (AWACS) program. This tracker was designed to evaluate application-specific Quality of Service (QoS) metrics to quantify its tracking services in a dynamic environment and to derive scheduling parameters directly from these QoS metrics to control tracker behavior. The prototype tracker was implemented on the MK7 operating system, which provided native value-based processor scheduling and a distributed thread programming abstraction. The prototype updates all of the tracked-object records when the system is not overloaded, and gracefully degrades when it is. The prototype has performed extremely well during demonstrations to AWACS operators and tracking system designers. Quantitative results are presented.

1 Introduction

Many currently deployed computer systems are insufficiently adaptive for the dynamic environments in which they operate. For instance, the AWACS Airborne Operational Control Program (AOCP), which includes the tracking system, has a specified maximum track-processing capacity. Since the tracker processes data in FIFO order, it fails to process later data if its processing capacity is exceeded. Since sensor reports generally come in the same order from one sweep to the next, it is likely that, under overload, sensor reports from a specific region will not be processed for several consecutive sweeps. This overload behavior can result in entire regions of the operator displays that do not get updated. This is a potentially serious problem because there is no inherent correlation between important regions of the sky and the arrival order of sensor reports.

This situation, while undesirable, is handled by skilled operators, who recognize that some data is not being processed and take remedial actions (e.g., reducing the gain of a sensor or designating portions of the sky that do not need to be processed). While these manual adaptations reduce the tracker workload, they are not ideal. Reducing sensor gain might cause smaller, threatening objects to go undetected.

Injecting more intelligence into the tracker could avoid such compromises. The US Air Force Research Laboratory at Rome (NY), The MITRE Corporation, and The Open Group Research Institute undertook a joint project to explore that approach. The project recently concluded after producing an Advanced Technology Demonstration (ATD) featuring a notional, adaptive AWACS tracker that can execute in a distributed configuration, with attendant scalability and fault tolerance benefits.

The ATD tracker "processes the right tracks at the right time" by appropriately managing the resources needed for track processing. The prototype updates all tracked-object records when it has sufficient resources, and gracefully degrades otherwise. A single underlying mechanism automatically provides this degradation, despite its manifestation as a succession of qualitative operational changes: First, more important tracks receive better service than less important tracks while all tracks continue to be maintained. Under severe, sustained, resource shortages, less important tracks are lost before more important tracks. (This is referred to as *dropping* tracks and is described more formally in Section 4.2.) Moreover, the tracker automatically delivers improved service whenever more resources become available—in essence, tracker performance gracefully degrades and gracefully improves without direct human intervention.

2 Adaptivity, Value-Based Scheduling, and Quality of Service

Military planners have observed that "you never fly the same mission twice," implying that flexibility and evolvability are important design characteristics. With that in mind, we employed a relatively straightforward approach to adaptivity for this project—decomposing an application (or a set of applications) into component computations, which are then assigned application-specific values reflecting their individual contributions to the overall mission. By scheduling computations to maximize the accrued application-specific value, the overall system can perform a given mission well.

In some cases, associating a value with a computation seems natural. For instance, in financial trading applications, the value of performing a computation might be a function of the market price and the production cost. More complex situations involve other factors, such as human safety, or deal with a non-monetary domain. The adaptive tracker described here deals with a domain that has not traditionally employed a monetary model, providing a worked example for such a domain.

This project selected a research operating system that provides a value-based scheduling policy to applications. This makes the mapping from the computation values to scheduling attributes trivial. On other operating systems, this mapping would typically translate the application-specific values into scheduling priorities, which have a much more restricted value domain. The mapping could be performed by either the application directly or by a middleware package.

While application decomposition with appropriate value assignments assists in effectively accomplishing a specified mission, application effectiveness could be further increased by employing feedback to directly drive its behavior. To do this, application-specific figures of merit (that we refer to as Quality-of-Service (QoS) metrics) are specified at design-time, and evaluated or estimated dynamically at run-time in order to monitor—and subsequently control—overall application operation. Section 4 discusses this in some depth for the AWACS ATD, where the adaptive tracker uses feedback based on QoS metrics for individual tracks to determine the allocation of processing resources for current track processing.

3 The Tracking Problem

Surveillance radar systems are an important class of real-time systems that have both civilian and military uses. These systems consist of components for sensor processing, tracking, and display. In this study we concentrated on the tracking component, which receives *sensor reports*—the output produced by the sensor-processing component—and uses them to detect objects and their movements [1]. Each object is typically represented by a *track record*, and the collection of all track records is commonly referred to as the *track file*. Sensor reports arrive at a tracker periodically, and each report describes a potential airborne object. The number of sensor reports can vary from radar sweep to radar sweep, and some sensor reports represent noise or clutter rather than planes or missiles. When new sensor reports arrive, a tracker correlates the information contained in the sensor reports with the current estimated track state to update the track records that represent the tracking system's estimate of the state of the airspace.

A typical tracker comprises *gating*, *clustering*, *data association*, and *prediction and smoothing* stages. Gating and clustering splits (*gates*) the problem data into mutually exclusive, collectively exhaustive subsets of sensor reports and track records, called *clusters*. Data association then matches the cluster's sensor reports with its track records. The final stage of a tracker—prediction and smoothing—computes the next position, velocity and other parameters for each object using its track history and the results of data association.

3.1 Tracking Software for the ATD

Advanced Technology Demonstrations are intended to transfer technology into an application setting so that it can be utilized and evaluated by the operational and acquisition communities. In order to focus on this transfer, the project used existing tracking software and terminology, ensuring that ATD evaluators would be familiar with the overall function and structure of the tracker.

The project adapted a tracker that processed information from multiple sensors, including multiple types of radar. That tracker performed a single-threaded computation: That is, a single thread in the tracker performed all of the tracking stages described above from receiving sensor reports to updating the track file.

3.2 Adaptive Tracking

The motivating problem was the (mis)behavior of the tracker under overload, which can be intentionally caused by an enemy. Since, under overload, all of the incoming data cannot be processed, the project's goal was to allow the tracker to do a better job of selecting a subset of incoming data that could feasibly be processed by allowing scheduling decisions that had previously been made at design time to be deferred until run time. There were two major changes: processing was divided into smaller-grained units of work that could be scheduled independently and concurrently; and value-based design principles were employed to determine appropriate scheduling parameters for those individual work units.

Multithreading and Concurrency Presumably, a single-threaded adaptive tracker could be constructed by properly ordering the association computations so that those that are most critical are attempted first. While that approach addresses the overload problem at hand, it suffers from the serious limitations described below, motivating us to design a multithreaded AWACS ATD tracker that utilizes a separate thread for each association computation to be performed.

First, a single-threaded tracker could not take advantage of available distributed resources, including both multiprocessors and other processing nodes.

Secondly, ordering the association computations in the application reduces or eliminates the possibility of taking advantage of certain operating system and middleware resource management facilities. For instance, the OS employed for the AWACS ATD offered a processor-scheduling policy that accepted explicit application time constraints for computations and performed the set of computations that maximized accrued value (in application-specific terms).

Application of Value-Based Design Principles Given the multithreaded design described above, with a separate thread assigned to perform each association computation, and an underlying scheduler that attempts to maximize accrued value, the adaptive tracker design problem is reduced to a straightforward question: Can scheduling parameters be selected for tracks and clusters that reflect the value associated with their processing?

Considerable effort was devoted to answering that question for the ATD. Fortunately, at about this time, the tracking community was independently encouraged to think in terms of "selling" track information to customers—that is, operators and decision makers. That point of view helped make the notion of establishing the application-specific value of a track quite natural. In fact, the project developed a set of QoS metrics for individual tracks. These per-track QoS metrics, described in Section 4.2, directly determine the scheduling parameters for a cluster, which, in turn, affect the future QoS metrics of each component track.

4 Value-Based Design

AWACS can perform a number of different missions: e.g., it can manage logistics such as refueling, perform air-traffic control, or carry out general surveillance. In order to maximize the depth of this initial work, we applied the principles of value-based design to a single mission. Because of its general utility and intuitive simplicity, a *surveillance* mission was chosen for the ATD.

When flying a surveillance mission, AWACS operators attempt to monitor all airborne objects in a large region. Once an object has been identified, the tracker follows its progress (i.e., "track" it). The more closely the tracker's estimate of an object's position and heading agree with reality, the better.

Moreover, once the tracker has identified a track, it should not "drop" it erroneously. A track is dropped if it is not updated for a number of input sensor cycles. While this is inevitable if no new sensor input is received for the track, the project focused on cases where sensor input is received, but is not processed. (A dropped track can be rediscovered; but this is a relatively costly operation and there is no assurance that any individual track will be reacquired.)

4.1 Adaptive Tracker Behaviors

A key step to producing an adaptive tracker was to develop a specification of its desired behavior—in particular, its behavior under overload. This is an application specification task, but was of interest to the project because it required specification of behaviors that were new to the application domain.

The specification of adaptive tracker behaviors occurred in two phases. The first phase described a high-level policy, but did not address tradeoffs that could arise when attempting to satisfy conflicting policy objectives. For example, the high-level adaptive tracker should preferentially process tracks that:

- are in danger of being dropped
- the user has identified as "important"
- have poor state (position and velocity) estimates
- are maneuvering
- potentially pose a high threat
- are moving at high speed

The second phase refined this high-level policy by addressing conflicting policy objectives. In general, project members could identify these conflicts, but needed expert assistance to determine the proper resolutions (i.e., tradeoffs).

4.2 QoS Metrics for Tracks

Policy refinement required the definition of QoS metrics. Where the high-level policy could usefully describe behaviors for more and less important tracks, or make a distinction between more and less accurate track positions, the implementation of that policy required precise specification of these terms.

There are a number of traditional metrics for tracker performance. Many of these do not address the central issues of the ATD. For example, the ability of the tracker to follow aircraft through maneuvers, while obviously a valuable capability, is primarily a function of factors such as a sensor's probability of detection, the sensor's revisit rate, and the association algorithm employed. The first two factors were out of our scope, and we have not yet implemented the dynamic selection of an association algorithm based on QoS metrics. Rather, the ATD focused on the decision of when to execute instances of known algorithms.

The project settled on three QoS metrics that could quantify track processing:

- Timeliness: the total elapsed time between the arrival of a sensor report and the update of the corresponding track file record.
- Track quality (TQ): a traditional measure of the amount of recent sensor data incorporated in the current track record. TQ is incremented or decremented after each scan and ranges between zero (lowest quality) and seven (highest quality). If TQ falls to zero, the track is dropped.
- Track accuracy: a measure of the uncertainty of the estimate of the track's position and velocity.

In addition to these QoS metrics, each track was dynamically assigned to one of two *importance classes*. A track could be deemed more important for a number of reasons, including designation by an operator, flying in an operator-designated region, or posing a significant threat to the AWACS platform.

For the sake of intellectual manageability and to simplify the task of producing the first prototype adaptive tracker, the QoS metrics and the track importance designation were coalesced into a small number of classes. As mentioned above, there were two track importance classes; in addition, there are three track quality classes and two track accuracy classes. (AWACS operators subsequently validated this level of granularity.)

4.3 Quantified Adaptations

Based on these classes, project members refined the high-level adaptive tracking policy by focusing on specific tradeoffs involving pairs of track QoS metrics or track importance. For instance, one ranking considered only the track importance and track quality of competing clusters. Several such pair-wise rankings were merged to form a total ordering of the cases (with numeric values) as a function of track quality, track accuracy, and track importance.

At that point, tracking experts helped quantify the relative values of the bins. We also examined several cluster scenarios "by hand" to assign overall values

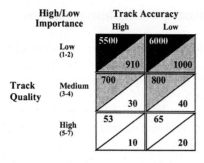
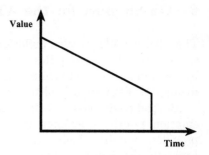

Fig. 1. Track Values as a Function of Track Quality-of-Service Metrics

Fig. 2. Shape of Time-Value Functions for Association Computations

and perform fine tuning. Figure 1 shows the set of valuations ("bins") that were employed for our demonstrations, which are described briefly in Section 7.

Once the bins had been assigned values, the translation to scheduling parameters was straightforward. The underlying operating system—The Open Group Research Institute's MK7—provides a value-based scheduler that accepts time constraints for specific computations. Each time constraint describes the application-specific value (in this case, the bin value of the cluster) of completing the designated computation as a function of time. For association computations, all of the time constraints had a similar shape (see Figure 2), which specified that "sooner was better than later."

5 Additional Design Benefits

Although scheduling and overload behavior were important aspects of the tracker design, there were other key issues as well.

Scalability — The relatively high computational cost of association processing justifies the investigation of distributed solutions, which seem particularly feasible since each association operates on a limited amount of data (a cluster's sensor reports and track file records) that can be easily assembled during gating and clustering. Consequently, not only is the ATD tracker multithreaded, but the association algorithms have been extracted and encapsulated in a separate object class. Association object instances (called *associators*) can—and have— been run on other nodes in a distributed tracking system. Moreover, there is provision in the tracker to select an associator on a cluster-by-cluster basis. This structure lays the groundwork for a scalable distributed tracker, where additional associators can be placed on newly added nodes for immediate use by the tracker.

Fault Tolerance — The distributed structure outlined above supports a straightforward form of fault tolerance. The tracker can clearly survive the failure of individual associators—as long as at least one associator survives. When associators fail, reducing system capacity, the basic adaptive, QoS-driven nature of the system results in the selection of a reasonable subset of work to perform.

6 OS Support for the ATD Prototype Implementation

The adaptive ATD tracker builds on several features of the underlying operating system, The Open Group Research Institute's MK7 [6][7], which provides a number of standards-based facilities as well as a set of unique capabilities designed to support distributed, real-time applications.

Beyond traditional real-time support (e.g., predictable execution times, preemption, priority scheduling, instrumentation, and low interrupt latency), MK7 provides a number of advanced features. The MK7 microkernel, which has been explicitly developed to support real-time applications, contains a scheduling framework that simultaneously supports priority-based and time-based scheduling policies. A value-based processor-scheduling policy called Best-Effort scheduling [4][7], which accepts application (and system) time constraints and schedules computations according to a heuristic that attempts to maximize total accrued value, was particularly useful for the ATD. Notably, this value-based scheduling policy and the threads it schedules co-exist with and interact with other parts of the OS and application, including synchronizers, preemptions, and "priority" modifications (e.g., priority inheritances and priority depressions).

MK7 threads are *distributed* (or migrating) threads[2][3], which move among the processes (i.e., MK *tasks*) of a distributed system by executing RPCs while carrying an environment that includes information like the thread's scheduling parameters, identity, and security credentials.

MK7 provides several standard interfaces—including a highly standards-compliant UNIX interface and the X Window System—that permit application programmers to reuse existing code. Both distributed threads and value-based scheduling are managed by an extended POSIX threads library. The extensions, while providing access to powerful facilities, are few in number and are generalizations of existing POSIX thread functions (e.g., through extensions of the POSIX thread cancellation interfaces)—thus simplifying the programmer's task considerably.

7 Prototype Results

The prototype performed extremely well during demonstrations to its target audience—AWACS operators and tracking system designers—most notably at the Boeing AWACS Prototype Demonstration Facility. Under "normal" (non-overload) conditions, the tracker handled all tracks as expected and delivered high QoS for all tracks. Overload conditions were simulated by artificially tightening the deadlines for the completion of association processing.

Figure 3 shows results of the tracker for a typical overload scenario. Batches of between ten and 14 sensor reports arrived at the tracker, with one-third of the tracks belonging to the more-important track class. The Association Capacity axis indicates the number of associations the tracker could usually perform in the specified processing time limit, and the Track Quality axis indicates the average TQ delivered for each track-importance class.

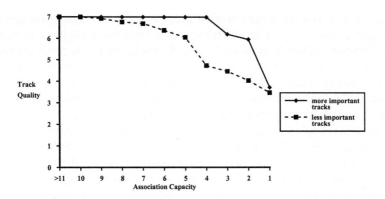

Fig. 3. Average Track Quality as a Function of Association Capacity

Figure 3 indicates that when the tracker can only process about 33% of the input, the prototype delivered essentially perfect TQ for the more important tracks, while delivering a reasonable (about 4.5) TQ for the less important tracks. (In some respects, this overload level can be likened to a system where the probability of detecting an airborne object is about 33%.) When the tracker was further constrained so that it could only process about 10% of the input, the prototype finally dropped some tracks—from the less important track class. No important tracks have been dropped during our demonstrations.

In addition, the demonstrations have shown that the tracker also adapts when new resources are added. In that case, which we demonstrate by loosening the time constraints on association processing, the prototype automatically delivers approximately the maximum achievable QoS.

Note that these results are not easily obtainable with static-priority tracking systems. In priority-based trackers, track priorities might reasonably be set according to track importance, where high importance implies high priority. In the prototype tracker, scheduling decisions are based on both importance and timeliness, and even relatively unimportant tracks can have very high application values in a surveillance mission—a situation that would not arise with straightforward priority-based scheduling.

8 Summary

The AWACS ATD project produced an adaptive, distributed tracker that was directly driven by Quality-of-Service metrics. Based on a novel design and incorporating knowledge from experts in the field, this tracker gracefully handles overloads, addressing a problem with currently deployed trackers and with trackers under development. The tracker was demonstrated to AWACS operators and tracker designers at Boeing in September 1998 and received supportive feedback, particularly regarding its behavior under overload and the operator interface.

This project and its demonstration provide another worked example in the area of value-based scheduling and further encourage our confidence in this technology. In addition, the derivation of the QoS metrics for tracks provided valuable insight into the nature and use of application-specific QoS metrics in a new application domain. Finally, the project produced a prototype tracker that can be used for further experimentation and can host future extensions, as well as be examined by tracker designers and implementers.

There are a number of open questions that should be addressed in the future. In this project, the value represented by a cluster of tracks and sensor reports was calculated by adding the values corresponding to each individual track in the cluster. While this is simple, has intuitive appeal, and produced good results in our tests, it might not be the best way to determine the value represented by a cluster. Moreover, there were several capabilities that were included in our design that have not yet been implemented—most notably the dynamic selection of an association algorithm for a cluster based on factors such as the amount of clutter, the number of maneuvering tracks, the computational cost of the algorithm, and the currently available processing resources. Also, AWACS program personnel have speculated that the overload adaptations in the ATD tracker could help manage the dramatically increased processing load associated with (next generation) multiple hypothesis trackers [5]. This seems credible and could broaden our expertise in the use of value-based scheduling in a new dimension—that is, the specification, in application-specific terms, of the benefit of evaluating each of the potentially large number of hypotheses associated with each track in the system. Finally, the tracker was designed to be able to perform different missions and to interact with other AWACS applications. Neither of these capabilities has been tested to date: The tracker has only performed a surveillance mission, and no other applications have been added. Exploring both of these areas should be fruitful, providing, among other things, an opportunity to explore hierarchical, value-based scheduling architectures.

References

1. Blackman, S.: Multiple-Target Tracking with Radar Applications. Artrech House, ISBN 0-89006-179-3 (1986)
2. Clark, R.K., Jensen, E.D., Reynolds, F.D.: An Architectural Overview of the Alpha Real-Time Distributed Kernel. Proc. of the USENIX Workshop on Micro-kernels and Other Kernel Architectures (1992) 127-146
3. Ford, B., Lepreau, J.: Evolving Mach 3.0 to a Migrating Thread Model. Proc. of the USENIX Winter 1994 Technical Conference (1994)
4. Locke, C.D.: Best-Effort Decision Making for Real-Time Scheduling. Ph.D. Thesis, Dept. of Electrical and Computer Engineering, Carnegie-Mellon University (1986)
5. Reid, D.B.: An Algorithm for Tracking Multiple Targets. IEEE Transactions on Automatic Control, Vol. AC-24 (1979) 843-854
6. Wells, D.: A Trusted, Scalable, Real-Time Operating System Environment. Dual-Use Technologies and Applications Conference Proceedings (1994) II:262-270
7. —: MK7.3 Release Notes. The Open Group Research Institute, Cambridge, MA (1997)

Non-preemptive Scheduling of Real-Time Threads on Multi-level-context Architectures

Jan Jonsson[1], Henrik Lönn[1], and Kang G. Shin[2]

[1] Dept. of Computer Engineering, Chalmers, S–412 96 Göteborg, Sweden
{janjo,hlonn}@ce.chalmers.se
[2] Dept. of Elec. Engr. and Computer Science, Univ. of Michigan, Ann Arbor
kgshin@eecs.umich.edu

Abstract. This paper addresses the problem of how to schedule periodic, real-time threads on a class of architectures referred to as *multi-level-context (MLC)* architectures. Examples of such architectures are real-time operating systems with support for user- or kernel-level threads, and multithreaded microprocessors endowed with on-chip contexts. A common feature of these architectures is that they provide support for the administration of threads within contexts at different levels of abstraction. Therefore, the cost for switching between threads will depend on the affinity of their corresponding contexts. The main contributions of this paper are to demonstrate (i) how the scheduling performance for off-line scheduling on MLC architectures can benefit from an integrated heuristic that is cognizant of both the time-criticality of a thread and the current context affinity; and (ii) how the predicted performance for on-line scheduling on MLC architectures can benefit from an off-line schedulability test that accounts for variations in the context affinity.

1 Introduction

Assessing the impact of architectural properties on the schedulability of a real-time application has become a very important issue in the real-time research community. Most modern microprocessor architectures are endowed with such mechanisms as caches and instruction pipelines whose temporal behavior are very difficult to predict *a priori*. Despite many recent successful attempts to model the behavior of these mechanisms [1,2], real-time scheduling research has not been able to keep pace with the progress in microprocessor design. This has led to an unfortunate gap between real-time scheduling theory and its practice.

An important remaining problem is to analyze the effects of *context switch* operations on the schedulability analysis. While traditional real-time schedulability tests [3] simply assumed that a context switch can be made at a negligible cost, more recent tests [4] are capable of incorporating a non-negligible context-switch cost. Unfortunately, these schedulability tests are based on the assumption that the cost for a context switch is constant and independent of the underlying thread-invocation pattern. This assumption does not hold for an important emerging class of architectures referred to as *multi-level-context*

(*MLC*) architectures. A common feature of these architectures is that they provide software and/or hardware support for the execution of multiple *threads* in the application. The administration of threads requires support for *contexts* at different levels of abstraction, that is, with different amounts of execution states to maintain. When there are more than one context level, the cost for switching between two threads will depend on the affinity of their contexts. Since the context-switch scenario cannot typically be predicted in advance, existing techniques for schedulability analysis can only be used with an MLC architecture at the expense of overly-pessimistic assumptions on the context-switch cost. This will dramatically reduce the usefulness of an off-line schedulability test.

Many existing real-time systems can be classified as MLC architectures, *e.g.*, real-time operating systems with support for user- or kernel-level threads [5]. Also, thread-level parallelism is believed to be the only alternative to maintain the performance growth of microprocessors [6]. Because one can predict an increasing demand for computing power in modern real-time applications [7], it is very likely that multithreaded architectures will also be used widely in future real-time systems. In order to analyze the real-time performance of such architectures, their degree of run-time predictability must be thoroughly assessed.

In this paper, we address the problem of non-preemptive scheduling of periodic, real-time threads on a uniprocessor MLC architecture. Non-preemptive scheduling is the natural choice for many systems since (i) uncontrolled preemption can cause extra context switches which may jeopardize the schedulability, (ii) exclusive access to shared resources is guaranteed which eliminates much of the need for synchronization and its associated overhead, and (iii) worst-case execution time analysis becomes easier since the instruction stream of a thread can execute without interference by other threads. Using a set of randomly-generated application thread sets and various context-configuration scenarios, we study the performance of a set of heuristic algorithms using the *success ratio* as the performance measure — that is, the ratio of the number of successfully-scheduled thread sets to the total number of evaluated thread sets.

Despite the growing interest in the MLC architectures, very few results on thread scheduling on them have been reported. The main intent of this paper is, therefore, to make the following two contributions:

C1. For real-time systems employing a time-driven thread dispatching policy, we demonstrate an integrated polynomial-time scheduling heuristic — cognizant of both time-criticality and the current context affinity — that significantly outperforms other heuristics when used with MLC architectures.

C2. For systems employing a priority-driven thread dispatching policy, we propose a sufficient schedulability test for the on-line EDF algorithm that makes far better predictions than existing tests when used with MLC architectures.

We start in Sect. 2 to elaborate on the problem of scheduling on MLC architectures, and then continue in Sect. 3 to introduce the assumed system models and the terminology used. In Sect. 4, we present an experimental evaluation of our proposed techniques. In Sect. 5, we discuss related work and give a brief description of future work. Finally, we summarize our findings in Sect. 6.

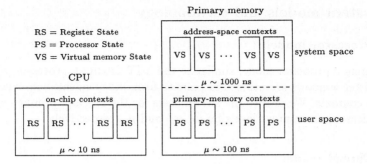

Fig. 1. Multi-Level-Context architecture.

2 Background and problem statement

In an MLC architecture, the schedulable entities are *threads*, streams of control that represent stretches of executable code in an application. Each thread executes within multiple levels of *contexts*. Each context is a locus of control that describes the current execution state of the thread at a certain abstraction level. When switching between two threads, a *context switch* must take place so as to exchange the active execution state. The cost for a context switch depends on the *context affinity* of the threads, which is defined as the amount of execution information shared between the threads.

Depending on the level of context abstraction, the cost for a context switch consists of different components. An *on-chip* context is typically found in multiple-context, multithreaded architectures where the registers are either duplicated in hardware [8,9], or partitioned in software [10], to hold multiple active threads. The cost for switching between two on-chip contexts on a pipelined processor is in the range of 5–10 clock cycles. The *primary-memory* context contains processor user registers (*e.g.*, in a thread control block). The total cost for these operations is typically in the range of 100–200 clock cycles for a RISC processor with a large number of user registers. An *address-space* context is typically found at the operating system level. The cost for switching between address-space contexts includes the cost for updating virtual-memory mechanisms (*e.g.*, TLB and MMU), operations that can take thousands of clock cycles to perform.

Fig. 1 illustrates how the context-switch cost μ relates to the different context levels in a typical MLC architecture. Apparently, depending on how the application threads are organized with respect to the contexts levels, the cost for a context switch can vary by as much as an order of a magnitude at run-time. Thus, depending on the way the application has been partitioned into threads, and the invocation pattern of the threads at run-time, the cost for switching between two threads will vary with the current context affinity of the threads, which, in turn, is a function of time. Unless such variations are taken into account during off-line analysis, an expensive context switch at run-time may cause a thread, and hence the application, to miss its deadline. The objective of this paper is to find effective real-time scheduling strategies that are robust with respect to such variations in the context-switch cost.

3 System models and terminology

3.1 Processor model

We assume a processor model similar to the MIT SPARCLE processor [8] where the register windows are utilized in pairs to form independent, non-overlapping on-chip contexts. We denote the switch costs for the on-chip, primary-memory, and address-space context levels by μ_ℓ, μ_g, and μ_a, respectively.

3.2 Thread model

We consider a real-time application that consists of a set $\mathcal{T} = \{\tau_i : 1 \leq i \leq n\}$ of threads. The threads are assumed to be part of tasks that represent major computations in the application. Each thread $\tau_i \in \mathcal{T}$ is characterized by a 5-tuple $\langle c_i, \phi_i, d_i, p_i, \vartheta_i \rangle$. The *worst-case execution time* c_i is an upper bound of the execution time of the thread and includes various architectural overheads such as the cost for cache misses and pipeline hazards, as well as the on-chip context-switch cost μ_ℓ. The *phasing* ϕ_i is the earliest time by which the first invocation of the thread will occur. The *relative deadline* d_i is the amount of time within which the thread must complete its execution. The *thread period* p_i is the time interval between two consecutive invocations of the thread. The *context* ϑ_i is the primary-memory state within which the thread will execute.

We organize the threads in *period classes*, with one class π_k for each distinct period used by threads in $\mathcal{T}$. We denote the collection of period classes by Π, the corresponding period of π_k by $p(\pi_k)$, and the $n_t(\pi_k)$ threads in π_k by $\tau_{\pi_k,1}, \tau_{\pi_k,2}, \ldots, \tau_{\pi_k,n_t(\pi_k)}$. It is also useful to collect the threads sharing the same primary-memory context in *context groups*, with one group $\theta_{\pi_k,r}$ for each distinct context used by threads in π_k. We denote the collection of $n_g(\pi_k)$ context groups in π_k by $\Theta_{\pi_k} = \{\theta_{\pi_k,1}, \theta_{\pi_k,2}, \ldots, \theta_{\pi_k,n_g(\pi_k)}\}$, the corresponding context of $\theta_{\pi_k,r}$ by $\vartheta(\theta_{\pi_k,r})$, and the number of threads in $\theta_{\pi_k,r}$ by $n_t(\theta_{\pi_k,r})$.

Precedence constraints between threads in $\mathcal{T}$ are represented as follows. If thread τ_j cannot begin its execution until thread τ_i has completed its execution, we write $\tau_i \prec \tau_j$. In this case τ_i is said to be a *predecessor* of τ_j, and, conversely, τ_j a *successor* of τ_i. A thread which has no predecessors is called an *input thread*, and a thread which has no successors is called an *output thread*. A *thread chain* is defined as an input thread followed by a series of parallel/serial threads and terminated by one output thread.

3.3 Execution model

The non-preemptive execution of application threads is administrated by a run-time dispatcher. When a thread is selected by the dispatcher, it begins its execution and runs without preemption until it voluntarily transfers control of the processor back to the dispatcher. To understand the behavior of the dispatcher, we start by describing the dynamic behavior of a thread τ_i. Let τ_i^k denote the kth invocation of τ_i, $k \in \mathbf{Z}^+$. The *absolute arrival time* $a_i^k = \phi_i + p_i(k-1)$ is the earliest time by which τ_i^k is allowed to start its execution. The *absolute deadline* $D_i^k = a_i^k + d_i$ is the latest time by which τ_i^k must complete its execution.

Now, assume that the dispatcher is in the process of selecting a new thread to schedule at time t. We then make the following definitions.

Definition: A thread τ_i is *ready* at time t if its immediate predecessors have finished their execution.

Definition: A thread τ_i is *released* at time t if it is ready, $a_i \leq t$, and the thread has received all necessary data from its immediate predecessors.

Definition: A thread is called a *continuation thread* if it is ready and belongs to the same context group as the most-recently-executed thread.

Definition: A thread is called a *completion thread* of period class π_k if it is executed last among all threads in π_k within one particular invocation.

Definition: A thread is said to be *returning* from period class π_j to period class π_k if it belongs to π_j, is the most-recently-executed thread, and the next thread that executes belongs to π_k.

Definition: A scheduling policy that allows itself to insert idle time slots instead of executing a ready thread is said to use *inserted idle time*.

4 Experimental evaluation

4.1 Quality assessment

Our primary performance measure is the *success ratio*, defined as follows. If a real-time scheduling strategy can aid in finding feasible schedules for x of the y thread sets considered, its success ratio is (x/y). We choose to study the delivered performance as a function of the **context-switch cost** (μ_g) and the **context affinity** (η). The former represents the cost for performing a switch to another primary-memory context and reflects two important characteristics of the system: (i) the number of on-chip registers that must be exchanged, and (ii) the granularity (in terms of execution time) of the application threads. The latter represents the probability that two arbitrarily-chosen threads in the generated thread set share the same primary-memory context. This parameter allows us to observe the effect of the number of primary-memory contexts in the system on scheduling performance. The affinity ranges from $\eta = 0.0$ (threads execute in separate contexts) to $\eta = 1.0$ (threads execute in a common context).

4.2 Architecture

Similar to the SPARCLE processor, we assume that there are four on-chip contexts. We also assume that there are eight primary-memory contexts. When a primary-memory context switch occurs, we assume that all on-chip contexts are reloaded and that all memory references hit in the cache during the context switch. The primary-memory context-switch cost ranges from 0% (no overhead) to 25% (significant overhead) of the mean thread execution time, c_{mean}. We also assume that all threads execute within one single address-space context. Any overhead associated with interrupt handling and thread dispatching in the run-time system is assumed to be negligible in our experiments.

4.3 Workload

To evaluate[1] the robustness of each heuristic in the presence of varying context-switch costs, we attempted to schedule 1024 randomly-generated sets of threads, each containing 40–50 threads organized in four chains of threads. Each chain of threads was then randomly mapped to one of the eight primary-memory contexts, as determined by the context affinity, η. The default value of η is 0.25. The chains were generated in such a way that the parallelism of the threads sharing one specific context could be fully exploited on the on-chip contexts. Individual thread execution times were chosen at random assuming a uniform distribution in the range of 15 to 25 μs, and a mean execution time $c_{mean} = 20$ μs.

All threads in a chain were assigned a common period, chosen at random (with uniform probabilities) from four period classes with the corresponding periods $p(\pi_1) = 0.8p$, $p(\pi_2) = 1.0p$, $p(\pi_3) = 1.2p$, and $p(\pi_4) = 1.6p$. The base period p was chosen so that the total utilization U of the processor was kept at approximately 85%. We assume for each thread τ_i that $\phi_i = 0$ and $d_i = p_i$.

4.4 Pre-run-time scheduling for a time-driven system

We first investigate the time-driven dispatching policy where run-time thread dispatching is performed using a table with explicit start and finish times of each thread. The table of thread invocations (the schedule) is constructed off-line. The off-line scheduling heuristics under investigation in this experiment are all variations of the *list scheduling* [12] strategy, which uses inserted idle time.

Fig. 2a plots the performance of the evaluated heuristics as a function of the primary-memory context-switch cost μ_g. The figure presents the results for five different heuristics. We first discuss the performance of two simple list-scheduling strategies: the earliest-deadline-first (EDF) and earliest-start-time (EST) heuristics. In these strategies, the next thread to schedule in the priority list is the one with the closest (absolute) deadline and the one that can begin execution soonest (accounting for any precedence or periodicity constraints), respectively. The plots indicate that the EDF heuristic does not perform well under the list-scheduling policy. In fact, EDF is not capable of scheduling more than 30% of all generated thread sets, regardless of context-switch cost and context affinity. Moreover, the success ratio drops as the context-switch cost increases owing to the fact that the deadline of a thread does not contain any information as to whether a context switch will be taken or not should the thread be scheduled next. In contrast, the plots for the EST heuristic indicate that EST performs better when the context-switch cost is significant, than when it is zero. In fact, the success ratio increases from 30% to approximately 50% when the context-switch cost increases from zero to 5%. The explanation for this behavior is as follows. As soon as the context-switch cost is non-negligible, the earliest start time of a thread indicates whether a context switch will be taken or not should the thread be scheduled. Hence, the list-scheduling algorithm can take advantage of this information when selecting the next thread to schedule.

---

[1] All modeling and simulations were performed within the GAST [11] framework.

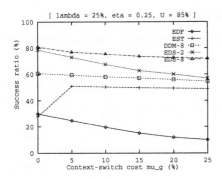 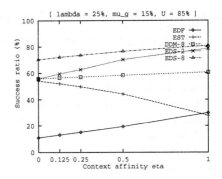

Fig. 2. Success ratio for time-driven dispatching as a function of a) the primary-memory context-switch cost μ_g assuming a context affinity $\eta = 0.25$; b) the context affinity η assuming a primary-memory context-switch cost $\mu_g = 15\%$.

In the dynamic deadline modification (DDM) technique [13], non-continuation threads have their original deadlines temporarily modified during thread selection so as to favor continuation threads. The temporary deadline of a non-continuation thread τ_i is $\hat{D}_i = D_i + k$, where D_i is the original deadline of the thread and k is an offset factor. In our experiments, we have used the DDM with an offset factor $k = 8$ (denoted in the figure by DDM-8). Fig. 2a clearly indicates that this approach is successful in increasing the success ratio.

An interesting choice of heuristic would be to integrate the time-criticality-cognizant traits of the EDF heuristic with the context-switch-cognizant traits of the EST heuristic. To this end, we define the *earliest-deadline-and-start-time* (EDS) heuristic, where the priority function for a thread τ_i is the weighted sum $D_i + w \times t_{est}$, where D_i is the (absolute) deadline of the thread, t_{est} is the earliest-start-time of the thread, and w is a weight factor. Fig. 2a shows the plots for the EDS heuristic with a weight factor $w = 2$ and 8 (denoted by EDS-2, and EDS-8, respectively). As indicated by these plots, the integrated approach gives a significant boost in performance. Initially, the success ratio increases to more than 75% for a negligible context-switch cost, and, as the context-switch cost increases, the plots show the potential of the integrated approach — in particular when the weight factor increases. For example, a weight factor $w = 8$ gives a very robust behavior as compared with the original EDF and EST heuristics[2].

Fig. 2b plots the performance of the evaluated heuristics as a function of the context affinity η. The plots indicate that EDS-8 is very robust with respect to variations in the context affinity. Also, note the anomalous behavior of EST as the context affinity increases. While the other heuristics benefit from an increase in the context affinity because the degree of sharing increases, EST performs worse as the earliest start time of a thread provides less and less information on whether a primary-memory context switch will be taken or not.

---

[2] Note that an increase in the weight factor not automatically implies an increase in performance. Too high a weight factor would give the "EST" part of the heuristic too strong an influence, which would cause the performance to reduce to that of original EST. We found that $w = 8$ through $w = 32$ gives the overall best performance.

4.5 Run-time scheduling for a priority-driven system

In this section, we explore a priority-driven dispatching scheme. Threads are kept in a list at run-time, ordered by their priorities. Whenever the dispatcher selects a thread to execute, it selects the one that has the highest priority in the list. Inserted idle time is not used for this dispatching scheme, which means that only released threads will be considered during thread selection. We assign thread priorities according to the earliest-deadline-first (EDF) policy.

Now, since we are using an on-line scheduling algorithm, it is necessary to verify, off-line, that all thread deadlines will be met at run-time with the employed dispatching policy. To this end, we use the schedulability test proposed by Zheng and Shin [14], modified to account for context-switch costs.

Theorem 1 (Zheng and Shin). *In the presence of non-real-time threads, a set of real-time threads* $\mathcal{T} = \{\tau_i : 1 \leq i \leq n\}$ *with* $d_i = p_i$ *are schedulable on a uniprocessor with a primary-memory context-switch cost* μ_g *under the non-preemptive EDF policy if the following conditions hold:*

1. $\sum_{\tau_i \in \mathcal{T}} (c_i + \mu_g)/p_i \leq 1.$

2. $\forall t \in S, \sum_{\tau_i \in \mathcal{T}} \lceil (t - p_i)/p_i \rceil^+ (c_i + \mu_g) + c_p \leq t,$
   where $S = \cup_{\tau_i \in \mathcal{T}} \{t : t \leq t_{max}, t = p_i + mp_i, m \in \mathcal{N}\},$
   $t_{max} = \max\{p_1, \ldots, p_n\}$

One obvious problem with the schedulability test in Theorem 1 is that it assumes that one primary-memory context-switch is taken for each thread invocation. This is not a reasonable assumption since, on an MLC architecture, multiple threads execute within shared contexts. Thus, many primary-memory context switches accounted for in Theorem 1 will, in fact, never be taken at run-time. Unfortunately, with the current run-time policy, it is difficult to calculate tight upper bounds on the number of context switches unless we completely predict the thread invocation pattern in advance. In order to make better predictions on the context-switch scenario, we introduce a degree of determinism by replacing the original EDF policy with a similar scheduling policy that we call the EDF* policy. The new scheduler behaves exactly like the original EDF scheduler except when choosing among released threads with the same deadline. Recall that the EDF rule allows for an arbitrary choice among threads with the same deadline. We will exploit this feature to enforce the precedence constraints on the threads. When there are multiple released threads with the nearest deadline, the EDF* policy will choose a continuation thread whenever one exists. This decision will be enforced at run-time by complementing the dynamic-priority policy with an additional static thread priority that will guarantee that continuation threads will be favored before other threads. The actual static-priority assignment is performed according to the CLASSIFY algorithm listed in Fig. 3. The main purpose of this algorithm is to guarantee that the execution of threads in the same period class gives rise to as few context switches as possible.

Algorithm CLASSIFY:

1. create a set Π of period classes: one class π_k for each distinct thread period;
2. create one local context group $\theta_{\pi_k,r}$ for each distinct thread context in π_k;
3. enumerate context groups within each period class according to significance;
4. initialize a variable z with the highest possible priority;
5. **for** { each $\pi_k \in \Pi$, shortest period first } **loop**
6.   **for** { each $\theta_{\pi_k,r} \in \pi_k$, most significant first } **loop**
7.   assign to all threads in $\theta_{\pi_k,r}$ a priority equal to z;
8.   decrease the priority variable z;
9.   **end loop**;
10. **end loop**;

Fig. 3. The period-based classification algorithm.

Now, to find a schedulability test that is less pessimistic than the one in Theorem 1, we analyze the period classes one by one and determine, for each period class π_k, the number of context switches caused by (i) threads being *invoked* in π_k, and (ii) threads *returning* from another period class to π_k. To this end, we use Lemma 1. The proof can be found in detail in [15].

Lemma 1. *Assuming that the non-preemptive EDF* policy and the* CLASSIFY *algorithm are used, an upper bound $n_c(\pi_k)$ on the total number of primary-memory context switches experienced while executing the threads in π_k during the time interval $p(\pi_k)$ can be found as follows.*

$$n_c(\pi_k) = \min\{n_t(\pi_k),\ n_g(\pi_k) + \sum_{\pi_j \in \Pi\ :\ p(\pi_j) < p(\pi_k)} \min\{\hat{n}_t(\pi_k, \pi_j),\ \lceil (p(\pi_k) - p(\pi_j))/p(\pi_j) \rceil\}\}$$

where $\hat{n}_t(\pi_k, \pi_j) = \sum_{\theta_{\pi_k,r} \in \Theta_{\pi_k}\ :\ \vartheta(\theta_{\pi_k,r}) \neq \vartheta(\theta_{\pi_j, n_g(\pi_j)})} n_t(\theta_{\pi_k,r})$

By means of Lemma 1, we now arrive at a new sufficient, but not necessary, schedulability test that is based on the assumption that the EDF* policy is employed at run-time. The proof can be found in [15].

Theorem 2 (Jonsson, Lönn, and Shin). *In the presence of non-real-time threads, a set of real-time threads $\mathcal{T} = \{\tau_i : 1 \leq i \leq n\}$ with $d_i = p_i$ are schedulable on a uniprocessor with a primary-memory context-switch cost μ_g under the non-preemptive EDF* policy if the following conditions hold:*

1. $\sum_{\tau_i \in \mathcal{T}} c_i/p_i + \sum_{\pi_k \in \Pi} n_c(\pi_k) \times \mu_g/p(\pi_k) \leq 1.$

2. $\forall t \in S,\ \sum_{\tau_i \in \mathcal{T}} \lceil (t - p_i)/p_i \rceil^+ c_i + \sum_{\pi_k \in \Pi} \lceil (t - p(\pi_k))/p(\pi_k) \rceil^+ n_c(\pi_k) \times \mu_g + c_p \leq t,$
   where $S = \cup_{\tau_i \in \mathcal{T}} \{t : t \leq t_{max},\ t = p_i + mp_i,\ m \in \mathcal{N}\},$
   $t_{max} = \max\{p_1, \ldots, p_n\}$

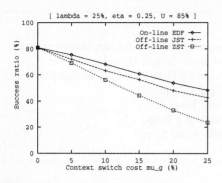 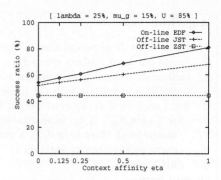

Fig. 4. Success ratio for dynamic-priority-driven dispatching as a function of a) the primary-memory context-switch cost μ_g assuming a context affinity $\eta = 0.25$; b) the context affinity η assuming a primary-memory context-switch cost $\mu_g = 15\%$.

Returning to the CLASSIFY algorithm in Fig. 3, we observe that the strategy used for enumerating the context groups within each period class is not specified in Step 3 of the algorithm. Recall that one of the strengths with the schedulability test in Theorem 2 is that it accounts for context affinity. More specifically, we observed that, given a certain period class π_k, only the context of the completion thread in period class π_j has to be checked against the threads in π_k to find the number of context switches caused by the completion thread returning from π_j to π_k. Now, in order to reduce the number of context switches, we must take advantage of the fact that only the completion thread has to be checked. This can be done by devising a good heuristic for enumerating the context classes within each period class. To this end, we make the following conjecture.

Conjecture 1. Assuming that the non-preemptive EDF* policy and the CLASSIFY algorithm are used, a good strategy for minimizing $n_c(\pi_k)$ is to guarantee that the completion thread is part of the context group that minimizes the following expression:

$$\theta_{\pi_k,r} \in \Theta_{\pi_k} : \sum_{\pi_j \in \Pi,\, \theta_{\pi_j,q} \in \Theta_{\pi_j}\,:\, p(\pi_j) > p(\pi_k)\,\wedge\, \vartheta(\theta_{\pi_j,q}) \neq \vartheta(\theta_{\pi_k,r})} n_t(\theta_{\pi_j,r})/p(\pi_j)$$

Fig. 4a plots the performance of the on-line EDF* algorithm as a function of the primary-memory context-switch cost μ_g. Also included in the plots is the modified Zheng/Shin schedulability test from Theorem 1 (denoted by ZST). The plots indicate that the EDF* policy performs much better than what can be predicted by ZST. For example, the performance of the on-line EDF* algorithm is more than 20 percentage points higher than the predicted one for $\mu_g \geq 20\%$. The reason for this poor correspondence between predicted and real performance is that ZST always assumes that a full primary-memory context switch will occur at a thread invocation. Since this strategy does not account for the variations in context-switch cost that can occur in an MLC architecture, the system will be poorly utilized if the context-switch cost is significant. In contrast, we can see that the predicted performance of the new schedulability test from Theorem 2 (denoted by JST) is significantly higher than that of ZST*. In fact, the increase in predicted performance can be as high as 20 percentage points in systems with significant context-switch cost. This increase should, of course, be attributed to the "intelligence" that was added through the EDF* policy, the CLASSIFY algorithm and Conjecture 1.

Looking at the scheduling performance as a function of the context affinity, we observe in Fig. 4b that the performance of JST follows the curve for EDF* and increases for higher values of context affinity. This shows that the combination of the CLASSIFY algorithm and Conjecture 1 works very well. No such behavior can be noticed for ZST, since it does not account for the context affinity.

5 Related work and discussion

Little has been reported in the literature on the problem of real-time scheduling on MLC architectures, despite its growing importance. Humphrey et al. [5] recognize the need to exploit fast context-switch operations in order to improve the performance in the Spring kernel but did not make any attempt to evaluate the impact of such a technique on the application schedulability. Fiske and Dally [9] evaluated the use of a priority-based thread selection mechanism in a multithreaded architecture with multiple on-chip register sets. However, their work only evaluated scheduling strategies whose objective is to minimize the average application completion time. The work that comes closest to ours is that by Cheng et al. [13]. They proposed the dynamic-deadline-modification (DDM) heuristic for a multiprocessor MLC architecture assuming arbitrary thread preemption. As shown thus far, our proposed EDS heuristic clearly outperforms the uniprocessor version of DDM with respect to all the parameters considered.

The proposed strategies for reducing the number of imposed context switches during schedulability analysis is just a first attempt to address the problem of scheduling on MLC architectures. One interesting topic for future work is to investigate whether it is possible to find good heuristics for enumerating the context groups, and whether there exists an optimal enumeration strategy that minimizes the total number of primary-memory context switches at run-time. Such a strategy would have to consider not only period times and contexts but also execution times and phasings of threads.

6 Conclusions

In this paper, we have proposed and evaluated scheduling heuristics suitable for multi-level-context (MLC) architectures, where the context-switch cost is a function of the thread invocation pattern. Our main contributions are: (i) an integrated off-line scheduling heuristic that accounts for both the time-criticality of a thread and the current context affinity, and a demonstration of how this heuristic deliver superior performance over traditionally-used scheduling heuristics; (ii) an improved schedulability test for the on-line EDF algorithm that accounts for the variations in context affinity that can occur in an MLC architecture.

References

1. C. A. Healy, D. B. Whalley, and M. G. Harmon, "Integrating the Timing Analysis of Pipelining and Instruction Caching," *Proc. of the IEEE Real-Time Systems Symposium*, Pisa, Italy, Dec. 5–7, 1995, pp. 288–297.
2. R. T. White, F. Mueller, C. A. Healy, D. B. Whalley, and M. G. Harmon, "Timing Analysis for Data Caches and Set-Associative Caches," *Proc. of the IEEE Real-Time Technology and Applications Symposium*, Montreal, Canada, June 9–11, 1997, pp. 192–202.
3. C. L. Liu and J. W. Layland, "Scheduling Algorithms for Multiprogramming in a Hard-Real-Time Environment," *Journal of the Association for Computing Machinery*, vol. 20, no. 1, pp. 46–61, Jan. 1973.
4. D. I. Katcher, H. Arakawa, and J. K. Strosnider, "Engineering and Analysis of Fixed Priority Schedulers," *IEEE Trans. on Software Engineering*, vol. 19, no. 9, pp. 920–934, Sept. 1993.
5. M. Humphrey, G. Wallace, and J. A. Stankovic, "Kernel-Level Threads for Dynamic, Hard Real-Time Environments," *Proc. of the IEEE Real-Time Systems Symposium*, Pisa, Italy, Dec. 5–7, 1995, pp. 38–48.
6. S. J. Eggers, J. S. Emer, H. M. Levy, J. L. Lo, R. L. Stamm, and D. M. Tullsen, "Simultaneous Multithreading: A Platform for Next-Generation Processors," *IEEE Micro*, vol. 17, no. 5, pp. 12–19, Sept./Oct. 1997.
7. K. Diefendorff and P. K. Dubey, "How Multimedia Workloads will Change Processor Design," *IEEE Computer*, vol. 30, no. 9, pp. 43–45, Sept. 1997.
8. A. Agarwal, J. Kubiatowicz, D. Kranz, B.-H. Lim, D. Yeung, G. D'Souza, and M. Parkin, "Sparcle: An Evolutionary Processor Design for Large-Scale Multiprocessors," *IEEE Micro*, vol. 13, no. 3, pp. 48–61, June 1993.
9. S. Fiske and W. J. Dally, "Thread Prioritization: A Thread Scheduling Mechanism for Multiple-Context Parallel Processors," *Proc. of the IEEE Symposium on High Performance Computer Architecture*, Raleigh, North Carolina, Jan. 22–25, 1995, pp. 210–221.
10. C. A. Waldspurger and W. E. Weihl, "Register Relocation: Flexible Contexts for Multithreading," *Proc. of the ACM Int'l Symposium on Computer Architecture*, San Diego, California, May 16–19, 1993, pp. 120–130.
11. J. Jonsson, "GAST: A Flexible and Extensible Tool for Evaluating Multiprocessor Assignment and Scheduling Techniques," *Proc. of the Int'l Conf. on Parallel Processing*, Minneapolis, Minnesota, Aug. 10–14, 1998, pp. 441–450.
12. C.-Y. Lee, J.-J. Hwang, Y.-C. Chow, and F. D. Anger, "Multiprocessor Scheduling with Interprocessor Communication Delays," *Operations Research Letters*, vol. 7, no. 3, pp. 141–147, June 1988.
13. B.-C. Cheng, A. D. Stoyenko, T. J. Marlowe, and S. Baruah, "A Scheduler Maximizing Maximum Tardiness for DSP Programs with Context Switch Overheads Considered," *Proc. of the Int'l Conf. on Signal Processing Applications & Technology*, Boston, Massachusetts, Oct. 7–10, 1996, pp. 771–775.
14. Q. Zheng and K. G. Shin, "On the Ability of Establishing Real-Time Channels in Point-to-Point Packet-Switched Networks," *IEEE Trans. on Communications*, vol. 42, no. 2/3/4, pp. 1096–1105, Feb./Mar./Apr. 1994.
15. J. Jonsson, H. Lönn, and K. G. Shin, "Non-Preemptive Scheduling of Real-Time Threads on Multi-Level-Context Architectures," Technical Report No. 98–6, Dept. of Computer Engineering, Chalmers University of Technology, S–412 96 Göteborg, Sweden, May 1998.

QoS Control and Adaptation in Distributed Multimedia Systems

Farid Naït-Abdesselam[1,2], Nazim Agoulmine[2]

[1] Advanced Communication Engineering Centre
Department of Electrical and Computer Engineering
The University of Western Ontario, London Ontario N6A 5B9, Canada.
Email: fnait@engga.uwo.ca.

[2] PR*i*SM Laboratory, University of Versailles
45 Avenue des Etats-Unis 78 000, Versailles. France.
Email: {naf, naz}@prism.uvsq.fr

Abstract. Presently, many distributed multimedia systems adapt to their changing environments and Quality of Service (QoS) requirements by exchanging control and feedback data between servers and clients. In order to realize a more flexible adaptation to QoS in distributed multimedia systems, this paper seeks to present a new approach, based on distributed software agents located in both network nodes and end-systems. By having a good knowledge of local resources at each component, and with their capabilities to communicate in order to share their knowledge, the distributed software agents can alleviate the major fluctuations in QoS, by providing load balancing and resource sharing between the competing connections. In order to show the feasibility of our active adaptation approach, simulations have been conducted to adapt a delay sensitive flow such as distributed interactive virtual environment. We have performed our evaluations for short and long range (i.e. self-similar) traffic patterns. Preliminary results show a viable system, which exhibits a smooth and noticeable improvement in perceptual QoS during a heavy loaded network. In addition, our results indicate that the network and the application have more to benefit from the algorithm, when the traffic exhibits long range dependence behavior.

1 Introduction

Recent years have seen great advances in computing and networking technology. At the network level, new high-speed technologies such as *Asynchronous Transfer Mode (ATM)* are actively being deployed in local, metropolitan and wide area network environments. These networks not only have the capability of transmitting information at high speed, but also have the potential to offer a wide range of *Quality of Service (QoS)* properties including bounds in delay, guarantees and isochronous communications.
Multimedia workstation technology for generating, processing and displaying streams of digital video and audio is also available in the marketplace with very high capacity storage systems and real-time video and audio codecs.
These technological developments are complemented by new user perspectives. New classes of distributed applications have been developed, such as Distance Learning, Desktop Video-Conferencing, Virtual Workshop and Video On Demand. These applications are characterized by their highly interactive nature and their significant use of multimedia information transfer. In these applications, communication requirements are extremely diverse and demand varying levels of service in terms of parameters such as latency, bandwidth, jitter, and loss rate [1]. These various applications share the need for *Quality of Service control and management* to ensure that the requirements of the users are satisfied [2]. However, even with service guarantees, fluctuations in the Quality of Service may occur because of shortage of resources, e.g., network congestion or switch failure.
In this work, we propose an adaptation approach that allows automatic recovery, from violations in Quality of Service, by deploying a new allocation policy to distribute the level of QoS, that

should be supported by all the components involved in meeting the end-to-end requirements of a multimedia connection. This approach, based on the mutual help between all the components involved in the support of the requested service, allows to reach a higher resource utilization in the underlying system, increase the probability of admitting a new connection in the network, and reduce the probability of violations in the negotiated Quality of Service.

The approach undertaken is based on distributed software agents, located on all the components involved in providing QoS guarantees [3]. We associate for each component of the Distributed Multimedia System an Agent called *QoS-Agent (QoSA)*. Hence, the Distributed Multimedia System can be seen as a multi-agent system, able to communicate and evaluate, , in the duration of the connection, the values of the used QoS parameters applicable to the resource utilization. Close cooperation between the agents is required to improve the Quality of Service of a given application, or react to violations that may occur. In our designs, we have taken into consideration the traffic volumes produced by the agents, and we have devised algorithmic approaches that reduce these traffic volumes. In order to assess the capabilities and behavior of the proposed scheme, we have applied it to a delay sensitive application such as video or virtual reality application, operating in a distributed environment. Our results have indicated that the proposed approach reduces the frequency of occurrence of violations significantly, it enables the system to recover very fast when a violation occurs, and improves the utilization of the underlying resources.

The remainder of this paper is organized as follows. Section two gives a short related work on QoS adaptation. Section three describes the agent-based approach for QoS adaptation, the inter-agents communication protocol and the exchanged. In section four, we describe our simulation results and analysis. Finally, in section five we present our conclusions.

2 Dynamic QoS Adaptation

In the literature there are many proposals to deal with QoS adaptation. Most of them highlight the need for QoS degradation or re-negotiation, in case of violations in the initially agreed Quality of Service [5][4][8]. For instance, [5] exploits the existing hierarchical coding techniques used by video coders to make a range of adaptation strategies, by selectively adding or removing coding layers in case of fluctuations in the network bandwidth. In [4][8], the approach undertaken is to adapt the application to the service provided by the network, by using information reports sent by receivers to detect the congestion state of the network, and then to react accordingly by regulating the sending rate. The approach considered in our work follows another avenue since it has the capabilities to keep, when it is possible, the agreed QoS provided to the users without any degradation in the service. The approach provides dynamic QoS adaptation, using dedicated active software agents, to balance resources between the components (nodes and end-systems) involved in the connections experiencing violations. A similar research work is presented in [7], which also uses the agent paradigm to deal with QoS adaptation in distributed multimedia systems. The major differences between [7] and our work are the following; the distributed architecture proposed in [7] uses a ring approach. When an agent detects congestion, it sends a violation message, along with the violation degree to the neighbouring network component. The agent of this component, forwards the violation message to the next in the ring, after attaching the maximum amount of resources, the component can provide. The process repeats, until the message returns to the node of the initiating agent. After that, the latter sends along the ring a message that includes the allocations. This approach requires the use of variable size messages. The recovery interval is twice the amount of the time that takes to go around the ring. Also, in case the messages are lost due to some overflow or network failure, no action will be taken. In addition, the recovering time of the approach is unbounded, and depends on the characteristics of the network topology (e.g. number of nodes in the ring, speed of links etc.) and the network load. This makes the method unable to take into consideration the actual QoS requirements of the applications. Our method is not based on the use of a ring topology. Violation messages are forwarded in both directions (forwards and backwards) and receiving agents reply immediately to the request. This increases the reliability and fault tolerance of the method (since loss of several reply messages will not halt the process). In addition to that, in our case, the agents have to respond within a limited amount of time and the message is only forwarded up to a fixed amount of hops depending on the load of the network. This maintains good relevance between derived solution and the present state of the connection or the network. Finally, our method uses timers, in order to limit the time required for transmission and recovery and imposes an upper bound on the waiting time to a certain upper bound. This guarantees the fast provision of effective solutions for recovery of violations.

The major motivation for dynamic QoS adaptation, using dedicated software agents in both network nodes and end-systems, is to explore the dynamic behavior of multimedia connections in order to improve resource utilization and increase the number of connections satisfied by the network. Dynamic QoS adaptation differs from the conventional management functions due to its real time characteristics. For instance, a connection which is experiencing degradation on QoS in a specific node would have to be recovered as soon as possible. For connections with Variable Bit Rate (VBR) pattern, QoS fluctuations can be so frequent that they should be handled in way that is transparent to the users; therefore, this action should be left to the QoS management system [6].

3 An Agent-based Approach for QoS Adaptation

This section presents our proposed approach for QoS adaptation based on software agents. The approach is based on distributed *QoS-Agents (QoSA)* located on all the components involved in providing QoS guarantees (Fig.1). Thus, the distributed multimedia system can be seen as a multi-agents system able to communicate and evaluate, at each instant, the values of the used QoS parameters applicable to the resources utilization.

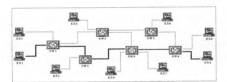

Fig. 1. Distributed Virtual Environment System Configuration.

The use of an agent-based infrastructure for Quality of Service adaptation seems to be a good choice for implementing several adaptation strategies, in which available resources need to be redistributed in order to maintain satisfaction of the end user requirements [7][10]. Such operations, in general, imply an intense (and sometimes complex) exchange of messages among the parties involved. The distributed software agents can therefore adequately support such kind of functions, and permit more flexibility in performing them.

Particularly interesting features, in performing this kind of operations, is the exploitation of agents' cooperation for optimization purposes. Each agent (QoSA) is able to collect information about nearest agents, and then complement the partial knowledge concerning the state of its own resources. As a result, each agent (QoSA), although autonomous, is influenced by, and can influence, the behavior of other agents.

When providing end-to-end QoS, each component involved in a specific multimedia connection has to contribute with certain guarantees on its own resources. If a component violates such guarantees, the associated (QoSA) will detect this and cooperation among involved agents will start in order to reassign resource reservations to different components. Agents whose resources are not fully utilized may reserve additional resources, in order to compensate for the violations of other agents.

3.1 Agents-based Distribution Scheme

We introduce here the principle of the Agents-based Distribution Scheme (ADS), which is proposed in this paper. The ADS is an adaptation scheme that helps to maintain the initially agreed Quality of Service for connections, even when one or more components involved in the configuration of interest failed to meet their commitment. It does this by re-allocating resources in a dynamic fashion and, in such a way that the end-to-end QoS requirements still unchanged. While the ADS principle can be used for any type of configurations, we focus in this work only on configurations involving a single stream of information.

When a QoS violation occurs in a particular component, the associated QoSA tries to find a local solution to the problem (e.g. modify locally the parameters and re-assign resources between all the connections) [11]. If it is not possible, this QoSA initiates a dialogue with other agents, involved in the same connection, and requests additional resources that can compensate for the current violation. If the overall system does not have enough resources to recover from the violation, the end-application is notified to make a self-adaptation [12], or to re-negotiate new Quality of Service parameters [9].

3.2 Agents Communication Protocol and Operations

In this section we give a short description of the operations that a QoSA performs, the inter-agents communication protocol, and the supported messages that are exchanged for the purpose of QoS adaptation and recovery.

Structure of the Messages
We have defined three services to support the interactions between agents: *QoS_Request* message, *Available_QoS* message, and *Allocate_QoS* message.

QoS_Request (Sender_Id, Connection_Id, Violation_Type, Requested_QoS, Hops)
This message is issued by the agent, which is experiencing violation in a specific connection. The message specifies the identification of the agent (Sender_Id), the link on which the violation occurred (Connection_Id), the type of violation (Violation_Type), as well as the extra QoS needed for QoS adaptation (Requested_QoS). Note that in this work, the extra QoS requested is in terms of delay and is expressed in ms. The Connection_Id and the Violation_Type uniquely identify each request for extra QoS, and it is local for each QoSA.

In order to inform the other neighboring agents involved in the connection, that segment of the connection going through the specific network element experiences violation, the agent sends the message in forward and backward directions to ask for extra QoS (See Fig.2). At this point we have to comment on the term neighboring. Every agent must solve and recover from the violation in a certain period of time. If this time period expires and the adaptation is not achieved, the service might be severely degraded. Therefore, the agent decides how far will send the QoS_Request message, and the decision is based on the application's requirements, and the characteristics of the connection. The Hops parameter in the message is used for that purpose, and it is set at the beginning equal to the maximum number of agents (components) that an agent, experiencing violation, should contact and ask for recovery. The value of this parameter is decremented by 1 every time another agent receives the message. The message will be either forwarded if the field is decremented to a non-zero value, or discarded if it is decremented to zero or has reached an end-system.

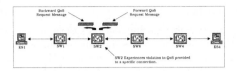

Fig. 2. QoS Request messages sent by the agent located in *SW2* to the participating agents in the connection.

Available_QoS (Sender_Id, Receiver_Id, Connection_Id, Violation_Type, Available_QoS)
The agents that have received a QoS_Request message from the agent experiencing violation issue this message, stating the amount of QoS that each of them can give in order to help the connection to recover from the violation (See Fig.3). In the parameters of this message, the amount of QoS that can be provided is included (Available_QoS). This helps the sender of the QoS_Request message to determine the exact amount of QoS that is needed, and it does so by applying the algorithm described and developed in [10].

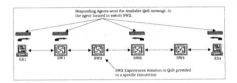

Fig. 3. Available QoS messages sent by all the agents willing to help recover the violation in the switch *SW2*.

Allocate_QoS (Sender_Id, Receiver_Id, Connection_Id, Violation_Type, Allocated_QoS)
The agent, that has previously detected the violation, will send this message to all the agents that have responded to the QoS_Request message, asking for a specific amount of QoS from each of them in order to resolve the violation (See Fig.4). Its request is uniquely identified to distinguish between requests originating from different nodes or different QoS types. This distinction can be made using the combination of Connection_Id and Violation_Type.

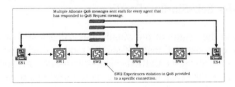

Fig. 4. Allocate QoS messages sent by the agent in the switch *SW2*, to the agents that have responded for recovery.

4 Simulation Results

In order to test our mechanisms within the multi-agents system, we ran several experiments using extensive simulations. The simulations are based on the Optimization Network Tool (OPNET) software. The simulated network configuration is similar to the one presented in the Fig.1. As controlled application, we took a model that best characterized a distributed interactive virtual application with frequent and timely sensitive traffic. The main QoS metric measured during the experiments was the end-to-end delay experienced by the traffic going between the two end-systems *ES1* and *ES4*. The traffic between these two end-systems goes through the switches *SW1*, *SW2*, *SW6*, and *SW4*, which constitutes the route of the controlled connection. The required value of this delay was set to 120 ms. As depicted in the Fig.1, there is also other traffic that is multiplexed in order to reach a high network load and to interleave the traffic considered, so that the packets are not all consecutive.

We have conducted evaluations for two different types of background traffic. The first follows the Poisson distribution. The second traffic model belongs to the traffic models that exhibit self-similarity and long-range dependence. The models of this traffic have been developed in our lab and are based on the use of alpha-stable distributions. We have shown [13] that these models are more accurate as compared to models proposed earlier (e.g. Fractional Brownian Motion [14] or MMPP [15]). Below we present and discuss our results.

4.1 Evaluation under Poisson Traffic

Our mechanisms, introduced to control the QoS metric delay and adapt to major fluctuations, have shown their major utility and their real effectiveness. The results are shown in Figs. 5 and 6. We have noticed during our simulations that the best results are reached when the agents probe the system every 0.01 ms (see Fig. 5 and Fig. 6). The results presented in Fig.5 are obtained by loading all the switches in our configuration (presented in Fig.1) to almost 100% of their capacity. In Fig.6, we present results obtained by loading the switches *SW2* and *SW4* at 75%, and all other switches at 100% of their capacity. Those different cases of load allowed us to test the accuracy of the mechanisms. As can be seen from Figs. 5 and 6, the proposed scheme obtains a good improvement in terms of end-to-end delay experienced by the traffic considered.

From the results reported in Fig.9(a), we also notice that the proposed algorithm reduces considerably the cell losses experienced by the aggregate traffic. The losses went down from almost 30% without implementing our mechanisms, to 2.5% when the mechanisms are used at their optimum rate. We can also see in Fig. 9(a) that our mechanisms show a much better effectiveness when the network is overloaded and when violations are more frequent. The violation of transit delay has been recovered in most of the cases for the considered service, and higher resources utilization is obtained in the underlying system.

4.2 Evaluation under Self-Similar Traffic

We present in this section the results of our simulations using a self-similar traffic model [13]. Recent studies [14] have demonstrated that self-similar models that show long range dependency are more appropriate for describing the traffic behaviour in today's networks. The results we have obtained were expectable, in the sense that we had more violations occurring than in the case of Poisson traffic.

The simulations we have conducted follow the same path as the ones presented in the previous section. We have loaded first all the switches to reach 100% on average of network load (results presented in Fig.7), and then we loaded *SW2* and *SW4* at 75% and all other switches at 100% (results presented in Fig.8).

In Fig.7 and Fig.8, we can see that under self-similar traffic, we reach very soon a violation in the expected end-to-end delay of our controlled connection and we keep having violations during the whole session. However, by applying the mechanisms implemented by our distributed software agents, the violations start to be reduced. By probing the resources every 0.01s, we reach always a better improvement, and a satisfaction of the end-to-end delay is obtained.

The plots shown in the Fig.9(b) give us an idea about how much we should pay in terms of cell losses, in order to keep a respectable end-to-end delay to our controlled connection. We can see that a remarkable improvement in reducing cell losses is obtained for the aggregate traffic going into the network, and at the same time we keep the quality of service of the connections at a higher level. By using the mechanisms implemented by our distributed software agents, the cell losses went down from almost 65% to 15%. This means that we pay less by using the new mechanisms to control the underlying resources.

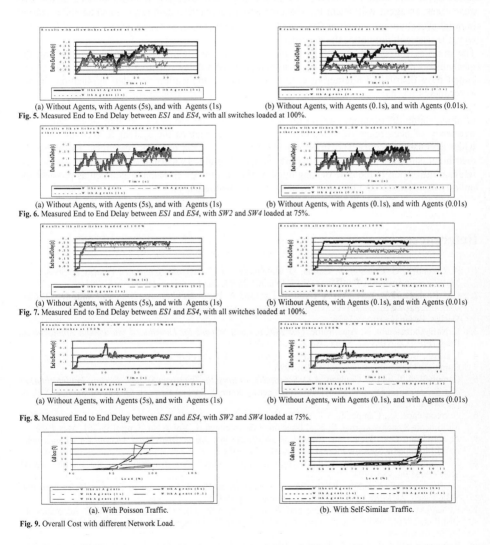

(a) Without Agents, with Agents (5s), and with Agents (1s)   (b) Without Agents, with Agents (0.1s), and with Agents (0.01s)
Fig. 5. Measured End to End Delay between *ES1* and *ES4*, with all switches loaded at 100%.

(a) Without Agents, with Agents (5s), and with Agents (1s)   (b) Without Agents, with Agents (0.1s), and with Agents (0.01s)
Fig. 6. Measured End to End Delay between *ES1* and *ES4*, with *SW2* and *SW4* loaded at 75%.

(a) Without Agents, with Agents (5s), and with Agents (1s)   (b) Without Agents, with Agents (0.1s), and with Agents (0.01s)
Fig. 7. Measured End to End Delay between *ES1* and *ES4*, with all switches loaded at 100%.

(a) Without Agents, with Agents (5s), and with Agents (1s)   (b) Without Agents, with Agents (0.1s), and with Agents (0.01s)
Fig. 8. Measured End to End Delay between *ES1* and *ES4*, with *SW2* and *SW4* loaded at 75%.

(a). With Poisson Traffic.   (b). With Self-Similar Traffic.
Fig. 9. Overall Cost with different Network Load.

5 Conclusion

In this paper, we have presented an approach to QoS adaptation in distributed multimedia applications. The approach is based on distributed software agents able to communicate and cooperate in order to achieve a better resource utilization, and then maintain the end-to-end QoS requirement of multimedia connections. We believe that the use of an agent-based system infrastructure for QoS adaptation is a good choice for implementing several adaptation strategies in which available resources need to be re-arranged and then keep the user satisfied with the service in use.

When providing end-to-end QoS, each component involved in the distributed multimedia application has to contribute with certain guarantees on its own resources. If a component violates such guarantees, an agent will detect the violation and a cooperation among involved agents is started in order to recover the violation. This can be done, for instance, by reassigning resources in such a way, the user will not notice the degradation.

The multi-agents system presented in this paper has shown his effectiveness by applying the mechanisms described to recover the delay violation of a delay sensitive traffic, such as one generated by a distributed interactive virtual environment application or video application. In our experiments we have considered two different types of background traffic. The first follows a Poisson distribution. The second traffic exhibits long-range dependence and is based on the use of alpha-stable distribution. Our results have indicated that the proposed approaches reduce the frequency of occurrence of violations significantly, it enables the system to recover very fast when a violation occurs, and improves the utilization of the network resources. However, the distributed software agents implementing the QoS-Weighted Distribution Algorithm, show their relevant utilization in the case of long burst traffic, as they try to balance as much as possible the underlying resources and therefore be able to consume the bursts.

References

1. A. Campbell, G. Coulson, F. Garcia, D. Hutchison and H. Leopold. *Integrated Quality of Service for Multimedia Communications*, Proc. IEEE INFOCOM'93, San Fransisco, USA, April 1993.
2. G.v. Bochmann, B. Kerherve, A. Hafid, P. Dini and A. Pons. *Architectural Design of Adaptive Distributed Multimedia Systems*, Proc. Of the IEEE International Workshop in Distributed Multimedia Systems Design, Berlin, Germany 1996.
3. F. Naït-Abdesselam, N. Agoulmine. *QoS Management Scheme for Distributed Multimedia Applications*, Proc. IFIP TC6 NIPS'97, Sofia , Bulgaria, Oct 1997.
4. I. Busse, Bernd Deffner, H. Schulzrinne. *Dynamic QoS Control of multimedia applications using RTP*. Computer Communications, vol19, Jan 1996.
5. Campbell, A.T. and G. Coulson. *A QoS Adaptive Multimedia Transport System: Design, Implementation and Experiences*, Distributed Systems Engineering Journal, Special Issue on Quality of Service, Vol. 4, pg. 48-58, April 1997.
6. Aurrecoechea, C., Campbell, A.T. and L. Hauw. *A Survey of QoS Architectures*, ACM/Springer Verlag Multimedia Systems Journal, Special Issue on QoS Architecture, Vol. 6 No. 3, pg. 138-151, May 1998.
7. A. Hafid and G.v. Bochmann. *Quality of Service Adaptation in Distributed Multimedia Applications*. ACM Multimedia Systems Journal, volume 6, issue 5, 1997
8. J-C. Bolot et al. Scalable feedback control for multicast video distribution in the internet. ACM SIGCOMM'94, Oct 1994.
9. H. Zhang and E. Knightly. RED-VBR: A Renegotiation-Based Approach to Support Delay-Sensitive VBR Video. ACM Multimedia Systems Journal, May 1997.
10. F. Naït-Abdesselam, N. Agoulmine and A. Kasiolas. Agents-based approach for QoS Adaptation in Distributed Multimedia Applications over ATM Networks. International Conference on ATM, Colmar, France, juin 1998.
11. R. Nagarajan, J.F. Kurose, D. Towsley, *Allocation of Local Quality of Service Constraints to meet End-to-End Requirements*, Proc. of IFIP Workshop on the Performance Analysis of ATM Systems, Martinique, Jan. 1993.

12. C. Diot, C. Huitema, T. Turletti. *Multimedia Application should be Adaptive*. HPCS Workshop. Mystic (CN), August 23-25, 1995
13. J. R. Gallardo, D. Makrakis and L. Orozco-Barbosa. *On the Use of Alpha-Stable Self-Similar Stochastic Processes for Modeling Traffic in Broadband Networks*, submitted to the IEEE Trans. on Communications.
14. S.Ben-Slimane and T. Le-Ngoc. *A Doubly Stochastic Poisson Model for Self-Similar Traffic*. Proc. IEEE ICC'95, Seattle, June 1995, pp. .
15. W.E. Leland, M.S. Taqqu, W. Willinger, and D.V. Wilson. *On the self-similar Nature of Ethernet traffic* (extended version), IEEE Trans. on Networking, no 1, pp 1-14, (1994).

Dependability Evaluation of Fault Tolerant Distributed Industrial Control Systems[1]

J.C. Campelo, P. Yuste, F. Rodríguez, P.J. Gil, J.J. Serrano

Department of Computer Engineering
Technical University of Valencia – Spain
{jcampelo, pyuste, prodrig, pgil, juanjo}@disca.upv.es

Abstract. Modern distributed industrial control systems need improvements in their dependability. In this paper we study the dependability of a fault tolerant distributed industrial control system designed in our university. This system is based on fault tolerant nodes interconnected by two communication networks. This paper begins showing the architecture of a single node in the distributed system. Reliability and safety results for this node are presented using a theoretical model based on Stochastic Activity Networks (SAN). Based on this architecture, the theoretical model of the distributed system is then presented; in order to evaluate the reliability and safety of the whole system models based on stochastic activity networks are used, and the results obtained using UltraSAN are presented.

1 Introduction

Distributed industrial control systems are becoming one of the most important research areas in embedded control applications. In this sense, control and supervision of those systems is accomplished through the co-operation of different nodes that are interconnected through industrial local area networks.

The fault tolerant systems group of the Technical University of Valencia, has developed a fault tolerant distributed industrial control system. In this sense, the fault tolerant architecture for the node of a distributed industrial control system is shown in this paper and a theoretical model of this architecture based on stochastic activity networks is used to obtain its reliability and safety. In order to study the distributed system, a hierarchical modeling approach is used, also with stochastic activity networks, based on the fault tolerant node model and joining several nodes to obtain a complex system.

This paper is organized as follows. After the introduction, an overview of the node of the distributed system is given. The structure and the model of the fault tolerant node are presented. In sections three and four the modelling approach of the distributed system and the results are presented. This paper ends with the conclusions obtained.

---

[1] This work is supported by the Spanish *CICYT* under project CICYT-TAP96-1090-C04-01

2 Node Architecture

The node architecture of the distributed industrial control system developed has as its main objective reliability and safety improvements with a low cost. This node (figure 1) consists of two communication networks, two microcontrollers, one for communication tasks and other for the execution of the control algorithms, and a dual-port RAM.

Commonly in industrial applications different nodes could accomplish similar functions. This is due to the fact that they can be connected to the same set of sensors and actuators. So, an important aspect is to store the state of each node in the dual-port RAM and to update it periodically. Thus, in case of failure in the application microcontroller, the communication microcontroller can send the state to other node that could continue its function. To do this, an important aspect is to reach a high coverage factor, mainly for the application microcontroller (the most complex component). In order to obtain a high coverage factor, the communication microcontroller also fulfils functions of watchdog processor. In this way, the communication microcontroller can detect errors in the other processor and inhibit possible outputs to the actuators and possible error messages to the network (this is known as fail-silent behaviour).

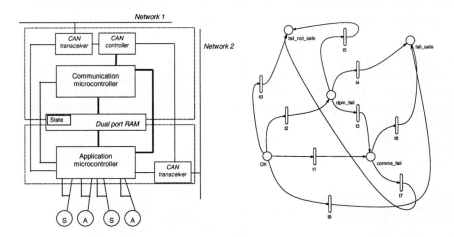

Fig. 1. Node Architecture   **Fig. 2.** SAN model

The SAN [1,2,3] model that represents a single node is shown in figure 2. In this model each place represents a state of the system and is the direct conversion of the Markov model [4].

3 Distributed System Model

Now, that the reliability and safety can be obtained solving the previous model, new questions arise about the reliability and safety of the whole distributed system. In order to study these aspects a distributed system model is presented.

We are assuming a distributed industrial control system composed by six fault tolerant nodes. In this system, due to the characteristics of the industrial process to control, nodes 3 and 4 and nodes 5 and 6, are connected to the same set of sensors and actuators. So, a failure in node 3 can be recovered by node 4 and vice-versa (when node 3 is in a fail-safe state it can send its state to node 4 and this one can continue both jobs). The same behaviour for nodes 5 and 6 also applies.

In order to model the distributed system, we are doing a composed model. As it can be seen in figure 3, the 6 single "node" and "connection" subnets compose this model. The connection subnet will be in charge of deciding which of the failures of the single nodes can be recovered by another node (node 3 can be recovered by node 4 and vice-versa and the same for nodes 5 and 6). In this model, node subnets are slightly different from the previous model. As the failure of a single node does not necessarily take the system to a failure state, there are no faulty states embedded in the subnet. Instead, the "fail safe" and "fail not safe" states are in the connection subnet (due to some UltraSAN limitations – instantaneous activities between common places - a more hierarchical model can not be done [5]). So, the node subnet only has the operational states representing the behaviour of the node. In figure 4 the node model can be seen. The places in this subnet will be common with the same places in the connection subnet. The connection subnet is shown in figure 5. In this subnet there are different kinds of places: first, places representing the inner state of each of the six nodes. These are common places that are both in the connection subnet and in the node subnets. There are the failure places, as in the previous model, two new places, "rg1" and "rg2" and the network places "networks_down", "OK_net1" and "OK_net2".

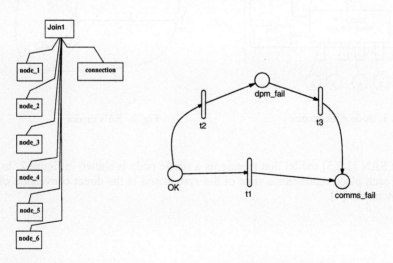

Fig. 3. Composed model   **Fig. 4.** Node subnet

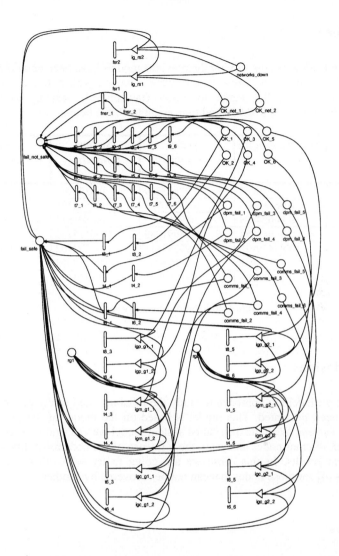

Fig. 5. SAN model: connection

The meaning of these new places is the following: the system has six nodes, two of them are independent and four are grouped in two clusters. Inside each cluster any node can recover any other node's failure and reconfigure itself to execute both tasks. In places *"rg1"* and *"rg2"* there is a number of marks equivalent to the number of failed nodes in cluster 1 and 2 respectively. Places *"OK_net1"* and *"OK_net2"* represent the correct operation of the two networks respectively and *"networks_down"* represents when it has more than one mark that both networks are out of service. As it can be seen in the connection subnet, USAN input gates are used to allow the relationships between the different node groups, to model the networks behaviour and the transitions to faulty states.

4 Results

Reliability and safety results solving the previous models have been obtained. In table 1 the results can be seen. These results were obtained with a failure rate of the application microcontroller equal to 1e-05 (faults/hour), varying coverage for transient faults from 0.75 to 0.999 and time equal to 10000 hours.

Transient faults coverage	Node reliability	Node safety	System reliability	System safety
0.75	0.9588465	0.97822869	0.8395765	0.8748919
0.80	0.96268653	0.98210725	0.8598195	0.8959124
0.85	0.96654194	0.9860013	0.8805511	0.91743846
0.90	0.97041281	0.98991089	0.90178286	0.9394824
0.95	0.97429918	0.99383611	0.92352702	0.96205666
0.99	0.97741949	0.99698756	0.94129954	0.98050646
0.999	0.97812293	0.99769802	0.9453453	0.98470625

Table 1. Reliability and safety results

5 Conclusions

In this paper a modelling approach to study the reliability and safety of a distributed system has been presented. This model is based on the model of the fault tolerant nodes in the system. With the distributed system model we can study the influence of different fault tolerant nodes analysing the influence of the node reliability and safety in the system results. In this sense we are going to try with other fault tolerant architectures [4] for those nodes that can not be recovered by another one.

References

1. Sanders, W.H., Obal, W.D.: Dependability evaluation using UltraSAN. 23th International Symposium on Fault Tolerant Computing.Toulouse, France (1993)
2. Sanders, W.H., Obal, W.D., Qureshi, M.A., Widjarnako, F.A.: UltraSAN version 3: architecture, features, and implementation. Technical report 95S02, Center for reliable and high-performance computing. University of Illinois at Urbana-Champaign (1995)
3. UltraSAN users manual. Center for Reliable and High Performance Computing, Coordinated Science Laboratory, University of Illinois at Urbana-Champaign
4. Campelo, J.C., Rodríguez, F., Gil, P.J., Serrano, J.J.: Dependability evaluation of fault tolerant architectures in distributed industrial control system. WFCS'97, 2nd IEEE International Workshop on Factory Communication Systems, Barcelona, Spain (1997)
5. Nelli, M., Bondavalli, A., Simonici, L.: Dependability modelling and analysis of complex control systems: an application to railway interlocking. EDDC-2, 2nd European Dependable Computing Conference. Taormina, Italy (1996)

An Approach for Measuring IP Security Performance in a Distributed Environment

Brett L. Chappell, David T. Marlow, Philip M. Irey IV, and Karen O'Donoghue

System Research and Technology Department
Combat Systems Branch
Naval Surface Warfare Center, Dahlgren Division
Dahlgren, Virginia 22448-5000
{chappellbl, marlowdt, ireypm, kodonog}@nswc.navy.mil

Abstract. The Navy needs to use Multi Level Security (MLS) techniques in an environment with increasing amount of real time computation brought about by increased automation requirements and new more complex operations. NSWC-DD has initiated testing of a security protocol based on the commercial standard, IPSEC, which is becoming available in Commercial Off The Shelf (COTS) computing products. IPSEC is viewed as a critical component towards providing MLS capabilities. Current implementations of IPSEC are implemented in software as part of the kernel system software. The system engineer must carefully develop security policies versus applying this technology in a brute force way. This paper describes the security issues, the IPSEC standard, testing performed at NSWC-DD and provides an approach to using this technology in the current resource constrained environment using today's COTS products.

1 Introduction

As the U.S. Navy develops new and upgraded shipboard systems for its surface combatants, reduction of total ownership costs has become an over-all driving concern. This has lead to a Navy-wide direction to reduce manning and use Commercial Off The Shelf (COTS) products. The reduced manning direction has lead to an over-all increase in the amount of computing needed, and the COTS mandate has lead to the use of commercially developed computers versus the previous use of computers which were designed to specification.

The Navy is expecting a great increase in the amount of computing required, not only due to the increased emphasis on automation as described, but also due to the increasing functionality that is being added to meet the needs of changing military requirements. For example, joint operations with nations that we were not previously allied with may be needed in the future. This presents the need to be able to quickly adjust to sharing (and to stop sharing) specific information with military units (e.g. ships) from a variety of countries. Thus, policies applied by the operational personnel must be able to enable (or disable) the sharing of specific information. In addition to the existing requirements of military security where information is classified (e.g. unclassified, confidential, secret,...) there are concerns of protecting all information from computer hackers who have the expertise with the military computers now that COTS computers are being used.

Currently, classified and mission critical applications reside on a physically separate platform from unclassified or non-mission critical applications, as shown in Figure 1. This approach is unacceptable for the future where increased computational requirements are forcing the use of information at a variety of classification levels by a large number of the shipboard computers. In addition, the network infrastructure used by mission critical computing units must have inherent redundancy in order to survive after battle damage or the failure of any one network component. Such network infrastructure cannot be replicated for each classification level with the cost constraints that are being imposed in all current and future developments. There are other reasons to migrate away from designs based on the physical separation of equipment. There is a shift throughout the Navy from the current "stove-pipe" systems to tightly integrated systems. Stove-pipe designs result in non-interoperable subsystems, duplication of resources, and difficulty in incorporating changes that are needed to meet new requirements.

The nature of the envisioned computing environment is such that application to application multilevel security (MLS) is required. Figure 2 illustrates a system which provides application to application MLS without the use of physical separation. The long term goal is to provide security services, such as confidentiality and authentication, to applications. In Figure 2, the dotted and dashed lines illustrate communication streams between applications executing at different security levels. This implies that security mechanisms must be in place to secure the data as it is transferred across the networking infrastructure.

Another concern is to provide the maximum flexibility for resource management in order to dynamically reconfigure the binding of computing tasks to physical computers even after events occur that result in battle damage where computers are lost. This implies that all applications must be capable of running on any computer. Thus it is desired that any computer be capable of executing both classified and unclassified applications simultaneously.

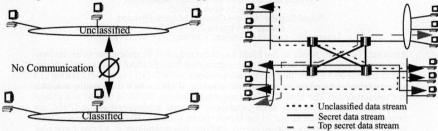

Figure 1. Infosec by physically separate networks

Figure 2. Application to application MLS

The IP Security Protocol (IPSEC) is a technology that may be used to provide security services at the network layer in a system such as the one shown in Figure 2. The protocol is an output of a working group within the Internet Engineering Task Force (IETF) whose charter is to provide security to the Internet Protocol (IP).[1, 2, 3] The focus of this paper is to develop a method in which the IPSEC protocol can be evaluated to determine its suitability for use in an MLS environment. The Naval Surface Laboratory (NRL) has provided an open source version of the IPSEC protocol which was used by NSWC-DD for experiments described in this paper.

2 IPSEC Overview

The primary services provided to the IP data packet by IPSEC are data confidentiality and authentication. In addition, IPSEC has been designed in such a way to minimize the risk of replay attacks and IP spoofing.

IP authentication allows the recipient of an IP packet to be certain that the contents of the headers and data field have not been modified. Also, the authentication service allows the recipient to be sure that the packet actually came from the host identified by the source IP address. IP confidentiality ensures that the data portion of the IP packet is not readable by unauthorized entities. Both the authentication and confidentiality services are achieved through the use of cryptographic techniques.

IPSEC uses a one way cryptographic hash function to provide authentication services.[2] The IPSEC protocol supports any number of one way hash functions. For interoperability, there is a function that is mandatory to implement. Currently, the mandatory function to implement is MD5, and the NRL implementation uses MD5 as the one way hash function.

In order to provide data confidentiality, IPSEC supports a number of encryption algorithms.[3] Currently, the Digital Encryption Standard (DES) is mandatory to implement for interoperability. The NRL implementation of IPSEC uses DES to provide the confidentiality services. DES is a symmetric key encryption algorithm, meaning that the same key is used to encrypt as well as decrypt the data. This implies that the sender and receiver must securely exchange the key. The IPSEC protocol includes a key exchange protocol [7], however, the NRL implementation used employs manual keying. A noteworthy event concerning DES occurred recently (mid-1998), where a system was built that is capable of cracking DES encrypted data by brute force in a matter of days.[4] The DES encryption algorithm will most likely be replaced by a more secure algorithm in the near future.

Before the IPSEC services can be invoked, a *security association (SA)* must to be established. A security association is a simplex connection that provides a specific IPSEC service to the traffic carried by it. An SA has three components: the type of service provided, either an Encapsulated Security Payload (ESP) service or Authentication service; the destination address of the channel; and a security parameters index (SPI). The ESP service provides data confidentiality for the data portion of an IP packet. The SPI is an integer number that is used by the end system to internally identify the communications stream. The three components together describe a unique SA. In order to communicate security between two systems using an IPSEC service, two SAs are required, one in each direction. Also, only one IPSEC service can be provided by a single SA, and this results in a separate SA for each IPSEC service desired.

2.1 IPSEC Implementation

A topic that deserves careful consideration is how and where to implement IPSEC. This section discusses some of the issues that should be considered when adding functionality to the IP layer.

IPSEC may be implemented by several different methods in a host. There are three common methods: 1) integration into the native IP stack of the operating system, 2) "bump-in-the-stack," meaning that IPSEC is implemented in software directly below the IP layer, and 3) "bump-in-the-wire," meaning that IPSEC is implemented in special hardware that acts as an external processor to the system.[1] The bump-in-the-wire implementation may be partially or completely separated from the resources of the computer. Figure 3 illustrates the three choices.

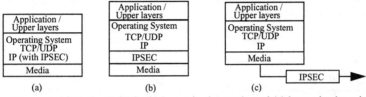

Figure 3. (a) Native IPSEC, (b) bump-in-the-stack, and (c) bump-in-the-wire

Each method of implementation has advantages and disadvantages. The first two methods listed are lower cost in terms of equipment as they are implemented in software. Integration of IPSEC into the native IP stack requires that source code for the operating system be available. Often, when dealing with COTS equipment, this code is either not available or prohibitively expensive. This method will most likely become the most popular method adopted by major OS vendors, such as Sun Microsystems or Microsoft. The second method, the bump-in-the-stack, is comparable to buying an enhancement to an operating system, possibly from a third-party vendor. If the operating system source code is not available, this may be a wise choice.

The third method may be the most expensive to produce after the design has been completed, since it requires additional hardware to be purchased with each computer system. However, specialized hardware implementations generally provide higher performance in processing cryptographic functions.

The NRL IPSEC implementation is a native IP stack implementation. Since the source code for some versions of Unix is freely obtainable (e.g., NetBSD), it is a suitable platform for an integrated implementation. However, a result of this type of implementation is that the cryptographic algorithms are implemented as system calls in the kernel. As will be noted later, this can have a significant impact on the proper operation of the kernel.

3 Testbed Configuration

The stand-alone test performed by NSWC-DD consisted of measuring the network performance of TCP connections between two Sun Sparcstation 20 workstations. The workstations have been configured with the NetBSD operating system, version 1.2. The operating system has been modified to integrate the alpha-five version of the IPSEC implementation from the Naval Research Laboratory (NRL) An in-house software package, the Communication Analysis and Simulation Tool (CAST), was integrated into the testbed by porting it to the NetBSD environment. CAST measures several different network performance related metrics, however, this paper will concentrate on two: end-to-end throughput and one-way latency.[5] Also, it is important to note that the NRL implementation of IPSEC has not been optimized for performance; it is a proof of concept that is still evolving.

3.1 IPSEC Configuration

There were three types of tests performed - one with no IPSEC service, one with IPSEC authentication, and one with IPSEC confidentiality. TCP connections between the sender and receiver were used for each of the three tests. Figure 4 illustrates the security associations required if two end systems desire to communicate using an IPSEC service. Specifically, the testbed configuration consisted of one Sparcstation 20, burke, connected to another Sparcstation 20, ollie, via 10 Mbps Ethernet and an Ethernet switch.

3.2 Network Time Protocol (NTP) Configuration

The CAST tool requires that the sender and receiver have synchronized clocks or that the offsets between the two clocks can be accurately measured. In addition, the frequency of the system clocks on the sender and receiver must change at the same rate. In this testbed, NTP version 3 was used to synchronize the system clocks. [6]

Figure 5 shows the configuration of NTP in this testbed. A GPS receiver is used to provide an accurate and stable notion of UTC time via an IRIG-B signal to an oscillator board in the time server. NTP was used to synchronize the system clock of the time server to the oscillator board and the system clocks of the two test machines to the system clock of the time server. For the purposes of this experiment, maintaining true UTC time is not required, however, this NTP configuration was available and used.

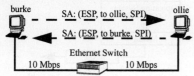

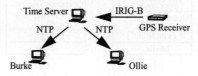

Figure 4. The testbed configuration with ESP security associations.

Figure 5. The NTP configuration

4 Analysis of results

The CAST tool has three phases of operation: 1) calculate clock offset, 2) calculate one-way latency, and 3) calculate end-to-end throughput. The calculated clock offset is used to compute both the one-way latency and end-to-end throughput phases.

End-to-end throughput is the amount of data that can be sent over the TCP connection in a time period. One-way latency is the amount of time that it takes to send data from one end of the connection to the other. Figures 6, 7, and 8 show the measured end-to-end throughput and one-way latency with no IPSEC being used, IPSEC authentication, and IPSEC confidentiality.

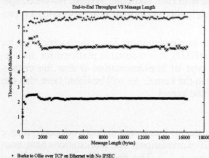

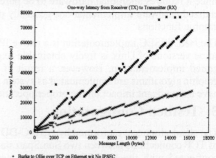

Figure 6. End-to-end throughput for the three tests from the sender to the receiver

Figure 7. One-way latencies for the three tests from the sender to the receiver

4.1 Effect on End-to-end Throughput and One-way Latency

Figures 6, 7, and 8 show that both the end-to-end throughput and one-way latency results are significantly impacted when using the IPSEC protocols.

The effect upon the end-to-end throughput is a direct relationship between the three test runs. The one-way latencies were measured from the sender to the receiver, and vice-versa, as shown in Figures 7 and 8. In both cases, the one-way latencies increase when using the IPSEC authen-

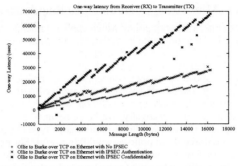

Figure 8. One-way latencies for the three tests from the receiver to the sender.

tication and the IPSEC confidentiality. In addition, as shown in Figures 9 and 10, the standard deviation in the latency measurements is increased when IPSEC is used, most notably when software encryption is employed.

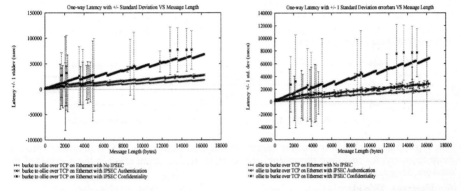

Figure 9. The standard deviation of the latencies between the sender and receiver

Figure 10. The standard deviation of the latencies between the receiver and the sender

5 Conclusions

A number of conclusion and lessons learned can be drawn from the results of this experiment. Based on the results, certain recommendations can be made on the use of the IPSEC protocols in future Navy systems.

First, it was shown that the IPSEC protocol can be used to secure communication between two end points. The beta release of the NRL code is a proof of concept that the protocol will work. Since many major operating system vendors have adopted IPSEC, such as Microsoft and a number of Unix vendors, it is a viable component of the solution to the application-to-application MLS requirement. However, it is important to realize that the total solution will be an integration of several components, such as IPSEC and operating system mechanisms.

Secondly, it was shown that the software implementation of IPSEC in the kernel significantly impacts network performance. In order to use such an implementation in an MLS environment as shown in Figure 2, a method must be developed to minimize the computational requirements. Specifically, real-time communications between two applications may not be possible while using the IPSEC services.

The method used to measure and analyze IPSEC performance is scalable beyond the single sender, single receiver example used throughout this paper. Since CAST is executed in the end system, this approach can used to evaluate IPSEC performance in a more distributed environment.

5.1 An Approach to Minimizing Computational Requirements used by IPSEC

As shown by the test results, a security policy which applies authentication and encryption on all information transferred is not appropriate in a resource constrained environment. The computational load of applying authentication and encryption is significant given today's COTS computing products. A straightforward approach to applying this technology with current products is to carefully choose the proper configuration of IPSEC that is well suited for the applications of interest. By trading off security with reduced performance in throughput and latency, a system engineer can work out a solution which balances the system's real-time requirements. In order to avoid unnecessary overhead, security policies need to be developed that describe which types of information flows require each type of protection. Also, a security management scheme to enforce these policies would be useful. For example, in most systems, the routine data being transferred is unclassified and does not require protection. In this situation, it is unnecessary and inefficient to encrypt and authenticate all data. It is possible that information exchanged within a given subsystem could be transferred with less IPSEC services than information exchanged between subsystems. Appropriate use of firewall technology may be needed in order to practically implement such an approach.

In the system shown in Figure 2, each system may have applications requiring multiple levels of service. The NRL implementation of IPSEC enables security associations (SAs) on a per-socket basis. Therefore, each application could potentially communicate with another application running on a different host using different data streams, each with a different security association via sockets, as shown in Figure 11.

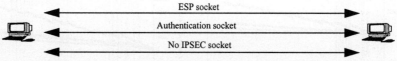

Figure 11. A simple approach to managing IPSEC security services

This type of approach will reduce the computational load required by IPSEC in order to provide more resources for real-time processing. However, it requires more effort in the analysis of the network design.

6 Future Work

During the testing of the NRL IPSEC implementation, observations were made about the system performance that are beyond the scope of this paper. Specifically, there is a significant impact upon basic system services, such as timekeeping and CPU scheduling, when software encryption is performed in the kernel. This impact is worthy of further exploration. In addition, an evaluation of the dynamic key management protocol of IPSEC is planned as the implementations mature.

7 References

1. Kent, S., and Atkinson, R., RFC2401, *Security Architecture for the Internet Protocol*, November, 1998.

2. Kent, S., and Atkinson, R., RFC2402, *IP Authentication Header*, November, 1998.

3. Kent, S., and Atkinson, R., RFC2406, *IP Encapsulated Security Payload*, November, 1998.

4. Gilmore, J., *Cracking DES: Secrets of Encryption Research, Wiretap Politics, and Chip Design*, Electronic Frontier Foundation, July, 1998.

5. Irey, P., Harrison, R., Marlow, D., *Techniques for LAN Performance Analysis in a Real-Time Environment*, Real-Time Systems - International Journal of Time Critical Computing Systems, Volume 14, Number 1, pp. 21-44, January, 1998.

6. Mills, D., *RFC-1305, Network Time Protocol (Version 3) Specification, Implementation, and Analysis*, March, 1992.

7. Maughan, D., Schertler, M., Schneider, M., and Turner, J., RFC2408, *Internet Security Association and Key Management Protocol*, November9, 1998.

An Environment for Generating Applications Involving Remote Manipulation of Parallel Machines

Luciano G. Fagundes, Rodrigo F. Mello, Célio E. Morón

Federal University of São Carlos – Department of Computer Science
São Carlos - SP - 13565-905 - BRAZIL
E-mail: {luciano, mello, celio}@dc.ufscar.br

Abstract. *This paper summarises the Visual Environment for the Development of Parallel Real-Time Systems and focuses on one of its tools, namely the Graphical User Interface Generator (GUIG). The aim of this tool is to facilitate the generation of a Graphical User Interface for applications developed using the Parallel Kernel Virtuoso (Virtuoso is a trademark of Eonic Systems - http://www.eonic.com). Roughly, the GUIG should be able to provide integration among the application being developed, the Internet and the graphical interface of the application. A first version of the Environment was released as Teaching Environment for Virtuoso (TEV), available for downloading from http://www.dc.ufscar.br/~tev/tev.html.*

1. Introduction

The Visual Environment for the Development of Parallel Real-Time Systems provides an integrated development environment for the construction of parallel real-time systems. This tool has been designed to facilitate the change from the sequential to the parallel paradigm. Parallel programming is being encouraged by the low cost of processors which makes possible the construction of powerful parallel systems, which are able to support high-performance applications. The new challenge is the software complexity and low control of computation. The sequential paradigm has forced inherent parallel problems to be mapped into sequential languages. Tools for sequential programming encapsulate the machine architecture, a feature that is not adequate for parallel systems because the programmer needs to know at least how many processors are available to execute his program. Some works in this direction can be found in [1,2,3,4,5].

The Visual Environment being described here is introduced as an alternative for approaching the problems mentioned above. This environment will be composed by a set of tools to support the final steps involved in the development of parallel real-time programs, one of the tools is the Graphical User Interface Generator.

When designing the Graphical User Interface Generator, the main concern was the communication speed on the Internet. On-line interaction may be spoiled due to the low speed of data transfer across the Internet. This point is being dealt with in two ways: first, we are using specific algorithms to maximise the network usage, and second, studying initiatives like the "Internet2" and the "Next Generation Internet", which research new technologies to build a new Internet, with speeds ranging from 100 to 1000 times higher than is available today. This tool will initially help in the construction of applications that use local high speed networks and, in a near future, with the increase of network speed it will be possible to create more complex Internet-based systems.

The tools composing the Visual Environment will be integrated through a common graphic interface, and therefore offer a high-level environment for the construction of parallel applications. This paper focuses in the Graphical User Interface Generator. The remaining of this paper is organised as follows: Section 2 describes the tools of the Visual Environment for the Development of Parallel Real-Time Systems; Section 3 discusses the implications of the migration from text-based I/O to Windows-based I/O, as well as the issues involved in the creation of Remote GUIs. Finally, a brief conclusion, that shows the most important issues of this article, is provided.

2. A Visual Environment for the Development of Parallel Real-Time Programs

The main feature of the Visual Environment [6, 7, 8] is to increase the productivity of the development of real-time systems. Visual programming represents an appropriate alternative since the graphic editor provides the programmer with an easy way to generate the main features of his project with the possibility of looking at low level details when necessary. The basic units of visual programming are the objects identified in the Kernel Virtuoso (Virtual Single Processor Programming System) [9], plus the visual components defined by the user. These abstractions help to minimise the maintenance costs while at the same time maximise the reuse of source code.

3. From Text-based I/O to Windows-based I/O and Remote Issues

Parallel machines usually do not have an operational system, they are rather linked to a host computer. The host computer has utilities that enable it to load applications in parallel machines. The interface between parallel computers and the host operating system is composed by those utilities. That interface carries out all input and output operations. The Kernel Virtuoso offers text-based I/O facilities; its utilities are called Host Server.

The Graphic User Interface Generator tool is aimed at supporting the creation of Graphical User Interfaces (GUIs) that will extend the actual architecture of user interaction with the application. It will migrate from text-based to WINDOWS-based systems. The environment described here has been developed for PCs; however, during system development all features emulate a parallel machine. When the user wants to execute in a real parallel machine, the system will provide a cross-compiler that will generate the specific binary code for the specific machine. Currently users need to have a parallel machine available at their facilities, or to use a text-based shell. The aim of the Visual Environment is to generate a system naturally distributed, so that users do not need to have their own parallel machines, but rather execute their programs remotely. It should be highlighted that by using the GUIs Generator, the user is able to develop all his work in a PC compatible, and he will run his applications in a real parallel machine only when he wants to do so.

3.1. Remote GUI Approach

This tool is aimed at supporting the generation of graphical user interfaces that can interact with remote parallel systems. The main idea behind the tool is to break the system into two distinct modules, one program that is cross compiled for a parallel machine (the server), and the other one that is generated using a windows-based language (the client). Both modules communicate through a network. The Graphical User Interface Generator is quite similar to the Parallel Program Generator [8]; that is, generates the GUIs using visual objects and connections.

The Parallel Programs Generators (PPG) [1,2,3,4,5] will be added with a new object: the GUI. This new object will have a graphical representation and a list of primitives.

At the low level, the GUI Server module is added to the parallel program generated by the PPG; this module is accessed to through an interface composed by C language primitives, in a way analogous to the kernel Virtuoso objects.

The Client module is composed by an external program. The chosen language has been the Java Language, but it could be any other language, provided that it supports network programming and has visual capabilities. The main objective of this is to manage user interaction and to show results from the parallel system.

The network programming level is transparent to the programmer; all connection management and network verification will be encapsulated into GUI primitives. In the GUI Generator it is possible to manipulate user events; however, this can only be done in the host machine, and not in the parallel machine. The interface between the GUI and the parallel machine will be carried out through local variables, which represent the state of the GUI Components (Buttons, Lists, Text Areas, etc). The parallel machine will need to verify the values of those variables and to take decisions about what to do with them.

The Remote GUI approach was chosen mainly because the environment was designed to support real-time systems, in particular those which do not need too much user interaction; that is, the idea is to have a GUI that can interact with the parallel system at run time. The generated real-time parallel system will not be tied up to the GUI, unless the programmer wants to do so. With this flexibility, the GUI will be executed only when an intervention becomes necessary in the parallel machine or when the system needs to obtain data. It should be noted that a parallel system which does not need data for its processing can execute in a stand-alone manner; however, if the programmer wants to generate a GUI, he will have available a user-friendly manner to manage his parallel real-time requirements. All this is possible using the GUI Generator.

The GUI generator is shown through the Parallel Program Generator. Figure 1 shows the GUI Generator. It is composed by a workspace where the GUI assembly happens. The component palette is at the left side; where components like Frames, Buttons, Check Boxes/Check Boxes Groups, Labels, Text Fields/Text Areas, Lists, Choices, Canvases, Scripts, etc are available. In addition, the GUI generator has another important window, the Object Inspector that manages the properties of the visual component.

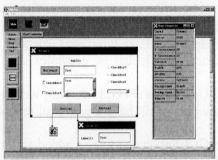

Fig 1. GUI Generator

The GUI assembly is carried out through dragging and dropping visual components, whereas the logic is specified through visual connections, or scripts in Java language.

This approach makes possible to explore two features that may be undermined by programmers of parallel machines, the processing power on the host machine, and the creation of user-friendly interfaces. As mentioned before, the host machine has been used mainly to manage I/O and to load the binary code into the parallel machine. With the creation of local scripts, the programmer gains in processing power which can be used for usual operations that do not need to compete with the tasks in the parallel machine. The parallel machine will be a real extension of the host, and not a system itself.

The GUI state will be verified by the parallel system through local variables. Each component has at least one variable that carries its state. The user can choose whether that information is sufficient or whether he needs a real image of the component.

The component image is interesting for data that is frequently used by the parallel system but one that needs more network usage to update all the data. The user should choose which components need a complete image and which not.

4. Conclusion

A significant limitation for the development of parallel real-time systems is the lack of adequate programming tools, mainly for supporting the final stages of the development life cycle. CASE tools represent an alternative in this direction, but require development to follow a single methodology from beginning to end. This work presented an environment to help with these problems, and a tool, Graphical User Interface Generator, that allows easy, and rapid, development of applications that use windows-based I/O and are naturally distributed.

The extensions presented here for the Visual Environment enable the user to create a GUI that allows remote interaction with parallel machines. Using the GUI generator it is possible to build systems that share remote machines using a user-friendly interface instead of the text-based interface currently being used. Finally, it is worth highlighting that the programmer only will need to supply the machine's address, all network programming will be generated by the Visual Environment.

References

[1] W. Cai, T.L. Pian, S. J. Turner. "A Framework for Visual Parallel Programming", In Proceedings of Aizu International Symposium on Parallel Algorithms/Architecture Synthesis, IEEE Computer Society Press, Japan, March 1995

[2] L. Schaefers, C. Scheidler, O. Kraemer-Fuhrmann, "TRAPPER - A Graphical Programming Environment for Industrial High-Performance Applications", Parle (Parallel and Languages Europe), Muenchen, June 1993, pp. 11.

[3] G. Dózsa, P. Kacsuk, T. Fadgyas, "Development of Graphical Parallel Programs in PVM Environments", In Proc. of 1st Austrian-Hungarian Workshop on Distributed and Parallel Systems, Miskolc, Hungria, October 1996, pp. 33-40.

[4] M. Aspnäs, R.J.R. Back, T. Långbacka, "Millipede - A Programming Environment Providing Visual Support for Parallel Programming", Reports on Computer Science & Mathematics, Åbo Akademi, Ser. A, N° 129, 1991.

[5] G. R. R. Justo, "PVMGraph: A Graphical Editor for the Design of PVM Programs", Technical Report, University of Westminster, May 1996.

[6] Célio E. Morón, José R. P. Ribeiro, Nilton C. da Silva, Towards a Rapid Prototyping by Linking Design, Implementation and Debugging in Real Time Parallel Programs *Systems,* Proceedings of the 9th IEEE International Workshop on Rapid System Prototyping, Leuven, Belgium, June 1998.

[7] Nilton C. da Silva, José R. P. Ribeiro, Célio E. Morón, Sérgio D. Zorzo, Unifying Implementation, Visualisation, Debugging and Validation of Temporal Requirements in Real-Time Parallel Programs, I Workshop on Real-Time System, in the 16º Brazilian Symposium of Computer Network, May 1998, (in Portuguese).

[8] José R. P. Ribeiro, Nilton C. da Silva, Célio E. Morón. *A Visual Environment for the Development of Parallel Real-Time Programs,* published on WPDRS'98 (Workshop on Parallel and Distributed Real-Time Systems), in LNCS # 1388 (Lectures Notes in Computer Science) from Springer-Verlag, Proceedings of the IPPS98 Workshop – IEEE, Orlando, Florida - USA, March, 30 to April, 3 / 1998.

[9] "Virtuoso - The Virtual Single Processor Programming System", User Manual, Version 3.11, EONIC SYSTEMS.

Real-Time Image Processing on a Focal Plane SIMD Array

Antonio Gentile[1], José L. Cruz-Rivera[2], D. Scott Wills[1], Leugim Bustelo[2],
José J. Figueroa[2], Javier E. Fonseca-Camacho[2], Wilfredo E. Lugo-Beauchamp[2],
Ricardo Olivieri[2], Marlyn Quiñones-Cerpa[2], Alexis H. Rivera-Ríos[2],
Iomar Vargas-Gonzáles[2], Michelle Viera-Vera[2]

[1] School of Electrical and Computer Engineering
Georgia Institute of Technology
Atlanta, Georgia 30332-0250
{gentile, scott.wills}@ee.gatech.edu

[2] Department of Electrical and Computer Engineering
University of Puerto Rico in Mayagüez
Mayagüez, Puerto Rico 00681-9042
jcruz@ece.uprm.edu

Abstract. Real-time image processing applications have tremendous computational workloads and I/O throughput requirements. Operation in mobile, portable devices poses stringent resource limitations (size, weight, and power). The SIMD Pixel Processor (SIMPil) has been designed at Georgia Tech to address these problems. In SIMPil, an image sensor array (focal plane) is integrated on top of a SIMD computing layer, where processing elements (PEs) are connected in a torus. A prototype processing element has been implemented in 0.8 μm CMOS technology. This paper evaluates the effectiveness of the SIMPil design on a set of important image applications. A target SIMPil system is described, which is capable of operating in the Tops/sec range in Gigascale technology.. Simulation results indicate sustained operation throughput in the range of 100-1000 Gops/sec. These results support the design choices and suggest that more complex, multistage applications can be implemented to execute at real-time frame rates.

1 Introduction

Large computational workloads and I/O requirements characterize real-time image processing applications. Typical applications may require 10-1000 Gops/s, which cannot be matched by today's general-purpose microprocessor capabilities (0.2-0.7 Gops/s) [1]. In addition, the traditional sequential access to image data further reduces the available performance. When these applications are to operate in mobile, portable devices, the above requirements couple with stringent resource limitations (size, weight, and power) to pose a formidable system design problem. Many SIMD systems [2], [3], [4] have been used for image processing before, but they achieve performance and generality at the expense of I/O coupling and physical size.

Advances in current semiconductor technology have enabled the integration of CMOS image sensors with digital processing circuitry [5]. In a focal plane imaging system, images are optically focussed into the sensor array, or *focal plane*, and are made available to the underlying processing engine in a single operation. Because large amount of data parallelism is inherent in image processing applications, SIMD (Single Instruction Multiple Data) architectures provide an ideal programming model [6]. The SIMD Pixel Processor (SIMPil) [7], is a focal plane imaging system, being designed at Georgia Tech, which incorporates a specialized SIMD architecture with an integrated array of image sensors.

This paper evaluates the effectiveness of the SIMPil design on a set of important image applications. The objective of this research is to verify the system capability to execute such applications within the inter-frame delay (15-30 ms). A target SIMPil system is described, which is capable of operating in the Tops/sec range in Gigascale technology. Several image processing applications, spanning from spatial and morphological filtering to image compression and analysis, have been implemented and simulated. Simulation results indicate sustained operation throughput in the range of 100-1000 Gops/sec. These results support the design choices and suggest that there is the potential for real-time execution of complex, multistage applications.

The rest of the paper is organized as follows. Section 2 describes the architecture of the SIMPil system and presents a target system for Gigascale integration. Section 3 describes the set of image processing applications implemented on SIMPil. Section 4 presents simulation results and performance evaluation. Finally, conclusions are offered in Section 5.

2   SIMPil System Architecture

The SIMD Pixel Processor (SIMPil) architecture consists of a mesh of SIMD processors on top of which an array of image sensors is integrated. A block diagram for a 16-bit implementation is illustrated in Fig. 1.

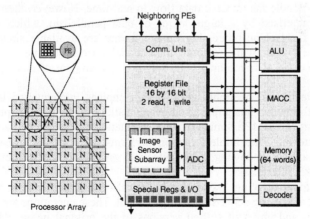

Fig. 1. Block diagram of the SIMPil SIMD system

Each processing element includes a RISC load/store datapath plus an interface to a 4×4 sensor subarray. A 16-bit datapath has been implemented which includes a 32-bit multiply-accumulator unit, a 16 word register file, and 64 words of local memory (the

ISA allows for up to 256 words). The SIMD execution model allows the entire image projected on many PEs to be acquired in a single cycle. Large arrays of SIMPil PEs can be simulated using the SIMPil Simulator, an instruction level simulator, running under Windows95™.

Early prototyping efforts have proved the feasibility of direct coupling of a simple processing core with a sensor device [7]. A 16 bit prototype of a SIMPil PE was designed in 0.8 µm CMOS process and fabricated through MOSIS. The prototypes were successfully tested and run at 25 MHz.

In the rest of the paper, reference is made to a future SIMPil system, integrating 4,096 PEs into a single monolithic device. Table 1 summarizes the most important system parameters [8]. This target system delivers a peak throughput of about 1.5 Tops/sec in a monolithic device, enabling image and video processing applications that are currently unapproachable using today's portable DSP technology.

Table 1. Technology and system parameters for the target SIMPil system

Target VLSI Technology	0.1 µm	Image Sensor Array Size	256×256
Target Clock Rate	500 MHz	Interconnection Network	Torus
Transistor per PE	35 K	Total No. of Transistors (est.)	143.4 M
Image Sensor per PE	4×4	Peak Operation Throughput	1.46 Tops/sec
System Size	64×64	Estimated Power	8.9 W
Chip Size	800 mm²		

3 SIMPil Application Suite

The SIMPil architecture is designed for image and video processing applications. In general, this class of applications is very computational intensive and requires high throughput to handle the massive data flow in real-time. However, these applications are also characterized by a large degree of data parallelism, which is maximally exploited by focal plane processing. Image frames are available simultaneously at each PE in the system, while retaining their spatial correlation. Image streams can be therefore processed at frame rate, with only nominal amount of memory required at each PE.

To evaluate the set of architectural design choices implemented in the SIMPil system, the following image-processing applications have been implemented and simulated using the SIMPil Simulator. Implementation details can be found in [9].

Spatial filtering. The implementation performs 2D convolution-based filtering. Operations such as shadowing, edge detection, and smoothing are executed using appropriate 3×3-filter masks.

Morphological filtering. Basic morphological operations (erosion, dilation) have been implemented using a 3×3 structuring element. These operations are implemented as intersection and union of shifted versions of the original image. More complex operations, such as opening, closing, inside edge detection, and skeletonization are then implemented by combining the two basic operations.

Vector Quantization. A full-search VQ encoding algorithm has been implemented on the SIMPil architecture. A 256-word codebook is used to achieve a 0.5 bit per pixel encoding of an 8-bpp gray-level image. Compression is achieved by subdividing the image into small blocks (e.g. 4x4 pixels) and finding the best match for each block among the available codewords. Two levels of parallelism are exploited in this implementation, via codebook distribution among the PEs and by processing multiple input blocks in parallel.

Wavelet decomposition. Discrete wavelet decomposition has been implemented for fingerprint compression and archival. Standard Daubechie's filters have been used to implement the low/high pass filters. A row-column scheme decomposes a gray-level image into 61 frequency bands.

Image rotation. A parallel rotation algorithm has been implemented to perform fast rotations of binary images. The rotation angle γ is first expressed as in (1)

$$\gamma = \alpha + n \cdot \frac{\pi}{2}, \quad \alpha \in \left[0, \frac{\pi}{2}\right], \quad \text{and } n = 0\ldots 3. \tag{1}$$

Rotations are then executed in two stages: a skew-based rotation of the angle α, and then a set of n fast ninety-degree rotations. This scheme is well suited for a SIMD implementation with regular communication patterns.

Image labeling. This implementation is based on a cluster analysis algorithm. It is used to classify objects in a binary image on the basis of object diameter. The objects are then labeled accordingly.

Quadtree region representation. This implementation operates on binary images to generate a quadtree representation. Quadtrees are based on the principle of recursive decomposition of space. The image is first decomposed in four equal-sized quadrants. If a quadrant is not uniform (entirely filled/empty), it is further decomposed in four more subquadrants. The decomposition stops when uniform quadrants are encountered, or the quadrant contains a single pixel.

These application kernels cover different aspects of the image-processing domain (image compression, image analysis, and image restoration), and they can be staged as components of larger full-featured applications. Larger applications, such as JPEG image encoding, region autofocus and region clustering are currently being implemented, integrating various subcomponents into single larger applications.

4 Results

The applications described in the previous section have been developed and simulated on the SIMPil Simulator. Table 2 summarizes application performances for all the applications implemented. A 64×64 system was used for all the implementations, with the only exception of the image labeling application, for which a 32x32 system was used. A standard NEWS interconnection network was used for all applications with the exception of VQ, for which a toroidal wrapped network was used. In the design of the SIMPil system, memory is traded off for higher throughput. Only a modest amount of memory is available per PE. This choice does not impair SIMPil performance. Both the values of memory used per PE and total system memory indicate that less than 100 memory words are required for most applications. The total system memory size does not exceed 1 MB in almost all cases.

System utilization is calculated as average concurrency, and it is relative to the total system size. Execution time and sustained throughput are computed with reference to the 500 MHz target platform described in section 2. For each application, the worst case scenario was used. Maximum size images were used in almost all cases. For the image labeling applications, a large number of small objects was used. For the image rotation by skew transforms, the total number of cycles refers to a rotation of 89°.

The simulations show that all applications execute in lower millisecond range, suggesting that they can combined in stages to form complex applications, which still execute at full frame rate (30-60 fps, or 15-30 ms). In most cases, large system utilization is recorded, which in turn provides for a large sustained throughput. The low system utilization for the ninety degree ring rotation algorithm is mainly due to the staged communication patterns used.

Table 2. Application performance comparison (SF: Spatial Filtering, IED: Inside Edge Detection, SKL: Skeletonization, VQ: Vector Quantization, WLT: Wavelet Decomposition, LBL: Image Labeling, QTREE: Quad Tree Decomposition, ROT Skew: Skew-based Rotation, ROT Ring: 90° Ring Rotation)

Application	Memory Size		Total Cycles	System Utilization (%)	Execution Time (μs)	Sustained Throughput (Gops/s)
	PE (16b word)	System (KB)				
SF	62	507.9	11,228	81.6 %	22.46	1,671.2
IED	20	163.8	747	94.2 %	1.49	1,929.2
SKL	20	163.8	14,169	99.3 %	28.34	2,033.7
VQ	51	417.8	81,774	91.7 %	163.55	1,878.0
WLT	56	458.7	23,300	68.5 %	46.60	1,402.9
LBL	86	176.1	170,281	43.9 %	340.56	224.8
QTREE	13	106.5	331,479	87.4 %	662.96	1,790.0
ROT Skew	134	1,097.7	580,636	57.5 %	1,161.27	1,177.6
ROT Ring	39	319.5	10,558	34.5 %	21.12	706.6

In general, the impact of communications across applications is contained within 10% of the total execution time. This illustrates the fact that most applications require limited communications and spend most of their time in local operations. In this, they mostly benefit from the focal plane input of the image data, which distributes the image to each PE in a single operation. A larger amount of communications characterizes the wavelet decomposition application (26%). In this case, a large number of data transfers are needed between different application stages, to rearrange data in preparation of the next stage. A different network organization could be used to reduce the amount of communications required. In general, a good balance between scalar and vector operations is shown by all applications (17-40%). The large degree of data parallelism in the application set is demonstrated by the larger slice of total execution time occupied by vector operations.

5 Conclusions

The demand for real time image processing applications for portable, battery-operated, hand-held devices has grown rapidly in recent years, and it has motivated the research on processing architectures that support focal plane data. The SIMPil processor has been designed to offer the required computational power and focal plane integration in a compact, monolithic device.

In this paper, a target SIMPil system for Gigascale technology is introduced, and the effectiveness of its design is evaluated for a set of important image processing applications. The application set is described, and performance analysis is given. Simulation results indicate sustained operation throughput in the range of 100-1000 Gops/sec and large system utilization for all implemented applications. Moreover, a balanced hardware resource allocation is shown. These results support the design choices made and suggest that more complex, multistage applications can be implemented on SIMPil to execute at real-time frame rates.

This work was supported in part by NSF, DARPA, and ARL.

References

1. D. S. Wills, J. M. Baker, H. H. Cat, S. M. Chai, L. Codrescu, J. L. Cruz-Rivera, J. C. Eble, A. Gentile, M. A. Hopper, W. S. Lacy, A. Lòpez-Lagunas, P. May, S. Smith, T. Taha, "Processing Architectures for Smart Pixel Systems", *IEEE Journal of Selected Topics in Quantum Electronics*, v.2 n.1, April 1996, p.24-34
2. K. E. Batcher, "Design of a massively parallel processor", *IEEE Transactions on Computers*, pp. 836-840, Sept. 1980
3. T. Blank, "MasPar MP-1 Architecture", *Proceedings of the 35th IEEE Computer Society International Conference*, San Francisco, California, pp.20-24, 1990
4. W. F. Wong, K. T. Lua, "A preliminary evaluation of a massively parallel processor: GAPP", *Microprocessing Microprogramming*, v.29, no. 1, pp. 53-62, 1990
5. E. Fossum, "Digital Camera System on a Chip," *IEEE Micro*, pp. 8-15, May 1998.
6. K. Diefendorff and R. Dubey. "How Multimedia Workloads Will Change Processor Design." IEEE Computer, Vol. 30, No. 9, September 1997, pp. 43-45.
7. H. H. Cat, A. Gentile, J. C. Eble, M. Lee, O. Vendier, Y. J. Joo, D. S. Wills, M. Brooke, N. M. Jokerst, A. S. Brown: "SIMPil: An OE Integrated SIMD Architecture for Focal Plane Processing Applications", *Proceedings of the Third IEEE International. Conference on Massively Parallel Processing using Optical Interconnection* (MPPOI-96), Maui, Hawaii, pp. 44-52, 1996
8. S. M Chai, A. Gentile, D. S. Wills, "Impact of power density limitation in Gigascale Integration for the SIMD Pixel Processor", to appear in *Proceedings of the 20th Anniversary Conference on Advanced Research in VLSI*, Atlanta, Georgia, 1999.
9. A. Gentile, J. L. Cruz-Rivera, D. S. Wills, L. Bustelo, J. J. Figueroa, J. E. Fonseca-Camacho, W. E. Lugo-Beauchamp, R. Olivieri, M. Quiñones-Cerpa, A. H. Rivera-Ríos, I. Vargas-Gonzáles, M. Viera-Vera, "Image Processing Applications on the SIMD Pixel Processor", *Technical Report*, GT-PICA-005-98, Pica Lab, Georgia Institute of Technology, Jun. 1998

Metrics for the Evaluation of Multicast Communications

Philip M. Irey IV and David T. Marlow

System Research and Technology Department, Combat Systems Branch
Naval Surface Warfare Center, Dahlgren Division, Dahlgren, Virginia 22448-5100
{ireypm, marlowdt}@nswc.navy.mil

Abstract. *In the distributed shipboard environment of interest to the United States Navy, there is an increasing interest in the use of multicast communications to reduce bandwidth consumption and to reduce latencies. Standardized metrics and tools are needed to evaluate the performance of multicast systems in this environment. This paper proposes a number of metrics and data collection and analysis techniques for assessing multicast communications performance. Of particular significance is a metric that correlates message reception among receivers and shows promise in analyzing topology related problems. The MCAST Tool Set (MTS), which uses the metrics and data collection techniques presented, is used to analyze several multicast application scenarios.*

1 Introduction

In the distributed shipboard environment, there are a number of data streams which must be received by multiple hosts. These data streams may be large in volume and sent often (e.g. 100 Hz gyro data). Multicast transfer techniques are expected to reduce aggregate bandwidth utilization and to reduce transfer latencies. As the number of recipients of these data streams grows in the shipboard environment, the need for utilizing multicast transfer instead of sequential data transfer to each recipient becomes obvious. Testing must be done to ensure that the performance of multicast implementations lives up to its expectations. A large body of work has been published on unicast metrics [1] and [2] which can provide a foundation for defining multicast metrics; however, multicast transfer breaks some of the fundamental assumptions of unicast metrics. The IETF has initiated efforts in defining metrics targeted at multicast exchanges [3] but little work has been done in this area.

2 Multicast Performance Metrics

This paper defines two types of metrics critical for assessing multicast performance: local metrics measured at a single sender or receiver and group metrics which represent an aggregate performance for all receivers in a multicast group. Local metrics are applicable to both unicast and multicast data transfer. Group metrics only apply to multicast transfer. All metrics defined are measured at the application to communications subsystem interface (e.g. the sockets API in a Unix environment) so delays introduced by transmission on the media, queueing delays, etc. are included in measurements based upon them.

2.1 Metric Notation

Metric names are defined with capital letters (e.g. *XYZ*). The subscript i is used to select a particular measurement from a set of measurements. The subscript j is used to select a particular host from a set of hosts. To represent a particular measurement on a particular host, a double subscript notation is used. For example, $LIAT_{j_i}$ represents $LIAT$ measurement number i on host number j. Statistics can also be applied to a set of measurements. The subscripts avg, max,

```
while (done==FALSE) {
    ts1=gettimeofday(ts1)
    send(A,++seq,ts1,data)
    ts2=gettimeofday(ts2)
    LIST[++LMS]=timediff(ts2,ts1)
    local_usleep(sleep_time)
}
```
FIGURE 1. Transmitter Pseudo-Code

and min represent average, maximum, and minimum values. The subscript *sdev* represents standard deviation. To denote a statistical measurement from a particular host in group metrics, a subscript is applied to the statistic subscript. For example, XYZ_{avg_j} represents the average value of the *XYZ* metric on receiver j. In the equations below, m represents the number of messages sent on a data stream and n represents the number of multicast receivers in a group.

2.2 Local Metrics

Local metrics are measured at either the transmitter or a receiver represented in the pseudo-code in Figures 1 and 2. Each message is sent to the address A which can address a unicast receiver or a multicast group of receivers. Each message sent contains a sequence number, seq, a timestamp, ts1, and data. Since the timestamp generated by the transmitter is used by receivers, it is assumed that either the clocks of the sender and all receivers are synchronized (e.g. NTP, GPS, etc.) or clock offsets can be computed [4].

2.2.1 LMS and LIST Metrics

The *LMS* (Local Messages Sent) and *LIST* (Local Inter-Send Time) metrics are both measured at the transmitter. All other local metrics presented below are measured at a receiver. *LMS* is the count of the number of messages sent to the address A by the transmitter. *LIST* metrics compute the time between successive message sends at the transmitter. Computing *LIST* is important because the measured value may not always correspond to the expected value [5]. *LIST* statistics are computed as shown in Equations 1-4

```
bytes_received=LMR=LMRL=LGN=0
expected_seq_num=0
tr1=gettimeofday()
first=TRUE
while (done==FALSE) {
  prev_tr1=tr1
  recv(A,seq,ts1,data)
  tr1=gettimeofday()
  if (first == TRUE) {
    start_time=ts1
    FIRST=FALSE
  }
  process_packet = FALSE
  if (seq == expected_seq_num) {
    process_packet=TRUE
  }
  else if (seq < expected_seq_num) {
    ++LMRL
  }
  else { /* seq > expected_seq_num */
    LGB[++LGN]= (expected_seq_num,seq)
    LGL[LGN]=seq-expected_seq
    process_packet=TRUE
  }
  if (process_packet) {
    LOWL[++LMR]=timediff(tr1,ts1)
    LIAT[LMR]=timediff(tr1,prev_tr1)
    expected_seq_num=seq+1
  }
  bytes_received=bytes_received+sizeof(data)
}
end_time=gettimeofday()
LAT=bytes_received/timediff(end_time,start_time)
```

FIGURE 2. Receiver Pseudo-Code

2.2.2 LMR, LMRL, LMRI, and LPMR Metrics

LMR_j (Local Messages Received) is the count of the number of messages received in sequence and those whose sequence number was greater than the sequence number expected for the next message. Messages received with a sequence number less than the expected one are counted in $LMRL_j$ (Local Messages Received Late). Messages which arrive out of order or are duplicate are classified as "late" using this criteria. $LMRI_j$ (Local Messages Received In-order) is defined to count only the messages which were received with a sequence number equal to the expected sequence number as shown in Equation 5. $LPMR_j$ (Local Percent Messages Received) computes the percentage of the messages sent by the transmitter which were received by a receiver as shown in Equation 6.

$$LIST_{avg} = \sum_{i=1}^{m} \frac{LIST_i}{m} \quad (1) \qquad LIST_{max} = \underset{\forall i}{MAX}\{LIST\} \quad (2)$$

$$LIST_{min} = \underset{\forall i}{MIN}\{LIST\} \quad (3) \qquad LIST_{sdev} = \sqrt{\left(\sum_{i=1}^{m}(LIST_i - LIST_{avg})^2\right)/(n-1)} \quad (4)$$

$$LMRI_j = LMR_j - LGN_j \quad (5) \qquad LPMR_j = LMR_j/LMS_j \quad (6)$$

2.2.3 Local One-Way Latency (LOWL) Metrics

LOWL metrics measure the one-way latency between the transmitter and a receiver as shown in Equations 7-10. These metrics compute the difference between ts1 and tr1. Messages which are late or lost do not contribute to these metrics since LMR_j is used.

$$LOWL_{avg_j} = \sum_{i=1}^{LMR_j} \frac{LOWL_{j_i}}{LMR_j} \quad (7) \qquad LOWL_{max_j} = \underset{\forall i}{MAX}\{LOWL_{j_i}\} \quad (8)$$

$$LOWL_{min_j} = \underset{\forall i}{MIN}\{LOWL_{j_i}\} \quad (9) \qquad LOWL_{sdev_j} = \sqrt{\left(\sum_{i=1}^{LMR_j}(LOWL_i - LOWL_{avg_j})^2\right)/(LMR_j - 1)} \quad (10)$$

2.2.4 Local Inter-arrival Time (LIAT) Metrics

LIAT metrics measure the time between the receipt of successive messages at a receiver. Since *LIAT* is computed using timestamps from the local receiver only, synchronized clocks or clock offset conversions are not necessary. Since $LIAT_1$ is always undefined, it is not used in computing

$$LIAT_{avg_j} = \sum_{i=2}^{LMR_j} LIAT_{j_i}/(LMR_j - 1) \quad (11)$$

$LIAT_{avg}$ as shown in Equation 11. *LIAT* statistics are computed as in Equations 8-10 except that i starts at 2 and the $LMR_j - 1$ term in Equation 10 should be replaced with the term $LMR_j - 2$. Messages which are late or lost do not contribute to these metrics since LMR_j is used.

2.2.5 Local Application-to-Application Throughput (LAT) Metric

The *LAT* metric is used to characterize the end-to-end throughput between the transmitter and a receiver. The start of the time interval begins with the receiver recording the value of ts1 in the first message received on the data stream in start_time. The receiver then records the time when the last message was received in end_time. LAT_j is the number of bytes received divided by the difference between these two timestamps.

2.2.6 Local Gap Boundaries (LGB) Set

LGB_j is the set of ordered pairs of the start and end points of gaps in the sequence space of received messages observed by receiver j. LGB_j is a set from which metrics are derived for host j. When a message is received with a sequence number which is not equal to expected_seq, one greater than the highest previously received in-order message, one of two scenarios has occurred: 1) the message has a sequence number greater than expected_seq in which case the ordered pair (expected_seq_num,seq-1) denoting the sequence space gap is recorded in LGB_j; or 2) the message has a sequence number less than expected_seq in which case the message is considered late and no action associated with LGB_j is taken and a sequence space gap is not recorded

2.2.7 LGN and LGL Metrics

The *LGN* (Local Gap Number) metric is a count of the sequence space gaps observed (i.e. messages received which had a sequence number greater than expected_seq). The *LGL* (Local Gap Length) metrics measure the size of sequence space gaps observed by a receiver. LGL_{j_k} for sequence space gap LGB_{j_k} is computed by subtracting the ending sequence number for gap k from the starting sequence number as shown in Equation 12. (Figure 2 shows (start, end)-tuples being recorded for LGB). LGL_{avg_j} is computed by iterating over k as shown in Equation 13. LGL_{max_j}, LGL_{min_j}, and LGL_{sdev_j} are computed similarly to Equations 2-4 except that they are computed over k.

$$LGL_{j_k} = LGB(end)_{j_k} - LGB(start)_{j_k} \quad (12) \qquad LGL_{avg_j} = \left(\sum_{k=1}^{LGC_j} LGL_{j_k}\right) / LGN_j \quad (13)$$

2.3 Group Multicast Metrics

Unlike local metrics which characterize the performance of the transmitter or a receiver, group metrics attempt to simultaneously characterize the performance of all receivers in a multicast group.

2.3.1 GOWL, GIAT, GAT, GGN, and GGL Metrics

GOWL (Group One-way Latency) metrics characterize the one-way latency performance of the group using *LOWL* measurements from each receiver. *GOWL* statistics can be computed as shown in Equations 14-17 which serve as prototypes for computing statistics on sets for *GIAT*,

$$GOWL_{avg} = \sum_{j=1}^{n} \frac{LOWL_{avg_j}}{n} \quad (14) \qquad GOWL_{max} = \underset{\forall j}{MAX}\{LOWL_{max_j}\} \quad (15)$$

$$GOWL_{min} = \underset{\forall j}{MIN}\{LOWL_{min_j}\} \quad (16) \qquad GOWL_{sdev} = \sqrt{\sum_{j=1}^{n}(LOWL_{avg_j} - GOWL_{avg})^2 / (n-1)} \quad (17)$$

GAT, *GGN* and *GGL* defined below. *GIAT* (Group Inter-arrival Time) metrics characterize the message inter-arrival performance of the group using *LIAT* measurements from each multicast receiver. *GAT* (Group Application-to-application Throughput) metrics characterize the application-to-application throughput performance of the group using *LAT* measurements from each multicast receiver. *GGN* (Group Gap Number) metrics characterize the number of message gaps observed by the group using *LGN* measurements from each multicast receiver. *GGL* (Group

Gap Length) metrics characterize the size of message gaps observed by the group using *LGL* measurements from each multicast receiver.

2.3.2 Group Reception Correlation (GRC) Metric

The *GRC* metric is used to measure the degree of correlation in the messages received by members of a group. To compute *GRC*, reception vectors are created for each group member. A reception vector records which messages on a data stream were received in sequence and which were not by a particular multicast receiver in the group. To compute the reception vector for multicast receiver j, $\vec{V}_j$, component i of $\vec{V}_j$ is set equal to one if the message with sequence number i was received in order and is set equal to zero otherwise. The information needed to populate the reception vectors is recorded in *LGB*.

A set of reception vectors, S, can be grouped into a reception matrix, R as shown in Equation 18. Each column j of R contains the reception vector for multicast receiver j. Each row i of R contains a Reception Report (RR) for all receivers for message i. The number of rows in R is always equal to m. The number of columns in R is equal to the number of reception vectors in S. The reception matrix in Equation 18 shows R constructed from S^* (the set which contains reception vectors for all multicast receivers in a group). R can be constructed for any subset, S, of S^*.

$$R = \begin{bmatrix} V_{11} & V_{12} & \cdots & V_{1n} \\ V_{21} & V_{22} & \cdots & V_{2n} \\ V_{31} & V_{32} & \cdots & V_{3n} \\ \cdots & \cdots & \cdots & \cdots \\ V_{m1} & V_{m2} & \cdots & V_{mn} \end{bmatrix} \quad (18)$$

GRC is computed on a reception matrix R as shown in Equation 21 which is derived from Equations 19 and 20. Equation 19 computes the Message Reception Correlation (*MRC*) for message i. In this equation, the sum of R_{ij} is the number of multicast receivers which received message i ($R_{ij}=1$) and n minus this sum is the number which did not ($R_{ij}=0$). The absolute value of the difference of these two quantities is then divided by n (the number of multicast receivers) which yields a value between 0 and 1. This value represents the degree of correlation among the multicast receivers in receiving message i. A degree of correlation equal to 1 indicates that all receivers in the group whose reception vectors are contained in R received that message or all of them did not receive that message. A degree of correlation equal to 0 indicates that half of the receivers received the message and the other half did not. The sum of the *MRC* values is computed and divided by m (the total number of messages which could have been received) to give the average degree of correlation for all messages, or *GRC*, as shown in Equation 20. The values of *GRC* range from 0 to 1.

$$MRC_i = \left| \left(\sum_{j=1}^{n} R_{ij} \right) - \left(n - \sum_{j=1}^{n} R_{ij} \right) \right| / n \quad (19) \qquad GRC = \sum_{i=1}^{m} \frac{MRC_i}{m} \quad (20)$$

3 Testing Considerations

The utility of the metrics defined in Sections 2.2 and 2.3 are dependent on the test environment. In the environment of interest in this paper, IP multicast data transmission is used which has unreliable semantics which can affect any measurements collected.

3.1 Correlation and Cross Checking of Group Receivers

The *GRC* computed for the group provides a measure of the consistency of data reception among the set of multicast receivers. If there is a low degree of correlation among the group members, it is unlikely that the receivers are making their measurements on the same set of samples. As an extreme example, suppose Host 1 receives all messages with odd sequence numbers and Host 2 receives all messages with even sequence numbers which were sent on the same datastream to a multicast group. In this case, Host 1 and Host 2 are making measurements on what can be viewed as two different sets of data. In this case, *GRC* is equal to zero. Non-gap group metrics should be viewed with caution in this case. Cross-checking of the data measurements can clarify what appears at first glance to be

$$GRC = \left(\sum_{i=1}^{m} \left| 2 \sum_{j=1}^{n} R_{ij} - n \right| \right) / (n \times m) \quad (21)$$

inconsistent results. For example, it is possible to measure a low value for *LOWL* and a high value for *LIAT* on a data stream. This seems inconsistent since the measured *LOWL* implies good network performance yet the *LIAT* implies poor network performance. Examining *LGC* might show a large number of gaps which would contribute to the large value for *LIAT*. *LOWL* is still low since the time interval used for that metric only uses timestamps associated with that message. The *LIAT* metric, on the other hand, uses a timestamp taken when a previous message was received on the data stream.

3.2 GRC Partitioning Algorithms

Two partitioning algorithms are defined to partition hosts into groups based on the *GRC* metric computed for those groups. First, a threshold value is specified. Next, groups of hosts whose *GRC* is greater than or equal to the threshold value are created. Loose Partitioning creates a set Π to partition the set of receiving hosts into subsets based on their RR's such that the GRC for each subset is greater than or equal to a threshold correlation, τ as shown in Equation 22 and 23. Strict partitioning creates a set $\widehat{\Pi}$ to partition the set of receiving hosts into subsets based on their RR's such that the GRC for each subset and all subsets of that subset is greater than or equal to a threshold correlation, τ as shown in Equations 24 and 25. It should be noted that there are 2^n subsets which can be generated from R (where $n = |R|$) so generating $\widehat{\Pi}$ can be an expensive operation.

$$P = \{s \subseteq R | GRC(s) \geq \tau\} \quad (22)$$

$$\Pi = \left\{ A \subseteq P | A \text{ is not a proper subset of any other element of } P \right\} \quad (23)$$

$$P = \{s \subseteq R | \forall T, T \subseteq s, GRC(T) \geq \tau\} \quad (24)$$

$$\widehat{\Pi} = \left\{ A \subseteq P | A \text{ is not a proper subset of any other element of } P \right\} \quad (25)$$

4 The MCAST Tool Set (MTS)

MTS was developed at the Naval Surface Warfare Center - Dahlgren Division to enable the performance analysis of UDP/IP multicast systems using the metrics defined in this paper. It used in Section 5 for instrumentation and analysis. There are four types of components in MTS: an instrumentation tool called MCAST (Multicast Communications Analysis and Simulation Tool), a test coordination tool, a data extraction tool, and several data analysis tools. Although these tools can be used individually, they are generally used together. (Contact the authors for information on obtaining MTS).

5 IP Multicast Performance Tests

Several tests were conducted to evaluate the performance of UDP over IP multicast using the metrics developed in Section 2. All testing was done using MTS on the testbed shown in Figure 3. The hosts are interconnected with 10/100/1000 Mbps Ethernet switches from Extreme Networks. It is important to note that multicast routing is not being used in the testbed. In this environment, all multicast traffic is forwarded across all collision domains.

5.1 Unicast Versus Multicast

As described in Section 2, local metrics are applicable to measuring both unicast and multicast performance. MCAST allows both unicast and multicast measurements to be made using these metrics. In these tests, Tide is the transmitter and Cheer is the receiver. First a unicast data stream is measured between the hosts. Next, a multicast data stream is measured between the hosts. The unicast and multicast *LET* and *LOWL* results are then compared.

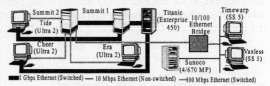

FIGURE 3. Multicast Testbed

5.1.1 100 Mbps Switched Ethernet

As can be seen in Figures 4 and 5, there is little if any difference between unicast and multicast transmission using 100 Mbps switched Ethernet with Tide as the transmitter and Cheer as the receiver.

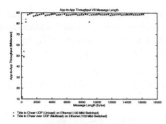

FIGURE 4. Unicast Versus Multicast LAT over 100 Mbit Switched Ethernet

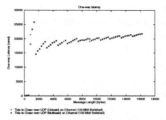

FIGURE 5. Unicast Versus Multicast LOWL over 100 Mbit Switched Ethernet

5.1.2 One Gigabit/Second (Gbps) Switched Ethernet

As shown in Figures 6 and 7, there is a noticeable performance difference in the *LAT* and *LOWL* metrics measured with Tide as the transmitter and Cheer as the receiver over Gigabit Ethernet. No satisfactory explanation for the performance differences observed between unicast and multicast transmission in the one Gbps tests has been found yet. Since no differences were observed in the 100 Mbps tests and differences were observed in the one Gbps tests, one might conclude that the multicast forwarding capabilities of the switch do not scale in the same way as the unicast forwarding at higher transmission rates or that architectural differences between the Summit 1 and Summit 2 switches used in the tests contribute to the performance differences observed. Further work in this area will attempt to test these theories and attempt to identify the source of the performance differences.

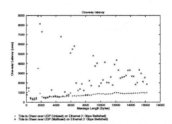

FIGURE 6. Unicast Versus Multicast LOWL over Gigabit Ethernet

5.2 Group Multicast Performance

When analyzing the behavior of a multicast group, the *GRC* metric is often a key in the analysis process. Figures 8, 9, and 10 show experiments, E1, E2, and E3, in which sunoco transmitted 10,000 messages to era, tide, timewarp, and vaxless. As shown in Figure 3, the transmitter (sunoco) uses a 10 Mbps shared Ethernet connection to communicate with the receiving hosts and thus can not exceed a transmission rate of 10 Mbps.

Figure 8 shows the results of E1 where sunoco sent messages of length 16384 bytes to the multicast group with a sleep_time equal to 0. As can be seen visually,

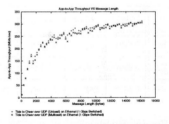

FIGURE 7. Unicast Versus Multicast LAT over Gigabit Ethernet

few messages were considered lost by any of the receivers. Table 1 shows a high degree of correlation (.9991) among the receiver set and that a high percentage of the messages were received (99.87%).

Figure 9 shows the results of E2 where sunoco sent messages of length 16384 bytes to the multicast group with a sleep_time equal to 0. In this scenario, unlike E1 and E3, a unicast background load of 6 Mbps (between two hosts not involved on the multicast test) was introduced on the 10 Mbps Ethernet. As can be seen, in Figure 9, a large number of messages were dropped. Table 1 shows a high degree of correlation among the receiver set (0.9989). It also shows that nearly all the messages which were considered lost were considered lost by all receivers (RR=0). In this environment, a message considered lost by all hosts probably indicates that the message was lost at the sender. It is likely the background load on the shared media to which the transmitter is attached caused large numbers of packet collisions which resulted in queue overflows on the transmitter.

Figure 10 shows the results of E3 where sunoco sent messages of length 256 bytes to the multicast group with a sleep_time equal to 0. For E3, Table 1 shows a much lower degree of correlation (.04498) than E1. This is also reflected in the Reception Reports (RR) in the table. Computing Π for E3 with $\tau = 0.5$ yields the set $\{\{R_1,R_2\},\{R_3,R_4\}\}$ where R_1,R_2,R_3,R_4 are the reception reports for Tide, Era, Vaxless, and Timewarp respectively. Consequently, the high performance machines, Tide and Era, correlate to the specified degree as do the low performance machines, Vaxless and Timewarp.

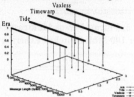

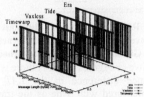

FIGURE 8. E1: Reception Vectors for Message 16384 Bytes in length (no background load)

FIGURE 9. E2: Reception Vectors for Messages 16384 bytes in length (6 Mbps background load)

6 Conclusions

The metrics and analysis techniques developed in this paper as well as their realization in MTS proved to be useful in analyzing the performance of multicast systems. The performance of complex multicast applications can be simulated and evaluated on the target hardware environment without the complexity of running the actual applications. Multicast performance variations among the set of receivers related to network topology, network utilization, and host performance were observed with MTS. In particular, the GRC metric proved to be extremely useful for analysis. Without the metrics defined here and their implementation in MTS, analyzing performance in a distributed multicast application would be very difficult.

TABLE 1. GRC Measurements

	GRC	RR=0	RR=N	0 < RR < N
E1	0.9991	0%	99.87%	.13%
E2	0.9989	1.81%	97.99%	0.2%
E3	0.4498	0%	21.06%	78.94%

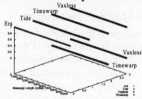

FIGURE 10. E3: Reception Vectors for Messages 256 bytes in length (no load)

7 Future Work

The work to date has focused on instrumenting IP Multicast in a non-routed environment. Future work in this area will examine a routed multicast environment and a switched VLAN environment. Further investigation is planned to investigate the Ethernet performance differences observed. Investigating applying multicast techniques and using metrics defined here in a network resource monitor which overcomes the scalability problems cited in [6] is planned.

References

1. Irey IV, Philip M., Harrison, Robert D., Marlow, David T., *Techniques for LAN Performance Analysis in a Real-Time Environment*, Real-Time Systems - International Journal of Time Critical Computing Systems, Volume 14, Number 1, pp. 21-44, January 1998.†
2. Paxon, V., Almes, G., Mahdavi, J., Mathis, M., *Internet Draft: Framework for IP Performance Metrics*, November 1997.
3. Dubray, K., *Internet-Draft: Terminology for IP Multicast Benchmarking*, July 1997.
4. Mills, D., *RFC-1305, Network Time Protocol (Version 3) Specification, Implementation and Analysis*, March 1992.
5. Irey IV, Philip M., Marlow, David T., Harrison, Robert D., *Distributing Time Sensitive Data in a COTS Shared Media Environment*, Joint Workshop on Parallel and Distributed Real-Time Systems, pp. 53-62, April 1997.†
6. Irey IV, Philip M., Hott, Robert W., Marlow, David T., *An Architecture for Network Resource Monitoring in a Distributed Environment*, Lecture Notes in Computer Science 1388, pages 1153-1163, Springer-Verlag, 1998†.

†Documents available on http://www.nswc.navy.mil/ITT/documents.

Distributing Periodic Workload Uniformly Across Time to Achieve Better Service Quality

Jaeyong Koh, Kihan Kim, and Heonshik Shin

Dept. of Computer Engineering, Seoul National University,
Shinlim-dong San 56-1, Gwanak-gu, Seoul 151-742, Korea.
{jyk, shinhs}@ce2.snu.ac.kr
http://cselab.snu.ac.kr/member/alcohol/jaykoh.html

Abstract. A multimedia system consists of substantial amount of continuous media workload scheduled periodically at deterministic time points. Phased scheduling is achieved by shifting *phase* or the lag between the invocation times of independent periodic tasks. A proper phase configuration distributes workload uniformly over time and reduces task interference that may otherwise result in jitter, deadline miss, and long response time. In our previous work, an algorithm is proposed that identifies optimal phase values for given periodic task set scheduled earliest deadline first[1]. We now present another phase identification algorithm and evaluate its effectiveness. The algorithm in this paper is less accurate but faster than the previous algorithm. In addition, the new approach is applicable to wider set of problems.

1 Introduction

Phased scheduling[1] has been proposed to deliver a better service quality of multimedia end system consisting of multiple periodic and aperiodic tasks. It is achieved by shifting the phase, or points of time at which mutually non-synchronized periodic tasks are scheduled for execution. In [1], we presented a phase optimization algorithm and evaluated the performance gain that results from the optimization. It is shown that phased scheduling substantially reduces deadline miss ratio and average response time of real-time tasks. However, our previous work is applicable only to Earliest Deadline First(*EDF*) scheduling discipline. Furthermore, it requires substantial amount of time for the optimization.

This paper presents an alternative approach to improve the system performance through phased scheduling. The new algorithm attempts to distribute the periodic workload evenly across the time span using artificial phases. This *time-wise* load balancing of periodic workload reduces the *jitter* or the variance in task response time. It improves worst-case performance by removing the *critical instant* where all the periodic tasks are released simultaneously. The proper phase also improves average performance such as average response time or deadline miss probability. The new algorithm is advantageous over our previous one in that it requires less amount of time to obtain the appropriate phase configuration. However, the obtained phase is not as accurate as in the previous algorithm which identifies the phase values optimal for the specific scheduling discipline, EDF.

2 A model of processor scheduling

We consider a single processor system where n periodic tasks $\tau_1, \ldots, \tau_i, \ldots, \tau_n$ are multi-tasked. The set of periodic tasks is denoted $\tau = \{\tau_1, \ldots, \tau_i, \ldots, \tau_n\}$. Each periodic task τ_i processes a continuous media stream and is a sequence of *task instances* produced with first release time or phase r_i and period P_i. A task instance processes each data unit of the media stream. The xth instance of task τ_i, denoted τ_{ix}, is invoked at the release time $e_{ix} = r_i + P_i(x-1)$ and is required to finish before the current period expires at time $e_{ix} + P_i[2]$. The computation time of task instances varies as the input data change. We assume, however, that the average computation time C_i for each task τ_i is known a priori. The *hyperperiod* is the least common multiple of all periods $P_1, \ldots, P_n$. Since the task set τ repeats an identical execution trace every hyperperiod, only one hyperperiod is examined to analyze the entire schedule. Finally, let $U_p = \sum_{i=1}^{n} C_i/P_i$, the utilization rate of periodic task set τ.

Aperiodic tasks may arrive in the system with unknown timing. Although their complete timing information is not available a priori, their average inter-arrival and computation times are assumed to be known. Let U_{ap} designate the utilization rate of aperiodic tasks, which equals the average computation time over the average inter-arrival time.

3 The phase improvement algorithm

Conventional schedulability analysis of real-time periodic tasks often assumes the critical instant where tasks are released all at once[3]. At this zero-phased instant, workload of the periodic tasks gets maximally condensed and the system is most likely to be overloaded. However, many periodic tasks are invoked by internal timer at deterministic time points. For these tasks, we may assign artificial phases to eliminate the condensation of workload. To identify an appropriate phase vector that improves service quality, we consider the regularity of workload distribution along time. Note that the zero phase implies the highly-irregular distribution of workload and degrades system performance. The new algorithm attempts to distribute the periodic workload evenly across the time span using artificial phases. Since the ratio of released workload to the elapsed time equals U_p within a sufficiently large time span, it is desirable that the periodic workload contained in any time interval $[t_1, t_2]$ equals $U_p(t_2 - t_1)$.

Periodic workload is inflicted into system at the release times e_{ix} of the task instances τ_{ix}. Let $< v_1, \ldots, v_j, \ldots, v_m >$ be the sorted list of the release times $\{e_{ix}\}$ within a hyperperiod. That is, v_j is the jth release time and m is the number of task instances within a hyperperiod. Relating to the sorted list $< v_1, \ldots, v_m >$, function ι is defined to map the index j of release time v_j onto the index i of periodic task τ_i released at v_j, i.e., $\iota(j) = i$ if and only if

$v_j = e_{ix}$. Similarly, for a function χ, we define $\chi(j) = x$ if and only if $v_j = e_{ix}$. Note that $v_j = e_{\iota(j)\chi(j)} = r_{\iota(j)} + P_{\iota(j)}(\chi(j) - 1)$ and $< v_1, \ldots, v_j, \ldots v_m > = < e_{\iota(1)\chi(1)}, \ldots, e_{\iota(j)\chi(j)}, \ldots, e_{\iota(m)\chi(m)} >$.

The workload released within the time interval $[0, v_j)$ for $j = 1, \ldots, m$ amounts to $\sum_{l=1}^{j-1} C_{\iota(l)}$. Uniform distribution of the workload implies that, within any time interval $[0, v_j)$, the ratio of total inflicted workload $\sum_{l=1}^{j-1} C_{\iota(l)}$ to the length of the interval v_j is identical across all $j = 2, \ldots, m$. Within a sufficiently large time span, the ratio $\sum_{l=1}^{j-1} C_{\iota(l)}/v_j$ converges to U_p. It follows that any interval lengths v_j should be as close to $\sum_{l=1}^{j-1} C_{\iota(l)}/U_p$ as possible for $j = 2, \ldots, m$. Let $v_j^0 = \sum_{l=1}^{j-1} C_{\iota(l)}/U_p$, the ideal length of the interval $[0, v_j)$. We intend to identify the phase values that approximate the interval lengths v_j as close to the ideal values v_j^0 as possible.

Without loss of generality, the value of each phase r_i is assumed to fall in the interval $[0, P_i)$. Since the phases are relative among themselves, we choose $r_n = 0$ as reference point to all the other phases. Let $\boldsymbol{r}_\tau$ be the n-dimensional vector space composed of possible phase values of task set τ, that is, $\boldsymbol{r}_\tau = (r_1, \ldots, r_i, \ldots, r_{n-1}, 0)$ for real numbers r_i such that $0 \leq r_i < P_i$. Finally, let $\boldsymbol{r}$ denote a phase vector belonging to the vector space $\boldsymbol{r}_\tau$. For some task sets, the precisely-uniform workload distribution is not possible. As an example, consider the task set $\{\tau_1 = (P_1 = 10, C_1 = 2, r_1 > 0), \tau_2 = (P_2 = 20, C_2 = 10, r_2 = 0)\}$. Ideally, the length of the interval $[0, v_2) = [0, r_1)$ should equal $C_2/U_p = 14.3$. However, since $r_1 < P_1 < 14.3$, it is impossible to set the interval length r_1 to the ideal value 14.3. The ideal workload distribution is not achievable for some periodic task sets as illustrated in this example. In this case, we try to keep the length of the interval $[0, v_j)$ as close to the ideal value v_j^0 as possible. The 'error' $v_j - v_j^0$ should be minimized for all $j = 2, \ldots, m$. More precisely, the phase configuration $\boldsymbol{r}$ is identified that minimizes the total amount of the variances $(v_j - v_j^0)^2$ or

$$\sum_{j=2}^{m}(v_j - v_j^0)^2 = \sum_{j=2}^{m}(r_{\iota(j)} + P_{\iota(j)}(\chi(j) - 1) - \sum_{l=1}^{j-1} C_{\iota(l)}/U_p)^2. \qquad (1)$$

Finding such a phase vector is a well-known convex quadratic programming problem of $\boldsymbol{r}$.

To formulate the objective function as shown in Formula (1), we first need to determine the sequence of release times $< v_1, \ldots, v_m >$. The sequence of release events $\{e_{ix}\}$ varies as the value of $\boldsymbol{r}$ changes within the phase vector space $\boldsymbol{r}_\tau$ as illustrated in Figure 1a-b. To consider each of the different release sequences separately, we partition the phase vector space $\boldsymbol{r}_\tau$ into subspaces in each of which the release sequence is uniquely determined. It is clear that the sequence of first $j \leq m$ release events $< v_1, \ldots, v_j >$ is obtained in the subspace $\{\boldsymbol{r} | \boldsymbol{r} \in \boldsymbol{r}_\tau, v_1 < \ldots < v_l < \ldots < v_j\}$ where v_l is a linear expression of a phase variable $v_l = r_{\iota(l)} + P_{\iota(l)}(\chi(l) - 1)$. Figure 1a illustrates two subspaces of $\boldsymbol{r}_\tau$ corresponding to two different release sequences $< e_{11}, e_{21} >$ and $< e_{21}, e_{11} >$ respectively.

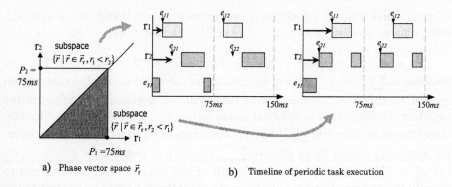

Fig. 1. Phase configuration and task release sequences illustrated for task set $\{(P_i, C_i)|(75, 21), (75, 27), (150, 16)\}$

Each release sequence $< v_1, \ldots, v_m >$ achieved in the subspace $\{r|r \in r_\tau, v_1 < \ldots < v_m\}$ is considered separately. [4] presents one method to generate all possible release sequences. Within each of the subspaces, the objective function $\sum_{j=2}^{m}(v_j - v_j^0)^2$ is obtained with the constraints $v_1 < \ldots < v_m$ and $0 \leq r_i < P_i$ for $i = 1, \ldots, n$. To that quadratic objective function and linear constraints, we apply the quadratic programming method to identify the phase vector r that achieves minimum $\sum_{j=2}^{m}(v_j - v_j^0)^2$ within each subspace. The best one of these local optima is then chosen to be the solution for our optimization. It should be noted that the complexity of quadratic programming in this case is $O(n^2)$ where n is the number of periodic tasks.

4 Performance evaluation

We show the impact of phased scheduling on deadline miss probability. In the experiment, periodic and aperiodic tasks are scheduled by proportional share discipline. We use the same set of periodic tasks as illustrated in Figure 1 except that their average computation times C_i are scaled up or down uniformly to achieve the desired utilization U_p. Three different periodic utilization rates are considered separately; $U_p = 60\%, 75\%,$ and 90%. The aperiodic load varies across the utilization unused by periodic tasks, i.e., $0 \leq U_{ap} \leq 1 - U_p$. The weight is assigned to each task, which equals the task's utilization rate. Both periodic and aperiodic tasks are considered *time-critical*: A task instance is accepted for execution only if it can finish by deadline. Otherwise it is discarded[5]. For more detailed description on experimental settings, refer to [4].

The deadline miss probabilities under the two phase configurations are compared in Figure 2. It is shown that the phased scheduling is most effective when the system is heavily loaded with periodic tasks. Note that the task set in Figure 1 has 75×75 enumerable integer phase vectors. Our previous algorithm[1] identifies the optimal phase vector by inspecting 66 phase configurations. The new algorithm proposed in this paper identifies the appropriate phase vector by applying quadratic programming 12 times.

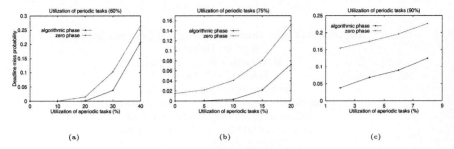

Fig. 2. Deadline miss probability

The algorithm presented in this paper is applicable to other scheduling disciplines and objective performances as well[4]. The phased scheduling shows similar effectiveness for various scheduling disciplines and performance metrics as well.

5 Conclusion

We have discussed a new phased scheduling algorithm which distributes periodic workload uniformly across the time span. The simulation results showed that this 'time-wise' load balancing of periodic workload significantly improves service quality. We are working on a phase identification algorithm with polynomial time complexity. Also, devising more efficient algorithms based on domain specific information such as scheduling discipline and objective performance remains further study.

References

1. J. Koh and H. Shin. Phased scheduling of continuous media tasks to improve quality of service. In *IEEE International Conference on Multimedia Computing and Systems*, pages 108–117, 1998.
2. R. Steinmetz. Analyzing the multimedia operating system. *IEEE Multimedia*, 2(1):68–84, Spring 1995.
3. C. L. Liu and J. W. Layland. Scheduling algorithms for multiprogramming in a hard-real environment. *Journal of the Association for Computing Machinery*, 20(1):46–61, 1973.
4. J. Koh and H. Shin. Distributing periodic workload uniformly across time to achieve better service quality. Technical Report http://cselab.snu.ac.kr/member /alcohol/rts.ps, Computer Engineering Dept., Seoul National University, 1998.
5. Karsten Schwan and Hongyi Zhou. Dynamic scheduling of hard real-time tasks and real-time threads. *IEEE Trans. on Software Engineering*, 18(8):736–748, Aug. 1992.

A Dynamic Fault-Tolerant Mesh Architecture

Jyh-Ming Huang[1] and Ted C. Yang[2]

[1] Department of Information Engineering and Computer Sciences
Feng-Chia University
100, Wen-Hwa Rd., Sea-Tween
Taichung 407, Taiwan
jmhuang@pine.iecs.fcu.edu.tw

[2] Computer & Communication Research Laboratories
Industrial Technology Research Institute
Bldg. 51, 195-11 Sec. 4, Chung Hsing Rd.
Chutung, Hsinchu 310, Taiwan
tedyang@ccl.itri.org.tw

Abstract. A desired mesh architecture, based on connected-cycle modules, is constructed. To enhance the reliability, multiple bus sets and spare nodes are dynamically inserted to construct modular blocks. Two reconfiguration schemes are associated, and can eliminate the spare substitution domino effect. Simulations show that both schemes provide for increase in reliability over the interstitial redundancy scheme[11] and the multi-level fault tolerance mesh(MFTM)[6], at the same redundant spare ratio. Especially, with global reconfiguration, the reliability improvement ratio per spare (RIPS) can be at least twice of that of the MFTM scheme. Furthermore, the lower port complexity in spare nodes as compared to those in both of the aforementioned schemes, and versatility in reconfiguration capability are two additional merits of our proposed architecture.

1 Introduction

In general, two alternative strategies for the design of fault-tolerant multiprocessor systems are usually considered[2]. One is to gracefully degrade the performance of the system[9], the other is the so-called structure fault tolerance. In systems with structure fault tolerance, rigid topology is maintained through redundant spare element replacements caused by the detection of faults[3][9] [10] [11] [12] [13]. There are two extreme approaches of utilizing the spare during the system reconfiguration phase: global[8][13] and local reconfiguration[8].

Many researchers have been investigating into the reconfiguration schemes based on processor arrays[1][7]. Singh proposed an area efficient fault tolerance scheme, named interstitial redundancy[11]. The main attractive features are short PE interconnections and high utilization of failure-free PE's. However, his scheme requires extensively redundant PE's with high number of ports. A multi-level fault-tolerant mesh design was proposed by Hwang[6]. When the redundancy level exceeds three, the design may not be practical because the area required for the interconnection of spare PEs may start dominating the area on the silicon, and the length of the interconnection links after reconfiguration may become objectionable.

A reliable cube-connected cycles structure, RECCC, was presented by Tzeng[12]. In that structure, buses and switches are inserted to serve as assistants for one-dimensional failed PE's reconfiguration, thus spare sharing can not be allowed between different dimensions. In addition, in some cases, when two faults are found within a window, spare substitution domino effect may take place.

To avoid the disadvantages of these schemes, we propose a compromised architecture to implement a dynamic fault tolerant scheme for mesh array. In our approach, four nodes are made into a connected-cycle module, and then multiple bus sets and soft switches are inserted for both inter-module normal connections and intra-module reconfiguration capability. To reduce the length of communication links after reconfiguration, spare nodes are inserted into the central position of a modular block.

2 Description of CCBM Structure

Our model, as shown in Fig. 1(a), consists of an m*n array of identical processing elements. For simplicity, we assume that m and n are all integer multiples of 2. To implement our architecture, we first connect four consecutive nodes in a counterclockwise direction to form a connected cycle, as shown in Fig. 1(b).

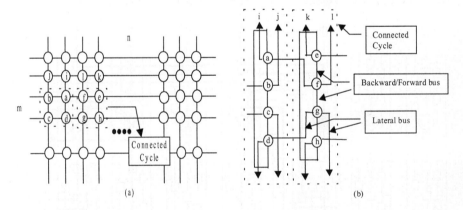

Fig. 1. (a) General model (b) The corresponding connections between two connected cycles

Fig. 2 depicts the detailed compact chip layout of the proposed fault-tolerant connected-cycle-based mesh(FT-CCBM), where spare nodes are denoted by darkened circles, and primary (active) nodes are represented by empty(white) circles filled with two numeric characters. The cb-i-bus, cf-i-bus, rl-i-bus and ll-i-bus indicates the ith cycle-connected backward bus, the ith cycle-connected forward bus, the ith right lateral-connected bus and the ith left lateral-connected bus, respectively. The cycle-connected buses are inserted to provide paths to spare, and the lateral-connected buses provide lateral connections to maintain the rigid FT-CCBM topology after reconfiguration. Switches on buses enable the buses to connect different pairs of nodes by appropriately making or breaking connections between bus segments, and between bus segments and node links. A switch can be set to one of the seven states shown in Fig. 3. To achieve multi-row fault tolerance, and avoid reconfiguration path conflict, multiple bus sets and extra switches located at the intersections of buses, and vertical reconfiguration buses that aside the spare connected cycle are needed.

Fig. 4 briefly shows the FT-CCBM structure of a conventional 2*n mesh with bus sets $i=2$. Each column represents a connected-cycle. Filled circles denote spares. By the number of given bus sets i, we can evenly divide the FT-CCBM into several *modular blocks,* such that each *modular block* consists of $2i^2$ primary nodes plus i spare nodes. *Modular blocks* aligned in a horizontal line form a *group*.

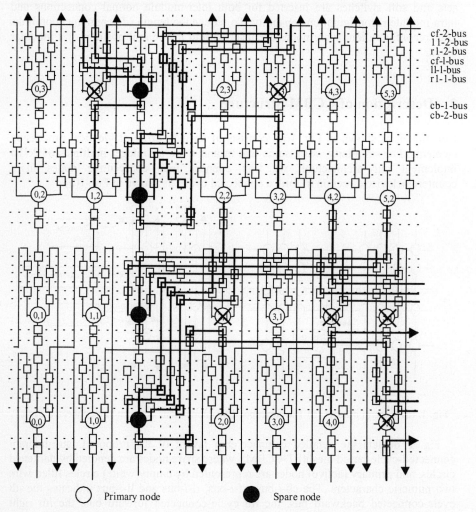

Fig. 2. A FT-CCBM compact chip layout and its reconfiguration scenarios

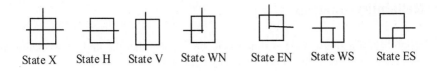

Fig. 3. Connecting states of a switch

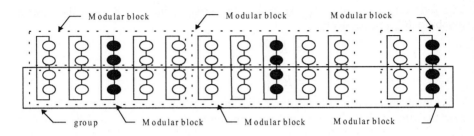

Fig. 4. The structure of 2*n FT-CCBM with i=2

3 Reconfiguration Schemes

Two schemes are proposed for system reconfiguration. In scheme-1, spare nodes can only replace faulty nodes in the same modular block. In scheme-2, in addition to local reconfiguration in scheme-1, partial global reconfiguration is also allowed if more than i nodes in a modular block are faulty.

The top half of Fig. 2 shows an example of scheme-1, with i=2 bus sets available. When a node(e.g. PE(1, 3)) in a modular block becomes faulty, scheme-1 first tries to replace the failed node with the spare node in the same row, by using the first bus set (i.e. cb-1-bus, cf-1-bus, rl-1-bus and ll-1-bus). If this is impossible, for example PE(3, 3), when the spare node in the same row has been used for a previous fault, then the second bus set along with the other row spare nodes are applied.

Scheme-2 is developed for overall system reliability improvement and better spare utilization. To achieve these goals, additional switches (as shown by the bolder boxes in Fig. 2) are needed. In this scheme, local reconfiguration is first performed as aforementioned. If the third faulty node is in the half modular block to the right(left) of the spare column, the available spares of the right(left) neighboring modular block will be borrowed to replace the faulty node. For instance, see the bottom half of Fig. 2, assume that the faults occur in PE(4, 1), PE(5, 0), PE(5, 1), and PE(2, 1) sequence, the faulty PE(4, 1) and PE(5, 0) are processed according to scheme-1, then when PE(5, 1) becomes faulty, because the spare nodes in the same modular block have all been used, the available spare in the left nearby modular block will be borrowed.

4 Reliability Analysis

The reliability of a single node at time t is $R_{pe}=e^{-\lambda t}$, given that the node is workable at time zero. For scheme-1, the block reliability R_{blk} can be expressed as:

$$R_{blk} = \left(\sum_{k=0}^{i} \binom{2i^2+i}{k} R_{pe}^{2i^2+i-k} (1-R_{pe})^k \right) \quad (1)$$

The following equations (2), (3) exploit the group reliability R_{g-1} and the system reliability R_{sys-1}, respectively.

$$R_{g-1} = R_{blk}^{\lfloor n/4i \rfloor} \quad (2)$$

$$R_{sys-1} = R_{g-1}^{\lfloor 2m/i \rfloor} \quad (3)$$

As for the analysis of scheme-2, we take an $i=2$ group as an example, and logically rearrange the modular block boundary as region B_0, B_1, B_2, ... , B_m, and B_r, shown in Fig. 5. We can individually compute the reliability R_{b0} of region B_0, and the reliabilities of regions B_1, B_2, ... , B_m, and B_r, denoted as R_{b1}, R_{b2}, ..., R_{bm}, and R_{br}, respectively. The group reliability R_{g-2} of scheme-2 is the product of all the region reliabilities. After the reliabilities of all groups are computed, the system reliability R_{sys-2}, equation (4), can be evaluated.

$$R_{b0} = \sum_{k=0}^{i} \binom{i^2+i}{k} R_{pe}^{i^2+i-k} (1-R_{pe})^k \quad \text{where } i \text{ is the number of spares.}$$

$R_{g-2} = R_{b0} * R_{b1} * R_{b2} * .... * R_{bm} * R_{br}$, and

$$R_{sys-2} = R_{g-2}^{N_g} \quad \text{where } N_g \text{ is the number of groups.} \quad (4)$$

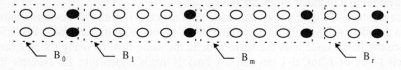

Fig. 5. A logical region view of scheme-2

5 Simulation and Comparison

A failure rate $\lambda_{pe}=0.1$ and some distinct values of the bus sets, $i=2,3,4,...$etc. are used to simulate on many different size FT-CCBM architecture. For simpilcity, however, only the simulation results for a 12*36 FT-CCBM are shown. Fig. 6 shows that the system reliability of scheme-2 is better than that of scheme-1 for the same number of bus sets. In addition, for a given redundancy ratio, maximum reliability can be achieved when the number of bus sets is 3 or 4, depending on whether a complete modular block is formed and whether spare nodes exist in the last region. Since when more bus sets are involved, the block redundant spare ratio will decrease swiftly, the system reliability will decrease if the number of bus sets exceeds 4.

We also made comparisons with the interstitial redundancy scheme[11] and the multi-level fault tolerant mesh(MFTM)[6]. Because the interstitial redundancy scheme performs only local reconfiguration and their redundant spare ratio is 1/4, therefore, we only compare its system reliability with our scheme-1. Results indicate our scheme always offers a much better reliability than the interstitial redundancy scheme.

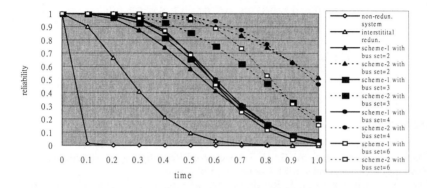

Fig. 6. System reliability of a 12*36 FT-CCBM

As for the MFTM, multi-level fault tolerance capability can be incorporated into the system. In order to have a fair comparison basis, we also evaluate the reliability improvement ratio per spare PE(RIPS) value as the MFTM did:

RIPS=(R_r-R_{non}) / The total number of spare PEs in the system,

where R_r and R_{non} are the reliabilities of the redundant and nonredundant system, respectively.

We compare our scheme-2 for systems with preferred bus sets=4 (denoted as FT-CCBM(2)) with a couple of two-level fault tolerance schemes of the MFTM, the first associates with k_1=1, k_2=1(denoted as MFTM(1,1)), and the second associates with k_1=2, k_2=1(denoted as MFTM(2,1)), where k_i is the number of spares in the ith level. Again, based on Fig. 7, our scheme in most cases, provides at least twice of the RIPS.

6 Conclusions

We have proposed a reliable structure with multi-row fault tolerance capability and spare substitution domino effect free as the two main merits. In addition, fewer ports in a spare node compared to both the interstitial redundancy scheme and the MFTM scheme, and versatile reconfiguration capabilities are two further merits of the proposed architecture.

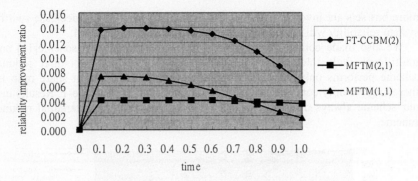

Fig. 7. The RIPS of a 12*36 mesh array with bus-sets=4

References

1. Chean, M., Fortes, Jose A. B.: A Taxonomy of Reconfiguration Techniques for Fault-Tolerant Processor Arrays. IEEE Computers, Vol. 23 (1990) 55-69
2. Chen, Y.Y., Upadhyaya, S. J.: Reliability, Reconfiguration, and Spare Allocation Issues in Binary-Tree Architectures Based on Multiple-Level Redundancy. IEEE Trans. on Computers, Vol. 42, no. 6 (1993) 713-723
3. Chen, Y.Y., Upadhyaya, S. J.: A comprehensive Reconfiguration Scheme for Fault Tolerant VLSI/WSI Array Processors. IEEE Trans. on Computers, Vol. 46, no. 12 (1997) 1363-1371
4. Chung, F.R.K., Leighton, F.T., Rosenberg, A.L.,: Diogenes: A Methodology for Designing Fault-Tolerant VLSI Processor Arrays. FTCS-13 (1983) 26-32
5. Huang, K.: Advanced Computer Architecture: Parallelism, Scalability, Programmability. Intl's Edition, McGraw-Hill (1993)
6. Hwang, I-S.: A High Reliability Two-Level Fault Tolerant Mesh Design and Applications. Journal of Chinese Institute of Engineers, Vol. 19, no. 5 (1996) 607-613
7. Grosspietsch, K.E.: Fault Tolerance in Highly Parallel Hardware Systems. IEEE Micro, 2 (1994) 60-68
8. Kuo, S.Y., Fuchs, W.K.: Reconfigurable Cube-Connected Cycles Architecture. Journal of Parallel And Distributed Computing, 9 (1990) 1-10
9. Raghavendra, C.S., Avizienis, A., Ercegovac, M.D.: Fault Tolerance in Binary Tree Architectures. IEEE Trans. on Computers, Vol. C-33, no. 6 (1984) 568-572
10. Singh, A.D.: A Reconfigurable Modular Fault-Tolerant Binary Tree Architecture. FTCS-17 (1987) 298-304
11. Singh, A.D.: Interstitial Redundancy: An area Efficient Fault Tolerance Scheme for Large Area VLSI Processor Arrays. IEEE Trans. on Computers, Vol. 37, no. 11 (1988) 1398-1410
12. Tzeng, N.F.: A Reliable Cube-Connected Cycles Structure. Journal of Parallel and Distributed Computing, 19 (1993) 64-71.
13. Wang, M., Culter, M., Su, S.: Reconfiguration of VLSI/WSI Mesh Arrays Processors with Two-Level Redundancy. IEEE Trans. on Computers, Vol. C-38, no. 4 (1989) 547-554

Evaluation of a Hybrid Real-Time Bus Scheduling Mechanism for CAN

Mohammad Ali Livani, Jörg Kaiser

University of Ulm, Department of Computer Structures, D-89069 Ulm, Germany
{Mohammad, Kaiser}@informatik.uni-ulm.de

Abstract. As a common resource, the CAN bus has to be shared by all computing nodes. Access to the bus has to be scheduled in a way that distributed computations meet their deadlines in spite of competition for the communication medium. The paper presents an evaluation of a resource scheduling scheme for CAN, which is based on time-slot reservation and dynamic priorities. The processing overhead as well as the schedulability parameters of the hybrid bus scheduling scheme are analyzed and compared with TDMA and the fixed priority assignment.

1 Introduction

The Controller Area Network (CAN) [8] is a field bus widely applied in industrial and automotive distributed real-time systems. One of the central issues in distributed real-time systems is scheduling the access to the communication medium so that distributed computations meet their deadlines. There exist several alternative approaches to solve this problem in CAN, which commonly exploit the priority-based bus arbitration mechanism of the CAN protocol[1].

The deadline-monotonic priority assignment [10] achieves meeting deadlines as ensured by an off-line feasibility test for a static system with periodic and sporadic tasks. By applying the dual priority scheme [4] a more flexible scheduling of hard and soft real-time communication is possible. The fixed priority assignment has been applied in the most common CAN-based communication systems implicitly, e.g. CAL [1], SDS [3], and DeviceNet [7]. The CANopen standard [2] defines a periodic communication scheme which is coordinated by a certain node (the SYNC master). Obviously, the SYNC master constitutes a single point of failure for the whole system. In [12] a combination of fixed and dynamic priority scheduling is approached. But this approach fails to schedule messages in a bus with 3 or more sender nodes due to a too short *time horizon*. The term *time horizon* is explained later in the paper.

In this paper, we first give a short presentation of a hybrid bus scheduling algorithm that combines mechanisms of TDMA and dynamic priority scheduling. A detailed description of the algorithm is given in [6]. In the rest of the paper the performance of the algorithm is evaluated. The evaluation consists of an analysis of

---

[1] Due to this arbitration technique, the identifier of a CAN-message serves as its priority (lower value = higher priority), and the bus itself acts as a priority-based dispatcher. The CAN arbitration mechanism requires that different nodes never start simultaneously sending messages with equal identifiers. This requirement must be satisfied by higher-level protocols.

the hybrid mechanism, and a comparison of the real-time communication schedulability under this scheme, TDMA, and the Deadline-Monotonic scheme.

2 The Hybrid Scheduling Mechanism

The bus scheduling scheme presented in this paper, is a combination of TDMA, dynamic priority, and user-defined fixed priority assignment. Unlike traditional TDMA schemes, our TDMA protocol uses dynamic priorities to ensure exclusive access rights of a sender during a reserved time-slot.

2.1 The Dynamic Priority Scheme

The dynamic priority scheme of the hybrid bus scheduling algorithm implements the Least-Laxity-First Scheme. Since CAN messages are non-preemptive, and their length varies in the same order of magnitude, the LLF scheme achieves a schedulability comparable with the Earliest-Deadline-First (EDF) scheduling, which is known to be optimal for the scheduling of a single resource. We chose LLF instead of EDF for its simpler implementation. In order to realize LLF in CAN, a mapping of the transmission laxity into the message priority has to be defined, such that the message with a shortest laxity gains the highest priority. Since the CAN identifier must provide uniqueness (cf. Footnote 1 on page 1) and information about the message subject, our algorithm uses only the most significant byte of the identifier as message priority.

Having a range $\{P_{min} .. P_{max}\}$ for the priority field, a laxity ΔL is mapped to a priority P, where $P = \lfloor \Delta L / \Delta tp \rfloor + P_{min}$ if $\Delta L < (P_{max}-P_{min})*\Delta tp$ and $P = P_{max}$ if $\Delta L \geq (P_{max}-P_{min})*\Delta tp$. The period Δtp is called the priority slot. Since any laxity value $\Delta L \geq (P_{max}-P_{min})*\Delta tp$ is mapped to P_{max}, the priority-based dispatcher (i.e. the CAN arbitration) cannot distinguish different laxity values greater or equal $(P_{max}-P_{min})*\Delta tp$. We denote $\Delta H = (P_{max}-P_{min})*\Delta tp$ as the *time horizon* of our LLF-scheduler. A correct deadline-based scheduling of messages from $N+1$ senders requires a time horizon greater or equal to the maximum transmission time of N arbitrary messages.

Given a maximum transmission time of ΔT_{max} (in CAN: 151 bits), an inter-frame space of 3 bits, a maximum failure rate of λ_{max}, and a maximum time loss of ΔT_{fail} per failure (in CAN: 168 bits), the time horizon ΔH must satisfy the condition: $\Delta H \geq N*(\Delta T_{max}+3) + \lceil \Delta H * \lambda_{max} \rceil * \Delta T_{fail}$. Thus $\Delta H \geq \dfrac{N*(\Delta T_{max}+3)}{1-\lambda_{max}*\Delta T_{fail}} + \Delta T_{fail}$ is a sufficient time horizon for correct deadline-based scheduling of real-time messages.

2.2 Scheduling Hard Real-time Messages by the Dynamic TDMA Scheme

Due to the high criticality of hard real-time messages, all their occurrences have to be predicted and respective transmission times have to be scheduled in advance. The scheduled transmission times must include the worst-case error handling delays and retransmission times under anticipated fault conditions. In [6] we have shown that every hard real-time message will be transmitted timely under anticipated fault conditions, if following requirements are fulfilled (see also Fig. 1):

(R1) for each hard real-time message (HRTM), an exclusive time-slot is reserved,
(R2) the length ΔT_h of the reserved time-slot of an HRTM h is greater or equal to the worst-case transmission time of h under worst-case failure occurrence,
(R3) the minimum gap ΔG_{min} between consecutive reserved time-slots is equal to the maximum time skew between any two non-faulty clocks in the system,
(R4) the latest ready time of an HRTM is at least ΔT_{fail} before its reserved time slot.
(R5) After its latest ready time, an HRTM has the highest priority in the system. This is satisfied by the condition: $P_{min}^{SRT} > P_{min}^{HRT} + \lfloor (\max\{\Delta T_h\} + \Delta T_{fail})/\Delta tp \rfloor$

The requirements are fulfilled by the hybrid scheduling algorithm presented in [6].

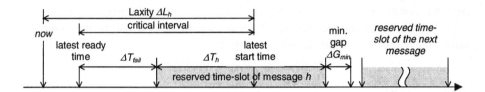

Fig. 1. The Critical Interval of a Hard Real-Time Message h

2.3 Scheduling Soft and Non-Real-Time messages

Soft real time messages have transmission deadlines which are considered by the system but no guarantees are given to meet them. Thus, the transmission resources for soft real-time messages are not reserved in advance. However, the Least-Laxity-First (LLF) scheduling scheme (cf. section 2.1) is used for optimal resource utilization. For non-real-time activities, messages are sent with low (user-defined) priorities. Different priority values for non-real-time messages can be used to express the user defined importance of a message.

3 Performance Analysis of the Algorithm

3.1 Computing overhead

The main additional overhead of the hybrid bus scheduling algorithm results from the necessary periodic modification of the dynamic priority of the ready-to-send message. In our current prototype, this task takes about 13 µ-seconds on a C167 microcontroller [9] at 20MHz. Since this is a periodic task which is triggered at each priority tick, the processor overhead can be calculated as *Overhead = 13µs / Δtp*. For example, in our current prototype Δtp has the same length (500µs) for both soft and hard real-time messages, resulting in a constant processing overhead of 2.6% during the transmission of any kind of real-time message. The processing overhead of the hybrid bus scheduling approach is similar to that of TDMA [5]. Both protocols need a globally synchronized time reference, and perform the correct bus scheduling using periodic tasks. In contrast, the fixed priority approach needs no additional processing effort for message scheduling.

3.2 Evaluation of the Schedulability

Since the dynamic TDMA scheme guarantees the timely transmission of every hard real-time message by reserving a time-slot, the maximum amount of schedulable hard real-time messages depends on the length of the time-slots and the length of the gaps between time-slots, which must be twice the maximum clock inaccuracy in the system. Table 1 contains examples for the maximum schedulability at a bus speed of 1 Mbit/s and assuming a maximum clock inaccuracy of ±20μs. For a message h with b bytes of data, the maximum length of the message including header and bit-stuffing is $L_h = 75 + \lfloor b * 9.6 \rfloor$. Under the assumption of f transmission failures due to bus/controller errors, the required minimum time-slot length is:

$$\Delta T_h = (L_h + 18) * f + L_h + 3 \ bit\text{-}times.$$

Table 1. Maximum Schedulability of HRTM at 1 Mbit/s under Various Fault Conditions and a Maximum Clock Inaccuracy of ±20 μsec

Data and fault model characteristics	Message length	Errors	Time loss due errors	Time-slot length	Schedulable messages
Zero bytes, single bus error	≤75 μs	1	≤93 μs	≥171 μs	≤4739
8 bytes, single bus error	≤151 μs	1	≤169 μs	≥323 μs	≤2754
Zero bytes, severe controller error	≤75 μs	16	≤1488 μs	≥1566 μs	≤622
8 bytes, severe controller error	≤151 μs	16	≤2704 μs	≥2858 μs	≤345
8 bytes, controller + single bus error	≤151 μs	17	≤2873 μs	≥3027 μs	≤326

Soft real-time messages are scheduled according to the Least-Laxity-First (LLF) scheme. Although LLF in CAN performs like EDF (which is an optimal scheduling policy), a soft real-time message will miss its deadline whenever the total waiting time (due to one non-preemptive low-priority transmission, more critical messages, and communication errors) exceeds its laxity.

The schedulability of hard real-time messages in our approach is similar to the schedulability in classic TDMA systems. In both cases, enough transmission time must be reserved to guarantee the successful transmission of the message under maximum number of tolerable consecutive faults. The advantage of our approach is that as soon as a hard real-time message is transmitted successfully, less critical messages may be transmitted in the rest of the reserved time-slot by LLF scheme. Hence, in a system with a mixture of hard and soft real-time communication, the total utilization of the bus bandwidth with the hybrid scheduling is considerably higher than with classic TDMA, especially at low error rates. However, while in classic TDMA systems a message may be transmitted in its reserved time-slot immediately, in our approach every message may be delayed after its ready time for up to ΔT_{max}, because the non-preemptive transmission of a less critical message may be already in progress.

To compare the schedulability of real-time messages with the Deadline-Monotonic scheme, we refer to the results of the analysis made by Tindell et al [11]. Although there is a feasibility test for the DM scheduling scheme, the timeliness of hard real-time communication is not monitored by the scheduling scheme. Hence, under exceptional fault and load conditions, system failures due to late hard real-time messages are possible. In contrast, both classic and dynamic TDMA approaches allow for monitoring the timeliness of the real-time messages, and detecting timing failures.

Another weak-point of the DM scheme is its inability to make difference between hard and soft real-time messages. If the DM scheme assigns a lower priority range to soft real-time messages, unnecessary soft timing failures are possible. If a soft real-time message s has a higher priority than a hard real-time message h, the higher priority of s may cause a late transmission of h, which is a fatal timing failure. This weak-point, however, can be eliminated by applying the dual priority scheduling [4].

References

[1] CiA: "CAN Application Layer (CAL) for Industrial Applications", *CiA Draft Standards 201..207, May 1993.*
[2] CiA: "CANopen Communication Profile for Industrial Systems" *CiA Draft Standard 301, Oct. 1996.*
[3] R.M. Crovella: "SDS: A CAN Protocol for Plant Floor Control", *Proc. 1st Int'l CAN Conference, 1994.*
[4] R. Davis: "Dual Priority Scheduling: A Means of Providing Flexibility in Hard Real-time Systems", *Report No. YCS230, University of York, UK, May 1994.*
[5] H. Kopetz and G. Grünsteidl, "TTP - A Time-Triggered Protocol for Fault-Tolerant Real-Time Systems", *Research Report No. 12/92, Institut für Technische Informatik, Technical University of Vienna, 1992.*
[6] M.A. Livani, J. Kaiser, W. Jia: "Scheduling Hard and Soft Real-Time Communication in Controller Area Network (CAN)", *23rd IFAC/IFIP Workshop on Real Time Programming, Shantou, China, June 1998.*
[7] D. Noonen, S. Siegel, P. Maloney: "DeviceNet Application Protocol", *Proc. 1st Int'l CAN Conference, 1994.*
[8] ROBERT BOSCH GmbH, "CAN Specification Version 2.0", *Sep. 1991.*
[9] Siemens AG: "C167, CMOS,User's Manual", *Published by Semiconductor Group, Siemens AG, 1996.*
[10] K. Tindell and A. Burns: "Guaranteeing Message Latencies on Control Area Network (CAN)", *Proceedings of the 1st International CAN Conference, 1994.*
[11] K. Tindell, A. Burns, A. Wellings: "Analysis of Hard Real-Time Communications", *J. Real-Time Systems 9(2), Sep. 1995.*
[12] K.M. Zuberi and K.G. Shin: "Non-Preemptive Scheduling of messages on Controller Area Network for Real-Time Control Applications", *Proc. Real-Time Techn. and Appl. Symposium, 1995.*

System Support for Migratory Continuous Media Applications in Distributed Real-Time Environments

Tatsuo Nakajima, Mamadou Tadiou Kone and Hiroyuki Aizu

Japan Advanced Institute of Science and Technology
1-1 Asahidai, Tatsunokuchi, Ishikawa, 923-12, JAPAN

Abstract. In this paper, we propose system support for building adaptive migratory continuous media applications in distributed real-time environments. In future distributed computing environments, various objects in homes and offices will embed computers, and various applications will be moved between these computers according to the location of these users. These environments are considered as one of typical future distributed real-time environments. However, since the computers may have dramatically different hardware and software configurations, the application cannot be moved without taking into account the configuration of the computer that the application is migrated. Therefore, migratory applications should be aware of environments that they are executed.

1 Introduction

In future distributed computing environments, users will expect to use their computing environments in any places even if they go out of their offices. In this case, it is convenient that their applications will be moved with them according to their location. For example, their intelligent agents that monitor their behaviors should give advice to them whenever they need helps. The users expect the agents to be executed in computers that exist near them. Traditionally, the scenario can be realized by carrying computers of respective users, and the agents can be executed on their computers.

The approach requires that the computers may execute heavy computation, and always connect to networks for acquiring information from servers. This means that users should carry powerful but heavy computers with them. The alternative approach is that the agents run on the computers near the agents' owners. In future, computers will be embedded in various objects that are placed everywhere, and we want to use these computers for executing applications for helping our work. In the environments, applications will be moved among computers that are embedded in objects existing near their users according to their location. The applications are called *migratory applications*[BC96, Car95].

These computing environments are considered as one of typical distributed real-time environments, and many techniques developed for distributed real-time environments such as real-time scheduling and synchronization can be used for building migratory applications. However, since computers may have dramatically different hardware and software configurations, these applications cannot be moved without taking into account the configurations of the computers that the applications are migrated. Therefore, migratory applications should be aware of environments in which the applications are executed, and should be adapted to respective computing environments to which the applications are moved[HKN96]. Also, these applications may need to process various types of media data such as video and audio for monitoring surrounding environments of the applications' users, or showing video and audio information that is interested by the users in a timely fashion. Therefore, environment-ware migratory continuous media applications are one of the most important classes of migratory applications in future computing environments.

In this paper, we present a system support for building environment-ware migratory continuous media applications. The work described in this paper is one of case studies showing what is required for supporting migratory applications. The experiences with building the system support clarify problems and requirements for realizing migratory applications in large scaled heterogeneous distributed real-time environments.

2 Issues for Supporting Environment-Aware Migratory Continuous Media Applications

For building environment-aware migratory continuous media applications, the following three issues should be taken into account.

- How do applications recognize computing environments in which the applications run, and how do they notify their changes to the applications ?
- How do continuous media applications should be adapted to respective computing environments ?
- How are applications migrated to other computers in the middle of their executions ?

Our system support consists of three components as shown in Figure 1, which answers the above respective issues, and it makes it possible to build environment-aware migratory continuous media applications in a systematic fashion.

The first component that answers the first issue is an environment server that monitors computing environments, and it notifies the changes in the environments to applications. The environment server provides primitives for accessing information about computing environments. Traditionally, respective information about computing environments requires to be accessed by different primitives. This makes the structures of environment-aware applications complicated. On the other hand, the uniform interface provided by the environment server enables us to build environment-aware applications in a systematic fashion.

The second component that answers the second issue is a continuous media toolkit that enables us to build continuous media applications in a highly configurable way. By using the toolkit, continuous media applications are constructed by composing several modules. The approach has two advantages. The first advantage is that the toolkit enables us to build continuous media applications with a little programming. The second approach is that the toolkit enables applications to change their configurations according to respective computing environments. The advantage makes it possible to build environment-aware continuous media applications systematically.

The last component that answers the third issue is the migration manager. The migration manager that makes continuous media application migratory has two functions. In the first function, the migration manager manages the reconstruction of an application on a computer that the application will be moved. In

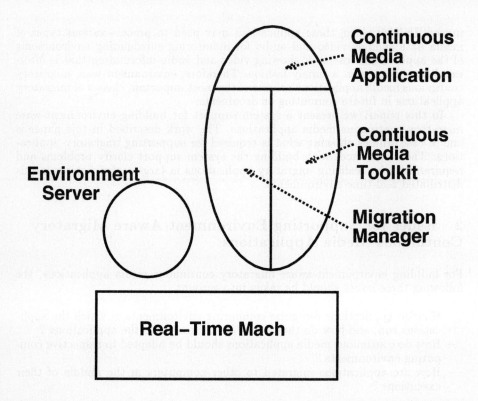

Fig. 1. Structure of Environment-Aware Migratory Continuous Media Application

the second function, it saves the states of applications, and restores the states in applications that are reconstructed on a new computer. In our approach, the amount of states of an application can be changed by taking into account the tradeoff between the time for migrating the application and the bandwidth of a network connected between two computers.

The above three components are currently implemented on Real-Time Mach[TNR90, NKAT93]. The environment server is implemented as a server, and the continuous media toolkit and the migration manager are implemented as libraries that are linked with applications.

3 Environment Server

The *environment server*[NAKS98] is important as a basic infrastructure for building environment-aware applications. In traditional operating systems, information about computing environments is managed in a very ad-hoc way, and applications require to access the information using different primitives for respective information. The approach makes the development of applications that are aware of various differences in computing environments very complicated. The *environment server* contains a database that manages environmental information in a uniform way. The interface of the environment server is well defined, and an application can access environmental information by using its uniform interface. This makes it possible to build environment-aware applications with a systematic framework.

4 Toolkit Support for Building Environment Aware Continuous Media Applications

In our toolkit, continuous media applications are defined as compositions of several modules. This makes it easy to build continuous media applications significantly. Also, the approach enables applications to change their configurations in the middle of the executions. The characteristic makes us to build environment-aware continuous media applications that can adapt to respective computing environments according to the configurations of computers on which the applications run.

Also, our toolkit hides many complicated programming for building continuous media applications such as real-time programming, media synchronization, and dynamic QOS control from programmers for making it easy to build complex continuous media applications[Nak97].

5 Migration Manager

The scripting language that is used for writing environment-aware continuous media applications is an extension of the TCL scripting language. Programmers can write continuous media applications using the script languages since the scripting language provides several primitives for calling the functions of our continuous media toolkit.

5.1 Protocols for Migrating Application

When a continuous media application detects that it should be migrated to other computers, it needs to negotiate to a computer that the application will be migrated. Let us assume that the computer executing the application currently is *computer A* and a new computer that will execute the application after the migration is *computer B*.

First, *computer A* sends a request to *computer B* for creating an application on *computer B*. In the current implementation, our continuous media application uses the same binary. Thus, *computer B* runs the binary when the request is received. We assume that the binary is created by compiling for respective machine architectures, and it is stored in respective computers that our migratory applications may be migrated.

After creating the application on *computer B*, the newly created application sends a request for acquiring a script from the application executed on *computer A*. Then, the application sends the script to *computer B*. After receiving the script program, the application running on *computer B* configures modules and streams according to the script program. Also, *Initialize* procedure in the script language is executed. The procedure includes necessary initializations for the application on *computer B* such as the registration of conditional statements to the environment server.

Next, the application running on *computer B* sends a request for saving the checkpoint of every module of the application running on *computer A* and transferring the checkpoint. On *computer A*, the checkpoint of an application is taken,

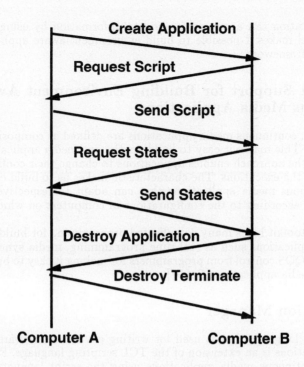

Fig. 2. Application Migration Protocol

and the checkpoint is transformed to the standard format by calling the *Checkpoint* procedure in the script. Then, the checkpoint is transfered to *computer B*. The checkpoint is transformed from the standard format according to the current configuration of the application on *computer B*, and restored in the application's modules by calling the *Restore* procedure in the script. Also, the states for respective streams and GUI are transfered and restored in the application on *computer B*. Lastly, the application running on *computer B* sends a destroy request to the application running on *computer A*. Then, the application on *computer A* executes *Finalize procedure* in the script for unregistering conditional statements in the environment server, and the application is destroyed on *computer A*. Finally, the application on *computer B* starts to be executed after the execution of the destroy request is terminated.

5.2 Reconfiguring Application

In the previous section, we describe how an application is migrated to other computers during the execution. This section describes how the application is adapted according to the configuration of computers that the application is migrated.

Figure 3 shows an example of a continuous media application that captures a video stream from a video camera, converts the video stream, and displays it to a display. Let us assume that the application is migrated between two computer A

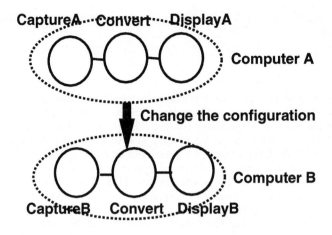

Fig. 3. Reconfiguration of Application

and B. Computer A has a high resolution video capture, and a window system. When the application is executed on computer A, it connects the capture A module that controls a high resolution video capture board with the convert module that converts a video stream. Also, the convert module is connected with the display A module that draws a video stream to a window. Now, let us assume that the application is migrated to computer B. Since computer B has a low resolution video capture card and a small display, the capture A module and the Display A module cannot be used on the computer B. Therefore, the application changes the configuration by replacing the connections as described in Figure 3 for taking into account the hardware configuration of computer B. After the computer B receives a script program, the application is reconstruct on the computer. Then, the display A module is replaced by the display B module, and the capture A module is replaced by the capture B module. On the other hand, if the application is migrated to computer A again, the configuration of the application is recovered to the original configuration.

In the migration manager, a script program manages the reconfiguration of an application when it is migrated to other computers. When the application is migrated to the computer whose configuration is drastically different from the original computer, the application receives a notification from the environment server, and the script program accesses the environment server for getting the configuration information of the new computer. In the example, the script program acquires information about a capture card and a display device of a new computer from the environment server, and the configuration of the modules is changed by executing the script program.

6 Agent Support

Migratory continuous media applications presented in this paper can be considered as mobile agents. In fact, our system provides several supports for implementing mobile agents. This section describes a brief overview of our mobile agent support.

6.1 Basic Model

In the design of the proposed architecture, four entities come into play: a virtual host, a virtual host interface, a place and a resource manager. A place, located inside a virtual host, is responsible for providing the adequate computation resources for the execution of an agent and grants it the desired QOS. A place confines the actions of the agent to a restricted environment for portability and security reasons. Each place is configured specially by an interface according to the requests submitted by the mobile agent. This interface is the *virtual host interface* which is also in charge of translating the subjective user defined QOS parameters into system compliant parameters. The virtual host interface receives mobile agents and translates their data (subjective QOS) into system defined QOS. With this data, it configures an adequate place inside a virtual host. Then, the mobile agent enters that place and uses the services provided there.

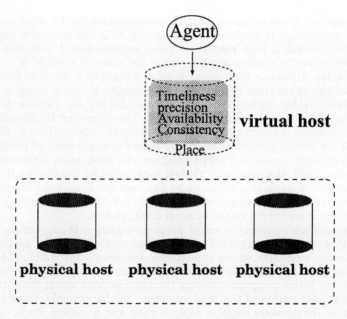

Fig. 4. A Virtual Host Spanning Multiple Machines

The virtual host interface separates the responsibilities of QOS negotiation and resources management. That is to say: the mobile agent system takes care of QOS negotiation at the user level while the virtual host deals with resource management at the machine level. The whole system acts more like a split-QOS server. In fact, we propose in this paper a protocol for QOS control that relies on the movements of mobile agents inside and between places. In order to satisfy its owner's QOS request, the mobile agent in our system implements an effective policy of QOS control inside places created on demand.

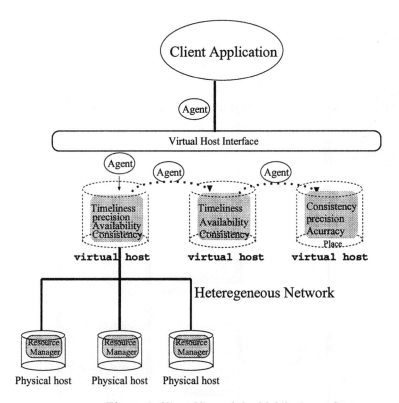

Fig. 5. A Client View of the Mobile Agent System

For example, when an application with strict time constraints like a video application makes a timeliness request together with accuracy, precision and consistency; the system gives precedence to the timeliness QOS and assigns it a high priority. In addition, if each of these QOS type is expressed in terms of "hard" or "soft" guaranty, then some tradeoff becomes necessary and the agent might visit several virtual hosts to achieve the best possible result.

6.2 QOS Assumption

In our system, we assume that a user targets a fixed number of QOS: timeliness for systems with time constraints, accuracy, precision, availability and consistency. We assume also that a number of host machines scattered across the network have the necessary resources to satisfy the user needs. Each place inside a virtual host in conjunction with the adequate resource managers located inside the host machine provides several types of QOS but not necessarily the entire set of QOS. In addition, in our context we assume that a resource reservation scheme exists at the level of each physical host machine. For matter of simplicity, here we do not deal with the case of multiple applications competing for the same resources.

6.3 Agents Movements

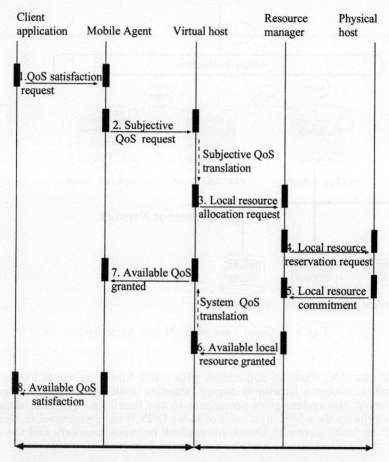

Fig. 6. The QOS Negotiation Process

Multiple QOS requests from users expressed as timeliness, accuracy, precision, availability and consistency may need several resources in order to be satisfied. In this case, the mobile agent in charge migrate with these requests to an appropriate virtual host. Before any operation, the subjective QOS is translated by the virtual host interface into system defined QOS. With this data, the interface configures a suitable place within the virtual host. Each virtual host spans, over the network, a number of resource managers and host machines that contain the needed resources. When a place is built, the mobile agent freely enters it. Then, it tries to secure the available QOS and migrate - if necessary - to other places to satisfy the complete set of requested QOS. In particular, when a strict time constraint is set by the client application, the mobile agent must assign a high priority to the timeliness parameter and lower priority to other QOS. Assuming that the execution environment is reliable, the agent migrates throughout the entire network, from virtual host to virtual host in order to satisfy all QOS requests.

6.4 Inter-Agents Communication

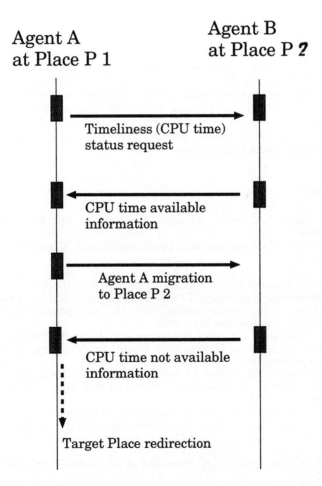

Fig. 7. Inter-Agents Communication

As a lot of agents migrate across the network on behalf of their owner to satisfy different QOS requests, they need to cooperate for optimization purposes. Agents communicate with one another by passing messages about new environmental conditions at their respective place. This way, an agent can have an impact on the computation of other agents. For example, when an agent, on behalf of the real time system user, plans to migrate to a place where timeliness (CPU time) used to be available, another agent staying at the targeted place knows that a failure of the local resource makes this QOS no longer satisfiable. So the latter agent, upon request, sends a message to the first agent which then changes its path.Also, the level of QOS available locally may not be sufficient and the agent in need looks elsewhere for better service. The inter-agent communication is achieved through asynchronous messages rather than Remote Procedure calls (RPC) for flexibility purposes. In this scheme, an agent calls a communication primitive and supplies the identity of the receiving agent and the message as arguments.

7 Conclusion

In this paper, we described system support for supporting environment-aware migratory continuous media applications, and presented three components that we have been developed. The first component is the environment server. The second component is the continuous media toolkit. The last component is the migration manager. The components enable us to build environment-aware migratory applications in a systematic way. We implemented a prototype version of the components on Real-Time Mach, and build an example program of environment-aware continuous media applications.

References

[BC96] K.A. Bharat, L. Cardelli, "Migratory Applications", SRC Research 138 Report, Digital, 1996.
[Car95] L. Cardelli, "A Language with Distributed Scope", Computer Systems, 8(1), MIT Press, 1995.
[HCK95] C. Harrison, D. Chess, and A. Kershenbaum, "Mobile Agent: Are they a good idea ?", Research Report RC19887, IBM Research Division, 1995.
[HKN96] A.Hokimoto, K.Kurihara, T.Nakajima, "An Approach for Constructing Mobile Applications using Service Proxies", The 16th International Conference on Distributed Computing Systems, May, 1996.
[NKAT93] T.Nakajima,T.Kitayama,H.Arakawa,and H.Tokuda, "Integrated Management of Priority Inversion in Real-Time Mach", *In Proceedings of the Real-Time System Symposium*, 1993
[Nak97] T. Nakajima, "A Toolkit for building Continuous Media Applications", In Proceedings of the 4th International Workshop on Real-Time Computing, Systems, and Applications, 1997.
[NAKS98] T. Nakajima, H. Aizu, M. Kobayashi, K. Shimamoto "Environment Server: A System Support for Adaptive Distributed Applications", In Proceedings of the International Conference on World Wide Computing and its Applications'98, 1998.
[TNR90] H. Tokuda, T. Nakajima, and P. Rao, "Real-Time Mach: Towards a Predictable Real-Time System", *In Proceeding of the USENIX 1st Mach Symposium*, October, 1990.
[Whi94] J.E. White, "Telescript Technology: The Foundation for the Electric Marketplace", General Magic Inc., 1994.

[WRB97] K.R. Wood, T. Richardson, F. Bennett, A. Harter, A. Hopper. "Global Teleporting with Java: Toward Ubiquitous Personalized Computing", *Computer* February 1997, Volume 30, Number 2, IEEE.

Dynamic Application Structuring on Heterogeneous, Distributed Systems[1]

Saurav Chatterjee

SRI International, Menlo Park, Ca 94025

saurav@erg.sri.com

http://www.erg.sri.com/projects/erdos

Abstract. The diversity of computers and networks within a distributed system makes these systems highly heterogeneous. System heterogeneity complicates the design of static applications that must meet quality-of-service (QoS) requirements. As part of the ERDoS[2] project, we have introduced a novel concept of dynamic application structuring where the system, at run time, chooses the best end-to-end implementation of an application, based on the system and resource attributes. Not only does this approach improve resource utilization and increase the total benefit to the user over that provided by the current static approaches, but it also is transparent to users and simplifies application development. We describe the models and mechanisms necessary for dynamic application structuring and use a set of multimedia applications to illustrate dynamic application structuring.

1 Introduction

Systems are becoming more and more heterogeneous. As the general use of faster and faster desktop computers increases, so does that of the relatively slower laptops and handheld machines. Similarly, the increased use of faster and faster networks (such as ATM[3] and gigabit Ethernets) is accompanied by the increased use of the much slower wireless networks and modems.

Designing quality-of-service (QoS)–based applications for this type of heterogeneous environment is difficult. Static applications that attempt to provide the same level of QoS irrespective of the attributes of the underlying system, are unsatisfactory because applications whose design assumes high-performance resource attributes are guaranteed to miss QoS on lower-performance resources. Applications whose design assumes a lowest common denominator can provide only minimal levels of QoS to everyone. It is impractical to design multiple versions of an application, each tailored for a particular system, because such applications can be used only by sophisticated users who have enough expertise to pick the best version. Second, the constant introduction of new and better resources requires continuous upgrades to such an application, which in turn requires a major development effort.

The SRI ERDoS[4] project has developed an innovative approach to this problem—at run time, the middleware dynamically creates customized applications and keep user's QoS requirements and the system and resource attributes. We have developed

---

[1] The work discussed in this paper is funded by DARPA/ITO under the Quorum program, Contract N66001-97-C-8525, and was performed by SRI International (SRI).
[2] ERDoS: End-to-End Resource Management of Distributed Systems.
[3] ATM: Asynchronous Transfer Mode.
[4] ERDoS: End-to-End Resource Management of Distributed Systems.

an application model that captures multiple ways to structure an end-to-end application; at run time, based on the current system and resource state, the ERDoS middleware chooses the structure that best meets the user's QoS requirements. This process is transparent to the user, enabling these applications to be run by almost anyone. ERDoS also dynamically creates distributed applications on the fly at run time, so that the development effort to create distributed applications is also minimized.

Our approach increases the probability of achieving a high level of user QoS in highly heterogeneous systems. Whereas a static application may fail because a particular resource is overloaded, our dynamic approach can find alternative application structures that use other resources but still provide the same end-to-end functionality and QoS. Furthermore, static approaches may reject an application if all system resources are highly utilized. Instead, our dynamic approach can utilize alternative application structures that provide partial QoS (which is better than no QoS).

For example, a handheld device may not have the computing power to decompress MPEG[5] video in real time. A static application that uses MPEG compression will not be able to run on this device; a sophisticated user may, at best, be able to request JPEG[6] compression instead of MPEG. ERDoS can take a even more sophisticated approach, one that is completely transparent to the user. ERDoS can autonomously utilize a proxy machine close to the handheld device to convert MPEG to JPEG, so that the data between the video source and the proxy can be MPEG compressed, while the data between the proxy and handheld device is JPEG compressed. Thus, ERDoS achieves the same level of QoS as a sophisticated user switching from MPEG to JPEG, but also achieves better network resource utilization, thereby enabling other applications to potentially achieve higher levels of QoS.

Another example illustrates graceful QoS degradation via dynamic application structuring. Suppose audio-video data is sent from a source to several users, one of whom is connected via a slow modem line. Attempting to transfer both audio and video over the modem line guarantees that timing requirements will be missed. Instead, ERDoS can dynamically choose an application structure that will filter out the video and transmit only the audio over the modem line.

The remainder of this paper describes dynamic application structuring in ERDoS. Section 2 provides our definition of QoS. Section 3 describes our models to capture the QoS behavior and requirements of applications and the system. Section 4 describes how our architecture aids in dynamic application structuring. In Sections 5 and 6 we describe how the applications are developed, instantiated, and executed within the ERDoS middleware. We summarize our work in Section 7.

---

[5] MPEG: Motion Picture Experts Group.
[6] JPEG: Joint Photographic Experts Group.

2 Definition of QoS

We define QoS via six orthogonal dimensions (*orthogonal* in the sense that a set of these dimensions can be used to capture QoS requirements). The six dimensions are shown in Figure 1. For any data, the QoS dimensions capture the following

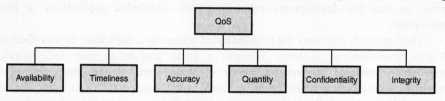

Fig. 1. ERDoS QoS Dimensions

information: what is the data (the *quantity* and *accuracy* QoS dimensions); when must it be available (*timeliness*); who has read (*confidentiality*) and write access (*integrity*); and during what percentage of time, within a time window, must these requirements be met (*availability*).

For example, a video application may demand a frame size of 640 × 480 pixels (quantity); a frame rate of 25 frames/s (timeliness); and jitter of 30 ms (timeliness).

3 Modeling a Distributed System

The determination of the best application structure requires an understanding of the trade-offs between different application structures, their resource usage, and the QoS they can achieve. This information is captured by means of our application, resource, and system models and benefit functions. The application model captures the multiple ways to structure an end-to-end application. The benefit function captures the relative importance of the different QoS parameters, from the user's perspective. The resource and system models capture system and resource attributes. Each is described below in detail.

3.1 Modeling Resources

Each Resource Model (RM) captures the QoS attributes and scheduling behavior of a single computing, communication, or storage resource [1]. We recently augmented our RM to include security attributes: specifically, how long it would take, in CPU cycles, for an intruder to break the encryption scheme or to modify secure data. For example, these attributes can be used to capture the security characteristics of a virtual private network (VPN). A VPN that provides DES encryption at the network level can specify confidentiality strength on the order of 2^{56} or 2^{168} for triple-DES encryption.

The System Model (SM) captures the set of resources constituting the system, the topological and administrative structure of the system, and policies and objectives governing each of these administrative domains [1]. Each system is modeled as a set of subsystems, similar to Internet domains. Each subsystem also has security metrics that specify the overall confidentiality and integrity strength within that subsystem.

3.2 Modeling Applications

ERDoS uses the Logical Application Stream Model (LASM) [1] to store the structure of an end-to-end application. At its simplest, a LASM is represented as a directed graph whose nodes correspond to Units of Work (UoWs) and whose edges dictate the order in which the UoWs execute. A UoW is an atomic module of work, on a single resource, that transforms the application's QoS. This decomposition of an application into UoWs that transform data and QoS is a fundamental concept of our application model. UoWs are the smallest components of applications that are used for allocation and routing. The LASM captures application structure at a *logical*, or system- and user-QoS-independent level.

Figure 2 illustrates a simple video-playback application modeled via the LASM. The boxes correspond to UoWs. Because the LASM is logical, there is no need to

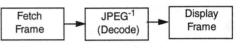

Fig. 2. Video LASM

explicitly identify I/O resources. This simple LASM facilitates portability and enables reallocation and rescheduling as the system state changes. We next describe an improvement upon this model, a recursive LASM that additionally enables ERDoS to dynamically structure applications. The recursive model is also ideal for modeling large-scale applications.

The structure of a recursive LASM is also represented as a directed graph, but the graph nodes now correspond to Logical Services (LSs). Graph edges now dictate the order in which the LSs execute. Each LS is realized by means of a Logical Realization of Service (LRoS), which is composed of either a set of child LS, executing in a specific order, or a single LUoW. LUoWs, as before, are not decomposable. The structure of an LRoS is identical to that of an LASM, in that it is represented as a directed graph, whose nodes also correspond to LSs. This model is recursive in that an LS is realized by an LRoS, which in turn is composed of a set of child LS, which in turn may each be realized by an LRoS. The recursion terminates when an LS is realized by a LUoW.

An example, a recursive video-playback LASM, is shown in Figure 3. In contrast to the example in Figure 2, the application developer does not need to specify a particular type of video decompression method at the top layer (Figure 3a); instead, the developer indicates the need for a "OutputVideoData" logical service.

Assume, for this example, that ERDoS finds five LRoSs for "OutputVideoData" (Figure 3b). The first three use video compression to compress the video; the last two filter out the video and only transmit audio. These last two are best when network bandwidth constraints make video transmission impractical. The first and the fourth provide no security, while the others do. The first is preferable within a secure intranet; the third is needed when the client is a simple appliance computer without enough computing power to decompress MPEG in real time; and the second is preferable under most other circumstances. With the third LRoS, ERDoS can utilize a proxy machine close to the client to convert MPEG to JPEG, so that the client can then decompress JPEG format in real time.

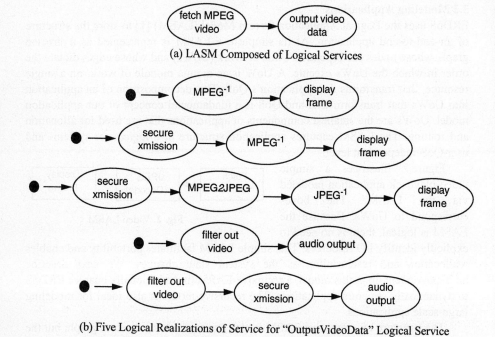

(a) LASM Composed of Logical Services

(b) Five Logical Realizations of Service for "OutputVideoData" Logical Service

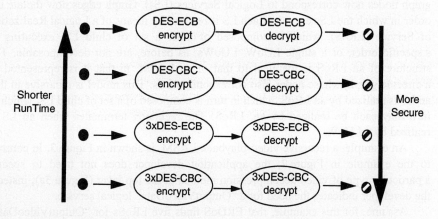

(c) Four Logical Realizations of Service for "Secure XMission" Logical Service

● Input Data Symbol
⊘ Output Data Symbol
◯ :Logical Service

Fig. 3. Recursive LASM for Video

Each "secure xmission" LS is furthermore realized by four LRoSs (Figure 3c). In this example, we use DES and triple-DES to encrypt and decrypt data. According to Schneier [2], these algorithms can be implemented with either CBC or ECB blocks. (CBC is approximately 15% slower than ECB but is more secure.)

This example illustrates how the LASM provides ERDoS the flexibility to structure applications in accordance with system state. Thus, the LASM specifies different ways to structure an application and ERDoS chooses the best combination at run time, based on a variety of factors including end-to-end QoS requirements and the state of the system.

Each UoW has multiple attributes [1], but a key attribute is its Resource Demand Model (RDM). The RDM specifies the resources required (e.g., CPU cycles, network bandwidth) to achieve different levels of QoS for the current UoW. This attribute is a key factor in determining which LRoSs to use.

3.3 Benefit Function

Each application can be structured in multiple ways. While many structures will achieve the same end-to-end QoS, some will not. We use benefit functions to specify relative importance to the user of the different QoS parameters for the application. ERDoS then uses this information is used to decide which structure will maximize QoS under the given resource constraints.

The benefit function is a n-dimensional function, where n is the number of QoS metrics of relevance to a specific application. The benefit function encodes the benefit the user receives as a function of QoS along the n dimensions.

The benefit function for an audio-video application typically has at least eight QoS metrics: frame jitter, frame size, frame rate, image clarity, audio quality, confidentiality, and integrity. While the QoS parameters are application specific, benefit is application content specific.

For example, consider two streaming video applications, one for a movie and another for a video conference between two staff members of a company discussing highly proprietary information. Clearly, confidentiality is paramount for the latter but not important for the former. On the other hand, a switch from audio-video to audio only minimally reduces benefit for the video conference, whereas the switch would greatly reduce benefit for the movie. This and other relative-importance relationships are captured using the benefit functions.

4 ERDoS Architecture

ERDoS does not store prelinked distributed applications. Instead, it dynamically creates distributed applications at instantiation time. It requires only that application developers create single-resource UoWs, which are strung together to create each distributed application. This process is completely autonomous and is transparent to users. In this section, we describe how the ERDoS architecture lends itself to the process.

ERDoS is a distributed resource management middleware composed of Resource Agents (RAs), Application Agents (AAs) and a System Manager (SM), as shown in Figure 4. Most resource management decisions are made at the SM, which is thus the central component of ERDoS. The SM also

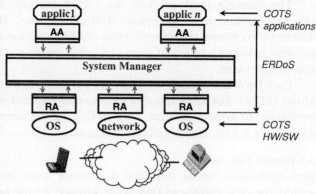

Fig. 4. ERDoS QoS Architecture

acts as a coordinator between the activities of the local resource agents and applications. The RAs (one per resource) and AAs (one per application) carry out the SM's commands and send it status updates. Each will be described in the following subsections.

Each AA is a wrapper around the application; it adds QoS monitoring and adaptation capabilities. The SM is a central entity that tracks the global picture of the system resources and applications. It controls resources via the RAs. A centralized system manager is inherently not scalable. We address this problem by using a hierarchical architecture for a distributed SM [3].

4.1 System Manager

The SM uses our adaptive resource management algorithms to efficiently manage the shared resources in the distributed system. Given end-to-end application specifications, the SM determines which resources to allocate to each application [4] and how to schedule all applications on shared resources [5]. It then conveys this information to the resource agents. The SM also degrades application QoS when needed. It determines which applications to degrade and along which QoS dimensions [6], and conveys this information to the application agents and the resource agents.

The SM coordinates the activities of all the applications in the distributed systems. It starts by deciding if an application can be admitted to the system, then instantiates the different UoWs that constitute the application at the different resources. The run time coordination between the different UoWs is also the SM's responsibility.

In large systems, a system can decompose into a set of subsystems, and the ERDoS SM can be decomposed into a set of subsystem managers. Each subsystem manger then manages all the applications that only utilize resources within that subsystem. Applications that span across multiple subsystems are handled by higher-level (parent) subsystem managers.

4.2 Resource Agents

The SM manages the shared resources of the distributed system by coordinating the activities of the local resource agents. RAs execute the SM's commands, monitor the QoS of each work module (the UoWs) on that resource, and notify the SM whenever a UoW misses its QoS. This architecture allows the RA to be resource specific, and shields the SM from the peculiarities of the heterogeneous resources.

Each RA coordinates the different UoWs executing on a resource and provides the control interface for the UoWs. When a new application arrives at the system, the SM instructs the RAs to instantiate a new application and a set of UoWs. All future control commands to the UoW from the SM are communicated via the RA. The RA provides the infrastructure for the inter-UoW communications, and the communication between the RA and the UoW. The UoW does not directly communicate with the SM. Also, the RAs do not communicate with each other. All coordination between RAs is managed via the system manager.[7]

Data flow between UoWs on the same resource are handled via a RAM disk. From a UoW's perspective, RAM disks behave like normal disks and export the standard file API: i.e., UoWs access them via standard methods available in the java.io classes such as *readLine()* or C/C++ methods such as *read()*. However, instead of storing the content on a physical disk, RAM disks store their content in memory, enabling very quick reading and writing. Our RAM disk code is embedded inside the RA. We chose RAM disks for I/O between UoWs because most off-the-shelf software expect files for input and output. Using RAM disks, these software modules can be made into UoWs without expensive and time-consuming reimplementation.

Data flow between UoWs on different machines is slightly more complex than the communication within a single machine. We handle inter-resource data flow by introducing two UoWs, one on the source side that sends data via TCP (or RTP over UDP) and the second on the receiver side that receives data. These UoWs, not visible at the application user level, are dynamically added by RAs as needed.

4.3 Application Agents

The SM coordinates the end-to-end application via the AAs. AAs translate application-specific, end-to-end QoS requirements into common, end-to-end QoS metrics [7] that ERDoS understands. For example, a video application's application-specific QoS parameters such as video frame size and clarity are translated into common volume and accuracy QoS parameters by the video application's agent. The use of AAs enables new applications, with application-specific QoS parameters, to execute on ERDoS without the need to modify the ERDoS SM.

An AA's functionality is distributed over each UoW that the application comprises. We implement these functionalities as application wrappers, where each application wrapper is a Java class or a C library. The individual UoW wrapper interfaces with the RAs. The AAs also allow the SM to dynamically reconfigure the application as the resource status changes.

---

[7] The majority of data traffic is between UoWs; therefore, the SM sets up individual data channels between these UoWs. Inter-RA and RA-SM communication is relatively rare in our architecture; hence, a central SM does not constitute a bottleneck.

5 Application Development and Instantiation in ERDoS

In this section, we describe the ERDoS development environment and our GUI-based tools, which easily store UoWs and specify LASMs, LRoSes, RMs, and SMs.

5.1 Application Developer's Environment

UoWs are Java or C/C++ methods that are written by programmers. To create a distributed application, the programmer must also specify all the UoWs that constitute a distributed application and must provide communication channels to transmit data between UoWs. A key limitation is that the programmer must a priori identify the resources on which each UoW will reside. Many distributed applications, such as LBL's *vic* [8], sidestep this problem by asking the application *user* to run the source and sink modules separately and to then specify the IP address and port number. This approach assumes that the user is sophisticated enough to supply this information and, even if this assumption is correct, is practical only when applications are simple (e.g., two UoWs on two machines). This approach clearly is not scalable when a complex, distributed application contains UoWs running on many machines.

The ERDoS development environment solves these problems. It requires programmers to create UoWs only by using standard C/C++ or Java development tools and to link with our application wrappers. The programmer then specifies the application structure in a *logical*, or system-independent, manner, using the System Development Tool (SDT) GUI. The SDT is based on our earlier System Engineering Workbench (SEW) [9], which was developed while the author was at Carnegie Mellon University. The ERDoS run-time environment, described in Section 6, then determines the structure of the application and the resource on which each UoW will execute, and creates communication channels to convey data between UoWs. This process is transparent to application users, so that almost everyone can use it. It is also transparent to application developers, significantly reducing development time.

The SDT has three GUI components—applications, resources and the system, and benefit functions. The GUI enables users to specify how *a-priori*-created UoWs are strung together to create end-to-end applications. The GUI enables *nonprogrammers* to create these end-to-end applications, because ERDoS does the actual interfacing between UoWs; the user has only to specify the structure of the application and the end-to-end QoS requirements of interest to that application. These applications are then placed in the Application Repository. This GUI can also be used to enter benefit function information, but that capability is not shown. The Resource Development GUI similarly is used to capture system state, including topology and resource attributes.

5.2 ERDoS Instantiation Environment

The application developer creates the component UoWs of the application and the LASM describing the application structure. This information is stored in the application repository for use by the users. The Instantiation GUI enables a user who wishes to run an application to choose it from the Application Repository. The user can also choose the content to be associated with the application. The SM then indexes into the Application Repository, finds the LASM for that application, and dynamically strings together UoWs to create an end-to-end application. QoS requirements are found from the BF associated with the content chosen by the user. The exact set of UoWs used to create the application is chosen by the application structuring algorithm. The SM then allocates resources to this application and schedules the application on these resources.

6 ERDoS Run-Time Environment

The ERDoS SM provides full resource management functionality, including allocation and scheduling, adaptation, and application structuring. After an application has been requested by the user, the SM instantiates it by invoking the appropriate UoWs at the different RAs. During run time, the SM coordinates the activities of the RAs and monitors the state of the system and the applications via the monitors inside the RAs. It also adapts to the changing system state and instructs the RAs to take corrective actions when faults occur. In the following discussion we describe the functions performed by the ERDoS system during the application's run time.

Figure 5 shows the ERDoS run-time infrastructure when a person in office A (at *desktop.officeA.com*) instantiates a audio-video conference with three other participants. The person at office B (at desktop.oficeB.com) is connected over the public Internet. The person at home.officeA.com is connected over a slow modem directly to his or her office machine. The person at laptop.officeC.com is connected over a virtual private network (VPN). All machines except the laptop are powerful desktop machines.

Based on the resource constraints and the application structure, the ERDoS SM determines that the network connection between desktop.officeA.com and desktop.officeB.com is insecure, and decides to use the LRoS that provides strong triple-DES-CBC encryption. Consequently, it instantiates seven UoWs: "fetch frame," "3xDES-CBC encrypt frame," "transmit frame over network via TCP," "receive frame over network via TCP," "3xDES-CBC decrypt frame," "decompress frame," and "display frame." Using this application structure, the SM instructs the appropriate RAs to instantiate the seven UoWs. The first three UoWs are instantiated by the RA on the source side ("fetch frame," "3xDES-CBC encrypt frame," and "transmit frame over network via TCP") and the remaining four by the RA on the destination side ("receive frame over network via TCP," "3xDES-CBC decrypt frame," "decompress frame," and "display frame").

The SM also determines that the network bandwidth between desktop.officeA.com and home.officeA.com is highly limited, and that timeliness-related QoS parameters will not be met if both audio and video are transmitted over the modem line. It also determines that the phone is moderately secure. Therefore, it chooses the appropriate LRoS that filters out video and then encrypts the audio. This procedure results in seven UoWs: "fetch frame," "filter out video," "DES-CBC encrypt audio," "transmit data over network via TCP," "receive data over network via TCP," "DES-CBC decrypt audio," and "output audio." For this application, the RAs instantiate four UoWs on the source side and three on the destination side.

Because the office A and office C are connected over a highly secure VPN, no additional encryption is necessary for the last application. However, the SM

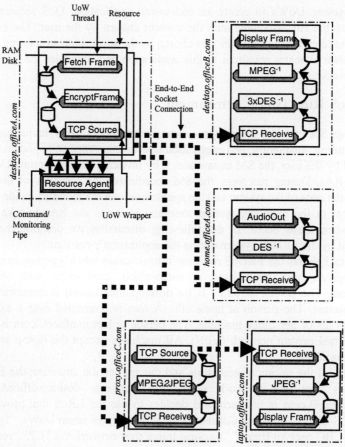

Fig. 5. ERDoS Infrastructure for Five Resources (Two Video Source Resources, Two Video Sink Resources, and a Proxy Resource) [The -1 superscript denotes a decompression or decryption UoW]. The System Manager is not shown. There also exists a RA on each machine, but only the RA on desktop.officeA.com is shown.

determines that laptop.officeC.com is not fast enough to decompress MPEG and still meet all timeliness parameters. To solve this problem, the SM uses a proxy machine that translates MPEG to JPEG video. This transformation enables the laptop user to view the video.

This dynamic structuring of an application, based on resource constraints, is a direct consequence of the modular and extensible architecture of ERDoS. The last application is now represented by a LASM consisting of eight UoWs. The SM instructs the appropriate RAs to instantiate eight UoWs. The first two are instantiated by the RAs on the source side ("fetch frame" and "transmit frame over network via TCP"). The next three ("receive frame over network via TCP," "MPEG to JPEG," and "transmit frame over network via TCP") are instantiated by the RA on the proxy machine (*proxy.org143.office.com*). The final three UoWs are instantiated by the RA on the destination side ("receive frame over network via TCP," "decompress frame," and "display frame"). The structure of each application is chosen by our allocation algorithm [4], which bases its decision on the available LRoSs, the system state, and the resource attributes.

7 Conclusion

Distributed systems are becoming increasingly heterogeneous. Developing a static application that can work in heterogeneous systems and still meet QoS requirements is increasingly difficult. In ERDoS, we introduce an innovative method to alleviate this problem by dynamically structuring applications at run time, based on system and resource attributes. Our approach increases the probability of achieving a high level of application QoS in highly heterogeneous systems. Whereas a static approach may fail because a particular resource becomes overloaded, our dynamic approach can find alternative application structures that use other resources but still provide the same end-to-end functionality and QoS. Furthermore, static approaches may reject an application if all system resources are highly utilized. Instead, our dynamic approach can utilize alternative application structures that provide partial QoS (which is better than no QoS at all).

References

1. Chatterjee, S., Sabata, B., Sydir, J., and Lawrence, T.: Modeling Applications for Adaptive QoS-based Resource Management. *Proc. 2nd IEEE High-Assurance System Engineering Workshop* (1997)
2. Schneier, B: Applied Cryptography, Protocols, Algorithms, and Source Code in C. 2nd ed. John Wiley and Sons, Inc. (1996)
3. Sabata, B., Chatterjee, S., Sydir, J., and Lawrence, T: Hierarchical Modeling of Systems for QoS-based Distributed Resource Management. Unpublished draft. SRI International, Menlo Park, California (1997)
4. Chatterjee, S.: A Quality of Service based Allocation and Routing Algorithm for Distributed, Heterogeneous Real Time Systems. *Proc. 17th IEEE Intl. Conf. Distributed Computing Systems (ICDCS97)*, Baltimore, Maryland (1997)
5. Chatterjee, S., and Strosnider, J.: Distributed Pipeline Scheduling: A Framework for Distributed, Heterogeneous Real-Time System Design. *Computer Journal* (British Computer Society), Vol 38(4) (1995)
6. Sabata, B, Chatterjee, S. and Sydir, J.: Dynamic Adaptation of Video for Transmission under Resource Constraints. In *Proc. 17th IEEE Intl. Conf. Image Processing (ICIP98)*, Chicago, Illinois (1998)
7. Sabata, B., Chatterjee, S., Davis, M., Sydir, J., and Lawrence, T.: Taxonomy for QoS Specifications. In *Proc. IEEE Computer Society 3rd International Workshop on Object-Oriented Real-Time Dependable Systems (WORDS '97)*, Newport Beach, California (1997)
8. McCanne, S., Jacobson, V.: vic: A Flexible Framework for Packet Video. *Proc. ACM Multimedia*, San Francisco, California (1995)
9. Chatterjee, S., Bradley, K., Madriz, J., Colquist, J., and Strosnider, J.: SEW: A Toolset for Design and Analysis of Distributed Real-Time Systems." In *Proc. 3rd IEEE Real-Time Technology and Applications Symposium (RTAS97)*, Montreal, Quebec (1997)
10. Jensen, E., Locke, C., and Tokuda, H.: A Time-driven Scheduling Model for Real-Time Operating Systems. In *Proc. IEEE Real-Time Systems Symposium* (1985)

Improving Support for Multimedia System Experimentation and Deployment*

Douglas Niehaus

Information and Telecommunication Technology Center
Electrical Engineering and Computer Science Department,
University of Kansas, Lawrence KS 66045, USA
niehaus@eecs.ukans.edu

Abstract. New applications with increasingly complex complex requirements present significant challenges to existing operating systems, motivating the design and implementation of new system abilities. Multimedia applications are excellent driving problems because they exhibit stringent temporal constraints and non-trivial interactions among applications within and across machine boundaries. This paper describes work done under several projects conducted by the author in the Information and Telecommunication Technology Center at the University of Kansas which had the need for real-time software in common. This common set of requirements motivated the development of real-time extensions to Linux capable of scheduling events with microsecond accuracy, and the creation of a facility called the Data Stream Kernel Interface capable of gathering performance data with sub-microsecond precision. This paper provides an overview of these new system abilities, how we have used them, and discusses their relevance to the design, implementation, and evaluation of multimedia systems.

1 Introduction

A new class of applications with increasingly complex requirements for an increasing range of system capabilities is emerging as the worldwide Internet continues to increase in size and complexity, and the power of the computing platforms in the homes of typical members of the public increases to a level that would have seemed a fantasy even a few years ago. Multimedia applications are excellent examples of this new class of programs because they exhibit most of the important characteristics of class members, including: increasingly stringent temporal constraints, significant interaction among other applications on the same machine, and significant interaction with network resources.

The challenges presented by such applications, which must be met by operating systems in the next few years, are likely to require significant changes in current system architectures. This is an important point that many researchers appreciate and a wide range of research projects are investigating different approaches to fundamental changes in system and computing architectures [1, 6, 9].

---
* This work was supported in part by grants from Sprint Corporation.

Investigators at Microsoft Research, for example, have recently written a position paper advocating their assertion that "it is time to reexamine the operating system's role in computing" [3]. While the work presented here is not as aggressive as a fundamental change in the operating system's role, our experience has indicated that several of the things we have done are useful.

The work discussed in this paper was done under several past and present research projects conducted by the author and his colleagues in the Information and Telecommunication Technology Center at the University of Kansas. Some projects addressed aspects of the design, implementation, and evaluation of next generation telecommunication networks. Other projects considered how emerging distributed applications might use these networks and how distributed software technology can be applied to the design, implementation, and deployment of these next generation networks. In the course of this work, an interesting set of results emerged which are relevant to the problem of improving support for experimentation and deployment of multimedia systems.

Specifically, we have spent the last several years working with the design, development, and evaluation of several software systems which must satisfy a variety of firm real-time execution constraints. We have concentrated on the use of commercial off-the-shelf (COTS) hardware and free, or at least inexpensive, software. Development of application and operating system software which must satisfy fine grain temporal constraints has also required us to significantly improve our ability to monitor events within all layers of the system in support of evaluating our work.

We believe that this work provides a useful approach to important components of experimental and deployment platforms for multimedia. First, it is inexpensive and open source, being based on the Linux OS, GNU software, and COTS hardware. This maximizes both the set of people who can use it and the ease of collaboration. Second, it is capable of scheduling and monitoring events at a significantly finer temporal resolution than most current systems, and certainly finer than current consumer systems. Third, continued development and use of this system has demonstrated that it provides these capabilities at a reasonable cost in overhead, and that it can be used for a wide range of interesting and innovative applications. Finally, preliminary tests support the idea that these capabilities can contribute to the support of distributed multimedia and the integration of network quality of service.

The rest of this paper first discusses system support for increased temporal resolution and scheduling applications with firm real-time constraints in Section 2, and then considers support we have created for evaluating the performance of such applications in Section 3. Section 4 describes several ways in which we have used these facilities in past and current research, and Section 5 presents some experimental results which demonstrate the viability of our approach. Section 6 points out where more detailed discussion of related work in several categories can be found. Finally, Section 7 briefly presents our conclusions and discusses future work.

2 System Support

Experimentation with systems supporting multimedia and other applications with more complex and demanding requirements for system support than conventional systems consider normal requires significant extension to the system support for controlling and monitoring events. We created the KU Real-Time (*KURT*) system in response to the need in several projects for a low cost system with microsecond temporal resolution in both timestamp data gathering and event scheduling. The first application of this ability was the ATM Reference Traffic system, which we use to help evaluate the performance of ATM networks, and which is discussed in Section 4. Later projects motivated us to expand the basic capabilities into a more generally applicable system which is the current version of KURT. This work is relevant for multimedia and quality of service researchers because it significantly increases the capabilities of a low cost and convenient experimental and deployment platform.

KURT is based on the freely available Linux operating system, modified in several important ways. First, by running the hardware timer as an aperiodic device we have increased the system's temporal resolution without significantly increasing the overhead of the software clock. We call this the *UTIME* component of KURT, which alone increases the utility of Linux for soft real-time applications.

The KURT system builds on top of UTIME, adding a number of features. KURT enables the system to switch between three modes: **normal mode**, in which it functions as a normal Linux system, **mixed real-time mode**, in which it executes designated real-time processes according to an explicit schedule, serving non-real-time processes when the schedule allows, and **focused real-time mode**, in which only real-time processes are run. Focused real-time mode is useful when KURT is being used as a dedicated real-time system satisfying the most stringent execution constraints. When real-time applications share a generic desktop workstation, KURT's mixed real-time mode is most often used, since it allows both real-time and non-real-time processes to run. When in either real-time mode, real-time processes can access any of the system services that are normally available to non-real-time processes. It is important to note that the use of different subsystems introduces different levels of scheduling distortion. Our current work includes analysis of how different subsystems create these scheduling distortions and the development of methods to control or eliminate them.

The basic approach to KURT implementation and its current scheduling resolution have been presented elsewhere. Srinivasan describes the original KURT design [14, 15], while Hill describes several modifications of KURT which improve its behavior in a number of ways [7, 8]. The essential points drawn from these sources, include: UTIME modifications increase KURT's temporal resolution to microseconds, events can be scheduled with an accuracy varying from tens to hundreds of microseconds depending on the Linux subsystems in use, and that KURT provides a powerful and flexible experimental platform. The rest of this section will discuss KURT modifications by Hill not addressed in the previous conference paper [14].

2.1 Scheduling Extensions

Kurt provides a simple and well structured interface for implementing different scheduling modules. Hill added two new modes to the KURT real-time process scheduling module to accommodate distributed real-time applications: *event-driven* and *buffered-periodic* [8]. Both of these scheduling modes were evaluated using a simple distributed multimedia system as described in Sections 3 and 4.

The event driven mode waits until a packet arrives, and then decides if the process waiting for the packet can be scheduled. KURT gives preference to processes that are already explicitly scheduled, but if no real-time process is already scheduled to run, the event-driven process is chosen for execution as though it had been explicitly scheduled. If there is already a running or pending explicitly scheduled process, the event-driven process is allowed to run in the next available slot in the real-time schedule.

The mode supporting periodic scheduling with buffering was created in response to the observation that a simple periodic process was able to find and process the intended frame a large percentage of the time, but that the arrival jitter of the network was too large to permit excellent performance. In this mode the system allows one or more frames associated with a periodic KURT process to accumulate before it begins execution. Our experiments in the LAN environment showed significant advantage to a buffer of only a single frame.

Atlas has also recently added a scheduling mode to KURT implementing her Statistical Rate Monotonic Scheduling (SRMS), which is a generalization of the classical RMS results of Liu and Layland for periodic tasks with highly variable execution times and statistical QoS requirements [2]. The SRMS scheduler is a simple, preemptive, fixed-priority scheduler. The paper describes the technical issues that arose while integrating SRMS into KURT Linux and presents the API developed. One of the issues cited, explicit scheduling of device driver bottom halves, is part of our current work to reduce scheduling distortion. We have not yet integrated the SRMS work into our generic KURT release but hope to do so in the near future.

2.2 Networking Extensions

The standard ATM implementation under Linux supports unspecified bit rate (UBR), or best effort, and constant bit rate (CBR) traffic. Data can be accessed directly as either individual cells, termed AAL0, or AAL5 packets. In addition, support is provided for IP over ATM and LANE. Hill extended the ATM implementation to include variable bit rate (VBR) traffic and to use raw AAL5 packets as the transport mechanism to minimize protocol processing overhead [8]. Multimedia data streams are usually compressed in some way, and thus typically exhibit variation in their transmission rate. VBR is thus the most commonly favored QoS for video streams. VBR service can be broken down further into two categories: real-time (rtVBR) and non-real-time (nrtVBR). Real-time VBR provides two QoS parameters: cell-delay variation (CDV) and cell transfer delay (CTD). CDV provides an upper bound on the variation in the delay cells experience while CTD provides an upper bound on the maximum end-to-end delay for cells.

Hill's modifications only support non-real-time VBR due to lack of support in the current version of the ATM switch control software. However, for the purposes of experimentation in a LAN environment, nrtVBR provided sufficient support for multimedia processes with periods in the tens of milliseconds range as arrival jitter was non-trivial but not excessive. Future work includes experiments with multimedia streams across WAN connections, which may reveal the advantage of more stringent rtVBR support.

2.3 Middleware Support

KURT also provides a good platform for our experimental work with CORBA middleware. Gopinath worked on the implementation and evaluation of real-time CORBA end-systems[5]. This work included the implementation of a simple real-time scheduling service on top of KURT, and simple real-time extensions to OmniORB[12]. We are currently using the KURT extensions, in conjunction with data collection capabilities discussed in Section 3, as part of a project creating tools for creating and conducting CORBA performance evaluation experiments[11]. We are extremely interested in the resulting capability to examine the influence of system support on overall ORB performance and to consider the design and implementation of explicit support for CORBA event services and the integration of kernel I/O and ORB level I/O management. This latter work is currently being done in the context of the OmniORB and the ACE ORB[13]. We intend to use a CORBA based version of our simple multimedia system as one of the driving applications as these efforts progress, which will provide for an interesting comparison between CORBA and non-CORBA based performance of essentially the same scenarios.

3 Experimentation Support

Many of the new facilities and possibilities for experimentation and support of applications discussed in Section 2 are interesting, promising, and initial results justify our current interest. The fact remains, however, that they are subjects of research because their ultimate utility, and thus their ultimate inclusion, in system of the future supporting multimedia and other sophisticated applications is as yet only potential. Further work, and further proof of their advantages, is required before potential can become reality.

Yet, the nature of the extended system capabilities means that the methods used to evaluate them must also be extended. For example, evaluating the KURT extensions to Linux requires the ability to gather data for performance evaluation with timestamps having sub-microsecond resolution. This is three or four orders of magnitude finer resolution than current commercial systems commonly provide. A more subtle difficulty is that evaluating management of events at increased temporal resolution is accompanied by a significant increase in the *volume* of performance data collected.

Data collection from internal levels of operating systems has traditionally been done using *ad hoc* and *data source specific* methods that are often of limited utility or unavailable to others interested in similar information. This represents a tremendous waste of time and effort since many problems are solved repeatedly by different investigators. We decided to reduce this waste of effort, while making available the broader range and greater volume of information we required, by creating a platform independent interface to support customizable collection of performance data from the operating system's internal subsystems. The interface which we have developed is called the Data Stream Kernel Interface (DSKI) [4]. The DSKI is currently implemented as a pseudo device driver on two machine architectures, PC hardware running Linux and DEC Alpha hardware running Digital Unix. The choice of the pseudo device driver interface was made to maximize convenience of porting the DSKI to other systems.

The DSKI provides both and *internal* and *external* interfaces. The external interface supports user programs gathering performance data from the operating system. The internal interface is used by kernel subsystems to register events and other information which may then be gathered by these user processes. An *event* is a time-stamped set of data generated when a thread of execution crosses the location of the event in the operating system executable code. The DSKI uses the Pentium Time Stamp Counter, or the equivalent register on the Alpha, to record timestamps at the resolution of the CPU clock. Each kernel subsystem, including device drivers, registers its events with the DSKI during initialization. Note that this works just as well for loadable device drivers.

The number of events within different components of the operating system which are of potential interest is huge, and thus presents a design challenge. We control events in several ways. First, they are conditionally compiled into the system so that their overhead can be eliminated in a production system. Even in an experimental system, each event compiled into the system may be enabled or disabled. This lowers the aggregate overhead since the overhead of a disabled event is $0.05 \mu s$ on a 133 MHz Pentium system.

Since the number of potentially interesting events associated with various subsystems is so large, we also provide for grouping the events into sets called *families*. Families of events may be enabled and disabled just as individual events are. The event families, and events within families, registered with the DSKI are presented to the user as a *name space* examined and manipulated using the *ioctl* system call of the DSKI. This is not elegant, but future work includes the relatively simple task of implementing a more graceful API on top of the current functionally complete one.

A process gathering performance data opens the DSKI, enables the sets of events in which it is interested, and then reads the resulting stream of events. Each enabled event is thus associated with at least one process, and perhaps

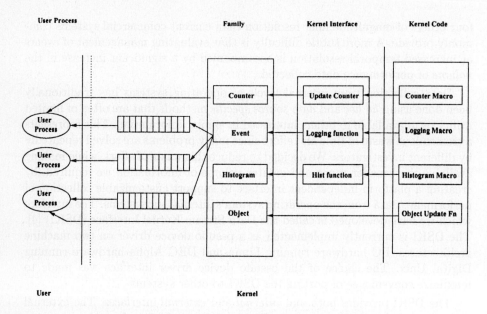

Fig. 1. The Data Stream Kernel Interface Architecture

more. The overhead of logging an event to a single event queue is $3\mu s$ with an additional $2\mu s$ for each additional event queue. In comparison, a simple *kprintf* used to output a timestamp and a long integer to the console required roughly $35\mu s$. More detailed evaluation of the DSKI overhead is given elsewhere [4].

Figure 1 illustrates the basic DSKI architecture. Note that events generated in the kernel flow through the DKSI and into event queues associated with processes. The figure also shows two other elements call the *counter* and the *object*. The *object* is simply a data structure updated by events in the kernel. A process gathering data receives a copy of the data structure's current state each time it chooses to read the DSKI stream attached to the object. The overhead of this approach is significantly lower than that of a datastream actively generating a series of event records, and it is an excellent way of gathering cumulative performance data. The *counter* is a special case of the object with even lower overhead, since the DSKI support for counters is optimized to increment the specified integer with minimal overhead. We have also recently created the *histogram* object which is a parameterized collection of counters. This has been particularly useful recently for gathering distributions of packet inter-arrival times and scheduling distortions.

Consider the differences among the DSKI data stream types applied to evaluation of the TCP/IP protocol stack. If we define a set of active events in the TCP/IP code where we want to gather detailed information, we create an active data stream which collects these events as they are generated by the kernel. We have done this many times and have been able to track individual packets from a socket down through the protocol stack on the sending machine and then up through the protocol stack to the socket on the receiving machine by collecting enough information (e.g. port number and byte sequence number) to uniquely identify the packet. However, if the data we want to collect is adequately represented by a passive object, for example aggregate TCP/IP throughput statistics, then using a passive data stream would incur less overhead. Since this information is not maintained by any specific object in the kernel, we define a data structure for this purpose which is updated by events to count the packets transmitted or received. Finally, if we are interested in the distribution of inter-arrival times of packets or the distribution of the number of packets transmitted or received during N-second periods, we can use the histogram object.

4 Example Driving Applications

Driving applications should, ideally, be software that is desired for its own sake which, at the same time, provides a significant level of practical exercise to a system under development. This section describes three driving applications that have been, and should continue to be, useful challenges to several projects.

The first significant driving application for KURT was the ATM Reference Traffic System (ARTS), which was the main focus of a project funding development of ATM network performance evaluation tools. ARTS can record and generate packet-level ATM traffic streams with microsecond accuracy. ARTS uses KURT's scheduling abilities in its traffic generation module by placing packet transmission events in an explicit KURT schedule [15]. As the application driving initial KURT development, ARTS was re-implemented several times, ending as a KURT real-time module operating in fixed scheduling mode. We avoid all but minor scheduling distortion when we are able to keep the entire packet transmission schedule in main memory, but on a 128MB machine this enables us to hold a schedule with approximately 3 million packet events. Using the disk removes all bounds on the schedule length, but does occasionally distort packet transmission events by delaying them until disk operations are complete. The need to evaluate the performance of ARTS with microsecond accuracy drove the development of the DSKI, as did the need to gather data to assist in optimizing performance and fixing bugs.

Hill extended and refined Srinivasan's work in the areas of Scheduling, I/O subsystems, and network quality of service discussed in Section 2 and in greater detail elsewhere [8, 7]. Hill's driving example was a distributed multimedia application with a simple client-server architecture. Thirty audio and video frames are produced every second, defining a 33 ms period for each frame. The accuracy of the frame period, the packet response latency, and the frame loss rate are obvious performance metrics. The streams of video and audio frames were handled by separate real-time processes, thus making the synchronization of the audio and video processes at the frame level a significant performance metric. The server was capable of generating multiple streams which were synchronized at the frame level.

KURT has also been used to support less obvious applications of real-time capabilities. Murthy used the basic KURT framework to implement a simple ATM switch, and to precisely control the rate at which it processed ATM cells. Similar modifications to host ATM code enabled him to control the rate at which ATM cell streams were produced and consumed. He then used this platform to conduct experiments testing the behavior of Available Bit Rate Service algorithms. His comparisons to the results of conventional simulations, and the execution time of those simulations, indicate that this approach can be used to conduct experiments significantly more quickly than by conventional means[10]. Even if the initial promise of this approach is not realized, Muthy's work shows that the KURT and DSKI provide a powerful and flexible platform for experimentation.

5 Illustrative Results

The results presented in this section illustrate and support the main message of this paper at two levels. First, they demonstrate that the software and methods described here are capable of supporting applications with the finer grain and more complex constraints and requirements typical of emerging multimedia applications. Second, that the evaluation technology and methods has been improved in a way commensurate with the increased capabilities of the systems being evaluated. In particular, the increased temporal resolution of the KURT supports both firm real-time execution of multimedia applications, and the DSKI's ability to gather performance data with microsecond accuracy.

The results presented in this section cannot cover all relevant and interesting aspects of the research described here. Interested readers must refer to the relevant cited works for more detailed results. Some of Hill's results evaluating the end-to-end quality of service provided by KURT and the the ATM LAN to the distributed multimedia application illustrate the relevant points well, and are drawn from his Master's thesis [8].

	Num. Clients	min ms	mean ms	max ms	Interval within which 'x'% of events fell			
					95%	99%	99.9%	99.99%
Std. Linux	1	10.036	32.989	113.947	0.419	2.011	22.447	78.935
Event-Driven	1	27.962	32.996	37.156	0.397	1.875	2.901	4.938
Periodic	1	30.229	32.997	35.769	0.543	2.011	2.397	2.673
Std. Linux	2	10.038	32.996	222.709	0.835	4.387	22.958	164.85
Event-Driven	2	13.215	33.004	59.541	0.585	2.348	17.249	24.815
Periodic	2	22.992	32.995	41.905	0.019	0.166	6.556	8.719
Std. Linux	4	10.038	32.997	246.677	1.283	6.394	30.247	172.375
Event-Driven	4	14.557	33.007	69.372	0.332	2.000	13.588	30.327
Periodic	4	20.529	32.998	45.453	0.181	1.317	2.663	12.131

Table 1. Confidence Intervals for Multimedia Client Period with Disk Activity

Table 1 provides information regarding how well the standard Linux scheduler, the event driven scheduler, and the periodic scheduler with buffering maintained the multimedia client's period which was nominally 33 ms. Note that the standard Linux scheduler was tested using the finer temporal resolution and DSKI support features of KURT since standard Linux did not provide the ability to gather this information. These data were taken when non real-time processes generating constant disk activity were active, creating unusually large scheduling distortion. The results show that KURT support maintained the period of the multimedia client significantly better than the standard Linux scheduler, as well as demonstrating that the DSKI supports gathering a large volume of data at a fine temporal resolution.

Table 2 presents the confidence intervals for the packet response latency. This data demonstrates that the performance of the event driven scheduler under KURT performs comparably to standard Linux with respect to this performance metric, which is reassuring but not surprising. The data for the periodic scheduling mode with buffering is interesting because it clearly shows the effect of the 1 packet buffer used by this method, since the latency is approximately 1.5 times the 33 ms packet inter-arrival time and multimedia client period.

	min μs	mean μs	max μs	Interval within which 'x'% of events fell			
				95%	99%	99.9%	99.99%
Std. Linux	139	156	10260	182	374	2469	5306
Event-Driven	124	143	9905	161	258	1842	5101
Periodic	46725	48341	54247	49691	49807	50524	53883

Table 2. Confidence Intervals for Packet Response Latency

The DSKI was also used to gather data about another performance metric; packet loss. This metric illustrated a significant difference between standard Linux and KURT support, especially when more than one multimedia stream was being processed. When only one stream was used, standard Linux lost 1 frame out of 3000 while the KURT methods lost 0. However, when 2 multimedia streams were used, standard Linux lost 107 of the 3000 while the event drive KURT method lost 1 and the buffered periodic method lost 0. When 4 streams were used, standard Linux lost 138 of 3000 frames, while the event driven KURT method lost 2 and the buffered periodic method lost 0.

Finally, Table 3 shows the time between when each of the audio and video processes began running to process the paired audio and video frames. The confidence intervals show the interval from the mean difference within which the given percentage of events fell. As the table shows, the buffered periodic real-time processes maintain the closest synchronization throughout the transfer. The maximum deviation experienced is only one quarter of the period. The event-driven process does not perform as well, experiencing some deviations of almost one full period. The standard Linux application experiences the most

	min ms	mean ms	max ms	Interval within which 'x'% of events fell			
				95%	99%	99.9%	99.99%
Std. Linux	0.114	21.374	52.197	6.935	7.895	8.915	15.715
Event-Driven	10.315	20.716	32.157	3.173	4.976	6.912	8.914
Periodic	3.125	5.165	8.525	0.596	0.937	1.375	2.956

Table 3. Confidence Intervals for Audio/Video Stream Synchronization

variance, with some individual variances approaching 2 full periods. Even with disk background activity, four real-time multimedia clients served by a single server process, running at 15 frames per second, remained fully synchronized after 90 minutes of continuous play. This shows the accuracy which the two real-time scheduling modes provides.

6 Related Work

The diversity of the work discussed in this paper brings with it a similar diversity of related work. Srinivasan provides an extensive discussion of how KURT relates to other real-time systems, as well as how the ARTS application compares to related ATM testing methods [14, 15]. Hill discusses a largely separate set of related real-time systems, as well as work in end-to-end quality of service [8]. The DSKI technical report discusses related work in performance data gathering[4], Gopinath discusses real-time support for ORBs[5], and Nimmagadda discusses ORB performance evaluation[11].

7 Conclusions and Future Work

This paper has presented work associated with several projects conducted by the author and his colleagues at the Information and Telecommunication Technology Center at the University of Kansas. The KURT extensions to Linux in support of a variety of firm real-time applications have shown that it can be effective, especially in conjunction with the ability to gather detailed performance data provided by the DSKI. Certainly, the development of these abilities is still relatively primitive, but the work discussed has shown that even modestly enhanced capabilities can make previously infeasible projects possible. The diverse and continuing application of KURT and the DSKI to current and future research projects will continue to drive their development. The two most interesting near-term extensions to KURT are support for explicit scheduling of device driver bottom halves, and integration of Atlas's SRMS and possibly other rate monotonic scheduling modules. The first should significantly decrease scheduling distortions, while the second would provide familiar and popular scheduling semantics.

References

1. T. Anderson, D. Culler, D. Patterson, and the NOW team, "A Case for Networks of Workstations: NOW," *IEEE Micro*, Feb 1995.
2. A. Atlas and A. Bestavros, "Design and Implementation of Statistical Rate Monotonic Scheduling in KURT Linux," Boston University Technical Report 98-013, September 1998.
3. W. Bolosky, R. Draves, R. Fitzgerald, C. Fraser, M. Jones, T. Knoblock, R. Rashid, "Operating System Directions for the Next Millenium," http://research.microsoft.com/sn/Millennium/mgoals.html
4. B. Buchanan, D. Niehaus, R. menon, S. Sheth, Y. Wijata, S. House "The Data Stream Kernel Interface," Technical Report ITTC-FY98-TR-11510-04, Information and Telecommunication Technology Center, Electrical Engineering and Computer Science Department, University of Kansas, 1998.
5. A. Gopinath, "Performance Measurement and Analysis of Real-Time CORBA Endsystems," Master's Thesis, Electrical Engineering and Computer Science Department, University of Kansas, 1998.
6. A. Grimshaw, W. Wulf, and the Legion Team. "The Legion Vision of a Worldwide Virtual Computer." *Communications of the ACM*, 40(1), January 1997.
7. R. Hill, B. Srinivasan, S. Pather, D. Niehaus "Temporal Resolution and Real-Time Extensions to Linux," Technical Report ITTC-FY98-TR-11510-03, Information and Telecommunication Technology Center, Electrical Engineering and Computer Science Department, University of Kansas, 1998.
8. R. Hill "Improving Linux Real-Time Support: Scheduling, I/O Subsystem, and Network Quality of Service Integration," Master's Thesis, Electrical Engineering and Computer Science Department, University of Kansas, 1998.
9. M. Jones, J. Barrera III, A. Forin, P. Leach, D. Rosu, and M. Rosu. An Overview of the Rialto Real-Time Architecture. In *Proceedings of the Seventh ACM SIGOPS European Workshop*, pages 249–256, September 1996.
10. S. Murthy "A Software Emulation and Evaluation of Available Bit Rate Service," Master's Thesis, Electrical Engineering and Computer Science Department, University of Kansas, 1998.
11. S. Nimmagadda, C. Liyanaarachchi, A. Gopinath, D. Niehaus, A. Kaushal, "Performance Patterns: Automated Scenario Based ORB Performance Evaluation," To Appear *COOTS '99*, May 1999.
12. The Olivetti & Oracle Research Laboratory. Omniorb2, http://www.orl.co.uk/omniorb/.
13. D. Schmidt, R. Bector, D. Levine, S. Mungee, and G. Parulkar. TAO: A Middleware Framework for Real-Time ORB Endsystems. In *IEEE Workshop on Middleware for Distributed Real-Time Systems and Services*, December 1997.
14. B. Srinivasan, S. Pather, R. Hill, F. Ansari, D. Niehaus "A Firm Real-Time System Implementation Using Commercial Off-The-Shelf Hardware and Free Software," In *Proceedings of the Real-Time Technology and Applications Symposium*, Denver, June 1998.
15. B. Srinivasan "A Firm Real-Time System Implementation Using Commercial Off-The-Shelf Hardware and Free Software," Master's Thesis, Electrical Engineering and Computer Science Department, University of Kansas, 1998.

Run-Time Systems for Parallel Programming

3rd RTSPP Workshop Proceedings

San Juan, Puerto Rico, USA, April 12, 1999

Organizing Committee

General Chair – Laxmikant V. Kale
Co Chair – Pete Beckman
Steering Chair – Matthew Haines
Program Chair – Ron Olsson

Program Committee

Henri Bal	Pete Beckman	Greg Benson
Luc Bougé	Matthew Haines	Laxmikant V. Kale
Koen Langendoen	David Lowenthal	Frank Müller
Ron Olsson	Raju Pandey	Alan Sussman

Preface

Runtime systems are critical to the implementation of parallel programming languages and libraries. They support the core functionality of programming models and the glue between such models and the underlying hardware and operating system. As such, runtime systems have a large impact on the performance and portability of parallel programming systems.

Despite the importance of runtime systems, there are few forums in which practitioners can exchange their ideas, and these are typically forums showcasing peripheral areas, such as languages, operating systems, and parallel computing. RTSPP provides a forum for bringing together runtime system designers from various backgrounds to discuss the state-of-the-art in designing and implementing runtime systems for parallel programming.

The RTSPP workshop will take place on April 12, 1999 in San Juan, Puerto Rico, USA, in conjunction with IPPS'99. This one-day workshop includes technical sessions of refereed papers and panel discussions. The 8 paper presentations were selected out of 17 submissions after a careful review process; each paper was reviewed by at least three members of the program committee. Based on the reviewers' comments, the authors revised their papers for inclusion in these workshop proceedings.

We thank the RTSPP Program Committee (see previous page) and the following additional people for taking part in the review process: Gene Fodor, Pete Keleher, Jean-Francois Méhaut, and Raymond Namyst. We also thank the Organizing Committee for RTSPP'97 and RTSPP'98 — Matt Haines, Greg Benson, and Koen Langendoen — for initiating this workshop, running it for the past two years, and helping us to organize it this year. Matt Haines was so helpful that we drafted him onto this year's Organizing Committee.

We hope that the Workshop will be interesting and exciting.

Ron Olsson
Laxmikant V. Kale
Pete Beckman
Matthew Haines

Efficient Communications in Multithreaded Runtime Systems

Luc Bougé, Jean-François Méhaut, and Raymond Namyst

Laboratoire de l'Informatique du Parallélisme, École normale supérieure de Lyon,
F-69364 Lyon, France.
{Luc.Bouge,Jean-Francois.Mehaut,Raymond.Namyst}@ens-lyon.fr

Abstract. Most of existing multithreaded environments have an implementation built on top of standard communication interfaces such as MPI which ensures a high level of portability. However, such interfaces do not meet the efficiency needs of RPC-like communications which are extensively used in multithreaded environments. We propose a new portable and efficient communication interface for RPC-based multithreaded environments, called MADELEINE. We describe its programming interface and its implementation on top of low-level network protocols such as VIA. We also report performance results that demonstrate the efficiency of our approach.

1 Introduction

For portability reasons, most existing multithreaded environments are often implemented on top of high-level standard communication interfaces such as PVM or MPI. A large subset of these environments are RPC-based environments because the provided functionalities rely implicitly or explicitly on a mechanism able to invoque remote services (*e.g.* RSR in Nexus [7], RPC in PM2 [12], Fork/Join in DTS [4]).

In fact existing communication interfaces do not adequately support the implementation of these environments [2], either because they are not portable (too low-level) or simply because they do not match the specific needs of these environments. Usually, implementations fall into the second category because most of existing environments emphasize portability (*e.g.* Chant [9], Athapascan [3]).

This situation, often been considered as a "natural" tradeoff between portability and efficiency, can be avoided. In this paper, we present a new portable communication interface (called MADELEINE) that provides both efficiency and high-level abstractions to RPC-based multithreaded environments. The software structure of MADELEINE is organized in two layers. The low-level layer (portability) isolates the code that needs to be adapted for each targeted network protocol. The high-level layer (API) provides advanced generic communication functionalities to optimize RPC operations.

MADELEINE is easily portable on low-level network protocols and is currently available on top of VIA [5], BIP [14], SCI [6], SBP [15], TCP, PVM [8] and MPI

[11]. Only a small constant overhead is introduced by MADELEINE on top of all these underlying procotols. Thus, its performance competes with low-level communications packages such as FM [13] or AM [16].

The remainder of this article is organized as follows. In the next section, we introduce the specific needs of RPC-based MTE as far as efficiency is concerned. The contribution of our work is presented in Section 3 where the MADELEINE communication interface is described and discussed. Section 4 describes the implementation of MADELEINE on top of several network protocols and shows some performance results. Section 5 presents related work. Finally, Section 6 summarizes the contribution of this work and lists the points we intend to address in the near future.

2 Zero-copy data transmissions in RPC-based Multithreaded Environments

A multithreaded environment is called "RPC-based" if most communications take place by means of remote invocations of services (also called Remote Procedure Calls). In this scheme of remote interaction, a client (usually a thread) sends a request to a server (usually a process) which executes a specified function (a service) to serve the request. According to the type of service executed, a response may be sent back to the client upon completion of the function. Many kinds of remote operations actually conform to this communication scheme and one can observe that most existing multithreaded environments provide RPC-based mechanisms [7, 9, 3, 4].

On a high speed network providing very low transmission latency (*e.g.* Myrinet [1]), any extra message copy introduced at some software level has a tremendous impact on performance [10]. Therefore, it is crucial that the underlying communication layer be able to ensure the delivery of messages with no intermediate extra copy of data. Many existing communication libraries actually provide such a functionality. However, the problem of realizing an RPC operation without any extra copy of data is more complex than realizing a zero-copy message transmission. Actually, when a RPC request is sent to a server, the server does not know the location where the data should be placed until it extracts preliminary information from the request message. Typically, the triggering of a RPC operation takes place in the following three steps on the server side:

1. The request type is identified (*i.e.* "Is this a migration or a remote put?") and some information is extracted from the network.
2. Then, actions are taken (for instance dynamic memory allocations) to prepare for the receipt of the data.
3. Finally, the data are extracted from the network and stored directly at the right location in memory.

As a result, implementing this scheme while avoiding extra copies of messages will require the client to send two messages (at least with a classical message passing interface): the first one to carry the *"header"* of the request and the second one to carry the regular data (the *"body"*). However, this approach does not yield good results if the underlying communication software is still buffering data on the receiving side. In this case, a single message would probably have been sufficient to carry all the information (header + body), since it could have been extracted in several steps.

3 The MADELEINE Communication Interface

To meet the efficieny and portability requirements of RPC-based multithreaded environments, we have designed a new communication interface called MADELEINE. Unlike environments like MPI or PVM, MADELEINE is not designed to be used directly by regular user-level applications. This interface is especially targeted at RPC-based multithreaded environments such as PM^2, Nexus or Chant. In this context, it is intended to bridge the gap between functionalities provided by low-level network protocols (such as BIP [14] or VIA [5]) and requirements of high-level abstractions (such as the RPC mechanism).

One important feature of MADELEINE is that it was designed to allow upper software layers (*i.e.* inside the multithreaded environment) to avoid extra copies of transmitted data. At the lowest level, this cooperation is realized using upcalls on the receiving side to allow upper layers to interactively participate in data emissions and extractions. At the user interface level, this cooperation is elegantly accomplished through the use of simple buffer management primitives.

Another main characteristic of MADELEINE is its "thread-awareness", which means that several threads may access its interface concurrently. Clearly, adequate synchronisation is enforced wherever necessary. For instance, only one thread is allowed to transmit data on a given communication link (in a given direction) at a time. Moreover, MADELEINE is able to properly schedule threads accessing to the network. For instance, polling operations may be grouped in order to avoid multiple threads simultaneously busy waiting on the same network interface.

The structure of MADELEINE is actually composed of two software layers: a generic user programming interface and a low-level portability interface whose implementation is network-dependent.

3.1 The Programming Interface

The MADELEINE programming interface provides a small set of primitives to allow the building of RPC-like communication schemes. Theses primitives look like classical message-passing-oriented primitives, and they actually only differ in the way the reception of data is handled. Basically, this interface provides primitives to send and receive *messages*, and several *packing* and *unpacking* primitives that allow the user to specify how data should be inserted into/extracted from messages.

Messages may consist of several pieces of data, located anywhere in userspace. They are assembled (resp. disassembled) incrementaly using *packing* (resp. *unpacking*) primitives. This way, a message can be built (resp. examined) at multiple software levels, which allows for instance the use of piggybacking techniques without losing efficiency. The following example illustrates this need. Let us consider a remote procedure call which takes an array of unpredictable size as a parameter. When the request reaches the destination node, the header is examined both by the multithreaded runtime (to allocate the appropriate thread stack and then to spawn the server thread) and by the user application (to allocate the memory where the array should be stored). This shows that several software layers may actually participate in the packing/unpacking of message data.

Sending messages Let us now describe how a user program prepares and sends a message to a remote node (Fig. 1). Such an operation starts by asking MADELEINE to allocate a new "message descriptor" for a given destination. Then, a series of *packing* operations are performed to "register" the differents part of the message. Finally, the emission operation is requested.

```
sendbuf_init(dest_node);

    sendbuf_pack ...
    sendbuf_pack ...

sendbuf_send();
```

Fig. 1. A typical send operation.

```
void handler()
{
    recvbuf_unpack ...
    recvbuf_unpack ...

    recvbuf_receive_data();
}

recvbuf_receive(handler);
```

Fig. 2. A typical receive operation.

The critical point of a send operation is obviously the series of *packing* calls. Such packing operations simply *virtually* append some data to a message under construction. Thus, data may not be copied, but just registered as part of the message.

The prototypes of these functions are very similar to the buffer management primitives provided by the PVM interface (*e.g.* **pvm_pkint**, **pvm_pkfloat**, etc.). For instance, the function used to append one or more integers to the current message has the following prototype:

      void sendbuf_pack_int(int mode, int *elems, int nb);

Among the parameters, **elems** is the address of an array of integers and **nb** represents the size of this array. The **mode** parameter plays an important role in the MADELEINE interface, because it determines the semantics of the operation.

It can be assigned different values, also called flags. These flags, defining the semantics of the **sendbuf_pack** operation, are mutually exclusive:

SND_SAFER When used, this flag indicates that MADELEINE should pack the data in a way that further modifications to the corresponding memory area should not impact on the message. In other words, an explicit "copy" is required. This is particularly mandatory if the data location is reused before the message is actually sent.

SND_LATER This flag indicates that MADELEINE should not consider accessing (*i.e.* copying or even sending) the value of the corresponding data until the **sendbuf_send** primitive is called. This means that any modification of these data between their packing and their sending will actually update the message contents. This is typically useful when a message is to be sent to various destinations with only minor variations. In this case, the data can be packed only once and their values can be updated between the send operations.

SND_CHEAPER This is the default flag. It allow MADELEINE to do its best to handle the data as efficiently as possible. The counterpart is that no assumption should be made about the way (and the time) MADELEINE will access the data. Thus, the corresponding data should be left unchanged until the send operation has completed. We describe in Section 4 how this can be implemented according to the underlying network protocol. Note that most part of data transmissions involved in parallel applications can accomodate the **SND_CHEAPER** semantics.

Of course, packing the same data multiple times with different semantics within the same message produces an erroneous program. There is no other particular restriction on the order or on the number of packing operations with respects to their semantics. The only constraint imposed by MADELEINE is that the sequence of packing operations will have to be "mirrored" on the receiving side, as seen in the next section.

Receiving messages Receiving a message is always done in three steps (Figure 2). First, the arrival of the message is detected (inside **recvbuf_receive**), which triggers the execution of a handler. Second, a number of unpacking operations (**recvbuf_unpack_***) are performed within the handler to specify where data should be stored in memory. Third, a "commit" operation is executed (**recvbuf_receive_data**) to force the completion of the data transmission.

The last two steps need further explanations. As the *packing* operations, the *unpacking* ones also require a special parameter which specifies *when* the data should be received. There are two possible values for this parameter:

RECV_EXPRESS This flag forces MADELEINE to guarantee that the corresponding data is immediately available upon the completion of the *unpacking* operation. Typically, this flag is mandatory when extracting data that will be used before the `recvbuf_receive_data` primitive is called. On some network protocols, this functionality may be available for free. On some others, it could penalize latency and bandwidth. That for, the user should extract data this way only when necessary. In the example described in section 2, the message *"header"* would have to be unpacked this way.

RECV_CHEAPER This is the default flag. It allows MADELEINE to defer the extraction of the corresponding data until the execution of `recvbuf_receive_data`. Thus, no assumption can be made about the exact moment at which the data will be extracted. Depending on the underlying network protocol, MADELEINE will do its best to minimize the overall message transmission time.

If combined with **SND_CHEAPER**, this flag always guarantees that the corresponding data is transmitted as efficiently as possible. On high-performance network protocols such as VIA or BIP for instance, data will effectively be transmitted without any extra copy.

For efficiency reasons, the way some data is to be extracted from a message should be known by MADELEINE on the sending side. That is, when some data is packed into a message, the programmer has to specify not only the sending mode, but also the receiving mode (for clarity, we omitted this detail in the previous section). Furthermore, the reverse is also true for the same reason: when unpacking some data out of a buffer, the sending mode has to be specified in addition to the receiving mode (by logically *or*-combining the flags). This constraint could have been avoided at the price of a significant performance drop (messages would need to be self-describing), so we did not retain this solution. Because MADELEINE internal messages are not self-describing, *packing* and *unpacking* operations have to be done in the same order (matching pack and unpack operations must have the same rank within each series).

Illustration We now describe a small example which illustrates the powerfulness of the MADELEINE interface. Let us say that we want to send a message containing two strings of arbitrary length (unpredictable on the receiving side). These strings are allocated by a function (`generate_str`) returning their address in memory. On the receiving side, it is necessary to get the size of each string (an integer) before extract the string themselves (destination memory has to be dynamically allocated). So the message will contain four elements (two integers and two strings) with a first constraint: each integer must be extracted **RECV_EXPRESS** before the corresponding string.

On the sending side, the code may look like the one in Figure 3. Note that the integers are packed using the **SND_SAFER** mode because the `size` variable is reused inside the loop.

Figure 4 shows the corresponding receiving code. Using the **RECV_CHEAPER** mode for the transmission of strings may allow MADELEINE to send both objects without extra copies in a single network transaction (with a gather/scatter mechanism). This is a major advantage over communication interfaces such as FM [13] that do not allow the user to express such a semantics. Thus, either a extra copy or an extra network packet would have to be sent in such a case.

```
char *data;
int i, size;

   sendbuf_init(dest_node);

   for(i=0; i<2; i++) {
      data = generate_str(); size = strlen(data)+1;
      sendbuf_pack_int(SND_SAFER | RECV_EXPRESS, &size, 1);
      sendbuf_pack_byte(SND_CHEAPER | RECV_CHEAPER, data, size);
   }

   sendbuf_send();
```

Fig. 3. Sending a sample message with the MADELEINE Programming Interface

3.2 The Portability Layer

The *Portability Layer* of MADELEINE is intended to provide a common interface to the various communication subsystems on which it may be ported. The design of this interface is critical because it has to be independent of any particular protocol while staying efficiently portable on top of any of them.

We chose an execution model that is inspired by the Active Messages model [16]. Thus, our model is essentially a message-passing protocol where exchanges are done between traditional processes. This means that although its implementation has to be thread-safe, the protocol does not need to be thread-aware. As a consequence, upper layers have to ensure that only one thread is processing a receive operation at a time.

Messages consist of a set of one or more vectors. Each vector is a contiguous area of memory and is defined by a pair (*address, size*) where the size is expressed in bytes. The first vector of a message has a particular semantics with respect to the receive operation: this vector contains the data that must be extracted prior to the rest of the message (*i.e.* the "header"). Once extracted, these data are immediately made available to the upper layers so that some application-level actions may be executed before the rest of the message is extracted.

The Figure 5 sketches a situation where process A is sending a message to process B. First, the message *header* (*i.e.* the first vector) is sent to process B. On its receipt, a *handler* is executed with the header as a parameter. This handler, which is defined by the upper layer, can inspect the header and take appropriate actions to prepare for full message extraction. Typically, the handler may compute the locations where data should be placed (step (3) in Figure 5). Process A is allowed to send the vectors of remaining data (i.e. the message body) when the handler has completed. Consequently, these data are directly stored at the right memory locations on their receipt.

We emphasize that this scenario is a conceptual description of the protocol. In fact, specific implementations may not strictly follow this communication scheme. We discuss more precisely the implementation issues in Section 4.

```
char *data[2];

void handler()
{
   int i, size;

   for(i=0; i<2; i++) {
      recvbuf_unpack_int(RECV_EXPRESS | SND_SAFER, &size, 1);
                        /* 'size' is available immediately */
      data[i] = malloc(size);
      recvbuf_unpack_byte(RECV_CHEAPER | SND_CHEAPER, data[i], size);
               /* Caution: data is *not* yet available here. */
   }
   recvbuf_receive_data();
}
int main_function()
{
   recvbuf_receive(handler); /* The current thread will block and  */
                        /* wait for the arrival of a message. */
   printf("I received the strings: %s and %s\n", data[0], data[1]);
}
```

Fig. 4. Receiving and inspecting a message with the MADELEINE Programming Interface

The Portability Interface The portability interface we propose is composed of only 8 function prototypes (Table 1) which represent the architecture-dependent primitives of MADELEINE. Two of them (init and exit) deal with the management of connections respectively at the beginning and at the end of the execution of an application. The remaining 6 primitives deal with communication.

Message sending As MADELEINE is intended to be used in a multithreaded context, operations that would ordinarily block the calling process (*e.g.* the sending of a message) should only block the calling thread. However, the portability layer has no knowledge about the thread package which is actually used in the upper layers.

Therefore, a sending operation is done in two steps. A call to the **send_post** primitive initiates the sending of a message and returns without blocking. The destination process and the message (an array of Unix standard **iovec** structures) are given as parameters. Then, further calls to **send_poll** have to be made until the primitive returns TRUE, which means that message transmission has completed on the sending side.

Message receiving During the receipt of a message, the first step is to receive the message header, which is done by a call to **recv_header_post(handler)**

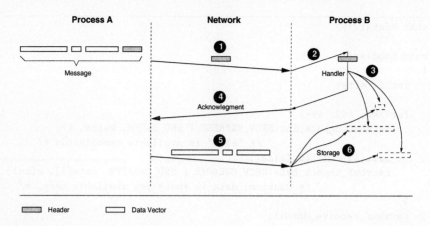

Fig. 5. Conceptual view of the data-exchange protocol in the portability layer.

Table 1. The MADELEINE portability interface.

Function	Operation
init(configSize, taskIDs)	Connections Setup
exit()	Connections Shutdown
send_post(dest, iovec, count)	Post the sending of msg to node 'dest'
send_poll(dest)	Check if msg sending is completed
recv_header_post(func)	Ready to receive a header
recv_header_poll()	Check if header received
recv_body_post(exped, iovec, count)	Ready to receive data
recv_body_poll(exped)	Check if data received

followed by one or more calls to recv_header_poll(). As soon as the header is completely received, the call-back function *handler* gets executed. Thus, the execution flow returns temporarily back to the upper layers.

The idea is that the handler has to inspect the header, to prepare for the receipt of the message body and to actually extract the data from the network. The second step typically leads to building an appropriate array of vectors. The last step is realized by making a call to recv_body_post (followed by subsequent calls to recv_body_poll) with the array as a parameter.

4 Implementation and Performance

The MADELEINE interface is designed to be portable, to provide efficient RPC-like communications and to minimize the overhead introduced over the underlying network protocol. We now illustrate these first two properties by describing the implementation and performance of MADELEINE on top of VIA, TCP. The overhead added by MADELEINE over the underlying protocol is small and constant. This point is detailed in [2].

4.1 High portability

Thanks to its modular structure, the MADELEINE communication library has been ported on a number of network protocols. In particular, it is available on top of low-level protocols such as VIA [5], BIP [14], SCI [6] and SBP [15]. It is also available on top of TCP/IP (for portability), PVM [8] and MPI [11][1]. This ability to adapt to very low-level protocols is the key to get the maximum performance of the underlying network. We now illustrate this point by presenting the implementation and performance of MADELEINE on top of VIA over a Fast-Ethernet network.

MADELEINE over the VIA Interface MADELEINE is one of the first high-level communication software available on top of VIA (Virtual Interface Architecture [5]), the interface proposed by Compaq, Intel and Microsoft for the management of high performance networks. The VI Architecture provides the user application with a protected, directly accessible interface to the network hardware: the *Virtual Interface* (VI). Once connected to another VI, a VI allows bi-directional point-to-point data transfer. Each VI contains a *Send Queue* and a *Receive Queue* where communication requests have to be posted. Such requests are described by *descriptors* that contain pointers to user data buffers. Data can be transmitted via regular message passing mechanisms or via remote direct memory access (read/write) operations. Before a descriptor is posted, all the data buffers it references should be explicitly *registered* in the VIA runtime.

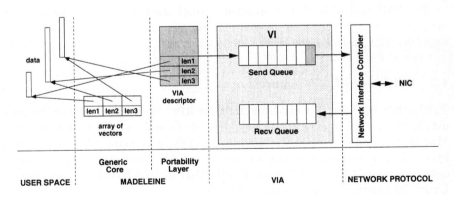

Fig. 6. Sending data in the VIA implementation of MADELEINE.

---

[1] Some vendors implementations of these libraries are very efficient (*e.g.* MPI-F on IBM SP2 or PVMe on Cray T3D).

The implementation of the portability layer on top of this VI architecture is easy and efficient, because gather/scatter functionalities are provided with the descriptor management. Figure 6 shows the typical execution path during the sending of a message *body* formed out of three vectors (see Section 3.2). The madeleine generic message management code first builds an array describing the message. This array is passed to the portability layer which is VIA specific in this case. A single VIA descriptor is then built by only copying the three entries of the array of vectors. At this point, the memory areas containing the user data are registered in VIA. Finally, the descriptor is posted in the Send Queue (which is an asynchronous operation) and the current thread can yield the processor until the operation is completed.

The implementation of a whole message sending is slightly more elaborate because VIA does not provide flow-control. Thus, before a descriptor like the previous one can be posted in a Send Queue, one has to make sure that a corresponding descriptor has already been posted in the destination Receive Queue. MADELEINE uses acknowledgment messages to control the posting of message bodies. As far as headers are concerned, a simple credit-based algorithm is used so that acknowledgement messages can be avoided in many situations.

We have used the M-VIA implementation of VIA (developped at Berkeley Lab, http://www.nersc.gov/research/FTG/via) on a Fast-Ethernet (100Mb/s) network to evaluate the advantage of using a low-level high-performance protocol in comparison to using TCP/IP[2]. Figure 7 shows the performance (latency) of raw message passing over MADELEINE using M-VIA and TCP as underlying protocols. That for, a ping-pong test was run between two Pentium133MHz PCs running Linux.

As one can see, the gap between the two curves is important, which confirms the interest of being portable on low-level protocols. We achieve a 37 μs latency with VIA (140 μs with TCP) and the bandwith for 64 KB messages is greater than 11.5 MB/s.

The PM^2 multithreaded environment, which used MADELEINE for communications, directly benefits from this performance. The PM^2 thread migration time over VIA on this same architecture is 150 μs.

4.2 Efficiency of RPC operations

To illustrate the adequacy of MADELEINE for RPC-based environments, we have measured the performance of processing remote procedure calls respectively over MADELEINE and over MPI. The parameter of the procedure call is an array of bytes which is transmitted without any extra copy. This experiment has been done on a pile of PCs (Intel PentiumPro 200MHz with 64 MB of memory running Linux) interconnected by a Myrinet network.

As discussed previously (Section 2), the only way of realizing such a remote invocation with MPI is to send two messages. The first one carries the header (containing the procedure Id and the array size) and the second one transports

---

[2] The TCP implementation of MADELEINE is described in the following section.

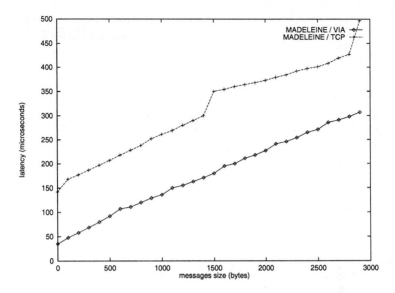

Fig. 7. Performance of MADELEINE on top of M-VIA and TCP over a 100Mb/s Ethernet network.

the array itself. With MADELEINE, we used the RECV_EXPRESS mode to pack the array size and the SND_CHEAPER+RECV_CHEAPER mode to pack the array itself (as in the example shown in Figure 4).

The next section compares the two approaches (MADELEINE vs. MPI) on top of the TCP/IP protocol. First, the TCP implementation of MADELEINE is described. Then, we report on some performance results.

MADELEINE over the TCP Protocol The implementation of MADELEINE over TCP, which provides flow control, is straightforward. The sending primitives of the portability layer (see Table 1) are just implemented by calls to the non-blocking **writev** Unix primitive. The message (header + body) is thus simply sent as a stream of bytes which is regulated by the internal TCP flow control algorithm. On the receiving side, the detection of the message availability (**recv_header_poll**) is realized through a call to the Unix **select** primitive. Then, the message header is extracted by means of a **read** operation. Finally, the message body can be extracted with a **readv** operation. As a result, no extra message nor extra copy is introduced by MADELEINE on top of TCP.

Figure 8 reports the times measured from the initialization of the communication on the source machine to the effective beginning of procedure execution on the destination machine. These times correspond to various array sizes as indicated on the abscissa.

As one can see, the gap in performance between MADELEINE and MPI is huge and proportional to message size (times obtained with MPI are approximately

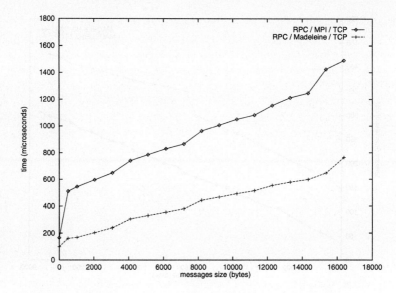

Fig. 8. Performance of Remote Procedure Calls over MADELEINE and over MPI. The underlying communication protocol is TCP/IP.

twice the times obtained with MADELEINE). This shows the better adequacy of the MADELEINE interface to support RPC operations on buffering protocols. Here, the performance gap is due to the number of TCP messages exchanged with the MPI version. In fact, MADELEINE needs only one TCP message to process the RPC, whereas the MPI interface requires the application to send two "user" messages. In addition, the internal flow-control algorithm of the MPI implementation may even trigger the sending of a third message from the receiver back to the sender. It is important to note that this gap is not in any way due to the "quality" of the MPI implementation. It simply reveals that the MPI interface lacks functionalities as far as RPC-like operations are concerned. Of course, experiments featuring other buffering protocols (such as SBP [15]) lead to similar results.

5 Related Work

Many communication libraries have been recently designed to provide portable interfaces and/or efficient implementations to build distributed applications. Among them, the FM (Illinois Fast Messages) [13] communication interface presents some similarities with the MADELEINE portability layer. FM is derived from the AM (Active Messages) communication software designed at Berkeley [16]. It could be considered as a "medium-level" communication layer in that sense that it can be ported on top of previously mentioned low-level ones. Its interface provides a very simple mechanism to send data to a receiving node that is notified upon arrival by the activation of a handler. The 2.0 release of this interface provides some gather/scatter features with the introduction of a "streaming message" concept that allows the sending and the receiving of data in several (not necessarily equal) pieces. This feature may allow an efficient implementation of zero-copy RPC operations as presented in section 2. However, this interface still provides the user with a low abstraction level: data insertion/extraction operations are always synchronous and put a strong constraint on network management. We have seen that MADELEINE does not have this drawback while providing a higher level of abstraction to the upper layers.

6 Conclusion and future work

Most existing multithreaded environments providing RPC-based functionalities are implemented on message-passing communication interfaces such as PVM or MPI for portability reasons. However, we have shown these interfaces are not adequate for this purpose because they lacks expressiveness as far as RPC operations are concerned. Thus, even if their implementation is efficient on a given network technology, applications that need RPC facilities may not be able to achieve much of the underlying hardware's performance.

We have proposed a new portable communication interface which better meets the needs of these environments. MADELEINE is composed of a programming interface providing high-level functionalities to upper-level software, and a portability interface hiding the network specifics. We have explained how these interfaces allows efficient (zero-copy) data transmissions during RPC operations.

We have implemented MADELEINE on top of several low-level network interfaces and have reported its performance on top of VIA and TCP/IP. These experiments have demonstrated the superiority of MADELEINE over classical message-passing interfaces. They also have proved that MADELEINE can be easily implemented on low-level protocols while achieving much of the underlying hardware's performance. MADELEINE is currently available on top of the following network interfaces : VIA, BIP, SBP, Dolphin-SCI, TCP, PVM and MPI. The implementation of an existing multithreaded environment (PM^2) has been successfully ported on top of MADELEINE. The performance of PM^2 on top of MADELEINE validates our work: with a BIP-based implementation on a PentiumPro 200MHz cluster connected by Myrinet, a null RPC takes 11 μs and a thread migration takes 65 μs.

In the near future, we intend to extend the MADELEINE portability interface in order to optimally exploit network protocols providing fixed-size preallocated buffers, such as SBP [15]. We also plan to implement a generic flow control algorithm in the high-level layer of MADELEINE, so that new portability layers would become even easier to implement. Finally, we plan to extend the programming interface in order to provide a better support for SCI networks.

References

1. BODEN, N., COHEN, D., FELDERMANN, R., KULAWIK, A., SEITZ, C., AND SU, W. Myrinet: A Gigabit per second Local Area Network. *IEEE-Micro 15-1* (feb 1995), 29–36. Available from http://www.myri.com/research/publications/Hot.ps.
2. BOUGÉ, L., MÉHAUT, J.-F., AND NAMYST, R. Madeleine: an efficient and portable communication interface for multithreaded environments. In *Proc. 1998 Int. Conf. Parallel Architectures and Compilation Techniques (PACT'98)* (ENST, Paris, France, Oct. 1998), IFIP WG 10.3 and IEEE, pp. 240–247.
3. BRIAT, J., GINZBURG, I., PASIN, M., AND PLATEAU, B. Athapascan runtime : Efficiency for irregular problems. In *Proceedings of the Europar'97 Conference* (Passau, Germany, 1997), Springer Verlag, pp. 590–599.
4. BUBECK, T., AND ROSENSTIEL, W. Verteiltes Rechnen mit DTS (Distributed Thread System). In *Proc. '94 SIPAR-Workshop on Parallel and Distributed Computing* (Suisse, Oct. 1994), M. Aguilar, Ed., Fribourg, pp. 65–68.
5. COMPAQ, INTEL, AND MICROSOFT. *Virtual Interface Architecture Specification*, December 1997. Version 1.0.
6. DOLPHIN INTERCONNECT SOLUTIONS INC. *PCI-SCI Adapter Programming Specification*, March 1997.
7. FOSTER, I., KESSELMAN, C., AND TUECKE, S. The Nexus approach to integrating multithreading and communication. *Journal on Parallel and Distributed Computing*, 37 (1996), 70–82.
8. GEIST, A., BEGUELIN, A., DONGARRA, J., JIANG, W., MANCHECK, R., AND SUNDERAM, V. *PVM: Parallel Virtual Machine. A User's Guide and Tutorial for Networked Parallel Computing*, 1994.
9. HAINES, M., CRONK, D., AND MEHROTRA, P. On the design of chant: A talking threads package. In *Proc. of Supercomputing'94* (Washington, November 1994), pp. 350–359.
10. LAURIA, M., AND CHIEN, A. MPI-FM: High performance MPI on workstation clusters. *Jounal on Parallel and Distributed Computing*, 40 (01) (1997), 4–18.
11. MESSAGE PASSING INTERFACE FORUM. *MPI: A Message-Passing Interface Standard*, March 1994. available from www.mpi-forum.org.
12. NAMYST, R., AND MEHAUT, J. PM^2: Parallel Multithreaded Machine. a computing environment for distributed architectures. In *ParCo'95 (PARallel COmputing)* (Sep 1995), Elsevier Science Publishers, pp. 279–285.
13. PAKIN, S., LAURIA, M., AND CHIEN, A. High Performance Messaging on Workstations : Illinois Fast Messages (FM) for Myrinet. In *Proc. of Supercomputing'95* (San Diego, California, December 1995). Available from http://www-csag.cs.uiuc.edu/papers/myrinet-fm-sc95.ps.
14. PRYLLI, L., AND TOURANCHEAU, B. BIP: A New Protocol designed for High-Performance Networking on Myrinet. In *Proc. of PC-NOW IPPS-SPDP98* (Orlando, USA, March 1998).
15. RUSSELL, R., AND HATCHER, P. Efficient kernel support for reliable communication. In *13th ACM Symposium on Applied Computing* (Atlanta, GA, February 1998). To appear.
16. VON EICKEN, T., CULLER, D., GOLDSTEIN, S., AND SCHAUSER, K. Active messages: a mechanism for integrated communication and computation. In *Proc. 19th Int'l Symposium on Computer Architecture* (May 1992).

Application Performance of a Linux Cluster Using Converse*

Laxmikant Kalé, Robert Brunner, James Phillips and Krishnan Varadarajan
kale@cs.uiuc.edu, {brunner,krishnan,jim}@ks.uiuc.edu

Department of Computer Science, and
Beckman Institute of Advanced Science and Technology
University of Illinois at Urbana-Champaign

Abstract. Clusters of PCs are an attractive platform for parallel applications because of their cost effectiveness. We have implemented an interoperable runtime system called Converse on a cluster of Linux PCs connected by an inexpensive switched Fast Ethernet. This paper presents our implementation and its performance evaluation. We consider the question of the performance impact of using inexpensive communication hardware on real applications, using a large production-quality molecular dynamics program, NAMD, that runs on this cluster.

1 Introduction

PCs, as commodity components, have always been fairly inexpensive. Over the past few years, PCs have gained significantly in performance even as their price continues to drop. As a result PCs make an attractive platform for computational programs. By connecting many PCs together in a cluster, one can build a "parallel machine" for a fraction of the cost of dedicated parallel machines such as the Cray T3E, SGI Origin 2000, or the ASCI Red (a one of a kind machine built by Intel, for the U.S. Department of Energy). The Beowulf project [12] exemplifies this cost effective approach to parallel computing. Several earlier parallel programming systems, such as Linda [1], Charm [2] and PVM [13] also supported clusters of workstations.

Such commodity clusters can either use commodity communication hardware, such as switched Fast Ethernet, or hardware designed especially for high-performance computing, such as Myrinet or switches based on the VI Architecture standard. Significant research has been carried out [10] in improving communication speeds using such dedicated hardware. Although such hardware improves communication speeds dramatically, at this point in time it is considerably more expensive. The large number of PCs on peoples' desks makes PC

---

* This work was supported by the National Institutes of Health (NIH PHS 5 P41 RR05969-04) and the National Science Foundation (NSF/GCAG BIR 93-18159 and NSF BIR 94-23827 EQ). James Phillips was supported by a Computational Science Graduate Fellowship from the United States Department of Energy.

clustering an attractive platform for parallel applications. As most of the compute power available on the desktop typically remains unused (especially after work hours, but even during the workday, since most user applications do not require much computational power) parallel applications can be deployed on such platforms without significant up-front costs. Such environments typically have only commodity communication hardware.

The questions that arise in this context are: Is a PC cluster with commodity communication hardware a viable parallel platform? What is the extent of communication performance loss with such hardware? And, most importantly, what is the impact on real applications of this low communication performance? We attempt to answer these questions in this paper.

As a part of a large collaborative research project, we are developing a parallel molecular dynamics program, called NAMD [6, 9]. Although the program runs on dedicated parallel machines at the national centers, we needed a platform we could use continuously. We have built a cluster of 32 Intel processors (16 2-way SMP nodes), connected by a 100 Mbps switched Fast Ethernet. We ported the Converse [5] run-time system to this platform. In this paper, we first describe the Converse port, and its raw communication performance. We then examine the impact of communication performance on a production application, NAMD.

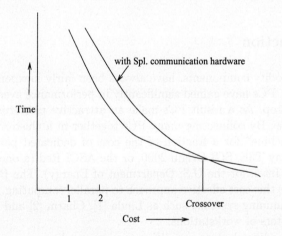

Fig. 1. Cost and Performance of systems with and without specialized communication hardware

The broader issue that this study raises is whether special purpose communication hardware is necessary for cost-effective performance on various applications. In general, the answer to this question depends on the number of processors to be used, and the characteristics of the application. Figure 1 shows a schematic plot of time-to-completion for an application versus the cost of the hardware, for (a) a simple system using commodity hardware (such as switched

Ethernet) and (b) a more expensive one using specialized communication hardware (such as Myrinet switches). On one processor, clearly both systems perform equally, but (b) costs more. On a very large number of processors, for most applications, (b) will perform better when comparing identical cost configurations. Somewhere in the middle, the curves cross over. The last section of our paper studies the factors that affect this crossover point.

2 Converse on the Linux cluster

Converse is a run-time framework that we have developed to simplify implementations of new parallel languages, and to support interoperability among them. Converse has a component based architecture, including modules for interprocess communication, user-level threads, prioritized scheduling, and load balancing. Using a common scheduler for entities across different parallel paradigms/languages, such as threads, handlers, and data-driven objects, Converse allows multiple language modules to interoperate concurrently. Its component architecture allows one to put together run-time systems for new languages quickly, without losing performance. Converse supports portability, and is available on major parallel machines and clusters of Unix workstations.

For the cluster of PCs, we adapted the workstations version of Converse. As we did not want to limit the number of PCs connected in a cluster, we use the connection-less UDP protocol. The user's message is divided into fixed-size packets. As UDP packet delivery is not guaranteed, we implement a window-based flow-control protocol. To deal with the fact that there are two SMP processors on each PC, we use a single socket per PC. A producer-consumer queue is used for each processor to store its incoming messages. Since there is exactly one producer and one consumer, this queue is implemented without using any locks. Messages between two processors within the PC are transmitted using the shared memory, avoiding the socket.

2.1 Communication Performance

To analyze the raw communication performance of our implementation, we used a simple ping-pong program. The results are shown in Figure 2. A short message takes about 250μsec. for its application-to-application delivery across the network. The incremental cost for each additional byte is approximately 130 nanoseconds, leading to the bandwidth of about 7.4 MB/sec (59 Mbits/sec). The communication bandwidth attained in this benchmark is also shown in the same Figure, as a function of message size. Figure 2 indicates that we attain half the peak bandwidth for 2 KB messages. Similar data for large messages is shown in Figure 3. We believe that our implementation can be further optimized to some extent, but we expect the attained bandwidth to be limited to about 10 MB/sec.

To compare this performance with a dedicated parallel machine, we present Converse communication performance on the ASCI Red machine in Figure 4. Latency of around 45μsec, and bandwidth of 50 MB/sec is attained.

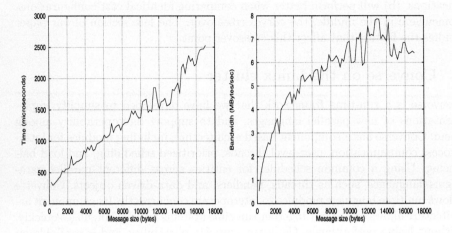

Fig. 2. Raw message time (one-way) and bandwidth for the ping-pong benchmark on the Linux cluster.

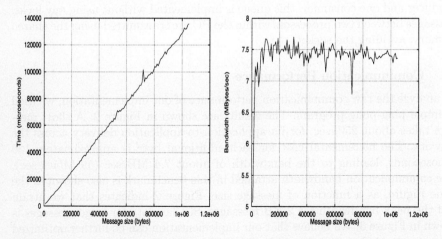

Fig. 3. Raw message time (one-way) and bandwidth for large messages on the Linux cluster.

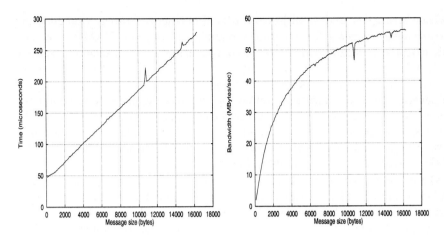

Fig. 4. Raw message time (one-way) and bandwidth for the ping-pong benchmark on the ASCI-Red.

On PC clusters, Myrinet based communication systems (such as FM [8], which was one of the earliest) provide a much higher performance compared with our switched Ethernet implementation. One of the fastest, for example, is BIP [10], which supports a less than 5μsec. latency, and 120 MB/sec bandwidth. So, while it is still important to try improving the performance on the commodity hardware, it is worthwhile asking whether this big a disparity in communication performance renders parallel applications infeasible. To answer this question, we turn to a study of a full-fledged application.

3 Parallel Molecular Dynamics: NAMD

Molecular dynamics programs simulate the motions of molecules by repeatedly computing the forces on all the atoms and numerically integrating the equations of motion. Our users are interested in using molecular dynamics simulation to study the behavior of biochemically significant molecules such as proteins, membranes and DNA. Typical simulations are composed of one thousand to one hundred thousand atoms, and require simulation times of hundreds of picoseconds. Since the numerical characteristics of such simulations mandate timesteps of the order of one femtosecond, a complete simulation requires several hundred thousand timesteps. Each timestep requires several CPU-seconds of computation, so sequential simulations would require weeks or months to complete. Parallel simulation is almost mandatory for such simulations, but parallelizing the integration algorithm is complicated since each timestep must be completed on every processor before the next timestep can begin. Therefore, efficient parallel simulation requires parallelizing operations within a single timestep which only requires a few seconds to compute sequentially.

NAMD is a parallel molecular dynamics program, designed to be scalable [6]. It is implemented using Charm++ [7] and Converse. It uses aggressive parallelization strategies, and an object-based load balancing strategy that relies on run-time measurements for accurate remapping of objects. NAMD is a relatively irregular program, with thousands of concurrent objects, each representing widely different computational load, and with complex communication patterns. It is one of the fastest production-quality parallel molecular dynamics program, and several scientific simulation studies of biomolecules have been conducted using it. NAMD relies on the message-driven communication model provided by Converse to adaptively overlap computation with communication. It also uses the multilingual capabilities of Converse to incorporate the DPMTA PVM-based full-electrostatics library from Duke University [11].

The current version of NAMD is optimized to run efficiently on dedicated parallel machines. This required optimizing the communication so that each timestep can be completed in less than a second on machines with hundreds of processors. The parallel molecular dynamics algorithm has the following outline:

1. Distribute atom positions from owning processors to several other processors which must compute atom forces.
2. Compute forces based on incoming atom positions.
3. Return forces to processors which own those atoms.
4. Update atom positions based on incoming forces.
5. Return to the first step.

NAMD uses a hybrid spatial/force decomposition scheme. Atoms are not owned by processors, but by objects called *patches*, which are distributed among the processors. Each patch is responsible for the atoms in a cubical region of the simulation space. Each patch sends requests to a set of *compute* objects, which compute the forces exerted on the atoms of that patch by other atoms in the simulation. Unlike in a pure spatial decomposition scheme, a compute object may reside on a different processor from its associated patches, in which case the atom positions from a remote patch are stored in a *proxy patch*. This minimizes communication for the case where several compute objects on a processor are accessing the patch data; the atom positions are sent to the processor only once per timestep, and the returned forces from each compute are combined and sent as a single message.

Allowing compute objects to reside on any processor, independently of where its patches reside, provides two important capabilities. First, NAMD can utilize parallel machines with more processors than patches. Second, compute objects can be shifted around to achieve good load balance. NAMD uses a run-time, measurement-based load balancing algorithm, that measures how much computation time each compute object consumes, and then periodically rebalances the compute objects to minimize idle time while simultaneously keeping communication overhead small.

NAMD is a production-quality application currently being used for several simulations. It is freely distributed. More details are available at the NAMD web site: http://www.ks.uiuc.edu/Research/namd/namd.html.

3.1 NAMD performance

NAMD has achieved its design goal of scalable performance on large parallel machines. For example, it yielded a speedup of over 180 using 220 processors. Efficient scalability to large numbers of processors required us to carefully optimize communication patterns in the program. These communication optimizations have had the unintended side effect of producing good speedups on smaller machines with relatively poor communication performance, such as the Linux/Fast Ethernet cluster.

Once Converse was ported to the cluster, NAMD was ported relatively effortlessly. Table 1 shows the performance of NAMD on the PC cluster (400MHz Pentium II), and compares it with its performance on two dedicated parallel machine, the Cray T3E (450MHz Alpha) and the Intel ASCI Red (200MHz Pentium Pro). Despite the irregular and dynamic nature of the parallel computation, the adaptive load balancing strategies and prioritized scheduling techniques in Converse lead to speed up of about 14.7 on 16 processors of this dedicated machine. The speedup using the cluster is about 12.2 on 16 processors, and 19.9 on 32 processors. Thus, in spite of the huge communication penalty, one is able to derive useful speedups on this cluster.

There are two separate conclusions one can derive from this experience. Firstly, it is clear that an existing network, not designed for parallel processing, can be used profitably for parallel applications. The second question is more subtle: is the investment in extra communication hardware worthwhile? Given NAMD performance on dedicated parallel machines, we expect the speedup would be around 28 on 32 PEs using specialized communication hardware. This represents about forty percent increase in performance over the Linux cluster performance on 32 processors. The current cost of additional communication hardware is between 25 to 40 percent of the cost of the PCs themselves. So, additional hardware may be a worthwhile investment. However, on 16 or fewer processors, additional hardware is clearly not cost effective. Even on 32 processors, planned optimizations will narrow the gap further. Another implication of this result is that although it is intellectually challenging to improve the raw communication cost, the effort will not lead to a commensurate performance improvement in real applications.

Furthermore, the execution times point out another advantage the clusters have over dedicated machines. The latest processor technology can be quickly integrated into a cluster, whereas a dedicated parallel machine is somewhat slower in incorporating newer processors. Although the ASCI Red used the fastest Intel processors available when it was designed, they are now less than half the speed of commonly available Linux machines.

4 A Controlled Benchmark

For our molecular dynamics program, we observed that beyond twenty processors, extra communication hardware becomes cost effective. However, we cannot

		Processors							
		1	2	4	8	16	20	24	30	32
T3E	Time			6.12	3.10	1.60	0.810				0.397
	Speedup		(1.97)	3.89	7.54	14.9				30.3
ASCI Red	Time	28.0	13.9	7.24	3.76	1.91				1.01
	Speedup	1.0	2.01	3.87	7.45	14.7				27.9
LINUX	Time	12.34	6.42	3.28	1.69	1.01	0.85	0.74	0.65	0.62
	Speedup	1.0	1.92	3.77	7.30	12.21	14.51	16.68	18.98	19.90

Table 1. Execution time (sec./timestep) for ER-ERE (36,573 atoms, 12Å cutoff).

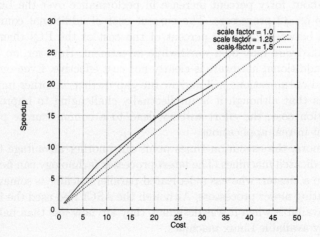

Fig. 5. NAMD Speedup, adjusted for communication cost.

generalize this observation to all application. The crossover point will depend on the characteristics of the particular application. More specifically, it depends on the communication load and how it scales with the number of processors in a given application.

To examine the effect of communication performance on the overall performance, and its relationship to the cost of the entire system, we studied a simple, well-controlled benchmark. The program used for the study is Jacobi relaxation using a five point stencil. The data for this program is a square array, which is partitioned across all the processors. Thus, each processor itself has a square piece of the array, assuming the total number of processors is a square. In each iteration, a processor sends the boundary elements (either a row or a column) to each neighboring processor. Thus, except for the processors on the boundary, each processor sends and receives four messages in each iteration. As the amount of computation per processor is proportional to the size of its partition, while the communication is proportional to the size of its boundary, we can control the computation to communication ratio by varying the size of the data array.

To compare the performance of this program on an architecture with fast communication hardware, we used the following procedure. First, we ran the program on the Cray T3E. To account for differences in absolute processor speed between the T3E and the Linux cluster, we compare speedups rather than absolute speeds. Thus, we are assuming that by using a more expensive communication hardware with Linux, we would be able to achieve speedups similar to those on a tightly coupled machine (T3E). To compare cost performance, we now plot the speedups as a function of cost. For concreteness, we assume that specialized communication hardware will cost fifty percent more than the cost of each processor with only commodity communications. The speedups for each problem size are plotted in Figure 6. Rather than plotting speedup against number of processors, the x-axis of these plots is *cost*, where *cost = # of processors * scale factor*. For example, the cost of two processor with commodity communications is 2.0, and the cost of two processor with specialized communications is 3.0.

For an application group deciding between building PC cluster with or without a high-cost communication network, can we draw any lessons from these experiments? Consider how the performance of Jacobi behaves in Figure 6 as the number of processors increases. A simple analysis shows that for 16 processors, each processor sends an average of 3 messages per iteration (48 messages total). For 25 processors, the above analysis gives almost the same number of messages, an average of 3.2 messages per iteration (80 messages total). For the grids examined, each message is only a few hundred bytes long. We know from the communication studies in earlier sections that for these messages, the α component (per message cost) dominates over the β component (per byte cost), so the communication time is determined by the number of messages. For a fixed grid size the computation scales perfectly with the number of processors. Therefore the computation to communication ratio decreases, making expen-

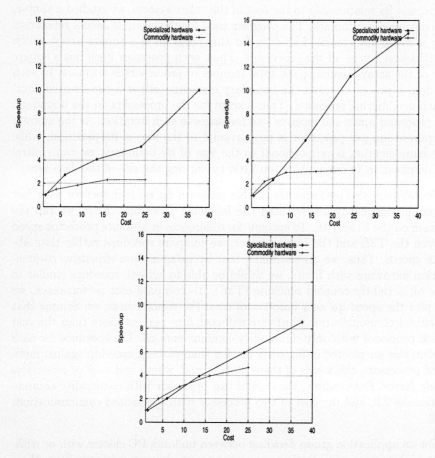

Fig. 6. Speedup versus cost for the Jacobi benchmark (Top left : 240×240 grid, Top right: 360×360 grid, Bottom: 480×480 grid)

sive communication hardware a more desirable alternative to larger numbers of processors.

Consider also the three problem sizes shown in Figure 6. Each problem size represents a different "application" with different computation to communication ratios. Clearly, the crossover point depends on the computation to communication ratio. As the grid size increases, the computation time increases. As described above, the communication time for these grid sizes is determined by the number of messages. Since the number of messages is fixed as the grid size changes, the communication time remains constant. Therefore the speed of communication hardware becomes a less significant factor in program performance, and the crossover point shifts to the right.

In general, one must study how the number of messages, m, and the number of bytes communicated per processor, n, varies as a function of the total number of processors used, p. Let us assume these characteristics are given by functions $m(p)$ and $n(p)$ respectively. Also, let α_1 and α_2 be the per message costs on the commodity and special-purpose hardware respectively. Similarly, let β_1 and β_2 be the per byte cost of communication.

For applications like Jacobi, the computation scales perfectly. So, if T_c is the computation time on 1 processors, the computation time on p processors will be T_c/p.

Let r be the fractional cost of communication hardware per processor, so that the communication hardware cost per processor $= r * $ (cost per node). Given a fixed budget, one has a choice of spending it on p_1 processors or $p_2 = p_1/(1+r)$ processors with expensive communication hardware.

Now that we are comparing equal cost configurations, the total time to completion, T_{total} for each can be calculated, as

$$T_{total} = \alpha_1 * m(p_1) + \beta_1 * n(p_1) + \frac{T_c}{p_1} \qquad (1)$$

and

$$T_{total} = \alpha_2 * m(p_2) + \beta_2 * n(p_2) + \frac{T_c}{p_2} \qquad (2)$$

$$= \alpha_2 * m(p_2) + \beta_2 * n(p_2) + \frac{T_c(1+r)}{p_1} \qquad (3)$$

Equations 1 and 3 clearly show the tradeoff involved: although the communication terms (α_2 and β_2 terms) are smaller for the expensive architecture, the computation time may increase because we can afford only fewer processors. The important point is that the switch-over critically depends on the nature of the functions $m(p)$ and $n(p)$, which can only be characterized through careful analysis of the particular application.

For application where the computation time itself does not scale well, due to load imbalances, critical path effects, or algorithmic overheads, a more refined analysis must be conducted to determine the crossover point. In such situations the crossover point will be pushed in favor of communication hardware, due to the higher efficiency attained on fewer processors.

5 Conclusion

A PC cluster is indeed a cost-effective platform for parallel applications. However, the question of whether to use special-purpose communication hardware (such as Myrinet) turns out to be somewhat complex. Our experience with NAMD, a production-quality molecular dynamics program, demonstrated that commodity hardware (such as switched 100Mb/sec Fast Ethernet) is adequate to provide good speedups up to 16 processors. Although the speedup on 32 processors was around 20 with the cluster, as compared with 28 on a dedicated parallel machine, one must take the cost of additional communication hardware into account. We showed a simple method for cost-performance analysis that can be used to decide whether to go for high-cost communication hardware. The crossover point beyond which it becomes profitable to deploy such hardware was seen to depend on specific communication-related characteristics of the application. Once the communication scaling behavior of the application is understood, one can simply plug in the appropriate values in expressions given in the paper, to determine which configuration will be appropriate for a given budget.

Success attained by the Avalon project [14], with a cluster of Alpha workstations connected by relatively inexpensive Fast Ethernet, indicates that for several applications one may be able to benefit by using commodity communication hardware. Of course, with the emergence of VI Architecture standard, and results such as those presented in this paper, coupled with growing popularity of clusters, the cost of special purpose hardware itself may drop substantially. However, this will be offset by further reductions in microprocessor costs. A careful cost analysis may still be needed in the future.

The Converse system itself is neutral with respect to communication technology. Converse runs on almost all available communication hardware, including ATM, shared Ethernet, switched Ethernet, and Myrinet. So, programs written using Converse will be able to exploit any emerging technology.

6 Acknowledgements

The computer time for the performance comparisons on the ASCI Red was provided by the Center for Simulation of Advanced Rockets at the University of Illinois at Urbana-Champaign, and the Accelerated Strategic Computing Initiative of the Department of Energy. The T3E time was provided by the Pittsburgh Supercomputing Center.

References

1. N. Carriero and D. Gelernter. Linda in Context. *Communications of the ACM*, 32(4), April 1989.
2. W. Fenton, B. Ramkumar, V. Saletore, A. Sinha, and L.V. Kalé. Supporting machine independent programming on diverse parallel architectures. In ICPP91 [3], pages 193–201.

3. *Proceedings of the 1991 International Conference on Parallel Processing*, St. Charles, IL, August 1991.
4. L. V. Kalé, Milind Bhandarkar, Robert Brunner, N. Krawetz, J. Phillips, and A. Shinozaki. NAMD: A case study in multilingual parallel programming. In *Proc. 10th International Workshop on Languages and Compilers for Parallel Computing*, pages 367–381, Minneapolis, Minnesota, August 1997.
5. L. V. Kalé, Milind Bhandarkar, Narain Jagathesan, Sanjeev Krishnan, and Joshua Yelon. Converse: An Interoperable Framework for Parallel Programming. In *Proceedings of the 10th International Parallel Processing Symposium*, pages 212–217, Honolulu, Hawaii, April 1996.
6. Laxmikant Kalé, Robert Skeel, Milind Bhandarkar, Robert Brunner, Attila Gursoy, Neal Krawetz, James Phillips, Aritomo Shinozaki, Krishnan Varadarajan, and Klaus Schulten. NAMD2: Greater scalability for parallel molecular dynamics. *J. Comp. Phys.*, 1998. In press.
7. L.V. Kale and S. Krishnan. Charm++: A portable concurrent object oriented system based on C++. In *Proceedings of the Conference on Object Oriented Programming Systems, Languages and Applications*, September 1993.
8. S. Pakin, M. Lauria, and A. Chien. High Performance Messaging on Workstations: Illinois Fast Messages (FM) for Myrinet. In *Proceedings of Supercomputing 1995*, dec 1995.
9. James C. Phillips, Robert Brunner, Aritomo Shinozaki, Milind Bhandarkar, Neal Krawetz, Laxmikant Kalé, Robert D. Skeel, and Klaus Schulten. Avoiding algorithmic obfuscation in a message-driven parallel MD code. In P. Deuflhard, J. Hermans, B. Leimkuhler, A. Mark, S. Reich, and R. D. Skeel, editors, *Computational Molecular Dynamics: Challenges, Methods, Ideas*, volume 4 of *Lecture Notes in Computational Science and Engineering*, pages 455–468. Springer-Verlag, November 1998.
10. L. Prylli and B. Tourancheau. BIP: A New Protocol Designed for High Performance Networking on Myrinet. *Lecture Notes in Computer Science*, 1388:472–??, 1998.
11. W. Rankin and J. Board. A portable distributed implementation of the parallel multipole tree algorithm. *IEEE Symposium on High Performance Distributed Computing*, 1995. [Duke University Technical Report 95-002].
12. T. Sterling, D. Savarese, D. J. Becker, J. E. Dorband, U. A. Ranawake, and C. V. Packer. BEOWULF : A parallel workstation for scientific computation. In *International Conference on Parallel Processing, Vol.1: Architecture*, pages 11–14, Boca Raton, USA, August 1995. CRC Press.
13. V.S. Sunderam. PVM: A Framework for Parallel Distributed Computing. *Concurrency: Practice and Experience*, 2(4), December 1990.
14. Michael S. Warren, Timothy C. Germann, Peter S. Lomdahl, David M. Beazley, and John K. Salmon. Avalon: An alpha/linux cluster achieves 10 gflops for $150k. Technical report, Los Alamos National Laboratories, http://loki-www.lanl.gov/papers/sc98, SC98 Gordon Bell Award Entry, 1998.

An Efficient and Transparent Thread Migration Scheme in the PM2 Runtime System

Gabriel Antoniu, Luc Bougé, and Raymond Namyst

LIP, ENS Lyon, 46, Allée d'Italie, 69364 Lyon Cedex 07, France.
Contact: {Gabriel.Antoniu,Luc.Bouge,Raymond.Namyst}@ens-lyon.fr.

Abstract. This paper describes a new *iso-address* approach to the dynamic allocation of data in a multithreaded runtime system with thread migration capability. The system guarantees that the migrated threads and their associated static data are relocated exactly at the same virtual address on the destination nodes, so that no post-migration processing is needed to keep pointers valid. In the experiments reported, a thread can be migrated in less than $75\mu s$.

1 Introduction

Multithreading has proven useful to implement massively parallel activities in distributed systems, since it provides an efficient way of overlapping communication and computation. When the application behavior is hardly predictable at compile time, *dynamic* load balancing becomes essential. It can be achieved by transparently migrating computation threads from the overloaded nodes to the underloaded ones. In the implementation described below, a thread can be migrated across the Myrinet network in less than 75 μs.

Migrating a thread usually means moving the thread stack, but sometimes may also mean moving the *static data* used by the thread. In this respect, several migration approaches have been implemented in the existing multithreaded systems, depending on the rationale underlying the use of thread migration. In Ariadne [8], threads are migrated to get close to the remote data they use. Static data never moves. On migration, the thread stack is relocated at a usually different address on the destination node, such that pointers need to be updated. As shown in 2, several problems cannot be solved by this approach. In Millipede [7], thread migration is directed by a load balancing module integrated in the system, whereas static data get moved only when they get accessed by remote threads. The threads and their data are always relocated at the same virtual addresses on all nodes. Yet, thread creation is expensive, therefore the number of concurrent threads is statically fixed at initialization. In both systems mentioned above, data are shared and can be accessed by more than one thread. UPVM [4] provides thread migration for PVM applications, in order to support load balancing. Threads have private heaps, for private dynamic allocations. Thread creation is expensive in this system, too, since it is carried out by means of a global synchronization. Besides, the size of the thread private heaps is fixed at thread creation: the amount of data that a thread can allocate is limited.

Our interest in iso-address allocation and migration stems from data-parallel compiling. Consequently, our study focuses on the case in which data are not shared: they belong to some unique thread and thus have to follow it on migration. Our iso-address allocator has been implemented in the PM2 multithreaded runtime system [10], which serves as a runtime support for two data-parallel compilers [1]. We target applications having to execute on *homogeneous* clusters of workstations or PCs interconnected by a high-speed network (e.g., Myrinet [9]).

This paper is structured as follows. In section 2, we give a quick description of PM2: a multithreaded runtime system providing thread migration. An overview of our iso-address approach is given in section 3 and some implementation details are presented in section 4. Section 5 shows some performance figures. Finally, section 6 summarizes our main results and points out what we intend to address in the near future.

2 PM2: a multithreaded runtime system with thread migration

PM2 is a multithreaded runtime system especially designed to serve as a runtime support for highly parallel irregular applications. In such applications, threads may need to start or terminate at arbitrary moments during the execution. At the same time, the system has to efficiently cope with a large number of concurrent threads. Therefore, PM2 provides very efficient primitives to handle these operations: creation, destruction and context switching. A distinctive feature of PM2 is thread migration. Since the execution of irregular applications may lead to severe load imbalances, thread migration can be used to support the implementation of load balancing policies based on dynamic activity redistribution.

In a PM2 application, there is a single (heavy) process running at each node and each such process may contain tens of thousands of threads. We often identify this container process with the node running it. At the simplest level, a PM2 thread is an execution flow managing a set of *resources*, i.e., its state descriptor and its private execution stack. The code to be executed by the threads is replicated on each node (SPMD approach) and is not part of the thread. Again, we emphasize that we do not consider the aspects of data sharing between threads in this paper, nor the problem of a thread using global process resources such as files, network interfaces, etc. In this setting, migrating a thread simply means moving the thread *resources* from the (heavy) process running on the local node to another (heavy) process located on some remote node. In PM2, the migration operation is carried out in three main steps:

1. The thread gets stopped (*frozen*) and its resources get copied to a communication buffer. The memory area storing the resources is set free.
2. The buffer contents get sent to the destination node through the network.
3. An adequate memory area is allocated on the destination node, the thread resources are copied into it, and the thread execution is resumed.

In PM2, any thread may "decide" to migrate to another node at any arbitrary point during its execution. It may also be *preemptively* migrated by another thread running on the same node. This latter property is essential, since it ensures that application threads may be *transparently* migrated across the nodes. Consequently, a *generic* module implemented *outside* the running application could balance the load by migrating the application threads. The threads are unaware of their being migrated and keep on running irrespective of their location.

```
Source code:
void p1()
{
  int x;

  x = 1;
  pm2_printf("value = %d\n", x);
  pm2_migrate(marcel_self(), 1);
  pm2_printf("value = %d\n", x);
}

Execution:
[node0] value = 1
[node1] value = 1
```

Fig. 1. Thread migration without pointers.

```
Source code:
void p2()
{
    int x;
    int *ptr = &x;

    x = 1;
    pm2_printf("value = %d\n", *ptr);
    pm2_migrate(marcel_self(), 1);
    pm2_printf("value = %d\n", *ptr);
}

Execution:
[node0] value = 1
Segmentation fault
```

Fig. 2. Thread migration in the presence of pointers to stack data

An example of thread migration is given on Figure 1. Assume that a thread running on node 0 calls procedure p1. The thread declares a local variable x, writes the value 1 to this variable, then prints it. Next, the thread migrates to node 1 and prints the value of the variable x again. At run time, we can see that the value 1 is displayed in both cases, before and after migration. The local variable x gets automatically moved to node 1, since it is stored in the thread stack.

A difficulty turns up as soon as a migrating thread makes use of pointers. Such a situation is illustrated on Figure 2. Here, the thread which calls p2 reads variable x through pointer ptr. After migration, there is no guarantee that variable x is still located at address ptr and the execution (most probably!) fails.

One way to tackle this problem is to update all references to stack data after migration, before the thread resumes its execution by adding some offset to all pointers. Two categories of pointers to stack data require such post-migration processing: the *implicit* pointers generated by the compiler in order to chain the stack frames and the *explicit* pointers used by the programmer. The former may be identified using some knowledge about the way they are generated by the compiler, whereas the latter need to be explicitly declared to the system, in order to enable their update after migration. Such an approach was implemented in the early versions of PM2, which provided primitives to register/unregister user-level pointers. When a thread moved to another node, all its registered pointers were updated (Figure 3).

```
Source code:
void p2()
{
  int x;
  int *ptr = &x;
  unsigned int key;
  key = pm2_register_pointer(&ptr);
  x = 1;
  pm2_printf("value = %d\n", *ptr);
  pm2_migrate(marcel_self(), 1);
  pm2_printf("value = %d\n", *ptr);
  pm2_unregister_pointer(key);
}

Execution:
[node0] value = 1
[node1] value = 1
```

Fig. 3. Thread migration with registered pointers

```
Source code:
void p3 ()
{
  int *t =
    (int *)malloc (100 * sizeof(int));

  t[10] = 1;
  pm2_printf("value = %d\n", t[10]);
  pm2_migrate(marcel_self(), 1);
  pm2_printf("value = %d\n", t[10]);
}

Execution:
[node0] value = 1
Segmentation fault
```

Fig. 4. Thread migration with pointers to heap data

Clearly, this approach does not extend to complex applications. Moreover it does not cope with resources located outside of the stack, such as heap data dynamically allocated by the malloc primitive of the C language. Figure 4 shows a thread which calls malloc to allocate some memory area, writes (potentially large) data into this area, migrates, and eventually tries to read at the same virtual address. The program obviously fails, since the allocated data has not been migrated.

One way to solve this problem consists in reallocating the data on the destination node. In this case, the programmer has to explicitly handle the data packing and unpacking, and to manage the pointer updating as the allocation address are usually different from the original one. As in the case of pointers to stack data, this approach cannot be used for arbitrarily complex applications making use of a large number of pointers to heap data. Moreover, this approach cannot cope with compiler-generated pointers in case optimization options are used, since such pointers are not registered and cannot be updated. Fundamental compiler optimizations such as using pointers instead of indices to scan large arrays are thus forbidden.

3 Our approach: the isomalloc memory allocator

3.1 General overview

A much better approach to the problem described in the previous section is to provide a mechanism able to guarantee that *both* the stack and the private, dynamically allocated data of a thread can be migrated and reallocated at the same virtual address on the destination node (*iso-address allocation and migration*). The idea is to *locally* allocate storage areas in a system-wide, *globally* consistent way. The allocation mechanism must guarantee that each range of virtual addresses at which memory has been mmapped at some node is kept free on all the other nodes. Such an approach has several advantages.

Simplicity The migration mechanism is simplified, because no post-migration pointer update is necessary any longer.
Transparency *Applications* may make free use of pointers without having to take into account possible problems related to thread migration. User-level pointers are always guaranteed to be safe.
Portability No *compiler* knowledge about the thread stack structure is required, since the stack contents remains exactly the same after migration. In particular, compiler-generated pointers are migration-safe, too. Consequently, any compiler may be used and compiler optimizations are allowed.
Preemptiveness Preemptive migration is possible, given that no assumption is made about the thread state at migration time.

The isomalloc allocation mechanism relies on a few basic principles. These rules ensure that each node may use its globally reserved memory without having to "inform" the other nodes. We thus avoid any synchronization when allocating memory to threads.

1. The physical execution environment is assumed to be *homogeneous* (same type of processor, same operating system). Moreover, all nodes have the same memory mapping: the same binary code is loaded on each of them at the same virtual address (so that no code needs getting moved upon migration). The (unique) process stack is also located at the same virtual address on all nodes.
2. On each node, all iso-address allocations take place within a special address range called *iso-address area*. We have located it between the process stack and the heap (Figure 5). This zone corresponds to the same virtual range on all nodes.
3. Separate ranges of virtual addresses within the iso-address area are *globally reserved* for each node, so that each address may be used by a single node at a time.
4. The *actual* memory allocation is carried out *locally*, within an address range belonging to the node on which the allocation request is made.

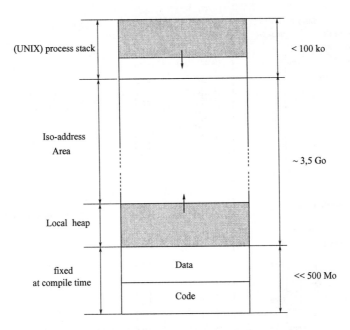

Fig. 5. All nodes have the same memory mapping. In particular, the iso-address area covers the same virtual address range on all nodes

3.2 The slot layer

In this improved view, a PM2 thread is an execution flow managing a set of *resources*, i.e., its state descriptor, its private execution stack, and a series a dynamically allocated sub-areas within the iso-address area. Let us introduce some terminology at this point for the sake of clarity. An *address slot* is a range of virtual addresses within the iso-address area. A slot is *free* if no memory has been mmapped at this address. Otherwise, it is *busy*, and we say that memory has been *allocated* in this slot. Then, data may be stored within this slot of virtual addresses. The **iso-address** discipline guarantees that a slot which is busy on a node is guaranteed to remain free on any other node.

Our goal is to design the management policy so as to avoid inter-node synchronization as far as possible and to remain compatible with the heap management mechanisms of the container (heavy) process. To manage slots in a consistent system-wide manner, it is convenient to give them a uniform size, very much like memory pages at the node level. The choice of this size is obviously crucial and we will discuss it later. We introduce again some terminology. At any point, exactly *one* agent, a node or a thread, is responsible for managing a given slot. It is the *owner* of the slot. The slots owned by a node or a thread are called its *private* slots. A slot owner is responsible for mmapping or unmmaping memory at this slot of addresses, and reading or writing data. Nobody but the owner is allowed to use the slot.

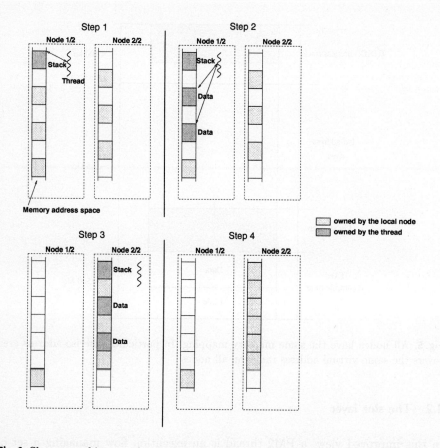

Fig. 6. Slot ownership may change due to migration. In this example, a thread is created and acquires a slot owned by the local node to store its stack (Step 1). The thread acquires other slots from the local node, to store its private data (Step 2). The thread migrates along with its slots (Step 3). The thread dies and its slots are acquired by the destination node (Step 4).

At initialization time, each slot is owned by a unique *node* and is free. When a thread is created, the local node *gives* the thread a slot to store its initial resources: this slot is from now on owned by the thread. When isomallocating data dynamically, a thread acquires additional slots from the local node. Notice that all this change of ownership do not require any synchronization between nodes whatsoever. A thread is associated with the list of its private slots where it stores its resources. On migration, these slots migrate along with the thread, which still owns them after the migration, though the memory is allocated at another node. At any point, a thread may release slots. They are then given to the node the thread is currently visiting. This node may be different from the node from which they have been acquired. On dying, a thread releases all the slots it currently owns. This *slot life cycle* is illustrated on Figure 6. Observe that the size of the slots and their initial distribution among nodes is completely irrelevant at this point. We will discuss the choice of this distribution later.

3.3 The block layer

Since our goal is to provide an allocation function compatible with the malloc C primitive, the isomalloc allocator has been refined so as to cope with arbitrarily-sized zones of memory. This leads to a new concept: the block. A *medium-sized* slot may contain multiple *small-sized* blocks.

Conversely, when large request are to be handled, a block may stretch over multiple contiguous slots. If the current local node owns the necessary number of *contiguous* slots, this allocation is carried out the same way as a simple, single-slot allocation. The set of contiguous slots is simply merged into a large slot. Otherwise, the node has to enter a negotiation with other nodes to *buy* from them the necessary set of *contiguous* slots. As such an operation involves synchronization and mutual exclusion, it is clearly much more expensive than "usual", local allocations. Everything has to be done to keep it exceptional. It is of course possible to increase the slot size defined at initialization. It is much more efficient to adjust the initial distribution of slots so as to favor the contiguity of the slots owned by nodes. We discuss these aspects further in Section 4.1.

3.4 The programming interface

The PM2 high-level programming interface provides two primitives by means of which threads may allocate (respectively release) memory in the iso-address area: pm2_isomalloc and pm2_isofree. These primitives have the same prototype as the classic C functions malloc and free:

```
void *pm2_isomalloc(size_t size);
void pm2_isofree(void *addr);
```

A thread must call pm2_isomalloc instead of malloc to allocate memory for private, non-shared data that are required to migrate with the thread. PM2 *guarantees* that all data stored at addresses returned by pm2_isomalloc follow the calling thread in case of migration. All addresses allocated by pm2_isomalloc have to be set free through a call to pm2_isofree. Using these primitives ensures that all references to the address areas handled by them remain valid and that accesses to the corresponding data are migration-safe. Migration is thus transparent and the migrating threads may use pointers in an arbitrary way.

An example of code using pm2_isomalloc is given in figure 7. Let us suppose that the procedure p4 is called by a thread running on node 0. The thread allocates memory blocks in the iso-address zone through successive calls to pm2_isomalloc and creates a linked list. Then, the thread begins to traverse the list while printing its elements. When the 101st element is reached, the thread migrates to node 1 and continues the traversal. As we can notice in figure 8, the first 100 list elements are displayed on node 0, whereas the next ones are displayed on node 1. All pointers in the list are still valid after migration, since PM2 guarantees that all blocks allocated by pm2_isomalloc migrate with the thread and keep the same virtual addresses.

```
#define NB_ELEMENTS 100000
#define NB_ITERATIONS 20000

typedef struct _item {int value;   struct _item *next;} item;

[...]
void p4() {
 int j; item *head, *ptr;

 /* Create a list. */
 head = NULL;
 for (j = 0; j < NB_ELEMENTS; j++) {
  ptr = (item *) pm2_isomalloc(sizeof(item));
  ptr->value = j * 2 + 1;    /* For example */
  ptr->next = head; head = ptr;
 }
 pm2_printf("I am thread %p\n", marcel_self());

[...]
 /* Print the list elements.  */
 j = 0; ptr = head;
 while(ptr != NULL) {
  if (j == 100) { /* Migrate! */
   pm2_printf("Initializing migration from node %d\n", pm2_self());
   pm2_migrate(marcel_self(), 1);
   pm2_printf("Arrived at node %d\n", pm2_self());
  }
  pm2_printf("Element %d = %d\n", j, ptr->value);
  ptr = ptr->next; j++;
 }
}
```

Fig. 7. Sample code using pm2_isomalloc. Procedure p4 is called by a thread initially running on node 0. After having allocated a few blocks in the iso-address area and constructed a linked list, the thread starts traversing the list. Arrived at element 100, the thread migrates to node 1 and continues the traversal.

4 Implementation details

4.1 Basic requirements

In order to implement our iso-address allocation strategy, we had to address the following points.

```
info%pm2load example1
[node0] I am thread eeff0020
[node0] Element 0 = 1
[node0] Element 1 = 3
[...]
[node0] Element 99 = 199
[node0] Initializing migration
        from node 0
[node1] Arrived at node 1
[node1] Element 100 = 201
[node1] Element 101 = 203
[node1] Element 102 = 205
```

```
info%pm2load example2
[node0] I am thread eeff0020
[node0] Element 0 = 1
[node0] Element 1 = 3
[...]
[node0] Element 99 = 199
[node0] Initializing migration
        from node 0
[node1] Arrived at node 1
[node1] Element 100 = -1797270816
[node1] Element 101 = 57654
Segmentation fault
```

Fig. 8. Execution trace for the code in Figure 7. The list traversal starts on node 0 and continues on node 1. Using malloc instead of pm2_isomalloc would result in a memory access error (Figure 9), since the list is not migrated with the thread in this case

Fig. 9. If the call to pm2_isomalloc is replaced by a call to malloc in the code given in figure 7, an error occurs when the thread tries to access its list after the migration

Iso-address area A specific part of the virtual space has to be dedicated to iso-address memory allocations on all nodes. To this purpose, we defined an *iso-address area* situated between the process stack and the heap (Figure 5). This is possible since all nodes are binary compatible and run by the same version of the operating system.

Global reservation, local allocation The iso-address area is divided into fixed-size virtual address slots, each of which is given to a unique node at initialization. To implement this *global reservation*, each node is provided with a private bitmap which identifies the slots owned by the node (see 4.2). The initial slot distribution pattern must ensure that no slot is shared by several nodes. On each node, actual, *local allocations* may only take place at the slots owned by the caller. Memory allocation is done using the mmap primitive, which allows for memory allocation at specified virtual addresses.

Slot distribution Initially, slots are distributed among the nodes according to some user-defined distribution pattern which may be chosen so as to meet the needs of the application. This choice should be made such that most allocations be local and negotiations are as seldom possible. In our current implementation, slots are assigned to nodes in a round-robin fashion: slot i belongs to node $i \mod p$ in a p-node configuration. This choice has been made for simplicity, but it behaves rather poorly for multi-slot allocations. Nothing prevents the user from choosing other distributions. For instance, instead of distributing single slots cyclically among the nodes, one may distribute series of contiguous slots (*block-cyclic* distribution). An extreme choice is to split the iso-adress area into p sub-areas, one for each node, but these scheme is not advisable if the heap of the container process needs to grow in unpredictable ways. Observe that nothing prevents the system from triggering at any point a *global negotiation* phase, where all nodes would simply exchange their (free) slots to maximize the contiguity,

Slot size As previously explained, the slot size was chosen so as to fit a thread stack and was fixed to 64 kB, that is 16 pages. Thus, thread creation is a local operation (i.e., no negotiation is needed) irrespective of the slot distribution, since a single slot is required. This is also valid for all allocations of blocks smaller than a slot. As for larger allocations, details are given in Section 4.4.

4.2 Managing slots

Each node keeps track of its private slots by means of a private bitmap. Each bit in this bitmap corresponds to a slot in the iso-address zone. Given that this zone is typically as large as 3.5 GB and that a slot corresponds to 64 kB, the size of such a bitmap amounts to 7 kB. In each bitmap, the bits are set to 1 if they correspond to slots owned by the local node, otherwise they are set to 0. If a bit is set to 1, the corresponding slot is free. If it is set to 0, the slot belongs either to another node (and it is necessarily free) or to some local or remote thread.

When a slot request is issued by a thread (for instance, when a thread is created or when it requires additional storage area), one of the slots owned by the local node is given to the thread and the corresponding bit is set to 0 in the local bitmap. The slot does not belong to the local node any more. When a slot is released by a thread (due to dynamic release or to thread death), the corresponding bit *in the current local bitmap* is set to 1. Observe that the bitmaps do not undergo any change on thread migration, since the migrating slots keep being owned by the thread and the corresponding bits keep their 0-value on all nodes. Notice also that, due to migration, a slot may be allocated on a node and released on another, so that the destination node may eventually acquire slots that it did not possess initially.

Threads manage their private slots in a double-linked list (Figure 10). This is in contrast with nodes which manage their private slots by means of a bitmap. Chaining the slots owned by a thread makes it much easier to manipulate them on migration. Actually, chaining is carried out by means of pointers stored in the slot headers. Given that the slot contents get copied at the same virtual address in case of migration, these pointers remain valid and the chaining is thus preserved. As with user-level pointers, no post-migration processing is necessary: an iso-address copy is enough.

4.3 Allocating blocks

In contrast to the traditional malloc/free primitives, which deal with dynamic allocations in a contiguous heap, pm2_isomalloc and pm2-isofree manage allocations of arbitrarily-sized blocks within a list of discontinuous slots. Each slot contains a double-linked list of free blocks. Blocks have headers storing their size, as well as pointers to the neighboring blocks in the list.

Block allocations are carried out as follows. When a thread requires some additional storage space, its slots are searched for a large enough free block. In

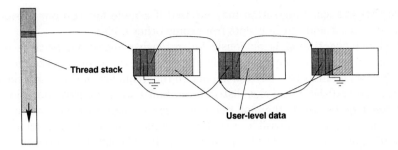

Fig. 10. Each thread keeps its private slots in a double-linked list

the current implementation, a first-fit strategy is used, but other strategies could be considered as well, especially if fragmentation is to be kept low. If no suitable block is found, a new free slot belonging to the current local node is acquired by the thread. It gets attached to its slot list. Then, a new block is allocated in this new slot. This scheme works for all requests for blocks smaller than the slot size, as long as the node owns at least one slot.

4.4 Coping with large-block allocations

To ensure the compatibility with malloc and free, our allocator can also cope with arbitrarily-sized block requests, larger than a slot. In order to satisfy such requests, the key point is to make up a larger slot out of n regular, contiguous slots and to allocate the block inside this new slot (where n is the smallest number of contiguous slots that would be necessary). For this purpose, the following steps are accomplished.

1. The slot bitmap of the local node is scanned, in order to find the necessary number of contiguous slots. A first-fit strategy is used. If this search is successful, the corresponding slots are given to the thread, which uses them to build up a large slot. This large slot gets attached to the slot list of the thread.
2. If the search fails, a global negotiation phase among all the nodes is launched. The initiating nodes behave as follows.
   (a) Enter a system-wide critical section. No other node is allowed to modify its slot bitmap within this section. (It may still run its code and allocate/free *blocks*, as long as no *slot* management is necessary.)
   (b) Gather the local bitmaps of all nodes.
   (c) Compute an *global or* taking all bitmaps as operands.
   (d) Search for the first series n contiguous available slots in this *global* bitmap and "buy" the non-local slots. It suffices to mark these slots are marked with "1" in the bitmap of the requesting node and "0" in the bitmap of their original owner node.
   (e) Send back the updated bitmaps to their respective nodes.
   (f) Exit the system-wide critical section.

Notice that the same algorithm may be used if a node has run out of slots. It simply enables a node to buy slots from some other nodes.

A global negotiation is obviously an expensive operation, because of the global communication required. It should therefore be kept as exceptional as possible. Two main factors have an impact on the frequency of these negotiations: the slot size and the initial slot distribution. Since all single-slot allocations are guaranteed to be local, the slot must be large enough to avoid multiple-slot allocations as much as possible. On the other hand, even for such allocations, negotiation may be avoided if the necessary number of contiguous slots are locally available. It is therefore important to choose a "good" initial slot distribution, in order to avoid negotiations even more. Observe that there is no restriction whatsoever on the initial distribution.

Notice also that the manipulation of the bitmaps on the local node may be completely arbitrary. It is in particular possible for the local node to to take advantage of a negotiation phase to "pre-buy" slots in prevision of foreseeable large allocation requests. It is also possible to completely restructure the slot distribution at the system level, for instance by grouping contiguous free slots as much as possible on the various nodes. The only requirement is that each slot present in the bitmaps must finally belong to exactly one node.

5 Performance and optimizations

We present here some results obtained on our PoPC cluster. Each node consists of a 200 MHz PentiumPro processor. The operating system is Linux 2.0.36. The nodes are interconnected by a Myrinet network from Myricom [9] accessed through the BIP low-level communication interface [12].

The time needed to migrate a thread with no static data between two nodes is less than 75 μs. It was measured by means of a thread ping-pong between two nodes. This time includes packing the thread resources, transferring them over the network, allocating the memory on the destination node and unpacking the resources. Notice that no post-migration processing whatsoever is needed thanks to our iso-address approach. This time should be compared to the 150 μs reported for the migration of a null thread in Active Threads [13]. This performance figure is partly due to the very efficient Madeleine communication layer used by PM2 [2].

Using the pm2_isomalloc function instead of the usual malloc induces a non-significant overhead for the requests of blocks larger than one slot, as shown in Figure 11. This overhead is mainly due to the negotiation automatically required by any multi-slot allocation when the slots are distributed in a round-robin way (which is the case in our experiment). This negotiation takes 255 μs in a 2-node configuration when using BIP/Myrinet. If the underlying architecture provides more than 2 nodes, another 165 μs should be added per extra node. Notice that, for large allocations, this overhead is small and rather insignificant compared to the total allocation time (see Figure 11, bottom). We can thus conclude that our approach scales well.

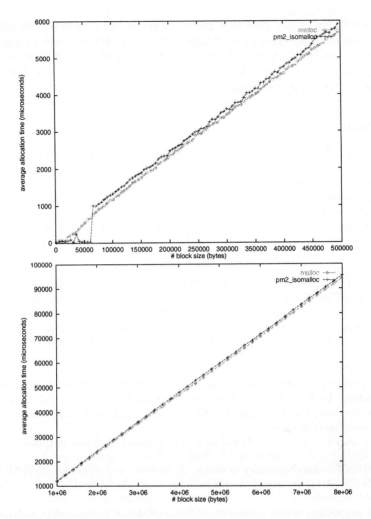

Fig. 11. Compared performance of malloc and pm2_isomalloc for respectively small and large requests in a 2-node configuration.

6 Conclusion and future work

To validate our approach, we have integrated the iso-address allocation primitives in the runtime libraries used by two data-parallel compilers [3, 6]. These compilers have been previously modified, in order to generate multithreaded code for PM2 [11, 1]. Thanks to our new allocator, the runtime code responsible for thread migration was significantly simplified. Given that pre- and post-migration processing were reduced, we could notice an improvement of our virtual processor migration time. We are currently working on these aspects.

A number of optimizations have been considered on top of the general scheme presented. Instead of unmmapping a slot each time it is released, we keep a number of mmapped empty slots in a process-wide cache. This saves the mmapping time at the next slot allocation. When migrating a *slot* attached to a thread, it is sufficient to send its internally allocated *blocks*. Additional details on the current implementation and a downloadable version can be found at http://www.ens-lyon.fr/~rnamyst/pm2.html.

References

1. L. Bougé, P. Hatcher, R. Namyst, and C. Perez. Multithreaded code generation for a HPF data-parallel compiler. In *Proc. 1998 Int. Conf. Parallel Architectures and Compilation Techniques (PACT'98)*, ENST, Paris, France, October 1998. Preliminary version available at ftp://ftp.lip.ens-lyon.fr/pub/LIP/Rapports/RR/RR1998/RR1998-43.ps.Z.
2. L. Bougé, J-F. Méhaut, and R. Namyst. Madeleine: an efficient and portable communication interface for multithreaded environments. In *Proc. 1998 Int. Conf. Parallel Architectures and Compilation Techniques (PACT'98)*, pages 240–247, ENST, Paris, France, October 1998. IFIP WG 10.3 and IEEE. Preliminary version avaiable at ftp://ftp.lip.ens-lyon.fr/pub/LIP/Rapports/RR/RR1998/RR1998-26.ps.Z.
3. Th. Brandes. *Adaptor (HPF compilation system), developped at GMD-SCAI*. Available at http://www.gmd.de/SCAI/lab/adaptor/adaptor_home.html.
4. J. Casas, R. Konuru, S. W. Otto, R. Prouty, and J. Walpole. Adaptive load migration systems for PVM. In *Proc. Supercomputing '94*, pages 390–399, Washington, D. C., November 1994. Available at http://www.mcs.vuw.ac.nz/~pmar/refs.html#R545.
5. D. Cronk, M. Haines, and P. Mehrotra. Thread migration in the presence of pointers. In *Proc. Mini-track on Multithreaded Systems, 30th Intl Conf. on System Sciences*, Hawaii, January 1997. Available at URL http://www.cs.uwyo.edu/~haines/research/chant.
6. P. J. Hatcher. UNH C*. Available at http://www.cs.unh.edu/pjh/vstar/cstar.html.
7. A. Itzkovitz, A. Schuster, and L. Shalev. Thread migration and its application in distributed shared memory systems. *J. Systems and Software*, 42(1):71–87, July 1998. Available at http://www.cs.technion.ac.il/Labs/Millipede/.
8. E. Mascarenhas and V. Rego. Ariadne: Architecture of a portable threads system supporting mobile processes. *Software: Practice & Experience*, 26(3):327–356, March 1996.
9. Myricom. Myrinet link and routing specification. Available at http://www.myri.com/myricom/document.html, 1995.
10. R. Namyst. *PM2: an environment for a portable design and an efficient execution of irregular parallel applications*. Phd thesis, Univ. Lille 1, France, January 1997. In French.
11. C. Perez. Load balancing HPF programs by migrating virtual processors. In *Second Int. Workshop on High-Level Progr. Models and Supportive Env. (HIPS'97)*, pages 85–92, April 1997.
12. B. Tourancheau and L. Prylli. BIP messages. Available at http://lhpca.univ-lyon1.fr/bip.html.
13. B. Weissman, B. Gomes, J. W. Quittek, and M. Holtkamp. Efficient fine-grain thread migration with Active Threads. In *Proceedings of IPPS/SPDP 1998*, Orlando, Florida, March 1998. Available at http://www.icsi.berkeley.edu/~sather/Publications/ipps98.html.

Communication-Intensive Parallel Applications and Non-dedicated Clusters of Workstations

Kritchalach Thitikamol and Peter Keleher

Department of Computer Science, University of Maryland,
College Park, MD 20742
{kritchal, keleher}@cs.umd.edu

Abstract. Time-sharing operating systems may delay application processing of incoming messages because other processes are scheduled when the messages arrive. In this paper, we present a simple adjustment to application polling behavior that reduces this effect when the parallel process competes with CPU-intensive sequential jobs. Our results showed that moderate improvement might be achieved. However, the effectiveness of the approach is highly application-dependent.

1. Introduction

Client-server applications often use busy loops to multiplex and retrieve incoming messages. Such loops might use UNIX `select()` calls to check for message arrival. If no message is pending, `select()` returns immediately and the applications may execute other work. The use of busy loops is often called message polling, in contrast to interrupt message delivery. Message polling is very simple to implement. However, in non-dedicated processors, time-sharing operating systems can inadvertently delay notification of incoming messages. This delay occurs because polling processes are inactive at the time their messages arrive. As a result, the message processing is delayed on both server and client sides and application performance is degraded.

In Section 2, we discuss such a performance problem in CVM [1], a software distributed-shared-memory system (DSM), executing on machines that have other active processes. Software DSMs are software systems that provide shared memory semantics across machines that support only message passing. CVM implements relaxed consistency models, supports multiple threads per processor, and allows thread migrations during application executions. In Section 3, we pinpoint the cause of the problem using simple client and server experiments and present our solution to the CVM problem. In Section 4, we then fully investigated implications from our results and the technique we used to reduce message delay. Finally, we presented our conclusion for similar polling structures and execution environments in Section 5.

Table 1. Characteristics of Sequential Processes.

Sequential Apps	Descriptions	Percent busy on an empty processor
grep	GNU grep program	15-20%
gcc	GNU gcc program	45-55%
infloop	Infinite loop program	100%

Previous studies of the message notification delay [2, 3] reduce its effect by applying special algorithms to re-schedule and synchronize incoming messages with polling processes. Many target message-passing parallel programs, in which synchronization delay is highly important. Significant improvements can be achieved with these approaches, but they often required modifications to the underlying operating system kernels. On the other hand, our technique modifies only the CVM application. In the worst case, our results also show significant performance improvement with our approach. The method basically allows CVM polling processes to avoid long context switches during the critical path of handling incoming messages. While the experiments we present here are specific to Linux 2.0.32 running on Pentium II's, we have confirmed that the problem exists on other operating systems, and similar approaches produce improvements.

2. CVM and Load Balancing

CVM maintains memory consistency by pulling shared data modifications from where they were created in a client-server style. CVM uses I/O signal handlers to notice incoming requests. However, when waiting for replies, CVM uses message polling to multiplex possible multiple replies and to schedule local threads.

We ran two applications, successive over-relaxation (SOR) and water molecular simulation application (Water) from Splash-2 [4], on nodes that had sequential jobs on them. The intent was to test CVM's thread reconfiguration mechanism for load balancing. Imbalance computation time observed in our experiment is a direct result from sequential jobs that compete for CPU cycles with CVM processes. By moving some of the computation to even their executions in thread granularity, CVM processes can avoid increasing global synchronization delay thus improves their overall performance in non-dedicated environment.

The machines we used are a cluster of Pentium-II 266MHz machines running Linux operating system and connected by Ethernet and Myrinet networks. The experiments basically included several runs of each application with different numbers of threads on four processors. We also ran one sequential process on the last processor. Table 1 shows characteristics of our sequential processes and their CPU-busy percentage on our Linux platform with no other processes. For each configuration execution, we measured its elapsed time after a specific sequential process started.

Table 2: Performance of CVM applications using 0-μsec timeout with sequential process on the last processor.

Appls	Active seq. processes	Elapsed time in second "8,8,8,8"	"10,10,10,2"
SOR	none	6.70	8.30
	grep	8.14	8.31
	gcc	10.66	10.17
	inf-loop	13.20	11.22
Water	none	25.88	30.78
	grep	28.58	30.95
	gcc	47.98	38.82
	inf-loop	51.70	44.97

Our expectation was that reconfiguring the mapping of threads to nodes would limit the degradation caused by the sequential job running on the last node.

Table 2 shows the performance results with 32 threads in total. The third and fourth columns show elapsed time from two configurations; "8,8,8,8" means 8 threads are mapped to each processor, and "10,10,10,2" means the first three processors have 10 threads and the last has only two threads. Both used our original busy-wait polling and are labeled 0-μsec in the table. The smaller number of threads on the last processor reflect competition for CPU cycles because of the sequential process. The results are disappointing, to say the least.

We also observed that even though we were balancing application computation, the results had high variability from run to run with different type of sequential process. We would expect the "10,10,10,2" to produce similar performance regardless of the guest processes because we expected that during the execution, barriers would force two threads on the last processor to wait until all 10-thread processors finished their computations. Also, the priority of sequential processes and CVM processes were no difference. In fact, the elapsed time of the last configuration SOR in Table 2 varied from 8.31, 10.17 to 11.22 seconds. These non-uniform results imply that there must be some other artifacts that are affecting barrier imbalance besides the application's computation.

3. Testing Message Polling with Sequential Processes

SOR's simplicity made it easy to rule out computation variability as the culprit. Hence, the load balance must be the result of changing computation costs. In order to verify that the problem was not an artifact of CVM's structure, we wrote simple client and server programs using UDP-sockets. Presumably, if we execute our server with the sequential processes, it would produce a relative degradation of communication performance similar to the first performance results in Table 2. More specifically, our client process continuously sends requests, blocks waiting replies, and then sends more requests. The server uses a non-blocking call to select() in order to implement a busy-wait loop. Calls to select() are made non-blocking by specifying a zero-microsecond (busy) delay as a parameter. During the experiments, we collected average message round trip time of 20,000 messages ping-ponging between client and server.

In order to explore other options that might affect our communication performance, we also tested our applications with constant timeouts passed to the select() calls. We tested 1, 2, 5, 10 and 20 μsec delays, and message sizes of one word and 8192 words.

Fig. 1 shows our client-server results, which we measured on Ethernet and Myrinet networks. The select() blocking periods are listed on the x-axis. The y-axis shows the average round trip time for each run. The results using one word messages are on the left and results using 8192-word messages are on the right. Lines in both charts represent results from different networks with different sequential processes in use shown below both charts. The "E" and "M" refer to Ethernet and Myrinet networks respectively.

The results with 0-μsec with any sequential processes confirmed our hypothesis. The average round trip time with simple busy loops degraded by a large amount in the presence of a sequential process. However, there was very little slowdown with nonzero μsec polling, regardless of the sequential processes. The busy-waiting produced better performance than non-zero μsec polling only when executing with no load. The average round trip time tended to get larger with sequential processes that generated higher percentage of CPU-busy time. The results from Ethernet and My-

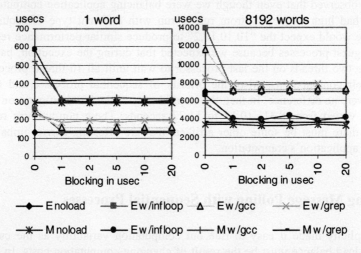

Fig. 1. The results from client-server experiment show changes in average message round trip time (*y-axis*) with different select() blocking periods (*x-axis*). The results using one word messages (*left*) and results using 8192-word messages (*right*) produce similar conclusion. Lines in both charts represent results from Ethernet (*E*) and Myrinet (*M*) networks with different sequential processes in use (*noload, infloop, gcc* and *grep*) shown at the bottom of the figure.

Table 3: Performance of CVM applications using 1-μsec timeout with a sequential process on the last processor and "10,10,10,2" mapping.

Appls	Active seq. processes	Elapsed time in second
SOR	none	8.31
	grep	8.44
	gcc	8.45
	inf-loop	8.44
Water	none	30.74
	grep	35.72
	gcc	35.97
	inf-loop	35.78

rinet networks showed similar patterns, but the Ethernet was faster for small messages and slower for large ones.

Based on these results, we re-ran the CVM experiments. Only the CVM process running on the node with the sequential job used 1-μsec timeouts instead of non-blocking `select()`. Table 3 shows their times in the last column. The results show slight degradation with little or no load, but significant improvements with high load, in comparison to "10,10,10,2" 0-μsec runs in Table 2.

Although we can conclude that message delay is likely the source of additional slowdown, we had yet to identify the exact mechanism of the performance problem. However, we hypothesized that the difference results from OS scheduler behavior.

4. Time Sharing Behaviors

Intuitively, it is easy to understand that message replies may be delayed because polling processes are suspended when the message arrives. The message cannot be seen by the application until the underlying operating system re-schedules the polling process. In order to verify that this is the case in our experiments, we re-ran the client-server experiments and recorded individual round trip times to see whether there were any abnormal message-delays with the infinite loop process. We selected the infinite loop program because it always uses up its time slice or quantum time and the scheduler behavior becomes fairly predictable. The client and server experiments were again repeated only with 0-μsec and 1-μsec blocking time in the `select()` call.

In addition, we modified the Linux kernel to report the current time counter of the server process. The time counter indicates how much time remains in the currently running process's time slice. Each time tick in time counter is about 10 milliseconds worth of execution time. The Linux priority-based scheduler preemptively makes a context switch when the current process's time counter becomes zero, or the process explicitly yields to another process often through UNIX system calls.

Fig. 2 shows individual round trip time of 150 messages listed on x-axis in the top chart and corresponding values of remaining Linux time counter of the server process at the bottom chart. The results confirmed that the 0-μsec server was switched out for full quantum time, approximately 0.2 second by Linux default, (because of the infinite loop program) more often than 1-μsec server. For example, the time counter of 0-μsec server went to zero more often than 1-μsec in Fig. 2. Note that the round trip time of the three spikes in the top chart was between 0.1 and 0.3 seconds, which is much higher than the normal round trip time of 0.029 second.

Although the numbers help us understand the source of the problem, we still need to understand exactly how the 1-μsec blocks change the scheduling behavior. It turns out that the scheduler periodically switches out the 0-μsec server because it exhausts its ticks. However, the 1-μsec block causes the kernel to temporarily yield the CPU to another process during the delay time. By watching the Linux scheduler during our 1-μsec server experiment, we observed 1505 context switches bouncing between server process and infinite loop process, comparing to only 21 context switches with 0-μsec. However, the 1-μsec switches are only for approximately the duration of the 1-μsec timeout. Additionally, the other process is usually the sequential job, which quickly uses up its ticks. The scheduler then adds new ticks to both the sequential job and all other jobs. This latter category includes the CVM process. Hence, the CVM process rarely exhausts its ticks, and rarely is swapped out for an entire time slice, resulting in better performance. Basically, the one microsecond timeout in our server experiments improves the chance of no full context switches forced by OS scheduler. The poor performance of 0-μsec server with a sequential process happened because it was inactive for full quantum time. The delay eventually adds up and creates abnormally large round trip times, especially in highly communicating processes.

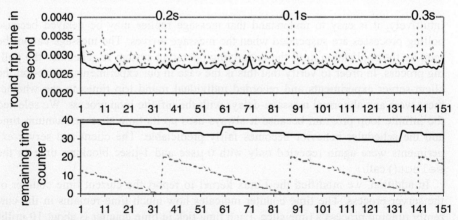

Fig. 2. In the top graph, 0-μsec server (*dotted line*) intermittently caused huge message round trip time while 1-μsec server (*normal line*) did not. The bottom graph showed that the 0-μsec server (*dotted line*) was swapped out more often than the 1-μsec server (*normal line*) as the OS counter of the 0-μsec server became zero frequently.

Although this behavior is clearly tied to specific implementations, we obtained similar results when ran the same experiments on an IBM-SP2 parallel machine running AIX 4.2. The AIX results also showed less variability in the "10,10,10,2" tests with 1-μsec blocks and sequential processes. The 1-μsec elapsed times were better with the infinite loop sequential program, but not GCC. Note that the sequential processes produced different CPU-busy percentages between AIX and Linux operating systems, except only the infinite loop program that maintains 100% busy.

We realized that the 1-μsec block was less effective with less resource-hungry sequential processes on both platforms, e.g. with grep program. In Linux, this is because time counters of lightly busy sequential processes become zero less often. The OS scheduler adds time counters only after a process exhausts its time, and it is likely that less intense sequential jobs use all of their quantum less frequently. Therefore, the blocking technique produces less improvement with less busy jobs. Fortunately, these jobs also produce less slow down to begin with.

Nonetheless, to avoid degradation with lightly busy processes, we suggest our approach be used in dynamic fashion. For example, CVM processes should be able to decide 0-μsec or 1-μsec blocking parameter based on CPU-busy percentage of adversary processes. Although we have not tested our approach with all available operating systems, our experience suggests that for 50-100% busy guest processes, 1-μsec blocking be worth trying.

Our further research reveals that modern operating systems, including Linux and AIX, similarly implement process scheduler with non-fixed priority. The scheduler basically recalculates running processes' priority value at each clock interrupt, which may not be equal to a period of time slice. Therefore, a process may lose its CPU control because its priority value is inferior to that of another dispatchable process. Consequently, we believe that the similar results from our technique can be obtained with other operating systems that use non-fixed priority scheduling. Ultimately, the idea is to avoid OS scheduler to penalize communication-intensive processes with a lot of full time-slice message delay, which have been proved to be expensive for parallel processing in non-dedicated environments.

5. Conclusions

We have shown that we can reduce the impact of competing sequential jobs by changing application code, without touching the operating system scheduler. Our technique consists of using blocking message polls when competing with CPU-intensive jobs. The result is that we reduce the chance of the parallel job being switched out for a long period of time. Its effectiveness, however, depends largely on the level of CPU usage of the local sequential processes. For best results, the technique should be adjusted dynamically. We confirmed the applicability of our method with simple client-server and CVM load balancing experiments on both Linux and AIX operating systems, and the results showed encouraging performance improvement.

References

1. P. Keleher. *The Relative Importance of Concurrent Writers and Weak Consistency Models.* in *Proceedings of the 16th International Conference on Distributed Computing Systems.* 1996 91-99. Hong Kong: IEEE.
2. A.C. Dusseau, R.H. Arpaci, and D.E. Culler. *Effective Distributed Scheduling of Parallel Workloads.* in *Sigmetrics'96 Conference on the Measurement and Modeling of Computer Systems.* 1996.
3. P.G. Sobalvarro, et al. *Dynamic coscheduling on workstation clusters.* in *Proceedings of the Workshop on Job Scheduling Strategies for Parallel Processing.* March 1998.
4. S.C. Woo, et al. *The SPLASH-2 Programs: Characterization and Methodological Considerations.* in *Proceedings of the 22nd Annual International Symposium on Computer Architecture.* June 1995 24--37. Santa Margherita Ligure, Italy.

A Framework for Adaptive Storage Input/Output on Computational Grids*

Huseyin Simitci, Daniel A. Reed, Ryan Fox, Mario Medina, James Oly, Nancy Tran, and Guoyi Wang

Department of Computer Science
University of Illinois
Urbana, Illinois 61801, USA
{simitci,reed,r-fox,c-medina,jamesoly,n-tran1,gwang2}@cs.uiuc.edu
WWW Home Page http://www-pablo.cs.uiuc.edu

Abstract. Emerging computational grids consist of distributed collections of heterogeneous sequential and parallel systems and irregular applications with complex, data dependent execution behavior and time varying resource demands. To provide adaptive input/output resource management for these systems, we are developing PPFS II, a portable parallel file system. PPFS II supports rule-based, closed loop and interactive control of input/output subsystems on both parallel and wide area distributed systems.

1 Introduction

Current users of homogeneous parallel systems often complain that it is difficult to achieve a high fraction of the theoretical performance peak. Distributed collections of heterogeneous systems (i.e., the computational grid), introduce new and more complex application tuning and performance optimization problems. Moreover, the sensitivity of performance to slight changes in application code, together with the large number of potential application performance problems (e.g., load balance, data locality, or input/output) and continually evolving system software, make application tuning complex and often counterintuitive.

Of the range of grid performance problems, the lack of high-performance input/output is one of the most debilitating. Not only is the performance gap between processors and secondary storage devices continuing to increase, the input/output performance of parallel systems is exceedingly sensitive to slight changes in application behavior. The time-varying availability of distributed resources and changing application demands of the computational grid exacerbates these problems. Hence, it is extraordinarily difficult to choose a single set of file

---

* This work was supported in part by the Defense Advanced Research Projects Agency under DARPA contracts DABT63-94-C0049 (SIO Initiative), F30602-96-C-0161, and DABT63-96-C-0027 by the National Science Foundation under grants NSF CDA 94-01124 and ASC 97-20202, and by the Department of Energy under contracts DOE B-341494, W-7405-ENG-48, and 1-B-333164.

system policies that yield high performance for all applications and resource demands.

Instead, by integrating dynamic performance instrumentation and malleable file system policies with adaptive control systems, a flexible runtime system could automatically configure resource management policies and parameters based on application request patterns and observed system performance. Such a dynamic system will allow applications and runtime libraries to adapt to changing hardware and software characteristics.

Recent experiences with flexible input/output policies [11] and adaptive runtime systems [18, 7] has shown that adapting to changing resource demands and availability can yield order of magnitude performance improvements. To provide a flexible testbed for high-performance input/output experiments, we have designed and implementation a prototype, second generation portable parallel file system, PPFS II, that includes real-time, adaptive policy control. PPFS II is designed to work atop either parallel systems or PC and workstation clusters.

The remainder of this paper is organized as follows. In §2, we outline the general design of PPFS II, followed in §3–§4 by a description of adaptive file caching and redundant disk striping. In §5, we describe *Autodriver*, a toolkit for interactive visualization of PPFS II behavior and steering of PPFS II policies. This is followed in §6 by a discussion of new input/output arrival traffic pattern forecasting techniques using ARIMA time series analysis and Hidden Markov models. Finally, §7 summarizes the current state of PPFS II and outlines directions for additional research.

2 PPFS II File System Architecture

Figure 1 illustrates the high-level components of our PPFS II prototype. Globus [7] provides communication and system interaction within a framework for dynamic systems on distributed computational grids. Atop Globus, we have implemented a toolkit for real-time adaptive control of distributed computations. This *Autopilot* toolkit provides a flexible set of performance sensors, decision procedures, and policy actuators to realize adaptive control of applications and resource management policies.

In PPFS II, performance sensors capture real-time performance data from distributed input/output components, decision procedures choose and configure input/output policies, and policy actuators implement policy decisions. All PPFS II sensors and actuators are registered with the *Autopilot* manager, which serves as a central place to publish and query available PPFS II components.

The PPFS II metadata manager organizes the distributed components of PPFS II files, noting data locations and access privileges. In turn, storage servers provide a uniform interface to native file systems available on the systems where they are executing. Decision server contains the fuzzy logic rule bases and control mechanisms for closed-loop adaptive control of the PPFS II file system. Finally, the Java-based *Autodriver* interface provides interactive monitoring and control.

Figure 1. PPFS II Distributed Grid Architecture.

Using the *Autopilot* toolkit, PPFS II integrates real-time access pattern classification software [12] and a fuzzy logic decision mechanism [18] for dynamically identifying access patterns and adaptively choosing file system policies. The fuzzy logic infrastructure accepts access pattern classifications and performance measures and uses a configurable rule base to choose file policies. Currently, rule bases for adaptive caching and striping are implemented.

Application clients access the file system via the Scalable I/O (SIO) Initiative's low-level API [4]. This API is intended as a portable interface for implementing higher level I/O libraries (e.g., MPI-IO [23]). Hence, the SIO API includes routines that can express complex parallel input/output patterns easily (e.g., strided accesses). It also includes a "hint" interface applications can use to specify both quantitative (i.e., offsets and sizes) and qualitative (.i.e., sequential or random) access patterns. For portability, PPFS II also supports standard UNIX file system calls through converters that are implemented atop the SIO API.

3 Adaptive File Caching

As discussed earlier, PPFS II relies on the *Autopilot* adaptive control infrastructure to implement an adaptive file caching scheme. Figure 2 shows the components of this adaptive caching system. Below, we describe this structure in more detail and discuss the performance benefits of adaptive caching.

3.1 PPFS II/Autopilot Interactions

PPFS II exploits *Autopilot*'s suite of configurable, lightweight sensors to capture quantitative and qualitative performance data for input/output policy decisions. To balance the communication cost of transmitting large volumes of raw performance data against local computation overhead, *Autopilot* allows one to attach data transformation functions to each sensor. These attached functions can compute simple statistics from raw data (e.g., sliding window averages) or

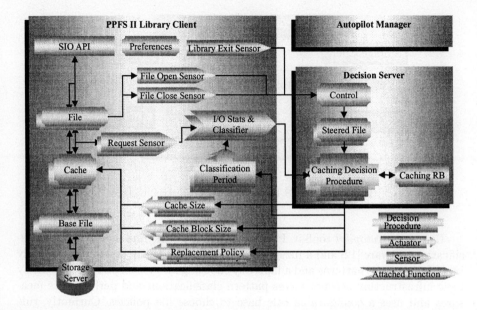

Figure 2. PPFS II Adaptive Caching System Components.

more complex transformations (e.g., computing qualitative access input/output patterns from request streams).

In PPFS II, two *Autopilot* sensors inform the decision server whenever a file is opened or closed by a remote client. Via this sensor notification, the decision server can choose caching and striping policies for the file's active lifetime.

In addition, each input/output request triggers a request sensor whose attached function calculates input/output statistics (e.g., request rates and sizes) and classifies the file access pattern in real-time. This classification exploits a trained neural network to classify access patterns as uniform or variable size, sequential or strided, and read-only, write-only, or read-update-write. Using the output of this sensor, the decision server can change caching and striping policies based on access pattern attributes and request rates.

The decision server controls the file system via four *Autopilot* actuators. These actuators accept remote inputs and can dynamically modify file system variables (e.g., cache size or cache block size) and to change cache management policies. Currently, the file system cache implements Least-Recently-Used (LRU), Most-Recently-Used (MRU) and random replacement policies, all of which can be changed during PPFS II execution. In summary, PPFS II implements a closed loop adaptive control system that dynamically tailors the file system cache's parameters to application's request patterns and observed input/output performance.

3.2 Adaptive Caching Experiments

To assess the performance benefits of adaptive input/output policy control, we have tested our PPFS II prototype in a variety of hardware and software configurations. These experiments verified the effectiveness of real-time performance monitoring and access pattern detection for policy selection and control with both single and multiple process benchmarks.

Below, we describe the results of one set of experiments conducted on a cluster of Linux PCs, each with three Ultra SCSI disks. To simplify explanation, we illustrate PPFS II's closed loop behavior using the interactive *Autodriver* interface, described in §5, to visualize and manipulate PPFS II behavior. In this example, shown in Figure 3, a benchmark program writes 128 KB blocks to a file striped over eight storage servers.

In the figure, the write response time sensor shows the application response times in microseconds. Initially, client caching is disabled by default. After activating caching via the cache mode actuator, file writes are buffered in the cache until it fills.

Using the cache block size actuator shown at the bottom of Figure 3, we increase the cache block size 512 KB. Subsequently, most of the application writes are buffered in the cache and written to secondary storage in groups large enough to benefit from striping across multiple disks. This significantly reduces average response time.

4 Adaptive, Redundant Disk Striping

Because disk capacities continue to rise faster than transfer rates, only input/output parallelism [19] can provide the bandwidth needed by high-performance parallel and and distributed applications. This technique, known as disk striping, is the fundamental idea behind striped file systems [8, 6] and disk arrays [15]. However, matching the distribution of data across storage devices to system utilization has proven critical to obtaining high performance.

Previously, Chen and Patterson [3] and Scheuerman *et al* [20] concluded that optimal striping units (i.e., the amount of continuous data written on each disk) were strongly dependent on the degree of resource contention. Intuitively, at low loads, aggressive striping trades throughput for reduced response times. At higher loads, less aggressive striping sacrifices response time for higher throughput.

Building on the results of our earlier parallel input/output analysis [12, 5], our ongoing PPFS II experiments have continued to explore the match of file striping to access policies and system loads. To understand the effects of file striping parameters on PPFS II performance, we have used a suite of SIO [4] benchmarks. As an example, Figure 4 shows the average response times of random, 128 KB requests, when eight clients read a 1 GB file.

The figure illustrates the complex interplay of inter-request delay, number of disks used for striping, and request size. There is more than two-fold performance differential based on stripe width. When one includes competing access

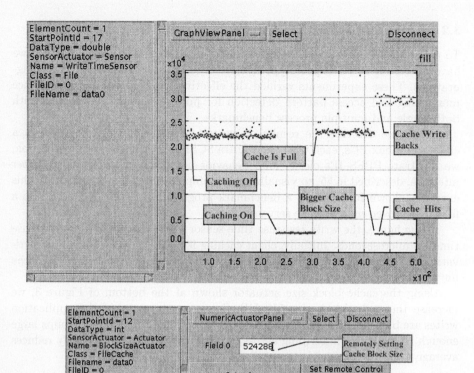

Figure 3. Annotated Write Response Time Control.

to multiple files, disks connected by variable latency, wide area networks, and multiple access patterns and request sizes, the complexity of the performance optimization problem becomes clear.

With these and other experiences, we have developed analytic queueing models of disk striping. This model considers request patterns, disk attributes, and network characteristics to predict the response times for striped requests. For detailed derivation and analysis of this model, see [21]. Based on this model, we have developed a fuzzy logic rule base for adaptive disk striping. This rule base distributes requests to multiple disks during periods of low system load to reduce the individual request response times. At high load, the rule base reduces the striping width to maximize system throughput.

These adaptive striping optimizations generate multiple redundant copies of the parallel files — each optimized for certain access and system characteristics. Simply put, redundant striping trades access disk capacity for higher input/output throughput.

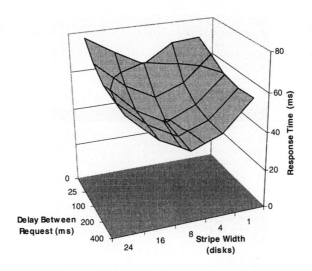

Figure 4. PPFS II Cluster I/O Elapsed Time (1 GB File Read)

5 Interactive PPFS II Control with Autodriver

Although PPFS II is primarily intended to operate as an adaptive, closed loop software system, it can also be steered interactively. These interactive controls allow file system designers to understand the behavior of PPFS II policies and to augment PPFS II software controls with user steering. *Autodriver* is our portable interface for desktop and mobile interaction with *Autopilot* and PPFS II sensors and actuators.

Written largely in Java to enhance portability, *Autodriver* queries *Autopilot* managers to obtain a list of available sensors and actuators, and allows users to select the system aspects they wish to visualize and control. *Autodriver* includes both textual and graphical views of connected sensors. Moreover, actuator values can be changed through either a generic value entry panel or via views customized for specific actuator types.

As an example, Figure 5 shows the *Autodriver* user interface. The window in the figure displays the sensors and actuators currently registered with the *Autopilot* manager. As sensors and actuators are created and destroyed, this list is automatically updated.

Selecting a sensor or actuator displays detailed information for that item in the lower part of the window. Double-clicking on a sensor or actuator creates a new window that shows, respectively, sensor data values or an input panel actuator control.

As a second example, the upper window in Figure 3 shows data for a PPFS II sensor named *WriteTimeSensor*. This view is a simple scrolling scatterplot of sensor values. One can zoom closer to focus on areas of particular interest or save

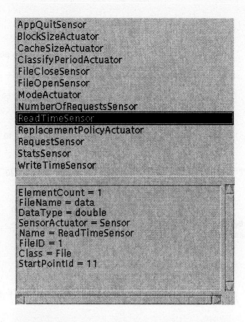

Figure 5. Autodriver Control Interface.

the data in a file for subsequent analysis. Finally, the lower window in Figure 3 illustrates remote actuator configuration and control.

Although Figure 3 illustrates use of *Autodriver* with our PPFS II file system prototype, *Autodriver* is domain independent. One can connect to any sensors and actuators registered with the *Autopilot* on either a local or remote system. By continuing to extend *Autodriver*'s Java interface, our goal is to make PPFS II and *Autopilot* sensors and actuators accessible anywhere and any time (e.g., via handheld devices and wireless networks).

6 New Access Pattern Classification Techniques

Currently, PPFS II uses artificial neural networks (ANNs) to automatically classify file access patterns during application execution. Building on the success of ANNs, we are exploring two complementary, more powerful approaches to access pattern classification based on time series analysis and hidden Markov models (HMMs). Each is described below.

6.1 Time Series Arrival Pattern Forecasting

To create ANNs, access pattern samples are used to train feedforward artificial neural networks [12]. Once trained, PPFS II read/write sensors invoke the ANN periodically for *qualitative* access pattern classification (e.g., sequential or strided).

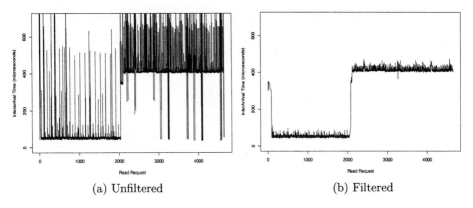

(a) Unfiltered (b) Filtered

Figure 6. PRISM Read Request Interarrival Times (Processor 0).

As a complementary approach to ANNs, Box-Jenkins ARIMA (Autoregressive Integrated Moving Average) time series analysis [2] can be used to create *quantitative* forecasts of future access patterns and trends. In particular, we are investigating ARIMA's ability to predict the *burstiness* of input/output requests. Such Quantitative forecasts of input/output arrival patterns would allow PPFS II to plan prefetching based on anticipated, rather than observed need — prefetching too late misses the window of opportunity for overlapping computations with disk input/output [16].

ARIMA is a stochastic data analysis and modeling methodology that can extract a wide variety of data patterns and seasonal trends, forecasting new behavior from past observations [1, 24]. As a test of ARIMA's applicability to input/output pattern forecasting, we have applied ARIMA to input/output trace data from the Intel Paragon XP/S at Caltech, captured using our Pablo instrumentation toolkit [17].

The Paragon XP/S PRISM application simulates the dynamics of turbulent flow using the three-dimensional Navier-Stokes equations [9, 22]. To study input/output request burstiness, we extracted the interarrival times of file read requests from the PRISM traces for processor zero. The request interarrival times for other processors exhibit similar behavior. Figure 6(a) shows the sequenced of interarrival times.

Inspection reveals that the data is very noisy, with many outliers that obscure the underlying data pattern.[1] After filtering with an approximate conditional mean robust filter, we obtain the data in Figure 6(b), which clearly shows two phase transitions, one near the beginning of execution and one a second near observation 2030. For simplicity's sake, we concentrate on the interval between the first and second transitions.

---

[1] We documented the wide variability of input/output times on the Intel Paragon XP/S in earlier studies [5].

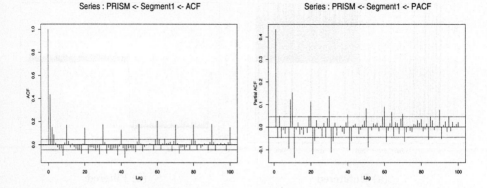

Figure 7. ACF and PACF of Filtered PRISM Series.

Fitting an ARIMA model to a time series consists of three basic steps: (a) model identification, (b) parameter estimation and diagnostic checking, and (c) forecasting. We briefly describe application of these steps to the filtered PRISM data. Space precludes a complete description of the ARIMA process; for additional details, see [24].

Model Identification and Parameter Estimation The two fundamental tools used by ARIMA to expose and identify data patterns are the autocorrelation function (ACF) and the partial autocorrelation function (PACF). ACF measures the linear relationships between pairs of observations, whereas PACF measures the conditional correlation by removing the effects of intervening observations.

Figure 7 shows the estimated ACF and PACF for the filtered PRISM series. The dotted band about zero represents a 95 percent confidence interval. In the ACF plot, significant spikes extend above the dotted band, appearing at regular intervals of ten requests. This correlation pattern indicates that the series is seasonal/periodic with a period of size 10.

The PRISM data also has a non-seasonal pattern, shown by the single large spike at lag one in the PACF plot, followed by a cutoff to insignificant values below the dotted band. Combining the seasonal and nonseasonal patterns gives an $ARIMA(1,0,0)(0,1,1)^{10}$ model. Additional parameter estimation yields $\phi_1 = 0.4964$ for the nonseasonal part and $\Theta_1 = 0.9930$ for the seasonal part.

Intuitively, these seasonal and non-seasonal patterns in the PRISM data are caused by loops and data-dependent control flow in the PRISM code. Seasonal data accrues from nested loops, whereas data-dependent control flow creates aperiodic behavior. The ARIMA model successfully captures both behaviors.

Model Forecasting Given an ARIMA model, it can be used to forecast request patterns by supplying the model with real-time performance data. To validate the forecast, we supplied the estimated ARIMA model with a shortened PRISM series with the final sixty input/output requests omitted.

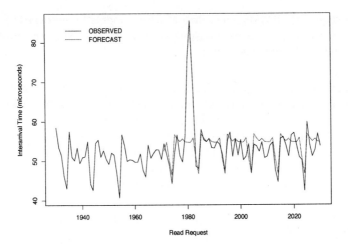

Figure 8. ARIMA Forecast Validation.

Figure 8 compares the forecasts for these sixty observations with the observed PRISM values. With one exception, the predictions are quite accurate, tracking both the seasonal and non-seasonal patterns.

Based on this encouraging match, we are working to develop on-line techniques for constructing ARIMA and other time series models. Via these on-line models, the PPFS II decision procedures can configure cache prefetching schemes that anticipate input/output request bursts, retrieving the desired disk blocks before the requests arrive.

6.2 Hidden Markov Model Access Pattern Forecasting

Recently, several groups have explored optimal caching and prefetching policies when parallel file access patterns are fully known [10]. However, such foreknowledge is often impossible to obtain. For instance, an application's access pattern may change across executions due to data dependencies or because differing system loads may change scheduling decisions for input/output tasks. Even if the access stream is known precisely, varying network latencies and loose client synchronization can change the ordering of requests at input/output servers. In these and similar situations, probabilistic access models can prove useful.

To explore the utility of probabilistic models, we have developed hidden Markov Model (HMM) access pattern classification techniques [14, 13]. In standard Markov models, the next labeled output state depends solely on the current state. Thus, a typical Markov model for file systems uses a state for each disk block, with state labels corresponding to block identifiers and transitions representing the probability that the source and destination states (blocks) are accessed in sequence. HMMs generalize Markov models by associating a probability with each state — a new output state is chosen using the probability distribution when the state is visited.

One can construct Markov models or HMMs from a trace of application input/output requests, either offline using an actual trace or incrementally during application execution. Once a Markov model is built for an application's request pattern, it can be used to control caching and prefetching decisions. For example, by computing high probability paths through the transition graph, prefetching algorithms can retrieve non-sequential groups of blocks that are most likely to be accessed in the future. With ancillary data on block interreference times, one can identify and remove from the file cache those blocks that will be referenced the longest in the future

Building on our preliminary experiences with HMM classification, we are adding HMMs to PPFS II. By enabling PPFS II to manage complex access patterns, these algorithms will allow PPFS II to adapt to changing conditions.

7 Conclusions and Future Work

PPFS II combines advanced techniques for adaptive control of distributed, parallel file systems. Our experiments with PC and workstation clusters have shown that automatic, rule-based adaptive control can yield significant performance improvements over statically configured input/output policies. However, these studies also revealed that optimal disk striping parameters (i.e., the striping unit and width) are strongly dependent on the interplay of application request rates and hardware configurations. Consequently, striping configurations, like caching and prefetching policies must be chosen with care if one is to maximize possible performance benefits.

To identify application input/output patterns, we continue to explore both time series and hidden Markov model (HMM) techniques. Our preliminary studies suggest that ARIMA time series analysis can accurately forecast an application's input/output patterns. However, input/output request models are currently constructed offline. We are investigating dynamic model identification methods that can adaptively derive new models using real-time performance data. Similarly, we believe HMMs can successfully build probabilistic models access pattern using previous executions. This generality will allow automatic classification of arbitrary access patterns.

Finally, the closed loop and interactive performance steering techniques developed for PPFS II are applicable to a broader range of resource domains than just input/output. We believe they can be profitably applied to task scheduling and network protocol configuration. Exploring application of adaptive techniques to these domains is the subject of ongoing work.

Acknowledgments

Special thanks to Ruth Aydt for her help with the *Autopilot* toolkit and to Douglas Simpson for his guidance in time series analysis. The sensor view applet presented here exploits the Java *PtPlot* package from UC-Berkeley.

References

1. ABRAHAM, B., AND LEDOLTER, J. *Statistical Method for Forecasting.* John Wiley and Sons, 1983.
2. BOX, G. E., AND JENKINS, G. M. *Time Series Analysis Forecasting and Control.* 2nd ed. San Francisco: Holden-day, 1976.
3. CHEN, P. M., AND PATTERSON, D. A. Maximizing Performance in a Striped Disk Array. In *Proceedings of the 17th Annual International Symposium on Computer Architecture* (1990), pp. 322–331.
4. CORBETT, P. F., PROST, J.-P., DEMETRIOU, C., GIBSON, G., RIEDEL, E., ZELENKA, J., CHEN, Y., FELTEN, E., LI, K., HARTMAN, J., PETERSON, L., BERSHAD, B., WOLMAN, A., AND AYDT, R. Proposal for a Common Parallel File System Programming Interface Version 1.0. http://www.cs.arizona.edu/sio/api1.0.ps, Nov. 1996.
5. CRANDALL, P. E., AYDT, R. A., CHIEN, A. A., AND REED, D. A. Characterization of a Suite of Input/Output Intensive Applications. In *Proceedings of Supercomputing '95* (Dec. 1995).
6. DIBBLE, P. C., SCOTT, M. L., AND ELLIS, C. S. Bridge: A High-Performance File System for Parallel Processors. In *Proc. 8th Int'l. Conf. on Distr. Computing Sys.* (San Jose, CA, jun 1988), pp. 154–161.
7. FOSTER, I., AND KESSELMAN, C. Globus: A Metacomputing Infrastructure Toolkit. *International Journal of Supercomputing Applications and High Performance Computing 11*, 2 (Summer 1997), 115–128.
8. HARTMAN, J. H., AND OUSTERHOUT, J. K. The Zebra Striped Network File System. *ACM Transactions on Computer Systems 13*, 3 (Aug. 1995), 274–310.
9. HENDERSON, R. D., AND KARNIADAKIS, G. E. Unstructured Spectral Element Methods for Simulation of Turbulent Flows. *Journal of Computational Physics 122(2)* (1995), 191–217.
10. KIMBREL, T., TOMKINS, A., PATTERSON, R. H., BERSHAD, B., CAO, P., FELTEN, E., GIBSON, G., AND KARLIN, A. R. A Trace-Driven Comparison of Algorithms for Parallel Prefetching and Caching. In *Proceedings of the 1996 Symposium on Operating Systems Design and Implementation* (1996), pp. 19–34.
11. MADHYASTHA, T. M., ELFORD, C. L., AND REED, D. A. Optimizing Input/Output Using Adaptive File System Policies. In *Proceedings of the Fifth Goddard Conference on Mass Storage Systems and Technologies* (Sept. 1996), pp. II:493–514.
12. MADHYASTHA, T. M., AND REED, D. A. Intelligent, Adaptive File System Policy Selection. In *Proceedings of the Sixth Symposium on the Frontiers of Massively Parallel Computation* (Oct 1996), IEEE Computer Society Press, pp. 172–179.
13. MADHYASTHA, T. M., AND REED, D. A. Exploiting Global Input/Output Access Pattern Classification. In *Proceedings of SC '97: High Performance Computing and Networking* (San Jose, Nov. 1997), IEEE Computer Society Press.

14. MADHYASTHA, T. M., AND REED, D. A. Input/Output Access Pattern Classification Using Hidden Markov Models. In *Proceedings of the Fifth Workshop on Input/Output in Parallel and Distributed Systems* (San Jose, CA, Nov. 1997), ACM Press, pp. 57–67.
15. PATTERSON, D., CHEN, P., GIBSON, G., AND KATZ, R. H. Introduction to Redundant Arrays of Inexpensive Disks (RAID). In *Proceedings of IEEE Compcon* (Spring 1989), pp. 112–117.
16. PATTERSON, R., GIBSON, G., GINTING, E., STODOLSKY, D., AND ZELENKA, J. Informed Prefetching and Caching. In *Proceedings of the 15th ACM Symposium on Operating Systems Principles, Copper Mountain, CO.* (December 1995), pp. 79–95.
17. REED, D., AYDT, R., NOE, R., ROTH, P. C., SHIELDS, K. A., SCHWARTZ, B., AND TAVERA, L. Scalable Performance Analysis: The Pablo Performance Analysis Environment. In *Proceedings of the Scalable Parallel Libraries Conference. IEEE Computer Society* (1993), pp. 104–113.
18. RIBLER, R. L., VETTER, J. S., SIMITCI, H., AND REED, D. A. Autopilot: Adaptive Control of Distributed Applications. In *Proceedings of HPDC 7* (July 1998).
19. SALEM, K., AND GARCIA-MOLINA, H. Disk Striping. In *Proceedings of the 2nd International Conference on Data Engineering* (Feb. 1986), ACM, pp. 336–342.
20. SCHEUERMANN, P., WEIKUM, G., AND ZABBACK, P. Data Partitioning and Load Balancing in Parallel Disk Systems. *The VLDB Journal 7*, 1 (Feb. 1998), 48–66.
21. SIMITCI, H., AND REED, D. A. Adaptive Disk Striping for Parallel Input/Output. In submitted for publication (1999).
22. SMIRNI, E., AND REED, D. Workload Characterization of Input Output Intensive Parallel Applications. In *Proceedings of the Conference on Modelling Techniques and Tools for Computer Performance Evaluation, Springer-Verlag Lecture Notes in Computer Science* (June 1997), vol. 1245, pp. 169–180.
23. THE MPI-IO COMMITTEE. MPI-IO: A Parallel File I/O Interface for MPI, April 1996. Version 0.5.
24. WEI, W. S. *Time Series Analysis Univariate and Multivariate Methods*. Addison-Wesley, 1990.

ARMCI: A Portable Remote Memory Copy Library for Distributed Array Libraries and Compiler Run-Time Systems

Jarek Nieplocha[1] and Bryan Carpenter[2]

[1]Pacific Northwest National Laboratory, Richland, WA 99352
<j_nieplocha@pnl.gov>
[2]Northeastern Parallel Architecture Center, Syracuse University
<dbc@npac.syr.edu>

Abstract. This paper introduces a new portable communication library called ARMCI. ARMCI provides one-sided communication capabilities for distributed array libraries and compiler run-time systems. It supports remote memory copy, accumulate, and synchronization operations optimized for non-contiguous data transfers including strided and generalized UNIX I/O vector interfaces. The library has been employed in the Global Arrays shared memory programming toolkit and Adlib, a Parallel Compiler Run-time Consortium run-time system.

1 Introduction

A portable lightweight remote memory copy is useful in parallel distributed-array libraries and compiler run-time systems. This functionality allows any process of a MIMD program executing in a distributed memory environment to copy data between local and remote memory. In contrast to the message-passing model used predominantly in these environments, remote memory copy does not require the explicit cooperation of the remote process whose memory is accessed. By decoupling synchronization between the process that needs the data and the process that owns the data from the actual data transfer, implementation of parallel algorithms that operate on distributed and irregular data structures can be greatly simplified and performance improved. In distributed array libraries, remote memory copy can be used for several purposes, including implementation of the object-based shared memory programming environment of Global Arrays [1] or ABC++ [2], and optimized implementation of collective operations on sections of distributed multidimensional arrays, like those in the PCRC Adlib run-time system [3] or the P++ [4] object oriented class library.

In scientific computing, distributed array libraries communicate sections or fragments of sparse or dense multidimensional arrays. With remote memory copy APIs (application programming interfaces) supporting only contiguous data, array sections are typically decomposed into contiguous blocks of data and transferred in multiple communication operations. This approach is quite inefficient on systems with high latency networks, as the communication subsystem would handle each contiguous portion of the data as a separate network message. The communication startup costs are incurred multiple times rather than just once. The problem is attributable to an inadequate API, which does not pass enough information about the intended data transfer and layout of user data to the communication subsystem. In principle, if a more descriptive communication interface is used, there are many ways communication libraries could optimize performance. For example: 1) minimize the number of underlying network packets by packing distinct blocks of data into as few packets as possible, 2) minimize the number of interrupts in interrupt-driven message-delivery systems, and 3) take advantage of any available shared memory optimizations (prefetching, poststoring, bulk data transfer) on shared memory systems. We illustrate the performance implication of communication interfaces with an example.

Consider a *get* operation that transfers a 2x100 section of a 10x300 Fortran array of double precision numbers on the IBM SP. The native remote memory copy operation *LAPI_Get* supports only contiguous data transfers [5,6]. Therefore, 100 calls are required to transfer data in 16-byte chunks. With 90µS latency for the first column and 30µS for the next ones (pipelining), the transfer rate is 0.5MB/s. For the same amount of contiguous data, the rate is 30MB/s. If the application used a communication interface supporting strided data layout, a communication library could send one LAPI active message specifying entire request. At the remote end the AM handler could copy data into a contiguous buffer and send all data in a single network packet. The data would then be copied to the destination. The achieved rate is 9MB/s. We argue that platform-specific optimizations such as these should be made available through a portable communication library, rather than being reproduced in each distributed array library or run-time system that requires strided memory copy.

In this paper, we describe design, implementation and experience with the ARMCI (Aggregate Remote Memory Copy Interface), a new portable remote memory copy library we have been developing for optimized communication in the context of distributed arrays. ARMCI aims to be fully portable and compatible with message-passing libraries such as MPI or PVM. Unlike most existing similar facilities, such as Cray SHMEM [7] or IBM LAPI [6], it focuses on the noncontiguous data transfers. ARMCI offers a simpler and lower-level model of one-sided communication than MPI-2 [8] (no epochs, windows, datatypes, Fortran-77 API, rather complicated progress rules, etc.) and targets a different user audience. In particular, ARMCI is meant to be used primarily by library implementors rather than application developers. Examples of libraries that ARMCI is aimed at include Global Arrays, P++/Overture [4,9], and the PCRC Adlib run-time system. We will discuss applications of ARMCI in Global Arrays and Adlib as particular examples. Performance benefits from using ARMCI are presented by comparing its performance to that of the native remote memory copy operation on the IBM SP.

2 ARMCI Model and Implementation

2.1 ARMCI Operations

ARMCI supports three classes of operations:
- data transfer operations including put, get and accumulate,
- synchronization operations—local and global fence and atomic read-modify-write, and

- utility operations for allocation and deallocation of memory (as a convenience to the user) and error handling.

The currently supported operations are listed in Table 1.

Table 1: ARMCI operations

Operation	Comment
ARMCI_Put, ARMCI_PutV, ARMCI_PutS	contiguous, vector and strided versions of put
ARMCI_Get, ARMCI_GetV, ARMCI_GetS	contiguous, vector and strided versions of get
ARMCI_Acc, ARMCI_AccV, ARMCI_AccS	contiguous, vector and strided versions of atomic accumulate
ARMCI_Fence	blocks until outstanding operations targeting specified process complete
ARMCI_AllFence	blocks until all outstanding operations issued by calling process complete
ARMCI_Rmw	atomic read-modify-write
ARMCI_Malloc	memory allocator, returns array of addresses for memory allocated by all processes
ARMCI_Free	frees memory allocated by ARMCI_Malloc
ARMCI_Poll	polling operation, not required for progress

The data transfer operations are available with two noncontiguous data formats:

1. *Generalized I/O vector.* This is the most general format intended for multiple sets of equally-sized data segments, moved between arbitrary local and remote memory locations. It extends the format used in the UNIX *readv/writev* operations by minimizing storage requirements in cases when multiple data segments have the same size. For example, operations that would map well to this format include scatter and gather. The format would also allow to transfer a triangular section of a 2-D array in one operation.

```
typedef struct {
  void *src_ptr_ar;
  void *dst_ptr_ar;
  int bytes;
  int ptr_ar_len;
} armci_giov_t;
```

For example, with the generalized I/O vector format a *put* operation that copies data to the process(or) *proc* memory has the following interface:

```
int ARMCI_PutV(armci_giov_t dscr_arr[], int len, int proc)
```

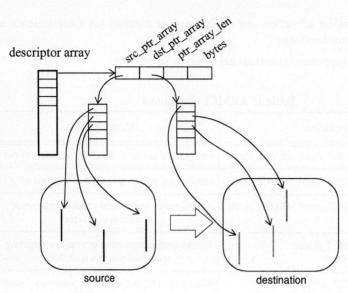

Figure 1: Descriptor array for generalized I/O vector format. Each array element describes transfer of a set of equally-sized (*bytes*) blocks of data.

The first argument is an array of size *arr_len*. Each array element specifies a set of equally-sized segments of data copied from the local memory to the memory at the remote process(or) *proc*.

2. *Strided*. This format is an optimization of the generalized I/O vector format for sections of dense multidimensional arrays. Instead of including addresses for all the segments, it specifies for source and destination only the address of the first segment in the set. The addresses of the other segments are computed using the stride information:

```
void *src_ptr;
void *dst_ptr;
int stride_levels;
int src_stride_arr[stride_levels];
int dst_stride_arr[stride_levels];
int count[stride_levels+1];
```

For example, with the strided format a *put* operation that copies data to the process(or) *proc* memory has the following interface:

```
int ARMCI_PutS(src_ptr, src_stride_ar, dst_ptr,
               dst_stride_ar, count, stride_levels, proc)
```

The first argument is an array of size *arr_len*. Each array element specifies a set of equally-sized segments of data to be copied from the local memory to the memory at the remote process(or) *proc*, see Figure 2.

For contiguous data, *put* and *get* operations are available as simple macros on top of their strided counterparts.

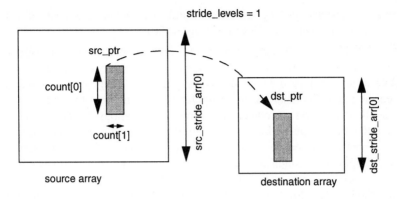

Figure 2: Strided format for 2-dimensional array sections (assuming Fortran layout)

2.2 Atomic Operations

At present time, ARMCI offers two atomic operations: *accumulate* and *read-modify-write*.

Accumulate is an important operation in many scientific codes. It combines local and remote data atomically $x = x + a*y$. For double precision data types, it can be thought of as an atomic version the BLAS *DAXPY* subroutine with x array located in remote memory. Unlike *MPI_Accumulate* in MPI-2, the ARMCI *accumulate* preserves the scale argument a available in *DAXPY*. For applications that need scaling, it has performance advantages on many modern microprocessors, where thanks to the *multiply-and-add* operation for floating point datatypes, the scaled addition is executed in the same time as addition operation alone. The ARMCI *accumulate* is available for integer, double precision, complex and double complex datatypes.

ARMCI_Rmw is another atomic operation in the library. It can be used to implement synchronization operations or as a shared counter in simple dynamic load balancing applications. The operation updates atomically a remote integral variable according to the specified operator and returns the old value. There are two operators available: *fetch-and-increment* and *swap*.

2.3 Progress and ordering

It is important for a communication library such as ARMCI to have straightforward and uniform progress rules on all supported platforms. They simplify development and performance analysis of applications and free the user from dealing with ambiguities of platform-specific implementations. On all platforms, the ARMCI operations are truly one-sided and complete regardless of the actions taken by the remote process(or). In particular, there is no need for remote process to make occasional communication calls or poll in order to assure that communication calls issued by other processes(ors) to this process(or) can complete. Although, for performance reasons, polling can be helpful (can avoid the cost of interrupt processing), it is not necessary to assure progress [10].

The ARMCI operations are ordered (complete in order they were issued) when referencing the same remote process(or). Operations issued to different processors can complete in an arbitrary order. Ordering simplifies the programming model and is required in many applications in computational chemistry (for example). Some systems allow ordering of otherwise unordered operations by providing a fence operation that blocks the calling process until the outstanding operation completes, so the next operation issued does not overtake it. This approach usually accomplishes more than applications might desire, and could have a negative performance impact on platforms where the copy operations are otherwise ordered. Usually, ordering can be accomplished with a lower overhead by using platform-specific means inside the communication library, rather than by a fence operation at the application level.

In ARMCI, when a *put* or *accumulate* operation completes, the data has been copied out of the calling process(or) memory but has not necessarily arrived to its destination. This is a local completion. A global completion of the outstanding put operations can be achieved by calling *ARMCI_Fence* or *ARMCI_AllFence*. *ARMCI_Fence* blocks the calling processor until all *put* operations issued to a specified remote process(or) complete at the destination. *ARMCI_AllFence* does the same for all outstanding *put* operations issued by the calling process(or), regardless of the destination.

2.4 Portability and implementation considerations

The ARMCI model specification does not describe or assume any particular implementation model (for example, threads). One of the primary design goals was to allow wide portability of the library. Another one was to allow an implementation to exploit the most efficient mechanisms available on a given platform, which might involve active messages [11], native put/get, shared memory, and/or threads.

For example, depending on whatever solution delivers the best performance, ARMCI *accumulate* might or might not be implemented using the "owner computes" rule. In particular, this rule is used on IBM SP where *accumulate* is executed inside the Active Message handler on the processor that owns the remote data whereas on the Cray T3E the requesting processor performs the calculations (by locking memory, copying data, accumulating, putting updated data back and unlocking) without interrupting the data owner.

On shared memory systems, ARMCI operations currently require the remote data to be located in shared memory. This restriction greatly simplifies the implementation and improves the performance by allowing the library to use simply an optimized memory copy rather than other more costly mechanisms for moving data between separate address spaces. This is acceptable in the distributed arrays libraries we considered since they are responsible for the memory allocation and can easily use *ARMCI_Malloc* (or other mechanisms such as using *mmap*, *shmget*, etc. directly) rather than the local memory allocation. However, the requirement can be lifted in future versions of ARMCI if compelling reasons are identified.

At present time, the library is implemented on top of existing native communication interfaces on:

- distributed-memory systems such as Cray T3E and IBM SP,
- shared memory systems such as SGI Origin and Cray J90, and Unix and Windows NT workstations.

On distributed-memory platforms, ARMCI uses the native remote memory copy (SHMEM on Cray T3E and LAPI on IBM SP). In addition, on IBM SP LAPI active messages and threads are employed to optimize performance of noncontiguous data transfers for all but very small and very large messages. As an example, in Figure 3 we demonstrate the implementation of the *strided get* operation on the IBM SP. Implementations of *strided put* and *accumulate* are somewhat more complex due to the additional level of latency optimization for short active messages and mutual exclusion in *accumulate*. The implementation of vector operations is close to that of their strided

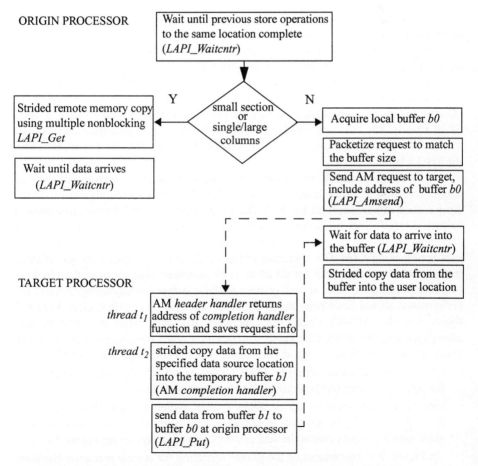

Figure 3: Implementation of the ARMCI strided get operation on the IBM SP

counterparts on the IBM SP. Mutual exclusion necessary to implement atomic operations is implemented with the Pthread mutexes on the IBM SP and the atomic swap on the Cray T3E.

The ARMCI ports on shared memory systems are based on the shared memory operations (Unicos shared memory on the Cray J90, System V on other Unix systems, and memory mapped files on Windows NT) and optimized remote memory copies for noncontiguous data transfers. On most of the platforms, the best performance of the memory copy is achieved with Fortran-77 compiler. Many implementations of the *memcpy* or *bcopy* operations in the standard C library are far from optimal even for well aligned contiguous data. Mutual exclusion is implemented using vendor-specific locks (on SGI, HP, Cray- J90), System V semaphores (Unix) or thread mutexes (NT).

The near-future porting plans for ARMCI include clusters of workstations, and Fujitsu distributed memory supercomputer on top of the Fujitsu MPlib low-level communication interface.

2.5 Performance

We demonstrate performance benefits of ARMCI by comparing it to performance of the native remote memory copy on the IBM SP in two contexts: 1) copying sections of 2-D arrays from remote to local memory and 2) gather operation for multiple double-precision elements. In both cases, we run the tests on four processors with processor 0 alternating its multiple requests between the other three processors which were sleeping at that time. This assured that the incoming requests were processed in the interrupt mode (worse case scenario). We performed these tests on a 512-node IBM SP at PNNL. The machine uses the 120MHz Power2 Super processor and 512MB main memory on each uniprocessor node, and the IBM high performance switch (with the TB-3 adapter).

In the first benchmark, we transferred square sections of 2-dimensional arrays of double precision numbers from remote to local memory (get operation) and varied the array section dimension from 1 to 512. The array section was transferred using either a strided *ARMCI_GetS* operation or multiple *LAPI_Get* operations that transferred contiguous blocks of data - columns in the array section. The latter implementation uses nonblocking *LAPI_Get* operation which allows to issue requests for all columns in the section and then wait for all of them to complete. Had we used the blocking operation waiting for data in each column to arrive before issuing another *get* call the performance would have been much worse. Even with this optimization, Figure 4 shows that the ARMCI strided *get* operation has very significant performance advantages over the native contiguous *get* operation when transferring array sections. There are two exceptions where the performances of the both approaches are identical:
- for small array sections where to *ARMCI_GetS* uses *LAPI_Get* before swithing to the Active Message (AM) protocol to optimize latency, and
- very large array sections where *ARMCI_GetS* switches again from the Active Message to *LAPI_Get* protocol to optimize bandwidth when the cost of the two extra memory copies makes the AM protocol less competitive, see Figure 3.

In Figure 5, performances of the *gather* operation for double precision elements is presented. In particular, the *gather* operation in this test fetches every third element

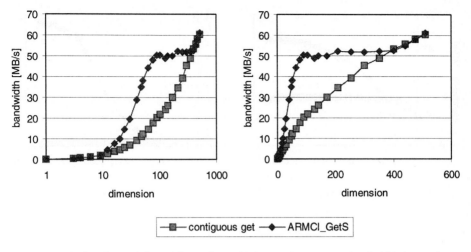

Figure 4: The log-linear (left) and linear-linear (right) graphs represent bandwidth for contiguous get operation and ARMCI *strided* get when moving square section of 2-D array.

of an array on a remote processor. For comparison, the operation is implemented on top an the ARMCI vector *get* operation, *ARMCI_GetV*, and multiple native *get* operations. In this case too we used nonblocking *LAPI_Get* followed by a single wait operation to improve performance of this implementation. Despite this optimization, the performance of ARMCI is by far better for all but very small requests for which the performances of the both approaches are identical. For small requests *ARMCI_GetV* is implemented using multiple nonblocking *LAPI_Get* operations up to the point where the Active Message based protocol becomes more competitive. This protocol is very similar to that for the strided get shown in Figure 3. The performance of *gather* implemented on top of the nonblocking native *get* operation can be explained by the pipelining effect [5] for multiple requests in LAPI. Only the first *get* request issued to a particular remote processor is processed in the interrupt mode, the following ones are received when processing has not been completed. Regardless of pipelining, only eight bytes of user data is send in each *LAPI_Get* message (single network packet) whereas in case of *ARMCI_GetV* all network packets sent by *LAPI_Put* inside AM handler are filled with user data. With the 1024-byte packets in the IBM SP network, the network utilization and performance of the *gather* operation on top of the native *get* operation are far from optimal.

3 Experience with ARMCI

3.1 Global Arrays

The Global Arrays (GA) [1] toolkit provides a shared-memory programming model in the context of 2-dimensional distributed arrays. GA has been the primary programming model for numerous applications, some as big as 500,000 lines of code, in quantum chemistry, molecular dynamics, financial calculations and other areas. As part of the DoE-2000 Advanced Computational Testing and Simulation (ACTS)

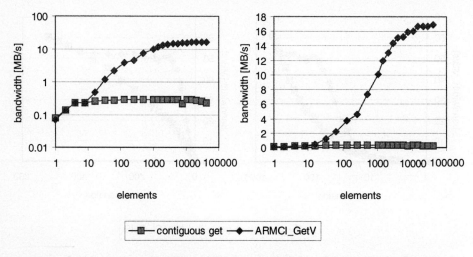

Figure 5: The log-log (left) and linear-log (right) graphs represent bandwidth in gather operation for double precision numbers as a function of the number of elements when implemented with contiguous get operation and the ARMCI *vector* get.

toolkit, GA is being extended to support higher dimensional-arrays. The toolkit was originally implemented directly on top of system-specific communication mechanisms (NX *hrecv*, MPL *rcvncall*, SHMEM, SGI *arena*, etc.). The original one-sided communication engine of GA had been closely tailored to the two-dimensional array semantics supported by the toolkit. This specificity hampered extensibility of the GA implementation. It became clear that a separation of the communication layer from the distributed array infrastructure is a much better approach. This was accomplished by restructuring the GA toolkit to use ARMCI hence making the GA implementation itself fully platform independent. ARMCI matches the GA model well and allows GA to support the arbitrary dimensional arrays efficiently. The overhead introduced by using an additional software layer (ARMCI) is small; for example on the Cray T3E *ga_get* latency has increased by less than 0.5μS when comparing to the original implementation of GA.

3.2 Adlib

The Adlib library was completed in the Parallel Compiler Runtime Consortium project [12]. It is a high-level runtime library designed to support translation of data-parallel languages [3]. Initial emphasis was on High Performance Fortran, and two experimental HPF translators used the library to manage their communications [13,14]. Currently the library is being used in the *HPspmd* project at NPAC [15]. It incorporates a built-in representation of a distributed array, and a library of communication and arithmetic operations acting on these arrays. The array model is a more general than GA, supporting general HPF-like distribution formats, and arbitrary regular sections. The existing Adlib communication library emphasizes collective communication rather than one-sided communication.

The kernel Adlib library is implemented directly on top of MPI. We are reimplementing parts of the collective communication library on top of ARMCI and expect to see improved performance on shared memory platforms. The layout information must now be accompanied by a locally-held table of pointers containing base addresses at which array segments are stored in each peer process (in message-passing implementations, only the local segment address is needed). Adlib is implemented in C++, and for now the extra fields are added by using inheritance with the kernel distributed array descriptor class (DAD) as base class. A representative selection of the Adlib communication schedule classes were reimplemented in terms of ARMCI. These included the *Remap* class, which implements copying of regular sections between distributed arrays, the *Gather* class, which implements copying of a whole array to a destination array indirectly subscripted by some other distributed arrays, and a few related classes.

The benchmark presented here is based on the *remap* operation. The particular example chosen abstracts the communication in the array assignment of the following HPF fragment

```
      real a(n, n), b(n, n)
!hpf$ distribute a(block, *) onto p
!hpf$ distribute b(*, block) onto p
      a = b
```

The source array is distributed in its first dimension and the destination array is distributed in its second dimension, so the assignment involves a data redistribution requiring an all-to-all communication. In practice the benchmark was coded directly in C++ (rather than Fortran) and run on four processors of the SGI Power Challenge. The timings for old and new versions are given in Figure 6. The initial implementation of

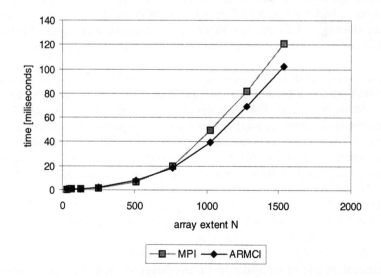

Figure 6: Timings for original MPI vs new ARMCI implementation of *remap*. The operation is a particular redistribution of an N by N array.

the operation on top of ARMCI version already shows a modest but significant 20% improvement over MPI for large arrays. For small arrays the MPI version was actually slightly faster due to extra barrier synchronizations still used in the initial ARMCI implementation of *remap* (for now these synchronizations were implemented naively in terms of message-passing). The MPI implementation is MPICH using the shared memory device, so the underlying transport is the the same in both cases.

In addition to improved performance of collective operations on shared memory platforms, adopting ARMCI as a component of the Adlib implementation will enable the next release of the library to incorporate one-sided *get* and *put* operations as part of its API—a notable gap in the existing functionality. In a slightly orthogonal but complementary development, we have produced an PCRC version of GA that internally uses a PCRC/Adlib array descriptor to maintain the global array distribution parameters. This provides a GA style programming model, but uses the Adlib array descriptor internally. The modularity of the GA implementation, enhanced by the introduction of the ARMCI layer, made changing the internal array descriptor a relatively straightforward task. An immediate benefit is that direct calls to the optimized Adlib collective communication library will be possible from the modified GA. This work was undertaken as part of an effort towards converging the two libraries.

4 Relation to Other Work

ARMCI differs from other systems in several key aspects. Unlike facilities that primarily support remote memory copies with contiguous data (for example SHMEM, LAPI, or the Active Message run-time-system for Split-C (which includes put/get operations) [16] it attempts to optimize noncontiguous data transfers, such as those involving sections of multidimensional arrays, and generalized scatter/gather operations. The Cray SHMEM library [7] available on the Cray and SGI platforms, and on clusters of PCs through HPVM [17], supports only put/get operations for contiguous data and scatter/gather operations for the basic datatype elements.

The Virtual Interface Architecture (VIA) [18] is a recent network architecture and interface standard that is relevant to ARMCI as a possible implementation target. VIA offers support for noncontiguous data transfers. However, its scatter/gather Remote Direct Memory Access (RDMA) operations can transfer data only between a noncontiguous locations in local memory and a contiguous remote memory area. For transfer of data between sections of local and remote multidimensional arrays, the VIA would require an additional copy to and from an intermediate contiguous buffer, which obviously has impact on the performance. This is unnecessary in ARMCI. ARMCI and VIA offer similar ordering of communication operations.

The MPI-2 one-sided communication specifications include remote memory copy operations such as *put* and *get* [8]. Noncontiguous data transfers are fully supported through the MPI derived data types. There are two models of one-sided communication in MPI-2: "active-target" and "passive-target". The MPI-2 one-sided communication and in particular its "active-target" version have been derived from message-passing, and its semantics include rather restrictive (for a remote memory copy) progress rules closer to MPI-1 than to existing remote memory copy interfaces like Cray SHMEM, IBM LAPI or Fujitsu MPlib. A version of MPI-2 one-sided communication called "passive-target" offers more relaxed progress rules and a simpler to use model than "active-target". However, it also introduces potential performance penalties by requiring locking before access to the remote memory ("window") and forbidding concurrent accesses to non-overlapping locations in a "window". ARMCI does not have similar restrictions and offers a simpler programming model than MPI-2. This makes it more appropriate for libraries and tools supporting some application domains, including computational chemistry.

5 Conclusions and Future Work

We introduced a new portable communication library targeting parallel distributed array libraries and compiler run-time systems. By supporting communication interfaces for noncontiguous data transfers ARMCI offers potential performance advantage over other existing systems for communication patterns used in many scientific applications. We presented its performance for the strided get and gather - type data transfers on the IBM SP where the ARMCI implementation outperformed by a large margin the native remote memory copy operations. ARMCI already demonstrated its value: we employed ARMCI to implement Global Arrays library and its new higher-dimensional array capabilities, and used it to optimize collective communication in the Adlib run-time system. The porting and tuning efforts will be continued. The future ports will include networks of workstations and other scalable systems. In addition, future extensions of the existing functionality will include nonblocking interfaces to *get* operations and new operations such as locks.

Acknowledgments

This work was supported through the U.S. Department of Energy by the Mathematical, Information, and Computational Science Division the Office of Computational and Technology Research DOE 2000 ACTS project, and used the Molecular Science Computing Facility in the Environmental Molecular Sciences Laboratory at the Pacific Northwest National Laboratory (PNNL). PNNL is operated by Battelle Memorial Institute for the U.S. Department of Energy under Contract DE-AC06-76RLO 1830.

References

1. J. Nieplocha, R. J. Harrison, and R. J. Littlefield. Global Arrays: A nonuniform memory access programming model for high-performance computers. *J. Supercomputing, 10:197–220, 1996.*
2. F.C. Eigler, W. Farell, S. D. Pullara, and G. V. Wilson, ABC++, in Parallel Programming using C++, G. Wilson and P. Lu, editors, 1996.
3. B. Carpenter, G. Zhang and Y. Wen, NPAC PCRC Runtime Kernel Definition, http://www.npac.syr.edu/projects/pcrc/kernel.html, 1997.
4. D. Quinlan and R. Parsons. A++/P++ array classes for architecture independent fnite difference computations. In Proceedings of the Second Annual Object-Oriented Numerics Conference, April 1994.

5. G. Shah, J. Nieplocha, J. Mirza, C. Kim, R. Harrison, R. K. Govindaraju, K. Gildea, P. DiNicola, and C. Bender, Performance and experience with LAPI: a new high-performance communication library for the IBM RS/6000 SP. Proceedings of the International Parallel Processing Symposium IPPS'98, pages 260-266, 1998.

6. IBM Corp., book chapter "Understanding and Using the Communications Low-Level Application Programming Interface (LAPI)" in IBM Parallel System Support Programs for AIX Administration Guide, GC23-3897-04, 1997. (available at http://ppdbooks.pok.ibm.com:80/cgi-bin/bookmgr/bookmgr.cmd/BOOKS/sspad230/9.1)

7. R. Barriuso, Allan Knies, SHMEM User's Guide, Cray Research Inc, SN-2516, 1994.

8. MPI Forum. MPI-2: Extension to message passing interface, U. Tennessee, July 18, 1997.

9. F. Bassetti, D. Brown, K. Davis, W. Henshaw, D. Quinlan, OVERTURE: An Object-Oriented Framework for High Performance Scientific Computing, Proc. SC98: High Performance Networking and Computing, IEEE Computer Society, 1998.

10. K Langendoen and J. Romein and R. Bhoedjang and H. Bal, Integrating polling, interrupts and thread management, Proc. Frontiers 96, pages 13-22, 1996.

11. T. von Eicken, D.E. Culler, S.C. Goldstein and K.E. Schauser, Active messages: A mechanism for integrated communications and computation, Proc. 19th Ann. Int. Symp. Comp. Arch., pp. 256-266, 1992.

12. Parallel Compiler Runtime Consortium, Common Runtime Support for High-Performance Parallel Languages, Supercomputing `93, IEEE CS Press, 1993.

13. J. Merlin, B. Carpenter and A. Hey, SHPF: a Subset High Performance Fortran compilation system, Fortran Journal, pp. 2-6, March, 1996.

14. G. Zhang, B. Carpenter, G. Fox, X. Li, X. Li and Y. Wen, PCRC-based HPF Compilation, 10th Internat. Wkshp. on Langs. and Compilers for Parallel Computing, LNCS 1366, Springer, 1997.

15. B. Carpenter, G. Fox, D. Leskiw, X. Li, Y. Wen and G. Zhang, Language Bindings for a Data-parallel Runtime, 3 Internat. Wkshp. on High-Level Parallel Programming Models and Supportive Envs., IEEE, 1998.

16. D. E. Culler, A. Dusseau, S. C. Goldstein, A. Krishnamurthy, S. Lumetta, T. von Eicken, and K. Yelick, Parallel Programming in Split-C, Proc. Supercomputing'93, 1993.

17. A. Chien, S. Pakin, M. Lauria, M. Buchanan, K. Hane, L. Giannini, and J. Prusakova. High Performance Virtual Machines HPVM: Clusters with supercomputing APIs and performance. *Proc. 8 SIAM Conf. Parallel Processing in Scientific Computations,* 1997.

18. Compaq Computer Corp., Intel Corp., Microsoft Corp., Virtual Interface Architecture Specification, Dec., 1997.

Multicast-Based Runtime System for Highly Efficient Causally Consistent Software-Only DSM

Thomas Seidmann

Simultan AG, Switzerland [**]
tseidmann@simultan.ch
http://www.simultan.ch/~thomas/

Abstract. This paper introduces the application of IP multicasting for enhancing of software-only DSM systems and, at the same time, simplification of the programming model by offering a simple memory consistency model. The described algorithm is the foundation of a runtime system implemented as filesystems for the Windows NT and FreeBSD operating systems.
Keywords: distributed shared memory, DSM, memory coherence protocol, IPv4/IPv6 multicasting, causal consistency, vector logical clock

1 Introduction

Software distributed shared memory (DSM) realized on a network of workstations connected through a conventional computer network has gained a lot of attention in both research and industry. On the other hand group communication, also known as multicasting, on IP networks is getting widespread over the Internet. The new IP generation, IPv6, relies heavily on the availability of multicasting. This paper presents the idea of an application of IP multicasting for the purpose of coherence traffic in software DSMs.

2 Related Work

Speight and Bennett reported in [5] about the use of multicasting in the Brazos project. There main differences between the Brazos approach and ours are the following:

1. DSM in Brazos is concentrated on scope consistency, that means a memory consistency model using synchronization variables; our approach is to provide causal consistent DSM.
2. Brazos relies implicitly on reliable multicast, whereas we do not.

The Orca distributed system [1] uses totally ordered group communication for a MRMW write-update coherence protocol. The protocol employs a centralized component - the sequencer - to achieve totally ordered multicast on top of the potentially unreliable IP multicast. In our design this centralized component is avoided.

[**] This research was supported in parts from the Slovak University of Technology in Bratislava under Grant VEGA 1/4289/97

3 Multicast-based Coherence Protocol for Causally Consistent DSM

The main requirements for the new multicast-based DSM coherence protocol are the following:

- Delivery of causal consistent memory.
- Use of IP multicast in its present (IPv4 and IPv6) shape, that means with no guarantee of ordering and reliability.
- Avoidance of central components like managers in Brazos or sequencers in Amoeba; the protocol should be completely distributed.

Although in the following text we use the term "page" denoting hereby the granularity of memory sharing (the first implementation is being developed for a page-based DSM), the protocol itself can be used with other types of DSM as well, for example with object-based DSM.

3.1 Components of the Memory Coherence Algorithm

The proposed protocol is basically a MRMW protocol with write-update using multicast transfer of diffs between the involved processes. The causality relationship between shared memory operations is achieved by means of **vector logical clocks** [3], which represent the basic building block in the algorithm. Every process maintains for every shared page a vector logical clock value - timestamp - consisting of:

- The process's own logical (Lamport, monotonic) clock value of the last write operation upon the page.
- Other known processes' logical timestamps of write operations upon the page.

This vector logical timestamp is transmitted within every message that concerns this particular page. In the first implementation of the DSM (see below) the vector logical timestamp is represented with a fixed array of logical timestamps, meaning the group of processes using the DSM is closed and must be known in advance. In a future enhancement, in order to overcome this limitation, the representation of vector logical timestamps will be changed to an associative array consisting of pairs $(PID, value)$, where PID denotes the process' global ID composed of the node's global ID (for example IP address) concatenated with the node local PID.

Every shared page in the DSM must be uniquely identified. This page ID is simply referred to as the page number. In the runtime system described in section 4 the page number can easily be calculated from the offset within the file representing an application's shared memory space.

Every process is responsible for keeping around the last m ($m >= 1$) diffs resulting from its own write accesses for every modified page.

3.2 Initialization of the Algorithm

When a process P wants to access a shared page which is not present locally, the page's vector logical timestamp is initialized with zero values. A new page initialized with zeroes is allocated from the operating system. The further action depends on the type of access to the page.

- If the access is a **write** access, it is performed and the element in the page's logical vector timestamp denoting the local process timestamp is advanced by one and a multicast message requesting the page is sent containing the page ID, the actual vector logical timestamp and the diff freshly produced by the write operation.
- In case of a **read** operation a multicast message denoting a request for the page is sent containing the page ID and the page's actual vector logical timestamp. The read operation returns zero without blocking the caller. Note that this kind of multicast message is also sent asynchronously to any accesses to the page upon expiration of a timer as a part of the mechanism for lost messages recovery; this mechanism is described in section 3.4.

The reaction of processes which receive the message specified above is described in section 3.3.

3.3 Scenarios in the Algorithm

Every process must perform some specific action upon the occurrence of two events: writing to a shared page and receipt of a multicast message with a requests and/or a diff for a page. These actions are described in the following paragraphs.

Write access to a shared page. Every process owning a copy of a shared page and performing a write access notifies all other processes of the performed operation by advancing its own logical clock for the page by one and sending out a multicast message containing the page ID, the page's current vector timestamp logical clock and the diff for the page produced by the write operation. The diff is recorded as specified in section 3.1, perhaps causing an older diff to be purged. This means that every write access must notify the runtime system, that is a context switch to kernel mode is performed, hereby increasing the cost of a write operation on a shared page.

Receipt of a multicast message with a request for a page. Any process keeping a copy of the requested page and whose vector logical timestamp of the is not smaller than the one carried with the message will reply to the message either with the missing diffs kept by the process, or a constructed diff containing the complete page if the diff isn't kept any more (the difference between the vector logical clock values is greater than m, the number of diffs kept by each process for pages modified by itself). The reply is sent in a multicast message and contains the page ID and the vector logical timestamp value of the page in addition to the diff.

Receipt of a multicast message with a diff for a page. This kind of messages are ordered according to the vector logical timestamp carried by them causing the causality relationship between accesses to the page to be respected. The use of vector logical clocks for determination of causal order of events has been described in the ISIS project (Cornell) [2]. Causally related write accesses are performed in their causal order, concurrent (not causally related) writes are performed in arbitrary order. For the purposes of ordering these messages a queue must be maintained by every process for every shared page it keeps locally. A diff to the page is ready to be applied when all diffs causally preceding it have already been applied. When a diff is applied to the contents of the page, the vector logical timestamp of the page is advanced by merging its original value with the timestamp of the diff, thus possibly making other diffs ready to be applied.

The queue if diffs yet to be applied reveals messages missed by the process, which have been seen by at least one process in the causal chain of accesses to the page. These missing diffs, which prevent a diff to be applied, are requested and replied with point-to-point messages.

3.4 Handling of Lost Messages

The previous paragraph showed us that the algorithm can cope with lost messages (both containing requests and/or diffs) as long as at least one of the other processes in the chain of causal accesses to a page has received the missing write update. However there is still a problem when this condition is not met, which is solved by a timer associated with each shared page. This timer is restarted every time a write update (a diff) for the page is received, and upon expiration of this timer a request message containing the page ID and its current vector logical timestamp is sent in a multicast message.

3.5 Summary of the Memory Coherence Protocol Algorithm

It can be shown, that the number of messages in case of no loss can be reduced by the use of multicasting compared to point-to-point communication from $O(N^2)$ to $O(N)$, which is the biggest advantage of the proposed protocol. The simple memory consistency offered by the application of the protocol can be viewed as a gain which has its prize in the rather expensive write operation – each one must be interrupted in order to formulate and send a write-update multicast packet. In case of a filesystem-form implementation of the runtime system (see next section) this means only one context switch between user and kernel mode, however.

4 Results

To examine and test the proposed DSM cache coherence protocol we've implemented it into the CVM system [4] and deployed it on a network of UNIX workstations communicating over an IPv6 network, both local and the 6bone [1]. Note that IPv6 itself is not a prerequisite, but merely a chosen characteristic of the used testbed. In the case of 6bone communication two sites were available, both with multicast routing enabled. The functionality of the protocol was verified on two well-known scientific applications, Barnes Hut (from the SPLASH suite) and FFT-3D.

The coherence-related communication is reduced compared to the sequential consistency mechanism contained in CVM in case of 5 processes by a factor in range between 4 and 5 in case of no loss of messages. Under conditions of 10% loss this factor drop down to 3 due to additional point-to-point messages.

The DSM protocol is being implemented in the form of specialized filesystems for the Windows NT and FreeBSD operating systems. The idea is appealing since it allows the use existing APIs for accessing the DSM instead of adding new ones. Both operating system families contain APIs for the filesystem-backed shared memory paradigm, namely the `MapViewOfFileEx()` family under Windows NT and `mmap()` family under the BSD dialects. The implemented filesystems have two significant features:

- The data contained by them are never serialized to stable storage, except by the applications themselves.
- Although the filesystems can be accessed with "normal" file access APIs, this is not the intended use.

The filesystem-form of implementations has the advantage of reducing the overhead of a write operation on a shared page, since all work related to memory coherence can be done in kernel mode and thus require only a single transition between user and kernel mode and back.

Clearly, various files opened via these filesystems correspond to different applications using the DSM. Every application is supposed to create a separate group for multicast communication.

5 Conclusions and Future Work

The results we have experienced so far make us believe of a right direction in our research. There are a couple of tasks we plan to take on in near future:

- **Formal specification and verification:** We are currently dealing with formal verification of the proposed (and other) protocols by means of the Z specification language.
- **Simulation of the memory coherence algorithm** under FreeBSD-current, which contains a framework called "dummynet" primarily intended for network simulations.
- **Overhead determination** of various effects of the proposed algorithm: group communication, additional context switches etc.
- **Open group of involved processes** as specified in section 3.1.

---

[1] experimental IPv6 network layered on top of the Internet

- **Group membership:** One application is using one group in the sense the Internet Group Membership Protocol (IGMP – in IPv4) or Multicast Listener Discovery (MLD – in IPv6). This behavior could be changed to use more groups and thus keep processes from receiving unnecessary interrupts. We try to elaborate some algorithm for this. An extreme solution that would be a separate group for each shared page.
- **Fault tolerance:** The current version of the DSM coherence protocol doesn't cope with the failure of nodes; we investigate this topic.
- **Security:** Since the some of the current IPsec protocols are not suitable for group communication (for example the ESP header), we investigate the means of achieving privacy for the transfer of coherence-based traffic.

References

1. Henri E. Bal, Raoul Bhoedjang, Rutger Hofman, Ceriel Jacobs, Koen Langendoen, Tim Ruehl, and M. Frans Kaashoek. Performance evaluation of the orca shared object system. *ACM Trans. on Computer Systems*, February 1998.
2. K. P. Birman. The process group approach to reliable distributed computing. *Commun. of the ACM*, 36:36–54, December 1993.
3. Randy Chow and Theodore Johnson. *Distributed Operating Systems & Algorithms*. Addison Wesley Longman, Inc., 1997.
4. Pete Keleher. Cvm: The coherent virtual machine. Technical report, University of Maryland, July 1997.
5. W. E. Speight and J. K. Bennett. Using multicast and multithreading to reduce communication in software dsm systems. In *Proc. of the 4th IEEE Symp. on High-Performance Computer Architecture (HPCA-4)*, pages 312–323, February 1998.

Adaptive DSM-Runtime Behavior via Speculative Data Distribution

Frank Mueller

Humboldt-Universität zu Berlin, Institut für Informatik, 10099 Berlin (Germany)
e-mail: mueller@informatik.hu-berlin.de   phone: (+49) (30) 2093-3011, fax:-3011

Abstract. Traditional distributed shared memory systems (DSM) suffer from the overhead of communication latencies and data fault handling. This work explores the benefits of reducing these overheads and evaluating the savings. Data fault handling can be reduced by employing a lock-oriented consistency protocol that synchronizes data when acquiring and releasing a lock, which eliminates subsequent faults on data accesses. Communication latencies can be reduced by speculatively replicating or moving data before receiving a request for such an action. Histories of DSM accesses are used to anticipate future data requests.

DSM-Threads implement an adaptive runtime history that speculatively distributes data based on access histories. These histories are kept locally on a node and are distinguished by each thread on a node. When regular access patterns on requests are recognized, the data will be speculatively distributed at the closest prior release of a lock. The benefits of this active data distribution and its potential for false speculation are evaluated for two consistency protocols. The data-oriented entry consistency protocol (Bershad *at. al.*) allows an even earlier data distribution at the closest prior release on the **same** lock. A simple multi-reader single-writer protocol, an improvement of Li/Hudak's work, uses the closest prior release of a speculatively **associated** lock for speculative data distribution and also eliminates data faults. In addition, a framework for speculative release consistency is formulated.

1 Introduction

In recent years, distributed systems have enjoyed increasing popularity. They provide a cost-effective means to exploit the existing processing power of networks of workstations. Shared-memory multi-processors (SMPs), on the other hand, do not scale well beyond a relatively small number of processors (typically no more than 40 CPUs) since the system bus becomes a bottleneck. Distributed systems have the potential to scale to a much larger number of processors if decentralized algorithms are employed that cater to the communication medium and topology. However, distributed systems may cause a number of problems.

First, the communication medium between processing nodes (workstations) often poses a bottleneck, although recent increases in throughput in FDDI, FastEther, ATM and SCI may alleviate the problem in part. Also, communication latencies are not reduced quite as significantly. Latency strongly impacts distributed systems since communication generally occurs as a result of a demand for resources from a remote node, and the node may idle until the request is served. In general, distributed applications must restrict communication to the absolute minimum to be competitive.

Another problem is rooted in the absence of a global state within a distributed system. Conventional techniques for concurrent programming cannot be applied anymore. Instead, distributed algorithms have to be developed based on explicit communication (message passing). But either such approaches use a centralized approach with servers as a potential bottleneck or decentralized approaches require new algorithms or extensive program restructuring. As a result, distributed algorithms are often hard to understand.

An alternative solution is provided by *distributed shared memory (DSM)* [13, 14]. DSM provides the logical view of a portion of an address space, which is shared between physically distributed processing nodes, *i.e.*, a global state. Address references can be distinguished between local memory accesses and DSM accesses, *i.e.*, an architecture with non-uniform memory access (NUMA) is created. In addition, the well-understood paradigm of concurrent programming can be used without any changes since the communication between nodes is transparent to the programmer. Notice that NUMA is experiencing a renaissance on the hardware level as well to cope with scalability problems of SMPs.

DSM systems have been studied for over a decade now. A number of consistency models have been proposed to reduce the amount of required messages at the expense of additional requirements that a programmer has to be aware of. Sequential consistency still provides the conventional programming of SMPs but has its limitations in terms of efficiency for DSM systems. It has been commonly implemented using the dynamic distributed manager paradigm at the granularity of pages [7]. We will refer to this implementation as sequential consistency for DSM. Entry consistency requires explicit association between locks and data that is either provided by a compiler or by the programmer with a granularity of arbitrary sized objects [1]. Release consistency is a weak model that allows multiple writes to a page at the same time, as long as the programmer ensures that the writes occur to distinct parts of the page between nodes. Implementations include *eager* release consistency where data is broadcasted at lock releases and *lazy* release consistency where data is requested upon demand when an attempt to acquire a lock is made [6]. These are the most common consistency models.

This paper explores new directions by adapting consistency models for speculative movement of resources. This approach should address two problems of DSM systems. First, access faults, required to trigger data requests, come at the price of interrupts. The increasing complexity in architecture, such as instruction-level parallelism, has a negative impact on the overhead of page faults since all pipelines need to be flushed. The increased complexity translates into additional overhead that may have been negligible in the past. Thus, avoiding page faults may become more important. Speculative data distribution can eliminate most of these page faults. Second, the number of messages in a DSM system often creates a bottleneck for the overall performance. By forwarding data speculatively ahead of any requests, the latency for data requests can be reduced and messages may be avoided.

The scope of this paper is limited to software DSM systems targeted at networks of workstations. The models for speculative data distribution are developed and implemented under *DSM-Threads* [9], a DSM system utilizing an interface similar to POSIX Threads (Pthreads) [19]. The aim of DSM-Threads is to provide an easy way for a programmer to migrate from a concurrent programming model with shared memory (Pthreads) to a distributed model with minimal changes of the application code. A major goal of DSM-Threads is the portability of the system itself followed by a requirement for decentralized runtime components and performance issues. The runtime system of DSM-Threads is implemented as a multi-threaded system over Pthreads on each node and copes without compiler or operating system modifications. Several data consistency models are supported to facilitate ports from Pthreads to DSM-Threads and address the requirements of advanced concepts for multiple writers. Heterogeneous systems are supported by handling different data representations at the communication level. The system assumes a point-to-point message-passing protocol that can be adapted to different communication media. It assumes that messages are slow compared to computation and that the size of messages does not have a major impact on message latencies.

The paper is structured as follows. After a discussion about related work, the basic concept of speculative data distribution are developed. Following this discussion, three specific models of speculation are proposed for common consistency models, namely speculative entry consistency, speculative sequential consistency and speculative release consistency. These models rely on the recognition of access pattern, as considered next. Thereafter, measurements within the DSM-Threads environment are presented and interpreted for speculative locking and speculative data distribution. After an outlook at future work, the work is summarized in the conclusions.

2 Related Work

Software DSM system have been realized for a number of consistency models mentioned before. Nevertheless, the support for speculative data distribution in the context of DSM has only recently been studied. Implementations of sequential consistency, such as by Li and Hudak [7], do not mention speculative approaches. The topic is neither been covered by the original work on entry consistency [1] nor for the associated lock protocols [12]. However, Keleher *et. al.* developed and evaluated a hybrid model as a compromise between eager and lazy release consistency [6]. Data is associated with locks based on heuristics and is speculatively forwarded to a growing number of interested nodes upon lock release. This model differs from our approach in that the latter goes beyond data speculation. We also consider speculative forwarding of lock ownership, varying data access patterns, on-line assessment of the benefits of speculation and suspension and resumption of speculation mechanisms. Recently, Karlsson and Stenström [4] as well as Bianchini *et. al.* [2] proposed prefetching schemes using histories of length two and sequential prefetching but still take page faults on already prefetched pages to ensure that prefetched data will be used. They report speedups ranging between 34% and slight performance degradations depending on the applications but most tested programs show good improvement. These studies differ from our approach since we extend speculation to the synchronization protocol, avoid page faults, incorporate asynchronous message delivery and support sequential and entry consistency. Schuster and Shalev [16] use access histories to steer migration of threads, which shows the range of applications for access histories. While access patterns have also been studied in the context of parallel languages, such as in HPF [15] or filaments [8], such systems rely on compiler support to generate code for data prefetching at user-specified directives. The work described in this paper aims at systems without compiler modifications. Previous work on branch prediction [17, 18] is closely related to the recognition of access patterns proposed in this work.

3 Speculative Data Distribution

Common consistency models for distributed shared memory systems are based on demand-driven protocols. In other words, if data is not accessible locally, it is explicitly requested from a remote node. Speculative data distribution, on the other hand, anticipates future access patterns based on the recorded history of access faults. This is depicted abstractly in Figure 1 where an multiple writes to data x are interleaved between nodes A and B. Instead of serving requests on demand by sending data, the data is forwarded to the node most likely to be accessing it ahead of any requests (see second write to x on node B). If this speculation of access sequences correctly anticipates the application behavior, then considerable performance savings will materialize given that both communication and interrupt overhead can be reduced. Should, however, the speculation not represent the actual program behavior, then the overhead of speculative movements adversely affects performance. Hence, there is a clear trade-off between the number of beneficial and the number of false speculations that can be tolerated.

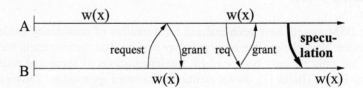

Fig. 1. Abstract Speculation Model for Access Histories

The problem of additional overhead of false speculations can be addressed by various means. One may require a later response from a destination node to confirm that speculative movement was beneficial. In order to keep communication overhead low, this confirmation should be piggybacked with the lock distribution as depicted in Figure 2. When the original node acquires the lock again, it will then be able to evaluate the effect of its previous speculation. Upon realizing that false speculation had occurred, future speculative movements can be inhibited. This confirmation protocol should be very effective since it does not result in any additional messages and the slight increase in size of lock messages is negligible for most communication protocols. An alternative option to confirmations within the runtime system are application-driven control interfaces that govern the use of speculation. This provision allows the programmer to specify access patterns for data areas with respect to program parts. Common access patterns, such as access by row, column, block or irregular, can drive the runtime protocol for speculations. However, this interface requires knowledge by the programmer about algorithms and execution environments that may not always be available.

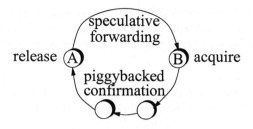

Fig. 2. Confirmation of Speculative Success

The consistency models for distributed shared memory and for distributed mutual exclusion are mostly implemented by synchronous message protocols. These protocols have to be extended to support speculative data distribution. In an asynchronous message-passing model, such as DSM-Threads, the protocols to implement consistency models are slightly more complex [10] and require a few additional considerations than their simpler synchronous counterparts. The main problem of speculative data forwarding is posed by race conditions that may occur in lock protocols and consistency protocols when data is requested while being speculatively forwarded as depicted in Figure 3. A data request may then be forwarded to its requester, a situation that conventional protocols do not anticipate, such that the protocol is broken. This situation should be handled by a silent consumption of data requests on the node that speculatively handed off the data. Such a behavior is safe for protocols that dynamically adjust their knowledge of ownership, such as the distributed dynamic manager protocol [7]. Advanced asynchronous protocols already handle such data races. For example, a token-based protocol for distributed mutual exclusion with priority support prevents these races in the described manner and is directly applicable to non-prioritized protocols as well [11]. The only restriction here is that in-order message handling is required for requests regarding the same data object. Another problem is posed by the requirement to process token/page grant messages when they have not been anticipated, *i.e.,* when they arrive unasked without prior issuing of a request.

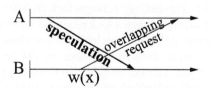

Fig. 3. Overlap between Speculation and Request

The integration of speculation into protocols implementing consistency models also requires a resolution at conflicts between demand-driven requests and speculative information. DSM-Threads always gives demand-driven requests precedence over speculative actions based on the motivation to keep response times of lock requests low and serve requests in a fair manner. If speculative forwarding had the upper hand, the performance may degrade due to false speculation. In the event of multiple pending requests, speculative information could be matched with the requests to serve a later request with a match ahead of earlier ones. However, such behavior may not only be undesirable, distributed data structures may even prevent them, which is the case for distributed lock protocols and distributed data managers.

The speculative distribution of data requires not only the recognition of access patterns but also the association of data with locks. This association can be provided by logging information about the data in the lock object. This simple method may already yield good improvements in performance. A change of the access pattern during program execution may, however, initially result in false speculation. This problem can be addressed by application-driven interfaces but such an approach requires specific knowledge by the programmer, as seen before. An alternative solution is motivated by the observation that a change of access patterns generally occurs between different program parts. If the source of a lock request was related to the program part, speculation could be inhibited when the execution crosses from one part into the next one. This could be achieved by associating the stack pointer and return address (program counter) with a lock upon a lock request. Such information would provide the means to prevent false speculation *after* a new program part is entered but it cannot prevent potentially false speculation at the last release of the lock in the *previous* part. Furthermore, obtaining the stack pointer and return address are machine-dependent operations that restrict portability. Finally, after one false speculation, the confirmation protocol as discussed above should already suffice to reduce the chance of consecutive false speculations.

3.1 Speculative Entry Consistency

Entry consistency associates data explicitly with locks, where the granularity of data is arbitrary (typically objects or abstract data structures). When a lock is transfered between nodes, the data is piggybacked. As a result, accesses to data protected by a lock never result in faults with the associated interrupts and communication. It is assumed data and locks are explicitly associated.

Speculative entry consistency improves upon the lock protocol of the consistency model. Commonly, locks are transfered upon request, *i.e.,* the protocol is demand driven. By logging the history of lock requests, lock movement can be anticipated if regular lock patterns exist. The analogy between lock requests and data accesses illuminates the potential for savings. When a lock request is received on a node that holds the mutual exclusive access (token) of the lock but has already released it, then the requester is recorded as a future speculative destination. The destination node is associated with the same lock object. This has no side-effect on other synchronization objects. Another round of an acquire and release on the same node results in speculative lock forwarding at the point of the release to the recorded destination node as depicted in Figure 4.

The enhanced locking model may be subject to false speculation, just as speculative data distribution, but such a situation may easily be detected by recording if the lock was acquired at the destination node before handing over the token to another node. Upon successful speculation, this protocol reduces the latency of lock operations (acquires) and the amount of messages sent since an acquire to a locally available lock does not result in remote requests anymore. Modifications to existing protocols in order to

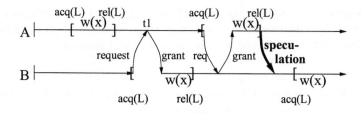

Fig. 4. Example for Speculative Entry Consistency

support speculation have to compensate for requests and speculative motion that occur in parallel as well as reception of a token without prior request. Barriers can be handled in a similar way by considering them as a sequence of a release followed by an acquire [6].

3.2 Speculative Sequential Consistency

Sequential consistency does not provide the luxury of *explicit* associations between data and locks. The model mandates a single-writer-multiple-reader protocol that operates at the granularity of pages. It also requires that the programmer enforces a partial order on memory accesses by using synchronization, reflecting the model of a physically shared-memory environment with multiple threads of control. The lock protocols can be extended to speculatively hand off mutual exclusive access as seen for speculative entry consistency. However, the lack of data association with locks implies that data accesses may still cause page faults in the presence of speculative locking.

Data association can be induced *implicitly* within the sequential consistency model by a reactive response to page faults. Figure 5 depicts the pseudo-code for activating speculations. Upon a page fault in node A, the page is associated with the previous

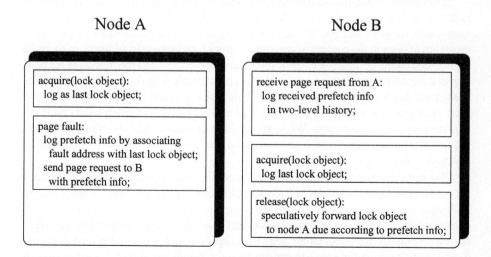

Fig. 5. Protocol for Speculative Sequential Consistency

lock. This information is sent to the owner of the page, node B. Upon reception, node B records this page-lock association in its history and responds with the requested page. At some last point in time, node B locks the same object again and accesses the page in question. Upon releasing the lock, the lock as well as the page are speculatively forwarded to node A if the prefetch information indicates such an action.

In general, a request is locally associated with the lock object of the most recently completed acquire operation but this information is recorded when a remote node receives a page request from the node that had a page fault. Upon the next release operation on the remote node, both the lock token and the data of the page will be speculatively forwarded, even without any remote requests as depicted in Figure 6. Thus, speculation exists at two levels. First, locks and pages are speculatively matched. Second, the association of future destination nodes with locks/pages supplements the speculations. This model has the advantage that the speculative destinations of locks and pages can be compared. In case of a match, the chance of false speculation is reduced. In case of a mismatch, the data may not be associated with the lock, although the local lock history could provide a previously released lock with a matching destination, which may be a likely source for association.

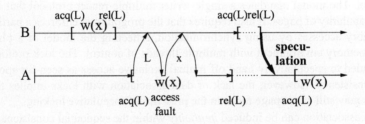

Fig. 6. Example for Speculative Sequential Consistency

Speculative sequential consistency provides potentially high benefits in terms of elimination of access faults and reduction in message traffic. Upon successful speculation, lock requests will not be issued, resulting in lower latencies of locks and fewer messages as seen in the context of speculative entry consistency. In addition, page faults are avoided on the destination node of speculative forwarding and the messages for page requests are saved. The latter savings may be considerable when both a read and a write fault occur for a remote write operation. This claim is based on the observation that there is no portable way to distinguish a read and a write access when a data fault occurs on a non-accessible page. While the faulting instruction may be disassembled to inquire whether the address fault lies within the read or within the written operands, the instruction decoding is highly unportable.

The portable method of distinguishing access modes on faults requires two faults for writes. The first fault results in a request for a read copy of the page. When a second fault occurs on a readable page, the page is supplied in write mode. Thus, the first request requires at least two messages for the read copy plus at least another two messages for the second one for invalidating other read copies, which was not depicted in Figure 6. Speculative sequential consistency can be quite beneficial in such a setting since it can save four page messages, one lock request message, two page faults and reduce the lock latency. False speculation may even be tolerated to some extend if the number of beneficial speculations is sufficient. However, the detection of false speculation with respect to data poses problems since page faults are avoided for successful speculation. Nonetheless, to assess the data/lock association one may perform the lock/data forwarding but, for the first time of speculative movement, leave the page protected to record the next read and write fault. Upon successful speculation, the page access will be enabled at the proper level (read or write) for subsequent speculations. Another option for write accesses will be discussed in the context of speculative release consistency.

3.3 Speculative Release Consistency

Keleher *et. al.* developed a so-called *lazy hybrid* version of release consistency, as mentioned in the related work [6]. The model is thought as an compromise between lazy release consistency and eager release consistency. A heuristic controls the creation of diffs (data differences resulting from write accesses) upon releases of locks and piggybacks them with lock grants. The destinations of such speculative data distribution are members of an *approximate* copy set that continually grows as processes declare interest in data. If a diff is rejected by the heuristic, lazy release consistency is used. A central manager is used to implement this protocol. The main application of this protocol are message-passing systems where the number of messages impacts performance more than size of messages. In comparison with the other variants of release consistency, the model only proves to be marginally better on the average. The reduced number of page faults only played a marginal role in their study.

The speculation realized by the hybrid model neglects opportunities for lock speculation, as explained in the context of speculative entry consistency. It also assumes centralized algorithms for managing pages. Thus, the speculative release consistency proposed here differs in a number of issues. First, speculative locks are utilized. Second, in the absence of a central manager, the approximate copy set has to be created at the root of page invalidation and is subsequently piggybacked with the lock grant message. Third, the approximate copy set can be updated by assessing the effectiveness of prefetching using the confirmation model discussed in the context of speculative sequential consistency. When speculative forwarding occurred without subsequent data access, members are removed from the copy set. This model reduces the overall amount of communication. It also shows that confirmations should be repeated from time to time to reflect a change of access patterns of a different program part in communication patterns. Finally, logging access patterns enables the speculation model discussed in this paper to selectively create diffs and grant locks. This will be discussed in more detail in the following.

3.4 Access Patterns

The collection and recognition of access patterns is a fundamental condition for speculative data distribution. Often, temporal locality exhibits simple patterns where each acquire was followed by an access miss, *e.g.* for sequential consistency and for speculative lock forwarding with entry consistency. Histories of depth one suffice to handle such simple patterns. We also experimented with histories of size two, where prefetching is delayed until an access is repeated. These histories of depth two have been successfully used in the context of branch prediction with certain variations [17]. Due to the similarity between control flow changes and lock requests along control flow paths, these histories are promising for lock protocols as well.

Another common pattern is given by local reuse of resources, for example when every second acquire is followed by an access fault. The latter case also provides opportunities for improvements, in particular for speculative release consistency. Here, not only lock forwarding but also diff creation can be delayed to the point of the second release of the lock. Again, similarities to branch prediction exist, in particular wrt. two-level prediction schemes [21]. However, the problem of DSM access histories requires that addresses of pages or locks are recorded while branch prediction records whether or not a branch was taken. In this sense, DSM access patterns may be closer to predicting the target address of indirect jumps, which is a problem that has also been handled successfully with two-level histories [18]. In the context of DSMs, the first-level history of depth n records the source nodes of the n most recent requests, often in a compressed format within k bits (see Figure 7). Such a *Target History Buffer (THB)* should be kept for each page or lock. (A more simplistic variant for branch prediction uses a single global THB but lacks precision, in particular for more complex access patterns.) The second-level history consists of a *predictor table (PT)* with entries for 2^k patterns indexed by a hashed value calculated from all THB entries. The PT simply contains at the index of the THB the last access recorded for this pattern, which is then used for the next prediction. Thus, the use of two-level histories requires that actual accesses be propagated to the predictor, which can be accomplished by piggybacking this knowledge together with page or lock migration.

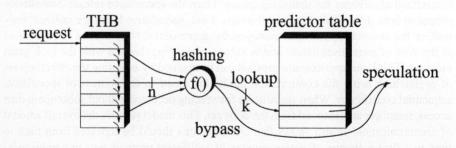

Fig. 7. Two-Level Access Histories

There are several options for using two-level histories within DSMs. One may either use the program counter the base address of a page (or a lock index) for requests. Alternatively, the fault address may be used. The difference between these options becomes clear when considering spatial locality in an SPMD model with synchronization at each iteration. When fault addresses are used, the addresses differ from iteration to iteration resulting in ever new indices for the PT. For column-oriented spatial locality, the base addresses of a page would always result in the same PT index yielding better predictions. Another way to handling this problem would be by considering only the most significant k bits of the fault address (and the least significant k bits for lock indices). For row-oriented spatial locality, on the other hand, the lower bits of the faulting address may be more telling assuming a uniform stride. To deal with both of these cases, the entire address should be saved in the THB and the hashing process can be augmented to recognize strides. For small strides (column indexing), the highest k bits are hashed while large strides may bypass the PT (see Figure 7) in order to prefetch the next page (in direction of the stride).

A problem is posed by irregular access patterns. While complicated patterns may still be recognized by two-level histories, a completely chaotic access behavior cannot be dealt with. First, such a situation has to be recognized. This can be accomplished by adding a two-bit saturating up-down counter to each PT entry (also similar to branch prediction [17]) that is incremented on successful speculation and decremented otherwise. Speculation will simply be inhibited when this counter is less than 2 (see Figure 8). This limits the amount of false speculation (with prefetching) to two instances and requires another two correct speculations (without prefetching) before prefetching is reactivated.

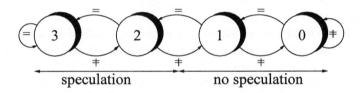

Fig. 8. DFA for a Two-Bit Saturating Up-Down Counter

In an SPMD programming model, a number of distributed processes may issue a request for the same page or lock almost at the same time. Sometimes, node A may be slightly faster than node B, the next time B may be faster. The resulting access patterns may seem irregular at first but the group of processes issuing requests within a certain time frame remains the same. Thus, two-level histories may also help out but it may take slightly longer before different patterns are stored in the corresponding PT-entries. In addition, it would help to delay transitive migrations to allow the current owner to utilize the forwarded information, thereby compensating for jitters of requests. Our system uses a short delay before it responds to further requests for this reason, which is similar to timed consistency but is often used in conjunction with consistency protocols that support data migration.

The two-level history scheme described here records the first access between synchronizations for implementations where subsequent faults are still forced before pages are validated (see rated work). The framework described in this paper avoids faults altogether after successful speculation but will only record the first two faults (for depth two histories or two-bit counters). It may also be useful to experiment with a combination of these options where subsequent faults are only forced on every ith iteration, depending on the stride. This would allow prefetching of the next page at page boundaries and re-confirmation of speculation.

Speculative Locking			Speculative Data Forwarding		
Speculation	No	Yes	Speculation	No	Yes
Messages	38	22	Messages	76	8
Local Locks	1	18	Page Faults	18	4

Table 1. Performance Evaluation

4 Measurements

Experiments were conducted with the aforementioned distributed shared-memory system called DSM-Threads. The user interface of the system closely resembles the POSIX Threads API [19]. It distributes threads remotely upon requests by spawning a remote process (if none exits) and executing the requested thread (instead of the main program). Each logical node is itself multi-threaded, *i.e.,* local threads can be executed in a physically shared memory. Nodes are identified by a tuple of host and process. The system supports asynchronous communication for message-passing systems, and the low-level message-passing implementations currently support TCP sockets and the SCI common message layer [3]. The lock protocol is an implementation along the lines of Naimi *et. al.* [12] although DSM-Threads supports asynchronous handling of the lock. The system also supports priority-based locks but this option was not used for the experiments to minimize the influence of more complex protocols [11]. Sequential consistency is realized by Li/Hudak's work [7] but uses an asynchronous variant of the protocol [10]. The system also provides entry consistency much alike Bershad *et. al.* [1]. The implementation of lazy release consistency following Keleher's work is still in progress [5]. All algorithms use decentralized distributed methods to prevent scalability problems due to software restrictions.

The DSM-Threads system runs under SunOS 4.1.x and the Linux operating systems at present. It utilizes POSIX threads at runtime and Gnu tools for compilation, assembling, linking and extraction of information from the symbol table [9]. The following experiments were conducted under Linux. A ping-pong test was designed to assess the potentials of speculation. Two logical nodes share data on the same page that is accessed for reading and writing. The accesses occur between an acquire and a release of the same lock object. Each node repeats the sequence of acquire, read accesses, write access and release 10 times. The runtime system was instrumented to report the number local locks (without communication) and the number of messages sent with respect to pages and locks. The amount of page faults could be inferred from these numbers.

The first experiment compares a lock protocol without speculation with a version that speculatively forwards locks, only. The left part of Table 1 depicts the number of messages recorded for locks within the program. Each acquire required a request message and a granting response without speculation, except for the first acquire since the token was locally available. With speculation, the first acquire on each node required two messages as before and resulted in a recording of the access pattern. All other acquires were preceded by speculative forwarding of the token at a release to the other node, resulting only in one message per release. The acquires proceeded without message latency as they could be granted locally. Excluding the start-up overhead, the number of messages can be reduced by 50% with this method.

The second experiment compares data migration without speculation with a version that speculatively forwards data at release operations. The right part of Table 1 depicts the number of messages recorded for data requests within the program. Lock requests are not depicted. Each access required a request message and a granting response without speculation, except for the first read and write accesses since the initial node had write access to the data. The remaining 9 iterations in the initial node and 10 iterations in the other node constituted of a read access (request and grant read) and a write access (invalidate copy set and acknowledge), for a total of $19*4 = 76$ messages. Each of the 19 instances resulted in a read fault followed by a write fault. With speculation, the first round of a read and write access pair on each node required four messages (as before) and resulted in recording the access pattern. All other accesses were preceded by speculative piggybacking of the data with the (speculative) token forwarding to the other node at the release point, resulting in no data messages at all. Notice that the overhead of the release message was already accounted for in the lock protocol. The data accesses on the destination node proceeded without message latency as they could be satisfied locally without a page fault. Thus, after a start-up overhead dependent on the number of nodes interested in the data, all data messages and page faults are eliminated. For n interested nodes, the initial overhead would amount to $4n$ messages and $2n$ faults. In the event of confirmation for speculations, an additional $2n$ faults would be required in the second round of access pairs.

The actual savings may depend on the communication medium and processing environment, which is common for DSM systems. If communication latencies are small relative to processing time, the observed savings can be achieved in practice. For longer latencies, chances increase that a lock request is issued before or while speculative forwarding occurs. In that case, the advantages of speculation may not materialize in full but the message performance would not be degraded. This applies mostly to speculative entry consistency since the lock protocol is the primary cause of messages. The book-keeping overhead in terms of local computation for speculation is relatively low.

However slow the communication system, speculation may still result in savings of messages although long latencies cause parallel request issues. In particular for speculative data distribution, the data will be supplied at the correct access level (read or write). In case of write accesses, the overhead of a prior read fault with its two messages can still be avoided. In conjunction with lock association, the full savings of speculative data distribution should still materialize for speculative sequential consistency.

5 Future Work

On-going work includes an implementation of lazy release consistency within DSM-Threads. A comparison with speculative release consistency would be interesting with respect to the competitiveness of speculative sequential consistency. Current work also includes the recognition of more complex access patterns, speculative confirmation during initial speculation and temporal re-confirmation of speculations. We are also considering to log access patterns on the faulting node and piggybacking the information with the lock token to reduce the chance of false associations between data and locks. Furthermore, it may be possible to avoid request messages (or delay them) if a demanding node can be ensured that speculation will provide him with the data right away. Finally, selected programs of the *de-facto* standard benchmark suite "Splash" are being adapted for DSM-Threads for a more comprehensive evaluation [20].

6 Conclusion

This work presented methods for speculative distribution of resources as a means to address the main short-comings of DSM systems, namely the amount of communication overhead and the cost of access faults. A number of models with speculative support are developed. Speculative entry consistency forwards lock tokens at the point of a release to another node based on past usage histories. Speculative sequential consistency not only utilizes a speculative lock protocol but allows data association with the most recently released lock. Future lock grants will then piggyback the associated data, thereby circumventing access faults at the destination node. Speculative release consistency extends Keleher's hybrid model to provide speculative lock forwarding, initial confirmation of the benefits of speculation and temporary re-confirmation of these benefits to detect changes in access patterns. Initial measurements show that the number of messages for the lock protocol can be almost reduced by half under ideal circumstances. The savings for speculative data distribution for sequential consistency are even more dramatic. After a start-up of a constant number of messages, all subsequent messages may be avoided by piggybacking the data with lock messages. Furthermore, all page faults may be avoided as well when the speculation is successful. The main problems of speculation are twofold. First, there exists a potential of false speculation, where resources are sent to a node that does not use them. The paper suggests two-level histories to recognize access patterns and use re-confirmation to avoid unnecessary communication. Second, large communication latencies relative the short processing intervals may allow request messages to be issued even though the requests will be granted by speculation. Speculation does not cause any penalty in this case, in fact, it may still reduce latencies but it cannot reduce the amount of messages if requests are sent in parallel with speculations. Overall, existing consistency protocols only require moderate changes to support speculation so that it should well be worth to incorporate these methods into existing DSM systems.

References

1. B. Bershad, M. Zekauskas, and W. Sawdon. The midway distributed shared memory system. In *COMPCON Conference Proceedings*, pages 528–537, 1993.
2. R. Biachini, R. Pinto, and C. Amorim. Data prefetching for software DSMs. In *Int. Conf. on Supercomputing*, 1998.
3. B. Herland and M. Eberl. A common design of PVM and MPI using the SCI interconnect. final report for task a 2.11 of ESPRIT project 23174, University of Bergen and Technical University of Munich, October 1997.

4. M. Karlsson and P. Stenström. Effectiveness of dynamic prefetching in multiple-writer distributed virtual shared memory systems. *J. Parallel Distrib. Comput.*, 43(7):79–93, July 1997.
5. P. Keleher, A. Cox, and W. Zwaenepoel. Lazy release consistency for software distributed shared memory. In *International Symposium on Computer Architecture*, pages 13–21, 1992.
6. Pete Keleher, Alan L. Cox, Sandhya Dwarkadas, and Willy Zwaenepoel. An evaluation of software-based release consistent protocols. *J. Parallel Distrib. Comput.*, 29:126–141, September 1995.
7. K. Li and P. Hudak. Memory coherence in shared virtual memory systems. *ACM Trans. Comput. Systems*, 7(4):321–359, November 1989.
8. D. K. Lowenthal, V. W. Freeh, and G. R. Andrews. Using fine-grain threads and run-time decision making in parallel computing. *J. Parallel Distrib. Comput.*, August 1996.
9. F. Mueller. Distributed shared-memory threads: DSM-threads. In *Workshop on Run-Time Systems for Parallel Programming*, pages 31–40, April 1997.
10. F. Mueller. On the design and implementation of DSM-threads. In *Proc. 1997 International Conference on Parallel and Distributed Processing Techniques and Applications*, pages 315–324, June 1997. (invited).
11. F. Mueller. Prioritized token-based mutual exclusion for distributed systems. In *International Parallel Processing Symposium*, pages 791–795, 1998.
12. M. Naimi, M. Trehel, and A. Arnold. A log(N) distributed mutual exclusion algorithm based on path reversal. *J. Parallel Distrib. Comput.*, 34(1):1–13, April 1996.
13. B. Nitzberg and V. Lo. Distributed shared memory: A survey of issues and algorithms. *IEEE Computer*, 24(8):52–60, August 1991.
14. J. Protic, M. Tomasevic, and V. Milutinovic. Distributed shared memory: Concepts and systems. *IEEE Parallel and Distributed Technology*, pages 63–79, Summer 1996.
15. Rice University, Houston, Texas. *High Performance Fortran Language Specification*, 2.0 edition, May 97.
16. A. Schuster and L. Shalev. Using remote access histories for thread scheduling in distributed shared memory systems. In *Int. Symposium on Distributed Computing*, pages 347–362. Springer LNCS 1499, September 1998.
17. J. E. Smith. A study of branch prediction strategies. In *Proceedings of the Eighth International Symposium on Computer Architecture*, pages 135–148, May 1981.
18. J. Stark and Y. Patt. Variable length path branch prediction. In *Architectural Support for Programming Languages and Operating Systems*, pages 170–179, 1998.
19. Technical Committee on Operating Systems and Application Environments of the IEEE. *Portable Operating System Interface (POSIX)—Part 1: System Application Program Interface (API)*, 1996. ANSI/IEEE Std 1003.1, 1995 Edition, including 1003.1c: Amendment 2: Threads Extension [C Language].
20. S. C. Woo, M. Ohara, E. Torrie, J. P. Singh, and A. Gupta. The SPLASH-2 programs: Characteriation and methodological considerations. In *Proceedings of the 22nd Annual International Symposium on Computer Architecture*, pages 24–37, New York, June 22–24 1995. ACM Press.
21. T.-Y. Yeh and Y. N. Patt. Alternative implementations of two-level adaptive branch prediction. In David Abramson and Jean-Luc Gaudiot, editors, *Proceedings of the 19th Annual International Symposium on Computer Architecture*, pages 124–135, Gold Coast, Australia, May 1992. ACM Press.

6th Reconfigurable Architectures Workshop

Hossam ElGindy – Program Committee Chair

School of Computer Science and Engineering
University of New South Wales
Sydney NSW 2052
hossam@cse.unsw.edu.au

Welcome to the sixth edition of the Reconfigurable Architectures Workshop. The workshop features 3 contributed paper sessions, 1 poster session and 1 panel on the status of reconfigurable architectures and computing. The 14 contributed presentations and 8 posters were chosen from 34 submissions by the program committee. The papers were chosen to cover the active research work in both theoretical and industrial/practical sides. We hope that the workshop will provide the environment for interaction and exchange between the two sides.

I would like to thank the organizing committee of IPPS/SPDP'99 for the opportunity to organize this workshop, the authors for their contributed manuscripts, and the program committee and referees for their effort in assessing the 34 submissions to the workshop. Finally I would like thank Viktor K. Prasanna for his tireless effort in getting all aspects of the workshop organization.

Workshop Chair

Viktor K. Prasanna, University of Southern California (USA)

Program Committee

    Jeff Arnold, Independent Consultant (USA)
    Michael Butts, Synopsys, Inc. (USA)
    Steve Casselman, Virtual Computer Corporation (USA)
    Bernard Courtois, Laboratory TIMA, Grenoble (France)
    Carl Ebeling, Univ. of Washington (USA)
    Hossam ElGindy, University of NSW (AUSTRALIA) - Chair
    Reiner Hartenstein, Univ. of Kaiserslautern (Germany)
    Brad Hutchings, Brigham Young Univ. (USA)
    Hyoung Joong Kim, Kangwon National Univ. (Korea)
    Fabrizio Lombardi, Northeastern University (USA)
    Wayne Luk, Imperial College (UK)
    Patrick Lysaght, Univ. of Strathclyde (Scotland)
    William H. Mangione-Smith, Univ. of California, Los Angeles (USA)
    Malgorzata Marek-Sadowska, Univ. of California, Santa Barbara (USA)
    Margaret Martonosi, Princeton Univ. (USA)

John T. McHenry, National Security Agency (USA)
Alessandro Mei, Univ. of Trento (Italy)
Martin Middendorf, Univ. of Karlsruhe (TH) (Germany)
George Milne, Univ. of South Australia (Australia)
Koji Nakano, Nagoya Institute of Technology (Japan)
Stephan Olariu, Old Dominion Univ. (USA)
Robert Parker, Information Sciences Institute East/USC (USA)
Jonathan Rose, Univ. of Toronto (Canada)
Hartmut Schmeck, Univ. of Karlsruhe (TH) (Germany)
Herman Schmit, Carnegie Mellon Univ. (USA)
Mark Shand, Compaq Systems Research Center (USA)
Jerry L. Trahan, Louisiana State Univ. (USA)
Ramachandran Vaidyanathan, Louisiana State Univ. (USA)

Publicity Chair

Kiran Bondalapati, University of Southern California (USA)

DEFACTO: A Design Environment for Adaptive Computing Technology

Kiran Bondalapati[†], Pedro Diniz[‡], Phillip Duncan[§], John Granacki[‡],
Mary Hall[‡], Rajeev Jain[‡], Heidi Ziegler[†]

Abstract. The lack of high-level design tools hampers the widespread adoption of adaptive computing systems. Application developers have to master a wide range of functions, from the high-level architecture design, to the timing of actual control and data signals. In this paper we describe DEFACTO, an end-to-end design environment aimed at bridging the gap in tools for adaptive computing by bringing together parallelizing compiler technology and synthesis techniques.

1 Introduction

Adaptive computing systems incorporating configurable computing units (CCUs) (*e.g.*, FPGAS) can offer significant performance advantages over conventional processors as they can be tailored to the particular computational needs of a given application (e.g., template-based matching, Monte Carlo simulation, and string matching algorithms). Unfortunately, mapping programs to adaptive computing systems is extremely cumbersome, demanding that software developers also assume the role of hardware designers. At present, developing applications on most such systems requires low-level VHDL coding, and complex management of communication and control. While a few tools for application developer are being designed, these have been narrowly focused on a single application or a specific configurable architecture [1] or require new programming languages and computing paradigms [2]. The absence of general-purpose, high-level programming tools for adaptive computing applications has hampered the widespread adoption of this technology; currently, this area is only accessible to a very small collection of specially trained individuals.

This paper describes DEFACTO, an end-to-end design environment for developing applications mapped to adaptive computing architectures. A user of DEFACTO develops an application in a high-level programming language such as C, possibly augmented with application-specific information. The system maps this application to an adaptive computing architecture that consists of multiple FPGAs as coprocessors to a conventional general-purpose processor.

DEFACTO leverages parallelizing compiler technology based on the Stanford SUIF compiler. While much existing compiler technology is directly applicable

---

† Department of Electrical Engineering-Systems, University of Southern California, Los Angeles, CA 90089, email: {kiran,hziegler}@usc.edu
‡ Information Sciences Institute, University of Southern California, 4676 Admiralty Way, Marina del Rey, CA 92092. email: {pedro,granacki,mhall,rajeev}@isi.edu
§ Angeles Design Systems. 2 N. First Street, Suite 400, San Jose, CA 95113. email: duncan@angeles.com

to this domain, adaptive computing environments present new challenges to a compiler, particularly the requirement of defining or selecting the functionality of the target architecture. Thus, a design environment for adaptive computing must also leverage CAD techniques to manage mapping configurable computations to actual hardware. DEFACTO combines compiler technology, CAD environments and techniques specially developed for adaptive computing in a single system.

The remainder of the paper is organized into three sections and a conclusion. In the next section, we present an overview of DEFACTO. Section 3 describes the system-level compilation based on the Stanford SUIF compiler. Section 4 presents the design manager, the tool that brings together the system-level compiler and commercial synthesis tools.

2 System Overview

The target architecture for DEFACTO consists of a single general-purpose processor (GPP) and multiple configurable computing units (CCUs). Each CCU can access its own memory and communicate with the GPP and other CCUs via data and control channels. In this architecture, the general-purpose processor is responsible for orchestrating the execution of the CCUs by managing the flow of data and control in the application execution. An architecture description language is used to describe the specific features of the communication channels (e.g., number of channels per CCU and their bandwidth) and topology of the connections between the CCUs; the amount and number of ports on each CCU and the CCU logic capacity. This architecture specification is used by DEFACTO to evaluate possible program partitioning strategies.

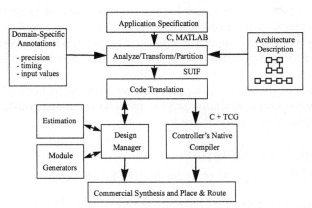

Fig. 1. DEFACTO Compilation Flow Outline

Figure 1 outlines the flow of information in the DEFACTO environment. The input for DEFACTO consists of an application specification written in a high-level language such as C or MATLAB and augmented with domain and application-specific annotations. These annotations include variable precision, arithmetic operation semantics and timing constraints.

The DEFACTO system-level compiler, represented by the Analyze/Transform/Partition box in Figure 1, first converts the application specification to SUIF's intermediate representation. In this translation, the system-level compiler preserves any application-specific pragmas the programmer has included. The system-level compiler then analyzes the input program for opportunities to exploit parallelism. Next the compiler generates an abstract representation of the parallelism and communication. This representation consists of the partitioning of tasks from the input program, specifying which execute on CCUs and which execute on the GPP, and capturing communication and control among tasks. When the compiler has derived a feasible program partitioning, it generates C source code that executes on the GPP and an HDL representation (i.e., a vendor-neutral hardware description language) of the portions of the computation that execute on the configurable computing units. While the C portion is translated to machine code by the GPP's native compiler, the portions of the computation that execute on the CCUs provides input to the design manager, and subsequently, commercial synthesis tools, to generate the appropriate bitstreams representing the configurations.

The code partitioning phase is iterative, whereby the system-level compiler interfaces with the design manager component. The design manager is responsible for determining that a given partition of the computation between the GPP and the CCUs is feasible, i.e., the resources allocated to the CCU actually fit into the hardware resources. For this purpose, the design manager makes use of two additional tools, the estimation and module generator components. The estimation tool is responsible for estimating the hardware resources a given computation requires while the module generator uses a set of parameterized library modules that implement predefined functions.

An overall optimization algorithm controls the flow of information in the DEFACTO system. The optimization goal is to derive a complete design that correctly implements the application within the constraints imposed by the architecture platform, minimizes overall execution time, avoids reconfiguration and meets any timing constraints specified by the programmer. The compiler iterates on its selection of partitions and program transformations until it has satisfied the hardware resource constraints and the application timing constraints subject to the semantics of the original program.

3 System-Level Compiler

The DEFACTO system-level compiler has the following goals:

- Identify computations that have the potential to yield performance benefits by executing on a CCU.
- Partition computation and control among the GPP and the CCUs.
- Manage data storage and communication within and across multiple CCUs.
- Identify opportunities for configuration reuse in time and space.

The compiler analysis produces an annotated hierarchical source program representation, which we call the task communication graph (TCG), specifying the parallel tasks and their associated communication. Subsequently, this representation is refined to partition the computation and control among CCUs and the GPP, and identify the communication and storage requirements. This refined representation is the input to the design manager. We now describe how the compiler identifies and exploits transformation opportunities using a simple code example depicted in Figure 2 (left).

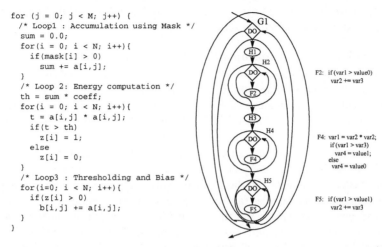

Fig. 2. Example program and its Hierarchical Task Communication Graph(TCG)

Figure 2 (right) shows the tasks in the TCG generated for the example code. At the highest level of the hierarchy, we have a single task that encompasses the entire computation. Next, task G1 accounts for the outer-loop body. At the next level, tasks H1 through H5 represent each of the statements (simple and compound) in the loop body of G1, namely two statements and three nested loops. At each of the loops, it is possible to derive functions representing the loops' computations as depicted by the functions F2, F4 and F5. The compiler analysis phase also annotates the TCG at each level of the representation with data dependence and privatization information. For example, at G1, the compiler will annotate that sum, z and th can be safely made private to each iteration of the loop since values in the variables do not flow from one iteration of G1 to another. These task-level data dependence and privatization annotations will help later phases of the compiler to relax the scheduling constraints on the execution of the computation. For this example, the code generation phase can explore parallelism or pipelining in the execution of the iterations of task G1.

3.1 Identifying Configurable Computing Computations

The compiler identifies parallel loops, including vector-style SIMD computations, more general parallel loops that follow multiple threads of control, and pipelined parallel loops using existing array data dependence analysis, privatization and reduction recognition techniques [3]. In addition to these parallelization analyses, the DEFACTO compiler can also exploit partial evaluation, constant folding and special arithmetic formats to generate specialized versions of a given loop body.

In the example, the loop corresponding to task H2 performs a sum reduction (the successive application of an associative operator over a set of values to produce a single value), producing the value in the sum variable. Because of the commutativity and associativity of the addition operator, the compiler can execute in any order the accumulations in the sum variable. Tasks H4 and H5 are easily parallelizable as they have no loop-carried dependences (any two iterations of these loops access mutually exclusive data locations).

3.2 Locality and Communication Requirements

Based on the representation and the data dependence information gathered for the program for a particular loop level in a nest, the compiler evaluates the cost associated with the movement of data and the possible improvement through the execution of the tasks on CCUs. For a multi-CCU architecture, we can use the data partitioning analysis to determine which partitions of the data can be accommodated that result in minimal communication and synchronization [5]. These data partitions are subject to the additional constraint that the corresponding tasks that access the data can be fully implemented on a single CCU.

The compiler has an abstract or logical view of communication channels that it can use to determine the impact of I/O bandwidth on partitioning and therefore on performance and size. For example, if the configurable machine provides for a bandwidth of 20 bits between CCUs with a transfer rate of 10 Mbps, the compiler can use these constraints to define the partitioning.

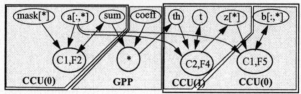

Fig. 3. Data and Computation Partition for the Example Code in Figure 2

For this example, it is possible to split the data in the b, z and a arrays by columns and privatize the variable sum, so that the iterations of the outer-most loop could be assigned to different CCUs in consecutive blocks and performed concurrently [4]. Figure 3 illustrates a possible data and computation partitioning for the code example. In this figure, we have represented each configuration C associated with a given set of hardware functions F.

3.3 Generating Control for the Partition

The final program implementation cannot operate correctly without a control mechanism. The control keeps track of which data is needed by a certain component and at what time that data must be available to a certain computation. Therefore, the system control has a dual role; it is both a communication and a computation coordinator. As a communication coordinator, the control takes the form of a finite-state machine (FSM) indicating data movement among the CCUs and the GPP. As a computation coordinator, the control takes is represented by another FSM indicating when CCU configuration and task computation occur. In effect, the control captures the overall task scheduling. The actual implementation of these FSMs is distributed among CCUs, initially in HDL format, and the GPP, in C code.

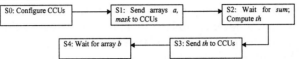

Fig. 4. GPP Control FSM Representation

As part of the partitioning process, the system-level compiler must annotate the TCG to indicate which tasks run on which CCU or GPP. The compiler then starts the automated generation of the control by using the TCG to generate corresponding FSMs for each of the CCUs and the GPP. The GPP FSM for task F1 is shown in Figure 4.

The GPP control is implemented using a set of C library routines that are architecture specific. The library consists of low-level interrupt handlers and communication and synchronization routines that are associated with the target architecture. To complete the system-level compiler's role in the control generation, it translates the control information for the GPP into calls such as send, receive and wait, using the C library, and inserts these into the GPP C code. The compiler also generates HDL for each of the CCUs' control FSMs. This HDL is used in the behavioral synthesis phase of the control generation.

3.4 Configuration Reuse Analysis

Because reconfiguration time for current reconfigurable devices is very slow (on the order of tens of milliseconds), minimizing the number of reconfigurations a given computation requires is the most important optimization. For the example code in Figure 2, it is possible to reuse a portion of the hardware that executes task F2 for task F5. Although the data inputs of these loop bodies are different, the functions are structurally the same. For the example, this information may be captured in the TCG by relabelling task F5 with task F2. Identifying common configuration reuse is only part of the solution. The compiler also uses the dependence analysis information to bring together (by statement reordering) portions of code that use common configurations. This strategy aims at minimizing the total number of configurations.

4 Design Manager

The design manager provides an abstraction of the hardware details to the system-level compiler and guides derivation of the final hardware design for computation, communication, storage, and control. In most cases, there will be a pre-defined configurable computing platform, where the available CCUs, storage elements (e.g., SRAM) and communication paths are all known a priori. In this case, the main responsibilities of the design manager are to map the computation, storage and communication from the high-level task abstractions to these pre-defined characteristics.

The design manager operates in two modes, designer and implementer. In the initial partitioning phase, the design manager assists the system-level compiler by estimating the feasibility of implementing specific tasks to be executed on a CCU. Working in tandem with the estimation tool, the design manager reduces the number of time-consuming iterations of logic synthesis and CCU place-and-route. If a feasible partitioning is not presented, the design manager provides feedback to the system-level compiler to guide it in further optimization of the design, possibly resulting in a new partitioning.

Once a feasible design is identified, the design manager enters into its second mode of operation, that of implementer. During this phase, the design manager interfaces with behavioral synthesis tools, providing them with both HDL and other target-specific inputs, culminating in the generation of the FPGA configuration bitstreams.

For example, for the TCG in Figure 2 we need to define the physical location, at each point in time, of the arrays a and z, how they are accessed, and how the data is communicated from storage to the appropriate CCU. We now address these mapping issues in a platform-independent manner.

4.1 Computation and Data Mapping

To provide feedback to the system-level compiler during the partitioning phase, the design manager collaborates with the estimation tool and (optionally) the module generator component. The design manager uses HDL associated with each task or group of tasks as input to a behavioral synthesis tool such as Monet to generate an RTL HDL description, and provides this HDL representation to the estimation tool. Or if the system-level compiler specifies an HDL module binding, the design manager can invoke the corresponding module representation in order to generate the RTL HDL. This representation will have captured both the details of the computation and data movement as specified in the TCG. In response to the design manager's request, the estimation tool returns both a timing and a sizing estimate. The design manager then returns these estimates to the compiler to help guide further efforts to find a partitioning that satisfies all of the constraints if the current one does not.

For our example, the compiler has partitioned the data into columns for the arrays a and b. The compiler has allocated array z to the memory of the first CCU. Figure 5 illustrates the conceptual partitioning of the computation and identifies which memories of which CCU hold which data The design manager captures this storage mapping in the RTL HDL for each CCU. If estimates are within the system-level constraints, the design manager generates the FPGA configuration bitstreams.

4.2 Estimation Tool

The design manager gives to the estimation tool a specific task in RTL HDL and expects the estimation tool to return three pieces of information: the estimated configurable logic block (CLB) count, timing, and threshold information specific to the FPGA technology. The estimates should also consider routing for a specified task placement. The off-line estimation manager maintains a table of useful estimates to allow for more expedient estimation.

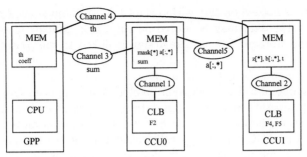

Fig. 5. Computation and Communication Mapping to 2 CCU Architecture

4.3 Communication Mapping

While the system-level compiler uses virtual communication channels, the design manager implements the physical communication channels required for data communication among the CCUs, GPP and memory elements. Data communication between the tasks must be mapped onto the predefined interconnect for the target architecture. While some of the channels can be implemented as part of the pipelined execution of the functions assigned to different CCUs, other channels require buffering. In the former case, the design manager must add hardware handshaking circuits to the configuration specification. In the latter, the design manager generates circuits that temporarily store data in local memory and later ship it to another CCU's memory.

For example, if there is a 20-bit channel between CCUs, and 4-bit words must be communicated between the CCUs according to the TCG specification, the design manager can implement a serial/parallel converter at the CCU I/O port to communicate 5 values in parallel. On the other hand, if 40-bit signals need to be communicated, the design manager implements a parallel/serial converter or multiplexer in the CCU I/O port. In this example, the columns of array a need to be sent to the CCUs for these later units to compute the updates to b. In addition the design manager can make use of the physical communication channels that exist between the adjacent units for maximum performance and rely on a GPP buffer-based scheme otherwise.

4.4 Storage Mapping

The system-level compiler specifies the data on which computation takes place as well as the communication of data. However, it does not specify how or where the data is stored. Mapping data onto existing memories and corresponding addresses on each of the CCUs, as well as creating storage within the CCUs is the responsibility of the design manager. Note that when the compiler generates a partitioning, the estimates must take into consideration I/O circuits, control logic and storage circuits that the design manager adds. Thus in an initial iteration, the compiler produces a best guess of a partitioning that should fit on the available CCUs and the design manager estimates the cost of the I/O, control and storage within the CCU to provide feedback on the feasibility.

The compiler specifies initial storage of data (GPP memory, result of a previous configuration on a CCU, etc.) as part of the data dependence annotations associated with its partitions. When the system control initiates the execution of the TCG, data is needed on chip within the CCU. Without some a priori defined models of how the data may be stored on chip, it is difficult to generalize the synthesis of the storage. The storage design is linked to the control design and the communication channel design. The communication channels feed the data to the CCU and transfer results from the CCU back to the GPP.

As an example of a design decision involved here, suppose that the tasks in a CCU need to operate on a matrix of data and tasks are executed sequentially on sub-matrices of the data (*e.g.*, an 8 × 8-byte window of an image). Further suppose that the TCG specifies parallel execution of the computation on all the data in each sub-matrix. In this case, it would be essential to access the sub-matrices sequentially from the external storage and store them on the CCU.

5 Conclusions

This paper has presented an overview of DEFACTO, a design environment for implementing applications for adaptive computing systems. The DEFACTO system uniquely combines parallelizing compiler technology with synthesis to automate the effective mapping of applications to reconfigurable computing platforms. As the project is still in its early stages, this paper has focused on the necessary components of such a system and their required functionality. The system-level compiler uses parallelization and locality analysis to partition the application between a general-purpose processor and CCUs. The design manager maps from the compiler's partitioning to the actual configurations on the FPGA, including the required communication, storage, and control for the CCUs and the GPP. A key distinguishing feature of DEFACTO is that it is designed to be architecture independent and retargetable.

Acknowledgements. This research has been supported by DARPA contract F30602-98-2-0113 and a Hughes Fellowship. The authors wish to thank John Villasenor and his students for their contributions to the design of the system.

References

1. D. Buell, J. Arnold and W. Kleinfelder, "Splash 2: FPGAs in a Custom Computing Machine," IEEE Symposium on FPGAs for Custom Computing Machines, Computer Society Press, Los Alamitos CA, 1996.
2. M. Gokhale and J. Stone, "NAPA C: Compiling for a Hybrid RISC/FPGA Architecture," IEEE Symposium on FPGAs for Custom Computing Machines, Computer Society Press, Los Alamitos CA, April, 1997.
3. Hall, M et al., "Maximizing Multiprocessor Performance with the SUIF Compiler," IEEE Computer, IEEE Computer Society Press, Los Alamitos CA, Dec. 1996.
4. Polychronopoulos, C. and Kuck D., "Guided-Self-Scheduling A Practical Scheduling Scheme for Parallel Computers," ACM Transactions on Computers, Vol. 12, No. 36., pp. 1425-1439, Dec. 1987.
5. Anderson J. and Lam, M., "Global Optimizations for Parallelism and Locality on Scalable Parallel Machines," In Proceedings of the ACM SIGPLAN Conference on Programming Language Design and Implementation (PLDI'93), pp. 112-125, ACM Press, NY, July 1993.

A Web–Based Multiuser Operating System for Reconfigurable Computing

Oliver Diessel, David Kearney, and Grant Wigley

School of Computer and Information Science
University of South Australia
Mawson Lakes SA 5095
{Oliver.Diessel, David.Kearney, Grant.Wigley}@unisa.edu.au

Abstract. Traditional reconfigurable computing platforms are designed to be used by a single user at a time, and are acknowledged to be difficult to design applications for. These factors limit the usefulness of such machines in education, where one might want to share such a machine and initially hide some of the technical difficulties so as to explore issues of greater value. We have developed a multitasking operating system to share our SPACE.2 coprocessing board among up to 8 simultaneous users. A suite of pre–configured tasks and a web based client allows novices to run reconfigurable computing applications. As users develop a knowledge of the FPGA design process they are able to make use of a more advanced PC client to build and upload their own designs. The development aims to increase access to the machine and generate interest in the further study of reconfigurable computing. We report on the design, our experience to date, and directions for further development.

1 Introduction

Definitions of reconfigurable computing are currently unclear. Rather than clarify the issue, we define usage of the term as it relates to this paper as describing the computations one performs on computers that include reconfigurable logic as part of the processing resource. The reconfigurable logic presently of interest is some form of Field Programmable Gate Array (FPGA) technology. Current reconfigurable machines may have the reconfigurable resource tightly coupled with a von Neumann processor integrated on a single chip, or loosely coupled to a von Neumann host via a general purpose bus such as PCI. While the coupling does influence performance in different application domains, the reason for employing reconfiguration is the same for both, namely, to speed up computations by implementing algorithms as circuits, thereby eliminating the von Neumann bottleneck and exploiting concurrency. In tightly coupled machines it may be viable to reconfigure on–chip logic at a fine grain of computation — perhaps even at the instruction level. However, the overheads of communicating over relatively low bandwidth buses demand that loosely coupled machines be reconfigured at a coarser grain. Applications on this latter class of machine are generally implemented as a front–end program executing on the sequential host, and a sequence of one or more configurations that are loaded and executed on a reconfigurable coprocessing board.

Loosely coupled reconfigurable coprocessing boards such as PAM [5], Splash 2 [2], and SPACE.2 [3] are single tasked devices despite usually being attached to hosts running multitasking operating systems. A possible reason for this is that most FPGAs are programmed in a slow configuration phase that establishes the circuitry for the whole chip at once. It is therefore easier to control performance and access if it is done for a single task at a time. However, the advent of dynamically reconfigurable FPGAs such as the Xilinx XC6200 and Atmel AT6K families, which allow part of the FPGA to be reconfigured while the rest of the chip continues to operate, has increased the interest in techniques and applications that exploit these chips' facility for multitasking.

Early work in this area was carried out on the loosely coupled DISC computer [6], which made use of a well–defined global context to allow the reconfigurable logic to be reconfigured and shared by multiple relocatable tasks during the run-time of an application. However, use of the DISC array is restricted to a single task at a time to avoid contention on globally shared control lines. More recently, the tightly coupled GARP processor [4] allows multiple configurations to be cached within the reconfigurable logic memory and supports time slicing on the controlling MIPS processor that is integrated with the reconfigurable logic array. Multiple users and multiple tasks thereby appear to be supported, but in fact only one configuration at a time may execute lest multiple configurations contend for control signals. Brebner described the issues involved in managing a virtual hardware resource [1]. He proposed decomposing reconfigurable computing applications into swappable logic units (SLUs), which describe circuits of fixed area and input/output (I/O) interfaces, so that multiple independent tasks might share a single FPGA. Brebner described two models for allocating the resources of the FPGA. The sea of accelerators model admits arbitrarily sized rectangular tasks and is suited to independent tasks of varying computational needs. On the other hand, the parallel harness model, which partitions the resource into fixed sizes, was thought to be more appropriate for cooperating tasks. We examine the implementation of a parallel harness model that partitions an array of FPGAs at the chip level. Our SLUs occupy an entire chip, and currently operate independently.

Our interest in developing a multiuser operating system for reconfigurable computing is motivated by the desire to increase accessibility to a SPACE.2 machine for CS and EE classes within a university environment. A second thrust of this work is to provide abstractions that simplify the programming and use of reconfigurable computers for novices. To this end, we have designed a web-based interface to SPACE.2 that allows multiple users to execute pre–configured applications at the same time. Users familiar with the FPGA design flow are also able to design and upload their own applications within a PC environment. Both are based on a simple multiuser operating system that partitions the FPGAs on a SPACE.2 board to allow up to 8 simultaneous users. This operating system provides the interface to web–based clients and manages the allocation of FPGAs within the SPACE.2 array to user tasks. We report on the development to date and the future direction of the project.

The remainder of this paper is organized as follows. Section 2 describes the architectural and applications development features of SPACE.2 that impact on the design of the multiuser operating system. Section 3 presents the design and reports on the development of the multiuser operating system and web–based clients. Our findings so far are reported on in Section 4. Section 4 also proposes remedies for overcoming the limitations experienced with the current design. The conclusions and directions for further work are presented in Section 5.

2 The SPACE.2 architecture

The SPACE.2 system consists of a DEC Alpha host into which one or more SPACE.2 processing boards are installed as 64 bit PCI localbus slaves [3]. The host processor communicates configuration and application data over the PCI bus to individual boards. Processing boards can operate independently of each other, but for particularly large applications they can also be connected together over a secondary backplane to form extensive arrays.

The compute surface of a SPACE.2 processing board consists of a 4×2 array of Xilinx XC6216 FPGAs. In addition, the board provides control logic for interfacing the configurable logic to the PCI bus, on–board RAM, and a clock module.

Limited connectivity to I/O blocks in the XC6216 makes a seamless array of gates spanning multiple FPGAs impossible. The devices are interconnected in a 2-dimensional mesh, with a pair of adjacent chips sharing 32 pads in the east-west direction and a column of 4 chips sharing 41 pads in the north-south direction. With appropriate pad configurations, opposing edges of the mesh may be tied together to create a toroidal structure. Access to the on–board RAM is available from the northern and southern edges of the FPGA array.

Several global signals are distributed throughout the array. Devices are programmed via a global 16 bit data bus. This bus is also available for FPGA register I/O. The appropriate chip is selected by the control logic, while mode, configuration, and state registers are selected through an 18 bit address bus. Three common clock frequencies are set in the clock module and distributed with a global clear signal to the FPGA array.

All software on the host interacts with the SPACE.2 processing boards through a character device driver under UNIX. A board is opened as a single–user file to allow a user to make exclusive use of the reconfigurable hardware. Once opened, a board's FPGA configuration and state can be read or written to by an application program.

2.1 Current design flow

SPACE.2 applications typically consist of two parts: a circuit design for configuring the FPGAs of a SPACE.2 board and a host program that manages I/O and controls the loading, clocking, and interrupt handling for the circuit. Applications initially therefore present a problem in hardware/software codesign.

When the hardware portion of the design has been identified, its design is elaborated in two stages. In the first stage, the logic of the required circuits is derived. Then, in the second stage, the logic is placed onto the FPGAs and its interconnections are routed.

The front–end or host program is responsible for loading the design onto the board, communicating with the running application, and responding to user interrupts. Configurations cannot be tested by simulation due to a lack of tools for simulating multi–chip designs. Circuits are therefore not tested until they are loaded onto SPACE.2. Debugging is then performed by reading the contents of registers while the circuit is live. This may necessitate modifying the original design to latch signals for debugging purposes.

Apart from the difficulty of recognizing opportunities for exploiting parallelism in problems, the current design flow inhibits experimentation with the SPACE.2 platform in several ways. Due to a lack of adequate abstractions, applications development requires the understanding and management of many low–level details. Moreover, we lack the tools to make this task easier. For example, designers need to partition tasks manually, and there is limited support for testing and verifying designs. In order to debug designs, a detailed knowledge of the system is needed. It should be noted that these problems are not unique to SPACE.2 but are generally accepted as barriers to the wider adoption of reconfigurable computing as a useful paradigm. One of the roles of our operating system is to hide as much detail as possible from the novice user.

3 The multiuser environment

In order to introduce reconfigurable computing to novice users and to develop their interests, we embarked on a project to develop two internet accessible interfaces and a multiuser operating system for the SPACE.2 machine. One of these interfaces provides a simple form for the user to select and run one of several pre–configured applications for the machine. This interface is available from common WWW browsers. As the user's knowledge of the FPGA design flow develops, they may upload their own designs from a network PC–based interface. Managing the requests of these two clients, the multiuser server establishes socket connections using TCP/IP, vets user authorizations, allocates FPGAs to users, loads user tasks, and handles user I/O to tasks.

3.1 Web–based demonstration platform

The web–based client consists of a CGI script and forms–based interface that contains a number of pre–configured designs. Users may choose to execute one of a number of simple applications, supply their own data, and obtain results upon completion of processing. The client thus abstracts the details of designing and running an application. With this interface we also hope to develop clear demonstrations of what the SPACE.2 board is capable of.

The client attempts to establish a connection with the server after the user selects an application and provides input data. In contrast to the PC–based client, no special authorization is required for web–based users. If an FPGA is available, the server provides a connection, and accepts an application in the form of a Xilinx CAL file one line at a time. This file is translated into global coordinates and loaded by the server. Thereafter the user input values are written to the appropriate registers. The simple applications developed so far return results to the user within a predefined number of clock periods. Event–driven applications are not yet catered for. After the results have been obtained, the FPGA resource is freed by the server, and the socket connection to the client is broken.

3.2 Support for applications design

Loading designs onto the SPACE.2 processing board and interacting with them is tricky since it involves lengthy sequences of control functions. When users begin to design their own applications, they should not be distracted by these details. We therefore developed a PC–based Tcl/Tk client to abstract away the complexity of loading and interacting with tasks. This client presents a mouse–driven graphical user interface that can be used from any PC on the internet that uses the Windows 95 operating system. Anybody may request a copy of the client by filling out a web–based form that is processed by the system's administrator in order to establish an entry for the user in the host's password file. The user is then provided with instructions on how to download and install the client by email.

The Tcl/Tk GUI interacts with a dynamic link library (DLL) that also resides on the PC for parsing requests between the client and the server. Use of the DLL provides communications independence between the client and server, thereby allowing either to be more easily replaced. The client provides an authentication procedure by sending a username and password to the server for checking against the host's password file. After a connection is established, the client sends the user's design to the server for translation and loading and sends the user's data for writing to the input registers. When requested to by the user, the server retrieves the application results from the output registers and passes them to the client for display.

Successful SPACE.2 designs must resolve several complex design issues. To overcome some of these problems we have chosen to constrain user designs in a number of ways: (1) We make use of the XC6200 family's support for bus–based register I/O. A "register" may be viewed as a virtual pins abstraction since the user need not worry about interfacing designs to particular pins. (2) The number of registers is limited to a maximum of eight for input and output respectively. Registers are 8 bits wide. The eight input registers are aligned in the leftmost column of an FPGA, and the eight output registers are aligned in the rightmost column. (3) The user design must be able to be fully loaded onto a single FPGA, thereby eliminating multi–chip partitioning problems. (4) The application must be designed to a fixed global clock speed.

Adding these constraints to applications limits the range of designs that can be created, but simplifies their complexity considerably.

3.3 Operating system design

Normally only one user at a time can gain access to a SPACE.2 board. However, with the possibility of several people wanting to gain access to the machine at the same time, a multiuser operating system is required. Currently, while a designer is using a SPACE.2 board, all other attempts to open the device driver are blocked by the kernel. The multiuser operating system removes the blocking semantics by adding an additional layer to the device driver. The system is modelled on a multiprocess server that accepts up to eight simultaneous socket connections — one for each chip.

The operating system initializes the socket connection code and establishes a file–based allocation table for the 8 FPGAs on the board. The server then listens on a designated port for connection requests that are initiated by the web or PC–based clients. When a user connects to a socket, the operating system checks whether the SPACE.2 board is available for use. This occurs in two stages. If the server does not yet have control of the SPACE.2 device driver, it attempts to gain control in order to determine whether any other user has dedicated access to the SPACE.2 board. After control of the device driver has been gained, the FPGA allocation table is checked for an available chip. If one is found, the server then forks off a process for the new user, and waits for further socket connection requests. If either stage fails, the request is refused and the client will wait until the user attempts to reconnect.

The system validates PC client users against the host's password file and determines the user's privileges (super or normal) before allocating an FPGA to the user. The user's process ID, name, and priveleges are stored in the allocation table for easy reference and to avoid contention. A super user has similar privileges to those of a UNIX root user. This user has access to all user commands as well as simple "house–keeping" functions such as altering the common system clock speed and removing users from the system. Normal users do not have access to these global commands. The list of super users is maintained in a separate file.

The multiuser operating system manages FPGA allocation and deallocation on a first come, first served (FCFS) basis. In the conventional single user operating environment, this function is explicitly handled by the user's front–end program. Once allocated, a chip is not relinquished until the socket connection is broken. The connection to a web–based client is broken after the results from a run have been obtained. A PC–based client is disconnected at the user's request.

The operating system is responsible for loading applications and handling I/O for the applications. A 3 digit code is used for communicating requests between the clients and the server. Requests consist of three parts: the code, followed by an integer representing the number of arguments to follow, and a list of arguments. The server must acknowledge each part of the request for the communication to proceed.

The application designs are uploaded in the form of a Xilinx CAL file, which can be loaded onto any XC6216 chip. The file is loaded up one line at a time and a special flag is sent to indicate EOF. Row and column addresses in this file are translated into the global SPACE.2 coordinate system so that the design is loaded onto the correct chip. Similarly, I/O requests are parsed so as to address the correct set of cells in the global coordinate system.

4 Performance review

4.1 What sorts of tasks work?

The interface constraints imposed by our clients limit the range of practical tasks that can be performed by the FPGAs. Nevertheless, the current environment does not preclude users from experimenting with reconfigurable computing. Indeed, we see it as an entry point for education, and believe there is ample scope for devising educational projects that fit within the constraints of the interface. Suitable pre–configured designs include simple bit manipulation tasks and more complex neural net classifiers and distributed multipliers. Moreover, it is a matter of designing alternative client interfaces to remove the restrictions on I/O since the multiuser server has the full capability of the device driver at its disposal.

Since all requests go through the original single user device driver, which services them sequentially, it is possible that performance will suffer. The server is therefore not practical for tasks that expect performance guarantees such as real–time tasks.

SPACE.2 is a platform designed for experimentation. The problem of providing a multiuser run–time environment for SPACE.2 requires solutions to numerous issues. These include: the design and compilation of suitable applications, the interplay of time– and space–sharing, what the granularity of swapping should be, and how the device driver and operating system kernel should be designed. We have just begun to explore some of these issues. There is considerable work to go on with.

4.2 Problems and potential solutions

The current implementation of the multiuser operating system suffers from several limitations. In this section we discuss these problems and propose what we perceive to be workable solutions.

The current Tcl/Tk client interface for user designed tasks limits the number of input and output registers to eight of each, and the width of these registers is set to 8 bits. Moreover, input is from the leftmost column and output is to the rightmost column of an FPGA. It may be desirable to have more registers, and that it be possible to interface with these at arbitrary locations, in particular, as an aid to debugging. A more flexible client interface would overcome this limitation.

Register I/O is acceptable for tasks involving low bandwidth transfers, but for high bandwidth streaming applications it causes overheads that could be avoided. Unfortunately for such applications, use of the on–board RAM is not supported in the current operating system. However, use of the RAM could be supported by applications that make use of the pair of chips at the northern edge of the board, which would leave the rest of the board free for tasks that make use of register I/O.

Another area where more support from the client interface may be necessary is dynamic reconfiguration. While the SPACE.2 board and device driver allow dynamic reconfiguration, the multiuser environment does not currently support it. Tasks are therefore limited in size to a single XC6216 chip. It should be possible to provide hooks and conditions for reconfiguration within the client interface. Alternatively, the operating system could be designed to interface with conventional front–end host programs that control the reconfiguration.

Future enhancements to the operating system should also consider more adaptive space–sharing schemes so as to allow allocating tasks in multiples of a single chip.

At present up to 8 simultaneous users are supported by partitioning the logic resource of a SPACE.2 board at the chip level. Any additional users are blocked from access to the board. To overcome this problem we propose implementing a time–sharing mechanism in the future. Time–sharing would place further constraints on tasks, such as ensuring that all intermediate results are latched before commencing a swap. However, having 8 chips available for swapping ensures a reasonable period over which to amortize the costs of swapping if they are carried out in a round–robin fashion. To reduce the overheads of swapping, it would be desirable to modify the board controller to enable on–board storage of configurations and state.

If time–sharing and swapping is to be supported, it may be necessary to investigate multithreading the device driver to reduce blocking, and to implement a scheduler to reduce the unpredictable delays that can result with the FCFS servicing of requests imposed by the current device driver.

Finally, the fixed global clock speed places an undue constraint on performance. It would be nice to find a solution to the problem of frequency–sharing the board — having multiple tasks of different clock speed sharing the board. To this end, we propose investigating supporting user programmable clock divider circuits within each chip.

5 Conclusions

We have described the initial design of a multiuser server for our SPACE.2 reconfigurable computing platform. This system partitions the array of eight FPGAs available on a SPACE.2 processing board among multiple users, one per user, so as to allow multiple independent tasks to execute simultaneously. This approach increases utilization and access to the system. However, our solution is rather coarse grained considering that the XC6216 chips used in the array are partially reconfigurable at the cell level. The resource is also under-utilized if less than 8 tasks are running. Moreover, the PCI bus and board device driver form a sequential channel for I/O to the array. Tasks are therefore loaded one at a time, and I/O to tasks is performed sequentially. A further limitation imposed by the board on the initial design is the need to share a common clock. To overcome these problems we intend supporting adaptive partitions and implementing a time-sharing scheme. We will investigate implementing programmable clock divider circuits on each chip to allow the board to be shared in the frequency domain. Further enhancements envisaged for the user interface are intended to facilitate the use of dynamic reconfiguration and support more flexible I/O.

Acknowledgments

The helpful comments and assistance provided by Matthew Altus, Bernard Gunther, Ahsan Hariz, Jan Machotka, Manh Phung, Joseph Pouliotis, and Karl Sellmann are gratefully acknowledged.

URL

Access to the server and documentation is available through links from the URL http://www.cis.unisa.edu.au/acrc/cs/rc/multios.

References

1. G. Brebner. A virtual hardware operating system for the Xilinx XC6200. In R. W. Hartenstein and M. Glesner, editors, *Field-Programmable Logic: Smart Applications, New Paradigms and Compilers, 6th International Workshop, FPL'96 Proceedings*, pages 327 – 336, Berlin, Germany, Sept. 1996. Springer-Verlag.
2. D. A. Buell, J. M. Arnold, and W. J. Kleinfelder, editors. *Splash 2: FPGAs in a Custom Computing Machine*. IEEE Computer Society, Los Alamitos, CA, 1996.
3. B. K. Gunther. SPACE 2 as a reconfigurable stream processor. In N. Sharda and A. Tam, editors, *Proceedings of PART'97 The 4th Australasian Conference on Parallel and Real-Time Systems*, pages 286 – 297, Singapore, Sept. 1997. Springer-Verlag.
4. J. R. Hauser and J. Wawrzynek. Garp: A MIPS processor with a reconfigurable coprocessor. In K. L. Pocek and J. M. Arnold, editors, *The 5th Annual IEEE Symposium on FPGAs for Custom Computing Machines (FCCM'97)*, pages 24 – 33, Los Alamitos, CA, Apr. 1997. IEEE Computer Society.
5. J. E. Vuillemin, P. Bertin, D. Roncin, M. Shand, H. H. Touati, and P. Boucard. Programmable active memories: Reconfigurable systems come of age. *IEEE Transactions on Very Large Scale Integration (VLSI) Systems*, 4(1):56 – 69, Mar. 1996.
6. M. J. Wirthlin and B. L. Hutchings. Sequencing run-time reconfigured hardware with software. In *FPGA'96 1996 ACM Fourth International Symposium on Field Programmable Gate Arrays*, pages 122 – 128, New York, NY, Feb. 1996. ACM.

Interconnect Synthesis for Reconfigurable Multi-FPGA Architectures *

Vinoo Srinivasan, Shankar Radhakrishnan, Ranga Vemuri, and Jeff Walrath

E-mail: {vsriniva, sradhakr, ranga, jwalrath}@ececs.uc.edu

DDEL, Department of ECECS, University of Cincinnati, OH 45221.

Abstract. Most reconfigurable multi-FPGA architectures have a programmable interconnection network that can be reconfigured to implement different interconnection patterns between the FPGAs and memory devices on the board. Partitioning tools for such architectures must produce the necessary pin-assignments and interconnect configuration stream that correctly implement the partitioned design. We call this process Interconnect Synthesis for reconfigurable architectures.

The primary contribution of this paper is a interconnection synthesis technique that is independent of the reconfigurable interconnection architecture. We have developed an automatic interconnect synthesis tool that is based on Boolean satisfiability. The target interconnection architecture is modeled in an architecture specification language. The desired interconnections among the FPGAs are specified as in the form of a netlist. The tool automatically generates the pin-assignments and the required values for the configuration-inputs in the architecture specification. We modeled several reconfigurable architectures and performed interconnect synthesis for varying number of desired nets. We provide experimental results that demonstrate the effectiveness of the interconnection synthesis technique.

1 Introduction

Modern multi-FPGA (Field Programmable Gate Array) reconfigurable boards host several FPGAs and memory devices that communicate through both dedicated and programmable interconnections [1–3]. The presence of programmable interconnections on the boards highly increases the utility of the board. However, the interconnection architectures largely vary from one board to another. This diversity can be handled by existing partitioning and synthesis tools, only with a considerable effort. More importantly, it is usually not straight forward to migrate between reconfigurable boards. The effort required in modifying existing CAD tools to handle a new RC (Reconfigurable Computer) interconnect architecture, is often comparable to that of developing a new CAD algorithm specific to that interconnect architecture.

Partitioning tools for RC architectures, in addition to partitioning computational tasks and logical memory segments, must also generate an interconnect configuration that establishes an appropriate connectivity across partitions. Typically, in programmable interconnection architectures, the pins of an FPGA cannot be treated as identical. Pin assignment plays an important role during

---

* This work is supported in part by the US Air Force, Research Laboratory, WPAFB, contract #F33615-97-C-1043.

partitioning for RC architectures. The functionality of the programmable interconnection network is a direct consequence of pin assignment. We define the *interconnect synthesis* problem for RC architectures to be the unified problem of generating a pin assignment and synthesizing the appropriate configuration stream for the programmable interconnection network, such that the interconnection requirement is met.

In this paper we present an interconnection synthesis technique that is independent of the target architecture. Interconnection synthesis is modeled as a Boolean satisfiability problem. The Boolean function to be satisfied is represented as a BDD (Binary Decision Diagram) [4, 5] and operations are performed on the BDD to generate the pin assignment and the configuration stream that satisfy it. Boolean satisfiability approaches have been used earlier for the conventional bi-level channel routing for ASICs [6]. More recently, Boolean satisfiability is used to model FPGA routing [7]. The authors present a methodology to automatically generate the necessary Boolean equation that determines the routability of a given region. Although we use Boolean satisfiability techniques, we differ from the above works because modeling pin assignment and configuration stream generation for interconnection architectures is different from FPGA or ASIC routing. Secondly, we have developed an automated environment for performing interconnect synthesis. In addition, we have formulated a novel interconnect architecture specification technique which helps straight-forward retargeting of the interconnect synthesis tool to various RC architectures.

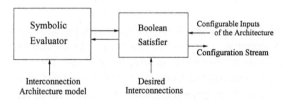

Fig. 1. Interconnect Synthesis Environment

A programmable interconnect network in a RC typically consists of a small number of such building block modules. Many instances of these building blocks are connected together, in a regular fashion, to form a reconfigurable interconnect fabric among the FPGAs and memories. In order to specify an RC interconnect architecture, we need two specification constructs: First, we should be able to specify the configuration behavior of the building blocks. Second, we should be able to specify a structural composition of instances of these building blocks. Given such specification, we need an elaboration mechanism that effectively extracts the symbolic Boolean equations that together represent the set of all allowable connections in the interconnect fabric. These equations can then be turned into the BDD representation and, in conjunction with the representation of the desired nets, be used to determine a viable configuration for the interconnect fabric if one exists.

Figure 1 shows the components that constitute the interconnect synthesis environment. We have used an RC architecture modeling language called PDL+ [8, 9] to specify interconnect architectures. ARC is an elaboration environment for PDL+ [10]. We use the symbolic evaluation capability in the ARC evaluator to generate the Boolean equations representing the allowable connections in the interconnect fabric. The Boolean satisfier tool, checks for the existence of a pin assignment and a corresponding configuration stream that makes the interconnection network achieve the desired interconnections. Although a detailed discussion of the semantics of PDL+ and ARC are beyond the scope of this paper, we provide enough information to make this presentation self contained. Remainder of this paper is organized as follows. The next section presents the previous work. Section 3 elaborates the interconnection network specification methodology. Section 4 details the working of the Boolean satisfier and its integration with the symbolic evaluator. Section 5, presents a technique for Boolean formulation of the pin assignment problem. Results of interconnect synthesis are presented in Section 6. Conclusions are drawn in the final section.

2 Previous Work

The problem of pin-assignment and inter-FPGA routing, in the presence of interconnection networks, has been investigated before. Hauck and Borriello [11] present a force-directed pin-assignment technique for multi-FPGA systems with fixed routing structure. The pin-assignment and placement of logic in the individual FPGAs are performed simultaneously to achieve optimal routing results. Mak and Wong [12] present a optimal board-level routing algorithm for emulation systems with multi-FPGAs. The interconnection architecture considered is a partial crossbar. The authors present an optimal $O(n^2)$ algorithm for routing two-terminal nets. However, they do not deal with multi-terminal nets, but prove it NP-complete.

Selvidge et al. [13] present TIERS, a topology independent pipelined routing algorithm for multi-FPGA systems. The desired interconnections and target interconnection topology are both represented as graphs and the tool establishes the necessary routing. The interesting feature of TIERS is that it time-multiplexes the interconnection resources thus increasing its utility. However the limitation is that the interconnection topology has only direct two-terminal nets between FPGAs. Topologies for programmable interconnections and multi-way connections are not considered.

Kim and Shin [14] studied the performance and efficiency of various interconnection architectures shuch as, direct interconnections, K-partite networks and partial cross bars. They present a performance driven partitioning algorithm where performance is based on the number of *hops* (FPGA boundary crossing) of a net. The interconnection architecture is fixed before partitioning and cannot be programmed according to the cut-set requirements. Multi-terminal connections are broken into two-terminal nets.

Khalid and Rose [15] present a new interconnection architecture called the HCGP (Hybrid Complete-Graph Partial-Crossbar). They show that HCGP is better than partial crossbar architectures. They present a routing algorithm that is customized for HCGP. The router minimizes the number of *hops* and keeps it less than three.

The difference between our work and those discussed above is that the interconnection synthesis technique presented in this paper works for any generic interconnection architecture. Any type of programmable architecture can be specified. The interconnection topology not fixed prior to partitioning. The necessary configuration information is produced as the result of interconnection synthesis. Both two-terminal and multi-terminal nets are allowed. The current model permits only two hops for routing each net.

3 Network Specification and Symbolic Evaluation

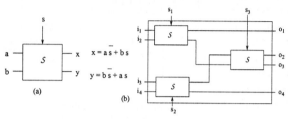

Fig. 2. Interconnection Network built using Switch-box Component

The interconnection network is a circuit that is composed of basic switch-box components of one or more types. These switch-box components determine the input-output mapping, based on certain *control* inputs. Figure 2(a) shows a simple switch-box component, S. When the control input (s) is zero, S transmits a to x and b to y and when the s is one, it transmits b to x and a to y. Figure 2(b) shows the interconnection network that instantiates three S switch-boxes.

```
module switch_box -- defining the model switch box module
   ports
      a, b, s, x, y : ioport;    rules -- rules for outputs x and y
      x'val = (a'val and not s'val) or (b'val and s'val);
      y'val = (b'val and not s'val) or (a'val and s'val);
end module;

module interconnection_network = board -- interconnect network instance
   ports -- ports of the interconnection network
      ioports = {i1, i2, i3, i4, s1, s2, s3, o1, o2, o3, o4};
   modules
      switch_boxes = {S1, S2, S3}; -- has three switch box instances
end module;
```

Fig. 3. Partial PDL+ Model of the Interconnection Network

Figure 3 shows a portion of the PDL+ model for the interconnection network. The design is composed of *modules* that are connected through *ports*. Ports have a single attribute called *val* of type Boolean. The module *switch_box* has *rules* to evaluate the attribute values of the output ports depending on the values of the inputs. The ARC system can perform symbolic evaluation of the various attributes in the specification model. The Boolean satisfier tool (Figure 1) communicates with the ARC runtime system through an API (Application Procedural Interface). Selective attribute values can be supplied by ARC system upon request from the Boolean satisfier.

4 Boolean Satisfiability Tool

The role of the Boolean satisfier is to generate values for the switches (configuration stream) that makes the interconnection network implement the desired interconnections. The desired interconnections are given to the Boolean satisfier in terms of a set of equivalence equations. Multi-terminal nets can also be specified using equivalence equations. The symbolic evaluator provides the boolean model of the interconnection architecture.

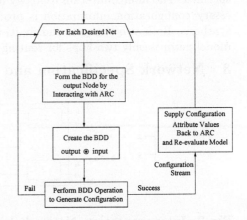

Fig. 4. Flow Diagram of the Boolean Satisfier Tool

The flow of the the Boolean satisfier tool is shown in Figure 4. For each desired net, the Boolean equation corresponding to the *val* attribute of the output port (or multiple output ports in the case of multi-terminal net) is extracted through the ARC-API. The Boolean equation is stored as a BDD. Binary Decision Diagrams (BDD) are efficient, graph-based canonical representations of Boolean functions [5,4]. We use the BDD library developed by the model checking group at the Carnegie Mellon University [16]. In order to reduce the memory utilization and BDD generation time, the library has modes that perform dynamic reordering during BDD operations. Similarly the equation for the *val* attribute input port is extracted. BDD operations are performed to generate the values for all control inputs that makes the equivalence of the input and the output TRUE.

As shown in Figure 4, when a feasible set of switch values (configuration stream) exists, these switch values are fed back into the ARC system and the board model is reevaluated with the new switch values. For the next desired net, the new reevaluated model is used. This process continues until all desired nets have been tried. In the next section, we present a methodology to model the pin assignment as a Boolean satisfiability problem.

5 Modeling Pin Assignment as Boolean Satisfiability

Due to the presence of the interconnection network, pins in the FPGA cannot be viewed identically. For example, the Wildforce board has four FPGAs (F1 .. F4). There are 36 direct connections between FPGAs, F1 to F2, F2 to F3, and F3 to F4. There are 9 nibbles (36 bits) going from each FPGA to the interconnection network. Only *corresponding* nibbles in each FPGA can be connected, i.e., nibble number i cannot be connected to nibble j of another FPGA if $i \neq j$. Thus assignment of nodes in the design to pins plays a crucial role in interconnect synthesis. The problem of mapping design pins to physical pins in the FPGA is called the *pin assignment problem*.

Given a design partitioned across multiple FPGAs, the desired connections is only available in terms of the design pins rather than the physical FPGA pins. For this reason, pin assignment must be captured into our Boolean formulation. Since a physical pin on the FPGA can be connected to any design pin, we introduce a virtual interconnection network between the design pins and FPGA pins.

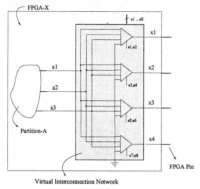

Fig. 5. Introduction of a Virtual Interconnection Network

Figure 5 shows the virtual interconnection network introduced for FPGA-X. This virtual interconnection network consists of four multiplexors. Each multiplexor can be independently controlled by two switches. Different values of switches implement different pin assignments. If the FPGA has N pins, then there are N multiplexors introduced to the design and correspondingly $O(N.log N)$ switches introduced. The design of the virtual interconnection network makes all the input lines to the virtual network symmetric. Without any loss in connectability, design pins can be mapped to any input of the virtual interconnection network. The network in Figure 5 is called "virtual" because this network is not synthesized as part of the design. Its presence is only to formulate the pin assignment into our Boolean satisfiability framework.

For multi-FPGA boards, pin assignment can be formulated as a Boolean satisfiability by introducing virtual interconnection networks for each FPGA. Thus the PDL+, for the interconnection architecture specification will include several virtual interconnection networks in addition to the existing interconnection network model. Now, the desired connections can be expressed in terms of design pins rather than FPGA pins. If a solution exists, the values of the switches that the Boolean satisfier tool produces will determine both pin assignment and interconnect network configuration.

6 Results

We have modeled the interconnection architecture of three boards – Wildforce [1] Gigaops, [2], and BorgII [3] and one commercial interconnection chip IQX160 [17]. We performed interconnect synthesis for various random desired nets on all four models. The Wildforce architecture has a complex interconnection network. Hence, for Wildforce, we present interconnect synthesis results with and without pin assignment. For Gigaops, BorgII, and IQX160, pin assignment exists at the time of interconnect synthesis. All experiments were conducted on an Sun Ultra Sparc 2 workstation, running at 296 MHz, with 128 MB of RAM.

The BorgII Architecture: BorgII [3] has two XC4000 FPGAs to perform logic, two for the purpose of interconnection. There are 38 programmable pins between one of the interconnect-FPGAs to both logic-FPGAs and 28 programmable pins between the other router-FPGA to the logic FPGAs. These are the only pins through which the FPGAs communicate. The PDL+ model for BorgII has 3 modules and 132 ports.

Table 6 presents the interconnect synthesis results for the BorgII architecture. Column two gives the number of desired nets. T_{bddgen} gives the total time spent in BDD generation. This includes the time taken for symbolic evaluation and generation of BDD that represents the Boolean equation corresponding to the desired net. T_{congen} is the time spent in performing the BDD operation that generates the required configuration. T_{congen} is a very small percentage of the total time. T_{avg} is the average run-time per desired net. Table 6 shows that the total execution time ($T_{bddgen} + T_{congen}$) increases with the number of desired nets. Total execution time is less than 10 seconds for the largest test case, having 66 desired nets. T_{avg} is 0.55 seconds for run #1 but gradually reduces to 0.14 for run #9. The reason for this is that the symbolic evaluation time reduces as the number of nets increase. This is because, for each possible net, the configuration values are fed back into the symbolic evaluator. The reevaluated model with the updated configuration values is smaller than the original model. Hence, the BDD generation times are likely to reduce as the number of calls to the symbolic evaluator increases.

Run #	# nets	T_{bddgen} (sec)	T_{congen} (μsec)	T_{avg} (sec)
1	2	1.105	3.86	0.55
2	4	2.255	7.72	0.56
3	8	2.697	15.44	0.34
4	16	3.632	30.88	0.23
5	24	4.085	46.32	0.17
6	33	4.935	98.26	0.15
7	33	4.868	86.68	0.15
8	66	8.868	126.63	0.13
9	66	9.268	130.48	0.14

Fig. 6. Results of BorgII

Run #	# nets	T_{bddgen} (sec)	T_{congen} (μsec)	T_{avg} (sec)
1	2	0.509	42.19	0.25
2	2	0.569	52.36	0.28
3	4	0.770	84.38	0.19
4	4	0.844	67.22	0.21
5	8	1.361	67.12	0.17
6	8	1.446	74.85	0.18
7	16	2.749	82.56	0.17
8	16	2.622	90.29	0.16
9	32	3.751	138.20	0.12
10	32	3.778	145.93	0.12

Fig. 7. Results of GigaOps

The GigaOps Architecture: The interconnect model for the GigaOps [2] architecture has eight 16-input 16-output switches that either transmit or cut-off the input from the output. The eight switches are modeled as two rings that establish interconnections between four MODs (Gigaops' multi-FPGA card). The PDL+ model for Gigaops has 9 modules and 168 ports. Table 6 presents the interconnection synthesis results for the Gigaops board. For case 1 (two desired nets), the total execution time is in the order of half a second and for thirty two desired nets the total execution time is about 4 seconds. T_{congen} is a very small percentage of the total time. The average execution time per net reduces as the number of desired net increases. This is similar to the observation in the case of BorgII results. T_{avg} ranges between a quarter of a second to about a tenth.

The IQX160 Interconnection Chip: IQX160 is a 160 pin FPID (Field Programmable Interconnect Device) chip that implements a cross bar type of interconnection network. Any of its 160 pins can be connected to any other pin or a set of pins. The PDL+ model for IQX160 has 1 module and 160 ports. Table 6 presents the interconnection synthesis results. For the IQX chip any connection is feasible. Results show that for most cases the average run-time per net is only a fraction of second. For the cases 9 and 10, T_{avg} is less than a tenth of a second.

The Wildforce Board Without Pin Assignment: We modeled the interconnection network to be composed of 9 switch-boxes. Each switch box establishes a connectivity among wires in the corresponding nibbles of the four FPGAs. The wild force model we use is a comprehensive model that has all the dedicated interconnections, in addition to the programmable interconnection network. The PDL+ model for the board has 64 modules and 1386 ports.

Table 6 presents the interconnection synthesis results of the Wildforce architecture. The interconnection synthesis time for the largest test case with 36 desired nets is about 5 seconds. The average synthesis time per desired net, varies between 0.22 seconds to 0.14 seconds. Unlike other boards there is not a sharp reduction in T_{avg} with the increase in number of desired nets. This is because, the Wildforce interconnection network has mutually disconnected components.

Run #	# nets	T_{bddgen} (sec)	T_{congen} (μsec)	T_{avg} (sec)
1	2	1.123	15.02	0.56
2	4	1.539	40.76	0.38
3	8	2.336	44.10	0.29
4	16	5.496	79.87	0.34
5	20	5.473	85.39	0.27
6	24	5.803	115.70	0.24
7	32	5.911	130.05	0.18
8	48	5.859	179.06	0.12
9	64	5.877	188.20	0.09
10	80	5.949	323.41	0.07

Fig. 8. Results for IQX160

Run #	# nets	T_{bddgen} (sec)	T_{congen} (msec)	T_{avg} (sec)
1	4	0.894	208.91	0.22
2	8	1.519	438.92	0.19
3	12	1.624	519.03	0.14
4	16	2.536	761.03	0.16
5	20	3.462	952.72	0.17
6	24	4.113	1107.34	0.17
7	28	4.579	1207.70	0.16
8	32	5.134	1392.17	0.16
9	36	5.196	1428.18	0.14

Fig. 9. WF without Pin Assignment

The Wildforce Board With Pin Assignment: For this experiment, a simple model of the Wildforce board is considered. We ignore direct connections between FPGAs for this experiment. Each FPGA has 9 lines connected to the interconnection network. Line i of any FPGA can only be connected to line i of any other FPGA. The PDL+ model for Wildforce architecture with modifications for pin assignment has 14 modules and 324 ports. Table 6 presents the results of interconnect synthesis with pin assignment for the Wildforce architecture. The test case 9 that has 36 desired nets, the run-time is about 155 seconds. The average time for interconnect synthesis per net is about 5 seconds. The synthesis time can be reduced with better Boolean formulations for the pin assignment.

Memory Usage: The ARC runtime system utilizes close to 12 MB of memory for all of the above experiments. Peak memory used by the Boolean satisfier for BorgII, Gigaops, IQX, and Wildforce (without pin assignment) experiments was approximately 70 KB. For Wildforce with pin assignment the memory used by the Boolean satisfier was 245 KB.

Run #	# nets	T_{bddgen} (sec)	T_{congen} (msec)	T_{avg} (sec)
1	4	18.351	72.10	4.59
2	8	28.798	93.53	3.59
3	12	60.698	146.65	5.06
4	16	71.727	167.27	4.48
5	20	111.586	203.24	5.58
6	24	123.947	205.91	5.16
7	28	140.536	200.31	5.02
8	32	152.691	217.84	4.77
9	36	155.283	268.18	4.31

Fig. 10. WF with Pin Assignment

7 Conclusions

This paper presented a novel technique for interconnection synthesis for reconfigurable multi-FPGA architectures based on Boolean satisfiability. The technique works for any generic interconnection architecture. We have developed a fully automatic interconnection synthesis tool that has a symbolic evaluator and a BDD based Boolean satisfier. Currently, the interconnect synthesis tool is used as part of SPARCS [18, 19], an integrated partitioning and synthesis system for adaptive reconfigurable computing systems. SPARCS provides automatic tools that perform temporal partitioning, spatial partitioning and high-level synthesis targeted toward dynamically reconfigurable multi-FPGA systems.

References

1. "Wildforce". *"WILDFORCE Reference Manual, Document #11849-0000"*. Annapolis Micro Systems, Inc.
2. "GigaOps". http://www.gigaops.com.
3. Pak K. Chan. "A Field-Programmable Prototyping Board: XC4000 User's Guide". Technical report, University of California, Santa Cruz, 1994.
4. R. E. Bryant. "Graph Based Algorithms for Boolean Function Manipulation". In *IEEE Trans. on Computers, C-35(8)*, pages 677–691, August 1986.
5. R. E. Bryant. "Binary Decision Diagrams and Beyond: Enabling Technologies for Formal Verification". In *Proc. ACM/IEEE ICCAD*, Nov. 1995.
6. S. Devadas. "Optimal layout via Boolean Satisfiability". In *Proc. ACM/IEEE ICCAD*, pages 294–297, 1989.
7. R. G. Wood, R. Rutenbar. "FPGA Routing and Routability Estimation via Boolean Satisfiability". In *IEEE Trans. on VLSI, vol. 6 No. 2*, pages 222–231, 1998.
8. Ranga Vemuri and Jeff Walrath. "Abstract models of reconfigurable computer architectures". In *SPIE 98*, Nov 1998.
9. Jeff Walrath and Ranga Vemuri. "A Performance Modeling and Analysis Environment for Reconfigurable Computers". In *Proc. ACM/IEEE ICCAD*, April 1998.
10. J. Walrath and R. Vemuri. "A Performance Modeling and Analysis Environment for Reconfigurable Computers". In *Proceedings of Parallel and Distributed Processing*, pages 19–24. Springer, March 1998.
11. Scott Hauck and Gaetano Borriello. "Pin Assignment for Multi-FPGA Systems". In *Proc. of FPGAs for Custom Computing Machines*, pages 11–13, April 1994.
12. Wai-Kei Mak and D.F. Wong. "On Optimal Board-Level Routing for FPGA based Logic Emulation". In *Proc. 32nd ACM/IEEE Design Automation Conference*, pages 552–556, 1995.
13. C. Selvidge, A. Agarwal, M. Dahl, J. Babb. "TIERS: Topology Independent Pipelined Routing and Scheduling for Virtual Wire Compilation". In *Proc. Int. Symp. on FPGAs*, pages 25–31, Feb. 1995.
14. Chunghee Kim and Hyunchul Shin. "A Performance-Driven Logic Emulation System: FPGA Network Design and Performance-Driven Partitioning". In *IEEE Trans. on CAD, vol. 15 No. 5*, pages 560–568, May 1996.
15. Mohammad Khalid and Jonathan Rose. "A Hybrid Complete-Graph Partial-Crossbar Routing Architecture for Multi-FPGA Systems". In *Proc. Int. Symp. on FPGAs*, pages 45–54, Feb. 1998.
16. http://www.cs.cmu.edu/ modelcheck/bdd.html.
17. *"IQX160 Starter Kit User's Guide"*. I-Cube, Inc.
18. I. Ouaiss, et.al. "An Integrated Partitioning and Synthesis System for Dynamically Reconfigurable Multi-FPGA Architectures". In *Proceedings of Parallel and Distributed Processing*, pages 31–36. Springer, March 1998.
19. S. Govindarajan, et.al. "An Effective Design Approach for Dynamically Reconfigurable Architectures". In *Int. Symposium on Field-Programmable Custom Computing Machines*, pages 312–313, April 1998.

Hardwired-Clusters Partial-Crossbar: A Hierarchical Routing Architecture for Multi-FPGA Systems

Mohammed A. S. Khalid[1] and Jonathan Rose

Department of Electrical and Computer Engineering
University of Toronto, Toronto, Ontario
CANADA M5S 3G4
{khalid, jayar}@eecg.toronto.edu

Abstract

Multi-FPGA systems (MFSs) are used as custom computing machines, logic emulators and rapid prototyping vehicles. A key aspect of these systems is their programmable routing architecture which is the manner in which wires, FPGAs and Field-Programmable Interconnect Devices (FPIDs) are connected. Several routing architectures for MFSs have been proposed [Arno92] [Butt92] [Hauc94] [Apti96] [Vuil96] [Babb97] and previous research has shown that the partial crossbar is one of the best existing architectures [Kim96] [Khal97]. Recently, the Hybrid Complete-Graph Partial-Crossbar Architecture (HCGP) was proposed [Khal98], which was shown to be superior to the Partial Crossbar. In this paper we propose a new routing architecture, called the **H**ardwired-**C**lusters **P**artial-Crossbar (HWCP) which is better suited for large MFSs implemented using multiple boards. The HWCP architecture is compared to the HCGP and Partial Crossbar and we show that it gives substantially better manufacturability. We compare the performance and cost of the HWCP, HCGP and Partial Crossbar architectures experimentally, by mapping a set of 15 large benchmark circuits into each architecture. We show that the HWCP architecture gives reasonably good cost and speed compared to the HCGP and Partial Crossbar architectures.

1 Introduction

Field-Programmable Gate Arrays (FPGAs) are widely used for implementing digital circuits because they offer moderately high levels of integration and rapid turnaround time [Brow92]. Multi-FPGA systems (MFSs), which are collections of FPGAs and memory joined by programmable connections, are used when the logic capacity of a single FPGA is insufficient, and when a quickly re-programmed system is desired.

The routing architecture of an MFS is the way in which the FPGAs, fixed wires, and programmable interconnect chips are connected. The routing architecture has a strong effect on the speed, cost and routability of the system. Many architectures have been proposed and built [Butt92] [Apti96] [Babb97] and some research work has been done to empirically evaluate and compare different architectures [Kim96] [Khal97]. These studies have shown that the partial crossbar is one of the best existing MFS architectures. Recently we proposed a new routing architecture for MFSs, called the HCGP [Khal98],

---

[1]Currently with Quickturn Design Systems, 55 West Trimble Road, San Jose, CA 95131-1013,USA

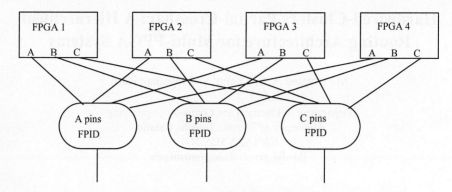

Figure 1 - The Partial Crossbar Architecture

that uses both hardwired and programmable connections to reduce cost and increase speed. We evaluated and compared the HCGP architecture and the partial crossbar architectures and showed that the HCGP architecture gives superior cost and speed.

The HCGP architecture gives excellent routability and speed for single board MFSs. The disadvantage of the HCGP, however, is that the wiring is very non-local and this becomes a major manufacturing problem when the FPGAs are spread across many boards - the number of connections required between the boards requires expensive, high pin-count connectors on each board. In this paper, we propose a new routing architecture called the Hardwired-Clusters Partial-Crossbar (HWCP), that lends itself to hierarchical implementations of large MFSs using FPGAs distributed across multiple boards. We show that the HWCP gives better manufacturability and reasonably good cost and speed compared to the HCGP and partial crossbar architectures.

2 Routing Architecture Description

In this Section we describe the partial crossbar, the HCGP and the HWCP architectures.

2.1 The Partial Crossbar Architecture

The partial crossbar architecture [Butt92] is used in logic emulators produced by Quickturn Design Systems [Quic96]. A partial crossbar using four FPGAs and three FPIDs is shown in Figure 1. The pins in each FPGA are divided into N subsets, where N is the number of FPIDs in the architecture. All the pins belonging to the same subset number in different FPGAs are connected to a single FPID. Note that any circuit I/Os will have to go through FPIDs to reach FPGA pins. For this purpose, a certain number of pins per FPID (50) are reserved for circuit I/Os.

The number of pins per subset (P_t) is a key architectural parameter that determines the number of FPIDs needed and the pin count of each FPID. A good value of P_t should require low cost, low pin count FPIDs. In this work we set $P_t = 17$. Our previous research [Khal97] has shown that, for real circuits, the routability and speed of the partial crossbar is not affected by the value of P_t used. But this is contingent upon using an intelligent inter-chip router that understands the architecture and routes each inter-FPGA net using

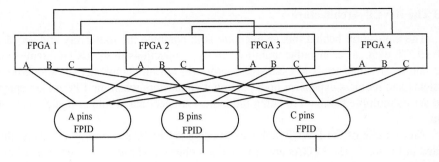

Figure 2 - The HCGP Architecture

only two hops to minimize the routing delay. However, a practical constraint is that we should avoid using P_t values that require expensive or even unavailable high pin count FPIDs.

2.2 The HCGP Architecture

The HCGP architecture for four FPGAs and three FPIDs is illustrated in Figure 2. The I/O pins in each FPGA are divided into two groups: hardwired connections and programmable connections. The pins in the first group connect to other FPGAs and the pins in the second group connect to FPIDs. The FPGAs are directly connected to each other using a complete graph topology, i.e. each FPGA is connected to every other FPGA. The connections between FPGAs are evenly distributed, i.e. the number of wires between every pair of FPGAs is the same. The FPGAs and FPIDs are connected in exactly the same manner as in a partial crossbar. As in the partial crossbar, any circuit I/Os will have to go through FPIDs to reach FPGA pins. For this purpose, a certain number of pins per FPID (50) are reserved for circuit I/Os.

A key architectural parameter in the HCGP architecture is the percentage of programmable connections, P_p. It is defined as the percentage of each FPGA's pins that are connected to FPIDs (the remainder are connected to other FPGAs). If P_p is too high it will lead to increased pin cost, if it is too low it will adversely affect routability. Our previous research [Khal98] has shown that $P_p = 60\%$ gives good routability at the minimum possible pin cost for the HCGP architecture. Therefore we use this value of P_p when comparing the HCGP to the other architectures.

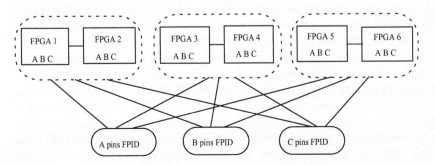

Figure 3 - The HWCP Architecture

2.3 The HWCP Architecture

The motivation behind this architecture is to combine the routability and speed of the HCGP with easier manufacturability. All the connections in the HCGP architecture are non-local which leads to excessive board-level wiring complexity for single-board systems and requires expensive high pin count connectors when the FPGAs are spread out across multiple boards. Providing more local connections would mitigate this problem.

An example of the HWCP architecture using six FPGAs and three FPIDs is illustrated in Figure 3. The FPGAs are grouped into clusters whose size is represented by a parameter called the cluster size (C_s). In Figure 3 $C_s = 2$, which implies three clusters. The pins in each FPGA are divided into two groups: hardwired and programmable connections. The pins in the first group connect to the other FPGAs within each cluster and the pins in the second group connect to FPIDs. All the FPGAs within each cluster are connected to each other in a complete graph topology. In Figure 3 the pins assigned for programmable connections are divided into three subsets A, B, and C. The pins from different FPGAs belonging to the same subset are connected using a single FPID. As in the HCGP architecture, the percentage of programmable connections (P_p) is a key parameter in HWCP.

Notice that the MFS size in the HWCP architecture is restricted to be a multiple of C_s. In this research the HWCP architecture was explored for C_s values 2, 3, and 4.

3 Comparison of the Manufacturability of the Three Architectures

In this Section we propose a metric for evaluating and comparing the manufacturability of MFSs and compare the architectures using this metric.

3.1 Manufacturability Metric: Connector Pin Count

When implementing an MFS on a printed circuit board, proper design practices must be observed to ensure correct operation of the MFS. The main problems that need to be addressed are interconnect delays, crosstalk, ringing and signal reflection, and ground bounce. These problems become more acute as the system speed increases. Since MFSs usually run at low to medium speeds, these problems can be easily tackled by using proper design practices.

There is one major problem encountered when implementing an MFS using multiple boards connected by a backplane. Each board may contain one or more FPGAs and FPIDs, e.g. each board in the TM-2 system has 2 FPGAs and 4 FPIDs [Lewi98]. If the number of pins needed for connecting each board to the backplane is very high, expensive high pin count connectors are required. This problem is severe in MFS architectures with non-local wiring like the partial crossbar and HCGP. In fact, it is difficult to obtain (commercially) a pin connector that uses more than 800 pins [Iers97]. Therefore we use the number of pins required on the backplane connector on each printed circuit board as a metric for measuring the manufacturability of an MFS routing architecture (when the FPGAs and FPIDs are distributed across multiple boards). We call this the *Connector Pin Count* (CPC for short). The lower the CPC value for a given size and type of architecture, the easier it will be to manufacture. Given the MFS size, values of the architecture parameters, the number of boards used, and the number of FPGAs and FPIDs used per board, it is fairly easy to calculate the CPC value.

MFS Size (#FPGAs)	Connector Pin Count (CPC)		
	HWCP	HCGP	PXB
8	61	106 (+74%)	96 (+57%)
12	81	138 (+70%)	128 (+58%)
16	89	149 (+67%)	144 (+62%)
20	97	160 (+65%)	156 (+61%)
24	101	167 (+65%)	160 (+58%)
28	101	170 (+68%)	168 (+66%)
32	105	174 (+66%)	168 (+60%)
Average	91	152 (+68%)	146 (+60%)

Table 1: CPC Values for the Three Architectures for Different MFS Sizes

3.2 Manufacturability Comparison

Here we compare the manufacturability of the three architectures by calculating the CPC values for different MFS sizes ranging from 8 FPGAs to 32 FPGAs. To calculate CPC, we assume that the number of FPIDs used is equal to the number of FPGAs for all MFS sizes. We also assume that each board contains 4 FPGAs and 4 FPIDs. It is assumed that the FPGA used is the Xilinx 4013E-1(PG223) which has 192 usable I/O pins. We assume $P_p = 60\%$ for HCGP and HWCP, $C_s = 4$ for HWCP, and $P_t = 17$ for the partial crossbar. These parameter values were chosen after careful experimentation (further details are given in [Khal98] and Section 5.1).

Table 1 shows the CPC values obtained for the three architectures for different MFS sizes. Column 1 shows the MFS size (number of FPGAs used), column two shows the CPC value obtained for the HWCP, the HCGP and the partial crossbar architectures. Note that the CPC value for the HCGP architecture is 68% more on average compared to the HWCP architecture across different MFS sizes considered. The CPC value for the partial crossbar architecture is 60% more on average compared to the HWCP. The clear conclusion is that the HWCP architecture is easier to manufacture compared to the HCGP and partial crossbar architectures.

In addition to providing better manufacturability, an MFS routing architecture should give minimum possible cost and provide good routability and speed performance. In the next Section we present the method used for experimentally evaluating and comparing MFS routing architectures. Although this method is targeted to only single-board systems, it still gives a good estimate of the cost, routability and speed of MFS routing architectures.

4 Experimental Overview

To evaluate the three routing architectures considered in this paper, we used the experimental procedure illustrated in Figure 4. Each benchmark circuit was partitioned, placed and routed into each architecture. Section 4.1 describes the general toolset used in this flow. The cost and speed metrics that we use to evaluate architectures are described in Section 4.2. A brief description of the 15 benchmark circuits used is given in Section 4.3.

4.1 General CAD Flow

The experimental procedure for mapping a circuit to an architecture is illustrated in Figure 4. We assume that the circuit is available as a technology mapped netlist of 4-LUTs and flip flops. First, the circuit is partitioned into a minimum number of sub-circuits using a multi-way partitioning tool that accepts as constraints the specific FPGA logic capacity and pin count. The FPGA used in our experiments is the Xilinx 4013E-1, which consists of 1152 4-LUTs and flip flops and 192 I/O pins. Multi-way partitioning is accomplished using a recursive bipartitioning procedure. The partitioning tool used is called *part*, which is based on the well known Fiduccia and Mattheyses partitioning algorithm with an extension for timing-driven pre-clustering. The output of the partitioning step is a netlist of connections between the sub-circuits.

The next step is the placement of each sub-circuit on a specific FPGA in the MFS. Given the sub-circuits and the netlist of interconnections, each sub-circuit is assigned to a specific FPGA in the MFS. The goal is to place highly connected sub-circuits into adjacent FPGAs (if the architecture has some notion of adjacency) so that the routing resources needed for inter-FPGA connections are minimized.

Given the chip-level interconnection netlist, the next step is to route each inter-FPGA net using the most suitable routing path.

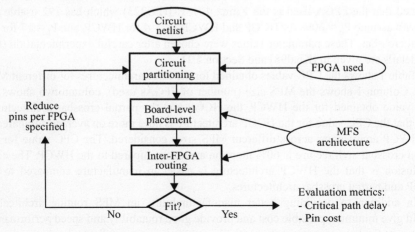

Figure 4 - Experimental Evaluation Procedure for MFSs

If the routing attempt fails, the partitioning step is repeated after reducing the number of I/O pins per FPGA specified to the partitioner. This usually increases the number of FPGAs needed, and helps routability by decreasing the pin demand from each FPGA, and providing more "route-through" pins in the FPGAs which facilitate routing.

We have developed a specific router for each of the architectures compared. (We had attempted to create a generic router but found that it had major problems with different aspects of each architecture [Khal98a].)

4.2 Evaluation Metrics

To compare the three routing architectures we implement benchmark circuits on each and contrast the pin cost and post-routing critical path delay, as described below.

4.2.1 Pin Cost

The cost of an MFS is likely a direct function of the number of FPGAs and FPIDs: If the routing architecture is inefficient, it will require more FPGAs and FPIDs to implement the same amount of logic as a more efficient MFS. While it is difficult to calculate the price of specific FPIDs and FPGAs, we assume that the total cost is proportional to the total number of pins on all of these devices. Since the exact number of FPGAs and FPIDs varies for each circuit implementation (in our procedure above, we allow the MFS to grow until routing is successful), we calculate, for each architecture, the total number of pins required to implement each circuit. We refer to this as the *pin cost* metric for the architecture.

4.2.2 Post-Routing Critical Path Delay

The speed of an MFS, for a given circuit, is determined by the critical path delay obtained after a circuit has been placed and routed at the inter-chip level. We call this the *post-routing critical path delay*. We have developed an MFS static timing analysis tool (MTA) for calculating the post routing critical path delay for a given circuit and MFS architecture. The operation and modeling used in the MTA are described in [Khal98] hence we do not discuss it here for the sake of brevity.

4.3 Benchmark Circuits

A total of fifteen large benchmark circuits were used in our experimental work. An extensive effort was expended to collect this suite of large benchmark circuits which ranged in size from 2000 to 6000 4-LUTs and flip flops. A detailed description of the benchmark circuits can be found in [Khal98].

5 Experimental Results

In this Section we determine the effect of varying the value of P_p and C_s on the routability and speed of the HWCP architecture and compare the HWCP, the partial crossbar and HCGP architectures.

5.1 HWCP Architecture: Analysis of P_p and C_s

The cluster size C_s and the percentage of programmable connections P_p, are important parameters in the HWCP architecture. In this section we explore the effect of P_p on the routability of the HWCP architecture for three values of C_s (2, 3, and 4). Our objective is to determine suitable values of C_s and P_p that give good routability at the minimum possible pin cost.

We mapped the 15 benchmark circuits to the HWCP architecture for P_p values 20, 30, 40, 50 and 60 for each of the following cases: (a) $C_s = 2$, (b) $C_s = 3$ and (c) $C_s = 4$. The best routability results for the HWCP architecture (with the lowest possible P_p value to minimize the pin cost) are obtained when $C_s = 4$ and $P_p = 60\%$. The percentage of circuits that routed for $P_p \leq 60\%$ were 20% for $C_s = 2$, 60% for $C_s = 3$ and 80% for $C_s = 4$.

The clear conclusion from these results is that $C_s = 4$ and $P_p = 60\%$ are suitable choices for achieving good routability at the minimum possible pin cost in the HWCP architecture. Therefore we will use these parameter values when comparing the HWCP to the other two architectures.

5.2 Pin Cost and Speed Comparison of the Three Architectures

The 15 benchmark circuits were mapped to the HWCP, the HCGP and the partial crossbar architectures using the experimental procedure described in Section 4.1. For the sake of brevity, we only summarize the experimental comparison results. A detailed comparison of these architectures can be found in [Khal99].

Across all the benchmark circuits, the partial crossbar needs 20% more pins on average, and as much as 25% more pins compared to the HCGP architecture. Clearly, the HCGP architecture is superior to the partial crossbar architecture in terms of the pin cost metric. This is because the HCGP exploits direct connections between FPGAs to save FPID pins that would have been needed to route certain nets in the partial crossbar.

The typical circuit delay is lower with the HCGP architecture: the HCGP gives significantly less delay (20% less on average) for twelve circuits compared to the partial crossbar and about the same delay for the rest of the circuits. The reason is that the HCGP utilizes fast and direct connections between FPGAs, whenever possible, to route delay-critical inter-FPGA nets.

The HWCP architecture failed to route three of the fifteen benchmark circuits. For the twelve circuits that routed on the HWCP, the pin cost is 17% more on average, and 33% in the worst case more compared to the HCGP architecture. The increase in pin cost is partly due to the fact that the HWCP required more FPGAs for some circuits to make the MFS size (measured by the number of FPGAs) a multiple of the cluster size ($C_s = 4$). The other reason is that the HCGP has superior routability compared to HWCP.

For the twelve circuits that routed on HWCP, the critical path delay is 15% more on average, and up to 29% more compared to the HCGP architecture. For the twelve circuits that routed, the average pin cost and speed results for the HWCP are comparable to the partial crossbar.

The results presented in this Section demonstrate that the HWCP architecture has the potential to provide good routability, speed, and manufacturability. Despite some routing failures, the routability results for HWCP are quite encouraging for the following reasons: first, we did not try higher cluster sizes (relative to the MFS size). The routability of the HWCP architecture improves with the increase in cluster size. Second, the partitioning and placement methods used for the HWCP architecture were not the best possible. Architecture-driven partitioning and placement methods may give better routability results.

6 Conclusions

In this paper we presented the Hardwired-Clusters Partial-Crossbar (HWCP), a new routing architecture suitable for large MFSs implemented using multiple boards. We show that the HWCP gives substantially better manufacturability compared to the HCGP and the partial crossbar architectures. Using an experimental approach, we evaluated and compared this architecture to the HCGP and the partial crossbar architectures and showed that it gives reasonably good pin cost and speed. To our knowledge, this is the first study of hierarchical architectures for board-level MFSs.

This research is a first step in exploring routing architectures for large MFSs that use hundreds of FPGAs spread out across multiple boards. For a more rigorous investigation of such hierarchical architectures, we will need extremely large benchmark circuits and appropriate mapping CAD tools.

References

[Apti96] Aptix Corporation, Product brief: The System Explorer MP4, 1996. Available on Aptix Web site: http://www.aptix.com.

[Arno92] J. M. Arnold, D. A. Buell, and E. G. Davis, "Splash 2," Proceedings of 4th Annual ACM Symposium on Parallel Algorithms and Architectures, pp. 316-322, 1992.

[Babb97] J. Babb et al, "Logic Emulation with Virtual Wires," IEEE Trans. on CAD, vol. 16, no. 6, pp. 609-626, June 1997.

[Brow92] S. Brown, R. Francis, J. Rose, and Z. Vranesic, Field Programmable Gate Arrays, Kluwer Academic Publishers, 1992.

[Butt92] M. Butts, J. Batcheller, and J. Varghese, "An Efficient Logic Emulation System," Proceedings of IEEE International Conference on Computer Design, pp. 138-141, 1992.

[Hauc94] S. Hauck, G. Boriello, C. Ebeling, "Mesh Routing Topologies for Multi-FPGA Systems", Proceedings of International Conference on Computer Design (ICCD'94), pp. 170-177, 1994.

[ICub97] I-Cube, Inc., The IQX Family Data Sheet, May 1997. Available at: www.icube.com.

[Khal97] M. A. S. Khalid and J. Rose, "Experimental Evaluation of Mesh and Partial Crossbar Routing Architectures for Multi-FPGA Systems," Proceedings of the Sixth IFIP International Workshop on Logic and Architecture Synthesis (IWLAS'97), pp. 119-127, 1997.

[Khal98] M. A. S. Khalid and J. Rose, "A Hybrid Complete-Graph Partial-Crossbar Routing Architecture for Multi-FPGA Systems," Proc. of 1998 Sixth ACM International Symposium on Field-Programmable Gate Arrays (FPGA'98), pp. 45-54, February 1998. This paper is available on the internet at "www.eecg.toronto.edu/~khalid/papers/fpga98.ps".

[Khal99] M. A. S. Khalid, Routing Architecture and Layout Synthesis for Multi-FPGA Systems, Ph. D. Thesis, University of Toronto, Toronto, Canada, 1999.

[Kim96] C. Kim, H. Shin, "A Performance-Driven Logic Emulation System: FPGA Network Design and Performance-Driven Partitioning," IEEE Trans. on CAD, vol. 15, no. 5, pp. 560-568, May 1996.

[Iers97] M. Van Ierssel, University of Toronto, Private Communication, 1997.

[Quic96] Quickturn Design Systems, Inc., System Realizer Data Sheet, 1996. Available on Quickturn Web site:http://www.quickturn.com.

[Vuil96] J. E. Vuillemin, P. Bertin, D. Roncin, M. Shand, H. Touati, and P. Boucard, "Programmable Active Memories: Reconfigurable Systems Come of Age," IEEE Transactions on VLSI, Vol 4, No. 1, pp. 56-69, March 1996.

[Xili97] Xilinx, Inc., Product Specification: XC4000E and XC4000X Series FPGAs, Version 1.2, June 16, 1997. Available on Xilinx Web site: www.xilinx.com.

Integrated Block-Processing and Design-Space Exploration in Temporal Partitioning for RTR Architectures *

Meenakshi Kaul
mkaul@ececs.uc.edu

Ranga Vemuri
Ranga.Vemuri@UC.EDU

Department of ECECS, University of Cincinnati, Cincinnati, OH 45221–0030

Abstract. *We present an automated temporal partitioning and design space exploration methodology that temporally partitions behavior specifications. We propose block-processing in the temporal partitioning framework for reducing the reconfiguration overhead for partitioned designs. Block-processing is a technique used traditionally in the area of parallel compilers, for increasing the computation speed by processing several inputs simultaneously. Block-processing technique has been integrated with task-level design space exploration to achieve designs that justify temporal partitioning of systems. An ILP-based methodology has been proposed to solve this problem. We present experimental results for the Discrete Cosine Transform (DCT).*

1 Introduction

The reconfiguration capability of SRAM-based FPGAs can be utilized to fit a large application onto an FPGA by partitioning the application *over time* into multiple segments. This division into temporal segments is called *temporal partitioning*. Such temporally partitioned applications are also called Run-Time Reconfigured (RTR) systems.

To overcome the effects of high reconfiguration overhead for existing reconfigurable hardware, we demonstrated in [8] how loop fission techniques can be used as a post-processing step after temporally partitioning a design to increase the throughput. In the current work, we extend our temporal partitioning technique to incorporate block-processing and design space exploration techniques and demonstrate how this integrated processing can be used to search for optimized temporally partitioned designs.

We assume the input specification to be a task graph [4], where each task consists of a set of operations. Depending on the resource/area constraint for the design, different implementations of the same task, which represent different area-time tradeoff points, can be contemplated. These different implementations are *design points* in the design space of a task. If a task is implemented with less resources then the operations in the task will be executed serially, thus increasing the latency of the task. On the other hand, an implementation with more resources reduces the latency but increases the area. Therefore, choosing the best design point for each task may not necessarily result in the best overall design for the specification. The optimal design point for a task in the context of optimizing the overall throughput of the design will depend on the architectural constraints and the dependency constraints among the tasks. In the subsequent discussion we will express the latency of a design point in terms of total execution time and not in number of clock cycles.

Block-Processing in Temporally Partitioned designs : In many application domains eg., Digital Signal Processing, computations are defined on very long streams of input data. In order to decrease overall execution time multiple inputs from the input data stream can be processed simultaneously. This approach known as *block-processing*, is used to increase the throughput of a system through the use of parallelism and pipelining in the area of parallel compilers [12] and VLSI processors [13]. We can

---

* This work is supported in part by the US Air Force, Wright Laboratory, WPAFB, under contract number F33615-97-C-1043.

apply the concept of block-processing to a single processor reconfigurable system to speedup the processing time.

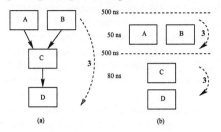

Fig. 1. Temporally partitioned design example

Figure 1 illustrates the use of block-processing to speed up computation in a temporally partitioned design. The task graph consists of 4 tasks A, B, C, D. It is partitioned into two temporal partitions as shown in Figure 1(b). The latency of temporal partition 1 is 50ns and of partition 2 is 80 ns. The reconfiguration time is 500 ns. The latency of the design is 50+500 +80+500 = 1130 ns. A single iteration of the task graph executes in 1130 ns. Now three iterations of this temporally partitioned design will take 3x1130 = 3390 ns. However if we perform block-processing by sequencing all 3 computations on each temporal partition, the time taken for the execution is (50+50+50)+500+(80+80+80)+500 = 1390 ns. Thus block-processing amortizes the reconfiguration overhead over the 3 inputs. Block-processing is possible only for applications that process a large stream of inputs. We represent such applications by a task graph having an implicit outer loop as shown in Figure 1(a). Note that block-processing is possible if there are no dependencies among the computations for different inputs. In compiler terminology this means there should be no loop-carried dependencies due to the implicit outer loop among different iterations of this loop. In this paper, we deal with applications for which no dependencies among computations is present. Most DSP applications fall in this category.

Integrating Design-Space Exploration and Block-Processing in Temporal Partitioning : For FPGA based architectural synthesis, the constraints of area in terms of CLBs (Configurable Logic Blocks)/FGs (Function Generators) and memory are to be met by the partitioned design. For the *spatial* partitioning problem increasing the number of partitions has the effect of increasing the overall area for the design, and directly affects the latency of the design. Increasing the area, generally increases the number of operations that can execute in parallel (if no dependency constraints exist) and thus decreases the latency of the design. However, for a temporal partitioning system increasing the number of partitions increases the area available for the design, but this increase is 'over time' and not 'over space'. This increase in number of partitions may or may not result in the reduction of the latency of the design.

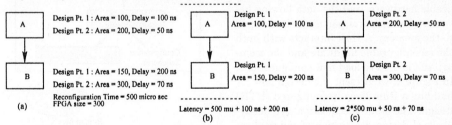

Fig. 2. Design space exploration

When the reconfiguration overhead is very large compared to the execution time of the task it is clear that minimizing the number of temporal partitions will achieve the *smallest latency* in the overall design. However, it is *not necessary* that the minimum latency design is the best solution. We illustrate this idea with an example. In the

Figure 2(a) a task graph is shown. Each task has two different design points on which it can be mapped. Two different solutions (b) and (c) are possible. If minimum latency solution is required then solution (b) will be chosen over solution (c) because the latency of (b) is 500.3 μ sec and latency of (c) is 1000.12 μ sec. Now, if we use (b) and (c) in the block-processing framework to process 5000 computations on each temporal partition. Then the execution time for solution (b) is 2000 μ seconds and for solution (c) is 1600 μ seconds. Therefore if we can integrate the knowledge about block-processing while design space exploration is being done then it is possible to choose more appropriate solutions.

The price paid for block-processing is the higher memory requirements for the reconfigured design. We call the number of data samples or inputs to be processed in each temporal partition to be the the *block-processing factor*, k. This is given by the user and is the minimum number of input data computations that this design will execute for typical runs of the application. The amount of block-processing is limited by the amount of memory available to store the intermediate results.

Currently many designers perform temporal partitioning manually [1] or the designer specifies the partitioning points to the tool [3]. Existing automated temporal partitioning techniques, extend existing scheduling techniques of high-level synthesis [2, 9] and focus on minimizing the number of partitions of the design. In [7], we presented a mathematical model for combined temporal partitioning and synthesis. In that approach, synthesis cost exploration is performed at an operation level in the task graph, and the number of alternative solutions explored becomes very large. It can be used to synthesize small-scale behavior specifications. Our current technique can also simultaneously handle multiple design constraints, eg., FPGA resources, on-board memories, and perform design exploration for large specifications.

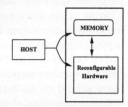

Fig. 3. RTR System Architecture

2 System Architecture and Design Flow

In Figure 3, the hardware architecture on which the Run-Time reconfigured design is to be mapped is shown. It consists of a reconfigurable hardware communicating with an external memory. Each temporal partition is mapped to the reconfigurable hardware, and the data flowing between two temporal partitions is mapped to the memory. A Host interacts with both the reconfigurable hardware and the memory and is used to load new configurations. In Figure 4, we present the design flow for building a Run-Time Reconfigured (RTR) design. The input specification, is a behavior level design description of the application to be implemented on the reconfigurable hardware. The input specification

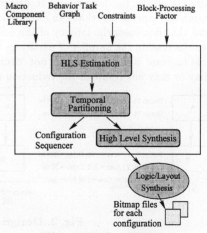

Fig. 4. Design Flow

consists of acyclic data flow task graph, with an outer implicit loop. The implicit loop signifies the successive items of input data that will be executed on this task graph.
Task Estimation: First, the behavior level estimation engine, part of a High Level

synthesis tool, DSS [10], generates multiple design points for each task separately based on the architecture and user constraints. The architecture constraints are the resources available on the reconfigurable hardware, the user constraints are the maximum clock-width for the design. The HLS tool makes use of a component library characterized for the particular reconfigurable device, to estimate the resource and delay.

Temporal Partitioning: Next, the temporal partitioning tool divides the task graph into multiple temporal segments, while mapping each task to its appropriate design point. We discuss the ILP formulation used to solve the multi-constraint temporal partitioning problem later in detail.

High Level Synthesis: A high level synthesis system is used to generate the RTL design for each temporal segment.

Logic/Layout Synthesis: We use commercial tools, for logic synthesis (Synplify tools from Synplicity) and layout synthesis (Xilinx M1 tools) to convert the RTL description of each configuration into bitmap files.

3 Temporal Partitioning

The inputs to our Temporal Partitioning system are - (1) Behavior specifications (2) Target Architecture Parameters (3) Block-processing Factor
In formal notation, the inputs are stated as -

T *set of tasks in the task graph.*
$t_i \rightarrow t_j$ *a directed edge between tasks, $t_i, t_j \in T$, exists in the task graph.*
$B(t_i, t_j)$ *number of data units to be communicated between tasks t_i and t_j.*
$B(env, t_j)$ *number of data units to be read by task t_j from the environment.*
$B(t_i, env)$ *number of data units to be written from task t_i to the environment.*
R_{max} *resource capacity of the reconfigurable processor.*
M_{max} *memory size of the RTR architecture.*
C_T *reconfiguration time of the reconfigurable processor.*
k *the block-processing factor for the design.*

The behavior specifications are in the form of a directed graph called the *Task Graph*. The vertices in the graph denote tasks, and the edges denote the dependency among tasks. Data communicated between two tasks, $B(t_i, t_j)$, will have to be stored in the on-board memory of the processor, if the two tasks connected by an edge are placed in different temporal partitions. The target architecture parameters specify the underlying resources and the reconfiguration time, C_T, for the device. Typically, resource capacity, R_{max}, is the combinational logic blocks/function generators on the FPGAs of the reconfigurable device. M_{max}, is the memory for storage of intermediate data available on the reconfigurable processor. k, the block-processing factor is the lower bound on the number of computations that this design will usually perform.

3.1 Preprocessing

Design Point Generation: Each task in the task graph is processed by a design space exploration and estimation tool which is part of a high level synthesis system. The high level synthesis tool generates a set of *design points* for each task. Each design point has an associated *module set* [11]. A module set, m, consists of the set of, possibly multiple, functional units used to implement the design point. Each design point is characterized by its area and latency. Each task t, will have a set of module sets, M_t, corresponding to the set of synthesized design points. We state this formally as -

M_t *set of module sets for a task $t \in T$.*
$R(m)$ *area for a design point using module set $m \in M_t$.*
$D(m)$ *latency of a design point using module set $m \in M_t$.*

Partition bounds Estimation: To find the number of partitions over which the temporal partitioning solution should be explored we calculate two bounds -

1. *MinAreaPartitions():* For calculating the lower bound on number of partitions N^l_{min}, we sum the *minimum* area module set, m, for each task. This value divided by the FPGA area will be the minimum number of partitions required to obtain a solution.

$N^l_{min} = \sum_{t \in T} R(m)/R_{max}, \quad m \in M_t$, m is the minimum area module set in M_t

2. *MaxAreaPartitions():* Ideally, we should be able to establish an upper bound on the number of partitions needed to be explored by the partitioner, if the maximum area design point for each task is chosen. However, we cannot establish, an upper bound on the maximum number of partitions. If a task is too large to fit in some temporal partition, it must go to a later partition. Then all the descendents of this task also cannot occupy the earlier temporal partition. This will leave some area on temporal partitions unoccupied due to dependency constraints, and the task graph will not fit, even though there is enough area left unoccupied on the partitions. We define, the minimum number of partitions, N^u_{min}, that need to be explored if *maximum* area design point for each task is mapped by the partitioner, to be -

$N^u_{min} = \sum_{t \in T} R(m)/R_{max}, \quad m \in M_t$, m is the maximum area module set in M_t

3.1.1 Partition Space Exploration

To explore better solutions for the temporal partitioning problem, we need to explore more than one partition bound. Finding the ideal partition size, N, is an iterative procedure, shown in Figure 5. We calculate the bounds on the number of partitions, N^l_{min} and N^u_{min}, as described earlier. We start the search at N^l_{min} and obtain an optimal solution, by forming and solving an ILP model of the temporal partitioning problem. The details of the model are described in Section 3.1.2. For the first ILP model there is no upper bound on constraint on the execution time of the design. The result of solving this model is a temporal partitioning solution for N partitions and the execution time D_a of the solution. We now relax N by 1, form and solve the ILP model again. This time however, since we are looking for a better solution than the one we have already achieved, D_a is the execution time constraint for the new ILP model. We continue to relax N and look for better solutions until the value of N reaches $N^u_{min} + \gamma$. Here γ is a user controlled parameter, called the *Partition Relaxation*, which defines the number of partitions beyond N^u_{min} that must be explored while searching for better solutions.

```
Algorithm Refine_Partition_Bound()
begin
    N^u_min ← MaxAreaPartitions()
    N^l_min ← MinAreaPartitions()
    N ← N^l_min  /* starting partition number */
    D_a ← Solve_ILP_Model()
    while D_a = 0 /* Partition bound was infeasible */
        N ← N + 1 /* next partition number */
        D_a ← Solve_ILP_Model()
    end while
    while N < N^u_min + γ and not TimeExpired()
        N ← N + 1 /* Relax N */
        D'_a ← Solve_ILP_Model()
        if D'_a ≠ 0 /* solution is feasible */
            D_a ← D'_a
        end if
    end while
    return(D_a) /* return with the last known best solution */
end Algorithm Refine_Partition_Bound
```

Fig. 5. Partition Refinement Procedure

3.1.2 ILP formulation for Design Space Exploration

We build the temporal partitioning model for the given tasks and their design points and the values of N and k. After linearization of the non-linear constraints, we solve it using a linear programming solver. To state formally the mathematical model we use the following definitions -

N bound on the number of partitions.
D_a constraint on the execution time of the design.
T_l set of tasks $t_i \in T$, where $\forall t_j \in T, \neg(t_i \to t_j)$, (leaf tasks of T).
T_r set of tasks $t_j \in T$, where $\forall t_i \in T, \neg(t_i \to t_j)$, (root tasks of T).
$t_i \xrightarrow{p} t_j$ a directed path from $t_i \in T$ to $t_j \in T$.
$P_{l\xrightarrow{p}r}$ $\{ t_i \xrightarrow{p} t_j \mid (t_i \in T_r) \wedge (t_j \in T_l)\}$, (set of paths from root tasks to leaf tasks).

Variables and Constraints Variable y_{tpm}, models partitioning and design point selection for a task. $w_{pt_1t_2}$, models data transfer requirement across partition boundaries. η, is the actual number of partitions finally used in the solution and will be less than or equal to N. d_p, models the execution time of a temporal partition.

$$y_{tpm} = \begin{cases} 1 \text{ if task } t \in T \text{ is placed in partition p}, 1 \leq p \leq N, \text{ using module set } m \in M_t \\ 0 \text{ otherwise} \end{cases}$$

$$w_{pt_1t_2} = \begin{cases} 1 \text{ if task } t_1 \text{ is placed in any partition } 1 \cdots p-1 \text{ and } t_2 \text{ is placed in any} \\ \quad p \cdots N \text{ and } t_1 \to t_2 \\ 1 \text{ if task } t_1 \text{ is placed in partition } p \text{ and } t_2 \text{ is placed in any } p+1 \cdots N \text{ and } t_1 \to t_2 \\ 0 \text{ otherwise} \end{cases}$$

η = Number of partitions actually used in solution.

d_p = execution time of partition p.

Variables $y_{tpm}, w_{pt_1t_2}$ are 0-1 variables, η is an integer variable and d_p can be integer or real depending on whether the latency values are integer or real.

Uniqueness Constraint: Each task should be placed in exactly one partition among the N temporal partitions, while selecting one among the various module sets for the task.

$$\forall t \in T \quad : \sum_{m \in M_t} \sum_{p=1}^{N} y_{tpm} = 1 \quad (1)$$

Temporal order Constraint: Because we are partitioning over time, a task t_1 on which another task t_2 is dependent cannot be placed in a later partition than the partition in which task t_2 is placed. It has to be placed either in the same partition as t_2 or in an earlier one.

$\forall t_2, \forall t_1 \to t_2, \forall p_2, 1 \leq p_2 \leq N-1$:

$$\sum_{m_1 \in M_{t_1}} \sum_{p_2 < p_1 \leq N} y_{t_1 p_1 m_1} + \sum_{m_2 \in M_{t_2}} y_{t_2 p_2 m_2} \leq 1 \quad (2)$$

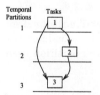

MODELLING EQUATIONS:
$W_{212}^* B(1,2) + W_{213}^* B(1,3) + W_{223}^* B(2,3) <= M_{max}$
$W_{312}^* B(1,2) + W_{313}^* B(1,3) + W_{323}^* B(2,3) <= M_{max}$
RESULT EQUATIONS:
$W_{212}^* B(1,2) + W_{213}^* B(1,3) + W_{223}^* B(2,3) <= M_{max}$
$W_{313} B(1,3) + W_{323}^* B(2,3) <= M_{max}$

Fig. 6. Memory Constraints

Memory Constraint: Data transfer across partition boundaries will occur due to two dependent tasks being placed in different temporal partitions. This intermediate data needs to be stored between partitions and should be less than the memory, M_{max}, of the reconfigurable processor. The variable $w_{pt_1t_2}$, if 1, signifies that t_1 and t_2 have a data dependency and are being placed across temporal partition p. Therefore the data being communicated between them, $B(t_1, t_2)$, will have to be stored in the memory of partition p. The sum of all the data being communicated across a partition should be less than the memory constraint of the partition.

$$\forall p, 1 \leq p \leq N \quad : \sum_{t \in T} \sum_{p \leq p_2 \leq N} y_{tp_2} * B(env, t) + \sum_{t \in T} \sum_{1 \leq p_3 \leq p} y_{tp_3} * B(t, env) + \sum_{t_2 \in T} \sum_{t_1 \to t_2} (w_{pt_1t_2} * B(t_1, t_2)) \leq M_{max} \quad (3)$$

Note that the variable $w_{pt_1t_2}$ has to model communication among tasks which are both on adjacent and non-adjacent temporal partitions. In Figure 6, we show how this variable models data transfer. We show in the figure the original equations used to model the constraints in the example for Temporal Partitions 2 and 3. The result equations show the variables which will be 1 in the mapping of tasks to partitions shown in the example and the constraint which has to be satisfied. $w_{pt_1t_2}$ are 0-1 non-linear terms constrained as -

$$\forall p, \ 1 \leq p \leq N, \ \forall t_2 \in T, \ \forall t_1 \to t_2, \ : w_{pt_1t_2} \geq \sum_{1 \leq p_1 < p} y_{t_1p_1} * \sum_{p \leq p_2 \leq N} y_{t_2p_2} \quad (4)$$

$$\forall p, \ 1 \leq p \leq N, \ \forall t_2 \in T, \ \forall t_1 \to t_2, \ : w_{pt_1t_2} \geq y_{t_1p} * \sum_{p+1 \leq p_2 \leq N} y_{t_2p_2} \quad (5)$$

Equations (4) and (5) are non-linear. We can use linearization techniques to transform the non-linear equations into linear ones, so that the model can be solved by a Linear Program solver. Linearization techniques have been used successfully before in [7] to solve the combined temporal partitioning and synthesis problem.

Resource Constraint: The sum of area costs of all the tasks mapped to a temporal partition must be less than the overall resource constraint of the reconfigurable processor. Typical FPGA resources include function generators, configurable logic blocks etc. Similar equations can be added if multiple resource types exist in the FPGA.

$$\forall p, \ 1 \leq p \leq N \ :$$
$$\sum_{m \in M_t} \sum_{t \in T} (y_{tpm} * R(m)) \leq R_{max} \quad (6)$$

Execution Time Constraint: The execution time of a partition will be the maximum execution time among all the paths of the task graph mapped to that partition. In Figure 7, we show

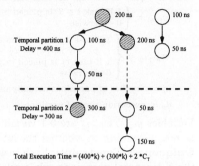

Fig. 7. Execution Time Estimation

how the execution time for a partition is determined. The final mapping of tasks to partitions, with the latency value for each task, is shown. In partition 1, three paths are mapped. The latency of this partition is the greatest latency along a path mapped to the partition, i.e., maximum among 350ns, 400ns, 150ns. The maximum latency in partition 2 is 300ns. If the block-processing factor is k, then the execution time of the partition is the latency multiplied to the block-processing factor. Formally the execution time of a temporal partition is given as -

$$\forall p, \ 1 \leq p \leq N, \ \forall (t_i \xrightarrow{p} t_j) \in P_{\substack{p \\ 1 \to r}} \ : \sum_{m \in M_t} \sum_{t \in t_i \xrightarrow{p} t_j} (y_{tpm} * D(m) * k) \leq d_p \quad (7)$$

All temporal partitions $1 \cdots N$ used in the formulation, may not be used in the final solution, if the tasks can fit in lesser number of partitions. To calculate the actual number of partitions used in the solution, we determine the highest numbered partition used by any leaf level task in the task graph by the following equation -

$$\forall t \in T_l \ : \sum_{m \in M_t} \sum_{p=1}^{N} (p * y_{tpm}) \leq \eta \quad (8)$$

Now the execution time constraint on the overall design can be stated in terms of equations (7) and (8) as -

$$\eta * C_T + \sum_{p=1}^{N} d_p \leq D_a \quad (9)$$

As discussed earlier, this constraint is used to search for a better solution as different partition bounds are being explored in Algorithm *Refine_Partition_Bound* in Figure 5.

Optimality Goal: The most optimal solution will be the design with the least execution time.

$$Minimize \ : \eta * C_T + \sum_{p=1}^{N} d_p \quad (10)$$

The solution of this ILP model gives us the optimum temporal partitioning for the give partition bound N, the block-processing factor k, and the set of design points for the tasks. If the amount of intermediate memory required to process k computations exceeds the memory constraint M_{max} of the architecture then the user needs to reduce k and temporally partition the design again.

4 Experimental Results

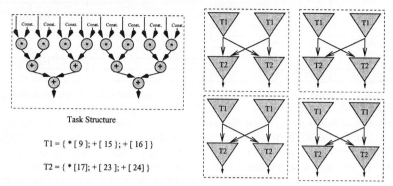

Fig. 8. Task graph for DCT

We performed temporal partitioning on the Discrete Cosine Transform (DCT) [16] which is the most computationally intensive part of the JPEG [17] algorithm. In this study, the DCT is a collection of 16 tasks, the structure of a task is shown in Figure 8. There are two kinds of task in the task graph, $T1$ and $T2$, whose structure is similar but whose operations have different bit widths. We obtained all the design points for each kind of tasks, by using estimation tools on the individual tasks. The functional units, area and latency costs for each is shown in Table 1.

Task	Desgn. Pt.	Characteristics					
		Area (CLBs)	Latency (ns)	$*_9$	$+_{16}$	$*_{16}$	$+_{24}$
T1	1	336	375	8	4		
	2	286	500	6	2		
	3	220	625	4	2		
	4	194	750	2	2		
	5	174	875	1	2		
T2	1	396	420			8	4
	2	356	560			6	2
	3	292	700			4	2
	4	276	840			2	2

Table 1. Design Points for DCT tasks

In Tables 2 - 4 we present the results of our temporal partitioning tool. In all the tables, R_{max} is the resource constraint of the FPGA, C_T the reconfiguration time, k the block-processing factor, and N the number of partitions onto which the design is partitioned. The latency of the final design (with the reconfiguration overhead), is shown in the column Latency. The execution time of the design for the k blocks of data is given in the column Design Execution Time. Partitioner Run Time in seconds is the time taken by our temporal partitioning tool to execute. All experiments were run using an ILP solver called CPLEX on an UltraSparc Machine running at 175 MHz. In Table 2, we present the result of temporal partitioning and design space exploration of the DCT for with and without block-processing factors. In Exp. 1, for a block-processing factor of 3,000, our temporal partitioning tool explores 3 temporal partitions for the design and results in a latency of 60,795 ns. In Exp. 2, with a block-processing factor of 1 (i.e., no computations are being sequenced), the tool gives a minimum latency design of 31,590 ns and uses just one temporal partition. This results in a statically configured design. Even though, the latency of the statically configured design in Exp. 2 is less than that of Exp. 1, this is not the best possible solution. This is because, if multiple computations are computed on both the static and RTR design, the RTR

Exp.	R_{max} (CLBs)	C_T (μs)	k	N	Latency (ns)	Design Execution Time	Partitioner Run Time (s)
1	4,000	30	3,000	1	31,590	4,800 μs	1
				2	60,795	2,445 μs	1
				3			Infeasible
2	4,000	30	1	1	31,590		1
3	2,304	30	3,000	2	61,590	4,830 μs	11
				3	91,215	3,735 μs	22
				4			Infeasible
4	2,304	30	1	2	61,590		57

Table 2. Results for combined design-space exploration and block-processing

design will outperform the static design. For executing 3,000 computations, the RTR design will take 2,445 μ sec, while the static design will take 4,800 μ sec. This is a 49% improvement of the RTR design over the static design. Exp. 3 and 4 were performed for different FPGA size of 2,304 CLBs. In Exp. 3, again with a block-processing factor of 3000, the optimal design takes 3 temporal partitions with the latency of the design being 91,215 ns. For Exp. 4, the optimal latency of the design is 61,590 ns. Again, the actual execution time of the design when the block-processing factor is considered while exploring the design space is superior. In all experiments in Table 2 the reconfiguration time considered is similar to the Xilinx 6000 series FPGAs.

The experiments in Table 2 illustrate that combining block-processing and design space exploration gives better temporal partitioning solutions. If the block-processing factor is equal to 1, then the temporal partitioning tool will tend to pick the design with minimum number of temporal partitions. If a relevant block-processing factor is given the tool will search for a design with more temporal partitions, because block-processing will amortize the effects of reconfiguration overhead. So, since we understand that the block-processing is necessary for good performance of a temporally partitioned design, we must integrate this idea early in the design process, while partitioning and design point selection is being performed.

Similar results will hold if the reconfiguration overheads are varied. In Table 3, we show results for different reconfiguration overheads. In Exp. 5 and 6, the reconfiguration overhead is in nano-seconds (similar to the reconfiguration overheads of context-switching FPGAs like the Time Multiplexed FPGA [14]). In Exp. 7 and 8, the reconfiguration overhead is in milli-seconds (similar to commercially available reconfigurable hardware [15]). As the reconfiguration overhead decreases we observe that for small values of k, the exploration process chooses more temporal partitions. However, for the reconfiguration overheads in milli-seconds even for values of k as large as 3,000 the temporal partitioner chooses designs with minimum temporal partitions.

Exp.	R_{max} (CLBs)	C_T	k	N	Latency (ns)	Design Execution Time	Partitioner Run Time (s)
5	2,304	30 ns	300	2	1,650	477.06 μs	17
				3	1,305	364.59 μs	19
6	2,304	30 ns	50	2	1,650	79.56 μs	25
				3	1,305	60.84 μs	7
7	2,304	3 ms	3,000	2	6,001,590	10,770 μs	80
8	2,304	3 ms	30,000	2	6,001,590	53,700 μs	29
				3	9001,215	45,450 μs	36

Table 3. Results for different reconfiguration overheads

In Table 4, we illustrate how design space exploration is beneficial. For same values of the block-processing factor k, we perform experiment with and without design space exploration. In Exp. 9, temporal partitioning is performed with only one design point for each task, the minimal area design point. In Exp. 10, all the design points are used. Again we observe that the tool chooses the most appropriate design points for the given constraints, when multiple design points are given to it, and results in a 27%

improvement of the design in Exp. 10. Therefore design space exploration must be integrated in the temporal partitioning process, rather than choosing the design point before temporal partitioning is performed.

Exp.	R_{max} (CLBs)	C_T (μs)	k	N	Latency (ns)	Design Execution Time	Partitioner Run Time (s)
9	2,304	30	3,000	2	31,715	5,145.6 μs	1
10	2,304	30	3,000	2	61,590	4,830 μs	22
				3	91,215	3,735 μs	204

Table 4. Results for design-space exploration

5 Conclusion

The algorithms presented in this paper are integrated in the SPARCS (Synthesis and Partitioning for Adaptive Reconfigurable Computing Systems) [5, 6] design environment being developed at the University of Cincinnati. SPARCS is an integrated design system for automatically partitioning and synthesizing designs for reconfigurable boards with multiple field-programmable devices (FPGAs). The SPARCS system contains a temporal partitioning tool, a spatial partitioning tool, and a high-level synthesis tool. For more details go to http://www.ececs.uc.edu/~ddel/projects/sparcs/sparcs.html.

References

1. R. D. Hudson, D. I. Lehn and P. M. Athanas, "A Run-Time Reconfigurable Engine for Image Interpolation", *IEEE Symposium on FPGAs for Custom Computing Machines, FCCM 1998, pp. 88-95.*
2. M. Vasiliko and D. Ait-Boudaoud, "Architectural Synthesis for Dynamically Reconfigurable Logic",*International Workshop on Field-Programmable Logic and Applications, FPL 1996, pp. 290-296.*
3. M. B. Gokhale and J. M. Stone, "NAPA C:Compiling for Hybrid RISC/FPGA Architectures", *IEEE Symposium on FPGAs for Custom Computing Machines, FCCM 1998, pp. 126-135.*
4. I. Ouaiss, S. Govindarajan, V. Srinivasan, M. Kaul, and R. Vemuri, "A Unified Specification Model of Concurrency and Coordination for Synthesis from VHDL", *International Conference on Information Systems, Analysis and Synthesis, ISAS 1998, pp. 771-778.*
5. I. Ouaiss, S. Govindarajan, V. Srinivasan, M. Kaul and R. Vemuri, "An Integrated Partitioning and Synthesis System for Dynamically Reconfigurable Multi-FPGA Architectures", *Reconfigurable Architectures Workshop in 12th International Parallel Processing Symposium and 9th Symposium on Parallel and Distributed Processing, IPPS/SPDP 1998, pp. 37-42.*
6. S. Govindarajan, I. Ouaiss, M. Kaul, V. Srinivasan and R. Vemuri, "An Effective Design Approach for Dynamically Reconfigurable Architectures", *IEEE Symposium on FPGAs for Custom Computing Machines, FCCM 1998, pp.312-313.*
7. M. Kaul and R. Vemuri, "Optimal Temporal Partitioning and Synthesis for Reconfigurable Architectures", *Design and Test in Europe, DATE 1998, pp. 389-396.*
8. M. Kaul, R. Vemuri, S. Govindarajan and I. Ouaiss, "An Automated Temporal Partitioning Tool for a class of DSP applications", *Workshop on Reconfigurable Computing in International Conference on Parallel Architectures and Compilation Techniques, PACT 1998, pp. 22-27.*
9. S. Trimberger, "Scheduling designs into a Time-Multiplexed FPGA", *ACM/SIGDA International Symposium on Field Programmable Gate Arrays, FPGA 1998, pp. 153-160.*
10. J. Roy, N. Kumar and R. Vemuri, "DSS: A Distributed High-Level Synthesis System for *VHDL* Specifications", *IEEE Design and Test of Computers, v9, n2, June 1992, pp. 18-32.*
11. R. Dutta et. al, "Distributed Design Space Exploration for High-Level Synthesis Systems", *29th Design Automation Conference, DAC 1992, pp. 644-650.*
12. M. Wolf, *High Performance Compilers for Parallel Computing*, Addison-Wesley Publishers, 1996.
13. S. Y. Kung, *VLSI Array Processors*, Prentice Hall 1988.
14. S. Trimberger, "A Time-Multiplexed FPGA", *IEEE Symposium on FPGAs for Custom Computing Machines, FCCM 1997, pp. 22-28.*
15. WILDFORCE Reference Manual, *Document #1189 - Release Notes, Annapolis Micro Systems, Inc..*
16. N. Narasimhan, V. Srinivasan, M. Vootukuru, J. Walrath, S. Govindarajan and R. Vemuri, "Rapid Prototyping of Reconfigurable Coprocessors", *International Conference on Application-Specific Systems, Architectures and Processors, 1996.*
17. G.K. Wallace, "The JPEG Still Picture Compression Standard", *ACM Communications, 1991.*

Improved Scaling Simulation of the General Reconfigurable Mesh

José Alberto Fernández-Zepeda, Ramachandran Vaidyanathan, and
Jerry L. Trahan

Department of Electrical & Computer Engineering
Louisiana State University, Baton Rouge, LA 70803, USA

Abstract. The reconfigurable mesh (R-Mesh) has drawn much interest in recent years, due in part to its ability to admit extremely fast algorithms for a large number of problems. For these algorithms to be useful in practice, the R-Mesh must be scalable; that is, any algorithm designed for a large R-Mesh should be able to run on a smaller R-Mesh without significant loss of efficiency. This amounts to designing a "scaling simulation" that simulates an arbitrary step of an $N \times N$ R-Mesh on a smaller $P \times P$ R-Mesh in $O\left(\frac{N^2}{P^2} f(N,P)\right)$ steps; $f(N,P)$ is a non-decreasing function representing the *simulation overhead*. The aim is to minimize this overhead, ideally to a constant.

In this paper, we present a scaling simulation for the general (unconstrained) R-Mesh. This simulation has an overhead of $\log N$ (smaller than the $\log P \log \frac{N}{P}$ overhead of the previous fastest scaling simulation), using a CREW LR-Mesh (a weaker version of the General R-Mesh) as the simulating model; prior simulations needed concurrent write.

1 Introduction

Reconfigurable bus-based models use dynamically alterable connections between processors to create various, changing, bus configurations. This allows efficient communication, and further, allows faster computation than on conventional "non-reconfigurable" models. This paper deals with the ability of a reconfigurable mesh (R-Mesh) [3, 6, 7, 10] to adapt an algorithm instance of an arbitrary size to run on a given smaller model size without significant loss of efficiency. This ability holds particular significance for flexibility in algorithm design and for running algorithms with various input sizes (designed for an arbitrary sized model) on an available model of given size. More formally, for any $P < N$, let a $P \times P$ R-Mesh simulate a step of an $N \times N$ R-Mesh, in $O\left(\frac{N^2}{P^2} f(N,P)\right)$ time, where $f(N,P)$ is a non-decreasing function that represents the *simulation overhead*. If the simulation overhead, $f(N,P)$, is a constant, then the model is said to have an *optimal scaling simulation*. On the other hand, if $f(N,P)$ depends only on P and is independent of N, then the model is said to have a *strong scaling simulation* [5]. The goal of designing a scaling simulation is to reduce

$f(N, P)$, ideally to a constant. Throughout this paper we assume the simulated (resp., simulating) R-Mesh to be of size $N \times N$ (resp., $P \times P$).

Reconfigurable models possess a large body of fast algorithms, yet only a handful of results exist for scaling these models. Previously, Maresca [8] established that the Polymorphic Processor Array (PPA) has an optimal scaling simulation. The PPA restricts the pattern of buses that can be created, severely curtailing the power of the model [14]. Ben-Asher et al. [1] proved that the LRN-Mesh (a restriction of the R-Mesh) also has an optimal scaling simulation. The LRN-Mesh admits only certain patterns of buses, making it unsuitable for fundamental problems such as graph connectivity [2]. Trahan and Vaidyanathan [13] have shown that, for certain restrictions of local connections, the RMBM (a reconfigurable bus-based model) has a strong scaling simulation. Recently, the authors [4,5] have proved that the FR-Mesh (another restriction of the R-Mesh) has a strong scaling simulation with a simulation overhead of $\log P$. These results impose restrictions on the model scaled (though, allowing increase in number of processors, the FR-Mesh is as "powerful" as the R-Mesh [12]).

The best results, so far, on scaling the general (unrestricted) R-Mesh are due to the authors [5] and Matias and Schuster [9]. We developed a scaling simulation for the unrestricted R-Mesh with a simulation overhead of $\log P \log \frac{N}{P}$. Matias and Schuster [9] designed a randomized scaling algorithm for the unrestricted R-Mesh simulated on the LRN-Mesh. This method has a constant (with high probability) simulation overhead, when $P \leq \frac{N}{\log N \log \log N}$; the simulating LRN-Mesh, however, uses the ARBITRARY rule to resolve concurrent writes, a rule not easily implementable on a bus.

In this paper, we construct a deterministic scaling simulation for the unrestricted R-Mesh, which improves on the best previous simulation overhead of $\log P \log \frac{N}{P}$. The key component of this scaling simulation is a novel method to convert a general R-Mesh algorithm to run on an LRN-Mesh of the same size. The conversion identifies structures that comprise a spanning forest of reduced bus configurations, then emulates these with linear buses following Euler tours of the trees. This approach has a $\log N$ simulation overhead. Further, the simulating LRN-Mesh uses only exclusive writes, whereas prior scaling simulations needed concurrent writes. We next extend the algorithm to a lower-overhead simulation of an FR-Mesh and to a simulation by a model using pipelined optical buses (PR-Mesh). Table 1 summarizes the results of this paper.

Table 1. Summary of results for the general R-Mesh and FR-Mesh scaling simulations

$N \times N$ Simulated Model	$P \times P$ Simulating Model	Simulation overhead
CRCW R-Mesh	CREW LRN-Mesh or PR-Mesh	$\log N$
CRCW FR-Mesh	CREW LRN-Mesh or PR-Mesh	$\log P$

Section 2 describes the R-Mesh and some of its versions. Sections 3.1 and 3.2 describe the scaling simulations of an R-Mesh and FR-Mesh using a CREW LRN-Mesh. Section 3.3 presents scaling simulations of an R-Mesh and FR-Mesh using an optical model. Finally, Section 4 identifies some open problems.

2 The Reconfigurable Mesh

An $R \times C$ *Reconfigurable Mesh* (*R-Mesh*) is a two-dimensional array of processors connected in an $R \times C$ grid. Each processor in the R-Mesh has direct "external connections" to adjacent processors through a set of four input/output ports. A processor can internally partition its set of ports so that ports in the same block of a partition are fused. These partitions, along with external connections between processors, define a global bus structure that connects the ports of processors. All ports that are part of the same bus are said to be in the same *component* of the R-Mesh. Figure 1 shows a component in bold. Also, Fig. 1 shows a 3×5 R-Mesh, depicting the fifteen possible port partitions of a processor. The R-Mesh is a synchronous model that may change its bus configurations at each step. It also assumes negligible delay on buses [7, 8, 10].

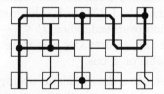

Fig. 1. Port partitions of an R-Mesh. Processor (0,0) is placed on the upper left corner.

The *LRN-Mesh* is a restricted version of the R-Mesh that allows processors to fuse only pairs of ports together or leave ports unfused. (That is, the LRN-Mesh does not permit the connections in processors (0,2), (0,4), (1,0), (1,1), and (2,2) of Fig. 1.) The *FR-Mesh* is another restriction of the R-Mesh that permits only the "fusing" and "cross-over" connections shown in processors (2,2) and (2,3), respectively, of Fig. 1. A component is *linear* iff it connects its ports only as allowed by the LRN-Mesh; otherwise the component is *non-linear*.

In most of this paper, we assume the concurrent read concurrent write (CRCW) model. Let the term *reading (writing) port* to refer to a port through which a processor is reading (writing). The R-Mesh will resolve concurrent writes by any of the following rules: COMMON (where all writing ports of a component must write the same value), COLLISION (where concurrent writes result in a distinguished collision symbol being written on the component), or COLLISION$^+$ (that behaves like the COMMON rule when all processors attempt to write the same value on a component, and like COLLISION otherwise). In this paper, we will also discuss the PRIORITY rule (in which the lowest indexed writing port of a component succeeds in writing) and the ARBITRARY rule (in which an arbitrary port succeeds in writing). PRIORITY and ARBITRARY rules are difficult to implement and will be used only as algorithmic tools. Regardless of the write rule used, the final value on the bus is called the *bus data*.

3 Scaling Simulation via LRN-Mesh

We now design a scaling simulation of the R-Mesh by an LRN-Mesh, where the sizes of both machines have the same order, and then (optimally) scale down the simulating LRN-Mesh. We will later refine this result so that the simulating LRN-Mesh uses only exclusive writes. The simulation allows the powerful and flexible algorithm design permitted by the CRCW model, while using the more feasible bus implementation of the CREW model.

3.1 Simulation Description

We will describe the simulation through the following phases:
1. Simulation of an $N \times N$ CRCW R-Mesh on a $2N \times 2N$ CRCW LRN-Mesh
2. Simulation of an $N \times N$ CRCW R-Mesh on a $P \times P$ CRCW LRN-Mesh
3. Simulation of an $N \times N$ CRCW R-Mesh on a $P \times P$ **CREW** LRN-Mesh

Phase 1. This phase establishes the following result.

- For any $P < N$, any step of an $N \times N$ COMMON, COLLISION, COLLISION$^+$, or PRIORITY CRCW R-Mesh can be simulated on a $2N \times 2N$ COMMON, COLLISION, COLLISION$^+$, or PRIORITY CRCW LRN-Mesh in $O(\log N)$ time.

We first describe the result for the COMMON write rule, then include the other write rules. We now construct a simulation of an $N \times N$ COMMON CRCW R-Mesh, $\mathcal{Q}$, on a $2N \times 2N$ COMMON CRCW LRN-Mesh, $\mathcal{Z}$, in $O(\log N)$ time.

A group of four processors of $\mathcal{Z}$ simulates a processor of $\mathcal{Q}$, where Fig. 2 shows selected configurations of a processor of $\mathcal{Q}$ and the corresponding configurations assumed by a group of $\mathcal{Z}$. Assign to each bus of $\mathcal{Z}$ a direction (in or out) as shown in Fig. 2; this assignment is mainly for ease of explanation and does not require the "more powerful" directional model [2]. Allow each processor in the group to write only to outgoing buses and read only from incoming buses.

Since an LRN-Mesh can simulate linear components directly, we focus our attention on non-linear components. Let the *non-linear graph* of $\mathcal{Q}$ be a graph in which, for each nonlinear component, the processors are nodes and the links between them are edges. The idea in this simulation is to iteratively grow a spanning tree for each component in the non-linear graph of $\mathcal{Q}$. During the spanning tree construction, we simultaneously construct an Euler tour of each tree. The Euler tour enables the LRN-Mesh to handle each tree as a linear bus. For clarity, we describe the algorithm acting on only one component on the non-linear graph.

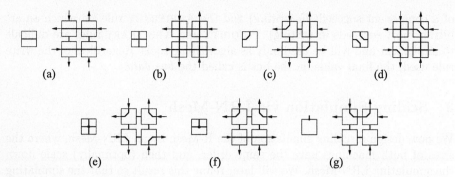

Fig. 2. Group configurations for processors with special types of connections: a-d) linear; e,f) non-linear; g) end-point bus.

We now describe the simulation. For brevity, we omit details.

Step 1 - Bus contraction: This step partially contracts the non-linear graph. It removes each chain that connects a leaf (degree 1 node) to an internal node (degree 3 or 4 node) to construct a *raked graph* (Fig. 3). Before removing these chains, $\mathcal{Z}$ simulates each corresponding bus segment independently and obtains their resulting bus data; these values will be used later during Step 7.

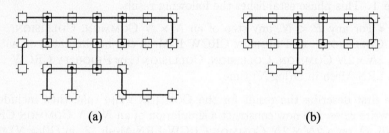

Fig. 3. a) Initial non-linear graph of the component (before the bus contraction of Step 1); b) raked graph (after Step 1).

Step 2 - Initiating spanning tree construction: This step further reduces the raked graph to a *distilled graph* and starts constructing a spanning tree. Let each internal node (degree 3 or 4) and leaf (degree 1) of the raked graph be a node of the distilled graph. Let a chain of degree-2 nodes in the raked graph be an edge in the distilled graph. Denote the corresponding groups of processors in $\mathcal{Z}$ as node groups and edge groups, respectively. Each node group chooses its neighboring node group with smallest index as its parent. Add the edge between each parent and child into a forest (that eventually will be a spanning tree). Denote as a *root group* a group with index smaller than any of its neighbors.

Step 3 - Initial Euler tour construction: This step constructs an Euler tour in each tree of the forest obtained in Step 2 using the configurations shown in

Fig. 2a-d for edge groups and Fig. 2e-g for node groups. The root group of each tree acts as a leader and broadcasts its index to all the node groups in its tree.

Repeat Steps 4 and 5 $2\log N$ times to complete a spanning forest of the distilled graph.

Step 4 - Grafting trees: This step combines trees in the forest by a *graft operation*, in which each tree chooses a neighboring tree with a smaller label and grafts onto it. This operation basically changes the internal connections of a node group in each of the two trees to incorporate the edge between them into the Euler tour.

Step 5 - Grafting rejected trees: It is possible that a tree has not been subject to a graft operation in Step 4, despite having a neighbor in another tree via an unselected edge. Force each such tree to graft onto some neighboring tree.

Step 6 - Adding unselected edges to the spanning tree: This step extends the spanning tree of the distilled graph to a spanning tree of the raked graph (Fig. 4).

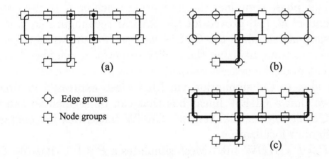

Fig. 4. a) Raked graph (after Step 1); b) Spanning tree of the distilled graph (before Step 6); c) Spanning tree of the raked graph (after Step 6).

Step 7 - Simulating bus communication: This step simulates the communication through the Euler tours of the spanning tree obtained in Step 6. Since an Euler tour connects all the ports of the same component to the same bus, the writer processors of that component can simulate the write rule of $\mathcal{Q}$. Finally, broadcast the resulting bus data to the bus segments contracted in Step 1. This step completes the simulation of the general R-Mesh by the LRN-Mesh. ∎

Example: Figure 4a shows a raked graph after the contraction step (Step 1). The bold bus in Fig. 4b represents the spanning tree of the distilled graph generated after the iterative process of Steps 4 and 5. Notice that all the node groups are part of the spanning tree, but not all of the edge groups. Figure 4c shows the spanning tree of the raked graph after Step 6.

All of the above steps run in constant time. The running time of the algorithm is due to the iterative process on Steps 4 and 5. After Step 3, it is possible to have $O(N^2)$ trees in a component. By grafting these trees in Steps 4 and 5, the number of trees reduces by at least half, so the algorithm runs in $O(\log N)$ iterations to complete the construction of the spanning tree. This algorithm readily adapts to the COLLISION and COLLISION$^+$ rules. When Q uses the PRIORITY rule, solve concurrent writes in Steps 1 and 7 using priority resolution in $O(\log N)$ time.

Phase 2. This phase uses the scaling simulation of Ben-Asher et al. [1] to scale the $N \times N$ COLLISION$^+$ CRCW LRN-Mesh down to a $P \times P$ COLLISION$^+$ CRCW LRN-Mesh. By extending this result to other write rules, we obtain the following.

- For any $P < N$, any step of an $N \times N$ COMMON, COLLISION, COLLISION$^+$, or PRIORITY CRCW R-Mesh can be simulated on a $P \times P$ COMMON, COLLISION, COLLISION$^+$, or PRIORITY CRCW LRN-Mesh in $O\left(\frac{N^2}{P^2} \log N\right)$ time.

Phase 3. This phase simplifies the simulating machine $\mathcal{R}$ to use only exclusive writes. Trahan et al. [14] proved for the RMBM reconfigurable model that the CREW version can simulate the CRCW version (for the write rules considered here) in constant time, utilizing the ability to perform neighbor localization (a procedure that finds the nearest marked processor to a reference processor). We extend this idea to the simulation via LRN-Mesh expressed in Phase 2. It is easy to prove that a CREW linear bus that can be segmented can simulate a COMMON, COLLISION, or COLLISION$^+$ CRCW linear bus in constant time by applying neighbor localization.

Since a $P \times P$ **CREW** LRN-Mesh simulates a $P \times P$ COMMON, COLLISION, or COLLISION$^+$ LRN-Mesh in constant time, then by using the result of Phase 2 we obtain the following theorem.

Theorem 1. *For any $P < N$, any step of an $N \times N$ COMMON, COLLISION, COLLISION$^+$, or PRIORITY CRCW R-Mesh can be simulated on a $P \times P$ **CREW** LRN-Mesh in $O\left(\frac{N^2}{P^2} \log N\right)$ time.* ∎

Although Matias and Schuster [9] also simulated the general R-Mesh via the LRN-Mesh, their simulation is randomized and quite different from the one proposed in this paper. Their simulation computes connected components in two stages. In the first stage, they obtained the connected components in each $P \times P$ sized window of Q. In the second stage, they completed the process, calculating the connected components on a graph with $O\left(\frac{N^2}{P}\right)$ nodes and $O\left(\frac{N^2}{P}\right)$ edges. The connected components algorithm in the second stage is the LRN-Mesh simulation of a randomized PRAM algorithm. On the other hand, our connected components algorithm (spanning forest construction) is deterministic and the input is a graph with $O(N^2)$ nodes and $O(N^2)$ edges. Another important difference is that, to attain the stated overhead in the simulation of Matias and

Schuster, the write rule for the simulating machine must be ARBITRARY, which is difficult to implement in a bus. Our simulation uses only exclusive writes.

3.2 Scaling Simulation of FR-Mesh using LRN-Mesh

The authors [5] previously developed a scaling simulation for the FR-Mesh. By applying Theorem 1 in some portions of this FR-Mesh scaling simulation, an LRN-Mesh can simulate an FR-Mesh with simulation overhead of $O(\log P)$.

Corollary 2. *For any $P < N$, any step of an $N \times N$ COMMON, COLLISION, COLLISION$^+$, or PRIORITY CRCW FR-Mesh can be simulated on a $P \times P$ CREW LRN-Mesh in $O\left(\frac{N^2}{P^2} \log P\right)$ time.*

3.3 Simulation of R-Mesh by PR-Mesh

The Pipelined Reconfigurable Mesh or PR-Mesh [12] is a special type of reconfigurable mesh that uses optical buses. It can configure its port connections to form linear buses as the LRN-Mesh can. Each optical bus can perform several one-to-one communications in constant time by using pipelining. An optical bus consists of two bus segments (upper and lower) connected at one of the ends, forming a directional U-shaped bus. Processors use the upper bus to write and the lower bus to read.

A $P \times P$ PR-Mesh can simulate each step of a $P \times P$ CREW LRN-Mesh with no cycles in constant time as follows. Scale the LRN-Mesh down to a $\frac{P}{2} \times \frac{P}{2}$ LRN-Mesh. Then, replicate the connections of this new LRN-Mesh on the PR-Mesh, creating a double bus structure to decide which of the two end-processors connects the upper and lower segments. Finally, use only one of these two buses to broadcast the written value. (This simulation corresponds to those in earlier papers [11, 12], though they did not explicitly address the leader election problem.) If the LRN-Mesh has cycles, the PR-Mesh can find a leader within the cycle in an additional $O(\log P)$ time, then break the cycle, and proceed as if the LRN-Mesh was acyclic.

Corollary 3. *For any $P < N$, any step of an $N \times N$ COMMON, COLLISION, COLLISION$^+$, or PRIORITY CRCW R-Mesh can be simulated on a $P \times P$ PR-Mesh in $O\left(\frac{N^2}{P^2} \log N\right)$ time.*

Corollary 4. *For any $P < N$, any step of an $N \times N$ COMMON, COLLISION, COLLISION$^+$, or PRIORITY CRCW FR-Mesh can be simulated on a $P \times P$ PR-Mesh in $O\left(\frac{N^2}{P^2} \log P\right)$ time.*

Remark: The $O(\log P)$ time to break cycles of the LRN-Mesh is an additive overhead in the simulation overhead of Corollaries 3 and 4 because the PR-Mesh performs this operation only once per each of the $\left(\frac{N^2}{P^2}\right)$ windows.

4 Open Questions

Open problems include investigating whether or not the General R-Mesh has an optimal scaling simulation or at least a strong scaling simulation. Another research direction is to find configurations (other than LRN-Mesh and FR-Mesh) that are useful computationally and have optimal or strong scaling simulations. Finally, is priority resolution in $o(\log P)$ time possible on a $P \times P$ FR-Mesh using COMMON or COLLISION rules? The answer to this could immediately improve simulation overheads for the R-Mesh and FR-Mesh.

Acknowledgments. This material is based upon work supported in part by the National Science Foundation under Grant No. CCR-9503882.

References

1. Ben-Asher, Y., Gordon, D., Schuster, A.: Efficient Self Simulation Algorithms for Reconfigurable Arrays. J. Parallel Distrib. Comput. **30** (1995) 1–22
2. Ben-Asher, Y., Lange, K.-J., Peleg, D., Schuster, A.: The Complexity of Reconfiguring Network Models. Info. and Comput. **121** (1995) 41–58
3. ElGindy, H., Wetherall, L.: A Simple Voronoi Diagram Algorithm for a Reconfigurable Mesh. IEEE Trans. Parallel Distrib. Systems **8** (1997) 1133–1142
4. Fernández-Zepeda, J.A., Trahan, J.L., Vaidyanathan, R.: Scaling the FR-Mesh under Different Concurrent Write Rules. World Multiconf. on Systemics, Cybernetics, and Informatics (1997) 437–444
5. Fernández-Zepeda, J.A., Vaidyanathan, R., Trahan, J.L.: Scaling Simulation of the Fusing-Restricted Reconfigurable Mesh. IEEE Trans. Parallel Distrib. Systems **9** (1998) 861–871
6. Jang, J.-w., Nigam, M., Prasanna, V. K., Sahni, S.: Constant Time Algorithms for Computational Geometry on the Reconfigurable Mesh. IEEE Trans. Parallel Distrib. Systems **8** (1997) 1–12
7. Li, H., Stout, Q. F.: Reconfigurable SIMD Massively Parallel Computers. IEEE Proceedings **79** (1991) 429–443
8. Maresca, M.: Polymorphic Processor Arrays. IEEE Trans. Parallel Distrib. Systems **4** (1993) 490–506
9. Matias, Y., Schuster, A.: Fast, Efficient Mutual and Self Simulations for Shared Memory and Reconfigurable Mesh. 7th IEEE Symp. Par. Distrib. Processing (1995) 238–246
10. Miller, R., Prasanna-Kumar, V. K., Reisis, D., Stout, Q.: Parallel Computations on Reconfigurable Meshes. IEEE Trans. Comput. **42** (1993) 678–692
11. Pavel, S., Akl, S. G.: On the Power of Arrays with Optical Pipelined Buses. Int'l. Conf. Par. Distr. Proc. Techniques and Appl. (1996) 1443–1454
12. Trahan, J. L., Bourgeois, A. G., Vaidyanathan, R.: Tighter and Broader Complexity Results for Reconfigurable Models. Parallel Processing Letters, special issue on Bus-based Architectures **8** (1998) 271–282
13. Trahan, J. L., Vaidyanathan, R.: Relative Scalability of the Reconfigurable Multiple Bus Machine. Proc. Workshop Reconfigurable Arch. and Algs. (1996)
14. Trahan, J. L., Vaidyanathan, R., Thiruchelvan, R. K.: On the Power of Segmenting and Fusing Buses. J. Parallel Distrib. Comput. **34** (1996) 82–94

Bit Summation on the Reconfigurable Mesh

Martin Middendorf *

Institut für Angewandte Informatik
und Formale Beschreibungsverfahren, Universität Karlsruhe,
D-76128 Karlsruhe, Germany
mmi@aifb.uni-karlsruhe.de

Abstract. The bit summation problem on reconfigurable meshes is studied for the case where the number of bits is as large as the number of processsors. It is shown that kn binary values can be computed on a reconfigurable mesh of size $k \times n$ in time $O(\log^* k + (\log n/\sqrt{k \log k}))$.

1 Introduction

Processor arrays with a dynamically reconfigurable bussytem promise interesting applications because they allow to solve several problems much faster than on most other parallel architectures. The reconfigurability of the bussytem allows to transfer computations from time to space, because the formation of a new bus structure may depend on local data and can therefore be used as part of the computation. Clearly, there is a tradeoff between the size of the mesh and the time that can be achieved to solve a problem. Therefore, many of the fast algorithms for reconfigurable meshes work only for instances that are small compared to the size of the mesh (e.g. constant time summation of n numbers on a mesh of size $n \cdot \log^3 n$ [10]) or for sparse instances (e.g. matrix multiplication where time depends on the degree of sparseness of the matrices [5]).

One of the basic problems on reconfigurable meshes studied by several authors is the summation of binary values where each processor stores at most one bit. Algorithms for this problem are used as building blocks to solve several other problems (e.g. summation of integers [1], matrix multiplication [10]).

There are two models of reconfigurable meshes that differ by the buswidth and the maximal length of the operands for the arithmetic and logic operations that can be performed by the processors. In the bit-model the width of the buses is only one bit and processors can perform only bit operations. Whereas in the word-model the width of buses is at least the logarithm of the number of processors and the processors can perform operations on operands of that length. The bit summation problem has been studied for both models.

For the bit-model Miller et al. [6] and Wang et al. [11] have shown that n binary values can be added in time $O(1)$ on an $n \times n$ bit-model reconfigurable mesh. This was improved by Jang et al. [4] who gave a $O(1)$ algorithm to compute

---

* This work was done during the authors stay at the Faculty of Computer Science, University of Dortmund, Germany

the sum of n^2 bits (one per PE) on a bit-model $n \log n \times n \log n$ reconfigurable mesh. Note, that in both cases the bits of the result value has to be stored in different processors.

For the word-model reconfigurable mesh the following results are known. Jang and Prasanna [2] showed that n bits can be summed in time $O(t)$ on an $2n^{1/t} \times n$ RG for $1 \leq t \leq \log n$. Olariu et al. [9] showed that prefix sums of n binary values can be computed in time $O((\log n/\log m) + 1)$ on an $m \times n$ reconfigurable mesh. Miyashita et al. (cited in [8]) showed how to solve the prefix-sum problem for n binary values in $O(\log \log n)$ time on an $(\log^2 n/(\log \log n)^2) \times n$ reconfigurable mesh. Nakano [8] has shown that the prefix-sums of n binary values can be computed in time $O(\log n/\sqrt{k \log k}) + \log \log n)$ on an $k \times n$ reconfigurable mesh. This result implies that the problem can be solved in time $O(\log \log n)$ on an $(\log^2 n/(\log \log n)^3) \times n$ mesh which improves the result of Miyashita et al.. Nakano [7] also showed that n bits can be summed up in time $O(\log n/\sqrt{k \log k} + 1)$ on an $k \times n$ reconfigurable mesh. The addition of binary values on reconfigurable meshes where the number of processors equals the size of the problem instance was studied by Jang et al. [3]. They showed that an array of $n \times n$ bits can be summed in time $O(\log^* n)$ on an $n \times n$ word-model reconfigurable mesh. If the size of the mesh is allowed to be $\log^2 n \times n^2$ then n^2 bits can be added in constant time on the word-model reconfigurable mesh (Jang et al. [3], Park et al. [10]).

In this paper we study the bit summation problem on word-model reconfigurable meshes for case that the number of bits is as large the number of processors. We generalise the result of Jang et al. [3] by considering $k \times n$ meshes. Moreover, our result partially improves the result of Nakano [7]. We show that kn binary values can be computed in time $O(\log^* k + (\log n/\sqrt{k \log k}))$ on a reconfigurable mesh of size $k \times n$ if $n \geq \log^3 k$. Compared to the result of Nakano we can handle larger problem instances while using the same number of processors and the same time if $(\log n/\sqrt{k \log k}) \geq \log^* k$. The basic idea of our algorithm is to use the method of Nakano first in small submeshes to obtain prefix-remainders and a small number of remaining bits in every submesh. Then, several submeshes are combined to a larger submesh. Again the method of Nakano is applied to obtain prefix-remainders now with respect to a larger modulus and even fewer remaining one bits in every submesh. This process is repeated until we end up with at most n remaining bits in the whole mesh. Then the algorithm of Nakano is applied directly on the whole mesh.

2 Model of Computation

The model of computation is an SIMD $k \times n$-array of processing elements (PE's) with dynamically reconfigurable buses as depicted in Figure 1. We assume that only linear buses can be formed, i.e. every processore can connect only pairs of its ports. Every PE can read from every bus it is connected to, but only one PE at a time can write the value of one of its registers on a bus, i.e. we have CREW-buses. If no PE writes on a bus then its value is 0.

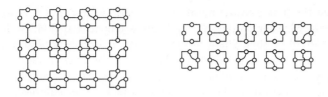

Fig. 1. Reconfigurable mesh (left) and possible connections of ports within a PE (right)

Every PE has a constant number of registers of length $\log n + \log k$. Every PE knows its row and column indices. Within one time step every PE can locally configure the bus, write to and/or read from one of the buses it is connected to, and perform some local computation. Signal propagation on buses is assumed to take constant time regardless of the number of switches on the bus — a standard assumption for this model of computation (e.g. [6]).

3 Nakano's Method

We first describe the algorithm of Nakano for computing the sum x of n binary values $a_1, a_2, \ldots, a_n$ on a reconfigurable mesh of size $k \times n$ ($k \geq 16$). For $k \leq 16$ his result holds trivially. The algorithm makes use of the result that prefix-remainders modulo m of n binary values can be computed in time $O(1)$ on an $(m+1) \times n$ reconfigurable mesh (e.g. Nakano et al. cited in [8]). The idea is to recursively apply the prefix-remainders algorithm using the fact that

$$a_1 + a_2 + \ldots + a_n = (a_1 + a_2 + \ldots + a_n) \bmod z + z \cdot (b_1 + b_2 + \ldots + b_n)$$

where $b_j = 1$ iff $a_j = 1 \wedge (a_1 + a_2 + \ldots + a_j) \bmod z = 0$. Instead of computing prefix-remainders with respect to some number z directly it is more space efficient to compute the prefix-remainders with respect to the prime factors of z. This is done in parallel for all prime factors of z using a submesh of size at least $(p+1) \times n$ for each prime factor p. Number z is chosen as the product of the first q prime numbers $p_1, p_2, \ldots, p_q$ for an integer q defined as follows: q is the maximal value such that $q \leq \lfloor \sqrt{k/\log k} \rfloor$ and $p_q \leq \lfloor \sqrt{k \log k} \rfloor - 1$. By the prime number theorem it can be shown that this definition of q allows to use for every prime factor p_i, $1 \leq i \leq q$ a submesh of size $\lfloor \sqrt{k \log k} \rfloor \times n$ and all this can be done on a mesh of size $k \times n$. By the Chinese Remainder theorem there exists exactly one $y \in [1:z]$ with $(a_1 + a_2 + \ldots + a_n) \bmod p_i = (a_1 + a_2 + \ldots + a_n) \bmod y$ for all $i \in [1:q]$. For this y it holds that $(a_1 + a_2 + \ldots + a_n) \bmod z = y$.

To determine the first q prime numbers each PE P_{ij}, $i < j$, $i \leq (k \log k)^{1/4}$, $j \leq \sqrt{k \log k}$ tests whether i divides j and sets a flag to 1 if it does. If for all $i < j$ i does not divide j then j is a prime number. It is easy to determine this in time $O(1)$ for all $j \leq n$ in parallel by ORing over the flags in every column. For identifying which prime number is the ith one, $1 \leq i \leq q$ a simple counting technique that works in time $O(1)$ can be applied in a $q \times p_q$ submesh.

To determine the number y for which $(a_1 + a_2 + \ldots + a_n) \bmod p_i = (a_1 + a_2 + \ldots + a_n) \bmod y$ holds for all $i \in [1:q]$ the prefix-remainders are computed for all p_i with respect to the sequence consisting of n one bits. Then it is easy to determine in time $O(1)$ in every column j, $j \in [1:z]$ whether $j \bmod p_i = (a_1 + a_2 + \ldots + a_n) \bmod p_i$ for every $i \in [1:q]$.

The computation of z as the product of the first q prime numbers is done by computing in every column j, $j \leq n$ of the mesh whether $j \bmod p_i = 0$ holds for every $i \in [1:q]$. The smallest such j equals z. Clearly this can be done in time $O(1)$. The whole algorithm of Nakano works as follows.

Algorithm Sum-1 (Nakano [7]):

1. Determine the first $q(n)$ prime numbers $p_1, p_2, \ldots, p_q$
2. Compute $z = p_1 \cdot p_2 \cdot \ldots \cdot p_q$
3. Initialise variables $i := 0$ and $x := 0$.
WHILE $OR(a_1, a_2, \ldots, a_n)$ DO
4. Set $i := i+1$ and compute $x_i := (a_1 + a_2 + \ldots + a_n) \bmod z$.
5. For $j \in [1:n]$ set $b_j := 1$ if $a_j = 1$, and $a_1 + \ldots + a_j \bmod z = 0$. Otherwise, set $b_j := 0$.
6. Set $a_j := b_j$ for $j \in [1:n]$.
7. $x := x + z^{i-1} \cdot x_i$

To analyse the running time of algorithm Sum-1 Nakano has shown the following lemma by using the prime number theorem.

Lemma 1. *[Nakano [7]] There exists a constant $c > 0$ and an integer k_0 such that for all $k \geq k_0$ $q/2 \geq c \cdot \sqrt{k/\log k}$ and $p_{q/2} \geq c \cdot \sqrt{k \log k}$.*

Since every of the steps (1) to (7) works in time $O(1)$ it remains to determine how often steps (4) to (7) are executed. Using Lemma 1 it can be shown that there exists a constant $c > 1$ such steps (4) to (7) are executed at most t times where t is the minimal integer with $(c^{\sqrt{k \log k}})^t \geq n$. Hence the running time is $O((\log n / \sqrt{k \log k}) + 1)$.

4 The Algorithm

In this section we describe our algorithm to compute the sum of kn bits $a_1, a_2, \ldots, a_{kn}$ on a $k \times n$ reconfigurable mesh. We assume that every PE stores exactly one of the given bits. We need the following definition: For an integer $m \geq 16$ let $q(m)$ be the maximal value such that $q(m) \leq \lfloor \sqrt{m/\log m} \rfloor$ and $p_{q(m)} \leq \lfloor \sqrt{m \log m} \rfloor - 1$.

To describe the general idea of the algorithm let us assume first that we have to sum only kn/k_0 bits which are stored in the PEs of every k_0th row for some integer $k_0 \geq 16$. The algorithm starts by partitioning the mesh into small horizontal slices each of size $k_0 \times n$. The bits are stored in the PEs of the first rows of the slices. Let $a_{m1}, a_{m2}, \ldots, a_{mn}$ be the bits in the mth slice, $m \in [1 : k/k_0]$. In every slice m the prefix remainders modulo $z_1 := p_1 \cdot \ldots \cdot p_{q(k_0)}$ of the bits in the first row of the slice are computed using the method of Nakano. After setting all those one bits a_{mj} to zero for which $(a_{m1} + a_{m2} + \ldots + a_{mj}) \bmod z_1 \neq 0$ holds there are at most n/z_1 remaining one bits in the first row of each slice. These bits will be added in the next steps. Now the mesh is partitioned into larger slices each containing $k_1 := z_1 \cdot k_0$ rows. All the at most n one bits in each new slice are sent to disjoint PEs of the first row of the slice. Again, prefix-remainders of the bits in the first row are computed in each slice, but this time modulo $z_2 := p_1 \cdot \ldots \cdot p_{q(k_1)}$. The algorithm proceeds recursively in the same manner each time using broader slices until there is only one slice left containing at most n one bits in the first row. In this situation the algorithm of Nakano can be applied directly.

One problem remains, namely, how to compute the (weighted) sum of all remainders which are obtained during each iteration for every slice. It is not clear how to do this using a larger mesh or more time (with respect to $Big - O$). The idea is to circumvent the problem by glueing together all slices used in one step of the algorithm so that they form one long band. In every iteration only one remainder is computed for the whole band.

Now we give a more detailed description of the algorithm called Sum-2. We first assume that we have a $k \times 3n$ mesh and kn bits that have to be summed. The bits are stored in the PEs of the n middle columns, i.e. in PEs P_{ij} with $1 \leq i \leq k$, $n + 1 \leq j \leq 2n$. We call the submesh containing the bits the *centre* and the submesh consisting of the first (last) n columns the *left* (*right*) *side*.

In the following we use the operation "the centre is configured into a band of size $m \times (kn/m)$". This operation is realized by partitioning the centre into slices of size $m \times n$ and connecting the last (first) PE of row i of slice s, for s odd (even), to the last (first) PE of row i of slice $s+1$, $i \in [1 : m]$, $s \in [1 : k/m - 1]$ (for ease of description we assume that m divides k). The sides are used to realize the connections. How this can be done is depicted in Figure 2.

The algorithm starts by configuring the centre into a band of size $(k_0 + 1) \times (kn/(k_0 + 1))$ for some constant $k_0 \geq 16$ (for ease of description we assume $k_0 + 1$ divides kn). For every row of the band the prefix-remainders of the bits in the row modulo k_0 are computed using the method of Nakano. Afterwards all bits are set to zero with the exception of those bits for which the prefix-sum modulo k_0 in the row up to the bit itself is zero. Then we compute the sum of the remainders modulo k_0 of the bits in a row that were obtained one for each row using an ordinary $O(\log k_0)$ time tree structured algorithm. The result is sent to PE P_{11} and stored there in variable x which will later hold the final result.

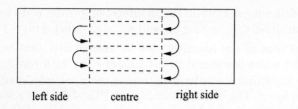 

Fig. 2. Forming a band by connecting slices. Right: detail showing connections between PEs (dark boxes) of two slices using the right side PEs (light boxes)

Observe, that there are at most n/k_0 remaining one bits in every row. The centre is partitioned into slices of size $k_0 \times n$ and the at most $k_0 \times (n/k_0) = n$ one bits in a slice are sent to disjoint PEs of the first row of that slice.

Now the algorithm determines all prime numbers $p_1, p_2 \ldots, p_{q(k)}$ as in the algorithm of Nakano. The centre is configured into a band of size $k_1 \times (kn/k_1)$ with $k_1 := k_0$. Then prefix-remainders modulo $z_1 := p_1 \cdot \ldots \cdot p_{q(k_1)}$ are computed in the band. Afterwards all one bits are set to zero with the exception of those one bits for which the prefix-sum in the band up to the bit itself equals 0 mod z_1. At most $kn/(z_1 \cdot k_0)$ one bits are remaining in the first row of the band. The computed remainder of the bits in the band modulo z_1 is sent to PE P_{11} where it is multiplied by k_0 and added to x.

Now, the centre is partitioned into broader slices containing $k_2 := z_1 \cdot k_0$ rows. All one bits in a slice are sent to disjoint PEs in the first row of the slice. The algorithm proceeds similar as above, i.e. the centre is configured into a band of size $k_2 \times (kn/k_2)$ with, the prefix remainders modulo $z_2 := p_1 \cdot \ldots \cdot p_{q(k_2)}$ are computed and so forth. This procedure is applied recursively on bands of increasing width and decreasing length. The recursion stops when the band has width n. Then the mesh contains at most n remaining one bits which are stored in different PEs of the first row and the algorithm of Nakano can be applied directly to compute the sum of the remaining bits. The obtained result is multiplied by $z_{r-1} \cdot z_{r-2} \cdot \ldots \cdot z_1 \cdot k_0$ and added to variable x in PE P_{11} which stores now the final result.

It remains to describe how the algorithm Sum-2 works on a mesh of size $k \times n$. The idea is to fold the sides that were used to realize the connections into the centre. To make this possible we perform the procedure — called Sum-Submesh — described above four times. The first time we sum only bits stored in PEs of the odd rows and odd columns reserving the other PEs for realizing the connections (see Figure 3). The other three times only bits in PEs of odd rows and even columns (respectively even rows and odd columns, even rows and even columns) are summed. As can be seen from Figure 3 we also need a "free" column before and behind the PEs and a "free" row below the PEs containing the bits that are summed in the procedure described above. Therefore algorithm Sum-2 starts by applying the algorithm of Nakano to sum the bits in the first and last columns and the last row so that they become free (Procedure Sum-Submesh is given later).

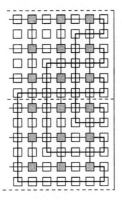

Fig. 3. Connections between two slices

Algorithm: Sum-2:

1. Compute the first prime numbers $p_1, p_2, \ldots, p_{q(k)}$ using a submesh of size $(k \log k)^{1/4} \times \sqrt{k \log k}$.
2. Compute the sum of the bits in the first and last columns, and the last row using algorithm Sum-1 three times.
3. If n and k are both even apply procedure Sum-Submesh four times to the following submeshes (the cases that n or k are odd are similar and omitted here): the submesh consisting of columns 1 to $n-1$ and rows 1 to k, the submesh consisting of columns 1 to n and rows 1 to k, the submesh consisting of columns 1 to $n-1$ and rows 2 to $k-1$, the submesh consisting of columns 1 to $n-1$ and rows 2 to $k-1$.

In the following procedure Sum-Submesh we assume that we have a mesh called centre of size $k \times n$ which contains the bits to be added and also "hidden" columns and rows of PEs that allow to configure the centre as a band. Recall that the prime numbers $p_1, \ldots, p_{q(k)}$ have already been computed by Sum-2 before procedure Sum-Submesh is invoked.

Procedure: Sum-Submesh:

1. Configure the centre as a band of size $(k_0 + 1) \times kn/(k_0 + 1)$.
2. FOR $i = 1$ TO $k_0 + 1$ DO
    Compute the prefix-remainders modulo k_0 of the bits in row i of the band. Let x_i be the remainder modulo k_0 of the bits in row i. Let $a_{i,1} \ldots, a_{i,kn/(k_0+1)}$ be the bits in row i of the band. Set $b_{ij} := 1$, $j \in [1 : kn/(k_0+1)]$ if $a_{ij} = 1$ and $a_{i1} + \ldots + a_{ij} \bmod k_0 = 0$ and, otherwise, set $b_{ij} := 0$.

3. Determine the sum y of all x_i that were computed in Step (2) by a simple $O(\log k_0)$ tree structured algorithm. Send the result to PE P_{11} and set $x := y$.
4. Partition the centre into slices of size $k_0 \times n$ and send all b_{ij}'s with value one into disjoint PEs of the first row of the slice as follows: In row m, $m \in [1 : k_0]$ there is at most one 1-bit b_{ij} in columns $1 + (l-1) \cdot k_0$ to $l \cdot k_0$ for each $l \in [1 : n/k_0]$. If there is such a 1-bit send it first in its row to the PE in column $m + (l-1) \cdot k_0$ and then along the column to the PE in the first row of the slice.
5. Set $r = 1$ and $k_1 = k_0$.

WHILE $k_r < k$ DO
6. Determine $z_r := p_1 \cdot p_2 \cdot \ldots \cdot p_{q(k_r)}$ and send it to PE P_{11}.
7. Configure a $k_r \times kn/k_r$ band. Compute the prefix remainders modulo z_r of the bits in the first row. Let $a_1, a_2, \ldots, a_{kn/k_r}$ be the bits in the first row of the band. For $i \in [1 : kn/k_r]$ set $b_i := 1$ if $a_i = 1$ and $a_1 + \ldots + a_i \bmod z_r = 0$ and otherwise, set $b_i := 0$. Send the obtained remainder modulo z_r off all bits in the first row to PE P_{11}, multiply it by $z_{r-1} \cdot \ldots \cdot z_1 \cdot k_1$ and add it to variable x.
8. Partition the centre into slices of size $k_{r+1} \times n$ where $k_{r+1} := k_r \cdot z_r$. Similar as in Step (4) send all b_i's which are one into disjoint PEs of the first row of it's slice.
9. Set $r := r + 1$
10. Compute the sum of the remaining n one bits using algorithm Sum-1. Send the result to PE P_{11}, multiply it by $z_r \cdot \ldots \cdot z_1 k_1$ and add it to variable x.

The correctness of algorithm Sum-2 is easy to show. Let us analyse the running time. Step (1) of Sum-2 takes time $O(1)$. Step (2) applies the algorithm Sum-1 and therefore takes time $O(\log n/\sqrt{k \log k} + 1)$. The final Step (3) invokes procedure Sum-Submesh four times. It remains to analyse the running time of Sum-Submesh. Step (1) of Sum-Submesh takes constant time. Step (2) takes time $O(k_0 + 1)$ which is $O(1)$ since k_0 is constant. Step (3) is done in time $O(\log k_0)$ which is constant. Each of the steps (4) - (9) can be done in time $O(1)$. Step (10) applies the algorithm Sum-1 and therefore needs time $O(\log n/\sqrt{k \log k} + 1)$. It remains to analyse how often the steps (6) - (9) are repeated. By Lemma 1 there exists a constant $c > 0$ and an integer $\hat{k}$ such that for all $k \geq \hat{k}$ we have $q/2 \geq c \cdot \sqrt{k/\log k}$ and $p_{q/2} \geq c \cdot \sqrt{k \log k}$. Thus, for $k_0 = k_1 \geq \hat{k}$ we have that $\log(p_1 \cdot p_2 \cdot \ldots \cdot p_{q(k_1)}) \geq c \cdot (q(k_1)/2) \cdot \log p_{q(k_1)/2} \geq c \cdot \sqrt{k_1/\log k_1} \cdot \log(c \cdot \sqrt{k_1 \log k_1}) \geq \frac{1}{2}c \cdot \sqrt{k_1 \log k_1}$. This implies that $z_1 = p_1 \cdot \ldots \cdot p_{k_1} \geq \hat{c}^{\sqrt{k_1 \log k_1}}$ for some constant $\hat{c}$. Hence $k_2 \geq k_1 \cdot \hat{c}^{\sqrt{k_1 \log k_1}}$. Iteratively one can show that $k_j \geq k_{j-1} \cdot \hat{c}^{\sqrt{k_{j-1} \log k_{j-1}}}$, $j \in [2 : r]$. Thus for k_1 large enough, but still constant, we have $k_j \geq k_{j-1}\hat{c}^{\sqrt{k_{j-1} \log k_{j-1}}} \geq 2^{\sqrt{k_{j-1}}}$. Thus $k_j \geq 2^{\sqrt{2^{\sqrt{k_{j-2}}}}}$. For k_{j-2} large enough we obtain $k_j \geq 2^{k_{j-2}}$. Thus for k_0 large enough we have $r/2 \leq \log^* k = O(\log^* k)$. Altogether we obtain that the running time of algorithm Sum-2 is $O(\log^* k + \log n/\sqrt{k \log k})$ and the following theorem is shown.

Theorem 1. *The sum of kn bits can be computed in $O(\log^* k + \log n/\sqrt{k \log k})$ time on a reconfigurable mesh of size $k \times n$ using only linear buses.*

5 Conclusion

We have shown that kn binary values can be computed on a reconfigurable mesh of size $k \times n$ in time $O(\log^* n + (\log n/\sqrt{k \log k}))$. This result partially improves a result of Nakano who has shown for a reconfigurable mesh of the same size as ours that the sum of n can be computed in time $O(\log n/\sqrt{k \log k}) + 1)$. That is we can add more bits on a reconfigurable mesh of the same size in the same time if $(\log n/\sqrt{k \log k}) \geq \log^* k$.

References

1. P. Fragopoulou, On the efficient summation of n numbers on an n-processor reconfigurable mesh. *Par. Proc. Lett.* 3: 71-78, 1993.
2. J. Jang, V.K. Prasanna. An optimal sorting algorithm on reconfigurable mesh. In: *Proc. Int. Parallel processing Symp.*, 130-137, 1992.
3. J.-W. Jang, H. Park, V.K. Prasanna. A fast algorithm for computing histogram on reconfigurable mesh. In: *Proc. of Frontiers of Massively parallel Computation '92*, 244-251, 1992.
4. J.-W. Jang, H. Park, V.K. Prasanna. A bit model of reconfigurable mesh. In :*Proc. Workshop on Reconfigurable Architectures*, Cancun, Mexico, 1994.
5. M. Middendorf, H. Schmeck, H. Schröder, G. Turner. Multiplication of matrices with different sparseness properties. to appear in: *VLSI Design*.
6. R. Miller, V.K. Prasanna-Kumar, D.I. Reisis, Q.F. Stout. Parallel Computations on Reconfigurable Meshes. *IEEE Trans. Comput.* 42: 678-692, 1993.
7. K. Nakano. Efficient summing algorithms for a reconfigurable mesh. In: *Proc. Workshop on Reconfigurable Architectures*, Cancun, Mexico, 1994.
8. K. Nakano. Prefix-sum algorithms on reconfigurable meshes. *Par. Process. Lett.*, 5: 23-35, 1995.
9. S. Olariu, J. L. Schwing, J. Zhang. Fundamental algorithms on reconfigurable meshes. Proc. 29th Allerton Conference on Communications, Control, and Computing, 811-820, 1991.
10. H. Park, H.J. Kim, V.K. Prasanna. An $O(1)$ time optimal algorithm for multiplying matrices on reconfigurable mesh. *Inf. Process. Lett.*, 47: 109-113, 1993.
11. B.F. Wang, G.H. Chen, H. Li. Configurational computation: A new computation method on processor arrays with reconfigurable bus systems. In: *Proc. 1st Workshop on Parallel Processing*, Tsing-Hua University, Taiwan, 1990.

Scalable Hardware-Algorithms for Binary Prefix Sums[*]

R. Lin[1], K. Nakano[2], S. Olariu[3], M. C. Pinotti[4], J. L. Schwing[5], and A. Y. Zomaya[6]

[1] Department of Computer Science, SUNY Geneseo, Geneseo, NY 14454, USA
[2] Department of Electrical and Computer Engineering, Nagoya Institute of Technology, Showa-ku, Nagoya 466-8555, Japan
[3] Department of Computer Science, Old Dominion University, Norfolk, VA 23529, USA
[4] I.E.I., C.N.R, Pisa, ITALY
[5] Department of Computer Science, Central Washington University, Ellensburg, WA 98926, USA
[6] Parallel Computing Research Lab, Dept. of Electrical and Electronic Eng, University of Western Australia, Perth, AUSTRALIA

Abstract. The main contribution of this work is to propose a number of broadcast-efficient VLSI architectures for computing the sum and the prefix sums of a w^k-bit, $k \geq 2$, binary sequence using, as basic building blocks, linear arrays of at most w^2 shift switches. An immediate consequence of this feature is that in our designs broadcasts are limited to buses of length at most w^2 making them eminently practical. Using our design, the sum of a w^k-bit binary sequence can be obtained in the time of $2k-2$ broadcasts, using $2w^{k-2} + O(w^{k-3})$ blocks, while the corresponding prefix sums can be computed in $3k-4$ broadcasts using $(k+2)w^{k-2} + O(kw^{k-3})$ blocks.

1 Introduction

Recent advances in VLSI have made it possible to implement algorithm-structured chips as building blocks for high-performance computing systems. Since computing binary prefix sums (BPS) is a fundamental computing problem [1, 4], it makes sense to endow general-purpose computer systems with a special-purpose parallel BPS device, invoked whenever its services are needed. Recently, Blelloch [1] argued convincingly that *scan operations* - that boil down to parallel prefix computation - should be considered primitive parallel operations and, whenever possible, implemented in hardware. In fact, scans have been implemented in microcode in the Thinking Machines CM-5 [10].

Given an n-bit binary sequence $a_1, a_2, \cdots, a_n$, it is customary to refer to $p_j = a_1 + a_2 + \cdots + a_j$ as the j-th *prefix sum*. The BPS problem is to compute all the prefix sums $p_1, p_2, \cdots, p_n$ of A. In this article, we address the problem of designing efficient and scalable hardware-algorithms for the BPS problem. We adopt, as the central theme of this effort, the recognition of the fact that the delay incurred by a signal propagating along a medium is, at best, linear in the distance traversed. Thus, our main design criterion is to keep buses as short as possible. In this context, the main contribution of this work is to show that we can use short buses in conjunction with the shift switching

---

[*] Work supported by NSF grants CCR-9522092 and MIP-9630870, by NASA grant NAS1-19858, by ONR grants N00014-95-1-0779 and N00014-97-1-0562 and by ARC grant 04/15/412/194.

technique introduced recently by Lin and Olariu [5] to design a scalable hardware-algorithm for BPS. As a byproduct, we also obtain a scalable hardware-algorithm for a parallel counter, that is, for computing the sum of a binary sequence (BS).

A number of algorithms for solving the BS and BPS problems on the reconfigurable mesh have been presented in the literature. For example, Olariu et al.[8] showed that an instance of size m of the BPS problem can be solved in $O(\frac{\log n}{\log m})$ time on a reconfigurable mesh with $2n \times (m+1)$, $(1 \leq m \leq n)$, processors. Later, Nakano [6, 7] presented $O(\frac{\log n}{\sqrt{m \log m}})$ time algorithms for the BP and BPS problems on a reconfigurable mesh with $n \times m$ processors. These algorithms exploit the intrinsic power of reconfiguration to achieve fast processing speed. Unfortunately, it has been argued recently that the reconfigurable mesh is unrealistic as the broadcasts involved occur on buses of arbitrary number of switches.

Our main contribution is to develop scalable hardware-algorithms for BS and BPS computation under the realistic assumption that blocks of size no larger than $w \times w^2$ are available. We will evaluate the performance of the hardware-algorithms in terms of the number of broadcasts executed on the blocks of size $w \times w^2$. Our designs compute the sum and the prefix sums of w^k bits in $2k-2$ and $3k-4$ broadcasts for any $k \geq 2$, respectively. Using pipelining, one can devise a hardware-algorithm to compute the sum of a kw^k-bit binary sequence in $3k + \lceil \log_w k \rceil - 3$ broadcasts using $2w^{k-2} + O(kw^{k-3})$ blocks. Using this design, the corresponding prefix sums can be computed in $4k + \lceil \log_w k \rceil - 5$ broadcasts using $(k + \lceil \log_w k \rceil + 2)w^{k-2} + O(kw^{k-3})$ blocks. Due to the stringent page limitation, we omit the description of the pipelining-scaled architecture.

2 Linear arrays of shift switches

An integer z will be represented *positionally* as $x_n x_{n-1} \cdots x_2 x_1$ in various ways as described below. Specifically, for every i, $(1 \leq i \leq n)$, *Binary:* $x_i = \lfloor \frac{z}{2^{i-1}} \rfloor \bmod 2$; *Unary:* $x_i = 1$ if $i = z+1$ and 0 otherwise; *Filled unary:* $x_i = 1$ if $1 \leq i \leq z$ and 0 otherwise, *Distributed:* x_i is either 0 or 1 and $\sum_{i=1}^{n} x_i = z$; *Base w:* Given an integer w, $(w \geq 2)$, $x_i = \lfloor \frac{z}{w^{i-1}} \rfloor \bmod w$; *Unary base w:* Exactly like the base w representation except that x_i is represented in unary form. The task of converting back and forth between these representation is straightforward and can be readily implemented in VLSI [2, 3, 9]

Fix a positive integer w, typically a small power of 2. At the heart of our designs lays a simple device that we call the $S(w)$. Referring to Figure 1, the input to the $S(w)$ consists of an integer b, $(0 \leq b \leq w-1)$, in unary form and of a bit a. The output is an integer c, $(0 \leq c \leq w-1)$, in unary form, along with a bit d, such that

$$c = (a+b) \bmod w \text{ and } d = \left\lfloor \frac{a+b}{w} \right\rfloor. \qquad (1)$$

We implement the $S(w)$ using the shift switch concept proposed in [5]. As illustrated in Figure 1, in this implementation, the $S(w)$ features w switching elements with the state changes controlled by a 1-bit state register. Suppose that $b_w b_{w-1} \ldots b_1$ and $c_w c_{w-1} \ldots c_1$ are the unary representation of b and c, respectively. The $S(w)$ cyclically

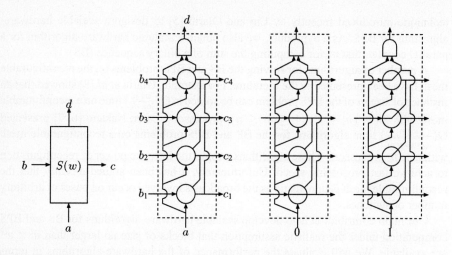

Fig. 1. *Illustrating the functional view of the $S(w)$ and its shift switch implementation*

shifts b if $a = 1$; otherwise b is not changed. Notice that this corresponds to the modulo w addition of a to the unary representation of b. Consequently, $d = 1$ if and only if $b_w = 1$ and $a = 1$. For this reason d is usually referred to as the *rotation bit*.

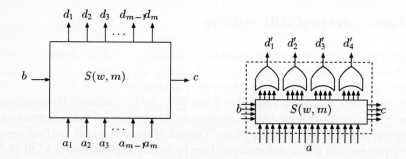

Fig. 2. *A functional view of the array $S(w, m)$ and of block $T(w, m)$*

In order to exploit the power and elegance of the $S(w)$, we construct a *linear array* $S(w, m)$ consisting of m $S(w)$s as illustrated in the left part of Figure 2.

Suppose that $b_w b_{w-1} \cdots b_1$ and $c_w c_{w-1} \cdots c_1$ are the unary representations of b and c, respectively. Clearly, (1) guarantees that

$$c = (a_1 + a_2 + \cdots + a_m + b) \bmod w \qquad (2)$$

and that
$$d = d_1 + d_2 + \cdots + d_m = \left\lfloor \frac{a_1 + a_2 + \cdots + a_m + b}{w} \right\rfloor. \qquad (3)$$

In the remainder of this work, when it comes to performing the sum or the prefix sums of a binary sequence $a_1, a_2, \cdots, a_m$, we will supply this sequence as the bit input of some a suitable linear array $S(w, m)$ and will insist that $b = 0$, or, equivalently, $b_w b_{w-1} \cdots b_2 b_1 = 00 \cdots 01$. It is very important to note that, in this case, (2) and (3) are computing, respectively,

$$c = (a_1 + a_2 + \cdots + a_m) \bmod w \text{ and } d = \left\lfloor \frac{a_1 + a_2 + \cdots + a_m}{w} \right\rfloor.$$

Note that the collection of rotation bits $d_1, d_2, \ldots, d_m$ of the $S(w, m)$ is sparse in the sense that among any group of consecutive w bits in $d_1, d_2, \ldots, d_m$, there is at most one 1. In other words, with m' standing for $\lceil \frac{m}{w} \rceil$, we have for every i, $(1 \leq i < m')$,

$$d_{(i-1)\cdot w+1} + d_{(i-1)\cdot w+2} + \cdots + d_{i\cdot w} = d_{(i-1)\cdot w+1} \vee d_{(i-1)\cdot w+2} \vee \cdots \vee d_{i\cdot w} \qquad (4)$$

and, similarly,

$$d_{(m'-1)\cdot w+1} + d_{(m'-1)\cdot w+2} + \cdots + d_{m'\cdot w} = d_{(m'-1)\cdot w+1} \vee \cdots \vee d_m. \qquad (5)$$

Now (4) and (5) imply that by using $\lceil \frac{m}{w} \rceil$ OR gates with fan-in w, $d_1 + d_2 + \cdots + d_m$ is converted to an equivalent $\lceil \frac{m}{w} \rceil$-bit distributed representation. Equipped with $\lceil \frac{m}{w} \rceil$ OR gates as above, the linear array $S(w, m)$ will be denoted by $T(w, m)$. We refer the reader to Figure 2 for an illustration. Consequently, we have the following result.

Lemma 1. *Given the w-bit unary representation of b and the m-bit distributed representation of a, a block $T(w, m)$ can compute the w-bit unary representation of $c = (a + b) \bmod w$ and the $\lceil \frac{m}{w} \rceil$-bit distributed representation of $d = \lfloor \frac{a+b}{w} \rfloor$ in one broadcast operation over buses of length m.*

3 Computing sums on short buses

Our design relies on a new block $U(w, w^k)$, $(k \geq 1)$, whose input is a w^k-bit binary sequence $a_1, a_2, \cdots, a_{w^k}$ and whose output is the unary base w representation $A_{k+1}A_k \cdots A_1$ of the sum $a_1 + a_2 + \cdots + a_{w^k}$. We note that A_{k+1} is 1 only if $a_1 = a_2 = \cdots = a_{w^k} = 1$. Recall that for every i, $(1 \leq i \leq k)$, $0 \leq A_i \leq w - 1$.

The block $U(w, w^k)$ is defined recursively as follows. $U(w, w^1)$ is just a $T(w, w)$. $U(w, w^2)$ is implemented using blocks $T(w, w)$ and $T(w, w^2)$: The w^2 input are supplied to $T(w, w^2)$. $T(w, w^2)$ outputs the LSD (Least Significant Digit) of the sum in unary base w representation, and the MSD (Most Significant Digit) of the sum in distributed representation. $T(w, w)$ is used to convert the distributed representation into the unary base w representation of the MSD of the sum.

Let $k \geq 3$ and assume that the block $U(w, w^{k-1})$ has already been constructed. We now detail the construction of the block $U(w, w^k)$ using w blocks $U(w, w^{k-1})$. The reader is referred to Figure 3 for an illustration of this construction for $w = 4$ and $k = 5$.

For an arbitrary i, $(1 \leq i \leq w)$, let $A_k^i A_{k-1}^i \cdots A_1^i$ denote the base w representation of the output of the i-th block $U(w, w^{k-1})$. We have

$$A_1 = (A_1^1 + A_1^2 + \cdots + A_1^w) \bmod w \text{ and } D_1 = \left\lfloor \frac{A_1^1 + A_1^2 + \cdots + A_1^w}{w} \right\rfloor.$$

An easy argument asserts that for every i, $(2 \leq i \leq k)$,

$$A_i = (A_i^1 + A_i^2 + \cdots + A_i^w + D_{i-1}) \bmod w \text{ and } D_i = \left\lfloor \frac{A_i^1 + A_i^2 + \cdots + A_i^w + D_{i-1}}{w} \right\rfloor.$$

Similarly, $A_{k+1} = D_k$. The equations above confirm that the design in Figure 3 computes the unary base w representation of the sum $a_1 + a_2 + \cdots + a_{w^k}$. Specifically, the rightmost $T(w, w^2)$ computes A_1. It receives the unary representation of 0 (i.e. $00 \cdots 01$) from the left and adds the filled unary representation of $A_1^1, A_1^2, \ldots, A_1^w$ to A_1^1. The output emerging from the right is the unary representation of A_1 and the rotation bits are exactly the distributed representation of D_1. The next $T(w, w^2)$ computes A_2 in the same manner. Continuing in the same fashion, $k - 1$ blocks $T(w, w^2)$ compute $A_1, A_2, \ldots, A_{k-1}$. Since $A_k^1, A_k^2, \ldots, A_k^w$ consist of 1 bit each and since D_{k-1} involves w bits, a block $T(w, 2w)$ is used to compute A_k. In this figure, the circles indicate circuitry that converts from unary to filled unary representation.

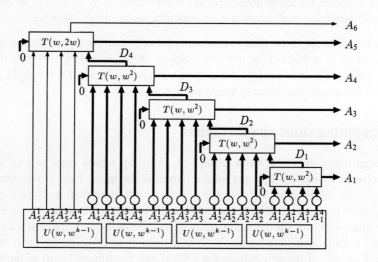

Fig. 3. *Illustrating the recursive construction of $U(w, w^k)$*

At this time it is appropriate to estimate the number of broadcasts involved in the above computation.

Lemma 2. *For every k, $(k \geq 2)$, and for every i, $(1 \leq i \leq k)$, the block $U(w, w^k)$ returns the unary base w representation of A_i at the end of $k + i - 2$ broadcasts and that of A_{k+1} at the end of $2k - 2$ broadcasts. Moreover, all the broadcasts take place on buses of length at most w^2.*

Proof. The proof is by induction on k. The basis is easy: the construction of $U(w, w^2)$ returns A_1 in one broadcast and A_2 and A_3 in two broadcasts. Thus, the lemma holds for $k = 2$. For the inductive step, assume that the lemma holds for $U(w, w^{k-1})$. Our construction of $U(w, w^k)$ guarantees that, after receiving $A_1^1, A_1^2, \ldots, A_1^w$, A_1 and D_1 are available in one further broadcast. Since $A_1^1, A_1^2, \ldots, A_1^w$ are computed in $k - 2$ broadcasts, it follows that A_1 is computed in $k - 1$ broadcasts.

Similarly, A_2 and D_2 are computed in one broadcast after receiving $A_2^1, A_2^2, \ldots, A_2^w$, and D_1. Since all of them are computed in $k - 1$ broadcasts, A_2 and D_2 are computed in k broadcasts. Continuing similarly, it is easy to see that A_i, $(1 \leq i \leq k)$, is computed in $k + i - 2$ broadcasts and that A_{k+1} is computed in $2k - 2$ broadcasts, as claimed.

We now estimate the number of blocks of type $T(w, m)$ with $m \leq w^2$ involved in the construction of $U(w, w^k)$.

Lemma 3. *A block $U(w, w^k)$, $(k \geq 2)$, involves $2w^{k-2} + O(w^{k-3})$ blocks $T(w, m)$ with $m \leq w^2$.*

Proof. Let $g(k)$ be the number of blocks used to construct $U(w, w^k)$. $U(w, w^2)$ uses 2 blocks $T(w, w)$ and $T(w, w^2)$. Thus, $g(2) = 2$. As illustrated in Figure 6, $U(w, w^k)$ uses w $U(w, w^{k-1})$s, $k-1$ $T(w, w^2)$s, and one $T(w, 2w)$. Hence, $g(k) = w \cdot g(k-1) + k$ holds for $k \geq 2$. By solving this recurrence we obtain $g(k) = 2w^{k-2} + O(w^{k-3})$, thus completing the proof of the lemma.

Now, Lemmas 2 and 3, combined, imply the following important result.

Theorem 4. *The sum of an w^k-bit binary sequence can be computed in $2k-2$ broadcasts using $2w^{k-2} + O(w^{k-3})$ blocks $T(w, m)$, with $m \leq w^2$.*

4 Computing prefix sums on short buses

The main goal of this section is to propose an efficient hardware-algorithm for BPS computation on short buses.

Observe that each block $U(w, w^k)$ has, essentially, the structure of a w-ary tree as illustrated in Figure 4. The tree has nodes at $k - 1$ levels from 2 to k. For an arbitrary node v at level h in this w-ary tree, let $a_{s+1}, a_{s+2}, \ldots, a_{s+w^h}$ be the input offered at the leaves of the subtree rooted at v and refer to Figure 5. Let $t(v)$ and $p(v)$ denote the *sum of v* and the *prefix sum of v* defined as follows: $t(v) = a_{s+1} + a_{s+2} + \cdots + a_{s+w^h}$, and $p(v) = a_1 + a_2 + \cdots + a_s$. The blocks corresponding to v can compute $t(v)$ locally, but not $p(v)$. Let $v_1, v_2, \ldots, v_w$ be the w children of v specified in left-to-right order. Clearly, for every j, $(1 \leq j \leq w)$, we have $p(v_j) = p(v) + t(v_1) + t(v_2) + \cdots + t(v_{j-1})$. Hence, by computing $p(v), t(v_1), t(v_2), \ldots, t(v_{w-1})$, we obtain the prefix sum of each of the children of v. This computation proceeds from the root down to the leaves. Once

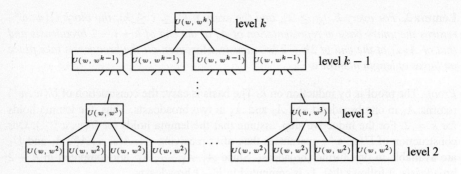

Fig. 4. *Illustrating the w-ary tree structure of $U(w, w^k)$ for $w = 4$*

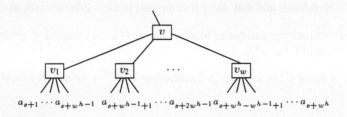

Fig. 5. *A node v and its w children $v_1, v_2, \ldots, v_w$*

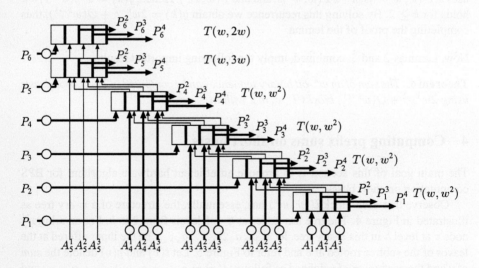

Fig. 6. *Illustrating a design for computing the prefix sums $P_k^j P_{k-1}^j \cdots P_1^j$*

this is done, by computing the local prefix sums at the leaves, we can get the resulting prefix sums of the input sequence.

We now show how this idea can be implemented efficiently. We begin by using a block $U(w, w^k)$ to compute the unary base w representation of $t(v)$ for every node v in the w-ary tree. For this purpose, let $P_{k+1} P_k P_{k-1} \cdots P_1$ be the unary base w representation of the prefix sum $p(v)$ of node v. As before, let $v_1, v_2, \ldots, v_w$ be the children of v, and assume that the subtree rooted at v has w^h input bits. Since the subtree rooted at each v_j has w^{h-1} input bits, let each $A_h^j A_{h-1}^j \cdots A_1^j$, $(1 \le j \le w)$, denote the unary base w representations of the sum $t(v_j)$. Note that the most significant digit A_h^j is either 0 or 1. Further, let $P_{k+1}^j P_k^j P_{k-1}^j \cdots P_1^j$, $(1 \le j \le w)$, be the unary base w representations the prefix sum $p(v_j)$ of v_j. Notice that, for all prefix sums except, perhaps, the last one, the most significant digits P_{k+1} and P_{k+1}^j are always 0 and will be ignored. Our design will have, therefore, to solve the following problem:

Input: The prefix sum $p(v) = P_k P_{k-1} \cdots P_1$ of v and the sum $t(v_j) = A_h^j A_{h-1}^j \cdots A_1^j$ of each child v_j, $(1 \le j \le w - 1)$ of v;
Output: The prefix sum $p(v_j) = P_k^j P_{k-1}^j \cdots P_1^j$ of each child v_j, $(2 \le j \le w)$, of v.

Note that we do not have to compute $p(v_1) = P_k^1 P_{k-1}^1 \cdots P_1^1$ because $p(v_1) = p(v)$. It is easy to confirm that, once this task is completed, the same task can be performed for the children $v_1, v_2, \ldots, v_w$ in order to compute the prefix sums of the grandchildren of v. By continuing in the same fashion until the leaves are reached, we get the final prefix sums. Figure 6 illustrates this task for $w = 4$, $h = 5$, and $k = 6$.

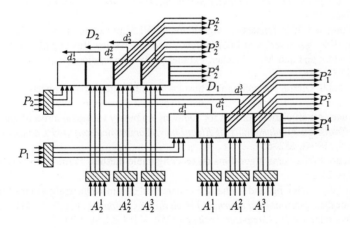

Fig. 7. *Illustrating the details of the prefix sums design of Figure 6*

Let us now estimate the number of broadcasts used by our design. Each child v_j of the root r of the w-ary tree corresponding to the $U(w, w^k)$ outputs its unary $t(v_j) = A_k^j A_{k-1}^j \cdots A_1^j$ in $2k - 4$ broadcasts. By Lemma 2 A_i^j, $(1 \le i \le k - 1)$, is obtained in $k + i - 3$ broadcasts, while A_k^j is returned in $2k - 4$ broadcasts. By the construction in Figure 6, P_i^j, $(1 \leq i \leq k)$, of the child v_i of the root r is computed in $k + i - 2$ broadcasts. Similarly, each of the P_i^j in the children of the root, i.e. nodes at level $k - 1$, is computed in $k + i - 1$ broadcasts. Finally, the unary base w representation of P_i^j in the leaves, i.e. nodes at level 0, is computed in $2k + i - 4$ broadcasts. Thus, we have the following result.

Lemma 5. *Our prefix sums design returns P_i^j, $(1 \leq i \leq k; 1 \leq j \leq w)$, in at most $2k + i - 4$ broadcasts.*

Thus, all the prefix sums can be computed in $3k - 4$ broadcasts. Next, we estimate the number of blocks used. Since each node uses k blocks, and the tree has $w^{k-2} + w^{k-3} + \cdots + w^0$ nodes, this construction adds $kw^{k-2} + O(kw^{k-3})$ blocks. Since by Theorem 4 $U(w, w^k)$ uses $2w^{k-2} + O(w^{k-3})$, a total number of $(k+2)w^{k-2} + O(kw^{k-3})$ blocks are used. Thus we have the following result.

Theorem 6. *The prefix sums of an w^k-bit binary sequence can be computed in $3k - 4$ broadcasts using $(k+2)w^{k-2} + O(kw^{k-3})$ blocks, with all the broadcasts taking place on buses of length at most w^2.*

References

1. G. Blelloch, Scans as primitives parallel operations, *IEEE Transactions on Computers*, C-18, (1989), 1526–1538.
2. J. J. F. Cavanaugh, *Digital Computer Arithmetic Design and Implementation*, McGraw-Hill, New York, 1984.
3. H. T. Kung and C. E. Leiserson, Algorithms for VLSI processor arrays, in C. Mead and L. Conway, Eds, *Introduction to VLSI Systems*, Addison-Wesley, Reading MA, 1980.
4. R. E. Ladner and and M. J. Fischer, Parallel prefix computation, *Journal of the ACM*, 27, (1980), 831–838.
5. R. Lin, and S. Olariu, Reconfigurable buses with shift switching – architectures and applications, *IEEE Transactions on Parallel and Distributed Systems*, 6, (1995), 93–102.
6. K. Nakano, An efficient algorithm for summing up binary values on a reconfigurable mesh, *IEICE Trans. on Fundamentals of Electronics, Communications and Computer Sciences*, E77-A, 4, (1994), 652–657.
7. K. Nakano, Prefix-sums algorithms on reconfigurable meshes, *Parallel Processing Letters*, **5**, (1995), 23-35.
8. S. Olariu, J. L. Schwing, and J. Zhang, Fundamental data movement algorithms for reconfigurable meshes, *International Journal of High Speed Computing*, 6, (1994), 311–323.
9. E. E. Swartzlander, Jr., *Computer Arithmetic* Vol. 1, IEEE CSP, 1990.
10. Thinking Machines Corporation, Connection machine parallel instruction set (PARIS), July 1986.

Configuration Sequencing with Self Configurable Binary Multipliers

Mathew Wojko
Dept. of Electrical & Computer Engineering
The University of Newcastle
Callaghan, NSW 2308, Australia
mwojko@ee.newcastle.edu.au

Hossam ElGindy
School of Computer Science and Engineering
University of New South Wales
Sydney, NSW 2052, Australia
hossam@cse.unsw.edu.au

Abstract.
In this paper we present a hardware design technique which utilises runtime reconfiguration for a particular class of applications. For a multiplication circuit implemented within an FPGA, a specific instance of multiplying by a constant provides a significant reduction of required logic when compared to the generic case when multiplying any two arbitrary values. The use of reconfiguration allows the specific constant value to be updated, such that at any time instance the constant multiplication value will be fixed, however over time this constant value can change via reconfiguration. Through investigation and manipulation of the sequence of required multiplication operations for given applications, sequences of multiplication operations can be obtained where one input changes at a rate slower than the other input. That is one input to the multiplier is fixed for a set number of cycles, hence allowing it to be configured in hardware as a constant and reconfigured at the periodicity of its change. Applications such as the IDEA encryption algorithm and every cycle Adaptive FIR filtering are presented which utilise this reconfiguration technique providing reduced logic implementations while not compromising the performance of the design.

1 Introduction

Since the introduction of reconfigurable computing as a viable computing alternative, several reconfigurable computing techniques have been suggested for applications which have been previously performed/executed by existing well known computing paradigms [2, 6, 10, 9, 13]. While reconfigurable computing defines its own computing paradigm, there are varying techniques which in general categorise its use and application. In this paper we provide a case for the use of Run-Time Reconfiguration (RTR) within FPGAs for high throughput data streaming applications requiring binary multiplication. We show how reconfiguration can be exploited without any effect on the operation of the application while at the same time providing reduced logic area implementations over those that do not use reconfiguration.

In many cases, reconfiguration can be viewed as providing an Area Time (AT) design alternative. Instead of implementing application specific circuitry, a configurable resource is provided that can be reconfigured over time to provide the specific functionality when required. This precludes an AT trade-off for application designs utilising reconfiguration over those that don't. Reconfiguration provides an AT advantage when the reduction in area coupled with the time to reconfigure provides better utilisation of the hardware resource.

For reconfigurable designs requiring r reconfiguration cycles every n cycles, the data throughput rate is reduced by the fraction $\frac{n}{n+r}$. While the hardware utilisation may be higher, the maximum potential throughput is reduced. Provided $r \leq n$, reconfiguration delay can be eliminated if two individual configurable resources are provided such that while one resource is being reconfigured, the other is active in operation until such time a new configuration is required. This process can be seen below in Figure 1. This process of switching between reconfigured contexts, termed *reconfigurable context switching* has been used previously in [4] and is a derivative of the DPGA concept and temporal pipelining [3]. The implication of such a proposal is that additional hardware resource is required for the second configuration. The overall hardware cost must not exceed that of any design techniques not using reconfiguration, since for both the maximum throughput rate is now achieved. Additionally, to sustain maximal throughput, applications must be able to effectively use one configurable resource while scheduling and performing the configuration of the second such that it can be seamlessly switched into operation.

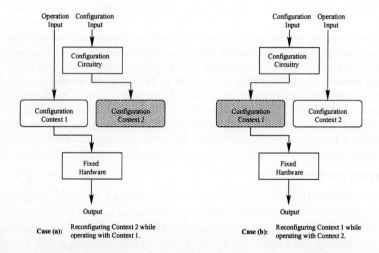

Fig. 1. Reconfiguring one context while the other is operating.

If these criteria can be satisfied then applications which can use the reconfigurable context switching technique to its advantage will comparatively provide a resource efficient technique when implemented. For this paper, we present a self configurable binary multiplier design comprising of two configuration resources which allow continuous processing on input data streams at the maximal throughput rate. Based on the number of cycles required to reconfigure, r, we then investigate the data flow for adaptive FIR filtering and the IDEA encryption algorithms with the objective of obtaining continuous sequences of constant value multiplications. This then allows such algorithms to utitilise the multiplier design presented. The result being an application design alternative which incorporates reconfiguration to allow the possibility for higher utilisation of the hardware resource over more traditional conventional design techniques.

2 Self-Configurable Multiplication Technique

The self-configurable multiplication technique can be applied to any FPGA allowing read/write RAM primitives mapped to the Logic Elements (LEs) of the architecture [14]. The self-configurable technique uses a lookup based distributed arithmetic approach [8] to perform the multiplication of two numbers. By storing one value in hardware Look-Up Tables (LUTs), the Distributed Arithmetic (DA) approach technique is frequently used for FPGA implementations since it provides significant hardware reduction. A comparison between a common general purpose multiplication technique and the constant coefficient DA approach is shown below in Figure 2 for 8-bit multiplication. The implementation difference is considerable since the cyclic convolution of inputs and top two stages of addition are replaced by two 12-bit wide, 4-bit addressable LUTs.

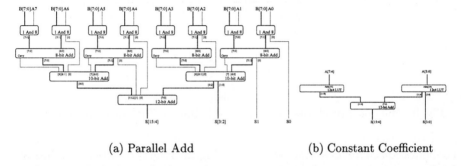

(a) Parallel Add (b) Constant Coefficient

Fig. 2. 8-bit Multiplier and Constant Coefficient Multiplier functional diagrams.

While the constant DA multiplier provides significant hardware reduction, as a general purpose multiplier its disadvantage is that one input remains constant. By use of reconfiguration we can update the LUT contents at run time providing a two input multiplier. Previous work has shown reconfigurable multiplier implementations on the XC6200 series FPGA [12, 1]. The reconfiguration of LUT content is performed through the SRAM configuration interface where the configuration data is precomputed. This interface has physical bit width and access time limitations. It is shown that a single LUT can be updated in $1.45 \mu s$ at 33MHz [1], which relates to 48 cycles for a design operating at this speed.

The self configurable multiplication technique eliminates this physical limitation by performing the LUT reconfiguration on chip. On new input, the configuration data is computed on chip and stored into the LUTs in parallel. No outside configuration intervention is required, hence describing the self configurable attribute of the multiplier. Reconfiguration is thus not physically limited and as a result faster. Typically, the required number of cycles for reconfiguring the logic content is that of the LUT addressable range.

Below, in Figure 3, a sixteen cycle 8-bit self configurable multiplier design is presented. Present are two 12-bit wide 4-bit addressable LUTs and a 12-bit adder which represent the basis of the 8-bit constant coefficient multiplier. The 4-bit counter and 12-bit accumulator represent the address and data generators respectively used to reconfigure the LUTs with new values. The two eight bit inputs A and B represent the multiplier inputs and the input line $CHANGE$ is used to signify that input B has changed and the multiplier LUTs must be reconfigured. The multiplier reconfiguration time is 16 cycles. Multipliers of less reconfiguration time can be implemented by reducing the LUT addressable size. A generic design technique is presented [14] where the required reconfiguration time is a design parameter such that multipliers with 8- and 4-cycle reconfiguration times can be implemented.

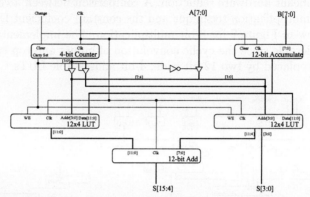

Fig. 3. Functional Block Diagram for an 8-bit 16-cycle Reconfigurable Multiplier.

Given this reconfigurable multiplier design, we extend it by increasing the configuration contexts to two. As shown in Figure 3 the two 12×4-bit addressable LUTs represent one configuration context. By providing another set of 12 × 4-bit LUTs we can guarantee continuous operation of the multiplier, i.e. while one context is operating the other is being reconfigured and vice versa, provided that $r \leq n$. In Tables 1(a), 1(b) and 1(c), we provide the hardware resource requirements (by using the Configurable Logic Block (CLB) count for implementations on Xilinx 4000 series FPGAs [7]) for 8-, 16- and 32-bit self configurable multipliers which use two configurable contexts. These values are compared to the implementation and associated CLB count for arbitrary value parallel add multipliers of the same bit size. Note that in every case the sixteen cycle reconfigurable multipliers require a lower CLB count than the parallel add technique, while the 8 cycle multipliers improve at 16-bits and up, and the four cycle multipliers provide no benefit.

Reconfig. Steps	# CLBs
16	44
8	74
4	142
0 (Parallel Add)	70

Reconfig. Steps	# CLBs
16	132
8	213
4	426
0 (Parallel Add)	273

Reconfig. Steps	# CLBs
16	455
8	765
4	1418
0 (Parallel Add)	1049

(a) 8-bit implementations   (b) 16-bit implementations   (c) 32-bit implementations

Table 1. 8-, 16- and 32-bit multiplier implementations on Xilinx XC4000 FPGAs.

3 IDEA Encryption

The IDEA encryption algorithm [11] presents a high throughput computationally intense algorithm that has been suggested for use as a benchmark for reconfigurable computing [5]. IDEA uses a 128-bit key and operates on 64-bit blocks of data. Each block is split into 4 16-bit sub-blocks and passed through eight successive rounds of computation where each round is as illustrated in Figure 4. For each round of encryption, the encryption key values vary since they are each defined as different 16-bit subsequences of the 128-bit key which remains fixed for the duration of the encryption process. All primitive operations within the IDEA algorithm are performed on 16-bit data.

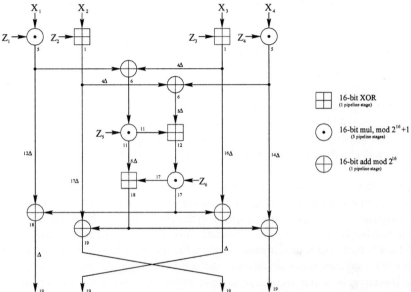

Fig. 4. One round of the IDEA Encryption algorithm.

Due to current logic capacity limitations of FPGAs, typical single chip implementations of the IDEA algorithm generally contain one instantiated single round computation unit [5]. Depending on the capacity of the FPGA this computation unit could be implemented as bit serial, bit parallel or some combination of both. The computation unit is equally time multiplexed between the eight successive rounds of required computation. With FPGA device logic capacities always increasing, parallel implementations of a single round of computation are compared with and without reconfiguration, i.e. with and without the use of the self configurable multiplier. To maximise throughput, each functional unit within the single round of computation is pipelined. Additional buffering for synchronisation between pipelined operations is used. The buffering infrastructure is set the same for implementations with and without reconfiguration. While the self configurable multiplier has a latency one cycle less than the general multiplier, we add an additional latency stage for buffer and control compatibility, and reuse. The buffering delays for synchronisation are implemented as displayed in Figure 4.

Between each round of computation, six new 16-bit sub-key sequences $Z_1 \ldots Z_6$ are selected from the 128-bit encryption key. These values remain constant for the duration of the round of computation. We observe in Figure 4 the pipeline latency to be 19 cycles. As a result, to achieve maximum data throughput from this architecture each round of computation is set to process 19 data values before the sub-key sequences are updated for the next round. In providing an implementation that uses the self configurable multiplier we identify that one input to the multiplier be constant for at least the number of required reconfiguration cycles, i.e. while one configuration context is being reconfigured, the other must be operating as a fixed coefficient multiplier. Since the single round latency is calculated at 19 cycles, using the 16-cycle self configurable multiplier will guarantee that the design performance will not be hampered by the time to reconfigure. New configuration contexts can be provided as they are required since each context has a total of 19 cycles to reconfigure its content.

Data streams are injected into the hardware implementation 19 blocks at a time. These 19 blocks of data ensure the pipeline is completely utilised and circulate through the single round computation unit eight times before exiting. At any time, the 19 blocks of data are stored within the 19 pipeline stages that exist within the implementation. For each round of computation, the 16-bit sub-key operators derived from the 128-bit key are constant. Each multiplier is configured to multiply by the constant 16-bit value for this time. During this time, the other multiplication configuration contexts are being configured for the next round of computation. However, we must ensure that the reconfiguration actions and switching of contexts are properly synchronised with the datapath latency within the implementation. Thus we must provide a sequencing of configurations such that correct operation is maintained with the data flow. Observing Figure 4, we see the multiply units situated in different latency positions. Relative to the time instance t_0 when data blocks first enter the round of computation, we provide configuration start times for each multiplication unit. For the two multiplication units multiplying by the constants Z_1 and Z_4, we begin reconfiguring their non-active configuration context at time $t_0 - 16$, and for the multiplication units multiplying by Z_5 and Z_6, we begin their reconfiguration at times $t_0 - 10$ and $t_0 - 4$ respectively. Once each context is configured it immediately becomes active reflecting the new constant value for the next round of computation.

Through the effective use of configuration sequencing we can guarantee that the constant value multiplications are scheduled at the right time to match the required data stream operations. We provide no negative impact on the performance of data flow, illustrating a hardware resource advantage over implementations that do not use reconfiguration. Preliminary implementation differences between designs which do and do not use reconfiguration have shown a CLB count saving of around 500. This comparative difference can be justified by referencing Table 1(b). Here we observe the CLB count difference between the 16 cycle and 0 cycle parallel add multiplier as 141. Given there are four multiplication units within the single round of IDEA computation, an implementation using reconfiguration should theoretically provide a savings of 564 CLBs.

4 Adaptive FIR Filtering

An FIR filter computes the dot product between a weighted coefficient vector and a finite series of time samples. Typically the coefficients are pre-computed for the filter and remain constant for the entire mode of operation. Samples stream through the filter such that the dot product result is computed every cycle from a window of samples equal to the coefficient vector length. The width of the sample window and coefficient vector is referred to as the number of taps the filter possesses. FIR implementations are well suited to FPGAs typically because the bit width of the operands is relatively small, i.e. between 8 and 16 bits, distributed arithmetic approaches can be used [8] for fixed coefficients, and pipelining and parallelism can be well exploited.

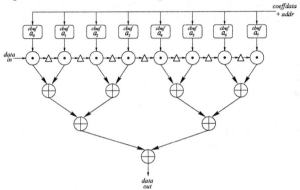

Fig. 5. 8 Tap FIR filter block diagram.

Adaptive FIR filtering allows the coefficient vector to be updated over time. The rate at which this vector can be updated is dependent on the application. Most adaptive FIR filters use multipliers which can multiply any two arbitrary values, and an interface which allows any coefficient values to be updated at any time. This is illustrated in Figure 5 with the addressable $coeff\_data$ interface and the coefficient buffers $cbuf$.

In order to utilise self configurable multipliers, we observe the data flow for the FIR filter. We see in Figure 5 that data flows into the filter from the left side and proceeds to the right where a delay stage is encountered between every tap (or multiplier). As a result we see every input sample present within the filter for a number of cycles equal to the number of taps. During this time, each sample is multiplied in order by the filter coefficients $a_0 \ldots a_7$. We observe the constant multiplication of $a_0 \ldots a_7$ by each input sample. If a self configurable multiplier with two configuration contexts is provided that can update a given context in the same or better time for which an input sample exists within the filter, then reconfiguration can be used to provide benefit. That is, the multiplier reconfiguration time r must be less than or equal to the number of taps for the filter. In the case of the example 8-tap FIR filter provided, an 8-cycle configurable multiplier is required. For a 16-tap Adaptive FIR filter, a 16-cycle configurable multiplier is required.

The design for an Adaptive FIR filter using self configurable multipliers sees changes to the internal data flow whereby the coefficients now propagate from multiplier to multiplier in the filter rather than the input samples. The operation of the filter is as follows. Filter coefficients are stored in the buffers labeled $cbuf$. These coefficients circulate through and around the filter multipliers such that each multiplier will repeatatively see the continuous sequence of all coefficients in order $a_0 \ldots a_7$. These coefficients can be updated at any time through the interface provided. The addressing mechanism for updating any particular coefficient must be aware of its current buffer location in order to update it correctly. Input samples stream in every cycle and are assigned to be configured within a multiplier. It is here that the configuration sequencing is critical to ensuring each input sample is multiplied by all coefficients in the order $a_0 \ldots a_7$. We must initiate the reconfiguration of each multiplier with an input sample such that its context switch occurs when coefficient a_0 is present. It is apparent that we are providing a circulating sequence of initiating reconfiguration of the multipliers within the filter which follows the circulation of coefficient values within. The initiation of configuration for each multiplier is offset r cycles before the arrival of coefficient a_0. The *data in* interface is broadcast to each multiplier, such that when the configuration of one multiplier is initiated it takes the sample value from the data in bus and configures its switch-able context to this value. In Figure 6 we see an example of the incorporation of self configurable multipliers in an an 8-tap Adaptive FIR filter.

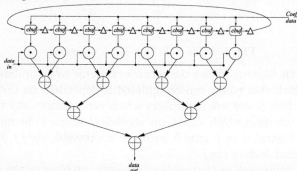

Fig. 6. 8 Tap Adaptive FIR filter block diagram using reconfiguration.

Provided that r for the multiplier is equal to or less than the number of the taps for the filter, and its CLB count is less than that of the arbitrary value parallel multiplier then a non performance compromising benefit will be achieved. Below in Table 2, we see several cases of adaptive FIR filters where the use of self configurable multipliers provides a reduced CLB count for implementations on the Xilinx XC4000 device series while not compromising performance. This includes the additional overhead for the updating of rotating coefficients and configuration sequencing control.

Filter Properties	#CLBs /w	#CLBs /wo	Ctrl o/head	Δ CLB Count	% Saving
16-tap 8-bit	1006	1344	78 CLBs	338	25.1
8-tap 16-bit	1976	2384	72 CLBs	408	17.1
16-tap 16-bit	2663	4777	142 CLBs	2114	44.3

Table 2. FIR implementations /w and /wo reconfiguration on Xilinx XC4000 FPGAs.

5 Conclusion

In this paper we have presented a case for the usefulness of RTR providing an alternative solution for traditional hardware implementations of given applications requiring multiplication. Considering the self configurable binary multiplication technique with two configuration contexts, application data flow analysis was required for its successful use. Two applications, IDEA encryption and adaptive FIR filtering were targeted. It was discovered that their data-flow could be sufficiently arranged such that one input to each multiplier in the application changed at a rate slower than the other. As a result, the self configurable multiplication technique was applied to these derived sequences of operations such that the reduced hardware resource requirements of the technique could be exploited. The results demonstrate a design alternative not impacting on achievable performance while providing an impetus to allow an area conscious advantage.

References

1. J. Burns A. Donlin J. Hoggs S. Singh M. de Wit. A dynamic reconfiguration runtime system. In *FPGAs for Custom Computing Machines*, April 1997.
2. A. DeHon. *Reconfigurable Architectures for General-Purpose Computing*. PhD thesis, Massachusetts Institute of Technology, 1996.
3. A. DeHon. DPGA utilization and application. In *FPGA '96 ACM/SIGDA Fourth International Symposium on FPGAs*, Montery CA, February 1996.
4. H. Eggers P. Lysaght H. Dick and G. McGregor. Fast reconfigurable crossbar switching in FPGAs. In *FPL'96*.
5. N. Weaver E. Caspi. Idea as a benchmark for reconfigurable computing. Technical Report Available from http://www.cs.berkeley.edu/projects/brass/projects.html, BRASS Research Group, University of Berkeley, December 1996.
6. J. R. Hauser and J. Wawrzynek. Garp: A mips processor with a reconfigurable coprocessor. In *Proc. FPGAs for Custom Computing Machines*, pages 12–21. IEEE, April 1997.
7. Xilinx Incorporated. Xilinx XC4000 data sheet, 1996.
8. Les Mintzer. FIR filters with field-programmable gate arrays. *Journal of VLSI Signal Processing*, 6(2):119–127, 1993.
9. C. Rupp, M. Landguth, T. Garverick, E. Gomersall, H. Holt, T. Arnold, and M. Gokhale. The NAPA adaptive processing architecture. In *Proc. FPGAs for Custom Computing Machines*, pages 28–37. IEEE, April 1998.
10. H. Schmit. Incremental reconfiguration for pipelined applications. In *Proc. FPGAs for Custom Computing Machines*, pages 47–55. IEEE, April 1997.
11. W. Stallings. *Network and Internetwork Security: Principles and Practice*. Prentice Hall, 1995.
12. B. Slous T. Kean, B. New. A multiplier for the XC6200. In *Sixth International Workshop on Field Programmable Logic and Appications*, 1996.
13. J. Vuillemin , P. Bertin , D. Roncin , M. Shand , H. Touati and P. Boucard. Programmable active memories: Reconfigurable systems come of age. *IEEE Transactions on VLSI Systems*, 4(1):56–69, March 1996.
14. M. Wojko and H. ElGindy. Self configurable binary multipliers for LUT addressable FPGAs. In Tam Shardi, editor, *Proceedings of PART'98*, Newcastle, New South Wales, Australia, September 1998. Springer.

Domain Specific Mapping for Solving Graph Problems on Reconfigurable Devices[*]

Andreas Dandalis, Alessandro Mei[**], and Viktor K. Prasanna

University of Southern California
{dandalis, prasanna, amei}@halcyon.usc.edu
http://maarc.usc.edu/

Abstract. Conventional mapping approaches to Reconfigurable Computing (RC) utilize CAD tools to perform the technology mapping of a high-level design. In comparison with the execution time on the hardware, extensive amount of time is spent for compilation by the CAD tools. However, the long compilation time is not always considered when evaluating the time performance of RC solutions. In this paper, we propose a domain specific mapping approach for solving graph problems. The key idea is to alleviate the intervention of the CAD tools at mapping time. High-level designs are synthesized with respect to the specific domain and are adapted to the input graph instance at run-time. The domain is defined by the algorithm and the reconfigurable target. The proposed approach leads to predictable RC solutions with superior time performance. The time performance metric includes both the mapping time and the execution time. For example, in the case of the single-source shortest path problem, the estimated run-time speed-up is 10^6 compared with the state-of-the-art. In comparison with software implementations, the estimated run-time speed-up is asymptotically 3.75 and can be improved by further optimization of the hardware design or improvement of the configuration time.

1 Introduction

Reconfigurable Computing (RC) solutions have shown superior execution times for several application domains (e.g. signal & image processing, genetic algorithms, graph algorithms, cryptography), compared with software and DSP based approaches. However, an efficient RC solution must achieve not only minimal execution time, but also minimal time for mapping onto the hardware [5, 7].

Conventional mapping approaches to RC (see Fig. 1) utilize CAD tools to generate hardware designs optimized with respect to execution time and area.

---

[*] This research was performed as part of the MAARC project. This work is supported by the DARPA Adaptive Computing Systems program under contract no. DABT63-96-C-0049 monitored by Fort Hauchuca.
[**] A. Mei is with the Department of Mathematics of the University of Trento, Italy. This work was performed while he was visiting the University of Southern California.

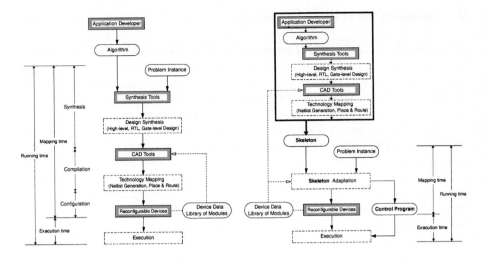

Fig. 1. Conventional (left) and Domain Specific (right) Mapping Approaches

The resulting mappings incur several overheads due to the dominant role of the CAD tools. The clock rate and the required area of the RC solution depend heavily on the CAD tools used and cannot be estimated reliably at compile time. Moreover, the technology mapping phase requires extensive compilation time. Usually, several hours of compilation time is required to achieve execution time in the range of *msec* [1,7]. In the case of mappings that are reused over time, compilation occurs only once and is not a performance bottleneck. But, for mappings that depend on the input problem instance, the mapping time cannot be ignored and often becomes a serious performance limitation.

In this paper we propose a novel RC mapping approach for solving graph problems. For each input graph instance, a new mapping is derived. The objective is to derive RC solutions with superior time performance. The time performance metric is the running time which is defined as the sum of the mapping time and the execution time. The key idea of the proposed approach is to reduce the intervention of the CAD tools at mapping time. A technology dependent design is synthesized based on the specific domain [algorithm,target]. The reconfigurable target is now visible to the application developer (see Fig. 1). At run-time, the derived design is adapted to the input graph instance. The proposed mapping approach eliminates the dominant role of the CAD tools and leads to "real-time" RC solutions. Furthermore, the time performance and the area requirements can be accurately estimated before compilation. This is particularly important in run-time environments where the parameters of the problem are not known *a priori* but time and area constraints must be satisfied.

In Section 2, the proposed mapping approach is briefly described. In Section 3, to illustrate our approach, we demonstrate a solution for the single-source shortest path problem. Finally, in Section 4, concluding remarks are made.

2 Domain Specific Mapping

A simple and natural model is assumed for application development. It consists of a host processor, an array of FPGAs, and external memory. The FPGAs are organized as a 2D-mesh. The memory stores the configuration data to be downloaded to the array and the data required during execution. The role of the memory is analogous to the role of the cache memory in a memory system in terms of providing a high-speed link between the host and the array. To illustrate our ideas, we consider an adaptive logic board from Virtual Computer Corporation to map our designs. This board is based on the XC6200 architecture.

In this paper, we consider graph problems as the application domain. Each graph instance leads to a different mapping. Thus, the mapping overhead cannot be ignored. Given a specific domain [algorithm,target], the objective is to derive a working implementation with superior time performance. The time performance metric is the running time which is defined as the sum of the mapping time and the execution time (see Fig. 1). To obtain the hardware implementation, an algorithm specific *skeleton* is synthesized based on the specific domain and is dynamically adapted to the input graph instance at run-time. The proposed mapping approach consists of three major steps (see Fig. 1):

1. **Skeleton design** For a given graph problem, a general structure (*skeleton*) is derived based on the characteristics of the specific domain. The *skeleton* consists of modules that correspond to elementary features of the graph (i.e. graph vertex). The modules are optimized hardware designs and their functionality is determined by the algorithm. Configurations for the modules and their interconnection are derived based on the target architecture.
   The interconnection of the modules is fixed and is defined to be general enough to capture the individual connectivity of different graph instances. Hence, the placement and routing of the modules are less optimized than in conventional CAD tools based approaches. The *skeleton* is derived before compilation and its derivation does not affect the running time. In addition, the *skeleton* exploits low-level hardware details of the reconfigurable target in terms of logic, placement, and routing.
2. **Adaptation to graph input instance** Functional and structural modifications are performed to the *skeleton* at run-time. Such modifications are dictated by the characteristics of the problem instance based on which the configuration of the final layout is derived.
   The functional modifications dynamically add or alter module logic to adapt the modules to the input data precision and problem size. The structural modifications shape the interconnection of the skeleton based on the characteristics of the problem instance.
   A software program (Control Program) is also derived to manage the execution in the FPGAs. This program schedules the operations and the on-chip data flow based on the computational requirements of the problem instance. In addition, it coordinates the data flow to/from the hardware implementation. Since the interconnection of the *skeleton* is well established in Step

1, the execution scheduling essentially corresponds to a software routing for the adapted *skeleton*.
3. **Configuration** Finally, the reconfigurable target is configured based on the adapted structure derived in Step 2. After the completion of the configuration, the control program is executed on the host to initiate and manage the execution on the hardware.

The proposed approach leads to RC solutions with superior time performance compared with conventional mapping approaches. Furthermore, in our approach, the *skeleton* mainly determines the clock rate and the area requirements. Hence, reliable time and area estimates are possible before compilation.

3 The Single-Source Shortest Path problem

To illustrate our ideas, we demonstrate a mapping scheme for the single-source shortest path problem. It is a classical combinatorial problem that arises in many optimization problems (e.g. problems of heuristic search, deterministic optimal control problems, data routing within a computer communication network) [2]. Given a weighted, directed graph and a source vertex, the problem is to find a shortest path from the source to every other vertex.

3.1 The Bellman-Ford Algorithm

For solving the single-source shortest path problem, we consider the Bellman-Ford algorithm. Figure 2 shows the pseudocode of the algorithm [3]. The edge weights can be negative. The complexity of the algorithm is $O(ne)$, where n is the number of vertices and e is the number of edges.

```
The Bellman-Ford algorithm
Initialize G (V,E)
    FOR each vertex i ε V
        DO label(i) ← ∞
    label(source) ← 0
Relax edges
    FOR k = 1.. n-1
        DO FOR each edge (i,j) ε E
            DO label(j) ← min {label(j), label(i) + w(i,j)}
Check for negative-weight cycles
    FOR each edge (i,j) ε E
        DO IF label(j) > label(i) + w(i,j)
            THEN return FALSE
    return TRUE
```

Problem size # vertices x # edges	# of iterations m* (average)
16 x 64	4.52
16 x 128	4.40
16 x 240	4.31
64 x 256	4.13
64 x 512	4.06
64 x 1024	4.03
128 x 512	5.32
128 x 1024	4.96
128 x 2048	4.97
256 x 1024	6.04
256 x 2048	5.75
256 x 4096	5.86
512 x 2048	7.02
512 x 4096	6.71
512 x 8192	6.76
1024 x 4096	7.40
1024 x 8192	7.77
1024 x 16384	7.80

Fig. 2. The Bellman-Ford Algorithm and experimental results for m^*

For graphs with no negative-weight cycles reachable from the source, the algorithm may converge in less than $n-1$ iterations [2]. The number of required iterations m^*, is the height of the shortest path tree of the input graph. This height is equal to the maximum number of edges in a shortest path from the source. In the worst case $m^* = n - 1$, where n is the number of the vertices.

We performed extensive software simulations to determine the relation between m^* and $n-1$ for graphs with no negative-weight cycles. Note that known RC solutions [1] always perform $n-1$ iterations of the algorithm, regardless the value of m^*. Figure 2 shows the experimental results for different problem sizes. For each problem size, $10^5 - 10^6$ graph instances were randomly generated. Then, the value of m^* for each graph instance was found and the average over all graph instances was calculated. For the considered problem sizes, the number of required iterations grows logarithmically as the number of vertices increases. For values of e/n smaller than those in the table in Fig. 2, m^* starts converging to $n-1$.

3.2 Mapping the Bellman-Ford algorithm

The skeleton The *skeleton* corresponds to a general graph $G(V,E)$ with n vertices and e edges. A weight $w(i,j)$ is assigned for each edge $(i,j) \in E$ (i.e. edge from vertex i to vertex j). The derived structure (see Fig. 3) consists of n modules connected in a pipelined fashion. An index $id = 0, 1, .., n-1$ and a *label* are uniquely associated with each module. Module i corresponds to vertex i. The weight of the edges is stored in the memory. No particular ordering of the weights is required. Each memory word consists of the weight $w(i,j)$ and the associated indices i and j.

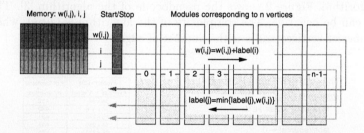

Fig. 3. The *skeleton* architecture for the Bellman-Ford algorithm

The Start/Stop module initiates execution on the hardware. An iteration corresponds to the e cycles needed to feed once the contents of the memory to the modules. The weights $w(i,j)$ are repetitively fed to the modules every e cycles. The algorithm terminates after m^* iterations. One extra iteration is required for the Start/Stop module to detect this termination. If no labels are modified during an iteration and $m^* \leq n$, the graph contains no negative-weight cycles reachable from the source and a solution exists. Otherwise, the graph contains a negative-weight cycle reachable from the source and no solution exists.

In each module (see Fig. 4), the values id and $label$ are stored and the relaxation of the corresponding edges is performed. In the upper part, the *label* is added to each incoming weight $w(i,j)$. The index i is compared with id to determine if the edge (i,j) is incident from vertex id. The weight $w(i,j)$ is updated only if $i = id$. In the lower part, the weight $w(i,j)$ is relaxed according to the *min* operation of the algorithm as shown in Figure 2. The index j is compared with id to determine if the edge (i,j) is incident to the vertex id. The *label* of the vertex id is updated only if $j = id$ and $w(i,j) < label$. When *label* is updated, a flag U is asserted.

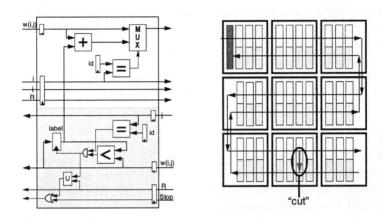

Fig. 4. The structure of the modules (left) and the placement of the *skeleton* into the FPGA array (right)

At the beginning of each iteration, the signal R is set to 1 by the Start/Stop module to reset all the flags. In addition, R resets a register that contains the signal *Stop* in module $n-1$. The signal *Stop* travels through the modules and samples all the flags. At the end of each iteration, the *Stop* signal is sampled by the Start/Stop module. If $Stop = 0$ and $m^* \leq n$, the execution terminates and a solution exists. Otherwise, if $Stop = 1$ after n iterations, no solution exists.

The *skeleton* placement and routing onto the FPGAs array (see Fig. 4) is simple and regular. The communication between consecutive modules is uniform and differs only at the boundaries of the array. Depending on the number of the required modules, a "cut" (Fig. 4) is formed that corresponds to the communication links of the last module (Fig. 3). During the adaptation of the *skeleton* the "cut" is formed and the labels in the allocated modules are initialized. The execution is managed by a control program executed on the host. This program controls the memory for feeding the required weights to the array. In addition, it initiates and terminates the execution via the Start/Stop module.

Area and running time estimates The above module was created based on the parametrized libraries for Xilinx 6200 series of FPGAs [9]. The footprint of each module was $(p + \lceil \log n \rceil) \times (4p + 2\lceil \log n \rceil + 10)$, where p denotes the number of bits in each weight/label, and n is the number of vertices. For $p = \log n = 16$, 4 modules can be placed in the largest device of the XC6200 family. The memory space required was $(p + 2\lceil \log n \rceil) \times e$ bits, where e is the number of edges. The needed memory-array bandwidth is $p + 2\log n$ bits/cycle to support the execution. To fully utilize the benefits of the FastMAP^{TM} interface, 140 MB/sec bandwidth is required. Under this assumption, the largest XC6200 device can be configured in 165 μsec using wildcards [8].

The algorithm terminates after $(m^* + 1) \times e + 2n$ cycles, where m^* is the number of required iterations for a given graph. One cycle corresponds to the clock period of the *skeleton*. The clock rate for the *skeleton* was estimated to be at least 15 MHz for p=16 bits, and at least 25 MHz for p=8 bits. In the clock rate analysis, all the overheads caused by the routing were considered. The clock rate was determined mainly by the carry-chain adder of the modules. By using a faster adder, improvements in the clock rate are possible. The mapping time was in the range of *msec*. The above mapping time analysis is based on the timings for the FastMAP^{TM} interface in the Xilinx 6200 series of FPGAs databook [8].

3.3 Performance Comparison

In [1], the shortest path problem is solved by using Dynamic Computation Structures (DCSs). The key characteristic of the solution is the mapping of each edge onto a physical wire. The experiments considered only problems with an average out-degree of 4 and a maximum in-degree of 8. For the instances considered, the compilation time was 4-16 hours assuming that a network of 10 workstations was available. Extensive time was spent for placement and routing. Hence, the resulting mapping time eliminated any gains achieved by fast execution time. To make fair comparisons with our solution, we assumed that the available bandwidth for configuring the array is 4 MB/sec as in [1]. Even though, the mapping time for our solution was estimated to be in the *msec* range (see Fig. 5).

	Check for negative-weight cycles	Mapping time	Execution time			Area requirements
			# of iterations	# of cycles	Clock rate	
Solution in [1]	NO	> 4 hours [+]	n-1	n-1	$\Omega(1/n^2)$	$\Omega(n^4)$
Our Solution	YES	~ 100 msec [++]	m*	(m*+1)e+2n	independent of n	O(n+e)

[+] a network of 10 workstations was used
[++] memory-array bandwidth 4MB/sec is assumed as in [1]

Fig. 5. Performance comparisons with the solution in [1]

Besides the mapping overhead, the mapping of edges into physical wires resulted in several limitations in [1], with respect to the clock rate and the area requirements. The clock rate depended on the longest wire which is $\Omega(n^2)$ in the worst case, where n is the number of vertices. This remark is supported by well-known theoretical results [6] which show that in the worst case, a graph takes $\Omega(n^4)$ area to be laid out, where n is the number of vertices, and that the longest wire is $\Omega(n^2)$ long. Therefore, as n increases, the execution time of their solution drops dramatically. For $n = 16, 64, 128$, the execution time was on the average 1.5-2 times faster than our approach while it became 1.3 times slower for $n = 256$. For larger n, the degradation of performance in [1] is expected to be more severe. Considering both the execution and the mapping time, the resulting speed-up comparing with the solution in [1] was 10^6.

Also, in [1], $n-1$ iterations were always executed and negative-weight cycles could not be detected. If checking for algorithm convergence and negative-weight cycles were included in the design, the resulting longest wire would increase further drastically affecting the clock rate and the excution time. Finally, the time performance and the area requirements in [1] are determined completely by the efficiency of the CAD tools and no reliable estimates can be made before compilation.

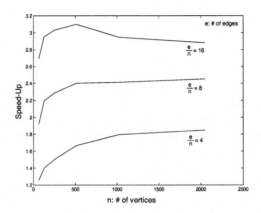

Fig. 6. Comparison of running time: our approach v.s. software implementation.

Area comparisons are difficult to make since different FPGAs were used in [1]. Furthermore, the considered graph instances in [1] were not indicative of the entire problem space since $e/n = 4$. For the considered instances in [1], one XC4013 FPGA was allocated per vertex but, as e/n increases, the area required grows rapidly. In our solution, $O(n)$ area for FPGAs and $O(e)$ memory were required. Moreover, our design is a modular design and can be easily adapted to different graph instances without complete redesign.

Software simulations were also performed to make time performance comparisons with uniprocessor-based solutions. The algorithm that was mapped onto the hardware was also implemented in C language. The software experiments were performed on a Sun ULTRA 1 with 64 MB of memory and a clock rate of 143 MHz. No limitations on the in/out-degree of the vertices were assumed. For each problem instance, $10^4 - 10^6$ graph instances were randomly generated and the average running time was calculated. The compilation time on the uniprocessor to obtain the executable was not considered in the comparisons. Moreover, the data were assumed to be in the memory before execution and no cache effects were considered. Under these assumptions, on the average, an edge was relaxed every 250 $nsec$.

For the hardware implementation, it was assumed that $p=16$ bits. The mapping time was proportional to the number of vertices of the input graph. Both the mapping and the execution time were considered in the comparisons. The achieved run-time speed-up was asymptotically 3.75. However, for the considered problem sizes (see Fig. 6), lower speed-up was observed. As e/n increases, the mapping time overhead is amortized over the corresponding execution time. Hence, shorter configuration time would result in convergence to the speed-up bound (3.75) for smaller e/n and n.

4 Conclusions

In this paper, a *domain specific* mapping approach was introduced to solve graph problems on FPGAs. Such problems depend on the input graph instance and constitute a suitable application domain to exploit reconfigurability. The proposed approach reduces the dominant role of CAD tools and leads to RC solutions with superior time performance. For example, for the single-source shortest path problem, a speed-up of 10^6 was shown compared with [1]. In comparison with software solutions, an asymptotic speed-up of 3.75 was also shown.

Future work includes more graph problems examples to further validate our mapping approach. In addition, we will focus on specific instances of NP-hard problems where the execution time is comparable to the corresponding mapping time provided by general purpose CAD tools based approaches. We believe that the proposed approach combined with a software/hardware co-design framework can efficiently attack specific instances of NP-hard problems.

References

1. J. Babb, M. Frank, and A. Agarwal, "Solving graph problems with dynamic computation structures", *SPIE '96: High-Speed Computing, Digital Signal Processing, and Filtering using Reconfigurable Logic*, 1996.
2. D. P. Bertsekas, J. N. Tsitsiklis, "Parallel and Distributed Computations: Numerical Methods", Athena Scientific, Belmont, Massachusetts, 1997.
3. T. H. Cormen, C. E. Leiserson, and R. L. Rivest, "Introduction to Algorithms", The Massachusetts Institute of Technology, 1990.
4. B. L. Hutchings, "Exploiting Reconfigurability Through Domain-Specific Systems", *Int. Workshop on Field Programmable Logic and Applications*, Sep. 1997.
5. A. Rashid, J. Leonard, and W. H. Mangione-Smith, "Dynamic Circuit Generation for Solving Specific Problem Instances of Boolean Satisfiability", *IEEE Symposium on Field-Programmable Custom Computing Machines*, Apr. 1998.
6. J. D. Ullman, "Computational Aspects of VLSI", Computer Science Press, Rockville, Maryland, 1984.
7. P. Zhong, M. Martonosi, P. Ashar, and S. Malik, "Accelerating Boolean Satisfiability with Configurable Hardware", *IEEE Symposium on Field-Programmable Custom Computing Machines*, Apr. 1998.
8. The XC6200 Field Programmable Gate Arrays Databook, http://www.xilinx.com/apps/6200.htm, April 1997.
9. Parameterized Library for XC6200, H.O.T. Works Ver. 1.1, Aug. 1997.

MorphoSys: a Reconfigurable Processor Targeted to High Performance Image Application

Guangming Lu [1], Ming-hau Lee [1], Hartej Singh [1], Nader Bagherzadeh [1], Fadi J. Kurdahi [1], Eliseu M. Filho [2]

[1] Department of Electrical and Computer Engineering
University of California, Irvine, CA 92715 USA
{glu, mlee, hsigh, nader, kurdahi}@ece.uci.edu

[2] Department of Systems and Computer Engineering
Federal University of Rio de Janeiro P.O.BOX 68511 21945-970 Rio de Janeiro, RJ, Brazil
eliseu@cos.ufrj.br

Abstract: This paper addresses the design idea of the MorphoSys Reconfigurable processor developed by the researchers in the UC, Irvine. With the demand to perform the multimedia operations efficiently, it is one of the directions that general processor needs to incorporate with some reconfigurable computing units, like FPGA. In MorphoSys project, we successfully propose a prototype to fulfill the above trend, which is comprised of a simplified general purpose MIPS-like RISC processor, called TinyRISC and 8x8 coarse grained reconfigurable cells, organized as SIMD architecture. MorphoSys is realized using 0.35um technology, and runs at 100Mhz with impressive performance enhancement compared with other architectures.

1. Introduction

With the demand for the multimedia application, it is extremely necessary to explore another way beyond the general purpose processor to speedup the operation of the multimedia operations, such as compression, decompression, encryption, decryption, pattern recognition, target recognition.

In addition to the MMX technology, where new ISAs are added in the general processor to speed up the multimedia operation, recently, some processor architectures have been proposed to provide multiple configuration contexts in chip to program the LUTs and crossbars, like DPGA[1], or incorporate the general purpose processor with reconfigurable computing unit, such as GARP[2], MATRIX[3], RaPiD[4], RAW[5]. Basically, this kind of new architecture can be sub-divided into two categories:
1) Fine-grained level reconfigurable unit, such as FPGA. Each bit is configurable.
2) Coarse-grained level reconfigurable unit, such as reconfigurable array processor. Only the functionality and connectivity of each processor can be programmed.

In this paper, we describe a new reconfigurable processor named MorphoSys (<u>Morpho</u>ing <u>Sy</u>stem), which is coarse level reconfigurable array processor. It is composed of a simplified version MIPS-like RISC, called TinyRISC, 8 by 8 Reconfigurable Cells (RC) Array, Frame Buffer, Context memory, DMA controller. We have developed VHDL version simulation environment and realized it by designing custom blocks, such as 8x8 RC Array, Frame Buffer, context memory, Datapath in the TinyRISC and synthesizing other control logic.

This paper is organized as follows: Section 2 gives the brief overview about the MorphoSys system, and each major component. Section 3 introduces the M1 (MorphoSys Version 1) simulation environment MorphoSim, and the programmability of M1 chip. Section 4 gives the performance analysis of M1 architecture. Section 5 explains how we realize the M1 in physical design.

2. MorphoSys System

Figure 1 shows the components of the M1. It is composed of TinyRISC, 8x8 RC Array, context memory, frame buffer, DMA controller, instruction cache, data cache and memory controller. The following section will briefly explain the function and architecture of each component.

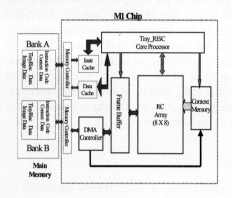

Figure 1. System level architecture for M1 chip

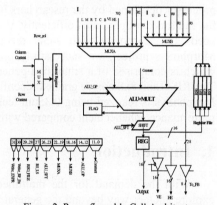

Figure 2. Reconfigurable Cell Architecture

2.1 RC Array
2.1.1 Architecture of Reconfigurable Cell

The reconfigurable cell (Figure 2) is the programmable unit of MorphoSys. It is coarse grained. The minimum programmable bitwidth is 16 bits. The RC Array (Figure 3) is consisted of 8x8 of identical RC. As showed in Figure 2, each RC comprises of an ALU-multiplier, a shift unit, and two multiplexers for ALU inputs. It has an output register, and a small register file. A context word, stored in the context register, defines the functionality of the RC. It also provides select bits to input multiplexers. In addition, the context word can also specify an immediate value (constant). One's counter (one part of ALU) is used to implement special functions, which require processing of binary image data, such as automatic target recognition (ATR).

2.1.2 Broadcast Context

This 32-bit register contains the context word to configure each RC. The programmability of the RC functionality and interconnection network is derived from the context word.

This context word is broadcast to RC in each row or column to achieve the data parallel operation. However, the RCs in different row or column may have different contexts applied to them. By switching row context broadcast to column context broadcast or vice versa, we can avoid data movement needed frequently in some applications, such as DCT. Meanwhile, it is also possible to enable only one specific row or column for operation in the RC Array. This feature is primarily useful in loading data into the RC Array. Since the context can be used selectively, and because the data bus limitations allow loading of only one column at a time, the same set of context words can be used repeatedly to load data into all the eight columns of the RC. This feature also allows irregular operations in the RC Array, for e.g. zigzag re-arrangement of array elements.

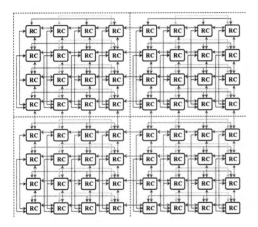

Figure 3. 8x8 RA Array

Figure 4: Express lane connectivity (between cells in same row, but adjacent quadrants, same for the column express lanes)

2.1.3 Interconnection Network

The RC interconnection network is comprised of two hierarchical levels.

Intra-quadrant (complete row/column) connectivity: The first layer of connectivity is within one quadrant (a quadrant is a 4 by 4 RC partition). In the current MorphoSys specification, the RC array has four quadrants. Within each quadrant, each cell can access the output of any other cell in its row and column, as shown in Figure 3.

Inter-quadrant (express lane and nearby quadrant) connectivity: Between each pair of adjacent quadrant, the nearby quadrant connectivity exists in both vertical the horizontal direction. At the higher or global level, there are connections between adjacent quadrants. These buses also called express lanes, run across rows as well as columns. Figure 4 shows two express lanes going in each direction across a row. Therefore, any cell in any quadrant can access any RC output in the same row/column in the adjacent quadrant. The express lanes greatly enhance global connectivity. Even irregular communication patterns, that otherwise require extensive interconnections, can be handled quite efficiently. For example, an eight-point butterfly is accomplished in only three cycles. The programmability of the connection is realized via the MUXA and MUXB in each RC, which is controlled by the configuration context. With these 2 levels connectivity, MorphoSys provides flexible interconnection to exchange data each other.

Data bus and Context bus: A 128-bit data bus from Frame Buffer to RC array is linked to column elements of the array. It provides two eight bit operands to each of the eight column cells. It is possible to load two operand data (Port A and Port B) in an entire column in one cycle. Eight cycles are required to load the entire RC array. The outputs of RC elements of each column are written back to frame buffer through Port A data bus. In order to perform row/column context broadcasting, there are 8-word-width context buses in both vertical and horizontal direction.

2.2 TinyRISC and MorphoSys Decoder

TinyRISC [6] is a simplified 32-bit MIPS RISC processor, which has four pipeline stages: fetch, decoder, execute, and write back. Since in this TinyRISC, it doesn't have the separate memory access pipeline stage, the memory address comes only directly from the register file. TinyRISC have 16 user accessible data registers and register number 0 is tied to zero. It uses a subset of conventional MIPS instruction set as well as some dedicated instructions used to control the RC Array, Context memory, frame buffer and DMA controller, etc.

MorphoSys Decoder, which is located at the decode stage, is dedicated to decode the MorphoSys instructions. Basically, It will either activate the DMAC to begin to transfer data between frame buffer, context memory with main memory, or let RC Array work through broadcasting configuration context. It establishes the communication between the general RISC processor, DMAC, Frame buffer, context memory and SIMD reconfigurable computing Units. Table 1 explains the some typical MorphoSys Instructions.

Table 1. Typical MorphoSys Instruction

LDCTXT	Load Context from Main Memory to Context memory.
LDFB	Load image data to frame buffer
STFB	Store the processed image data back to Main Memory
DBCBC	Column context broadcast, get image data from both banks
DBCBR	Row context broadcast, get image data from both banks
SBCB	Context broadcast, get one image data from a certain bank in frame buffer
CBCAST	Context broadcast, no data from frame buffer
WFBI	Write image data back to frame buffer with immediate frame buffer address
WFB	Write image back to the frame buffer, frame buffer address is from register
RCRISC	Write one data from RC Array to TinyRISC

2.2.3 Frame Buffer, Context Memory and DMAC

The frame buffer is consisted of 2 sets. Each set has 2 banks. Each bank has the 64x8 bytes storage. Bank A will provide the operand A for the RC Array, while Bank B will provide the operand B for the RC Array. The key feature of the Frame buffer is that once the starting address is given, it will activate 2 rows and read out the consecutive 8 bytes in order to provide 8 operands for the 8 RC's in a certain column. Meanwhile, when one set is reading/writing data to/from RC, the other set can load/write data from/to main memory.

Context memory provides the SIMD-like instructions for 8x8 RC Array. All of the RCs in each column/row share the same context. Due to similarity of jobs each RC performs, we decide to centralize the context memory in each RC into column context and row context. They provide the context for column broadcasting and row broadcasting respectively. Obviously, the alternative solution is to make each RC have its context memory. So, each RC can perform different operation. But, it is the tradeoff between flexibility and chip size. Currently, we reserve 16 contexts for each column and row, since it will cover most of the contexts for MPEG2 application [7] mapping to M1 chip. Each context has 32 bits. The behavior and connectivity of RC is controlled by the context.

Basically, the DMAC handles all of the data movements involving RC Array with outside memory. It provides 3 atomic operations: from memory to Frame buffer, from memory to context memory and from frame buffer to memory. In fact, it is another state machine to handle the communication protocol between main memory and in-chip memory.

3. Simulation and Programmability of M1

MorphoSim is the VHDL version simulator for the MorphoSys reconfigurable computing processor. Using this simulator, it becomes feasible to verify the mapping algorithm, and validate the physical design. We have developed a compiler based on the SUIF compiler [8]. So, we can program M1 chip either on the C level (with in-line code) or on the assembly language level.

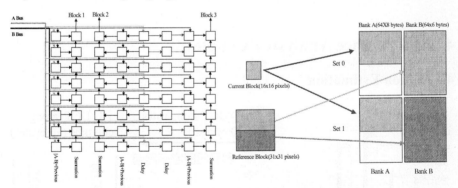

Figure 5. The Motion Estimation Mapping

Figure 6. Data Storage in Frame buffer for Motion Estimation

The following is one segment of C program example of Motion Estimation, which is compiled by the MorphoSys compiler, and generate the executable code for the M1 reconfigurable processor. In this mapping case, the RC Array can handle 3 blocks each time, generate the total sum of difference for each pixel, and send these 3 results back to the TinyRISC. TinyRISC will compare these 3 values with the previous minimum value, get the smallest one, then calculate the Motion Vector, finally, store it in the register file. This Motion Vector will be used again in the Motion compensation in the MPEG2 application [9]. The data storage in the Frame buffer is visualized in the Figure 6. All the instruction with prefix TR_ will be executed in the RC Array. The rest of the instruction will be executed in the TinyRISC. Current version can't support automatic job partition between TinyRISC and RC Array. It should have the interference from users. But, this is an important sub-project of our current researches.

```
ME() {
   int H, V, minvalue, MVX, MVY, d1, d2, d3,  reg,  c1,bankstep, base;
# d1 get the data from Col#1; d2 get the data from Col#2; get the data from Col#2;
# set the minvalue as the biggest postitive number;
   bankstep = 4;    base = 0;   reg = 0; minvalue = 0x7FFFFFFF;
   for (H=0; H<16; H++) {
      for ( V=0; V<16; V=V+3) {
#####    The following instructions will be performed in RC Array
#               reg,  set,  all, base,   ctx, Bank_B
         TR_dbcbc( reg,   0, 1,    0,     0,  0*bankstep);
         TR_dbcbc( reg,   0, 1,    1,     1,  1*bankstep);
            . . .

         TR_dbcbc( reg,   0, 1,   15,    2,  15*bankstep );
         TR_dbcbc( reg,   0, 1,    0,    3,   0);
         TR_dbcbc( reg,   0, 1,    1,    4,   0);
### Get the data generated in the previous iteration.
         TR_rcrisc (d1, 1);
         TR_rcrisc (d2, 2);
         TR_rcrisc (d3, 7);
#### choose the smallest among d1, d2,d3 and minvalue, then calculate the motion
Vectors, done in TinyRISC
         minvalue = min(d1, d2, d3, minvalue);
##       update the corresponding motion vector

            . . .
      } }
}
```

4. Performance Analysis Examples

4.1 Motion Estimation

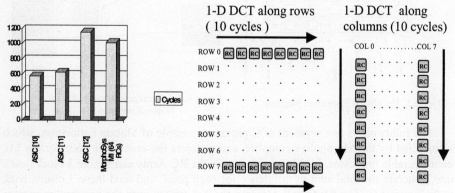

Figure 7: Performance Comparison for Motion Estimation

Figure 8: Computation of 2-D DCT across rows/columns (without data transposing)

The following comparison is based on 8x8 current block moving around on a reference block (search area) with 8-pixel displacement. Currently, full search algorithm is employed in the MorphoSys Motion Estimation mapping [17]. It is compared with three ASIC architectures implemented in [10], [11], [12]. The ASIC architectures have same processing units with MorphoSys. The figure 7 shows that the MorphoSys is comparable to the cycles required by the ASIC designs. Pentium MMX takes almost 29000 cycles for the same task, which is almost thirty times more than MorphoSys.

4.2 Two Dimension DCT

The Figure 8 shows the mapping of 8x8 pixels block to 8x8 RC Array. We broadcast the Horizontal context and Vertical contexts to perform Vertical DCT and Horizontal DCT. MorphoSys requires 21 cycles to complete 2-D DCT (or IDCT) on 8x8 block of pixel data [17]. This is in contrast to 240 cycles required by Pentium MMX ^{TM} [13]. REMARC [14] takes 54 cycles to implement the IDCT, even though it uses 64 nano-processors. The relative performance figures for MorphoSys and other implementations are given in Figure 9.

4.3 ATR Application

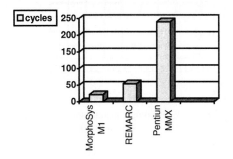
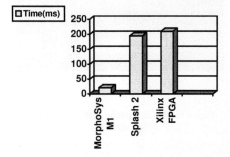

Figure 9: DCT/IDCT Performance Comparison (cycles)

Figure 10: ATR Performance Comparison (ms)

Automatic Target Recognition (ATR) is to automatically detect, classify, recognize and identify an object. Under the same condition, for 16 pairs of target templates, Xilinx XC4010 FPGA [15] needs 210 ms processing time, Splash 2 [16] needs 195 ms, MorphoSys needs 21 ms (@ 100 Mhz) (Figure 10) [17].

5. Layout Realization of M1

We have finished the physical design cache and Register file in the TinyRISC, Frame buffer, individual RC using Magic. For the other pure logic circuits, we will use Synopsys to synthesize them, and use the AutoCells (from Mentor) to place standard cells and route. Since the rich connectivity and symmetrical property in 8x8 RC array, it is hard for commercial Router to handle the routing for 8x8 Array properly. Automatic router will completely destroy the regularity, and make the clock H tree unbalanced. Therefore, we developed L language script file to finish the regular routing, and build the balanced clock tree (Figure 11). Inside RC, we only use Metal 1 and Metal 2 for internal routing, and reserve the Metal 3 and Metal 4 for connectivity among RC. Mentor Graphics MicroPlan and MicroRoute will handle the final routing on the top level, as showed in the Figure 1, including standard-cell blocks, such as DMAC, and custom design blocks, such as cache, RC Array. The final floor plan will be illustrated in the figure 12. We use the delay information gotten from MicroPlan to do back-annotation and use Lsim to do post-layout simulation.

6. Conclusions and future work

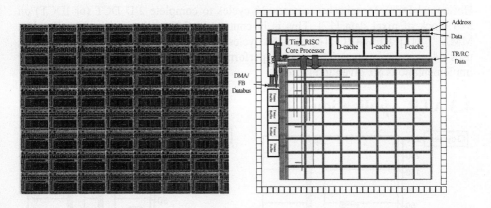

Figure 11. Regular routing of 8x8 RC Array (12mmx10mm)

Figure 12. Final floor plan for the M1 chip (core 14mmx14mm).

In this paper, We have depicted the architecture and functionality of each main component of the MorphoSys and evaluated the performance of applications, such as Motion Estimation, DCT, ATR. Meanwhile, we have presented the simulation environment -- Morphosim and MorphoSys compiler, and will finally realize it in physical layout. We present a feasible way to integrate general-purpose microprocessor with an Array of reconfigurable cells. We can freely increase the size of the reconfigurable array based on the availability of the die area of the chip and performance requirement. MorphoSys is designed to be an independent system. But for M1, it has to rely on a host to communication with other peripheral devices. In current prototype, we download image data, executable code, context data through standard PCI bus, and upload the processed image data back to host to visualize it. The PCB design for M1 test chip is under development. Meanwhile, we will continue to enhance the current compiler to support the automatic job partition between TinyRISC and 8x8 RC Array.

7. Acknowledgments

This research is funded by the Defense and Advanced Research Projects Agency (DARPA) of the Department of Defense under the contract number F-33615-97-C-1126.

References:

1) E. Tau, D. Chen, I. Eslick, J. Brown, A. DeHon, "A First Generation DPGA Implementation". *FPD'95 Third Canadian Workshop of Field-Programmable Devices, May 1995.*
2) J. R. Hauser and J. Wawrzynek, "Garp: A MIPS Processor with a Reconfigurable Co-processor," *Proc. of the IEEE Symposium on FPGAs for Custom Computing Machines, 1997.*
3) E. Mirsky and A. DeHon, "MATRIX: A Reconfigurable Computing Architecture with Configurable Instruction Distribution and Deployable Resources," *Proceedings of IEEE Symposium on FPGAs for Custom Computing Machines,* 1996, pp.157-66.
4) C. Ebeling, D. Cronquist, and P. Franklin, "Configurable Computing: The Catalyst for High-Performance Architectures," *Proeedings of IEEE Internatinoal Conference on Application-specific Systems, Architectures and Processors,* July 1997, pp. 364-72.
5) J. Babb, M. Frank, V. Lee, E. Waingold, R. Barua, M. Taylor, J. Kim, S. Devabhaktuni, A. Agrawal, "The RAW Benchmark Suite: computation structures for general-purpose computing," *Proc. IEEE Symposium on Field-Programmable Custom Computing Machines,* FCCM 97, 1997, pp. 134-43.
6) A. Abnous, C. Christensen, J. Gray, J. Lenell, A. Naylor and N. Bagherzaheh, "Design and implementation of the TinyRISC microprocessor" *Microprocessors and Microsystems. Vol.16, No.4, pp.187-94, 1992.*
7) Stanford PVRG-MPEG by anonymous ftp from havefun.stanford.edu:/pub/mpeg/MPEGv1.2.tar.Z.
8) SUIF Compiler system, The Stanford SUIF Compiler Group, *http://suif.stanford.edu*
9) MPEG2 Standard. http://www.mpeg.org.
10) C. Hsieh, T. Lin, "VLSI Architecture For Block Matching Motion Estimation Algorithm" *IEEE Transaction on CSVT, Vol. 2, June, 1992.*
11) S.H. Name, J.S. Baek, T.Y.Lee, M.K. Lee, "A VLSI Design For Full Search Block Matching Motion Estimation" *Proceedings of IEEE ASIC Conference, Rochester, NY, Set 1994.*
12) K-M Yang, M-T Sun, and L.Wu, "A Family of VLSI Designs for Motion Compensation Block Matching Algorithm" *IEEE Transaction on Circuits and Systems. Vol 36. No.10, Oct 89.*
13) Intel Application Notes for Pentium MMX. *http://developer.intel.com/drg/mmx/appnotes.*
14) T. Miyamori and K. Olukotun, "A Quantitative Analysis of Reconfigurable Coprocessors for Multimedia Applications," *Proceedings of IEEE Symposium on Field-Programmable Custom Computing Machines,* April 1998
15) J. Villasenor, B. Schoner, K. Chia, C. Zapata, H. J. Kim, C. Jones, S. Lansing, and B. Mangione-Smith "Configurable Computing Solutions for Automatic Target Recognition" *Proceedings of IEEE Workshop on FPGAs For Custom Computing Machines, April 1996.*
16) M. Rencher and B.L. Hutchings, "Automatic Target Recognition on SPLASH 2" Proceedings of IEEE Symposium on FPGAs for Custom Computing Machines, April 1997.
17) H. Singh, M. Lee, G. Lu, F. Kurdahi, N. Bagherzadeh, T. Lang, R. Heaton, E. Filho, "MorphoSys: An Integrated Re-configurable Architecture" *Proceeding of the NATO Symposium on Concepts and Integration, April, 1998.*

An Efficient Implementation Method of Fractal Image Compression on Dynamically Reconfigurable Architecture

Hidehisa Nagano, Akihiro Matsuura, and Akira Nagoya

NTT Communication Science Laboratories
2-4 Hikaridai, Seika-cho, Soraku-gun, Kyoto, 619-0237, JAPAN
{nagano, matsuura, nagoya}@cslab.kecl.ntt.co.jp

Abstract. This paper proposes a method for implementing fractal image compression on dynamically reconfigurable architecture. In the encoding of this compression, metric computations among image blocks are the most time consuming. In our method, processing elements (PEs) configured for each image block perform these computations in a pipeline manner. By configuring PEs, we can reduce the number of adders, which are the main computing elements, by half even in the worst case. This reduction increases the number of PEs that work in parallel. In addition, dynamic reconfigurability of hardware is employed to omit useless metric computations. Experimental results show that the resources for implementing the PEs are reduced to 60 to 70% and the omission of useless metric computations reduces the encoding time to 10 to 55%.

1 Introduction

Due to recent advances in programmable logic devices such as Field Programmable Gate Arrays (FPGAs), reconfigurable computing has become a realistic computing paradigm [1]. Some reconfigurable devices have begun to support dynamic and partial reconfiguration [2][3], and recent research has made use of such reconfigurability to accelerate solving problems [4]. In this paper, we focus on the acceleration of fractal image compression using dynamically reconfigurable computing.

Fractal image compression [5] is expected to be a promising compression method for digital images in terms of its high compression ratios and fidelity. However, due to the unacceptable encoding time, use of this compression method is limited to areas where the encoding time is not critical. Therefore, acceleration of the encoding is indispensable to a variety of uses. In fractal-based image encoding, an image is divided into sub-blocks and the root mean square metrics among them are computed. These metric computations carry the burden in encoding the entire image because numerous metric computations for the sub-blocks must be done.

In our method, processing elements (PEs) are devoted to the corresponding sub-blocks. Each PE computes the metrics between the corresponding sub-blocks and the candidate sub-blocks in the pipeline. In each step of the pipeline, one metric computation is executed on a PE. In addition, plural PEs implemented on the reconfigurable architecture concurrently perform the metric computations.

Our method has two advantages owing to the reconfigurability of the hardware. One advantage is the structure of PEs. Since each PE is reconfigured for a sub-block, a number of multiplications in the metric computations can be regarded as constant multiplications. This makes it possible to use shifts and adders instead of variable-by-variable general-purpose multipliers. By using constant multiplications, the number of adders needed for PEs is reduced, even in the worst case, to half that of PEs in the straightforward implementation. This reduces the reconfigurable resources needed for the PE, and consequently increases the number of PEs that work in parallel. The second advantage is the omission of useless metric computations. We can accelerate the encoding by omitting metric computations that do not influence the fidelity of the decoded image. Such omission can be achieved by the partial reconfiguration of the hardware.

Experimental results show that the resources needed to implement the PEs can be reduced to approximately 60 to 70% of the straightforward implementation, even in the worst case. Results also show that omitting useless computations reduces the encoding time to 10 to 55%. Thus, our method utilizing reconfigurability significantly accelerates the encoding.

This paper is organized as follows. Sect. 2 briefly describes fractal image compression. In Sect. 3, we present a dynamically reconfigurable architecture of the encoder for fractal image compression. In Sect. 4, experimental results are shown and analyzed. Sect. 5 concludes this paper.

2 Fractal Image Compression

In this section, we describe the outline of fractal image compression. A fractal-based image encoding scheme was first promoted by M. Barnsley [6]. The fractal image compression for grey-scale images was developed by A. Jacquin. For a detailed survey and extensions of fractal image compression, we refer readers to a paper by Jacpuin [7].

2.1 Iterated Function System (IFS)

Here, we briefly present the underlying mathematical principles of fractal image compression based on a theory of IFS [8][9]. An IFS consists of a collection of contractive affine transformations $\{w_i : M \to M | 1 \leq i \leq m\}$, where w_i maps a metric space M to itself. A transformation w is said to be contractive when there is a constant number $s (0 \leq s < 1)$ and $d(w(\mu_1), w(\mu_2)) \leq s\, d(\mu_1, \mu_2)$ holds for any points $\mu_1, \mu_2 \in M$, where $d(\mu_1, \mu_2)$ denotes the metric between μ_1 and μ_2. The collection of transformations defines a map $W(S) = \bigcup_{i=1}^{m} w_i(S)$, where $S \subset M$. Note that the map W is applied not to the set of points in M but to the set of sub-sets in M. Let W be a contractive map on a compact metric space of images with the bounded intensity, then the Contractive Mapping Fixed-Point Theorem asserts that there is one special image x_W, called the attractor, with the following properties.

1. $W(x_W) = x_W$.
2. $x_W = \lim_{n \to \infty} W^n(S_0)$, which is independent of the choice of an initial image S_0.

Fractal image compression relies on this theorem. Suppose the attractor of a contractive map W is the original image to be compressed. From property 2, the original image is obtained from any initial image by applying W. The goal of fractal image compression is to find the contractive map W whose attractor is sufficiently close to the original image and to store the parameters of W instead of the intensity values.

2.2 Image Compression Using IFS

The fractal image compression algorithm we focus on is based on the quad-tree decomposition scheme [5][8]. This scheme was developed for grey-scale images.

Let $R_1, R_2, \ldots, R_p$ be non-overlapping sub-squares of $B \times B$ pixels in an image. These sub-squares are called *range blocks*. The union of the set of range blocks covers the entire image. All $2B \times 2B$ sub-squares, $D_1, D_2, \ldots, D_q$, are called *domain blocks*, where all of the over-lapping sub-squares of this size are taken into account. The collection of all domain blocks is called *the domain pool*. The map W in Sect. 2.1 maps sub-sets of the domain pool to range blocks.

For each R_k, we search for the domain block that approximates R_k with the smallest metric out of the domain pool. A range block that is approximated by an acceptable metric is said to be *covered*. A range block for which acceptable domain blocks are not found is said to be *uncovered*. Note that before calculating metrics, domain blocks are resized to the same size as range blocks by averaging the intensity values of 2×2 pixels. Let the pixel intensities of the resized domain block be $a_1, a_2, \ldots, a_n$ and let those of the range block be $b_1, b_2, \ldots, b_n$. The metric $R(D_j, R_k)$ between a domain block D_j and a range block R_k is defined as $R(D_j, R_k) = \sum_{i=1}^{n}(s a_i + o - b_i)^2$, where s is the scaling factor and o is the offset factor. Differentiating R by s and o, s and o that minimize R are calculated as

$$s = \frac{n \sum_{i=1}^{n} a_i b_i - \sum_{i=1}^{n} a_i \sum_{i=1}^{n} b_i}{n \sum_{i=1}^{n} a_i^2 - (\sum_{i=1}^{n} a_i)^2}, \quad o = \frac{\sum_{i=1}^{n} b_i - s \sum_{i=1}^{n} a_i}{n}. \quad (1)$$

In this case, $R(D_j, R_k)$ is calculated as

$$R(D_j, R_k) = \frac{1}{n}(n \sum_{i=1}^{n} b_i^2 - (\sum_{i=1}^{n} b_i)^2) - \frac{s^2}{n}(n \sum_{i=1}^{n} a_i^2 - (\sum_{i=1}^{n} a_i)^2). \quad (2)$$

The entire fractal image compression scheme is shown in Fig. 1. We call this algorithm EXAUS. The parameter *tolerance* settles the compression ratio and the fidelity. If a range block is uncovered, that is, if *tolerance* is violated for all domain blocks, the range block is divided into four smaller range blocks and the best-match domains for these smaller range blocks are searched. In this case, domain blocks with half the size in the side length are evaluated. To increase the probability of finding good range-domain matches, the domain pool may include sub-squares obtained by rotating domain blocks $90°$, $180°$, $270°$ and $360°$ and flipping them vertically (eight transformations). These sub-blocks are also called domain blocks. After finding the best matches for all range blocks, s and o and the number specifying the best-match domain block are stored for each range block instead of their pixel intensities; as such, the encoding is completed. For s and o, the necessary bit widths to be preserved are empirically known to be 5 and 7, respectively. On the other hand, for decoding the image, the corresponding transformations are repeated. This iteration is started from any initial image. Iterations of 10 to 20 times are sufficient to decode an image with good fidelity.

With regard to the computing time, the iterations in the decoding are not so troublesome. However, the encoding time is often not acceptable due to the many iterations of complex metric computations. For example, in encoding a 256×256 pixel image, the number of range blocks of 8×8 pixels is 1024 and the number of domain blocks of 16×16 pixels is 58081 (not including the domain blocks obtained by rotating and flipping). Therefore, the total number of iterations is about 60 million. For reducing the encoding time, some schemes that try to reduce the number of range-domain comparisons have been proposed [7]. In these schemes, only range-domain comparisons among sub-blocks with the same characteristics are evaluated. Unfortunately, the encoding times using these schemes are still unacceptable.

In the next section, we propose a method for implementing fractal image compression using parallel computation on a reconfigurable hardware. Our method has two advantageous features for reducing the encoding time. The first one concerns accelerating the inside of the innermost loop of EXAUS by using pipeline processing. In each step of the pipeline, one iteration is executed on a PE. Looking at the equations (1)–(2), we can observe that many computations, $\sum_{i=1}^{n} a_i$, $\sum_{i=1}^{n} a_i^2$, $(\sum_{i=1}^{n} a_i)^2$, $\sum_{i=1}^{n} b_i$, $\sum_{i=1}^{n} b_i^2$, $(\sum_{i=1}^{n} b_i)^2$, $(n\sum_{i=1}^{n} a_i^2 - (\sum_{i=1}^{n} a_i)^2)$ and $(n\sum_{i=1}^{n} b_i^2 - (\sum_{i=1}^{n} b_i)^2)$, can be performed in advance out of the innermost loop and have to be done only once for each domain block and range block. Therefore, inevitable computations inside of the innermost loop are other computations for s and R and comparisons. More precisely, the indispensable burden is the part $\sum_{i=1}^{n} a_i b_i$. The method presented in this paper lightens this burden by executing it in a pipeline. The second feature concerns reducing the number of iterations of metric computations. In EXAUS, the metric computations for all pairs of range and domain blocks are done. With regard to the fidelity of the decoded image, it is not necessary to evaluate all range-domain matches. Our scheme omits these useless computations. This omission is owing to the dynamic and partial reconfigurability of the hardware. In the next section, we describe these features in detail.

```
min_R = tolerance;
For (j = 1 ; j ≤ num_range ; j++) {
    For (k = 1 ; k ≤ num_domain ; k++) {
        compute s;
        if (0 ≤ s < 1.0)
            if (R(D_j, R_k) < min_R)
                { min_R = R(D_j, R_k); best_domain[j] = k; }
    }
    if (min_R == tolerance)
        set R_j uncovered and divide it into 4 smaller blocks;
}
```

Fig. 1. Pseudocode of EXAUS.

3 Encoder Architecture

In this section, we propose an encoder of fractal image compression implemented on dynamically reconfigurable hardware. In this paper we assume that the dynamically reconfigurable hardware is constructed from fine-grain programmable logic elements (LEs) with a register and programmable wires among them such as Xilinx XC6200 series FPGAs [2] or Atmel AT40K FPGAs [3]. We also suppose that these LEs and wires can be configured by setting configuration data dynamically and partially. Some kind of architectures of LEs have been proposed such as Look-Up-Table (LUT) based architectures, multiplexer based architectures and others. However, generalizing these LEs as only programmable logic function units with limited inputs and a output, for example, 3 or 4 inputs, we consider the encoder architecture independent from the architectures of LEs and realized as wired logic of LEs.

3.1 A Pipelined PE Network

Fig. 2 shows the overview of the proposed encoder implemented on dynamically reconfigurable hardware. The PE performs computations inside the innermost loop of EXAUS. Each PE is devoted to a given range block and finds the best-match domain among all domain blocks that are given through the data path in pipeline. In each step of the pipeline, one iteration of the inside of the innermost loop is executed on each PE. After finding the best match, the PE informs the best-match domain to the control unit. Then, the PE is reconfigured for another range block with the configuration data given

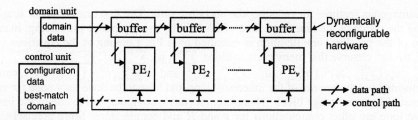

Fig. 2. Overview of the Encoder.

by the control unit. These PEs are reconfigured until no uncovered range block exists. All PEs are connected by the data path and perform the calculations in a pipeline manner while transferring the domain data. The domain unit inputs the domain data into the PE network until all range blocks are checked against all domain blocks. A buffer transfers domain data to the PE connected with it and the neighboring buffer through the data path without the pipeline stole during the reconfiguration of that PE.

3.2 Dynamically Reconfigurable PE

Here, we describe the PE architecture in detail. A PE performs the calculations inside of the innermost loop of EXAUS. As stated in Sect. 2.2, the computation of $\sum_{i=1}^{n} a_i b_i$ is the burden for computing R. Therefore, it is desirable that $\sum_{i=1}^{n} a_i b_i$ is calculated in parallel. The basic implementation is shown in Fig. 3. The parameters $a_1, a_2, \ldots, a_n$ are enter the PE in parallel and $\sum_{i=1}^{n} a_i b_i$ is calculated in the pipeline. The hardware component for calculating $\sum_{i=1}^{n} a_i b_i$ is denoted by P1. The hardware component for calculating s, o and R is denoted by P2. P2 finds the minimum R and stores the identity number of the domain block that provides it. These computations are also performed in the pipeline. In each step of the pipeline, one iteration of the inside of the innermost loop is completed on a PE. Precomputable parameters for the domain block, such as $\sum_{i=1}^{n} a_i$, $\sum_{i=1}^{n} a_i^2$ and others, are given to the PEs through the data path in parallel cooperating with the pipeline. The other precomputable parameters for the range block are stored in P2. In order to implement P1, we need n multipliers and $n-1$ adders. On the other hand, the computations performed in P2 are three multiplications, three comparisons, and one division. Since the hardware resources needed for P2 and the buffer are much smaller than those for P1, it is important to reduce the hardware resources for P1 for total resource reduction. By reducing the resources for a PE, we can implement many PEs at the same time and increase the number of range blocks processed in parallel.

Now, we describe an efficient implementation method using reconfigurability. Note that the multipliers in P1 in Fig. 3 are implemented with adders and shifts. Since shifts are implemented by wiring on reconfigurable devices, it is important to reduce the number of adders. Therefore, we implement P1 as follows. Let the integer b_i be presented as $b_{i,l} b_{i,l-1} \cdots b_{i,1}$ in the binary presentation, where $b_{i,j}$ is 0 or 1, and let the bit width of a_i's and b_i's be l. Then, multiplication $a_i b_i$ is presented as $a_i b_i = \{(a_i b_{i,l}) \ll (l-1)\} + \{(a_i b_{i,l-2}) \ll (l-2)\} + \cdots + \{a_i b_{i,1}\}$, where $a \ll j$ denotes the j-bit shift of a to the left. The multiplication is implemented by $l-1$ adders. In this case, the number of adders for computing $\sum_{i=1}^{n} a_i b_i$ at P1 is

$$(l-1)n + (n-1) = ln - 1. \qquad (3)$$

On the other hand, in the case of reconfiguring a PE for a range block, we can make use of the fact that $b_{i,j}$ is a constant. Namely, we can get rid of the part $(a_i b_{i,j})$ if $b_{i,j}$ is 0.

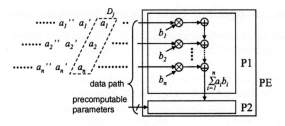

Fig. 3. PE architecture.

First, we rearrange $\sum_{i=1}^{n} a_i b_i$ as follows.

$$\sum_{i=1}^{n} a_i b_i = \{(a_1 b_{1,1} + a_2 b_{2,1} + \cdots + a_n b_{n,1})\}$$
$$+ \{(a_1 b_{1,2} + a_2 b_{2,2} + \cdots + a_n b_{n,2}) \ll 1\}$$
$$\cdots$$
$$+ \{(a_1 b_{1,j} + a_2 b_{2,j} + \cdots + a_n b_{n,j}) \ll (j-1)\}$$
$$\cdots$$
$$+ \{(a_1 b_{1,l} + a_2 b_{2,l} + \cdots + a_n b_{n,l}) \ll (l-1)\}. \quad (4)$$

The term of the j-th row of (4),

$$(a_1 b_{1,j} + a_2 b_{2,j} + \cdots + a_n b_{n,j}), \quad (5)$$

can be computed in a different way by inverting $b_{i,j}$'s for all i, $1 \leq i \leq n$, as follows.

$$\sum_{i=1}^{n} a_i - (a_1 \overline{b_{1,j}} + a_2 \overline{b_{2,j}} + \cdots + a_n \overline{b_{n,j}}). \quad (6)$$

If the number of non-zero bits among $b_{1,j}, b_{2,j}, \ldots, b_{n,j}$ is more than $\frac{n}{2}$, we configure P1 using (6) instead of (5). This is done for all j. Then, the number of adders needed to implement P1 is

$$\frac{nl}{2} - 1 \quad (7)$$

in the worst case. Therefore, the number of adders is reduced at least by half. The worst case happens when the number of non-zero bits of $b_{1,j}, b_{2,j}, \ldots, b_{n,j}$ is $\frac{n}{2}$ or $\frac{n}{2} + 1$ for all $j (1 \leq j \leq l)$. In practice, since b_i's $(1 \leq i \leq n)$ are intensities of a sub-block of a digital image, $b_{i,j}$'s $(1 \leq i \leq n)$ tend to be the same, especially when j is close to l. Therefore, the worst case rarely happens and the number of adders is expected to be much smaller than that of the worst case. In Sect. 4, the scheme's experimental results show the reduction of adders and resources for the PE.

$\sum_{i=1}^{n} a_i b_i$ discussed here can be implemented by distributed arithmetic (DA) using memories [10], if LEs are LUT based and can be used as memories. This implementation can be space efficient. However, the space efficiency is terribly affected by the number of inputs, M, of LEs (i.e., the number of content bits possible for a LUT). When M is 3, the DA implementation is not comparable even with the our worst case implementation for practical n and l. In addition, the DA implementation is limited to LUT based architectures. Therefore, we would not adopt the DA implementation.

3.3 A Compression Algorithm Using Dynamic Reconfiguration

Next, we describe the improvement of EXAUS. We use the dynamic partial reconfigurability of the pipelined PE network stated in Sect. 3.1. In this PE network, each PE can be reconfigured without the pipeline stole since buffers transfer the domain data. Therefore, all PEs do not have to be reconfigured at the same time. In EXAUS, all domain blocks are checked for each range block. However, by choosing the appropriate *tolerance* and finding only one range-domain match that accepts the *tolerance* for each range block, we can obtain a decoded image with sufficiently good fidelity.

Generally, the fidelity of the decoded image is evaluated with the signal-to-noise ratio (*SNR*) defined as $SNR = 10\log_{10}(dr(S)^2/d_e(S,S'))$ (dB). Here, $dr(S)$ is the dynamic range of the original image S and $d_e(S,S')$ is the metric between S and the decoded image S'. $d_e(S,S')$ is defined as $d_e(S,S') = \sum_{i=1}^{t}(\mu_i - \mu_i')^2$, where μ_i's and μ_i's are pixel intensities of S and S'. An *SNR* of approximately 28 to 32 dB is sufficient for fine fidelity.

Now, we present the improved algorithm that omits the exhaustive range-domain matches in Fig. 4. We call this algorithm OMIT. This improvement can be implemented on hardware by dynamically and partially reconfiguring each PE as soon as the first acceptable range-domain match is found.

```
For (j = 1 ; j ≤ num_range ; j++) {
    flag = 0;
    For (k = 1 ; k ≤ num_domain ; k++) {
        compute s;
        if (0 ≤ s < 1.0)
            if (R(D_j, R_k) < tolerance)
                { best_domain[j] = k; flag = 1; break; }
    }
    if (flag == 0)
        set R_j uncovered and divide it into 4 smaller blocks;
}
```

Fig. 4. Pseudocode of OMIT.

4 Experiments

4.1 Resources for PE

First, we evaluate the number of adders needed to implement a P1 and the resources for PEs. Table 1 shows the number of adders needed to implement P1 when the range size is 4×4 pixels and 8×8 pixels for two benchmark images, "Girl" (256×256 pixels, 8 b/p) of SIDBA and "Lenna" (480×512 pixels, 8 b/p). NO-RECONF denotes the number of adders needed to implement a P1 without using reconfigurability. WORST of RECONF

Table 1. Number of adders.

| Range size | NO- | RECONF | | |
| (pixels) | RECONF | WORST | AVERAGE | |
			Girl	Lenna
8×8	511	255 (49.9%)	150 (29.3%)	136 (26.6%)
4×4	127	63 (49.6%)	36 (28.3%)	29 (22.8%)

Table 2. Number of LEs for a PE

Range size (pixels)	NO-RECONF	RECONF		
		WORST	AVERAGE	
			Girl	Lenna
8×8	10532	6548 (62.2%)	4674 (44.4%)	4642 (44.1%)
4×4	3535	2591 (73.3%)	2109 (59.7%)	2057 (58.1%)

denotes the number of adders using reconfigurability in the worst case and its ratio to that of NO-RECONF. The values of WORST are calculated by (7) and those of NO-RECONF is calculated by (3). AVERAGE of RECONF denotes the number of adders in the average case using reconfigurability. Table 1 shows that the number of adders is significantly reduced by utilizing reconfigurability.

Table 2 shows the total number of LEs needed for the PE. Again, NO-RECONF denotes the number of LEs for a PE implemented without using reconfigurability. WORST of RECONF denotes the number of LEs in the worst case using reconfigurability and its ratio to the number of NO-RECONF. AVERAGE of RECONF denotes these average numbers using reconfigurability. Here, we assume that an LE is an LUT with four inputs and an output. Carry save adders are used for the adders for P1. The final summation and the carry vector provided by these carry save adders is added by a carry look ahead adder at the end. Therefore, one carry look ahead adder is implemented in P1. We evaluated the number of LEs by counting 1-bit full adders needed for carry save adders of P1 and assuming that a 1-bit full adder is implemented with two LEs. Altera FPGA mapping tool MAX+Plus II was used to evaluate the number of LEs for P2 and the carry look ahead adder. The number of LEs for P2 is 1556 when the range-size is 8×8, and 1343 when the range-size is 4×4.

Table 2 indicates that the value of WORST is approximately 60% of that of NO-RECONF when the range-size is 8×8. When the range-size is 4×4, the ratio is approximately 70%. These reductions are important for implementing many PEs under the resource limitation to accelerate the encoding. Although further investigations for dynamic allocation of PEs are needed, the average case suggests that dynamically changing the number of PEs and making full use of limited resources is promising.

4.2 Efficiency of The Improved Algorithm

Here, to confirm the efficiency of OMIT, we evaluate the number of iterations of the inside of the innermost loop by computer simulation. The numbers of iteration for EXAUS and OMIT are compared under the same SNRs for "Girl" and "Lenna." SNRs are evaluated with the decoded images obtained after 20 iterations of the mapping. The image was divided into 8×8-size range blocks at the beginning. Uncovered range blocks were divided into four 4×4-size range blocks. Domain blocks obtained by rotating and flipping are not included in the domain pool. We evaluated the summation of iterations for 8×8 and 4×4-size range blocks.

In each step in our pipelined PE network, a domain block is inputted to the network. Therefore, the number we evaluated indicates the encoding time when one PE is implemented. Fig. 5 shows the evaluated values for "Girl" and "Lenna." For "Girl," the values for OMIT are from 55% to 70% of EXAUS when the SNR is from 28 to 30 dB. For "Lenna," these ratios are from 10% to 30%. Note that for "Lenna," no 8×8-size range block is divided in EXAUS when the SNR is 30 dB. Therefore, the SNR does not fall to less than 30 dB, and the number of iterations cannot be reduced more. The ratios for "Lenna" are given by comparing the values of OMIT and the minimum value of

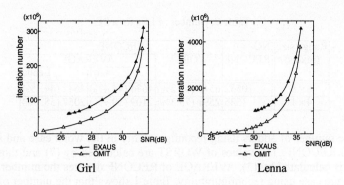

Fig. 5. Number of iterations

EXAUS. As a result, the improvement of the algorithm reduces the encoding times for these images to 10 to 55% with fine fidelity.

5 Conclusion

In this paper, we have proposed a method for implementing fractal image compression on a dynamically reconfigurable architecture. In our method, PEs are specialized for image blocks and concurrently perform the computations in a pipeline manner. Furthermore, plural PEs work in parallel. By using the reconfigurability, the reconfigurable resources needed for each PE is reduced to 60% to 70% of that of the straightforward implementation in the worst case. This reduction is quite important for implementing many PEs under the resource limitation in order to accelerate the encoding. In addition, an improved fractal image compression algorithm omitting useless computations, which can be achieved by dynamic and partial reconfigurability, can reduce the encoding time to 10 to 55%. As a result, by applying these utilization of the reconfigurability to the proposed pipelined PE network, the encoding time of the PE network can be significantly reduced under the reconfigurable resource limitations.

In future works, we plan to implement the presented encoder and confirm the effectiveness of the presented PE network in practice.

References

1. T. Miyazaki, "Reconfigurable Systems: A Survey," *Proc. of ASP-DAC '98*, pp. 447–457, Feb. 1998.
2. Xilinx, *XC6200 Field Programmable Gate Arrays*, Apr. 1997.
3. Atmel, *AT40K FPGAs*, Dec, 1997.
4. J. R. Koza, F. H. Bennett III, J. L. Hutchings, S. L. Bade, M. A. Keane, and D. Andre, "Evolving Computer Programs using Rapidly Reconfigurable Field-Programmable Gate Arrays," *Proc. of FPGA '98*, pp. 209–219, Feb. 1998.
5. Y. Fisher (Ed.), *Fractal Image Compression: Theory and Application*, Springer, 1996.
6. M. F. Barnsley, V. Ervin, D. Hardin, and J. Lancaster, "Solution of an Inverse Problem for Fractals and Other Sets," *Proc. of Natl. Acad. Sci USA*, 83:1975–1977, Apr. 1986.
7. A. E. Jacquin, "Fractal Image Coding: A Review," *Proc. of the IEEE*, vol. 81, no. 10, pp. 1451–1465, Oct. 1993.
8. Y. Fisher, *Fractal Image Compression*, SIGGRAPH Course Notes, 1992.
9. M. F. Barnsley, *Fractals Everywhere*, AK Peters, 1993.
10. S. A. White, "Applications of Distributed Arithmetic to Digital Signal Processing: A Tutorial Review," *IEEE ASSP Magazine*, pp. 4–19, July. 1989.

Plastic Cell Architecture: A Dynamically Reconfigurable Hardware-based Computer

Hiroshi Nakada, Kiyoshi Oguri, Norbert Imlig, Minoru Inamori, Ryusuke Konishi, Hideyuki Ito, Kouichi Nagami, and Tsunemichi Shiozawa

NTT Optical Network Systems Laboratories
1-1 Hikarinooka Yokosuka-shi Kanagawa-ken 239-0847 Japan
{nakada, oguri, imlig, ina, ryusuke, hi, nagami, shiozawa}@exa.onlab.ntt.co.jp

Abstract. This paper describes a dynamically reconfigurable hardware-based computer called the Plastic Cell Architecture (PCA). PCA consists of dual-structured sea-of-cells that consist of a built-in part and a plastic part. The built-in part forms 'cellular automata' while the plastic part looks like an SRAM-based FPGA. We detail the design flow especially how to implement logic onto the processing element. The advantages of PCA, considering VLSI implementation and several application examples, are also described.

1 Introduction

During the past few years, 'system on silicon' which can realize a total information processing system on one chip has been widely discussed with the assumption of the future dense VLSI era. Most of the systems proposed, however, are just the integration of conventional elements onto one large silicon chip, where down sizing and performance improvement will be achieved as a natural consequence. Given the inherent limitations of this idea, an entire new approach to next generation computer systems is needed that will make the best use of 'system on silicon'.

In general, there are two approaches to execute process at higher speed on a programmable system; one is reducing the clock cycle (or shortening the critical paths) in the hardware, the other is to extract and utilize parallelism in the process. In other words, the former approach challenges the 'time domain' while the latter approach tackles the 'space domain'. The former approach has common in most of the processor-based systems that execute software described by the computer language. On the other hand, many combinations of FPGA-based systems and Hardware Description Language (HDL) have been proposed in recent years, and these takes the latter approach. Unfortunately, FPGAs have not yet become a viable alternative to general purpose processors.

We have proposed both types of reconfigurable systems for programmable telecommunication systems, namely, FPGA-type [1], and processor type [2]. Moreover, we also engaged in developing an HDL-based CAD system [3]. We revisited the loosely-coupled relationship between the HDL and reconfigurable devices again given our experience and the next generation in VLSI technology.

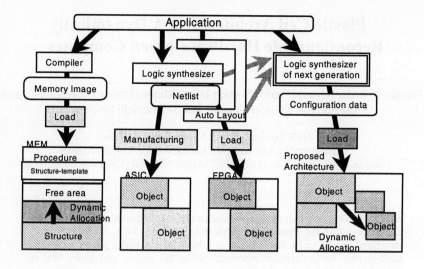

Fig. 1. The common and different points of implementation of application

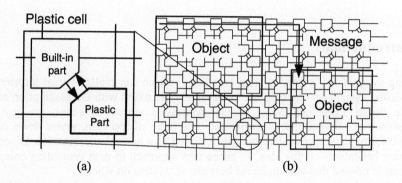

Fig. 2. (a) Plastic cell (b) Cell array with two hardware-objects communicating by message passing

Figure 1 compares implementing functionality in either software or hardware. While software is compiled into a memory image, hardware is synthesized into a static circuit image. Starting from this netlist representation a mask layout is generated for ASIC production in a fabrication or a configuration bit-stream for downloading into an FPGA. The major difference lies in the runtime behavior. Software supports dynamic memory allocation. On the other hand, dynamic reconfiguration is basically not supported by hardware. Since many algorithms operate on dynamically generated complex data structures, software was traditionally thought to be the only way to implement such functionality. Algorithms implemented as wired logic can also be realized by memory. Until the appearance of the FPGA, nobody thought about exploiting this dynamic nature of wired logic.

Our news approach to an original processing architecture has two aspects, namely, (i) HDL which provides the semantics of dynamic 'generation' and 'deletion' of the processing element, and (ii) hardware which realizes the above 'generation' and 'deletion' operation dynamically. We started this project from the basic hardware architecture.

This paper mainly describes the hardware mechanism and lower design process of the new reconfigurable computer, called the Plastic Cell Architecture (PCA). Section 2 describes the dual structure of PCA in detail. Section 3 mentions how to map logic to processing element of PCA. This involves a kind of VLSI layout problem. The advantage of PCA with considerations for VLSI implementation and applications are described in section 4. Finally, we conclude this paper by mentioning several related works and future study.

2 Architecture

2.1 Dual Structure

In order to allow dynamic runtime reconfiguration after the initial state has been set, a system must be composed of two distinct elements: a variable part as the place holder of the new configuration and a fixed part responsible for setting the former. In the example of a CPU-memory configuration, the memory corresponds to the variable part and the CPU to the fixed part. The same is true for FPGAs. Their configuration SRAM memory corresponds to the variable part while the attached CPU can be regarded as the fixed part.

Though our proposed architecture is also composed of two parts, the main differences from the above examples are as follows:
 - Both parts are simple, and they are interconnected within a single LSI chip.
 - The system consists of a huge mesh of above connected pairs (pairs of fixed and variable parts).

Figure 2.a shows the 'plastic cell' of our architecture. We call the variable part the 'plastic part' and the fixed part the 'built-in part'. Both parts are connected. In order to exploit parallelism, the cells are arranged in an orthogonal array as indicated Fig.2.b. The corresponding parts of adjacent cells are also interconnected.

The functions of the built-in part are fixed and consist of plastic part reconfiguration, data I/O from/to plastic part, and mutual communication with the adjacent plastic parts. The array of built-in parts forms a network of 'cellular automata' [4]. A cellular automation is a processing model in which a collection of identical state machines exchange information with their neighbors by a simple set of rules. A plastic part is composed of several look-up table (LUT) memories, which can achieve various logic operations and a bypass function from one adjacent plastic part to another adjacent plastic part. This part can be reconfigured independently of other subsystems in the chip. The plastic part can also organize the memory object for data storage.

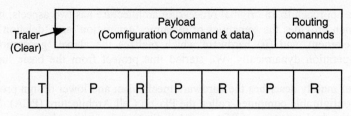

Fig. 3. Series of messages for inter-object communication (Two types)

2.2 Object and Communication

In a PCA, a processing module is called an 'object', its dynamic generation/deletion triggers the operation of another 'object'. For example, the output of an object '8-bit ALU' requires generation of an object '4-bit comparator', and so on. Objects are formed by several plastic parts, and generation, deletion, and data communication procedures are realized simply by message passing via built-in parts.

In order to dynamically allocate new objects the following steps are necessary:
1. Free area is searched and reserved
2. A new object is generated by injecting its behavior into the plastic part
3. The newly formed object is enabled for runtime operation
4. The communication path between built-in and plastic part is opened to allow message passing
5. Operation of the hardware object, communication with other objects
6. After receiving a release message, the area is freed for new allocation

Objects and generated objects have a mother-child relationship. Thus, only the mother object can send a release message and free the resources of an allocated child.

In order to enable proper configuration, the route of each message should be fixed. In fact, we employ exact routing rather than adaptive message routing. A series of messages is formed into a variable length packet as shown by Fig.3. Each packet is composed of a series of commands and data, which are categorized by three types of frames: routing command, payload, and trailer, which are processed by the built-in part. More detailed discussion for communication protocol was found in another presentations [4],[5],[6].

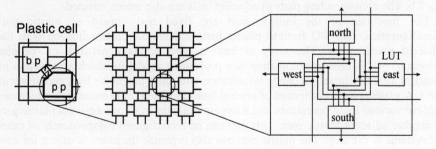

Fig. 4. Example of a plastic part architecture

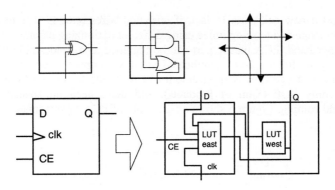

Fig. 5. Examples of logic circuit elements with plastic part

2.3 Plastic Part and Asynchronous Circuit

Figure 4 shows an example of a plastic part architecture. Each plastic part consists of an array of 'basic cells'. A basic cell consists of four 4-input LUTs as shown in the figure. The most important feature is its regular structure. In general FPGAs, each cell has many components such as flip-flops, selectors, and transistor switches. On the other hand, a plastic part has only LUTs and wire. Only these components enable any binary logic operation for any direction of input/output signals, data branch, data transfer, and data storage. These examples are shown in Fig.5. This structure is called the 'sea of LUTs'.

One of the most remarkable points is that there is no dedicated flip-flop in the plastic part, that is, the flip-flop is constructed by combining and looping back of LUTs. Obviously it is not practical to use a global clock. Consequently, we regard all the logic circuit on the plastic part as asynchronous circuit, which has no clock signal. This is suitable for asynchronous pipeline architecture on for the built-in part.

3 Design Flow of Circuit on Plastic Part

In order to construct an object on PCA's plastic parts, the object is expanded into logic circuits described by netlists. At that time, a layout CAD technique is required for implementing netlists onto cells automatically and efficiently. As stated above, plastic part is a kind of SRAM type FPGA. In general, conventional FPGA layout system consists of following stages:

Stage 1: Technology mapping for technology independent logic circuits into LUTs.
Stage 2: Placement of each LUT.
Stage 3: Routing each wire between LUTs.

Usually, above each stage is optimized independently because general FPGA is so inhomogeneous that it is necessary to relocate one or more placed/routed modules in order to insert a new cell/wire. In fact, if unrouted wires remain, it is necessary to back an upper stage and to re-optimize using different optimizing parameters.

On the other hand, PCA plastic part can be regarded as a 'relocatable PLA'. This feature is very useful to place complicated circuits compactly such as finite state machine. If PLA design flow is applicable for PCA layout, the output format of logic synthesis is only SOP (Sum of Products), and there need no special stage for placement and routing. Figure 6 shows an example of PCA layout image using PLA design.

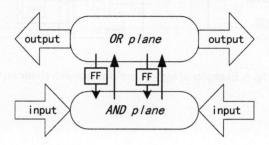

Fig. 6. Example of PCA layout image using PLA design scheme

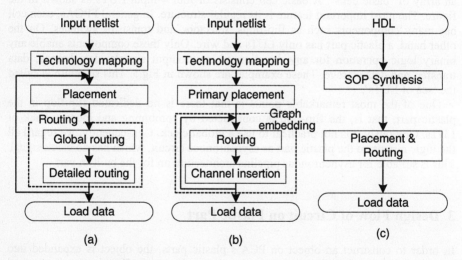

Fig. 7. (a) Conventional FPGA design flow (b) PCA plastic part layout design using conventional ASIC design technique (c) PCA plastic part layout design using PLA design scheme

Figure 7 compares PCA layout design flow to the usual FPGA design flow. As shown in Fig.7.b, conventional ASIC design technique is still useful for some cases, such as inter module communications or uneven distributed flip-flop circuits. Even in these cases, plastic part layout algorithm is simpler than usual FPGA case. For example, a classical LSI layout model (grid model) [8] for formulation and Dijkstra method [9] for routing algorithm are directly applicable for PCA plastic part layout.

4 Discussion

4.1 Advantage of PCA - VLSI Implementation -

In this section, we will discuss the advantages of PCA with regard to future VLSI implementation. As stated above, it is expected that memory fabrication technique can be applied to PCA because of its simple and regular structure. Figure 8 compares the number of transistors per chip of several types of VLSIs. In this figure, the Pentium II, fabricated using the newest 0.35micron technology, is composed of only 7.5 million transistors [10], while DRAM consists of 64 million transistors. On the other hand, the newest Xilinx FPGA contains 25 million transistors if 0.25micron technology is applied [10]. According to the SIA roadmap [11], the densest MPU using 0.25micron technology will place 11 million transistors in a chip. This shows that the regular memory structure significantly increases chip density. Of course, this comparison is not exactly fair because the object that should be compared is the actual logic circuit on the chip.

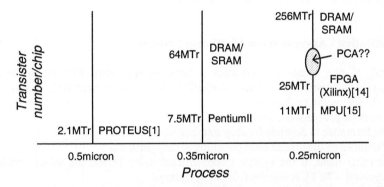

Fig. 8. Recent VLSI density trends [10],[11]

Table I Forecast of future VLSI chip density from SIA roadmap 1997 edition [11]

Year	1997	1999	2001	2003	2005	2009	2012
Process (nm)	250	180	150	130	100	70	50
DRAM (bit/chip)	267M	1.07G	1.7G	4.29G	17.2G	68.7G	275G
MPU (transistors/chip)	11M	21M	40M	76M	200M	520M	1.40B

Though above discussion itself is not so meaningful, consider the trend in VLSI technology. Table 1 is a forecast of future VLSI chip density [11]. It is noticeable that the density of memory will be quadruple every 3 years from now on, while MPU density will fair to double every 3 years. That is, while the difference in density between memory and MPU is about 24 times in 1997, the difference will increase to 56 times in 2003, and 200 times in 2012. Since even the regular FPGA can realize higher density than MPU, the PCA, which has a pure memory structure in its plastic parts, will strongly benefit from this memory technology trend.

For this reason, it is expected that the scale of logic in a PCA will quickly catch up with that of an MPU given the trend in process technology. Let us consider 32-bit array multiplier for example. In general, a 32-bit array multiplier costs about 1000 logic gates because about 10 logic gates are required for a 1-bit full adder. This would require about 51 thousand transistors with ASIC realization if one gate costs four transistors. Suppose the same function is implemented on a PCA, it would costs about 2000 basic cells for a 32-bit array multiplier because a 1-bit full adder can be constructed with 2 basic cells including wiring. In total, about 1.3 million transistors are required for PCA implementation assuming 640 transistors per basic cell. Though PCA implementation wastes about 25 times more transistors than ASIC implementation, it differs by only 2.5 times in terms of area if the PCA VLSI is 10 times denser than the ASIC. Furthermore, if the difference of density widens over 25 times, PCA array multiplier would be smaller than the ASIC version. While this comparison is not directly useful because of built-in part area, the area difference of both realizations will converge in the near future.

4.2 Application - Asynchronous Multi-chip Systems -

In general, if one wants to construct a large system with FPGAs, multi-chip architecture is indispensable because of one FPGA has insufficient area. In this case, there are the following problems:
(P1) Pin neck problem occurs between chips,
(P2) Performance is degrade by chip interconnection,
(P3) Partitioning the circuit onto the chips is very difficult.
On the other hand, if the system is constructed using PCAs, the above problems will be resolved. The PCA has the following features:
(F1) Chip has pins only between built-in parts,
(F2) Communication between objects (or chips) is asynchronous and multiplexed,
(F3) Homogeneous and non-hierarchical structure,

Feature (F1) will solve problem (P1). Problem (P2) will be eliminated by setting high performance paths between chips. This is not so difficult given feature (F2) because communication between chips is the same as the communication between objects in a chip. Furthermore, architecture partitioning will not be a problem because feature (F3) is covers the chip level and also the multi-chip (or system) level.

Telecommunication systems are one example of an application that fully utilizes the above features. Most public telecommunication systems realize low layered high speed processing by synchronous circuits while IP packets or ATM cells are asynchronously multiplexed in the communication channel. This architecture demands complicated buffer structure. If we select the appropriate frame format for inter-object communication, PCA may become a key device for programmable telecommunication systems. The authors are now considering a telecommunication system model based on PCA.

5 Conclusion

A new dynamically reconfigurable hardware-based computer (PCA) was introduced in this paper. PCA, which is based on built-in and plastic parts, is a cross between an FPGA and cellular automata. A von Neumann architecture is obtained if only one cell part is extracted. The memory corresponds to the plastic part and the built-in part can be compared with a mini CPU engine.

Some runtime reconfigurable SRAM based FPGAs have recently entered the market. However, a CPU is required for configuration. A more decentralized configuration approach is proposed in the wormhole system [12] [13]. The difference between their approach and ours is the fine granularity of the PCA. A highly fault-tolerant cellular architecture that mimics nature and biological concepts was proposed in [14]. It is widely known that many algorithms can be speeded up if memory can realize useful logic functions as well as data storage. This is the case with the plastic part of our cell.

We are now simulating the architecture in order to balance performance, resource, and VLSI implementation issues. The enhancement of the HDL for 'generation' and 'deletion' operation is also being introduced.

References

1. Ohta, N., Nakada, H., Yamada, K., Tsutsui, A., Miyazaki T.: PROTEUS: Programmable Hardware for Telecommunication Systems. Proc. ICCD '94 (1994) 178–183
2. Inamori, M., Ishii, K., Tsutsui, A., Shirakawa, K., Nakada, H.: A News Processor Architecture for Digital Signal Transport Systems. Proc. ICCD '97 (1997) 157–162
3. Nakamura, Y., Ogu.ri, K., Nagoya, A.: Synthesis from Pure Behavioral Descriptions. In Camposano, R., Wolf, W. (eds.): High-Level VLSI Synthesis. Kluwer Academic Publishers (1991) 205–229
4. von Neumann, J.: The Theory of Self-Reproducing Automata. Univ. of Illinois Press (1966)
5. Nagami, K., Oguri, K., Shiozawa, T., Ito, H., Konishi, R.: Plastic Cell Architecture for general Purpose Reconfigurable Computing. Proc. FCCM '98 (1998)
6. Ito, H., Oguri, K., Nagami, K., Konishi, R., Shiozawa, T.: Plastic Cell Architecture for General Purpose Reconfigurable Computing. Proc. RSP"98 (1998)
7. Oguri, K., Imlig, N., Ito, H., Nagami, K., Konishi, R., Shiozawa, T.: General Purpose Computer Architecture Based on Fully Programmable Logic. Proc. ICES '98 (1998)
8. Ullman, J.D.: Computational Aspects of VLSI. Rockville, MD: Computer Science Press (1983)
9. Aho, A.V., Hopcroft, J.E., Ullman, J.D.: Data Structures and Algorithms. Addison-Wesley Publishing Co., Inc. (1983)
10. Xilinx Inc.: XC4000XV Product Highlights. http://www.xilinx.com/products/xc4000xv.htm (1998)
11. Semiconductor Industry Association (SIA).: The National Technology Roadmap for Semiconductors, 1997 Edition. (1997)
12. Bittner, R., Athanas, P., Musgrove, M.: Colt: An Experiment in Wormhole Run-time Reconfiguration. Proc. SPIE Photonics East `96 (1996)
13. Luk,, W., Shirazi, N., Cheung, P.Y.K.: Compilation Tools for Run-Time Reconfigurable Designs. Proc. FCCM'97 (1997) 56–65
14. Nussbaum, P., Marchal, P., Piguet, Ch.: Functional Organisms Growing on Silicon. Proc. ICES'96 (1996) 139–151

Leonardo and Discipulus Simplex:
An Autonomous, Evolvable Six-Legged Walking Robot

Gilles Ritter, Jean-Michel Puiatti, and Eduardo Sanchez

Logic Systems Laboratory, Swiss Federal Institute of Technology,
CH – 1015 Lausanne, Switzerland
E-mail: {name.surname}@di.epfl.ch

Abstract

Evolutionary systems based on genetic algorithms (GAs) are common nowadays. One of the recent uses of such systems is in the burgeoning field of evolvable hardware which involves, among others, the use of FPGAs as a platform on which evolution takes place. In this paper, we describe the implementation of Discipulus Simplex, an autonomous evolvable system that controls the walking of our six-legged robot Leonardo. Discipulus Simplex is based on GAs and logic system reconfiguration, and is implemented into a single FPGA. As a result, we obtained a fully autonomous robot which is able to learn to walk with the aid of on-line evolvable hardware without any processors or off-line computation.

1 Introduction

The idea of applying the biological principle of natural evolution to artificial systems, introduced more than four decades ago, has seen impressive growth in the past few years. Usually grouped under the term *evolutionary algorithms* or *evolutionary computation*, we find the domains of genetic algorithms, evolution strategies, evolutionary programming, and genetic programming [1]. As a generic example of artificial evolution, we consider genetic algorithms.

A genetic algorithm is an iterative procedure that involves a (usually) constant-size population of individuals, each represented by a finite string of symbols, *the genome*, encoding a possible solution in a given problem space. This space, referred to as the *search space*, comprises all possible solutions to the problem at hand. The algorithm sets out with an initial population of individuals that is generated at random or heuristically. In every evolutionary step, known as a *generation*, the individuals in the current population are *decoded* and *evaluated* according to some predefined quality criterion, referred to as the *fitness*, or *fitness function*. To form a new population (the next generation), individuals are *selected* according to their fitness, and then transformed via genetically-inspired operators, of which the most well known are *crossover* ("mixing" two or more genomes to form novel offspring) and *mutation* (randomly flipping bits in the genomes). Iterating this evaluation-selection-crossover-mutation procedure, the genetic algorithm may eventually find an acceptable solution, i.e., one with high fitness.

One of the recent uses of evolutionary algorithms is in the burgeoning field of *evolvable hardware*, which involves, among others, the use of FPGAs as a platform on which evolution takes place [2,3]. In this paper, taking advantage of this new methodology, we use an autonomous evolvable system to control the walking of a six-legged robot.

The learning of walking behavior in legged robots is a well-known problem in the autonomous robotic field, because it is non-trivial to solve due to the complexity of the movement synchronization. Several approaches have been used to solve this problem, such as the use of reinforcement learning [4] or subsumption [5]. The proposed solutions are generally bio-inspired and are implemented with one or more processors which can be embedded in the robot or off-line. We believe it is interesting to have a circuit that controls the motions of the robot and can modify its functionality in order to find the right behavior. Thompson *and al.* use such an approach [2] (pp. 156-162), exploiting an evolvable system in order to control the motion of a two-wheeled robot. The resulting system (robot and evolvable hardware) was quite large and for practical reasons they had to simulate the world in which the robot evolved.

In our approach we want to avoid the use of processors and of off-line computations generally needed to solve the walk problem. We use a small autonomous 6-legged robot – Leonardo [6] – as a platform, and we develop *Discipulus Simplex*, an autonomous evolvable hardware walking control. Discipulus Simplex controls the motion of Leonardo's legs, and by way of evolution changes its behavior in order to obtain correct walking behavior. The entire hardware fits within a single FPGA. The result is small autonomous 6-legged robot (24cm x 20cm, weighting 1 kg) controlled by an autonomous evolving system implemented in an FPGA. The most important point is that this system evolves by itself with no processor or external aid.

The remainder of this paper is organized as follows. Section 2 describes briefly the mechanics and electronics of the robot Leonardo. Section 3 describes Discipulus Simplex, the evolvable system that realizes the learning of the walking behavior. Finally, in Section 4 we present some conclusion.

2 *Leonardo:* **An Autonomous Six-Legged Robot**

Leonardo [6] is an autonomous six-legged robot developed as an experimental platform for bio-inspired algorithms. The main features of this robot are its low cost and its small size. These goals were attained by the use of efficient and simple mechanic and electronic parts.

Leonardo's mechanics combine originality, simplicity, and efficiency. The robot has 13 degrees of freedom: 2 degrees of freedom (elevation and propulsion) in each of the 6 legs, and 1 degree of freedom in the body. The latter is the most original mechanical part of the robot and allows the robot to make efficient turns. The body articulation is shown in Figure 1(a). Figure 1(b) shows a front view of one of Leonardo's legs. Each leg has two servo-motors that control the elevation and the propulsion respectively; furthermore, lateral motions (i.e., a third pseudo-degree of freedom) are allowed by the introduction of an elastic joint. The sensorial part is composed of two simple contacts that indicate whether or not a leg is touching the ground or an obstacle. New sensors can be easily incorporated through the extension ports which are provided in the electronic part of the robot.

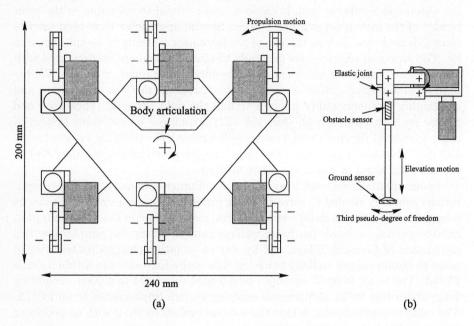

Fig. 1. Leonardo's mechanics: top view of the body (a), front view of a leg (b)

The robot is controlled through a main board, currently existing in two different versions: (1) a processor-based card; and (2) an FPGA-based board which is used for this paper. The processor-based controller is derived from the Khepera robot hardware [7].

The FPGA-based board has been developed specially for this experiment and is composed only of an FPGA (Xilinx XC4036EX), configuration ROM memory, a stabilized power supply that can be driven by an accumulator or an external source of power, and a clock. The FPGA directly controls the servo-motors and the sensors, replacing the micro-controller and the co-processor of the preceding board. Figure 2 depicts Leonardo with the FPGA-based board.

3 *Discipulus Simplex:* An Evolvable Logic System used for Walking Control

The objective for this evolvable hardware system (Discipulus Simplex) is to allow the robot to learn how to walk by itself. To implement this task we use one single

Fig. 2. Leonardo (six-legged robot) and Discipulus Simplex (evolvable hardware walking control)

FPGA which drives each leg of the robot and also performs the learning process. This constitutes an autonomous hardware system which can evolve and drive a robot without any help from processors, while remaining very simple (only a single FPGA chip is needed).

The learning process is realized with a genetic algorithm. The principle is to hold a population of individuals stored in the memory system which encodes a specific walk for the robot. The individual encodes the walk for the robot through a reconfigurable state machine which generates the sequence of movements for the steps. Once executed, the genetic algorithm generates a new population of individuals with average fitness usually better than the previous generation. This continues until a good individual is found for the walking behavior.

Three main modules are thus needed in our system. The first is the reconfigurable logic system for the walking control: it has to change its behavior during the learning process, in accordance with the walking performance. To configure this evolvable state machine we use a genome (individual), encoded by a bit-stream, so as to generate the sequence of movements. The genome with the greater fitness in the current population is provided to the evolvable state machine by the genetic algorithm. The second module performs the learning process: it is the genetic algorithm processor (GAP) responsible for the genetic operations (selection, crossover and mutation) on the population of genomes. Finally, the third module is the fitness module, responsible for the evaluation of an individual's performance. Figure 3 shows the complete evolvable hardware system, implemented in the FPGA, with the three main modules.

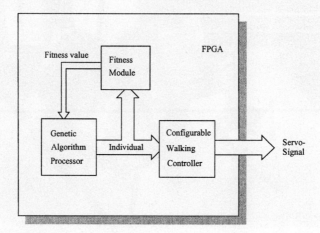

Fig. 3. Global view of Discipulus Simplex: the evolvable hardware system used for evolving walking behavior.

3.1 Reconfigurable state machine as walking control

The walk of the robot is controlled by a state machine which is able to modify its behavior through reconfiguration. This is the part of the system that evolves during the learning process. The behavior of the state machine is encoded in a bit-stream (i.e,, genome) which is provided by the learning process.

A genome encodes two steps of the walk. In each step there are six subparts, one for each leg. This means that the six parts are used and decoded at the same time during the walk. Finally, inside the six parts there are three bits which encode the movement of the leg during the step. The first bit codes whether the leg first goes up or down. The second bit codes whether the leg goes forward or backward. The last bit codes whether the leg goes up or down after the horizontal move. In all, one individual is composed of 36 bits, giving rise to a search space of size $2^{36} = 68$ billion possibilities.

The evolvable walking controller, including all these functionalities, is separated into two distinct modules. The main module is the reconfigurable state machine which is configured by the individual and generates the sequence of movements. The second module generates the signals for the servo-motor of each leg. Figure 4 shows the evolvable walking controller module and its submodules.

There are two servo-controls for each leg which generate PWM (Pulse Width Modulation) signals for the servo-motors from the position given by the parameterizable state machine.

3.2 Genetic algorithm processor (GAP)

To perform the learning process we use a genetic algorithm. The algorithm generates a population of individuals (genomes) and its goal is to continually improve

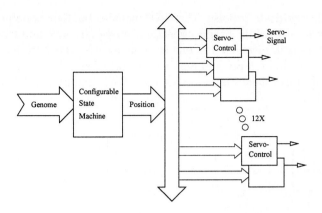

Fig. 4. Evolvable walking controller: module breakdown

the population's walking behavior. During the genetic process, we have to evaluate the performance of each individual. This performance measure is called the *fitness value*.

To evaluate fitness, the first idea is to get information on how the robot walks directly from the robot. For example, with a measure of the distance traveled in a given unit of time and a measure of the equilibrium of the robot with the help of sensors on his legs. But the problem is that the robot has a dynamic constraint and needs to try a genome for about five seconds to execute the walk. This time is too long to be used in our case. Therefore, we had to define a fitness function only in terms of logic computations. We could choose to define fitness by comparing an individual with a known solution; this would, however, diminish our work's interest. We chose to define the fitness with some high-level rules about the dynamics of the robot. After tests and simulations, we retained three rules which give good results, without knowledge of the solution:

1. The first physical rule involves the equilibrium of the robot. For example, if the robot has three legs raised on the same side, it will stumble and fall, resulting in a bad fitness value.
2. The second rule tests the symmetry between the two steps. If a leg goes forward in the first step, it should go backward in the next step. This can be deduced from observation of the walk of animals.
3. The last rule concerns the coherence of the movement of a leg. For example, it is not good for a leg to be down before it goes forward, the robot will go backward. The leg has to be up before going forward. On the other hand, the leg has to be down before doing a propulsion movement (going backward).

These rules are interesting in that they do not include knowledge of the solution genome. They arise from physical considerations, so they can be applied to other robots with other leg configurations.

The genetic algorithm is implemented in a main module called GAP (Genetic Algorithm Processor), which contains different modules that perform each a part of the genetic algorithm process. The GAP includes the four principal operators for the genetic algorithm: fitness, selection, crossover, and mutation. Each of these operators is implemented in one module, so they can work independently. The GAP also includes a random number generator, an initialisation unit, and a population of individuals. Figure 5 shows the GAP modules with their different components.

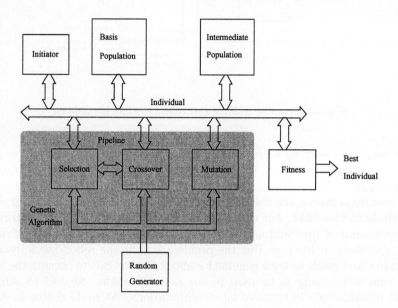

Fig. 5. Genetic Algorithm Processor.

When the GAP starts to work, the initialization module generates a new population of individuals using pseudo-random numbers. Once done, it starts the genetic algorithm cycle.

The four principal operators run in a fixed order. From the initial population the fitness operator is applied, then selection, then crossover, and finally mutation. To decrease computation time by a factor of about two, we ran the selection and crossover operators in a pipeline. During these operations, however, the selection operator needs to read in the population and the crossover operator needs to write the new individuals in an intermediate population. This is why we used two populations of individuals.

There are several possibilities for implementing each genetic operator [8]. In our case, the choices were constrained by the logic system. It is difficult to implement complex functions in a logic system. So we have to choose an operator which can be implemented with low computational cost.

The first operator which runs every time is the random number generator. It generates a new pseudo-random number for all genetic operators at each clock cycle. It is implemented as a one-dimensional cellular machine (XOR system).

It does not depend on the execution of the genetic algorithm, in order to render the evolutionary process less data-dependent.

The implementation choice made for the selection module was that of tournament selection [8] because it does not use real numbers and divisions which are difficult to implement in logic systems. This operator randomly draws two individuals from the population. A threshold defines the probability that the better individual will be selected.

For the crossover operator, the single-point crossover method is used. This method takes two individuals in the population and randomly selects a crossover position in the genome. The two genomes are cut at the crossover point and the part after the point are swapped, creating two new genomes. A threshold define how many crossover operations are performed on the population.

The mutation operator is single-bit mutation. This method randomly flips a bit in an individual's genome.

The last operator to define is the fitness operator. This operator evaluates the performance of each individual, using the high-level rules described above.

3.3 Implementation and tests

The evolvable logic system was synthesized for a Xilinx XC4036ex FPGA, from a VHDL language description. The use of VHDL is interesting because it allows to define parameters such as selection threshold, crossover threshold, population size, etc. So, it is possible to parameterize the entire logic system and it is easy to modify it.

The GAP implements the operators described above: selection, crossover, mutation, and fitness evalutation. The different parameters used for the GAP are:

- Population size: 32 individuals.
- Genome size: 36 bits.
- Selection threshold: 0.8.
- Crossover threshold: 0.7.
- Number of mutations: 15 bits (over 1152 bits).
- Frequency: 1 MHz.

Using the three rules for fitness evaluation, the maximum fitness does not necessarily correspond to the best walk known for the robot. However, the walking behavior found with the maximum fitness respecting all these rules is nonetheless good. To evolve the maximum fitness it needs an average of about 2000 generations. This number is small, if the speed of the GAP is considered. For example, if we had to test all the 68 billion possibilities for the genome, we would need about 19 hours at 1 MHz to obtain the best genome. With this system, the avarage time needed is only about 10 minutes.

The complete system implemented in the XC4036ex FPGA uses 96 percent of the available CLBs, i.e. 1244 CLBs. It represents around 34500 logic gates.

4 Conclusion

In this paper, we have investigated whether evolvable hardware systems can be used for the learning of the walking behavior of a six-legged robot. Using our six-legged robot Leonardo as a platform, we built an evolvable hardware walking control – called Discipulus Simplex – based on GAs and reconfiguration. Discipulus Simplex is implemented into a single FPGA which evolves on-line and is able to learn to walk. As a result, we obtained a fully autonomous robot which is able to learn to walk with the aid of on-line evolvable hardware without any processors or off-line computation.

This realization validates the use of evolvable hardware into autonomous systems, but still is a first step in the direction of a complex autonomous evolvable system. In future work, we will take advantage of the computational power provided by the GAP, and use the same kind of evolvable system in order to solve problems which deal with bigger genomes (i.e., more complex reconfigurable systems) and where the final solution is not known.

5 Acknowledgments

The authors want to thank Moshe Sipper and Gianluca Tempesti for their helpful comments and contributions, and André Badertscher for photographs.

References

1. Z. Michalewicz, *Genetic Algorithms + Data Structures = Evolution Programs*. Heidelberg: Springer-Verlag, third ed., 1996.
2. E. Sanchez and M. Tomassini, eds., *Towards Evolvable Hardware*, vol. 1062 of *Lecture Notes in Computer Science*. Heidelberg: Springer-Verlag, 1996.
3. M. Sipper, E. Sanchez, D. Mange, M. Tomassini, A. Pérez-Uribe, and A. Stauffer, "A phylogenetic, ontogenetic, and epigenetic view of bio-inspired hardware systems," *IEEE Transactions on Evolutionary Computation*, vol. 1, pp. 83–97, April 1997.
4. W. Ilg and K. Berns, "A learning architecture based on reinforcement learning for adaptive control of the walking machine lauron," *Robotics and Autonomous Systems*, vol. 15, pp. 321–334, 1995.
5. P. Maes and R. Brooks, "Learning to coordinate behaviors," in *Proceedings of the Eighth National Conference on Artificial Intelligence*, (Boston, MA), pp. 796–802, August 1990.
6. J.-M. Puiatti, "Robot 6 pattes," tech. rep., LAMI – Swiss Federal Institute of Technology at Lausanne, 1995.
7. F. Mondada, E. Franzi, and P. Ienne, "Mobile robot miniaturization: A tool for investigation in control algorithms," in *Proceedings of the Third International Symposium on Experimental Robotics* (T. Yoshikawa and F. Miyazaki, eds.), (Kyoto, Japan), pp. 501–513, 1993.
8. M. Mitchell, *An Introduction to Genetic Algorithms*. Cambridge, MA: MIT Press, 1996.

Reusable Internal Hardware Templates

Ka÷an Agun and Morris Chang
Illinois Institute of Technology
Computer Science Department
{agunsal | chang}@charlie.iit.edu

Abstract. This paper describes the framework of internal hardware templates. These reusable templates can be instantiated, inside the FPGA, to the required precision. Thus, the resource utilization of the target RCMs can be improved. Moreover, the configuration time can be eliminated after the first use of the template. The detail design is presented.

1. Introduction

After the recent development of Reconfigurable Computing Machines (RCMs), researchers have started to explore efficient algorithms and design methodologies to enhance the resource utilization. One of the software design techniques, design reuse, may help to achieve higher efficiency. Similar to the templates in C++, for example, hardware templates can be instantiated to the precision required by the applications. This way, only the necessary resources are used in the target RCMs. Moreover, this improves the portability among different RCMs. By using a library of templates, hardware system designs are independent from underlying technologies (e.g. FPGA architectures, HDLs). Reusability design technique makes these templates more powerful for system development.

In this paper we will introduce new design approaches to provide reusable hardware designs for RCMs. Currently, hardware templates are instantiated with parameters during the compilation time. In this paper, we propose to build *internal hardware templates* which will be instantiated internally by the hardware itself.

2. Reusability

Reuse, among the topics of software engineering, is one of the approaches to control complexity of software programs. Reusability is simply the re-application of source code and the most effective technique to achieve efficiency by reducing development time. HDLs can get the benefit of the same approach. Increasing the usage of hardware functions among the hardware tasks and among the different RCMs is crucial to the system design of RCMs. This can be done through building a set of parameterized templates. But reuse must be applied systematically to get its potential for improvement of RCMs. Ideally they are embedded in the design principles.

Generic model libraries based on hardware templates have been used in ASIC design. Researchers use application algorithms rather than the hardware design methodology for their hardware templates to increase reuse and reduce development time. Hardware templates are good ways to represent reusable hardware functions through generic and parameterized design methodology. Hardware templates can be classified into two categories. They are:

External Hardware Templates

These types of templates will be instantiated during compilation with parameters. External hardware templates are generic and parameterized hardware functions.

Internal Hardware Templates

The base hardware design will be compiled by HDL tools, but instantiation can be done in hardware independent from the design tools. Therefore it requires hardware support. This approach provides high performance with the cooperation of design tools and the FPGA architecture. The detail design is presented in next section.

3. A Design Alternative: Internal Hardware Templates

Considering a bit-slice design of n-bit adder, the smallest component is a one-bit adder which includes 3 inputs: a, b, cin and 2 outputs: sum and $cout$. Any size of adder can be instantiated from this design by passing parameter n. This adder can be implemented through modern Hardware Design Languages (e.g. VHDL). This design can be mapped into target FPGA with configuration data. The configuration of this adder consists of n one-bit adder logic functions, connections among these one-bit adders, and control information that shows how to place and route the configuration in the FPGA.

Configurable Logic Block (CLB) or Logic Element (LE) of FPGA implements these 4 input and 2 output logic functions. The architectures of these logic blocks are mostly based on Look Up Tables (LUT) which are small memories and multiplexers with SRAM. In our example, one CLB based on LUT can perform one-bit adder. Figure 1 shows the structure of the FPGA based on LUT and a logic function.

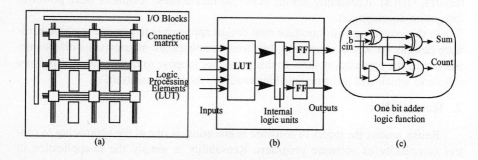

Figure. 1. Basic FPGA structure.

The idea is to divide a design into identical smallest components, so that several of these can form any size of the design. Configuration of the component and the rules of connections among them which is a kind of formula will form an *internal hardware template* to be downloaded into FPGA. The internal hardware template can be stored in the FPGA SRAM cache. Configuration controller will conduct all the operations to instantiate the templates internally in the FPGA. Without considering the technology of FPGAs, rolling of one-bit adder in the FPGA is much faster than downloading n bit adder configuration. Because the configuration controller allows to configure all selected cells at once (Figure 2). This design improvement will reduce the configuration overhead as well.

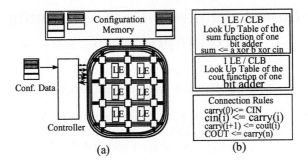

Figure. 2. A FPGA Architecture and Internal Hardware Template Example

Configuration time for a single FPGA can be a significant overhead in RCMs. In some recent FPGA architectures (e.g. XC6200 from Xilinx), it is possible to configure multiple CLBs at the same time. This reduces the configuration time drastically. However, the routing configuration is still required during the download. Moreover, for each different configuration (say 8-bit adder or 10-bit adder), different routing configuration data is required to be downloaded.

We propose to embed routing algorithms into the FPGA. This feature provides exact unrolling of internal templates inside the device as a part of controller. A routing configuration table will provide enough information to the controller to complete hardware design. Symmetrical architecture of the FPGA routing matrixes decreases the complexity of routing configuration.

This design decision minimizes the communication cost by eliminating configuration time. Having the routing algorithms inside the chip provides more portability to design tools. After download, instantiating the templates is completely platform independent. Sending a command with a parameter can instantiate any size of the hardware design.

4. Conclusion

Although there have been several development in the area of reconfigurable computing, the technology of the FPGAs and CPLDs are needed to be improved to provide better solutions for the wide-range applications. Developing hardware technology will reduce the complexity of designs by providing new features for synthesis tools. Internal hardware templates proposed in this paper is one of them. With this design approach, it will be easy to manage FPGA or CPLD devices. Operating Systems may use these programmable devices as a part of their resources, if the operating complexity is reduced and a global language is provided.

An On-Line Arithmetic-Based Reconfigurable Neuroprocessor

Jean-Luc Beuchat and Eduardo Sanchez

Logic Systems Laboratory, Swiss Federal Institute of Technology,
CH – 1015 Lausanne, Switzerland
E-mail: {name.surname}@di.epfl.ch

Abstract. Artificial neural networks can solve complex problems such as time series prediction, handwritten pattern recognition or speech processing. Though software simulations are essential when one sets about to study a new algorithm, they cannot always fulfill real-time criteria required by some practical applications. Consequently, hardware implementations are of crucial import. The appearance of fast reconfigurable FPGA circuits brings about new paths for the design of neuroprocessors. All arithmetic operations are carried out with on-line operators. This short paper briefly describes reconfigurable FPGA-based neural networks and gives an introduction to on-line arithmetic.

1 Introduction

The brain is a "computer" which exhibits characteristics such as robustness, distributed memory and calculation, interpretation of imprecise or noisy sensorial information. The design of electronic devices demonstrating such characteristics would be very interesting from the engineering point of view. Applications of these circuits include artificial vision, autonomous robotics or speech processing. Artificial neural networks constitute a possible way to realize such devices. The appearance of fast reconfigurable Field-Programmable Gate Arrays (FPGA) offers new paths for neuroprocessor implementation. However, FPGAs are too small to implement a large fully connected neural network and its learning rule. We exploit two philosophies to solve this problem:

❑ All arithmetic operations are carried out with on-line operators. Section 2 describes the advantages of this methodology.
❑ A neural algorithm is split into several steps executed sequentially, each of which is associated with a specific FPGA configuration. Such an approach leads to an optimal use of hardware resources. The efficiency of this method has been proven by J. G. Eldredge [1]. The reconfiguration paradigm also allows the implementation of multiple algorithms on the same hardware [2].

2 On-line Arithmetic

A neural network contains numerous connections, thus requiring a substantial amount of wires. As parallel arithmetic requires large buses, it is not well suited for such implementations. Bit-serial communications seem to be an effective solution.

Bit-serial data transmission may begin with the least significant bit (LSB) or with the most significant bit (MSB). Though the LSB paradigm seems more natural (the digits of the result are generated from right to left when we carry out an addition or a multiplication using the "paper and pencil" method), it does not allow the design of algorithms for division or square root. In general, the implementation of complex serial operations makes use of LSB and MSB operators. Let us consider the example depicted by figure 1. All digits of $(a + b) + (c \cdot d)$ must be known before the computation of the square root.

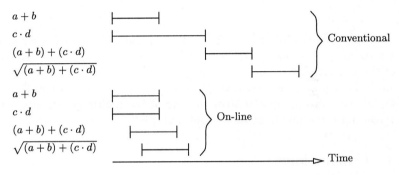

Fig. 1. *Computing $\sqrt{(a+b) + (c \cdot d)}$.*

The on-line mode, introduced in [3], *performs all computations in MSB mode thanks to a redundant number representation*[1]. It allows to overlap computation, leading so to a lower processing time (figure 1).

Usually, a number $a \in \mathbb{R}$ is written in radix r as $\sum_{i=-\infty}^{\infty} a_i r^{-i}$, where $a_i \in \mathcal{D}_r = \{0, 1, \ldots, r-1\}$. $\mathcal{D}_r$ is the digit set. On-line algorithms described in the literature use Avizienis' signed-digit systems [4], where numbers are represented in radix r with digits belonging to $\mathcal{D}_r = \{-a, -a+1, \ldots, a-1, a\}$, $a \leq r - 1$.

Our neuroprocessor exploits on-line arithmetic in radix 2, i.e. the digit set is $\{-1, 0, 1\}$. It uses a bit-level representation of the digits, called *Borrow-Save*, defined as follows: two bits, a_i^+ and a_i^-, represent the ith digit of a, such that $a_i = a_i^+ - a_i^-$. Consequently, digit 1 is encoded by $(1,0)$, digit -1 is encoded by $(0,1)$, while digit 0 is represented by $(0,0)$ or $(1,1)$.

Many on-line operators are described in the literature. Each of them is characterized by a delay δ, which indicates the number of clock cycles required to compute the MSB of the result. Figure 2a depicts the schedule diagram of an operator of delay 2. An input signal i is provided at time $t = 0$. Two clock cycles are required to elaborate the MSB of the output signal o. A new bit of the result is then produced at each clock cycle.

---

[1] In a redundant system, additions can be performed without carry propagation. This property allows the design of MSB adders.

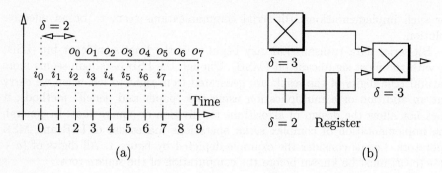

Fig. 2. *(a) Delay of an on-line operator. (b) Pipeline with on-line operators.*

Suppose now that the result of a first operator Γ is needed for further computations. The second operation may begin as soon as Γ has generated the MSB of the partial result. Figure 2b shows how to chain different on-line operators. The addition and the first multiplication are carried out in parallel. As the delay of the adder is smaller than the one of the multiplier, we use a register to synchronize the inputs of the second multiplier.

3 Conclusions

We have exploited the reconfiguration principles and on-line arithmetic to implement a neuroprocessor with on-chip training. The reconfigurable approach allows an optimal use of hardware resources and on-line arithmetic leads to a space-efficient implementation of data transmission. This approach is effective to implement large neural networks on a single FPGA, which is for instance intersting for autonomous robotic applications.

References

1. J.G. Eldredge and B.L. Hutchings. Density Enhancement of a Neural Network Using FPGAs and Run-Time Reconfiguration. In Duncan A. Buell and Kenneth L. Pocek, editors, *Proceedings of the IEEE Workshop on FPGAs for Custom Computing Machines*, pages 180–188. IEEE Computer Society Press, 1994.
2. Jean-Luc Beuchat and Eduardo Sanchez. A Reconfigurable Neuroprocessor with On-chip Pruning. In L. Niklasson, M. Bodén, and T. Ziemke, editors, *ICANN 98*, Perspectives in Neural Computing, pages 1159–1164. Springer, 1998.
3. Kishor S. Trivedi and Milos D. Ercegovac. On-line Algorithms for Division and Multiplication. *IEEE Transaction on Computers*, C-26(7), July 1977.
4. Algirdas Avizienis. Signed-Digit Number Representations for Fast Parallel Arithmetic. *IRE Transactions on Electronic Computers*, 10, 1961.

The Re-Configurable Delay-Insensitive FLYSIG[1] Architecture[2]

Wolfram Hardt, Achim Rettberg, Bernd Kleinjohann

Cooperative Computing & Communication Laboratory
Siemens AG & University Paderborn
Fürstenallee 11, D-33 102 Paderborn, Germany
e-mail: {hardt, achcad, bernd}@c-lab.de

Abstract: This article describes a new re-configurable architecture design suitable for high performance signal processing applications. The main benefits beside re-configurability are high computation performance, robustness against technology changes, and short design time. Therefore four design paradigms are combined: re-configurable functionality, data-flow orientation, delay-insensitive implementation, and pipelined bit-serial operators. Based on this paradigms a re-configurable high performance processor for cyclic execution of real-time critical algorithms, i.e. a specialized application domain has been designed. Comparable to FPGA based designs operators and interconnections are configurable. In contrast to FPGAs the basic configurable parts are much more complex. This ensures feasibility of complex designs and high complexity The implementation follows a bottom up approach while the concept has been specified top-down.

Overview

Advances in semiconductor technology have opened new dimensions of complexity for integrated circuits. Full custom designs are considered more frequently, if high computation performance is needed. Because time to market shortens more and more it is important to provide hardware prototypes in an early design stage. Re-configurable components as FPGAs have been proven very useful for functional validation. If real-time characteristics and chip area have to be considered, validation based on a re-configurable prototype is only possible if the prototype architecture and the final chip architecture are similar. In general this can not be guaranteed by FPGA components

---

[1] data**FL**ow oriented dela**Y**-insensitive **SIG**nal processing
[2]. The authors would like to acknowledge the support provided by Deutsche Forschungsgemeinschaft DFG, project SPP RP.

especially if architectures are adapted to specific application domains. We describe an approach providing a re-configurable architecture maintaining the target architecture for the signal processing application domain. Our approach is based on four well known design paradigms brought together for the first time:

- Re-configurable functionality ,
- data-flow oriented architecture design,
- delay-insensitive implementation style, and
- deeply pipelined, bit-serial operator implementation.

Typical applications are real-time control and processing of data-streams, e.g. encoding, decoding, and filters for audio, video, or sensor data. The main benefits of our design approach are re-configurability of operators and interconnection, high performance computation, robustness against technology changes within large limits, and cheap implementation in terms of chip area and design time. The first paradigm, configurable functionality, is well known from FPGA technology. We have added this idea to our approach to support rapid prototyping within short design cycles. The described target architecture is modified by introducing configurable operators and configurable operator interconnection. The result is a prototyping architecture which is in principle identical with the target architecture except for the amount of operators and interconnection possibilities. The concrete target architecture for a successfully prototyped design can be derived easily from the prototyping architecture. The second paradigm, data-flow oriented architecture design [5], adapts the architecture to the class of algorithms in view. The matching of algorithms and architecture is honored by high performance. The third paradigm, delay-insensitive implementation style [1], ensures the robustness against technology changes. We decided to use dual-rail encoding [2] of all signals, so a maximum of robustness can be achieved. Due to the overhead from dual-rail encoding the third paradigm, deeply pipelined bit-serial operator implementation, is involved. The implementation costs of this kind of operators are very low. By pipelining stage parallelism is introduced which leads to high throughput rates. More details are given in [3]. Reasonable improvements in terms of computation performance and reduction of design time are the results from maintaining the structure of the application for the prototype architecture as well as for the target architecture. The architecture concept is given in Fig.1.

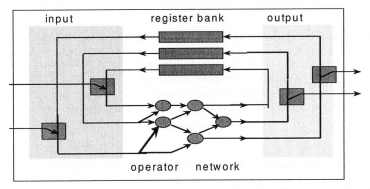

Fig. 1. Data-flow architecture used for implementation of cyclic executed algorithms

To provide re-configurability the two shaded blocks of the FLYSIG architecture (Fig. 1) are implemented by cross bars. This allows the re-configuration of interconnection. In addition the operators of the operator network can be implemented re-configurable. This is especially interesting for operations with similar implementations. We have implemented this architecture for different operator networks. Due to the high flexibility several applications could be mapped onto the same chip. Our approach allows timing analysis and exact estimation of the final chip size during the prototyping design phase. First results show also that the achieved sampling rate (for up to 32 bit data width) is much higher than FPGA technology has provided up to now.

References

[1] H. Bisseling, H. Eemers, M. Kamps, and A. Peeters. Designing delay-insensitive circuits. Technical report, IVO, Eindhoven University of Technology, Sept. 1990.
[2] M. Dean, T. Williams, and D. Dill. Efficient self-timing with level-encoded 2-phase dual-rail (LEDR). In C. H. Sequin, editor, *Advanced Research in VLSI*. MIT Press, 1991.
[3] W. Hardt and B. Kleinjohann. Flysig: Dataflow Oriented Delay-Insensitive Processor for Rapid Prototyping of Signal Processing. In *9. IEEE International Workshop on Rapid System Prototyping*. IEEE Computer Society Press, 1998.
[4] J. Sparso, J. Staunstrup, and M. Dantzer-Soerensen. Design of delay-insensitive circuits using multi-ring structures. In G. Musgrave, editor, *EURO-DAC '92*. IEEE Computer Society Press, 1992.
[5] H. Terada, M. Iwata, S. Miyata, and S. Komori. Superpipelined dynamic data-driven VLSI processors. In *Advanced Topics in Dataflow Computing and Multithreading*, pages 75-85. IEEE Computer Society Press, 1995.

Digital Signal Processing with General Purpose Microprocessors, DSP and Reconfigurable Logic

Steffen Köhler, Sergej Sawitzki, Achim Gratz, and Rainer G. Spallek

Institute of Computer Engineering
Dresden University of Technology
D-01062 Dresden, Germany
e-Mail: {stk,sawitzki,gratz,rgs}@ite.inf.tu-dresden.de

Abstract. This paper compares selected digital signal processing algorithms on a variety of computing platforms in terms of achievable performance and cost. The experiments were carried out on a standard PC platform, DSP, a RISC microcontroller and on Xilinx XC4013XL FPGA. Our results confirm that general purpose microprocessors are not well suited to these tasks. Both DSP and FPGA achieve higher performance and/or better cost/performance ratios at the expense of lesser generality and a more complicated development cycle. The porting of the algorithms to DSP and FPGA requires about the same amount of work, whereby the cost/performance ratio of the reconfigurable FPGA solution is very attractive.

1 Introduction

Digital signal processing still requires special hardware solutions to achieve an acceptable level of performance. This fact is underlined by the rapid expansion of the DSP market, especially in the embedded domain, and confirmed by the results achieved for general purpose microprocessors in this paper. If the reconfigurability of FPGA is employed, very elegant and cost effective solutions are possible for applications that are otherwise only accessible by high-end DSP. The reconfigurable FPGA is more flexible than the DSP and especially interesting for exploration of newly developed algorithms that are not yet supported by special features of DSP. While dynamic reconfiguration may extend the reach of the FPGA solution even further, the added complexity seems to disfavour that approach. For PC I/O-cards with integrated signal processing, reconfiguration at startup seems to suffice for many conceivable applications.

2 Hardware Equipment

Our experiments were conducted on a standard PC platform with four different system configurations as described below.
General Purpose CPU. Two conventional PC systems were used to determine the performance of general purpose processors (AMD K6 CPU clocked at 200 MHz, Dual Intel Pentium II clocked at 350 MHz. All test programs were ran under the Linux operating system with algorithms coded in C (singlethreaded).

Table 1. DSP Algorithm Run Times (in milliseconds) and Speedups

	FFT (512pt) rt/sp		LPC Filter rt/sp		Viterbi Decoder rt/sp	
AMD K6 200	1.2	8.33	0.00100	10.00	6.95	1.00
Pentium II	0.5	20.00	0.00059	16.94	3.20	2.17
DSP 56302	0.6	16.67	0.00030	33.33	1.40	4.96
TMS 320C50	2.2	4.55	0.00100	10.00	3.00	2.32
ARM stand-alone	10.0	1.00	0.01000	1.00	4.80	1.45
ARM with FPGA	1.4	7.14	0.00060	16.67	0.27	25.74

Digital Signal Processors (DSP). We used a PCI extension board with a Motorola DSP 56302 (24 bit fixed point DSP) operating at 85 MHz [1] and an ISA extension board with Texas Instruments TMS 320C50 clocked at 40 MHz (16 bit fixed point DSP) [3]. Both boards were plugged into the K6-based PC which was acting as a host running the Linux operating system and accessing the DSP via native drivers. The algorithms were completely rewritten in assembly language for each DSP.

RISC Microcontroller and FPGA. We used an additional PCI extension card containing an 32 bit ARM microcontroller produced by Atmel and clocked at 8 MHz and two Xilinx XC4013XL FPGA operating at the same clock frequency [2]. In the first experiment the algorithms were compiled with the native ARM compiler and ran on the microcontroller alone. In the second experiment the microcontroller was used as a host to program the FPGA and exchange data with them while the cores of the algorithms were implemented in Handel-C and downloaded to one of the FPGA. To simplify the synchronization, the ARM waited for the FPGA to complete it's respective operation. The development environment was ran under Microsoft Windows 95.

3 Benchmark Suite and Experimental Results

We considered a range of applications which usually serve as benchmarks for DSP vendors and extracted the algorithmic cores. Note that XC4013 devices consist of 576 cells each, and 113 of them (on average) are consumed by the host interface.

Fast Fourier Transform [4]. A one-dimensional FFT was ran for an input set of 512 points. We limited all computations to 16-bit wide operands due to the hardware available. In the case of FPGA the butterfly operation was implemented in hardware consuming 526 CLBs.

LPC Synthesis Filter [6]. Voice synthesis using the LPC10 compression algorithm served as a second test example. We used the FPGA to implement a 10-stage lattice filter (471 CLBs) to accelerate the computations.

Viterbi Decoder [5]. We implemented a $\frac{1}{3}$-rate convolutional decoding algorithm. In the reconfigurable case the Trellis look-up operation was executed in the FPGA while the state matrix and Trellis graph were stored in the SRAM on the FPGA board. 349 CLBs were used to implement the corresponding circuit.

Table 1 summarizes the results of our measurements. For the PC platforms, best timings are shown. It is impossible to eliminate the wide fluctuations in timing val-

Table 2. Cost/Performance Trade-Off

Circuit	Price ($)	Performance/$ (arb. units)	
		achieved	achievable
AMD K6 200	65	99	—
Pentium II	250	52	—
DSP 56302	30	610	—
TMS 320C50	10	560	—
XC4013XL-3	60	280	1120

ues of the general purpose architectures due to interrupt handling and related cache conflicts, which degrade the sustainable performance significantly and make real-time digital signal processing with the host CPU an elusive goal on the PC platform.

4 Conclusions

Although it seems that DSP provide better performance in two of three cases, as Table 1 suggests, the picture changes if the post layout timing data of the designs are considered. Note that Xilinx has announced a new family of XC4000E compatible devices that are claimed to be twice as fast as the currently available parts. Unfortunately, the limitations of the development board do not allow to take advantage of the maximum speed obtainable with even the slower (and cheaper) devices on the FPGA board. The trade-off between cost and performance is another interesting point of view. Setting the circuit prices in relation to the performance data in Table 1 shows that FPGA provide the most performance per dollar on average (at least in best case), see Table 2. This comparison can be continued at the system level with the same result.

Our experience with the Handel-C compiler has shown that typical DSP algorithms can be ported in a reasonable amount of time (2–3 weeks) with good results. To achieve good performance with a DSP, the algorithms often have to be re-coded in assembler, which is a time-consuming process and requires a lot of special knowledge about the DSP used. This suggests that time-to-market for the DSP and the FPGA solution is similar and the better cost/performance ratio of the FPGA solution is very attractive.

References

1. Motorola Inc. DSP56302 24-Bit Digital Signal Processor User's Manual. Austin, TX, 1996
2. Snook, M. ASPIRE Development Board Datasheet. Advanced RISC Machines Ltd., 1998
3. Texas Instruments Inc. TMS320C5x User's Guide. Houston, TX, 1993
4. Mitra, S.K. and Kaiser, J.F. (ed.) Handbook for Digital Signal Processing. John Wiley & Sons, Inc., New York, 1993
5. Benedetto, S., Biglieri, E., and Castellani, V. Digital Transmission Theory. Prentice-Hall, p. 407, 1987
6. Tremain, E.T. The Government Standard Linear Predictive Coding Algorithm: LPC-10. Speech Technology Magazine, pp. 40-49, 1982

Solving Satisfiability Problems on FPGAs Using Experimental Unit Propagation Heuristic

Takayuki Suyama, Makoto Yokoo, and Akira Nagoya

NTT Communication Science Laboratories
2-4 Hikaridai Seika-cho Soraku-gun, Kyoto, JAPAN
{suyama, yokoo, nagoya}@cslab.kecl.ntt.co.jp

Abstract. This paper presents new results on an approach for solving satisfiability problems (SAT), that is, creating a logic circuit that is specialized to solve each problem instance on Field Programmable Gate Arrays (FPGAs). This approach has become feasible due to recent advances in Reconfigurable Computing.
We develop an algorithm that is suitable for a logic circuit implementation. This algorithm is basically equivalent to the Davis-Putnam procedure with Experimental Unit Propagation. The required hardware resources for the algorithm are less than those of MOM's heuristics.

1 Introduction

Recent reconfigurable hardware technologies enable users to rapidly create logic circuits specialized to solve each problem instance. We have chosen satisfiability problems to examine the effectiveness of this approach. In this paper, we present a new method of implementing an efficient algorithm with a powerful dynamic variable ordering heuristic on a logic circuit. This algorithm is basically equivalent to the Davis-Putnam procedure with Experimental Unit Propagation [2].

2 Design Flow

Figure 1 shows the logic synthesis flow of a logic circuit that solves a specific SAT problem.

3 Problem Definition

A satisfiability problem (SAT) for propositional formulas in conjunctive normal form can be defined as follows. A boolean *variable* x_i is a variable that takes the value true or false (represented as 1 or 0, respectively). We call the value assignment of one variable a *literal*. A *clause* is a disjunction of literals, e.g., $(x_1 + \overline{x_2} + x_3)$. Given a set of clauses $C_1, C_2, \ldots, C_m$ and variables $x_1, x_2, \ldots, x_n$, the satisfiability problem is to determine if the formula $C_1 \cdot C_2 \cdot \ldots \cdot C_m$ is satisfiable, i.e., to determine whether an assignment of values to the variables exists so that the above formula is true. This problem was the first computational task shown to be NP-complete.

4 Algorithm

The algorithm used in this paper is basically equivalent to the Davis-Putnam procedure that introduces a dynamic variable ordering with *Experimental Unit Propagation* (EUP) that is a heuristic used for the branching. The outline of the algorithm can be described as follows.

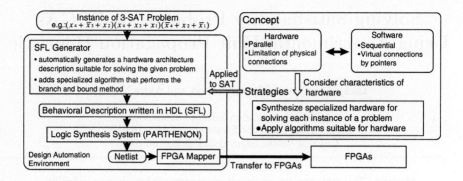

Fig. 1. Logic synthesis flow for SAT problems

Let I be the input SAT problem instance, and L be the working list containing problem instances, where L is initialized as $\{I\}$. A problem instance is represented as a pair of a set of value assignments and a set of clauses that are not yet satisfied.

1. If L is empty, then I is unsatisfiable, so stop the algorithm. Otherwise, select the first element S from L and remove S from L.
2. If S contains an empty clause, then S is unsatisfiable, go to 1.
3. If all clauses in S are satisfied, print the value assignments of S as one solution, go to 1.
4. If S contains a unit clause (a clause with only one variable), then set this variable to the value which satisfies the clause, simplify the clauses of S and go to 2.
5. **Branching:** In order to select a variable which should be assigned at the next step, count what time chain reactions of unit propagation occur when assuming each unassigned variable is set at 0 and 1. If an unsatisfied clause exists when the variable is set at 0, set it to 1, and vice versa. If this does not occur, the variable which has the largest counting number is set to 0. Go to 1.

The algorithm can be straightforwardly represented as a finite state machine.

5 Evaluation

We first evaluate the efficiency of the developed algorithm by software simulation. We use hard random 3-SAT problems as examples. The number of clauses divided by the number of variables, which is called *clause density*, is 4.3. In Figure 2, we show the log-scale plot of the average required time of over 100 problems, assuming the clock rate is 1MHz. We can see from Figure 2 that the search tree of MOM's[3] grows at the rate of $O(2^{n/17.3})$, where n is the number of variables. On the other hand, the search tree growing at the rate of the new method (EUP) is $O(2^{n/20.0})$. This shows that the new method is more efficient with a larger number of variables.

We use an ALTERA FLEX10K250 FPGA chip to implement the algorithm. We have implemented a particularly difficult 3-SAT AIM benchmark problem[1], "aim-100-2_0-no-1.cnf", which consists of 100 variables, 200 clauses. The number of Logic Cells (LCs) is 9,464 for implementing this problem. The utilized rate of LCs is 77%.

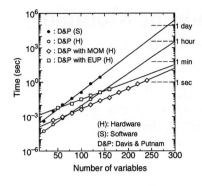 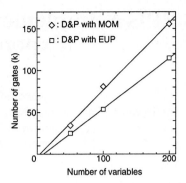

Fig. 2. Required time at 1MHz on hard random 3-SAT problems

Fig. 3. Required gates for "aim-<#variables>-1_6-yes1-1.cnf"

A logic circuit simulator shows this circuit is capable of running at a clock rate of 12.07MHz.

Figure 3 shows the number of required gates for "aim-{50,100,200}-1_6-yes1-1.cnf". This shows the number of gates when the circuit is organized by primitive gates. Note that these circuits are initial ones synthesized from HDL descriptions. If these circuits are optimized, the number of gates can be reduced more with keeping the same trend. The required gates of EUP are about 30% less than that of MOM. This is because almost all the circuits of branching of EUP can be shared with the other steps. On the other hand, the MOM algorithm requires additional circuits for branching.

6 Conclusions

This paper presented new results on solving SAT using FPGAs. We developed an algorithm which is suitable for implementation on a logic circuit. This algorithm is basically equivalent to the Davis-Putnam procedure that introduces Experimental Unit Propagation. In this approach, a logic circuit specific to each problem instance is created on FPGAs. The number of gates required to implement the proposed method is 30% less than that of MOM. In addition, the order of the proposed method is better than that of MOM. We have actually implemented benchmark problems with 100 variables and it can run at 12MHz. In the future, we will refine the implementation of the algorithm on FPGAs, and perform various evaluations on implemented logic circuits.

References

1. Y. Asahiro, K. Iwama, and E. Miyano. Random generation of test instances with controlled attributes. In *Proceedings of the DIMACS Challenge II Workshop*, 1993.
2. C. M. Li and Anbulagan. Heuristics based on unit propagation for satisfiability problems. In *Proc. of 15th International Joint Conference on Artificial Intelligence*, pages 366–371, 1997.
3. T. Suyama, M. Yokoo, and H. Sawada. Solving satisfiability problems using logic synthesis and reconfigurable hardware. In *Proc. of the 31st Annual Hawaii International Conference on System Sciences Vol. VII*, pages 179–186, 1998.

FPGA Implementation of Modular Exponentiation

Alexander Tiountchik
Institute of Mathematics, National Academy of Sciences of Belarus,
11 Surganova st, Minsk 220072, Belarus
e-mail: aat@im.bas-net.by

Elena Trichina
Advanced Computing Research Centre, University of South Australia,
Mawson Lakes, SA 5095, Australia
e-mail: elena.trichina@unisa.edu.au

Abstract. An efficient implementations of the main building block in the RSA cryptographic scheme is achieved by mapping a bit-level systolic array for modular exponentiation onto Xilinx FPGAs. One XC6000 chip, or 4 Kgates accommodates 132-bit long integers. 16 Kgates is required for modular exponentiation of 512 bit keys, with the estimated bit rate 800 Kb/sec.

1 Systolic Array for Modular Exponentiation

The design of public key cryptography hardware is an active area of research because the speed of cryptographical schemes is a serious bottleneck in many applications. A cheap and flexible modular exponentiation hardware accelerator can be achieved using Field Programmable Gate Arrays. FPGA design presented in this paper is based on an efficient systolic array for a modular exponentiation such that the whole exponentiation procedure can be carried out entirely by the single systolic unit without use of global memory. This procedure is based on a Montgomery multiplication [2], and uses a high-to-low binary method of exponentiation.

Let A, B be elements of $\mathbf{Z}_m$, where $\mathbf{Z}_m$ is the set of integers between 0 and $m-1$. Let h be an integer coprime to m, and $h > m$. Montgomery multiplication (MM) is an operation $A \overset{h,m}{\otimes} B = A \cdot B \cdot h^{-1} \bmod m$. Several algorithms suitable for hardware implementation of MM are known [1, 4, 7]. In this paper, as a starting point we use the algorithm described and analysed in [7]. Let numbers A, B and m be written with radix 2: $A = \sum_{i=0}^{N-1} a_i \cdot 2^i$, $B = \sum_{i=0}^{M} b_i \cdot 2^i$, $m = \sum_{i=0}^{M-1} m_i \cdot 2^i$, where $a_i, b_i, m_i \in \mathbf{GF}(2)$, N and M are the numbers of digits in A and m, respectively. B satisfies condition $B < 2m$, and has at most $M+1$ digits. Extend a definition of A with an extra zero digit $a_N = 0$. The algorithm for MM is given below (1).

$s := 0;$
For $i := 0$ **to** N **do**
Begin
$$u_i := \big((s_0 + a_i * b_0) * w^1\big) \bmod 2 \qquad (1)$$
$$s := (s + a_i * B + u_i * m) \mathrm{div} 2$$
End

Using B instead of A results in calculating $B \overset{h,m}{\otimes} B = (\overset{h,m}{\otimes} B)^2$, called M-squaring. A fast way to compute $B^n \bmod m$ is by reducing the computation to a sequence of modular squarings and multiplications [5]. Let $[n_0 \ldots n_k]$ be a binary representation of n, i.e., $n = n_0 + 2n_1 + + \cdots + 2^k n_k$, $n_j \in \mathbf{GF}(2)$, $k = \lfloor \log_2 n \rfloor$, $n_k = 1$. Let β denote a partial product. We start out with $\beta = B$ and run from n_{k-1} to n_0 as follows: if $n_j = 0$, then $\beta := \beta^2$; if $n_j = 1$, then $\beta := \beta^2 * B$. Thus, we need at most $2k$ operations to compute B^n.

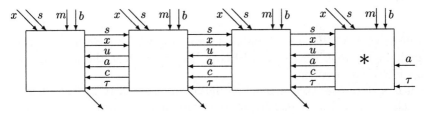

Fig. 1. Linear Systolic Array for Montgomery Exponentiation; M = 3.

A linear systolic array for high-to-low modular exponentiation is presented in Fig. 1. Details of its systematic design can be found in [6]. The systolic array comprises $M + 1$ processing elements (PE), each PE is able to operate in two modes, multiplication and squaring. To control the operation modes, a sequence of one-bit control signals τ is fed into the rightmost PE and propagated through the array. If $\tau = 0$ the PE implements an operation of M-multiplication, if $\tau = 1$, M-squaring. The order in which control signals are input is determined by the binary representation of n. Each vertex is associated with the operation

$$s_{j-1}^{(i+1)} + 2 \cdot c_{out} := s_j^{(i)} + a_i \cdot b_j + u_i \cdot m_j + c_{in},$$

where $s_j^{(i)}$ denotes the j-th digit of the i-th partial product of s, c_{out} and c_{in} are the output and input carries. A starred vertex, i.e., a vertex marked with "*", performs calculations of $u_i := ((s_0 + a_i * b_0) * w) \bmod 2$ besides an ordinary operation.

To perform M-squaring we need only the b_j's inputs. This eliminates the rightmost a-input entirely. To deliver all b_js to the starred vertex, we have to pump them through the graph using additional arcs x_js. Vertex operations are to be slightly modified to provide propagation of these digits: each non-starred vertex just transmits its x-input data to x-output, while when arriving at the rightmost vertex, these data are "reflected" and propagated to the left as if they were ordinary a_i's input data. The operations for M-squaring can be specified exactly as above (apart from x's transmissions) for main vertices; the starred vertex computes $u_i := ((s_0 + x_i * b_0) * w) \mod 2$, assigns $a_i := x_i$, and performs an operation as above (with $c_i n = 0$).

A timing function that provides a correct order of operations is $t(v) = 2i + j$ [6]. The total running time is thus at most $(4\lfloor \log_2 n \rfloor + 1)M + 8\lfloor \log_2 n \rfloor$ time units.

2 Mapping a Systolic Array onto FPGAs

To map a systolic array onto FPGAs, one has to:

- find suitable logic designs for a FPGA realisation of individual PEs and compile these designs into blocks of FPGA cells;
- specify logic designs that combine these blocks into an array and compile it;
- optimise the design.

A straightforward realization of the first two steps leads to a successful routing for designs with 67 PEs. Our "trial and error" experience with optimisation of the mapping systolic arrays onto FPGAs has shown that the following ideas ensure 100% improvement.

– It is wise to base your manual optimisation on the XACTStep6000 system proposals because the XACTStep6000 system does it with the objective of successful routing; a complex set of criteria takes into account a number of parameters which are known only to the system that optimises the allocation from the point of view of all these criteria.

– Using a hierarchy of modules facilitates the design process. However, an automatic routing may spread gates that belong to different levels of the hierarchy rather far apart. The more levels of the hierarchy the less control one may exercise over the overall design, so try to minimise the number of levels.

– If the outputs of some gates are to be stored in registers, these gates and registers should not be at different hierarchical levels, because a gate–register pair may occupy only one cell but if the gate is embedded in a module, while the register is outside of this module, they inevitably will be placed in different cells, and often rather far apart. Thus, if registers are to be used to store output data of a module, it is desirable to insert these registers inside the module.

– When a systolic array cannot be mapped in a straightforward fashion onto a FPGA chip (e.g., a long and narrow line of PEs has to be mapped onto a square array of logic cells), it is advisable to partition this array into blocks of PEs with respect to the width of the board, so that every block can be allocated on a chip in a form of a border-to-border straight line. In our case, one block comprises 13 PEs. Pack these blocks in a zig-zag "snake" to fill in the whole 64×64 sea of cells on a chip.

– To ensure regularity and locality of interconnections between blocks, and between PEs in different blocks, create for each PE a few types of logic designs with identical functionality, but with input/output gates representing some suitable permutation of the gates in the original design. Use these designs to build "mirror images" of the original block under rotation and reflection. These images must be used at every second row of the zig-zag and where a "snake"' turns.

3 Summary

We presented a new implementation of modular exponentiation based on Montgomery multiplication on fine–grained FPGAs. With hand–crafted optimisation we managed to embed a modular exponentiation of 132-bit long integers into one Xilinx XC6000 chip, which is to our knowledge one of the best fine-grained FPGA designs for a modular exponentiation reported so far. 3,000 out of 4,096 gates are used for computations and registers, providing 75% density.

Reported in this paper hardware implementation relies on configurability of FPGAs, but does not use run-time reprogrammability or/and SRAM memory. This makes our design simpler and easy to implement. The price to pay is that more chips are needed to implement RSA with longer keys. 512-bit keys need four XC6000 chips connected in a pipeline fashion, or 16 Kgates. Taking into account the total running time, we can estimate the bit rate for a clock frequency of 25 MHz being approximately 800 Kb/sec for 512 bit keys, which is comparable with the rate reported in a fundamental paper of Shand and Vuillemin [5], and an order of magnitude better than that one in [1] and [3].

References

1. K. Iwamura, T. Matsumoto and H. Imai, Modular Exponentiation Using Montgomery Method and the Systolic Array, IEICE Technical Report, vol. 92, no. 134, ISEC92-7, 1992, pp.49–54.
2. P. L. Montgomery, Modular multiplication without trial division. *Mathematics of Computations*, 1985 (44) 519–521.
3. H. Orup, E. Svendsen, E. And, VICTOR an efficient RSA hardware implementation. In: *Eurocrypt 90*, LNCS, vol. 473 (1991) 245–252
4. J. Sauerbrey, A Modular Exponentiation Unit Based on Systolic Arrays, in *Advances in Cryptology – AUSCRYPT'93*, Springer-Verlag, LNCS, vol. 718 (1993) 505–516.
5. M. Shand, J. Vuillemin, Fast Implementation of of RSA Cryptography. In *Proc. of the 11th IEEE Symposium on Computer Arithmetics*, 1993. pp.252–259.
6. A. A. Tiountchik, Systolic modular exponentiation via Montgomery algorithm. J. *Electronics Letters*, 1998 (34).
7. C. D. Walter, Systolic Modular Multiplication. *IEEE Trans. on Comput.*, 1993 (42) 376–378.

Workshop on
Java for Parallel and Distributed Computing

Denis Caromel	Serge Chaumette	Geoffrey Fox
INRIA	LaBRI	NPAC
Nice Univ.	Bordeaux 1 Univ.	Syracuse Univ.

IPPS/SPDP'98 that was held in Orlando, Florida, in April 1998, showed that Java is becoming more and more important for the IPPS/SPDP community, i.e. for parallel and distributed computing. Many of the papers that were presented in Orlando referred to Java. Furthermore the attendance of the tutorial on Parallel and Distributed Computing Using Java that was given there clearly demonstrated the widespread interest.

One of the aims of this workshop is therefore to bring together the IPPS/SPDP community around Java, and to provide an opportunity to share experience and views of current trends and activities in the domain. It focuses on Java for parallel and distributed computing and supportive environments.

The papers that were submitted to the workshop were of excellent quality and the number of submissions reflected the big amount of work which is being done in this area. We believe that the work that was done by the referees and the program committee lead to the selection of papers significant of the current research trends in the domain. These papers address three main topics:

1. Optimizations in Distributed Object Computing.

Java seems to be perfectly suited for distributed and parallel computing. Nevertheless it still suffers from a lack of efficiency especially when RMI or object serialisation is used.

2. Models and Libraries for Distribution.

To attract the users of non-Java parallel and distributed programming libraries to Java, it is necessary to provide them with good quality Java implementations of such libraries. To be widely adopted, these implementations must furthermore offer new features, by taking advantage of some of Java high level functionalities.

3. Metacomputing Environements and Principles.

Metacomputing is probably one of the big issues of the Java technology. Such kind of computing requires both appropriate environments and libraries but also adequate models.

We decided to keep some space for a **special session on Java Grande Forum** and for open discussions. The goal of the Java Grande Forum is to develop community consensus and recommendations for either changes to Java or establishment of standards (frameworks) for Grande libraries and services. These language changes or frameworks are designed to realize the best ever Grande programming environment. The forum holds meetings linking the large scale computing community, Sun MicroSystems and the commodity computing participants. Membership of the forum is open and further details may be found at http://www.javagrande.org.

January 1999

Denis Caromel
Serge Chaumette
Geoffrey Fox

Workshop Organization

Program Co-Chairs

Denis Caromel	INRIA, University of Nice Sophia Antipolis, Nice, FRANCE
Serge Chaumette	LaBRI, University Bordeaux 1, Bordeaux, FRANCE
Geoffrey Fox	NPAC, Syracuse University, Syracuse, NY, USA

Program Commitee

Jack Dongarra	University of Tenessee, USA
Ian Foster	Argonne National Laboratory, USA
Doug Lea	State University of New York at Oswego, USA
Vladimir Getov	University of Westminster, London, UK
Dennis Gannon	Indiana University, USA
Siamak Hassanzadeh	Sun Microsystems Computer Corp., USA
George K. Thiruvathukal	Loyola University, USA
Michael Philippsen	University of Karlsruhe, GERMANY
David Walker	Cardiff University, UK

More Efficient Object Serialization

Michael Philippsen and Bernhard Haumacher

University of Karlsruhe, Germany
phlipp@ira.uka.de and hauma@ira.uka.de
http://wwwipd.ira.uka.de/JavaParty/

Abstract. In current Java implementations, Remote Method Invocation is too slow for high performance computing. Since Java's object serialization often takes 25%–50% of the time needed for a remote invocation, an essential step towards a fast RMI is to reduce the cost of serialization.

The paper presents a more efficient object serialization in detail and discusses several show-stoppers we have identified in Sun's official serialization design and implementation. We demonstrate that for high performance computing some of the official serialization's generality can and should be traded for speed. Our serialization is written in Java, can be used as a drop-in replacement, and reduces the serialization overhead by 81% to 97%.

1 Introduction

From the activities of the Java Grande Forum [2, 9] and from early comparative studies [1] it is obvious that there is growing interest in using Java for high-performance applications. Among other needs, these applications frequently demand a parallel computing infrastructure. Although Java offers appropriate mechanisms to implement Internet scale client/server applications, Java's remote method invocation (RMI) is too slow for environments with low latency and high bandwidth networks, e.g., clusters of workstations, IBM SP/2, and SGI Origin. Explicit socket communication – the only reasonable alternative to RMI – is not object-oriented, too low level, and too error-prone to be seriously considered by a broad user community.

Since Java's current object serialization often takes at least 25% and up to 50% of the time needed for a remote invocation, an essential step towards a fast RMI is to reduce the cost of serialization as far as possible.

1.1 Basics of Object Serialization

Object serialization [8] is a significant functionality needed by Java's RMI implementation. When a method is called from a remote JVM, method arguments are either passed by reference or by copy. For objects and primitive type values that are passed by copy, object serialization is needed: The objects are turned into a byte array representation, including all their primitive type instance variables and including the complete graph of objects to which all their non-primitive instance variables refer. This wire format is unpacked at the recipient and turned back into a deep copy of the graph of argument objects. The serialization can copy even cyclic graphs from the caller's JVM to the callee's JVM; the serialization keeps and monitors a hash-table of objects that have already been packed to avoid repetition and infinite loops. For every single remote method invocation, this table has to be reset, since part of the objects' state might have been modified.

In general, programmers do not implement marshaling and unmarshaling. Instead, they rely on Java's reflection mechanisms (dynamic type introspection) to automatically derive an appropriate byte array representation. However, programmers can provide routines (`writeObject` or `writeExternal`) that do more specific operations. These routines are invoked by the serialization mechanism instead of introspection. Similar routines must be provided for the recipient.

From the method declaration, the target JVM of a remote invocation knows the declared types of all objects expected as arguments. However, the *concrete* type of individual arguments can be any subtype thereof. Hence, the byte stream representation needs to be augmented with type information.

Any implementation of remote method invocation that passes objects by value will have to implement the above functionality.

In the remainder of this paper, the term "serialization" refers to the functionality of writing and reading byte array representations in general. The official implementation of serialization that is available in the JDK is called "JDK-serialization". Our implementation is called "UKA-serialization".

1.2 Cost of JDK-Serialization

We have done measurements demonstrating that Java's current object serialization takes 25%–50% of the cost of a remote method invocation.

μs per object	32 int		4 int, 2 null		tree(15)	
RMI `ping(obj)`	4620	100%	3220	100%	6170	100%
Socket `send(obj)`	3290	71%	1820	56%	4720	76%
`Serialize(obj)`	1756	38%	765	24%	3081	50%

Table 1. Ping times (μs) of RMI (=100%), socket communication, and just JDK-serialization. The argument object `obj` has either 32 `int` values, 4 `int` values plus 2 `null` pointers, or it is a balanced binary tree of 15 objects each of which holds 4 `int`s.

On two Suns (300 MHz Sun Ultra 10 (Sparcs IIi), running Solaris 2.6) we have used JDK 1.2beta3 (JIT enabled) on some RMI benchmarks.[1] Table 1 gives

---

[1] The JDK-serialization offers two wire protocols. Although protocol 2 is default, RMI uses protocol 1 because it is slightly faster. Our comparisons use protocol 1 as well.

the results. For three different types of objects we measured the time of a remote invocation of `void ping(obj)`. In addition, we have taken the time needed for the communication over existing Java socket connections. This includes serialization. Finally, we measured the time needed for the JDK-serialization of the argument. It can easily be seen that the serialization takes at least 25% of the time. The cost of serialization grows with growing object structures up to 50% in our measurements. Benchmarks conducted by the Manta team [11] show similar results.

1.3 Organization of this paper

The rest of the paper is organized as follows. Section 2 summarizes the related work. Section 3 looks at specific performance problems of the JDK-serialization and discusses benchmark results that quantify them. Section 4 sketches the design of the UKA-serialization, our more efficient drop-in replacement for the JDK-serialization. In section 5 we give a detailed performance analysis of the UKA-serialization.

2 Related Work

Some groups have published their ideas on improving the JDK-serialization.

- At Illinois University, an improved remote method invocation has been implemented that is based on an alternative object serialization, see [10]. The authors experimented with explicit routines to write and read an object's instance variables.
  In our work, we use explicit routines as well but show that close interaction with the buffer management can further improve the performance.
- Henri Bal's group at Amsterdam is currently working on the compiler project *Manta* [11]. Manta has an efficient remote method invocation (35 μs for a remote null invocation, i.e., without serialization) on a cluster of workstations connected by Myrinet. To achieve this result, Manta compiles a subset of Java to native code. The implementation of serialization involves the automatic generation of marshaling routines which avoid dynamic inspection of the object structure and make use of the fact that Manta knows the layout of the objects in memory. The paper does not mention performance numbers on serialization of general objects, i.e., graphs.
  Similar to this work, we use explicit marshaling routines and recommend that the JNI (Java native interface) exploits its knowledge of memory layout. However, our work sticks to Java, avoids native code, and can handle polymorphism and subtyping in arguments.
- There are other approaches to Java computing on clusters, where object serialization is not an issue. For example, in *Java/DSM* [12] a JVM is implemented on top of Treadmarks [4]. Since no explicit communication is necessary and because all communication is handled by the underlying DSM, no serialization is necessary.

However, although this approach has the conceptual advantage of being transparent, there are no performance numbers available to us.
- An orthogonal approach is to avoid object serialization by means of object caching. Objects that are not sent will not cause any serialization overhead. See [5] for the discussion of a prototype implementation of this idea and some performance numbers.

Whereas the impact of serialization performance on Grande applications is obvious, object serialization will as well become relevant for the Corba world in the future. While Corba currently cannot pass objects by copy, OMG [6] is working with Sun and others to add value classes. Future versions of IIOP (Internet inter-ORB protocol) are likely to apply object serialization techniques [7].

In addition to parallel computing, object serialization is a critical issue in the area of persistent object storage. Since in that area the life time of objects is much longer than in high-performance computing, object recovery needs to be feasible even from newer versions of the JDK and from programs that do not have access to the appropriate ByteCode files. Therefore, this area needs other optimization strategies than the ones discussed below.

3 Improving JDK-Serialization Performance

The JDK-serialization has mainly been designed to implement persistent objects and to send and receive objects in typical client-server applications. These goals have strongly influenced some design and implementation decisions.

3.1 Explicit Marshaling instead of Reflection

For regular users of object serialization, it is a nice feature that the JDK offers general code that can do the marshaling and unmarshaling automatically by dynamic inspection of the object. For high performance computing however, better performance can be achieved.

Explicit marshaling is much faster when JDK-serialization is used, as a comparison of the full bars from Figure 1 (~ 346 μs and ~ 1410 μs) and Figure 2 (~ 120 μs and ~ 490 μs) shows. In Figure 1, dynamic type introspection is used, whereas in Figure 2, the programmer has provided explicit marshaling and unmarshaling routines. Explicit marshaling is still much faster when all applicable optimization of the UKA-serialization are switched on, as a comparison of sizes of the white boxes show.

The following sections 3.2 to 3.4 will discuss the patterned areas of the bars of Figure 2. The UKA-serialization can avoid all of them and even achieves better latency hiding.

3.2 Slim Encoding of Type Information

Persistent objects that have been stored to disk must be readable even if the ByteCode that was originally used to instantiate the object is no longer available.

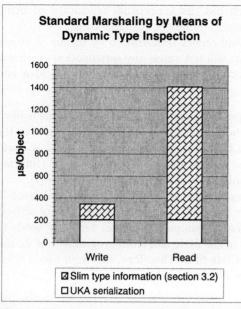

Fig. 1. JDK-serialization with dynamic type inspection and the effect of all applicable optimizations of the UKA-serialization.

For example, a new version of the class might have a different layout of its instance variables. To cope with these situations, the complete type description is included in the stream of bytes that represents the state of an object being serialized.

For parallel Java programs on clusters of workstations and DMPs this is not required. The life time of all objects is shorter than the runtime of the job. When objects are being communicated it is safe to assume that all nodes have access to the same ByteCode, for example through a common file system. Hence, there is no need to completely encode and decode the type information in the byte stream and to transmit that information over the network.

The UKA-serialization uses a textual encoding of class names and package prefixes. Even shorter representations are possible. Simplifying type information has improved the performance of serialization significantly. See the patterned areas of the bars of Figures 1 and 2 that are marked 3.2.

3.3 Two types of Reset

To achieve copy semantics, every new method invocation has to start with a fresh hash-table so that objects that have been transmitted earlier will be re-transmitted with their current state.[2] The current RMI implementation achieves

---

[2] As we have mentioned in the Related Work section, caching techniques could be used to often avoid retransmission of objects whose states have not changed.

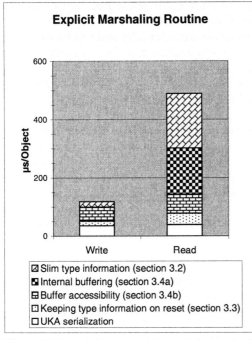

Fig. 2. JDK-serialization with explicit marshaling and the effect of all applicable optimizations of the UKA-serialization.

The full bars show the times needed by the JDK-serialization to write/read an object with 32 int values with explicit marshaling routines.

The patterned areas show the individual savings due to the optimizations discussed in sections 3.2 to 3.4. Slim type encoding has more effect when dynamic type inspection is used (see Figure 1); but techniques 3.3 and 3.4 can only be used with explicit marshaling.

By switching on all optimizations in the UKA-serialization, only the times of the lower white boxes remain. These are much smaller than what can be achieved with dynamic type inspection.

that effect by creating a new JDK-serialization object for every method invocation. An alternative implementation could probably call the serialization's `reset` method instead.

The problem with both approaches is that they not only clear the information on objects that have already been transmitted. But in addition, they clear all the information on types.

To alleviate this problem, the UKA-serialization offers a new `reset` routine that only clears the object hash-table but leaves the information on types unchanged. The dotted area of the bars of Figure 2 that are marked with 3.3 show how much improvement the UKA-serialization can achieve by providing a second reset routine. (Unfortunately, the current official RMI implementation does not use it.)

3.4 Better Buffering

The JDK-serialization has two problems with respect to buffering.

a) **External versus Internal Buffering.** On the side of the recipient, the JDK-serialization does not implement buffering strategies itself. Instead, it uses buffered stream implementations (on top of TCP/IP sockets). The stream's buffering is general and does not know anything about the byte representation of objects. Hence, its buffering is not driven by the number of bytes that are needed to marshal an object.

The UKA-serialization handles the buffering internally and can therefore exploit knowledge about an object's wire representation. The optimized buffering strategy reads all bytes of an object at once.

b) **Private versus Public Buffers.** Because of the external buffering used by the JDK-serialization, programmers cannot directly write into these buffers. Instead, they are required to use special `write` routines.

UKA-serialization on the other hand implements the necessary buffering itself. Hence, there is no longer a need for this additional layer of method invocations. By making the buffer public, explicit marshaling routines can write their data immediately into the buffer. Here, we trade the modularity of the original design for improved speed.

Because it is cumbersome to deal with buffer overflow conditions in the explicit marshaling routines, we are working on a tool that automatically generates proper routines.

The patterned area marked 3.4a in Figure 2 shows the effect of external buffering. The patterned area marked 3.4b indicate the additional gain that can be achieved by having the explicit marshaling routines write to and read from the buffer directly.

3.5 Reflection Enhancements

Although we haven't implemented it in the UKA-serialization because of our pure-Java approach, some benchmarks clearly indicate that the JNI (Java native interface) should be extended to provide a routine that can copy all primitive-type instance variables of an object into a buffer at once with a single method call. For example, class `Class` could be extended to return an object of a new class `ClassInfo`:

```
ClassInfo getClassInfo(Field[] fields);
```

The object of type `ClassInfo` then provides two routines that do the copying to/from the communication buffer.

```
int   toByteArray(Object obj,    int objectoffset,
                  byte[] buffer, int bufferoffset);
int fromByteArray(Object obj,    int objectoffset,
                  byte[] buffer, int bufferoffset);
```

The first routine copies the bytes that represent all the instance variables into the communication buffer (ideally on the network interface board), starting at the given buffer offset. The first `objectoffset` bytes are left out. The routine returns the number of bytes that have actually been copied. Hence, if the communication buffer is too small to hold all bytes, the routine must be called again, with modified offsets.

Some experiments indicate that the effect of accessible buffers, see Figure 2(3.4a), would increase if such routines were made available in the JNI.

3.6 Handling of Floats and Doubles

In scientific applications, floats and arrays of floats are used frequently (the same holds for doubles). It is essential that these data types are packed and unpacked efficiently.

The conversion of these primitive data types into a machine-independent byte representation is (on most machines) a matter of a type cast. However, in the JDK-serialization, the type cast is implemented in a native method called via JNI (`Float.floatToIntBits(float)`) and hence requires various time consuming operations for check-pointing and state recovery upon JNI entry and JNI exit. We therefore recommend that JIT-builders inline this method and avoid crossing the JNI barrier.

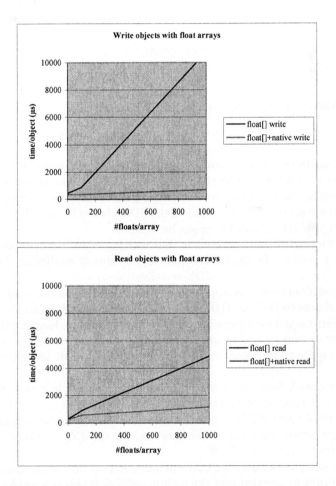

Fig. 3. Serialization of float arrays (same benchmark setup).

Moreover, the JDK-serialization of float arrays (and double arrays) currently invokes the above-mentioned JNI-routine *for every single array element.* We have implemented fast native handling of whole arrays with dramatic improvements, as shown in Figure 3. This, however, cannot be done in pure Java and is left for JVM vendors to fix.

4 Design

An important characteristic of the UKA-serialization is that it only improves the performance for objects that are equipped with explicit marshaling and unmarshaling routines discussed in section 3. We call these objects *UKA-aware* objects. For UKA-unaware objects, the UKA-serialization does not help. Instead, standard JDK-serialization is used. Therefore, the JDK-serialization code must – in some way or another – still be present in any design of the UKA-serialization.

In the sections below, we discuss in a increasingly detailed way why straightforward approaches fail and why subclassing the JDK-serialization imposes major problems. Section 4.5 then shows a design that works.

4.1 CLASSPATH approach fails

The necessary availability of standard JDK-serialization code rules out a design that is based on CLASSPATH modifications.

The straightforward approach to develop a drop-in replacement for the JDK-serialization is to implement all the improvements directly in a copy of the existing serialization classes (`java.io.ObjectOutputStream` and `...InputStream`). The resulting classes must then shadow the JDK classes in the CLASSPATH so that the original classes will no longer be loaded.

The advantage of this approach is that existing code that uses serialization functionality need not be changed in any way. By simply modifying the CLASSPATH, one can switch from the JDK-serialization to a drop-in serialization.

The disadvantage of this approach is that it is not maintainable. Unfortunately, the source code of the JDK-serialization keeps changing significantly from version to version and even from beta release to beta release. Keeping the drop-in implementation current and re-implementing all the improvements in a changing code base is quite a lot of work, especially since existing JDK-serialization needs to survive.

Another straightforward CLASSPATH approach fails: it is impossible to simply rename the JDK-serialization classes and put UKA-classes with the original names into the CLASSPATH. This idea does not work, since the renamed classes can no longer access some native routines because the JNI encodes class names into the names of native methods.

Since for early versions of the UKA-serialization we have suffered under quick release turn-over we decided that the maintainability problem is more significant than the advantages gained by this approach. Therefore, UKA-serialization is designed as subclasses of JDK-serialization classes.

4.2 Consequences of Subclassing the JDK-Serialization

Designing the UKA-serialization by subclassing the JDK-serialization causes two general disadvantages.

First, existing code that uses serialization functionality has to be modified in two ways: (a) the UKA-subclass needs to be instantiated wherever a JDK-parent-class has been created before. Additionally (b), every existing user-defined subclass of a JDK-class needs to become a subclass of the corresponding UKA-class, i.e., the UKA-classes need to be properly inserted into the inheritance hierarchy. These modifications are sufficient since the UKA-serialization objects are type compatible with the standard ones due to the subclass relationship.

Even if the source of existing code is not available, the class files can be retrofitted to work with the UKA-serialization. Our retrofitting tool modifies the class file's constant table accordingly. After retrofitting, a precompiled class creates instances of the new serialization instead of the original one.

Using the retrofitting trick we were able to use the UKA-serialization in combination with RMI although most of the RMI source code is not part of the JDK distribution.[3]

The second general disadvantage is that the security manager must be set to allow object serialization by a subclass implementation. There is no way to avoid a check by the security manager because it is done in the constructor of JDK's `ObjectOutputStream`.

For using the UKA-serialization from RMI, this is not a big problem, since the RMI security manager allows serialization by subclasses anyway.

4.3 Problems when Subclassing the JDK-serialization

Unfortunately, after subclassing the JDK-serialization, the standard implementation can no longer be used. This is due to a very restrictive design that prevents reuse.

Since `writeObject` is final in `ObjectOutputStream` it cannot be overridden in a subclass. The API provides an alternative, namely a hook method called `writeObjectOverride` that is transparently invoked in case a private boolean flag (`enableSubclassImplementation`) is set to true. This flag is true only if the parameter-less standard constructor of the JDK-serialization is used for creation, i.e., only if the serialization is implemented in a subclass. The standard constructor however does (intentionally) not properly initialize `ObjectOutputStream`'s data structures and thus prevents using the original serialization implementation.[4]

There are two approaches to cope with that problem. The first approach uses delegation in addition to subclassing. Although the existing code of the JDK-serialization is not touched, the necessary code gets quite complicated. Moreover,

---

[3] Only three RMI classes needed retrofitting, namely `java.rmi.MarshalledObject`, `sun.rmi.server.MarshalInputStream`, and ...`MarshalOutputStream`.

[4] The reason for this design is that it allows the security manager to check permissions. However, the same checks could be done with other designs as well.

for certain constellations of `instanceof`-usage this approach does not work at all. See section 4.4 for the details.

The implementation of the UKA-serialization does not use the delegation approach. Instead we moderately and maintainably changed the existing JDK-serialization classes to enable reuse. This is more "dirty" but results in a cleaner overall design. See section 4.5.

4.4 Subclassing plus Delegation

The only way out without touching the implementation of the JDK-serialization is to allocate an additional `ObjectOutputStream` delegate object within the UKA-serialization. Its `writeObject()` method is invoked whenever a UKA-unaware object is serialized. Since the delegate object can be created lazily, it does not introduce any overhead unless UKA-unaware objects are serialized.

Subclassing plus delegation has two disadvantages. First, it is not as simple as it appears. But more importantly, it does not work correctly under all circumstances.

With respect to simplicity it must be noted, that for the delegate object another subclass of the JDK-serialization is needed for cases where existing code itself is using subclasses of the JDK-serialization. (RMI for example does it.) In addition to the hook method mentioned above, the JDK-serialization has several other dummy routines which can be overridden in subclasses. Therefore, if a standard JDK-serialization stream would be used as delegate, it's dummy routines would be called instead of the user-provided implementations. To solve this problem, the delegate is a subclass of the JDK-serialization and provides implementations for *all* methods that can be overridden in the JDK-implementation. The purpose of the additional methods is to forward the call back to the UKA-serialization and hence to the implementation provided by the user's subclass.

There is no guarantee, that subclassing plus delegation works correct in cases where existing code itself is using subclasses of the JDK-serialization and where UKA-unaware objects provide explicit marshaling routines that use the `instanceof` operator to find out the specific type of a current serialization object. (RMI for example does it.) Since the objects are UKA-unaware they are handled by the delegate. Therefore, the `instanceof` operator does no longer signal type compatibility to the serialization subclasses provided by the code.

Since RMI does exactly this (there are subclasses of the JDK-serialization, and the code uses explicit marshaling routines that check the type of a serialization stream object), subclassing plus delegation does not work correctly with RMI. Especially painful are problems with the distributed garbage collector that are hard to track down due to their indeterministic nature. (However, subclassing plus delegation does work correctly with "well-behaved" users of serialization functionality.)

4.5 Subclassing plus Source Modification

We now present an approach that works. Being based on subclassing the JDK-serialization, it has the general disadvantages discussed in section 4.2.

The idea is to moderately modify the source code of existing JDK-serialization classes to enable reusing the existing functionality from subclasses. The modification is kept small enough to not affect maintainability.

Three simple changes are sufficient in every JDK release: First, the code of ObjectOutputStream's regular constructor is copied into an additional initialization method _init(OutputStream). Second, to switch between the subclass and the standard serialization, the access modifier of the above-mentioned flag enableSubclassImplementation is relaxed from private to protected.[5] And third, the final modifier of writeObject(Object) is removed to override the method directly and to save an unnecessary call of the hook method.

Since these modifications are simple they can easily be applied to updated versions of the JDK without too much thought. Because no additional serialization object is introduced, the UKA-serialization works fine, even with RMI.

We hope that Sun will incorporate the ideas of the UKA-serialization in future releases of the JDK so that this "dirty" source code modification will not be necessary for ever.

5 Results

We have designed the UKA-serialization according to the approach sketched in section 4.5. The implementation includes the optimizations that are discussed in sections 3.1 to 3.4. Further optimizations would need help from the JVM vendors.

5.1 Improvement of Serialization

Section 3 has discussed the individual problems of the JDK-serialization and has provided solutions. Because of all the details, the quantitative result might have passed unnoticed.

Instead of 1410 μs for reading (JDK-serialization of a certain type of object by means of dynamic type inspection) the UKA-serialization takes just 39 μs, which amounts to an improvement of about 97%. For writing, the improvement is in the order of 90%. See the bold line of Table 2. Similar improvements occur for different types of objects. The second column of Table 2 shows the measurements (in μs) for an object with four ints and two null pointers. The last column presents the measurements for a balanced binary tree of 15 objects each of

---

[5] Javasoft recently released a beta version of RMI-IIOP. This software uses an extended serialization that is compatible with Corba's IIOP. This extension faces the same problems as ours. But instead of modifying the access modifier, the authors provide a native library routine to toggle the flag. We consider this to be even more "dirty" than our approach since it is no longer platform independent.

which holds four `ints`. Note that reading and writing now often take about the same time and therefore allow for balanced overlapping and hence better object transfer times. Overlapping is only relevant for large object graphs that are sent in several packets.

μs per object	32 int w	r	4 int, 2 null w	r	tree(15) w	r
JDK serialization	346	1410	169	596	1192	1889
UKA-serialization	35	39	19	28	201	354
improvement %	**90**	**97**	**89**	**95**	**83**	**81**
explicit marshaling	228	920	81	308	396	647
slim type encoding	19	187	16	159	72	213
internal buffering	0	159	6	19	0	330
buffer accessibility	48	65	30	49	502	291
two types of reset	16	40	17	33	21	54

Table 2. Improvements for several types of objects.

While for flat objects about 90% or more of the serialization overhead can be avoided, for the tree of objects the improvement is in the range of 80%. This is due to the fact that the work needed to deal with potential cycles in the graph of objects cannot be reduced significantly.

The lower half of the table shows the contributions of the individual improvement techniques.

5.2 Improvement of RMI

To use the UKA-serialization with RMI, we have applied the retrofitting tool to Sun's RMI class files. Wherever a regular `ObjectOutputStream` has been created in the original code, the upgraded versions used the corresponding UKA-objects. Due to latency hiding effects the improvement (36%–51%) is even better than expected. See Table 3.

μs per object	32 int		4 int, 2 null		tree(15)	
`RMI ping(obj)`	4193	100%	2890	100%	5330	100%
`Socket send(obj)`	2790	67%	1541	53%	3990	75%
`Serialize(obj)`	1756	42%	765	26%	3081	58%
RMI with UKA	2043		1840		2900	%
improvement %		**51 %**		**36 %**		**46 %**

Table 3. Ping times (μs) of RMI, socket communication, and JDK-serialization.

6 Conclusion

The UKA-serialization demonstrates that it is possible to implement object serialization in Java with improved performance by at least a factor 5. Some additional help from the JVM vendors could speed things up even further. The UKA-serialization which is pure Java and freely available [3] can be used as a drop-in replacement for the JDK-serialization. A retrofitting tool is provided that upgrades existing class files to work with the UKA-serialization classes. When used with RMI, our synthetic benchmarks improve by 36% to 51%. Benchmark results with full applications will follow.

In the future, we will work on a tool that automatically generates the explicit marshaling routines that right now must be provided by hand. Moreover, we are attacking the remaining cost of a remote method invocation by using high speed networks and by optimizing the RMI layer.

Acknowledgments

We would like to thank the JavaParty team. Matthias Jacob did excellent work on the performance comparison (Java versus Fortran) on a geophysics application. Christian Nester and Matthias Gimbel suffered through the beta-testing. The Java Grande Forum and Siamak Hassanzadeh from Sun Microsystems provided the opportunity and some financial support to find and discuss shortcomings of the JDK-serialization.

References

1. Matthias Jacob, Michael Philippsen, and Martin Karrenbach. Large-scale parallel geophysical algorithms in Java: A feasibility study. *Concurrency: Practice and Experience*, 10(11–13):1143–1154, September–November 1998.
2. Java Grande Forum. http://www.javagrande.org.
3. JavaParty. http://wwwipd.ira.uka.de/JavaParty/.
4. P. Keleher, A. L. Cox, and W. Zwaenepoel. Treadmarks: Distributed shared memory on standard workstations and operating systems. In *Proc. 1994 Winter Usenix Conf.*, pages 115–131, January 1994.
5. Vijaykumar Krishnaswamy, Dan Walther, Sumeer Bhola, Ethendranath Bommaiah, George Riley, Brad Topol, and Mustaque Ahamad. Efficient implementations of Java Remote Method Invocation (RMI). In *Proc. of the 4th USENIX Conference on Object-Oriented Technologies and Systems (COOTS'98)*, 1998.
6. OMG. http://www.omg.org.
7. OMG. *Objects by Value Specification*, January 1998. ftp://ftp.omg.org/pub/docs/orbos/98-01-18.pdf.
8. Sun Microsystems Inc., Mountain View, CA. *Java Object Serialization Specification*, November 1998. ftp://ftp.javasoft.com/docs/jdk1.2/serial-spec-JDK1.2.pdf.
9. George K. Thiruvathukal, Fabian Breg, Ronald Boisvert, Joseph Darcy, Geoffrey C. Fox, Dennis Gannon, Siamak Hassanzadeh, Jose Moreira, Michael Philippsen, Roldan Pozo, and Marc Snir (editors). Java Grande Forum Report: Making Java

work for high-end computing. In *Supercomputing'98: International Conference on High Performance Computing and Communications*, Orlando, Florida, November 7-13, 1998. panel handout.
10. George K. Thiruvathukal, Lovely S. Thomas, and Andy T. Korczynski. Reflective remote method invocation. *Concurrency: Practice and Experience*, 10(11-13):911-926, September-November 1998.
11. Ronald Veldema, Rob van Nieuwport, Jason Maassen, Henri E. Bal, and Aske Plaat. Efficient remote method invocation. Technical Report IR-450, Vrije Universiteit, Amsterdam, The Netherlands, September 1998.
12. Weimin Yu and Alan Cox. Java/DSM: A platform for heterogeneous computing. *Concurrency: Practice and Experience*, 9(11):1213-1224, November 1997.

A Customizable Implementation of RMI for High Performance Computing*

Fabian Breg and Dennis Gannon

Department of Computer Science, Indiana University

Abstract. This paper describes an implementation of Java's Remote Method Invocation (RMI) that is designed to run on top of the Globus high performance computing protocol. The primary contribution of this work is to illustrate how the object serialization mechanism used by RMI can be extended so that it becomes more configurable. This allows the implementation of object serialization protocols that are more efficient than the default or that are compatible with other distributed object models like HPC++, which is based on C++. Both issues are important when RMI is to be used in scientific computing.

1 Introduction

One of the main reasons of the popularity of the Java programming language is its support for distributed computing. Java's API for sockets, URLs and other networking facilities is much simpler than what is offered by other programming languages, like C and C++, which is why Java is likely to be adopted in scientific computing. These constructs, however, are still too tedious to use for generic distributed applications. Java's Remote Method Invocation (RMI) [18] was designed to make distributed application programming as simple as possible, by hiding the remoteness of distributed objects as much as possible.

RMI is a framework that allows objects to invoke methods on remote objects (i.e. objects residing in a different Java Virtual Machine, JVM for short) in the same way as methods of local objects (in the same JVM) are invoked. The syntax of a remote method invocation is similar to a method invocation on a local object, but the semantics for parameter passing are different. While remote objects are still passed by reference, non-remote objects are passed by copy instead (local method invocations pass all objects by reference). Also, references to remote objects must be obtained from a special entity called the registry. RMI does, however, provide a convenient framework for distributed computing.

As noted by others (see Sect. 2), Java RMI has a number of shortcomings when it comes to performance and flexibility. The first contribution of this work is to overcome these shortcomings improving both flexibility and performance of an important bottleneck of RMI: object serialization. To this end, we reimplemented RMI from scratch, retaining the original specification as much as possible. We

---

* This research has been supported by a contract with the Dept. of Energy DOE2000 project.

have only slightly altered the API to implement a flexible framework for object serialization. Since we only add methods to the API, our implementation can still handle most applications written for Java RMI.

The second contribution of this work is to investigate into the interoperability of RMI with another remote method invocation framework: HPC++ [2]. We will only briefly outline our approach to obtaining interoperability. A more in-depth study into the issues involved has been conducted in a previous article [3].

We are planning on using our implementation of RMI in future versions of our Distributed Problem Solving Environment Component Architecture Toolkit [5]. In this project, we will have a component architecture toolkit implemented in Java that communicates with distributed components written in HPC++.

The rest of this paper is organized as follows. In Sect. 2 we will point out some related research into RMI conducted by others. Section 3 gives a brief overview of the Java RMI framework. Section 4 gives an overview of some important design and implementation issues of our RMI implementation. Section 5 focuses on the object serialization and compares a number of object serialization protocols. Section 6 concludes this paper.

2 Related Work

One of the goals of the **Java Grande Forum** [8] is to investigate into methods to optimize Java RMI performance. In a short note presented to this forum, Philippsen and Haumacher [11] critiqued the performance of Java RMI and offered a number of suggestions to improve both its performance and flexibility. They conclude that a more open approach would encourage the widespread use of RMI in the scientific computing area.

The same is suggested by Thiruvathukal and others in [15]. They implemented a more open version of the Java RMI framework, which also provided some additional functionality, such as dynamically obtaining the interface of remote objects. Their API, however, is quite different from the original. Our goal was to make minimal changes to the official specification, while still providing some openness.

As part of the **Ninja** project [9] at the University of California, Berkeley, **NinjaRMI** [17] has been developed. This project adds some interesting features that are not (yet) available in Sun's RMI version. They do not address object serialization or interoperability with other languages.

Another implementation of RMI is provided by Raje and others. **ARMI** offers the possibility of performing asynchronous method invocations on remote objects. Their implementation is built on top of the original Java RMI and thus suffers from the same performance drawbacks.

In [16] Veldema and others implemented their own version of RMI for use in their **Albatross** [1] project. Their goal was to optimize RMI for homogeneous cluster computers and they offer a programming model based on **JavaParty** [12]. In addition, they translate Java applications to native code, thereby giving up the cross-platform portability. Nevertheless, their approach is interesting for scientific computing, because very low latency and high bandwidth is obtained with their implementation.

There are a number of distributed object frameworks like RMI available. **CORBA** [6] provides a distributed object model for interoperability between different platforms. Besides remote method invocation, **Voyager** [10] also provides the notion of active migratable entities, called agents. Hirano and others compared the performance of some of these different distributed object technologies in [7], including **HORB** [13], their own distributed object technology. They showed that indeed RMI exhibits worse performance than other distributed object technologies and adds significant overhead to the use of sockets.

3   Remote Method Invocation

Java **Remote Method Invocation (RMI)** [14] is a framework designed to simplify distributed object oriented computing in Java. An overview of the Java RMI model is shown in Fig. 1.

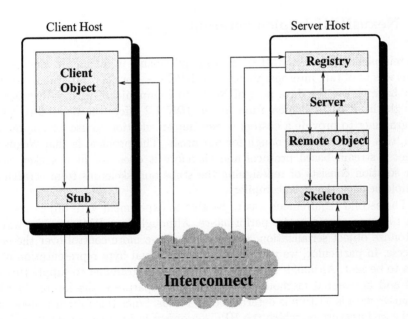

Fig. 1. RMI overview

RMI is essentially a client-server model in which the client talks to the remote object through a proxy called a **stub**. The stub itself talks to an active entity on the server side called a **skeleton**, which in turn calls the appropriate method on the remote object (which is local to the skeleton). The result is communicated back through the skeleton and the stub to the client. The stubs and skeleton are generated by a special compiler, `rmic`, and are completely hidden from the programmer.[1]

To obtain a reference to a remote object, the client contacts the **registry** on the remote host, providing it the name of the desired remote object. The remote object must previously have been bound in the registry by a server process. The remote reference that is passed from server via the registry to the client is actually the stub object itself.

To pass non-remote objects as a parameter of a remote method invocation, the notion of **object serialization** has been introduced in Java 1.1. Object serialization transforms an object or even a graph of objects into a sequence of bytes that can be sent over the network, allowing the other side to reconstruct the original object (graph). Note that the RMI parameter passing mechanism for non-remote objects is pass-by-copy, which is different from the parameter passing semantics for method invocations on local objects. From a programmer's point of view, remote objects are passed by reference. What is actually transfered is the stub object (which itself is passed by copy).

4 NexusRMI Implementation

We reimplemented Java RMI with two requirements in mind. The first requirement was to obtain interoperability with HPC++. For this, our stubs and skeletons have to use the Java port of Nexus to communicate with each other. Although the RMI implementation in the JDK 1.2 offers the possibility for the programmer to provide a custom socket implementation to use for communication, this is not flexible enough for our needs. The problem is that NexusJava is not a stream based protocol and therefore it does not fit a socket model. Our solution consists of redesigning the stubs and skeletons from scratch and reimplementing the `rmic` compiler.

The second requirement was to be able to experiment with object serialization to try to obtain better performance. Although Java RMI provides ways to customize object serialization, we needed to have more control over the entire process. In particular, we need to specify the actual byte representation of the data to be sent. Although HPC++ requires the programmer to supply the marshal and unmarshal methods for each object separately, the format in which primitive data is written is determined by Nexus. Since this format is dependent on the architecture on which the HPC++ object is running, we must deal with all possible formats, which is why we use the NexusJava framework for reading and writing primitives. Also, offering the possibility to specify the object serialization protocol allows us to implement various optimizations. Examples of this are described in Sect. 5.

---

[1] In an RMI application, the remote reference is declared to be of the remote interface type. The generated stub is defined to implement this interface, allowing the remote reference to be a reference to the stub.

This section describes the design and implementation of NexusRMI and how both requirements are met by it. In Sect. 5 we evaluate the performance of object serialization in our framework.

4.1 NexusRMI Transport Protocol

To understand the underlying communication protocol of NexusRMI, we will first introduce the NexusJava [4] model of communication. An overview of NexusJava is given in Fig. 2. In Nexus, a **node** or a host can contain multiple **contexts**, which in turn can contain multiple **threads**. A thread can communicate with an object in another context by issuing a **remote service request** on a **startpoint**. The startpoint will forward this request to the **endpoint** connected to it. The endpoint will then invoke the appropriate method on its associated **user object**.

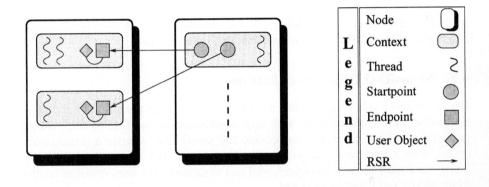

Fig. 2. NexusJava model

Figure 3 shows the different abstractions of NexusJava in our RMI implementation (refer to the NexusJava overview figure for a legend of the symbols used). Basically, a skeleton object contains an endpoint, with the actual remote object associated to it as a user object. The registry can be considered just another remote object and thus has a similar implementation. If the client invokes a remote method, the stub issues a remote service request on the startpoint connected to the appropriate skeleton endpoint. No reply message is implicit in a remote service request. The result value of the remote method is sent to client through a separate remote service request. To this end, the client itself contains an endpoint and it sends a startpoint to itself as part of the remote method invocation request message. A separate handler thread at the client extracts the result and passes it to the thread that initiated the remote method invocation.

When a remote object is exported, i.e. when the stub/skeleton pair is created, containing the necessary startpoints and endpoints. The stubs are registered in the registry when the server binds the remote object and are passed to the client

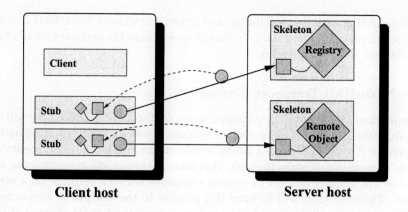

Fig. 3. NexusRMI communication system

when the client performs a lookup for the remote object. A startpoint to the registry is obtained through NexusJava's `attach()` method.

4.2 NexusRMI Object Serialization

Besides obtaining interoperability with HPC++, another purpose of our implementation was to experiment with object serialization in RMI to improve its performance. The intent was to provide the possibility to write a serialization protocol on a per remote object basis. The idea is that a remote object may know what kind of data to expect, and can thus make the best decision on the exact object serialization protocol.

To be able to specify the object serialization protocol on a per remote object basis, we had to alter the RMI specification slightly. When a remote object is exported, i.e. when the stubs and skeleton are created, we pass the serialization protocol with it, so that both the stubs and the skeleton know how to communicate with each other. A remote object can be exported by invoking `UnicastRemoteObject.exportObject()` method. If, however, the remote object extends `UnicastRemoteObject`, it is exported implicitly in the constructor. The NexusRMI interface of `UnicastRemoteObject` is shown in Fig. 4.[2]

The `Serialize` class is the superclass of objects implementing a serialization protocol. Its definition is shown in Fig. 5. Data is written into a `PutBuffer`[3] and read from a `GetBuffer`. The codebase is used to dynamically download classes from the specified site [14, sect 3.8]. Whenever a class has to be loaded that is not locally available the URL is passed to the `RMIClassLoader`, which than downloads the class. The `initWrite()` method initializes the object for

---

[2] Our implementation of RMI is mainly based on the JDK 1.1 version.
[3] The implementation of the `PutBuffer` class has been adapted for this project to allow it to dynamically expand as needed. This behavior is needed since we cannot simply determine the total buffer size requirements for an arbitrary graph of objects.

```
public class UnicastRemoteObject extends RemoteServer
{
    protected UnicastRemoteObject()              throws RemoteException;
    protected UnicastRemoteObject(Serialize ser) throws RemoteException;
    public Object clone()                  throws CloneNotSupportedException;
    public static RemoteStub exportObject(Remote obj)
                                                 throws RemoteException;
    public static RemoteStub exportObject(Remote obj, Serialize ser)
                                                 throws RemoteException;
}
```

Fig. 4. UnicastRemoteObject

writing. The initRead() method initializes the object for reading. It is given the buffer to read from and the URL to dynamically load classes from. When all data has been written, the buffer with the serialized data can be obtained by invoking getBuffer(). The next six methods implement the reading and writing of HPC++ specific data, which are not to be altered by the custom serialization protocol. The rest of the methods implement the reading and writing of primitives and references. The actual implementations in this class are no-ops.

To implement a custom serialization protocol, the Serialize class has to be extended and the desired read and write methods need to be implemented. Only those methods actually needed have to be overridden. The next section will show some implementations of object serialization protocols that can be used in our framework.

Serialization of primitive types is fairly simple to implement since NexusJava provides methods to write primitives and arrays of primitives. To serialize non-array references, we need to be able to access the complete internal state of such references. Using the JDK 1.1, it is not possible to access the values of private and protected fields of an object. Furthermore, to reconstruct an object at the receiving side, we need a uniform way to create any object. Since the only way to construct an object is by invoking a constructor and the signature of a constructor is not guaranteed to be fixed, we need some other way of performing object serialization.[4]

Our approach is to add methods to a serializable class, which traverses all fields of an instance of the class and reads or writes these fields from or to a buffer. We implemented a compiler which automatically adds such methods to classes. If a parameterless constructor is not available in the class, the compiler will also add one for us so we can create instances of serializable objects in a uniform way.[5] Since we add our methods to the original Java source code, this

---

[4] In JDK 1.2, because of its more flexible security implementation, it is possible to access all fields of an object through reflection. However, we still have the problem of creating an uninitialized instance.

[5] This method is not completely safe, since an existing parameterless constructor may produce undesired side-effects.

```
public class Serialize
{
  protected Nexus nexus;
  protected PutBuffer outbuf;
  protected GetBuffer inbuf;
  protected URL codebase;

  public Serialize();
  public void initWrite();
  public void initRead(Object buffer, String codebase);
  public Object getBuffer();

  public final void writeHPCxxHeaderStartpoint(Startpoint sp);
  public final void writeHPCxxHeaderOffset(int off);
  public final void writeHPCxxHeaderHandlerid(int id);
  public final Startpoint readHPCxxHeaderStartpoint();
  public final int readHPCxxHeaderOffset();
  public final int readHPCxxHeaderHandlerid();

  public void writeboolean(boolean b);
  public void writebyte(byte b);
  public void writechar(char c);
  public void writeshort(short s);
  public void writeint(int i);
  public void writelong(long l);
  public void writefloat(float f);
  public void writedouble(double d);
  public void writeObject(Object o);
  public void writeFinalObject(Class type, Object o);

  public boolean readboolean();
  public byte readbyte();
  public char readchar();
  public short readshort();
  public int readint();
  public long readlong();
  public float readfloat();
  public double readdouble();
  public Object readObject();
  public Object readFinalObject(Class type);
}
```

Fig. 5. Serialize

method only works for serializable classes for which we have access to the source code.

5 Experiments with NexusRMI

In this section, we give some examples of object serialization protocol implementations that we have experimented with in our NexusRMI framework. We will show that, with the facilities offered in NexusRMI, a wide range of protocol semantics can be implemented, from semantics close to Java's own to semantics that can improve the performance of specific applications.

Section 5.1 describes the `DefaultSerialization` protocol. This protocol allows the serialization and deserialization of all primitive and reference types.[6] It offers the possibility of dynamically downloading classes from the server host and it will handle graphs of objects correctly, constructing a logically equivalent graph of objects on the receiving size. This protocol is close to the protocol used in Java RMI [7] and serves as the default serialization protocol in NexusRMI.

The `RelaxedSerialization` protocol described in Sect. 5.2 relaxes the default serialization semantics a little by removing the dynamic class loading facility and the check for previously serialized objects. This means that a graph of objects can still be passed as a parameter or return value, but only if it satisfies certain restrictions explained later. In that sense, this protocol is potentially unsafe, but may give better performance.

Finally, we present the `SpecificSerialization` protocol that can only handle objects very specific to the serialization protocol. This is of course a very restricted protocol, but this approach may further improve the performance for a specific application. This protocol is described in Sect. 5.3.

The benchmark that we will use to compare the performance of the various approaches is one that sends back and forth an array of Complex number objects, each containing two doubles. We experimented with different types of object serialization protocols as will be described in Sect. 5.4.

5.1 The `DefaultSerialization` Protocol

In this section we will describe a general serialization protocol that can handle any object given to it. Reading and writing primitive types is done by delegating this task directly to Nexus' buffer implementations. Reading and writing references to objects is done by invoking the methods added to these object by our compiler. For array references, a distinction is made between arrays of primitives and arrays of references. The serialization of the former is done by the Nexus library, the latter is serialized by serializing each reference separately. Some Objects that do not have the serialization methods added (because source code was not available) are serialized in an ad hoc manner.

---

[6] provided that source code is available.
[7] class versioning is not yet implemented.

By default, complete type information has to be included to be able to create an instance of the right class. We provide additional methods for (de)serializing parameters of a final class, since these do not need complete type information (the stub compiler can obtain the type information statically). Furthermore, care has to be taken to make sure that multiple references to the same object does not cause the creation of multiple objects at the receiver. Related to this is detecting cycles in the object graph so that the object serialization does not end up in an infinite loop.

5.2 The RelaxedSerialization Protocol

A majority of the data that we want to serialize has a 'flat' structure for which the check for multiple references to an object is not needed. By removing this check, we should obtain better performance in case many objects are serialized.

Another source of overhead may be the dynamic loading of classes from the server host. Even if classes are available locally, just invoking the RMIClassLoader instead of loading classes directly by an invocation of Class.forName() may yield some overhead.

Both optimizations were incorporated in RelaxedSerialization, the second serialization protocol that we experiment with in this paper. Its implementation is similar to the DefaultSerialization protocol, except for the optimizations described above.

5.3 The SpecificSerialization Protocol

If parameters and return values for a particular remote method can only take values of a restricted set of types, it is possible to optimize object serialization significantly. Tests for types of objects that are never encountered could be removed completely. Also, the internal state of some objects could be accessed more efficiently than they are accessed in the DefaultSerialization and the RelaxedSerialization protocol.

To see how much the performance of RMI could be improved by using a more efficient serialization protocol, we performed some experiments with an array of Complex number objects, each containing two (primitive) double values that are publicly accessible. We wrote a little program that just sends these objects back and forth plus an accompanying object serialization protocol that can only handle these types of objects. The only methods that had to be provided were the writeObject() and readObject(). Since all fields of the object are public, these two methods directly access these fields to obtain higher performance.

The next section describes our experiments and the results obtained.

5.4 Object Serialization Performance

In this section we will compare the performance of the three serialization protocols described in the previous sections. These serialization protocol implementations all use the serialization primitives offered by NexusJava. In essence, object serialization in NexusRMI can be divided into three layers as shown in Fig. 6. The `DataConversion` layer takes care of writing primitive values in appropriate form in a byte array. The `PutBuffer/GetBuffer` layer offers facilities for managing outgoing and incoming buffers respectively. The `NexusRMI` layer is in fact any custom implemented serialization protocol. Only serialization protocols implemented in the `NexusRMI` layer can be 'plugged' into NexusRMI.

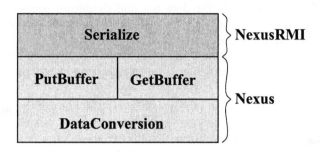

Fig. 6. Overview of serialization layers

The performance of our serialization protocols is measured by invoking the `writeObject()` and `readObject()` methods offered by the `NexusRMI` layer. To assess which parts of object serialization cause the most overhead, we will also measure object serialization performance by directly invoking the the methods from `GetBuffer` and `PutBuffer` and by directly invoking the methods from `DataConversion`. These methods cannot be 'plugged' into NexusRMI, however, since the interface required by NexusRMI is bypassed.

The first alternative method uses the `PutBuffer` implementation to write the simple internal state directly, and the `GetBuffer` implementation to read the internal state directly. Comparing the performance of this method with the performance of the specialized object serialization protocol of Sect. 5.3 gives us a good indication of the overhead incurred by the object serialization framework of NexusRMI.

In the other methods, we bypass the `PutBuffer` and `GetBuffer` implementations and use the `DataConversion` object implementation provided by Nexus-Java to directly write into a byte array. We first measure the serialization and deserialization performance, including the overhead caused by allocating the buffer when writing, and allocating the data structure (the Complex number object array and each complex number individually) when reading. Lastly, we measure the (de)serialization performance when using the `DataConversion` object, without including the memory allocation overhead.

We assessed the performance of object serialization by calculating the **write rate** and **read rate** separately. We measured the time T (in seconds) needed to serialize a particular object of size K (in Mbits). The write rate is than defined to be K/T (in Mbits per second or mbps). We also calculated the read rate by dividing the size of the object by the time it takes to deserialize it. The performance of the different object serialization protocols on a uniprocessor 300 MHz UltraSPARC II is summarized in Fig. 7.

As can be seen, the upper limit for object serialization for complex number objects is approximately 40 mbps (megabits per second). For lower number of objects, the resolution of the used timer is too low, which causes the measured time to be 0 msec, causing the object serialization performance to become infinity. The allocation of the buffer does not add significant overhead, while our implementation of the `PutBuffer` class does add significant overhead. The object serialization framework of our NexusRMI package does not add much overhead when using the special serialization protocol for complex numbers. Finally, adding generality to the object serialization performance causes the performance to drop some more.

Object deserialization does suffer from overhead due to the allocation of the data. In this case, each complex number must be allocated on the heap separately. The implementation of the `GetBuffer` class does not add significant overhead, which was to be expected, since it is just a wrapper around bytearray. Contrast this with the `PutBuffer` implementation, which had to provide the dynamic expansion of the buffer. Also on the deserialization side, our object serialization framework does not add much overhead in itself, except when the protocol gets more general.

To compare our serialization protocols in the context of RMI we ran a simple RMI program that sends back and forth an array of complex number objects. We tested the three serialization protocols that are implemented according to the NexusRMI serialization framework as well as the default JavaRMI for comparison. We used a dual processor UltraSparc to perform the test in order to keep actual transfer time at a minimum, and still allow client and server to run in parallel. The results are shown in Fig. 8.

As can be seen, the default serialization protocol achieves the same order of performance as Java RMI for larger message sizes. Note, however, that Java RMI's object serialization semantics is still stronger than ours. Also, the implementation of the RMI system is different. Where we rely on compiler generated class traversal routines to access the internal state of an object, Java RMI uses reflection. On the other hand, Java RMI uses TCP to communicate, while we use the Java port of the Nexus library, which uses TCP as its underlying communication mechanism. As noted in [3], using NexusJava also inhibits the overlapping of computation and serialization, something that is done in Java RMI.

Relaxing protocol semantics improves RMI performance, although not dramatically, which could be expected after investigating the object serialization protocol performance in isolation as we did earlier.

The specialized serialization protocol improves performance even further. The performance thus obtained is, however, still disappointing. The reason for this poor performance is probably a combination of reasons also mentioned in [11], where the most important would be the fact that each array element has to be separately serialized. The serialization of the actual primitive data is done by

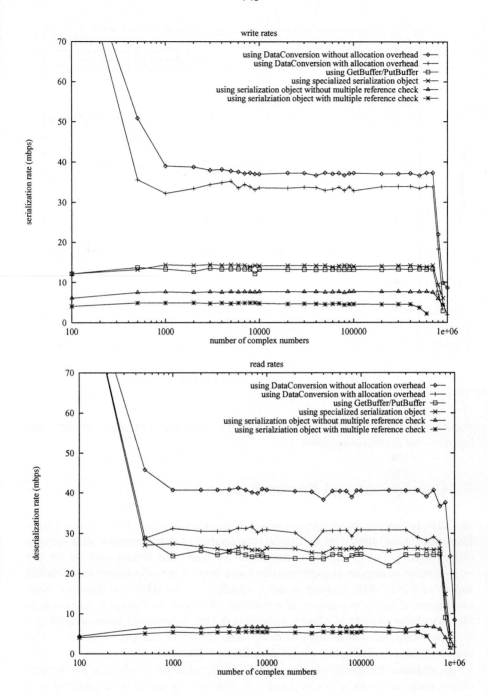

Fig. 7. Write and read rates on a uniprocessor UltraSPARC II

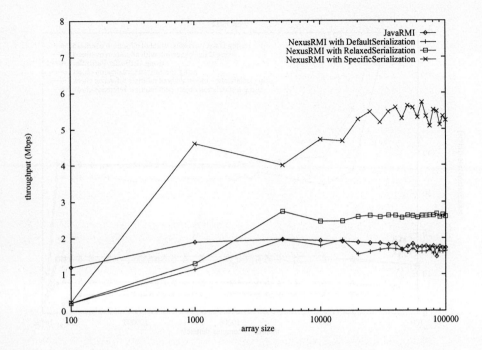

Fig. 8. Performance of RMI implementations

explicitly converting each data item to a byte representation and copying these bytes to the destination buffer.

6 Conclusions and Future Research

We have presented an implementation of the Java Remote Method Invocation framework that allows us to almost transparently communicate with remote object written in the HPC++ language. We have also shown how we let the implementor of remote objects exercise more control over the object serialization aspect of RMI, while making minimal changes to the RMI specification. Next we measured the performance of a number of object serialization protocols to identify those aspects of object serialization that prevent us from obtaining a high performance. These tests, however, also show that the performance can be improved by restricting the object serialization protocol.

In addition to specifying the object serialization protocol on a per remote object basis, one may also wish to specify the underlying network protocol on a per remote object basis. This would allow remote objects to easily communicate with objects that use a different transport layer. The issue here is to design a common transport interface that can be efficiently implemented by any other existing communication library.

Java RMI was designed as a very general framework for remote objects to communicate. Although this is very convenient for most application programmers, when dealing with performance demanding applications, it is generally useful to sacrifice some generality in favor of efficiency. We have shown that this is possible with RMI too, while retaining the convenience for applications that do not really need the performance.

References

1. Albatross project, 1998. http://www.cs.vu.nl/albatross.
2. P. Beckman, D. Gannon, and E. Johnson. Portable Parallel Programming in HPC++. 1996.
3. F. Breg, S. Diwan, J. Villacis, J. Balasubramanian, E. Akman, and D. Gannon. Java RMI Performance and Object Model Interoperability: Experiments with Java/HPC++. *Concurrency: Practice and Experience*, 10:(to appear), 1998.
4. I. Foster, G.K. Thiruvathukal, and S. Tuecke. Technologies for ubiquitous supercomputing: a Java interface to the Nexus communication system. *Concurrency: Practice and Experience*, 9(6):465–475, jun 1997.
5. D. Gannon, R. Bramley, T. Stuckey, J. Villacis, J. Balasubramanian, E. Akman, F. Breg, S. Diwan, and M. Govindaraju. Developing Component Architectures for Distributed Scientific Problem Solving. *IEEE Computational Science & Engineering*, 5(2):50–63, 1998.
6. Object Management Group. The Common Object Request Broker: Architecture and Specification, jul 1995.
7. S. Hirano and H. Yasu, Y. Igarashi. Performance Evaluation of Popular Distributed Object Technologies for Java. *Concurrency: Practice and Experience*, 10:(to appear), 1998.
8. Java Grande Forum. http://www.javagrande.org/.
9. Ninja project, 1998. http://ninja.cs.berkeley.edu/.
10. Objectspace. Objectspace Voyager Core Package Technical Overview, 1997.
11. M. Philippsen and B. Haumacher. Bandwidth, Latency, and other Problems of RMI and Serialization. 1998.
12. M. Philippsen and M. Zenger. JavaParty - Transparent Remote Objects in Java. *Concurrency: Practice and Experience*, 9(11):1225–1242, nov 1997.
13. H Satoshi. HORB: Distributed Execution of Java Programs. 1997.
14. Sun Microsystems. *Java(TM) Remote Method Invocation Specification*, oct 1997. revision 1.42 jdk1.2Beta1.
15. G.K. Thiruvathukal, L.S. Thomas, and A.T. Korczynski. Reflective Remote Method Invocation. *Concurrency: Practice and Experience*, 10:(to appear), 1998.
16. R. Veldema, R. van Nieuwpoort, J. Maassen, H.E. Bal, and A. Plaat. Efficient Remote Method Invocation. Technical Report IR-450, Vrije Universiteit, Amsterdam, sep 1998.
17. M. Welsh. Ninjarmi, 1998. http://www.cs.berkeley.edu/~mdw/proj/ninja/.
18. A. Wollrath, J. Waldo, and R. Riggs. Java-Centric Distributed Computing. *IEEE Micro*, 17(3):44–53, may/jun 1997.

mpiJava: An Object-Oriented Java Interface to MPI

Mark Baker[1], Bryan Carpenter[2], Geoffrey Fox[2],
Sung Hoon Ko[2] and Sang Lim[2]

[1] School of Computer Science
University of Portsmouth,
Southsea, Hants,
UK, PO4 8JF
Mark.Baker@port.ac.uk

[2] NPAC at Syracuse University
Syracuse, New York,
NY 13244, USA
{dbc,gcf,shko,slim}@npac.syr.edu

Abstract. A basic prerequisite for parallel programming is a good communication API. The recent interest in using Java for scientific and engineering application has led to several international efforts to produce a message passing interface to support parallel computation. In this paper we describe and then discuss the syntax, functionality and performance of one such interface, mpiJava, an object-oriented Java interface to MPI. We first discuss the design of the mpiJava API and the issues associated with its development. We then move on to briefly outline the steps necessary to 'port' mpiJava onto a range of operating systems, including Windows NT, Linux and Solaris. In the second part of the paper we present and then discuss some performance measurements made of communications bandwidth and latency to compare mpiJava on these systems. Finally, we summarise our experiences and then briefly mention work that we plan to undertake.

1 Introduction

It is generally recognised that the vast majority of scientific and engineering applications are written in either C, C++ or Fortran. The recent popularity of Java has led to it being seriously considered as a good language to develop scientific and engineering applications, and in particular for parallel computing[1][2][3][4]. Sun's claims, on behalf of Java, that it is simple, efficient and platform-neutral - a natural language for network programming - makes it attractive to scientific programmers who wish to harness the collective computational power of parallel platforms as well as networks of workstations or PCs, with interconnections ranging from LANs to the Internet. The attractiveness of Java for scientific computing is being encouraged by bodies like Java Grande[5]. The Java Grande forum has been set up to co-ordinate the communities efforts to standardise many aspects of Java and so ensure that its future development makes it more appropriate for scientific programmers.

Developers of parallel applications generally use the Single Program Multiple Data (SPMD) model of parallel computing, wherein a group of processes cooperate by executing identical program images on local data values. A programmer using the SPMD model has a choice of explicit or implicit means to move data between the cooperating processes. Today, the normal explicit means is via message passing and the implicit means is via data-parallel languages, such as HPF. Although using implicit means to develop parallel applications is generally thought to be easier, explicit message passing is more often used. The reasons for this are beyond the scope of this paper, but currently developers can produce more efficient and effective parallel applications using message passing.

A basic prerequisite for message passing is a good communication API. Java comes with various ready-made packages for communication, notably an interface to BSD sockets, and the Remote Method Invocation (RMI) mechanism. Both these communication models are optimised for client-server programming, whereas the parallel computing world is mainly concerned with 'symmetric' communication, occurring in groups of interacting peers.

This symmetric model of communication is captured in the successful Message Passing Interface standard (MPI), established a few years ago[6]. MPI directly supports SPMD model of parallel computing. Reliable point-to-point communication is provided through a shared, group-wide communicator, instead of socket pairs. MPI allows numerous blocking, non-blocking, buffered or synchronous communication modes. It also provides a library of true collective operations (broadcast is the most trivial example). An extended standard, MPI 2, allows for dynamic process creation and access to memory in remote processes.

The existing MPI standards specify language bindings for Fortran, C and C++. In this article we discuss a binding of MPI 1.1 for Java, and describe an implementation using Java wrappers to invoke C MPI calls through the Java Native Interface[8]. The software is publically available from:

http://www.npac.syr.edu/projects/pcrc/mpiJava

1.1 Related work

Early work by two of the current authors on Java MPI bindings is reported in[9][10]. In these papers we compared various approaches to parallel programming in Java, including sockets and MPI programming. A comparable approach to creating full Java

MPI interfaces is used by JCI[4], the *Java-to-C interface generator*. In JCI, Java wrappers are automatically generated from the C MPI header. This eases the implementation work, but does not lead to a fully object-oriented API.

MPIJ is a completely Java-based implementation of MPI which runs as part of the Distributed Object Group Metacomputing Architecture[12] (DOGMA). Being closely modelled on the MPI-2 C++ bindings, MPIJ implements a large subset of MPI functionality. MPIJ communication uses native marshalling of primitive Java types. This technique allows MPIJ to achieve communications speeds comparable with native MPI implementations. jmpi[13] is an MPI environment built upon JPVM[14], a Java-based implementation of PVM. jmpi, offers a full Java API to MPI 1.1 as well as features such as thread safety and multiple communication end-points per task. jmpi is a pure Java MPI environment, but is complicated by the need to call JPVM methods. Another recently announced Java MPI interface, called JavaWMPI[15], is built upon the Windows MPI environment WMPI[20]. JavaWMPI has a very similar in structure to mpiJava, but the syntax of the interface is less object-oriented and a procedural method is used to perform polymorphism between Java datatypes.

MPI Software Technology, Inc. has also announced their intention to deliver a commercial Java interface to MPI called JMPI[16]. Java implementations of the related PVM message-passing environment have been reported by Yalamanchilli et. al.[17] and The MPI Forum[6]. Many of the above-mentioned groups, as part of the Java Grande forum activities, have recently published a position paper[18] in an attempt to standardise on a single API.

1.2 Overview of this article

First we outline the mpiJava API and describe various special issues that arise in Java. Implications of object serialization are also explored briefly as are the difficulties due to the lack of true multidimensional arrays in Java.

This discussion is followed by a description of an implementation of the proposed Java binding through a set of wrappers that use the JNI to call existing MPI implementations. The virtues and problems of this implementation strategy are discussed, and results of tests and benchmarks on Solaris, Windows NT and Linux are presented.

2 Introduction to the mpiJava API

The MPI standard is explicitly object-based. The C and Fortran bindings rely on 'opaque objects' that can be manipulated only by acquiring object handles from constructor functions, and passing the handles to suitable functions in the library. The C++ binding specified in the MPI 2 standard collects these objects into suitable class hierarchies and defines most of the library functions as class member functions. The mpiJava API follows this model, lifting the structure of its class hierarchy directly from the C++ binding. The major classes of mpiJava are illustrated in Figure 1.

The class MPI only has static members. It acts as a module containing global services, such as initialization of MPI, and many global constants including the default communicator COMM_WORLD.

The most important class in the package is the communicator class Comm. All communication functions in mpiJava are members of Comm or its subclasses. As usual in MPI, a communicator stands for a 'collective object' logically shared

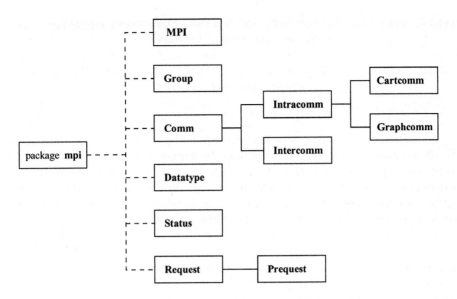

Fig. 1. Principal classes of mpiJava

by a group of processors. The processes communicate, typically by addressing messages to their peers through the common communicator.

Another class that is important for the discussion below is the Datatype class. This describes the type of the elements in the message buffers passed to send, receive, and all other communication functions. Various basic datatypes are predefined in the package. These mainly correspond to the primitive types of Java, shown in Figure 2.

MPI datatype	Java datatype
MPI.BYTE	byte
MPI.CHAR	char
MPI.SHORT	short
MPI.BOOLEAN	boolean
MPI.INT	int
MPI.LONG	long
MPI.FLOAT	float
MPI.DOUBLE	double
MPI.PACKED	

Fig. 2. Basic datatypes of mpiJava

The standard send and receive operations of MPI are members of Comm with interfaces:

```
public void Send(Object buf, int offset, int count, Datatype
                 datatype, int dest, int tag)

public Status Recv(Object buf, int offset, int count, Datatype
                   datatype, int source, int tag)
```

In both cases the actual argument corresponding to buf must be a Java array. In the current implementation they must be arrays with elements of primitive type. By implication they must be one-dimensional arrays, because Java 'multidimensional arrays' are really arrays of arrays. In these and all other mpiJava calls, the buffer array argument is followed by an offset that specifies the element of in array where the message actually starts.

2.1 Special features of the Java binding

The mpiJava API is modelled as closely as practical on the C++ binding defined in the MPI 2.0 standard (currently we only support the MPI 1.1 subset). A number of changes to argument lists are forced by of the restriction that arguments cannot be passed by reference in Java. In general outputs of mpiJava methods come through the result value of the function. In many cases MPI functions return more than one value. This is dealt with in mpiJava in various ways. Sometimes an MPI function initializes some elements in an array and also returns a count of the number of elements modified. In Java we typically return an array result, omitting the count. The count can be obtained subsequently from the length member of the array. Sometimes an MPI function initializes an object conditionally and returns a separate flag to say if the operation succeeded. In Java we return an object handle which is null if the operation fails. Occasionally an extra field is added to an existing MPI class to hold extra results - for example the Status class has an extra field, index, initialized by functions like Waitany. Rarely none of these methods work and we resort to defining auxilliary classes to hold multiple results from a particular function. In another change to C++, we often omit array size arguments, because they can be picked up within the wrapper by reading the length member of the array argument.

As a result of these changes mpiJava argument lists are often more concise than the corresponding C or C++ argument lists.

Normally in mpiJava, MPI destructors are called by the Java finalize method for the class. This is invoked automatically by the Java garbage collector. For most classes, therefore, no binding of the MPI_class_FREE function appears in the Java API. Exceptions are Comm and Request, which do have explicit Free members. In those cases the MPI operation could have observable side-effects (beyond simply freeing resources), so their execution is left under direct control of the programmer.

```
// Simple program

import mpi.*;

class Hello {
  static public void main(String[] args){
    MPI.Init(args);
    int myrank = MPI.COMM_WORLD.Rank();
    if(myrank == 0)}
      char [] message = "Hello, there".toCharArray() ;
      MPI.COMM_WORLD.Send(message,0,message.length, MPI.CHAR, 1, 99);
    } else {
      char [] message = new char [20] ;
      MPI.COMM_WORLD.Recv(message, 0, 20, MPI.CHAR, 0, 99) ;
      System.out.println("received:" + new String(message) + ":");

    }
    MPI.Finalize();
  }
}
```

Fig. 3. Minimal mpiJava program (run in two processes)

2.2 Derived datatype vs Object serialization

In MPI new *derived types* of class Datatype can be created using suitable library functions. The derived types allow one to treat contiguous, strided, or indirectly indexed segments of program arrays as individual message elements. The corresponding array subsections can then be communicated in a single function call, potentially exploiting any special hardware or software the platform provides for exchanging scattered data between user space and the communication system.

Currently mpiJava provides all the derived datatype constructors of standard MPI, with one limitation: it places significant restrictions on its binding of MPI_TYPE_STRUCT. In C or Fortran this function can be used to describe an entity combining fields of different primitive (or derived) type. Because of the assumption that buffers are one-dimensional arrays with elements of primitive type, mpiJava imposes a restriction that all the types combined by its Datatype.Struct member must have the same *base type*, which must agree with the element type of the buffer array. Also mpiJava does not provide an analogue of MPI_BOTTOM buffer address, or the MPI_ADDRESS function for finding offsets relative to this absolute member base. In C or Fortran these functions allow buffers to include fields from separately declared variables or arrays, but the mechanism does not sit well with the pointer-free Java language model.

Approaches based on the MPI derived datatype model do not seem to be the best way to alleviate this restriction. A better option is probably to exploit the run-time type information already provided in Java objects. We are developing a version of mpiJava that adds one new predefined datatype:

MPI.Object

A message buffer can then be an array of any serializable Java objects. The objects are serialized automatically in the wrapper of send operations, and unserialized at their destination.

The absence of true multi-dimensional arrays in Java limits another use of derived data types. In MPI the MPI_TYPE_VECTOR function creates a derived datatype representing a strided section of an array. In C or Fortran this strided section can be identified with a section of a multi-dimensional array. It could describe, say, an edge of the local patch of a two-dimensional distributed array. In Java there is no equivalence between a multi-dimensional array and a contiguous patch of memory, or a one-dimensional array. The programmer may choose to linearize all multi-dimensional arrays in the algorithm, representing them as one-dimensional arrays with suitable index expressions. In this case derived datatypes can be used to send and receive sections of the array. Alternatively the programmer may use Java arrays of arrays to represent multi-dimensional arrays. This simplifies the index arithmetic in the program. Sections of the array are then explicitly copied to one-dimensional buffers for communication. The latter option seems to be more popular with programmers.

Although, for reasons of conformance of with MPI standards, we expect to continue supporting derived datatypes in mpiJava, their value in the Java domain is less clear-cut than in C or Fortran. Allowing serializable objects as buffer elements is probably a more powerful facility.

3 mpiJava implementations on PC Platforms

3.1 Introduction

As Java is a platform-neutral language there is much interest in 'porting' mpiJava to PC-based systems, in particular Windows NT and Linux. To 'port' mpiJava onto a PC environment it is necessary to have a native MPI library, a version of the Java Development Toolkit (JDK) and a C compiler. mpiJava consists of two main parts: the MPI Java classes and the C stubs that binds the MPI Java classes to the underlying native MPI implementation. We create these C stubs using JNI - the means by which Java can call and pass parameters to and from a native API. Figure 4 provides a simple schematic view of the software layers involved.

To 'port' mpiJava onto a new platform, generally two steps need to be undertaken:

- Create a native library out of the compiled JNI C stubs.
- Compile MPI Java in class libraries - ensuring that the correctly named stub library is loaded by the Java System.loadLibrary("stublib") call in the main source file MPI.java.

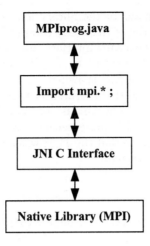

Fig. 4. Software Layers

The development and testing of mpiJava was undertaken on various Sun and SGI UNIX platforms using MPICH. As interfacing Java to MPI is not always trivial, in earlier implementation we often saw low-level conflicts between the Java runtime and the interrupt mechanisms used in the MPI implementations. The situation is improving as the JDK matures, in particular version 1.2β allows the use of *green* or native threads, which have eliminated the interrupt problem that we encountered with earlier releases of the JDK. mpiJava is now stable on NT platforms using WMPI and JDK 1.1 or later as well as UNIX platforms using MPICH and JDK 1.2β.

3.2 Windows NT

To test mpiJava under Windows NT we had the choice of a number of MPI implementations to pick from[19]. We chose WMPI from the Instituto Supererior de Engenharia de Coimbra, Portugal. WMPI is a full implementation of MPI for Microsoft Win32 platforms. WMPI is based on MPICH and includes a p4[21] device standard. P4 provides the communication internals and a startup mechanism. The WMPI package is a set of libraries (for Borland C++, Microsoft Visual C++ and Microsoft Visual FORTRAN). The release of WMPI provides libraries, header files, examples and daemons for remote start-up. WMPI can co-exist and interact with MPICH/ch_p4 in a cluster of mixed UNIX and Win32 platforms. WMPI is still under development and is freely available.

mpiJava under WMPI: To create a release of mpiJava for WMPI the following steps were undertaken:

- *Step 1* - Compile the mpiJava JNI C interface into a Win32 Dynamic Link Library (mpiJava.dll).

- *Step 2* - Modify the name of the library loaded by the mpiJava interface (MPI.java) to that of the of the newly compiled library.
- *Step 3* - Compile the Java MPI interface into class libaries.
- *Step 4* - Create a JNI interface to WMPI. This was necessary as under WMPI a master process is first spawned. Its purpose is to first read in a job configuration file and use the information within it to set up and run the actual MPI processes. An idiosyncrasy of WMPI is that all MPI processes must have a file name with the extension *.EXE*. This led to the need to produce a JNI to WMPI so that the JVM was loaded and the 'main' method of mpiJava Java class started.

3.3 Linux

At the time of writing this paper, our attempts to 'port' mpiJava to Linux are in progress. We are currently experiencing problems similar to those encountered during our early attempts to create the interface on Solaris, mentioned in section 3.1 . Sun's releases the JDK for Solaris and NT platforms first. On other platforms, such as Linux, it is necessary for developers to 'port' the JDK. The most recent release of the JDK for Linux is 1.1.7 and this is version is known to be the cause of our problems. It is anticipated that JDK 1.2 will be available for Linux shortly and that we will be able to report our experiences with Linux at the IPPS/SPDP 99 workshop in April 1999.

3.4 Functionality Tests

An integral part of the development of this project was to produce or translate a number of basic MPI test codes to mpiJava. An obvious starting point was the C test suite originally developed by IBM . This suite had been modified to comply fully with the MPI standard and to be compatible with the MPICH. The suite consists of fifty-seven C programs that test the following MPI calls and data types; collective operations, communicators, data types, environmental inquiries, groups, point to point and virtual topologies. These codes were all translated to mpiJava.

Under WMPI and Solaris-MPICH these codes were run either as multiple processes on a single machine (Shared Memory mode - SM) or as multiple processes running on separate machines (Distributed Memory mode - DM). Under WMPI and Solaris-MPICH all the codes ran in both modes without alterations. Our experiences using Linux-MPICH will be reported when JDK 1.2 is available for Linux.

4 Simple Communications Performance Measurement

4.1 Introduction

At this early stage of our project we have decided to restrict performance measurements to those that will give some indication of the basic inter-processor communications performance. The actual computational performance of each process is felt to be dependent on the local JVM and associated technologies used by specific vendors to increase the performance of Java.

4.2 PingPong Communications Performance Tests

In this program increasing sized messages are sent back and forth between processes - this is commonly called PingPong. This benchmark is based on standard blocking MPI_Send/MPI_Recv. PingPong provides information about latency of MPI_Send/MPI_Recv and uni-directional bandwidth. To ensure that anomalies in message timings are minimised the PingPong is repeated many times for each message size. The codes used for these tests were those developed by Baker and Grassl[23]. The three existing codes (MPI-C, MPI-Fortran and Winsock-C) were used for comparison and we implemented an mpiJava version for our purposes.

The main problem encountered running the PingPong code was that under WMPI on Win32 MPI_Wtime() had been implemented with a millisecond resolution. It was necessary to adapt each of the codes to use an alternative timer with microsecond (μs) resolution. The performance tests shown in the next section were run on two similar systems:

- Two dual processor (P6 200 MHz) NT 4 workstations each with 128 MBytes of DRAM.
- Two dual processor (UltraSparc 200 MHz) Solaris workstations with 256 MBytes of DRAM.

Both systems were connected via 10BaseT Ethernet and the tests were carried out when there was little network activity and on quiet machines.

4.3 Message Startup Latencies

	Wsock	WMPI-C	WMPI-J	MPICH-C	MPICH-J	Linux-C	Linux-J
SM	144.8 μs	67.2 μs	161.4 μs	148.7 μs	374.6 μs	- μs	- μs
DM	244.9 μs	623.9 μs	689.7 μs	679.1 μs	961.2 μs	- μs	- μs

Table 1. Time for 1 Byte Messages

In Table 1 we show the transmission time in microseconds (s) to send a 1 byte message in each of the environments tested. In SM the mpiJava wrapper adds an extra 94μs (140%) and 226μs (152%) compared to WMPI and MPICH C respectively. In DM the mpiJava wrapper adds an extra 66μs (11%) and 282μs (42%) compared to WMPI and MPICH C respectively. The Wsock figures are those for a WinSock implementation of PingPong benchmark using TCP. Linux results will be presented during the conference workshop.

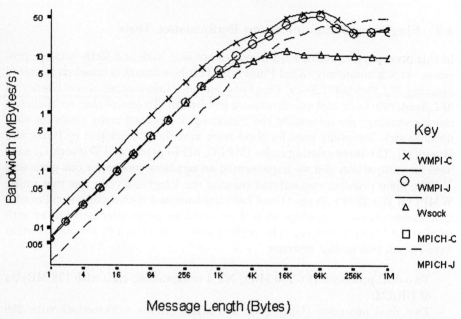

Fig. 5. PingPong Results in Shared Memory (SM) mode

4.4 Results in Shared Memory Mode (Figure 5)

The `mpiJava` curve mirrors that of C with an almost constant offset up to 8K, thereafter the curves converge meeting at 256K. Under MPICH, the curves for C and `mpiJava` mirror each other in a similar fashion to those under WMPI, again there is a constant offset and convergence at around 256K.

Under WMPI the peak bandwidth of C is around 65 MBytes/s and `mpiJava` is 54 MBytes/s. The peaks occur at around 64K. Under MPICH the bandwidth is flattening out, but still increasing for C and `mpiJava`, at the 1M. The actual rate measured at this point is about 50 MBytes/s.

Clearly the WMPI C code perform best of those tested. The performance of `mpiJava` in SM under WMPI is good - it exhibits a fairly constant overhead of 95μs up to 2K, thereafter it converges with the C curve. The performance the C code under MPICH is slightly surprising as the NT and Solaris platforms used for these tests had similar specifications. It is assumed that the performance reflects the usage of MPICH rather than a native version of MPI for Solaris. Even so, the MPICH results for `mpiJava` show that it exhibits reasonable performance.

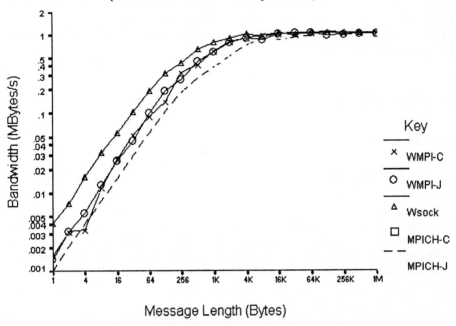

Fig. 6. PingPong Results in Distributed Memory (DM) mode

4.5 Results in Distributed Memory Mode (Figure 6)

In DM the differences between the MPI codes is not as pronounced as seen in SM. Under WMPI the C and mpiJava codes display very similar performance characteristics throughout the range tested. Under MPICH, there is distinct performance difference between C and mpiJava, However the difference is much smaller than in SM and the curves converge at the 4K. All curves peak at about 1 MByte/s, which is about 90% of the maximum attainable on 10 Mbps Ethernet link.

4.6 Overall Results Discussion

In both SM and DM modes mpiJava adds a fairly constant overhead compared to normal native MPI. In an environment like WMPI, which has been optimised for NT, the actual overheads of using mpiJava are relatively small at around 100ms. Under MPICH the situation is not quite so good, here the use of mpiJava introduces an extra overheads of between 250 - 300μs.

It should be noted that these results compare codes running directly under the operating system with those running in the JVM. For example, according to a single 200 MHz PentiumPro will achieve in excess of 62 Mflop/s on a Fortran version of LinPack. A test of the Java LinPack code gave a peak performance of 22 Mflop/s for the same processor running the JVM. The difference in performance will account for much of the additional overhead that mpiJava imposes on C MPI codes. From this it can be deduced that the quality and performance of JVM on each platform will have the greatest effect on the usefulness of mpiJava for scientific computation.

5 Conclusions

5.1 Overall Summary

We have discussed the design and development of mpiJava - a pure Java interface to MPI. We have also highlighted the benefits of a fully object-oriented Java API compared to those currently available. Our performance tests have shown that mpiJava should fulfil the needs of MPI programmers not only in terms of functionality but also in terms of good performance when compared to similar C MPI programs. Unfortunately, at the time of submission of this paper we have been unable to test mpiJava under Linux, but we believe that the problems we have encountered we be overcome soon and we will be able to present our finding during the Workshop in April 1999. Overall, however, we feel that we have implemented a well designed, functional and efficient Java interface to MPI.

5.2 Particular Conclusions

- mpiJava provides a fully functional and efficient Java interface to MPI.
- Our performance tests have shown that, in terms of communications speeds, WMPI on NT out performs MPICH on Solaris.
- When used for distributed computing the current implementation of mpiJava does not impose a huge overhead on-top of the native MPI interface.
- We have discovered some of the limitation in the usage of JNI. In particular with MPICH where we had problems with UNIX signals. We are hopeful that these problems will disappear when we start using JDK 1.2 and native threads.
- Our performance tests indicate that much of the additional latency that mpiJava imposes is due to the relatively poor performance of the JVM rather than the impact of messages traversing additional software layers.
- The syntax of mpiJava is easy to understand and use, thus making it relatively simple for programmers with either a Java or Scientific background to take up.
- We believe that mpiJava will also provide a popular means for teaching students the fundamentals of parallel programming with MPI.

5.3 Future Work

We plan to continue to improve mpiJava with further Java features, such as object serialization, and also add-in the functionality that has been proposed in MPI-2. We intend to 'port' mpiJava to a multitude of new MPI environments, including LAM, Sun MPI and Globus. We are also planning a pure-Java MPI environment which does not rely on native MPI services.

References

1. *Parallel Compiler Runtime Consortium, HPCC and Java* - a report by the Parallel Compiler Runtime Consortium, http://www.npac.syr.edu/users/gcf/hpjava3.html, May 1996.
2. G.C. Fox, editor, *Java for Computational Science and Engineering - Simulation and Modelling II*, volume 9(11) of Concurrency: Practice and Experience, November 1997.
3. G.C. Fox, editor, *ACM 1998 Workshop on Java for High-Performance Network Computing*, Palo Alto, February 1998, Concurrency: Practice and Experience, 1998.
4. V. Getov, S. Flynn-Hummel, and S. Mintchev. *High-Performance parallel programming in Java: Exploiting native libraries*, In ACM 1998 Workshop on Java for High-Performance Network Computing. Palo Alto, February 1998, Concurrency: Practice and Experience, 1998. To appear.
5. Java Grande Forum - http://www.javagrande.org/
6. *Message Passing Interface Forum. MPI: A Message-Passing Interface Standard*, University of Tennessee, Knoxville, TN, June 1995. http://www.mcs.anl.gov/mpi
7. G.C. Fox, editor, *Java for Computational Science and Engineering - Simulation and Modelling*, volume 9(6) of Concurrency: Practice and Experience, June 1997.
8. R. Gordon, *Essential JNI: Java Native Interface*, Prentice Hall, 1998.
9. D.B. Carpenter, Y. Chang, G.C. Fox, D. Leskiw, and X. Li, *Experiments with HPJava* Concurrency: Practice and Experience, 9(6):633, 1997.
10. D.B. Carpenter, Y. Chang, G.C. Fox, and X. Li, *Java as a language for scientific parallel programming*, In 10th International Workshop on Languages and Compilers for Parallel Computing, volume 1366 of LNCS, pages 340–354, 1997.
11. S. Mintchev and V. Getov, *Towards portable message passing in Java: Binding MPI*, Recent Advances in MPI and PVM, Editors, M. Bubak, J. Dongarra and J. Wasniewski, volume 1332 of LNCS pages 135 - 142, Springer Verlag, 1997.
12. DOGMA - http://zodiac.cs.byu.edu/DOGMA/
13. K. Dincer and K. Ozbas, *jmpi and a Performance Instrumentation Analysis and Visualization Tool for jmpi*, 1st UK Workshop on Java for High Performance Network Computing, Southampton, UK, September 1998.
14. A.J. Ferrari, *JPVM: Network parallel computing in Java*, In ACM 1998 Workshop on Java for High-Performance Network Computing. Palo Alto, February 1998, Concurrency: Practice and Experience, 1998. To appear.
15. P. Martin, L.M. Silva and J.G. Silva, *A Java Interface to MPI*, Proceeding of the 5th European PVM/MPI Users' Group Meeting, Liverpool UK, September 1998 WMPI - http://dsg.dei.uc.pt/w32mpi/
16. G. Crawford III, Y. Dandass, and A. Skjellum, *The JMPI commercial message passing environment and specification: Requirements, design, motivations, strategies, and target users*, http://www.mpi-softtech.com/publications

17. N. Yalamanchilli and W. Cohen, *Communication performance of Java based parallel virtual machines*, In ACM 1998 Workshop on Java for High-Performance Network Computing. Palo Alto, February 1998, Concurrency: Practice and Experience, 1998. To appear.
18. B. Carpenter, V. Getov, G. Judd, T. Skjellum and G. Fox, *MPI for Java - Position Document and Draft API Specification*, November 1998 - http://www.npac.syr.edu/projects/pcrc/mpiJava
19. M.A. Baker and G.C. Fox, *MPI on NT: A Preliminary Evaluation of the Available Environments*, 12th IPPS & 9th SPDP, LNCS, Jose Rolim (Ed.), Parallel and Distributed Computing, Springer Verlag, Heidelberg, Germany. ISBN 3-540 64359-1, April 1998.
20. WMPI - http://dsg.dei.uc.pt/w32mpi/
21. R. Butler and E. Lusk, *Monitors, messages, and clusters: The p4 parallel programming system*, Parallel Computing, 20:547-564, April 1994.
22. IBM Test Suite - ftp://info.mcs.anl.gov/pub/mpi/mpi-test/ibmtsuite.tar
23. PingPong Benchmarks - http://www.dcs.port.ac.uk/~mab/TOPIC/
24. LinPack
    http://performance.netlib.org/performance/html/linpack.data.col0.html
25. Java LinPack - http://www.netlib.org/benchmark/linpackjava/

An Adaptive, Fault-Tolerant Implementation of BSP for Java-Based Volunteer Computing Systems

Luis F. G. Sarmenta

Massachusetts Institute of Technology
Laboratory for Computer Science
545 Technology Square
Cambridge, MA 02139
lfgs@cag.lcs.mit.edu
http://www.cag.lcs.mit.edu/bayanihan/

Abstract. In recent years, there has been a surge of interest in Java-based *volunteer computing* systems, which aim to make it possible to build very large parallel computing networks very quickly by enabling users to join a parallel computation by simply visiting a web page and running a Java applet on a standard browser. A key research issue in implementing such systems is that of choosing an appropriate programming model. While traditional models such as MPI-like message-passing can and have been ported to Java-based systems, they are not generally well-suited to the heterogeneous and dynamic structure of volunteer computing systems, where nodes can join and leave a computation at any time. In this paper, we present an implementation of the Bulk Synchronous Parallel (BSP) model, which provides programmers with familiar message-passing and remote memory primitives while remaining flexible enough to be used in dynamic environments. We show how we have implemented this model using the Bayanihan software framework to enable programmers to port the growing base of BSP-based parallel applications to Java while achieving adaptive parallelism and protection against both the random faults and intentional sabotage that are possible in volunteer computing systems.

1 Introduction

Volunteer computing is a form of distributed computing that seeks to make it as easy as possible for as many people as possible to donate computer time towards solving a parallel problem. By maximizing potential worker pool size and minimizing setup time, volunteer computing makes it possible to build very large networks of workstations very quickly, and creates many exciting new possibilities, including not only global *true volunteer* systems, such as `distributed.net` [4] (which was used to crack the RC5-56 challenge), but also local *forced volunteer* systems for use within institutions that have so far lacked the necessary expertise and time to pool their existing workstations to provide inexpensive supercomputing facilities for their computational needs [17].

In the past two years, there has been a particular surge of interest in *Java-based* volunteer computing systems, which would allow Internet users to join a parallel computation by simply visiting a web page and letting their standard web browser run a Java applet. Compared to existing network-of-workstations (NOW) software such as PVM [10] or MPI [12], such volunteer systems would be orders-of-magnitude easier to setup. Instead of having to spend weeks setting up accounts and installing software on potentially thousands of machines, one can setup an institution-wide NOW literally overnight by simply asking employees to point their browsers to a particular intranet site and to leave their browsers running before they go home or while they work. Moreover, Java's platform-independence allows programmers to worry less about generating code for many different architectures, and concentrate more on the applications themselves.

While *using* such Java-based systems should be very easy, however, *developing* the infrastructure for them is not as straightforward. A major research problem today is that of choosing an appropriate programming model. By enabling anyone to volunteer any kind of computer at any time, Java-based – and particularly *browser*-based – volunteer systems create a heterogeneous and dynamic environment for which currently popular parallel programming models such as shared memory and MPI-style message-passing are not well-suited. In basic MPI, for example, programs are written assuming that the number of physical processors is known and fixed; there is little or no support for situations where nodes leave or join the system at unexpected times. In general, a good programming model for volunteer computing systems has to be *adaptively parallel* and *fault-tolerant*. It cannot depend on static information on the number of nodes in the system and their individual processing speeds. At the same time, it must be able to tolerate data loss or corruption due not only to random hardware and software faults, but also to malicious *saboteurs* pretending to be volunteers but submitting erroneous results.

In this paper, we present a programming model based on the *Bulk Synchronous Parallel* (BSP) model [23, 20], and show how it can be used in Java-based volunteer computing systems. We begin by presenting the BSP programming model, and then show how we have implemented it using the Bayanihan volunteer computing framework [17–19], taking advantage of Bayanihan's existing adaptive parallelism and fault-tolerance mechanisms. We also describe the new programming interface and show how it can be used to write realistic parallel applications. We conclude by presenting some current results, and discussing related, ongoing, and future work.

2 The BSP Programming Model

In BSP, a parallel program runs across a set of virtual processors (called *processes* to distinguish them from physical processors), and executes as a sequence of parallel *supersteps* separated by barrier synchronizations. Each superstep is composed of three ordered phases, as shown in Fig. 1: 1) a *local computation* phase, where each process can perform computation using local data values and issue communication requests such as remote memory reads and writes; 2) a *global communication* phase, where data is exchanged between processes according to the requests made during the local computation phase; and 3) a *barrier synchronization*, which waits for all data transfers to complete and makes the transferred data available to the processes for use in the next superstep.

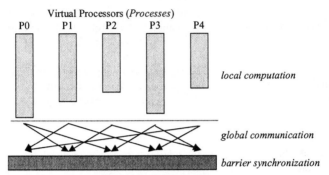

Fig. 1. A BSP superstep.

Figure 2 shows pseudo-code for a simple program where each process passes data to its right neighbor and performs a little local computation. As shown, BSP is very similar to conventional SPMD/MPMD (*single/multiple program, multiple data*)-based programming models such as MPI, and is at least as flexible, having both remote memory (e.g., bsp_put()) and message-passing (e.g., bsp_send() and bsp_recv()) capabilities. The *timing* of communication operations, however, is different – since the global communication phase does not happen until all processes finish the local computation phase, the effects of BSP communication operations are not felt until the *next* superstep. In line 6 of Fig. 2, for example, s gets the *unchanged* local value of d. Similarly, in line 9, the call to bsp_recv() returns null or generates an exception.

This postponing of communications to the end of a superstep is the key idea in BSP. It removes the need to support *non*-barrier synchronizations between processes, and guarantee that processes within a superstep are *mutually independent*. This makes BSP easier to implement on different architectures than traditional models, and makes BSP programs easier to write and to analyze mathematically. For example, since the timing of BSP communications make circular data dependencies between BSP processes impossible, there is no risk of deadlock or livelock in a BSP program. Also, the separation of the computation, communication, and synchronization phases allows one to compute time bounds and predict performance using relatively simple mathematical equations [20].

Because of these advantages, a growing number of people are now applying BSP to a wide variety of realistic parallel applications [5], ranging from small general-purpose numeric libraries (e.g., for matrix operations), to large computational research applications used in both academia and industry (e.g.,

```
1   void shift2()
2   { right = ( bsp_pid()+1 ) % bsp_nprocs();
3     s = 0; d = bsp_pid();
4     bsp_register( "d",&d );  // register variable "d"
                               // begin superstep 0
5     bsp_put( right,"d",d );  //   put my pid to right's d *at next sync*
6     s = d;                   //   s <- current d (i.e., my pid )
7     bsp_sync();              // synchronize
                               // begin superstep 1; d is now left's pid
8     bsp_send( right, d );    //   send d to right * at next sync *
9     s += bsp_recv();         //   error! nothing to receive yet!
10    bsp_sync();              // synchronize
                               // begin superstep 2; msg is now in recv q.
11    d = bsp_recv();          //   receive d sent in line 7
12    s += d;                  //   s <- s + 2nd-left's pid
13  }
```

Fig. 2. BSP pseudo-code for a two-step cyclic shift.

computational fluid dynamics, plasma simulation, etc.). By implementing BSP in Java, we hope to enable programmers to port these applications to Java, and make it easier to do realistic research using volunteer computing systems.

3 Implementing the BSP Model

3.1 The Bayanihan Master-Worker Based Implementation

BSP's structure makes BSP relatively easy to implement in Java-based volunteer computing systems. Since processes are independent within supersteps, we can package the work to be done in a superstep as separate work packets and farm them out to volunteer worker nodes using a master-worker system. Figure 3 shows how we have implemented BSP using the existing master-worker model of the Bayanihan framework [19]. As shown, BSP processes are implemented as **BSPWork** objects and placed in a *work pool* on the server. Volunteer *worker clients* each have a *work engine* which runs in a loop, repeatedly making remote method calls via HORB [14] (a distributed object library similar to Sun's RMI [22]) to its corresponding *advocate* on the server side to request more work. Each advocate forwards these calls to the *work manager*, which looks into the work pool, and returns the next available uncompleted work object.

When the work engine receives the new work object, it calls the work object's **bsp_run()** method, which starts the local computation phase of the superstep indicated by the **restorestep** field. As **bsp_run()** executes, communication requests are logged in **MsgQueue** fields, and the values of shared variables are saved in the **bspVars** field, a hash table that stores objects under **String**-type names. Execution continues until **bsp_sync()** is called, at which point **restorestep** is incremented and control returns to the work engine. The work engine then sends

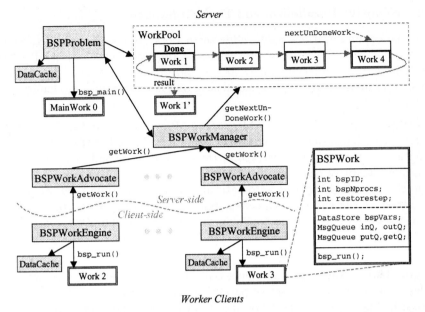

Fig. 3. Implementing BSP on top of the Bayanihan master-worker framework.

the whole work object – including **bspVars**, the **MsgQueue**'s, and **restorestep** – back to the server, which in turn marks the original work object "done", and stores the returned work object as its *result*.

When all work objects are done, the work manager notifies the *problem* object, and the global communication and barrier synchronization phases begin. The **BSPProblem** object goes through the result work objects, and performs all the global communication operations by moving data from one work object to another. The remote memory access operations (**bsp_get()** and **bsp_put()**) are done by reading and writing to the work objects' **bspVars** fields. Similarly, the message-passing-style **bsp_send()** operation is performed by enqueuing the message data into the destination node's **inQ** field, so that in the local computation phase of the next superstep, they can be retrieved by calling **bsp_recv()**. After all the communication operations have been performed, each work object in the work pool is replaced by its updated version, and control returns to the work manager. The next superstep begins as the work manager starts giving workers new work objects wherein **restorestep** indicates the new superstep, **bspVars** contains updated variable values, and **inQ** contains new incoming messages.

In addition to the work objects in the work pool, the server also has a special *main work* object, which is like other work objects, except that it is not farmed out to workers, but is run on the server itself. In place of **bsp_run()**, the server calls the **bsp_main()** method. As described in Sect. 4.1, the main work object can be used for such tasks as *spawning* new work objects, coordinating other work objects, collecting and displaying results, and interacting with users.

3.2 Adaptive Parallelism and Fault-Tolerance

Since no communications occur while these work packets are running, we can rerun work packets or run several copies of the same work object at a time, resolving any redundancies safely at the end of the superstep. This makes it possible to implement adaptive parallelism and fault-tolerance mechanisms.

Checkpointing and Process Migration. Previous research in C-based BSP systems has shown that the call to `bsp_sync()` at the end of the superstep provides a convenient place to *checkpoint* the program state and migrate processes to achieve adaptive parallelism and crash-tolerance [15, 13]. In our implementation, checkpointing happens automatically as each work object returns itself to the server at the end of a superstep. Since the returned work object includes all necessary process state in `bspVars`, the `MsgQueue`'s, and other non-transient fields, it can easily be saved and restored later or migrated to another machine.

Eager Scheduling. The underlying Bayanihan master-worker framework provides a simple form of adaptive parallelism called *eager scheduling* [3, 19]. As shown in Fig. 3 (Sect. 3.1), work objects in the work pool are stored in a circular list, with a pointer keeping track of the next available uncompleted work. As workers call `getWork()`, the pointer moves forward, assigning different work objects to different workers. Faster workers will tend to call `getWork()` more often, and naturally get a bigger share of the total work. This gives us a simple but effective form of dynamic load balancing. Moreover, since the list is circular, previously-assigned but uncompleted work can be reassigned to other workers. This "eager" behavior guarantees that slow workers do not cause bottlenecks – fast workers with nothing left to do will simply bypass slow ones, redoing work themselves if necessary. (Redundant results are simply ignored.) It also provides a basic form of crash-tolerance. If a worker crashes or quits, and leaves its work undone, for example, it is alright because the work will eventually be reassigned to another worker. In this way, computation can go on as long as at least one worker is still alive. In fact, even if all the workers crash or quit, the computation can continue as soon as a new worker becomes available.

Fault and Sabotage Tolerance. To be useful on a large-scale, volunteer computing systems must be able to tolerate not only random data faults, but also intentional *sabotage* from malicious volunteers producing erroneous data. While achieving such fault-tolerance is difficult in most parallel programming models, where data can move arbitrarily between processes, it is relatively easy in master-worker systems, where result data are neatly packaged into independent packets that can be checked, compared, and recomputed, if necessary.

At present, we have implemented and are studying two approaches to fault- and sabotage-tolerance for master-worker systems: *majority voting* and *spot-checking*. In *majority voting*, the work manager waits for an agreeing majority of at least m out of r results from different workers to be received before a work object is considered done. If r results are received without reaching a majority agreement, all the r results are invalidated, and the work object is redone. In *spot-checking*, the work manager precomputes a randomly selected work object before the start of a new superstep, and returns this *spotter work* with probability p at each call to getWork(). Then, if a worker is sent a spotter work and does not return the expected result, it is *blacklisted* as untrustable, and the work manager *backtracks* through all the current results, invalidating any results dependent on results from the offending worker. Spot-checking may be less reliable in cases where saboteurs do not *always* submit bad data, but is much more time-efficient, since it only takes an extra p fraction of the time instead of taking m times longer as voting does.

Early experiments have demonstrated the potential effectiveness of these techniques [19]. Spot-checking, tested with backtracking but without blacklisting, was particularly promising, maintaining error rates of only a few percent, even in cases where *over half* of the volunteers were saboteurs. Furthermore, the high rate at which saboteurs were being caught suggests that these error rates will be even smaller with blacklisting enabled. While still not zero, these low error rates may be enough for some applications where a few bad answers may be acceptable or can be screened out. These include *image rendering*, where a few scattered bad pixels would be acceptable, some *statistical computations*, where outliers can be detected and either ignored or double-checked, *genetic algorithms*, where bad results are naturally screened out by the system, and others. For other applications that assume 100% reliability, however, we are exploring ways of *combining* voting, spot-checking, backtracking, and blacklisting, as well as developing new mechanisms, to reduce the error rate as much as possible.

Any progress made in developing mechanisms for master-worker systems in general can be applied automatically to BSP programs by virtue of our master-worker based implementation of BSP. Programmers need only define an appropriate hasSameValue() method for each data class used in their work objects so that two work objects can be compared just like ordinary result objects by comparing their bspVars and MsgQueue fields.

4 Writing BSP programs in Java

To write a Bayanihan BSP program, one simply extends the BSPWork base class and defines an appropriate bsp_run() method containing the BSP program code. No changes to the Java language or Java virtual machine (JVM) are required. A Bayanihan program can run on any JDK 1.0-compliant browser or JVM.

4.1 The Bayanihan BSP Methods

The BSPWork class provides several methods which subclasses inherit and can call within bsp_run() in the same way one would call library functions in C. These are based on BSPLib [5], a programming library for C and Fortran currently being used in most BSP applications, but have been modified to account for language and style differences between Java and C. These methods are shown in Table 1, subdivided by function into the following categories (in order): *inquiry*, *message passing*, *direct remote memory access*, and *synchronization*.

Table 1. Basic Bayanihan BSP methods

method	meaning
int bsp_pid()	my BSP process identifier
int bsp_nprocs()	number of virtual processes
double bsp_time()	local time in seconds
void bsp_send(BSPMessage msg)	send prepared message
void bsp_send(int dest, int tag, Object data)	create tagged message, and send to *dest*
BSPMessage bsp_recv()	receive message
int bsp_qsize()	number of incoming messages
BSPMessage bsp_peek()	preview next message
int bsp_peekTag()	preview next message tag
void bsp_save(String name, Object data)	save data under *name*
Object bsp_restore(String name)	restore data under *name*
Object bsp_unregister(String name)	remove data under *name*
void bsp_put(int dest, String destName, Object data)	write data to *destName* variable of *dest*
Object bsp_get(int src, String srcName, String destName)	read *srcName* variable of *src* to local *destName*
boolean bsp_sync()	sync; possibly checkpoint and migrate
boolean bsp_step()	mark beginning of superstep
boolean bsp_cont()	mark continuation of superstep

Message-Passing. The message-passing methods are similar in functionality to those found in systems such as MPI [12] or PVM [10], except that, as noted earlier, messages sent during a superstep can only be received by the recipient in the *next* superstep. In the present implementation, messages are queued at the recipient in arbitrary order, but can be identified by tag using the bsp_peekTag() method, or by calling the getBSPTag() method of the BSPMessage object returned by bsp_recv(). The data itself can be retrieved as an object by calling the getData() method of the BSPMessage. Any data object that implements the java.io.Serializable interface or has a HORB proxy (generated by the HORB compiler) can be passed in a message. Note that Java's object-oriented features and platform-independence make sending and receiving messages much easier than in C-based systems such as MPI or PVM, where one must explicitly decompose complex messages into its primitive components.

Direct Remote Memory Access. Bayanihan BSP also provides the direct remote memory access (DRMA) methods bsp_put() and bsp_get(), which allow processes to write to and read from remote variables by specifying the owner's process ID and the String-type name of the desired variable.

The bsp_save() method registers a variable into the bspVars table. This allows it to be saved at a checkpoint, and to be read remotely using bsp_get(). A variable remains registered until it is removed with bsp_unregister(), and its contents can be modified freely by its owner using accessor methods. If the object itself is replaced, however, then bsp_save() must be called again to update the reference stored in bspVars. A variable can also be registered by processes other than its owner by using the bsp_put() method. This technique can be used by main work objects to initialize worker variables.

The bsp_restore() method restores the value of a variable so it can be used within a superstep. Variables that are used in all or most supersteps can be restored in the initVars() method, which is always called before bsp_run(). This guarantees that the variables are always restored and available at the beginning of a superstep. If an attempt is made to restore an unregistered name, bsp_restore() returns null. Typically, variables restored with bsp_restore() are declared as transient fields since their values are already stored in bspVars and do not need to be serialized independently.

Like the message-passing methods, the DRMA methods do not take effect until the next superstep. Thus, for example, a bsp_get() called during a superstep will only update the destination variable in the next superstep. The value received by bsp_get() is the value of the source variable at the end of the local computation phase, before any puts or gets are done. Similarly, the value sent by bsp_put() is the value of the data object at the end of the local computation phase. This may be different from the value of the data object at the time bsp_put() was called, unless the data is first *cloned* before bsp_put() is called.

Synchronization and Checkpointing. Calls to bsp_sync() not only indicate barrier synchronizations, but also mark potential checkpoint locations where a virtual process may be paused, saved, and moved to a different physical processor. While pausing and moving processes is easily done as described in Sect. 3.1, resuming execution is harder. In C, resuming execution can be done relatively easily by either using system-specific low-level mechanisms (as done in [13]), or using a pre-compiler to annotate the source code with a jump table using switch and goto statements [21]. Unfortunately, these techniques are impossible in Java, where there are no explicit goto's – not even at the virtual machine level.

Thus, instead of using direct jumps, we use if statements to *skip* over already-executed code. Figures 4 and 5 show an example. In general, the code for a superstep should be enclosed in an if (bsp_step()) block and ended with a call to bsp_sync(), as shown in step 0. As a syntactic shortcut, an if (bsp_sync()) statement can be used to end a superstep and start the next one at the same time, as shown at the end of step 1. In cases where function calls, for, while, if-else, or other special constructs prevent a superstep's code from being enclosed in one block, one can use an if (bsp_cont()) statement to continue the superstep on the other side of the offending statement as shown.

```
void bsp_run()                          step 2+n, part b;
{ step 0;                               foo(); // func with sync
  bsp_sync();                           step 3+n, part b;
  step 1;                               bsp_sync();
  bsp_sync();                         } // end bsp_run()
  step 2, part a;
  for ( int i=0; i<n; i++ )             void foo()
  { step 2+i, part b;                   { // cont. current step
    bsp_sync();                           step 2+n, part c;
    step 2+i+1, part a;                   bsp_sync();
  }                                       step 3+n, part a;
  // code continued in next col. ...    } // end foo();
```

Fig. 4. BSP pseudo-code without code-skipping.

```
void bsp_run()                          if ( bsp_cont() )
    throws MigrateMeException           { step 2+n, part b
{ if ( bsp_step() )                     }
  { step 0                              foo(); // func with sync
  }                                     if ( bsp_cont() )
  bsp_sync();                           { step 3+n, part b
  if ( bsp_step() ) // explicit         }
  { step 1                              bsp_sync();
  }                                   } // end bsp_run()
  if ( bsp_sync() ) // shortcut
  { step 2, part a                      void foo() throws
  }                                         MigrateMeException
  for ( int i=0; i<n; i++ )             { if ( bsp_cont() )
  { if ( bsp_cont() )                   { // cont. current step
    { step 2+i, part b                    step 2+n, part c
    }                                   }
    if ( bsp_sync() )                   if ( bsp_sync() )
    { step 2+i+1, part a                { step 3+n, part a
    }                                   }
  }                                   } // end foo()
  // code continued in next col.
}
```

Fig. 5. Transformed BSP pseudo-code with code skipping.

Figure 6 shows how these methods are implemented. When restarting a checkpointed `BSPWork` object, `bsp_run()` is called with `curstep` initialized to 0, and `restorestep` set to the step to be resumed. The `bsp_step()` and `bsp_cont()` functions, called at the beginning of each code block, return `false` while `curstep` has not yet reached `restorestep`, allowing blocks that have already been executed to be skipped. At the end of each skipped superstep, the call to `bsp_sync()` increments `curstep` so that `curstep` eventually reaches `restorestep`, and the desired superstep's block is allowed to execute. The next call to `bsp_sync()` then ends the superstep by updating `restorestep` and throwing an exception which causes `bsp_run()` to return control to the work engine.

```
protected boolean bsp_sync()            protected boolean bsp_step()
         throws MigrateMeException      { return ( curstep>=restorestep );
{ if ( curstep++ >= restorestep )       }
  { restorestep = curstep;
    throw new MigrateMeException();
  }                                     protected boolean bsp_cont()
  return bsp_step();                    { return bsp_step();
}                                       }
```

Fig. 6. Code for `bsp_sync()`, `bsp_step()`, `bsp_cont()`.

As demonstrated in Fig. 5, these methods can be used even when `bsp_sync()` calls occur in loops, `if-else` blocks, and function bodies. Moreover, the code transformation rules are simple enough to be applied manually without using pre-compilers. In many cases, all a programmer needs to do is enclose the superstep code in braces, and add an `if` in front of `bsp_sync()`. As shown here, and more clearly in the realistic sample code in Sect. 4.2, the resulting code is still reasonably readable. In fact, the indented blocks may even help emphasize the divisions between supersteps. This simplicity offers a significant advantage in convenience over other programming interfaces that require language changes and sophisticated pre-compilers, since it does not require programmers to generate and keep track of extra files, and allows them to more easily use off-the-shelf integrated development and build environments. Furthermore, if more readability is desired, one can implement a parallel language with nothing more than a C-style preprocessor by using `#define` macros to replace `if (bsp_step())` with `parbegin`, `bsp_sync();` with `parend`, and `if (bsp_cont())` with `parcont`.

As far as we know, the only significant disadvantage of this code-skipping technique is that resuming execution inside or after an unbounded or very long loop may take unnecessary extra time because the skipping has to go through all previous iterations of the loop before reaching the current superstep. We are currently working on a solution to this problem, which may take the form of a special version of `bsp_cont()` which updates loop variables and `curstep` without having to iterate through the loop.

BSPMainWork. A Bayanihan BSP program is started by giving the name of a subclass of `BSPMainWork` to the `BSPProblem` object. `BSPMainWork` is a subclass of `BSPWork` that provides extra methods for server-side control of a computation, as shown in Table 2. The `BSPProblem` object assigns the `BSPMainWork` the reserved ID of 0, and calls its `bsp_main()` method, which is then expected to spawn work objects of the desired class. As the program runs, the main work object runs together with the other processes, and can communicate with them using BSP methods. In addition to spawning new work, the main work object can be used for coordinating other work objects, for collecting and sending results to interested *watcher clients* [19], and for receiving and reacting to user requests. This is similar to the common programming practice in SPMD/MPMD systems like PVM, MPI, and BSPLib of having a "master" or "console" process that handles data distribution and coordinates the operation of all the other processes.

Table 2. Basic `BSPMainWork` methods

method	meaning
int bsp_spawn(String class, int n)	spawn n BSPWork's of class *class*
int bsp_spawn(BSPWork w, int n)	spawn n BSPWork's of same class as w
void bsp_postResult(Result res)	post result for watchers
Object bsp_getRequest()	get user input from watchers
boolean isMain()	am I the main work?

To write an application, one can write separate `BSPMainWork` and `BSPWork` classes, or a single `BSPMainWork` class with separate `bsp_main()` and `bsp_run()` methods. The latter is useful when the main work object has the same fields as the other work objects, and has the added advantage of keeping all the code in one file. Since `bsp_main()` defaults to calling `bsp_run()`, one can also write a single `BSPMainWork` class with a single `bsp_run()` method containing code for *both* the main work and the other processes. This allows SPMD-style programming, and is useful in cases where the main work code is very similar to the worker code, or where it is useful to see the master and worker codes for each superstep in the same place. The `isMain()` function can be used within `bsp_run()` to let the main work process differentiate itself.

4.2 Sample Code

Figure 7 shows a matrix multiplication example demonstrating the use of the Bayanihan BSP programming interface. Here, we use a single `BSPMainWork` subclass with separate `bsp_main()` and `bsp_run()` methods. The `initVars()` method is called in both the worker and main work objects before running `bsp_run()` and `bsp_main()`. The algorithm used is based on one of the MPI example programs in [12], where each process i from 1 to n is given a copy of the matrix B and row $i-1$ of A, and computes row $i-1$ of the product C.

```java
public class BSPMatMultMain extends BSPMainWork
{ transient Matrix A, B, C;
  // ... some code omitted ...

  // called before bsp_main() and bsp_run()
  public void initVars()  // initialize B and C
  { B = (Matrix)bsp_restore( "B" );
    C = (Matrix)bsp_restore( "C" );
  }

  // run by main work on server
  public void bsp_main() throws TaskMigratedException
  { if ( bsp_step() )        // *** superstep 0 ***
    { // create n-by-n matrices, spawn, send a and B
      A = createSampleMatrix( n ); B = createSampleMatrix( n );
      bsp_spawn( this, n );    // spawn n workers of this type
      for ( int i=1; i <= n; i++ )
      { bsp_put( i, "B", B );                // write B
        bsp_send( i, i-1, A.getRow(i-1) ); // send row of A
      }
    }
    if ( bsp_sync() )        // *** superstep 1 ***
    { // workers compute C; main does nothing
    }
    if ( bsp_sync() )        // *** superstep 2 ***
    { // collect and print results on server
      BSPMessage msg;
      C = new Matrix( n );
      while( (msg = bsp_recv()) != null )   // get result rows
      { C.setRow( msg.getBSPTag(), (Row)msg.getData() );
      }
      trace( "C = \n" + C );  // print C
      bsp_save( "C", C );     // allow workers to get C
    }
    if ( bsp_sync() )        // *** superstep 3 ***
    { // workers print C
    }
  }

  // run by worker clients
  public void bsp_run() throws TaskMigratedException
  { if ( bsp_step() )        // *** superstep 1 ***
    { // restored local var B contains matrix B put
      BSPMessage msg = bsp_recv();           // get row
      if ( msg != null )
      { int i = msg.getBSPTag();             // get row #
        Row a = (Row)msg.getData();          // get row data
        Row c = Matrix.rowMatMult( a, B );   // compute
        bsp_send( MAIN, i, c );              // send row back
      }
      bsp_unregister( "B" ); // we don't need B anymore
    }
    if ( bsp_sync() )        // *** superstep 2 ***
    { // main collects results, get it
      bsp_get( MAIN, "C", "C" ); // get C for next step
    }
    if ( bsp_sync() )        // *** superstep 3 ***
    { // local var C now contains result
      trace( "C = \n" + C )      // print C
    }
  }
}
```

Fig. 7. Bayanihan BSP code for matrix multiplication.

5 Results

By hiding the previously-exposed implementation details of barrier synchronization and process communication from the programmer, the new BSP programming interface makes it much easier to write Bayanihan applications than before. At present, we have successfully used it not only to rewrite existing applications in a more readable and maintainable SPMD style, but also to implement new programs and algorithms such as *Jacobi iteration*, whose more complex communication patterns were difficult to express in the old model.

Unfortunately, while it is now possible to *write* a much wider variety of parallel programs with Bayanihan, such programs may not necessarily yield useful speedups yet since the underlying implementation is still based on the master-worker model. For example, even though one can now use peer-to-peer style communication patterns in one's code, these communications are still sequentialized at the implementation level since they are all performed by the central server. Also, since all work objects are currently checkpointed at the end of each superstep (i.e., workers send back their *entire* process state to the server at each bsp_sync()), algorithms that keep large amounts of local state between supersteps will perform poorly as this state would be unnecessarily transmitted back and forth between the server and the worker machines.

Figure 8 shows speedup results from some tests using 11 200 MHz Pentium PCs running Windows NT 4.0 on a 10Mbit Ethernet network, with Sun's JDK 1.1.7 JVM on the server and Netscape 4.03 on the 10 clients. The Mandelbrot test is a BSP version of the original Bayanihan Mandelbrot demo running on a 400x400 pixel array divided into 256 square blocks, with an average depth of 2048 iterations/pixel. The matrix multiplication test is a variation of the sample code in Sect. 4.2 where the worker processes get 10 rows of A at a time to improve granularity, and do not get C from the main work. It was run using 700x700 `float` matrices. The Jacobi iteration test solves Poisson's equation on a 500x500 `double` array decomposed into a 10x10 2D grid, with an exchange of border cells and a barrier synchronization done at the end of each iteration.

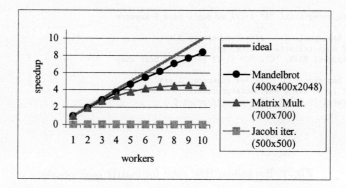

Fig. 8. Speedup measurements (relative to sequential version).

As shown, the "embarassingly parallel" Mandelbrot demo performed and scaled reasonably well. In contrast, the much finer-grain Jacobi iteration test performed very poorly, being at best about 40 times slower than a pure sequential version. This is not so surprising, though, given that it takes less time for the server to compute an element than to send it to another processor. Although speedups are theoretically still possible if the array is very large and workers are allowed to keep "dirty" local state for a long time, we cannot expect to get any speedup in this case since work objects send all their elements back to the server at the end of each iteration. On a more positive note, the matrix multiplication test, which employs a master-worker-like communication pattern (although written as a message-passing program), performed much better than the Jacobi iteration test, though not as well as the Mandelbrot test. The limiting factor in this case was the time needed to send the matrix B to the worker clients. Since we currently do not support multicast protocols, the server has to send B to workers individually. Thus, the broadcast time grows linearly with the number of worker clients, and eventually limits speedup. In the future, it may be possible to improve performance by using a more communication-efficient algorithm. However, most such algorithms keep local state between supersteps, and will thus probably be inefficient in the current implementation.

These results show that at present, Bayanihan BSP is still best used with coarse-grain applications having master-worker style communication patterns, and more work needs to be done to improve performance and scalability in other types of applications. Meanwhile, however, simply allowing programmers to write much more complex parallel programs than before is already a major result in itself, and represents a significant step towards making Java-based volunteer computing systems useful for realistic applications.

6 Current Issues and Future Work

At present, we are looking into ways of improving the performance of Bayanihan BSP applications. Some ideas we are studying include:

Caching. Implementing a form of software-based caching can improve performance and scalability in several ways. In the matrix multiplication program in Sect. 4.2, for example, the matrix B is sent to all *virtual* processes. Without caching, this would cause B to be sent over the communication network potentially thousands of times. By adding a cache to each work engine as shown in Fig. 3 in Sect. 3.1, we only need to send B once for each *physical* processor. We have successfully used such a caching scheme in the experiment presented in Sect. 5, and are currently generalizing it to apply to other situations as well.

Less frequent checkpointing and lazy migration. As noted, checkpointing and migrating processes at *all* calls to bsp_sync() can lead to unnecessarily poor performance in programs where large amounts of local state persist between supersteps. We can improve efficiency by having bsp_sync() normally send back communication requests only, and only occasionally send back the entire process state. We can also implement a *lazy* migration scheme that does not migrate a virtual process unless it is causing the system to slow down unnecessarily.

Peer-to-peer communication. Using the server for the global communication phase makes it easy for worker applets to communicate with each other despite the Java applet security restrictions that limit an applet to communicating only with its source host. Unfortunately, it can also be a serious and unnecessary sequential bottleneck in cases where such restrictions can be removed, such as within a trusted intranet. For these situations, we can implement peer-to-peer communication by modifying the work engine to send and accept communication requests to and from other work engines directly. We can still distribute the resulting worker code as an applet by either *signing* it, or asking volunteers to manually turn off their browser's applet network restrictions (Microsoft Internet Explorer 4.0, Sun's Hot Java browser, and Sun's Java plug-in [22], which works with Netscape, allow this). Implementing peer-to-peer communications should greatly improve the performance of parallel algorithms that depend on parallelizing communications as well as computation.

Fault-tolerance across supersteps. At present, our crash-tolerance and fault-tolerance mechanisms only work within a superstep. It is not clear yet how they can be extended if checkpoints are not performed each superstep. In such cases, a crash or fault may force a rollback to the previous checkpoint, possibly wasting work unnecessarily. It would be useful to develop a way to recover from a fault without having to do a full rollback.

Programming interface improvements. Aside from performance improvements, we are also studying some programming interface improvements such as the loop version of bsp_cont() mentioned in Sect. 4.1, tag-based retrieval of incoming messages, and saving and restoration of BSP variables in recursive functions.

7 Related Work

In recent years, an increasingly growing number of Java-based parallel computing systems have emerged [8, 1, 2]. Among these are systems with interfaces based on PVM, such as JPVM [9], and MPI, such as DOGMA [7]. Some of these, such as DOGMA and JavaParty [16], have already been successfully used in realistic parallel applications. Unfortunately, however, most of these systems use command-line Java *applications*, and thus still require some time and user expertise to set up. Making them use Java *applets* that can be run in a browser would be difficult or impossible not only because they require peer-to-peer communication not normally available to applets, but also because the semantics of the conventional message-passing models they use make it difficult to implement the adaptive load-balancing and fault-tolerance mechanisms required in the dynamic environment of applet-based volunteer systems.

A number of applet-based systems have also emerged, including Charlotte [3], Javelin [6], DAMPP [24], some RC5 cracking applications [11], and the original Bayanihan [17]. These are much easier to use than application-based systems since they allow a volunteer to join a computation by simply visiting a web page. Unfortunately, however, applet security restrictions and the need for adaptive parallelism and crash-tolerance seem to have so far limited applet-based systems to using ad hoc master-worker-based programming models that drastically limit the types of programs that one can write. Among applet-based systems, the most general and programmable one so far seems to be Charlotte [3]. Charlotte has a clean programming interface that provides transparent cache-coherent distributed shared memory between browser-based worker applets, while supporting adaptive parallelism in the form of eager scheduling. Its execution model is very similar to BSP in that computation is also divided by barrier synchronizations into superstep-like parallel blocks, and changes to distributed memory are not felt by other processors until after the next barrier. In addition, Charlotte has a fully-developed distributed caching mechanism that is still lacking in Bayanihan. Charlotte, however, requires defining a separate class for each superstep, and does not have message-passing functions.

The Bayanihan BSP interface bridges the two worlds of application- and applet-based parallel Java systems by allowing programmers to write programs using the flexible and easy-to-use remote memory and message-passing style found in peer-to-peer systems, while at the same time allowing these programs to be run in much more volunteer-friendly applet-based environments.

8 Summary and Conclusion

In this paper, we have shown how the BSP programming model can be implemented and used in Java-based volunteer computing systems. Our implementation makes use of existing adaptive parallelism and fault-tolerance mechanisms, such as eager scheduling, majority voting, and spot-checking, already developed in our Bayanihan volunteer computing framework, while providing a new programming interface with familiar and powerful methods for remote memory and message passing operations. At present, the performance and scalability of our implementation is limited in some applications, but is expected to improve with future work on mechanisms for lazy checkpointing and migration, caching, and peer-to-peer communication. Meanwhile, the BSP programming interface we have developed represents a significant improvement in programmability and flexibility over current programming interfaces for volunteer-based systems, and enables programmers to start porting and writing more realistic research applications for use with volunteer computing systems.

References

1. *Proc. ACM 1997 Workshop on Java for Science and Engineering Computation.* Las Vegas (1997) http://www.npac.syr.edu/users/gcf/03/javaforcse/acmspecissue/latestpapers.html

2. *Proc. ACM 1998 Workshop on Java for High-Performance Network Computing.* Palo Alto (1998). http://www.cs.ucsb.edu/conferences/java98/
3. Baratloo, A., Karaul, M., Kedem, Z., Wyckoff, P.: Charlotte: Metacomputing on the Web. Proc. of the 9th International Conference on Parallel and Distributed Computing Systems. (1996) http://cs.nyu.edu/milan/charlotte/
4. Beberg, A. L., Lawson, J., McNett, D.: http://www.distributed.net/
5. BSP Worldwide Home Page. http://www.bsp-worldwide.org/
6. Cappello, P., Christiansen, B. O., Ionescu, M. F.,, Neary, M. O., Schauser, K. E., Wu, D.: Javelin: Internet-Based Parallel Computing Using Java. ACM Workshop on Java for Science and Engineering Computation. (1997) http://www.cs.ucsb.edu/research/superweb/
7. DOGMA Home Page. http://zodiac.cs.byu.edu/DOGMA/
8. Fox, G.C. ed.: Special Issue on Java for Computational Science and Engineering – Simulation and Modeling. *Concurrency: Practice and Experience.* **9**(6) (1997).
9. Ferrari, A.: JPVM: Network Parallel Computing in Java. *Proc. ACM 1998 Workshop on Java for High-Performance Network Computing.* Palo Alto (1998). http://www.cs.virginia.edu/~ajf2j/jpvm.html
10. Geist, A., Beguelin, A., Dongarra, J., Jiang, W., Manchek, R., Sunderam, V.: PVM: Parallel Virtual Machine: A User's Guide and Tutorial for Networked Parallelism. MIT Press. (1994) http://www.netlib.org/pvm3/book/pvm-book.html
11. Gladychev, P., Patel, A., O'Mahony, D.: Cracking RC5 with Java applets. *Proc. ACM Workshop on Java for High-Performance Network Computing* (1998).
12. Gropp, W., Lusk, E., Skjellum, A.: Using MPI. MIT Press. (1994)
13. Hill, J.M.D., Donaldson, S.R., Lanfear, T.: Process Migration and Fault Tolerance of BSPlib Programs Running on Networks of Workstations. *Proc. Euro-Par'98. LNCS*, Vol. 1470. Springer, Berlin (1998) 80-91
14. Hirano, S.: HORB: Distributed Execution of Java Programs. *Proc. WWCA'97. LNCS*, Vol. 1274. Springer, Berlin(1997) 29-42 http://www.horb.org/
15. Nibhanupudi, M.V., Szymanski, B.K.: Adaptive Parallelism in the Bulk Synchronous Parallel model. *Proc. Euro-Par'96. LNCS*, Vol. 1124. Springer (1996)
16. Philippsen, M., Zenger, M.: JavaParty – Transparent Remote Objects in Java, in: *Concurrency: Practice and Experience.* **9**(11)(1997) 1125-1242
17. Sarmenta, L.F.G.: Bayanihan: Web-Based Volunteer Computing Using Java. *Proc. WWCA'98. LNCS*, Vol. 1368. Springer, Berlin (1998) 444-461
18. Sarmenta, L.F.G., Hirano, S., Ward, S.A.: Towards Bayanihan: Building an Extensible Framework for Volunteer Computing Using Java. *Proc. ACM Workshop on Java for High-Performance Network Computing* (1998).
19. Sarmenta, L.F.G., Hirano, S.: Building and Studying Web-Based Volunteer Computing Systems Using Java. *Future Generation Computer Systems*, Special Issue on Metacomputing (1999) http://www.cag.lcs.mit.edu/bayanihan/
20. Skillicorn, D., Hill, J.M.D., McColl, W.F.: Questions and answers about BSP. *Scientific Programming.* **6**(3) (1997) 249-274
21. Strumpen, V., Ramkumar, B.: Portable Checkpointing for Heterogeneous Architectures. *Fault-Tolerant Parallel and Distributed Systems.* Kluwer Academic Press (1998) 73-92
22. Sun Microsystems: Java Home Page. http://java.sun.com/
23. Valiant, L.G.: A bridging model for parallel computation. *Communications of the ACM.* 33(8) (1997) 103-111
24. Vanhelsuwe, L.: Create your own supercomputer with Java. JavaWorld. (Jan. 1997) http://www.javaworld.com/jw-01-1997/jw-01-dampp.ibd.html

High Performance Computing for the Masses

Mark Clement, Quinn Snell, and Glenn Judd

Computer Science Department
Brigham Young University
Provo, Utah 84602-6576
(801) 378-7608
{clement, snell, glenn}@cs.byu.edu
http://ccc.cs.byu.edu

Abstract. Recent advances in software and hardware for clustered computing have allowed scientists and computing specialists to take advantage of commodity processors in solving challenging computational problems. The setup, management and coding involved in parallel programming along with the challenges of heterogeneous computing machinery prevent most non-technical users from taking advantage of compute resources that may be available to them. This research demonstrates a Java based system that allows a naive user to make effective use of local resources for parallel computing. The DOGMA system provides a "point-and-click" interface that manages idle workstations, dedicated clusters and remote computational resources so that they can be used for parallel computing. Just as the "web browser" enabled use of the Internet by the "Masses", we see simplified user interfaces to parallel processing as being critical to widespread use. This paper describes many of the barriers to widespread use and shows how they are addressed in this research.

1 Introduction

Supercomputers and clusters of computers have long been used by parallel processing experts to solve computationally demanding problems. Users from other fields often find parallel application development and use too complex and confusing. Five years ago the Internet was a realm for a few technical people who could deal with remembering Internet addresses and obscure command line driven programs. Internet Web Browsers changed the face of the Internet and allowed non-technical people access to a wealth of information. This research intends to change the face of parallel processing in much the same fashion.

Imagine a typical evening at a major university or corporation, between the hours of midnight and 8am. The offices are all closed and thousands of computers are sitting idle. At the same time, researchers in several departments are waiting for results from applications that have been running for months on a single machine. Several avenues of research are unexplored because it takes so long to complete computations on a single machine. These applications could take advantage of a large number of processors if they were parallelized and the computing resources were available. This research proposes a way to make idle computing resources and dedicated clusters available for non-computer-experts (the masses) to use.

Barriers to widespread use of parallel computing include:

- Difficult installation of system software on cluster nodes. It is often necessary to have a development environment and extensive system understanding in order to compile and link applications for use on a compute node. Recently the creation of distributed/parallel programs has been eased somewhat by developments such as MPI[8], PVM[9], and CORBA[15]. These standards have allowed parallel/distributed programs to be written that will function on several different platforms in both tightly and loosely coupled environments. If you are using MPI or PVM, the ".rhosts" file must be correctly configured and other system resources must be manipulated in order to utilize the system. The names of available machines must be known and configured into the application or execution environment. The DOGMA system described here relies only on a screen saver and an installation of Java on processors. No other compilation or configuration must be performed. Since most workstations will have Java installed for the web browser, all necessary configuration can generally be performed with an access to a web page.
- Heterogeneous computing environments can also complicate the use of clusters. Runtime libraries, differing processor architectures, obscure security configurations and networking connectivity can discourage naive users from attempting configuration. Separate binary directories must be maintained for each processor architecture, and applications must be compiled several times in order to create binaries for all candidate architectures. In many cases it is difficult or impossible to determine the cause of these problems from system error messages. Because the DOGMA system is based on Java, most of these problems are not present. Any machine with a Java Virtual Machine can run Java Byte code so only one version of executables must be maintained.
- Selecting which cluster nodes will be involved in the computation can also be difficult for non-computer specialists. The DOGMA system includes scheduling software to determine optimal placement. The user need not participate in determining which processors, or how many processors should be involved in the computation
- Users have traditionally not wanted other users to have access to their machines because of the security risks involved. DOGMA provides two answers to this problem. The security measures inherent in the Java sandbox protect the machine and local file system by not allowing access and encrypting messages. Secondly, the DOGMA system allows a user to set up a group of machines that can only be used by his own applications. The researcher would then be protected from any outside use of his machines.

Although this work is still in progress, we have received enthusiastic support from researchers in Statistics, Zoology and Rendering groups. Each of these groups can use large quantities of processor cycles, but is prevented from making progress in their research area because of the limitations of a single processor.

These researchers are aware of efforts in their fields to use, parallel processing, but do not have the technical expertise to set up clustered systems. It is difficult for them to justify the expense involved in developing expertise in their groups when the benefits are hard to quantify. With the widespread use of high performance sequential processors, many new applications have been developed that have demanding processing requirements. Clusters present a powerful option for these applications if current barriers to implementation can be overcome.

2 Applications

Several application areas have been targeted for initial work with DOGMA. These applications require months or years to complete on available machines. Three statistical applications are being implemented to analyze large insurance databases. Current applications are unable to process more than approximately 30,000 individual records. Several million records are available which can not be analyzed by current technology. Although the Statistics department is reluctant to allow an installation of PVM or MPI, the are enthusiastic about using the DOGMA screen saver. Most system administrators are convinced that the Java sandbox will protect their machines from modification and their installation procedure does not have to be modified to allow DOGMA to run.

Phylogenetic Inference requires significant computational resources. Researchers in the Zoology department at Brigham Young University (BYU) are currently working with tens of individual DNA sequences. Analysis requires months of compute time even on these small data sets. They would like to experiment with hundreds of sequences and require faster turn-around time in order to effectively test new algorithms. Researchers speculate that they may be able to determine which cancer drugs to use for particular patients by examining DNA. They are also involved in analyzing AIDS virus DNA to determine epidemiological information. In both cases the research is being stymied by the lack of computing power.

The College of Fine Arts is currently operating a laboratory of high-end graphics workstations for use in animation and rendering. The workstations are often not available for student use due to the long rendering times when the machines cannot be used in an interactive mode. The DOGMA rendering application will free up these machines for modeling and visual development. The rendering stage will be performed by a DOGMA application in much less time, thus allowing students faster turn-around time and more appropriate use of resources. The resulting animation sequences will be of higher quality because the students will have more opportunities for interactive refinement.

Once the DOGMA system is being used widely, our research can focus on optimizations and more efficient resource allocation and utilization. These algorithms are important and this will be the first opportunity to investigate large-scale use of idle computational resources. Once we have completed evaluations at BYU, we feel confident that widespread use will continue throughout the Internet in these specific application areas. Additional applications should be possible once the "Masses" view parallel processing as an attainable technology.

3 DOGMA

The DOGMA system has been designed to take advantage of idle computing cycles as well as dedicated clusters of computers. Figure 1 shows a screen shot of a DOGMA rendering application. The user selects the application type and the specific application from the tree diagram shown on the left. When the application is selected, the user clicks to select the input file shown in the lower left corner. The application requirements are then shown in the lower right box. The application is run by clicking on the execute button. At this point the DOGMA system launches the program on available nodes. If additional machines become idle, their screen savers may contact the DOGMA system and become involved in the computation. The only requirement placed upon the user is some familiarity with the application and its inputs. Other applications may ask the user to simply cut and paste input files into the DOGMA interface or access input files on a web server and via a URL.

DOGMA is built using Java Remote Method Invocation (RMI) as a foundation. RMI allows Java objects to interact with other Java objects residing in separate Java Virtual Machines (JVM's); however, the programmer is required to handle issues such as object location at a low level. DOGMA's runtime system is known as the Distributed Java Machine (DJM) layer. It allows several JVM's to act as one distributed machine while freeing the programmer from needing to worry about the location of objects, the names of nodes, the location of nodes, etc. This layer also allows application binary code to be served via an arbitrary number of http servers, and eliminates the need for applications to be installed on local file systems.

DOGMA has a decentralized architecture divided into units known as "configurations". Each configuration is managed by a "configuration manager" which is capable of linking to other configuration managers in a hierarchical fashion. Configuration managers may optionally allow other configuration managers to join them and use their local resources. For instance in Figure 2 configuration manager A can utilize the resources of B but the reverse is not true as indicated by the unidirectional arrow. However, configuration manager B and C can both utilize each other's resources as indicated by the bi-directional arrow. Configuration managers periodically exchange information such as total number of nodes and total number of nodes. This information then propagates through the hierarchy.

As an example, consider Figure 2. An arrow from one configuration manager to another indicates that the source configuration manager may run remote jobs on the destination configuration manager. Note that configuration managers never see the global view. So from configuration manager A's point of view there is only one configuration manager attached to it, but this configuration manager

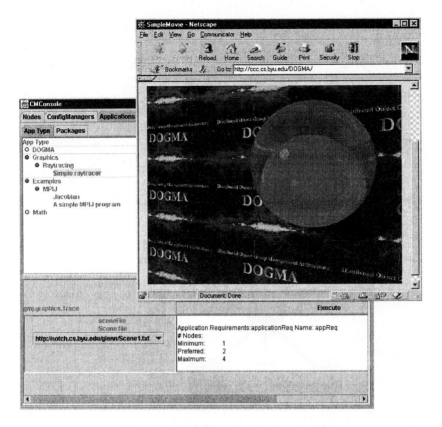

Fig. 1. DOGMA Application Interface

has 35 nodes which is the sum of all nodes downstream in the configuration graph.

Within each configuration three types of nodes may contact the configuration manager to join the system:

- Dedicated nodes - these nodes have access to DOGMA source code and are launched from the configuration manager's system console.
- Browser-based nodes - these nodes join the system by "browsing in" to a specific web-page served by the configuration manager.
- Screen-saver based nodes - these are nodes which have the DOGMA screen saver installed on them, and join the system automatically when they are idle.

Also within each configuration are "code servers" which are http servers used to distribute binary code for a given set of applications; they allow DOGMA to function with no shared file system. Optionally, "data servers" may be added to the configuration to serve application data files.

When nodes join a configuration they sign a "contract" with the configuration manager informing the configuration manager of the duration for which they will

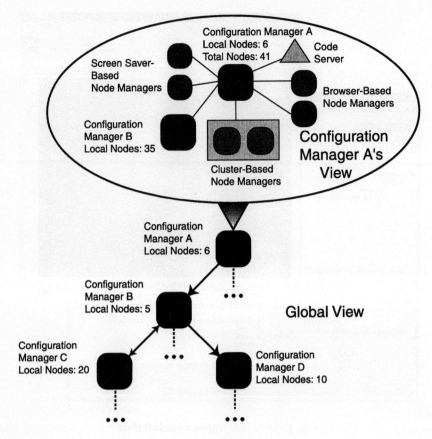

Fig. 2. DOGMA Architecture

stay in the system. This prevents the configuration manager from scheduling long running jobs on transitory nodes. Nodes also inform the configuration manager of certain attributes such as processing power.

To improve application performance and enhance usability, applications have an associated definition file which informs DOGMA of the application's system requirements and startup arguments. Application requirements may consist of the number of nodes, processing power required, networking connection required, whether or not the application allows a varying number of nodes, how long the application expects to run, etc.. Startup arguments free the user from the need to remember which arguments are required by which application, and can be simple text entries, a list of choices, etc..

When an application is run, the system attempts to match the application with a set of nodes which meet the application's requirements. If it cannot find a set of nodes locally, it may (depending on the application requirements) look remotely for nodes to run the application on. The information exchanged periodically by configuration managers helps them make intelligent decisions about where to remotely run a job.

4 Writing DOGMA Code

DOGMA is programmed using one of three programming API's: MPIJ, Distributed Object Groups, or Dynamic Object Groups. MPIJ is DOGMA's implementation of the MPI standard (we are working with other members of the Java Grande Forum on a formal proposal for MPI bindings). Distributed Object Groups allow programs to be written which will automatically partition and reassemble data. Dynamic Object Groups allow master/slave style programs to be written in a very simple fashion which can scale to large numbers of nodes where the number of nodes is allowed to vary at runtime.

4.1 Writing DOGMA Code

Writing Distributed Object Group programs is fairly simple. As in RMI, an interface file must be written defining which methods in a given object (or object group) may be invoked remotely. Upon group method invocation, arguments of type Partition or one of its subclasses are automatically partitioned by the system while other arguments are simply sent to every node in the group. For example:

```
interface SkelElement extends Element {
        public OneDArrayFloatPartition
        addFloatToArray(
        OneDArrayFloatPartition myPartition,
        float floatToAdd)
        throws RemoteException;
}
```

In this case, addFloatToArray is defined to accept a Partition data structure, myPartition, that will contain only the processor's required subset of the data. Group elements then extend the element base class for the data configuration scheme to create the corresponding implementation of the distributed object. A master (a class which has a distributed object group as a member) may then instantiate a distributed object group as follows:

```
skelGroup =
        new SkelElementGroup("my.pkg.SkelElementImpl",
        minNodes, preferredNodes, maxNodes,
        new OneDArrayController(),
        DistObjectGroup.CCFG_NONE);
```

This example uses the OneDArrayController to allocate a one-dimensional array and automatically perform the data distribution. Methods may then be invoked on the entire group with their arguments being partitioned if they are of type Partition or one of its subclasses. As the invocation is asynchronous, a handle to the invocation is returned:

handle = dog.addFloatToArray(data, 7.0f);

This call invokes the addFloatToArray method on all available processors, sending each processor a piece of the array. The programmer then uses the handle to request the results of the method. The programmer may request that the results be automatically reassembled, or that an array of results be returned. The statement

partition = (OneDArrayFloatPartition)
       handle.getAssembledReturn();

causes the results to automatically be reassembled into the array **partition**.
Group element communication is similar to other parallel systems. Note that objects may be sent as easily as primitive data as in the following statement. The object **histogram** is broadcast to all participating objects.

sendToAll(histogram);
Object[] otherHistogram = receiveFromAll();

The Distributed Object Group paradigm presented here provides a high level abstraction for parallelism that also achieves high performance.

4.2 DOGMA MPI

DOGMA's Java MPI implementation is based on the Java Grande Forum proposal for Java MPI bindings. MPIJ is a pure Java MPI implementation conforming to these standards. Other implementations use the Java Native Interface to link to native MPI implementations. However, as will be shown in later sections, Java communication performance is equivalent to that of native socket communication.

```
import mpij.*;
public class SampleMPIApplication extends MPIApplication {
    public void MPIMain(String[] args) {
    MPI.init();
    myNode = MPI.Comm_WORLD.rank();
    totalNodes = MPI.Comm_WORLD.size();
    // broadcast an array from node 0 to everyone in COMM_WORLD
    double[] d;
```

```
        d = new double[100];
        MPI.COMM_WORLD.bcast(d, 0, d.length, MPI.DOUBLE, 0);
        ...
    }
}
```

The above code fragment is a simple example of the MPIJ bindings. It allocates an array and broadcasts it to all members of the process group. This code sample is very similar to a C++ MPI implemented code. It is fairly easy to port existing MPI code to this system.

A DOGMA program can be written specifically for the use of distributed object groups or for message passing using the Java MPI calls. It is also possible to combine the two paradigms. Porting of MPI programs from C to Java has proven to be very simple, and the performance obtained is promising. A finite element program was recently ported from C to Java using our MPI calls. The only changes necessary were in the method calling conventions of Java and in the allocation of arrays.

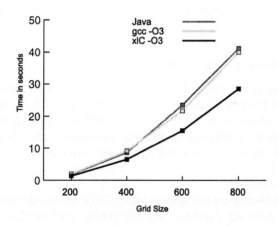

Fig. 3. Java Performance Comparison

5 DOGMA Performance

Despite the advances in source code level compatibility, traditional distributed programming environments require machine specific compilation and operating system customization. This makes the inclusion of a new architecture into an existing distributed computing platform non-trivial. With the advent of Java, code distribution is now relatively simple. Unfortunately, many have come to view Java as incapable of the performance required for scientific applications. However, Java just-in-time compiler performance is quickly approaching that of optimized C. Figure 3 shows the performance of a finite element algorithm on a single node of an SP2. The performance of the Java code is nearly indistinguishable from the fully optimized gnu gcc compiled executable. The custom IBM xlc compiler achieves higher performance, but scheduled improvements in JIT compiler technology promise to close this gap. The DOGMA system provides the portability of Java as well as high performance for heterogeneous parallel computing.

DOGMA is portable across all Java virtual machines that fully support the JDK version 1.1.3 or greater. We have successfully run applications on large groups of heterogeneous nodes with no source code modification. The applications include: matrix multiplication, image FFT, image histogram equalization, finite differences, and ray tracing to name a few. The number of nodes included in the system has ranged from 1 to 40 (including several dual-processor machines) with the nodes residing in 4 different domains. The system has also incorporated both browser-based nodes, which were able to execute applications and donate CPU power, as well as nodes running the DOGMA screen saver to donate idle CPU time to DOGMA. In addition, DOGMA is capable of seamlessly harnessing the power of multiprocessor machines by assigning multiple objects to the machine. Local objects communicate via direct method invocation rather than through RMI.

5.1 Computation Performance

The results shown previously in Figure 3 demonstrate similar findings for a single node of the IBM SP-2. The graph shows execution times for a finite differences code with varying size grids. In all grid sizes, JIT compiled code is comparable to the performance of fully optimized gcc compiled code. There is still room for improvement, as is demonstrated by the performance of xlC compiled code. Because of dynamic runtime optimizations, many predict superior performance as JIT compilers mature.

The previous results are not always typical. JIT compiled code that is heavily object oriented does not currently provide similar performance. JIT compilers need more improvements handling object oriented code, but clearly they have the potential for high performance computing. The Hotspot JIT compiler from Sun promises to solve some of these performance issues by using adaptive compilation techniques pioneered by the Self project at Stanford [4, 11].

5.2 Communication Performance

High communication performance is essential in a supercomputing environment. In Java, communication between nodes can take place via two mechanisms: sockets and Remote Method Invocation (RMI). The DOGMA environment uses both schemes. Group method invocation is accomplished via RMI, and inter-node message passing using the MPI library is performed using Java sockets.

Most messages in scientific codes consist of floating point data that is either double or single precision. Java communication is based on the InputStream and OutputStream classes; neither of which support communication with data types other than bytes. Therefore, in order to send an array of floating point data, the data must be converted into a byte array. Java is a strongly typed language and, as such, does not allow type casting of pointers. Unlike C programs which allow a floating point array to be directly treated as an array of characters, each Java data element must be type cast and copied into a byte array to be transmitted. This technique is termed data marshaling, and adds significant overhead.

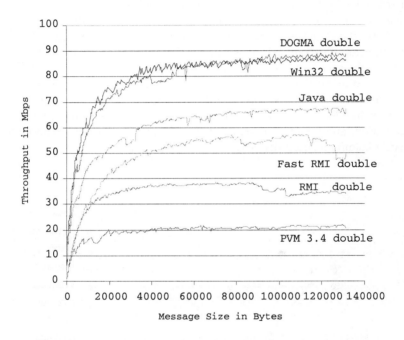

Fig. 4. Native communication versus Java for floating point data.

Figure 4 clearly shows the overhead due to data marshaling. However, the graph also demonstrates that significant performance increases are possible by performing the data marshaling in native code. The "Fast RMI double" line on the graph was obtained using native data marshaling. This simple routine is linked to the Java code using the Java Native Interface (JNI). This routine could be directly added to the Java virtual machine or incorporated into the InputStream and OutputStream classes to allow for communication of other data types. The graph also shows the comparison of message passing performance between two DOGMA nodes versus communication between two PVM nodes. Both tests were run on the same two workstations. PVM uses native sockets as a communication medium, while DOGMA communication is based on Java sockets and the JNI marshaling methods. Amazingly, DOGMA communication performance is better than PVM.

Even DOGMA latency (160 microseconds) is much better than that of a PVM message (1147 microseconds). Such a result implies that communication intensive applications will fare better in the DOGMA environment. This hypothesis will be demonstrated later in this paper.

5.3 Aggregate Performance

The above results are meaningless unless they translate into aggregate cluster performance for parallel applications. In this section, we demonstrate DOGMA application performance in comparison with PVM and MPI applications. In each case, the target applications were coded using the same algorithm for each environment. For comparison purposes, the applications were also run on the cluster using mpich on the Linux operating system. We also ran the applications using the WinMPICH message passing system. However, the system is still very unstable, and the results were very poor. Thus we chose the Linux system as our base-line. The applications of choice were: 2 dimensional FFT, Gaussian elimination, and a finite differences code. The applications were chosen for their varying communication demands.

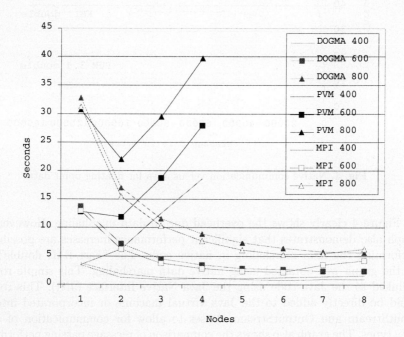

Fig. 5. Performance Comparison for Gaussian Elimination

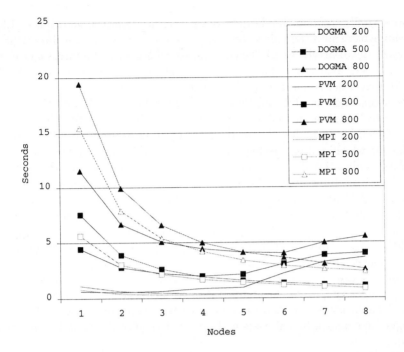

Fig. 6. Performance Comparison for Finite Element Calculation.

The application performance graphs in Figures 5 and 6 show the comparative performance results for the Gaussian elimination and finite element calculations. Each of the applications require successively less communication. As the communication demands decrease, the computational performance of the individual nodes is more important. In these cases, DOGMA application performance is better than PVM performance for larger numbers of nodes, but does not exceed Linux MPI application performance. It does, however, approach that level of performance.

6 Related Work

Java's platform independent/Internet-centric nature make it natural for use in distributed computing. This has lead several groups to begin research on distributed/parallel programming with Java. The following section reviews these other avenues of research and differentiates DOGMA from these existing systems.

The JavaPVM[19], JPVM[7] and IceT[10] message passing schemes do not provide automatic data distribution nor do they provide the object group method invocation found in DOGMA. DOGMA supports traditional message passing via an MPI implementation, but features object-group oriented programming as its preferred paradigm. Linda derivatives such as Jada[18], JavaSpaces[12] and Javelin[5] have not been able to achieve comparable performance. Java/DSM[20], Spar[13], JavaParty[17] and Charlotte[3] provide a shared address space, while DOGMA seeks to make the developer aware of the distributed nature of the program in order to increase performance.

Systems such as JET[16], SuperWeb[1]/Javelin[5], IceT, and Charlotte are similar to DOGMA in that they incorporate Web browser functionality; however, DOGMA also adds cluster based computing in order to utilize groups of nodes for which a user has login privilege. DOGMA also makes use of signed applets to allow for direct communication between browser-based nodes rather than requiring browser-based nodes to communicate only with a web server. In addition, DOGMA adds screen saver based browsing to allow nodes to automatically donate idle CPU time rather than requiring explicit user donation of CPU time.

More recent implementations[14] such as mpiJava[2] link native MPI implementations to Java via the Java Native Interface. These efforts are also concentrating on pure Java implementations. Condor[6] is a distributed batch system that provides for a hierarchical integration of clusters of workstations. It is suited well to batch processing and has extensive load balancing features. DOGMA is designed for non-technical users and does not require the installation any software.

DOGMA seeks to do more than simply provide a Java version of a message passing standard. Rather, DOGMA has three main goals:

- The creation of a distributed computing platform which can harness the power of clusters of computers as well as anonymous idle computers in both Internet and intranet settings.
- Providing mechanisms for easily writing programs capable of efficiently using the power of such a system.
- Providing a point and click interface that non-technical users can use. DOGMA is usable by people who don't feel comfortable with writing any code or performing any task other than accessing a web page.

7 Conclusions

The DOGMA system allows idle machines to be incorporated into supercomputing clusters irrespective of their CPU or operating system. Since use of these clusters is much less expensive than traditional supercomputing platforms, more researchers and average users can have access to high performance compute resources. Applications such as phylogenetic inference, statistical analysis and large database indexing (which often require much more compute power than is available on a single system) can be approached using DOGMA clusters.

Though there is much future research to be done, DOGMA has proven to be very capable of enabling clusters of computers and individual anonymous computers to act as a single powerful networked computing entity. DOGMA volunteer facilities also provide an excellent means to increase workstation utilization. Object migration is currently being incorporated to allow for increased performance, and useful inclusion of transitory nodes in a large number of applications. Other issues such as security and fault tolerance enhancements are also being examined.

The availability of significant CPU resources will allow more users to solve complex optimization and simulation problems in a reasonable amount of time. The DOGMA system provides a significant improvement over existing parallel environments. The point-and-click interface will open up new scientific frontiers for researchers in other disciplines.

See the DOGMA homepage at for more current information: http://ccc.cs.byu.edu/DOGMA/.

References

1. A. D. Alexandrov, M. Ibel, K. E. Schauser, and C. J. Scheiman. Superweb: Towards a global-based parallel computing infrastructure. In *Proceedings of the 11th International Parallel Processing Symposium*, April 1997.
2. M. Baker, B. Carpenter, S. H. Ko, and X. Li. mpijava: A java interface to mpi. In *First UK Workshop on Java for High Performance Network Computing, Europar*, September 1998.
3. A. Baratloo, M. Karaul, Z. Kedem, and P. Wyckoff. Charlotte: Metacomputing on the web. In *Proceedings of the ISCA International Conference on Parallel and Distributed Computing*, 1996.
4. C. Chambers. *The Design and Implementation of the Self Compiler, an Optimizing Compiler for Object-Oriented Programming Languages*. PhD thesis, Stanford University, 1992.
5. B. O. Christiansen, P. Cappello, M. F. Ionescu, M. O. Neary, K. E. Schauser, and D. Wu. Javelin: Internet-based parallel computing using java. In *ACM 1997 PPoPP Workshop on Java for Science and Engineering Computation*, June 1997.
6. D. Epema, M. Livny, R. van Dantzig, X. Evers, and J. Pruyne. A worldwide flock of condors : Load sharing among workstation clusters. *Journal on Future Generations of Computer Systems*, 12, 1996.
7. A. Ferrari. Jpvm. Technical report, http://www.cs.virginia.edu/~ajf2j/jpvm.html, 1997.
8. M. Forum. Mpi: A message-passing interface standard. Technical report, University of Tennessee, June 1995.
9. A. Geist, A. Beguelin, J. Dongarra, W. Jiang, R. Manchek, and V. Sunderam. PVM 3 user's guide and reference manual. Technical Report ORNL/TM-12187, Oak Ridge National Laboratory, September 1994.
10. P. A. Gray and V. S. Sunderam. IceT: Distributed computing and java. In *Proceedings of the ACM 1997 PPoPP Workshop on Java for Science and Engineering Computation*, June 1997.
11. U. Hoelzle. *Adaptive Optimization for Self: Reconciling High Performance with Exploratory Programming*. PhD thesis, Stanford University, 1992.
12. Javasoft. Javaspaces. Technical report, http://chatsubo.javasoft.com/javaspaces/, 1997.
13. H. J. S. Kees van Reeuwijk, Arjan J.C. van Gemund. Spar: A programming language for semi-automatic compilation of parallel programs. In *ACM 1997 PPoPP Workshop on Java for Science and Engineering Computation*, June 1997.
14. S. Mintchev and V. Getov. *Recent Advances in PVM and MPI*. Springer Verlag, 1997.

15. OMG. The common object request broker: Architecture and specification. *2.0 ed.*, July 1997.
16. H. Pedroso, L. M. Silva, and J. G. Silva. Web-based metacomputing with jet. In *Proceedings of the ACM 1997 PPoPP Workshop on Java for Science and Engineering Computation*, June 1997.
17. M. Philippsen and M. Zenger. Javaparty - transparent remote objects in java. In *Proceedings of the ACM 1997 PPoPP Workshop on Java for Science and Engineering Computation*, June 1997.
18. D. Rossi. Jada: Multiple object spaces for java. Technical report, http://www.cs.unibo.it/~rossi/jada/, 1996.
19. D. Thurman. Javapvm. Technical report, http://homer.isye.gatech.edu/chmsr/JavaPVM.html/, 1997.
20. W. Yu and A. Cox. Java/DSM: A platform for heterogeneous computing. In *Proceedings of the ACM 1997 PPoPP Workshop on Java for Science and Engineering Computation*, June 1997.

Process Networks as a High-Level Notation for Metacomputing

Darren Webb, Andrew Wendelborn, and Kevin Maciunas
{darren,andrew,kevin}@cs.adelaide.edu.au

Department of Computer Science, University of Adelaide
South Australian 5005, Australia
Fax: +61 8 8303 4366, Tel +61 8 8303 4726

Abstract. Our work involves the development of a prototype Geographical Information System (GIS) as an example of the use of process networks as a well-defined high-level semantic model for the composition of GIS operations. Our Java-based implementation of this prototype is known as PAGIS (Process network Architecture for GIS).

Our process networks consist of a set of nodes and edges connecting those nodes assembled as a Directed Acyclic Graph (DAG). In our prototype, nodes represent services and edges represent the flow of data (in this case sub-processed imagery) between services. Services are pre-defined operations that can be performed on imagery, presently selected from the Generic Mapping Tools (GMT) library. In order to control the start and end-point of the DAG, we define an input node (the original image) and an output node (the result image).

To exploit potential parallelism, we extend our idea of a process network to a distributed process network, where each service may be processed on different computers. A single server coordinates computation and computation is performed by any number of workers. The server and workers together can beseen as a metacomputer. The server takes a process network from a client and distributes work to the workers. Each worker applies for work and decides if it is capable of performing the work offered. In this way, scheduling is essentially dynamic, and computation can be performed without client intervention.

1 Introduction

In this paper, we introduce the concepts of process networks, distributed process networks, and metacomputing, and the application of these concepts in the development of a prototype Geographical Information System (GIS). Process networks provide a well-defined high-level semantic model for the composition of processing steps, here used in a GIS.

PAGIS, our prototype GIS, provides a sophisticated user interface for domain experts to apply transformations to selected satellite imagery. The user selects and orders transformations in a process network from their local computer. This process network is then submitted to a metacomputer for computation across geographically distributed high-performance computers. Scheduling is dynamic so the user requires no intervention in how processes are distributed. This factor is especially important for systems such as a GIS, as domain experts are generally not interested in learning parallel programming paradigms.

Section 2 of this paper defines and explains important characteristics of process networks and introduces distributed process networks. Section 3 is a brief discussion of metacomputing and metacomputing issues. Section 4 provides an overview of the PAGIS GIS, and Section 5 discusses some future work.

2 Process Networks and Distributed Process Networks

2.1 A Semantic Model for Composing Computations

The Kahn model of process networks [6] can be used to semantically represent transformations to be applied to data. Process networks are directed acyclic graphs where a node represents a process and an edge represents the flow of data from one process to another. Kahn defines a process to be a mapping from one or more input streams to one or more output streams. Concurrent processes communicate only through directed first in - first out streams of tokens with unbounded capacity such that each token is produced exactly once, and consumed exactly once. Production of tokens is non-blocking, while consumption from an empty stream is blocking.

Stevens, et al [8] present a Java implementation for dataflow networks. In this case, a network is defined to be a directed graph, comprising a set of threads connected by a set of unidirectional first in - first out queues. Each thread executes an infinitely executing sequential program. To enable each thread to run in finite memory each edge in the graph represents a bounded queue that nodes can read and write data tokens. A thread blocks if the queue is empty on a get or full on a put. A master thread executes to detect deadlock in the network - solved by increasing the size of the offending queues.

We propose a process network model that is a hybrid of Kahn's process network and the dataflow network presented by Stevens, et al. Our interpretation of a process network is a directed acyclic graph where nodes represent conceivably infinitely executing sequential lightweight processes, and edges represent bounded queues. Our implementation blocks if the queue is empty on a get or full on a put, but does not require a master thread to resolve deadlock as the process network is acyclic. The only time deadlock can occur is when the all queues are empty, which means there is no input into the network.

There are several properties that make process networks desirable as a computational model. The following are basic properties noted in [3]:

1. Each process is a sequential program that consumes tokens from its input queues and produces tokens to its output queues.
2. Each queue has one source and one destination.
3. The network has no global data.

4. Each process is blocked if it tries to read a communication channel with insufficient data. The read proceeds when the channel acquires sufficient data.
5. Writing to a communication channel is generally not blocking, with each queue capable of storing an unlimited amount of data. We utilize a blocking write to enable our process networks to operate in bounded memory.

Given these basic properties, more advanced properties can be derived. These are:

Concurrency. This is safe in process networks because the process network has no global state, and hence concurrent computation of the network nodes will not affect the state of other nodes. More important to a programmer is concurrency is implicit, exploited by overlapping communication with computation and concurrently executing independent processes.

Figure 1 shows a simple UNIX script that takes $2*(s1+s2+s3)$, where $s1$, $s2$ and $s3$ are the time taken calculating the three results. Concurrently executing these processes and caching their results, shown diagrammatically in Fig. 2, takes just $s1+s2+s3$, a parallel speedup of approximately two.

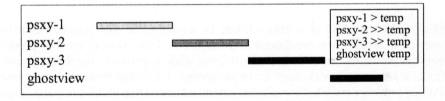

Fig. 1. Computation time of a trivial UNIX script

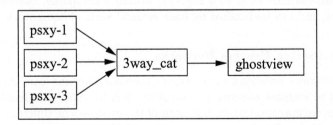

Fig. 2. The pipeline in Fig. 1 as a process network

Scheduling. In an information system such as PAGIS, it is important that scheduling be handled transparently, as domain experts composing a computation should not be required to understand how the components of a computation are scheduled. Process networks satisfy this criterion. All processes in our process networks are started independently; those processes that have no data to consume block. A process is eligible to be scheduled if it wishes to read data from an input port, and data are available on that input. In this way, scheduling is data-driven and dynamic. Scheduling in PAGIS is based on the dynamic ordering of process input and output operations. It may be possible to regard network operators as dataflow nodes, with firing rules that specify the data items required for an individual firing of the node. The scheduler could then operate by scheduling such activations in a data-driven, demand-driven, or hybrid, manner.

Determinacy. A concurrent system requires some guarantee of consistent behavior across implementations. Our process networks are acyclic directed graphs that can be shown to never deadlock. Deadlock due to lack of storage is the only source of deadlock in cyclic networks. Cyclic process networks can be defined (and executed by the PAGIS scheduler), but we have not used them so far in our application. Kahn has shown that some cyclic networks may deadlock in a unique state regardless of scheduling method [6]. In a non-deadlocking process network, method of implementation and order of computation do not affect the result.

Hierarchy. A node of a network can be expressed in any suitable notation provide it satisfies the conditions of Section 2.1. Thus, we can use the process network notation, or a cached result from such a network, itself for a node. Such a process network may be implemented differently from the outer-level PAGIS process network; for example, a node allocated to a multi-computer might implement process networks in a fine-grained manner. We are exploring issues of notational support for hierarchical networks, and of interfacing between different implementation techniques (at node boundaries). Note also that such process networks can (compared to the PAGIS model of process network) more readily exploit computations local to a high-performance computer, because latencies are smaller, and can be masked by finer-grained units of computation.

2.2 Distributing Process Networks

Process networks executed on a single-processor machine can exploit multi-processing to emulate concurrent execution, but to take full advantage of any parallelism, we may need to execute each of the processes on different processors. We can do this by distributing the processes of the process network such that we have a distributed process network. Distributed process networks can be used to provide an abstraction over potential parallelism, such that the user is implicitly programming a computation that is executed in parallel. The five processes in Fig. 2 can be executed on different machines, with results communicated over a high-speed network. In this way, each process is executed in parallel, successfully exploiting parallelism available in the problem.

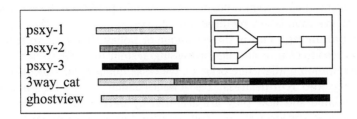

Fig. 3. Computation time of the process network in Fig. 2

2.3 Summary

A GIS requires a simple mechanism for the construction of generic computations by domain experts. Process networks provide us with an elementary model for representing complex computations in a GIS. It is a simple yet powerful semantic model for representing the composition of certain computations and higher-order functions.

Concurrency in process networks is implicit, achieved through concurrent computation of independent processes, and overlapping communication between processes with computation of others. To derive parallelism, we extend the concept of a process network to a distributed process network where processes are executed across distributed computers.

As computations are implicitly concurrent and scheduling is dynamic, user intervention in such matters is not required. These aspects make process networks beneficial for many applications, especially for metacomputing. We now proceed to show that our process networks are also a useful abstraction over the resources of a metacomputing system.

3 Metacomputing

3.1 An Overview

A metacomputer consists of computers ranging from workstations to supercomputers connected by high-bandwidth local area networks and an environment that enables users to take advantage of the massive computation power available through this connectivity. The metacomputing environment gives the user the illusion of a single powerful computer. In this way, complex computations can be programmed without requirement for special hardware or operating systems. There are a number of difficulties that arise when designing and implementing such systems. These issues include:

1. **Transparency.** This is especially important in a heterogeneous system and is generally solved by installing an interpreter on each node or by using a standard cross-platform language.
2. **Synchronization and Scheduling.** These issues define when parts of a larger computation should begin. These should preferably be transparent so user intervention is not required.
3. **Load balancing.** This is the technique or policy that aims to spread tasks among resources in a parallel processing system in order to avoid some resources standing idle while others have tasks queuing for execution.
4. **Security and Robustness.** These issues define policies for what resources are available and how they are accessed by particular users, and how the system should recover when a computation terminates prematurely.

Metacomputers are intended only for computationally intensive applications. Applications that are not computationally intensive can be disadvantaged when processed on a metacomputer, as the cost of communication and synchronization between co-operating processors may outweigh the benefits of concurrent processing.

3.2 Metacomputing and Distributed Process Networks

Without easy-to-use and robust software to simplify the virtual computer environment, the virtual computer will be too complex for most users [7]. Process networks, as a semantic model for composing computation, can provide the simplification required so users can compute efficiently and effectively. The user interface can be made very simple, requiring only a mechanism for graphical construction of a graph such that the user can compose networks of processes and the flow of data between them. In this way, the user can be offered an easy-to-use, seamless computational environment. The process network model can also facilitate high-performance through implicit parallelism, in cases (such as the abovementioned computationally intensive applications) where nodes can mask large communication latencies by overlapping intensive computation.

Distributed process networks provide a simple model for distributing work in a program. The nodes of our process networks can be easily mapped to metacomputer nodes. Edges of the process network represent communication between nodes of the metacomputer.

3.3 Summary

A metacomputer is an abstraction for the transparent grouping of heterogeneous compute resources. The abstraction is provided by a simple user interface that facilitates the effective and efficient use of these compute resources as a single homogeneous system. The distributed process network model allows us to easily distribute work to processing nodes in a metacomputer system. Java is a suitable mechanism for performing the distribution as it assists us to effortlessly communicate between distributed heterogeneous machines.

4 PAGIS: A Prototype GIS

4.1 Background and Overview

A Geographical Information System (GIS) is intended to aid domain experts such as meteorologists and geologists make accurate and quantitative decisions by providing tools for comparing, manipulating, and processing data about geographic areas. Such a system can help interpret what is shown by the image because it relates the image to known geographical areas and features. Images can help a domain experts interpretation and understanding of features of an area, showing aspects of the areas that are not noticeable or measurable at ground level.

Our project involves the development of a prototype GIS for processing geostationary satellite data obtained from the Japan Meteorological Agency's (JMA) [4] GMS-5 satellite. GMS-5 is a geostationary satellite positioned at approximately 140 °E at the equator with an altitude of approximately 35,800 kilometers. GMS-5 provides more than 24 full-hemisphere multi-channel images per day with each image 2291x2291 pixel-resolution. These images require approximately 204MB of storage capacity each day. The Distributed High-Performance Computing project (DHPC) [1] currently maintains a GMS-5 image repository capable of caching recent image data on a 100GB StorageWorks RAID with older data stored automatically by a 1.2TB StorageTek tape silo. We utilize this repository as the source of data for our prototype GIS.

Processing satellite data represents a time-consuming application we use to model real-world problems and test non-trivial computational techniques on distributed high performance computers. In our GIS application, we use the Generic Mapping Tools (GMT) [2] library to perform manipulation of the satellite imagery. This forms the basis of our "computational engine". GMT is a library of approximately 60 utilities for the manipulation and display of geographic data. These utilities enable the domain expert to perform mathematical operations on data sets, project these data sets onto new coordinate systems, and draw the resulting images as postscript or one of a few raster formats. GMT was chosen as out computational engine because of its modular and atomic design. Each utility serves a specific function, but together provide a powerful suite of utilities for a GIS.

We utilize the Java programming language to tie together our model, data and computational engine. The fact that Java is a distributed, interpreted, architecture neutral, portable, multithreaded, and dynamic language makes it highly desirable for the implementation of a metacomputer system such as PAGIS. Communication between distributed machines is possible using Java's RMI (Remote Method Invocation), essentially an evolution of RPC (Remote Procedure Call). RMI uses object serialization to enable Java objects to be transmitted between address spaces. Objects transmitted using the object serialization system are passed by copy to the remote address space [7].

There are drawbacks in using Java. Java is an interpreted language, and as such can not take advantage of architecture-dependent code optimizations. In addition, code interpretation is performed on a Java Virtual Machine (JVM), which needs to be available for all architectures in the metacomputer. In our case, our Thinking Machine's CM-5 does not have a native JVM that can take advantage of its 128 nodes. This reduces the ability to use certain machines in the metacomputer system optimally. Java is, however, a step in the right direction for distributed systems.

4.2 An Introduction to PAGIS

PAGIS is a simple yet sophisticated proof-of-concept metacomputer system that uses process networks as a semantic model for the composition of complex tasks in a GIS. Users are able to select and view a start image, compose a process network, submit the process network to a server for computation, then display the results. These actions are abstracted by a sophisticated interface that facilitates the composition and submission of a process network to a metacomputer for computation.

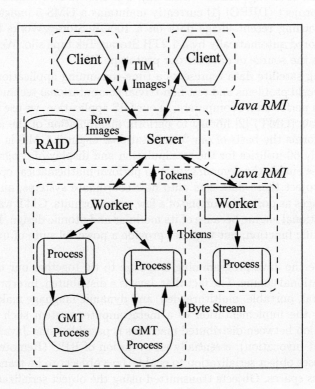

Fig. 4. The PAGIS system architecture, with dashed rectangles used to represent processing at potentially different nodes

The PAGIS metacomputer system represents an end-user-oriented approach to computation, rather than a program-development oriented architecture. In an end-user-oriented system, the user effectively programs applications to be executed on the metacomputer. By utilizing distributed process networks, issues of scheduling and parallelism are abstracted so the process of deciding how computations are to be performed is drastically simplified.

The end-user approach is facilitated by a service-based architecture. The domain expert is able to construct computations consisting of services to be applied to data. One service could be the conversion of satellite data from one format to another, another might be georectification. Services need to be sufficiently coarse-grained that the user is not swamped with a plethora of services, but fine enough that the system be effective.

The PAGIS system comprises three main levels: the Client, Server, and Worker. The correlation between these three levels is shown in Fig. 4, and discussed in detail in the following sub-sections.

4.3 The Server

The server is the central point of communication within the PAGIS system. It can be regarded as the front-end to our metacomputer system. It consists of a collection of Java classes used to coordinate various requests from any number of clients and workers.

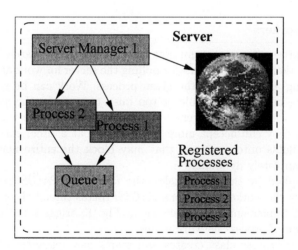

Fig. 5. The PAGIS Server

Each process network has a corresponding server manager object on the server. The server manager is a thread object that contains references to all the processes and edges that make up the process network, as well as references to the original and result images. This relationship is shown in Fig. 5. This architecture allows the server to be multithreaded such that it is capable of concurrently handling multiple process networks from any number of clients.

When a server manager thread is started, it parses a textual representation of the process network to a collection of *PAGISProcessID* objects representing nodes, and *Queue* objects representing edges. The processes are registered with the server as available for distribution to the workers, and the thread suspends itself.

We derived *get()* and *put()* methods to manipulate *Queue* objects from [3]. The server channels all requests to put and get data from the remote process object to the appropriate *Queue* object on the server. This requires the server to find the appropriate server manager the *PAGISProcessID* object belongs to, and the server manager to find the appropriate *Queue* object to modify. *Queue* objects are currently kept on the server but they could be kept on one of the communicating workers to reduce the impact of network latency. We are currently investigating a number of ways this could be done.

When a process signals its termination, the appropriate server manager thread is resumed. The server manager thread checks for process network completion and suspends itself if there are processes still active. If the process network is complete, the thread performs some clean-up operations, then notifies the server of its completion. Once all the processes have complete, the client is able to download the result image. The results of the process network are cached for later access and reuse.

4.4 The Workers

A worker represents a metacomputer node, sitting transparently behind the front-end. Workers are responsible for polling the server for work and, when available, accepting and executing the given process. Work can be rejected should the processing node be incapable or too busy to perform the work. This is a simple policy used for transparency and load balancing. The workers are multithreaded so they can accept more than one job at a time, ensuring that the server never holds onto processes that may block the entire network, keeping scheduling completely dynamic.

Work supplied by the server takes the form of a *PAGISProcessID* object that is used by the worker to create a *PAGISProcess* thread object. This object, whose thread implementation is shown in Fig. 6, triggers the execution and communication of process network nodes.

The *PAGISProcess* class creates and starts new instances of the *ProcessWriter* and *ProcessReader* threads. These objects are threads so the *ProcessWriter* may continue if the *ProcessReader* blocks on a read. The *ProcessWriter* thread, shown in Fig. 7, reads data from the server and writes that data to the process. The *ProcessReader* thread shares a similar structure except it tries to read results from the process, blocking if none are available, and writing this data to output queues on the server. This is a source of parallelism, where data is sent to the server while processing continues. The node continues consuming

```
public class PAGISProcess extends Thread {
  Server s; PAGISProcessID pid;
  public Service(Server s,PAGISProcessID pid) {
    this.s = s; this.pid = pid;
  }
  void run() {
    Process p = start_process();
    (new ProcessWriter(p.getOutputStream(),s,pid)).start();
    (new ProcessReader(p.getInputStream(),s,pid)).start();
    p.waitFor();
  }
}
```

Fig. 6. The PAGISProcess class

```
public class ProcessWriter extends Thread {
  OutputStream w; Server s; PAGISProcessID pid;
  ProcessWriter(OutputStream w, Server s, PAGISProcessID pid) {
    this.w = w; this.s = s; this.pid = pid;
  }
  void run() {
    Packet p;
    while ((p = s.server_consume(pid)) != null)
      process_write(w,p);
    w.close();
  }
}
```

Fig. 7. A thread for reading data from a server and then writing to a process

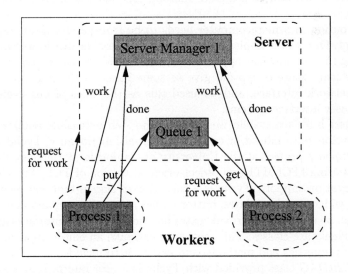

Fig. 8. Processes communicating with the server

and producing data until the process is complete. When the node has completed processing, the worker signals to the server that the process is done. This ensures the server knows the progress of the network.

Figure 8 presents an example of two workers communicating with the server through their server manager. The workers request *PAGISProcessID* objects from the server and obtain *Process1* or *Process2*. The worker then creates new *PAGISProcess* objects and their threads are started. The two processes communicate through a shared *Queue* on the server. The software currently does not take advantage of locality, so two communicating processes on the same node still communicate through the server.

Once the processes have completed, they notify the server manager of their completion, and the threads terminate. From this example, scheduling can be seen to be dynamic, dependent only on the availability of data, and pipelined parallelism is implicit, requiring no intervention by user or server to obtain.

4.5 The Client

The client is at the user end of the metacomputer system. The client consists of two pieces of software: the user interface and communication system. Our user interface design is influenced by a system proposed by Roger Davis, an interpreter capable of constructing arbitrarily complex process network topologies [10]. This interpreter is known as the Icon Processor or *ikp*. *ikp* allows the user to graphically build process networks. Users can also coordinate processes to run on remote hosts, allowing the user to distribute the processes across a number of machines. Each remote machine must be running an *ikp* daemon, in order to perform user authentication procedures similar to rlogin and rsh. The user must be aware of which remote hosts are running this daemon in order to distribute processes.

Distribution in a metacomputer needs to be transparent. *ikp* is deficient in this aspect. Our server implementation described earlier provides transparency by making the remote hosts apply for work and decide if work can be performed. The graphical nature of *ikp* did give us some ideas on the requirements for a process network interface, and we used this as the basis of our evaluation of different user interface systems.

The user interface system we use is Tycho - an extensible application that allows developers to inherit or compose essential functionality. Tycho was built for the Ptolemy project at the University of California at Berkeley [3]. It was developed using ITCL/ITK, an object-oriented version of TCL/TK with C and Java integration support. Tycho is a collection of classes that represent various graphical widgets such as slates, buttons, lines (or textboxes), and menus. These objects can be extended to provide extra functionality. User interfaces are classes just as widgets are classes. This means we can take an editor application class, inherit its properties and methods, and extend it to include extra functionality.

The *EditDAG* class provided with Tycho is a user interface that allows the user to construct directed acyclic graphs. These graphs are constructed by creating a start node, selecting that node, then creating a connecting node. A *DAG* class handles cycle checking on the fly so users can never build an illegal graph. The same class allows us to access and manipulate a Tycho Information Model (TIM) that textually represents the graph designed by the user. This TIM is used as input to our metacomputer.

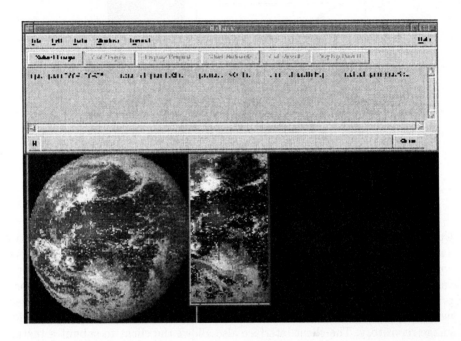

Fig. 9. The PAGIS graphical interface showing georectification of GMS-5 imagery

Our extension of the *EditDAG* interface involved adding a toolbar of standard commands and a menu of standard services. The user uses the menu to compose a network of services, and the toolbar to select and view original images, submit the TIM to the server, and select and view result images. The PAGIS user interface is demonstrated in Fig. 9.

As the server is written in Java, we need to utilize the Tycho-Java interface so that Tycho callbacks correspond to communications with the server. We have a *UserSide* class that performs method invocations on the server and some rudimentary graphical operations. (This class is also used by a simple Java AWT interface we developed that enables us to test the user-side software independent of the Tycho interface.) The *TclPAGIS* class implements the *TychoJava* interface. Calls to this class result in corresponding calls to the *UserSide* class. This separates any Tycho-related code from our communications code. Figure 10 describes the relationship between these entities.

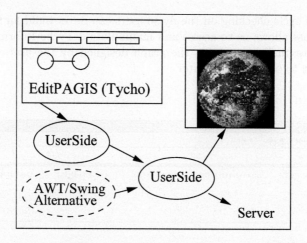

Fig. 10. PAGIS Client architecture

4.6 Communication Mechanisms

Communication is achieved with the use of Java RMI. The server implements two Java interfaces: one for client-server communication, another for worker-server communication. These interfaces act as an API for communication between the various entities. The interfaces provide access to relevant functions of the appropriate server manager. The client-server interface allows the client to create a new server manager instance and select an image to be retrieved from the DHPC image repository. The client interface also allows the client to submit a textual representation of the process network, start the process network, request the status of the network, and retrieve the result image. The worker-server interface provides workers with the ability to apply for and retrieve work, get and put data to the queues, and signal the completion of work.

4.7 Summary

The use of process networks has given us a simple solution to several of the metacomputing issues raised such as transparency, scheduling, and ease of use. Process networks offer implicit parallelism, dynamic scheduling, determinacy, and trivial work distribution. We are able to offer a high level of transparency to the user as the model implicitly manages issues such as parallelism, scheduling, determinacy, and distribution.

The PAGIS server software controls communication between any number of clients and metacomputer nodes. Each metacomputer node requests *PAGISProcessID* objects from the server and, if accepted, creates a *PAGISProcess* object that executes the desired process. The process consumes tokens from and produces tokens to *Queue* objects on the server. The *PAGISProcess* objects represent process network nodes and the *Queue* objects represent edges. The client enables domain experts to graphically compose computations, without needing to address issues such as parallelism, scheduling and distribution.

We are considering issues of load balancing. Process-level load balancing can be implemented by enabling workers to reject jobs based on their inability to perform them. Fine-grained load balancing is not appropriate within our coarse-grained computational engine, but a finer-grained model will be possible within an implementation of hierarchical networks as mentioned in Section 2.1. A simple implementation for hierarchy may simply pre-process the TIM and substitute (non-recursive) sub-networks with their TIM representation. An interesting issue is that of caching sub-network results. Tycho lacks support for building hierarchical networks so we are also currently examining other user interfaces, such as Ptolemy, that are capable of this. We hope that PtolemyII's version of Tycho will provide direct support[9]. We are also experimenting with Swing and 'Drag and Drop' to implement our own client as part of the DISCWorld infrastructure. Robustness and security have also not been implemented.

The nodes that constitute our process networks are selected from a configuration file. This list can be altered so any generic process can be executed; thus, the system is not limited to a GIS.

5 Discussion

A performance analysis of PAGIS is in progress. Our main performance concern currently is Java process IO. The Java implementations we have used for Sun Sparc and Dec Alpha appear to buffer output from processes by default. This means that calls to read data from a process block until the data is fully buffered, meanwhile the opportunity for latency hiding is lost. The run time of a process network has hence become $T+l*t$ where T is the run-time of a process network on a single machine, l is average message latency, and t is the number of tokens transmitted during execution of the process network. This is obviously suboptimal, and is our current focus of attention.

The use of RMI from a prototyping view-point was highly successful. RMI has enabled us to easily manipulate the server configuration and communication protocol. We hope to implement a version in the near future that transmits raw tokens over sockets to isolate the effect of using RMI. In the mean time, an obvious communications improvement, as highlighted in Section 4.3, will be removing our centralized queue system. Communications between two nodes currently requires two remote method invocations, one *put* and one *get* from the shared *Queue*. Remote method invocations can be halved just by moving the queue to the sending or receiving node.

It is worth noting that the architecture used is not the only possibility. PAGIS shows how a simple semantic model can be used to build metacomputing applications. The deterministic nature of our process network implementation means we can safely use other architectures preserving process network semantics to get the same results, and likely more efficiently. An example of this is the future use of PAGIS in the DISCWorld infrastructure. DISCWorld itself is a metacomputing system that executes services available from other nodes or resource administrators that have chosen to provide and advertise these services. DISCWorld provides a common integration environment for clients to access these services and developers to make them available [12]. Our hope is that PAGIS may be integrated with DISCWorld to utilise this integration environment as well as DISCWorld's distribution and scheduling mechanism.

We are also investigating the possible use of a Java package for concurrent and distributed programming developed by [11] called Java//. Java// abstracts over the Java thread model and RMI to provide an API for the implementation of concurrent and distributed systems. This system natually supports latency hiding because of its asynchronous will provide us with a transparent notion of active object, abstracting the distinction between local and remote nodes. Also, Java//'s asynchronous model naturally supports latency hiding.

Acknowledgements

This work was carried out under the Distributed High-Performance Computing Infrastructure (DHPC-I) project of the On Line Data Archives (OLDA) program of the Advanced Computational Systems (ACSys) Cooperative Research Centre (CRC), funded by the Research Data Networks CRC. ACSys and RDN are funded under the Australian Commonwealth Government's CRC Program.

References

1. Distributed High Performance Computing Project, University of Adelaide. See: http://www.dhpc.adelaide.edu.au
2. "Free Software Helps Map and Display Data", Paul Wessel and Walter H. F. Smith, EOS Trans. AGU, 72, p441, 445-446, 1991.
3. "The Tycho User Interface System", Christopher Hylands, Edward A. Lee, and H. John Reeckie, University of California at Berkeley, Berkeley CA 94720.
4. Japanese Meteorological Agency. See: http://www.jma.gov.jp
5. JavaSoft Java Studio. See: http://www.javasoft.com/studio
6. "The Semantics of a Simple Language for Parallel Programming", G. Kahn, Proceedings of IFIP Congress 74.
7. "RMI - Remote Method Invocation API Documentation", Sun Microsystems Inc., 1996.
8. "Implementation of Process Networks in Java", Richard S. Stevens, Marlene Wan, Peggy Laramie, Thomas M. Parks, and Edward A. Lee, 10 July 1997.
9. Ptolemy II. See: http://ptolemy.eecs.berkeley.edu/ptolemyII
10. "Iconic Interface Processing in a Scientific Environment", Roger Davis, Sun Expert, 1, p80-86, June 1990.
11. "A Java Framework for Seamless Sequential, Multi-threaded, and Distributed Programming", Denis Caromel and Julien Vayssière, INRIA Sophia Antipolis. See: http://www.inria.fr/sloop/javall
12. "DISCWorld: An Environment for Service-Based Metacomputing", Hawick, James, Silis, Grove, Kerry, Mathew, Coddington, Patten, Hercus, and Vaughan. Future Generation Computer Systems Journal, Special Issue on Metacomputing, to appear. See: http://www.dhpc.adelaide.edu.au/reports/042

Developing Parallel Applications Using the JAVAPORTS Environment

Demetris G. Galatopoullos   Elias S. Manolakos *

Electrical and Computer Engineering Department
Northeastern University, Boston, MA 02115, USA

Abstract. The JAVAPORTS system is an environment that facilitates the rapid development of modular, reusable, Java-based parallel and distributed applications for networked machines with heterogeneous properties. The main goals of the JAVAPORTS system are to provide developers with: (i) the capability to quickly generate reusable software components (code templates) for the concurrent tasks of an application; (ii) a Java interface allowing anonymous message passing among concurrent tasks, while keeping the details of the coordination code hidden; (iii) tools that make it easy to define, assemble and reconfigure concurrent applications on clusters using pre-existing and/or new software components. In this paper we provide an overview of the current state of the system placing more emphasis on the tools that support parallel applications development and deployment.

1 Introduction

The foremost purpose of the JavaPorts system [1] is to provide a user-friendly environment for targeting concurrent applications on clusters of workstations. In recent years, the *cluster* topology has brought to surface a powerful alternative to expensive large-scale multiprocessor machines for developing coarse-grain parallel applications. With the latest advances in interconnection networks, the spare CPU power of such clusters achieves performance comparable to those obtained by supercomputers for compute-intensive problems.

With the introduction of the Java programming language [2], the multi-platform integration of applications running on heterogeneous workstations became a reality. Our system takes advantage of the object-oriented nature of Java in order to provide an environment where programmers may develop, map and easily configure parallel and distributed applications without having extensive knowledge of concurrent or object-oriented programming. This goal is accomplished through the separation of the details of processors coordination from the user-defined computational part and the ability to create and update software components for a variety of cluster architectures.

The developer builds software components for concurrent Java applications using the JAVAPORTS Application Configuration Tool (Section 3). The tool offers

---

* e-mail: demetris{elias}@cdsp.neu.edu

a complete software cycle process with incremental phases. A *Task Graph* is first specified using a configuration language and/or a graphical user interface. *Tasks* in this graph are assigned to compute *Nodes* (machines) and may communicate via *Ports*. A Port in the JAVAPORTS system context, is a software abstraction that serves as a mailbox to a task which may write to (or read from) it in order to exchange messages with another peer task [1]. Ports ensure the safe and reliable point-to-point connectivity between interconnected tasks.

The JAVAPORTS system will automatically create Java program *templates* for each one of the defined tasks using the information extracted from the Task Graph. The part of the code necessary for creating and registering the ports is automatically inserted by the system. The JavaPorts Application Configuration Tool may be called multiple times each time the configuration file is modified by the user without affecting the user-added code. This scheme allows for the correct incremental development of reusable software components.

There are several new and on-going research projects around the world aiming at exploiting or extending the services of Java to provide frameworks that may facilitate parallel applications development. It would be impossible to account for all of them in the given space, so we just mention here the **Java//** [3, 4] and **Do!** [5] projects that employ *reification* and *reflection* to support transparent remote objects and high level synchronization mechanisms. The **JavaParty** [6] is a package that supports transparent object communication by employing a run time manager responsible for contacting local managers and distributing objects declared as `remote`. (JavaParty minimally extends Java in this respect). The **Javelin** [7] and **WedFlow** [8] projects try to expand the limits of clusters beyond local subnets, thus targeting large scale heterogeneous distributed computing. For example, **Javelin** is composed of three major parts: the broker, the client and the host. Processes on any node may assume the role of a client or a host. A client may request resources and a host is the process which may have and be willing to offer them to the client. Both parties register their intentions with a broker, who takes care of any negotiations and performs the assignment of tasks. Message passing is a major bottleneck in **Javelin** because the messages destined for remote nodes traverse the slow TCP or UDP stacks utilized by the Internet. However, for compute-intensive applications, such as parallel raytracing, **Javelin** has demonstrated good performance.

Notionally, the major difference between our design and the above projects is that the programmer visualizes a parallel application as an assembly of concurrent tasks with well-defined boundaries and interface mechanisms. This view is inspired by the way complex digital systems are designed with the aid of modern hardware description languages, such as **VHDL** [9]. The work distribution of a parallel application is described through a simple configuration language while allowing the user to have absolute control over "what-goes-where". After the application configuration file is completed and compiled, the JAVAPORTS system will automatically create reusable software task templates. Tasks will be able to exchange information with their peer tasks through ports using bi-directional asynchronous and anonymous point-to-point communication. The JAVAPORTS system is able to handle all possible types of data and therefore the message scheme is general and user-defined.

The rest of the paper is organized as follows. In the next section we briefly describe the main elements of the JAVAPORTS system that include the Task Graph and its allocation, the structure of the Java code templates generated by the system and the interface used for transparent inter-task communications. A more detailed description of the JAVAPORTS system is offered in [1]. Then in section 3 we outline the application design flow. In section 4 we present some preliminary experimental results using the system. Our findings are summarized in section 5 where we also point to work in progress.

2 The JAVAPORTS system

Each task in the JAVAPORTS system is an Ideal Worker which simply carries out a certain job without being concerned about how the input data arrives at its ports or where the output data goes to after it leaves its ports. Moreover, each task is unaware of the identity of its peer tasks (anonymous communications) [10, 11]. The synchronization among concerned ports is handled by the JAVAPORTS system classes and it is transparent to the programmer. However, the definition of the connectivity among computing entities is user-defined. Once this definition is complete, the JAVAPORTS system will return to the user a Java code template for each defined task. These templates will be used by the programmer in order to complete the implementation of the specific application.

2.1 Task Graph

The programmer may visualize the mapping of the work partitions onto the nodes of the cluster by using the Task Graph. The topology of this graph is captured into a configuration file created and edited by the programmer. The JAVAPORTS system will make use of the parameters set in the configuration file in order to generate or update the task templates.

A sample Task Graph is shown in Figure 1, where boxes represent tasks and dotted boxes machines (cluster nodes) respectively. This Task Graph depicts a star configuration with machine M1 assuming the role of the central node. There are four user-defined tasks, one allocated on machine M1 (task T1), two on machine M2 (tasks T2 and T3) and one on M3 (task T4). The task which is local to M1, utilizes three ports in order to communicate and the same is true for tasks T2, T3 and T4 which are remote to T1.

We have defined a set of commands available to the user which are sufficient for a complete description of the mapping of the application onto the cluster:

- define machine machine_var = ``machine_dns_name'' machine_type_flag
With this command the user will be able to define a machine in the cluster. The machine_var is a user-defined label. The machine_dns_name is the machine IP name as identified by the Internet domain or the cluster's local domain server.

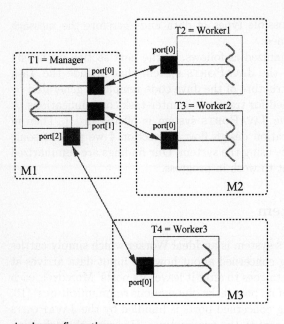

```
begin configuration
  begin definitions
    define    machine    M1= "corea.cdsp.neu.edu" master
    define    machine    M2= "walker.cdsp.neu.edu"
    define    machine    M3= "hawk.cdsp.neu.edu"
    define    task       T1= "Manager"    numofports=3
    define    task       T2= "Worker1"    numofports=1
    define    task       T3= "Worker2"    numofports=1
    define    task       T4= "Worker3"    numofports=1
  end definitions
  begin allocations
    allocate  T1   M1
    allocate  T2   M2
    allocate  T3   M2
    allocate  T4   M3
  end allocations
  begin connections
    connect   M1.T1.P[0]    M2.T2.P[0]
    connect   M1.T1.P[1]    M2.T3.P[0]
    connect   M1.T1.P[2]    M3.T4.P[0]
  end connections
end configuration
```

Fig. 1. Task Graph and the corresponding configuration file.

The machine_type_flag is used to specify which node will act as the application "master" i.e. used to launch the distributed application on the JAVAPORTS system.

- define task task_var = ''task_name'' numofports = number_of_ports

The user may define a task with this command. The task_var is a user defined label for each task. The task_name is a mnemonic string that will be used to name the corresponding Java template file generated by the JAVAPORTS system. The numofports parameter is an integer denoting the number of defined ports for the task.

- allocate task_variable machine_variable

Provided that a certain task and machine have already been defined, this command will assign the task to the specified machine for execution.

- connect machine_var.task_var.P[port_index]
  machine_var.task_var.P[port_index]

The connect command is used to specify port-to-port connections between two tasks. All port indices are zero-based. For example, if a task has three (3) ports defined, it can read/write to local ports with names P[0], P[1], P[2] .

- begin/end section_name

These commands are required in order to create readable and easy to debug configuration files. The boundaries of each section and of the whole configuration file should be identified by a begin and an end statement.

In Figure 1 we provide a task graph and its corresponding description. The user may alter the allocation of tasks to nodes, without having to re-write any code he or she added in an existing template, and have the parallel algorithm executed on a different cluster setup. The JAVAPORTS system will perform all the code template updates needed to run the concurrent tasks in the new cluster setup, even if all tasks are assigned to the same node (machine). We are currently developing a graphical tool for defining and modifying task graphs that will simplify even further the application development process.

2.2 User Program Templates

For illustration purposes we depict the templates generated for tasks T1 (Manager) and T3 (Worker2) in Figures 2 and 3 respectively. A more detailed discussion is provided in [1]. The automatically generated JAVAPORTS system code is denoted with bold letters. The rest of the code is user-added and remains un-altered after possible application configuration modifications. In this part, the Manager task first forms a data message that is sent to Worker2 and then continues with some local computation. The worker task waits until it receives the data message and then produces some results that are sent back to the Manager.

After the compilation of the configuration file, the JAVAPORTS system provides the user with pre-defined templates in order to complete the specific application implementation. Specialized scripts will automatically dispatch the user executable files to their correct destination nodes.

public class Manager

```
  // User-added variables
  Object ResultHandle;

  run()
  {
     Init InitObject = new Init();
     InitObject.configure(port);

     // User-added algorithm implementation starts here

     // Form the data message and send it out
     Message msg = new Message(UserData);
     port[1].AsyncWrite(msg,key2);
                    ⋮
     // Code for any local processing may go here
                    ⋮
     // Get references to expected results
     ResultHandle = port[1].AsyncRead(key2);
                    ⋮
     //Code to retrieve results may go here
  }
  public static void main()
  {
     Runnable ManagerThread = new Manager();
     Thread manager = new Thread(ManagerThread);
     manager.start();
  }
```

Fig. 2. Template for the Manager task T1.

Port objects communicate among them using the Remote Method Invocation (RMI) [12] package. Each port is registered locally and its counterpartner is looked up remotely. Such operations have proven to be time consuming and therefore they are only performed during the initialization phase of each task.

The JAVAPORTS system does not constrain the user from exploiting additional concurrency in an attempt to improve the performance of an application. In fact, concurrency is encouraged since multiple tasks may be executed on the

```
public class Worker2

    // User-added variables
    Object ListHandle, DataHandle;

    run()
    {
        Init InitObject = new Init()
        InitObject.configure(port)

        // Get a reference to the object where the incoming data message is stored
        ListHandle = port[0].AsyncRead(key2);

        // Use it to wait, if needed, until the data has arrived
        DataHandle = ListHandle.PortDataReady();
                          ⋮
        // The algorithm that uses the incoming data and produces results goes here
                          ⋮
        // Prepare a message with the results
        Message msg = new Message(ResultData);
                          ⋮
        // Write back the results
        port[0].AsyncWrite(msg,key2)
    }
public static void main()
{
    Runnable Worker2Thread = new Worker2();
    Thread worker2 = new Thread(Worker2Thread);
    worker2.start();
}
```

Fig. 3. Template for the Worker2 task T3.

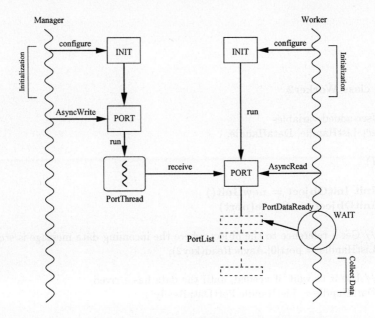

Fig. 4. A message communication using JAVAPORTS.

same node by treating the local destination ports in a similar fashion as the remote ports. If the appropriate hardware resources become available concurrent tasks may be executed in parallel, otherwise they will be time-sharing the same CPU. Such capability eliminates the considerable development costs for re-engineering the application code each time the allocation of software components to machines is altered for some reason.

2.3 The PORTS interface

Using the interface concept of the Java language in order to deal with multiple inheritance, we have introduced an allowable set of operations on a port [1]. These operations may be properly applied on a port object in order to realize the messages exchange with remote ports. We briefly outline these operations.

- AsyncWrite(Msg, MsgKey): It will write the MSG message object to a port along with a key for personalizing the message.
- AsyncRead(MsgKey): It signifies the willingness of the calling task to access a certain message object expected at this port. It returns a reference to a port list object where the expected message will be deposited upon arrival.
- PortDataReady(): It is an accessor method to the objects of the port list. When a message has arrived at a port this method returns a reference to the object where the incoming data resides; otherwise it waits until the message arrives.

The communication protocol between two JAVAPORTS task threads residing on different address spaces is outlined in moderate detail below. We only describe the communication operation in one direction, from the Manager (sending thread) to the Worker (receiving thread), since the operation in the reverse direction is symmetrically identical. A snapshot of such operation is illustrated in Figure 4.

The Manager task thread issues an `AsyncWrite` to its associated port with a user-defined key. This key will identify the data on the remote side. The `AsyncWrite` method spawns a secondary thread, the `PortThread`, and returns. This thread will be responsible for transporting the data to the remote port asynchronously and without blocking the Manager task thread. The `PortThread` calls the `receive` method of the remote port object. This method makes use of the RMI serialization mechanism in order to transfer the message to the remote node. The incoming data will be inserted in the appropriate object element of the remote port list.

On the receiving node, the Worker task thread issues an `AsyncRead` to its associated port using the corresponding key of the expected incoming data. The method will return a reference to the specific object in the port list where the incoming message will be deposited upon arrival. At the point in time when the Worker thread needs this data, it calls the method `PortDataReady` on the specified list object and then enters the wait state. The `receive` method, which the Manager `PortThread` calls during the transportation of data, inserts the incoming data to the corresponding list object and then notifies all the waiting threads in the address space of the receiving node. Once the Worker thread is notified, it exits the wait state and checks whether the incoming data has the correct key. If not it re-enters the wait state, otherwise it retrieves the data.

3 The JAVAPORTS Application Development Environment

3.1 Application Design Flow

In this section we describe a procedure that facilitates incremental development of Java-based parallel applications in the JAVAPORTS environment. Each phase of the procedure corresponds to a step in the application development cycle. Upon completion of each step the user may checkpoint the correctness of the current portion of the application before advancing to the subsequent step.

The procedure makes use of the JAVAPORTS Application Configuration Tool (JACT). JACT translates the user's task graph specifications (captured in a configuration file) into Java code template files for the specified tasks. At a later stage in the design, these files will be completed by adding the needed user code that implements the specific targeted algorithm. All phases of the application design flow are depicted in Figure 5 and described below:

- PHASE 1: DEFINE A TASK GRAPH.

This is the initial stage of the development cycle. The user may use any text

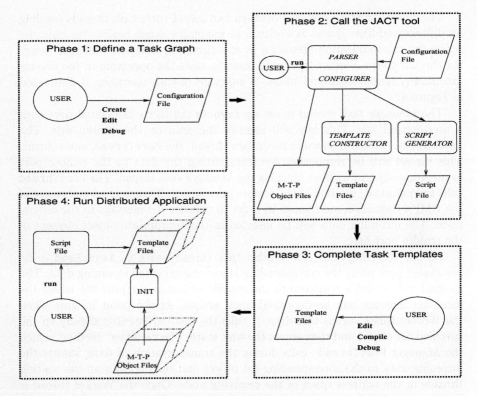

Fig. 5. Application Design Flow.

editor in order to create and save the configuration file. The name of the file is user-defined and should be provided as an input to JACT in Phase 2.

• PHASE 2: CALL THE JAVAPORTS APPLICATION CONFIGURATION TOOL. During this phase, the user will make calls to the JACT, in order to compile the configuration file and subsequently create the necessary templates for each defined task. The JACT is primarily composed of a Configurer, a Parser, a Template Constructor and a Script Generator.

The Configurer is the central module of the JACT and controls the operation of the rest of the modules in the tool. First the Configurer calls the Parser which compiles the configuration file and alerts the user for syntax or configuration errors. The syntax errors are due to incorrect usage of the configuration language commands. JACT will return informative messages about such errors to guide the user towards fixing them. Configuration errors occur when the user violates certain rules, for example when there are missing machine or task definitions, or the M-T-P property of the system does not hold and thus an incorrect scenario is specified (see for more details section 3.2). Once the user configuration file has been successfully parsed, the Configurer will automatically store the information needed for each machine into a separate object file. These files will be later used by the JAVAPORTS system in order to create and register the necessary ports per task.

At this point, the Configurer will call the Template Constructor module which is responsible for generating or updating the templates files. Each template file corresponds to a task with the same name, specified by the user in the configuration file.

As a last step, the Configurer will call the Script Generator to create a startup script that can be used to launch the distributed application from a single (master) node. As an example, the simple script generated automatically for the star topology described by Figure 1 is given below:

```
setenv CLASSPATH $HOME/public-html/codebase
rsh walker setenv CLASSPATH $HOME/public-html/codebase
rsh hawk setenv CLASSPATH $HOME/public-html/codebase
rsh walker.cdsp.neu.edu java JACT.Worker1 walker
rsh walker.cdsp.neu.edu java JACT.Worker2 walker
rsh hawk.cdsp.neu.edu java JACT.Worker3 hawk
java JACT.Manager corea
```

- PHASE 3: COMPLETE THE TASK TEMPLATE PROGRAMS.

During this stage the user will complete the template programs by adding the necessary Java code into each template file in order to implement the desired algorithm. These templates will be compiled using the Java standard compilers. The Java Virtual Machine running on the current workstation will inform the user about any syntax errors. The user will not be aware of potential runtime errors in the algorithm's implementation until the next stage of the design.

- PHASE 4: RUNNING THE DISTRIBUTED APPLICATION.

This is the final stage of the design flow. The user will reach this stage once the completed template files have been successfully compiled. The user may start the JAVAPORTS system and run the current application by calling the startup script which was created in Phase 2. In a UNIX environment the script will be already pre-compiled using the `source` system command. The user may simply type the name of the script at the shell prompt in order to run it. The script file will initiate each task on its corresponding machine. Subsequently each task will instantiate an object of class `Init` that will read the necessary configuration information from the corresponding M-T-P Object file and will update the ports configuration and their connections. At this stage the user is likely to be faced with runtime errors due to network communication failures among machines, null objects, deadlocks etc. The JAVAPORTS system is designed to utilize the rich set of exceptions that the Java programming language supports and inform the user of such problems when they are detected.

3.2 The M-T-P Tree files

The Configurer of JACT stores the information acquired from the configuration file into Tree Structures Object files. This information will be used to

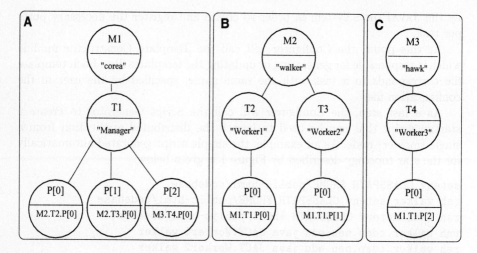

Fig. 6. M-T-P Object Files.

achieve port configuration in a manner transparent to the user. The Machine-Task-Port triplets are organized in tree data structures. The root of a tree is always a machine; the first level nodes are tasks and the leaf nodes are ports. (Hence the name M-T-P tree).

Each node at any level of the tree carries certain properties. A machine root node includes the `machine_variable` assigned to a specific machine by the user at configuration time as well as the `machine_dns_name`. A task node stores the `task_var` and `task_name` assigned to it and a port node stores the connection data comprised of the M-T-P triplet of the matching port. Figure 6 displays the three Object files which the JACT will create for the Task Graph depicted in Figure 1. The Object File A corresponds to machine M1, Object File B to machine M2 and C to machine M3 respectively. The task nodes represent the tasks defined in the corresponding configuration file and the leaves the ports which connect tasks. The connection data for each port as well as the information data for the machines and tasks are shown inside each node (circle).

JACT performs all necessary checking during parsing so that the data stored onto each tree represents legal and correct configurations. The M-T-P policy requires that each component of the tree be added through the configuration file. Furthermore, the root of a tree should always be a machine node, the tasks always reside at the first level nodes and all ports will always be at leaf nodes. If no tasks are defined for a machine, a tree file will be created for it including only the root node carrying the machine information. Every time a configuration file is altered JACT should be called to parse it and update the tree object files, as needed.

The JACT tool makes use of the object serialization and de-serialization procedures provided by the Java language. During the serialization process, all data types and graphs of the M-T-P objects are flattened and written to an output stream which in turn is stored persistently into a file. During the de-serialization process, the INIT object of each task process will open the corresponding file, read the object from the input stream and reconstitute it to its original form. The INIT object will extract all the necessary information from the tree which belongs to the current task in order to create, register with RMI and connect the corresponding ports.

4 Experimental Results

In this section we outline some experiments conducted using the JAVAPORTS system and present only few indicative performance results. A more detailed evaluation is outside the scope of this paper and will appear elsewhere [13]. As a first illustrative example we have chosen the matrix-vector multiplication problem applied to a variety of two-node configurations, since the same scenario was also chosen by other researchers (see [3] and [14]). The multiplicand is assumed to be a 1000-by-1000 square matrix and the multiplier an 1000-by-1 column vector. The multiplication is conducted by two concurrent tasks (a manager and a worker) allocated on two nodes (Sparc-4 workstations by Sun Microsystems running Solaris version 2.5.1 and belonging to a local 10Mb/s Ethernet subnet). The part of the multiplicand matrix needed by the worker task resided on the remote node before the commencement of the experiment. The communication overhead, i.e. the time it takes for the multiplier vector to be sent from the manager to the remote worker task as well as the time it takes for the partial product vector to be returned back from the worker to the manager has been measured and taken into consideration in the timing analysis.

In the two-node scenario, the number of rows of the multiplicand matrix allocated on each one of the two nodes was varied from 0 to 1000. The manager task starts first, sends the multiplier vector to the remote worker, then performs its local computation and finally collects back the partial result returned by the worker and assembles the overall product vector. In the left panel of Figure 7, the solid line curve shows the total time needed by the manager task to complete as a function of the number of rows allocated to the remote worker task. The dashed line curves show the times needed for the computation of the local partial product in the manager and the time the manager should wait subsequently for the remote partial product vector to arrive from the worker task. In essence, the decomposition of the total manager time to its major sequential components verified that as the work allocated to the remote worker increases the manager local computation time decreases and the waiting time (for worker results) increases. From the analysis of the results we observe that the minimum overall time is achieved close to the point where the rows of the matrix are split evenly among the two machines. This minimum time corresponds to 62% of the time needed to solve the problem (1000 rows) using a single thread Java program in one machine. Therefore, the achieved speedup by using two concurrent tasks in two nodes is about 1.6, a very good result considering that the network is a standard 10Mb/s shared Ethernet and the problem (message) size not too large.

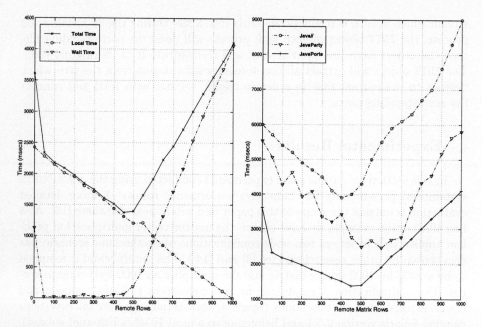

Fig. 7. *Left panel:* Time of the manager vs. rows allocated in the remote worker in a two-node scenario; *Right panel:* Comparison of three approaches using the same matrix-vector multiplication example.

For the two-node scenario, the JAVAPORTS performance was also compared to that of two other major designs, the Java// [3] and the JavaParty [6]. Similar to our system, the JavaParty application was ran on two Sparc-4 nodes whereas the Java// experiment made use of two UltraSparc stations according to [3]. To be able to obtain the JavaParty times, a single thread Java program was written. Following the guidelines of the JavaParty group, the worker task was declared as remote, thus allowing the system to migrate the worker task object to the remote node. As it can be seen from the plot in the right panel of Figure 7, the minimum time was achieved at approximately the same number of remote rows for all three designs. Among them, the JAVAPORTS design exhibited the best performance across the full range of rows allocated to the remote worker node.

5 Conclusions

We have provided an overview of the JAVAPORTS project, an environment for flexible and modular concurrent programming on cluster architectures. The design encourages reusability by allowing the developer to build parallel applications in a natural and modular manner. The components generated automatically by the system may be manipulated and (without modifying any part of the user code) they can be utilized to run the same concurrent application in a new cluster setup. The user-defined message object may be altered to customize the needs of transferring any type of data across the network. By giving the developer such capabilities, the JAVAPORTS environment may be customized for specific client-server applications where a large variety of services may be available. We managed to keep the task of tracking remote objects and marshalling of data between different address spaces transparent to the developer. The user is not aware of how the data is communicated among nodes in the cluster.

Currently there is a considerable amount of on-going work. The completion of the graphical tool for allowing the developer to enter and modify a task graph in the JAVAPORTS system is our top priority. This tool will provide the user with a visual sense of how the application is configured to run on the cluster. This is an important aid in parallel computing where partitioning data and functionality among nodes is much easier to understand when it is depicted visually. In addition we are improving the configuration language which provides a text based approach to the allocation of tasks to nodes. Experienced users may find this entry point more convenient for the development and customization of their application.

We are also targeting large scale applications using a variety of commonly employed task graph patterns (star, pipeline, mesh, cube). In addition to compute-intensive computations we are considering other application domains, such as distributed network management, distributed simulation and networked embedded systems that may take advantage of the easy configuration capabilities of the JAVAPORTS system. Some of these applications are selected to allow us to fully test and optimize specific aspects of the JAVAPORTS design before it is released for general use.

References

1. D. Galatopoullos and E. S. Manolakos. Parallel Processing in heterogeneous cluster architectures using Javaports. *ACM Crossroads*, (1999, to appear).
2. K. Arnold and J. Gosling. *The Java Programming Language*, (1996) Addison Wesley Longman, Inc.
3. D. Caromel, W. Klauser, and J. Vayssiere. Towards Seamless Computing and Metacomputing in Java. *Concurrency Practice and Experience*, **10** (1998) (11–13):1043–1061.
4. Proactive PDC - Java//. http://www.inria.fr/sloop/javall/index.html, (1998).
5. P. Launay and J. Pazat. Generation of distributed parallel Java programs. Technical Report 1171, IRISA, France, (1998).
6. M. Philippsen and M. Zenger. Javaparty - Transparent Remote Objects in Java. *Concurrency: Practice and Experience*, **9** (1997) (11):1225–1242.
7. B.O. Christiansen et al. Javelin:Internet Based Parallel Computing Using Java. *Concurrency: Practice and Experience*, **9** (1997) (11):1139–1160.
8. Dimple Bhatia et al. WebFlow–A Visual Programming Paradigm for Web/Java Based Coarse Grain Distributed Computing. *Concurrency: Practice and Experience*, **9** (1997) (6):555–577.
9. Z. Navabi. *VHDL: Analysis and Modeling of Digital Systems*. McGraw Hill, (1998).
10. F. Arbab. The IWIM model for coordination of concurrent activities. *Coordination '96, Lecture Notes on Computer Science*, **1061**, (1996).

11. F. Arbab, C.L. Blom, F.J.Burger, and C.T.H.Everaas. Reusability of Coordination Programs. Technical Report CS-TR9621, Centrum voor Wiskunde en Informatica, The Netherlands, (1996).
12. Remote Method Invocation Specification. Sun Microsystems, Inc., (1996).
13. D. Galatopoullos and E. S. Manolakos. JavaPorts: An environment to facilitate parallel computing on a heterogeneous cluster of workstations. *Informatica*, (1998, submitted).
14. R.R. Raje, J.I. William, and M. Boyles. An Asynchronous Remote Method Invocation (ARMI) mechanism for Java. *Concurrency: Practice and Experience*, **9** (1997) (11): 1207-1211.

3rd Workshop on Optics and Computer Science
Message from the Program Chairs

Advances in semiconductor technologies coupled with progress in parallel processing and multi-computing are placing stringent requirements on inter-system and intra-system communications. Demands for high density, high bandwidth, and low power interconnections are already present in a wide variety of computing and switching applications including, for example, multiprocessing and parallel computing (simulations of real problems, monitoring of parallel programming, etc), and enhanced digital telecommunications services (broadcast TV, video on demand, video conferencing, wireless communication, etc.). Furthermore, with advances in silicon and Ga-As technologies, processor speed will soon reach the gigahertz (GHz) range. Thus, the communication technology is becoming and will remain a potential bottleneck in many systems. This dictates that significant progress needs to be made in the traditional metal-based interconnects, and/or that new interconnect technologies such as optics are introduced in these systems.

Optical means are now widely used in telecommunication networks and the evolution of optical and optoelectronic technologies tends to show that they could be successfully introduced in shorter distance interconnection systems such as parallel computers. These technologies offer a wide range of techniques that can be used in interconnection systems. However, introducing optics in interconnect systems also means that specific problems have yet to be solved while some unique features of the technology must be taken into account in order to design optimal systems. Such problems and features include device characteristics, network topologies, packaging issues, compatibility with silicon processors, system level modeling, and so on.

The purpose of this workshop is to provide a good opportunity for the optics, architecture, communication, and algorithms research communities to get together for a fruitful cross-fertilization and exchange of ideas. The goal is to bring the optical interconnects research into the mainstream research in parallel processing, while at the same time to provide the parallel processing community with a more comprehensive understanding of the advantages and limitations of optics as applied to high-speed communications. In addition, we intend to assemble a group of major research contributors to the field of optical interconnects for assessing its current status and identifying future directions.

Each manuscript was reviewed by two or three program committee members. The quality of the accepted papers already speaks about the intelligent and diligent work of the authors, the reviewers, and the program committee members. We hope that WOCS will be successful again in 2000.

January 1999

Yi Pan, Program Chair
Keqin Li, Vice Program Chair

WOCS'99 Workshop Organization

Program Committee

Yi Pan (Chair)	University of Dayton, USA
Keqin Li (Vice Chair)	State University of New York, USA
Selim G. Akl	Queen's University, Canada
Mohammed Atiquzzaman	University of Dayton, USA
Pierre Chavel	Institut d'Optique Theorique et Appliquee, France
Hyeong-Ah Choi	George Washington University, USA
Patrick Dowd	University of Maryland, USA
Hossam ElGindy	The University of Newcastle, Australia
Joseph W. Goodman	Stanford University, USA
Mounir Hamdi	Hong Kong Univ. of Science & Technology, HK
Ahmed Louri	University of Arizona, USA
Philippe J. Marchand	University of California at San Diego, USA
Rami Melhem	University of Pittsburgh, USA
Timothy Pinkston	University of Southern California, USA
Chunming Qiao	State University of New York at Buffalo, USA
Sanguthevar Rajasekaran	University of Florida, USA
Sartaj Sahni	University of Florida, USA
Hong Shen	Griffith University, Australia
Ted Szymanski	McGill University, Canada
Hugo Thienpont	Vrije Universiteit Brussel, Belgium
Jerry L. Trahan	Louisiana State University, USA
Ramachandran Vaidyanathan	Louisiana State University, USA
Yuanyuan Yang	University of Vermont, USA
Si Qing Zheng	University of Texas at Dallas, USA

Steering Committee

Afonso Ferreira (Chair)	Laboratoire d'Informatique du Parallelisme, ENS Lyon, France
Pierre Chavel	Institut d'Optique Theorique et Appliquee, France
Timothy Drabik	Georgia Tech Lorraine, Metz, France, and Georgia Tech., Atlanta, USA
Sadik C. Esener	University of California, San Diego, USA
Paul Spirakis	Computer Technology Institute, Patras, Greece

Permutation Routing in All-Optical Product Networks

Weifa Liang[1]   Xiaojun Shen[2]

[1] Department of Computer Science
Australian National University
Canberra, ACT 0200, Australia
[2] Computer Science Telecommunications Program
University of Missouri - Kansas City
Kansas City, MO 64110, USA

Abstract. In this paper we study permutation routing techniques for all-optical networks. Firstly, we show some lower bounds on the number of wavelengths needed for implementing any permutation on an all-optical network in terms of bisection of the network. Secondly, we study permutation routing on product networks by giving a lower bound on the number of wavelengths needed, and presenting permutation routing algorithms for the wavelength non-conversion and conversion models, respectively. Finally, we investigate permutation routing on a cube-connected-cycles network by showing that the number of wavelengths needed for implementing any permutation in one round is $\lfloor 2 \log n \rfloor$, which improves on a previously known general result for bounded degree graphs by a factor of $O(\log^3 n)$ for this special case.

1 Introduction

In this paper we address designing routing algorithms for an emerging new generation of networks known as *all-optical network* [1,10,16,24]. This kind of network offers the possibility of interconnecting hundreds to thousands of users, covering local to wide area, and providing capacities exceeding substantially those a conventional network can provide. The network promises data transmission rates several orders of magnitudes higher than current electronic networks. The key to high speed in the network is to maintain the signal in optical form rather than electronic form. The high bandwidth of the fiber-optic links is utilized through *Wavelength-Division Multiplexing* (WDM) technology which supports the propagation of multiple laser beams through a single fiber-optic link provided that each laser beam uses a distinct optical wavelength. Each laser beam, viewed as a carrier of signal, provides transmission rates of 2.5 to 10 Gbps, thus the significance of WDM in very high speed networking is paramount. The major applications of the network are in video conferencing, scientific visualization and real-time medical imaging, supercomputing and distributed computing [24,25]. A comprehensive overview of the physical theory and applications of this technology can be found in the books by Green [10] and McAulay [16]. Some studies [1,18,23] model the all-optical network as an *undirected* graph, and thus the routing paths for implementing requests on the network are also undirected. However, it has since become apparent that optical amplifiers placed on fiber-optic links are directed devices. Following [5,9,17], we will model the all-optical network as a *directed symmetric* graph, and each routing path on it is a directed path.

The Model. An all-optical network consists of vertices (nodes, stations, processors, etc), interconnected by point-to-point fiber-optic links. Each fiber-optic link supports a given number of wavelengths. The vertex may be occupied either by terminals, switches, or both. *Terminals* send and receive signals. *Switches* direct the input signals to one or more of the output links. Several types of optical switches exist. The *elementary switch* is capable of directing the signals coming along each of its input links to one or more of the output ones. The elementary switch cannot, however, differentiate between different wavelengths coming along the same link. Rather, the entire signal is directed to the same output(s) [6,1,23]. The *generalized switch*, on the other hand, is capable of switching incoming signals based on their wavelengths [1,23]. Using acousto-optic filters, the switch splits the incoming signals to different streams associated with various wavelengths, and may direct them to separate outputs. In both cases, on any given link different messages must use different wavelengths. Unless otherwise specified, we adopt generalized switches for our routing algorithms. Besides there being a difference in the use of switches, there is a difference regarding the wavelength assignment for routing paths. In most cases, we only allow the assignment of a unique wavelength to each routing path. We call this WDM model the *Wavelength Non-Conversion Model*. If we allow each routing path to be assigned different wavelengths for its different segments, we call it a *Wavelength Conversion Model* (called the λ-routing model in [7]).

Previous Related Work. Optical routing in an arbitrary undirected network G was considered by Raghavan and Upfal [23]. They proved an $\Omega(1/\beta^2)$ lower bound on the number of wavelengths needed to implement any permutation in one round, where β is the edge expansion of G (which is defined later). They also presented an algorithm which implements any permutation on bounded degree graphs using $O(\log^2 n/\log^2 \lambda)$ wavelengths within one round with high probability, where n is the number of nodes in G and λ is the second largest eigenvalue (in absolute value) of the transition matrix of the standard random walk on G. For degree d arrays, they presented an algorithm with an $O(dn^{1/d}/\log n)$ worst case performance. Aumann and Rabani [5] presented a near optimal implementation algorithm for bounded degree networks in one round. Their algorithm needs $O(\log^2 n/\beta^2)$ wavelengths. For any bounded dimension array, any given number of wavelengths, and an instance I, Aumann and Rabani [5] suggested an algorithm which realizes I using at most $O(\log n \log |I| T_{opt}(I))$ rounds, where $T_{opt}(I)$ is the minimum number of rounds necessary to implement I. Later, Kleinberg and Tardos [13] obtained an improved bound of $O(\log n)$ on the approximation of the number of rounds required to implement I. Recently, Rabani [22] further improved the result in [13] to $O(poly \ \log \log n T_{opt}(I))$. Pankaj [18] considered the permutation routing issue in hypercube based networks. For hypercube, shuffle-exchange, and De Bruijn networks, he showed that routing can be achieved with $O(\log^2 n)$ wavelengths. Aggrawal et al [1] showed that $O(\log n)$ wavelengths are sufficient for the routing in this case. Aumannn and Rabani [5] demonstrated that routing on hypercubes can be finished with a constant number of wavelengths. Gu and Tamaki [11] further showed that two wavelengths for directed symmetric hypercubes and eight wavelengths for undirected hypercubes are sufficient. Pankaj [18,19] proved an $\Omega(\log n)$ lower bound on the number of wavelengths needed for routing a permutation in one round on a bounded degree network. Barry and Humblet [6,7] gave bounds for routing in passive (switchless) and λ-networks. An almost matching upper bound is presented later in [1]. Peiris and Sasaki [20] considered bounds for elementary switches. The connection between packet routing and optical routing is discussed in [1]. The integral multicommodity flow problem related to optical routing has been discussed in [5,3,4]. A comprehensive survey for optical routing is in [9].

Our Results. In this paper we first present some lower bounds on the number of wavelengths needed for implementing any permutation on all-optical networks in terms of the bisections of the networks. We then consider the product networks, which can be decomposed into a *direct product* of two networks. Many well known networks such as hypercubes, meshes, tori, etc, are product networks. We first show a lower bound on the number of wavelengths needed for implementing any permutation on product networks in one round, then present permutation routing algorithms for product networks, based on the wavelength non-conversion and conversion models respectively. Finally we study the permutation issue on cube-connected-cycles networks by implementing any permutation in one round using only $\lfloor 2 \log n \rfloor$ wavelengths. This improves on a general result for bounded degree networks in [5] by a factor of $O(\log^3 n)$ for this special case.

2 Preliminaries

We define some basic concepts involved in this paper.

Definition 1. An all-optical network can be modeled as a *directed symmetric* graph $G(V, E)$ with $|V| = n$ and $|E| = m$, where for each pair of vertices u and v, if a directed edge $\langle u, v \rangle \in E$, then $\langle v, u \rangle \in E$ too.

Definition 2. A *request* is an ordered pair of vertices (u, v) in G which corresponds to a message to be sent from u to v. u is the *source* of v and v is the *destination* of u.

Definition 3. An *instance* I is a collection of requests. If $I = \{(i, \pi(i)) \mid i \in V\}$, where π is a permutation of vertices in G.

Definition 4. Let $P(x, y)$ denote a directed path in G from x to y. A *routing* for an instance I is a set of directed paths $\mathcal{R} = \{P(x, y) \mid (x, y) \in I\}$. An instance I is *implemented* by assigning wavelengths to the routing paths and setting the switches accordingly.

Definition 5. The *conflict graph*, associated with a permutation routing

$$\mathcal{R} = \{P(i, \pi(i)) \mid i \in V, \ i \text{ is a source and } \pi(i) \text{ is the destination of } i\}$$

on a directed or undirected graph $G(V, E)$, is an undirected graph $G_{\mathcal{R},\pi} = (V', E')$, where each directed (undirected) routing path in $\mathcal{R}$ is a vertex of V' and there is an edge in E' if the two corresponding paths in $\mathcal{R}$ share at least a common directed (or undirected) edge of G.

Definition 6. The *edge-expansion* $\beta(G)$ of $G(V, E)$ is the minimum, over all subsets S of vertices, $|S| \leq n/2$, of the ratio of the number of edges leaving S to the size of S ($\subset V$).

Definition 7. A *bisection* of a graph $G(V, E)$ is defined as follows: Let P be a partition which partitions V into two disjoint subsets V_1 and V_2 such that $|V_1| = \lceil |V|/2 \rceil$ and $|V_2| = \lfloor |V|/2 \rfloor$. The *bisection problem* is to find such a partition (V_1, V_2) that $|C|$ is minimized, where $C = \{(i, j) \mid i \in V_2, \ j \in V_1, \text{ and } (i, j) \in E\}$. Let $c(G) = |C|$. The bisection concept for undirected graphs can be extended to directed graphs. For the directed version, define $C = \{\langle i, j \rangle \mid i \in V_2, \ j \in V_1, \text{ and } \langle i, j \rangle \in E\}$.

Definition 8. The *congestion* of a permutation π on $G(V, E)$ is defined as follows. Let $\mathcal{R}$ be a set of routing paths for π on G. Define $C(e, \mathcal{R}) = \{P(i, \pi(i)) \mid i \in V, \ e \in P(i, \pi(i)), \text{ and } P(i, \pi(i)) \in \mathcal{R}\}$. Then, the congestion problem for π on G is to find a routing $\mathcal{R}$ such that $\max_{e \in E}\{|C(e, \mathcal{R})|\}$ is minimized. Denote by

$$congest(G, \pi) = \min_{\mathcal{R}} \max_{e \in E} \{|C(e, \mathcal{R})|\}.$$

Let Π be the set of all permutations on G. Then, the congestion of G, $congest(G)$, is defined as

$$congest(G) = \max_{\pi \in \Pi} \{congest(G, \pi)\}.$$

3 Lower Bounds

Theorem 9. *For any all-optical network $G(V, E)$, let $c(G)$ be the number of edges in a bisection of G, $congest(G)$ be the congestion of G, and $w_{\min}$ be the number of wavelengths needed to implement any permutation on G in one round. Then*

$$w_{\min} \geq congest(G) \geq \frac{n-1}{2c(G)}.$$

Proof. Following the congestion definition and the optical routing rule that different signals through a single fiber-optic link must be assigned different wavelengths, it is easy to see that $w_{\min} \geq congest(G)$.

Let (V_1, V_2) be a bisection of G and $|V_1| \geq |V_2|$. Assume that there is a permutation π which permutes all vertices in V_2 to V_1. Then the congestion of G for π is $congest(G, \pi) \geq \lfloor n/2 \rfloor / c(G) \geq \frac{n-1}{2c(G)}$. Since $congest(G) \geq congest(G, \pi)$, the theorem then follows. □

Theorem 9 always holds no matter whether the WDM model is the wavelength conversion model or not. From this theorem we have the following corollaries.

Corollary 10. *In a chain L_n of n vertices, the number of wavelengths for implementing any permutation on it in one round is at least $\lfloor n/2 \rfloor$.*

Corollary 11. *In a ring R_n of n vertices, the number of wavelengths for implementing any permutation on it in one round is at least $\frac{n-1}{4}$.*

Corollary 12. *Let M be $l \times h$ mesh with $n = hl$ vertices. Suppose $l \leq h$. The number of wavelengths for implementing any permutation on M in one round is at least $\frac{\sqrt{n}}{2} - 1$.*

4 Product Networks

Define the *direct product* of two graphs $G = (V_1, E_1)$ and $H = (V_2, E_2)$, denoted by $G \times H$, as follows. The vertex set of $G \times H$ is the Cartesian product $V_1 \times V_2$. There is an edge between vertex (v_1, v_2) and vertex (v_1', v_2') in $G \times H$ when $v_1 = v_1'$ and $(v_2, v_2') \in E_2$, or $v_2 = v_2'$ and $(v_1, v_1') \in E_1$. G and H are called the *factors* of $G \times H$. Notice that $G \times H$ consists of $|V_2|$ copies of G connected by $|V_1|$ copies of H, arranged in a grid-like fashion. Each copy of H is called a *row* and each copy of G is called a *column*. An edge in $G \times H$ between (v_1, u) and (v_2, u) is called a *G-edge* if $(v_1, v_2) \in E_1$, and an edge of $G \times H$ between (v, u_1) and (v, u_2) is called an *H-edge* if $(u_1, u_2) \in E_2$. Now let $W = G \times H$ and π be a permutation on W. Assume the factors G and H of W are equipped with routing algorithms. We are interested in designing a routing algorithm for W by using the routing algorithms for its factors as *subroutines*.

4.1 Lower bounds on the number of wavelengths

Lemma 13. *Let $G \times H$ be a directed symmetric product network. Let $c(X)$ be the number of the edges in a bisection of X and $V(X)$ be the vertex set of X. Suppose that d_G and d_H are the maximum in-degrees (out-degrees) of G and H. Then we have the following lower bounds for $c(G \times H)$:*

1. *$\min\{|V(H)|c(G), |V(G)|c(H)\}$ if both $|V(G)|$ and $|V(H)|$ are even;*
2. *$\min\{|V(G)|c(H) + \lfloor |V(G)|/2 \rfloor d_H + c(G), |V(H)|c(G)\}$ if $|V(G)|$ is even and $|V(H)|$ is odd;*
3. *$\min\{|V(H)|c(G) + \lfloor |V(H)|/2 \rfloor d_G + c(H), |V(G)|c(H)\}$ if $|V(G)|$ is odd and $|V(H)|$ is even;*
4. *$\min\{|V(G)|c(H) + \lfloor |V(G)|/2 \rfloor d_H + c(G), |V(H)|c(G) + \lfloor |V(G)|/2 \rfloor d_G + c(H)\}$ otherwise (both $|V(G)|$ and $|V(H)|$ are odd).*

Proof. We first consider the case where $p = |V(G)|$ and $q = |V(H)|$ are even. Let $V(G) = \{v_1, v_2, \ldots, v_p\}$ be the vertex set of G. Suppose the bisection of G is $(V_1(G), V_2(G))$, where $V_1(G) = \{v_1, v_2, \ldots, v_{p/2}\}$ and $V_2(G) = \{v_{p/2+1}, \ldots, v_p\}$.

We replace each vertex of G by graph H, and make the corresponding connection between two vertices in different copies of H. This results in the graph W. Now we partition the vertex set of W into two disjoint subsets of equal size according to the bisection of G. Thus, the number of edges of W in this partition is $|V(H)|c(G)$. Since W is a symmetric network, there is another partition which satisfies the above partition condition. The number of edges in this latter partition is $|V(G)|c(H)$. Since $c(G \times H)$ is a partition that has the minimum number of edges, $c(G \times H) \leq \min\{|V(H)|c(G), |V(G)|c(H)\}$.

We then consider the case where $p = |V(G)|$ is even and $q = |V(H)|$ is odd. We use only G-edges of W for the partition. Then, the number of edges of W in this partition is $|V(H)|c(G)$ by the above discussion. In the rest we consider using both H- and G-edges for the bisection of W. Let $V(H) = \{u_1, u_2, \ldots u_q\}$. Suppose that a bisection of H is $(V_1(H), V_2(H))$, where $V_1(H) = \{u_1, u_2, \ldots, u_{\lceil q/2 \rceil}\}$ and $V_2(H) = \{u_{\lceil q/2 \rceil + 1}, \ldots, u_q\}$. It is clear that $|V_1(H)| - |V_2(H)| = 1$ because q is odd. Now we give another partition $(V_1'(H), V_2'(H))$ of H, where $V_1'(H) = V_2(H) \cup \{v\}$ and $V_2'(H) = V_1(H) - \{v\}$. The number of H-edges in this new partition is no more than $c(H) + d_H(v) \leq c(H) + d_H$. We replace each vertex of H by graph G, and make the corresponding connection between two vertices in different copies of G. This leads to the graph W. Partition the vertex set of W into two disjoint subsets such that: the size difference is at most one according to a bisection of H; and the vertices in the copy of G corresponding to v of H are partitioned into two disjoint subsets by a bisection of G. Thus, the number of edges of W in this partition is no more than $\lceil |V(G)|/2 \rceil c(H) + \lfloor |V(G)|/2 \rfloor (c(H) + d_H(v)) + c(G) \leq |V(G)|c(H) + \lfloor |V(G)|/2 \rfloor d_H + c(G)$. Therefore $c(G \times H) \leq \min\{|V(H)|c(G), |V(G)|c(H) + \lfloor |V(G)|/2 \rfloor d_H + c(G)\}$. The other two cases can be shown similarly, omitted. □

Having Lemma 13, the following lower bounds on the number of wavelengths are derived easily.

Theorem 14. Let $W = G \times H$ be a directed symmetric product network. Let $c(X)$ be the number of edges in a bisection of X. Suppose that d_G and d_H are the maximum in-degrees (out-degrees) of G and H. Then, a lower bound for the minimum number of wavelengths $w_{\min}(W)$ for implementing any permutation on W in one round is:

1. $\max\{\frac{|V(G)|}{2c(G)} - 1, \frac{|V(H)|}{2c(H)} - 1\}$ if both $|V(G)|$ and $|V(H)|$ are even;
2. $\max\{\frac{|V(H)|}{2c(H) + d_H + 2c(G)/|V(G)|} - 1, \frac{|V(G)|}{2c(G)} - 1\}$ if $|V(G)|$ is even and $|V(H)|$ is odd;
3. $\max\{\frac{|V(G)|}{2c(G) + d_G + 2c(H)/|V(H)|} - 1, \frac{|V(H)|}{2c(H)} - 1\}$ if $|V(G)|$ is odd and $|V(H)|$ is even;
4. $\max\{\frac{|V(G)|}{2c(G) + d_G + 2c(H)/|V(H)|} - 1, \frac{|V(H)|}{2c(H) + d_H + 2c(G)/|V(G)|} - 1\}$ otherwise (both $|V(G)|$ and $|V(H)|$ are odd).

Proof. From Theorem 9, it is clear that $w_{\min}(W) \geq \frac{|V(G)||V(H)|}{2c(G \times H)} - 1$. Use of the lower bounds for $c(G \times H)$ from Lemma 13 in this inequality gives the theorem.
□

4.2 A routing algorithm on the packet-passing model

The permutation routing on a direct product network W has been addressed for the packet-passing model [2,8]. Baumslag and Annextein [8] presented an efficient algorithm for permutation routing on that model which consists of the following three phases. 1. Route some set of permutations on the copies of G; 2. Route some set of permutations on the copies of H; 3. Route some set of permutations on the copies of G. Since the product network is a symmetric network, the above three phases can be applied alternatively, i.e., by first routing on H, followed by G, and followed by H.

Now, consider the following naive routing method. First route each source in a column to its destination row in the column. Then route each source in a row to its destination column in the row. This fails to be an edge congestion-free algorithm because there may be several sources in the same column that have their destinations in a single row (causing congestion at the intersection of the column and the row). By using an initial extra phase, the above congestion problem can be solved. That is, we first "rearrange" each column so that, after the rearrangement, each row consists of a set of sources whose destinations are all in distinct columns. After this, a permutation of each row is required to get each source to its correct column after the rearrangement. Once all of the sources are in their correct columns, a final permutation of each column suffices to get each source to its correct destination. This final phase is indeed a permutation since all destinations of sources are distinct. Thus, the aim of the first phase is to find a set of sources P_R, once per column, such that every source in P_R has its destination in a distinct column for each row R.

To this end, a bipartite graph $G_B(X, Y, E_B)$ is constructed as follows. Let X and Y represent the set of columns of W. There is an edge between $x_i \in X$ and $y_j \in Y$ for each source in column i whose destination is in column j. Since π is a permutation, it follows that G_B is a regular, bipartite multigraph. Thus, G_B can be decomposed into a set of edge disjoint perfect matchings. Note that the sources involved in a single perfect matching all have their destinations in distinct columns. Therefore, for each row R, use all of the sources that correspond to a single perfect matching for the set P_R. Each of these sets P_R, is "lifted" to row R during the first phase of the algorithm. Since each source is involved in precisely one perfect matching, the mapping of sources in a column during the first phase is indeed a permutation of the column.

4.3 Routings on the wavelength conversion model

The algorithm in Section 4.2 can be expressed in a different way. That is, π is decomposed into three permutations σ_i, $i = 1, 2, 3$, where σ_1 and σ_3 are permutations in columns of W and σ_2 is a permutation in rows of W. For example, let v be a source in W at the position of row i_1 and column j_1, and $\pi(v) = u$ be the destination of v at the position row i_2 and column j_2. Suppose v is "lifted" to the position of row i'_1 and column j_1 in the first phase. Then,

$\sigma_1(i_1, j_1) = (i'_1, j_1)$, followed by the second phase of $\sigma_2\sigma_1(i_1, j_1) = \sigma_2(i'_1, j_1) = (i'_1, j_2)$, and followed by the third phase of $\sigma_3\sigma_2\sigma_1(i_1, j_1) = \sigma(i'_1, j_2) = (i_2, j_2)$.

We now present a permutation routing algorithm on W for π on the wavelength conversion model, based on the above observation. We start with the following major theorem.

Theorem 15. *Given permutation routing algorithms for networks G and H, there is a routing algorithm for the product network $G \times H$. The number of wavelengths for any permutation in one round is at most $\max\{2w(G), w(H)\}$ if $w(G) \leq w(H)$; or $\max\{2w(H), w(G)\}$ otherwise, where $w(X)$ represents the number of wavelengths needed to implement any permutation on network X in one round.*

Proof. The routing algorithm by Baumslag and Annexstein is for the packet-passing model, which is a totally different model from our WDM model. Hence, some modifications to their algorithm are necessary. As we can see, implementing the permutation routing on W for π can be decomposed into three permutations σ_i, $1 \leq i \leq 3$.

Without loss of generality, we assume $w(G) \leq w(H)$. Following the three phases of the above algorithm, the permutation σ_1 can be implemented with $w(G)$ wavelengths (in other words, all routing paths can be colored with $w(G)$ colors); the permutation σ_2 can be implemented with $w(H)$ wavelengths (all routing paths can be colored with $w(H)$ colors, and the colors for σ_1 can be re-used here); the permutation σ_3 can be implemented with $w(G)$ wavelengths (all routing paths can be colored with $w(G)$ colors, and the colors for σ_1 cannot be used here). Now, for a request $(i, \pi(i))$, let (u_1, v_1) be the position of i in W and (u_2, v_2) be the position of $\pi(i)$ in W. Then, the routing path L_i for request $(i, \pi(i))$ consists of three routing segments $L_{i,1}$, $L_{i,2}$ and $L_{i,3}$ which correspond to σ_1, σ_2, and σ_3, where $L_{i,1}$ only contains the vertices in column v_1 of W and consists of G-edges; $L_{i,2}$ only contains the vertices in row u'_1 of W and consists of H-edges, assuming that $\sigma_1(i)$ is at row u'_1 and column v_1 in W; and $L_{i,3}$ only contains the vertices in column v_2 of W, and G-edges.

Any path L_i can further be made a simple path L'_i by deleting the cycles it contains. Let $L'_{i,j}$ be the corresponding segment of $L_{i,j}$, $1 \leq j \leq 3$. L'_i then is assigned three different wavelengths for each of the segments of L'_i, which are the wavelengths for $L_{i,j}$ originally, $j = 1, 2, 3$. That is, each L'_i for a request $(i, \pi(i))$ can be implemented with at most three wavelengths. But, the wavelengths used for $L_{i,1}$ cannot be used for $L_{i,3}$. So, at least $2w(G)$ wavelengths are needed. In summary, implementing any permutation π on W in one round can be done with $\max\{2w(G), w(H)\}$ wavelengths. □

Following Theorem 15, the following corollaries can be derived directly.

Corollary 16. *There is an algorithm for implementing any permutation in a directed symmetric hypercube H_q of $n = 2^q$ vertices with 2 wavelengths.*

Proof. Since $H_q = K_2 \times H_{q-1}$ and $w(K_2) = 1$, by Theorem 15, $w(H_q) = \max\{2w(K_2), w(H_{q-1})\}$. Furthermore, $H_{q-1} = K_2 \times H_{q-2}$, and it is easy to show that $w(H_q) = 2$ by induction on q. □

Corollary 17. *For any directed symmetric $l \times h$ mesh M with $l \leq h$ and $n = lh$, there is an algorithm for implementing any permutation on M in one round. The number of wavelengths used by this algorithm is at most $\max\{l, \lfloor h/2 \rfloor\}$.*

Proof. Assume $l \leq h$ and $M = L_l \times L_h$, where L_i represents a chain of i vertices. Obviously $w(L_i) = \lfloor i/2 \rfloor$. By Theorem 15, $w(M) = \max\{2w(L_l), w(L_h)\} \leq \max\{l, \lfloor h/2 \rfloor\}$. □

Corollary 18. *There is an algorithm for implementing any permutation in a directed symmetric $\sqrt{n/2} \times \sqrt{2n}$ mesh with n vertices in one round. The number of wavelengths used by this algorithm is at most $\sqrt{n/2}$ which is almost optimal.*

Proof. Let M be the $\sqrt{n/2} \times \sqrt{2n}$ mesh. By Corollary 17, $w(M) = \max\{2w(L_{\sqrt{n/2}}), w(L_{\sqrt{2n}})\} = \sqrt{n/2}$, and by Corollary 12, the lower bound of the number of wavelengths for any permutation on M is $\Omega(\sqrt{n})$, so, this bound is almost tight. □

4.4 Routings on the wavelength non-conversion model

We now study the permutation routing on W for the wavelength non-conversion model in which each routing path is assigned a unique wavelength. We only consider the case of $w(G) \leq w(H)$. The case of $w(H) \leq w(G)$ is similar, omitted.

Following the proof of Theorem 15, a routing path L'_i for every $i \in V(W)$ is obtained, and L'_i can be further divided into three segments $L'_{i,k}$, $k = 1, 2, 3$. Each of the segments has been assigned a wavelength (a color). Let γ_j be the color (wavelength) of $L'_{i,k}$. Then $(\gamma_1, \gamma_2, \gamma_3)$ is the ordered color tuple of L'_i. We treat $(\gamma_1, \gamma_2, \gamma_3)$ as a coordinate point in a 3-dimensional Cartesian coordinate system. Assume that each coordinate point in the system has been assigned a unique label, i.e., L'_i is assigned the wavelength numbered by the label of $(\gamma_1, \gamma_2, \gamma_3)$. Then, the total number of coordinate points for all routing paths on W in a permutation is $w(G) \times w(H) \times w(G) = w(G)^2 w(H)$.

Now, consider two routing paths L'_i and L'_j, $i \neq j$. Let L'_i be colored with $(\gamma_1, \gamma_2, \gamma_3)$ and L'_j be colored with $(\beta_1, \beta_2, \beta_3)$. If there exists a k such that $\gamma_k \neq \beta_k$, $1 \leq k \leq 3$, then the wavelengths assigned for L'_i and L'_j are different. Otherwise, if $\gamma_k = \beta_k$ for all k, $1 \leq k \leq 3$, then L'_i and L'_j are assigned the same wavelength. However, this wavelength assignment may not be incorrect because there still exist some routing paths sharing common edges which are assigned the same wavelength. We illustrate this situation by an example (see Figure 1). Assume that the given permutation is π. Consider two routing paths L'_i and L'_j, where L'_i starts from i, goes through y', $\pi(j)$, x, x', and ends at $\pi(i)$, L'_j starts from j, goes through x', $\pi(i)$, y, y', and ends at $\pi(j)$. Clearly L'_i and L'_j share two common segments which are from y' to $\pi(j)$ and from x to $\pi(i)$ even though they have the same color tuple. So, they cannot be assigned the same wavelength on the WDM model. To cope with this situation, the following approach is applied. Let $l_{\max}(G)$ be the number of edges in the longest routing path on G for any permutation. Define $\mathcal{R}_l = \{L'_i \mid L'_i \text{ is labeled by } l \text{ in the coordinate system}\}$.

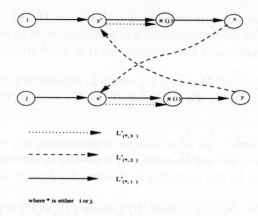

Fig. 1. An example.

Obviously the set of routing paths for π is $\mathcal{R} = \bigcup_{l=1}^{w(G)^2 w(H)} \mathcal{R}_l$. For each $\mathcal{R}_l$, an auxiliary graph $G_l = (V_l, E_l)$ that is a subgraph of the conflict graph on W, is constructed as follows. Every vertex in V_l corresponds to an element in $\mathcal{R}_l$. There is an edge between two vertices if the two routing paths share at least one common edge. Then we have

Lemma 19. *Let $G_l = (V_l, E_l)$ be defined as above, then the maximum degree of G_l is $l_{\max}(G)$.*

Proof. Let L'_i and L'_j be the corresponding routing paths of two vertices in G_l. We know that L'_k consists of three segments $L'_{k,1}$, $L'_{k,2}$, and $L'_{k,3}$, $k = i$ or $k = j$. By the definition, $L'_{i,p}$ and $L'_{j,p}$ are edge disjoint for all p, $1 \leq p \leq 3$. But $L'_{i,1}$ and $L'_{j,3}$ (similarly $L'_{j,1}$ and $L'_{i,3}$) may not be edge disjoint, see Fig. 1. Since the number of edges in any routing path is no greater than $l_{\max}(G)$, there are at most $l_{\max}(G)$ other routing paths sharing common edges with L'_i. Therefore, the maximum degree of G_l is $l_{\max}(G)$. □

By Lemma 19, all vertices in G_l can be colored with $l_{\max}(G) + 1$ colors such that the adjacent vertices are colored with different colors. This coloring can be done in polynomial time by a greedy approach. This means, for each $\mathcal{R}_l$, we can assign $l_{\max}(G) + 1$ wavelengths for the routing paths in it such that those paths sharing common edges are assigned different wavelengths. There are $w(G)^2 w(H)$ different $\mathcal{R}_l$s. Therefore, the total number of wavelengths required for any permutation is $(l_{\max}(G) + 1) w(G)^2 w(H)$, which is formally described as follows.

Theorem 20. *Given permutation routing algorithms for networks G and H, there is a permutation routing algorithm for the product network $W = G \times H$. The number of wavelengths for any permutation on W in one round is $(l_{\max}(G) + 1) w(G)^2 w(H)$ if $w(G) \leq w(H)$; or $(l_{\max}(H) + 1) w(H)^2 w(G)$ otherwise, where*

$w(X)$ represents the number of wavelengths needed to implement any permutation on network X in one round, and $l_{\max}(X)$ is the number of edges of the longest routing path on X for any permutation.

Let $|V(G)| = p$, $|V(H)| = q$, and $n = pq$. Assume that both p and q are functions of n. If both $w(H)$ and $w(G)$ are linear functions of the vertex sizes of G and H, for example, $w(G) = ap$ and $w(H) = bq$ where a and b are constants with $0 < a, b < 1$. Without loss of generality, we further assume that $w(G) \leq w(H)$. Then, it needs $(l_{\max}(G) + 1)w(G)^2 w(H) = (l_{\max}(G) + 1)(a^2 b)pn = cn^{1+\alpha} > n$ wavelengths where $p = n^\alpha$. As a matter of the fact, any permutation can be implemented on an arbitrary all-optical network in one round if n wavelengths are available. To cope with this case, we present another permutation routing algorithm. We start with the following perfect matching lemma.

Lemma 21. *[15] Let $G_B(X, Y, E)$ be a bipartite graph such that for every subset S of X, we have $|N(S)| \geq |S|$, where $N(S)$ is the subset of Y that are adjacent to vertices in S. Then G_B has a perfect matching of size $\min\{|X|, |Y|\}$.*

Suppose $p \leq q$. Our idea comes from Youssef [26]. That is, at each time, we select p sources and their destinations such that these sources belong to distinct rows and their destinations belong to distinct columns. Such sources can be found using perfect matching in a bipartite graph $G_B = (X, Y, E)$ where X is the set of rows and Y is the set of columns. There is an edge connecting $x \in X$ and $y \in Y$ if there is a source in row x whose destination is in column y. Clearly G_B is a bipartite multigraph, the degree of every vertex of G_B in X is q, and the degree of every vertex of G_B in Y is p. Since any subset $S \subseteq X$, $|N(S)| \geq |S|$. Thus, there is a perfect matching in G_B by Lemma 21. By deleting this matching, we can find the next perfect matching in the remaining graph, and so on. As a result, G_B is decomposed into q edge disjoint perfect matchings. Since the routing paths in a perfect matching are edge disjoint, they can be assigned the same wavelength. So, we have

Lemma 22. *Given a directed symmetric network $G \times H$ with $|V(G)| = p$, $|V(H)| = q$ and $p \leq q$, there is an algorithm for implementing any permutation on $G \times H$ in one round if q wavelengths are available.*

If we allow $\max\{w(G), w(H)\}$ wavelengths to assign every fiber-optic link of $G \times H$, we have

Theorem 23. *Let $G \times H$ be a directed symmetric network. On the wavelength non-conversion model, if there are permutation algorithms for implementing any permutation on G and H in one round with $w(G)$ and $w(H)$ wavelengths respectively, then there is a permutation algorithm for implementing any permutation on $G \times H$ in $\frac{|V(H)|}{\max\{w(H), w(G)\}}$ ($\leq 2c(H)+1$) rounds with $\max\{w(H), w(G)\}$ wavelengths if $|V(G)| \leq |V(H)|$, or in $\frac{|V(G)|}{\max\{w(H), w(G)\}}$ ($\leq 2c(G)+1$) rounds with $\max\{w(H), w(G)\}$ wavelengths, where $w(G)$ and $w(H)$ are the linear functions of their sizes and $c(X)$ is the number of edges in a bisection of X.*

Proof. We only consider the case $|V(G)| \leq |V(H)|$. The analogous case $|V(H)| < |V(G)|$ is omitted. By Theorem 9, $w(H) \geq \frac{|V(H)|-1}{2c(H)}$. So, $|V(H)| \leq 2w(H)c(H) + 1$. According to Lemma 22, in order to implement any permutation on $G \times H$ with $\max\{w(H), w(G)\}$ wavelengths, the number of rounds needed is at most

$$\frac{|V(H)|}{\max\{w(H), w(G)\}} \leq \frac{2w(H)c(H) + 1}{\max\{w(H), w(G)\}}.$$

That is, the number of rounds is at most $|V(H)|/w(H) \leq \frac{2w(H)c(H)+1}{w(H)} \leq 2c(H) + 1$ if $w(G) \leq w(H)$; or $|V(H)|/w(G) \leq \frac{2w(H)c(H)+1}{w(G)} \leq 2c(H) + 1$ otherwise. □

5 Routings for Specific All-Optical Networks

In this section we study permutation routings on some specific all-optical networks. Before we proceed, we introduce Lemma 24 by Gu and Tamaki [11].

Lemma 24. *Let H_q be a directed symmetric hypercube. Then any permutation on it can be implemented in one round with two wavelengths.*

We now consider the well known *cube-connected-cycles* (CCC) network [21,14], which is constructed from the r-dimensional hypercube by replacing each vertex of the the hypercube with a cycle of r vertices in the CCC. The resulting graph has $n = r2^r$ vertices with degree 3. By modifying the labeling scheme of the hypercube, we can represent each vertex by a pair $\langle w, i \rangle$ where i ($1 \leq i \leq r$) is the position of the vertex within its cycle and w (any r-bit binary string) is the label of the vertex in the hypercube that corresponds to the cycle. Then two vertices $\langle w, i \rangle$ and $\langle w', i' \rangle$ are linked by an edge in the CCC if and only if either (1) $w = w'$ and $i - i' \equiv \pm 1 \mod r$, or (2) $i = i'$ and w differs from w' in precisely the ith bit. Edges of the first type are called *cycle edges*, while edges of the second type are referred to as *hypercube edges*. For the CCC, we have the following theorem.

Theorem 25. *Let $G(V, E)$ be a directed symmetric CCC network defined as above. Then any permutation on G can be implemented either in $\lfloor \log n \rfloor$ rounds using 2 wavelengths, or in one round using $\lfloor 2 \log n \rfloor$ wavelengths on the wavelength non-conversion model.*

Proof. The basic idea is to choose a vertex from each cycle as a source such that no two sources have their destinations in the same cycle. Route these chosen sources to their destinations on the supergraph, which can be implemented in one round using 2 wavelengths by Lemma 24. Thus, any permutation on the CCC can be implemented in r rounds. Since $r = \log n - \log r \leq \lfloor \log n \rfloor$, $\lfloor \log n \rfloor$ rounds suffices.

It remains to show how to choose the sources for each round such that none of the destinations of two sources are in the same cycle. The approach proceeds as follows. First we construct a bipartite graph $G(X, Y, E_{XY})$ where X and Y

represent the sets of 2^r cycles of G. For $v \in X$ and $u \in Y$, if a source in v will route its message to a destination in u, then an undirected edge (u,v) is added to E_{XY}. Clearly, $G(X,Y,E_{XY})$ is a regular bipartite multigraph with degree r. By Hall's theorem, $G(X,Y,E_{XY})$ can be decomposed into r edge disjoint perfect matchings, and each perfect matching can be found in polynomial time. The corresponding routing paths for each perfect matching can be implemented using two wavelengths by Lemma 24. Therefore, any permutation in the CCC can be implemented with either $\lfloor \log n \rfloor$ rounds using 2 wavelengths, or one round using $2 \times r \leq \lfloor 2 \log n \rfloor$ wavelengths. □

Note that the CCC is a regular network with out-degree (in-degree) 3. Aumann and Rabani [5] showed that $O(\log^2 n/\beta^2(G))$ wavelengths suffice for implementing any permutation on a network G of constant degree in one round, where $\beta(G)$ is the edge-expansion of G. The CCC is a network of degree three, and $\beta(\mathrm{CCC}) \leq \frac{2^{r-1}}{2^{r-1}r} = 1/r$. For this special network, our permutation routing algorithm implements any permutation in one round with $2 \log n$ wavelengths, which improves the number of wavelengths used in [5] by a factor of $O(\log^3 n)$.

Next we generalize the CCC further to the following network $G(V,E)$ with $n = 2^r \times s$ vertices satisfying $s \geq r$. This network is a hypercube supergraph of 2^r supervertices, and each supervertex is a cycle of s vertices. Clearly CCC is a special case of this generalized network with $s = r$. Now we see that the number of wavelengths needed for implementing any permutation on G depends on the number of vertices of degree three. For example, if $2^r = \sqrt{n}$, then $s = \sqrt{n}$. The network G contains $2^r \times r = \frac{\sqrt{n} \log n}{2}$ vertices of degree three, and $n - \frac{\sqrt{n} \log n}{2}$ vertices of degree two. Thus, implementing any permutation on G using the above approach requires either $\sqrt{n}$ rounds if there are 2 wavelengths available, or one round if there are $2\sqrt{n}$ wavelengths available. Thus, we have

Lemma 26. *Let $G(V,E)$ be a directed symmetric, generalized CCC network defined as above, with $s \geq r$. Then any permutation on G can be implemented either in s rounds using 2 wavelengths, or in one round using $2s$ wavelengths on the wavelength non-conversion model. The number of vertices of degree three in G is $r2^r$, and the number of vertices of degree two is $n - r2^r$.*

6 Conclusions

In this paper we have shown some lower bounds on the number of wavelengths needed for implementing any permutation on all-optical networks in terms of bisection. We have also shown a lower bound on the number of wavelengths required on the product networks, and presented permutation routing algorithms for it, based on the wavelength non-conversion and conversion models respectively. We finally considered the permutation issue on a cube-connected-cycles network. The number of wavelengths needed for implementing any permutation on this network in one round is $\lfloor 2 \log n \rfloor$, which improves on a general result for bounded degree networks in [5] by a factor of $O(\log^3 n)$ for this special case.

References

1. A. Aggarwal, A. Bar-Noy, D. Coppersmith, R. Ramaswami, B. Schieber and M. Sudan. Efficient routing in optical networks. *J. of the ACM* **46**:973–1001, 1996.
2. N. Alon, F. R. K. Chung, and R. L. Graham. Routing permutations on graphs via matchings. *SIAM J. Discrete Math.* **7**:516–530, 1994.
3. B. Awerbuch, Y. Azar, and S. Plotkin. Throughput competitive on-line routing. *Proc. of 34th IEEE Symp. on Found. of Computer Science*, 1993, 32–40.
4. B. Awerbuch, Y. Bartal, A. Fiat and A. Rosén. Competitive non-preemptive call control. *Proc. of 5th ACM-SIAM Symp. on Discrete Algorithms*, 1994, 312–320.
5. Y. Aumann and Y. Rabani. Improved bounds for all optical routing. *Proc. of 6th ACM-SIAM Symp. on Discrete Algorithms*, 1995, 567–576.
6. R. A. Barry and P. A. Humblet. Bounds on the number of wavelengths needed in WDM networks. *In LEOS'92 Summer Topical Mtg. Digest* 1992, 114–127.
7. R. A. Barry and P. A. Humblet. On the number of wavelengths and switches in all-optical networks. *IEEE Trans. Commun.* **42**:583–591, 1994.
8. M. Baumslag and F. Annexstein. A unified framework for off-line permutation routing in parallel networks. *Math. Systems Theory* **24**:233–251, 1991.
9. B. Beauquier, J-C Bermond, L. Cargano, P. Hell, S. Pérnces, and U. Vaccaro. Graph problems arising from wavelength-routing in all-optical networks. *Proc. of 2nd Workshop on Optics and Computer Science*, IPPS'97, April, 1997.
10. P. E. Green. *Fiber- Optic Communication Networks*. Prentice Hall, 1992.
11. Q -P Gu and H. Tamaki. Routing a permutation in the hypercube by two sets of edge-disjoint paths. *Proc. 10th IPPS.*, IEEE CS Press, 561–576, 1996.
12. J. M. Kleinberg. *Approximation Algorithms for Disjoint Paths Problems*. Ph.D dissertation, Dept. of EECS, MIT, Cambridge, Mass., 1996.
13. J. Kleinberg and E. Tardos. Disjoint paths in densely embedded graphs. *Proc. of 36th Annual IEEE Symposium on Foundations of Computer Science*, 1995, 52–61.
14. F. T. Leighton. *Introduction to Parallel Algorithms and Architectures: Arrays · Trees · Hypercubes*. Morgan Kaufmann Publishers, CA, 1992.
15. C. L. Liu. *Introduction to Combinatorial Mathematics*. McGraw-Hill, NY, 1968.
16. A. D. McAulay. *Optical Computer Architectures*. John Wiley, 1991.
17. M. Mihail, C. Kaklamanis, and S. Rao. Efficient access to optical bandwidth. *Proc. of 36th IEEE Symp. on Founda. of Computer Science*, 1995, 548–557.
18. R. K. Pankaj. *Architectures for Linear Lightwave Networks*. Ph.D dissertation. Dept. of EECS, MIT, Cambridge, Mass., 1992.
19. R. K. Pankaj and R. G. Gallager. Wavelength requirements of all-optical networks. *IEEE/ACM Trans. Networking* **3**:269–280, 1995.
20. G. R. Pieris and G. H. Sasaki. A linear lightwave Beneš network. *IEEE/ACM Trans. Networking* **1**:441–445, 1993.
21. F. Preparata and J. Vuillemin. The cube-connected cycles: A versatile network for parallel computation. *Comm. ACM* **24**:300–309, 1981.
22. Y Rabani. Path coloring on the mesh. *Proc. of 37th Annual IEEE Symposium on Foundations of Computer Science*, Oct., 1996, 400–409.
23. P. Raghavan and E. Upfal. Efficient routing in all-optical networks. *Proc. of 26th Annual ACM Symposium on Theory of Computing*, May, 1994, 134–143.
24. R. Ramaswami. Multi-wavelength lightwave networks for computer communication. *IEEE Communications Magazine* **31**:78–88, 1993.
25. R. J. Vitter and D. H. C. Du. Distributed computing with high-speed optical networks. *IEEE Computer* **26**:8–18, 1993.
26. A. Youssef. Off-line permutation routing on circuit-switched fixed-routing networks. *Networks* **23**:441–448: 1993.

NWCache: Optimizing Disk Accesses via an Optical Network/Write Cache Hybrid [*]

Enrique V. Carrera and Ricardo Bianchini

COPPE Systems Engineering
Federal University of Rio de Janeiro
Rio de Janeiro, Brazil 21945-970
{vinicio,ricardo}@cos.ufrj.br

Abstract. In this paper we propose a simple extension to the I/O architecture of scalable multiprocessors that optimizes page swap-outs significantly. More specifically, we propose the use of an optical ring network for I/O operations that not only transfers swapped-out pages between the local memories and the disks, but also acts as a system-wide write cache. In order to evaluate our proposal, we use detailed execution-driven simulations of several out-of-core parallel applications running on an 8-node scalable multiprocessor. Our results demonstrate that the NWCache provides consistent performance improvements, coming mostly from faster page swap-outs, victim caching, and reduced contention. Based on these results, our main conclusion is that the NWCache is highly efficient for most out-of-core parallel applications.

1 Introduction

Applications frequently access far more data than can fit in main memory. Reducing disk access overhead is the most serious performance concern for these out-of-core applications. For this reason, programmers of these applications typically code them with explicit I/O calls to the operating system. However, writing applications with explicit I/O has several disadvantages [14]: programming often becomes a very difficult task [21]; I/O system calls involve data copying overheads from user to system-level buffers and vice-versa; and the resulting code is not always portable (performance-wise) between machine configurations with different memory and/or I/O resources, such as different amounts of memory or I/O latency. In contrast with the explicit I/O style of programming, we advocate that out-of-core applications should be based solely on the virtual memory mechanism and that disk access overheads should be alleviated by the underlying system. Essentially, our preference for virtual memory-based I/O is analogous to favoring shared memory instead of message passing as a more appropriate parallel programming model.

Basically, virtual memory-based I/O involves reading pages to memory and writing (or swapping) pages out to disk. Page reads can usually be dealt with efficiently by dynamically prefetching data to main memory (or to the disk controller cache) ahead of its use, e.g. [13, 10]. In cases where dynamic prefetching is not effective by itself,

---

[*] This research was supported by Brazilian CNPq.

prefetching can be improved with compiler involvement [14] or with future-access hints [9]. Page swap-outs are more difficult to optimize however, even though these writes happen away from the critical path of the computation. The performance problem with page swap-outs is that they are often very bursty and, as a result, the operating system must at all times reserve a relatively large number of free pages frames to avoid having to stall the processor waiting for swap-out operations to complete. In fact, the more effective the page prefetching technique is, the greater is the number of free page frames the operating system must reserve. This situation is especially problematic for scalable multiprocessors where not all nodes are I/O-enabled (e.g. [1, 2]), since both disk latency and bandwidth limitations delay page swap-outs.

In this paper, we propose a simple extension to the I/O architecture of scalable multiprocessors that uses an optical network to optimize page swap-outs significantly. Our proposal is based on the observation that the extremely high bandwidth of optical media provides data storage on the network itself and, thus, these networks can be transformed into fast-access (temporary) data storage devices. Specifically, we propose NWCache, an optical ring network/write cache hybrid that not only transfers swapped-out pages between the local memories and the disks of a scalable multiprocessor, but also acts as a system-wide write cache for these pages. When there is room in the disk cache, pages are copied from the NWCache to the cache such that the pages swapped out by a node are copied together. The NWCache does not require changes to the multiprocessor hardware and only requires two simple modifications to the operating system's virtual memory management code.

The NWCache has several performance benefits: it provides a staging area where swapped-out pages can reside until the disk is free; it increases the possibility of combining several writes to disk; it acts as a victim cache for pages that are swapped out and subsequently accessed by the same or a different processor; and it reduces the data traffic across the multiprocessor's interconnection network and memory buses.

In order to assess how the performance benefits of NWCache affect overall performance, we use detailed execution-driven simulations of several out-of-core parallel applications running on an 8-node scalable cache-coherent multiprocessor with 4 I/O-enabled nodes. We consider the two extremes in terms of the page prefetching techniques: optimal and naive prefetching. Under the optimal prefetching strategy, our results demonstrate that the NWCache improves swap-out times by 1 to 3 orders of magnitude. The NWCache benefits are not as significant under the naive prefetching strategy, but are still considerable. Overall, the NWCache provides execution time improvements of up to 64% under optimal prefetching and up to 42% under naive prefetching. Our results also show that a standard multiprocessor often requires a huge amount of disk controller cache capacity to approach the performance of our system.

2 Background

Wavelength Division Multiplexing networks. The maximum bandwidth achievable over an optic fiber is on the order of Tbits/s. However, due to the fact that the hardware associated with the end points of an optical communication system is usually of an electronic nature, transmission rates are currently limited to the Gbits/s level. In order to approach the full potential of optical communication systems, multiplexing techniques must be utilized. WDM is one such multiplexing technique. With WDM, several independent communication channels can be implemented on the same fiber. Optical networks that use this multiplexing technique are called WDM networks. Due to the rapid development of the technology used in its implementation, WDM has become one of the most popular multiplexing techniques. The NWCache architecture focuses on WDM technology due to its immediate availability, but there is nothing in our proposal that is strictly dependent on this specific type of multiplexing.

Delay line memories. Given that light travels at a constant and finite propagation speed on the optical fiber (approximately 2.1E+8 meters/s), a fixed amount of time elapses between when a datum enters and leaves an optical fiber. In effect, the fiber acts as a delay line. Connecting the ends of the fiber to each other (and regenerating the signal periodically), the fiber becomes a delay line memory [18], since the data sent to the fiber will remain there until overwritten. An optical delay line memory exhibits characteristics that are hard to achieve by other types of delay lines. For instance, due to the high bandwidth of optical fibers, it is possible to store a reasonable amount of memory in just a few meters of fiber (e.g. at 10 Gbits/s, about 5 Kbits can be stored on one 100 meters-long WDM channel).

3 The Optical Network/Write Cache Hybrid

3.1 Traditional Multiprocessor Architecture and Virtual Memory Management

We consider a traditional scalable cache-coherent multiprocessor, where processors are connected via a wormhole-routed mesh network. As seen in figure 1, each node is the system includes one processor (labeled "μP"), a translation lookaside buffer ("TLB"), a coalescing write buffer ("WB"), first ("L1") and second-level ("L2") caches, local memory ("LM"), and a network interface ("NI"). Each I/O-enabled node also includes one disk and its controller connected to the I/O bus. The multiprocessor can be extended with a NWCache simply by plugging the NWCache interface ("NWC") into the I/O bus of its nodes. The disk controller of the I/O-enabled nodes can then be plugged to the NWCache interface. This interface connects the node to the optical ring and filters some of the accesses to the disk, as detailed below.

The only part of the multiprocessor operating system we must consider is its virtual memory management code. Again, we assume a standard strategy here. More specifically, our base system implements a single machine-wide page table, each entry of which is accessed by the different processors with mutual exclusion. Every time the access rights for a page are downgraded, a TLB-shootdown operation takes place, where all other processors are interrupted and must delete their TLB entries for the page.

The operating system maintains a minimum set of free page frames per node of the multiprocessor. On a page fault, the operating system sends a request for the page[2] to the corresponding disk across the interconnection network. (We assume that pages are stored in groups of 32 consecutive pages. The parallel file system assigns each of these

---

[2] For simplicity of presentation, from now on we will not differentiate a virtual memory page from a disk block.

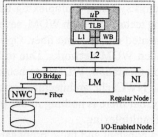
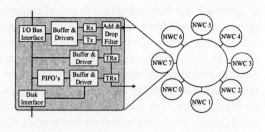

Fig. 1. Overview of Node. **Fig. 2.** Overview of NWCache System.

groups to a different disk in round-robin fashion.) For each request, the disk controller reads the page from its cache (cache hit) or disk (cache miss).

When the page faulted on arrives from its disk, the global page table is updated, allowing all other processors to access (and cache) the page data remotely. If the arrival of this page reduces the number of free page frames on the node below the minimum, the operating system uses LRU to pick a page to be replaced. If the page is dirty, a page swap-out operation is started. Otherwise, the page frame is simply turned free.

Page prefetching is not within the scope of this work. Thus, we consider the extreme prefetching scenarios: optimal prefetching and naive prefetching. Optimal prefetching attempts to approximate the performance achieved by highly-sophisticated compilers [14] or application hints [9] that can prefetch data from disks to disk caches or to memory. Our idealized technique then assumes that all page requests can be satisfied directly from the disk cache; i.e. all disk read accesses are performed in the background of page read requests.

Under the naive-prefetching scenario, the only prefetching that takes places occurs on a disk cache miss, when the controller fills its cache with pages sequentially following the missing page. We believe this technique is very naive for two reasons: a) files in our system are stripped across several disks; and b) several of our applications do not fetch pages sequentially.

A page that is swapped out of memory is sent to the corresponding disk cache across the network. The disk controller responds to this message with an ACK, if it was able to place the page in its cache. Writes are given preference over prefetches in the cache. The ACK allows the swapping node to re-use the space occupied by the page in memory. The disk controller responds with a NACK, if there was no space left in its cache (i.e. the disk controller's cache is full of swap-outs). The controller records the NACK into a FIFO queue. When room becomes available in the controller's cache, the controller sends a OK message to the requesting node, which prompts it to re-send the page.

3.2 The Optical Ring

Overview. Figure 2 overviews the optical ring that implements the NWCache. The ring is only used to transfer pages that were recently swapped out of memory by the different multiprocessor nodes to the disks, while "storing" these pages on the network itself. All other traffic flows through the regular interconnection network.

The swapped-out pages are continually sent around the ring's WDM channels, referred to as cache channels, in one direction. In essence, the ring acts as a write cache, where a page that is swapped out of main memory resides until enough room for it exists in its disk controller's cache. If the page is requested again while still stored in the NWCache, it can be re-mapped to main memory and taken off the ring.

The storage capacity of the ring is completely independent of the individual or combined local memory sizes. The storage capacity of the ring is simply proportional to the number of available channels, and the channels' bandwidth and length. More specifically, the capacity of the ring is given by: $capacity\_in\_bits = (num\_channels \times fiber\_length \times transmission\_rate) \div speed\_of\_light$, where $speed\_of\_light = 2 \times 10^8 m/s$.

Ring management. Each cache channel transfers and stores the pages swapped out by a particular node. A page can be swapped out to the NWCache if there is room available on the node's cache channel. An actual swap out to the NWCache allows the swapping node to re-use the space occupied by the page in memory right away. At the time of swapping, the node must set a bit (the Ring bit) associated with the page table entry for the page, indicating that the page is stored on the NWCache. The swapping node must also send a message to the NWCache interface of the corresponding I/O-enabled node. This message includes the number of the swapping node and the number of the page swapped out. The remote NWCache interface then saves this information in a FIFO queue associated with the corresponding cache channel.

Every time the disk controller attached to the interface has room for an extra page in its cache, the interface starts snooping the most heavily loaded cache channel and copies as many pages as possible from it to the disk cache. After the page is sent to the disk controller cache, the ACK is sent to the node that originally swapped the page out. The ACK is taken by the node to mean that it can now re-use the space occupied by the page on the ring and prompts it to reset the Ring bit associated with the page.

Two important characteristics of how pages are copied from the NWCache to the disk cache increase the spatial locality of the different writes in the cache: a) pages are copied in the same order as they were originally swapped out; and b) the interface only starts snooping another channel after the swap-outs on the current channel have been exhausted. When a node swaps consecutive pages out, these two characteristics allow several writes to be batched to disk in a single operation.

Pages can be re-mapped into memory straight from the NWCache itself. On a page fault, the faulting node checks if the Ring bit for the page is set. If not, the page fault proceeds just as described in the previous section. Otherwise, the faulting node uses the last virtual-to-physical translation for the page to determine the node that last swapped the page out. Then the faulting node can simply snoop the page off the correct cache channel. Moreover, the faulting node must send a message to the NWCache interface of the I/O-enabled node responsible for the page informing the interface of the page and cache channel numbers. This message tells the remote NWCache interface that the page does not have to be written to disk, since there is again a copy of it in main memory. The remote interface then takes the page number off of the cache channel's FIFO queue and sends the ACK to the node that originally swapped the page out to the NWCache.

Note that the NWCache does not suffer from coherence problems, since we do not allow more than one copy of a page beyond the disk controller's boundary. The single copy can be in main memory or on the NWCache.

Software cost. The software cost of the NWCache approach is negligible (provided that the kernel code is available, of course). The operating system code must be changed to include the Ring bits and to drive the NWCache interface.

Hardware cost. The electronic hardware cost of the NWCache is restricted to the I/O bus and disk interfaces, the FIFOs, and the buffers and drivers that interface the electronic and optical parts of the NWCache interface. The optical hardware requirements of the NWCache are also minimal. The NWCache interface at each node can read any of the cache channels, but can only write to the cache channel associated with the node and, thus, does not require arbitration. The NWCache interface regenerates, reshapes, and reclocks this writable cache channel. To accomplish this functionality, the interface requires two tunable receivers, one fixed transmitter, and one fixed receiver, as shown in figure 2. One of the tunable receivers is responsible for reading the pages to be written to disk from the NWCache, while the other tunable receiver is used to search the NWCache for a page that has been faulted on locally. The fixed transmitter is used to insert new data on the writable cache channel. In conjunction with this transmitter, the fixed receiver is used to re-circulate the data on the writable cache channel. Thus, the optical hardware cost for the ring is then only $4 \times n$ optical components, where n is the number of nodes in the multiprocessor. The number of cache channels is n. This cost is pretty low for small to medium-scale multiprocessors, even for today's optical technology costs. As aforementioned, commercial WDM products can be found with more than one hundred channels. Mass production of optical components and further advances in optical technology will soon make these costs acceptable even for large-scale multiprocessors.

Note that, even though our optical ring acts as a cache for disk data, the ring does not guarantee non-volatility (as some disk controller caches do), much in the same way as using idle node memory for swap-outs (e.g. [3]). This is not a serious problem however, since these approaches optimize the virtual memory management for applications that do not involve stringent reliability constraints, such as scientific applications.

4 Methodology and Workload

We use a detailed execution-driven simulator (based on the MINT front-end [20]) of a DASH-like [12] cache-coherent multiprocessor based on Release Consistency [4]. Memory, I/O, and network contention are fully modeled. The simulation of the multiprocessor operating system is limited to the part that really matters for our purposes: virtual memory management.

Our most important simulation parameters and their default values are listed in table 1. The values of the other parameters are comparable to those of modern systems, but have an insignificant effect on our results. The disk cache and main memory sizes we simulate were purposely kept small, because simulation time limitations prevent us from using real life input sizes. In fact, we reduced optical ring and disk cache storage

Parameter	Value
Number of Nodes	8
Number of I/O-Enabled Nodes	4
Page Size	4 KBytes
TLB Miss Latency	100 pcycles
TLB Shootdown Latency	500 pcycles
Interrupt Latency	400 pcycles
Memory Size per Node	256 KBytes
Memory Bus Transfer Rate	800 MBytes/sec
I/O Bus Transfer Rate	300 MBytes/sec
Network Link Transfer Rate	200 MBytes/sec
WDM Channels on Optical Ring	8
Optical Ring Round-Trip Latency	52 usecs
Optical Ring Transfer Rate	1.25 GBytes/sec
Storage Capacity on Optical Ring	512 KBytes
Optical Storage per WDM Channel	64 KBytes
Disk Controller Cache Size	16 KBytes
Min Seek Latency	2 msec
Max Seek Latency	22 msecs
Rotational Latency	4 msec
Disk Transfer Rate	20 MBytes/sec

Table 1. Simulator Parameters and Their Default Values – 1 pcycle = 5 nsecs.

capacities by a factor of 32 and main memory sizes by a factor of 256 with respect to their sizes in real systems. Our goal with these reductions was to produce roughly the same swap-out traffic in our simulations as in real systems.

Our choice for amount of optical storage deserves further remarks. Increasing the size of the NWCache can be accomplished by lengthening the optical fiber (and thus increasing ring round-trip latency) and/or using more cache channels. With current optical technology the capacity of the NWCache could be increased a few fold, but certainly not by a factor of 32. Nevertheless, we believe that in the near future this magnitude increase in size will be possible. In fact, our capacity assumptions can be considered highly conservative with respect to the future potential of optics, especially if we consider multiplexing techniques such as OTDM (Optical Time Division Multiplexing) which will potentially support 5000 channels [15].

Our application workload consists of seven parallel programs: Em3d, FFT, Gauss, LU, Mg, Radix, and SOR. Table 2 lists the applications, their input parameters, and the total data sizes (in MBytes) the inputs lead to. All applications *mmap* their files for both reading and writing[3] and access them via the standard virtual memory mechanism.

---

[3] The UNIX *mmap* call forces the user to specify the largest possible size of a writable file. This was not a problem for us, since it was always possible to determine the exact final size of each writable file for our applications. Nevertheless, we feel that the UNIX *mmap* call is much too restrictive for a pure virtual memory-based style of programming.

Program	Description	Input Size	Data (MB)
Em3d	Electromagnetic wave propagation	32 K nodes, 5% remote, 10 iters	2.5
FFT	1D Fast Fourier Transform	64 K points	3.1
Gauss	Unblocked Gaussian Elimination	570 × 512 doubles	2.3
LU	Blocked LU factorization	576 × 576 doubles	2.7
Mg	3D Poisson solver using multigrid techs	32 × 32 × 64 floats, 10 iters	2.4
Radix	Integer Radix sort	320 K keys, radix 1024	2.6
SOR	Successive Over-Relaxation	640 × 512 floats, 10 iters	2.6

Table 2. Application Description and Main Input Parameters.

5 Experimental Results

In this section we evaluate the performance of the NWCache comparing it against the performance of a standard multiprocessor. Before doing this, however, it is important to determine the best minimum number of page frames for each combination of prefetching technique and multiprocessor. We performed experiments where we varied this minimum number for each of the applications in our suite. In the presence of the NWCache, the vast majority of our applications achieved their best performance with a minimum of only 2 free page frames, regardless of the prefetching strategy.

The best configuration for the standard multiprocessor is not obvious. Under optimal prefetching, four of our applications (Gauss, LU, Radix, and SOR) favor large numbers (> 12) of free page frames, while two of them (Em3d and MG) achieve their best performance with only 2 free frames. The other application, Radix, requires 8 free frames for best performance. On the other hand, under naive prefetching, all applications except SOR favor small numbers (2 or 4) of free page frames. Thus, we picked 12 and 4 frames as the best minimum numbers of free frames under optimal and naive prefetching, respectively. All the results in this subsection will be presented for these configurations.

We are interested in assessing the extent to which the benefits provided by the NWCache actually produce performance improvements. As we have mentioned in the introduction, the NWCache has several performance benefits: it provides a staging area where swapped-out pages can reside until the disk is free; it increases the possibility of combining several writes to disk; it acts as a victim cache for pages that are swapped out and subsequently accessed by the same or a different processor; and it reduces the data traffic across the multiprocessor's interconnection network and memory buses. We now look at statistics related to each of these benefits in turn.

Write staging. Tables 3 and 4 show the average time (in pcycles) it takes to swap a page out of memory under the optimal and naive prefetching strategies, respectively. The tables show that swap-out times are 1 to 3 orders of magnitude lower when the NWCache is used. The main reason for this result is that the NWCache effectively increases the amount of disk cache from the viewpoint of the memory. A swap out is only delayed in the presence of NWCache when the swapping node's cache channel fills up. In contrast, when the NWCache is not assumed, swap outs are much more frequently delayed due to lack of space in the disk cache. As expected, the tables also show that

Application	Standard	NWCache
em3d	49.2	1.8
fft	86.6	3.1
gauss	30.9	1.0
lu	39.6	2.0
mg	33.1	0.6
radix	48.4	2.7
sor	31.8	1.3

Table 3. Average Swap-Out Times (in Mpcycles) under Optimal Prefetching.

Application	Standard	NWCache
em3d	180.4	2.8
fft	318.1	31.8
gauss	789.8	86.3
lu	455.0	24.3
mg	150.8	19.2
radix	1776.9	2.8
sor	819.4	12.5

Table 4. Average Swap-Out Times (in Kpcycles) under Naive Prefetching.

Application	Standard	NWCache	Increase
em3d	1.11	1.12	1%
fft	1.20	1.39	16%
gauss	1.06	1.07	1%
lu	1.13	1.24	10%
mg	1.11	1.16	5%
radix	1.08	1.12	4%
sor	1.46	2.30	58%

Table 5. Average Write Combining Under Optimal Prefetching.

Application	Standard	NWCache	Increase
em3d	1.10	1.10	0%
fft	1.35	1.38	2%
gauss	1.03	1.04	1%
lu	1.05	1.05	0%
mg	1.05	1.11	6%
radix	1.05	1.07	2%
sor	1.18	1.37	16%

Table 6. Average Write Combining Under Naive Prefetching.

swap-out times are much higher under the optimal prefetching technique than under its naive counterpart. This is a consequence of the fact that under optimal prefetching the reduced page read times effectively cluster the swap-outs in time, increasing contention.

Write combining. Given the way swap-outs are copied from the NWCache to the disk cache, the locality of the writes in the disk cache is often increased. When consecutive pages can be found in consecutive slots of the disk controller's cache, the writes of these pages can be combined in a single disk write access. The data in tables 5 and 6 confirm this claim. The tables present the average number of swap-outs that are combined in each disk write operation; maximum possible combining factor is 4, the maximum number of pages that can fit in a disk controller cache. Results show that increases in write combining are only moderate under the naive prefetching strategy, but can be significant under the optimal prefetching strategy. Again, the temporal clustering of swap-outs under optimal prefetching is responsible for this result. It becomes more common for the disk controller to find consecutive writes in its cache at the same time.

Victim caching. Table 7 presents the page read hit rates in the NWCache under the optimal and naive prefetching techniques. The table shows that hit rates are slightly higher under optimal prefetching than under naive prefetching, again due to the temporal characteristics of the page swap-outs under the two techniques. In addition, these results show that hit rates can be as high as 50+% (Gauss and MG) or as low as 10-% (Em3d). These results are a combination of two factors: the size of the memory working set and the degree of data sharing in the applications. Gauss, MG, and Em3d exhibit a significant amount of sharing, but only Gauss and MG have working sets that can (almost) fit in the combined memory/NWCache size. The other applications achieve hit rates in the 10 to 25% range.

Application	Naive	Optimal
em3d	8.5	10.0
fft	9.8	13.0
gauss	49.9	58.3
lu	13.5	19.5
mg	41.1	59.1
radix	17.2	22.6
sor	25.8	24.1

Table 7. NWCache Hit Rates Under Different Prefetching Techniques.

Application	Standard	NWCache	Reduction
em3d	13.4	9.7	28%
fft	25.9	19.6	24%
gauss	16.7	10.4	38%
lu	21.5	20.3	6%
mg	19.1	6.7	63%
radix	12.6	9.2	27%
sor	14.3	10.2	29%

Table 8. Average Page Fault Latency (in Kpcycles) for Disk Cache Hits Under Naive Prefetching.

The beneficial effect of victim caching is more pronounced under naive prefetching, since read fault overheads represent a significant fraction of the applications' execution times. Furthermore, pages can be read only slightly faster from the NWCache than the disk controller's cache, which reduces the potential gains of victim caching under optimal prefetching.

Contention. The NWCache alleviates contention for two reasons: a) page swap-outs are not transferred across the interconnection network; and b) page reads that hit in the NWCache are transferred neither across the network nor across the I/O node's memory bus (whenever the request for the corresponding I/O node can be aborted in time). In order to evaluate the benefit of the NWCache in terms of contention alleviation, we collected statistics on the average latency of a page read from the disk controller's cache under the naive prefetching technique. A comparison of the statistics from the standard multiprocessor against those of the NWCache-capable multiprocessor provides a rough estimate of the amount of contention that is eliminated.

Table 8 displays these data in thousands of processor cycles. The table shows that the NWCache reduced disk cache hit latencies by as much as 63%. For most applications, reductions range from 24 to 38%. Given that it takes about 6K pcycles to read a page from a disk cache in the total absence of contention, we can see that the contention reductions provided by the NWCache are always significant. For instance, the average disk cache hit takes about 21K pcycles for LU running on the standard multiprocessor. Out of these cycles, about 15K are due to contention of several forms. In the presence of the NWCache, the number of cycles due to contention in LU is reduced to about 14K, which means a 7% reduction. At the other extreme, consider the reduction in disk cache hit latency achieved by MG, 63%. Out of the 19K cycles that MG takes to read pages from disk caches on the standard multiprocessor, about 13K are due to contention. In the presence of the NWCache, the number of cycles due to contention in MG is reduced to about 700, which means a 95% reduction in contention.

Under optimal prefetching the NWCache is not as successful at alleviating contention, since there is usually not enough time to prevent a page transfer through the network and the I/O node bus when the page read request hits in the NWCache.

The results we presented thus far confirm that the NWCache indeed benefits performance significantly in several ways. The greatest performance gains come from very fast swap-outs, victim caching, and contention reduction.

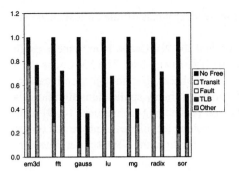

Fig. 3. Performance of Standard MP (left) and NWCache MP (right) Under Optimal Prefetching.

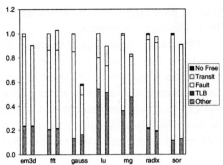

Fig. 4. Performance of Standard MP (left) and NWCache MP (right) Under Naive Prefetching.

Overall performance. Figures 3 and 4 show the normalized execution times of each of our applications under optimal and naive prefetching, respectively. From top to bottom, each bar in the graphs is divided into: the stall time resulting from the lack of free pages frames ("NoFree"); the overhead of waiting for another node to finish bringing a page into memory ("Transit"); the overhead of page faults ("Fault"); the overhead of TLB misses and TLB shootdown ("TLB"); and the components of execution time that are not related to virtual memory management ("Others"), including processor busy, cache miss, and synchronization times.

Figure 3 shows that under the optimal prefetching strategy, NoFree overheads are always very significant for the standard multiprocessor, especially for Gauss and SOR. The operating system frequently runs out of free frames on the standard multiprocessor due to the fact that page reads are completed relatively quickly, while page swap-outs are very time-consuming. When the multiprocessor is upgraded with the NWCache, NoFree times are reduced significantly as a result of its much faster page swap-outs.

The figure also demonstrates that for several applications the time taken on non-virtual memory operations is significantly reduced in the presence of the NWCache. These reductions come from improvements in the cost of remote data accesses and better synchronization behavior, both of them a result of the significant reduction of the traffic through the memory system (network and memories). Overall, we see that the NWCache provides performance improvements that average 41% and range from 23% (Em3d) to 60 and 64% (MG and Gauss), when optimal prefetching is assumed. In fact, improvements are greater than 28% in all cases, except Em3d.

The performance results when naive prefetching is assumed are quite different. Under this technique, execution times are dominated by page fault latencies, since disk cache hit rates are never greater than 15%. The page fault latencies then provide the much needed time for swap-outs to complete. As a result, NoFree times almost vanish, diminishing the importance of the fast swap-outs in the NWCache architecture.

Under naive prefetching, the addition of the NWCache to the multiprocessor improves performance from 3 (Radix) to 42% (Gauss) for all applications except FFT, which degrades by 3%. NWCache-related improvements come from reasonable reductions in page fault latencies, which result from reading pages off the optical cache and from alleviating contention.

Discussion. In summary, we have shown that the NWCache is extremely useful when prefetching is effective, mostly as a result of fast swap-outs. The NWCache is not as efficient when prefetching is either ineffective or absent, even though victim caching and contention alleviation improve the performance of many applications significantly. We expect results for realistic and sophisticated prefetching techniques [14, 9] to lie between these two extremes. In addition, as prefetching techniques improve and optical technology develops, we will see greater gains coming from the NWCache architecture.

6 Related Work

As far as we are aware, the NWCache is the only one of its kind; an optical network/write cache hybrid has never before been proposed and/or evaluated. A few research areas are related to our proposal however, such as the use of optic fiber as delay line memory and optimizations to disk write operations.

Delay line memories. Delay line memories have been implemented in optical communication systems [11] and in all-optical computers [8]. Optical delay line memories have not been used as write caches and have not been applied to multiprocessors before, even though optical networks have been a part of several parallel computer designs, e.g. [5, 6]. The only aspects of optical communication that these designs exploit are its high bandwidth and the ability to broadcast to a large number of nodes. Besides these aspects, the NWCache architecture also exploits the data storage potential of optics.

Disk write operations. Several researchers have previously set out to improve the performance of write operations for various types of disk subsystems. These efforts include work on improving the small write performance of RAIDs (e.g. [19]), using non-volatile RAM as a write cache (e.g. [17]), logging writes and then writing them to disk sequentially (e.g. [16]), using idle or underloaded node memory for storing page swap-outs [3], and using a log disk to cache writes directed to the actual data disk [7]. The two latter types of work are the closest to ours.

Storing page swap-outs in another node's memory is only appropriate for workstation networks where one or more nodes may be idle or underloaded at each point in time. The same technique could not be applied in our computing environment, a multiprocessor running an out-of-core application, since processors are always part of the computation and usually do not have spare memory they could use to help each other.

The storage architecture proposed in [7], the Disk Caching Disk (DCD), places the log disk between the RAM-based disk cache and the data disk, effectively creating an extra level of buffering of writes. New data to be written to disk is then stored in the RAM cache and later written sequentially on the log disk. Overwriting or reading a block requires moving around the log disk to find the corresponding block. When the data disk is idle, the data is copied from the log disk to the data disk. This scheme improves performance as it reduces seek and rotational latencies significantly when writing new data to the log disk and, as a result, frees space in the RAM cache more rapidly. Overwriting or reading involves seek and rotational latencies that are comparable to those of accesses to the data disk.

Just like the DCD, the NWCache also attempts to improve the write performance by creating an extra level of buffering. However, the NWCache places this buffer between main memory and the disk caches and thus does not require any modifications to standard disk controllers. In addition, overwriting or reading data from the NWCache is as efficient as writing new data to the NWCache. Another advantage of the NWCache is that it creates an exclusive path for disk writes to reach the disk controllers. Nevertheless, the technology used to implement the additional buffering in the DCD allows it much more buffer space than our optics-based buffer.

7 Conclusions

In this paper we proposed a novel architectural concept: an optical network/write buffer hybrid for page swap-outs called NWCache. The most important advantages of the NWCache are its fast swap-outs, its victim caching aspect, and its ability to alleviate contention for memory system resources. Through a large set of detailed simulations we showed that an NWCache-equipped multiprocessor can easily outperform a traditional multiprocessor for out-of-core applications; performance differences in favor of our system can be as large as 64% and depend on the type of disk data prefetching used. Based on these results and on the continually-decreasing cost of optical components, our main conclusion is that the NWCache architecture is highly efficient and should definitely be considered by multiprocessor designers.

Acknowledgements

The authors would like to thank Luiz Monnerat, Cristiana Seidel, Rodrigo dos Santos, and Lauro Whately for comments that helped improve the presentation of the paper.

References

1. A. Agarwal, R. Bianchini, D. Chaiken, K. Johnson, D. Kranz, J. Kubiatowicz, B.-H. Lim, K. Mackenzie, and D. Yeung. The MIT Alewife Machine: Architecture and Performance. In *Proceedings of the 22nd International Symposium on Computer Architecture*, June 1995.
2. R. Alverson, D. Callahan, D. Cummings, B. Koblenz, A. Porterfield, and B. Smith. The Tera Computer System. In *Proceedings of the 1990 International Conference on Supercomputing*, July 1990.
3. E. Felten and J. Zahorjan. Issues in the Implementation of a Remote Memory Paging System. Technical Report 91-03-09, Department of Computer Science and Engineering, University of Washington, March 1991.
4. K. Gharachorloo, D. Lenoski, J. Laudon, P. Gibbons, A. Gupta, and J. L. Hennessy. Memory Consistency and Event Ordering in Scalable Shared-Memory Multiprocessors. In *Proceedings of the 17th Annual International Symposium on Computer Architecture*, pages 15–26, May 1990.
5. K. Ghose, R. K. Horsell, and N. Singhvi. Hybrid Multiprocessing in OPTIMUL: A Multiprocessor for Distributed and Shared Memory Multiprocessing with WDM Optical Fiber Interconnections. In *Proceedings of the 1994 International Conference on Parallel Processing*, August 1994.

6. J.-H. Ha and T. M. Pinkston. SPEED DMON: Cache Coherence on an Optical Multichannel Interconnect Architecture. *Journal of Parallel and Distributed Computing*, 41(1):78–91, 1997.
7. Y. Hu and Q. Yang. DCD – Disk Caching Disk: A New Approach for Boosting I/O Performance. In *Proceedings of the 23rd International Symposium on Computer Architecture*, pages 169–177, May 1996.
8. H. F. Jordan, V. P. Heuring, and R. J. Feuerstein. Optoelectronic Time-of-Flight Design and the Demonstration of an All-Optical, Stored Program. *Proceedings of IEEE. Special issue on Optical Computing*, 82(11), November 1994.
9. T. Kimbrel et al. A Trace-Driven Comparison of Algorithms for Parallel Prefetching and Caching. In *Proceedings of the 2nd USENIX Symposium on Operating Systems Design and Implementation*, October 1996.
10. D. Kotz and C. Ellis. Practical Prefetching Techniques for Multiprocessor File Systems. *Journal of Distributed and Parallel Databases*, 1(1):33–51, January 1993.
11. R. Langenhorst et al. Fiber Loop Optical Buffer. *Journal of Lightwave Technology*, 14(3):324–335, March 1996.
12. D. Lenoski, J. Laudon, T. Joe, D. Nakahira, L. Stevens, A. Gupta, and J. Hennessy. The DASH Prototype: Logic Overhead and Performance. *IEEE Transactions on Parallel and Distributed Systems*, 4(1):41–61, January 1993.
13. K. McKusick, W. Joy, S. Leffler, and R. Fabry. A Fast File System for UNIX. *ACM Transactions on Computer Systems*, 2(3):181–197, August 1984.
14. T. Mowry, A. Demke, and O. Krieger. Automatic Compiler-Inserted I/O Prefetching for Out-Of-Core Applications. In *Proceedings of the 2nd USENIX Symposium on Operating Systems Design and Implementation*, October 1996.
15. A. G. Nowatzyk and P. R. Prucnal. Are Crossbars Really Dead? The Case for Optical Multiprocessor Interconnect Systems. In *Proceedings of the 22nd International Symposium on Computer Architecture*, pages 106–115, June 1995.
16. M. Rosenblum and J. Ousterhout. The Design and Implementation of a Log-Structured File System. *ACM Transactions on Computer Systems*, 10(2):26–52, February 1992.
17. C. Ruemmler and J. Wilkes. UNIX Disk Access Patterns. In *Proceedings of the Winter 1993 USENIX Conference*, January 1993.
18. D. B. Sarrazin, H. F. Jordan, and V. P. Heuring. Fiber Optic Delay Line Memory. *Applied Optics*, 29(5):627–637, February 1990.
19. D. Stodolsky, M. Holland, W. Courtright III, and G. Gibson. Parity Logging Disk Arrays. *ACM Transactions on Computer Systems*, 12(3):206–235, August 1994.
20. J. E. Veenstra and R. J. Fowler. MINT: A Front End for Efficient Simulation of Shared-Memory Multiprocessors. In *Proceedings of the 2nd International Workshop on Modeling, Analysis and Simulation of Computer and Telecommunication Systems*, January 1994.
21. D. Womble, D. Greenberg, R. Riesen, and D. Lewis. Out of Core, Out of Mind: Practical Parallel I/O. In *Proceedings of the Scalable Parallel Libraries Conference*, October 1993.

NetCache: A Network/Cache Hybrid for Multiprocessors [*]

Enrique V. Carrera and Ricardo Bianchini

COPPE Systems Engineering
Federal University of Rio de Janeiro
Rio de Janeiro, Brazil 21945-970
{vinicio,ricardo}@cos.ufrj.br

Abstract. In this paper we propose the use of an optical network not only as the communication medium, but also as a system-wide cache for the shared data in a multiprocessor. More specifically, the basic idea of our novel network/cache hybrid (and associated coherence protocol), called NetCache, is to use an optical ring network on which some amount of recently-accessed shared data is continually sent around. These data are organized as a cache shared by all processors. We use detailed execution-driven simulations of a dozen applications to evaluate a multiprocessor based on our NetCache architecture. We compare a 16-node multiprocessor with a third-level NetCache against two highly-efficient systems based on the DMON and LambdaNet optical interconnects. Our results demonstrate that the NetCache multiprocessor outperforms the DMON and LambdaNet systems by as much as 99 and 79%, respectively. Based on these results, our main conclusion is that the NetCache is highly efficient for most applications.

1 Introduction

The gap between processor and memory speeds continues to widen; memory accesses usually take hundreds of processor cycles to complete, especially in scalable shared-memory multiprocessors. Caches are generally considered the best technique for tolerating the high latency of memory accesses. Microprocessor chips include caches so as to avoid processor stalls due to the unavailability of data or instructions. Modern computer systems use off-chip caches to reduce the average cost of memory accesses. Unfortunately, caches are often somewhat smaller than their associated processors' working sets, meaning that a potentially large percentage of memory references might still end up reaching the memory modules. This scenario is particularly damaging to performance when memory references are directed to remote memories through the scalable interconnection network. In this case, the performance of the network is critical.

The vast majority of multiprocessors use electronic interconnection networks. However, relatively recent advances in optical technology have prompted several studies on the use of optical networks in multiprocessors. Optical fibers exhibit extremely high bandwidth and can be multiplexed to provide a large number of independent communication channels. These characteristics can be exploited to improve performance

---

[*] This research was supported by Brazilian CNPq.

(significantly) by simply replacing a traditional multiprocessor network with an optical equivalent. However, we believe that optical networks can be exploited much more effectively. The extremely high bandwidth of optical media provides data storage in the network itself and, thus, these networks can be transformed into fast-access data storage devices.

Based on this observation and on the need to reduce the average latency of memory accesses in parallel computers, in this paper we propose the use of an optical network not only as the communication medium, but also as a system-wide cache for the shared data in a multiprocessor. More specifically, the basic idea of our novel network/cache hybrid (and associated coherence protocol), called NetCache, is to use an optical ring network on which some amount of recently-accessed shared data is continually sent around. These data are organized as a cache shared by all processors. Most aspects of traditional caches, such as hit/miss rates, capacity, storage units, and associativity, also apply to the NetCache.

The functionality provided by the NetCache suggests that a multiprocessor based on it should deliver better performance than previously-proposed optical network-based multiprocessors. To verify this hypothesis, we use detailed execution-driven simulations of a dozen applications running on a 16-node multiprocessor with a third-level shared cache, and compare it against highly-efficient systems based on the DMON [7] and LambdaNet [6] optical interconnects. Our results demonstrate that the NetCache multiprocessor performs at least as well as the DMON-based system for our applications; the performance differences in favor of the NetCache system can be as significant as 99%. The NetCache system also compares favorably against the LambdaNet multiprocessor. For nine of our applications, the running time advantage of the NetCache machine ranges from 7% for applications with little data reuse in the shared cache to 79% for applications with significant data reuse. For the other applications, the two systems perform similarly.

Given that optical technology will likely provide a better cost/performance ratio than electronics in the future, we conclude that designers should definitely consider the NetCache architecture as a practical approach to both low-latency communication and memory latency tolerance.

2 Background

2.1 Optical Communication Systems

Two main topics in optical communication systems bear a direct relationship to the ideas proposed in this paper: Wavelength Division Multiplexing (WDM) networks and delay line memories.

WDM networks. The maximum bandwidth achievable over an optic fiber is on the order of Tbits/s. However, due to the fact that the hardware associated with the end points of an optical communication system is usually of an electronic nature, transmission rates are currently limited to the Gbits/s level. In order to approach the full potential of optical communication systems, multiplexing techniques must be utilized. WDM is one such multiplexing technique. With WDM, several independent communication channels can be implemented on the same fiber. Optical networks that use this

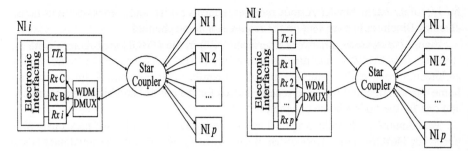

Fig. 1. Overview of DMON. **Fig. 2.** Overview of LambdaNet.

multiplexing technique are called WDM networks. Due to the rapid development of the technology used in its implementation, WDM has become one of the most popular multiplexing techniques. The NetCache architecture focuses on WDM technology due to its immediate availability, but there is nothing in our proposal that is strictly dependent on this specific type of multiplexing.

Delay line memories. Given that light travels at a constant and finite propagation speed on the optical fiber (approximately 2.1E+8 meters/s), a fixed amount of time elapses between when a datum enters and leaves an optical fiber. In effect, the fiber acts as a delay line. Connecting the ends of the fiber to each other (and regenerating the signal periodically), the fiber becomes a delay line memory [11], since the data sent to the fiber will remain there until overwritten. An optical delay line memory exhibits characteristics that are hard to achieve by other types of delay lines. For instance, due to the high bandwidth of optical fibers, it is possible to store a reasonable amount of memory in just a few meters of fiber (e.g. at 10 Gbits/s, about 5 Kbits can be stored on one 100 meters-long WDM channel).

2.2 DMON

The Decoupled Multichannel Optical Network (DMON) is an interesting WDM network that has been proposed by Ha and Pinkston in [7]. The network divides its $p + 2$ (where p is the number of nodes in the system) channels into two groups: one group is used for broadcasting, while the other is used for point-to-point communication between nodes. The first group is formed by two channels shared by all nodes in the system, the control channel and the broadcast channel. The other p channels, called home channels, belong in the second group of channels.

The control channel is used for distributed arbitration of all other channels through a reservation scheme [3]. The control channel itself is multiplexed using the TDMA (Time Division Multiple Access) protocol [3]. The broadcast channel is used for broadcasting global events, such as coherence and synchronization operations, while home channels are used only for memory block request and reply operations. Each node can transmit on any home channel, but can only receive from a single home channel. Each node acts as "home" (the node responsible for providing up-to-date copies of the blocks) for $1/p$ of the cache blocks. A node receives requests for its home blocks from its home channel. Block replies are sent on the requester's home channel.

Figure 1 overviews a network interface ("NI") in the DMON architecture, with its receivers (labeled "Rx") and tunable transmitter ("TTx"). As seen in the figure, in this architecture each node requires a tunable transmitter (for the control, broadcast, and home channels), and three fixed receivers (two for the control and broadcast channels, and one for the node's home channel).

We compare performance against an update-based protocol [1] we previously proposed for DMON. This protocol has been shown to outperform the invalidate-based scheme proposed for DMON in [7]. Our protocol is very simple since all writes to shared data are sent to their home nodes, through coalescing write buffers. Our update protocol also includes support for handling critical races that are caused by DMON's decoupling of the read and write network paths. The protocol treats these races by buffering the updates received during the pending read operation and applies them to the block right after the read is completed.

Given that a single broadcast channel would not be able to deal gracefully with the heavy update traffic involved in a large set of applications, we extended DMON with an extra broadcast channel for transferring updates. A node can transmit on only one of the coherence channels, which is determined as a function of the node's identification, but can receive from both of these channels. In addition, we extended DMON with two fixed transmitters (one for the control channel and the other for the coherence channel on which the node is allowed to transmit) to avoid incurring the overhead of re-tuning the tunable transmitter all the time. Besides the extra channel (and associated receivers) and the additional fixed transmitters, the hardware of the modified DMON network is the same as presented in figure 1. Thus, the overall hardware cost of this modified DMON architecture in terms of optical components is then $7 \times p$.

2.3 LambdaNet

The LambdaNet architecture has been proposed by Goodman *et al.* in [6]. The network allocates a WDM channel for each node; the node transmits on this channel and all other nodes have fixed receivers on it. In this organization each node uses one fixed transmitter and p fixed receivers, as shown in figure 2. The overall hardware cost of the LambdaNet is then $p^2 + p$.

No arbitration is necessary for accessing transmission channels. Each node simultaneously receives all the traffic of the entire network, with a subsequent selection, by electronic circuits, of the traffic destined for the node. This scheme thus allows channels to be used for either point-to-point or broadcast communication.

Differently from DMON, the LambdaNet was not proposed with an associated coherence protocol. The LambdaNet-based multiprocessor we study in this paper uses a write-update cache coherence protocol, where write and synchronization transactions are broadcast to nodes, while the read traffic uses point-to-point communication between requesters and home nodes. Just as the update-based protocol we proposed for DMON, the memory modules are kept up-to-date at all time. Again, in order to reduce the write traffic to home nodes, we assume coalescing write buffers.

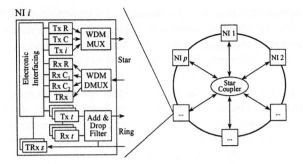

Fig. 3. Overview of NetCache Architecture.

The combination of the LambdaNet and the coherence protocol we suggest for it represents a performance upper bound for multiprocessors that do not cache data on the network, since the update-based protocol avoids coherence-related misses, the LambdaNet channels do not require any medium access protocol, and the LambdaNet hardware does not require the tunning of transmitters or receivers.

3 The Network/Cache Hybrid

3.1 Overview

Each node in a NetCache-based multiprocessor is extremely simple. In fact, all of the node's hardware components are pretty conventional, except for the network (NetCache) interface. More specifically, the node includes one processor, a coalescing write buffer, first and second-level caches, local memory, and the NetCache interface connected to the memory bus. The interface connects the node to two subnetworks: a star coupler WDM subnetwork and a WDM ring subnetwork. Figure 3 overviews the NetCache architecture, including its interfaces and subnetworks. The star coupler subnetwork is the OPTNET network [1], while the ring subnetwork is unlike any other previously-proposed optical network. The following sections describe each of these components in detail.

3.2 Star Coupler (Sub)Network

Just as in DMON, the star coupler WDM subnetwork in the NetCache divides the channels into two groups: one group for broadcast-type traffic and another one for direct point-to-point communication. Three channels, a request channel and two coherence channels, are assigned to the first group, while the other p channels, called home channels, are assigned to the second group.

The request channel is used for requesting memory blocks. The response to such a request is sent by the block's home node (the node responsible for providing up-to-date copies of the block) on its corresponding home channel. The coherence channels are used for broadcasting coherence and synchronization transactions. Just like the control channel in DMON, the request channel uses TDMA for medium access control. The access to the coherence channels, on the other hand, is controlled with TDMA with variable time slots [3]. Differently from DMON, home channels do not require arbitration, since only the home node can transmit on the node's home channel.

As seen in figure 3, each node in the star coupler subnetwork requires three fixed transmitters (one for the request channel, one for the home channel, and the last for one of the coherence channels), three fixed receivers (for the broadcast channels), and one tunnable receiver labeled "*TRx*" (for the home channels). The hardware cost for this subnetwork is then $7 \times p$ optical components.

3.3 Ring (Sub)Network

The ring subnetwork is the most interesting aspect of the NetCache architecture. The ring is used to store some amount of recently-accessed shared data, which are continually sent around the ring's WDM channels, referred to as cache channels, in one direction. These data are organized as cache blocks shared by all nodes of the machine. In essence, the ring becomes an extra level of caching that can store data from any node. However, the ring does not respect the inclusion property of cache hierarchies, i.e. the data stored by the ring is not necessarily a superset of the data stored in the upper cache levels. More specifically, the ring's storage capacity in bits is given by: $capacity\_in\_bits = (num\_channels \times fiber\_length \times transmission\_rate) \div speed\_of\_light\_on\_fiber$, where $speed\_of\_light\_on\_fiber = 2 \times 10^8 m/s$.

The ring works as follows. Each cache channel stores the cached blocks of a particular home node. Assuming a total of c cache channels, each home node has $t = c/p$ cache channels for its blocks. Since cache channels and blocks are associated with home nodes in interleaved, round-robin fashion, the cache channel assigned to a certain block can be determined by the following expression: $block\_address \bmod c$. A particular block can be placed anywhere within a cache channel.

Just as in a regular cache, our shared cache has an address tag on each block frame (shared cache line). The tags contain part of the shared cache blocks' addresses and consume a (usually extremely small) fraction of the storage capacity of the ring. The access to each block frame by different processors is strictly sequential as determined by the flow of data on the cache channel. The network interface accesses a block via a shift register that gets replicas of the bits flowing around the ring. Each shift register is as wide as a whole block frame. When a shift register is completely filled with a block, the hardware moves the block to another register, the access register, which can then be manipulated. In our base simulations, it takes 10 cycles to fill a shift register. This is the amount of time the network interface has to check the block's tag and possibly to move the block's data before the access register is overwritten.

The NetCache interface at each node can read any of the cache channels, but can only write to the t cache channels associated to the node. In addition, the NetCache interface regenerates, reshapes, and reclocks these t cache channels. To accomplish this functionality, the interface requires two tunable receivers, and t fixed transmitter and receiver sets, as shown in figure 3. One of the tunable receivers is for the cache channel used last and the other is for pre-tuning to the next channel. The t fixed transmitters are used to insert new data on any of the cache channels associated with the node. In conjunction with these transmitters, the t fixed receivers are used to re-circulate the data on the cache channels associated with the node. Thus, the hardware cost for this subnetwork is then $2 \times p + 2 \times c$ optical components.

Taking both subnetworks into account, the overall hardware cost of the NetCache architecture is $9 \times p + 2 \times c$. In our experiments we set $c = 8 \times p$, leading to a total of $25 \times p$ optical components. This cost is a factor of 3.6 greater than that of DMON, but is linear on p, as opposed to quadratic on p as the LambdaNet's.

3.4 Coherence Protocol

The coherence protocol is an integral part of the NetCache architecture. The NetCache protocol is based on update coherence, supported by both broadcasting and point-to-point communication. The update traffic flows through the coherence channels, while the data blocks are sent across the home and cache channels. The request channel carries all the memory read requests and update acknowledgements. The description that follows details the protocol in terms of the actions taken on read and write accesses.

Reads. On a read access, the memory hierarchy is traversed from top to bottom, so that the required word can be found as quickly as possible. A miss in the second-level cache blocks the processor and is handled differently depending on the type of data read. In case the requested block is private or maps to the local memory, the read is treated by the memory, which returns the block to the processor.

If the block is shared and maps to another home node, the miss causes an access to the NetCache. The request is sent to the corresponding node through the request channel, and the tunable receiver in the star coupler subnetwork is tuned to the correct home channel. In parallel with this, the requesting node tunes one of its tunable receivers in the ring subnetwork to the corresponding cache channel. The requesting node then waits for the block to be received from either the home channel or the cache channel, reads the block, and returns it to the second-level cache.

When a request arrives at the home node, the home checks if the block is already in any of its cache channels. In order to manage this information, the NetCache interface keeps a hash table storing the address of its blocks that are currently cached on the ring. If the block is already cached, the home node will simply disregard the request; the block will eventually be received by the requesting node. If the block is not currently cached, the home node will read it from memory and place it on the correct cache channel, replacing one of the blocks already there if necessary. (Replacements are random and do not require write backs, since the memory and the cache are always kept up-to-date.) In addition to returning the requested block through a cache channel, the NetCache sends the block through the home node's home channel.

It is important to note that our protocol starts read transactions on both the star coupler and ring subnetworks so that a read miss in the shared cache takes no longer than a direct access to remote memory. If reads were only started on the ring subnetwork, shared cache misses would take half a roundtrip longer (on average) to satisfy than a direct remote memory access.

Writes. Our multiprocessor architecture implements the release consistency memory model [4]. Consecutive writes to the same cache block are coalesced in the write buffer. Coalesced writes to a private block are sent directly to the local memory through the first and second-level caches. Coalesced writes to a shared block are always sent to one of the coherence channels in the form of an update, again through the local caches. An update only carries the words that were actually modified in each block.

Each update must be acknowledged by the corresponding block's home node before another update by the same node can be issued, just so the memory modules do not require excessively long input queues (i.e. update acks are used simply as a flow control measure). The other nodes that cache the block simply update their local caches accordingly upon receiving the update. When the home node sees the update, the home inserts it into the memory's FIFO queue, and sends an ack through the request channel. The update acks usually do not overload the request channel, since an ack is a short message (much shorter than multi-word updates) that fits into a single request slot.

After having inserted the update in its queue, the home node checks whether the updated block is present in a cache channel. If so, besides updating its memory and local caches, the home node updates the cache channel. If the block is not present in a cache channel, the home node will not include it.

There are two types of critical races in our coherence protocol that must be taken care of. The first occurs when a coherence operation is seen for a block that has a pending read. Similarly to the update protocol we proposed for the DMON network, our coherence protocol treats this type of race by buffering updates and later combining them with the block received from memory. The second type of race occurs because there is a window of time between when an update is broadcast and when the copy of the block in the shared cache is actually updated. During this timing window, a node might miss on the block, read it from the shared cache and, thus, never see the update. To avoid this problem, each network interface includes a small FIFO queue that buffers the block addresses of some previous updates. Any shared cache access to a block represented in the queue must be delayed until the address of the block leaves the queue, which guarantees that when the read is issued, the copy of the block in the shared cache will be up-to-date. The management of the queue is pretty simple. The entry in front of the queue can be dropped when it has resided in the queue for the same time as two roundtrips around the ring subnetwork, i.e. the maximum amount of time it could take a home node to update the copy of the block in the shared cache. Hence, the maximum size of the queue is the maximum number of updates that can be issued during this period; in our simulations, this number is 54.

4 Methodology and Workload

We simulate multiprocessors based on the NetCache, DMON and LambdaNet interconnects. We use a detailed execution-driven simulator (based on the MINT front-end [12]) of 16-node multiprocessors. Each node of the simulated machines contains a single 200-MHz processor, a 16-entry write buffer, a 4-KByte direct-mapped first-level data cache with 32-byte cache blocks, a 16-KByte direct-mapped second-level data cache with 64-byte cache blocks, local memory, and a network interface. Note that these cache sizes are purposedly kept small, because simulation time limitations prevent us from using real life input sizes.

Operation	Latency
Shared cache read hit	
1. 1st-level tag check	1
2. 2nd-level tag check	4
3. Avg. shared cache delay	25
4. NI to 2nd-level cache	16
Total shared cache hit	46
Shared cache read miss	
1. 1st-level tag check	1
2. 2nd-level tag check	4
3. Avg. TDMA delay	8
4. Memory request	1*
5. Flight	1
6. Memory read	76+
7. Block transfer	11
8. Flight	1
9. NI to 2nd-level cache	16
Total shared cache miss	119

Table 1. Read Latency Breakdowns (in pcycles) for NetCache – 1 pcycle = 5 nsecs.

Operation	Latency	
	LambdaNet	DMON
2nd-level read miss		
1. 1st-level tag check	1	1
2. 2nd-level tag check	4	4
3. Avg. TDMA delay	–	8
4. Reservation	–	1*
5. Tuning delay	–	4
6. Memory request	1*	2
7. Flight	1	1
8. Memory read	76+	76+
9. Avg. TDMA delay	–	8
10. Reservation	–	1*
11. Block transfer	11*	12
12. Flight	1	1
13. NI to 2nd-level	16	16
Total 2nd-level miss	111	135

Table 2. Read Latency Breakdowns (in pcycles) for DMON and LambdaNet – 1 pcycle = 5 nsecs.

Shared data are interleaved across the memories at the block level. All instructions and first-level cache read hits are assumed to take 1 processor cycle (pcycle). First-level read misses stall the processor until the read request is satisfied. A second-level read hit takes 12 pcycles to complete. Writes go into the write buffer and take 1 pcycle, unless the write buffer is full, in which case the processor stalls until an entry becomes free. Reads are allowed to bypass writes that are queued in the write buffers. A memory module can provide the first two words requested 12 pcycles after the request is issued. Other words are delivered at a rate of 2 words per 8 pcycles. Memory and network contention are fully modeled.

In the coherence protocols we simulate only the secondary cache is updated when an update arrives at a node; the copy of the block in the first-level cache is invalidated. In addition, in order to reduce the write traffic, our multiprocessors use coalescing write buffers for all protocol implementations. A coalesced update only carries the words that were actually modified in each block. Our protocol implementations assume a release-consistent memory model [4].

The optical transmission rate we simulate is 10 Gbits/s. In the NetCache simulations we assume 128 cache channels. The length of the fiber is approximately 45 meters. These parameters lead to a ring roundtrip latency of 40 cycles and a storage capacity of 32 KBytes of data. The shared cache line size is 64 bytes. The NetCache storage capacity we simulate is compatible with current technology, but is highly conservative with respect to the future potential of optics, especially if we consider multiplexing techniques such as OTDM (Optical Time Division Multiplexing) which will potentially support 5000 channels [10].

Program	Description	Input Size
CG	Conjugate Gradient kernel	1400 × 1400 doubles, 78148 non-0s
Em3d	Electromagnetic wave propagation	8 K nodes, 5% remote, 10 iterations
FFT	1D Fast Fourier Transform	16 K points
Gauss	Unblocked Gaussian Elimination	256 × 256 floats
LU	Blocked LU factorization	512 × 512 floats
Mg	3D Poisson solver using multigrid techniques	24 × 24 × 64 floats, 6 iterations
Ocean	Large-scale ocean movement simulation	66 × 66 grid
Radix	Integer Radix sort	512 K keys, radix 1024
Raytrace	Parallel ray tracer	teapot
SOR	Successive Over-Relaxation	256 × 256 floats, 100 iterations
Water	Simulation of water molecules, spatial alloc.	512 molecules, 4 timesteps
WF	Warshall-Floyd shortest paths algorithm	384 vertices, (i,j) w/ 50% chance

Table 3. Application Description and Main Input Parameters.

The latency of shared cache reads in the NetCache architecture is broken down into its base components in table 1. The latency of 2nd-level read misses in the LambdaNet and DMON-based systems is broken down in table 2. In the interest of saving space and given that most or all coherence transaction overhead is usually hidden from the processor by the write buffer, we do not present the latency breakdown for these transactions. Please refer to [1] for further details. All numbers in tables 1 and 2 are in pcycles and assume channel and memory contention-free scenarios. The values marked with "*" and "+" are the ones that may be increased by network and memory contention/serialization, respectively.

Our application workload consists of twelve parallel programs: CG, Em3d, FFT, Gauss, LU, Mg, Ocean, Radix, Raytrace, SOR, Water, and WF. Table 3 lists the applications and their inputs.

5 Experimental Results

5.1 Overall Performance

Except for CG, LU, and WF, all our applications exhibit good speedups (> 10) on a 16-node NetCache-based multiprocessor with a 32-KByte shared cache. Figure 4 shows the running times of our applications. For each application we show, from left to right, the NetCache, LambdaNet, and DMON-U results, normalized to those of NetCache.

A comparison between the running time of the LambdaNet and DMON-U systems shows that the former multiprocessor performs at least as well as the latter for all applications, but performance differences are always relatively small; 6% on average. This relatively small difference might seem surprising at first, given that the read latency is by far the largest overhead in all our applications (except WF) and the contention-free 2nd-level read miss latency in the DMON-U system is 22% higher than in the LambdaNet system. However, the LambdaNet multiprocessor is usually more prone to

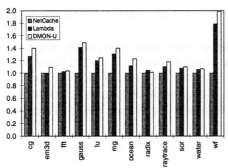
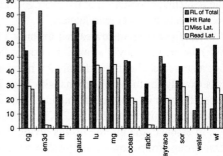

Fig. 4. Running Times. **Fig. 5.** Read Latencies and Hit Rates.

contention effects than the DMON-U and NetCache systems, due to two characteristics of the former system: a) its read and write transactions are not decoupled; and b) its absence of serialization points for updates from different nodes leads to an enormous bandwidth demand for updates. Contention effects are then the reason why performance differences in favor of the LambdaNet system are not more significant.

The NetCache system compares favorably against the DMON-U multiprocessor. The performance advantage of the NetCache system averages 27% for our applications. Except for FFT and Radix, where the performance difference between these two systems is negligible, the advantage of the NetCache ranges from 7% for Water to 99% for WF, averaging 32%. This significant performance difference can be credited to a large extent to the ability of the shared cache to reduce the average read latency, since their contention-free 2nd-level read miss latencies only differ by 13% and contention affects the two systems in similar ways.

Figure 4 demonstrates that a comparison between the NetCache and LambdaNet multiprocessors also clearly favors our system. Our running times are 20% lower on average for our application suite. The performance of the two systems is equivalent for 3 applications: Em3d, FFT, and Radix. For the other 9 applications, the performance advantage of the NetCache multiprocessor averages around 26%, ranging from about 7% for SOR and Water to 41 and 79% for Gauss and WF, respectively. Disregarding the 5 applications for which the performance difference between the two systems is smaller than 10%, the average NetCache advantage over the LambdaNet goes up to 31%. Again, the main reason for these favorable results is the ability of the NetCache architecture to reduce the average read latency, when a non-negligible percentage of the 2nd-level read misses hit in the shared cache.

5.2 Effectiveness of Data Caching

Figure 5 concentrates our data on the effectiveness of data caching in the NetCache architecture, again assuming 16 processor nodes. For each application we show a group of 4 bars. The leftmost bar represents the read latency as a percentage of the total execution time for a NetCache multiprocessor without a shared cache. The other bars represent our system with a 32-KByte shared cache: from left to right, the shared cache hit rate, the percentage reduction of the average 2nd-level read miss latency, and the percentage reduction in the total read latency.

The figure demonstrates that for only 3 applications in our suite (Radix, Water, and WF) the read latency is a small fraction of the overall running time of the machine without a shared cache; the read latency fraction is significant for the other 9 applications. It is exactly for the applications with large read latency fractions that a shared cache is most likely to improve performance. However, as seen in the figure, not all applications exhibit high hit rates when a 32-KByte shared cache is assumed.

In essence, the hit rate results divide our applications into 3 groups as follows. The first group (from now on referred to as *Low-reuse*) includes the applications with insignificant data reuse in the shared cache. In this group are Em3d, FFT, and Radix, for which less than 32% of the 2nd-level read misses hit in the shared cache. As a result, the reductions in 2nd-level read miss latency and total read latency are almost negligible for these applications. The second group of applications (*High-reuse*) exhibit significant data reuse in the shared cache. Three applications belong in this group: Gauss, LU, and Mg. Their hit rates are very high, around 70%, leading to reductions in 2nd-level read miss latency and total read latency of at least 35%. The other 6 applications (CG, Ocean, Raytrace, SOR, Water, and WF) form the third group (*Moderate-reuse*), with intermediate hit rates and latency reductions in the 20 to 30% range.

Overall, our caching results suggest that the NetCache multiprocessor could not significantly outperform the LambdaNet system for Em3d, FFT, Radix, and Water; these applications simply do not benefit significantly from our shared cache, either because their hit rates are low or because the latency of reads is just not a performance issue. On the other hand, the results suggest that WF should not benefit from the NetCache architecture either, which is not confirmed by our simulations. The reason for this (apparently) surprising result is that the read latency reduction promoted by the NetCache has an important side-effect: it improves the synchronization behavior of 7 applications in our suite (CG, LU, Mg, Ocean, Radix, Water, and WF). For WF in particular, a 32-KByte shared cache improves the synchronization overhead by 56% by alleviating the load imbalance exposed during this application's barrier synchronizations.

5.3 Comparison to Other Systems

To end this section we compare the NetCache architecture qualitatively to yet another set of systems.

One might claim that it should be more profitable to use a simpler optical network (such as DMON or our star coupler subnetwork) and increase the size of 2nd-level caches than to use a full-blown NetCache. Even though this arrangement does not enjoy the *shared* cache properties of the NetCache architecture, it is certainly true that larger 2nd-level caches could improve the performance of certain applications enough to outperform a NetCache system with small 2nd-level caches. However, the amount of additional cache space necessary for this to occur can be quite significant. Furthermore, even on a system with extremely large 2nd-level caches, most applications should still profit from a shared cache at a lower level, especially when the remote memory accesses of a processor can be prevented by accesses issued by other processors (as in Gauss). These effects should become ever more pronounced, as optical technology costs decrease and problem sizes, memory latencies (in terms of pcycles) and optical transmission rates increase.

In the same vein, it is possible to conceive an arrangement where the additional electronic cache memory is put on the memory side of a simpler optical network. These memory caches would be shared by all nodes of the machine, but would add a new level to the hierarchy, a level which would effectively slow down the memory accesses that could not be satisfied by the memory caches. The NetCache also adds a level to the memory hierarchy, but this level does not slow down the accesses that miss in it. Thus, even in this scenario, the system could profit from a NetCache, especially as optical costs decrease and latencies and transmission rates increase.

6 Related Work

As far as we are aware, the NetCache architecture is the only one of its kind; a network/cache hybrid has never before been proposed and/or evaluated for multiprocessors. A few research areas are related to our proposal however, such as the use of optic fiber as delay line memory and the use of optical networks in parallel machines. Delay line memories have been implemented in optical communication systems [9] and in all-optical computers [8]. Optical delay line memories have not been used as caches and have not been applied to multiprocessors before, even though optical networks have been a part of several parallel computer designs, e.g. [5, 7].

Ha and Pinkston [7] have proposed the DMON network. Our performance analysis has shown that the NetCache multiprocessor consistently outperforms the DMON-U system. However, DMON-based systems have an advantage over the NetCache architecture as presented here: they allow multiple outstanding read requests, while the NetCache architecture does not. This limitation results from the star coupler subnetwork (OPTNET) having a single tunable receiver that must be tuned to a single home channel on a read access. We have eliminated this limitation by transforming a read request/reply sequence into a pair of request/reply sequences [2]. This extension only affects a read request that is issued while other requests are outstanding and, thus, its additional overhead should not affect most applications.

7 Conclusions

In this paper we proposed a novel architectural concept: an optical network/cache hybrid (and associated coherence protocol) called NetCache. Through a large set of detailed simulations we showed that NetCache-based multiprocessors easily outperform other optical network-based systems, especially for the applications with a large amount of data reuse in the NetCache. In addition, we evaluated the effectiveness of our shared cache in detail, showing that several applications can benefit from it. Based on these results and on the continually-decreasing cost of optical components, our main conclusion is that the NetCache architecture is highly efficient and should definitely be considered by multiprocessor designers.

Acknowledgements

The authors would like to thank Luiz Monnerat, Cristiana Seidel, Rodrigo dos Santos, and Lauro Whately for comments that helped improve the presentation of the paper. We would also like to thank Timothy Pinkston, Joon-Ho Ha, and Fredrik Dahlgren for their careful evaluation of our work and for discussions that helped improve this paper.

References

1. E. V. Carrera and R. Bianchini. OPTNET: A Cost-Effective Optical Network for Multiprocessors. In *Proceedings of the International Conference on Supercomputing '98*, July 1998.
2. E. V. Carrera and R. Bianchini. Designing and Evaluating a Cost-Effective Optical Network for Multiprocessors. November 1998. Submitted for publication.
3. P. W. Dowd and J. Chu. Photonic Architectures for Distributed Shared Memory Multiprocessors. In *Proceedings of the 1st International Workshop on Massively Parallel Processing using Optical Interconnections*, pages 151–161, April 1994.
4. K. Gharachorloo, D. Lenoski, J. Laudon, P. Gibbons, A. Gupta, and J. L. Hennessy. Memory Consistency and Event Ordering in Scalable Shared-Memory Multiprocessors. In *Proceedings of the 17th Annual International Symposium on Computer Architecture*, pages 15–26, May 1990.
5. K. Ghose, R. K. Horsell, and N. Singhvi. Hybrid Multiprocessing in OPTIMUL: A Multiprocessor for Distributed and Shared Memory Multiprocessing with WDM Optical Fiber Interconnections. In *Proceedings of the 1994 International Conference on Parallel Processing*, August 1994.
6. M. S. Goodman *et al.* The LAMBDANET Multiwavelength Network: Architecture, Applications, and Demonstrations. *IEEE Journal on Selected Areas in Communications*, 8(6):995–1004, August 1990.
7. J.-H. Ha and T. M. Pinkston. SPEED DMON: Cache Coherence on an Optical Multichannel Interconnect Architecture. *Journal of Parallel and Distributed Computing*, 41(1):78–91, 1997.
8. H. F. Jordan, V. P. Heuring, and R. J. Feuerstein. Optoelectronic Time-of-Flight Design and the Demonstration of an All-Optical, Stored Program. *Proceedings of IEEE. Special issue on Optical Computing*, 82(11), November 1994.
9. R. Langenhorst *et al.* Fiber Loop Optical Buffer. *Journal of Lightwave Technology*, 14(3):324–335, March 1996.
10. A. G. Nowatzyk and P. R. Prucnal. Are Crossbars Really Dead? The Case for Optical Multiprocessor Interconnect Systems. In *Proceedings of the 22nd International Symposium on Computer Architecture*, pages 106–115, June 1995.
11. D. B. Sarrazin, H. F. Jordan, and V. P. Heuring. Fiber Optic Delay Line Memory. *Applied Optics*, 29(5):627–637, February 1990.
12. J. E. Veenstra and R. J. Fowler. MINT: A Front End for Efficient Simulation of Shared-Memory Multiprocessors. In *Proceedings of the 2nd International Workshop on Modeling, Analysis and Simulation of Computer and Telecommunication Systems (MASCOTS '94)*, January 1994.

A Multi-Wavelength Optical Content-Addressable Parallel Processor (MW-OCAPP) for High-Speed Parallel Relational Database Processing: Architectural Concepts and Preliminary Experimental System

Peng Yin Choo[1], Abram Detofsky[2], Ahmed Louri[3]

[1] Department of Electrical and Computer Engineering, The University of Arizona,
Tucson, AZ 85721,
choop@bigdog.engr.arizona.edu

[2] Department of Electrical and Computer Engineering, The University of Arizona,
Tucson, AZ 85721,
detofsky@ece.arizona.edu

[3] Department of Electrical and Computer Engineering, The University of Arizona,
Tucson, AZ 85721,
louri@ece.arizona.edu

Abstract. This paper presents a novel architecture for parallel database processing called Multi-Wavelength Optical Content Addressable Parallel Processor (MW-OCAPP). MW-OCAPP is designed to provide efficient parallel retrieval and processing of data by moving the bulk of database operations from electronics to optics. It combines a parallel model of computation with the many degrees of processing freedom that light provides. MW-OCAPP uses a polarization and wavelength-encoding scheme to achieve a high level of parallelism. Distinctive features of the proposed architecture include (1) the use of a multiple-wavelength encoding scheme to enhance processing parallelism, (2) multiple-comparand word-parallel and bit-parallel magnitude comparison with an execution-time independent of the data size or word size, (3) the implementation of a suite of eleven database primitives, and (4) multi-comparand two-dimensional data processing. The MW-OCAPP architecture realizes eleven relational database primitives: difference, intersection, union, conditional selection, maximum, minimum, join, product, projection, division and update. Most of these operations execute in constant time independent of the data size. This paper outlines the architectural concepts and motivation behind MW-OCAPP's design, as well as describes the architecture required for implementing the equality and magnitude comparison processing cores. Additionally, a physical demonstration of the multi-wavelength equality operation is presented.

1.0 Introduction

Databases are emerging as the most important ingredients in information systems. They have penetrated all fields of the human endeavor and are no longer limited to business-oriented processing. Database systems are becoming the backbone of (1) business, (2) computer-aided design and manufacturing, (3) medicine and pharmaceuticals, (4) geographic information systems, (5) defense-related systems, (6) multimedia (text, image, voice, video, and regular data) information systems, among many others.

Searching, retrieving, sorting, updating, and modifying non-numeric data such as databases can be significantly improved by the use of content-addressable memory (CAM) instead of location-addressable memory [1,2,3,4]. CAM-based processing is not only more akin to the way database users address their data (in parallel), but it is also faster than location-addressing schemes since the overhead cost of address computations is completely eliminated. Using the CAM model, the absolute address location of each object has no logical significance; all access to data is by content. Therefore, the CAM model is a naturally parallel form of database representation and manipulation with potential benefits in simplicity of expression (programming), storage capacity, and speed of execution. Unfortunately, large computers based on CAM have not been realized due to the difficulty and high cost of implementing them in conventional electronic technology. Until now, CAMs have only been included in computers systems as small auxiliary units [4,5,6]. Optics can alleviate the cell complexity of CAM-based storage cells by migrating their interconnects into the third dimension [7]. The high degree of connectivity available in free-space and fiber-based optical interconnects and the ease with which optical signals can be expanded (which allows for signal broadcasting) and combined (which allows for signal funneling) can also be exploited to solve interconnection problems.

A novel architecture called the Optical Content Addressable Parallel Processor (OCAPP) was demonstrated that implemented a limited set of high-speed database operations [2,8,9,10]. The architecture combined a parallel associative model of computation with high-speed optical technology. The implemented operations were divided into parallel equivalence and relative magnitude searches, and were accomplished using an intensity and polarization-encoding scheme [11]. The parallel algorithms used in OCAPP can be considered one-dimensional such that a single comparand (search argument) is matched against an entire database in constant time ($O(1)$). However, if n number of comparands are to be compared (a two-dimensional operation), the execution time increases from $O(1)$ to $O(n)$. Multi-comparand comparison can be accomplished in constant time by stacking replicates of optical units, but this is a very inefficient use of space and resources. The implementation of efficient two-dimensional database operations requires that an additional property of light be exploited.

A new architecture is now proposed that we call the Multi-Wavelength Optical Content Addressable Parallel Processor (MW-OCAPP). It utilizes polarization-division multiplexing introduced in OCAPP combined with wavelength-division multiplexing to achieve an even higher degree of parallelism and system integration. The ability to propagate multiple-wavelength light-planes through the same space without mutual interference allows for true two-dimensional operations to take place. This enables the processing of multiple data arguments at the same time and within the same space, thus greatly increasing the parallelism of the processor over single-wavelength designs. Additionally, utilizing these additional degrees of freedom allows for a more compact system to be created, simplifying the manufacturing process and relaxing the alignment and power constraints.

2.0 Overview of the MW-OCAPP system architecture

Fig. 1 illustrates a structural organization of MW-OCAPP's processing model. The optical processing logic of MW-OCAPP consists of six modules: an optical Selection Unit, an optical Match-Compare Unit, an optical Equality Unit, an optical Magnitude Comparison Unit, an optical Relational Operations Unit, and an optical Output Unit. The MW-OCAPP architecture is designed in such a way as to implement a total of eleven database primitives. Most of these execute in a time-span that is independent of the problem size. MW-OCAPP can realize: difference, intersection, union, conditional selection, join, maximum, minimum, product, projection, division and update.

The inputs to MW-OCAPP are the comparand array (CA) and relational array (RA). Each row (tuple) of the CA is polarization logic encoded with different wavelengths by a multiwavelength source array and an electronically-addressable spatial light modulator. This form of encoding allows for the superposition and parallel processing of multiple comparands as they propagate through the Match-Compare and Equality units. The Selection Unit produces the Selection Register (SR), a light-plane that holds the multiple tuples in the Comparand Array to be matched. The optical Match-Compare Unit produces the Match-Compare Register (MCR), a wavelength/polarization encoded light plane that holds the locations of all the matched and mismatched bits. The Magnitude Comparison Unit takes the CA and RA and performs a magnitude comparison (greater-than and less-than) between CA and RA tuples, and outputs less-than (LR) and greater-than (GR) registers. The optical Equality Unit takes the MCR and produces an output called the Equality Register (ER) that represents the intersection locations of the CA and RA tuples. The ER, LR and GR light-planes pass through the optical Relational Operation Processing Unit where they are operated on to produce the Relational Operation Registers (RORs). Lastly, the ROR and MCR light-planes are routed through the optical Output Unit that interfaces with a host computer.

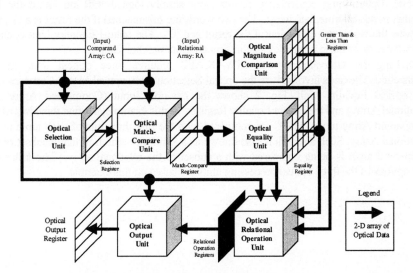

Fig. 1. MW-OCAPP Schematic Organization

3.0 MW-OCAPP Optical Implementation

3.1 Data Encoding

MW-OCAPP uses several methods for encoding a data-plane on a light-plane. Binary patterns are represented by spatially distributed orthogonally-polarized locations on a 2-D pixilated grid [11]. Logical '1' is defined as vertically polarized light and logical '0' as horizontally polarized light. The presence or absence of light (intensity threshold) within a light-plane indicates the selection or de-selection of tuples or attributes in the system. Individual tuples are differentiated from one another by polarization encoding each on a unique wavelength.

3.2 Implementation of Equivalence Operation

The equality operation is one of the most basic operations that a database machine must perform. The equality operation is rooted in the XOR concept:

$$[CA_{(m-1)}CA_{(m-2)}\ldots\ldots CA_{(1)}CA_{(0)}] \oplus [RA_{(m-1)}RA_{(m-2)}\ldots\ldots RA_{(1)}RA_{(0)}] =$$
$$\{ [CA_{(m-1)} \oplus RA_{(m-1)}] \ [CA_{(m-2)} \oplus RA_{(m-2)}] \ldots\ldots [CA_{(1)} \oplus RA_{(1)}] \ [CA_{(0)} \oplus RA_{(0)}] \} \quad (1)$$

where 'm' represents the bit position in the words to be compared and the symbol '$\oplus$' represents the logical XOR operation. The Comparand Array (CA) is the data array that resides in the comparand registers and Relational Array (RA) is the main database that occupies the main associative memory of the CAM. If a CA word and an RA word are a match, formula (1) will produce a resultant word containing only logical '0' bits:

if $\quad CA_{(m-1)} = RA_{(m-1)}, CA_{(m-2)} = RA_{(m-2)}, \ldots, CA_{(1)} = RA_{(1)}$ and $CA_{(0)} = RA_{(0)}$, then

$$[CA_{(m-1)}CA_{(m-2)}\ldots CA_{(1)}CA_{(0)}] \oplus [RA_{(m-1)}RA_{(m-2)}\ldots RA_{(1)}RA_{(0)}] = [\text{'}0\text{'}_{(m-1)} \text{'}0\text{'}_{(m-2)} \ldots \text{'}0\text{'}_{(1)} \text{'}0\text{'}_{(0)}] \quad (2)$$

If the CA and RA do not match, the resultant XOR word will be a mixture of '0' and '1' bits. Determining equivalency results from simply logical 'OR'ing all of the bits together in this intermediate word. The two words are mismatched if the result is a '1', and likewise the words are equivalent if the result is a '0'. The optical Equality Unit operates under these same basic principles, just in a much more parallel fashion.

Testing for equivalency with MW-OCAPP requires three optical units in the architecture. The units involved are the optical Selection Unit, optical Match-Compare Unit and optical Equality Unit. Fig. 2 shows the example input (Comparand Array and Relational Array) and the output Equality Register. In this example, the first three bits of the Comparand Array tuple "1100" are compared with the first three bits of each tuple in the Relational Array. Such a search is also known as a similarity search in that a subset of the comparand tuple is being compared with a subset of the relational array. The following descriptions of the three optical processing units are based on this example.

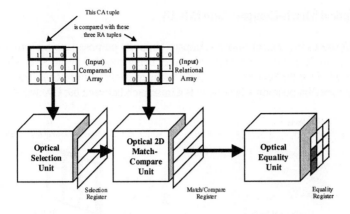

Fig. 2. Schematic organization for the Equivalency operation. The bold rectangles illustrate a similarity-search example (search for a match within a masked field).

3.2.1 Optical Selection Unit

The optical Selection Unit is illustrated in Fig. 3. Its purpose is to encode a pixilated two-dimensional optical wavefront with the comparand array (CA) to be processed. The "rows" in the wavefront, each encoded on a unique wavelength, represent tuples in the CA. Polarization encoding of the desired data pattern is employed to differentiate the binary states of each of the pixilated "bits".

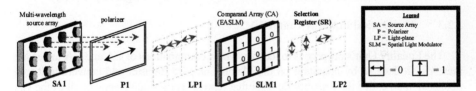

Fig. 3. Optical Selection Unit

The optical Selection Unit begins with a multi-wavelength source array (SA1) in which each row (which corresponds to a separate tuple) radiates at a different wavelength. Only a single wavelength is required, so only the first row is selected to radiate. Since only the first three bit positions in row one are involved in this operation, the source in the last column is de-selected. Therefore, SA1 indirectly functions like a masking register in the CAM. This wavefront passes through a horizontally-oriented polarizer (P1) to reset all of the bit positions to the '0' logical state (LP1). Light-plane LP1 impinges on an electronically-addressable spatial light modulator (SLM1) which polarization-encodes the light passing through it with the bit pattern 1-1-0. The resultant light-plane, LP2, is called the Selection Register (SR) and represents the optically-encoded version of the comparand array.

3.2.2 Optical Match-Compare Unit (MCU)

Fig. 4 illustrates the optical Match-Compare Unit. Its purpose is to bit-wise exclusive-or (XOR) each tuple in the comparand array (after being encoded in the Selection Unit and passed to the MCU as the SR) with each tuple in the relational array (RA). This produces a logical '1' at every bit position where there is a mismatch between the CA and RA.

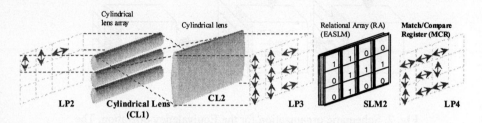

Fig. 4. Optical Match-Compare Unit

The encoded light plane from the optical Selection Unit (LP2) passes through a lens array which spreads each of the rows (LP3) with different wavelengths over the full surface of an EASLM (SLM2). Light plane LP3 passes through SLM2 encoded with the relational array (RA) to be searched. The EASLM rotates the polarization(s) of the incident light according to the logic states of its pixels, effectively generating the result of the logical XOR operation in LP4. Light plane LP4 is called the Match-Compare Register (MCR) and contains all of the bit match and mismatch locations of each of the CA and RA tuple combinations (designated by horizontally and vertically-polarized light respectively).

3.2.3 The Optical Equality Unit (EU)

The Equality Unit's purpose is to identify which combinations of CA and RA tuples are matches. It does so by operating on the Match-Compare Register and converting it to a pixelated map called the Equality Register that represents the equivalency of all of the CA and RA tuple combinations.

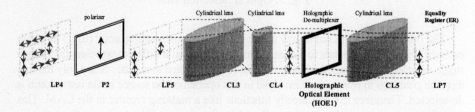

Fig. 5. Optical Equality Unit

Fig. 5 illustrates the optical Equality Unit. The MCR (LP4) enters the Equality Unit and passes through a vertically-oriented polarizer (P2) to form LP5. LP5 contains an illuminated pixel corresponding to all bit mismatch positions. LP5 is funneled down to a single column by CL3 and CL4. This single column (LP6) must now be wavelength demultiplexed into a plane that has a pixel count width equal to the number of tuples in the CA. This can be accomplished using a holographic element (HOE1) that is fabricated to deflect light at an angle that is a function of its wavelength. Cylindrical lens (CL5) collimates the light exiting HOE1 and produces the Equality Register (LP7).

Decoding this Equality Register light-plane is fairly simple. The light-plane is a two-dimensional representation of the intersection of the CA and RA. If 'n' represents the number of tuples in the CA and 'm' represents the number of tuples in the RA, then the ER must consist of $m \times n$ pixels. It is encoded in "negative logic" meaning that non-illuminated pixels correspond to exact matches. For an m by n ER grid, $pixel_{mn}$ is illuminated such that tuple RA_m is not-equal-to tuple CA_n.

3.3 Magnitude Comparison Operation

Magnitude comparison is the second fundamental operation that MW-OCAPP implements. Operations of this type include the greater-than, less-than, in-bound, out-of-bound, and extremum tests. The magnitude comparison functional schematic organization is shown in Fig. 6. The input to the optical Magnitude Comparison Unit is the Equality Register from the Equality Unit, and the output is the Less-than (LR) and Greater-than (GR) registers [4]. The algorithm that MW-OCAPP uses to perform this operation can be broken down into four steps: (1) compute and store the Comparand Rank Comparison Register (CRCR) comparing the CA with a rank table; (2) compute and store the Relational Rank Comparison Register (RRCR) comparing the RA with a rank table; (3) compute and store the Equivalency Register (EQR) comparing the CA and RA; (4) compute and output the Less-than (LR) and Greater-than (GR) registers. Tasks 1-3 are processed by the optical Selection, Match-Compare and Equality units, and the results (the Equality Registers) are stored sequentially in the Optical Buffer Subunit. Step 4 of the algorithm requires that these three registers (termed the Equivalency Register (EQR), the Comparand Rank Comparison Register (CRCR), and the Relational Rank Comparison Register (RRCR)) be presented simultaneously to both the Rank Thresholding Subunit and the LR Extraction Subunit for final processing.

The Optical Buffer Subunit which performs the temporary storage of the EQR, CRCR and RRCR can be realized using an integrated "smart-pixel array" (SPA) [12]. The subunit consists of a single detector array matched to the dimensions of the Equality Register, and three polychromatic emitter arrays of the same dimensions. Hybrid source-detector modules can be realized using flip-chip bonding [13].

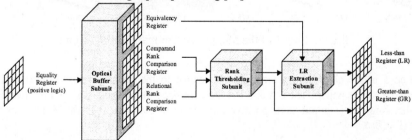

Fig. 6. Magnitude Comparison Functional Schematic Organization

In the following discussion, a single comparand bit pattern (1-1-0) is compared with a single relational bit pattern (1-1-1). Fig. 7 illustrates the Selection, Match-Compare and Equality Units in series that are configured to perform Step 1 in the magnitude comparison algorithm. The hardware processes the comparand word by testing it for equivalency against a ranked binary look-up table. The lookup table (SLM2) is an EASLM encoded with sequential binary words from 0 to $2^m - 1$ (one word per row) arranged in ascending order from top to bottom. The number of entries in the binary look-up table, 2^m, is determined by the maximum wordlength m that the unit is processing. This "equality" operation reduces the input word to a single pixel whose vertical position in the light-plane indicates its absolute magnitude. Light-plane LP7 is converted to positive-logic (swapping illuminated pixels for non-illuminated ones) using trivial hardware and is stored as the CRCR (LP8) in the Optical Buffer Subunit (Fig. 8a).

Step 2 in the magnitude comparison algorithm essentially is identical to Step 1 except that the Relational Array is loaded into SLM1 instead of the Comparand Array. Light-plane LP7 produced in this step in the algorithm is stored in its positive-logic form (LP9) as the RRCR in the Optical Buffer Subunit (Fig. 8b).

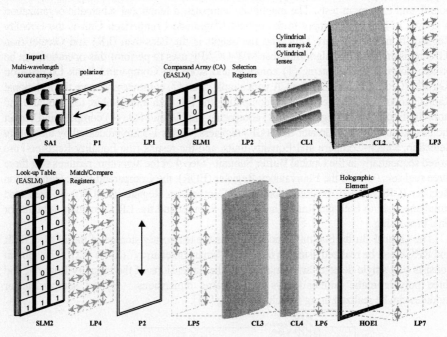

Fig. 7. Step 1 of the Magnitude Comparison Algorithm

Step 3 in the algorithm is most closely related to the equality process described in section 4.3 of this paper. The CA and RA are compared in the usual way and the resultant Equality Register is stored in its positive-logic form in the Optical Buffer Subunit as the EQR (Fig. 8c).

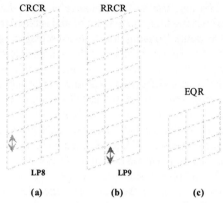

Fig. 8. (a) The Comparand Rank Comparison Register (CRCR); (b) The Relational Rank Comparison Register (RRCR); (c) The Equivalency Register (EQR)

Step 4 in the magnitude comparison algorithm directs the Optical Buffer Subunit to present the EQR, CRCR and RRCR registers simultaneously to the Rank Thresholding and LR Extraction Subunits for final processing[14]. This process is illustrated below.

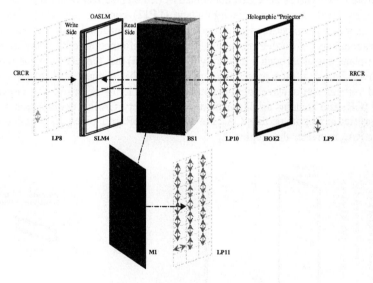

Fig. 9. The Rank Threshold Subunit (part 1)

Fig. 9 illustrates the first part of the Rank Thresholding Subunit (RTS). The RRCR (LP9) encounters a holographic "projector" element (HOE2) with a very specific fan-out function. A pixel in the ith row of the incident light-plane must be projected to every row r from $r = 1$ to $r = i - 1$. This function can most easily be realized by creating an array of discrete sub-holograms over the surface of the holographic plate. Each sub-hologram has a specific fan-out function (varying from top to bottom) and is tuned to a specific wavelength of interest (varying left to right). Relating this to the discussion, the illuminated pixel of LP9 occupies row eight. The fan-out function of the holographic element at this location produces a light-plane (LP10) in which rows one through seven are illuminated. Meanwhile, the CRCR light-plane (LP8) impinges on the write side of an OASLM (SLM4) which rotates the polarization states of LP10 that reflects off of its read-side. This light-plane (LP11) is steered out of the system by beamsplitter (BS1) and mirror (M1) combination. LP11 contains a horizontally-polarized light pixel for every CA/RA tuple pair such that the RA tuple exceeds the CA tuple in magnitude. In this example, it can be seen from the single horizontally-polarized pixel in LP11 that the RA tuple, 1-1-1, is greater than the CA tuple, 1-1-0. The pixel residing in the first column of LP11 indicates that it corresponds to the first comparand of the CA. The specific wavelength of this plane determines which tuple in the RA it corresponds to. In this case, there is only one RA tuple to be compared, so there can be only one wavelength. If a second CA tuple, 0-1-1, were also compared with the existing RA tuple, there would exist in LP11 a horizontally-polarized pixel in the 4th row of column two. Since RA tuples are differentiated by wavelength at this stage, adding additional RA tuples would translate into overlapping but independent light-planes coexisting in LP11. This light-plane must now be manipulated such that it can easily be sensed by a detector array at the output.

Fig. 10 illustrates the second part of the Rank Threshold Unit (RTU). The output (LP11) from Part 1 passes through a horizontally-oriented polarizer (P5) which allows only the modified pixels to pass. This light-plane (LP12) contains the relative magnitude comparison information between all of the tuples in the comparand array and all the tuples in the relational array. Further processing allows this information to be easily interpreted. The light-plane passes through a lens array (CL4 & CL5) that funnels the plane down to a single row. This row then passes through a holographic element such that the light is deflected vertically at a wavelength-dependent angle. LP14 is termed the Greater-than Register (GR) and contains an illuminated pixel for every CA/RA combination such that RA_m is greater in magnitude than CA_n. For an m by n GR grid, $pixel_{mn}$ is illuminated such that tuple RA_m is greater-than tuple CA_n.

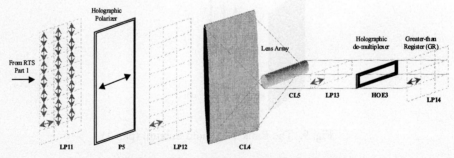

Fig. 10. The Rank Threshold Unit (part 2)

The magnitude comparison architecture has been used thus far to compare a single-comparand 1-1-0 with a single tuple in the relational array, 1-1-1, and produce the example GR. Using the same algorithm, it can be shown that the GR shown in Fig. 11(c) results from the RTU operating on the CA and RA in Fig. 11(a) and Fig. 11(b). Extracting the Less-than register from the GR requires logical bit-wise NOR'ing the GR with the positive-logic version of the Equality Register shown in Fig. 11(d). The resulting light-plane shown in Fig. 11(e) is the Less-than register (LR).

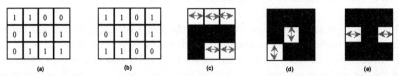

Fig. 11. Magnitude Comparison of the CA (a) and RA (b). Each row in the CA and RA is a separate tuple. The greater-than register (c) is generated by the Rank Thresholding subunit operating on (a) and (b). The LR Extraction subunit takes the logical NOR of (c) with the positive-logic version of the equality register (d) to produce the less-than register (e)

Fig. 12 illustrates the LR Extraction Subunit required to realize this NOR'ing process. Both the GR (LP8) and the ER (LP9) enter LR Extraction Subunit and are OR'ed together using beamsplitter BS1 to produce LP10. LP10 impinges on the write side of an OA-SLM (SLM3). At the same time, a horizonatally-polarized light-plane is launched from SA2. It reflects off of OA-SLM (SLM3) and has the polarization states of its pixels modified according to the illumination pattern on its read side. All pixels that had their polarization states rotated 90 degrees will be blocked by polarizer P4. Light-plane LP11 exiting the module is the logic-reversed version of LP10, otherwise known at the Less-than register (LR).

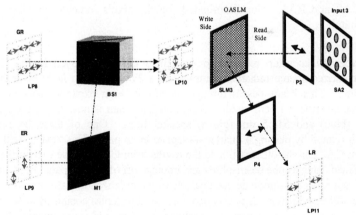

Fig. 12. LR Extraction subunit. Extraction of the less-than register (LR) using the equality register (ER) and the greater-than register (GR) as inputs.

Decoding of the LR is identical to that of the GR and ER. The plane's width in pixels is equal to the number of tuples in the comparand array and the height is equal to the number of tuples in the relational array. For the LR, an illuminated pixel indicates that a specific tuple in the relational array is less-than a specific tuple in the comparand array. For an m by n LR grid, pixel$_{mn}$ is illuminated such that tuple RA$_m$ is less-than tuple CA$_n$.

3.4 Physical Demonstration of a Preliminary Version of MW-OCAPP

This section presents the physical demonstration of MW-OCAPP that has been done thus far. In this preliminary demonstration, the Equality Register that contains the "exact-match" information between each of the tuples in the RA and CA is generated. Fig. 13 shows the bit patterns contained in the example RA and CA as well as the expected Equality Register that results from the operation. The Comparand Array (CA) contains two tuples, "0-0-0-0" and "1-1-1-1". These are compared against four tuples in the Relational Array (RA).

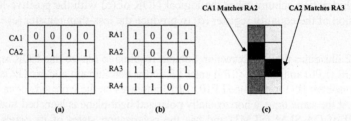

Fig. 13. The experimental setup matches two tuples found in (a) the Comparand Array (CA) with the four tuples found in (b) the Relational Array (RA). The Equality Register that results (c) indicates that there is a match between CA1 and RA2 as well as CA2 and RA3. Non-illuminated (black) pixels indicate an exact match.

The experimental setup required to implement this is displayed in Fig. 14. The vertically-polarized source radiation comes from a 2W argon ion laser in its multi-line configuration. This beam passes through an afocal beam expander lens system that broadens the beam to a 1.5cm width. This multi-line beam is now filtered to extract the 488.0 nm (blue) and 514.5 nm (green) spectral lines. One of these beams has its polarization rotated by placing a quarter waveplate in its path. This "color" will hold the results from CA2 and the other will hold the results from CA1. These two purified beams are recombined using a cube beamsplitter and impinge off of a FLC spatial light modulator. This SLM is a reflection-mode device that rotates the polarization states by 90 degrees of each of the addressed pixels. It is encoded with the four tuples contained in the relational array. This passes through a vertically-oriented polarizer and some imaging lenses that focus the plane down to a single column on the screen. A holographic grating is placed between the lens system and the screen in order to spatially separate the two color channels into individual columns.

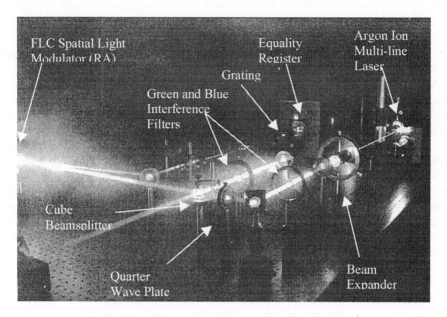

Fig. 14. Demonstration system photograph

Fig. 15 shows the Equality Register at several points in the imaging system. Fig. 15a is an image of the light plane just prior to its passage through the imaging optics. Both the green and blue channels completely overlap. Fig. 15b shows the Equality Register at the focal plane. Notice that there is a dark pixel in the second row of the first column and in the third row of the second column. Both of these locations correspond correctly to tuple "matches" predicted in Fig. 13.

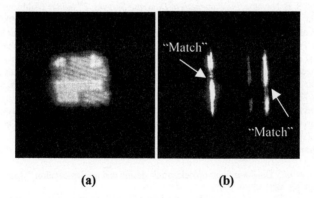

(a)　　　　　　　(b)

Fig. 15. The Equality Register at (a) just prior to wavelength demultiplexing, and (b) at best focus. The blurring is caused by the low shutter speed required to image the registers combined with the page-swap refreshing of the FLC SLM

4 Discussion/Conclusions

In this paper, an optical associative processing system called MW-OCAPP is presented. It harnesses a unique method of wavelength and polarization-multiplexing of dataplanes to achieve a high level of parallelism. Optical implementation is made possible by exploiting the non-interactive behavior of coincident light-planes of differing wavelength. This architecture offers database systems constant-time parallel equality and magnitude comparison of multiple comparands with multiple tuples in a relational array.

When compared with conventional iterative word-parallel magnitude comparison schemes where w is the wordlength and n is the number of comparands to be operated on, MW-OCAPP's magnitude comparison algorithm offers a speedup of wn. This performance suggests a substantial performance improvement over previous designs in database operations such as sorting which typically use the magnitude comparison operation repetitively. Additionally, MW-OCAPP's multi-comparand approach offers a speedup of m over conventional word-parallel CAM-based systems, where m is the number of comparands to be compared and n is the number of tuples in the database.

This research was supported by National Science Foundation grant MIP-9505872.

References

[1] S.Y. Su, *DATABASE COMPUTERS Principles, Architectures, and Techniques*. New York: McGraw-Hill, 1988.
[2] L. Chisvin and R. J. Duckworth, "Content-Addressable and Associative Memory: Alternatives to the Ubiquitous RAM," IEEE Computer, pp 51-64, July 1989.
[3] P. B. Berra, A. Ghaffor, P.A. Mitkas, S. J. Marchinkowski and M. Guizani, "The Impact of Optics on Data and Knowledge Base Systems," IEEE Transactions on Knowledge and Data Engineering, Vol. 1, No 1, pp 111-132, March 1989.
[4] Ahmed Louri, "Optical content-addressable parallel processor: architecture, algorithms, and design concepts," *Applied Optics*, Vol. 31, No. 17, pp. 3241-3258, 1992.
[5] C. J. Date, *An Introduction to Database Systems* (Addison-Wesley, Mass., 1986).
[6] R. Elmasri and S. B. Navathe, *Fundamentals of Database Systems, 2nd editon* (Addison-Wesley, 1994)
[7] K. Giboney, et. al., "The ideal light source for datanets," *IEEE Spectrum*, February 1998, pp. 43-53.
[8] Ahmed Louri and James Hatch Jr., "An Optical content-Addressable Parallel Processor for High-Speed Database Processing," in Applied Optics, vol. 33, no. 35, pp 8153 – 8164, December 10, 1994.
[9] Ahmed Louri and James Hatch Jr., "An Optical Content-Addressable Parallel Processor for High-Speed Database Processing: Theoretical concepts and Experimental Results," in *IEEE Computer*, IEEE Computer Society, Special Issue on Associative Processors, vol. 27, no. 11, pp. 65 – 72, November 1994.
[10] Ahmed Louri and James Hatch Jr., "Optical Implementation of a Single-Iteration Thresholding Algorithm with Applications to Parallel Database/Knowledge-base Processing," Optics Letters, vol. 18, no. 12, pp. 992-994, June 15, 1993.
[11] A. W. Lohmann, "Polarization and optical logic," Applied Optics, vol. 25, pp. 1594-1597, Mar. 1990.
[12] R. D. Snyder *et. al.*, "Database filter: optoelectronic design and implementation," *Applied Optics*, Vol. 36, No. 20, pp. 4881-4889, 1997.
[13] J. Neff, "Free-space optical interconnection for digital systems," in Optoelectronic Packaging, A. Mickelson, N. Basavanhally, Y. Lee, editors, John Wiley and Sons, New York, 1997.
[14] A. Detofsky, P. Choo, A. Louri, "Optical implementation of a multi-comparand bit-parallel magnitude comparison algorithm using wavelength and polarization-division multiplexing with application to parallel database processing," Optics Letters, Vol. 23, No 17, pp. 1372-1374, September 1, 1998.

Optimal Scheduling Algorithms in WDM Optical Passive Star Networks

Hongjin Yeh, Kyubum Wee, and Manpyo Hong

College of Information and Communication, Ajou University, 442-749 Suwon, Korea
{hjyeh, kbwee, mphong}@madang.ajou.ac.kr

Abstract. All-to-all broadcast scheduling problems are considered in WDM optical passive star networks where k wavelengths are available in the network. It is assumed that each node has exactly one tunable transmitter and one fixed tuned receiver. All transmitters can tune to k different wavelengths, and have the same tuning delay δ to tune from one wavelength to another. In this paper, we take δ to be a nonnegative integer which can be expressed in units of packet durations. When all-to-all broadcasts are scheduled periodically in the network, the lower bounds are established on the minimum cycle length depending on whether each node sends packets to itself or not. And then, we present optimal scheduling algorithms in both cases for arbitrary number of wavelengths and for arbitrary value of the tuning delay. [1]

1 Introduction

In optical communications, one of the most popular packet switched network architecture is the passive star network because of its simplicity; packets may be transmitted directly from source to destination without any routing process. Additionally, WDM(Wavelength Division Multiplexing) is used to overcome the bottleneck of communication bandwidth due to optoelectronic components.[1] It is well known that tuning delay of transmitter/receiver can affect performance of the network. One of critical issues is how to minimize the effect of tuning delay for scheduling static or dynamic traffic patterns.

Let's consider WDM optical passive star networks with N nodes and k wavelength channels available in the network. All-to-all broadcast scheduling algorithms on WDM passive star networks may be classified into five categories depending on the number and tunability of transmitters/receivers for each node:

– STT-SFR, SFT-STR, MFT-SFR, SFT-MFR
– MTT-SFR, SFT-MTR
– STT-STR, MFT-MFR, STT-MFR, MFT-STR
– MTT-STR, MFT-MTR, STT-MTR, MTT-MFR
– MTT-MTR

---
[1] This research was supported by the Institute of Information Technology Assessment under Grant 97-G-0574.

where S: single, M: multiple, T: tunable, F: fixed, T: transmitter, R: receiver.

We assummed that each node has exactly one tunable transmitter (e.g., laser) and one fixed tuned receiver (e.g., filter). The network is asynchronous so that transmitters may send a packet or tune its wavelength at anytime. All transmitters can tune to k different wavelengths, and have the same tuning delay δ to tune from one wavelength to another. In this paper, we take δ to be an arbitrary nonnegative integer which can be expressed in units of packet durations. We also normalize the unit time such that all packets can be transmitted in the unit time so that the packet size is ignored for transmissions.

A special case of the general scheduling traffic patterns is the all-to-all broadcast in which every node has a single packet to be transmitted to the other. The problem is, given values k and δ, how one can schedule all-to-all broadcasts using k wavelengths within a repeating cycle of minimum length. A transmission schedule is a timetable that shows each packet is transmitted at which time using which wavelength. In order to tranfer a packet, the transmitter should tune to the wavelength of the receiver. A transmitter cannot transmit packets while it is tuning to another wavelength. Hence it needs to transmit all the packets that use the same wavelength without interruption if possible, and only then tune to another wavelength. Also a wavelength should be used many times, since the number of wavelengths used in WDM is limited compared to the number of nodes. Thus, obtaining an optimal schedule amounts to finding how to overlap tansmissions of packets and tuning delays as much as possible.

There are two kinds of all-to-all broadcast scheduling problems depending on whether each node sends packets to itself or not. In case each node sends single packet to itself, N^2 packets should be transmitted for each all-to-all broadcast. Pieris and Sasaki [5] proposed a scheduling algorithm under the assumtion that k divides N, and showed that the schedule length of their algorithm falls between $\max\{\delta + \frac{N^2}{k}, N\sqrt{\delta}\}$ and $\max\{\delta + \frac{N^2}{k}, k\delta + N - \frac{N}{k} + \frac{N^2}{k^2}\}$. Later it was shown that the schedule of [5] is actually optimal by Choi, Choi, and Azizoglu[2].

For the case where k does not divide N, they also proposed optimal scheduling algorithms for three different cases depending on the range of δ except for a very narrow range where their solution is within the bound of $\frac{13}{12}$ of the optimal schedule length [3]. In the case where each node does not send packets to itself, the number of packets to be sent for each all-to-all broadcast is $N(N-1)$. Lee, Choi, and Choi[4] showed that $\max\{\frac{N(N-1)}{k}, k\delta+N-1\}$ is a lower bound for the cycle lengths of schedules and proposed an algorithm for obtaining a transmission schedule that meets this lower bound under the assumption that k divides N. Chang, Yeh, and Wee[6] proposed an algorithm for obtaining a schedule where each node receives packets with a regular interval, and showed that the optimal schedule length in that case is $\max\{C(N-1), \lceil \frac{\delta}{C-1}+\frac{1}{C-1}\lceil \frac{C-1}{k}\rceil\rceil\}$, where $C = \frac{N}{k}$ is an integer.

In this paper, we present scheduling algorithms that finds optimal schedules for arbitrary values of N, k, and δ.

2 Lower Bounds to Schedule Cycle Lengths

Let us consider WDM optical passive star networks with N nodes $n_0, n_1, \cdots, n_{N-1}$, where are k wavelengths $w_0, w_1, \cdots, w_{k-1}$ available in the network. We take arbitrary number of wavelengths k and arbitrary tuning delay δ for a given network composed of N nodes.

In this paper, it is assumed that all-to-all broadcasts are repeated periodically after initialization of the network. The problem is to schedule packet transmissions for each node pairs so that all nodes transmit once to every node within a repeating cycle as short as possible. The transmission schedules are called *optimal* if their cycle lengths meet a lower bound as shown in the following. All-to-all broadcast scheduling problems can be divided into two categories; one is the case that every node transfers packets to all nodes including itself, another is the case that each node transfers packets to the other nodes except itself.

Firstly, let us talk about the case that every node transmits once to itself for a cycle. For a given $2 \le k \ll N$, let R_i denote the set of nodes in which their receivers' wavelengths are identically fixed to w_i for $0 \le i \le k-1$. To maximize sharing all the wavelengths, let's define $R_i = \{n_j | j \equiv i \pmod{k}, 0 \le j \le N-1\}$. Then, $|R_i| = \lceil \frac{N}{k} \rceil$ or $|R_i| = \lfloor \frac{N}{k} \rfloor = \lceil \frac{N}{k} \rceil - 1$. Specially, it is noted that $\max_{0 \le i \le k-1} |R_i| = |R_0| = \lceil \frac{N}{k} \rceil$ by the definition.

Theorem 1. *For values k and δ, the cycle length of any all-to-all broadcast schedule is at least*

$$\max\left\{ \left\lceil \frac{N}{k} \right\rceil N, k\delta + N \right\}$$

where each of N^2 node pairs has single packet to be transmitted for a cycle.

Proof. Each node must receive N packets via only one wavelength fixed to its receiver. These packets are transmitted from all the nodes in the network including itself. Since more than two packets can not be transmitted via the same wavelength at a time, the cycle length is at least the number of packets, $|R_i|N$ for $0 \le i \le k-1$, to be transmitted using a wavelength w_i for a cycle. Thus, $|R_0| = \lceil \frac{N}{k} \rceil$ means that the cycle length is at least $\lceil \frac{N}{k} \rceil N$.

On the other hand, every node transmits N packets using k different wavelengths for a cycle. If we take into account the fact that all-to-all broadcasts are repeated periodically, each node has to tune its transmitter k times for a cycle. Thus, as for a node, the cycle length is at least the sum of total tuning delay for a cycle and transmission time for N packets i.e., $k\delta + N$.

Another case is that nodes do not transfer any packets to themselves through all-to-all broadcast. That is, the number of packets transmitted to a set of nodes R_i using the wavelength w_i could be different if the transmitting node is an element of R_i for $0 \le i \le k-1$. Thus, the lower bound to optimal schedule length is slightly changed.

Theorem 2. *For values k and δ, the cycle length of any all-to-all broadcast schedule is at least*

$$\max\{\left\lceil\frac{N}{k}\right\rceil(N-1), k\delta + N - 1\}$$

where nodes do not tranfer any packets to themselves; each of $N(N-1)$ node pairs has single packet to be transmitted for a cycle.

Proof. The proof is in the same manner as theorem 1.

In the next section, we present optimal scheduling algorithms for the latter which may be more complicated than the former.

3 Optimal Scheduling Algorithms

The main ideas to construct an optimal scheduling algorithm can be derived from two important system parameters, *the number of wavelengths* and *the tuning delay*, that affect performance of WDM optical passive star networks. The more wavelength channels are available in the network, the more packets can be transmitted simultaneously. On the other hand, tuning delay is dependent on practical devices and would limit the network throughput; while nodes spend time to tune from one wavelength to another, they have to delay packet transmissions. Therefore, transmission schedules compromise the following two contradictory conditions:

- As for a node, packet transmissions must be scheduled as soon as possible after every tuning delay for a cycle.
- As for a wavelength, packet transmissions must be scheduled as soon as possible from one node to another.

Remark In the previous research, it would be better that the packets transmitted via the same wavelength in a node are scheduled all together before tuning the transmitter to another wavelength. It should be noted, in the case of repeating all-to-all broadcasts periodically, that the cycle length includes the time for tuning to the wavelength used for the beginning of the next broadcast. That is, a wavelength looks like to be used one more time in a cycle with viewpoint of one stage of repeating all-to-all broadcasts.

In our proposed transmission schedules, the order of wavelengths used by each node is $w_j, w_{j-1}, \cdots, w_1, w_o, w_{k-1}, \cdots, w_{j+1}$, where $0 \leq j \leq p \bmod k$. That is, the first group to transmit packets to is R_j with $|R_j| = \lceil\frac{N}{k}\rceil$. The order of nodes using the same wavelength is $n_0, n_1, \cdots, n_{k-1}$, or a circular shift of this order. Without loss of generality, we can assume that the order of wavelengths used by each node is $w_0, w_{k-1}, \cdots, w_1$, since we know that $|R_0| = \lceil\frac{N}{k}\rceil$. Also in the following we notate w_0 and w_k interchangeably. The same comment applies to R_0 and R_k.

Let $N \times k$ matrix M that represents relations between a node n_p for $0 \le p \le N-1$ and the set of nodes R_{k-i} containig n_p for $0 \le i \le k-1$. Then, the matrix $M = (m_{pi})$ is defined by

$$m_{pi} = \begin{cases} 1 \text{ if } p + i \equiv 0 \pmod{k} \\ 0 \text{ otherwise} \end{cases}$$

for $0 \le p \le N - 1$ and $0 \le i \le k - 1$. Note that the indices of m_{pi} starts from zero. For example, in the case of $N = 8$ and $k = 3$, the matrix M is the following:

$$M = \begin{pmatrix} 1 & 0 & 0 \\ 0 & 0 & 1 \\ 0 & 1 & 0 \\ 1 & 0 & 0 \\ 0 & 0 & 1 \\ 0 & 1 & 0 \\ 1 & 0 & 0 \\ 0 & 0 & 1 \end{pmatrix}$$

Thus, $m_{12} = 1$ means that the receiver of the node n_1 is fixed tuned to the wavelength $w_{3-2} = w_1$.

Lemma 1. *For $N \times k$ matrix $M = (m_{pi})$,*

$$\sum_{j=0}^{i} m_{pj} - \sum_{j=0}^{i-1} m_{(p+1)j} \ge 0$$

for $0 \le p \le N - 1$ and $0 \le i \le k - 1$.

Proof. By the definition of the matrix M, the values $\sum_{j=0}^{i} m_{pj}$ and $\sum_{j=0}^{i-1} m_{(p+1)j}$ are always 0 or 1. If $\sum_{j=0}^{i} m_{pj} = 1$, $\sum_{j=0}^{i} m_{pj} - \sum_{j=0}^{i-1} m_{(p+1)j} \ge 0$ because the value $\sum_{j=0}^{i-1} m_{(p+1)j}$ is at most 1.

Otherwise, i.e., If $\sum_{j=0}^{i} m_{pj} = 0$, $p + j \not\equiv 0 \pmod{k}$ for $0 \le j \le i$. Since $(p+1) + (j-1) \not\equiv 0 \pmod{k}$ for $1 \le j \le i$ implies $(p+1) + j \not\equiv 0 \pmod{k}$ for $0 \le j \le i - 1$, $\sum_{j=0}^{i-1} m_{(p+1)j} = 0$. Thus, $\sum_{j=0}^{i} m_{pj} - \sum_{j=0}^{i-1} m_{(p+1)j} = 0$. Therefore, given inequality always holds.

Each node n_p starts transmitting packets to the nodes in group R_{i-1} using wavelength w_{i-1} right after finishing transmitting packets to the nodes in group R_i using wavelength w_i delayed only by tuning delay δ. The time s_p when node n_p transmits the very first packet using wavelength w_k (i.e., w_0) is set as follows:

$$s_p = \left\lceil \frac{N}{k} \right\rceil p - \left\lceil \frac{p}{k} \right\rceil + 1$$

Algorithm 1 shows the initial transmission schedules compromising tuning delay and lack of the wavelengths. In the algorithm, the statement $T(n_p, t) = n_q$ means that node n_p transmits a packet to node n_q at time t. The wavelength used in this case is $w_{q \bmod k}$. In the algorithm, the cycle length of optimal schedules L_{opt} is defined by $\max\{\lceil \frac{N}{k} \rceil (N-1), k\delta + N - 1\}$.

Algorithm 1 *Initial Schedules of an All-to-all Broadcast*

1: For $p = 0$ to $N - 1$ do
2:      $t \leftarrow s_p$
3:      For $r = k$ to 1 do
4:         $i \leftarrow r \bmod k$
5:         For $j = 0$ to $|R_i| - 1$ do
6:            If $(p \neq jk + i)$
7:               $T(n_p, t) \leftarrow n_{jk+i}$
8:               $t \leftarrow t + 1$
9:         $t \leftarrow t + \delta$

Figure 1 shows the transmission schedule constructed by Algorithm 1 when $N = 8$, $k = 3$, and $\delta = 5$. The numbers on the leftmost column represent the nodes, and time goes on from left to right.

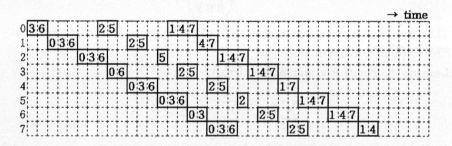

Fig. 1. An example of initial schedules

Lemma 2. *In the schedules generated by Algorithm 1, no two nodes use the same wavelength at the same time.*

Proof. We only have to check that two adjacent nodes n_p and n_{p+1} do not use the same wavelength simultaneously. Let α be the time when node n_p finishes transmitting packets using wavelength w_{k-i}, and β be the time when node n_{p+1} finishes tuning its transmitter to start packet transmissions using wavelength w_{k-i}. Then α and β are expressed as follows:

$$\alpha = s_p + \text{the number of packets transmitted} + \text{total tuning delay}$$

$$= \left(\left\lceil \frac{N}{k} \right\rceil p - \left\lceil \frac{p}{k} \right\rceil + 1\right) + \sum_{j=0}^{i}(|R_{k-j}| - m_{pj}) + i\delta$$

Similarly,

$$\beta = \left(\left\lceil \frac{N}{k} \right\rceil (p+1) - \left\lceil \frac{p+1}{k} \right\rceil + 1\right) + \sum_{j=0}^{i-1}(|R_{k-j}| - m_{(p+1)j}) + i\delta$$

Hence

$$\beta - \alpha = \left\lceil \frac{N}{k} \right\rceil - \left\lceil \frac{p+1}{k} \right\rceil + \left\lceil \frac{p}{k} \right\rceil - |R_{k-i}| + \sum_{j=0}^{i} m_{pj} - \sum_{j=0}^{i-1} m_{(p+1)j}$$

$$\geq \left\lceil \frac{p}{k} \right\rceil - \left\lceil \frac{p+1}{k} \right\rceil + \sum_{j=0}^{i} m_{pj} - \sum_{j=0}^{i-1} m_{(p+1)j}$$

since $|R_{k-i}| \leq \left\lceil \frac{N}{k} \right\rceil$. Recall that $\sum_{j=0}^{i} m_{pj} - \sum_{j=0}^{i-1} m_{(p+1)j} \geq 0$ by Lemma 1.

Hence $\beta - \alpha \geq 0$ because $\left\lceil \frac{p}{k} \right\rceil - \left\lceil \frac{p+1}{k} \right\rceil = 0$ unless p is a multiple of k. Now, we have to consider the case when $p \equiv 0 \pmod{k}$. In this case, $\left\lceil \frac{p}{k} \right\rceil - \left\lceil \frac{p+1}{k} \right\rceil = -1$. By the definition of matrix M, $p + 0 \equiv 0 \pmod{k}$ implies that $m_{p0} = 1$ is unique nonzero element in p-th row. Similarly, $(p+1) + (k-1) \equiv 0 \pmod{k}$ implies that $m_{(p+1)(k-1)} = 1$ is unique nonzero element in $(p+1)$-th row. Thus, $\sum_{j=0}^{i} m_{pj} - \sum_{j=0}^{i-1} m_{(p+1)j} = 1 - 0 = 1$ for $0 \leq i \leq k-1$. Therefore, $\beta - \alpha \geq 0$ always holds for any value p.

Now let us denote the time for each node n_p to transmit all the packets for an all-to-all broadcast as L_p. Note that the tuning delay for next all-to-all broadcast is not considered.

Lemma 3. *In the schedules generated by Algorithm 1, $L_p = (k-1)\delta + N - 1$ for any p in $0 \leq p \leq N - 1$.*

Proof. Lemma 2 guarantees that each node does not have to wait for using a wavelength that is being used by another node. In other words, each node can start transmitting packets as soon as tuning to the wavelength is completed. By the way, every node n_p transmits $N - 1$ packets. And every node n_p needs to tune $k - 1$ times. Hence, $L_p = (k-1)\delta + N - 1$ for any p.

Since all L_p are identical for $0 \leq p \leq N - 1$, let us denote L_p simply as L. Now we claim that if we repeat the transmission schedule constructed by Algorithm 1 by many times, any time span of L_{opt} starting from any position of the repeated schedule is an optimal schedule. Algorithm 2 finds an optimal schedule. Algorithm 2 differs from Algorithm 1 in that time t is replaced by $t - L_{opt}$ if t is later than L_{opt}.

For example, Figure 2 shows the repetitions of the transmission schedule of Figure 1. Note that in Figure 2, $L_{opt} = 22$ is calculated with the values N, k and δ. If we take out any portion of Figure 2 spanning 22 time units, that constitutes an optimal schedule.

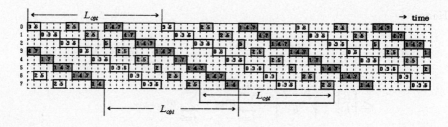

Fig. 2. An example of repeating initial schedules

Algorithm 2 *Optimal Schedules for an All-to-all Broadcast*
1:   For $p = 0$ to $N - 1$ do
2:       $t \leftarrow s_p$
3:       For $r = k$ to 1 do
4:           $i \leftarrow r \bmod k$
5:           For $j = 0$ to $|R_r| - 1$ do
6:               If $(p \neq jk + i)$
7:                   If $(t > L_{opt})$ $T(n_p, t - L_{opt}) \leftarrow n_{jk+i}$
8:                   else $T(n_p, t) \leftarrow n_{jk+i}$
9:                   $t \leftarrow t + 1$
10:          $t \leftarrow t + \delta$

Fig. 3. An example of optimal schedules with $N(N-1)$ packets

Theorem 3. *For arbitrary values k and δ, the cycle length of transmission schedules by Algorithm 2 is optimal.*

Proof. All we have to do is to show that $L_{opt} - L$ is no less than tuning delay δ. Recall that $L = (k-1)\delta + (N-1)$ by Lemma 3. If $L_{opt} = \lceil \frac{N}{k} \rceil (N-1)$ then

$$\lceil \frac{N}{k} \rceil (N-1) \geq k\delta + N - 1 \Leftrightarrow \lceil \frac{N}{k} \rceil (N-1) - \{(k-1)\delta + (N-1)\} \geq \delta$$
$$\Leftrightarrow L_{opt} - L \geq \delta$$

Otherwise, $L_{opt} = k\delta + N - 1$ implies

$$L_{opt} - L = k\delta + N - 1 - \{(k-1)\delta - (N-1)\} = \delta$$

Remark In the above example, $L_{opt} = 3 \times 5 + 8 - 1 = 22$ is affected by the value $\delta = 5$. Our algorithm also works in case that the value δ is small enough to determine L_{opt} only with values N and k. Figure 4 shows the optimal schedule generated by Algorithm 2 when $\delta = 4$ with the same values $N = 8$ and $k = 3$. In such case, $L_{opt} = \lceil\frac{8}{3}\rceil(8-1) = 21$.

Fig. 4. An example of optimal schedules for $\delta = 4$

Remark Another scheduling problem for all-to-all broadcasts takes place when every node transmits packets to itself. Our algorithms can be applicable to this problem if the line 6 in Algorithm 2 is deleted. In fact, each node has the same number of packets that are transmitted using the same wavelength. This property makes the proposed algorithms more simple. The proof of optimality in transmission schedules is omitted. The following example is depicted to compare two versions of optimal schedules when $N = 8, k = 3$ and $\delta = 5$.

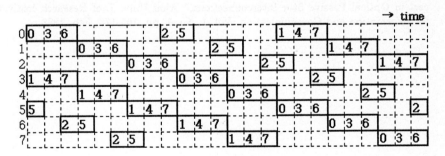

Fig. 5. An example of optimal schedules with N^2 packets

4 Conclusions

We presented optimal scheduling algorithms for all-to-all broadcasts repeated periodically in WDM optical passive star networks where each node has one tunable transmitter and one fixed receiver. Our algorithms work for arbitrary values k and δ. There are two versions: one for the case where nodes send packets to themselves and the other case where they do not. Both versions find optimal schedules.

As long as we use electrical buffers, it is important to guarantee enough time to electrically process received packets which are usually of very big size in optical networks. Hence, we are currently working on optimal scheduling algorithms with additional condition that each node receives packets at a regular interval. On the other hand, finding optimal schedules for the other classes as mentioned in the Introduction, particulary the case where both transmitters and receivers are tunable, would be an interesting future research topic.

References

1. C. A. Brackett, "Dense Wavelength Division Multiplexing Networks: Principles and Applications," IEEE J. on Selected Areas in Communications. Vol. 8, No. 6, pp. 948-964, Aug. 1990.
2. H. Choi, H-A. Choi, and M. Azizoglu, "Efficient Scheduling of Transmissions in Optical Broadcast Networks," IEEE/ACM Trans. on Networking, Vol. 4, No. 6, pp. 913-920, Dec. 1996.
3. H. Choi, H-A. Choi, and M. Azizoglu, "On the All-to-All Broadcast Problem in Optical Network," Proceedings of Infocom '97, 1997.
4. S-K. Lee, A. D. Oh, H. Choi, H-A. Choi, "Optimal Transmission Schedules in TWDM Optical Passive Star Networks," Discrete Applied Mathematics, Vol. 75, No. 1, pp. 81-91, Jan. 1997.
5. G. R. Pieris and G. H. Sasaki, "Scheduling Transmissions in WDM Broadcast-and-select Networks," IEEE/ACM Trans. on Networking, Vol. 2, No. 2, pp. 105-110, April, 1994.
6. S. Chang, H-J. Yeh, K. Wee, "Regular Transmission Schedules for All-to-All Broadcast in Optical Passive Star Interconnections," Ajou Univ. J. of Research Institut for Information and Communication, Vol. 1, No. 1, pp. 165-173, Feb. 1998.

OTIS-Based Multi-Hop Multi-OPS Lightwave Networks

David Coudert[1], Afonso Ferreira[1], and Xavier Muñoz[2]

[1] Project SLOOP, CNRS-I3S-INRIA, BP 93, F-06902 Sophia-Antipolis, France
{dcoudert,afferrei}@sophia.inria.fr
[2] UPC, 08024 Barcelona, Spain xml@mat.upc.es

Abstract. Advances in optical technology, such as low loss Optical Passive Star couplers (OPS) and the possibility of building tunable optical transmitters and receivers have increased the interest for multiprocessor architectures based on lightwave networks because of the vast bandwidth available. Many research have been done at both technological and theoretical level. An essential effort has to be done in linking those results. In this paper we propose optical designs for two multi-OPS networks: the single-hop POPS network and the multi-hop stack-Kautz network; using the Optical Transpose Interconnecting System (OTIS) architecture, from the Optoelectronic Computing Group of UCSD. In order to achieve our result, we also provide the optical design of a generalization of the Kautz digraph, using OTIS.

1 Introduction

The advances in optical technology, such as micro-lenses, low energy loss optical passive star couplers (OPS) [14, 20], optical transmission lines allowing more than 1 G-bits per second with less power consumption than electrical wires [12] as well as tunable optical transmitters and receivers [8, 21, 23], have increased the interest for optical interconnection networks for multiprocessor systems [4] because of their large bandwidth.

Our work focuses on multi-OPS optical network topologies and design. Although many results exist in the literature at both technological and theoretical level, an essential effort has to be done in linking those results together. In this paper, we study the optical design of several graph-theoretical topologies using the Optical Transpose Interconnecting System (OTIS) architecture, from the Optoelectronic Computing Group of UCSD [19].

Optical systems are usually divided in two classes, according to the number of intermediate processors a message has to visit before delivery. In a *single-hop* network, the nodes communicate with each other in only one step [20]. Such topologies require either a large number of transceivers per node, or rapidly tunable transmitters and receivers. In a *multi-hop* topology [21], there is no direct path between all pairs of nodes: a communication should use intermediate nodes to reach the destination. This allows the use of a small number of statically tuned transmitters and receivers, but on the other hand the processing of the information by the intermediate nodes causes a loss of speed. Moreover, OPS-based networks can be further classified according to the number of optical couplers used (See p. 209–233 of [4]), being single-OPS [8, 21] or multi-OPS [7, 9]. A great deal of research effort have been concentrated on single-hop *single-OPS* topologies [10, 22, 25]. However, *multi-OPS* networks seem more viable and cost-effective under current optical technology [9, 11].

Optical networks topologies can be studied using graph-theoretical tools, as graphs and digraphs play an important role in the analysis and synthesis of *point-to-point* networks [2, 24]. With respect to *one-to-many* networks, where messages sent by the processors can be broadcast to all outputs of the OPS couplers, they can be modeled by hypergraphs, a generalization of graphs, where edges may connect more than two nodes [1]. In this paper, we address implementation issues of three optical interconnection networks:

- Graphs of Imase and Itoh, a generalization of Kautz graphs allowing point-to-point communications [15].
- The single-hop multi-OPS *POPS* network, which allows one-to-many communications at every communication step [9].
- The multi-hop multi-OPS *stack-Kautz* network, which also allows one-to-many communications at every communication step [11].

Graphs of Imase and Itoh are very interesting for the realization of an interconnection network due to their large number of nodes, N, for small constant degree, d, and low diameter, $\lceil \log_d N \rceil$. The POPS network is based on complete graphs and on the concept of *groups* of processors. The stack-Kautz is based on the Kautz graphs [18] and on the stack-graphs, a useful model to manipulate multi-OPS networks [7].

This paper is organized as follows. We start by recalling, in Sec. 2, the results from the literature upon which we constructed ours. In particular, we present the Optical Transpose Interconnecting System (OTIS) architecture from [19], the optical passive star, and the single-hop multi-OPS Partitioned Optical Passive Star network (POPS) network from [9]. We also recall the stack-graphs from [7], the definition of the Kautz graph from [18], generalized with the graph of Imase and Itoh from [15], and close presenting the multi-hop multi-OPS stack-Kautz network from [11]. In Section 3, we propose the implementation of some building blocks using the OTIS architecture: first, the optical interconnections between a group of processors and its corresponding OPS couplers, and then an optical interconnection network having the topology of a graph of Imase and Itoh. Finally, Section 4 presents the application of these building blocks to the optical design of the POPS network and of the stack-Kautz network. We close the paper with some concluding remarks and directions for further research.

2 Preliminaries

In this section, we recall the basic features of the OTIS architecture and OPS couplers. We also show the model proposed in [7] for studying multi-OPS networks with stack-graphs. Then, we exemplify the use of this model on the POPS network. Finally, we recall the definition of the Kautz graph, the Imase and Itoh's graph and of the stack-Kautz.

2.1 OTIS

The **Optical Transpose Interconnection System** (OTIS) architecture, was first proposed in [19]. $OTIS(G,T)$ is an optical system which allows point-to-point (1-to-1) communications from G groups of size T onto T groups of size G. This architecture connects the transmitter of processor (i,j), $0 \leq i \leq G-1$, $0 \leq j \leq T-1$, to the receiver of processor $(T-1-j, G-1-i)$.

Optical interconnections in the OTIS architecture are realized with two planes of lenses [5] in a free optical space as shown in Fig. 1.

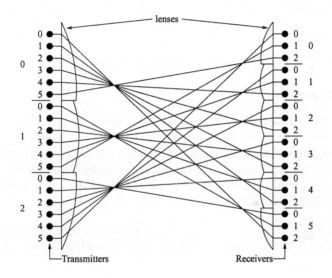

Fig. 1. $OTIS(3,6)$.

The OTIS architecture was used in [24] to realize interconnections networks such as hypercubes, 4-D meshes, mesh-of-trees and butterfly for multi-processors systems. It was shown that, in such electronic interconnection networks (in which connections are realized with electronic wires), a set of wires can be replaced with pairs of transmitters and receivers connected using the OTIS architecture. This is interesting in terms of speed, power consumption [12] and space reduction.

2.2 Optical passive stars

An **optical passive star coupler** is a single-hop one-to-many optical transmission device. An OPS(s, z) has s inputs and z outputs. In the case where s equals z, the OPS is said to be of degree s (see Fig. 2, left). When one of the input processors sends a message through an OPS coupler, the s output processors have access to it. An OPS coupler is a **passive** optical system, i.e. it requires no power source. It is composed of an optical multiplexer followed by an optical fiber or a free optical space and a beam-splitter that divides the incoming light signal into s equal signals of a s-th of the incoming optical power. Note that only one optical beam has to be guided through the network (see Fig. 2, right). A practical realization of an OPS coupler using a hologram [14] at the outputs, is described in [6]. Throughout this paper, we will deal with **single-wavelength** OPS couplers of degree s. Consequently, only one processor can send an optical signal through it per time step.

Fig. 2. A degree 4 optical passive star coupler.

2.3 Stack-graphs

We saw in the introduction that one-to-many networks (e.g., OPS based networks) are better modeled by hypergraphs. Let us define a special kind of directed hypergraphs, called *stack-graphs*, which have the property of being very easy to deal with. Informally, they can be obtained by piling up copies of a directed graph and subsequently viewing each stack of arcs as a hyperarc [7]. A formal definition is as follows.

Definition 1. *[7] Let $G = (V, A)$ be a directed graph. The **stack-graph** $\varsigma(s, G) = (V_\varsigma, A_\varsigma)$ is as follows, s being called **stacking-factor** of the stack-graph.*

1. *The set of nodes V_ς of $\varsigma(s, G)$ is $V_\varsigma = \{0, \cdots, s-1\} \times V$, $s \geq 1$.*
2. *Let π be the projection function defined from V_ς onto V such that $\pi((i, v)) = v$, for $0 \leq i < s$ and $v \in V$.*
3. *The set of hyperarcs A_ς of $\varsigma(s, G)$ is then $A_\varsigma \stackrel{def}{=} \{a_\varsigma = (\pi^{-1}(u), \pi^{-1}(v)) | (u, v) \in A\}$.*

Therefore, an OPS coupler of degree s can be modeled by a hyperarc linking two sets of s nodes, meaning that the nodes of one can only send messages through the hyperarc while the other set can only receive messages through the same hyperarc. Figure 3 shows an OPS coupler modeled by a hyperarc.

2.4 POPS

The **Partitioned Optical Passive Star network** $POPS(t, g)$, introduced in [9], is composed of $N = tg$ processors and g^2 OPS couplers of degree t. The

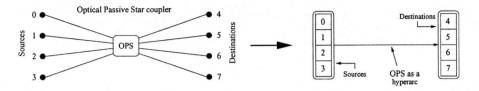

Fig. 3. Modeling an OPS by a hyperarc.

processors are divided into g groups of size t (see Fig. 4). Each OPS coupler is labeled by a pair of integers (i,j), $0 \le i, j < g$. The input of the OPS (i,j) is connected to the i-th group of processors, and the output to the j-th group of processors. The POPS is a single-hop multi-OPS network.

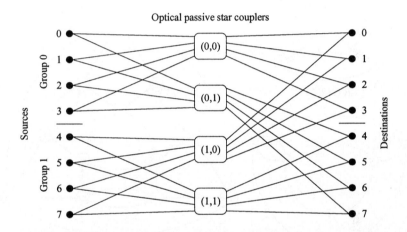

Fig. 4. Partitioned Optical Passive Star POPS(4,2) with 8 nodes.

Since an OPS coupler is modeled as a hyperarc, the POPS network $POPS(d, g)$ can be modeled as a stack-K_g^+ (or $\varsigma(d, K_g^+)$, for short) of stacking-factor d, where K_g^+ is the complete digraph with loops[1] and with g nodes and g^2 arcs (see Fig. 5), as proposed in [3].

2.5 Kautz graph

The Kautz graph was first defined in [18] as follows.

Definition 2. *[18] The **directed Kautz graph** $KG(d, k)$ of degree d and diameter k is the digraph defined as follows (see Fig. 6).*

1. *A vertex is labeled with a word of length k, $(x_1, \cdots, x_k)$, on the alphabet $\Sigma = \{0, \cdots, d\}$, $|\Sigma| = d + 1$, in which $x_i \ne x_{i+1}$, for $1 \le i \le k - 1$.*

---
[1] A loop is an arc from a node to itself.

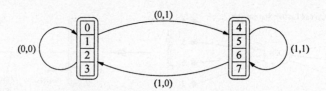

Fig. 5. $POPS(4,2)$ modeled as $\varsigma(4, K_2^+)$.

2. There is an arc from a vertex $x = (x_1, \cdots, x_k)$ to all vertices y such that $y = (x_2, \cdots, x_k, z)$, $z \in \Sigma$, $z \neq x_k$.

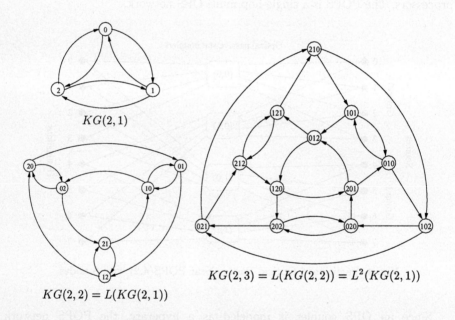

Fig. 6. Three line digraph iterations of the Kautz graph $KG(d, k)$.

Another definition of direct Kautz graph, in terms of *line digraph iteration*, was proposed in [13]. They show that $KG(d, 1)$ is the complete digraph K_{d+1} and that $KG(d, k) = L^{k-1}(K_{d+1})$, where $L(G)$ is the line digraph of a digraph G. Figure 6 shows three of these iterations.

The Kautz graph $KG(d, k)$ has $N = d^{k-1}(d+1)$ nodes, constant degree d and diameter k ($k \leq \log_d N$). It is both Eulerian and Hamiltonian and optimal with respect to the number of nodes if $d > 2$ [18]. As an example, $KG(5, 4)$ has $N = 3750$ nodes, degree 5 and diameter 4.

Notice that routing on the Kautz graph is very simple, since a shortest path routing algorithm (every path is of length at most k) is induced by the label of the nodes. It can be extended to generate a path of length at most $k+2$ which survives $d-1$ link or node faults [17].

2.6 Imase and Itoh

Kautz graphs are very interesting in terms of number of nodes for fixed degrees and diameters. However, its definition does not yield graphs of any size. Graphs of Imase and Itoh have thus been introduced in [15] as a Kautz graph generalization in order to obtain graphs of every size.

Imase and Itoh's graphs are defined using a congruence relation, as follows.

Definition 3. *[15] The* **graph of Imase and Itoh** $II(d,n)$, *of degree d with n nodes, is the directed graph in which:*

1. *Nodes are integers modulo n.*
2. *There is an arc from the node u to all nodes v such that $v \equiv (-du - \alpha)$ mod n, $1 \leq \alpha \leq d$.*

Figure 10, left, shows an example of Imase and Itoh's graph with $II(3,12)$. It is drawn such that the left nodes and the right nodes represents the same nodes. As $II(3,12)$ is a directed graph, the left nodes are connected to the outgoing arcs and the right nodes to the incoming arcs.

It has been shown in [15] that the diameter of $II(d,n)$ is $\lceil \log_d n \rceil$ and in [16] that the graph $II(d,n)$ is the Kautz graph $KG(d,k)$ when $n = d^{k-1}(d+1)$ (for example, in Fig. 10, left, $II(3,12)$ is $KG(3,2)$).

2.7 Stack-Kautz

We now dispose of a good model for multi-OPS networks (the stack-graphs) and also of a graph having good properties to build a multi-hop network (the Kautz graph). Hence, we can recall the multi-hop multi-OPS architecture based on the *stack-Kautz*.

In order to define the optical interconnection network called stack-Kautz, introduced in [11], we use the Kautz graph with loops $KG^+(d,k)$, where every node has a loop and hence degree $d+1$.

Thus, we can define the stack-Kautz graph as follows.

Definition 4. *[11] The* **stack-Kautz graph** $SK(s,d,k)$ *is the stack-graph of stacking-factor s, degree $d+1$ and diameter k, $\varsigma(s, KG^+(d,k))$ (see Fig. 7).*

The *stack-Kautz* network has the topology of the stack-Kautz graph $SK(s,d,k)$ and $N = sd^{k-1}(d+1)$ nodes. Each node is a processor labeled by a pair (x,y) where x is the label of the stack in $KG(d,k)$ and y is an integer $0 \leq y < s$, i.e., x is the label of a processor group and y is the label of a processor in this group. Since the stack-Kautz network inherits most of the properties of the Kautz graph, like shortest path routing, fault tolerance and others, in a previous work, we showed that it is a good candidate for the topology of an OPS-based lightwave network [11].

We note that the definition of stack-Kautz network can be trivially extended to the stack-Imase-Itoh network.

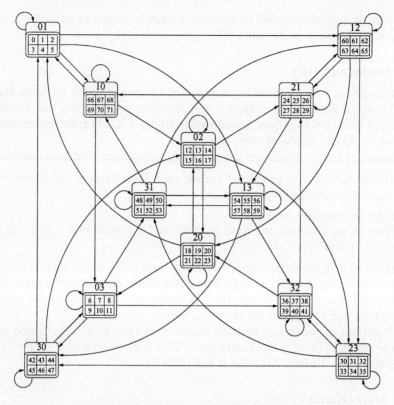

Fig. 7. Stack-Kautz network $SK(6, 3, 2)$.

3 Implementing some building blocks with OTIS

In this section, we explain how the OTIS architecture can be used both to connect a group of processors to the inputs of OPS couplers, and to connect the outputs of OPS couplers to a group of processors, and we show that the optical interconnections of an Imase and Itoh's based network can be simply realized using the OTIS architecture.

3.1 Creating groups of processors

We can realize the optical interconnections between the t processors of a group and the inputs of g OPS couplers (each processor being connected to the g OPS couplers), in a simple way using one $OTIS(t, g)$ architecture plus g optical multiplexers (input part of an OPS coupler). Figure 8 shows how to connect the transmitters of a group of 6 processors to 4 optical multiplexers, using $OTIS(6, 4)$.

Analogously, we can also realize the optical interconnections between the outputs of g OPS couplers and the t processors of a group, using one $OTIS(g, t)$ ar-

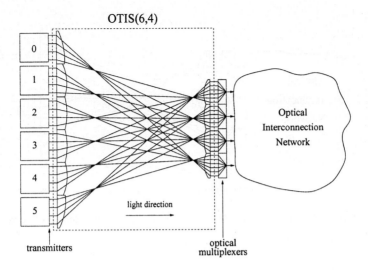

Fig. 8. Interconnections between a group of 6 processors and the input of 4 OPS couplers.

chitecture plus g beam-splitters. Figure 9 shows how to connect 3 beam-splitters to the receivers of a group of 5 processors, using $OTIS(3,5)$.

To build specific topologies, the output optical multiplexers are to be connected to its corresponding input beam-splitters, through the topology of the Optical Interconnection Network.

3.2 Optical implementation of the graph of Imase and Itoh

Proposition 1. *The optical interconnections of the graph of Imase and Itoh $II(d,n)$ of degree d with n nodes can be perfectly realized with the OTIS architecture $OTIS(d,n)$.*

Proof. The OTIS architecture $OTIS(d,n)$ has dn inputs (d groups of size n) and dn outputs (n groups of size d). The inputs $e = (i,j)$, $0 \leq i < d$, $0 \leq j < n$, are connected to the outputs $s = (n - j - 1, d - i - 1)$.

The graph of Imase and Itoh $II(d,n)$ of degree d with n nodes connect a node u to the nodes $v \equiv (-du - \alpha) \mod n$, $1 \leq \alpha \leq d$.

We associate d inputs of $OTIS(d,n)$ to each node of $II(d,n)$, such that the input $e = (i,j)$ is associated to the node $u = \lfloor \frac{ni+j}{d} \rfloor$. In other words, a node u is associated to inputs $e_{du+\alpha-1} = (\lfloor \frac{du+\alpha-1}{n} \rfloor, du + \alpha - 1 - \lfloor \frac{du+\alpha-1}{n} \rfloor n)$, $1 \leq \alpha \leq d$.

We also associate d outputs of $OTIS(d,n)$ to each node v of $II(d,n)$, an output $s = (n - j - 1, d - i - 1)$ being associated to node $v = n - j - 1$, and a node v being associated to outputs $s = (v, d - \alpha)$, $1 \leq \alpha \leq d$.

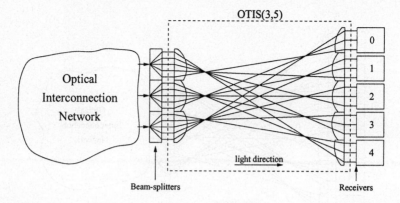

Fig. 9. Interconnections between the outputs of 3 OPS couplers and a group of 5 processors.

Due to the OTIS architecture, a node u is then connected to nodes

$v_\alpha = n - \left(du + \alpha - 1 - \lfloor \frac{du+\alpha-1}{n} \rfloor n\right) - 1$, $1 \leq \alpha \leq d$,
$v_\alpha = n\left(1 + \lfloor \frac{du+\alpha-1}{n} \rfloor\right) - du - \alpha$, $1 \leq \alpha \leq d$,
$v_\alpha \equiv (-du - \alpha) \mod n$, $1 \leq \alpha \leq d$.

The neighborhood of u, obtained by using the OTIS architecture is exactly the neighborhood of u in the graph of Imase and Itoh. Finally, by associating inputs of the OTIS architecture $OTIS(d, n)$ to nodes of the graph of Imase and Itoh $II(d, n)$ as we did, we have perfectly realized the interconnections of $II(d, n)$ using the OTIS architecture $OTIS(d, n)$.

Figure 10 shows how the optical interconnections of the graph of Imase and Itoh are realized with the OTIS architecture.

Corollary 1. *The Kautz graph $KG(d, k)$ is the graph of Imase and Itoh $II(d, d^{k-1}(d+1))$. Hence, we can realize the optical interconnections of $KG(d, k)$ using $OTIS(d, d^{k-1}(d+1))$.*

4 Implementing multi-OPS networks with OTIS

We saw in the previous section that the OTIS architecture can be used to realize the optical interconnections of the graph of Imase and Itoh and consequently of the Kautz graph. It was also shown how to connect the processors of a group with its corresponding OPS couplers. In this section, we show how to build the POPS and the stack-Kautz networks using these building blocks

4.1 POPS with OTIS

$POPS(t, g)$ has g groups of t processors and g^2 OPS couplers. We have explain how to realize the optical interconnections between the inputs and the outputs

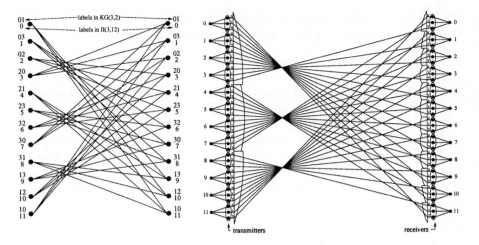

Fig. 10. $II(3,12)$ with $OTIS(3,12)$.

of t processors of a group through g OPS couplers using two $OTIS(t,g)$ architectures. We also have modeled the $POPS(t,g)$ network as a stack-K_g^+ of stacking factor t. As $OTIS(g,g)$ realize the optical interconnections of K_g^+, we can realize all the optical interconnections of the $POPS(t,g)$ network using two $OTIS(t,g)$ and one $OTIS(g,g)$, as shown in Fig. 11.

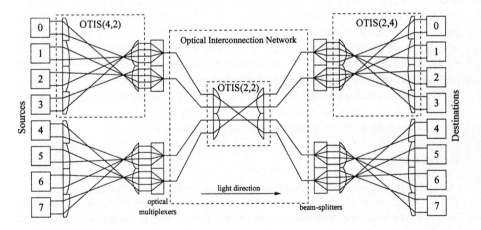

Fig. 11. Optical interconnections of $POPS(4,2)$ using the OTIS architecture.

4.2 Stack-Kautz with OTIS

The stack-Kautz network $SK(s,d,k)$ has $d^{k-1}(d+1)$ groups of s processors and $d^{k-1}(d+1)^2$ OPS couplers of degree s. Each processor has degree $d+1$.

The groups: We have to explain how to connect a group of processors to its corresponding OPS couplers. Using $d^{k-1}(d+1)$ $OTIS(s,d+1)$ plus $d^{k-1}(d+1)$

$OTIS(d+1, s)$ architectures, we can connect all the groups of s processors to its corresponding $d+1$ optical multiplexers and $d+1$ beam-splitters.

The Optical Interconnection Network: Now, we have to realize the optical interconnection network which connects the optical multiplexers to its corresponding beam-splitters. By Corollary 1, that the optical interconnections of a Kautz network $KG(d, k)$ can be realized using one $OTIS(d, d^{k-1}(d+1))$. The stack-Kautz network $SK(s, d, k)$ is based on the Kautz graph $KG(d, k)$. Hence, we can use one $OTIS(d, d^{k-1}(d+1))$ to realize the optical interconnections between the optical multiplexers and its corresponding beam-splitters. We remark that the loops are not taken into account in the optical interconnection network and we consider that they are connected using an appropriate technique (e.g., optical fiber).

An example: Figure 12 shows how those interconnections are realized for $SK(6,3,2)$ (modeled in Fig. 7) using 12 $OTIS(6,4)$, 12 $OTIS(4,6)$, 48 optical multiplexers, 48 beam-splitters and one $OTIS(3, 12)$. $SK(6,3,2)$ has 72 processors (12 groups of 6 processors) of degree 4, connected in a network of diameter 2.

5 Conclusion

In this paper, we showed the optical design of several graph-theoretical network topologies using available optical technology. Our work advances the interplay between graph theory and optics technologies. Indeed, the key of our work was the relationship it established between the OTIS architecture and the graph of Imase and Itoh. Also important was the proposed optical design of the concept of a group of processors, since such a concept seems to be central in multi-OPS communication networks. It is of interest to study the optical design of other multi-OPS networks using the tools developed in this paper.

Finally, a corollary of our results is that the OTIS architecture can be viewed as the graph of Imase and Itoh. Therefore, properties of existing OTIS-based networks can be studied using the properties of such a graph.

References

1. C. Berge. *Hypergraphes: Combinatoire des ensembles finis.* Bordas, 1987.
2. J.-C. Bermond, C. Delorme, and J. J. Quisquater. Strategies for interconnection networks: some methods from graph theory. *Graphs and Combinatorics*, 5:107–123, 1989.
3. P. Berthomé and A. Ferreira. Improved embeddings in POPS networks through stack-graph models. In *Third Internationnal Workshop on Massively Parallel Processing using Optical Interconnections*, pages 130–135. IEEE Press, July 1996.
4. P. Berthomé and A. Ferreira, editors. *Optical Interconnections and Parallel Processing: Trends at the Interface.* Kluwer Academic, 1997.
5. M. Blume, G. Marsen, P. Marchand, and S. Esener. Optical Transpose Interconnection System for Vertical Emitters. *OSA Topical Meeting on Optics in Computing*, Lake Tahoe, March 1997.

6. M. Blume, F. McCormick, P. Marchand, and S. Esener. Array interconnect systems based on lenslets and CGH. Technical Report 2537-22, SPIE International Symposium on Optical Science, Engineering and Instrumentation, San Diego, 1995.
7. H. Bourdin, A. Ferreira, and K. Marcus. A performance comparison between graph and hypergraph topologies for passive star WDM lightwave networks. *Computer Networks and ISDN Systems*, 30:805–819, 1998.
8. C. Brackett. Dense Wavelength Division Multiplexing Networks: Principles and Applications. *IEEE Journal on Selected Areas in Communication*, 8:947–964, 1990.
9. D. Chiarulli, S. Levitan, R. Melhem, J. Teza, and G. Gravenstreter. Partitioned Optical Passive Star (POPS) Topologies for Multiprocessor Interconnection Networks with Distributed Control. *IEEE Journal of Lightwave Technology*, 14(7):1601–1612, 1996.
10. I. Chlamtac and A. Fumagalli. Quadro-star: A high performance optical wdm star network. *IEEE Transactions on Communications*, 42(8):2582–2591, aug 1994.
11. D. Coudert, A. Ferreira, and X. Muñoz. Multiprocessor Architectures Using Multihops Multi-OPS Lightwave Networks and Distributed Control. In *IEEE International Parallel Processing Symposium*, pages 151–155. IEEE Press, 1998.
12. M. Feldman, S. Esener, C. Guest, and S. Lee. Comparison between electrical and free-space optical interconnects based on power and speed considerations. *Applied Optics*, 27(9):1742–1751, May 1988.
13. M.A. Fiol, J.L.A. Yebra, and I. Alegre. Line digraphs iterations and the (d,k) digraph problem. *IEEE Trans. on Computers C-33, 400-403*, 1984.
14. D. Gardner, P. Harvey, L. Hendrick, P. Marchand, and S. Esene. Photorefractive Beamsplitter For Free Space Optical Interconnection systems. *Applied Optics*, (37):6178–6181, September 1998.
15. M. Imase and M. Itoh. Design to Minimize Diameter on Building-Block Network. *IEEE Transactions on Computers*, C-30(6):439–442, June 1981.
16. M. Imase and M. Itoh. A Design for Directed Graphs with Minimum Diameter. *IEEE Transactions on Computers*, C-32(8):782–784, August 1983.
17. M. Imase, T. Soneoka, and K. Okada. A fault-tolerant processor interconnection network. *Systems and Computers in Japan*, 17(8), 1986.
18. W.H. Kautz. Bounds on directed (d,k) graphs. Theory of cellular logic networks and machines. *AFCRL-68-0668 Final report, 20-28*, 1968.
19. G. Marsen, P. Marchand, P. Harvey, and S. Esener. Optical transpose interconnection system architectures. *Optics Letters*, 18(13):1083–1085, July 1993.
20. B. Mukherjee. WDM-based local lightwave networks part I: Single-hop systems. *IEEE Networks*, pages 12–27, may 1992.
21. B. Mukherjee. WDM-based local lightwave networks part II: Multi-hop systems. *IEEE Networks*, pages 20–32, July 1992.
22. K.N. Sivarajan and R. Ramaswami. Lightwave Networks Based on de Bruijn Graphs. *IEEE/ACM Transactions on Networking*, 2(1):70–79, apr 1994.
23. F. Sugihwo, M. Larson, and J. Harris. Low Threshold Continuously Tunable Vertical-Cavity-Surface-Emitting-Lasers with 19.1nm Wavelength Range. *Applied Physics Letters*, 70:547, 1997.
24. F. Zane, P. Marchand, R. Paturi, and S. Esener. Scalable Network Architectures Using The Optical Transpose Interconnection System. In *Massively Parallel Processing using Optical Interconnections*, pages 114–121. IEEE Press, Oct. 1996.
25. Z. Zhang and A.S. Acampora. Performance analysis of multihop lightwave networks with hot potato routing and distance-age-priorities. *IEEE Transactions on Communications*, 42(8):2571–2581, aug 1994.

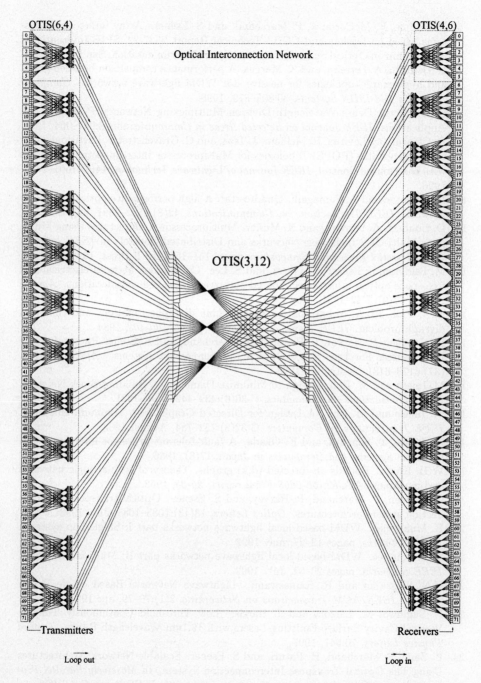

Fig. 12. Optical interconnections of $SK(6,3,2)$ using the OTIS architecture.

Solving Graph Theory Problems Using Reconfigurable Pipelined Optical Buses

Keqin Li[1], Yi Pan[2], and Mounir Hamdi[3]

[1] Dept. of Math & Computer Sci., State Univ. of New York, New Paltz, NY 12561
li@mcs.newpaltz.edu
[2] Dept. of Computer Science, University of Dayton, Dayton, OH 45469
pan@cps.udayton.edu
[3] Dept. of Computer Sci., Hong Kong Univ. of Sci. & Tech., Kowloon, Hong Kong
hamdi@cs.ust.hk

Abstract. We solve a number of important and interesting problems from graph theory on a linear array with a reconfigurable pipelined optical bus system. Our algorithms are based on fast matrix multiplication and extreme value finding algorithms, and are currently the fastest algorithms. We also distinguish the two cases where weights have bounded/unbounded magnitude and precision.

1 Introduction

It has been recognized that many important and interesting graph theory problems can be solved based on matrix multiplication. A representative work is a parallel matrix multiplication algorithm on hypercubes, which has time complexity $O(\log N)$ on an N^3-processor hypercube for multiplying two $N \times N$ matrices [3]. By using this algorithm, many graph problems can be solved in $O((\log N)^2)$ time on an N^3-processor hypercube. Since the summation of N values takes $O(\log N)$ time even on a completely connected network, further reduction in the time complexity seems difficult due to the limited communication capability of static networks used in distributed memory multicomputers.

The performance of parallel algorithms on shared memory multiprocessors, e.g., PRAMs, can be much better. For instance, on a CRCW PRAM, a boolean matrix product can be calculated in constant time by using $O(N^3)$ processors. This implies that the transitive closure and many related problems of directed graphs can be solved in $O(\log N)$ time by using N^3 processors [5]. Since a general matrix multiplication takes $O(\log N)$ time even on a PRAM, the all-pairs shortest problem requires $O((\log N)^2)$ time, no matter how many processors are used.

Using reconfigurable buses, a number of breakthrough have been made. For instance, a constant time algorithm for matrix multiplication on a reconfigurable mesh was reported in [18]. However, the number of processors used is as many as N^4. It is also known that the transitive closure as well as other related problems can be solved in constant time on a reconfigurable mesh [20]. However, the algorithm only works for undirected graphs, while the problem is defined on directed graphs.

The recent advances in optical interconnection networks have inspired a great interest in developing new parallel algorithms for classic problems [9]. Pipelined optical buses can support massive volume of data transfer simultaneously and realize various communication patterns. An optical bus can also implement some global operations such as calculating the summation and finding the extreme values of N data items in constant time. Furthermore, an optical interconnection can be reconfigured into many subsystems which can be used simultaneously to solve subproblems. The reader is referred to [10, 16] for more detailed discussion on these issues.

In this paper, we solve graph theory problems on the LARPBS (linear arrays with reconfigurable pipelined bus system) computing model. (LARPBS was first proposed in [15, 16], and a number of algorithms have been developed on LARPBS [4, 7, 8, 10–12, 14, 15, 17].) We use matrix multiplication algorithms as subroutines. We show that when the weights are real values with bounded magnitude and precision, the all-pairs shortest paths problem for weighted directed graphs can be solved in $O(\log N)$ time by using $O(N^3)$ processors. If the weights have unbounded magnitude and precision, the problem can be solved in $O(\log N)$ time with high probability by using $O(N^3)$ processors. The transitive closure problem for directed graphs can be solved in $O(\log N)$ time by using $O(N^3/\log N)$ processors. Other related problems are also discussed. These algorithms are currently the fastest.

2 Reconfigurable Pipelined Optical Buses

A pipelined optical bus system uses optical waveguides instead of electrical signals to transfer messages among electronic processors. In addition to the high propagation speed of light, there are two important properties of optical pulse transmission on an optical bus, namely, unidirectional propagation and predictable propagation delay. These advantages of using waveguides enable synchronized concurrent accesses of an optical bus in a pipelined fashion [2, 6]. Such pipelined optical bus systems can support a massive volume of communications simultaneously, and are particularly appropriate for applications that involve intensive regular or irregular communication and data movement operations such as permutation, one-to-one communication, broadcasting, multicasting, multiple multicasting, extraction and compression. It has been shown that by using the coincident pulse addressing technique, all these primitive operations take $O(1)$ bus cycles, where the bus cycle length is the end-to-end message transmission time over a bus [10, 16]. (*Remark.* To avoid controversy, let us emphasize that in this paper, by "$O(f(p))$ time" we mean $O(f(p))$ bus cycles for communication plus $O(f(p))$ time for local computation.)

In addition to supporting fast communications, an optical bus itself can be used as a computing device for global aggregation. It was proven in [10, 16] that by using N processors, the summation of N integers or reals with bounded magnitude and precision, the prefix sums of N binary values, the logical-or and logical-and of N Boolean values can be calculated in constant number of bus cycles.

A linear array with a reconfigurable pipelined bus system (LARPBS) consists of N processors P_1, P_2, ..., P_N connected by a pipelined optical bus. In addition to the tremendous communication capabilities, an LARPBS can also be partitioned into $k \geq 2$ independent subarrays $LARPBS_1$, $LARPBS_2$, ..., $LARPBS_k$, such that $LARPBS_j$ contains processors $P_{i_{j-1}+1}$, $P_{i_{j-1}+2}$, ..., P_{i_j}, where $0 = i_0 < i_1 < i_2 \cdots < i_k = N$. The subarrays can operate as regular linear arrays with pipelined optical bus systems, and all subarrays can be used independently for different computations without interference (see [16] for an elaborated exposition).

The above basic communication, data movement, and aggregation operations provide an algorithmic view on parallel computing using optical buses, and also allow us to develop, specify, and analyze parallel algorithms by ignoring optical and engineering details. These powerful primitives that support massive parallel communications plus the reconfigurability of optical buses make the LARPBS computing model very attractive in solving problems that are both computation and communication intensive.

3   Matrix Multiplication

The problem of matrix multiplication can be defined in a fairly general way. Let S be a set of data with two binary operators $\oplus$ and $\otimes$. Given two $N \times N$ matrices $A = (a_{ij})$ and $B = (b_{jk})$, where $a_{ij} \in S$, and $b_{jk} \in S$, for all $1 \leq i, j, k \leq N$, the product $C = AB = (c_{ik})$ is defined as

$$c_{ik} = (a_{i1} \otimes b_{1k}) \oplus (a_{i2} \otimes b_{2k}) \oplus \cdots \oplus (a_{iN} \otimes b_{Nk}),$$

for all $1 \leq i, k \leq N$.

Several methods that parallelize the standard matrix multiplication algorithm have been developed on LARPBS by using communication capabilities of optical buses [10]. As a matter of fact, we can establish the following general result.

Lemma 1. *The product of two $N \times N$ matrices can be calculated in $O(T)$ time by using $N^2 M$ processors, assuming that the aggregation $x_1 \oplus x_2 \oplus \cdots \oplus x_N$ can be calculated in $O(T)$ time by using $M \geq 1$ processors.* ∎

The Appendix gives the implementation details of the generic matrix multiplication algorithm of Lemma 1. Compared with the results in [10], Lemma 1 is more general in two ways. First, while only numerical and logical data are considered in [10], Lemma 1 is applicable to arbitrary data set S and operations $\oplus$ and $\otimes$. Second, Lemma 1 covers a wide range of processor complexity. For example, if S contains real numbers whose magnitude and precision are bounded, and $\otimes$ and $\oplus$ are numerical multiplication and addition, then the summation of N reals can be calculated in $O(N/M)$ time by M processors, where $1 \leq M \leq N$, which implies that matrix multiplication can be performed in $O(N/M)$ time by $N^2 M$ processors for all $1 \leq M \leq N$. In particular, matrix multiplication can be performed

- in $O(N)$ time by N^2 processors;
- in $O(1)$ time by N^3 processors.

These results were developed in [10].

For boolean matrices, where S is the set of truth values, and $\otimes$ is the logical-and, and $\oplus$ is the logical-or, a different approach has been adopted. The method is to parallelize the Four Russians' Algorithm, with the following performance [7].

Lemma 2. *The product of two $N \times N$ boolean matrices can be calculated in $O(1)$ time by using $O(N^3/\log N)$ processors.* ∎

4 Finding Extreme Values

One instance of $\oplus$ is to find the minimum of N data. This operation is used in several graph problems.

Let S be the set of real values. The following result is obvious, since even the radix sorting algorithm can be implemented on an LARPBS in constant time by using N processors when the magnitude and precision are bounded [16].

Lemma 3A. *The minimum value of N data with bounded magnitude and precision can be found in $O(1)$ time by using N processors.* ∎

When the reals have unbounded magnitude or precision, different approaches are required. One approach is to use more processors. It is obvious that by using N^2 processors, the minimum can be found in constant time by making all the possible comparisons [17]. The method can be easily generalized to the following, using the same method in PRAM [19].

Lemma 3B. *The minimum value of N data with unbounded magnitude and precision can be found in $O(1)$ time by using $N^{1+\delta}$ processors, where $\delta > 0$ is any small constant.* ∎

By using the above method in Lemma 3B as a subroutine, the well known doubly logarithmic-depth tree algorithm [5] has been implemented on LARPBS, that can find the minimum of N data in $O(\log \log N)$ time by using N processors [16]. The number of processors can easily be reduced by a factor of $O(\log \log N)$.

Lemma 3C. *The minimum value of N data items with unbounded magnitude and precision can be found in $O(\log \log N)$ time, by using $N/\log \log N$ processors.* ∎

The third method to handle general real values with unbounded magnitude and precision is to use randomization, as shown by the following lemma [17, 19].

Lemma 3D. *The minimum value of N data with unbounded magnitude and precision can be found in $O(1)$ time with high probability (i.e., with probability $1 - O(1/N^\alpha)$ for some constant $\alpha > 0$) by using N processors.* ∎

By Lemma 1 and Lemmas 3A-3D, we have

Theorem 4. *When $\oplus$ is the "min" operation, the product of two $N \times N$ matrices can be calculated in $O(1)$ time by using N^3 processors, if the matrix entries are of bounded magnitude and precision. For matrix entries of unbounded magnitude and precision, the problem can be solved*

- *in $O(1)$ time by using $N^{3+\delta}$ processors;*
- *in $O(\log \log N)$ time by using $N^3 / \log \log N$ processors;*
- *in $O(1)$ time with high probability by using N^3 processors.* ∎

5 Repeated Squaring

It turns out that to solve graph theory problems, we need to calculate the Nth power of an $N \times N$ matrix A. This can be obtained by $\lceil \log N \rceil$ successive squaring, i.e., calculating A^2, A^4, A^8, and so on. Such a computation increases the time complexities in Theorem 4 by a factor of $O(\log N)$.

Theorem 5. *When $\oplus$ is the "min" operation, the Nth power of an $N \times N$ matrix can be calculated in $O(\log N)$ time by using N^3 processors, if the matrix entries are of bounded magnitude and precision. For matrix entries of unbounded magnitude and precision, the problem can be solved*

- *in $O(\log N)$ time by using $N^{3+\delta}$ processors;*
- *in $O(\log N \log \log N)$ time by using $N^3 / \log \log N$ processors;*
- *in $O(\log N)$ time with high probability by using N^3 processors.* ∎

The last claim in Theorem 5 needs more explanation. In one matrix multiplication, there are N^2 simultaneous computation of minimum values using the Monte Carlo method in Lemma 3D. Since the algorithm in Lemma 3D has failure probability $O(1/N^\alpha)$, the failure probability of one matrix multiplication is $O(N^2/N^\alpha) = O(1/N^{\alpha-2})$. Since this Monte Carlo matrix multiplication is performed for $\lceil \log(N-1) \rceil$ times, the success probability of the all-pairs shortest paths computation is

$$\left(1 - O\left(\frac{1}{N^{\alpha-2}}\right)\right)^{\log N} = 1 - O\left(\frac{\log N}{N^{\alpha-2}}\right) = 1 - O\left(\frac{1}{N^{\alpha-2-\epsilon}}\right),$$

for any $\epsilon > 0$. The above argument implies that we need α (in Lemma 3D) to be no less than, say, 3. Fortunately, this can be easily achieved because a Monte Carlo algorithm which runs in $O(T(N))$ time with probability of success $1 - O(1/N^\alpha)$ for some constant $\alpha > 0$ can be turned into a Monte Carlo algorithm which runs in $O(T(N))$ time with probability of success $1 - O(1/N^\beta)$ for any large constant $\beta > 0$ by running the algorithm for $\lceil \beta/\alpha \rceil$ consecutive times and choosing a one that succeeds without increasing the time complexity.

6 All-Pairs Shortest Paths

Let $G = (V, E, W)$ be a weighted directed graph, where $V = \{v_1, v_2, ..., v_N\}$ is a set of N vertices, E is a set of arcs, and $W = (w_{ij})$ is an $N \times N$ weight matrix, i.e., w_{ij} is the distance from v_i to v_j if $(v_i, v_j) \in E$, and $w_{ij} = \infty$ if $(v_i, v_j) \notin E$. We assume that the weights are real numbers, whose magnitude and precision can be bounded or unbounded. The all-pairs shortest paths problem is to find $D = (d_{ij})$, an $N \times N$ matrix, where d_{ij} is the length of the shortest path from v_i to v_j along arcs in E.

Define $D^{(k)} = (d_{ij}^{(k)})$ be an $N \times N$ matrix, where $d_{ij}^{(k)}$ is the length of the shortest path from v_i to v_j that goes through at most $(k-1)$ intermediate vertices. It is clear that $D^{(1)} = W$, and $D^{(k)}$ can be obtained from $D^{(k/2)}$ by

$$d_{ij}^{(k)} = \min_{l}(d_{il}^{(k/2)} + d_{lj}^{(k/2)}),$$

where $k > 1$, and $1 \leq i, j, l \leq N$. Such a computation can be treated as a matrix multiplication problem $D^{(k)} = D^{(k/2)} D^{(k/2)}$, where $\otimes$ is the numerical addition operation $+$, and $\oplus$ is the "min" operation. It is also clear that $D = D^{(N-1)}$, so that we can apply Theorem 5.

Theorem 6. *The all-pairs shortest paths problem for a weighted directed graph with N vertices can be solved in $O(\log N)$ time by using N^3 processors, if the weights are of bounded magnitude and precision. For weights of unbounded magnitude and precision, the problem can be solved*

- *in $O(\log N)$ time by using $N^{3+\delta}$ processors;*
- *in $O(\log N \log \log N)$ time by using $N^3 / \log \log N$ processors;*
- *in $O(\log N)$ time with high probability by using N^3 processors.* ∎

6.1 Related Problems

It is well known that [1,3] there are many other interesting graph theory problems very closely related to the all-pairs shortest paths problem in the sense that the solution to the all-pairs shortest paths problem can be used to easily assemble the solutions to these problems. The following corollary gives a list of such problems, and the reader is referred to [3] for more details of these problems. All these problems can be solved on LARPBS with the time and processor complexities specified in Theorem 6.

Corollary 7. *All the following problems on a weighted directed graph with N vertices can be solved with the same time and processor complexities in Theorem 6: radius, diameter, and centers, bridges, median and median length, shortest path spanning tree, breadth-first spanning tree, minimum depth spanning tree, least median spanning tree, max gain, topological sort and critical paths.* ∎

7 Minimum Weight Spanning Tree

Let $G = (V, E, W)$ be a weighted undirected graph, where $V = \{v_1, v_2, ..., v_N\}$ is a set of N vertices, E is a set of edges, and $W = (w_{ij})$ is an $N \times N$ weight matrix, i.e., $w_{ij} = w_{ji}$ is the cost of the edge $\{v_i, v_j\} \in E$, and $w_{ij} = \infty$ if $\{v_i, v_j\} \notin E$. It is assumed that the edge costs are distinct, with ties broken using the lexicographical order. The minimum weight spanning tree problem is to find the unique spanning tree of G such that the sum of costs of the edges in the tree is minimized.

It was shown in [13] that the minimum weight spanning tree problem can be solved in the same way as that of the all-pairs shortest paths problem. Let the cost of a path be the highest cost of the edges on the path. Define $C^{(k)} = (c_{ij}^{(k)})$ to be an $N \times N$ matrix, where $c_{ij}^{(k)}$ is the shortest path from v_i to v_j that passes through at most $(k-1)$ intermediate vertices. Then, we have $c_{ij}^{(1)} = w_{ij}$, and

$$c_{ij}^{(k)} = \min_l(\max(c_{il}^{(k/2)}, c_{lj}^{(k/2)})),$$

for all $k > 1$, and $1 \le i, j, l \le N$. Such a computation can be treated as a matrix multiplication problem $C^{(k)} = C^{(k/2)} C^{(k/2)}$, where $\otimes$ is the "max" operation, and $\oplus$ is the "min" operation. Once $C^{(N-1)}$ is obtained, it is easy to determine the tree edges, namely, $\{v_i, v_j\}$ is in the minimum weight spanning tree if and only if $c_{ij}^{(N-1)} = w_{ij}$.

Theorem 8. *The minimum weight spanning tree problem for a weighted undirected graph with N vertices can be solved in $O(\log N)$ time by using N^3 processors, if the weights are of bounded magnitude and precision. For weights of unbounded magnitude and precision, the problem can be solved*

- *in $O(\log N)$ time by using $N^{3+\delta}$ processors;*
- *in $O(\log N \log \log N)$ time by using $N^3 / \log \log N$ processors;*
- *in $O(\log N)$ time with high probability by using N^3 processors.* ∎

The spanning tree problem for undirected graphs is a special case of the minimum weight spanning tree problem in the sense that $w_{ij} = w_{ji} = 1$ for an edge $\{v_i, v_j\} \in E$. Since all the weights are of bounded magnitude, we have

Theorem 9. *The spanning tree problem for an undirected graph with N vertices can be solved in $O(\log N)$ time by using N^3 processors.* ∎

8 Transitive Closure

Let $G = (V, E)$ be a directed graph, and A_G be the adjacency matrix of G, which is an $N \times N$ boolean matrix. Let A_G^* be the adjacency matrix of G's transitive closure. By applying Lemma 2, the following result was shown in [7].

Theorem 10. *The transitive closure of a directed graph with N vertices can be found in $O(\log N)$ time by using $O(N^3 / \log N)$ processors.* ∎

9 Strong Components

A strong component of a directed graph $G = (V, E)$ is a subgraph $G' = (V', E')$ of G such that there is path from every vertex in V' to every other vertex in V' along arcs in E', and G' is maximal, i.e., G' is not a subgraph of another strong component of G.

To find the strong components of G, we first calculate $A_G^* = (a_{ij}^*)$, where $a_{ij}^* = 1$ if there is a path from v_i to v_j, and $a_{ij}^* = 0$ otherwise. Then, v_i and v_j are in the same component if and only if $a_{ij}^* = a_{ji}^* = 1$, for all $1 \le i, j \le N$. Based on A_G^*, we construct $C = (c_{ij})$, where $c_{ij} = 1$ if v_i and v_j are in the same component, and $c_{ij} = 0$ otherwise. If the strong components are represented in such a way that every vertex v_i remembers the vertex $v_{s(i)}$ with the smallest index $s(i)$ in the same component, then $s(i)$ is the minimum j such that $c_{ij} = 1$, where $1 \le j \le N$. The construction of C and the finding of the $s(i)$'s can be performed on an N^2-processor LARPBS in $O(1)$ time.

Theorem 11. *The strong components of a directed graph with N vertices can be found in $O(\log N)$ time by using $O(N^3/\log N)$ processors.* ∎

The connected component problem for undirected graphs is just a special case of the strong component problem for directed graphs.

Theorem 12. *The connected components of an undirected graph with N vertices can be found in $O(\log N)$ time by using $O(N^3/\log N)$ processors.* ∎

10 Summary and Conclusions

We have considered fast parallel algorithms on the model of linear array with a reconfigurable pipelined bus system for the following important graph theory problems:

- All-pairs shortest paths;
- Radius, diameter, and centers;
- Bridges;
- Median and median length;
- Shortest path spanning tree;
- Breadth-first spanning tree;
- Minimum depth spanning tree;
- Least median spanning tree;
- Max gain;
- Topological sort and critical paths;
- Minimum weight spanning tree;
- Spanning tree;
- Transitive closure;
- Strong components;
- Connected components.

Our algorithms are based on fast matrix multiplication and extreme value finding algorithms, and are currently the fastest algorithms.

Appendix. Implementation Details

We now present the implementation details of the generic matrix multiplication algorithm of Lemma 1. The algorithm calculates the product $C = AB = (c_{ik})$ of two $N \times N$ matrices $A = (a_{ij})$ and $B = (b_{jk})$ in $O(T)$ time by using $N^2 M$ processors, assuming that the aggregation $x_1 \oplus x_2 \oplus \cdots \oplus x_N$ can be calculated in $O(T)$ time by using M processors.

For convenience, we label the $N^2 M$ processors using triplets (i, j, k), where $1 \leq i, j \leq N$, and $1 \leq k \leq M$. Processors $P(i, j, k)$ are ordered in the linear array using the lexicographical order. Let $P(i, j, *)$ denote a group of consecutive processors $P(i, j, 1), P(i, j, 2), ..., P(i, j, M)$. It is noticed that the original system can be reconfigured into N^2 subsystems, namely, the $P(i, j, *)$'s. Each processor $P(i, j, k)$ has three registers $A(i, j, k)$, $B(i, j, k)$, and $C(i, j, k)$.

We start with the case where $M \geq N$. The input and output data layout are specified as follows. Initially, elements a_{ij} and b_{ji} are stored in registers $A(1, i, j)$ and $B(1, i, j)$ respectively, for all $1 \leq i, j \leq N$. If we partition A into N row vectors $A_1, A_2, ..., A_N$, and B into N column vectors $B_1, B_2, ..., B_N$,

$$A = \begin{bmatrix} A_1 \\ A_2 \\ \vdots \\ A_N \end{bmatrix}, \quad B = [B_1, B_2, ..., B_N],$$

then the initial data distribution is as follows:

$$\begin{array}{cccc} P(1,1,*) & P(1,2,*) & \cdots & P(1,N,*) \\ (A_1, B_1) & (A_2, B_2) & \cdots & (A_N, B_N) \end{array}$$

where other processors are not shown here for simplicity, since they do not carry any data initially. When the computation is done, c_{ij} is found in $C(1, i, j)$. If we divide C into N row vectors $C_1, C_2, ..., C_N$,

$$C = \begin{bmatrix} C_1 \\ C_2 \\ \vdots \\ C_N \end{bmatrix},$$

then the output layout is

$$\begin{array}{cccc} P(1,1,*) & P(1,2,*) & \cdots & P(1,N,*) \\ (C_1) & (C_2) & \cdots & (C_N) \end{array}$$

Such an arrangement makes it easy for C to be used as an input to another computation, e.g., repeated squaring.

The algorithm proceeds as follows. In Step (1), we change the placement of matrix A in such a way that element a_{ij} is stored in $A(i, 1, j)$, for all $1 \leq$

A Generic Matrix Multiplication Algorithm on LARPBS

```
for 1 ≤ i,j ≤ N do in parallel // Step (1). One-to-one communication.
    A(i,1,j) ← A(1,i,j)
endfor

for 1 ≤ i,k ≤ N do in parallel // Step (2). Multiple multicasting.
    A(i,2,k),A(i,3,k),...,A(i,N,k) ← A(i,1,k)
endfor
for 1 ≤ j,k ≤ N do in parallel
    B(2,j,k),B(3,j,k),...,B(N,j,k) ← B(1,j,k)
endfor

for 1 ≤ i,j,k ≤ N do in parallel // Step (3). Local computation.
    C(i,j,k) ← A(i,j,k) ⊗ B(i,j,k)
endfor

for 1 ≤ i,j ≤ N do in parallel // Step (4). Aggregation.
    C(i,j,1) ← C(i,j,1) ⊕ C(i,j,2) ⊕ ··· ⊕ C(i,j,N)
endfor

for 1 ≤ i ≤ N, 2 ≤ j ≤ N, do in parallel // Step (5). One-to-one communi.
    C(1,i,j) ← C(i,j,1)
endfor
```

$i,j \leq N$. This can be accomplished by a one-to-one communication. After such replacement, we have the following data distribution,

$$
\begin{array}{cccc}
P(1,1,*) & P(1,2,*) & \cdots & P(1,N,*) \\
(A_1,B_1) & (A_2,B_2) & \cdots & (A_N,B_N) \\
\\
P(2,1,*) & P(2,2,*) & \cdots & P(2,N,*) \\
(A_2,-) & (-,-) & \cdots & (-,-) \\
\\
\vdots & \vdots & \ddots & \vdots \\
\\
P(N,1,*) & P(N,2,*) & \cdots & P(N,N,*) \\
(A_N,-) & (-,-) & \cdots & (-,-)
\end{array}
$$

where the linear array of $N^2 M$ processors are logically arranged as a two dimensional $N \times N$ array, and each element in the array stands for a group of M processors $P(i,j,*)$. The symbol "−" means that the A and B registers are still undefined.

In Step (2), we distribute the rows of A and columns of B to the right processors, such that processors $P(i,j,*)$ hold A_i and B_j, for all $1 \leq i,j \leq N$. This can be performed using multiple multicasting operations. After multicasting, the data distribution is as follows:

$$
\begin{array}{cccc}
P(1,1,*) & P(1,2,*) & \cdots & P(1,N,*) \\
(A_1,B_1) & (A_1,B_2) & \cdots & (A_1,B_N) \\
\\
P(2,1,*) & P(2,2,*) & \cdots & P(2,N,*) \\
(A_2,B_1) & (A_2,B_2) & \cdots & (A_2,B_N) \\
\\
\vdots & \vdots & \ddots & \vdots \\
\\
P(N,1,*) & P(N,2,*) & \cdots & P(N,N,*) \\
(A_N,B_1) & (A_N,B_2) & \cdots & (A_N,B_N)
\end{array}
$$

At this point, processors are ready to compute. In Step (3), $P(i,j,k)$ calculates $a_{ik} \otimes b_{kj}$.

Then, in Step (4), the values $\texttt{C}(i,j,k)$, $1 \le k \le N$, are aggregated by the M processors in $P(i,j,*)$, for all $1 \le i,j \le N$, and the result c_{ij} is in $\texttt{C}(i,j,1)$. Here, for the purpose of multiple aggregations, the original system is reconfigured into N^2 subsystems $P(i,j,*)$'s, for $1 \le i,j \le N$. After calculation, the data distribution is as follows:

$$
\begin{array}{cccc}
P(1,1,1) & P(1,2,1) & \cdots & P(1,N,1) \\
(c_{11}) & (c_{12}) & \cdots & (c_{1N}) \\
\\
P(2,1,1) & P(2,2,1) & \cdots & P(2,N,1) \\
(c_{21}) & (c_{22}) & \cdots & (c_{2N}) \\
\\
\vdots & \vdots & \ddots & \vdots \\
\\
P(N,1,1) & P(N,2,1) & \cdots & P(N,N,1) \\
(c_{N1}) & (c_{N2}) & \cdots & (c_{NN})
\end{array}
$$

Finally, in Step (5), one more data movement via one-to-one communication brings the c_{ij}'s to the right processors.

It is clear that Steps (1), (2), (3), and (5) are simple local computations or primitive communication operations, and hence, take $O(1)$ time. Step (4) requires $O(T)$ time. Thus, the entire algorithm can be executed in $O(T)$ time.

In general, when $M \ge 1$ and $M \le N$, to store a vector in a group $P(i,j,*)$ of consecutive processors $P(i,j,1)$, $P(i,j,2)$, ..., $P(i,j,M)$, we can divide a vector of length N into sub-vectors of length $\lceil N/M \rceil$ such that each processor $P(i,j,k)$ is assigned a sub-vector. This increases the time of Steps (1), (2), (3), and (5) by a factor of $O(N/M)$. Since $T = \Omega(N/M)$, the complete algorithm still takes $O(T)$ time.

Acknowledgments

Keqin Li was supported by National Aeronautics and Space Administration and the Research Foundation of State University of New York through NASA/University Joint Venture in Space Science Program under Grant NAG8-1313 (1996-1999). Yi Pan was supported by the National Science Foundation under Grants CCR-9211621 and CCR-9503882, the Air Force Avionics Laboratory, Wright Laboratory, Dayton, Ohio, under Grant F33615-C-2218, and an Ohio Board of Regents Investment Fund Competition Grant. Mounir Hamdi was partially supported by the Hong Kong Research Grant Council under the Grant HKUST6026/97E.

References

1. S.G. Akl, *Parallel Computation: Models and Methods*, Prentice-Hall, Upper Saddle River, New Jersey, 1997.
2. D. Chiarulli, R. Melhem, and S. Levitan, "Using coincident optical pulses for parallel memory addressing," *IEEE Computer*, vol. 30, pp. 48-57, 1987.
3. E. Dekel, D. Nassimi, and S. Sahni, "Parallel matrix and graph algorithms," *SIAM Journal on Computing*, vol.10, pp.657-673, 1981.
4. M. Hamdi, C. Qiao, Y. Pan, and J. Tong, "Communication-efficient sorting algorithms on reconfigurable array of processors with slotted optical buses," to appear in *Journal of Parallel and Distributed Computing*.
5. J. JáJá, *An Introduction to Parallel Algorithms*, Addison-Wesley, 1992.
6. S. Levitan, D. Chiarulli, and R. Melhem, "Coincident pulse techniques for multiprocessor interconnection structures," *Applied Optics*, vol. 29, pp. 2024-2039, 1990.
7. K. Li, "Constant time boolean matrix multiplication on a linear array with a reconfigurable pipelined bus system," *Journal of Supercomputing*, vol.11, no.4, pp.391-403, 1997.
8. K. Li and V.Y. Pan, "Parallel matrix multiplication on a linear array with a reconfigurable pipelined bus system," *Proceedings of IPPS/SPDP '99*, San Juan, Puerto Rico, April 12-16, 1999.
9. K. Li, Y. Pan, and S.-Q. Zheng, eds., *Parallel Computing Using Optical Interconnections*, Kluwer Academic Publishers, Boston, Massachusetts, 1998.
10. K. Li, Y. Pan, and S.-Q. Zheng, "Fast and processor efficient parallel matrix multiplication algorithms on a linear array with a reconfigurable pipelined bus system," *IEEE Transactions on Parallel and Distributed Systems*, vol. 9, no. 8, pp. 705-720, August 1998.
11. K. Li, Y. Pan, and S.-Q. Zheng, "Fast and efficient parallel matrix computations on a linear array with a reconfigurable pipelined optical bus system," in *High Performance Computing Systems and Applications*, J. Schaeffer ed., pp. 363-380, Kluwer Academic Publishers, Boston, Massachusetts, 1998.
12. K. Li, Y. Pan, and S.-Q. Zheng, "Efficient deterministic and probabilistic simulations of PRAMs on a linear array with a reconfigurable pipelined bus system," to appear in *Journal of Supercomputing*.
13. B.M. Maggs and S.A. Plotkin, "Minimum-cost spanning tree as a path-finding problem," *Information Processing Letters*, vol.26, pp.291-293, 1988.
14. Y. Pan and M. Hamdi, "Efficient computation of singular value decomposition on arrays with pipelined optical buses," *Journal of Network and Computer Applications*, vol.19, pp.235-248, July 1996.

15. Y. Pan, M. Hamdi, and K. Li, "Efficient and scalable quicksort on a linear array with a reconfigurable pipelined bus system," *Future Generation Computer Systems*, vol. 13, no. 6, pp. 501-513, June 1998.
16. Y. Pan and K. Li, "Linear array with a reconfigurable pipelined bus system – concepts and applications," *Journal of Information Sciences*, vol. 106, no. 3-4, pp. 237-258, May 1998.
17. Y. Pan, K. Li, and S.-Q. Zheng, "Fast nearest neighbor algorithms on a linear array with a reconfigurable pipelined bus system," *Journal of Parallel Algorithms and Applications*, vol. 13, pp. 1-25, 1998.
18. H. Park, H.J. Kim, and V.K. Prasanna, "An $O(1)$ time optimal algorithm for multiplying matrices on reconfigurable mesh," *Information Processing Letters*, vol.47, pp.109-113, 1993.
19. S. Rajasekaran and S. Sen, "Random sampling techniques and parallel algorithm design," in *Synthesis of Parallel Algorithms*, J.H. Reif, ed., pp.411-451, Morgan Kaufmann, 1993.
20. B.-F. Wang and G.-H. Chen, "Constant time algorithms for the transitive closure and some related graph problems on processor arrays with reconfigurable bus systems," *IEEE Transactions on Parallel and Distributed Systems*, vol.1, pp.500-507, 1990.

High Speed, High Capacity Bused Interconnects Using Optical Slab Waveguides

Martin Feldman, Ramachandran Vaidyanathan, and Ahmed El-Amawy

Department of Electrical & Computer Engineering
Louisiana State University, Baton Rouge, LA 70803, USA
{feldman,vaidy,amawy}@ee.lsu.edu

Abstract. In this paper we consider the use of optical slab waveguides as buses in a parallel computing environment. We show that slab buses can connect to many more elements than conventional electrical or fiber optic buses. We also introduce a novel multiplexing scheme called mode division multiplexing that vastly increases the number of independent channels that a single slab can support. We show that optical slab waveguides have, in principle, capacities of over a million independent channels (distinguished by about 1000 "out-of-plane modes" and about 1000 wavelengths) in a single physical medium, with each channel capable of sustaining a load of over 1000. This becomes comparable to the high capacity of a free space optical system, but with the ability to broadcast each light source to many physically separated locations. Preliminary experiments on the "sawtooth slab bus" point to the feasibility of practical slab buses. We also present a bus arbitration example that uses the high capacity and loading of slab buses to achieve sublogarithmic arbitration time.

1 Introduction

Buses are essential components in most computer systems. Their uses range from buses within processors and host buses on a board, to backplane buses connecting boards and global shared buses in multiprocessor systems.

As interconnects, buses have several desirable features. Being shared resources, they can be used more efficiently and flexibly than dedicated connections. Bus-based systems support broadcasting easily and naturally, have reasonable costs, and admit simple layouts. Besides traditional multiple bus systems [3, 11, 14, 24, 26], buses have been used in a variety of computing structures including enhanced meshes [2, 4, 22], reconfigurable networks [8, 12, 17, 19] and in the context of emulating interconnection functions and topologies [6, 7, 13, 15, 16].

As the number of connections to a bus increases, the following problems arise:

Loading: In this paper we will use the term *loading* loosely to mean the maximum number of connections that can be made to a bus without significantly degrading the quality of the signal on the bus.

Connections to an *electrical* bus cause capacitive loading that limits the rate at which the signal switches state reliably (i.e., bus clock rate). In addition, cross talk problems arise from the close proximity of high frequency signals in adjacent buses. With existing technology an electrical bus operating at a few hundred MHz, can only support about 30 connections. A fiber optic bus can connect to a somewhat larger number of elements (about 100), still inadequate for moderately large systems.

Arbitration cost (for asynchronous systems): As the number of masters connected to a bus increases, so does the likelihood of multiple masters attempting to use the bus simultaneously. Such conflicts have to be resolved by an arbitration mechanism, whose complexity (cost, delay) depends directly on the number of bus masters.

In this paper we propose buses that use *optical slab waveguides*. Unlike optical fibers that have a single (or a few) transmission modes, slabs can support several thousands of modes. To a large extent, slab buses can overcome both of the above problems of traditional buses. We will show that, in principle, a single slab waveguide can admit over a million independent channels, each capable of supporting a loading of over 1000. This high capacity of slab buses can also be used to devise a fast bus arbitration scheme. Traditional (single) bus-based systems are also limited by bus bandwidth. Multiple bus systems, capable of overcoming this limitation are easily implemented on a single slab waveguide, whose independent channels can be treated as individual buses.

Much previous work on optical buses has centered on the adaptation of components used for optical fiber communication. This is understandable since optical fiber communications is an established technology. However, many of the considerations important to long haul optical fiber communications are not relevant to the communication between proximate processors. In order to operate over long distances at high frequencies, optical fibers are designed to have minimal attenuation, less than 1 db/kilometer, and minimal dispersion. Clearly, the low attenuation and dispersion required for transmission over many kilometers is totally unnecessary between processors at most a few feet apart. Moreover, the low dispersion is usually obtained by restricting the fibers to single mode operation. This requires the use of laser diode transmitters for efficient light coupling to the fiber. In addition, if the signals from several fiber inputs are passively coupled or fed into a single fiber (as in a bus), most of the available power will be reflected. This is because the optical power that is carried by a single mode fiber with input from a single source is determined by the brightness of the source. This power cannot be increased by the addition of other sources. Not only does this seriously reduce the power available in many optical bus configurations, it also couples some of the power back into the laser that emitted the light. This unwanted feedback can cause instabilities in the operation of the laser.

Free space optical systems have also been studied as interconnects for parallel computing [9, 18, 21]. In a typical free space optical system (Fig. 1), lenses (not shown) image an array of light emitting diodes or lasers on a matching array of photodetectors.

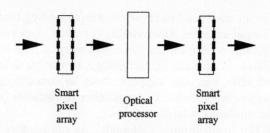

Fig. 1. Structure of a typical massively parallel free space optical system.

They are capable of carrying an enormous amount of data. In addition, with "smart pixels" some processing can be done locally between each photodiode and its corresponding light emitter. Processing can also be done in additional optical components such as liquid crystal arrays or acoustooptic modulators inserted in the optical path. However, free space optical systems have limited applications as buses, where light from each of the input sources is broadcast to all of the detectors.

This paper reports some preliminary results with slab waveguides as buses. In the next section we will show why a single mode fiber optic bus can support a loading of only about 100. Slab buses, on the other hand, are shown to be capable of connecting to over 1000 elements per channel (Section 3). In Section 4 we introduce a novel multiplexing scheme called mode division multiplexing that is possible only on slab waveguides. This technique can be used in conjunction with other conventional methods such as wavelength division multiplexing to increase the number of independent channels in the bus by a factor of about 1000. In Section 5 we briefly outline some results from preliminary experiments on slab buses. In Section 6 we show how the high capacity of slab buses can be used in a fixed priority-based arbitration scheme that resolves conflict among N w-bit processors in $O\left(\frac{\log N}{\log w}\right)$ steps. The fact that this method can be used to resolve conflicts in access to any shared resource may be of independent interest. Finally in Section 7 we summarize our results and make some concluding remarks.

2 Loading Limitation of Fiber Buses

A typical optical fiber bus consists of an U-shaped optical fiber connected by directional couplers to light sources (on one end of the bus) and detectors (on the other) [5, 10, 20] (see Fig. 2). Normally, directional couplers are used to transfer power from one fiber to the other, while the direction of information flow along the fibers remains the same. The fraction of power transferred between the fibers (*coupling factor*) is a parameter of the directional coupler. To ensure that all detectors receive the same amount of power (regardless of the transmitting light source) different couplers need different coupling factors. Moreover, if one unit

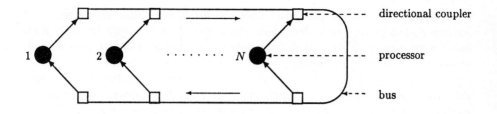

Fig. 2. Fiber optic bus with directional couplers.

of power is transmitted by a source, each of the N detectors only receive $\frac{1}{N^2}$ units of power, as shown below.

For the following analysis we assume a single mode fiber bus and neglect transmission and insertion losses. Number the processors from 1 to N (left to right) as shown in Fig. 2. Let the coupling factors for the light source and detector of processor i be α_i and β_i, respectively. That is, if processor i transmits one unit of power through its light source, then α_i units are transferred to the bus. If processor i is not transmitting and the bus is carrying one unit of power, then α_i units of power are siphoned off from the bus. Likewise at the detector end, if one unit of power appears on the bus, then β_i units of power is transferred across the coupler to the detector of processor i. Clearly, $\alpha_1 = \beta_1 = 1$ as the first processor does not require a coupler.

Let each light source transmit one unit of power. A unit of power transmitted by the first source reduces to $(1 - \alpha_2)(1 - \alpha_3) \cdots (1 - \alpha_i)$ just after the ith source. Since the power available to the detectors must be independent of the transmitting source, $\alpha_i = (1-\alpha_2)(1-\alpha_3) \cdots (1-\alpha_i)$, for each $i > 1$. This implies that $\alpha_i = \frac{1}{i}$. In a similar manner it can be shown that $\beta_i = \frac{1}{i}$. The power on the bus, just after the Nth light source is $\alpha_N = \frac{1}{n}$. The detector of processor i receives $\frac{1}{N}(1 - \beta_N)(1 - \beta_{N-1}) \cdots (1 - \beta_{i+1})\beta_i = \frac{1}{N^2}$ units of power.

This inefficient power distribution limits the loading, or the maximum number of detectors that may be connected to the bus. Let N_f denote this maximum for a fiber bus. Also let λ (in microns) and τ (in nsecs) denote the wavelength of the light and the detection time, respectively. Assuming a 100% transmission efficiency, it can be shown that P, the number of photons per detector per mW of incident energy is given by

$$P = 5.04 \times 10^6 \frac{\tau \lambda}{(N_f)^2}.$$

With $\tau = 1$ nsec (corresponding to a Gigabit data rate) and $N_f = 100$, P ranges from about 300 photons in red light ($\lambda \approx 0.6$ microns) to about 500 photons at the limiting wavelength for silicon photodetectors (about 1.0 microns). These numbers are close to the minimum required for detection with ample error margins, especially in the presence of background noise and with more realistic transmission rates. Thus, the maximum loading for fiber buses is about 100.

Though the analysis presented above is for a particular arrangement of sources and detectors on a fiber bus, the essential ideas apply to other fiber bus configurations as well (such as couplers arranged in a binary tree with N leaves with each coupler having a 50% coupling factor). This is because the reciprocal property of directional couplers is a fundamental limitation: it implies that the power carried by a fiber cannot be extended indefinitely simply by adding more inputs. To do so would increase the power per mode, and hence the brightness in the fiber without limit, which would violate the second law of thermodynamics.

3 Slab Waveguides

The term *slab* is used in this paper to characterize waveguides for which the dimensions of the cross section are much larger than the wavelength of the light. Slab waveguides have many modes in which the light may be transmitted. The term *fiber* is used for waveguides that permit only a single (or a few) modes.

It can be shown [25, page 312] that the number of modes, m, in the plane of a slab of width d is given by

$$m = \frac{2d\sqrt{(n^2-1)}}{\lambda},$$

where n is the index of refraction of the slab in air, and λ is the wavelength of the light. For the slabs considered here, with d equal to a few mm, m is on the order of several thousands. In contrast with optical fibers, the total power carried by a slab can be proportional to the number of sources (as long as there are at least as many modes as sources).

In a slab bus with a carefully designed geometry, it is possible to divide the power from a laser diode or LED equally among all detectors (assuming no transmission and insertion loss). That is, if the light source transmits one unit of power, then each of N detectors receives $\frac{1}{N}$ units of power. This directly translates to a larger loading, N_s. Under the same assumptions used above to calculate the loading of a fiber bus, the loading, N_s, of a slab bus would approximately be equal to $(N_f)^2$; recall that N_f is the loading of a fiber bus. With even a conservative 10% transmission efficiency, N_s is around 1000, which corresponds to approximately 10 times as many photons per unit time, an order of magnitude improvement over fiber buses, in the number of photons and hence the number of detectors.

It should be pointed out that the parameters used here are for adequately bright light sources, such as laser diodes, and sufficiently sensitive detectors, for example avalanche diodes, that are well within the present state of the art.

Several slab geometries with various properties are possible. The slabs considered in this paper have a "sawtooth" shape (Fig. 3). In these waveguides light is injected and removed from ports spaced along the side. These locations could correspond to the locations of different circuit boards. There is a minimum distance between the last transmitter and the first receiver, so that no element is shadowed. The light is nominally divided uniformly between the receivers, with minimal loss and reflection. The sawtooth shape may also be folded, so that each card in a rack may have both a transmitter and a receiver.

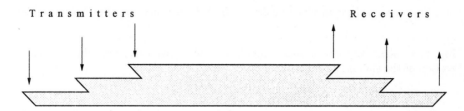

Fig. 3. The sawtooth slab waveguide; in this figure, the *plane of the slab* is the plane of this page; a view of this plane will be called the *side view*. The direction perpendicular to the plane of the slab will be termed the *top view*.

For our discussion, we classify the transmission modes of the light propagating in the slab into two groups.

In-plane modes (Fig. 4(a)), or modes in which the propagation of light is parallel to the plane of the slab but in general not collimated within that plane.

Out-of-plane modes (Fig. 4(b)), in which the light is collimated, but its propagation is not necessarily parallel to the plane of the slab.

Of course, most modes have components which are both in-plane and out-of-plane.

The in-plane multimode capacity of the slab waveguide is useful in mode mixing, so that the energy from each of the transmitters is uniformly distributed among the receivers. It is also important that the slab waveguide have minimal delay, dispersion, and echo. Echo is defined as light which does not reach a detector by a direct path, but returns after a relatively long delay, possibly interfering with the signal from the next transmission on the same channel. Although we have not yet measured echo in sawtooth guides, we have observed that the polarization of the propagating light is preserved. This implies that conventional polarization techniques, e.g. double passes through a quarter wave plate, could be used to minimize echoes, if necessary.

4 Mode Division Multiplexing

In this section we describe a novel multiplexing scheme called *mode division multiplexing (MDM)*, that significantly increases the number of channels that a single slab can support. We will show that more than 1000 independent channels can coexist on a slab. Since each of these channels can utilize other multiplexing schemes, MDM can increase the number of channels available through these schemes by a factor of 1000.

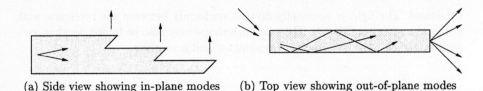

(a) Side view showing in-plane modes   (b) Top view showing out-of-plane modes

Fig. 4. In-plane and out-of-plane modes of a slab waveguide; for simplicity only a portion of the slab is shown.

In this section the in-plane modes and the out-of-plane modes are treated very differently: the in-plane modes are well mixed to insure uniform illumination at the detectors, while the out-of-plane modes are well separated to provide multiple communication channels (Fig. 5).

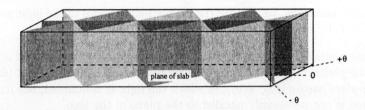

Fig. 5. A section of the slab, showing three out-of-plane modes. These modes are well separated as they propagate in different angles when viewed in a direction perpendicular to the plane of the slab (top view). In the plane of the slab itself, the modes are well mixed.

If light is propagated through the slab waveguide parallel to the plane of the slab, then the light leaving the slab will also be collimated in a direction parallel to the plane of the slab. This is true at each of the slab's exit ports. Additional modes may also be introduced in the slab, in which the light is again collimated, but at various angles to the plane of the slab. The light from these modes exits the slab in a parallel beam at an angle, θ_{air}, to the plane of the slab that is given by

$$\sin \theta_{\text{air}} = n \sin \theta_{\text{slab}}, \qquad (1)$$

where n is the index of refraction and θ_{slab} is the angle of the beam within the slab. Because it may undergo either an odd or an even number of reflections, each mode exits at both positive and negative angles, even though it may have been excited by light which only entered the slab at either a positive or a negative angle. These properties have two important consequences: First, light introduced at various angles may be detected separately by different detectors, so that multiple channels of communication are possible within the same waveguide. Second, light may be intentionally introduced at both positive and negative

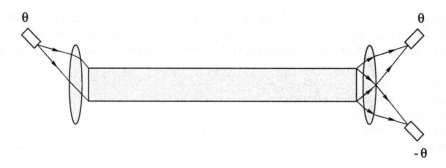

Fig. 6. A single out-of-plane mode entering and exiting a slab. The figure shows a top view of the slab and, for clarity, only one input port and one output port have been shown.

angles by different transmitters, and then be detected by the same detector. This permits operation of the two transmitters in an "OR" configuration, if that is desired. Note that a mirror may be used to fold the light exiting the slab, so that although it exits at both positive and negative angles a single detector is all that is required for each mode.

The principle of mode division multiplexing is illustrated in Fig. 6, which shows the light from a laser diode being collimated into a parallel beam and filling the slab. Note that Fig. 6 shows only one input port and one output port; similar effects occur at all input and output ports. It is important that the light not be collimated in the in-plane direction, in order to insure good mixing and uniform illumination of the detectors. The collimation could be performed with a cylindrical lens whose axis is perpendicular to the plane of the laser diodes. Such a lens is most simply formed by shaping the input face of the slab.

The light leaving the slab is focused by a second lens, either spherical or cylindrical, to a spot of size about $\frac{\lambda f}{t}$, where f is the focal length of the second lens and t is the thickness of the slab.

Additional laser diodes may also be arranged at angles θ_N to the plane of the slab (Fig. 7). In this figure the input lens has been placed a focal length away

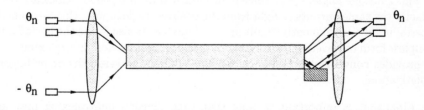

Fig. 7. Multiple out-of-plane modes entering and exiting a slab. The figure shows a top view of the slab. For clarity, only light entering from one input port and leaving through a one exit port of the slab have been shown.

from the LED's and also a focal length away from the slab. This permits the LED's to emit light perpendicular to their plane, facilitating their fabrication on a single chip, while also collimating their light in the slab. The light from each of these additional laser diodes propagates within the slab at an angle given by equation (1). It then exits at angles of $\pm\theta_N$ and is focused by the output lens into two lines. The images are lines because the light is not collimated in the orthogonal direction. As indicated in Fig. 7, a mirror can be used to overlap the two images for improved light collection; it is important to use all the light because the light can potentially be shared between a large number of detectors.

If we take the total useful range of the focused output light as $\pm f$, then the number of resolved spots is $\frac{f}{(\lambda f/t)} = \frac{t}{\lambda}$. The number of useful modes is somewhat less, because of the necessity of providing clearance between channels and because of possible degradation of the modes during transmission through the slab. Nevertheless, for a slab a few mm thick and visible light the number of distinct out-of-plane modes is on the order of 1000. (Note that in addition to these out-of-plane modes, the slab may have several thousands of in-plane modes that are used to support a large loading.)

Wavelength division multiplexing may also be introduced, by using sources at different wavelengths. The different wavelength signals may then be directed to different detectors, for example by using dichoric mirrors, which reflect some wavelengths while transmitting others. In principle this can be done at each of the output ports, and for each of the propagating modes, so that the total number of channels at each exit port would be the product of the number of wavelengths and the number of modes. A more interesting concept is to use a dispersive device, such as an efficient blazed diffraction grating, to direct the different wavelengths to corresponding detectors (Fig. 8). This could also be done in conjunction with mode division multiplexing, again greatly expanding the total number of channels available for communication. In principle, the light signals at different wavelengths could be spatially separated into more than 1000 different channels, resulting in an array of 1000 × 1000 elements. Each case requires a detector for each channel, but these may be arranged in an array on a single chip (or a few chips).

It is interesting that inputs may also be arranged at negative angles, as shown in Fig. 7 for one angle, $-\theta_N$. Since light launched from these locations arrives at exactly the same outputs as light from the positive angle inputs, this presents the possibility of a permanently "built in" OR function. It also forms the basis for the simplest form of wavelength division and polarization division multiplexing, since it provides convenient input ports for two different wavelengths or orthogonal polarizations.

Finally, it is important to note that time division multiplexing may also be used, further extending the capacity of the slab mode. This is especially significant in its possible application as a pipelined bus, a widely studied model of computation [10, 20].

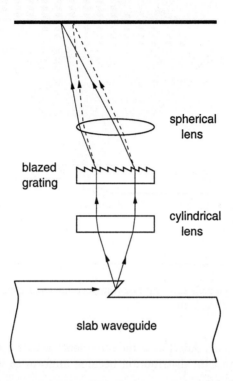

Fig. 8. Using a blazed grating to demultiplex a WDM signal. The figure shows the spatial separation of two wavelengths (indicated by solid and dotted lines)

5 Experimental Results

Preliminary experiments have been performed with sawtooth optical slabs machined from Plexiglas and polished with 1500 grit sandpaper. The slabs were 3 mm thick, and had 8 input ports and 8 output ports, generally arranged as in Fig. 3. The cross sectional area of each port was 3×3 mm^2, and the ports were spaced 25 mm apart. The separation between the input ports and the output ports was varied between 200 mm and 800 mm.

The source of light was a red light emitting diode (LED) operated in a constant current mode. The detector was a photodiode operating in the photovoltaic mode into a resistive load under conditions which limited the output voltage to 50 mV to insure linearity. No optics was used to match the LED to the input ports, or the photodiode to the output ports. A digital voltmeter was used to perform dc measurements of the photocurrent.

We observed an insertion loss of approximately 10 db, largely attributable to coupling losses at the input and output ports. The uniformity between pairs of input and output ports was approximately 1.5 db, full range between the best

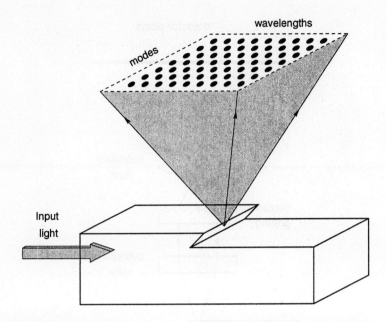

Fig. 9. A portion of the slab showing two dimensional separation of modes and wavelengths at a "tooth" of the slab.

pair and the worst pair. Additional measurements across a cross section of the slab showed that the light distribution became uniform to within a factor of 2 after just a few cm.

In a second experiment light from a HeNe laser beam at 6328 Å was passed through a 25 mm × 300 mm × 3 mm Plexiglas slab. The beam did not fill the slab. The angle by which the light was inclined to the plane of the slab was varied up to approximately 60°. Fig. 10 is an oscilloscope trace which shows the angular distribution of the intensity in the exit beam at an angle to the slab of 18°. The angular distribution of a similar beam which did not pass through the slab is also shown. Although the exact shape of the profile is not yet understood in detail, it is certainly comparable to the profile from the comparison beam.

6 Bus Arbitration

The discussion in the previous sections indicate that a single slab waveguide has the potential to provide over a million independent channels (corresponding to 1000 out-of-plane modes and 1000 colors used in wavelength division multiplexing). Moreover, each of these channels could support over 1000 loads. Although we do not expect such an enormous capacity to be utilized on a single bus, at least initially, a slab waveguide can be used to implement multiple bus systems with much larger loads than previously possible. This is comparable to the high capacity of a free space optical system, but with the ability to broadcast each light source to many physically separated locations.

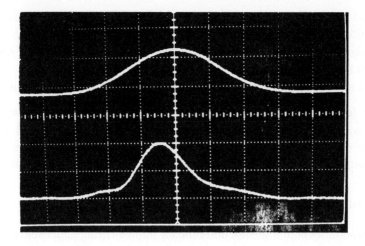

Fig. 10. Angular distribution of reference beam (top trace) and beam through slab (bottom trace). The full scale is 8 mrads

As an example of what slab waveguides can offer in the parallel processing context, we take up the topic of bus arbitration. In this section we show how a the bus arbitration scheme of [1, 23] can be adapted to exploit the high capacity and loading of slab buses.

Consider a bus with N processors numbered $0, 1, \cdots, N-1$. Without loss of generality, assume that $N = 2^k$, for some integer k. For any integer b (where $2 \leq b \leq k$), the k-bit processor addresses can be expressed in base 2^b, simply by grouping the k address bits into $\lceil \frac{k}{b} \rceil$ groups, each with at most b contiguous bits. For convenience, assume that $\frac{k}{b}$ is an integer. Thus the address of processor i can be written in base 2^b as $a_{i,1}, a_{i,2}, \cdots, a_{i,\frac{k}{b}}$, where for each $1 \leq j \leq \frac{k}{b}$, $a_{i,j}$ is the jth b-bit digit in the base 2^b representation of the address of processor i.

The aim is to determine the highest indexed processor that wants to write on the bus. Suppose we have $2^b - 1$ (one bit) buses available for the arbitration process. Let these buses be indexed $1, 2, \cdots, 2^b - 1$. We will also assume that each processor has word size w and can, in constant time, determine the leftmost (or rightmost) one, if any, in a w-bit word. (This operation can be built into the processor at a cost comparable to that of a w-bit adder.) For our purpose, $b \geq w$.

At the first step, each processor wanting to write on the bus participates in the arbitration algorithm. Let processor i be one such processor. If $a_{i,1} > 0$, then this processor indicates its presence by writing a 1 on bus $a_{i,1}$. Several processors may be writing on this bus, but they will all be writing a 1. Next, each processor i that wants to access the bus reads from all $2^b - 1$ arbitration buses, and determines if there are any positions larger than $a_{i,1}$ for which a processor wants the bus. That is, is there a higher priority processor vying for the bus? If so, processor i withdraws from the rest of the arbitration process. If $a_{i,1} = 0$, then processor i checks to see if any processor has written on any arbitration bus. Processors that find no higher priority processor vying for the bus continue to subsequent iteration. This step requires, $O(\frac{2^b}{w})$ time.

By applying this method to all $\frac{k}{b} = \frac{\log N}{b}$ digits of the addresses, the highest indexed processor attempting to write on the bus can be selected.

The entire procedure runs in $O\left(\frac{2^b \log N}{wb}\right)$ steps. If $w = \Theta(2^b)$, then the time needed is $O\left(\frac{\log N}{\log w}\right)$. This is an improvement over the conventional $O(\log N)$ time priority resolution (that proceeds one address bit at a time) and is achieved without extra hardware resources. The large number of channels available on a single physical slab admits a large value of b, a hence a lower arbitration time.

7 Conclusions

Slab waveguides hold immense potential for implementing high capacity buses for connecting proximate processors. Preliminary results indicate that a single slab waveguide has, in principle, the potential to provide over a million independent channels (corresponding to 1000 out-of-plane modes and 1000 colors used in wavelength division multiplexing). Moreover, each of these channels can support around 1000 loads. This is comparable to the large capacity of a free space optical system, but with the ability to broadcast each light source to many physically separated locations.

More modest systems, with smaller waveguides and less complex optics, are quite practical with existing technology and still would provide enormous advantages in bused interconnections. The general question of how best to exploit features of slab buses in a computing environment is open.

Acknowledgments

The authors wish to thank Diedre Reese and Yih Chyi Chong for assisting in the experimental work, and James Breedlove for constructing the optical slabs.

References

1. Alnuweiri, H.: A Fast Reconfigurable Network for Graph Connectivity and Transitive Closure. Parallel Processing Letters **4** (1994) 105–115
2. Bokka, V., Gurla, H., Olariu, S., Schwing, L., Wilson, L.: Time-Optimal Domain Specific Querying on Enhanced Meshes. IEEE Trans. Parallel & Distributed Systems **8** (1997) 13–24
3. Chen, W.-T., Shue, J.-P.: Performance Analysis of a multiple Bus Interconnection Network with Hierarchical Requesting Model. IEEE Trans. Computers **40** (1991) 834–842
4. Chen, Y., Chen, W., Chen, G., Sheu, J.: Designing Efficient Parallel Algorithms on Mesh Connected Computers with Multiple Broadcasting. IEEE Trans. Parallel & Distributed Systems **1** (1990) 241–246
5. Chiarulli, D.M., Levitan S., Melhem, R.G.: Optical Bus Control for Distributed Multiprocessors. J. Parallel Distributed Computing **10** (1990) 45–54

6. Dharmasena, H.P., Vaidyanathan, R.: An Optimal Multiple Bus Network for Fan-in Algorithms. Proc. Int. Conf. on Parallel Processing (1997) 100–103
7. Dighe, O.M., Vaidyanathan R., Zheng, S.Q.: The Bus-Connected Ringed Tree: A Versatile Interconnection Network. J. Parallel Distributed Computing **33** (1996) 189–196
8. Fernández-Zepeda, J.A., Vaidyanathan R., Trahan, J.L.: Scaling Simulation of the Fusing-Restricted Reconfigurable Mesh. IEEE Trans. Parallel & Distributed Systems **9** (1998) 861–871
9. Gourlay, J., Yang, T.-Y., Dines, J., Walker, A.: Development of Free-Space Digital Optics in Computing. IEEE Computer **31** (Feb. 1998) 38–44
10. Guo, Z., Melhem, R.G.: Embedding Binary X-Trees and Pyramids in Processor Arrays with Spanning Buses. IEEE Trans. Parallel & Distributed Systems **5** (1994) 664–672
11. Jiang, H., Smith, K.C.: PPMB: A Partial-Multiple-Bus Multiprocessor Architecture with Improved Cost Effectiveness. IEEE Trans. Computers **41** (1992) 361–366
12. Jang, J.-w., Nigam, M., Prasanna, V.K., Sahni, S.: Constant Time Algorithms for Computational Geometry on the Reconfigurable Mesh. IEEE Trans. Parallel & Distributed Systems **8** (1997) 1–12
13. Kamath, S.T., Vaidyanathan, R.: Running Weak Hypercube Algorithms on Multiple Bus Networks. Proc. ISCA Int. Conf. on Parallel & Distributed Computing Systems (1997) 217–222
14. Karim, Md.N.: Design and Analysis of Reduced Connection Multiple Bus Systems: A Probabilistic Approach. Ph.D. dissertation, Louisiana State University, Baton Rouge, 1996
15. Kulasinghe, P., El-Amawy, A.: On the Complexity of Bussed Interconnections. IEEE Trans. Computers **44** (1995) 1248–1251
16. Kulasinghe, P., El-Amawy, A.: Optimal Realizations of Sets of Interconnection Functions on Synchronous Multiple Bus Systems. IEEE Trans. Computers **45** (1996) 964–969
17. Miller, R., Prasanna-Kumar, V.K., Reisis, D., Stout, Q.: Parallel Computations on Reconfigurable Meshes. IEEE Trans. Computers **42** (1993) 678–692
18. Mitkas, P.A., Betzos G.A., Irakliotis, L.J.: Optical Processing Paradigms for Electronic Computers. IEEE Computer **31** (Feb. 1998) 45–51
19. Nakano, K.: A Bibliography of Published Papers on Dynamically Reconfigurable Architectures. Parallel Processing Letters **5** (1995) 111–124
20. Qiao, C., Melhem, R.G.: Time-Division Optical Communications in Multiprocessor Arrays. IEEE Trans. Computers **42** (1993) 577–590
21. Raksapatcharawong, M., Pinkston, T.M.: An Optical Interconnect Model for k-ary n-cube Wormhole Networks. Proc. Int. Parallel Processing Symp. (1996) 666–672
22. Serrano, M.J., Parhami, B.: Optimal Architectures and Algorithms for Mesh-Connected Parallel Computers with Separable Row/Column Buses. IEEE Trans. Parallel & Distributed Systems **4** (1993) 1073–1080
23. Trahan, J.L., Vaidyanathan R., Subbaraman, C.P.: Constant Time Graph Algorithms on the Reconfigurable Multiple Bus Machine. J. Parallel Distributed Computing **46** (1997) 1–14
24. Wilkinson, B.: On Crossbar Switch and Multiple Bus Interconnection Networks with Overlapping Connectivity. IEEE Trans. Computers 41 (1992) 738–746
25. Wilson, J., Hawkes, J.F.B.: Optoelectronics. Prentice Hall, 1989.
26. Yang, Q., Bhuyan, L.N.: Analysis of Packet-Switched Multiple-Bus Multiprocessor Systems. IEEE Trans. Computers **40** (1991) 352–357

A New Architecture for Multihop Optical Networks

A. Jaekel[1], S. Bandyopadhyay[1] and A. Sengupta[2]

[1] School of Computer Science, University of Windsor
Windsor, Ontario N9B 3P4

[2] Dept. of Computer Science, University of South Carolina
Columbia, SC 29208

Abstract. Multihop lightwave networks are becoming increasingly popular in optical networks. It is attractive to consider regular graphs as the logical topology for a multihop network, due to its low diameter. Standard regular topologies are defined only for networks with the number of nodes satisfying some rigid criteria and are not directly usable for multihop networks. Only a few recent proposals (e.g., GEMNET [10]) are regular and yet allow the number of nodes to have any arbitrary value. These networks have one major problem - node addition requires a major redefinition of the network. We present a new logical topology which has a low diameter but is not regular. Adding new nodes to a network based on this topology involves a relatively small number of edge definition/redefinitions. In this paper we have described our new topology, a routing scheme that ensures a low diameter and an algorithm for adding nodes to the network.

1 Introduction

Optical networks have become one of the emerging technologies for computer communication and provides low-loss transmission over a frequency range of about 25THZ. Signals can be transmitted over large distances at high speeds before amplification or regeneration is needed. Wavelength division multiplexing(WDM) allows the simultaneous data transmission at different carrier frequencies over the same optical fiber. A WDM network can offer lightpath service where a lightpath is a circuit switched connection between two nodes in the network and it is set up by assigning a dedicated wavelength to it on each link in its path [1]. We are looking at networks that provide permanent lightpaths which are set up at the time the network is initially deployed or is subsequently updated by addition of new nodes. Local area optical WDM networks may be categorized as single-hop or multi-hop networks. A single hop network allows communication between any source-destination pair using a single lightpath from the source to the destination. In a multihop network [2], to communicate from a source to a destination, it is necessary to use a composite lightpath consisting of a number of lightpaths. In other words, to communicate from a source node u to a destination node v, we have to select nodes $x_1, x_2, ..., x_{k-1}$ such that there is a lightpath L_0 from node u to node x_1, a lightpath L_1 from node x_1 to node x_2, ..., a lightpath L_{k-1} from node x_{k-1} to node v. Our composite lightpath L for communication from u to v consists of the lightpaths $L_0, L_1, ..., L_{k-1}$ and, in order to use this composite lightpath to send some data from u to v, we have to first send the data from node u to node x_1 using lightpath L_0. This data will be buffered at node x_1 using electronic media until the lightpath L_1 from node x_1 to node x_2 is available. When L_1 is available, the data will be communicated to node x_2 using lightpath L_1. This process continues until the data is finally communicated from node x_{k-1} to destination node v. In this case we say that this communication required k lightpaths and hence k *hops*. In optical networks, communication using a single lightpath is relatively fast and the process of buffering at intermediate nodes using electronic media is relatively slow. Each stage of electronic buffering at intermediate nodes x_1, x_2, ..., x_{k-1} adds an additional delay in the communication and limits the speed at which data communication can be made from the source node u to the destination node v. Thus a network which requires, on an average, a lower number of hops to go from any source to any destination is preferable.

The physical topology of an optical network defines how the nodes in the network consisting of computers (also called *endnodes*) and optical routers are connected using optical fibers. A popular technique for implementing multihop networks is the use of multistar networks where each endnode broadcasts to and receives optical signals from a passive optical coupler. To define a lightpath from node x to node y, the node x must broadcast, to the passive optical coupler, at a wavelength $\lambda_{x,y}$ and the node y must be capable of receiving this optical signal by tuning its receiver to the same wavelength $\lambda_{x,y}$. This scheme allows us to define a lightpath between any pair of nodes. The logical topology of a network defines how different endnodes are connected by lightpaths. It is convenient to represent a logical topology by a directed graph (or digraph) $\mathbf{G} = (\mathbf{V}, \mathbf{E})$ where $\mathbf{V}$ is the set of nodes and $\mathbf{E}$ is the set of the edges. Each node of $\mathbf{G}$ represents an endnode of the network and each edge (denoted by $x \rightarrow y$) represents a lightpath from end node x to endnode y. In the graph representation of the logical topology of a multihop network, for communication from u to v, we can represent the composite lightpath L consisting of the lightpaths $L_0, L_1, \ldots, L_{k-1}$ by the graph-theoretic path $u \rightarrow x_1 \rightarrow x_2 \rightarrow \ldots \rightarrow x_{k-1} \rightarrow v$.

If we use the shortest length path for all communication between source-destination pairs, the maximum hop distance is the *diameter* of $\mathbf{G}$. It is desirable that the logical topology has a low average hop distance. A rough rule of thumb is that networks with small diameters have a low average hop distance. To ensure a high throughput and low cost, a logical topology should be such that

- $\mathbf{G}$ has a small diameter
- $\mathbf{G}$ has a low average hop distance.
- The indegree and outdegree of each node should be low. The last requirement is useful to reduce the number of optical transmitters and receivers.

These conditions may be contradictory since a graph with a low degree usually implies an increase in the diameter of the digraph and hence the throughput rate. Several different logical topologies have been proposed in the literature. An excellent review of multihop networks is presented in [3].

Both regular and irregular structures have been studied for multihop structures [4],

[5], [6], [7], [8], [9]. Existing regular topologies(e.g., shuffle nets, de Bruijn graphs, torus, hypercubes) are known to have simple routing algorithms and avoid the need of complex routing tables. Since the diameter of a digraph with n nodes and maximum outdegree d is of $O(log_d n)$, most of the topologies attempt to reduce the diameter to $O(log_d n)$. One problem with these network topologies is that the number of nodes in the network must be given by some well-defined formula involving network parameters. This makes it impossible to add nodes to such networks in any meaningful way and the network is not scalable. This problem has been avoided in the GEMNET architecture [3], [10], which generalizes the shufflenet [6] to allow any number of nodes. A similar idea of generalizing the Kautz graph has been studied in [11] showing a better diameter and network throughput than GEMNET. Both these topologies are given by regular digraphs.

One simple and widely used topology that has been studied for optical networks is the bidirectional ring network. In such networks, each node has two incoming lightpaths and two outgoing lightpaths. In terms of the graph model, each node has one outgoing edge to and one incoming edge from the preceding and the following node in the network. Adding a new node to such a ring network involves redefining a fixed number of edges and can be repeated indefinitely. This is an attempt to develop a topology which has the advantages of a ring network with respect to scalability and the advantages of a regular topology with respect to low diameter.

In this paper we introduce a new scalable topology for multihop networks where the graph is not, in general, regular. The major advantage of our scheme is that, as a new node is added to the network, most of the existing edges of the logical topology are not changed, implying that the routing schemes between the existing nodes need little modification. The diameter of a network with n nodes is $O(log_d n)$ and we need to redefine only $O(d)$ edges. In section 2, we describe the proposed topology and derive its pertinent properties. In section 3 we present a routing scheme for the proposed topology and establish that the diameter is $O(log_d n)$. We have conclude with a critical summary in section 4.

2 Scalable Topology for Multihop Networks

In this section we will define the interconnection topology of the network by defining a digraph **G**. In doing this we need two integers - n and d, where n represents the number of nodes in the network and $d \leq n$, is used to determine the maximum indegree/outdegree of a node. As mentioned earlier, the digraph is not regular - the indegree and outdegree of a node varies from 1 to $d+1$.

2.1 Definitions and notations

Let Z_p be the set of all p-digit strings choosing digits from $Z = \{0, 1, 2, ..., d-1\}$ and let any string of Z_p be denoted by $x_0 x_1 ... x_{p-1}$. We divide Z_p into $p+1$ sets $S_0, S_1, ..., S_p$ such that all strings in Z_p having x_j as the left-most occurrence of 0 is included in S_j, $0 \leq j < p$ and all strings with no occurrence of 0 (i.e. $x_j \neq 0$, $0 \leq j < p$) are included in S_p. We now define an ordering relation between every pair of strings in Z_p. Each string in S_i is smaller than each string in S_j if $i < j$. For two strings $\sigma_1, \sigma_2 \in S_j$, $0 \leq j \leq p$, if $\sigma_1 = x_0 x_1 ... x_{p-1}$, $\sigma_2 = y_0 y_1 ... y_{p-1}$ and t is the smallest integer such that $x_t \neq y_t$ then $\sigma_1 < \sigma_2$ if $x_t < y_t$.

Let k be the integer such that $d^k < n < d^{k+1}$. We note that, for a string in Z_{k+1}, $|S_{k+1}| = (d-1)^{k+1}$ and $|S_j| = (d-1)^j d^{k-j}$, $0 \leq j \leq k$.

Definition 1. Let σ_1 be a string of Z_p, where $\sigma_1 = x_0 x_1 ... x_i x_{i+1} ... x_{p-1} \in S_i$, $0 \leq i < p$, so that i is the left-most digit which is 0.

For $0 \leq i < p-1$: We will call the string σ_2 obtained by interchanging the digits in the ith and the (i+1)th position in σ_1, the *image* of σ_1. Clearly, if $x_{i+1} = 0$, σ_1 and σ_2 represent the same string. Otherwise, $\sigma_2 \in S_{i+1}$. An *image* is not defined for any string in group S_p or in group S_{p-1}.

We consider a network of n nodes, where $d^k < n < d^{k+1}$. We will represent each node of the interconnection topology by a distinct string $x_0 x_1 ... x_k$ of Z_{k+1}. As $d^k < n < d^{k+1}$, all strings of Z_{k+1} will not be used to represent the nodes in **G**. We will use the n smallest strings from Z_{k+1} to represent the nodes of **G**. We will use S_M to denote the set containing the largest string. We will use the term *used* string to denote a string of Z_{k+1} which has been already used to represent some node in **G**. We will call all other strings of Z_{k+1} as *unused* strings. We will use a node and its string representation interchangeably.

Definition 2. We consider a string $\sigma = x_0 x_1 ... x_k \in Z_{k+1}$. We will call a string $x_0 x_1 ... x_{i-1} (x_j x_{j+1} ... x_k)$ as a *prefix(suffix)* of the string $\sigma = x_0 x_1 ... x_k$.

Definition 3. A *pivot* in a string $\sigma = x_0 x_1 ... x_k$ is the left-most 0 in the string. If $\sigma \in S_j$, $0 \leq j \leq k$, then x_j is the pivot for σ. A pivot is not defined for a string $\sigma \in S_{k+1}$.

Properties 1-3 given below are for strings used to denote nodes in a network with n nodes so that these strings are all in Z_{k+1}.

Property 1. All strings of S_0 are used strings.

Property 2. If $\sigma \in S_j$ is an used string, then all strings of $S_0, S_1, ..., S_{j-1}$ are also used strings.

Property 3. If $\sigma_1 = 0x_1...x_k$, $x_1 \neq 0$ and σ_2, the image of σ_1, is an unused string, then all strings of the form $x_1x_2...x_kj$, $0 \le j \le d-1$ are unused strings.

The proofs for Properties 1 - 3 are trivial and are omitted.

2.2 Proposed Interconnection Topology

We now define the edge set of the digraph **G**. Let any node u in **G** be represented by $u = x_0x_1...x_k$. The outgoing edges from node u can be of four types and are defined as follows:

Type A: There is an edge $u = x_0x_1x_2...x_k \to x_1x_2...x_kj$ whenever $x_1x_2...x_kj$ is an used string, for some $j \in Z$,

Type B: There is an edge from $u \to v$, where v be the image of u, as defined earlier, whenever the following conditions hold:

    a) $u \in S_i$, $0 \le i < k-1$

    b) v represents an used string

    c) $u \neq v$ and

    d) a type A edge $u \to v$ does not exist in the network.

Type C: There is an edge $u = 0x_1x_2...x_k \to 0y_1...y_k$, where $x_1x_2...x_k, y_1y_2...y_k \in Z_k$ and $y_1y_2...y_k$ is the image of $x_1x_2...x_k$ whenever the following conditions hold:

    b) $x_1 \neq 0$ and $x_10x_2...x_k$, the image of u, is an unused string

or

    c) $x_1 = 0$

Type D: There is an edge $0x_1x_2...x_k \to 0x_2...x_kj$ for all $j \in Z$ whenever the following conditions hold:

    a) $x_1 \neq 0$ and

    b) $x_10x_2...x_k$, the image of u, is an unused string

We note that if $u \in S_j$, $0 < j \le k$ node $v = x_1x_2...x_kj$ always exists (from property 2, since $v \in S_{j-1}$).

As an example, we show a network for $d = 2, k = 2$ with 5 and 6 nodes respectively in figure 1(a) and (b). We have used a solid line for an edge of the Type A, a line of dots for edges of Type B and a line of dashes and dots for edges of Type D. In this small network there are no edges of Type C. We note that Type D edges are the only ones that may need to be redefined as more nodes are added to the network. the edge from 010 to 100 satisfies the condition for both an edge of the type $x_0 x_1 x_2 ... x_k \to x_1 x_2 ... x_k j$ and an edge of the type $0 x_1 x_2 ... x_k \to x_1 0 x_2 ... x_k$.

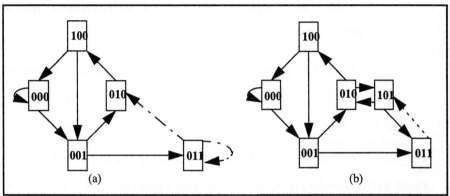

Fig. 1. Interconnection topology with $d = 2$, $k = 2$ with (a) $n = 5$ and (b) $n = 6$ nodes.

2.3 Limits on Nodal Degree

In this section, we derive the upper limits for the indegree and the outdegree of each node in the network. We will show that, by not enforcing regularity, we can easily achieve scalability. As we add new nodes to the network, minor modifications of the edges in the logical topology suffice, in contrast to large number of changes in the edge-set as required by other proposed methods.

Theorem 1: In the proposed topology, each node has an outdegree of up to $d+1$.

Proof: Let u be a node in the network given by $x_0 x_1 ... x_k \in S_j$ and let w be the image of u, if an image can be defined for the node. We consider the following three cases:

i) $0 < j \le k$: For every v given by $x_1 x_2 ... x_k t$ for all t, $0 \le t \le d-1$ is an used string since $v \in S_{j-1}$. Therefore the edge $u \to v$ exists in the network. There are the only d such edges from u. In addition there may be one outgoing edge $u \to w$ from u to its image w. Hence, u has a minimum outdegree d and maximum outdegree $d+1$.

ii) $j = 0$: According to our topology defined above, u will have an edge to $x_1 x_2 ... x_k j$ whenever $x_1 x_2 ... x_k j$ is an used string for some $j \in Z$. We have two subcases to consider:

- If p ($p > 0$) of the strings $x_1 x_2 ... x_k j$ are used strings, for some j, $0 \leq j < d$ and w is also an used string, then u has edges to all the p nodes with used strings of the form $x_1 x_2 ... x_k j$ and to w. Hence u has outdegree $p + 1$. So, u has an outdegree of at least 1 and at most $d+1$.

- Otherwise, if w is an unused string, then all strings of the form $x_1 x_2 ... x_k j$ are unused strings (**Property 3**) and u has d outgoing edges of Type D to nodes of the form $0 x_2 x_3 ... x_k j$, $0 \leq j < d$. and one outgoing edge of Type C. Hence u has outdegree $d+1$.

iii) $j = k+1$: If p of the strings $x_1 x_2 ... x_k j$ are used strings, for some j, $0 \leq j < d$, then u has outdegree of p. We note that $x_1 x_2 ... x_k 0 \in S_k$ is an used string. Therefore $1 \leq p \leq d$, and u has an outdegree of at least 1 and at most d, since an image of u is not defined in this case.

Theorem 2. In the proposed topology, each node has an indegree of up to $d+1$.

Proof: Let us consider the indegree of any node v given by $y_0 y_1 ... y_k \in S_j$. As described in 2.2, there may be four types of edges to a node v as follows:

- An edge (Type A) $t y_0 y_1 ... y_{k-1} \to y_0 y_1 ... y_k$ whenever $t y_0 y_1 ... y_{k-1}$ is an used string, for some $t \in Z$. There may be at most d edges of this type to v.
- There may be an edge (Type B) $u \to v$, where v is the image of u.
- If $y_0 = 0$, v may have one incoming edge of Type C.
- If $y_0 = 0$ and $t y_0 y_1 ... y_{k-1}$ is an unused string for some $t \in Z$, there is an edge (Type D) $0 t y_1 ... y_{k-1} \to y_0 y_1 ... y_k$. There may be at most d such edges to v.

We have to consider 2 cases, j = 0 and j > 0.

If j > 0, v can have at most one incoming edge $u \to v$, where v is the image of u and up to d edges of the type $t y_0 y_1 ... y_{k-1} \to y_0 y_1 ... y_k$. Therefore, the maximum possible indegree of v is d+1.

If j = 0, an edge of the type $0 t y_1 ... y_{k-1} \to y_0 y_1 ... y_k$ exists if and only if the corresponding edge of $t y_0 y_1 ... y_{k-1} \to y_0 y_1 ... y_k$ (Type A) does not exist in the network. Therefore, there is always a total of exactly d incoming edges to v of these two types. Node $v \in S_0$ may also have one incoming edge of Type C. Therefore, the maximum indegree is $d+1$.

2.4 Node Addition to an Existing Network

In this section we consider the changes in the logical topology that should occur when a new node is added to the network. We show that at most $O(d)$ edge changes in G would suffice when a new node is added to the network. When a multistar implementation is considered, this means $O(d)$ retunings of transmitters and receivers, whereas for a wavelength routed network, this means redefinition of $O(d)$ lightpaths. In contrast, other proposed topologies [10], [11] require $O(nd)$ number of edge modifications. As discussed in the previous section, the nodes are assigned the smallest strings defined earlier. When adding a new node u, we will assign the smallest unused string to the newly added node. Let the smallest unused string be $x_0 x_1 ... x_k \in S_j$. To add a node we have to consider the following two cases:

Case i) $0 < j \leq k$: Let $v = x_1 x_2 ... x_k t$, $0 \leq t \leq d-1$, then $v \in S_{j-1}$ and must exist in the network. Let $w_0 = 0 x_0 x_1 ... x_{k-1}$ then $w_0 \in S_0$ and is guaranteed to exist in the network. Let w_c be a node in the network, such that u is the image of w_c then $w_c \in S_{j-1}$ and is guaranteed to exist in the network.

- We add an edge $u \rightarrow v$ for each node $v = x_1 x_2 ... x_k t$ in the network.
- We add a new edge $w_0 = 0 x_0 x_1 ... x_{k-1} \rightarrow u$ to the network
- If $w_c \neq w_0$, we add an edge $w_c \rightarrow u$ to the network.
- If $u \in S_1$, we delete the edge $w_c \rightarrow v$ for each node $v = 0 x_2 ... x_k t$ in the network.

Every string $t x_0 x_1 ... x_{k-1}$, $1 \leq t \leq d-1$, $w \in S_{j+1}$ and is an unused string. Therefore w_0 and w_c are the only predecessors of u. Also, if w is the image of u, then $w \in S_{j+1}$ and does not exist in the network.

Case ii) $j = k+1$: Let $v = x_1 x_2 ... x_k t$, $0 \leq t \leq p-1$ be used strings in the network. Also, let $w = t x_0 x_1 ... x_{k-1}$, $0 \leq t \leq q-1$ be used strings

- We add a new edge $w \rightarrow u$ to the network for each $w = t x_0 x_1 ... x_{k-1}$, $0 \leq t \leq q-1$.
- We add a new edge $u \rightarrow v$ to the network for each $v = x_1 x_2 ... x_k t$, $0 \leq t \leq p-1$.

We note that $w_0 = 0x_0x_1...x_{k-1} \in S_0$ is an used string. Therefore, there is at least one w such that $w \to u$ exists. Similarly, $x_1x_2...x_k0 \in S_k$ is an used string. Therefore, there is at least one v such that $u \to v$ exists.

Fig. 2.(a) shows again the network with 6 nodes given in Figure 1. We choose the smallest unused string $u = 110$ to represent the new node being inserted. The following new edges are added to the network.

i) The node u will have outgoing edges of Type A to nodes 100 and 101.

ii) The node 110 is the image of node 101. Hence there will be a Type B edge from 101 to 110.

iii) Finally, there will be a Type A edge from 011 to 110.

The final network with 7 nodes is shown in Fig. 2.(b).

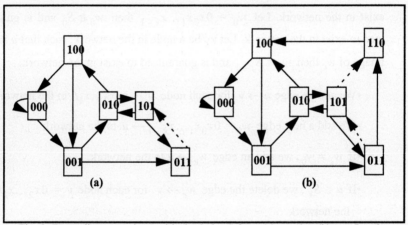

Fig. 2. Expanding a topology with $d = 2$, $k = 2$ from (a) $n = 6$ to (b) $n = 7$ nodes.

3 Routing strategy

In this section, we present the routing scheme in the proposed topology from any source node S to any destination node D. Let S be given by the string $X = x_0x_1...x_k \in S_j$ and D be given by the string $Y = y_0y_1...y_k \in S_t$. In general, a path from S to D may require a number of hops, where each hop represents a traversal of an edge in the network. We will first informally describe the routing scheme and then illustrate it through an example. A formal proof of correctness is omitted due to lack of space. The outline of the routing algorithm is given below.

```
Find_path

(digits_to_be_inserted, pivot_position) ←
    Find-best-suffix-prefix(source, destination);
current-node  ←  source
path  ←  List consisting of source only

While (current-node != destination)
(current-node,path,pivot-position) ←
    Insert-digits(path,digits_to_be_inserted,pivot-position);
(current-node, path, Pivot-position) ← Move-pivot(path, pivot-position);
EndWhile;
return path;
```

In this section, we will explain the three main functions, *Find-best-suffix-prefix*, *Insert-digits* and *Move-pivot* through an example. The details of the functions are omitted due to lack of space.

The function *Find-best-suffix-prefix* first finds the length (L1) of the longest set of digits from source string X which can also be used as digits in destination string Y. The suffix of X of length L1 must contain the same digits as the corresponding prefix of Y of length L1 and the ordering of the digits may differ only in the position of the *pivot*. This function determines the position of this pivot (pivot position) in the source string and the list of new digits that must be shifted in (list of digits to be inserted) to obtain the destination string.

The function *Insert-digits*, calculates the next node in the path by traversing an edge of Type A or Type D from the current node and inserting the next digit from the list of digits to be inserted. This is done only if the following conditions are satisfied.
    a. the destination has not been reached
    b. the next node calculated by *Insert-digits* exists in the network
    c. there is at least one more digit to be inserted and
    d. the left-most digit of the current node will not be used as part of destination string Y.

The function *Move-pivot* calculates the next node by traversing an edge of Type B from the current node. This essentially causes the pivot to move one digit to the right. This may be necessary for one of two reasons:
    i) the next node calculated in *Insert-digits* does not exist or
    ii) it is necessary to move the pivot to its proper location with respect to the destination string Y.

The next node calculated in this step is used as part of the path only if it exists in the network.

Example. We consider the topology of Fig. 2(a). Let, the source node be S=001 and the destination node be D=011. We can find a path from S to D, by following edges of Type A from each node to the next and inserting successive digits of the destination. The final path is given by P_1= 001 → 010 → 101 → 011. Since each node in the path exists in the network, P1 is a valid path.

We now consider a second example where S=011 and D=000. We try to follow a Type A edge from node 011 to node 110. But 110 is an *unused* string and the corresponding node does not exist in the network. Therefore, we must first follow an edge of Type B from 011 to its image 101. The path from node 101 to node 000 can be constructed using only Type A edges as in the previous example. The final path is given by P_2= 011 → 101 → 010 → 100 → 000.

Theorem 3. The function *Find-path* will always perform correctly if there is no node u in the network such that $u \in S_{k+1}$.

We omit the proof, due to lack of space.

Theorem 4. The function *Find-path* will find a path of at most 2k+1 hops, if there is no node u in the network such that $u \in S_{k+1}$.

Proof.

Part 1: The maximum possible number of digits in list of digits to be inserted is k+1. So, Insert-digits introduces at most k+1 hops traversing edges of Type A or Type D.

Part 2: We consider a source string $X = x_0 x_1 ... x_k$ with m+1 0's. Let the leftmost 0 occur in the ith position, the next 0 in the $(i+p_1)$th position, the next in the $(i+p_2)$th position and so on until the last 0 in the $(i+p_m)$th position. Each of these 0's may become the pivot in the function move-pivot. Using the jth 0 as the pivot, we can traverse at most $(i+p_{j+1})-(i+p_j)-1$ edges of type B. Therefore, the maximum number of Type B edges traversed by *Move-pivot* is

$(i+p_1)-i-1 + (i+p_2)-(i+p_1)-1 + (i+p_3)-(i+p_2)-1 ... + k-(i+p_m)-1 = k-(m+1)$

Therefore, the maximum number of hops = $(k+1) + (k-m-1) = 2k-m = 2k$, for m=0.

Theorem 5. If a network contains all nodes in $S_0, S_1, ..., S_k$ then

• there exists an edge $S \to \gamma = x_1 x_2 ... x_k 0$ and

• $\sigma(\gamma, D)$ represents a path from γ to D of length that cannot exceed k+1.

Proof. Since the network contains all nodes in $S_0, S_1, ..., S_k$, $\gamma \in S_j$ for some j, $j \leq k$ and must exist. Our topology (section 2.2) ensures that the edge $S \to \gamma$ exists. The path given below consists only strings belonging to groups S_i, $0 \leq i \leq k$ and hence are used strings:

$x_1 x_2 ... x_k 0 \to x_2 ... x_k 0 y_0 \to x_3 ... x_k 0 y_0 y_1 \to ... \to y_0 y_1 ... y_k$

The number of edges in the above path is $k+1$, hence the proof.
Theorem 6: The diameter of the network in $2k+1$.

If there is no node u in the network such that $u \in S_{k+1}$, then *Find-path* will return a path of $2k$ or less hops (**Theorem 4**).

If there is a node u in the network such that $u \in S_{k+1}$, then **Theorem 5** can be used to find a path of $k+1$ hops from S to D, hence the proof.

4 Conclusions

In this paper we have introduced a new graph as a logical network for multihop networks. We have shown that our network has an attractive average hop distance compared to existing networks. The main advantage of our approach is the fact that we can very easily add new nodes to the network. This means that the perturbation of the network in terms of redefining edges in the network is very small in our architecture. The routing scheme in our network is very simple and avoids the use of routing tables.

References

[1] R. Ramaswami and K. N. Sivarajan, "Opical Networks: A Practical Perspective", Morgan-Kaufmann, 1998.
[2] B. Mukherjee, "Optical Communication Networks', McGraw-Hill, 1997.
[3] B. Mukherjee, "WDM-based local lightwave networks part II: Multihop systems," *IEEE Network*, vol. 6, pp. 20-32, July 1992.
[4] K. Sivarajan and R. Ramaswami, "Lightwave Networks Based on de Bruijn Graphs," *IEEE/ACM Transactions on Networking, Vol. 2, No. 1*, pp. 70-79, Feb 1994.
[5] K. Sivarajan and R. Ramaswami, "Multihop Networks Based on de Bruijn Graphs," *IEEE INFOCOM '91*, pp. 1001-1011, Apr. 1991.
[6] M. Hluchyj and M. Karol, "ShuffleNet: An application of generalized perfect shuffles to multihop lightwave networks," *IEEE/OSA Journal of Lightwave Technology*, vol. 9, pp.1386-1397, Oct. 1991.
[7] B. Li and A. Ganz, "Virtual topologies for WDM star LANs: The regular structure approach," *IEEE INFOCOM '92*, pp.2134-2143, May 1992.
[8] N. Maxemchuk, "Routing in the Manhattan street network," *IEEE Trans. on Communications*, vol. 35, pp. 503-512, May 1987.
[9] P. Dowd, "Wavelength division multiple access channel hypercube processor interconnection," *IEEE Trans. on Computers*, 1992.
[10] J. Innes, S. Banerjee and B. Mukherjee, "GEMNET : A generalized shuffle exchange based regular, scalable and modular multihop network based on WDM lightwave technology", IEEE/ACM Trans. Networking, Vol 3, No 4, Aug 1995.
[11] A. Venkateswaran and A. Sengupta, "On a scalable topology for Lightwave networks", Proc IEEE INFOCOM'96, 1996.

Pipelined Versus Non-pipelined Traffic Scheduling in Unidirectional WDM Rings*

Xijun Zhang[1] and Chunming Qiao[2]

[1] Department of Electrical Engineering (EE)
University at Buffalo (SUNY), Buffalo, NY 14260
xz2@eng.buffalo.edu

[2] Departments of Computer Science and Engg. (CSE) and EE
University at Buffalo (SUNY), Buffalo, NY 14260
qiao@computer.org

Abstract. Two traffic scheduling schemes, namely, non-pipelined and pipelined scheduling, are considered for uniform traffic in unidirectional WDM rings. Their performances in terms of schedule length and throughput are compared. Optimal or near optimal schedules are obtained for a given number of wavelengths, and a given number of transceivers per node as well as their tuning time.

1 Introduction

Due to the lack of sophisticated optical logic devices, communications in all-optical Wavelength Division Multiplexed (WDM) networks are normally carried out in a circuit-switching fashion to avoid conversions of data between electronic and optical domains at intermediate nodes. In addition, in order to reduce the control overhead involved in dynamically establishing connections, one may schedule pre-determined communication patterns. This basically leads to hybrid Time/Wavelength Division Multiplexed access wherein a connection will be scheduled on a particular wavelength (assuming no wavelength conversion at any intermediate node) during a particular time slot.

A typical communication pattern is uniform, all-to-all traffic, which requires each node in a network to send a unique message to every other node. Such a pattern arises in many important parallel computing algorithms. Furthermore, it is usually more beneficial to use a highly tuned schedule for the uniform traffic pattern than attempting to find a schedule for a non-uniform but dense traffic pattern [1, 2]. The schedule length is an important performance metric as it affects, for example, the throughput that can be achieved in the network.

The problem of traffic scheduling in conventional networks has been studied extensively (see, for example, [1–8]). Traffic scheduling in broadcast-and-select WDM networks based on star-couplers has also received much attention (see,

---

* This research is supported in part by a grant from NSF under contract number MIP-9409864 and ANIR-9801778.

for example, [9–12]). In this paper, we study the problem of scheduling traffic in WDM rings.

Our work (including those in [13–15]) differs from other closely related work (including those in [16, 17]), in that we consider WDM rings with a limited number of wavelengths per link, and/or a limited number of transceivers per node. The traffic scheduling problem addressed in this work is unique (and more challenging) also because the tuning time of each transceiver will be considered. Furthermore, we will consider pipelined transmissions, which is made possible by two unique properties of optical transmissions, namely, unidirectional propagation and predictable propagation delay [18, 19].

The rest of this paper is organized as follows. In Section 2, we give a general description of the WDM ring network to be considered as well as the traffic scheduling problem. Non-pipelined and pipelined transmission scheduling schemes are studied in Section 3 and Section 4, respectively. Numerical results on their performance in terms of schedule length and throughput are given in Section 5. Finally, we conclude the paper in Section 6.

2 Problem Description

We consider a unidirectional (e.g. clockwise) WDM ring having K wavelength channels on each link and N nodes. At each node, there are T fully tunable transceivers whose maximum tuning time is Δ seconds. Since the transceivers can be pre-tuned, we will ignore the *initial* tuning time hereafter and accordingly, consider the effect of Δ only when $T < K$. We assume that there is no wavelength conversion. For simplicity, it is also assumed that the propagation delay between any two adjacent nodes is d seconds, and the network is globally synchronized. We will study the scheduling problem for all-to-all uniform traffic, in which each node has one message of M bits to be sent to each of the other $N-1$ nodes.

Due to limited network resources, not all the messages can be sent/received simultaneously. Accordingly, our objective is to determine a priori a schedule which includes all the necessary information (such as the time to transmit and receive, and the wavelength to be used). The schedule length, denoted by L_s, is the time needed for all the messages to be received.

In this paper, we will consider two traffic scheduling schemes, namely, non-pipelined and pipelined, as illustrated in Figure 1. More specifically, assume that the transmission rate on each wavelength is B bits per second. In non-pipelined transmission, when it is node i's turn to transmit to another node s hops away, ($s = 4$ in Figure 1), it transmits the entire message. Accordingly, $\frac{M}{B} + s \cdot d$ seconds are needed for the message to be received. During this period, those intermediate nodes cannot transmit on the same wavelength On the other hand, in pipelined transmission, all nodes can transmit on the same wavelength simultaneously. In order to avoid collision, each node only sends a packet of p bits at a time, where $\frac{p}{B} \leq d$, and hence, needs to send q packets for each message, where $p \cdot q = M$.

In the remainder of the paper, we will examine non-pipelined and pipelined traffic scheduling schemes in more detail, determine optimal or near optimal

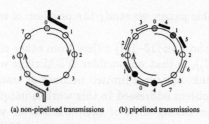

Fig. 1. (a) Non-pipelined and (b) Pipelined transmission (the number on each message/packet denotes the destination node).

schedules for each scheme, and compare the performance of the two schemes in terms of the schedule length as well as the network throughput.

3 Non-pipelined Scheduling

The maximum propagation time over all connections in the N-node unidirectional ring is $(N-1) \cdot d$, which will be denoted by t_p. We will study non-pipelined scheduling under the assumption that M is large relative to t_p. A non-pipelined schedule can be regarded as a sequence of *rounds* (i.e. super time-slots). In each round, it is desirable to schedule as many all-optical connections as possible, one for each message to be transmitted, in order to fully utilize the I/O capacity of the transceivers, the bandwidth of the wavelengths, or both. A message is transmitted over a connection only at the beginning of each round, and a node will not send a new message on the same wavelength until the previous message from this node has been received. To facilitate synchronization, the duration of each round, R, is made equal to the sum of the message transmission time and the maximum propagation delay, i.e. $R = \frac{M}{B} + t_p$. Accordingly, if the schedule requires r rounds, then the schedule length L_s is the sum of $r \cdot R$ and any tuning time.

Although message transmissions occur only at the beginning of each round, the tuning of a transmitter (and receiver) may start immediately after a message is transmitted (and received, respectively), and hence can overlap with message propagation. More specifically, given the tuning time Δ, we may define *tuning delay* $\delta = 0$ if $\Delta \leq t_p$ and $\delta = \Delta - t_p$ if otherwise.

3.1 Negligible Tuning Delay

We start with the simple case where $\delta = 0$. To facilitate our presentation, a connection is said to have *stride s* if a message is to be transmitted to a node s hops away. Two connections with common end nodes (e.g. one from i to j and the other from j to i) have complementary strides, s and $N - s$, and thus form a *circle* and are scheduled in the same round to fully utilize the bandwidth of one wavelength.

Multiple circles are scheduled in the same round to fully utilize the bandwidth of multiple wavelengths. The maximum number of circles that can be scheduled in a round, however, is limited by both the number of wavelengths, K, and the number of transceivers at each node, T. Accordingly, the *theoretical lower bound* (TLB) on the schedule length depends on both K and T. Specifically, the TLB imposed by T is $TLB_T = \lceil \frac{N-1}{T} \rceil \cdot R$, and the TLB imposed by K is $TLB_K = \lceil \frac{\frac{N(N-1)}{2}}{K} \rceil \cdot R$ [15]. The overall TLB is thus $TLB(K, T, \delta = 0) = max\{TLB_K, TLB_T\}$.

Single transceiver per node

We first consider the case where $T = 1$. Assuming N is odd, one method to enumerate all the connections is shown in Figure 2, in which a straight line with an arrow from node i to node j is used to represent the circle involving nodes i and j. Circles are grouped into *virtual topologies (VTs)* such that each VT contains only parallel lines as shown in the figure. This way, there are $\frac{N-1}{2}$ circles in each VT, and every node except one is involved in one circle in a VT. There are N VTs in total, and they have similar patterns but contain different connections. In general, rotating the connections in VT_i by $j \geq 1$ nodes clockwise results in $VT_{mod(i+j, N)}$ (where $mod()$ is the modular function, which will be omitted hereafter in the subscript of all VT's).

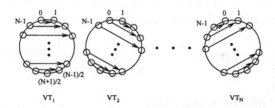

Fig. 2. Virtual topologies representing all the connections when N is odd.

Since $T = 1$, two circles can be scheduled in the same round only if they do not involve common nodes. Given that each node is involved in at most one circle in any VT, all the circles in one VT can be scheduled in one round as long as K is large enough. If K is less than the number of circles in each VT, the circles in one VT have to be scheduled in multiple rounds. Specifically, any K of the $\frac{N-1}{2}$ circles in a VT can be scheduled in one round. Let $\frac{N-1}{2} = mK + y$. After $m = \lfloor \frac{\frac{N-1}{2}}{K} \rfloor$ rounds (and a total of $m \cdot N$ rounds for the N VTs), there will be $y = mod(\frac{N-1}{2}, K) < K$ circles left in each VT. When $y > 0$, in order to fully utilize the bandwidth, the circles left in different VTs can also be scheduled in the same round if they do not involve common nodes.

Let x be the number of circles in each VT, $y = mod(x, K)$, and z be the number of VTs. An upper bound on the number of rounds needed has been obtained as [15],

$$f(x,z,K) = \begin{cases} \frac{x}{K} \cdot z & x \geq K, y = 0 \\ \lfloor \frac{x}{K} \rfloor z + \lceil \frac{\lceil \frac{x}{y} \rceil y}{K} \rceil mod(z,y) + \lceil \frac{\lfloor \frac{x}{y} \rfloor y}{K} \rceil [y - mod(z,y)] & x \geq K, y > 0 \\ z & x < K \end{cases} \quad (1)$$

Accordingly, the upper bound on the schedule length when $T = 1$ for an odd N is $UB(K, T = 1, \delta = 0) = f(\frac{N-1}{2}, N, K) \cdot R$. The case where N is even is similar except that we have to consider connections having even strides and odd strides (or circles formed by such connections) separately.

Note that the circles left can be packed *recursively* using a procedure which we call "recursive packing procedure"[15]. The schedule length achieved using the recursive packing procedure will be less than the upper bound given above.

Multiple transceivers per node

When $T > 1$ transceivers are used at each node, more circles may be scheduled in each round. A straightforward extension of the scheduling method proposed for the case of $T = 1$ is to combine T VTs to form a *super*-VT such that the I/O capacity provided by the T transceivers per node can be fully utilized by scheduling the circles in the super-VT in the same round when K is large enough. Note that N may not be divided evenly by T, in which case, the remaining $mod(N, T)$ VTs can be considered separately. As an example, when N is odd (the case when N is even is similar), there are $\frac{N-1}{2} \cdot T$ circles within each super-VT, and there are $\lfloor \frac{N}{T} \rfloor$ such super-VTs. Therefore,

$$UB(K, T, \delta = 0) = [f(\frac{N-1}{2} \cdot T, \lfloor \frac{N}{T} \rfloor, K) + \lceil \frac{\frac{N-1}{2} \cdot mod(N,T)}{K} \rceil] \cdot R \quad (2)$$

3.2 Non-negligible Tuning Delay

In this section, we consider the effects of non-negligible tuning delay (i.e. $\delta > 0$) on the schedule length for the simplest case where N is odd and $T = 1$. We derive an upper bound under the most conservative assumption that any two consecutive transmissions by a node are always scheduled on different wavelengths. In other words, we assume that a node always has to wait for δ seconds before any transmission (including the first one).

Recall that after m rounds, there are y circles left in each VT, and the only possible value of y that is larger than $\frac{N}{4}$ is $y = \frac{N-1}{2}$. In this case, each VT has to be scheduled separately in one round. Since the I/O capacity is fully utilized in each round by each VT, there has to be a delay of δ before each round (except the first round with pre-tuned transceivers). Accordingly, the upper bound on the schedule length is $UB(K, T = 1, \delta) = N \cdot R + (N - 1) \cdot \delta$.

When $y = 0$, there are exactly $m \cdot K$ circles in each VT, and m rounds are needed to schedule them. Let VT_i^h, where $1 \leq h \leq m$, represent the circles scheduled in round h for VT_i (see Figure 3(a) for an illustration of VT_1^1, VT_1^2, VT_2^1 and VT_2^2). Note that $VT_1^1, VT_1^2, \cdots, VT_1^m$ can be scheduled with no tuning delay. However, since VT_2^1 is obtained by rotating VT_1^1 by one node clockwise, it shares one common node with VT_1^2 (and some common nodes with VT_1^1, but no common node with VT_1^{1+k} if $k \geq 2$). This means that VT_2^1 cannot be scheduled until δ seconds after VT_1^2. If $\delta \leq (m-2)R$, the tuning delay can be hidden by overlapping it with the rounds in which VT_1^3 through VT_1^m are scheduled (see Figure 3(b)). However, if $\delta > (m-2)R$, the round for VT_2^1 has to be delayed for $\delta - (m-2)R$ seconds after the round for VT_1^m (see Figure 3(c)).

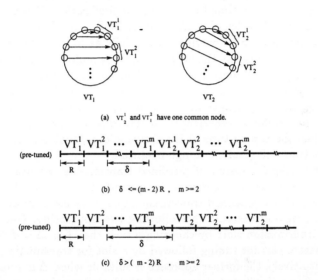

Fig. 3. Non-pipelined scheduling with non-negligible tuning delay.

Summarizing the above discussion, an upper bound on the schedule length when $C = mK$, where $m \geq 2$, is:

$$UB(K, T=1, \delta) = \begin{cases} N \cdot m \cdot R & \delta \leq (m-2)R, m \geq 2 \\ N \cdot m \cdot R + (N-1) \cdot [\delta - (m-2)R] & \delta > (m-2)R, m \geq 2 \end{cases} \quad (3)$$

When $0 < y \leq \frac{N}{4}$, the first $m \cdot K$ circles in each VT can be scheduled in the same way as described above. Afterwards, a similar method can be applied to schedule the $N \cdot y$ circles left. By overlapping tuning of transceivers with as many rounds as possible, the effects of tuning delay can be reduced or even eliminated.

4 Pipelined Scheduling

Pipelined transmission allows a batch of N packets, one from each node, to be transmitted on the same wavelength at the same time. These N packets are sent over connections with the same stride s (e.g. $s = 4$ in Figure 1(b)) such that they will be received simultaneously. When multiple batches are to be transmitted on the same wavelength, the transmission of the second or later batch can start as soon as the previous batch is being received. Accordingly, when calculating the time needed to send and receive multiple batches of packets on the same wavelength, one may include the transmission time of the first batch ($\frac{P}{B}$), and omit that of all subsequent batches. Hereafter, in order to simplify our presentation, we will focus on the so-called *sub-schedule length* and the corresponding *sub-schedules*. The former, denoted by L'_s, is the time for each node to send to (and receive from) every one of the other $N-1$ nodes *a single packet* (instead of q packets), excluding the time to transmit the first packet. Based on the above discussion, the schedule length, L_s, is equal to $\frac{P}{B} + q \cdot L'_s$ (may be shorter if several sub-schedules can be overlapped partially as to be discussed later).

In pipelined transmission, the behavior of every node is the same at any given time. Accordingly, we will only need to consider how to schedule transmissions at one node. In addition, even though a node cannot transmit (or receive) another packet on the same wavelength in $s \cdot d$ seconds, the transmitter (or receiver) is free to tune to a different wavelength to transmit (or receive) another packet. However, when the tuning time Δ is relatively large, tuning a transceiver once every packet time will result in a high overhead.

In the following discussion of pipelined transmission, we use the notion of a time slot, whose duration is d seconds, instead of the notion of a round used before. We will study pipelined transmission under the assumption that $\Delta \leq d$. In particular, we will describe the following two approaches. One uses mixed tuning and transmission slots by choosing p such that $\frac{P}{B} + \Delta = d$. The other is to use a separate slot for tuning followed by a slot for transmitting a packet of $d \cdot B$ bits. Intuitively, the former approach is suitable when Δ is small compared to d. When Δ is a significant portion of d, the latter may be more suitable.

4.1 Mixed Tuning and Transmission Slots

In this subsection, we study the first approach whereby tuning and packet transmission can take place in one slot. When using this approach, the actual value of Δ will have no effect on the sub-schedule length (although the larger the Δ, the smaller the p and the larger the q, and accordingly, the longer the schedule length and the lower the throughput). We will first study the case where $T = K$, then extend the discussion to two other cases where $2 \leq T < K$ and $T = 1$, respectively.

The case where $T = K$

When $T = K$, each transceiver can be fixed at a distinct wavelength, and thus no tuning is needed except for initialization. Since the total time needed for a node to send one packet to every other node is $\frac{N(N-1)}{2}$ slots on a single wavelength, the TLB on the sub-schedule length is $TLB_p(T=K) = \lceil \frac{\frac{N(N-1)}{2}}{K} \rceil \cdot d$, which may be achieved if all packet transmissions can be uniformly distributed among K wavelengths.

Our idea for approaching the TLB_p is to group the transmissions into *phases* with a fixed number of slots, L_p. More specifically, on each wavelength, up to two transmissions with strides s and $L_p - s$ can be scheduled in each phase. Figure 4 shows an example schedule with $L_p = N = 17$ slots, where each row corresponds to a wavelength, and each arrowed line segment corresponds to a connection with a specific stride. In this example, $TLB_p(T = K = 4) = 34 \cdot d$, which is achieved in two phases since the sub-schedule length is $L'_s = 2 \cdot L_p \cdot d = 34 \cdot d$.

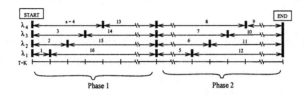

Fig. 4. An example of pipelined scheduling ($N = 17, T = K = 4$).

Note that in some cases, combining connections with strides s and $N - s$ to form phases with $L_p = N$ may not be appropriate. In general, the appropriate value of L_p depends on N and K as to be discussed next. In order to find an efficient sub-schedule, we first determine how big K needs to be.

Theorem 1. *At most $\lceil \frac{N}{2} \rceil$ wavelengths are needed for scheduling pipelined transmissions when $T = K$.*

Proof. Since the largest s is $N - 1$, the sub-schedule length cannot be shorter than $N - 1$ slots. Without loss of generality, assume that the connection with stride $N - 1$ is scheduled on wavelength λ_1. Let (s_1, s_2, λ_i) denote that two connections with strides s_1 and s_2 are scheduled on wavelength λ_i, where $2 \leq i \leq K$. A straightforward way to schedule the remaining $N - 2$ connections is $(1, N - 2, \lambda_2), (2, N - 3, \lambda_3), \cdots$, and so on. This sub-schedule requires $\lceil \frac{N}{2} \rceil$ wavelengths in total, and the sub-schedule length achieved is $(N - 1) \cdot d$, the shortest schedule possible. □

Note that the proof above also implies an optimal sub-schedule for the case where $K = \lceil \frac{N}{2} \rceil$, which is accomplished in one phase with a length of $L_p = N - 1$ slots. When $K < \lceil \frac{N}{2} \rceil$, let $N - 1 = 2mK + \mu$, where $m = \lfloor \frac{N-1}{2K} \rfloor$ and $\mu = mod(N - 1, 2K)$. It turns out that the sub-schedule length L'_s will be minimized if each phase has a length of $L_p = N + \mu$ slots[14]. The resulting sub-schedule is illustrated in Figure 5, where each number inside the squares denotes the stride of a connection, and an arrow indicates the direction at which the strides increase.

From Figure 5, it is clear that when $T = K$, an upper bound on the sub-schedule length is,

$$UB_p(T = K) = [m(N + \mu) + \mu] \cdot d \qquad (4)$$

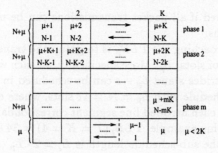

Fig. 5. One heuristic sub-schedule in pipelined transmission.

and the upper bound on the schedule length is thus simply $\frac{P}{B} + q \cdot UB_p(T = K)$.

In the next two subsections, we will only consider $K < \lceil \frac{N}{2} \rceil$ because when $K = \lceil \frac{N}{2} \rceil$, $m = 0$ and the $N - 1$ connections can always be scheduled in the same way as the last μ connections, as to be described next.

The case where $2 \leq T < K$

When $2 \leq T < K$, in order to use all the wavelengths available, a transmitter has to transmit on different wavelengths in an *interleaved* way. Specifically, it has to transmit one packet on one wavelength at the beginning of one slot, and then another packet on another wavelength at the beginning of another slot, and so on. Since each transmitter interleaves among $\lceil \frac{K}{T} \rceil$ wavelengths, $\lceil \frac{K}{T} \rceil - 1$ more slots are needed to send a packet on each of the K wavelengths comparing with the case where $T = K$. Intuitively,

$$TLB_p(2 \leq T < K) = TLB_p(T = K) + (\lceil \frac{K}{T} \rceil - 1) \cdot d \qquad (5)$$

We now consider how to modify the sub-schedule in Figure 5 to approach the above TLB_p. First, we divide the K wavelengths into $\lceil \frac{K}{T} \rceil$ groups such that the ith group, where $1 \leq i < \lceil \frac{K}{T} \rceil$, contains the following T wavelengths, $\lambda_{K-(i-1)T}$, $\lambda_{K-(i-1)T-1}$, $\cdots$, λ_{K-iT+1}, and the last group (i.e. the $\lceil \frac{K}{T} \rceil$-th group) contains only $mod(K,T)$ wavelengths, namely, $\lambda_{mod(K,T)}$, $\lambda_{mod(K,T)-1}$, $\cdots$, λ_2, λ_1. Let the T transmitters transmit on the first group of wavelengths at the beginning of each phase, and then on the second group, and so on. Note that there will be two transmissions of complementary strides on each wavelength in each phase. The first transmissions in phase 1 on the wavelengths in group i ($1 \leq i \leq \lceil \frac{K}{T} \rceil$) will be delayed for $i - 1$ slots relative to the case where $T = K$ (see Figure 6(a)). If we denote by $D_T(k)$ the delay of the first transmission on λ_k in phase 1, where $1 \leq k \leq K$, we have, $D_T(k) = \lceil \frac{K-k+1}{T} \rceil - 1$.

In fact, the first m phases in the sub-schedule in Figure 5 can be simply modified by delaying every transmission on λ_k by $D_T(k)$ slots. This is possible, or in other words, the resulting sub-schedule is valid, because of the following theorem (whose proof is omitted).

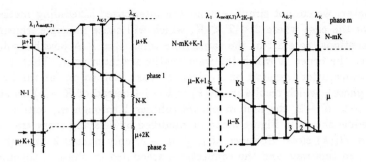

Fig. 6. (a) Phase 1 when $2 \leq T < K$ (left) and (b) Schedule of the remaining μ connections when $2 \leq T < K$ and $K < \mu < 2K$ (right).

Theorem 2. *In the modified sub-schedule, at most T packets are transmitted and received simultaneously by a node in the first m phases.*

We can schedule the remaining μ (where $0 < \mu < 2K$) connections, in the order of decreasing strides, as follows. When $K < \mu < 2K$, we schedule the first K of them on λ_K through λ_1, and the remaining $\mu - K$ of them on $\lambda_{2K-\mu}$ through λ_{K-1} (see Figure 6(b)). In this way, the connection that ends last is the one with stride $\mu - K$, which is scheduled on $\lambda_{2K-\mu}$ with a delay of $D_T(2K - \mu)$ slots. When $0 < \mu \leq K$, we schedule the μ connections on λ_μ through λ_1. Accordingly, the connection with stride μ ends last on λ_μ with a delay of $D_T(\mu)$ slots. The overall delay relative to the case where $T = K$ for different μ is thus:

$$D_T = \begin{cases} D_T(2K - \mu) & K < \mu < 2K \\ D_T(\mu) & 0 < \mu \leq K \\ D_T(1) & \mu = 0 \end{cases} \qquad (6)$$

and based on Eq. (4), the modified sub-schedule will have a length of

$$UB_p(2 \leq T < K) = [m(N + \mu) + \mu + D_T] \cdot d \qquad (7)$$

Note that in the above sub-schedule, when $\mu = 0$, transmission on the first group of wavelengths (containing λ_K) starts and ends earlier than that on the last group (containing λ_1) by $D_T(1)$. Accordingly, when the sub-schedule needs to be repeated (e.g. in order for each node to send multiple packets to every other node), two consecutive sub-schedules can overlap with each other partially just as the two phases do in Figure 6(a). As a result, the schedule length is only $D_T(1)$ (instead of $q \cdot D_T(1)$) longer than the schedule length when $T = K$. Even when $\mu > 0$, it is possible for two consecutive sub-schedules to overlap partially, and hence, the overall schedule length when $T < K$ will be increased by less than $q \cdot D_T$ when compared to the schedule length when $T = K$.

The case when $T = 1$

We may first derive the following by extending Eq. (5),

$$TLB_p(T = 1) = TLB_p(T = K) + (K - 1) \cdot d \qquad (8)$$

However, we cannot simply extend the interleaved scheduling method proposed for the case where $2 \leq T < K$, which will cause multiple transmissions (and receptions) to start at the same time when $T = 1$. A possible solution is to reverse the order at which the wavelengths are interleaved. That is, we will let the transmitter transmit on λ_1 first, λ_2 next, and so on (at the beginning of each phase), which results in $D_1(k) = k - 1$ (instead of $K - k$). In short, as long as $\mu > K - 2$, the first m phases (phases 1 through m) outlined in the sub-schedule shown in Figure 5 can be modified by delaying every transmission on λ_k by $D_1(k)$ slots. However, if $\mu \leq K - 2$, only phases 2 through m can be modified this way, and the connections scheduled in phase 1, together with the remaining μ connections, will be scheduled using a heuristic method. One such heuristic method, for example, is to consider the connections in the order of decreasing stride, and try to schedule each connection as soon as possible.

4.2 Using Separate Tuning and Transmission Slots

When $T < K$ and the tuning time Δ is a large portion of d, it makes more sense to use a separate slot for tuning of transceivers. Obviously, the (sub-)schedule length will be longer than that when $\Delta = 0$. However, the (sub-)schedule length is the same for any $0 < \Delta < d$ as long as a dedicated slot is used for tuning.

The heuristic method mentioned above for the case when $T = 1$ can be modified to ensure that each transceiver has a slot for tuning before transmitting/receiving a packet on a different wavelength. Note that, when adopting this approach, it is implied that not every slot for packet transmissions has to be preceded by a tuning slot. In addition, the ratio of time used for transmission over that used for tuning could be larger. Accordingly, a better performance than that obtained by using mixed tuning and transmission slots may be obtained as to be shown next.

5 Numerical Results

In this section, we present some results on the schedule length (L_s) and throughput, which is $\frac{N(N-1) \cdot M}{L_s}$, achieved in non-pipelined and pipelined scheduling. By default, we assume that $B = 10$ Gbps (equivalent to OC-192) and $M = 100$ Mbits (accordingly, the default *message time* is $\frac{M}{B} = 10ms$). In addition, $N = 19$ and $d = 50\mu s$ (accordingly, the distance between two adjacent nodes is 10 km, and the maximum delay t_p is $900\mu s$).

Figure 7(a) shows the schedule lengths achieved, as a function of T and K, with all other parameters set to their default values (note that, the results are applicable to any $\Delta \leq 900\mu s$ since δ will be 0). As can be seen, in most cases, the TLB is achieved, and can be closely approached otherwise. In addition, as long as $K \leq \frac{N}{2}$, there is no need to use more than $T = 1$ transceivers per node since the schedule performance will not be improved.

The throughputs in non-pipelined transmission scheduling when $T = 1$ and $\Delta = 5\mu s$ is shown in Figure 7(b) for different values of M and d. Curve (1) shows

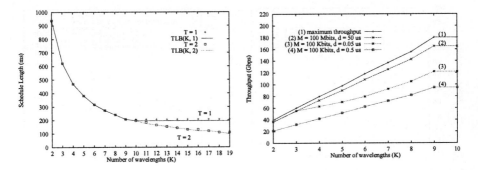

Fig. 7. Non-pipelined scheduling: (a) Schedule lengths achieved using the recursive packing procedure when $\Delta = 0$; (b) Throughput achieved for different values of M and d when $T = 1$ and $\Delta = 5\mu s$.

the maximum throughput that would be achieved if both the propagation delay and tuning delay were ignored. When propagation delay is considered, one needs to have a large ratio of message time over t_p (or $\frac{M}{B} \gg t_p$) in order to achieve a high throughput. For example, curve (2) shows that, with the default M and d, and accordingly, the ratio of message time over t_p is $\frac{100}{9}$ and $\delta = 0$ (since $\Delta < t_p$), a high throughput can be achieved. In fact, our results (not shown) indicate that the same throughput can be achieved when both M and d (and thus the message time and t_p, respectively) are reduced proportionally, e.g. by 100 times (i.e. to $M = 1$ Mbits and $d = 0.5\mu s$), as long as δ remains 0. However, having a large ratio of message time over t_p is not sufficient for achieving a high throughput. For example, if we reduce the two by too much, e.g. 1000 times, that is, to $d = 0.05\mu s$ (and $M = 100$ kbits), we have $\delta = 4.1\mu s$ and consequently, the throughput will be much lower as shown by curve (3). Note that, if we keep the same M (e.g. to 100 kbits) but increase d to $0.5\mu s$ to make $\delta = 0$, the ratio decreases and consequently, an even lower throughput will be achieved (see curve (4)) since the propagation delay is increased.

Figure 8(a) shows the schedule lengths achieved using pipelined scheduling with the same default settings of the parameters as in Figure 7(a). Since $\Delta = 0$, we have set $p = d \cdot B$ and $q = \frac{M}{p}$. Based on the discussion in Section 4, we have limited the number of wavelengths to be less than $\lceil \frac{N}{2} \rceil$. Note that, only $TLB_p(T = K)$ is shown in Figure 8(a) since the TLB on the schedule length for any given T is almost the same when $K \leq \lceil \frac{N}{2} \rceil$ (however, unlike in non-pipelined scheduling, increasing the number of transceivers, e.g. from 1 to 2, will reduce the schedule length in pipelined scheduling). The results show that the TLB is closely approached. In addition, pipelined scheduling results in a shorter schedule length than non-pipelined scheduling.

The throughputs achieved in pipelined scheduling for various Δ (and T and K) with the default settings of all other parameters (and an appropriate p and q) are shown in Figure 8(b). For the purpose of making comparisons, non-pipelined throughput for any T (up to $K = 10$) and any $0 \leq \Delta \leq d$ (or any $\Delta \leq t_p$) is also

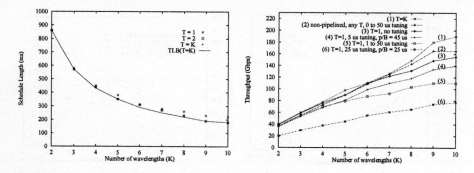

Fig. 8. Pipelined scheduling: (a) Schedule length when $\Delta = 0$; (b) Throughput for various Δ values.

shown as curve (2). Note that this curve is the same as curve (2) of Figure 7(b), even though the latter is for $T = 1$ and $\Delta = 5\mu s$.

As can be seen from curve (1), when $T = K$, (and hence with any tuning time), pipelined scheduling achieves a higher throughput. Note that, based on Figure 8(a) and related discussion, it is also conceivable that with a large T and a small Δ, pipelined scheduling can achieve a higher throughout than non-pipelined scheduling. In fact, curve (3) shows that even when $T = 1$ (as long as $\Delta = 0$), the pipelined throughput can be higher when $K \leq 6$.

On the other hand, curves (4) through (6) show that pipelined throughput can be lower than non-pipelined throughput when T is small (e.g. $T = 1$) and Δ is non-zero. These results also show that in such cases, the approach corresponding to curve (4), which combines tuning ($5\mu s$) and packet transmission ($45\mu s$) in one slot ($50\mu s$), is better when Δ is small, but the other approach which uses a separate slot for tuning is better when Δ is large (e.g. $25\mu s$).

6 Conclusion

In this paper, we have studied the problem of scheduling non-pipelined and pipelined transmissions for uniform traffic in unidirectional WDM rings, taking into consideration of a limited number of wavelengths, K, a limited number of transceivers per node, T, and their tuning time Δ. We have determined the TLBs on the schedule length, and proposed scheduling methods which can achieve optimal or near optimal schedules for various cases. We have also found that in general, pipelined scheduling is more suitable if T is equal (or close) to K or Δ is small relative to the propagation delay between the two adjacent nodes (which limits the size of the packets in pipelined transmission), but otherwise, non-pipelined scheduling is more suitable especially if message size (M) is large relative to t_p, the maximal propagation delay in the ring network. Our results can also be used to determine the amount of resources (e.g. K and T) for a given throughput requirement to achieve cost-effectiveness.

References

1. S. Hinrichs, C. Kosak, D. R. O'Hallaron, T. M. Stricker, and R. Take, "An architecture for optimal all-to-all personalized communication," in *Proc. Sixth Annual ACM Symposium on Parallel Algorithms and Architecture (SPAA)*, pp. 310–319, 1994.
2. S. H. Bokhari, "Multiphase complete exchange: A theoretical analysis," *IEEE Transactions on Computers* 45(2), pp. 220–229, 1996.
3. C.-T. Ho and M. T. Raghunath, "Efficient communication primitives on hypercubes," in *Proc. Sixth Distributed Memory Concurrent Computers*, pp. 390–397, 1991.
4. D. Scott, "Efficient all-to-all communication patterns in hypercube and mesh topologies," in *Proc. Sixth Distributed Memory Concurrent Computers*, pp. 398–403, 1991.
5. S. Johnsson and C.-T. Ho, "Optimal broadcasting and personalized communication in hypercubes," *IEEE Transactions on Computers* 38(9), pp. 1249–1268, 1989.
6. Y.-J. Suh and S. Yalamanchili, "All-to-all personalized exchange in two-dimension and 3-dimension tori," in *10th International Parallel Processing Symposium*, 1996.
7. L. Tassiulas and J. Joung, "Performance measures and scheduling policies in ring networks," *IEEE/ACM Transactions on Networking* 3(5), pp. 576–584, 1995.
8. E. Varvarigos and D. P. Bertsekas, "Communication algorithm for isotropic tasks in hypercubes and wraparound meshes," *Parallel Computing* 18, pp. 1233–1257, 1992.
9. G. R. Pieris and G. H. Sasaki, "Scheduling transmissions in WDM broadcast-and-select networks," *IEEE/ACM Transactions on Networking* 2(2), pp. 105–110, 1994.
10. H. Choi, H.-A. Choi, and M. Azizoglu, "Optimum transmission scheduling in optical broadcast networks," in *Proc. Int'l Conference on Communication*, pp. 266–270, 1995.
11. M. S. Borella and B. Mukherjee, "Efficient scheduling of nonuniform packet traffic in a WDM/TDM local lightwave network with arbitrary transceiver tuning latencies," in *Proceedings of IEEE Infocom*, pp. 129–137, 1995.
12. G. N. Rouskas and V. Sivaraman, "On the design of optimal TWDM schedules for broadcast WDM networks with arbitrary transceivers tuning latencies," in *Proceedings of IEEE Infocom*, pp. 1217–1224, 1996.
13. C. Qiao, X. Zhang, and L. Zhou, "Scheduling all-to-all connections in WDM rings," in *SPIE Proceedings, All Optical Communication Systems: Architecture, Control and Network Issues*, pp. 218–229, November 1996.
14. X. Zhang and C. Qiao, "Pipelined transmission scheduling in all-optical TDM/WDM rings," in *IC3N'97*, pp. 144–149, 1997.
15. X. Zhang and C. Qiao, "Scheduling in unidirectional WDM rings and its extensions," in *SPIE Proceedings, All Optical Communication Systems: Architecture, Control and Network Issues III*, pp. 208–219, 1997.
16. A. Elrefaie, "Multiwavelength survivable ring network architectures," in *Proc. Int'l Conference on Communication*, pp. 1245–1251, 1993.
17. J.-C. Bermond, L. Gargano, S. Perennes, A. A. Rescigno, and U. Vaccaro, "Efficient collective communication in optical networks," in *International Colloquium on Automata, Languages, and Programming*, pp. 574–585, 1996.
18. C. Qiao and R. G. Melhem, "Time-division optical communications in multiprocessor arrays," *IEEE Transactions on Computers* 42(5), pp. 577–590, 1993.
19. R. G. Melhem, D. Chiarulli, and S. P. Levitan, "Space multiplexing of waveguides in optically interconnected multiprocessor systems," *The Computer Journal* 32(4), pp. 362–369, 1989.

Irregular '99
Sixth International Workshop on Solving Irregularly Structured Problems in Parallel

Program Chair: Tao Yang, UC Santa Barbara

Steering Committee:

Afonso Ferreira, CNRS-I3S-INRIA Sophia Antipolis
José Rolim, University of Geneva

Program Committee:

Daniel Andresen, Kansas State University
Scott Baden, UC San Diego
Soumen Chakrabarti, IBM Almaden Research Center
Siddhartha Chatterjee, University of North Carolina
Ricardo Correa, DC/UFC, Brazil
Michel Cosnard, LORIA, France
Geoffrey Fox, Syracuse University
Apostolos Gerasoulis, Rutgers University
Howard Ho, IBM Almaden Research Center
Oscar Ibarra, UC Santa Barbara
Vipin Kumar, University of Minnesota
Esmond Ng, Lawrence Berkeley National Lab
Keshav Pingali, Cornell University
Sanjay Ranka, University of Florida
Joel Saltz, University of Maryland
Horst Simon, NERSC, Lawrence Berkeley Lab
Denis Trystram, IMAG, France
Shanghua Teng, University of Illinois
Herry Wijshoff, University of Leiden
Tao Yang, UC Santa Barbara

Invited Speakers:

Guang R. Gao, University of Delaware
Esmond G. Ng, Lawrence Berkeley National Lab

Foreword

The Sixth International Workshop on Solving Irregularly Structured Problems in Parallel (*Irregular '99*) is an annual workshop addressing issues related to deriving efficient parallel solutions for unstructured problems, with particular emphasis on the inter-cooperation between theoreticians and practitioners of the field. Irregular'99 is the sixth in the series, after Geneva, Lyon, Santa Barbara, Paderborn, and Berkeley.

The number of papers submitted was 26 and 13 papers have been selected for presentation by the Program Committee on the basis of referee reports. The final scientific program of Irregular '99 consists of four sessions and two invited talks.

We wish to thank all of the authors who responded to the call for papers and our invited speakers. Thank the members of the Program Committee for reviewing and selecting papers, Kai Shen, Martin Simons, Hong Tang, Creto Vidal, and Huican Zhu for their help in paper reviewing. We also would like to thank the IPPS'99 conference co-chair José Rolim and members of the steering committee for helping us in organizing this workshop.

January 1999   Tao Yang

Irregular'99 Final Program
April 16, 1999

8:30am-9:15am

Invited Talk by Esmond G. Ng (Lawrence Berkeley National Lab): Incomplete Cholesky Parallel Preconditioners with Selective Inversion (Joint work with Padma Raghavan, Univ. of Tennessee)

9:15am-10:30am Partitioning

Self-avoiding walks over adaptive unstructured grids, Gerd Heber, Rupak Biswas, and Guang R. Gao, Univ. of Delaware and NASA Ames Research Center.

A Graph Based Method for Generating the Fiedler Vector of Irregular Problems, Michael Holzrichter and Suely Oliveira, Univ. of Iowa

Hybridizing Nested Dissection and Halo Approximate Minimum Degree for Efficient Sparse Matrix Ordering, Francois Pellegrini, Jean Roman LaBRI, Universite Bordeaux I, Patrick Amestoy Enseeiht, France

10:30am -11:00am Coffee break

11:00am -12:15pm Systems Support I

ParaPART: Parallel Mesh Partitioning Tool for Distributed Systems, Jian Chen and Valerie E. Taylor, Northwestern University

Sparse computations with PEI, Frederique Voisin, Guy-Rene Perrin, Universite Louis Pasteur, France

Optimizing Irregular HPF Applications using Halos, Siegfried Benkner, C&C Research Laboratories, NEC Europe Ltd.

1:30pm-2:15pm

Invited Talk by Guang R. Gao (University of Delaware): From EARTH to HTMT: An Evolution of A Multitheaded Architecture Model

2:15pm-3:30pm Systems Support II

Irregular Parallel Algorithms in Java, Brian Blount, Siddhartha Chatterjee, University of North Carolina; Michael Philippsen, Univ. of Karlsruhe

A Simple Framework to Calculate the Reaching Definition of Array References and Its Use in Subscript Array Analysis, Yuan Lin and David Padua, University of Illinois at Urbana-Champaign

Dynamic Process Composition and Communication Patterns in Irregularly Structured Applications, C.T.H. Everaars, B. Koren and F. Arbab, Centre for Mathematics and Computer Science, Netherland.

3:30pm-4:00pm Coffee break

4:00pm-5:30pm Graph Algorithms and Applications

Scalable Parallelization of Harmonic Balance Simulation, David Rhodes, Army CECOM/RDEC, Apostolos Gerasoulis, Rutgers University

Towards an effective Task Clustering Heuristic for LogP Machines, Cristina Boeres, Aline Nascimento and Vinod Rebello, UFF, Brazil

A parallel range minima algorithm for coarse grained multicomputers, Henrique Mongelli and Siang Wun Song, University of Sao Paulo, Brazil

Deterministic Branch-and-Bound on Distributed Memory Machines, Kieran T. Herley, University College Cork, Andrea Pietracaprina, Geppino Pucci, Universita' di Padova

Self-Avoiding Walks over Adaptive Unstructured Grids

Gerd Heber[1], Rupak Biswas[2], and Guang R. Gao[1]

[1] CAPSL, University of Delaware, 140 Evans Hall, Newark, DE 19716, USA
{heber,ggao}@capsl.udel.edu
http://www.capsl.udel.edu
[2] MRJ/NASA Ames Research Center, MS T27A-1, Moffett Field, CA 94035, USA
rbiswas@nas.nasa.gov
http://science.nas.nasa.gov/~rbiswas

Abstract. In this paper, we present self-avoiding walks as a novel technique to "linearize" a triangular mesh. Unlike space-filling curves which are based on a geometric embedding, our algorithm is combinatorial since it uses the mesh connectivity only. We also show how the concept can be easily modified for adaptive grids that are generated in a hierarchical manner based on a set of simple rules, and made amenable for efficient parallelization. The proposed approach should be very useful in the runtime partitioning and load balancing of adaptive unstructured grids.

1 Introduction

Advances in adaptive software and methodology notwithstanding, parallel computational strategies will be an essential ingredient in solving complex, real-life problems. However, parallel computers are easily programmed with regular data structures; so the development of efficient parallel adaptive algorithms for unstructured grids poses a serious challenge. An efficient parallelization of these unstructured adaptive methods is rather difficult, primarily due to the load imbalance created by the dynamically-changing nonuniform grid and the irregular data access pattern. Nonetheless, it is generally believed that unstructured adaptive-grid techniques will constitute a significant fraction of future high-performance supercomputing. Mesh adaptation and dynamic load balancing must therefore be accomplished rapidly and efficiently, so as not to cause a significant overhead to the numerical simulation.

Serialization techniques play an important role when using a finite element method (FEM) over adaptive unstructured grids. A numbering of the unknowns allows for a matrix-vector notation of the underlying algebraic equations. Special numbering techniques (Cuthill-McKee, frontal methods) have been developed to optimize memory usage and locality of the algorithms. In addition, the runtime support for decomposing dynamic adaptive grids is often based on a linear representation of the grid hierarchy in the form of a *space-filling curve*. Several researchers have demonstrated the successful application of techniques based on space-filling curves to N-body simulations, graph partitioning, and other graph-related problems [4, 6–9].

The general idea of a space-filling curve is a serialization (or linearization) of visiting points in a higher-dimensional space. A standard method for the construction is to embed the object under study into a regular environment (where the standard space-filling curves live). However, this approach introduces an artificial structure in that the entire construction depends on the embedding. Furthermore, this type of a linear representation ignores the combinatorial structure of the mesh which drives the formulation of operators between the finite element spaces. The question arises whether one could reduce the different requirements for a serialization in an adaptive FEM over unstructured grids to a common denominator.

In this paper, we present a new approach to constructing a *self-avoiding walk* through a triangular mesh. Unlike space-filling curves which are based on a geometric embedding, our algorithm is combinatorial since it uses the mesh connectivity only. We also show how the concept can be easily modified for adaptive grids (that are generated in a hierarchical manner based on a set of simple rules) and made amenable for efficient parallelization (using a novel indexing scheme).

2 Self-Avoiding Walks

A self-avoiding walk (SAW) over an arbitrary unstructured two-dimensional mesh $\mathcal{M}$ is an enumeration of all the triangles of $\mathcal{M}$ such that two successive triangles share an edge or a vertex. Note that a SAW visits each triangle exactly once, entering it over an edge or a vertex, and exiting over another edge or vertex. In the cases where the SAW jumps over vertices, we imply that the triangles following one another in the enumeration do not share an edge. As simple examples show, the requirement that triangles following one another in a SAW must share an edge is too strong. In general, SAWs do not exist under such conditions (see Fig. 1 for a simple example where jumping over a vertex is necessary to visit every triangle). It is important to note that a SAW is not a Hamiltonian path.

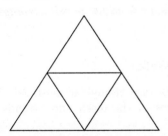
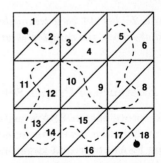

Fig. 1. A simple example that shows the need for SAWs to jump over vertices

Fig. 2. An example of a PSAW over a triangular mesh

SAWs can be used to improve the parallel efficiency of several irregular algorithms, in particular issues related to locality and load balancing. However, we need special classes of SAWs that will facilitate dynamic load balancing with good locality in terms of runtime memory access and interprocessor communication. We thus consider a special class of SAWs called *proper self-avoiding walks* (PSAW) where jumping twice over the same vertex is forbidden (see Fig. 2 for an example).

2.1 Definitions

Let us first fix some notation. For a SAW ω over a triangular mesh $\mathcal{M}$, we denote the i-th triangle in ω by $\omega(i)$ and the length of ω by $\#\omega$. Note that $\#\omega$ is also the number of triangles in $\mathcal{M}$. Formally, the properness of a SAW can be stated as follows:

Definition 1. *A self-avoiding walk (SAW) ω is called* <u>proper</u> $\iff_{\text{def}}$

$$\omega(i-1) \cap \omega(i) \neq \omega(i) \cap \omega(i+1) \qquad \forall i \in \{2, \ldots, \#\omega - 1\}. \tag{1}$$

If two triangles t_1, t_2 share an edge, we write $t_1 \mid t_2$. If $\omega(i) \mid \omega(i+1)$, we write $\omega(i) \vdash \omega(i+1)$ to indicate that ω enters $\omega(i+1)$ from $\omega(i)$ over an edge. If $\omega(i) \nmid \omega(i+1)$, we write $\omega(i) \curvearrowright \omega(i+1)$ to indicate that ω jumps into $\omega(i+1)$ from $\omega(i)$ over a vertex.

In the following, we make use of two technical results.

Lemma 1. *Let t, t_1, t_2, t_3 be triangles in a mesh $\mathcal{M}$ and*

$$t_i \mid t \qquad \forall i \in \{1, 2, 3\}. \tag{2}$$

Then, either

$$t_1 \cap t_2 \cap t_3 = \emptyset \tag{3}$$

or $\mathcal{M}$ is isomorphic to a tetrahedron. □

Note that if two triangles in a mesh share two vertices, they *must* share the corresponding edge.

Lemma 2. *Let $\mathcal{M}$ be a triangular mesh. Then, there exists a triangle τ in $\mathcal{M}$ such that $\mathcal{M}-\tau$, the complex obtained by removing τ from $\mathcal{M}$, is still a triangular mesh.* □

2.2 Existence of Proper Self-Avoiding Walks

We can prove that a PSAW exists for any arbitrary triangular mesh $\mathcal{M}$. The proof is inductive over the number of triangles in the mesh and extends an existing PSAW over to a larger mesh.

Proposition 1. *There exists a proper self-avoiding walk (PSAW) for an arbitrary triangular mesh $\mathcal{M}$.*

Proof. Assume that PSAWs exist for meshes with n triangles. Let $\mathcal{M}$ be a mesh with $n+1$ triangles.

Let τ be a triangle in $\mathcal{M}$ such that $\mathcal{M} - \tau$, the mesh consisting of all the triangles in $\mathcal{M}$ except τ, is a mesh (see Lemma 2). This mesh has n triangles and hence, by our induction assumption, a PSAW $\omega_{\mathcal{M}-\tau}$ exists for $\mathcal{M} - \tau$. We show that $\omega_{\mathcal{M}-\tau}$ can be extended to another PSAW $\omega_{\mathcal{M}}$ for $\mathcal{M}$. We call $\omega_{\mathcal{M}}$ a *proper extension* of $\omega_{\mathcal{M}-\tau}$.

Let t be an triangle of $\mathcal{M} - \tau$ sharing an edge with τ (i.e. $t \mid \tau$ holds). For an appropriate $i \in \{1, \ldots, n\}$, there holds

$$t = \omega_{\mathcal{M}-\tau}(i). \tag{4}$$

The discussion can be split up naturally into four cases, depending on whether $\omega_{\mathcal{M}-\tau}$ enters t through an edge or a vertex and leaves through an edge or a vertex, respectively.

We omit the subscript $\mathcal{M} - \tau$ from $\omega_{\mathcal{M}-\tau}$ in the following for the sake of clarity. Also, if we do not know whether the transition from $\omega(i)$ to $\omega(i+1)$ goes over an edge or a vertex, we write $\omega(i) \to \omega(i+1)$.

In the figures below, parts of $\omega_{\mathcal{M}-\tau}$ are drawn as solid lines whereas the modifications leading to $\omega_{\mathcal{M}}$ are drawn as dashed lines. Triangles are indexed by their position in the walk (e.g. $\omega(i)$ is denoted by i).

Case I: $\boxed{\omega(i-1) \vdash \omega(i) \vdash \omega(i+1)}$

Figure 3 illustrates the modifications necessary if $\omega_{\mathcal{M}-\tau}$ enters and leaves t through edges.

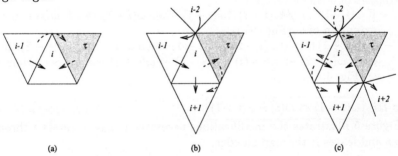

Fig. 3. Walk entering and exiting the triangle adjacent to τ via edges

- If $\omega(i-1) \cap \tau \neq \omega(i-2) \cap \omega(i-1)$, then $\omega(i-1) \to \tau \vdash \omega(i)$ is a proper extension of ω (Fig. 3(a)).
- If $\omega(i-1) \cap \tau = \omega(i-2) \cap \omega(i-1)$, then $\omega(i-2) \curvearrowright \omega(i-1)$.
  - If $\omega(i+1) \cap \tau \neq \omega(i+1) \cap \omega(i+2)$, then $\omega(i) \vdash \tau \to \omega(i+1)$ is a proper extension of ω (Fig. 3(b)).
  - If $\omega(i+1) \cap \tau = \omega(i+1) \cap \omega(i+2)$, then $\omega(i+1) \curvearrowright \omega(i+2)$.
    * If $\omega(i-1) \cap \omega(i+1) \neq \omega(i+1) \cap \omega(i+2)$, then $\omega(i-2) \to \tau \vdash \omega(i) \vdash \omega(i-1) \to \omega(i+1)$ is a proper extension of ω (Fig. 3(c)).

* Let $\omega(i-1) \cap \omega(i+1) = \omega(i+1) \cap \omega(i+2)$. According to Lemma 1, either $\mathcal{M}$ is tetrahedral or $\omega(i-1) \cap \omega(i+1) \cap \tau = \emptyset$. If $\mathcal{M}$ is tetrahedral, there is nothing to prove. For the other case, we have $\emptyset = \omega(i-1) \cap \omega(i+1) \cap \tau = \omega(i+1) \cap \omega(i+2) \cap \tau = \omega(i+1) \cap \tau = \omega(i+1) \cap \omega(i+2)$. This is contradicts our assumption $\omega(i+1) \curvearrowright \omega(i+2)$ which means $\omega(i+1) \cap \omega(i+2) \neq \emptyset$. Therefore, if $\mathcal{M}$ is not tetrahedral, the case $\omega(i-1) \cap \omega(i+1) = \omega(i+1) \cap \omega(i+2)$ is not possible.

Case II: $\boxed{\omega(i-1) \vdash \omega(i) \curvearrowright \omega(i+1)}$

Figure 4 illustrates the modifications necessary if $\omega_{\mathcal{M}-\tau}$ enters t through an edge and leaves it through a vertex.

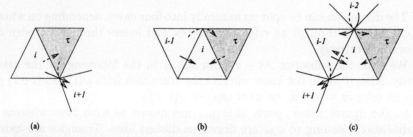

Fig. 4. Walk entering the triangle adjacent to τ via an edge and exiting via a vertex

- If $\omega(i) \cap \omega(i+1) \subset \omega(i) \cap \tau$, then $\omega(i) \vdash \tau \to \omega(i+1)$ is a proper extension of ω (Fig. 4(a)).
- Suppose $\omega(i) \cap \omega(i+1) \not\subset \omega(i) \cap \tau$. (Note that $\omega(i) \cap \omega(i+1)$ definitely consists of a single vertex!)
    - If $\omega(i-1) \cap \tau \neq \omega(i-2) \cap \omega(i-1)$, then $\omega(i-1) \to \tau \vdash \omega(i)$ is a proper extension of ω (Fig. 4(b)).
    - If $\omega(i-1) \cap \tau = \omega(i-2) \cap \omega(i-1)$, then $\omega(i-2) \curvearrowright \omega(i-1)$. In that case, $\omega(i-2) \to \tau \vdash \omega(i) \vdash \omega(i-1) \to \omega(i+1)$ is a proper extension of ω (Fig. 4(c)).

Case III: $\boxed{\omega(i-1) \curvearrowright \omega(i) \vdash \omega(i+1)}$

Figure 5 illustrates the modifications necessary if $\omega_{\mathcal{M}-\tau}$ enters t through a vertex and leaves it through an edge.

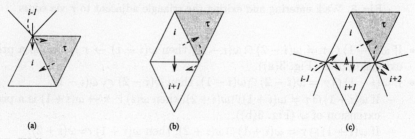

Fig. 5. Walk entering the triangle adjacent to τ via a vertex and exiting via an edge

- If $\omega(i-1) \cap \omega(i) \subset \omega(i) \cap \tau$, then $\omega(i-1) \to \tau \vdash \omega(i)$ is a proper extension of ω (Fig. 5(a)).
- Suppose $\omega(i-1) \cap \omega(i) \not\subset \omega(i) \cap \tau$.
  - If $\omega(i+1) \cap \tau \neq \omega(i+1) \cap \omega(i+2)$, then $\omega(i) \vdash \tau \to \omega(i+1)$ is a proper extension of ω (Fig. 5(b)).
  - If $\omega(i+1) \cap \tau = \omega(i+1) \cap \omega(i+2)$, then $\omega(i+1) \curvearrowright \omega(i+2)$. In that case, $\omega(i-1) \to \omega(i+1) \vdash \omega(i) \vdash \tau \to \omega(i+2)$ is a proper extension of ω (Fig. 5(c)).

Case IV: $\boxed{\omega(i-1) \curvearrowright \omega(i) \curvearrowright \omega(i+1)}$

Figure 6 illustrates the modifications necessary if $\omega_{\mathcal{M}-\tau}$ enters and leaves t through vertices.

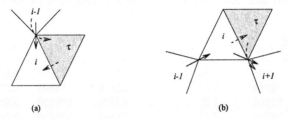

Fig. 6. Walk entering and exiting the triangle adjacent to τ via vertices

- If $\omega(i-1) \cap \omega(i) \subset \omega(i) \cap \tau$, then $\omega(i-1) \to \tau \vdash \omega(i)$ is a proper extension of ω (Fig. 6(a)).
- If $\omega(i-1) \cap \omega(i) \not\subset \omega(i) \cap \tau$, then $\omega(i) \vdash \tau \to \omega(i+1)$ is a proper extension of ω (Fig. 6(b)).

Since we have been able to properly extend $\omega_{\mathcal{M}-\tau}$ in all cases, our proposition is proved. □

A closer look at the proof of Proposition 1 and an auxiliary construction allow us to prove the following stronger result:

Proposition 2. *Let $\mathcal{M}$ a triangular mesh and $t_1, t_2 \in \mathcal{M}$ be two arbitrary triangles. Then, there exists a proper self-avoiding walk (PSAW) ω with $\omega(1) = t_1$ and $\omega(\#\mathcal{M}) = t_2$.* □

The proof of Proposition 1 actually provides a set of elementary rules which can be used to construct a PSAW: starting from any random triangle of $\mathcal{M}$, choose new triangles sharing an edge with the current submesh and extend the existing PSAW over the new triangle. In the process of deriving the existence of PSAWs, a natural mathematical framework emerges which can be easily adapted for other special classes of SAWs.

The complexity of this algorithm is $O(n \log n)$, where n is the number of triangles in the mesh. This complexity is obtained as follows. Since the triangles are organized in a red-black tree, a search requires $O(\log n)$ time (the edge-triangle incidence relation can also be organized in a red-black tree). Since a search has to be performed for each triangle in the mesh, the overall complexity of our algorithm is $O(n \log n)$.

We have implemented a sequential algorithm for the generation of a PSAW using the set and map containers from the C++ Standard Template Library. Both containers are implemented with red-black trees; however, one may also think about an implementation using hash tables. The results presented in Tables 1 and 2 were obtained for some sample meshes on a 110 MHz microSPARC II using GNU's g++ compiler (version 2.7.2.3) with the -O flag. Note that the times in the tables below include the setup time for the mesh and the edge-triangle incidence structure.

Table 1. Runtimes for various mesh sizes

Triangles	Edges	Time (s)
2048	3136	0.30
4096	6240	0.57
8192	12416	1.31
16384	24768	2.71
32768	49408	5.54
65536	98688	11.82
131072	197120	24.07

Table 2. Runtimes for various triangle/edge ratios

Triangles	Edges	Ratio	Time (s)
65536	131073	0.500	12.85
65536	114690	0.571	12.27
65536	106500	0.615	12.21
65536	102408	0.640	12.15
65536	100368	0.653	12.13
65536	99360	0.660	12.02
65536	98880	0.663	11.99
65536	98688	0.664	11.82

The results in Table 1 indicate the dependence of the runtime on the number of triangles (or edges). According to our description of the algorithm, one would suspect that the runtime may also depend on the triangle/edge ratio. The results in Table 2 explore this possibility; however, the numbers clearly show that the dependence of the execution time on the triangle/edge ratio is negligible and confirm our complexity estimate for the algorithm.

A standard parallelization of the deterministic construction algorithm would be based on the observation that if the extensions of an existing PSAW did not interfere with one another, then they could be done in parallel. Such an approach is feasible in dealing with single meshes when no additional structural information is available, or with families of meshes without any hierarchical structure, although scalability would be a major concern.

3 Adaptive Triangular Grids

As is evident from the discussion in the previous section, a parallel algorithm for constructing a PSAW is non-trivial, and the walk has to be completely rebuilt after each mesh adaptation. However, PSAWs provide enough flexibility to design a *local* algorithm for *hierarchical* adaptive changes in an unstructured mesh. In contrast to a general adaptation scheme that produces just another unstructured mesh, a hierarchical strategy pays particular attention to the initial triangulation and uses a set of simple coarsening and refinement rules:

- When a triangle is refined, it is either subdivided into two (green) or four (red) smaller triangles (Fig. 7).
- All triangles with exactly two bisected edges have their third edge also bisected; thus, such triangles are red.
- A green triangle cannot be further subdivided; instead, the previous subdivision is discarded and red subdivision is first applied to the ancestor triangle (Fig. 8).

Such a policy also assures that, with repeated refinement, the quality of the adapted mesh will not deteriorate drastically from that of the initial mesh. This strategy is similar to that described in [1] for tetrahedral meshes.

    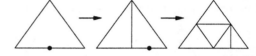

Fig. 7. Green (left) and red (right) refinement of a triangle

Fig. 8. Special policy for refining a green triangle

Such a scheme exploits the structure of the adaptation hierarchy to simplify the mesh labeling process. Our goal is to modify (or adapt) the labeling only in regions where the mesh has been altered but not to rebuild the entire indexing. This indexing idea has been extensively developed in [3]. It is a combination of coarse and local schemes. The coarse scheme labels the objects of the initial mesh so that the incidence relations can be easily derived from the labels. The local scheme exploits the regularity and the finiteness of the refinement rules to produce names for the mesh objects at subsequent refinement levels. The coarse and local schemes are combined by taking the union of the Cartesian products of the coarse mesh objects with their corresponding local schemes. Ambiguities are resolved by using a normal form of the index.

The key features of this scheme are:
- Each object is assigned a global name that is independent of any architectural considerations or implementation choices.
- Combinatorial information is translated into simple arithmetic.
- It is well-behaved under adaptive refinement and no artificial synchronization/serialization is introduced.
- It can be extended (with appropriate modifications) to three dimensions [2].

Our basic strategy for such hierarchical adaptive meshes is to apply the general algorithm only once to the initial mesh to construct a coarse PSAW, and then to reuse the walk after each adaptation. This requirement leads us to the concept of *constrained proper self-avoiding walks* (CPSAW) that exploits the regularity of the refinement rules. In a way, a PSAW is global since it is related to the initial coarse mesh, while a CPSAW is local since it is restricted to a triangle of the coarse mesh. A PSAW leaves a trace or footprint on each coarse triangle which is then translated into a boundary value problem for extending the walk over the adapted mesh. In addition, the extensions for different coarse triangles can be decoupled; thus, the construction process of a CPSAW is inherently parallel. Formal mathematical definitions of a constraint and the compatibility of PSAWs with constraints are given in [5].

Figure 9 depicts an example of how a CPSAW is generated from an existing PSAW when the underlying mesh is adapted. Figure 9(a) shows a portion of a PSAW over four triangles of an initial coarse mesh. The entry and exit points are respectively marked as α and β. These four triangles are then refined as shown in Fig. 9(b). The entry and exit constraints for each of the four original triangles are shown in Fig. 9(c) (an exploded view is used for clarity). Figure 9(d) shows the extension of the original PSAW to the higher level of refinement. Note that it is a direct extension dictated by the constraints. This CPSAW can often be improved by replacing certain jumps over vertices by jumps over edges when adjacent triangles in the walk share an edge (Fig. 9(e)).

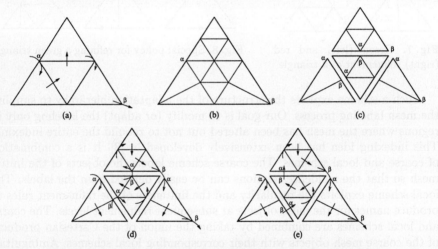

Fig. 9. Extending a PSAW to a CPSAW over an adapted mesh

The regularity of the local picture allows us to prove the existence of CPSAWs; full details are given in [5]. Moreover, CPSAWs are also appropriate for multigrid methods since they "know" the refinement level of the adapted mesh. The process of generating a CPSAW from an initial PSAW after mesh adaptation is embarrassingly parallel, since each processor is responsible for only its own chunk of the walk. However, load balancing is still an important issue that must be addressed when the coarse mesh is locally adapted.

4   Conclusions

In this paper, we have developed a theoretical basis for the study of self-avoiding walks over two-dimensional adaptive unstructured grids. We described an algorithm of complexity $O(n \log n)$ for the construction of proper self-avoiding walks over arbitrary triangular meshes and reported results of a sequential implementation. We discussed parallelization issues and suggested a significant improvement for hierarchical adaptive unstructured grids. For this situation, we have also proved the existence of constrained proper self-avoiding walks.

The algorithms presented allow a straightforward generalization to tetrahedral meshes; however, there is some flexibility because of additional freedom in three dimensions. The application of this methodology to partitioning and load balancing holds a lot of promise and is currently being investigated. However, modifications to improve locality in the sense of a generalized frontal method requires a more detailed study of the relationships between more specialized constraints on walks and cache memory behavior.

Note that our basic algorithm can be improved in many ways. For instance, we completely ignored additional information like nodal degrees. Also, given the existence of PSAWs, an acceleration technique might be favorable. For example, walks have a long tradition in the application of Monte Carlo methods to study long-chain polymer molecules. An investigation of these alternate approaches is beyond the scope of this paper but could be the focus of future work.

Acknowledgements

The work of the first and third authors was partially supported by NSF under Grant Numbers NSF-CISE-9726388 and NSF-MIPS-9707125. The work of the second author was supported by NASA under Contract Number NAS 2-14303 with MRJ Technology Solutions.

References

1. Biswas, R., Strawn, R.C.: Mesh Quality Control for Multiply-Refined Tetrahedral Grids. Appl. Numer. Math. **20** (1996) 337–348
2. Gerlach, J.: Application of Natural Indexing to Adaptive Multilevel Methods for Linear Triangular Elements. RWCP Technical Report 97-010 (1997)
3. Gerlach, J., Heber, G.: Fundamentals of Natural Indexing for Simplex Finite Elements in Two and Three Dimensions. RWCP Technical Report 97-008 (1997)
4. Griebel, M., Zumbusch, G.: Hash-Storage Techniques for Adaptive Multilevel Solvers and their Domain Decomposition Parallelization. AMS Contemp. Math. Series **218** (1998) 279–286
5. Heber, G., Biswas, R., Gao, G.R.: Self-Avoiding Walks over Two-Dimensional Adaptive Unstructured Grids. NAS Technical Report NAS-98-007 (1998)
6. Ou, C.-W., Ranka, S., Fox, G.: Fast and Parallel Mapping Algorithms for Irregular Problems. J. of Supercomp. **10** (1995) 119–140
7. Parashar, M., Browne, J.C.: On Partitioning Dynamic Adaptive Grid Hierarchies. 29th Hawaii Intl. Conf. on System Sciences (1996) 604–613
8. Pilkington, J.R., Baden, S.B.: Dynamic Partitioning of Non-Uniform Structured Workloads with Spacefilling Curves. IEEE Trans. Par. Dist. Sys. **7** (1996) 288–300
9. Salmon, J., Warren, M.S.: Parallel, Out-of-Core Methods for Fast Evaluation of Long-Range Interactions. 8th SIAM Conf. on Par. Proc. for Sci. Comput. (1997)

A Graph Based Method for Generating the Fiedler Vector of Irregular Problems[1]

Michael Holzrichter[1] and Suely Oliveira[2]

[1] Texas A&M University, College Station, TX,77843-3112
[2] The University of Iowa, Iowa City, 52242, USA,
oliveira@cs.uiowa.edu,
WWW home page: http://www.cs.uiowa.edu

Abstract. In this paper we present new algorithms for spectral graph partitioning. Previously, the best partitioning methods were based on a combination of Combinatorial algorithms and application of the Lanczos method when the graph allows this method to be cheap enough. Our new algorithms are purely spectral. They calculate the Fiedler vector of the original graph and use the information about the problem in the form of a preconditioner for the graph Laplacian. In addition, we use a favorable subspace for starting the Davidson algorithm and reordering of variables for locality of memory references.

1 Introduction

Many algorithms have been developed to partition a graph into k parts such that the number of edges cut is small. This problem arises in many areas including finding fill-in reducing permutations of sparse matrices and mapping irregular data structures to nodes of a parallel computer.

Kernighan and Lin developed an effective combinatorial method based on swapping vertices [12]. Multilevel extensions of the Kernighan-Lin algorithm have proven effective for graphs with large numbers of vertices [11]. In recent years spectral approaches have received significant attention [15]. Like combinatorial methods, multilevel versions of spectral methods have proven effective for large graphs [1]. The previous multilevel spectral algorithms of [1]is based on the application of spectral algorithms at various graph levels. Previous (multilevel) combinatorial algorithms coarsen a graph until it has a small number of nodes and can be partioned more cheaply by well known techniques (which may include spectral partitioning). After a partition is found for the coarsest graph, the partition is then successively interpolated onto finer graphs and refined [10, 11]. Our focus is on finding a partition of the graph using *purely spectral techniques*. We use the graphical structure of the problem to develop a multilevel spectral partitioning algorithm applied to the original graph. Our algorithm is based on a well known algorithm for the calculation of the Fiedler vector: the

---

[1] This research was supported by NSF grant ASC-9528912, and currently by NSF/DARPA DMS-9874015

Davidson algorithm. The structure of the problem is incorporated in the form of a graphical preconditioner to the Davidson algorithm.

2 Spectral Methods

The graph Laplacian of graph G is $L = D - A$, where A is G's adjacency matrix and D is a diagonal matrix where $d_{i,i}$ equals to the degree of vertex v_i of the graph. One property of L is that its smallest eigenvalue is 0 and the corresponding eigenvector is $(1, 1, \ldots, 1)$. If G is connected, all other eigenvalues are greater than 0.

Fiedler [5, 6] explored the properties of the eigenvector associated with the second smallest eigenvalue. (These are now known as the "Fiedler vector" and "Fiedler value," respectively.) Spectral methods partition G based on the Fiedler vector of L. The reason why the Fiedler vector is useful for partitioning is explained in more detail in the next section.

3 Bisection as a Discrete Optimization Problem

It is well known that for any vector x

$$x^T L x = \sum_{(i,j) \in E} (x_i - x_j)^2,$$

holds (see for example Pothen et al. [15]).

Note that there is one term in the sum for each edge in the graph. Consider a vector x whose construction is based upon a partition of the graph into subgraphs P_1 and P_2. Assign +1 to x_i if vertex v_i is in partition P_1 and -1 if v_i is in partition P_2. Using this assignment of values to x_i and x_j, if an edge connects two vertices in the same partition then $x_i = x_j$ and the corresponding term in the Dirichlet sum will be 0. The only non-zero terms in the Dirichlet sum are those corresponding to edges with end points in separate partitions. Since the Dirichlet sum has one term for each edge and the only non-zero terms are those corresponding to edges between P_1 and P_2, it follows that $x^T L x = 4 *$ (number of edges between P_1 and P_2).

An x which minimizes the above expression corresponds to a partition which minimizes the number of edges between the partitions. The graph partitioning problem has been transformed into a discrete optimization problem with the goal of

- minimizing $\frac{1}{4} x^T L x$
- such that
    1. $e^T x = 0$ where $e = (1, 1, \ldots, 1)^T$
    2. $x^T x = n$.
    3. $x_i = \pm 1$

Condition (1) stipulates that the number of vertices in each partition be equal, and condition (2) stipulates that every vertex be assigned to one of the partitions. If we remove condition (3) the above problem can be solved using Lagrange multipliers.

We seek to minimize $f(x)$ subject to $g_1(x) = 0$ and $g_2(x) = 0$. This involves finding Lagrange multipliers λ_1 and λ_2 such that

$$\nabla f(x) - \lambda_1 \nabla g_1(x) - \lambda_2 \nabla g_2(x) = 0.$$

For this discrete optimization problem, $f(x) = \frac{1}{2}x^T L x$, $g_1(x) = e^T x$, and $g_2(x) = \frac{1}{2}(x^T x - n)$. The solution must satisfy

$$Lx - \lambda_1 e - \lambda_2 x = 0.$$

That is, $(L - \lambda_2 I)x = \lambda_1 e$. Premultiplying by e^T gives

$$e^T L x - \lambda_2 e^T x = e^T e \lambda_1 = n \lambda_1.$$

But $e^T L = 0$, $e^T x = 0$ so $\lambda_1 = 0$. Thus

$$(L - \lambda_2 I)x = 0$$

and x is an eigenvector of L.

The above development has shown how finding a partition of a graph which minimizes the number of edges cut can be transformed into an eigenvalue problem involving the graph Laplacian. This is the foundation of spectral methods. Theory exists to show the effectiveness of this approach. The work of Guattery and Miller [7] present examples of graphs for which spectral partitioning does not work well. Nevertheless, spectral partition methods work well on bounded-degree planar graphs and finite element meshes [17]. Chan et al. [3] show that spectral partitioning is optimal in the sense that the partition vector induced by it is the closest partition vector to the second eigenvector, in any l_s norm where $s \geq 1$. Nested dissection is predicated upon finding a good partition of a graph. Simon and Teng [16] examined the quality of the p-way partition produced by recursive bisection when p is a power of two. They show that recursive bisection in some cases produces a p-way partition which is much worse than the optimal p-way partition. However, if one is willing to relax the problem slightly, they show that recursive bisection will find a good partition.

4 Our Approach

We developed techniques which use a multilevel representation of a graph to accelerate the computation of its Fiedler vector. The new techniques

- provide a framework for a multilevel preconditioner
- obtain a favorable initial starting point for the eigensolver
- reorder the data structures to improve locality of memory references.

The multilevel representation of the input graph G_0 consists of a series of graphs $\{G_0, G_1, \ldots, G_n\}$ obtained by successive coarsening. Coarsening G_i is accomplished by finding a maximum matching of its vertices and then combining each pair of matched vertices to form a new vertex for G_{i+1}; unmatched vertices are replicated in G_{i+1} without modification. Connectivity of G_i is maintained by connecting two vertices in G_{i+1} with an edge if, and only if, their constituent vertices in G_i were connected by an edge. The coarsening process concludes when a graph G_n is obtained with sufficiently few vertices.

We used the Davidson algorithm [4, 2, 14] as our eigensolver. The Davidson algorithm is a subspace algorithm which iteratively builds a sequence of nested subspaces. A Rayleigh-Ritz process finds a vector in each subspace which approximates the desired eigenvector. If the Ritz vector is not sufficiently close to an eigenvector then the subspace is augmented by adding a new dimension and the process repeats. The Davidson algorithm allows the incorporation of a preconditioner. We used the structure of the graph in the development of our Davidson preconditioner. More details about this algorithm is given in [8, 9].

4.1 Multilevel Preconditioner

We will refer to our new preconditioned Davidson Algorithm as PDA. The preconditioner approximates the inverse of the graph Laplacian matrix. It operates in a manner similar to multigrid methods for solving discretizations of PDE's [13]. However, our preconditioner differs from these methods in that we do not rely on obtaining coarser problems by decimating a regular discretization. Our method works with irregular or unknown discretizations because it uses the coarse graphs for the multilevel framework. PDA can be considered as our main contribution to the current research on partitioning algorithms. The next two subsections decrib other features of our algorithms. These additional features aimed to spedd up each interation of PDA. The numerical results of Sec. 5 will shown that we were successful in improving purely spectral partitioning algorithms.

4.2 Favorable Initial Subspace

The second main strategy utilizes the multilevel representation of G_0 to construct a favorable initial subspace for the Davidson Algorithm. We call this strategy the Nested Davidson Algorithm (NDA).

The Nested Davidson Algorithm works by running the Davidson algorithm on the graphs in $\{G_0, G_1, \ldots, G_n\}$ in order from G_n to G_0. The k basis vectors spanning the subspace for G_{i+1} are used to construct l basis vectors for the initial subspace for G_i. The l new basis vectors are constructed from linear combinations of the basis vectors for G_{i+1} by interpolating them onto G_i and orthonormalizing. Once the initial basis vectors for G_i have been constructed, the Davidson algorithm is run on G_i and the process repeats for G_{i-1}.

4.3 Locality of Memory References

No assumption is made about the regularity of the input graphs. Furthermore the manner in which the coarse graphs are constructed results in data structures with irregular storage in memory. The irregular storage of data structures has the potential of reducing locality of memory accesses and thereby reducing the effectiveness of cache memories.

We developed a method called Multilevel Graph Reordering (MGR) which uses the multilevel representation of G_0 to reorder the data structures in memory to improve locality of reference. Intuitively, we permute the graph Laplacian matrices to increase the concentration of non-zero elements along the diagonal. This improved locality of reference during relaxation operations which represented a major portion of the time required to compute the Fiedler vector.

The relabeling of the graph is accomplished by imposing a tree on the vertices of the graphs $\{G_0, G_1, \ldots, G_n\}$. This tree was traversed in a depth-first manner. The vertices were relabeled in the order in which they were visited. After the relabelling was complete the data structures were rearranged in memory such that the data structures for the ith vertex are stored at lower memory addresses than the data structures for the $i+1^{\text{st}}$ vertex.

An example of reordering by MGR is shown in Figures 1 and 2. The vertices are shown as well as the tree overlain on the vertices. (The edges of the graphs are not shown.) If a vertex lies to the right of another in the figure, then its datastructures occupy higher memory addresses. Notice that after reordering, the vertices with a common parent are placed next to each other in memory. This indicates that they are connected by an edge and will be referenced together during relaxation operations. Relaxation operations represented a large fraction of the total work done by our algorithms. This reordering has a positive effect of locality of reference during such operations.

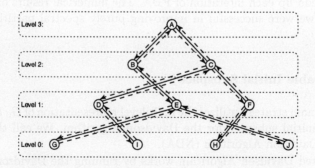

Fig. 1. Storage of data structures before reordering

The algorithms described above are complementary and can be used concurrently while computing the Fiedler vector. We call the resulting algorithm

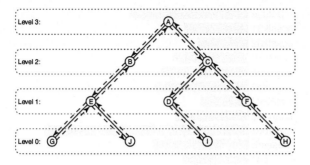

Fig. 2. Storage of data structures after reordering

when PDA, NDA and MGR are combined: the Multilevel Davidson Algorithm (MDA).

5 Numerical Results

The performance of our new algorithms were tested by computing the Fiedler vector of graphs extracted from large, sparse matrices. These matrices ranged in size from approximately $10,000 \times 10,000$ to more $200,000 \times 200,000$. The number of edges ranged from approximately 300,000 to more than 12 million. The matrices come from a variety of application domains, for example, finite element models and a matrix from financial portfolio optimization.

Figure 3 shows the ratio of the time taken by MDA when using PDA, NDA and MGR concurrently to the time taken by the Lanczos algorithm to compute the Fiedler vector. The ratio for most graphs was between 0.2 and 0.5 indicating that MDA usually computed the Fiedler vector two to five times as fast as the Lanczos algorithm. For some graphs MDA was even faster. For the **gearbox** graph MDA was only slightly faster.

Our studies indicate that the two main factors in MDA's improved performance shown in Figure 3 are the preconditioner of PDA and the favorable initial subspace provided by NDA.

We have observed that NDA does indeed find favorable subspaces. The Fiedler vector had a very small component perpendicular to the initial subspace.

We were able to measure the effect of MGR on locality of memory references. Our studies found that MGR on the average increased the number of times a primary cache line was reused by 18 percent indicating improved locality by the pattern of memory accesses.

6 Conclusions

We have developed new algorithms which utilize the multilevel representation of a graph to accelerate the computation of the Fiedler vector for that graph. The

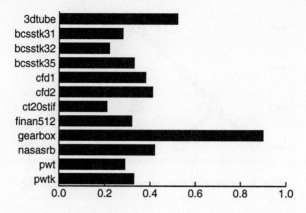

Fig. 3. Ratio of Davidson to Lanczos execution time

Preconditioned Davidson Algorithm uses the graph structure of the problem as as a framework for a multilevel preconditioner. A favorable initial subspace for the Davidson algorithm is provided by the Nested Davidson algorithm. The Multilevel Graph Reordering algorithm improves the locality of memory references by finding a permutation of the graph Laplacian matrices which concentrates the non-zero entries along the diagonal and reordering the data structures in memory accordingly. More numerical results can be found in [9]. The full algorithms for the methods described in this paper are available in [8]. PDA can be considered as our main contribution to spectral partitioning of irregular graphs.

References

1. S. Barnard and H. Simon. A fast multilevel implementation of recursive spectral bisection for partitioning unstructured problems. In *Proceedings of the Sixth SIAM Conference on Parallel Processing for Scientific Computing*, Norfolk, Virginia, 1993. SIAM, SIAM.
2. L. Borges and S. Oliveira. A parallel Davidson-type algorithm for several eigenvalues. *Journal of Computational Physics*, (144):763–770, August 1998.
3. T. Chan, P. Ciarlet Jr., and W. K. Szeto. On the optimality of the median cut spectral bisection graph partitioning method. *SIAM Journal on Computing*, 18(3):943–948, 1997.
4. E. Davidson. The iterative calculation of a few of the lowest eigenvalues and corresponding eigenvectors of large real-symmetric matrices. *Journal of Computational Physics*, 17:87–94, 1975.
5. M. Fiedler. Algebraic connectivity of graphs. *Czechoslovak Mathematical Journal*, 23:298–305, 1973.
6. M. Fiedler. A property of eigenvectors of non-negative symmetric matrices and its application to graph theory. *Czechoslovak Mathematical Journal*, 25:619–632, 1975.
7. S. Guattery and G. L. Miller. On the quality of spectral separators. *SIAM Journal on Matrix Analysis and Applications*, 19:701–719, 1998.

8. M. Holzrichter and S. Oliveira. New spectral graph partitioning algorithms. submitted.
9. M. Holzrichter and S. Oliveira. New graph partitioning algorithms. 1998. The University of Iowa TR-120.
10. G. Karypis and V. Kumar. Multilevel k-way partitioning scheme for irregular garphs. to appear in the Journal of Parallel and Distributed Computing.
11. G. Karypis and V. Kumar. A fast and high quality multilevel scheme for partitioning irregular graphs. *SIAM Journal on Scientific Computing*, 20(1):359–392, 1999.
12. B. Kernighan and S. Lin. An efficient heuristic procedure for partitioning graphs. *The Bell System Technical Journal*, 49:291–307, February 1970.
13. S. McCormick. *Multilevel Adaptive Methods for Partial Differential Equations*. Society for Industrial and Applied Mathematics, Philadelphia, Pennsylvania, 1989.
14. S. Oliveira. A convergence proof of an iterative subspace method for eigenvalues problem. In F. Cucker and M. Shub, editors, *Foundations of Computational Mathematics Selected Papers*, pages 316–325. Springer, January 1997.
15. Alex Pothen, Horst D. Simon, and Kang-Pu Liou. Partitioning sparse matrices with eigenvectors of graphs. *SIAM J. Matrix Anal. Appl.*, 11(3):430–452, 1990. Sparse matrices (Gleneden Beach, OR, 1989).
16. H. D. Simon and S. H. Teng. How good is recursive bisection. *SIAM Journal on Scientific Computing*, 18(5):1436–1445, July 1997.
17. D. Spielman and S. H. Teng. Spectral partitioning works: planar graphs and finite element meshes. In *37th Annual Symposium Foundations of Computer Science*, Burlington, Vermont, October 1996. IEEE, IEEE Press.

Hybridizing Nested Dissection and Halo Approximate Minimum Degree for Efficient Sparse Matrix Ordering*

François Pellegrini[1], Jean Roman[1], and Patrick Amestoy[2]

[1] LaBRI, UMR CNRS 5800, Université Bordeaux I & ENSERB
351, cours de la Libération, F-33405 Talence, France
{pelegrin|roman}@labri.u-bordeaux.fr
[2] ENSEEIHT-IRIT
2, rue Charles Camichel, 31071 Toulouse Cédex 7, France
Patrick.Amestoy@enseeiht.fr

Abstract. Minimum Degree and Nested Dissection are the two most popular reordering schemes used to reduce fill-in and operation count when factoring and solving sparse matrices. Most of the state-of-the-art ordering packages hybridize these methods by performing incomplete Nested Dissection and ordering by Minimum Degree the subgraphs associated with the leaves of the separation tree, but to date only loose couplings have been achieved, resulting in poorer performance than could have been expected. This paper presents a tight coupling of the Nested Dissection and Halo Approximate Minimum Degree algorithms, which allows the Minimum Degree algorithm to use exact degrees on the boundaries of the subgraphs passed to it, and to yield back not only the ordering of the nodes of the subgraph, but also the amalgamated assembly subtrees, for efficient block computations.
Experimental results show the performance improvement, both in terms of fill-in reduction and concurrency during numerical factorization.

1 Introduction

When solving large sparse linear systems of the form $Ax = b$, it is common to precede the numerical factorization by a symmetric reordering. This reordering is chosen in such a way that pivoting down the diagonal in order on the resulting permuted matrix PAP^T produces much less fill-in and work than computing the factors of A by pivoting down the diagonal in the original order (the fill-in is the set of zero entries in A that become non-zero in the factored matrix).

The two most classically-used reordering methods are Minimum Degree and Nested Dissection. The Minimum Degree algorithm [19] is a local heuristic that performs its pivot selection by selecting from the graph a node of minimum degree. The Nested Dissection algorithm [8] is a global heuristic recursive algorithm which computes a vertex set S that separates the graph into two parts A and B, ordering S last. It then proceeds recursively on parts A and B until their

---

* This work is supported by the French *Commissariat à l'Énergie Atomique* CEA/CESTA under contract No. 7V1555AC, and by the GDR ARP of the CNRS.

sizes become smaller than some threshold value. This ordering guarantees that no non zero term can appear in the factorization process between unknowns of A and unknowns of B.

The Minimum Degree algorithm is known to be a very fast and general purpose algorithm, and has received much attention over the last three decades (see for example [1, 9, 16]). However, the algorithm is intrinsically sequential, and very little can be theoretically proven about its efficiency.

On the other hand, many theoretical results have been carried out on Nested Dissection ordering [5, 15], and its divide and conquer nature makes it easily parallelizable. In practice, Nested Dissection produces orderings which, both in terms of fill-in and operation count, compare well to the ones obtained with Minimum Degree [11, 13, 17]. Moreover, the elimination trees induced by nested dissection are broader, shorter, and better balanced, and therefore exhibit much more concurrency in the context of parallel Cholesky factorization [4, 6, 7, 11, 17, 18, and included references].

Due to their complementary nature, several schemes have been proposed to hybridize the two methods [12, 14, 17]. However, to our knowledge, only loose couplings have been achieved: incomplete Nested Dissection is performed on the graph to order, and the resulting subgraphs are passed to some Minimum Degree algorithm. This results in the fact that the Minimum Degree algorithm does not have exact degree values for all of the boundary vertices of the subgraphs, leading to a misbehavior of the vertex selection process.

In this paper, we propose a tight coupling of the Nested Dissection and Minimum Degree algorithms, that allows each of them to take advantage of the information computed by the other. First, the Nested Dissection algorithm provides exact degree values for the boundary vertices of the subgraphs passed to the Minimum Degree algorithm (called *Halo* Minimum Degree since it has a partial visibility of the neighborhood of the subgraph). Second, the Minimum Degree algorithm returns the assembly tree that it computes for each subgraph, thus allowing for supervariable amalgamation, in order to obtain column-blocks of a size suitable for BLAS3 block computations.

The rest of the paper is organized as follows. Section 2 describes the metrics we use to evaluate the quality of orderings. Section 3 presents the principle of our Halo Minimum Degree algorithm, and evaluates its efficiency. Section 4 outlines the amalgamation process that is carried out on the assembly trees, and validates its interest in the context of block computations. In the concluding section, we show that relatively high speed-ups can be achieved during numerical factorization by using our ordering scheme.

2 Metrics

The quality of orderings is evaluated with respect to several criteria. The first one, called NNZ, is the overall number of non zero terms to store in the factored reordered matrix, including extra logical zeros that must be stored when block storage is used. The second one, OPC, is the operation count, that is, the number of arithmetic operations required to factor the matrix. To comply with existing RISC processor technology, the operation count that we consider in this paper accounts for all operations (additions, subtractions, multiplications, divisions) required by sequential Cholesky factorization, except square roots; it is equal to $\sum_c n_c^2$, where n_c is the number of non-zeros of column c of the factored matrix, diagonal included. The third criterion is the size, in integer words, of the secondary storage used to index the array of non-zeros. It is defined as SS_b when block storage is used, and SS_c when column compressed storage is used.

All of the algorithms described in this paper have been integrated into version 3.3 of the SCOTCH static mapping and sparse matrix ordering software package [17] developed at the LaBRI, which has been used for all of our tests. All the experiments were run on a 332MHz PowerPC 604E-based RS6000 machine with 512 Mb of main memory. The parallel experiments reported in the conclusion were run on an IBM-SP2 with 66 MHz Power2 thin nodes having 128 MBytes of physical memory and 512 MBytes of virtual memory each.

Table 1. Description of our test problems. NNZ_A is the number of extra-diagonal terms in the triangular matrix after it has been symmetrized, whenever necessary.

Name	Columns	NNZ_A	Description
144	144649	1074393	3D finite element mesh
598A	110971	741934	3D finite element mesh
AATKEN	42659	88237	
AUTO	448695	3314611	3D finite element mesh
BCSSTK29	13992	302748	3D stiffness matrix
BCSSTK30	28924	1007284	3D stiffness matrix
BCSSTK31	35588	572914	3D stiffness matrix
BCSSTK32	44609	985046	3D stiffness matrix
BMW3_2	227362	5530634	3D stiffness matrix
BMW7ST_1	141347	3599160	3D stiffness matrix
BRACK2	62631	366559	3D finite element mesh
CRANKSEG1	52804	5280703	3D stiffness matrix
CRANKSEG2	63838	7042510	3D stiffness matrix
M14B	214765	3358036	3D finite element mesh
OCEAN	143437	409593	3D finite element mesh
OILPAN	73752	1761718	3D stiffness matrix
ROTOR	99617	662431	3D finite element mesh
TOOTH	78136	452591	3D finite element mesh

3 Halo Minimum Degree Ordering

In the classical hybridizations of Nested Dissection and Minimum Degree, incomplete Nested Dissection is performed on the original graph, and the subgraphs at the leaves of the separation tree are ordered using a minimum degree heuristic. The switching from Nested Dissection to Minimum Degree is controlled by a criterion which is typically a threshold on the size of the subgraphs. It is for instance the case for ON-METIS 4.0 [14], which switches from Nested Dissection to Multiple Minimum Degree at a threshold varying between 100 and 200 vertices, depending on the structure of the original graphs.

The problem in this approach is that the degrees of the boundary vertices of the subgraphs passed to the Minimum Degree algorithm are much smaller than they would be if the original graph were ordered using Minimum Degree only. Consequently, the Minimum Degree algorithm will tend to order boundary vertices first, creating a "skin" of low-degree vertices that will favor fill-in between separator vertices and interior vertices, and across neighbors of the boundary vertices.

To avoid such behavior, the Minimum Degree algorithm must work on subgraphs that have real vertex degrees on their boundary vertices, to compute good orderings of all of the subgraph vertices. In order to do that, our Nested Dissection algorithm adds a topological description of the immediate surroundings of the subgraph with respect to the original graph (see figure 1). This additional information, called *halo*, is then exploited by our modified Approximate Minimum Degree [1] algorithm to reoder subgraph nodes only. The assembly tree (that is, the tree of supervariables) of both the subgraph and halo nodes is then built, for future use (see section 4). We refer to this modified approximate minimum degree as **Halo-AMD**, or **HAMD**.

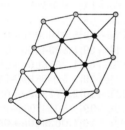 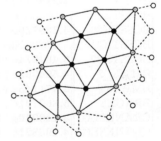

a. Without halo.   b. With halo.

Fig. 1. Knowledge of subgraph vertex degrees by the Minimum Degree algorithm without and with halo. Black vertices are subgraph internal vertices, grey vertices are subgraph boundary vertices, and white vertices are halo vertices, which are seen by the Halo Minimum Degree algorithm but do not belong to the subgraph.

Since it is likely that, in most cases (and especially for finite-element graphs), all of the vertices of a given separator will be linked by paths of smaller reordered numbers in the separated subsets, leading to completely filled diagonal blocks, we order separators using a breadth-first traversal of each separator subgraph from a pseudo-peripherial node [10], which proves to reduce the number of extra-diagonal blocks. Hendrickson and Rothberg report [12] that performing Minimum Degree to reorder last the subgraph induced by all of the separators found in the Nested Dissection process can lead to significant NNZ and OPC improvement, especially for stiffness matrices. This hybridization type could also be investigated using our HAMD algorithm, but might be restricted to sequential solving, since breaking the separator hierarchy is likely to reduce concurrency in the elimination tree.

To experiment with the HAMD algorithm, we have run our orderer on several test graphs (see table 1) with three distinct strategies. In the first one, we perform graph compression [3, 12] whenever necessary (referred to as CP), then multilevel nested dissection (ND) with a subgraph size threshold of 120 to match ON-METIS's threshold, and finally the Approximate Minimum Degree of [1] (AMD, see table 3). In the second strategy, we replace the AMD algorithm by our halo counterpart, HAMD (table 2). In the third one, AMD is used to reorder the whole graph (see table 4).

Table 2. Ordering results with CP+ND+HAMD, with threshold 120. Bold means best over tables 2 to 5.

Name	NNZ	OPC	SSb
144	**4.714472e+07**	**5.663156e+10**	1.961818e+06
598A	**2.604285e+07**	**1.877590e+10**	1.491780e+06
AATKEN	3.742874e+07	1.891992e+11	**6.657630e+05**
AUTO	**2.214524e+08**	**4.557806e+11**	6.515154e+06
BCSSTK29	**1.580533e+06**	**3.347719e+08**	3.683400e+04
BCSSTK30	**4.339720e+06**	**1.185672e+09**	**6.619000e+04**
BCSSTK31	4.389614e+06	1.234546e+09	**1.893440e+05**
BCSSTK32	**5.484974e+06**	1.292770e+09	**1.251660e+05**
BMW3_2	**4.564966e+07**	3.205625e+10	5.046610e+05
BMW7ST_1	**2.471148e+07**	1.133335e+10	**2.777360e+05**
BRACK2	**5.869791e+06**	**1.803800e+09**	8.194700e+05
CRANKSEG1	**3.148010e+07**	**3.004003e+10**	**1.062940e+05**
CRANKSEG2	**4.196821e+07**	4.598694e+10	**1.231160e+05**
M14B	**6.262970e+07**	**6.113363e+10**	2.926878e+06
OCEAN	2.047171e+07	1.301932e+10	2.093677e+06
OILPAN	9.470356e+06	3.333825e+09	**1.016020e+05**
ROTOR	**1.569899e+07**	**9.277418e+09**	1.342506e+06
TOOTH	**1.041380e+07**	**6.288746e+09**	1.026880e+06

Using HAMD instead of AMD for ordering the subgraphs reduces both the NNZ and the OPC (see tables 2 and 3). For a threshold value of 120, gains for our test graphs range from 0 to 10 percent, with an average of about 5 percent.

When the threshold value increases (see figures 2.a and figures 2.b), gains tend to increase, as more space is left for HAMD to outperform AMD, and then eventually decrease to zero since AMD and HAMD are equivalent on the whole graph. However, for most graphs in our collection, the NNZ and OPC computed by CP+ND+HAMD increase along with the threshold. Therefore, the threshold should be kept small.

Table 3. Ordering results with CP+ND+AMD, with threshold 120, compared to CP+ND+HAMD (table 2). Bold means better than CP+ND+HAMD.

Name	NNZ	%	OPC	%	SSb
144	4.768610e+07	1.15	5.672072e+10	0.16	**1.830258e+06**
598A	2.641262e+07	1.42	1.883037e+10	0.29	**1.439660e+06**
AATKEN	3.760711e+07	0.48	1.892114e+11	0.01	9.088100e+05
AUTO	2.228161e+08	0.62	4.560077e+11	0.05	**5.899395e+06**
BCSSTK29	1.705363e+06	7.90	3.559825e+08	6.34	**2.372500e+04**
BCSSTK30	4.714555e+06	8.64	1.295689e+09	9.28	6.655200e+04
BCSSTK31	4.676407e+06	6.53	1.291574e+09	4.62	1.897200e+05
BCSSTK32	5.958303e+06	8.63	1.410264e+09	9.09	1.265890e+05
BMW3_2	4.896745e+07	7.27	3.307174e+10	3.17	**5.039480e+05**
BMW7ST_1	2.687398e+07	8.75	1.200949e+10	5.97	2.866670e+05
BRACK2	6.056172e+06	3.18	1.822081e+09	1.01	**7.729910e+05**
CRANKSEG1	3.271575e+07	3.93	3.114353e+10	3.67	1.109120e+05
CRANKSEG2	4.372140e+07	4.18	4.760978e+10	3.53	1.277210e+05
M14B	6.349348e+07	1.38	6.128062e+10	0.24	**2.709220e+06**
OCEAN	2.072061e+07	1.22	1.304039e+10	0.16	**1.918925e+06**
OILPAN	1.041493e+07	9.97	3.543230e+09	6.28	1.085550e+05
ROTOR	1.608042e+07	2.43	9.322391e+09	0.48	**1.338132e+06**
TOOTH	1.064755e+07	2.24	6.311166e+09	0.36	**9.697280e+05**

Table 4. Ordering results with (H)AMD alone, compared to CP+ND+HAMD (table 2). Bold means better than CP+ND+HAMD.

Name	NNZ	%	OPC	%	SSb
144	9.393420e+07	99.25	2.406120e+11	324.87	2.963011e+06
598A	4.592155e+07	76.33	6.224629e+10	231.52	2.293154e+06
AATKEN	**2.589400e+07**	-30.82	**1.202562e+11**	-36.44	9.855260e+05
AUTO	5.386292e+08	143.23	2.701255e+12	492.67	1.044489e+07
BCSSTK29	1.786234e+06	13.01	4.714500e+08	40.83	3.819000e+04
BCSSTK30	**3.582124e+06**	-11.24	**9.446337e+08**	-20.33	7.488900e+04
BCSSTK31	5.568769e+06	26.86	2.909362e+09	135.66	2.123520e+05
BCSSTK32	**4.989134e+06**	-9.04	**9.535496e+08**	-26.24	1.404360e+05
BMW3_2	5.071095e+07	11.09	4.721008e+10	47.27	5.406270e+05
BMW7ST_1	2.613003e+07	5.74	1.466054e+10	29.36	2.983190e+05
BRACK2	7.286548e+06	6.47	3.057435e+09	69.50	8.804010e+05
CRANKSEG1	4.036152e+07	24.14	5.256665e+10	74.99	1.396250e+05
CRANKSEG2	5.866411e+07	39.78	9.587950e+10	108.49	1.685880e+05
M14B	1.105011e+08	76.44	1.880238e+11	207.56	4.030059e+06
OCEAN	3.306994e+07	61.54	3.557219e+10	173.23	2.322347e+06
OILPAN	1.013450e+07	7.01	4.183709e+09	25.49	1.104820e+05
ROTOR	2.475696e+07	57.70	2.491224e+10	168.53	1.615861e+06
TOOTH	1.613290e+07	54.92	1.810155e+10	187.84	1.136381e+06

Table 5. Ordering results with ON-METIS 4.0, compared to CP+ND+HAMD (table 2). Bold means better than CP+ND+HAMD.

Name	NNZ	%	OPC	%	SSc
144	5.006260e+07	6.19	6.288269e+10	11.04	3.020012e+06
598A	2.670144e+07	2.53	1.945231e+10	3.60	2.119415e+06
AATKEN	**3.297272e+07**	-11.91	**1.545334e+11**	-18.32	5.789739e+06
AUTO	2.358388e+08	6.50	5.140633e+11	12.79	1.023613e+07
BCSSTK29	1.699841e+06	7.55	3.502626e+08	4.63	1.753670e+05
BCSSTK30	4.530240e+06	4.39	1.216158e+09	2.57	3.120820e+05
BCSSTK31	**4.386990e+06**	-0.06	**1.178024e+09**	-4.58	4.363570e+05
BCSSTK32	5.605579e+06	2.20	**1.230208e+09**	-4.84	5.073400e+05
BMW3_2	4.593307e+07	0.62	**2.815020e+10**	-12.18	2.619580e+06
BMW7ST_1	2.558782e+07	3.55	**1.097226e+10**	-3.19	1.600835e+06
BRACK2	6.025575e+06	2.65	1.851941e+09	2.67	9.413400e+05
CRANKSEG1	3.343857e+07	6.22	3.284621e+10	9.34	7.667890e+05
CRANKSEG2	4.320028e+07	2.94	**4.521963e+10**	-1.67	9.279760e+05
M14B	6.652286e+07	6.22	6.791640e+10	11.09	4.399879e+06
OCEAN	**1.950496e+07**	-4.72	**1.178200e+10**	-9.50	2.247596e+06
OILPAN	**9.064734e+06**	-4.28	**2.750608e+09**	-17.49	6.773740e+05
ROTOR	1.627034e+07	3.64	**9.460553e+09**	1.97	1.702922e+06
TOOTH	1.094157e+07	5.07	7.153447e+09	13.75	1.247352e+06

Two classes of graphs arise in all of these experiments. The first one, containing finite-element-like meshes, is very stable with respect to threshold (see figure 2.a). The second one, which includes stiffness matrices, has a very unstable behavior for small threshold values (see figure 2.b). However, for every given threshold, the ordering produced by HAMD always outperforms the ordering produced by AMD. Due to the small threshold used, the orderings computed by our CP+ND+HAMD algorithm are only slightly better on average than the ones computed by ON-METIS 4.0, although the latter does not use halo information (see table 5). This is especially true for stiffness matrices.

These experiments also confirm the current overall superiority of hybrid ordering methods against pure Minimum Degree methods (see table 4); concurring results have been obtained by other authors [11, 12, 13]. As a concluding remark, let us note that using block secondary storage almost always leads to much smaller secondary structures than when using column compressed storage (see tables 2 and 5).

4 Amalgamating supervariables

In order to perform efficient factorization, BLAS3-based block computations must be used with blocks of sufficient sizes. Therefore, it is necessary to amalgamate columns so as to obtain large enough blocks while keeping fill-in low.

Since, in order for the HAMD algorithm to behave well, the switching between the ND and HAMD methods must occur when subgraphs are large enough, the

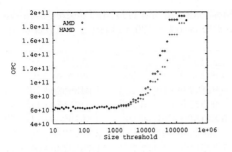 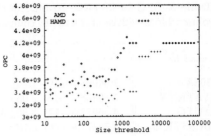

a. For graph M14B.   b. For graph OILPAN.

Fig. 2. Values of OPC obtained with CP+ND+AMD and CP+ND+HAMD for various threshold values, for graphs M14B and OILPAN.

amalgamation process will only take place for the columns ordered by HAMD, as separators are of sufficient sizes. The amalgamation criterion that we have chosen is based on three parameters cmin, cmax, and frat: a leaf of the assembly tree is merged to its father either if it has less columns than some threshold value cmin or if it induces less fill-in than a user-defined fraction of the surface of the merged block frat, provided that the merged block has less columns than some threshold value cmax.

The partition of the original graph into supervariables is achieved by merging the partition of the separators and the partition of the supervariables amalgamated in each subgraph. One can therefore perform an efficient block symbolic factorization in quasi-linear space and time complexities (the linearity is proven in the case of complete Nested Dissection [5]).

To show the impact of supervariable amalgamation, we have run our orderer with strategy CP+ND+HAMD, for size thresholds of 120 and 1111 (chosen to be about ten times the first one), respectively with no amalgamation at all (cmax = 0) and with a minimum desired column block size of cmin = 16; OPC results are summarized in table 6.

The amalgamation process induces a slight increase of NNZ and OPC (less than 15 and 5 percent respectively on average for large graphs), with the benefit of a dramatic reduction of the size of the secondary storage (of more than 70 percent on average). This reduction is not important in terms of memory occupation, since the amount of secondary storage is more than one order of magnitude less than NNZ, but it shows well that efficient amalgamation can be performed only locally on the subgraphs, at a cost much lower than if global amalgamation were performed in a post-processing phase.

The impact of amalgamation on NNZ and OPC is equivalent for the two threshold values, i.e. for column block sizes of about 1 to 10 percent of the subgraph sizes. This leaves some freedom to set amalgamation parameters according to the desired BLAS effects, independently of the Nested Dissection threshold.

Table 6. OPC for CP+ND+HAMD, for thresholds 120 and 1111, and desired minimum column block widths of 0 and 16, compared to case $t = 120$ and $cmax = 0$ (table 3).

Name	t=120, cmin=16	%	t=1111, cmax=0	%	t=1111, cmin=16	%
144	5.697046e+10	0.60	5.861587e+10	3.50	5.898574e+10	4.16
598A	1.900982e+10	1.25	1.918044e+10	2.15	1.942980e+10	3.48
AATKEN	1.892588e+11	0.03	1.905714e+11	0.73	1.910815e+11	0.99
AUTO	4.569721e+11	0.36	**4.549330e+11**	-0.19	4.561968e+11	0.09
BCSSTK29	3.554720e+08	6.18	3.404095e+08	1.68	3.591577e+08	7.28
BCSSTK30	1.478795e+09	24.72	1.396119e+09	17.75	1.704818e+09	43.78
BCSSTK31	1.434832e+09	16.22	1.400022e+09	13.40	1.619635e+09	31.19
BCSSTK32	1.787947e+09	38.30	1.353213e+09	4.68	1.884575e+09	45.78
BMW3_2	3.639430e+10	13.53	3.110278e+10	-2.97	3.565307e+10	11.22
BMW7ST_1	1.407066e+10	24.15	1.201388e+10	6.00	1.495933e+10	31.99
BRACK2	1.888913e+09	4.72	1.931003e+09	7.05	2.024382e+09	12.23
CRANKSEG1	3.293365e+10	9.63	4.055242e+10	34.99	4.407524e+10	46.72
CRANKSEG2	5.046633e+10	9.74	6.102002e+10	32.69	6.636002e+10	44.30
M14B	6.164424e+10	0.84	6.214652e+10	1.66	6.270303e+10	2.57
OCEAN	1.321423e+10	1.50	**1.271274e+10**	-2.35	1.295883e+10	-0.46
OILPAN	5.130055e+09	53.88	3.431957e+09	2.94	5.380494e+09	61.39
ROTOR	9.440305e+09	1.76	9.893124e+09	6.64	1.006988e+10	8.54
TOOTH	6.388041e+09	1.58	6.339282e+09	0.80	6.461720e+09	2.75

5 Conclusion and perspectives

In this paper, we have presented a tight coupling of the Nested Dissection and Approximate Minimum Degree algorithms which always improves the quality of the produced orderings.

Table 7. Elapsed time (in seconds) for the numerical factorization phase of the MUMPS code for matrix OILPAN. Uniprocessor timings are based on CPU time.

Procs	AMD	CP+ND+AMD	CP+ND+HAMD
1	41	40	37
8	9.9	8.1	6.9
16	7.6	5.5	4.9
24	7.2	4.9	4.4

This work is still in progress, and finding exactly the right criterion to switch between the two methods is still an open question. As a matter of fact, we have evidenced that performing Nested Dissection too far limits the benefits of the Halo-AMD algorithm. On the opposite, switching too early limits the potentiality of concurrency in the elimination trees.

As a concluding remark, we present in table 7 the influence of the ordering scheme on the performance of the numerical factorization. The multifrontal code MUMPS (see [2], partially supported by the PARASOL project, EU ESPRIT IV LTR project 20160) was used to obtain these results on an IBM-SP2. Our ordering scheme provides well balanced assembly trees (the maximum speed-ups obtained with AMD only and with CP+ND+HAMD are respectively 5.69 and 8.41).

References

[1] P. Amestoy, T. Davis, and I. Duff. An approximate minimum degree ordering algorithm. *SIAM J. Matrix Anal. and Appl.*, 17:886–905, 1996.
[2] P. Amestoy, I. Duff, and J.-Y. L'Excellent. Multifrontal parallel distributed symmetric and unsymmetric solvers. *to appear in special issue of Comput. Methods in Appl. Mech. Eng. on domain decomposition and parallel computing*, 1998.
[3] C. Ashcraft. Compressed graphs and the minimum degree algorithm. *SIAM J. Sci. Comput.*, 16(6):1404–1411, 1995.
[4] C. Ashcraft, S. Eisenstat, J. W.-H. Liu, and A. Sherman. A comparison of three column based distributed sparse factorization schemes. In *Proc. Fifth SIAM Conf. on Parallel Processing for Scientific Computing*, 1991.
[5] P. Charrier and J. Roman. Algorithmique et calculs de complexité pour un solveur de type dissections emboîtées. *Numerische Mathematik*, 55:463–476, 1989.
[6] G. A. Geist and E. G.-Y. Ng. Task scheduling for parallel sparse Cholesky factorization. *International Journal of Parallel Programming*, 18(4):291–314, 1989.
[7] A. George, M. T. Heath, J. W.-H. Liu, and E. G.-Y. Ng. Sparse Cholesky factorization on a local memory multiprocessor. *SIAM Journal on Scientific and Statistical Computing*, 9:327–340, 1988.
[8] A. George and J. W.-H. Liu. *Computer solution of large sparse positive definite systems*. Prentice Hall, 1981.
[9] A. George and J. W.-H. Liu. The evolution of the minimum degree ordering algorithm. *SIAM Review*, 31:1–19, 1989.
[10] N. E. Gibbs, W. G. Poole, and P. K. Stockmeyer. A comparison of several bandwidth and profile reduction algorithms. *ACM Trans. Math. Soft.*, 2:322–330, 1976.
[11] A. Gupta, G. Karypis, and V. Kumar. Scalable parallel algorithms for sparse linear systems. In *Proc. Stratagem'96, Sophia-Antipolis*, pages 97–110, July 1996.
[12] B. Hendrickson and E. Rothberg. Improving the runtime and quality of nested dissection ordering. *SIAM J. Sci. Comput.*, 20(2):468–489, 1998.
[13] G. Karypis and V. Kumar. A fast and high quality multilevel scheme for partitioning irregular graphs. TR 95-035, University of Minnesota, June 1995.
[14] G. Karypis and V. Kumar. METIS – *A Software Package for Partitioning Unstructured Graphs, Partitioning Meshes, and Computing Fill-Reducing Orderings of Sparse Matrices – Version 4.0*. University of Minnesota, September 1998.
[15] R. J. Lipton, D. J. Rose, and R. E. Tarjan. Generalized nested dissection. *SIAM Journal of Numerical Analysis*, 16(2):346–358, April 1979.
[16] J. W.-H. Liu. Modification of the minimum-degree algorithm by multiple elimination. *ACM Trans. Math. Software*, 11(2):141–153, 1985.
[17] F. Pellegrini and J. Roman. Sparse matrix ordering with SCOTCH. In *Proceedings of HPCN'97, Vienna, LNCS 1225*, pages 370–378, April 1997.
[18] R. Schreiber. Scalability of sparse direct solvers. Technical Report TR 92.13, RIACS, NASA Ames Research Center, May 1992.
[19] W. F. Tinney and J. W. Walker. Direct solutions of sparse network equations by optimally ordered triangular factorization. *J. Proc. IEEE*, 55:1801–1809, 1967.

ParaPART: Parallel Mesh Partitioning Tool for Distributed Systems

Jian Chen   Valerie E. Taylor
Department of Electrical and Computer Engineering
Northwestern University
Evanston, IL 60208
{jchen, taylor}@ece.nwu.edu

Abstract. In this paper, we present ParaPART, a parallel version of a mesh partitioning tool, called PART, for distributed systems. PART takes into consideration the heterogeneities in processor performance, network performance and application computational complexities to achieve a balanced estimate of execution time across the processors in the distributed system. Simulated annealing is used in PART to perform the backtracking search for desired partitions. ParaPART significantly improves performance of PART by using the asynchronous multiple Markov chain approach of parallel simulated annealing. ParaPART is used to partition six irregular meshes into 8, 16, and 100 subdomains using up to 64 client processors on an IBM SP2 machine. The results show superlinear speedup in most cases and nearly perfect speedup for the rest. Using the partitions from ParaPART, we ran an explicit, 2-D finite element code on two geographically distributed IBM SP machines. Results indicate that ParaPART produces results consistent with PART. The execution time was reduced by 12% as compared with partitions that consider only processor performance; this is significant given the theoretical upper bound of 15% reduction.

1 Introduction

High performance distributed computing is rapidly becoming a reality. Nationwide high speed networks such as vBNS [16] are becoming widely available to interconnect high speed computers at different sites. Projects such as Globus [9] and Legion [13] are developing software infrastructure for computations that integrate distributed computational and informational resources. Further, a National Technology Grid is being constructed to provide access to advanced computational capabilities regardless of the location of the users and resources [10]. The distributed nature of this kind of computing environment calls for consideration of heterogeneities in computational and communication resources when partitioning meshes. In this paper, we present ParaPART, a parallel version of a mesh partitioning tool, called PART, for distributed systems.

In [5], we presented an automatic mesh partitioning tool, PART, for distributed systems. PART takes into consideration the heterogeneities in processor performance, network performance and application computational complexities to achieve a balanced estimate of execution time across the processors in a distributed system. We addressed four major issues on mesh partitioning for distributed systems in [6] and demonstrated significant increase in efficiency by using partitions from PART. In this paper, we present ParaPART, a parallel version of PART, which significantly improves performance. Simulated annealing is used in PART to perform the backtracking search for desired partitions. However, it is well known that simulated annealing is computationally intensive. In ParaPART, we use the asynchronous multiple Markov chain approach of parallel simulated annealing [14] for which the simulated annealing process is considered a Markov chain. A server processor keeps track of the best solution and each client processor performs an independent simulated annealing with a different seed. Periodically, the client processors exchange solutions with the server and either continues with its own solution or with a better solution obtained from the server.

ParaPART is used to partition six irregular meshes into 8, 16, and 100 subdomains using up to 64 client processors on an IBM SP2 machine. The results show superlinear speedup in most cases and nearly perfect speedup for the rest. Using mesh partitions from ParaPART, we ran an explicit, 2-D finite element code on two geographically distributed IBM SP machines. We used Globus [9] software for communication between the two SPs. The results indicate that ParaPART produces partitions consistent with PART. The execution time was reduced by 12% as compared with partitions that consider only processor performance; this is significant given the theoretical upper bound of 15% reduction.

The remainder of the paper is organized as follows: Section 2 is related work. Section 3 describes ParaPART. Section 4 is experimental results. Section 5 gives conclusion.

2 Related Work

Currently, there are five widely used mesh partitioning tools; the tools are METIS [17, 18], HARP [26], CHACO [15], TOP/DOMDEC [25], and JOSTLE [28]. These tools use one or more of the following partitioning algorithms: Greedy [8], Kernighan-Lin [19], Recursive Coordinate Bisection [2], Recursive Spectral Bisection [24], Recursive Inertia Partition [23], Recursive Graph Bisection [11], and Simulated Annealing. All of the aforementioned tools have the goal of generating equal partitions with minimum cuts. Some tools, such as METIS, considers processor performance heterogeneity with an easy modification to the input file. To achieve good performance on a distributed system, however, the heterogeneities in both network and processor performance must be taken into consideration. In addition, computational complexity of the application as well as difference element types in the mesh must also be considered. PART differs from existing tools in that it takes into consideration all of these heterogeneities.

PART uses simulated annealing to partition the mesh. Figure 1 shows the serial version of simulated annealing algorithm. This algorithm uses the Metropolis criteria (line 8 to 13 in Figure 1) to accept or reject moves. The moves that reduce the cost function are accepted; the moves that increase the cost function

```
1. Get an initial solution S
2. Get an initial temperature T > 0
3. While stopping criteria not met {
4.    M = number of moves per temperature
5.    for m = 1 to M {
6.       Generate a random move
7.       Evaluate changes in cost function: ΔE
8        if (ΔE < 0) {
9           accept this move, and update solution S
10.      } else {
11.         accept with probability P = e^(-ΔE/T)
12.         update solution S if accepted
13.      }
14.   } /*end for loop*/
15.   Set T = rT (reduce temperature)
16.} /*end while loop */
17.Return S
```

Fig. 1. Simulated annealing.

may be accepted with probability $e^{-\frac{\Delta E}{T}}$, thereby avoiding being trapped in local minima. This probability decreases when the temperature is lowered. Simulated Annealing is computationally intensive, therefore, a parallel version of simulated annealing is used for ParaPART. There are three major classes of parallel simulated annealing [12]: serial-like [20, 27], parallel moves [1], and multiple Markov chains [3, 14, 22]. Serial like algorithms essentially break up each move into subtasks and parallelize the subtasks (parallelizing line 6 and 7 in Figure 1). For the parallel moves algorithms, each processor generates and evaluates moves independently; cost function calculation may be inaccurate since processors are not aware of moves by other processors. Periodic updates are normally used to address the effect of cost function error. Parallel moves algorithms essentially parallelize the for loop (line 5 to 14) in Figure 1. For the multiple Markov chains algorithm, multiple simulated annealing processes are started on various processors with *different* random seeds. Processors periodically exchange solutions and the best is selected and given to all the processors to continue their annealing processes. In [3], multiple Markov chain approach was shown to be most effective for VLSI cell placement. For this reason, ParaPART uses the multiple Markov chain approach. The details of the use of the multiple Markov chain approach is described below.

3 ParaPART Description

ParaPART is a parallel version of PART, which entails three steps. The first step generates a coarse partitioning for the distributed systems. Each group gets a subdomain that is proportional to its number of processors, the performance of the processors, and the computational complexity of the application. A group is a collection of processors with the same performance and interconnected via a local network. Hence computational cost is balanced across all the groups. In the second step, the subdomain that is assigned to each group from step 1 is partitioned among its processors. This coarsening step is performed using METIS. Within each group, parallel simulated annealing is used to balance the execution time by considering variance in network performance. Processors that communicate remotely with processors in other groups will have reduced computational load to compensate for the longer communication time. The third step addresses the global optimization, taking into consideration differences in the local interconnect performance of various groups. Again, the goal is to minimize the variance of the execution time across all processors. In this step, elements on the boundaries of group partitions are moved according to the execution time variance between neighboring groups. For the case when a significant number of elements is moved between the groups in step 3, the second step is executed again to equalize the execution time in a group given the new computational load. For the experiments conducted thus far, step 3 has not been needed. A more detailed description of PART is in [5].

ParaPART replaces the simulated annealing in PART with the asynchronous multiple Markov chain approach of parallel simulated annealing [14]. Given P processors, a straightforward implementation of the multiple Markov chain approach would be initiating simulated annealing on each of the P processors with a different seed. Each processor performs moves independently and then finally the best solution from those computed by all processors is selected. In this approach, however, simulated annealing is essentially performed P times which may result a better solution but not speedup.

To achieve speedup, P processors perform an independent simulated annealing with a different seed, but each processor performs only N/P moves (N is the number of moves performed by the simulated annealing at each temperature). Processors exchange solutions at the end of each temperature. The exchange of data occurs synchronously or asynchronously. In the synchronous multiple Markov chain approach, the processors periodically exchange solutions with each other. In the asynchronous approach, the client processors exchange solutions with a server processor. It has been reported that the synchronous approach is more easily trapped in a local optima than the asynchronous [14], therefore ParaPART uses the asynchronous approach. During solution exchange, if the client solution is better, the server processor is updated with the better solution; if the server solution is better, the client gets updated with the better solution and continue from there. Processor exchanges solution with the server processor at the end of each temperature.

To ensure that each subdomain is connected, we check for disconnected components at the end of ParaPART. If any subdomain has disconnected components, the parallel simulated annealing is repeated with a different random seed. This process continues until there is no disconnected subdomains or the number

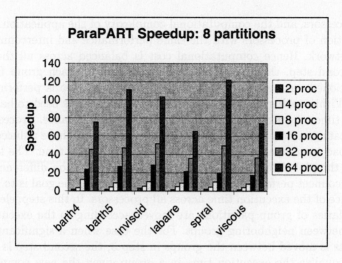

Fig. 2. ParaPART speedup for 8 partitions.

of trials exceed three times. A warning message is given in the output if there are disconnected subdomains.

4 Experimental Results

In this section, we present the results from two different experiments. The first experiment focuses on the speedup of ParaPART. The second experiment focuses on the quality of the partitions generated with ParaPART.

4.1 Speedup Results

Table 1. ParaPART execution time (seconds): 8 partitions.

# of proc.	barth4	barth5	inviscid	labarre	spiral	viscous
1 proc	158.4	244.4	248.0	163.3	183.0	214.6
2 proc	91.2	133.5	76.2	112.3	88.5	142.2
4 proc	44.4	54.7	46.6	32.4	41.2	53.7
8 proc	12.6	27.8	24.2	18.8	18.3	27.6
16 proc	6.7	9.1	8.8	7.6	9.3	12.6
32 proc	3.5	5.2	4.8	4.6	3.7	5.9
64 proc	2.1	2.2	2.4	2.5	1.5	2.9

ParaPART is used to partition six 2D irregular meshes with triangular elements: barth4 (11451 elem.), barth5 (30269 elem.), inviscid (6928 elem.), labarre (14971

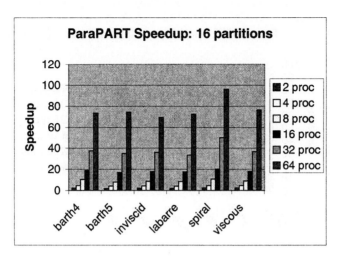

Fig. 3. ParaPART speedup for 16 partitions.

elem.), spiral (1992 elem.), and viscous (18369 elem.). The cost function of Para-PART is the estimate of the execution time of the WHAMS2D. The running time of using ParaPART to partition the six irregular meshes into 8, 16, and 100 subdomains are given in Table 1, 2, and 3, respectively. It is assumed that the subdomains will be executed on a distributed system consisting of two IBM SPs, with equal number of processors but different processor performance. Further, the machines are interconnected via vBNS for which the performance of the network is given in Table 4 (discussed in Section 4.2). In each table, column 1 is the number of client processors used by ParaPART, and columns 2 to 6 are the running time of ParaPART in seconds for the different meshes. The solution quality of using two or more client processors is within 5% of that of using one client processor. In this case, the solution quality is the estimate of the execution time of WHAMS2D.

Table 2. ParaPART execution time (seconds): 16 partitions.

# of proc.	barth4	barth5	inviscid	labarre	spiral	viscous
1 proc	772.4	781.6	729.6	717.0	615.9	749.6
2 proc	340.2	416.5	340.5	375.6	260.1	355.8
4 proc	168.4	183.5	168.5	176.3	135.1	160.9
8 proc	75.7	98.2	83.8	85.7	57.7	84.2
16 proc	40.7	46.0	41.0	40.8	30.7	41.7
32 proc	20.6	22.4	20.3	21.5	12.3	20.5
64 proc	10.5	10.6	10.6	9.9	6.4	9.8

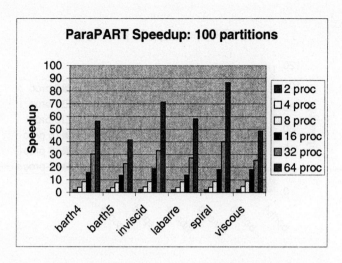

Fig. 4. ParaPART speedup for 100 partitions.

Table 3. ParaPART execution time (seconds): 100 partitions.

# of proc.	barth4	barth5	inviscid	labarre	spiral	viscous
1 proc	16212.5	17688.9	13759.2	16953.1	5433.4	18993.9
2 proc	8603.3	8997.9	6142.1	8626.6	2607.8	8260.7
4 proc	3982.2	4666.4	3082.3	4273.5	1304.4	3974.0
8 proc	1910.5	2318.6	1610.7	2007.8	630.3	2140.2
16 proc	1025.9	1310.7	728.1	1234.7	299.9	1056.4
32 proc	535.4	777.9	415.6	620.8	135.5	744.3
64 proc	288.3	426.3	192.8	291.2	62.7	391.7

Figure 2, 3, and 4 are graphical representations of the speedup of ParaPART relative to one client processor. The figures show that when the meshes are partitioned into 8 or 16 subdomains, superlinear speedup occurs in all cases when the number of client processors is more than 4. When the meshes are partitioned into 100 subdomains, superlinear speedup occurs only in the cases of two smallest meshes, spiral and inviscid. Other cases show slightly less than perfect speedup. This superlinear speedup is attributed to the use of multiple client processors conducting a search, for which all the processors benefit from the results. Once a good solution is found by any one of the clients, this information is given to other clients quickly, thereby reducing the effort of continuing to search for a solution. The superlinear speedup results are consistent with that reported in [21].

4.2 Quality of Partition

Our testbed includes two IBM SP machines, one located at Argonne National Laboratory (ANL) and the other located at the San Diego Supercomputing Center (SDSC). These two machines are connected by vBNS (very high speed Backbone Network Service). We used Globus [9] software for communication between the two SPs. Macro benchmarks were used to determine the network and processor performance. The results of the network performance analysis are given in Table 4. Further, experiments were conducted to determine that the SDSC SP processors nodes were 1.6 times as fast as the ANL ones. Our target application is an explicit, 2-D finite element code, WHAMS2D [4]. The computational complexity of WHAMS2D is linear with the size of the problem.

Table 4. Values of α and β for the different networks.

ANL SP Vulcan Switch	$\alpha_1 = 0.00069s$	$\beta_1 = 0.000093s/KB$
SDSC SP Vulcan Switch	$\alpha_2 = 0.00068s$	$\beta_2 = 0.000090s/KB$
vBNS	$\alpha_3 = 0.059s$	$\beta_3 = 0.0023\ s/KB$

Using the partitions generated by ParaPART as input, the WHAMS2D application is executed on the ANL and SDSC IBM SPs. We used 4 processors from ANL SP and 4 processor from SDSC SP due to limitations of coscheduling of resources on the two sites. The execution of WHAMS2D is given in Table 5. Column one identifies the irregular meshes. Column two is the execution time resulting from the partitions from ParaPART. Column three is the execution time resulting from METIS which takes computing power into consideration (each processor's compute power is used as one of the inputs to the METIS program). The results in Table 5 show that the execution time is reduced by up to 12% as compared to METIS; this is consistent with partitions generated with PART. The reduction of 12% is significant given that we derived in [7] that the best reduction by taking network performance into consideration is 15%. This reduction comes from the fact that even on a high speed network such as the vBNS, the message start up cost on remote and local communication still has a large difference.

5 Conclusion

In this paper, we demonstrate the use of a parallel mesh partitioning tool, ParaPART, for distributed systems. The novel part of ParaPART is that it uses the asynchronous multiple Markov chain approach of parallel simulated annealing for mesh partitioning. ParaPART is used to partition six irregular meshes into 8, 16, and 100 subdomains using up to 64 client processors on an IBM SP2 machine.

Table 5. Execution time (seconds) using the vBNS on 8 processors: 4 at ANL, 4 at SDSC.

Mesh	ParaPART	Proc. Perf. (METIS)
barth5	242.0	266.0
viscous	150.0	172.0
labarre	138.0	146.0
barth4	111.0	120.0

Results show superlinear speedup in most cases and nearly perfect speedup for the rest.

We used Globus software to run an explicit, 2-D finite element code using mesh partitions from ParaPART. Our testbed includes two geographically distributed IBM SP machines. Experimental results indicate up to 12% reduction in execution time compared with using partitions that only take processor performance into consideration; this is consistent with the results from using PART. This reduction comes from the fact that even on a high speed network such as the vBNS, the message start up cost on remote and local communication still has a large difference.

References

1. P. Banerjee, M. H. Jones, and J. S. Sargent. Parallel simulated annealing algorithms for cell placement on hypercube multiprocessors. *IEEE Trans. on Parallel and Distributed Systems*, 1(1), January 1990.
2. M. J. Berger and S. H. Bokhari. A partitioning strategy for nonuniform problems on multiprocessors. *IEEE Trans. Comput.*, C-36:570–580, May 1987.
3. J. A. Chandy, S. Kim, B. Ramkumar, S. Parkes, and P. Banerjee. An evaluation of parallel simulated annealing strategies with application to standard cell placement. *IEEE Trans. on Comp. Aid. Design of Int. Cir. and Sys.*, 16(4), April 1997.
4. H. C. Chen, H. Gao, and S. Sarma. WHAMS3D project progress report PR-2. Technical Report 1112, University of Illinois CSRD, 1991.
5. J. Chen and V. E. Taylor. Part: A partitioning tool for efficient use of distributed systems. In *Proceedings of the 11th International Conference on Application Specific Systems, Architectures and Processors*, pages 328–337, Zurich, Switzerland, July 1997.
6. J. Chen and V. E. Taylor. Mesh partitioning for distributed systems. In *Proceedings of the Seventh IEEE International Symposium on High Performance Distributed Computing*, Chicago, IL, July 1998.
7. J. Chen and V. E. Taylor. Mesh partitioning for distributed systems: Exploring optimal number of partitions with local and remote communication. In *Proceedings of Ninth SIAM Conference on Parallel Processing for Scientic Computing*, San Antonio, TX, March 22-24 1999. to appear.
8. C. Farhat. A simple and efficient automatic fem domain decomposer. *Computers and Structures*, 28(5):579–602, 1988.

9. I. Foster, J. Geisler, W. Gropp, N. Karonis, E. Lusk, G. Thiruvathukal, and S. Tuecke. A wide-area implementation of the Message Passing Interface. *Parallel Computing*, 1998. to appear.
10. I. Foster and C. Kesselman. Computational grids. In I. Foster and C. Kesselman, editors, *The Grid: Blueprint for a New Computing Infrastructure*, pages 15–52. Morgan Kaufmann, San Francisco, California, 1986.
11. A. George and J. Liu. *Computer Solution of Large Sparse Positive Definite Systems*. Prentice-Hall, Englewood Cliffs, New Jersey, 1981.
12. D. R. Greening. Parallel simulated annealing techniques. *Physica*, D(42):293–306, 1990.
13. A. S. Grimshaw, W. A. Wulf, and the Legion team. The legion vision of a worldwide virtual computer. *Communications of the ACM*, 40(1), January 1997.
14. G. Hasteer and P. Banerjee. Simulated annealing based parallel state assignment of finite state machines. *Journal of Parallel and Distributed Computing*, 43, 1997.
15. B. Hendrickson and R. Leland. The chaco user's guide. Technical Report SAND93-2339, Sandia National Laboratory, 1993.
16. J. Jamison and R. Wilder. vbns: The internet fast lane for research and education. *IEEE Communications Magazine*, January 1997.
17. G. Karypis and V. Kumar. A fast and high quality multilevel scheme for partitioning irregular graphs. Technical report, Department of Computer Science TR95-035, University of Minnesota, 1995.
18. G. Karypis and V. Kumar. Multilevel k-way partitioning scheme for irregular graphs. Technical report, Department of Computer Science TR95-064,University of Minnesota, 1995.
19. B. Kernighan and S. Lin. An efficient heuristic procedure for partitioning graphs. *Bell System Technical Journal*, 29:291–307, 1970.
20. S. A. Kravitz and R. A. Rutenbar. Placement by simulated annealing on a multiprocessor. *IEEE Trans. Comput. Aided Design*, CAD-6:534–549, July 1987.
21. V. Kumar, A. Grama, A. Gupta, and G. Karypis. *Introduction to Parallel Computing: Design and Analysis of Algorithms*. Benjamin/Cummings, 1994.
22. S.Y. Lee and K.G. Lee. Asynchronous communication of multiple markov chain in parallel simulated annealing. In *Proc. of Int. Conf. Parallel Processing*, volume III, pages 169–176, St. Charles, IL, August 1992.
23. B. Nour-Omid, A. Raefsky, and G. Lyzenga. Solving finite element equations on concurrent computers. In A. K Noor, editor, *Parallel Computations and Their Impact on Mechanics*, pages 209–227. ASME, New York, 1986.
24. H. D. Simon. Partitioning of unstructured problems for parallel processing. *Computing Systems in Engineering*, 2(2/3):135–148, 1991.
25. H. D. Simon and C. Farhat. Top/domdec: a software tool for mesh partitioning and parallel processing. Technical report, Report RNR-93-011, NASA, July 1993.
26. H. D. Simon, A. Sohn, and R. Biswas. Harp: A fast spectral partitioner. In *Proceedings of the Ninth ACM Symposium on Parallel Algorithms and Architectures*, Newport, Rhode Island, June 22-25 1997.
27. A. Sohn. Parallel n-ary speculative computation of simulated annealing. *IEEE Trans. on Parallel and Distributed Systems*, 6(10), October 1995.
28. C. Walshaw, M. Cross, S. Johnson, and M. Everett. *JOSTLE: Partitioning of Unstructured Meshes for Massively Parallel Machines*, In N. Satofuka et al, editor, *Parallel Computational Fluid Dynamics: New Algorithms and Applications*. Elsevier, Amsterdam, 1995.

Sparse Computations with PEI

Frédérique Voisin, Guy-René Perrin

ICPS, Université Louis Pasteur, Strasbourg,
Pôle API, Boulevard Sébastien Brant,
F-67400 Illkirch, France
voisin,perrin@icps.u-strasbg.fr

1 Introduction

A wide range of research work on the static analysis of programs forms the foundation of parallelization techniques which improve the efficiency of codes: loop nest rewriting, directives to the compiler to distribute data or operations, etc. These techniques are based on geometric transformations either of the iteration space or of the index domains of arrays.

This geometric approach entails an abstract manipulation of array indices, to define and transform the data dependences in the program. This requires to be able to express, compute and modify the placement of the data and operations in an abstract discrete reference domain. Then, the programming activity may refer to a very small set of primitive issues to construct, transform or compile programs. PEI is a program notation and includes such issues.

In this paper we focus on the way PEI deals with sparse computations. For any matrix operation defined by a program, a sparse computation is a transformed code which adapts computations efficiently in order to respect an optimal storage of sparse matrices [2]. Such a storage differs from the natural memory access since the location of any non-zero element of the matrix has to be determined. Due to the reference geometrical domain in PEI we show how a dense program can be transformed to meet some optimal memory storage.

2 Definition of the Formalism PEI

2.1 An Introduction to PEI Programming

The theory PEI was defined [8, 9] in order to provide a mathematical notation to describe and reason on programs and their implementation. A program can be specified as a relation between some *multisets* of value items, roughly speaking its *inputs* and *outputs*. Of course, programming may imply to put these items in a convenient organized directory, depending on the problem terms. In scientific computations for example, items such as arrays are functions on indices: the index set, that is the reference domain, is a part of some Z^n. In PEI such a multiset of value items mapped on a discrete reference domain is called a *data field*.

For example the multiset of integral items $\{1, -2, 3, 1\}$ can be expressed as a data field, say A, each element of which is recognized by an index in Z (e.g. from

0 to 3). Of course this multiset may be expressed as an other data field, say M, which places the items on points $(i,j) \in \mathbb{Z}^2$ such that $0 \le i,j < 2$. These two data fields A and M are considered equivalent in PEI since they express the same multiset. More formally, there exists a bijection from the first arrangement onto the second one, e.g. $\sigma(i) = (i \bmod 2, i \text{ div } 2)$. We note this by the equation:

where
$$\begin{array}{l} \texttt{M = align :: A} \\ \texttt{align} = \lambda i \mid (0 \le i \le 3) . (i \bmod 2, i \text{ div } 2) \end{array}$$

Any PEI program is composed of such unoriented equations. Here is an example of PEI program:

Example 1. For any i in $[1..n]$ compute $x_i = \sum_{j=1}^{j=n} a_{i,j} \times v_j$.

Any x_i can be defined by a recurrence on the domain $\{j \mid 1 \le j \le n\}$ of $\mathbb{Z}$ by:

$$\begin{cases} s_{i,1} = a_{i,1} \times v_1 & (1) \\ s_{i,j} = s_{i,j-1} + (a_{i,j} \times v_j) & 1 < j \le n \quad (2) \\ x_i = s_{i,n} & (3) \end{cases}$$

where the $s_{i,j}$ are intermediate results. The recurrence equation (2) emphasizes uniform dependencies $(0,1)$ for the calculation of $s_{i,j}$.

In PEI, this definition can be expressed in the following way. Matrix A is expressed as a data field A mapped on the domain $\{(i,j) \mid 1 \le i,j \le n\}$. Vectors v and x are expressed as data fields V and X mapped on $\{i \mid 1 \le i \le n\}$: this means that for any i, the values v_i and x_i are located at point i. Of course, any program which lies on another mapping for A, V or X defines an equivalent program, provided its operations result in an equivalent data field.

The recurrence defining the variable s suggests to map the data field S in $\mathbb{Z}^2$. Since the v_j are used by the $s_{i,j}$, for any i, the values of V are mapped in $\mathbb{Z}^2$ too, by a kind of data *Strip-Mining* on data field B. Its values are then broadcasted to *localize* the right values onto the right locations in order to compute the *recurrence* steps. Figure 1 shows a PEI program and its geometrical illustration. Each equation is commented below :

- the first equation defines the mapping of the input data field A. Function **matrix**, defined on $\mathbb{Z}^2$, applies a change of basis (notation ::) which is the identity on the domain $[1..n] \times [1..n]$. It means that the values are placed onto a square $[1..n] \times [1..n]$,
- the second equation implicitly defines B. The function **align**, defined on $\mathbb{Z}^2$, applies a change of basis which expresses that the projection of its argument B, supposed to be a row, on a one-dimensional domain, is equal to V. So, V is *aligned* on a two-dimensional domain. Notice that functions, such as **align**, define the context of the program: they are expressed as λ-expressions, in which the separator "|" allows to define the domain of the function, and the "." begins the function body,

- in the third equation, the operation /&/ builds the pairs of value items of A and B ◁ spread) which are placed at the same location: the last expression applies a *geometrical operation* (notation ◁) which broadcasts the values of the "first row" of data field B in the direction $(1,0)$ in Z^2,
- the data field S results from the application of a so-called *functional operation* (notation ▷) on two data fields. The first one results of a geometrical operation on S which expresses the dependency $(0,1)$ in Z^2,
- the last equation defines the data field X, by a change of basis: it projects the part of S mapped on $\{(i,n) \mid 1 \leq i \leq n\}$ onto Z. So, the result X is one-dimensional.

```
MatVec: (A,V) ↦ X {                      matrix  = λ(i,j) |(1≤i≤n,1≤j≤n).(i,j)
  matrix :: A = A                        project = λ(i,j) |(1≤i≤n,j=n).i
  align :: B = V                         shift   = λ(i,j) |(1≤i≤n,1<j≤n).(i,j-1)
  P = prod ▷ (A /&/ (B ◁ spread))        last    = λ(i,j) |(1≤i≤n,j=n).(i,j)
  S = add  ▷ (P /&/ (S ◁ shift))         spread  = λ(i,j) |(1≤i≤n,1≤j≤n).(1,j)
  X = project :: (S ◁ last)              add     = id # λ(a·b).a+b
}                                        prod    = λ(a·b).a×b
align = λ(i,j) |(i=1,1≤j≤n).j
```

align   spread   shift and add   last

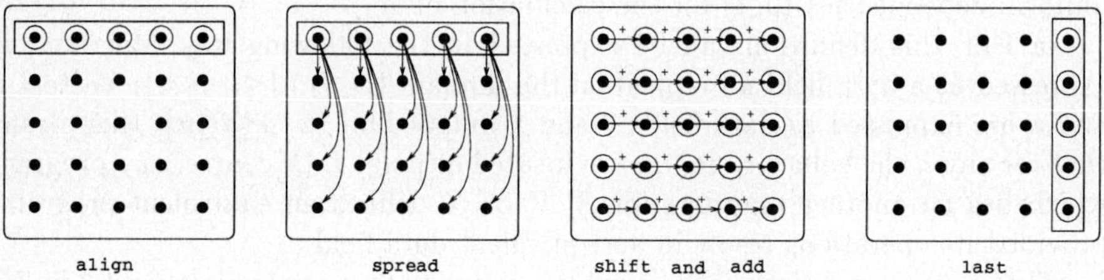

Fig. 1. Matrix-Vector multiplication program in PEI and its geometrical illustration

2.2 Syntactic Issues

The previous example allows to understand that data fields are main features in PEI. Besides data fields, partial functions define operations on these objects. In the following A, B, X, etc. denote data fields, whereas f, g, etc. denote functions. The PEI notation for partial functions is derived from the lambda-calculus: any function f of domain $dom(f) = \{x \mid P(x)\}$ is denoted as $\lambda x \mid P(x).f(x)$. Moreover a function f defined on disjunctive sub-domains is denoted as a partition $f_1 \# f_2$ of functions.

Expressions are defined by applying operations on data fields. PEI defines one internal operation on the data field set, called *superimposition*: it is denoted

as /&/ and builds the sequences of values of its arguments[1]. Three external operations are defined and associate a data field with a function:
- the *functional operation*, denoted as $\triangleright$, applies a function f (which may be a partial function) on value items of a data field X (notation f $\triangleright$ X),
- the *change of basis* (denoted as ::). Let h be a bijection, h::X defines a data field that maps the value items of X onto another discrete reference domain,
- the *geometrical operation*, denoted as $\triangleleft$, moves the value items of a data field on its domain: the data field X $\triangleleft$ g is such that the item mapped on some index z, $z \in dom(\mathtt{g})$, "comes from" X at index $\mathtt{g}(z)$.

2.3 Semantics and Program Transformations

The semantics of PEI is founded on the notion of discrete domain associated with a multiset. We call *drawing* of a multiset M of values in V, a partial function v from some Z^n in V whose image is M.

Such a drawing should define a natural functional interpretation $[\![X]\!] = v$ of a data field X. As we have told it at the beginning of this section, any other drawing can be deduced by applying any bijection, say h. As soon as its domain $dom(h)$ contains $dom(v)$, its interpretation should be also equal to $v \circ h^{-1}$. This is not sound and we have to consider a data field X as the abstraction of any drawing of a given multiset indeed. Here is the definition: a *data field* X is a pair $(v : \sigma)$, composed of a drawing v of a multiset M and of a bijection σ such that $dom(v) \subset dom(\sigma)$.

This definition induces a sound interpretation of a data field X = $(v : \sigma)$ as the function $[\![X]\!] = v \circ \sigma^{-1}$. It founds the semantics of operations on data fields.

The semantics allows to define transformations of programs in PEI, as soon as a data field is substituted for any equivalent one in an expression, or by applying some *algebraic law* on the operations. Here is a few examples of such laws (a detailed list is given in [10]). Assuming right and left expressions are valid expressions:

```
f1 ▷ (f2 ▷ X)     = (f1 ∘ f2) ▷ X        h::(X1/&/ X2) = (h::X1)/&/ (h::X2)
X ◁ (g1 ∘ g2)     = (X ◁ g1) ◁ g2        h::(f ▷ X)    = f ▷ (h::X)
(X1/&/ X2) ◁ g =  (X1 ◁ g)/&/ (X2 ◁ g)   h::(X ◁ g)    = (h::X) ◁ h ∘ g ∘ h⁻¹
```

3 PEI and Sparse Computations

PEI can deal with sparse computations by adapting a code in order to respect an optimal storage of sparse data structures. Such a storage differs from the natural memory access since the location of any non-zero element of the matrix has to be determined. In terms of PEI features this means that the optimal storage and the dense array are two equivalent data fields: the change of basis from one data field onto the other one defines the way the sparse matrix

---

[1] We use two operators on sequences: an associative constructor denoted as "." and the function "id" which is the identity on sequences of one element.

is stored. So, transforming the code consists in applying this change of basis to the dense program.

Let us consider again the example of a matrix-vector product, assuming the matrix is composed of a few bands parallel to the diagonal (see Fig. 2 for example, where non-zero elements are located at indices such that either $i = j+2$, or $i = j$, or $i = j - 2$).

The storage scheme we choose to use is MSR, a modified version of Compressed Sparse Row (CSR, described in [1]), that requires two arrays A and JA. A stores the non-zero values : first the diagonal values, then the other non-zero values as they appear in the matrix row by row. JA stores in its first positions the pointers to the beginning of each line in A, and then the column indices corresponding to the non-zero elements in A (see Fig. 2).

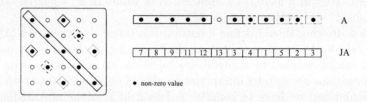

Fig. 2. Sparse matrix in dense representation and MSR storage

3.1 Program Transformation in PEI

The non-zero value items of the matrix and the vector value items must belong to the domain of the change of basis to be applied in PEI. Therefore, a first transformation step consists for example of aligning the vector with the main diagonal: MSR storage is then well-adapted. This alignment can be defined by rewriting spread as (see Fig. 3):

spread = pivot ∘ diagonal

where

pivot   = $\lambda(i,j)$ $|(1 \le i \le n, j=i) . (1,j)$
diagonal = $\lambda(i,j)$ $|(1 \le i \le n, 1 \le j \le n) . (j,j)$

we obtain the new program in PEI:

```
{
matrix :: A = A
align :: B = V
P = prod ▷ ( A /&/ ( (B ◁ pivot) ◁ diagonal ) )
...
}
```

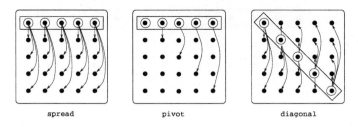

Fig. 3. Geometrical functions spread, pivot and diagonal

The change of basis itself (Fig. 2) is defined by the following function:

gather = $\lambda(i,j) \mid (1 \leq i \leq 2, 3 \leq j \leq 4, j=i+2) . (n+i+1)$
    # $\lambda(i,j) \mid (3 \leq i \leq n-2, 5 \leq j \leq n, j=i+2) . (n+2i-1)$
    # $\lambda(i,j) \mid (1 \leq i \leq n, 1 \leq j \leq n, i=j) . (i)$
    # $\lambda(i,j) \mid (3 \leq i \leq n-2, 1 \leq j \leq n-4, j=i-2) . (n+2i-2)$
    # $\lambda(i,j) \mid (n-1 \leq i \leq n, n-3 \leq j \leq n-2, j=i-2) .$

Its inverse is :

gather^{-1} = $\lambda(i) \mid (1 \leq i \leq n) . (i, i))$
    # $\lambda(i) \mid (n+2 \leq i \leq n+3) . (i-n-1, i-n+1)$
    # $\lambda(i) \mid (n+4 \leq i \leq 3n-5), (i-n) \bmod 2 = 0) .$
       $((i-n-1)/2+2), (i-n-1)/2)$
    # $\lambda(i) \mid (n+4 \leq i \leq 3n-5), (i-n) \bmod 2 = 1) .$
       $((i-n-1)/2+1), (i-n-1)/2+3)$
    # $\lambda(i) \mid (3n-4 \leq i \leq 3n-3) . (i-2n+3, i-2n+1)$

Applying the change of basis to P leads to this definition:

gather::P = prod ▷ (gather::A /&/ gather::((B ◁ pivot) ◁ diagonal))
     = prod ▷ (gather::A /&/
        (gather::(B ◁ pivot)) ◁ gather ∘ diagonal ∘ gather^{-1}))

Let

P' = gather::P
A' = gather::A
T' = gather::(B ◁ pivot)

We can write P' as P' = prod ▷ (A'/&/(T' ◁ spread)) where the new function spread, illustrated in Fig. 4 is defined below

spread = gather ∘ diagonal ∘ gather^{-1}
   = $\lambda(i) \mid (1 \leq i \leq n) . (i)$
    # $\lambda(i) \mid (n+2 \leq i \leq n+3) . (i-n+1)$
    # $\lambda(i) \mid (n+4 \leq i \leq 3n-5, (i-n) \bmod 2 = 0) . ((i-n-1)/2)$
    # $\lambda(i) \mid (n+4 \leq i \leq 3n-5, (i-n) \bmod 2 = 1) . ((i-n-1)/2+3)$
    # $\lambda(i) \mid (3n-4 \leq i \leq 3n-3) . (i-2n+1)$

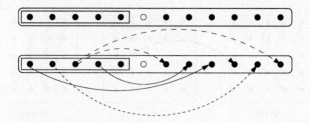

Fig. 4. New geometrical functions `align` and `spread`

In order to apply the change of basis `gather` in the equations, the definition of S is constrained on the domain of `gather` by geometrical operation `sparse` and folding:

$$\begin{aligned}
\texttt{sparse} = \quad & \lambda(i,j) \ | \ (3 \leq i \leq n, 1 \leq j \leq n, i=j+2) \\
\# \ & \lambda(i,j) \ | \ (1 \leq i \leq n, 1 \leq j \leq n, i=j) \\
\# \ & \lambda(i,j) \ | \ (1 \leq i \leq n-2, 1 \leq j \leq n, i=j-2)
\end{aligned}$$

S ◁ sparse = add ▷ (P ◁ sparse /&/ (S ◁ shift ∘ shift) ◁ sparse)
            = add ▷ (P ◁ sparse /&/ (S ◁ sparse) ◁ sparse_shift)

where

$$\begin{aligned}
\texttt{sparse\_shift} = \ & \texttt{shift} \circ \texttt{shift} \circ \texttt{sparse} \\
= \quad & \lambda(i,j) \ | \ (3 \leq i \leq n, 3 \leq j \leq n, i=j) \,.\, (i, j-2) \\
\# \ & \lambda(i,j) \ | \ (1 \leq i \leq n-2, 3 \leq j \leq n, i=j-2) \,.\, (i, j-2)
\end{aligned}$$

Applying the change of basis `gather` to the summation results in the new geometrical operation `shift` (see Fig. 5):

$$\begin{aligned}
\texttt{shift} = \ & \texttt{gather} \circ \texttt{sparse\_shift} \circ \texttt{gather}^{-1} \\
= \quad & \lambda(i) \ | \ (n+2 \leq i \leq n+3) \,.\, (i-n-1) \\
\# \ & \lambda(i) \ | \ (n+4 \leq i \leq 3n-5, (i-n) \bmod 2 = 1) \,.\, ((i-n-1)/2 + 1) \\
\# \ & \lambda(i) \ | \ (3 \leq i \leq n-2) \,.\, (n+2i-2) \\
\# \ & \lambda(i) \ | \ (n-1 \leq i \leq n) \,.\, (2n+i-3)
\end{aligned}$$

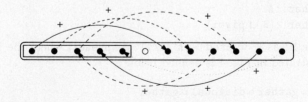

Fig. 5. New geometrical function `shift`

And

$$\begin{aligned}\text{\tt last} = \text{\tt gather} &\circ \text{\tt sparse\_last} \circ \text{\tt gather}^{-1}\\
= \quad &\lambda(i) \mid (n+1 \leq i \leq n+2)\\
\# \ &\lambda(i) \mid (n+3 \leq i \leq 3n-5, (i-n) \bmod 2 = 1)\\
\# \ &\lambda(i) \mid (n-1 \leq i \leq n)\end{aligned}$$

Fig. 6. New geometrical function last

where

$$\begin{aligned}\text{\tt sparse\_last} = \quad &\lambda(i,j) \mid (1 \leq i \leq n-2, j=i+2)\\
\# \ &\lambda(i,j) \mid (n-1 \leq i \leq n, j=i)\end{aligned}$$

This completes the program transformations in PEI, and defines a sparse solution. It can be implemented in terms of classical arrays in MSR storage.

3.2 Concrete Implementations

PEI expression leads easily the definition of the arrays in MSR storage. For example, let $\text{\tt second} = \lambda(i,j).(j)$, we clearly have :

$$JA(i) = \text{\tt second} \circ \text{\tt gather}^{-1}(i) \text{ for } i > n+1$$

Note that PEI does not require to implement the first $n+1$ elements of JA because references to rows are useless since the vector is explicitly aligned with the computation points by the change of basis.

4 Conclusion

The presented technique shows the interest of using PEI for program transformation to address sparse computations. Our approach is similar to Bik and Wijshoff's [3]: we both obtain the sparse formulation by taking the dense code and restricting the computation to the non-zero elements. However, PEI allows to express all kind of storages. Sparse General Pattern [7], Compress Diagonal Storage, Jagged Diagonal Storage or other storages described in [1] can be considered by applying convenient change of basis in PEI. The example considered here is based on a tridiagonal matrix which is a special case of sparse matrix. Unfortunately sparse matrices do not always have this special shape and the main difficulty is then to define the functions. It can be achieved by reading the matrices for example from a Harwell-Boeing formatted file [4]. In fact, the non-zero elements of the matrix can be stored in any chosen order; the vector alignment with the matrix is then done through the application of the change of basis to the whole dense program. It provides a theoretical framework in which the changes applied to the dense program are proven to be correct by composition of functions using formal calculus, independent from routines written in a given language.

We can compare our approach to Pingali's approach [6] where sparse data structures are viewed as database relations. The evaluation of relational queries provides a way to select the non-zero elements and to align the data so that computation is performed on the selected points. In PEI, these operations are expressed by functions.

At last, PEI aims at providing parallel code. Instead of scanning the sparse data-structure to perform computation, we consider it as a parallel data and we perform parallel computation after proper data alignment has been done. Therefore, PEI can be used in order to build a sparse computation library in the data parallel programming model, for example expressed in HPF. The definition of the library requires the definition of a PEI to HPF translator. Such a software tool is still under development [5].

References

1. R. Barrett, M. Berry, T. F. Chan, J. Demmel, J. Donato, J. Dongarra, V. Eijkhout, R. Pozo, C. Romine, and H. Van der Vorst. *Templates for the Solution of Linear Systems: Building Blocks for Iterative Methods, 2nd Edition*. SIAM, Philadelphia, PA, 1994.
2. A. Bik and H. Wijshoff. Advanced compiler optimizations for sparse computations. *J. of Parallel and Distributed Computing*, 31:14–24, 1995.
3. Aart J. C. Bik and Harry A. G. Wijshoff. Automatic data structure selection and transformation for sparse matrix computations:. *IEEE Transactions on Parallel and Distributed Systems*, 7(2):109–126, February 1996. also on LNCS.
4. I. S. Duff, R. G. Grimes, and J. G. Lewis. User's guide for Harwell-Boeing sparse matrix test problems collection. Tech. Report RAL-92-086, Computing and Information Systems Department, Rutherford Appleton Laboratory, Didcot, UK, 1992.
5. Stéphane Genaud. On deriving HPF code from PEI programs. Technical Report RR97-04, ICPS, September 1997.
6. V. Kotlyar, K. Pingali, and P. Stodghill. A relational approach to the compilation of sparse matrix programs. *Lecture Notes in Computer Science*, 1300:318–, 1997.
7. Serge Petiton and Nahid Emad. *The data parallel programming model: foundations, HPF realization, and scientific applications*, chapter A data parallel scientific computing introduction, pages 45–64. Springer-Verlag Inc., 1996.
8. E. Violard and G.-R. Perrin. PEI : a language and its refinement calculus for parallel programming. *Parallel Computing*, 18:1167–1184, 1992.
9. E. Violard and G.-R. Perrin. PEI : a single unifying model to design parallel programs. *PARLE'93, LNCS*, 694:500–516, 1993.
10. E. Violard and G.-R. Perrin. Reduction in PEI. *CONPAR'94, LNCS 854*, pages 112–123, 1994.

Optimizing Irregular HPF Applications Using Halos

Siegfried Benkner

C&C Research Laboratories
NEC Europe Ltd.
Rathausallee 10, D-53757 St. Augustin, Germany

Abstract. This paper presents language features for High Performance Fortran (HPF) to specify non-local access patterns of distributed arrays, called halos, and to control the communication associated with these non-local accesses. Using these features crucial optimization techniques required for an efficient parallelization of irregular applications may be applied. The information provided by halos is utilized by the compiler and runtime system to optimize the management of distributed arrays and the computation of communication schedules. High-level communication primitives for halos enable the programmer to avoid redundant communication, to reuse communication schedules, and to hide communication overheads by overlapping communication with computation. Performance results of a kernel from a crash simulation code on the NEC Cenju-4, the IBM SP2, and on the NEC SX-4 show that by using the proposed extensions a performance close to hand-coded message-passing codes can be achieved for irregular problems.

1 Introduction

Many technical and scientific applications utilize irregular grids or unstructured meshes in order to model properties of the real world. Implementing these models in a programming language like Fortran results in codes where the main computations are performed within loops based on indirect (vector-subscripted) array accesses. An efficient parallelization of such codes with a high-level data-parallel paradigm like High Performance Fortran (HPF) is a challenging task. Since the access patterns of distributed arrays are hidden within dynamically defined indirection arrays and thus cannot be analyzed at compile time, the major tasks of HPF compilers, which include work distribution, local memory allocation, local address generation, and communication generation have to be deferred to the runtime. Most of the runtime parallelization techniques are based on the inspector/executor strategy [12, 13]. With this approach the compiler generates for each parallelizable loop preprocessing code, called inspector, which at runtime analyzes the access patterns of distributed arrays, and, besides a number of other tasks, derives the required communication schedules for implementing non-local data accesses by means of message-passing. The executor usually comprises a communication phase, based on the derived schedules, followed by a local computation phase executing a transformed version of the original loop.

The inspector phase may be very time-consuming and, in the case of a complex loop, may by far exceed the sequential execution time of the loop. Therefore, such schemes may be applied with reasonable efficiency only in those cases where the information obtained by the inspector is reused over many subsequent executions of a loop as long as the access patterns and distributions of arrays do not change. Since the detection of invariant inspector phases by means of compile-time or runtime analysis techniques [8, 12] is restricted, recently language extensions [2, 4] have been proposed that allow the user to assert that communication schedules associated with irregular loops are invariant, either for the whole program run or within certain computational phases, and thus may be reused.

In this paper we present language extensions for HPF which allow the user to specify non-local access patterns of distributed arrays, called *halos*, and to control the communication associated with these non-local accesses. The provision of halos enables the compiler to apply crucial compile-time and/or runtime optimization techniques required for an efficient parallelization of irregular applications. In particular, halos alleviate the local management of distributed data, the local address generation and the computation and reuse of communication schedules. Moreover, the presented features enable the programmer to avoid redundant communication, to aggregate communication, and to hide communication overheads by overlapping communication with computation.

This paper is structured as follows: In Section 2 we introduce the concept of halos and describe how halos may be specified in an HPF program. In Section 3 we present high-level communication statements for halos. In Section 4 a brief discussion of implementation aspects and performance results on three parallel machines are presented. Finally, Section 5 summarizes the paper and discusses some related work. We assume that the reader is familiar with the basic concepts of HPF [7].

2 Halos

The (super) set of non-local elements of a distributed array that may be accessed during the execution of the program by a particular processor may be specified by the user by defining a *halo*. A halo is an irregular, possibly non-contiguous area describing, based on global indices, the non-local accesses to distributed arrays. A halo for a distributed array A is specified by means of the *halo-directive* which has the following form:

!HPF+ HALO (*halo-spec-list*) ON *halo-proc-ref* :: A

A *halo-spec* is a one-dimensional integer array expression that contains exactly one *halo-dummy variable* which ranges from 1 to the number of processors in the corresponding processor array dimension. This *halo-dummy variable* must occur in exactly one dimension of the associated processor array reference *halo-proc-ref*. By evaluating the *halo-spec* for each index of the corresponding processor array dimension, the set of indices of array A that may result in non-local data accesses during execution of the program is obtained. If a dimension of an array is not distributed, then the corresponding *halo-spec* must be a "*".

The *halo-directive* is a specification directive and thus all expressions appearing in a *halo-spec* must be *specification expressions*. However, in most irregular applications the distribution of arrays is dependent on runtime information and thus has to be determined dynamically. In this case halos may be specified by means of the HALO attribute within the REDISTRIBUTE directive. This feature allows recomputing or changing the halo of an array whenever its distribution is changed.

Figure 1, which is an excerpt from an HPF+[4] kernel of the IFS weather prediction code [1], illustrates the use of a halo for a dynamically distributed array. In this example we assume that both the mapping array used as argument to the INDIRECT distribution format and the halo array are generated by a partitioner and read in at runtime.

An alternative method to specify halos is based on using array valued functions. With this approach the halo is computed by means of an EXTRINSIC HPF_LOCAL function, which returns the halo indices in a one-dimensional integer array. In order to compute the halo of a distributed array by means of a function, information about the distribution of the array is usually required. This can be achieved by passing the array that is being distributed as argument to the halo-function and using the distribution inquiry functions of the HPF Local Library, like for example LOCAL_LINDEX and LOCAL_UINDEX, which return the lower-bound and upper-bound of the local section an a particular processor (see [7] page 242).

Specifying a halo for a distributed array constitutes an assertion that all non-local accesses are restricted to the halo. Halos do not change the semantics of data distribution directives and the concept of *ownership*. Array elements that are stored in the halo on a particular processor are not *owned* by that processor. Consequently, the semantics of the ON HOME clause, which may be used to specify the distribution of the iteration space of irregular loops is not influenced by halos.

Usually HPF compilers extend the local section of a distributed array that has to be allocated on a particular processor in order to store copies of non-local data elements moved in from other processors. This additional storage is referred to as *extension area*. The main advantage of such an approach, compared to a method where non-local data is stored in separate buffers, is that a uniform local addressing scheme for accessing local and non-local data can be utilized. For regular codes the size of the required extension area can usually be determined by the compiler by analyzing the access patterns of distributed arrays. In the case of indirect array accesses, however, the amount of storage required for non-local data cannot be determined statically. Therefore, many compilers determine the size of the extension area based on simple heuristics, for example, as some fraction of the local section. If it turns out at runtime that the extension area is too small, the array has to be reallocated with a larger extension area and the local addressing scheme has to be modified accordingly, which may cause severe overheads.

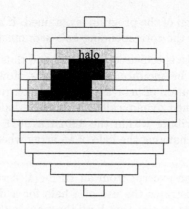

```
        TYPE AOA
            INTEGER, POINTER, DIMENSION(:) :: IDX
        END TYPE AOA
        TYPE (AOA), ALLOCATABLE :: SLHALO(:)

!HPF$ PROCESSORS :: R(NUMBER_OF_PROCESSORS())

        REAL, DIMENSION(NGP,NFL)           :: ZSL
!HPF+ DYNAMIC, RANGE(INDIRECT,*), HALO :: ZSL
        ...
        ALLOCATE(SLHALO(NUMBER_OF_PROCESSORS()))
        READ(...) MAP                    ! mapping array for grid points

        DO I = 1, NUMBER_OF_PROCESSORS()
            READ(...) N                  ! number of non-local elements on R(I)
            ALLOCATE(SLHALO(I)%IDX(N))
            READ(...) SLHALO(I)%IDX(:) ! indices of non-local elements on R(I)
        END DO

!HPF+ REDISTRIBUTE(INDIRECT(MAP),*), HALO(SLHALO(I)%IDX,*) ON R(I) :: ZSL
```

Fig. 1. Specifying Halos. In this example, the first dimension of array ZSL represents the reduced grid used in the semi-Lagrangian interpolation method of the IFS weather prediction code [1], and the second dimension stores physical quantities at each grid point. In order to store the reduced grid without wasting memory it is linearized along the latitudes. As a result of this linearization, an INDIRECT distribution is required. The figure sketches the local section of a particular processor (black) and the required halo (gray). Both the mapping array MAP and the halo array SLHALO, which describes on each processor the non-local indices with respect to the first dimension of ZSL, are read before the array is distributed by means of the REDISTRIBUTE statement. The array SLHALO is an array of arrays defined by means of derived types, where SLHALO(I)%IDX contains the non-local indices that need to be accessed on processor R(I).

The information provided by halos is utilized to allocate memory for non-local data by extending the local section on each processor according to the size of the halo. Moreover, the communication schedule for gathering the non-local data described by a halo from their owner processors can be determined at the time the array is allocated. This schedule is then stored in the distribution descriptor of the array and can be applied whenever it is necessary to update the halo. The main advantage of such a scheme is that the same schedule can be used in different parts of a program instead of determining a communication schedule by means of an inspector at each individual loop. Since the halo must represent the super-set of all non-local data accesses, it can never happen that the array has to be reallocated with a larger extension area.

Halos can be considered as a generalization of *shadows*, which have been included in the Approved Extensions of HPF-2. By means of the SHADOW directive the user may specify how much storage will be required for storing copies of non-local data accessed in nearest neighbor stencil operations. The shadow directive has the form SHADOW(ls:hs) where ls and hs (which must be scalar integer specification expressions) specify the widths of the shadow edges at the low end and high end of a distributed array dimension. Shadows are usually specified only in the context of BLOCK distributions, whereas halos may be specified for any distribution, including GEN_BLOCK and INDIRECT distributions. As currently defined in HPF-2, shadows are not applicable in the context of dynamically distributed arrays.

3 High-level Communication Primitives for Halos

In order to enable the user to control the communication associated with halos without having to deal with the details of message-passing, high-level communication primitives are provided.

By means of the UPDATE_HALO directive the user may enforce that the halo of all processors is updated by receiving the actual values from the owner processor. Although updating of halos could be managed automatically, it may be impossible for the compiler to avoid redundant updates of halos. If, for example, an array with a halo is accessed in subsequent loops without writing to elements that are contained in the halo of some processor, it is not necessary to update the halo at each individual loop. In the case of indirect data accesses or complicated control structures, such an optimization can not be performed automatically and thus needs to be triggered by the user.

The halo of a distributed array may not only be used to read copies of non-local data elements but also to perform non-local writes of array elements. This may be the case if particular assignments are not processed according to the owner computes rule, for example, if the iteration space of a loop is distributed explicitly by the user by means of the ON HOME clause. If processors write to non-local array elements in their halo in the context of array reduction operations, then the results of these operations have to be sent to the owners of the corresponding array elements and accumulated. For this purpose the REDUCE_HALO directive is provided. In a REDUCE_HALO operation the flow of messages is in the opposite direction of an UPDATE_HALO operation. A processor which is the *owner* of an array element that is contained in the halo of other processors, receives all the corresponding halo elements from these processors and accumulates them according to the reduction operation specified in the REDUCE_HALO directive.

Similar to asynchronous I/O as defined in the HPF-2 Approved Extensions, the UPDATE_HALO and REDUCE_HALO directives may be combined with the ASYNCHRONOUS attribute and the WAIT statement in order to hide the communication overheads by overlapping the update or reduction of the halo with computation. Updating the halo of an array asynchronously and utilizing halos in the context of reductions is shown in Figure 2.

4 Implementation and Performance Results

In this section we give a brief overview of the implementation of the described features in the Vienna Fortran Compiler (VFC) [3], a source-to-source compilation system that transforms HPF+ [4] programs into explicitly parallel Fortran 90 SPMD message-passing programs with embedded calls to an MPI based runtime library. VFC supports block, cyclic, generalized block, indirect distributions, basic alignments, and dynamic data distributions. For the parallelization of irregular loops the INDEPENDENT directive, the ON HOME clause, and the REUSE clause are provided. The REUSE clause [2] allows the user to assert that the access patterns of distributed arrays are invariant and thus the corresponding communication schedules may be reused once they have been determined by means of an inspector or based on the HALO attribute.

VFC transforms all distributed arrays such that they are dynamically allocated in the target program. For each distributed array calls to runtime library procedures are generated that take as arguments the shape of the array and the distribution and, if specified, the halo information. These functions evaluate at runtime the distribution and compute the local section that has to be allocated on the processor that called the function. If an array has a halo, the local section is extended accordingly. In this case, the communication schedules required for updating the halo are computed and stored in the distribution descriptor of the array.

The parallelization of INDEPENDENT loops with indirect accesses to distributed arrays is based on the inspector/executor paradigm [12, 13]. The required runtime support is based on an extended version of the CHAOS/PARTI library [6]. The first part of the inspector determines a distribution of the loop iteration space, usually by evaluating the ON HOME clause. The remainder of the inspector code consists of a series of loops that trace the accesses to distributed arrays on a dimension-by-dimension basis. The resulting traces are passed to runtime procedures that determine the required communication schedules, the size of the extension area for storing copies of non-local data, and a corresponding local addressing scheme.

```
!HPF$ PROCESSORS R(NUMBER_OF_PROCESSORS())
      REAL          :: X(3,NN), F(6,NN)
      INTEGER       :: IX(4,NE)
!HPF+ DYNAMIC, HALO :: X, F
!HPF$ DYNAMIC       :: IX
!HPF$ ALIGN F(*,I) WITH X(*,I)

      INTERFACE
         EXTRINSIC(HPF_LOCAL) FUNCTION halo_function(I, BS, IX)
            INTEGER, POINTER :: halo_function(:)
            INTEGER          :: I, BS(:), IX(:,:)
            ...
         END FUNCTION halo_function
      END INTERFACE
      ...

!HPF+ REDISTRIBUTE(*,GEN_BLOCK(BS)),                            &
!HPF+ HALO (halo_function(I, BS, IX)) ON R(I) :: X
      ...
!HPF+ UPDATE_HALO, ASYNCHRONOUS(ID=I1) :: X
      ...     ! some computation that does not involve X
      F = 0.
!HPF+ WAIT (ID=I1)
      ...

!HPF$ INDEPENDENT, ON HOME (IX(1,I)), REDUCTION(F)
      DO I = 1, NE
         F(:,IX(:,I)) = F(:,IX(:,I)) + ...
      END DO
      ...
!HPF+ REDUCE_HALO(+) :: F
```

Fig. 2. Update of halos. This example is an excerpt of a crash simulation kernel [5]. The code employs a finite element mesh based on 4-node shell elements as sketched in Figure 3. In the code the mesh is presented by means of the arrays X and F which store the coordinates and forces of each node, and the array IX which stores the 4 node numbers of each element. At the time the REDISTRIBUTE statement is executed, the halo of X is determined on each processor by evaluating the array valued function halo_function. Since F is aligned with X it also gets redistributed with the same halo as X. The UPDATE_HALO operation ensures that all copies of a particular element of X that are in the halo of some processor have the same value as on the owner processor. The communication to update the halo of X is done asynchronously and overlapped with some computation that does not involve X. The INDEPENDENT loop performs a sum-scatter reduction operation to add the elemental forces back to the forces stored at the nodes. The loop can be executed without communication due to the halo of F. In Figure 3, the local part of this operation is indicated by the solid arrows. The execution of the REDUCE_HALO directive accumulates all halo elements of F on the corresponding owner processors. For example, processor 1 receives the local values of F at node n3 from processor 2 and processor 3, and adds them to its own value of F at node n3.

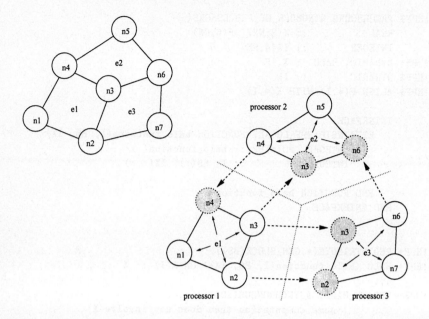

Fig. 3. Finite Element Mesh. This figure sketches the finite element mesh utilized by the crash simulation kernel. The mesh is based on 4-node shell elements In the figure three elements are shown, each of which is mapped to a processor as depicted in the right-hand side. The nodes at the boundary between the processors, have to be accessed on several processors. For example, node n3 is owned by processor 1 but needs to be accessed also on the other processors. Thus n3 is contained in the halo of processor 2 and 3. The local (owned) nodes L(p) and the halos H(p) on processor p are as follows: L(1) = {n1, n2, n3}, H(1) = {n4}; L(2) = {n4, n5}, H(2) = {n3, n6}; L(3) = {n6, n7}, H(3) ={n2, n3}. The dotted arrows indicate the flow of messages in the case of an UPDATE_HALO operation. The solid arrows indicate the local part of the sum-scatter reduction operation to add the elemental forces back to the forces stored at the nodes. The communication required during the REDUCE_HALO operation is in the oposite direction of the dotted arrows.

If a distributed array has a halo, then the computation of communication schedules is not performed within the inspector, since in this case a schedule has already been determined at the time the array has been allocated. Moreover, if the compiler can determine that the access patterns of a distributed array are invariant, or if the user asserts this by means of the REUSE clause, then the inspector code is guarded such that it is executed only when the loop is reached for the first time.

In Table 1 performance results of a crash simulation kernel [5] (see Figure 2 and 3) and an equivalent hand-coded MPI/F77 version on a NEC Cenju-4 distributed memory machine with 32 VR10000 processors, a NEC SX-4 shared memory vector supercomputer with 8 processors, and on an IBM SP2L with 32 RS/6000 processors are presented.

	NEC Cenju-4			IBM SP2			NEC SX-4		
	MPI	HPF+	HPF-2	MPI	HPF+	HPF-2	MPI	HPF+	HPF-2
1 processor	524	592	8721	904	1320	16953	33	42	4414
2 processors	263	293	4318	457	676	8578	18	23	2235
4 processors	129	144	2217	235	344	4512	8.7	4.4	1125
8 processors	63	70	1143	128	192	2938	4.4	8.2	579
16 processors	31	36	619	79	106	1462	-		
32 processors	16	19	369	48	73	990	-		

Table 1. Elapsed times (secs) of main computational loop of crash simulation kernel.

The kernel is based on an explicit time-marching scheme realized as a sequential loop from within procedures to calculate the shell-element forces and to update the velocities and displacements are called. In our evaluation a mesh with 102720 nodes and 102400 elements was used and 1000 time-steps were performed.

Table 1 shows the elapsed times of the hand-coded MPI kernel, the times obtained with the HPF+ kernel that utilized the described HALO features (except asynchronous updates) and the REUSE clause, and the times for a variant using HPF-2 features only. The HPF kernels were parallelized with VFC 1.5 and the generated SPMD programs were compiled with the Fortran 90 compiler of the target machine.

The huge gap between the two HPF variants is due to the fact that in the HPF-2 kernel communication schedules are computed with an inspector in every iteration, wheres in the HPF+ kernel the communication schedules required for X and F are computed once, based on the HALO attribute, at the time the arrays are allocated/distributed and reused over all time-steps.

5 Conclusions

In this paper we presented language extensions for HPF that allow the user to specify non-local access patterns for distributed arrays and to control the communication associated with these non-local accesses. The concept of halos allows the user to specify in addition to the distribution of arrays a description of the non-local data elements that may have to be accessed at runtime. This information allows computing the communication schedules for implementing non-local data accesses at the time the distribution of an array is evaluated and reusing these schedules in different parts of a program. By means of high-level communication primitives the user can control and optimize the communication associated with halos without having to deal with the details of message-passing. The performance results show that by using these extensions a performance close to hand-coded message-passing codes can be achieved for irregular codes.

Similar extensions have been proposed in HPF/JA [11], including *explicit shadows*, where the user may indicate that computation should be performed on the local section of a distributed array and on the allocated shadow edges by means of an extended on-clause. The REFLECT directive is provided to control the update of shadow edges. HPF/JA includes also language features for communication schedule reuse.

In contrast to the static concept of shadows as provided by the Approved Extensions of HPF-2, halos may also be used in the context of dynamically distributed arrays. The halo of an array may be changed whenever the array has to be redistributed.

References

1. S. Barros, D. Dent, L. Isaksen, G. Robinson, G. Mozdzynski, and F. Wollenweber. The IFS model: A parallel production weather code. Parallel Computing 21, 1995.
2. S. Benkner, P. Mehrotra, J. Van Rosendale, and H. Zima. High-Level Management of Communication Schedules in HPF-Like Languages. ACM Proceedings International Conference on Supercomputing, Melbourne, Australia, July 1998.
3. S. Benkner, K. Sanjari, V. Sipkova, and B. Velkov. Parallelizing Irregular Applications with the Vienna HPF+ Compiler VFC. Proceedings International Conference on High Performance Computing and Networking (HPCN'98), Amsterdam, April 1998, Lecture Notes in Computer Science, Vol. 1401, pp. 816-827, Springer Verlag.
4. S. Benkner. HPF+: High Performance Fortran for Advanced Industrial Applications. Proceedings International Conference on High Performance Computing and Networking (HPCN'98), Amsterdam, April 1998, Lecture Notes in Computer Science, Vol. 1401, pp. 797-808, Springer Verlag.
5. J. Clinckemaillie, B. Elsner, G. Lonsdale, S. Meliciani, S. Vlachoutsis, F. de Bruyne, M. Holzner. Performance Issues of the Parallel PAM-CRASH Code. The International Journal of Supercomputing Applications and High-Performance Computing, Vol.11, No.1, pp.3-11, Spring 1997.
6. R. Das, Y.-S. Hwang, M. Uysal, J. Saltz, and A. Sussman. Applying the CHAOS/PARTI Library to Irregular Problems in Computational Chemistry and Computational Aerodynamics, Proceedings of the 1993 Scalable Parallel Libraries Conference, pp. 45-56, IEEE Computer Society Press, October 1993.
7. High Performance Fortran Forum. *High Performance Fortran Language Specification 2.0*, Rice University, January, 1997.
8. M. W. Hall, S. Hirandani, K. Kennedy, and C.-W. Tseng. Interprocedural compilation of Fortran D for MIMD distributed-memory machines. Proceedings of Supercomputing (SC92), Minneapolis, November, 1992.
9. Message Passing Interface Forum. MPI: A Message-Passing Interface Standard Vers. 1.1, June 1995.
10. R. von Hanxleden, K. Kennedy, C. Koelbel, R. Das, and J. Saltz. Compiler Analysis for Irregular Problems in Fortran D. Proceedings of the 5th Workshop on Languages and Compilers for Parallel Computing, New Haven, August 1992.
11. JAHPF Homepage: http://www.tokyo.rist.or.jp/~shunchan/index-e.html
12. R. Ponnusamy, J. Saltz, and A. Choudhary. Runtime Compilation Techniques for Data Partitioning and Communication Schedule Reuse. Proceedings Supercomputing '93, pp. 361-370, 1993.
13. J. Saltz and K. Crowley and R. Mirchandaney and H. Berryman. Run-Time Scheduling and Execution of Loops on Message Passing Machines. Journal of Parallel and Distributed Computing, 8(4), pp. 303-312, April 1990.

From EARTH to HTMT: An Evolution of a Multiheaded Architecture Model

Guang R. Gao

University of Delaware

Abstract. In this talk, we discuss the issues and challenges solving irregularly structured problems in parallel from the angle of system architectures and support To this end, multithreaded architecture models and systems provide an new opportunity for meeting such challenges.

We begin by a brief review on the evolution of multithreaded models and architectures — in particular the EARTH (Efficient Architecture For Running Threads) architecture model We outline the program execution model and architecture issues for multithreaded processing elements suitable for use as the node processor facing the challenges of the irregular applications. These include the question of choosing appropriate program execution model, the organization of the processing element to achieve good utilization and speedup in the presence of irregular data and control flow patterns, the support for fine-grain interprocessor communication and dynamic load balancing, and related compiling issues. We have implemented the EARTH architecture model on a number of experimental high-performance multiprocessor platforms and we will present some experiment results on the effectiveness of the EARTH architecture on irregular applications.

We briefly present the conception and evolution of a US petaflop supercomputer architecture project — the Hybrid Technology Multithreaded Architecture project. The program execution model and architecture overview of the HTMT machine will be discussed.

Irregular Parallel Algorithms in Java*

Brian Blount[1], Siddhartha Chatterjee[1], and Michael Philippsen[2]

[1] The University of North Carolina, Chapel Hill, NC 27599-3175, USA.
[2] The University of Karlsruhe, 76128 Karlsruhe, Germany.

Abstract. The nested data-parallel programming model supports the design and implementation of irregular parallel algorithms. This paper describes work in progress to incorporate nested data parallelism into the object model of Java by developing a library of collection classes and adding a `forall` statement to the language. The collection classes provide parallel implementations of operations on the collections. The `forall` statement allows operations over the elements of a collection to be expressed in parallel. We distinguish between `shape` and `data` components in the collection classes, and use this distinction to simplify algorithm expression and to improve performance. We present initial performance data on two benchmarks with irregular algorithms, EM3d and Convex Hull, and on several microbenchmark programs.

1 Introduction

Software technology for scalable parallel computers lags behind the rapid developments in the hardware for such systems. Current programming languages on these systems often expose architecture-specific details to the programmer, making it difficult to develop scalable and portable software. Software standards have emerged for dense array codes in the form of data-parallel languages such as High Performance Fortran [16] (HPF) and message passing libraries such as MPI [28]. However, many irregular algorithms, such as those used in sparse matrix solvers and geometric problems, can be difficult to express in these systems.

Scientific applications continue to grow in the sophistication of the data structures and algorithms they use. Techniques such as adaptive unstructured meshes [4] in computational fluid dynamics (CFD), hierarchical n-body simulations [10, 17, 22, 26], and supernodal sparse Cholesky [2] and LU [21] factorization are difficult to write in an architecture-independent manner in traditional performance-oriented languages. One programming model capable of expressing such irregular algorithms is *nested data parallelism* [5]. While data parallelism allows the expression of parallelism through operations over the elements of a collection, nested data parallelism allows the elements of a collection to themselves be collections. This allows parallelism to be expressed on complex structures, such as graphs, trees, and sparse matrices.

We are investigating the incorporation of nested data parallelism into the object model of Java [1]. We choose Java for several reasons. First, there is a growing interest

---

* This research is supported in part by the National Science Foundation under grant CCR-9711438.

in Java for scientific computing [14, 15, 19]. Second, Java programs may utilize efficient numerical libraries through `native` methods. Third, the object-oriented features of Java are useful in developing modular and reusable software components. Finally, Java is portable; a Java program can run on any machine that implements a Java Virtual Machine (JVM).

We enhance Java in two ways to support nested data parallelism: we add a library of collection classes and a `forall` statement. The collection classes provide parallel operations on the collections themselves (aggregate parallelism), and the `forall` statement allows the expression of parallelism on operations over the elements of a collection (elementwise parallelism). To support the nested data-parallel model, members of collections can themselves be collections, and the `forall` statements may be nested. Each collection class consists of three members: a *data* object, a *shape* object, and a *map* object. The encapsulation of these separate components allows for cleaner expression of operations on the collections and enables program optimizations.

The remainder of the paper is organized as follows. Section 2 describes our framework. Section 3 presents performance results. Section 4 discusses related work. Section 5 presents conclusions and future work.

2 Nested Data Parallelism Framework

We introduce two features into Java to support nested data parallelism: a library of collection classes and a `forall` statement. The collection classes provide parallel implementations of aggregate operations, such as reductions and parallel prefix [5], and also provide functionality similar to the collection classes in JDK 1.2 [29]. The `forall` statement allows operations over the elements of a collection to be expressed in parallel. To support nested data parallelism, collections may be members of other collections and `forall` statements may be nested.

A pre-compiler [24] translates the `forall` statements into multi-threaded Java code. Because the translated code uses threads as the source of parallelism, we prohibit the explicit use of threads in our extended language. The transformed code uses a package of classes to manage the threads at runtime. The library of collection classes also uses this package to implement its parallel aggregate operations. We derived our translation from one previously presented by one of the authors [24]. Our current version differs in two ways. First, it supports nested `forall` statements more efficiently. Second, it uses Java as a target language instead of Pizza [23]. The original translation utilized the closure mechanism of Pizza, which we replace with *inner classes* [1]. We present the translation by examining the simple example of a `forall` statement that initializes an array.

```
1    forall (int i = 0; i < n; i++) a[i] = i;
```

Our approach transforms the above example into the code shown below. We use a work pile and worker threads. Worker threads start at program invocation and continuously perform any work placed on the work pile. An (anonymous) inner class encapsulates the work of a `forall` loop in its `exec` method, as shown in lines 2–8. The inner class is a subclass of `ForallWork`, which contains four variables that are initialized in the constructor (line 2): the upper and lower bounds of the work it must perform (lb and ub), the step size of the loop (step), and a handle to its ForallController (fac), a runtime support class. The exec method first checks in line 4 to see if the work object should be split. This is performed by the ForallController and is the method by which multiple work objects are created. If the object is split, the ForallController places the two resulting objects on the work pile. If the object is not split, the exec method performs the work in line 6. Finally, in line 7, the work object informs the ForallController that it is complete. Line 1 of the code instantiates a ForallController object. Line 2 creates a work object and line 9 places it on the work pile. When all of the work objects are complete, the controller's finalBarrier method in line 10 returns.

```
1    ForallController _fac = new ForallController();
2    ForallWork _work = new ForallWork(0, n, 1, _fac) {
3      public void exec() {
4        if (!this.fac.split(this))
5          for(int i = this.lb; i < this.ub; i += this.step)
6            a[i] = i;
7        this.fac.workDone();
8      } };
9    WorkPile.doWork(_fac, _work);
10   _fac.finalBarrier();
```

The translation above ignores exceptions that might be thrown inside the body of the forall, e.g., if $a.length < n$ in the example above. If this occurs, all work objects should cease execution and the exception should be thrown in the method that the forall statement occurred in. This is handled by maintaining information in the ForallController. The version of exec below handles exceptions and replaces lines 3–8 in the code block above. Any exception thrown inside the loop body is caught and stored in lines 6–8. All the work threads are signaled to stop execution through the setting of breakFlag. The finalBarrier method throws the stored exception when all work objects are complete.

```
1    public void exec() {
2      if (!this.fac.breakFlag) {
3        if (!this.fac.split(this)) {
4          for(int i = this.lb; i < this.ub; i += this.step) {
5            if (this.fac.breakFlag) break;
6            try { a[i] = i; } catch(Throwable _ex) {
7              this.fac.caughtException = _ex;
8              this.fac.breakFlag = true;
9    } } } }
10     this.fac.workDone();
11   }
```

Note that the semantics for a thrown exception are different for a forall loop than for a for loop. If an exception is thrown inside iteration i of a for loop, all iterations before i are guaranteed to have completed, and all iterations after i are guaranteed to have never begun execution. If an exception is thrown inside iteration i of a forall loop, there are no guarantees regarding which iterations have executed. This may complicate handling exceptions for a programmer.

Our translation supports nested forall statements differently than the original. In the original, the finalBarrier method simply slept, waiting on other worker threads to complete work. When nested forall statements occur, all the worker threads may enter finalBarrier methods, leaving no worker threads to execute any work. At this point, a low-priority watchdog thread recognized the situation and spawned more worker threads. This approach has the disadvantage that the executing program may eventually have more worker threads than originally intended. Our translation solves this problem by modifying the finalBarrier method. If a thread enters this method and cannot proceed, it executes work off of the work pile until it can proceed.

The collections within our library resemble collections from previous work [8] and consist of three components: a data object, a shape object, and a map object. The data object is a one-dimensional array containing all the elements of a given collection in some arbitrary order. The shape object specifies the "shape" of the collection by defining a domain on which the collection can be indexed. The map object maps values from this index domain to locations in the data array. As an example, consider a matrix collection with r rows and c columns. Its data array is of length n, where $n \geq r \times c$. Its shape object is of type MatrixShape, and contains r and c as data. Its map function maps $\{(i,j) : 0 \leq i < r, 0 \leq j < c\}$ to $\{k : 0 \leq k < n\}$. Each component is a first-class object and can be manipulated and shared arbitrarily. The collections rely on sharing of data and performing actions "lazily" on index domains whenever possible.

Each collection class supports parallel aggregate operations, such as reductions and prefix sum [5]. The implementation of these parallel operations is similar to the translation scheme for the forall statement and follows a uniform strategy: encapsulate work in a work object, and encapsulate the state of the operation in a controller object. As an example, the reduction operation sum uses two classes, ReduceSumController and ReduceSumWork. A ReduceSumWork object encapsulates the work of the sum by performing the sum operation on a portion of the collection. Similar to a ForallController, a ReduceSumController object maintains the state of the sum operation. The state includes two pieces of information: whether the sum operation is complete, and the reduced value computed by each work object. Once all the work objects are complete, the controller uses the values the work objects calculated to determine the final sum.

3 Performance

We wrote two benchmarks, EM3d and Convex Hull, and several microbenchmarks to evaluate the performance of our implementation of nested data parallelism in Java. The microbenchmarks fit into two categories: programs that measure the performance of Java's concurrency primitives, and benchmarks that measure the performance of the parallel operations added to the language. Table 1 lists the environments we used for our measurements.

	UltraSparc 170 (US170)	UltraSparc (US)	HyperSparc (HS)	Sun Enterprise 3000 (SE)	SGI Onyx 2 (SGI)
CPU	Ultra-I	Ultra-IIi	HyperSPARC	UltraSPARC	MIPS R8000
Processors	1	1	4	4	4
Clock Rate	167 MHz	300 MHz	100 MHz	170 MHz	195 MHz
Memory	96 MB	248 MB	160 MB	384 MB	2176 MB
OS	Solaris 2.5.1	Solaris 2.6	Solaris 2.6	Solaris 2.5.1	IRIX 6.4
JVM	1.1.6, 1.2beta4-K	1.1.6, 1.2beta4p	1.1.6, 1.2beta4p	1.1.6, 1.2beta4-K	3.1 (Sun 1.1.5)

Table 1. Testing Environment (1.2beta4p stands for 1.2beta4production). JDK 1.2beta4 always performed better than 1.1.6, so we do not present the results for JDK 1.1.6. All JVMs are JIT-enabled. We present results for native threads, implemented by the operating system, and for green threads, implemented by the JVM.

Our translation of nested data parallelism realizes its concurrency through Java's thread interface. There are three operations that may introduce overhead: thread instantiation, lock use, and thread scheduling. Our implementation amortizes the first overhead by instantiating threads only once, at program invocation. The second overhead depends on the implementation of locks, and is discussed below. Both the operating system and Java's scheduling primitives (such as yield, wait, and notify) influence the third overhead. Since our model relies heavily on wait and notify, we need to examine their costs further.

Java provides locking through its synchronized keyword, which can modify a method or a block of code. Synchronized code obtains a lock before it begins execution and releases a lock when it terminates execution. Many of the work pile's methods are synchronized because the worker threads access a central work pile. To quantify the cost of obtaining a lock, we compare the cost of calling a method that is synchronized with the cost of calling a method that is not synchronized. Both methods take no arguments, return no values, and perform a trivial amount of work. The results are shown in Table 2. The locking mechanism introduces significant overhead. SGI's JVM performs better than Sun's, but its overhead is still large. Several projects have examined reducing this cost [3, 20], and JDK 1.2 shows improvement over JDK 1.1.6. However, this overhead may require some redesign of our runtime classes to reduce the number of synchronized methods.

We used three microbenchmarks to measure the performance of the parallel operations added to the language. The first program executes a single forall statement with 64 parallel iterates. Each iterate performs a for loop with 100,000 sequential iterations. The second program performs a plus_reduce operation on a vector of 2.5 million integers. The third program performs a plus_scan operation on a vector of 1 million integers.

The top portion of Table 3 presents the performance of these three programs. Parallel execution of the forall statement introduced an overhead of 10% to 40%, because of object creation and synchronized methods introduced by the transformation. The parallel plus_reduce operation introduced a 15% to 60% overhead. The greater

Machine	JVM	Native Threads				Green Threads			
		Sync (μs)	No Sync (μs)	Overhead (μs)	Ratio	Sync (μs)	No Sync (μs)	Overhead (μs)	Ratio
US 170	1.2b4k	1.38	0.07	1.31	19.71	1.52	0.07	1.45	21.71
US	1.2b4p	0.62	0.01	0.61	62.00	—	—	—	—
HS	1.2b4p	0.85	0.03	0.82	28.33	—	—	—	—
SE	1.2b4k	1.52	0.06	1.46	25.33	1.50	0.07	1.43	21.43
SGI	3.1	1.13	0.22	0.91	5.14	0.58	0.21	0.37	2.76

Table 2. Overhead of synchronized methods. Both the synchronized method and the non-synchronized method were called 2048 times and the average cost was taken. We placed a small amount of work inside the methods to verify that the method calls were not optimized away. Performance is similar for synchronized blocks.

overhead may be due to the fact the reduce controller performs more work than the forall controller. The overhead of the parallel plus_scan operation was approximately 300%. This is because the parallel operation performs a three step algorithm, and each work object is placed on the work pile twice. The implementation of this operation must be re-evaluated to decrease the overhead. On machines with one processor, there were no parallel improvements. On the machines with four processors, near-linear speed up was seen for most operations. Some further speedup was seen with additional threads. The optimal number of threads varies among the machines, and requires further investigation.

Several peculiarities are seen in the data. On the Enterprise, the operating system scheduled the threads sequentially for the forall microbenchmark, so no speedup was seen. We are currently investigating the reason for this anomaly. Also, multiple threads did not provide speedup on the SGI until 4 or 8 threads were used. This is probably due to the scheduling algorithm of the operating system. While our implementation should be efficient independent of the scheduling algorithm used, it does depend on the minimum expectation of threads being scheduled on multiple processors when run on a multiple processor machine.

We have thus far implemented two irregular algorithms in our extended Java language: EM3d and Planar Convex Hull. The EM3d kernel [13] (EM3d) models the propagation of electromagnetic waves through objects in three dimensions. We include it in our benchmark suite because of its popularity and the availability of a number of implementations of it on various systems. The implementation of this algorithm in our extended language consists of two nested forall statements and an aggregate plus_reduce operation. The planar convex hull problem (PCH) is as follows: given n points in the plane, determine the points that lie on the perimeter of the smallest convex region containing them. The problem has many applications ranging from computer graphics to statistics. We implement the quickhull algorithm [6] for this problem. Our program uses both forall statements and aggregate operations.

The bottom of Table 3 presents the results for these two algorithms. EM3d executed 100 iterations on a graph with 4000 H nodes and 4000 E nodes, where each node had 30 neighbors. PCH executed 100 iterations on a set of 100000 points contained inside

Program	Machine	JVM	Sequential Version (s)	Parallel Version Number of Threads				
				1 (s)	2 (s)	4 (s)	8 (s)	16 (s)
Forall	US170	1.2b4k	0.45	0.62	0.68	0.66	0.65	0.66
	US	1.2b4p	0.25	0.27	0.29	0.31	0.32	0.32
	HS	1.2b4p	0.61	0.74	**0.37**	**0.19**	**0.23**	**0.23**
	SE	1.2b4k	0.44	0.61	0.58	0.58	0.57	0.57
	SGI	3.1	1.74	1.79	1.79	**0.89**	**0.61**	**0.56**
Plus-Reduce	US170	1.2b4k	1.08	1.71	1.74	1.77	1.80	1.74
	US	1.2b4p	0.23	0.30	0.34	0.35	0.34	0.35
	HS	1.2b4p	0.86	1.01	**0.51**	**0.26**	**0.30**	**0.31**
	SE	1.2b4k	1.04	1.64	**0.92**	**0.55**	**0.49**	**0.51**
	SGI	3.1	0.90	1.26	1.41	**0.80**	**0.53**	**0.52**
Plus-Scan	US170	1.2b4k	0.49	1.50	1.56	1.61	1.79	1.61
	SE	1.2b4k	0.49	1.44	0.75	0.52	0.50	0.50
	SGI	3.1	0.41	1.21	1.31	1.27	0.57	0.48
EM3d	US170	1.2b4k	1.21	**1.09**	1.22	1.22	1.27	1.23
	US	1.2b4p	0.26	0.38	0.28	0.28	**0.23**	**0.24**
	HS	1.2b4p	0.53	0.56	**0.29**	**0.17**	**0.20**	**0.21**
	SE	1.2b4k	1.07	**1.03**	**0.60**	**0.45**	**0.43**	**0.44**
	SGI	3.1	1.05	1.11	1.24	1.33	1.58	1.57
PCH	US170	1.2b4k	2.80	2.91	3.08	2.95	3.02	3.02
	SE	1.2b4k	2.77	2.85	**2.37**	**2.03**	**2.04**	**2.19**
	SGI	3.1	2.86	2.97	3.09	3.02	8.33	3.23

Table 3. Performance of microbenchmark and benchmark programs. Each program executed 100 iterations, and the average cost is shown. Times are in seconds. Results where parallel performance is better than sequential performance are shown in boldface.

a circle of radius 10000. Of the 100000 points, 63 fell on the hull. Parallel execution of the two programs introduces almost no overhead because the amount of work in the algorithm hides the overhead introduced in parallel execution. The EM3d performance resembles the performance of the `forall` benchmark. Because the inner forall loop is small (only 30 iterations), we run it sequentially. The same applies to its reduce operation. The PCH benchmark shows little or no speedup on any of the machines. This may be because much of the parallelism in the algorithm is not exploited. The algorithm recursively calls itself twice. The two calls are independent and may be executed in parallel. Because there is no way to express this easily in the language, the two calls are run sequentially.

4 Related Work

NESL [7] is a functional language designed to support nested data parallelism. It is a strongly typed, polymorphic language with a type inference system, which makes it possible to write generic functions that work on any type of sequence. Because NESL is purely functional, it suffers the usual problems when updating subsets of aggregate variables, and it makes it impossible for users to manage memory use. Also, NESL's runtime system is incompatible with other languages.

The Amelia [27] project introduces nested data parallelism into C++. We use many of the same techniques as the Amelia project, but do not suffer the problems with unrestricted pointer semantics of C++. The pC++ [11] language defines a new programming model called the "distributed collection model." This model is not quite data-parallel and does not support nested data parallelism. Since collections may not be members of other collections, the data types available to a programmer are somewhat limited. The ICC++ [12] language is an extension to C++ that supports concurrency through *conc* blocks and collection objects. The implementation uses advanced compiler optimizations and a custom runtime system to gain high performance. Any two statements within a *conc* block may be executed concurrently if there are no data dependences between the two.

High Performance Fortran (HPF) [16] defines a collection-oriented parallel language that supports one collection type: the multi-dimensional array. Parallelism in HPF programs comes through the use of a `forall` construct and special parallel operations on arrays, such as array permutation and reduction. The language is not oriented towards nested data parallelism, although pointer types can be used to support nested collections [25], and certain routines in the standard HPF library can be used to support nested data-parallel operations.

Titanium [30] is a dialect of Java designed for SPMD parallel programming. While Titanium uses Java as a base, it adds many extensions to the language, such as immutable classes, parallel constructs, and multi-dimensional arrays. However, Titanium provides no direct support for nested data parallelism and does not use Java as a target language. Hummel *et al.* [18] explore mechanisms and propose an API for writing SPMD programs in Java, targeting a parallel Java runtime system on the IBM POWER-parallel System SP machine. Their focus is on flat array-based data parallelism.

5 Conclusions and Future Work

This paper presents a method to incorporate nested data parallelism into Java. It describes a scheme for implementing collection classes, and a method for implementing `forall` statements and aggregate operations using inner classes and threads. The microbenchmarks examined demonstrate that these implementations can achieve good performance. The two irregular algorithms demonstrate that our model is well-suited for expressing most types of parallelism.

We are pursuing several future directions for this work. First, we are investigating two ways to reduce the overhead of the parallel operations. First, we want to modify the implementation of the `prefix sum` operation to reduce its overhead to the level of the other operations. Second, we may use multiple work piles to reduce the number of worker threads waiting on the one work pile. This model would provide one work pile for each worker thread, and the work piles could perform work stealing similar to the method implemented in Cilk [9].

Second, we are experimenting with various techniques for determining the optimal granularity for work objects. Our current strategy simply defines the optimal granularity as a constant number of iterations, and divides the work objects until they contain fewer than this number of iterations. A more reasonable approach would factor both number of iterations and amount of work performed in an iteration in determining optimal granularity.

Third, we are considering adding another parallel construct to our language: the spawn keyword. This would allow a limited amount of task parallelism, such as the recursive calls to quickhull described above, to be expressed easily. A forall statement is appropriate to express parallelism over tasks when the number of tasks is unknown. However, when the number of parallel tasks is known, such as the recursive calls in quicksort, quickhull, and Strassen's matrix multiplication, the spawn keyword provides a much simpler mechanism to express the parallelism. This will have substantial ramifications for semantics and implementations that need to be carefully examined.

Finally, we are examining other irregular computations, such as conjugate gradient, supernodal sparse Cholesky and LU factorizations, and CFD problems using adaptive unstructured meshes. These algorithms will provide a better suite of test cases to evaluate the expressiveness and performance of our framework.

References

1. K. Arnold and J. Gosling. *The Java^{TM} Programming Language*. The Java^{TM} Series. Addison-Wesley Publishing Company, 1996.
2. C. C. Ashcraft. The domain/segment partition for the factorization of sparse symmetric positive definite matrices. Enginering Computing & Analysis Technical Report ECA-TR-148, Boeing Computer Services, Seattle, WA, Nov. 1990.
3. D. F. Bacon et al. Thin locks: featherweight synchronization for Java. In *Proc. PLDI'98*, pages 258–268, 1998.
4. R. Biswas and R. C. Strawn. A new procedure for dynamic adaption of three-dimensional unstructured grids. *Applied Numerical Mathematics*, 13:437–452, 1994.
5. G. E. Blelloch. *Vector Models for Data-Parallel Computing*. The MIT Press, Cambridge, MA, 1990.
6. G. E. Blelloch. NESL: A nested data-parallel language (version 2.6). Technical Report CMU-CS-93-129, School of Computer Science, Carnegie Mellon University, Pittsburgh, PA, Apr. 1993. Updated version of CMU-CS-92-103, January 1992.
7. G. E. Blelloch et al. Implementation of a portable nested data-parallel language. *JPDC*, 21(1):4–14, Apr. 1994.
8. B. L. Blount and S. Chatterjee. An evaluation of Java for numerical computing. In *Proc. ISCOPE'98*, Dec. 1998. LNCS 1505, pp. 35-46, Springer Verlag.
9. R. D. Blumofe et al. Cilk: An efficient multithreaded runtime system. In *Proc. PPoPP'95*, pages 207–216, Santa Barbara, CA, July 1995. ACM.
10. J. A. Board Jr. et al. Scalable variants of multipole-accelerated algorithms for molecular dynamics applications. Technical Report TR94-006, Department of Electrical Engineering, Duke University, Durham, NC, 1994.
11. F. Bodin et al. Implementing a parallel C++ runtime system for scalable parallel systems. In *Proc. SC'93*, pages 588–597, November 1993.
12. A. A. Chien and J. Dolby. ICC++: A C++ dialect for high-performance parallel computation. In *Proc. ISOTAS'96*, Mar. 1996.

13. D. E. Culler et al. Parallel programming in Split-C. In *Proc. SC'93*, pages 262–273, Nov. 1993.
14. G. C. Fox. Java for high performance scientific and engineering computing. http://www.npac.syr.edu/projects/javaforcse/.
15. D. B. Gannon. High Performance Java. http://www.extreme.indiana.edu/hpJava/index.html.
16. High Performance Fortran Forum. High Performance Fortran language specification. *Scientific Programming*, 2(1–2):1–170, 1993.
17. Y. Hu, S. L. Johnsson, and S.-H. Teng. High Performance FORTRAN for Highly Irregular Problems. In *Proc. PPoPP'97*, pages 13-24, June 1997.
18. S. F. Hummel, T. Ngo, and H. Srinivasan. SPMD programming in Java. *Concurrency: Practice and Experience*, 9(6):621–631, June 1997. Special issue on Java for computational science and engineering—simulation and modeling.
19. Java Grande Forum. The Java Grande Forum charter document. http://www.npac.syr.edu/javagrande/jgfcharter.html.
20. A. Krall and M. Probst. Monitors and Exceptions: How to implement Java efficiently. *Concurrency: Practice and Experience*, 10(11):837–850, Sept. 1998.
21. X. S. Li. *Sparse Gaussian Elimination on High Performance Computers*. PhD thesis, Department of Computer Science, University of California at Berkeley, Berkeley, CA, Sept. 1996. Available as technical report CSD-96-919.
22. L. S. Nyland, J. F. Prins, and J. H. Reif. A data-parallel implementation of the fast multipole algorithm. In *Proc. DAGS'93*, pages 111–122, Hanover, NH, June 1993.
23. M. Odersky and P. Wadler. Pizza into Java: Translating theory into practice. In *Proc. POPL'97*, Jan. 1997.
24. M. Philippsen. Data parallelism in Java. In J. Schaefer, editor, *High Performance Computing Systems and Applications*. Kluwer Academic Publishers, Boston, Dordrecht, London, 1998.
25. J. Prins, S. Chatterjee, and M. Simons. Expressing irregular computations in modern Fortran dialects. In D. O'Hallaron, editor, *Languages, Compilers, and Run-Time Systems for Scalable Computers*, pages 1–16. Springer, 1998. LNCS 1511.
26. K. E. Schmidt and M. A. Lee. Implementing the fast multipole algorithm in three dimensions. *Journal of Statistical Physics*, 63(5/6), 1991.
27. T. J. Sheffler and S. Chatterjee. An object-oriented approach to nested data parallelism. In *Proceedings of the Fifth Symposium on the Frontiers of Massively Parallel Computation*, pages 203–210, McLean, VA, Feb. 1995.
28. M. Snir et al. *MPI: The Complete Reference*. MIT Press, 1996.
29. The Java collections framework. http://java.sun.com/products/jdk/1.2/docs/api/index.html.
30. K. Yelick et al. Titanium: A high-performance Java dialect. *Concurrency: Practice and Experience*, 10(11):825–836, Sept. 1998.

A Simple Framework to Calculate the Reaching Definition of Array References and Its Use in Subscript Array Analysis*

Yuan Lin and David Padua

Department of Computer Science, University of Illinois at Urbana-Champaign
{yuanlin,padua}@uiuc.edu

Abstract. The property analysis of subscript arrays can be used to facilitate the automatic detection of parallelism in sparse/irregular programs that use indirectly accessed arrays. In order for property analysis to work, array reaching definition information is needed. In this paper, we present a framework to efficiently calculate the array reaching definition. This method is designed to handle the common program patterns in real programs. We use some available techniques as the building components, such as data dependence tests and array summary set representations and operations. Our method is more efficient as well as more flexible than the existing techniques.

1 Introduction

Program restructuring, including automatic parallelization, has been a useful alternative to manual program optimization for regular, dense computations [2]. However, program restructuring techniques for sparse, irregular problems are not well understood. Compilers usually rely on data dependence tests to enable transformations. The effectiveness of the data dependence test is determined by its ability to accurately analyze array subscripts in loops. However, in sparse/irregular programs, indirect addressing via indices stored in auxiliary arrays[1] is often used. The array subscript can be complex and sometimes its value is difficult or impossible to know at compile time. For this reason, automatic program restructuring is usually believed to be less effective at exploiting implicit parallelism in sparse codes than in their dense counterparts. However, a recent study of a collection of sparse and irregular programs [8] has shown that the use of subscript arrays often follows a few common patterns. Based on these patterns, program restructuring techniques can be extended to handle sparse and irregular programs.

---

* This work is supported in part by Army contract DABT63-95-C-0097; Army contract N66001-97-C-8532; NSF contract MIP-9619351; and a Partnership Award from IBM. This work is not necessarily representative of the positions or policies of the Army or Government.
[1] In this paper, we call the array that appears in the subscript of other arrays the *subscript array* and the indirectly accessed array the *host array*.

For example, in the sparse matrix computations based on the Compressed Column Storage(CCS) or Compressed Row Storage(CRS) format, the host array is divided into several segments, and subscript arrays are used to store the offset pointer and the length for each segment. Figure 1.(a) shows an example, where `offset()` points to the starting position of each segment whose length is given in `length()`. Figure 1.(b) shows a common loop pattern using the `offset()` and `length()` arrays. The loop traverses the host array segment by segment. Figure 1.(c) shows a common pattern used to define `offset()`. For the code in this example, the compiler first can perform *array property analysis*[8] on the code in Fig.1.(c) to find that `offset()` has a *regular distance* of `length()` and then use this information in the data dependence analysis for array `data()` in Fig.1.(b) to find that loop do_200 is parallel. Similar examples can be presented for other important compiler transformations, including those for locality enhancement and communication optimizations.

Besides regular distance, the properties of subscript arrays, such as monotonicity and injectivity, as well as having maximum/minimum values and constant values, have been found useful [3]. We can derive the properties and infer the max/min/constant values from where the subscript array is defined and then use the information to analyze the loops that access the host array. We call this process *property analysis of subscript arrays,* or simply *subscript array analysis.*

For the subscript array analysis to be correct, we must make sure that the definitions can reach the use. For the example in Fig.1, we want to know whether the values of all the elements of `offset()` read in statement s3 are provided by statement s1 and s2. In order to get this information, we need *array reaching definition* analysis.

The array reaching definition problem can be described as, "Given an array read occurrence, find all the assignment statements that may provide the value accessed by the read occurrence." Since different executions of a statement may reference different elements of the same array, the array reaching definition problem may need to identify the instance use/def pairs. The original problem is called the *statement-wise array reaching definition problem* and the latter is called the *instance-wise array reaching definition problem.* Since, in real programs, array regions defined by a single statement tend to have the same property, the statement-wise array reaching definition information is precise enough for our purposes.

In this paper, we present a simple framework to calculate the array reaching definitions at the statement level. Such a simple framework is useful because, in real programs, most array accesses are simple and, in most cases, there is an order between the array accesses. We can use traditional data dependence tests to find this order. The reaching definitions are then computed by using set operations. As a result, our method performs the set operation only once for each occurrence rather than repeatedly on each loop level as done by other methods [5, 6]. Our method, in most cases, is more efficient and more flexible.

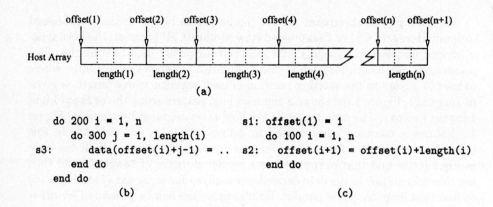

Fig. 1. Examples of Offset and Length Subscript Array

2 A Brief Discussion of The New Method

Any framework that computes the array reaching definition has to solve, either explicitly or implicitly, two problems. The first one is Access Order. Suppose two statements S_1 and S_2 write some array elements that are read by another statement S_3. It is necessary to know which statement writes the array elements last. If some writes of S_2 come after some writes of S_1, then those definitions generated by S_1 may be killed by S_2. The second problem is Coverage. Suppose that, in the program region being analyzed, there are several statements writing the array elements read by another statement S_r. If they do not write all the array elements read by S_r, then a definition may come from a statement outside the program region. In our method, we call each read or write of an array in the program text an *occurrence* of that array. We treat as a unit all the array elements that can be accessed by each array occurrence. For example, in the codes in Fig.2.(a), there are four array occurrences: O_1, O_2, O_3 and O_4. The first three are read occurrences and the last one is a write occurrence. When executed, they access $\{a(10:90)\}$, $\{a(1:50)\}$, $\{a(51:100)\}$ and $\{a(1,100)\}$, respectively.

The coverage problem can be solved by using the set subtract operation. For instance, in the previous example, because $(\{[1:100]\} - \{[51:100]\}) - \{[1:50]\} = \phi$, $s2$ and $s3$ provide all the definitions used in $s4$.

The set operation has to be applied by taking into account the access order. The previous example illustrates one of the two cases: the *straight line coverage* case. The other case is the *cross iteration coverage*. In the code in Fig.2.(b), an array element written by $s1$ in one iteration will be killed by $s2$ in the following iteration. Since the element is used by $s3$ two iterations later, the definitions in $s1$ do not reach $s3$. In other words, the coverage is performed across the iterations. The access order in this code also is clear. All the elements that can be accessed

```
              do l=10,90
s1:     a(l)=...  - 01                                              do i=1, m
        end do                                                        do j=1, n
        do i=1, 50              do i=1, 100         s1:     a(1,j-2)=...       - 01
s2:     a(i)=...  - 02   s1:    a(i)=...    - 01    s2:     a(i,j)=...         - 02
        end do           s2:    a(i-1)=...  - 02    s3:     a(i-1,j+1)=...     - 03
        do j=51, 100     s3:    ...=a(i-2)  - 03    s4:     ...=a(i-3,6*i+j)   - 04
s3:     a(j)=...  - 03          end do                                end do
        end do                                                      end do
        do k=1, 100
s4:     ...=a(k)  - 04
        end do
            (a)                      (b)                                (c)
```

Fig. 2. Some code samples

by both $s1$ and $s2$ first are accessed by $s1$ and then by $s2$. Thus, $a[1:98]$ is defined first by $s1$ and then by $s2$.

The basic idea of our method is to (1) identify all the array elements that an occurrence can access, (2) determine the execution order of the occurrences, and (3) use set operations to derive the reaching definitions. A simplified high-level algorithm of our method is shown in Fig.3. Given an array occurrence O of an array $a()$, we use $\mathcal{R}(O)$ to denote the set of the elements of $a()$ that are accessed by O during the execution of the program. The $<_R$ order is explained in the next section.

Input: A read occurrence R, and m write occurrences $W_1, W_2, ..., W_m$ that $W_i <_R R$, $1 \leq i \leq m$. The m write occurrences are sorted such that $W_i <_R W_j$ if $i < j$.

Output: A subset S of the m write occurrences that provide reaching definitions for the read occurrence R, and a set T of the array elements read at R, but whose definitions do not come from any of the m write occurrences.

Procedure:
$$S = \phi, T = \mathcal{R}(R) \ ;$$
for i=m downto 1 do
    if $T \cap \mathcal{R}(W_i) \neq \phi$ then
        $S = S \cup \{W_i\}$;
        $T = T - \mathcal{R}(W_i)$;
    end if
    if $T == \phi$ then return (S, T) ;
end for
return (S, T)

Fig. 3. A simplified high-level algorithm of our method

3 Access Order

3.1 Access Order

The algorithm in Fig.3 requires an order between the write occurrences and the read occurrence, as well as an order between the write occurrences. These orders guarantee that we can obtain the reaching definitions by performing the set operations (intersection, union, and subtraction) on the set of elements these occurrences can access.

Given a read occurrence R of an array and a write occurrence W of the same array, we say that $W <_R R$ if any element in $\mathcal{R}(R) \cap \mathcal{R}(W)$ is written at least once by W before it is first read by R. When $W <_R R$, $\mathcal{R}(R) - \mathcal{R}(W)$ gives all the array elements that are read at R but are not previously defined at W.

Given two write occurrences W_1 and W_2 and one read occurrence R of an array, we say $W_1 <_R W_2$ if

1. $W_1 <_R R$, and
2. $W_2 <_R R$, and
3. for each element in $\mathcal{R}(R) \cap \mathcal{R}(W_1) \cap \mathcal{R}(W_2)$, there is a write of this element by W_1 after any write of this element by W_2 and before any read by R.

When $W_1 <_R W_2$, $(\mathcal{R}(R) - \mathcal{R}(W_1)) \cap \mathcal{R}(W_2)$ gives the set of elements that are defined by W_2 and read by R but not killed by W_1.

3.2 Access Distance

We define the *access time* of an array element by an occurrence O to be the time during the program execution when the statement containing O is executed and the array element is accessed. The access time can be represented by a set of tuples $(iter, pos)$, where $iter$ is the iteration vector of the enclosing loops and pos is the text position of the statement in the program. It is a set because an array element may be accessed multiple times by an occurrence.

Suppose occurrence O_1 and O_2 can access some common array elements. Then, the *access distance* between O_1 and O_2 is the difference between the access times of the common array elements by these two occurrences. The access distance is represented by $(< d_1, d_2, ..., d_n >, (pos1, pos2))$, where d_i represents the difference of the corresponding elements in the iteration vectors. The value of d_i can be either a constant number or a $*$. It is a constant number when the corresponding elements in all the differences of iteration vectors are equal; otherwise, it is a $*$.

Given three occurrence O_1, O_2, and O_3 of an array, we say $O_1 \ll_{O_3} O_2$, if

1. there is a path from statement S_1 which contains O_1 to statement S_3 which contains O_3, and there is a path from statement S_2 which contains O_2 to statement S_3 in the control flow graph, and
2. S_1 dominates S_2, and
3. S_1 and S_2 are not enclosed in any common loop that does not enclose S_3.

Given two write occurrences W_1 and W_2 and one read occurrence R of an array, suppose the access distance between W_1 and R is $D_1 = (diter_1, (pos_{W_1}, pos_R))$ and the access distance between W_2 and R is $D_2 = (diter_2, (pos_{W_2}, pos_R))$. We say that $D_1 \prec_R D_2$ if

1. $diter_1$ is lexicographically less than $diter_2$, or
2. $diter_1 = diter_2$, and they contain no $*$ element, and $W_1 \ll_R W_2$.

We say $D_1 \asymp_R D_2$ if $diter_1 = diter_2$ and they contain no $*$ element, and pos_{W_1} is the same as pos_{W_2}. We say D_1 and D_2 is not comparable when neither $D_1 \prec_R D_2$, $D_2 \prec_R D_1$, nor $D_1 \asymp_R D_2$ holds.

Suppose $d = (diter, (pos_{O_1}, pos_{O_2}))$ is the access distance between O_1 and O_2. We say $0 \prec d$ if $diter$ is lexicographically bigger than vector 0, or $diter$ equals vector 0 and pos_{O_1} precedes pos_{O_2}.

Proposition 1. *Given a write occurrence W and a read occurrence R that can access some common array elements, let d be the access distance between W and R. If $0 < d$, then $W <_R R$.*

Proposition 2. *Given two write occurrences W_1 and W_2 and a read occurrence R that can access some common array elements, let D_1 be the access distance between W_1 and R and D_2 be the access distance between W_2 and R. If $D_1 \prec_R D_2$, then $W_1 <_R W_2$.*

3.3 Sorting the Write Occurrences

Now, given m write occurrences and a read occurrence, we can calculate the access distance from each write occurrence to the read occurrences and then sort the write occurrences according to the access distance. For example, the four occurrences in Fig.2.(c) can be ordered as O_2, O_1, O_3, and O_4.

Because two access distances may not be comparable, the result of the sorting may be a DAG.

3.4 Calculating the Access Distance

The access distance can be calculated in a way similar to that of the dependence distance. We will not present any detailed method here, but rather point out a way to extend existing methods.

Most methods that can calculate the dependence distance require that the subscript expression have only one loop index, such as the occurence O_1 in Fig.4. In real programs, however, many subscript expressions do not fit this constraint, especially when arrays are *reshaped* across procedure boundaries. Fortunately, as some researchers [9, 11] have indicated, the affine subscript expressions often have interleaved access patterns and can be *delinearized* into several independent subscripts with each subscript having the form $c_1(i + c_2) + c_3$. In this case, a dependence test such as [9] can be used to calculate the dependence distance. For example, occurrences O_2, O_3, and O_4 in Fig.4 can be delinearized to $b'(i+8, j)$, $b'(i, j)$, and $b'(i+1, j-1)$, respectively, with b' declared as $b'(20, *)$. Hence, it is easy to see that O_2 is executed before O_4.

```
          do i=1, 10                           do i=1, 10
            do j=1, 100                          do j=1, 100
01:           a(i,j) =                             a(i,j) = ...
02:           b(i+20*j+8) = ...                    b'(i+8,j) =
03:                      = b(i+20*j)   ===>                  = b'(i,j)
04:           b(i+20*j-19) = ...                   b'(i+1,j-1) = ...
            end do                               end do
          end do                               end do
```

Fig. 4. Subscript Delinearization

4 Coverage

4.1 Summary Set Representation

Once the execution order of the write occurrences is clear, the compiler can check the coverage to eliminate from the final reaching definitions the occurrences whose definitions are killed by other occurrences. Our method represents the array regions by set. Several schemes for set representations have been proposed, including convex regions [13], data access descriptors [1, 11] and regular sections [7]. Any one of these can be used in our framework.

Here we use regular sections [6] to illustrate the basic idea. A regular section of an array $A()$ is denoted by $(A(s_1, s_2, ..., s_m), accuracy)$, where m is the dimension of $A()$, $s_i (i = 1, ..., m)$ is a section in the form of $(l : u)$, and l, u are symbolic expressions. $(l : u)$ represents all integer values between l and u, including l and u. The value of $accuracy$ can either be MAY or $MUST$. MAY means $A(s_1, s_2, ..., s_m)$ is a superset of the real section, while $MUST$ means it is a subset. The regular section, which represents the array region an occurrence can access in a loop, can be calculated by *expanding* the loop index across the range of the loop [6, 11]. For example, in the loop in Fig.2.(c), the region of array $a()$ written by O_3 is $(a(0 : m - 1, 2 : n + 1), MUST)$, and it is $(a(4 : m + 3, 7 : 6m + n), MAY)$ by O_4.

4.2 Coverage Check

The major operation on the summary set is the *coverage check*, which checks if the set of elements written by an occurrence contains those written by other occurrences and computes difference as the uncovered region. The basic operation in coverage check is the set subtraction operation.

The result of the set subtraction often is an approximation of the real result because the subtraction operation usually is not closed in the set representation. We, therefore, want the uncovered region to be the superset of the real one. To be conservative, we compute the MAY regions for read occurrences and $MUST$ regions for write occurrences.

5 Putting It All Together

The overall algorithm is shown in Fig.5.

Input: A read occurrence R in a program section.
Output: 1. The set S of the write occurrences in the program section that provide the reaching definitions.
2. The array region $RegOut$ that contains elements read in R but may not be defined in the program section.
Procedure:
(1) Find in the program section all the write occurrences on which R is flow dependent. Assume they are $W_1, W_2, ..., W_n$;
(2) Summarize the MAY region $\mathcal{R}(R)$ for the read occurrence R and the $MUST$ region $\mathcal{R}(W_i)$ for each write occurrence W_i, $1 \leq i \leq n$;
(3) Calculate the access distance from each write occurrence to the read occurrence;
(4) Sort the write occurrences according to the access distance and store in T the resulting DAG ;
(5) Let $S = \{W_i | W_i$ is executed before R, and not $(W_i <_R R)\}$;
(6) Do the coverage check by using T and R as the input and store the result in $(S', RegOut)$;
(7) $S = S \cup S'$;
(8) return $(S, RegOut)$.

Fig. 5. Putting It All Together

6 Related Work

The existing array dataflow analysis techniques can be categorized according to the granularity of array information they can get, instance-wise or statement-wise.

In order to get the instance-wise reaching definition, the techniques in the first category usually use expensive techniques. Feautrier [4] uses a parametric integer programming method. Maydan provides a faster technique for many common situations. However, when handling multiple writes, both methods can grow exponentially. Pugh and Wonnacott [12] model the problem as verifying Presburger formulas. They extend the Omega test to answer questions in a subclass of Presburger arithmetic. Another method that also uses the Omega test is proposed by Maslov [10]. Although these methods can derive very precise reaching definition information, we have found that, for analysis of the programs we have studied, coarse grain statement-wise information suffices.

Statement-wise techniques [5, 6, 14] in the second category do not analyze array elements individually. Instead, these techniques work on sets of array elements accessed by each statement in the program. These sets usually have a regular shape. Simple set operations, such as union, intersection, and difference, also are performed on such units.

The method discussed in this paper belongs to the statement-wise category. One difference between our method and the others is that we check the coverage directly, while the other methods follow the general procedure proposed in [5]. This procedure calculates the upward exposed use set and downward exposed write set repeatedly while program structures, such as intervals, are being built. Ideally, this approach should be able to get a more precise result than our simple technique. However, on the forms used to represent sets in practice, the set operations usually are not closed and the representation itself often is an approximation of the real array region. Consequently, the advantage of their method is lost in practice. In the cases where the set operation is exact, the occurrences usually have single dependence distance. Our framework is based on this fact and uses a simple method to achieve the same result as the method in [5] in most real cases we have found. Another difference is that most of the other methods were designed to solve a specific array data flow problem, such as array privatization, whereas our method was designed to compute the array reaching definition in general. The array reaching definition not only can be used to test array privatization, but also can be used for many other purposes, such as the subscript array analysis discussed early in this paper.

7 Conclusion

Indirectly accessed arrays are used intensively in sparse and irregular programs and make the programs difficult to parallelize by parallelizing compilers. Our previous study [8] has shown that property analysis of subscript arrays can be used to facilitate the automatic detection of parallelism in this case. In order for array property analysis to work, we need array data flow analysis to derive the array reaching definition information.

In this paper, we presented a simple framework to calculate the statement-wise array reaching definitions. We found that, in real programs, most array occurrences have single or partial single dependence distance. We use this distance to establish an order between these occurrences. Based on this order, the array reaching definition can be computed by using set operations in a simple way. This technique and subscript array analysis will allow many automatic parallelization methods to be extended to handle sparse and irregular programs.

References

1. V. Balasundaram and K. Kennedy. A technique for summarizing data access and its use in parallelism-enhancing transformations. In *Proceedings of the ACM SIGPLAN'89 Conference on Programming Language Design and Implementation (PLDI)*, Portland, OR, June 1989.
2. William Blume, Rudolf Eigenmann, Keith Faigin, John Grout, Jay Hoeflinger, David Padua, Paul Petersen, William Pottenger, Lawrence Rauchwerger, Peng

Tu, and Stephen Weatherford. Polaris: Improving the effectiveness of parallelizing compilers. In Keshav Pingali, Uptal Banerjee, David Gelernter, Alex Nicolau, and David Padua, editors, *Proceedings of the 7th International Workshop on Languages and Compilers for Parallel Computing*, Lecture Notes in Computer Science, pages 141–154, Ithaca, New York, August 8–10, 1994. Springer-Verlag.

3. Rudolf Eigenmann and William Blume. An effectiveness study of parallelizing compiler techniques. In *Proceedings of the 1991 International Conference on Parallel Processing*, volume II, Software, pages II-17–II-25, Boca Raton, FL, August 1991. CRC Press.
4. P. Feautrier. Dataflow analysis of scalar and array references. *International Journal of Parallel Programming*, 20(1):23–52, February 1991.
5. Thomas Gross and Peter Steenkiste. Structured dataflow analysis for arrays and its use in an optimizing compiler. *Software Practice and Experience*, 20(2):133–155, February 1990.
6. Junjie Gu, Zhiyuan Li, and Gyungho Lee. Symbolic array dataflow analysis for array privatization and program parallelization. In Sidney Karin, editor, *Proceedings of the 1995 ACM/IEEE Supercomputing Conference, December 3–8, 1995, San Diego Convention Center, San Diego, CA, USA*, New York, NY 10036, USA, 1995. ACM Press and IEEE Computer Society Press.
7. P. Havlak. *Interprocedural Symbolic analysis*. PhD thesis, Rice University, May 1994.
8. Y. Lin and D. Padua. On the automatic parallelization of sparse and irregular fortran programs. In *Proc. of 4th Workshop on Languages, Compilers, and Runtime Systems for Scalable Computers (LCR98)*, volume 1511 of *Lecture Notes in Computer Science*, pages 41–56. Springer-Verlag, Pittsburgh, PA, 1998.
9. Vadim Maslov. Delinearization: An efficient way to break multiloop dependence equations. In *Proceedings of the Conference on Programming Language Design and Implementation (PLDI)*, volume 27, pages 152–161, New York, NY, July 1992. ACM Press.
10. Vadim Maslov. Lazy array data-flow dependence analysis. In ACM, editor, *Proceedings of 21st Annual ACM SIGACT-SIGPLAN Symposium on Principles of Programming Languages*, pages 311–325, New York, NY, USA, ???? 1994. ACM Press.
11. Yunheung Paek, Jay Hoeflinger, and David Padua. Simplification of array access patterns for compiler optimizations. In *Proceedings of the ACM SIGPLAN'98 Conference on Programming Language Design and Implementation (PLDI)*, pages 60–71, Montreal, Canada, 17–19 June 1998.
12. William Pugh and David Wonnacott. An exact method for analysis of value-based array data dependences. In *Proceedings of the Sixth Annual Workshop on rogramming Languages and Compilers for Parallel Computing*, December 93.
13. R. Triolet, F. Irigoin, and P. Feautrier. Direct parallelization of call statements. In *Proceedings of the SIGPLAN'86 Symposium on Compiler Construction*, pages 176–185, Palo Alto, CA, July 1986.
14. Peng Tu and David Padua. Automatic array privatization. In Uptal Banerjee, David Gelernter, Alex Nicolau, and David Padua, editors, *Proceedings of the 6th International Workshop on Languages and Compilers for Parallel Computing*, Lecture Notes in Computer Science, pages 500–521, Portland, Oregon, August 12–14, 1993. Intel Corp. and the Portland Group, Inc., Springer-Verlag.

Dynamic Process Composition and Communication Patterns in Irregularly Structured Applications [*]

C.T.H. Everaars, B. Koren and F. Arbab
email: Kees.Everaars@cwi.nl, Barry.Koren@cwi.nl and Farhad.Arbab@cwi.nl

CWI, P.O. Box 94079, 1090 GB Amsterdam, The Netherlands, Telefax 020-5924199
(+31 20 5924199)

Keywords parallel languages, parallel computing, distributed computing, coordination languages, models of communication, irregular communications patterns, unstructured process composition, software renovation, multi-grid methods, sparse-grid methods, computational fluid dynamics, three-dimensional flow problems.

> **Abstract.** In this paper we describe one experiment in which a new coordination language, called **MANIFOLD**, is used to restructure an existing sequential Fortran code from computational fluid dynamic (CFD), into a parallel application. **MANIFOLD** is a coordination language developed at CWI (Centrum voor Wiskunde en Informatica) in the Netherlands. It is very well suited for applications involving dynamic process creation and dynamically changing (ir)regular communication patterns among sets of independent concurrent cooperating processes. With a simple, but generic, master/worker protocol, written in the **MANIFOLD** language, we are able to reuse the existing code again, without rethinking or rewriting it. The performance evaluation of a standard 3D CFD problem shows that **MANIFOLD** performs very well.

1 Introduction

A workable approach for modernization of existing software into parallel/distributed applications is through coarse-grain restructuring. If, for instance, entire subroutines of legacy code can be plugged into the new structure, the investment required for the re-discovery of the details of what they do can be spared. The resulting renovated software can then take advantage of the improved performance offered by modern parallel/distributed computing environments, without rethinking or rewriting the bulk of their existing code. The necessary communications between the different partners in such a new concurrent system can have different forms. In some cases, the channel structures representing the communication patters between the different partners, are regular and the numbers of

---

[*] Partial funding for this project is provided by the National Computing Facilities Foundation (NCF), under project number NRG 98.04.

the partners is fixed (structured static communication). In other cases, the communication patterns are irregular and the number of partners changes over time (unstructured dynamic communication). There are many different languages and programming tools available that can be used to implement this kind of communications, representing very different approaches to parallel programming. Normally, languages like Compositional C++, High Performance Fortran, Fortran M, Concurrent C(++) or tools like MPI, PVM, and PARMACS are used (see [1] for some critical notes on these languages and tools). There is, however, a promising novel approach: the application of *coordination* languages [2–4].

In this paper we describe one experiment in which a new coordination language, called MANIFOLD, was used to restructure an existing Fortran 77 program into a parallel application. MANIFOLD is a coordination language developed at CWI (Centrum voor Wiskunde en Informatica) in the Netherlands. It is very well suited for applications involving dynamic process creation and dynamically changing (ir)regular communication patterns among sets of independent concurrent cooperating processes [5,6]. Programming in MANIFOLD is a game of dynamically creating process instances and (re)connecting the ports of some processes via streams (asynchronous channels), in reaction to observed event occurrences. This style reflects the way one programmer might discuss his interprocess communication application with another programmer on telephone (let process *a* connect process *b* with process *c* so that *c* can get its input; when process *b* receives event *e*, broadcast by process *c*, react on that by doing this and that; etc.). As in this telephone analogy, processes in MANIFOLD do not explicitly send to or receive messages from other processes. Processes in MANIFOLD are treated as black-box workers that can only read or write through the openings (called ports) in their own bounding walls. It is always a third party - a coordinator process called a manager - that is responsible for setting up the communication channel (in MANIFOLD called a stream) between the output port of one process and the input port of another process, so that data can flow through it. This setting up of the communication links from the *outside* is very typical for MANIFOLD and has several advantages. An important advantage is that it results in a clear separation between the modules responsible for computation and the modules responsible for coordination, and thus strengthens the modularity and enhances the re-usability of both types of modules (see [1,6,7]).

The MANIFOLD system runs on multiple platforms and consists of a compiler, a run-time system library, a number of utility programs, and libraries of built-in and predefined processes of general interest. Presently, it runs on IBM RS60000 AIX, IBM SP1/2, Solaris, Linux, Cray, and SGI IRIX [1].

The original Fortran 77 code in our experiment was developed at CWI by a group of researchers in the department of Numerical Mathematics, within the framework of the BRITE-EURAM Aeronautics R&D Programme of the European Union. The Fortran code consist of a number of subroutines (about 8000 lines) that manipulate a common date structure. It implements their multi-

---

[1] For more information, refer to our html pages located at
http://www.cwi.nl/ farhad/manifold.html.

```
program SEQ_CODE
begin

Preamble:
  - Some initialization work
  - Some initial sequential computations

Heavy computational job:
  for i = 1 to N
    - Heavy computations that can't be done in parallel
    - Heavy computations that can in principle be done in parallel
  endfor

Postamble:
  - Some final sequential computations
  - Printing of results

end
```

Fig. 1. The schema of the sequential code

grid solution algorithm for the Euler equations representing three-dimensional, steady, compressible flows. They found their full-grid-of-grids approach to be effective (good convergence rates) but inefficient (long computing times). As a remedy, they looked for methods to restructure their code to run on multi-processor machines and/or to distribute their computation over clusters of workstations.

Clearly, the details of the computational algorithms used in the original program are too voluminous to reproduce here, and such computational detail is essentially irrelevant for our restructuring. Thus we refer for a detailed description of the software to the last four chapters of reference [8], the official report on the BRITE-EURAM project, and instead use a simplified pseudo program in this paper that has the same logical design and structure as the original program. In section 2, we present this simplified pseudo program and give its parallel counterpart. Next, in section 3, we describe how we implement our parallel version using the coordination language **MANIFOLD**. In section 4, we show performance results for the standard test case of an ONERA M6 half-wing in transonic flight. Finally, the conclusion of the paper is in section 5.

2 The Simplified Pseudo Code and its Parallel Version

The simplified pseudo code as distilled from the original program is shown in figure 1. The heavy computations that, in principle, can be done in parallel represent the original Fortran version's pre- or post- Gauss-Seidel relaxations on all the cells of a certain grid [9]. Because the relaxation subroutine reads and writes data concerning its own grid only, the relaxations can in principle be done in parallel for all the grids to be visited at a certain grid level. In figure 2, we show the parallel version of the simplified pseudo code. There, we create inside a loop a worker-pool consisting of a number of workers to which we delegate the relaxations of the different grids. Note that the number of workers is dependent on the index i of the loop. When the workers can run as separate processes using different processors on a multi-processor hardware, then we have the desired parallel structure.

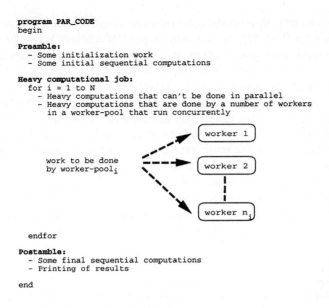

```
program PAR_CODE
begin

   Preamble:
      - Some initialization work
      - Some initial sequential computations

   Heavy computational job:
      for i = 1 to N
         - Heavy computations that can't be done in parallel
         - Heavy computations that are done by a number of workers
           in a worker-pool that run concurrently




      endfor

   Postamble:
      - Some final sequential computations
      - Printing of results

end
```

Fig. 2. The schema of the parallel code

3 The Parallel Implementation using Manifold

We can describe the parallel schema of figure 2 in a kind of master/worker protocol in which the master performs all the computations of the sequential code except the relaxations which are done by the workers.

```
1  // protocolMW.m
2
3  #define IDLE terminated(void)
4
5  export manifold ProtocolMW(manifold Master, manifold Worker,
                              event create_worker, event ready)
6  {
7    auto process master is Master.
8
9    begin: (master, IDLE).
10
11   create_worker: {
12     process worker is Worker.
13
14     begin: (&worker -> master -> worker, IDLE).
15   }.
16
17   ready: halt.
18 }
```

In **MANIFOLD** we can do this in a general way, as shown in the code above (for details see reference [9]). There, the master and the worker are parameters of the protocol implemented in the **MANIFOLD** language as a separate coordinator process (line 5). In this protocol we describe only how instances of the master and worker process definitions communicate with each other. For the protocol it is irrelevant to know what kind of computations are performed in the master and the worker.

In the coordinator process we create and activate the master process that embodies all computation except the relaxations (line 7). Each time the master arrives at the pre- or post-relaxation, he raises an event to signal the coordinator to create a worker (line 12). In this way, a pool of workers is set to work for the master whereby each pool contains its own number of workers (see figure 2). Before a worker can really work, it should know on which grid it must perform the relaxation. Because the master has this information available, the coordinator sets up a communication channel (called a stream in **MANIFOLD**) between the master and the worker so that the master can send this information to the worker (see the arrow between the master and the worker on line 14 which represents a stream). Also, the coordinator must inform the master of the identification of the worker so that it can activate the worker (see the arrow between the reference of the worker denoted by &worker and the master). When all the workers of a certain worker-pool are created and activated in this way, the master waits until the workers are done with the relaxation and are prepared to terminate. After this rendezvous, the master continues its work (i.e., the index i of the loop in figure 2 is incremented by one) until it again arrives at a point where it wants to use a pool of workers to delegate the relaxations to.

Note that it is not necessary to have a streams from workers to the master through which the workers send the results of their relaxations back to the master. The reason for this is that the relaxation work of the different workers, running as different **MANIFOLD** processes, can run as threads (light-weight processes) [10]. in the same operating-system-level (heavy weight) process, and thus can share the same global data space. Therefore, the restructuring we present here is not suitable for distributed memory computing. Nevertheless, the restructured program we present here *does* improve the performance of the application as we will see in the next section. For a description of the distributed memory restructuring see http://www.cwi.nl/~farhad/CWICoordina.html.

Having implemented the master/worker protocol in **MANIFOLD** in a general way, the only thing we need to do is to use this protocol by replacing its formal parameters by actual values which are processes. The actual master and worker manifolds are easy to implement as atomic processes written in C. These C functions then call the original Fortran code (8000 lines) to do the real work [9].

4 Performance Results

A number of experiments were conducted to obtain concrete numerical data to measure the effective speed-up of our parallelization All experiments were run on an SGI Challenge L with four 200 MHZ IP19 processors, each with a MIPS R4400 processor chip as CPU and a MIPS R4010 floating point chip for FPU. This 32-bit machine has 256 megabytes of main memory, 16 kilobytes of instruction cache, 16 kilobytes of data cache, and 4 megabytes of secondary unified instruction/data cache. This machine runs under IRIX 5.3, is on a network, and is used as a server for computing and interactive jobs. Other SGI machines on this network function as file servers.

Computations were done for both the sparse- and the semi-sparse-grid approaches. For the sparse-grid approach, the finest grid levels considered are: 1,

Table 1. Average elapsed times (hours:minutes:seconds) for sparse- and semi-sparse-grid-of-grids approaches.

	level	sequential	parallel
	1	11.24	5.84
sparse	2	1:37.42	34.06
	3	9:15.56	2:47.40
	2	50.43	27.33
semi-sparse	4	18:02.10	5:58.72
	6	4:36:08.54	1:14:44.04

2 and 3; for the semi-sparse-grid approach, the finest grid levels are: 2, 4 and 6. The higher the grid level the heavier the computational work. The results of our performance measurements for both the sparse- and the semi-sparse-grid approaches are summarized in table 1, which shows the elapsed times versus the grid level. All experiments were done during quiet periods of the system, but, as in any real contemporary computing environment, it could not be guaranteed that we were the only user. Furthermore, such unpredictable effects as network traffic and file server delays, etc., could not be eliminated and are reflected in our results. To even out such "random" perturbations, we ran the two versions of the application on each of the three levels close to each other in real time. This has been done for each version of the application, five times on each level. From the raw numbers obtained from these experiments we discarded the best and the worst performances and computed the averages of the other three which are shown in table 1.

From the results, it clearly appears that the **MANIFOLD** version takes good advantage of the parallelism offered by the four processors of the machine. The underlying thread facility in our implementation of **MANIFOLD** on the SGI IRIX operating system allows each thread to run on any available processor. For the sparse-grid and the semi-sparse-grid applications, the **MANIFOLD**-code times are about 3.25 and 3.75 times smaller, respectively, than the sequential-code times. So, in both cases we have obtained a nearly linear speed-up.

The dynamic creation of workers in different work pools for the sparse- and the semi-sparse-grid versions, are respectively shown in Figures 3 and 4. From

Table 2. Work pool and worker statistics.

application	level	n_p	$(n_w)_{max}$	$(n_w)_{total}$
	1	6	3	10
sparse	2	18	6	50
	3	42	10	170
	2	18	3	38
semi-sparse	4	82	7	336
	6	268	12	1838

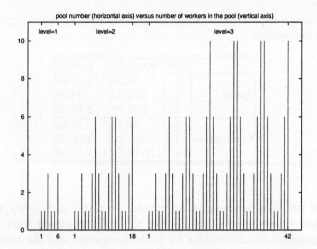

Fig. 3. Different pools of workers created during the sparse-grid parallel applications.

Figure 3 we see that for level=1, 6 pools of workers were created with their corresponding synchronization points each with 1, 1, 3, 1, 1 and 3 workers on board, respectively. This makes the total number of worker processes for this application 10. For level=2 there are 18 pools with a total of 50 workers, and for level=3 these numbers are 42 and 170, respectively. The numbers for both the sparse- and the semi-sparse-grid applications are summarized in Table 2. Here, n_p denotes the number of pools, $(n_w)_{\max}$ the maximum number of workers in a pool and $(n_w)_{\text{total}}$ the total number of workers in the application. Note that in the semi-sparse-grid application in level 6 the average elapsed time went from 4 hours and 36 minutes down to 1 hour and 14 minutes by introducing 268 worker-pools in which a total of 1838 workers, as independently running processes, did their relaxation work.

5 Conclusions

Parallelizing existing sequential applications is often considered to be a complicated and challenging task. Mostly it requires thorough knowledge about both the application to be parallelized and the development system to be used. Our experiment using **MANIFOLD** to restructure existing Fortran code into a parallel application indicates that this coordination language is well-suited for this kind of work.

The highly modular structure of the resulting application and the ability to use existing computational subroutines of the sequential Fortran program are remarkable. The unique property of **MANIFOLD** that enables such high degree of modularity is inherited from its underlying IWIM (*Idealized Worker Idealized Manager*) model in which communication is set up from the *outside* [6]. The core relevant concept in the IWIM model of communication is isolation of the

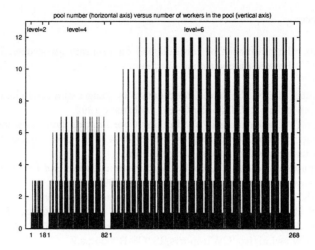

Fig. 4. Different pools of workers created during the semi-sparse-grid parallel applications.

computational responsibilities from communication and coordination concerns, into separate, pure computation modules and pure coordination modules. This is why the MANIFOLD modules in our example can coordinate the already existing computational Fortran subroutines, without any change.

An added bonus of pure coordination modules is their re-usability: the same MANIFOLD modules developed for one application may be used in other parallel applications with the same or similar cooperation protocol, regardless of the fact that the two applications may perform completely different computations (the sparse-grid and semi-sparse-grid application use the same protocol manifold, see also reference [7] for this notion of re-usability).

The performance evaluation of our test problem shows that MANIFOLD performs very well. Encouraged by the good results of this pilot study and the practical value it has for the partners who already use the sequential sparse-grid software, we now intend to develop a distributed restructuring of this sequential application. Also the distributed restructuring essentially consists of picking out the computation subroutines in the original Fortran 77 code, and gluing them together with coordination modules written in MANIFOLD. Again no rewriting of, or other changes to, these subroutines are necessary and we can reorganize according to a master/worker protocol. The additional work we have to do now, is to arrange for the MANIFOLD coordinators to send and receive the proper segments of the global data structure, which in the parallel version were available through shared memory, via streams. We can thereby implement the MANIFOLD glue modules (as separately compiled programs!) in such a way that their MANIFOLD code can run in distributed, as well as parallel environments.

References

1. F. Arbab. The influence of coordination on program structure. In *Proceedings of the 30th Hawaii International Conference on System Sciences*. IEEE, January 1997.
2. D. Gelernter and N. Carriero. Coordination languages and their significance. *Communication of the ACM*, 35(2):97–107, February 1992.
3. F. Arbab, P. Ciancarini, and C. Hankin. Coordination languages for parallel programming. *Parallel Computing*, 24(7):989–1004, July 1998. special issue on Coordination languages for parallel programming.
4. G.A. Papadopoulos and F. Arbab. *Coordination Models and Languages*, volume 46 of *Advances in Computers*. Academic Press, 1998.
5. F. Arbab. Coordination of massively concurrent activities. Technical Report CS-R9565, Centrum voor Wiskunde en Informatica, Kruislaan 413, 1098 SJ Amsterdam, The Netherlands, November 1995. Available on-line http://www.cwi.nl/ftp/CWIreports/IS/CS-R9565.ps.Z.
6. F. Arbab. The IWIM model for coordination of concurrent activities. In Paolo Ciancarini and Chris Hankin, editors, *Coordination Languages and Models*, volume 1061 of *Lecture Notes in Computer Science*, pages 34–56. Springer-Verlag, April 1996.
7. F. Arbab, C.L. Blom, F.J. Burger, and C.T.H. Everaars. Reusable coordinator modules for massively concurrent applications. *Software: Practice and Experience*, 28(7):703–735, June 1998. Extended version.
8. H. Deconinck and eds. B. Koren. *Euler and Navier-Stokes Solvers Using Multi-Dimensional Upwind Schemes and Multigrid Acceleration*. Notes on Numerical Fluid Mechanics 57. Vieweg, Braunschweig, 1997.
9. C.T.H. Everaars and B. Koren. Using coordination to parallelize sparse-grid methods for 3D CFD problems. *Parallel Computing*, 24(7):1081–1106, July 1998. special issue on Coordination languages for parallel programming.
10. Bradford Nicols, Dick Buttlar, and Jacqueline Proulx Farrell. *Pthreads Programming*. O'Reilly & Associates, Inc., Cebastopol, CA, 1996.

Scalable Parallelization of Harmonic Balance Simulation

David L. Rhodes[1] and Apostolos Gerasoulis[2]

[1] US Army CECOM/RDEC, Fort Monmouth, NJ 07703, USA
d.l.rhodes@ieee.org
[2] Rutgers University, Piscataway, NJ 08854, USA
gerasoul@cs.rutgers.edu

Abstract. A new approach to parallelizing harmonic balance simulation is presented. The technique leverages circuit substructure to expose potential parallelism in the form of a directed, acyclic graph (dag) of computations. This dag is then allocated and scheduled using various linear clustering techniques. The result is a highly scalable and efficient approach to harmonic balance simulation. Two large examples, one from the integrated circuit regime and another from the communication regime, executed on three different parallel computers are used to demonstrate the efficacy of the approach.

1 Introduction

The harmonic balance simulation method is widely used in the arena of large-signal, steady-state analysis of nonlinear circuits. It is often times applied to high frequency electronics since it directly accommodates frequency domain models [9, 10, 4]. Direct support for frequency domain models allow harmonic balance to be preferable in situations where distributed elements or frequency-dependent effects are important, for example, where accurate dispersive transmission line analysis (*e.g.* due to skin-effects, dielectric properties) is necessary. Such conditions are typical in radio-frequency (rf) and microwave circuits and increasingly arise in high-speed, deep sub-micron integrated circuits (IC) as well.

Nonlinear circuit simulations, of any type, are computationally intensive and therefore there have been several efforts to speed them via the use of parallel computation. Since linear matrix operations lie at the foundation of many nonlinear circuit simulation techniques, including harmonic balance, it is important to consider a variety of previous circuit simulation parallelization efforts. For example, several non-harmonic balance analog circuit simulators have been speed-up by parallelization of their inherent linear matrix operations [12, 3, 8]. A parallelization of the (linear part of the) harmonic balance method has been demonstrated [14]. While this effort provided 97% processor efficiency this approach—as well as the others cited—all display limited *scalability*.

Scalability is a measure of the ability of the parallelization technique to *efficiently* make use of an increasing number of available processing elements (PEs).

In the ideal, speed-up scales linearly with an increasing number of PEs since *potential* computational power is increasing linearly. Thus, a method which maintains close to linear speed-up over a wide range of available PEs is called scalable. Since we are interested in an efficient technique for today's parallel processors which contain tens to hundreds of PEs, scalability is an important concern.

The method in [14] parallelized the entire analysis of the linear portion of the harmonic balance method on a per-frequency basis, and thus the scalability of this approach is completely limited by the number of frequencies required in the analysis—adding PEs beyond the number of frequencies for the particular input file causes no additional speed up. Alternatively, methods which parallelize the underlying mathematics as it arises in circuit simulation have not, as of yet, achieved reasonable scalabilty. For example, the results in [12, 3] show efficiencies that are rapidly falling off even for ten PE computing systems. A peak efficiency of about 38% for 8 processors (about 3×) has been shown, but speed up then *decreases* beyond 8 processors [8].

A different parallel approach which leverages *circuit substructure*, as we do here, was demonstrated for nonlinear transient domain analysis using the waveform relaxation technique [17]. This approach yielded somewhat better results (efficiencies) than the mathematically oriented approaches but also shows signs of scalability limitations. For example, efficiencies of about 30–60% are shown for ten processors (about 6× speed up), while the method here achieves about 40% efficiency on 64 processors (*i.e.* 25× speed-up) and about 25% efficiency for 128 processors (32× speed-up). This is for the harmonic balance technique, of course, and not waveform relaxation as in [17].

2 Harmonic Balance Technique

While an overview of the harmonic balance method is given with the aid of Figure 1, please see the references for a full treatment of the technique [9, 10, 4]. The harmonic balance technique divides the circuit or system description into two portions: *linear* and *nonlinear*, interconnected at the *interface* (a set of circuit nodes). The 'balance' then entails iterative adjustment (using an optimization procedure) of the harmonic voltages at the interface until the harmonic currents as given by the linear and nonlinear sides 'agree'. This is done for a fixed set of frequency points, or harmonics, as dictated by the input parameters (the term frequency will be used henceforth for consistency). Sometimes only a few frequency points are used, say for nonlinear amplifier studies, while in other cases hundreds to thousands of frequency points are used, *e.g.* when complete time domain information is important. The 'analysis' on each side of the interface is quite different, but in each case we are interested in computing the frequency-domain currents at the interface given the frequency-domain voltages.

During the balancing iterations, each of the linear- and nonlinear-sides must compute frequency-domain currents given frequency-domain voltages at the interface and at external ports. For the **linear-side**, this is computed as a matrix multiplication (at each frequency) of the frequency-domain voltages with the

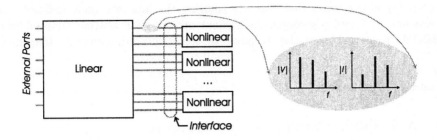

Fig. 1. Visualization of the harmonic balance method

equivalent ('reduced') *admittance matrix* of the linear portion of the circuit. Thus, the linear portion of the circuit may be reduced to its equivalent admittance matrix at each frequency point *once* and then subsequent 'analyses' during the balance process are merely matrix multiplications.

In order to compute the equivalent admittance matrix (at each frequency) of the linear side, circuit nodes that are internal to the linear side and not connected to either of the interface or external ports must be reduced. As it turns out from Kirchoff's currect/voltage laws (KCL and KVL respectively), Gaussian elimination becomes the basic step of such an equivalent reduction. Assuming the typical matrix notation [9, 10, 4], an internal node, k, may be equivalently reduced in the admittance matrix using the expression:

$$Y_{i,j}^f = Y_{i,j}^f - \frac{Y_{i,m}^f Y_{m,j}^f}{Y_{k,k}^f} \qquad \forall m \neq k \quad \text{and} \quad \forall i,j,f \qquad (1)$$

where $\mathbf{Y}^f$ is the admittance matrix for a single frequency point f. Assuming that the reduction is to a constant number of nodes, this reduction can be seen to be of *cubic* order; *i.e.*, as linear circuit nodes are added, computational time for the reduction to a fixed-sized admittance matrix increases in a cubic fashion.

On the **nonlinear-side**, each nonlinear model appears connected individually at the interface. Figure 1 only shows example models which have three nodes (as typical of transistors), but any number greater than or equal to two is possible. As is the case for the linear side, nonlinear model evaluation requires computation of frequency-domain currents given the frequency-domain voltages; these models typically make use of Fourier domain transform techniques to allow time-domain modeling.

The overall current at each interface node (at each frequency) is the sum of current contributions from both the linear and nonlinear sides. In accordance with KCL, this current should be zero. Thus an appropriate error function widely used in harmonic balance uses the L2 norm:

$$error = \left(\sum_{j \in \text{interface nodes}} \sum_{k \in \text{frequency}} I_{j,k}^2 \right)^{1/2} \qquad (2)$$

where $I_{j,k}$ is the sum of all currents entering the interface node indexed j at frequency k. Finally, a conjugate-gradient optimization routine adapted from MINPACK [1] performs the 'balance' by iteratively adjusting the frequency-domain voltages until an acceptably small current *error* is reached. Of course, at each frequency-domain voltage revision, the linear and nonlinear evaluations described must be used.

3 A Scalable Parallelization Technique

The harmonic balance algorithm, irrespective of parallelization, is then:

Form linear matrix and reduce to nodes of external ports and nonlinear models
Guess harmonic voltages at interface
forever
    Compute harmonic currents from linear side (matrix multiplication)
    Compute harmonic currents from nonlinear side (model evaluation)
    Evaluate *error* (Eq. 2)
    if *error* is acceptably small **then done**
    Update harmonic voltage guess (via conjugate-gradient optimization)

Within this analysis framework, several parallelization techniques are possible. In general, since only a few components are interconnected in typical circuits, even as the circuit itself becomes large, the admittance matrices—or other expressions of circuit equations—tend to be *sparse* [15]. Even with this sparseness characteristic, parallelization of the matrix solutions in other circuit simulators has met with limited scalability [12,3]. On the linear side only, parallelization of the entire process of model evaluations, matrix fill and reduction was shown effective, but that approach has scalability limited to the number of frequencies to be analyzed [14].

At first, it might appear that the harmonic balance iterations will dominate computation time as these are looped. However, as mentioned earlier the linear portion of the computation, represented entirely on the first line of the algorithm, actually dominates many harmonic balance calculations. As the circuit grows, linear fill and reduction tend towards cubic order, while the balance remains close to linear. This isn't always true of course, but many rf/microwave circuits do not exhibit strong interaction among nonlinear elements (unlike digital switching circuits)—although there are certainly exceptions. Thus, in many cases the linear part of the analysis dominates.

Obviously, a critical element for any parallelization technique is that sufficient parallelism be *exposed*, with a granularity that is not too small. **Granularity** may be loosely defined as the ratio of useful computation to inter-PE communication time. It is therefore a function of the particular multi-processor computing resource, even for a fixed problem—thus a technique is needed which both exposes a good deal of parallelism with a granularity useful for the computing resource at hand. The method here is just such a technique, leveraging circuit substructure to expose medium grain parallelism.

An introductory example will help clarify the general technique; first consider the linear part of harmonic balance by the example circuit, composed using subcircuits, as shown in Figure 2. In Figure 2.A, the circuit A has an instance of circuit $B1$ and $B2$ in it. In turn, $B1$ has instances of $C1$, $C2$ and $C3$ in it while $B2$ contains $C4$ and $C5$. The dark circles are circuit nodes, shown with their node labels.

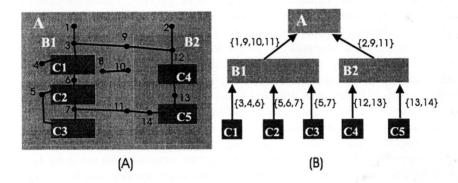

Fig. 2. (A) Linear subcircuit structure; (B) computational structure resulting from the method, where arrows represent data-dependencies

A typical analysis method would fully elaborate the circuit, forming a very large, sparse matrix [15] for the entire circuit description including all circuit nodes in all subcircuits. Alternatively, each subcircuit's internal nodes can be reduced prior to fill into the next level; with this technique, a full elaboration never exists. In this sense, such a method takes advantage of the sparsity present in circuit descriptions but without use of explicit sparse matrix techniques. Figure 2.B shows the computational flow of this approach for the example in Figure 2.A. The computation represented by each box is: (i) **matrix fill** in which requires admittance matrix evaluation of local circuit element models and use of subcircuit admittance matrices in the fill, and (ii) **matrix reduction** which then reduces internal circuit nodes to allow presentation of the reduced admittance matrix to the next level. Note that the boxes in Figure 2.B represent the *computation* required for this subcircuit and the arc represents the passing of admittance information to the next hierarchical level. Arcs are labeled using the node names for the equivalent, reduced subcircuit as it must appear at the higher level. Since subcircuit matrix information is required prior to computation, this approach gives rise to a directed, acyclic graph (dag), actually a special form called an **in-tree**, of computations.

Note that the linear matrix reduction computation shown in Figure 2.B is for a single frequency, each frequency point in the analysis requires an independent computation, so the whole linear computation becomes a **forest** of in-trees. Inclusion of nonlinear models within the hierarchy is handled by 'carrying' the nodes attached to any nonlinear model 'up' the hierarchy. Such a procedure results in exactly the form Figure 1 while leaving internal linear-only node reduction in hierarchical form as in Figure 2.

For nontrivial circuits, this method exposes a good deal of parallelism, which exists across siblings for the in-tree at each frequency and across the trees themselves. The method in [14] can be viewed as a subset of this approach in that it parallelizes across entire frequency trees only. Since all communicated results are only needed by one consumer (the graph is an in-tree), a message-passing approach is appropriate. The current implementation is message-passing based and uses the Message Passing Interface (MPI) standard [7, 16].

Even though linear computation is the dominant and more interesting part of the overall technique, next consider opportunities for parallelization in the balance portion of the method. First, the ('**forever**') loops themselves are serially dependent—*i.e.* for some iteration i, the frequency-domain voltage update is done in loop $i-1$, while the update itself cannot be done until the ith *error* is computed. Thus, only parallelization within a single step is possible (assuming that the basic algorithm itself is not changed). Within this loop the nonlinear current computations dominate—the linear current computations are small matrix multiplications (at each frequency), and the error evaluation and frequency-domain voltage updates are also small calculations. The optimization update, a non-sparse, second-order, conjugate-gradient matrix computation is also not readily parallelizable and also is of small size. This matrix size is exactly

$$2 \times \textit{number-frequencies} \times \textit{number-nodes-in-interface} \qquad (3)$$

where the factor of 2 arises because the frequency-domain voltages are complex-valued numbers. More importantly, this calculation does not usually tend to be a significant contributor to the iteration computation time.

Fortunately, the nonlinear current computation is readily parallelizable, by parallelizing the evaluation of each nonlinear device, as can be visualized in Figure 1. Thus, the non-linear analysis becomes a straightforward parallelization of the nonlinear model evaluations; the next section describes the background and specific technique for parallelizing the forest of in-trees that arises from the linear computation.

4 Allocation and Scheduling

For the linear part of the method, the forest of in-trees must be allocated and scheduled on the number of PEs to be used—this resource constrained scheduling problem is readily recognized as NP-hard/complete [5]. A further complication is that computation times (runtimes) are not known. Therefore an allocation/scheduling approach which *does not require* graph node runtimes is used. It is interesting to note that node computation times for realistic inputs do in fact vary widely—by two to three orders of magnitude—giving rise to a very 'irregular' scheduling problem.

Alternatively, approaches which require runtimes could be considered: (i) a quick runtime estimation function could be developed, or (ii) since runtimes are usually not highly dependent on frequency, the in-trees could be 're-scheduled' after the first frequency point. Of course, any scheduling time must be included in the net parallel efficiency achieved—meaning that only 'low-order' methods are potential candidates. Since good results have been acheived without the use of runtimes, as discussed next, and since the scheduling time for even low-order schedulers would have significant impact, these approaches were rejected.

The method makes use of a **linear clustering** [6] applied to the in-tree structure. Linear clusters are just single dependency chains of the in-tree, each of these is then statically allocated to a PE by assigning each in order to a PE, modulo the number of PEs—this allocation is a *wrap allocation of linear clusters*. Within this static allocation, local scheduling of each PE is dynamic, each PE loops first from the lowest to highest frequency and then from lowest (tree) level to highest looking for a 'ready' computation.

The first results obtained using this method revealed that load-balancing remained problematic. Obviously there is no assurance that a balanced allocation will result from this, and in fact without node runtimes or at least estimates, no allocation method could make such a guarantee. To remedy this situation, **rotated allocation** across 'frequency' in-trees is used. That is, a single in-tree is clustered and wrap allocated, but then this static assignment is 'rotated' across frequencies (modulo the number of PEs). The culmination of these approaches is the final allocation/scheduling technique and that for which results are subsequently shown. Note that communication overheads observed in execution traces (not shown for space reasons) are consistent with other investigations [18, 2, 11].

5 Examples and Results

In order to validate and demonstrate the approach, three different large-scale, parallel processors were used, with up to 128 PEs, an order of magnitude higher than presented in previous circuit simulators [3, 17]. These are: Cray's T3E, IBM's SP2, SGI's Origin-2000. Vendor supplied MPI routines were used in each case. Every effort has been made to ensure that the runtimes presented are accurate, however, there are a few factors that cannot be controlled. Since these machines are multi-user, various types of interference are possible including processor multi-tasking on some machines and communication contention. Nevertheless, multiple executions were done on 'quiet' machines and the data presented is felt to be correct. Two examples are executed on each of the processors with varying numbers of PEs.

The first example comes from the monolithic, microwave integrated circuit (MMIC) domain. It is a linear circuit hierarchically composed with 7 parametric circuit descriptions which, when elaborated, contains 169 subcircuit instances. It is analyzed at 20 frequency points, this gives 3,380 computational nodes total in the in-tree forest when elaborated across both circuit hierarchy and frequency. Figure 3 shows the results when using wrap clustering, (frequency index) rotating allocation, and dynamic local scheduling for the Cray T3E for up to 128 PEs. As can be seen, runtime is consistently improving all the way up to 128 PEs and, importantly, shows **reaching 32× speed up** (Figure 3.B). This appears to be the **highest reported speed up** for any type of circuit simulation. Figure 4.A and Figure 4.B show similar results for the IBM SP2 for up to 64 PEs and the SGI Origin-2000 for up to 32 PEs. All graphs are plotted in log-log form since the ideal runtime curve becomes linear with a slope of -1, *i.e.* ideal runtime is $x/number\text{-}of\text{-}PEs$ where x is a single PE runtime.

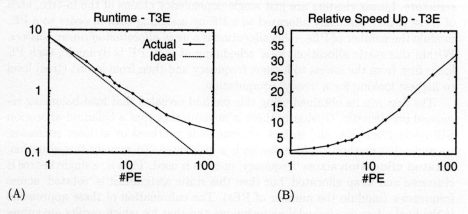

Fig. 3. Ideal and actual runtimes (A) and resulting relative speed-up (B) for the mmic circuit on the Cray T3E

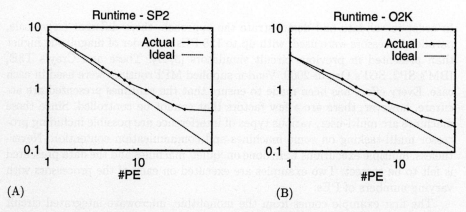

Fig. 4. Runtime data for mmic circuit for the IBM SP2 (A) and SGI Origin 2000 (B)

The second example file represents a set of 8 radios arranged in a circular fashion; each radio has a nonlinear transmitter and receiver, as well as filtering circuitry and an antenna model. A simple dispersive (loss dependent on frequency) model for the atmosphere is included. When elaborated, the circuit contains 16 nonlinear elements—a set of 6 frequency (harmonically related) points are analyzed. Figure 5.A shows the results for the linear part of the method only for this input. The parallelization here is not as good as for mmic on the T3E as shown in Figure 3 but consistent improvement up to 32 PEs is shown. Figure 5.B shows the results for the whole analysis. By examining the runtime difference for a single PE in Figures 5.A and 5.B, it can be seen that the nonlinear balance time is about 15 seconds of the 100 seconds run (this clearly illustrates the earlier statement that linear analysis time tends to dominate a harmonic balance simulation). As can be seen in Figure 5.B, *overall* parallel efficiency is not as good, but this is partly because the circuit only has 16 nonlinear elements, hence for any number of PEs over this amount the nonlinear portion of the method cannot be sped up.

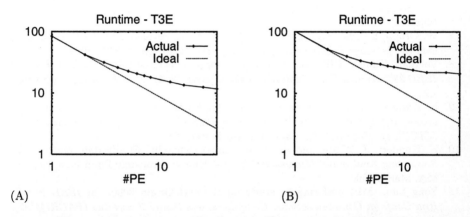

Fig. 5. Ideal and actual runtimes for linear portion only (A) and full analysis (B) for the cosite file executed on the Cray T3E

6 Concluding Remarks

A novel approach to the scalable parallelization of a circuit simulation problem has been developed. This includes a new approach to exposing parallelism as well as application-domain specific methods for allocation and scheduling. Measured speed-ups for three different parallel computers demonstrate the efficacy of the approach.

The authors would like to thank Professor Shirley Browne of the University of Tennessee Knoxville for assistance in obtaining execution trace results using the VAMPIR tool [13], the DoD High Performance Computing Modernization Office's CEN Computational Technology Area and ARPA under Order Number F425 (via ONR contract N6600197C8534) for partial project support.

References

1. http://netlib.att.com/netlib/minpack/.
2. Ghieth A. Adandah and Edward S. Davidson. Modeling the communication performance of the IBM SP2. In *Proc. of 10th IEEE Int. Parallel Processing Symp.*, pages 249–57, 15–19 April 1996. Honolulu, HI.
3. C. Chen and Y. Hu. A practical scheduling algorithm for parallel LU factorization in circuit simulation. In *Proc. of IEEE Int. Symp. on Circuits and Systems*, pages 1788–91, 8–11 May 1989. Portland, OR.
4. Benjamin Epstein, Stewart Perlow, David Rhodes, J. Schepps, M. Ettenberg, and R. Barton. Large-signal MESFET characterization using harmonic balance. In *IEEE Microwave Theory and Techniques-S Digest*, pages 1045–8, paper NN–2, May 1988. New York, NY.
5. Michael R. Garry and David S. Johnson. *Computers and Intractability: A Guide to the Theory of NP-Completeness*. W.H. Freeman and Company, New York, NY, 1979.
6. A. Gerasoulis and T. Yang. On the granularity and clustering of directed acyclic task graphs. *IEEE Trans. on Parallel and Distributed Systems*, 4(6):686–701, June 1993.
7. W. Gropp, E. Lusk, and A. Skjellum. *Using MPI: Portable Parallel Programming with the Message Passing Interface*. MIT Press, Cambridge, MA, 1994.
8. Volkhard Klinger. DiPaCS: a new concept for parallel circuit simulation. In *Proc. of IEEE 28th Annual Simulation Symposium*, pages 32–41, 9–13 April 1995. Phoenix, AZ.
9. K. Kundert and A. Sangiovanni-Vincentelli. Finding the steady-state response of analog and microwave circuits. In *IEEE 1988 Custom Integrated Circuits Conf. (CICC)*, 16–19 May 1988. paper 6.1, Rochester, NY.
10. K. Kundert, J. White, and A. Sangiovanni-Vincentelli. *Steady-state Methods for Simulating Analog and Microwave Circuits*. Kluwer Academic Publishers, Boston, MA, March 1990.
11. Yong Luo. MPI performance study on the SGI Origin 2000. In *IEEE Pacific Rim Conf. on Communications, Computers, and Signal Processing (PACRIM'97)*, pages 269–72, 20–22 August 1997. Victoria, CA.
12. K. Mayaram, P. Yang, J. Chern, R. Burch, L. Arledge, and P. Cox. A parallel block-diagonal preconditioned conjugate-gradient solution algorithm for circuit and device simulations. In *IEEE Int. Conf. on Computer-Aided Design (ICCAD'90)*, pages 446–9, 11–15 Nov 1990. Santa Clara, CA.
13. PALLAS GmbH, Brühl, Germany. *Vampir User's Manual, Release 1.1*.
14. D. Rhodes and B. Perlman. Parallel computation for microwave circuit simulation. *IEEE Trans. on Microwave Theory and Techniques*, 45(5):587–92, May 1997.
15. V. Rizzoli, F. Mastri, F. Sgallari, and V. Frontini. The exploitation of sparse-matrix techniques in conjunction with the piecewise harmonic balance method for nonlinear microwave circuit analysis. In *IEEE Microwave Theory and Techniques-S Digest*, volume 3, pages 1295–8, paper OO–5, May 1990. Dallas, TX.
16. M. Snir, S. Otto, S. Huss-Lederman, D. Walker, and J. Dongarra. *MPI: The Complete Reference*. MIT Press, Cambridge, MA, 1996.
17. E. Xia and R. Saleh. Parallel waveform-newton algorithms for circuit simulation. *IEEE Trans. on Computer-Aided Design*, 11(4):432–42, April 1992.
18. Zhiwei Xu and Kai Hwang. Modeling communication overhead: MPI and MPL performance on the IBM SP2. *IEEE Parallel and Distributed Technology: Systems and Applications*, 4(1):9–23, 1996.

Towards an Effective Task Clustering Heuristic for LogP Machines

Cristina Boeres, Aline Nascimento* and Vinod E. F. Rebello

Instituto de Computação, Universidade Federal Fluminense (UFF)
Niterói, RJ, Brazil
e-mail: {boeres,aline,vefr}@pgcc.uff.br

Abstract. This paper describes a task scheduling algorithm, based on a *LogP*-type model, for allocating arbitrary task graphs to fully connected networks of processors. This problem is known to be NP-complete even under the delay model (a special case under the *LogP* model). The strategy exploits the replication and clustering of tasks to minimise the ill effects of communication overhead on the makespan. The quality of the schedules produced by this *LogP*-based algorithm, initially under delay model conditions, is compared with that of other good delay model-based approaches.

1 Introduction

The effective scheduling of a parallel program onto the processors of a parallel machine is crucial for achieving high performance. One common program representation is a *directed acyclic graph* (*DAG*) whose nodes denote *tasks* and whose edges denote data dependencies (and hence communication, if the tasks are mapped to distinct processors). These program communication costs can be reduced by clustering or grouping tasks together to share processors. Until recently, the standard communication model in the scheduling community has been the *delay model*, where the sole architectural parameter for the communication system is the *latency* or *delay*, i.e. the transit time for each message [15].

Clearly, the majority of scheduling algorithms perform some sort of *clustering* (i.e. grouping two or more tasks onto the same processor). However in this paper, the term *cluster algorithms* refer specifically to the class of algorithms which initially consider each task as a cluster (allocated to a unique processor) and then merge clusters (tasks) if the makespan can be reduced.

DSC [17] is a clustering algorithm that finds optimal 1-linear schedules (without replicating tasks) for *DAG*s with granularity γ which are within a factor of $1 + \frac{1}{\gamma}$ of the true optimal schedule. Under this delay model heuristic, an optimal schedule can be found in polynomial time for fork, join, and coarse-grained (inverted) tree *DAG*s. Note that Jung et al. concluded that the best schedule produced by an algorithm that ignores replication may be worse than

---

* The author is supported by a postgraduate scholarship from CAPES, Ministério de Educação e Cultura (MEC), Brazil.

the true optimal schedule by a multiplicative factor which is a function of the communication delay [10].

If recomputation is allowed, there are cases where coarse-grained *DAG*s can be scheduled (truly) optimally in polynomial time [13, 14]. Palis *et al.* [14] proposed a clustering algorithm *PLW* which produces a schedule with a makespan at most $(1+\varepsilon)$ times the makespan of the true optimal, where the granularity of the *DAG* is at least $\frac{1-\varepsilon}{\varepsilon}$. Therefore, for any granularity of graph, *PLW* produces a schedule whose makespan is at most twice the optimal.

Unfortunately, research has shown that the delay model is not that realistic since it assumes certain properties (e.g. the ability to *multicast*, i.e. to send a number of different messages to different destinations simultaneously; and that communication can completely overlap with computation of tasks) that may not hold in practice – the CPU generally has to manage or at least initiate each communication [6]. The dominant cost of communication in today's architectures is that of crossing the network boundary. This is a cost which cannot be modelled as a latency and thus implies that new classes of scheduling heuristics are required to generate efficient schedules for realistic abstractions of today's parallel computers [4].

These architectural characteristics have motivated new (and now widely accepted) parallel programming models such as *LogP* [6] and *BSP* [16]. The *LogP model*, for example, is an MIMD message-passing model with four architectural parameters: the latency, L; the overhead, o, the CPU penalty incurred by a communication action; the gap, g, a lower bound on the interval between two successive communications; and the number of processors, P. Although the *BSP model* takes a less prescriptive view of communication actions, it has been shown that the two models are similar particularly terms of communication [4, 16].

In comparison with the delay model, identifying scheduling strategies that specify such *LogP*-type characteristics in their communication models is difficult. Only recently have a few scheduling algorithms appeared in the literature. Zimmerman *et al.* extended the work of [11, 17] to propose a clustering algorithm which produces optimal k-linear schedules for tree-like graphs when $o \geq g$ [18]. Boeres *et al.* [2, 3, 4] proposed a task replication strategy for scheduling arbitrary UET-UDC (unit execution task/unit data communication) *DAG*s on machines with a bounded number of processors under both the *LogP* and *BSP* models. A novel feature being the use of task replication and tight clustering to support the bundling of messages as a technique to reduce communication costs.

The purpose of this work is to continue to study the problem of scheduling arbitrary task graphs under *LogP*-type models, by building on the results of existing delay model-based task replication clustering algorithms. In the following section, we present a replication-based task clustering algorithm for scheduling general or arbitrary *DAG*s on *LogP*-type machines (although an unbounded number of processors is assumed). The specific architectural parameters considered have been adopted from the *Cost and Latency Augmented DAG* (CLAUD) model [5] (developed independently of *LogP*). In the CLAUD model, an application is represented by a *DAG G* and the set of m identical processors by $\mathcal{P}$.

The overheads associated with the *sending* and *receiving* of a message are denoted by λ_s and λ_r, respectively, and the communication latency or delay by τ. The CLAUD model very naturally models *LogP*, given the following parameter relationships: $L = \tau$, $o = \frac{\lambda_s + \lambda_r}{2}$, $g = \lambda_s$ and $P = m$.

The delay model is a special case in under a *LogP*-type model where the overheads are zero. Therefore, any proposed *LogP*-type scheduling algorithm still has to produce good schedules under delay model conditions. This is the focus of this paper and Section 3 compares the makespans produced by the new algorithm under delay model conditions against two other well known delay model clustering algorithms – *DSC* and *PLW*. Section 4 draws some conclusions and outlines future work regarding scheduling for *LogP*-type machines. (Throughout this paper, the term *delay model conditions* is used to refer to situation where the overheads are zero and the term *LogP conditions* used when the overheads are nonzero.)

2 A New Heuristic

A parallel application is represented by a DAG $G = (V, E, \varepsilon, \omega)$, where V is the set of tasks, E is the precedence relation among them, $\varepsilon(v_i)$ is the execution cost of task $v_i \in V$ and $\omega(v_i, v_j)$ is the weight associated to the edge $(v_i, v_j) \in E$ representing the amount of data units transmitted from task v_i to v_j.

For the duration of each communication overhead the processor is effectively *blocked* unable to execute other tasks in V or even to send or receive other messages. Consequently, any scheduling algorithm must, in some sense, view these sending and receiving overheads also as "tasks" to be executed by processors.

The new clustering heuristic (BNR) attempts to minimise the makespan by trying to construct a cluster for each task v_i so that v_i starts executing at the earliest possible time. Cluster v_i will contain the *owner task* v_i and, determined by a cost function, copies of some of its ancestors (i.e. BNR employs task replication). Like most clustering algorithms, BNR consists of two phases: the first determines which tasks should be included and their order within a cluster; the second, identifies the clusters required to implement the input DAG G and maps each necessary cluster to a unique processor. Even though an unbounded number of processors is assumed, BNR does try to minimise the number of processors required as long as the makespan is not adversely affected.

2.1 Properties for *LogP* scheduling

In order to schedule tasks under a *LogP*-type model, BNR applies a couple of new *general scheduling restrictions* with regard to properties of clusters (one in each phase). In the first phase, only the owner task of a cluster may send data to a task in another cluster. This restriction does not adversely affect any makespan since, by definition, an owner task will not start execution later than any copy of itself. Also, if a non-owner task u in cluster $C(v_i)$ were to communicate with a task in another cluster, the processor to which $C(v_i)$ is allocated would incur a send overhead after the execution of u which in turn might force task v_i to start at a time later than its earliest possible. The second phase uses the restriction that each cluster can have only one successor. If two or more clusters share the same predecessor cluster, then this phase will assign a unique copy of the predecessor cluster to each successor irrespective of whether or not the data being sent to each successor is identical. This removes the need to incur the multiple send overheads at the end of the predecessor cluster which unnecessarily delay the execution of successor clusters.

These general restrictions are new in the sense that they do not need to be applied by scheduling strategies which are used exclusively for delay model conditions (because of the assumptions made under the delay model). The purpose of these restrictions is to aid in the minimisation of the makespan, however they do incur the penalty of increasing the number of clusters required and thus the number of processors needed. Where the number of processors used is viewed as a reflection on the cost of implementing the schedule, a post-pass optimisation can be applied to relax the above restrictions and remove clusters which now become redundant (i.e. those clusters whose removal will not increase the makespan). Note that Löwe et al. [7] suggests that the problem of determining the optimal amount of replication is effectively NP-complete for the *LogP* model.

2.2 The first phase

For each task $v_i \in V$ (where i is assigned topologically), BNR constructs a cluster $C(v_i)$ (with v_i being its owner) from the ancestors of v_i. The term $iancs(C(v_i))$ defines the set of immediate ancestors of tasks already in $C(v_i)$ which themselves are not in $C(v_i)$, i.e. $iancs(C(v_i)) = \{u_j \mid (u_j, t_l) \in E \wedge u_j \notin C(v_i) \wedge t_l \in C(v_i)\}$. A task $u_j \in iancs(C(v_i))$ is incorporated into cluster $C(v_i)$ if v_i's start time can be reduced. Thus, BNR includes u_j in $C(v_i)$ based on comparing the start time of v_i under the following two situations: (a) when $u_j \notin C(v_i)$; and (b) when $u_j \in C(v_i)$.

The pseudocode for BNR can be seen in Algorithm 3. The strategy visits each task $v_i \in G$, constructing cluster $C(v_i)$ and calculating v_i's *earliest schedule time* $e_s(v_i)$ (which is defined as the start time of v_i in its own cluster $C(v_i)$) as described in the manner below. Note that we assume the definition of the *earliest start time* of task v_i to mean the optimal or earliest possible start time of v_i [15]. Our objective is to attempt to find a schedule for each $v_i \in V$ such that $e_s(v_i) = e(v_i)$.

Let $C(v_i) = \langle t_1, \ldots, t_l, \ldots t_k, v_i \rangle$ be tasks of cluster v_i in execution order. These tasks are not listed in non-decreasing order of their earliest start time nor earliest schedule time but rather in an order determined by the sequence in which tasks were included in the cluster and the position of their immediate successor task at the time of their inclusion. The *critical cluster path cost*, $m(C(v_i))$, is the earliest time that v_i can start **due** to the ancestor tasks of v_i in $C(v_i)$. This is calculated simply by ignoring the communications of all tasks $\in iancs(C(v_i))$ to $C(v_i)$, as shown in lines 2 and 3 of Algorithm 1. $et(t_p, v_i)$ is the end time of this copy of task t_p in $C(v_i)$ and which is effectively never scheduled earlier than the original (owner) task t_p in $C(t_p)$(line 3). The end time of t_1, the first task, $(et(t_1, v_i) = e_s(t_1) + \varepsilon(t_1))$ is based on its earliest schedule time even though its ancestors (when it has some) are not (yet) in the cluster. The reason for this is that BNR, in some sense, bases this value on the likely cost in future iterations, i.e. best, of course, being limited by the task's earliest schedule time.

Algorithm 1 : *critical-cluster-path* $(C(v_i))$;
1 let $C(v_i) = \langle t_1, t_2, \ldots, t_k \rangle$; $et(t_1, v_i) := e_s(t_1) + \varepsilon(t_1)$;
2 for $p = 2, \ldots, k$ do
3    $et(t_p, v_i) := \max\{et(t_{p-1}, v_i), e_s(t_p)\} + \varepsilon(t_p)$;
4 $m(C(v_i)) := et(t_k, v_i)$;

Algorithm 2 : *critical-ancs-path* $(v_i, C(v_i))$
1 $maxc := 0$; define $iancs(C(v_i))$; let $C(v_i) = \langle t_1, t_2, \ldots, t_k \rangle$;
2 if $(iancs(C(v_i)) \neq \emptyset)$ then
3   for all $u_j \in iancs(C(v_i))$ do
4   for each $(u_j, t_l) \in E$ such that $t_l \in C(v_i)$ do
5        $c(u_j, t_l, v_i) := e_s(u_j) + \varepsilon(u_j) + \lambda_s + L \times \omega(u_j, t_l) + \lambda_r$;
6   for $p = l, \ldots, k$ do
           $c(u_j, t_l, v_i) := \max\{c(u_j, t_l, v_i), e_s(t_p)\} + \varepsilon(t_p)$;
         end for each
7   if $(maxc < c(u_j, t_l, v_i))$ then
8          $maxc := c(u_j, t_l, v_i)$; $mtask := u_j$;
         end if
       end for all

Task u_j is a possible candidate to be incorporated into cluster $C(v_i)$ if $u_j \in iancs(C(v_i))$. Under **situation** (a), for each task u_j, BNR calculates the earliest time that v_i can start **due solely** to the chosen task u_j (known as the *critical ancestor path cost* $c(u_j, t_l, v_i)$) to be the sum of: the execution of u_j $(e_s(u_j) + \varepsilon(u_j))$; the sending overhead (λ_s) (due to the *LogP* general scheduling restrictions, there is no need to determine the number of overheads and gaps g incurred, since BNR only permits a cluster to send one message at the expense of needing multiple copies of the same cluster); the communication delay $(L \times \omega(u_j, t_l))$; the receiving overhead (λ_r); and the computation time of tasks t_l to t_k (i.e. the critical cluster path cost for task t_l to t_k). The task u_j whose $c(u_j, t_l, v_i)$ is the largest is the immediate ancestor of cluster $C(v_i)$ who most delays task v_i. This function can be seen in Algorithm 2 which assigns *mtasks* and *maxc* to the critical immediate ancestor and its critical ancestor path cost, respectively (line 8).

If *maxc* is greater than the critical cluster path cost, $m(C(v_i))$, than task *mtask* becomes the chosen candidate for inclusion into cluster $C(v_i)$. If this is not the case, the process stops and cluster $C(v_i)$ is now complete. This forms the condition of the main loop in the *construct-cluster* function as shown in lines 5 to 11 of Algorithm 3. Before committing $mtask$ to $C(v_i)$, it is necessary to compare the cost $maxc$ with situation (b) – the critical cluster path cost of the cluster $C(v_i) \cup \{mtask\}$. If $maxc$ is less than this new cost, task $mtask$ is not included (line 11) and the formation of cluster $C(v_i)$ is finished. Otherwise, task $mtask$ is included, critical ancestor path costs are calculated for the new set of tasks in $iancs(C(v_i))$ (line 9), the start time of v_i is updated (line 10) and the loop is repeated. Note that it is not necessary for all of the ancestors of v_i to be visited when constructing the cluster $C(v_i)$.

Algorithm 3 : *construct-cluster* (v)
1 if $pred(v) = \emptyset$ then $e_s(v) := 0$;
2 else
3    $(maxc, mtask) := critical\text{-}ancs\text{-}path(v, C(v))$;
4    $m(C(v)) := 0$; $e_s(v) := \max\{m(C(v)), maxc\}$;
5   while $(m(C(v)) < maxc)$;
6      $C(v) := C(v) \cup \{mtask\}$; $m'(C(v)) := m(C(v))$;
7      $m(C(v)) := critical\text{-}cluster\text{-}path(C(v))$;
8   if $(m(C(v)) \leq maxc)$ then
9        $(maxc, mtask) := critical\text{-}ancs\text{-}path(mtask, C(v))$;
10       $e_s(v) := \max\{m(C(v)), maxc\}$;
      else
11       $e_s(v) := \max\{m'(C(v)), maxc\}$; $C(v) := C(v) - \{mtask\}$;
    end while
  end if

2.3 The second phase

The second phase of the approach is to determine which of the generated clusters are needed to implement the schedule. This can be achieved by simply tracing the cluster dependencies starting with the sinks. Note that multiple copies of clusters may be required. These schedule clusters are then allocated to processors, at worst one cluster per processor. In some cases, there is scope for optimisation: clusters whose execution times do not overlap can be mapped to a single processor; depending on the communication parameters, the number of copies of a cluster could be reduced without affecting the makespan, e.g. clusters have the ability to multicast under the delay model.

2.4 Algorithm analysis

At first glance, BNR appears to be similar to that of the PLW algorithm of Palis et al. [14] ignoring the differences for $LogP$ scheduling. Both of these algorithms consist of two phases, utilise task replication to form clusters (one for each task in G), even growing them by adding a single task at a time. However, the key difference is the use of the *earliest schedule time* to determine the start time of owner tasks (i.e. only for the original tasks in G and not their copies) rather than the *earliest start time* for all tasks including their copies. (Note that the earliest start time is a bound [14] since it may be impossible to actually schedule a task at this time [15]). A second difference is the execution order of tasks within a cluster. Note that PLW orders tasks in nondecreasing order of their earliest start time. The effect of these differences are highlighted by the results presented in the next section.

The complexity of PLW is $\mathcal{O}(n^2 \log n + ne)$, where $n = |V|$ and $e = |E|$. For BNR, the *critical-cluster-path* function is clearly $\mathcal{O}(n)$. Examining function *critical-ancs-path*, $\mathcal{O}(e)$ edges could emanate from tasks in $iancs(v_i)$. Given k, the number of tasks currently in cluster $C(v_i)$, there are at most $(n-k)$ tasks in $iancs(C(v_i))$. Consequently for the worst case number of edges, as the number of tasks in $C(v_i)$ grows, so the number of edges emanating from $iancs(C(v_i))$ diminishes. The complexity of *critical-ancs-path* is then $\mathcal{O}(ek)$ and the whole strategy $\mathcal{O}(n^2 ek)$.

3 Results

The results produced by BNR have been compared with the clustering heuristic PLW [14] and DSC [17] using a benchmark suite of DAGs which includes out-trees (OT), in-trees (IT), diamond DAGs (D), a set of randomly generated DAGs (R) and irregular DAGs (I) taken from the literature.

A few of the experimental results obtained are shown in Tables 1, 2 and 3. The subscript of the DAGs represent the number of tasks in the respective graphs. Each heuristic produced a schedule with a (predicted) makespan M and an actual makespan S (only one value appears in the column if M and S are the same). The time for the actual makespan of the schedule was derived from a *parallel machine simulator* [4]. The required number of processors is P.

The first set of experiments focus on the delay model and fine-grained DAGs for the following reasons. The best scheduling algorithms are based on the delay model since it is the standard scheduling model. The delay model is in fact a special case under the $LogP$ model ($g = o = 0$) where tasks can exploit the property of multicasting. Although the scheduling restrictions cause $LogP$ strategies, such as the one proposed in this work, to appear greedy with respect to the number of processors required, it is important to show that the makespans produced and processors used are comparable to those of delay model approaches under their model. Scheduling approaches such as DSC and PLW have already been shown to generate good results (optimal or near-optimal) for various types of DAGs. For this experimental analysis, the graphs were assigned unit execution and unit communication costs and the latency L was set to 1, 2, 5 and 10 units for different tests.

In the 196 experiments carried out in the first set, the schedules produced by BNR were never worse than those of DSC or PLW. BNR was better than DSC in 164 cases and equal in 32, and in 166 and 30 cases, respectively, against PLW. The performances of both DSC and PLW progressively worsen in com-

Table 1. Results for communication latency (L) equal to 1 and 2 units.

	Communication Cost = 1						Communication Cost = 2					
	DSC		PLW		BNR		DSC		PLW		BNR	
DAG	M/S	P	M/S	P	M/S	P	M/S	P	M/S	P	M/S	P
OT_{63}	11	32	8	46	6	32	11	21	8	36	6	32
OT_{511}	17	256	11/12	344	9	256	17	171	13	308	9	256
IT_{63}	11	32	11	21	11	16	11	21	13	21	11	21
IT_{511}	17	256	17	341	17	256	17	171	19/21	341	17	171
D_{64}	22	8	23/28	32	22	6	30	5	28/37	26	28	8
D_{400}	58	20	59/76	200	58	14	78	11	70/97	164	76	14
I_{41} [11]	16	13	16	22	15	10	21	9	18/22	20	17	13
R_{13}	8	5	9	7	7	5	9	3	9	7	8	4
R_{124}	8	67	8/9	76	7	64	10	62	11	73	9	65
R_{310}	18	110	21/22	189	16	105	24/25	106	26/29	173	20	121
R_{510}	20	170	23/25	327	17	184	25	155	31/34	295	21	214

Table 2. Results for communication latency (L) equal to 5 and 10.

	Communication Cost = 5						Communication Cost = 10					
	DSC		PLW		BNR		DSC		PLW		BNR	
DAG	M/S	P	M/S	P	M/S	P	M/S	P	M/S	P	M/S	P
OT_{63}	19	13	6	32	6	32	29/32	16	6	32	6	32
OT_{511}	31/33	121	14	260	9	256	50/53	115	9	256	9	256
IT_{63}	19	13	19	21	17	12	29	16	24	9	22	11
IT_{511}	31	121	29/31	169	30	108	50/49	115	41	73	36	88
D_{64}	55	6	39/56	20	36	11	61/60	3	54/62	17	45	9
D_{400}	163	18	95/160	97	104	35	230/225	15	137/229	66	135	52
I_{41} [11]	31	5	28/26	19	23	8	52	4	34/32	8	29	8
R_{13}	13	1	12	5	11	3	13	1	12	2	12	2
R_{124}	14/16	54	16/17	61	11	60	23	53	20/21	53	15	56
R_{310}	43/45	98	41	131	29	140	73/75	95	75/79	111	42	138
R_{510}	46/44	128	51/55	226	32	229	80	116	71/82	185	47	244

parison with BNR as the DAGs effectively become more fine-grained due to the increasing latency value. DSC produces schedules which are, on average, 12.2%, 16%, 32.1% and 39.1% worse than those of BNR for each of the respective latency costs. For PLW these values are 20.6%, 23.9%, 25% and 25.3%. DSC generally utilises the fewest processors since it does not replicate tasks. BNR, on the whole, uses fewer processors than PLW (especially for small latencies), but where this is not true the compensation is the better makespan.

A second set of experimental results is presented in Table 3. This table contains a comparison of makespans produced by the three strategies for a group of documented irregular, non-unit cost graphs used by various researchers. BNR continues to perform well.

	DSC		PLW			BNR	
DAG	M/S	P	M/S	P	M/S	P	
I_{7a} [8]	9	2	8	5	8	3	
I_{7b} [9]	8	3	11/12	5	8	3	
I_{10} [14]	30	3	27	3	26	3	
I_{13} [1]	301	7	275	8	246	8	
I_{18} [12]	530†/550	5	490/480	8	370	9	

† Note that Reference [12] reports a schedule time of 460 on 6 processors.

Table 3. Results for non-unit (task and edge) cost irregular DAGs ($L = 1$).

4 Conclusions and future work

Based on the results obtained so far, BNR compares favourably against traditional clustering-based scheduling heuristic such as DSC and PLW which are dedicated exclusively to the delay model. Clearly, results of further experiments using graphs with a more varied range of granularities and connectivities are needed to complete the practical evaluation of the algorithm. The main objective is, however, to analyse the heuristic's behaviour under $LogP$ conditions, comparing results under these conditions with other $LogP$ scheduling algorithms. In particular, we wish to investigate MSA [2] to ascertain the benefits of its technique of bundling messages in a $LogP$ environment. In terms of obtaining theoretical bounds on makespan optimality under $LogP$ conditions, further work needs to be done although we suspect that under the delay model the bound is no worse than that of PLW.

Acknowledgments

The authors would like thank Tao Yang and Jing-Chiou Liou for providing the DSC and PLW algorithms, respectively, and for their help and advice.

References

1. I. Ahmad and Y-K Kwok. A new approach to scheduling parallel programs using task duplication. In K.C. Tai, editor, *International Conference on Parallel Processing*, volume 2, pages 47–51, Aug 1994.
2. C. Boeres and V. E. F. Rebello. A versatile cost modelling approach for multicomputer task scheduling. *Parallel Computing*, 1999. (to appear).
3. C. Boeres and V.E.F. Rebello. Versatile task scheduling of binary trees for realistic machines. In C. Lengauer, M. Griebl, and S. Gorlatch, editors, *The Proceedings of the Third International Euro-Par Conference on Parallel Processing (Euro-Par'97)*, LNCS 1300, pages 913–921, Passau, Germany, August 1997. Springer-Verlag.
4. C. Boeres, V.E.F. Rebello, and D. Skillicorn. Static scheduling using task replication for LogP and BSP models. In D. Pritchard and J. Reeve, editors, *The*

Proceedings of the Fourth International Euro-Par Conference on Parallel Processing (Euro-Par'98), LNCS 1470, pages 337–346, Southampton, September 1998. Springer-Verlag.

5. G. Chochia, C. Boeres, M. Norman, and P. Thanisch. Analysis of multicomputer schedules in a cost and latency model of communication. In *Proceedings of the 3rd Workshop on Abstract Machine Models for Parallel and Distributed Computing*, Leeds, UK., April 1996. IOS press.
6. D. Culler, R. Karp, D. Patterson, A. Sahay, K.E. Schauser, E. Santos, R. Subramonian, and T. von Eicken. LogP: Towards a realistic model of parallel computation. In *Proceedings of the 4th ACM SIGPLAN Symposium on Principles and Practice of Parallel Programming*, San Diego, CA, May 1993.
7. J. Eisenbiergler, W. Lowe, and A. Wehrenpfennig. On the optimization by redundancy using an extended LogP model. In *Proceeding of the International Conference on Advances in Parallel and Distributed Computing (APDC'97)*, pages 149–155. IEEE Comp. Soc. Press, 1997.
8. A. Gerasoulis and T. Yang. A comparison of clustering heuristics for scheduling directed acyclic graphs on multiprocessors. *Journal of Parallel and Distributed Computing*, 16:276–291, 1992.
9. J-J. Hwang, Y-C. Chow, F.D. Anger, and C-Y. Lee. Scheduling precedence graphs in systems with interprocessor communication times. *SIAM J. Comput.*, 18(2):244–257, 1989.
10. H. Jung, L. Kirousis, and P. Spirakis. Lower bounds and efficient algorithms for multiprocessor scheduling of DAGs with communication delays. In *Proc. ACM Symposium on Parallel Algorithms and Architectures*, pages 254–264, 1989.
11. S.J. Kim and J.C. Browne. A general approach to mapping of parallel computations upon multiprocessor architectures. In *Proceedings of the 3rd International Conference on Parallel Processing*, pages 1–8, 1988.
12. Y. K. Kwok and I. Ahmad. Dynamic critical-path scheduling: An effective technique for allocating tasks graphs to multiprocessors. *IEEE Transactions on Parallel and Distributed Systems*, 7(5):505–521, May 1996.
13. W. Lowe and W. Zimmermann. On finding optimal clusterings of task graphs. In *Aizu International Symposium on Parallel Algorithm and Architecture Synthesis*, pages 241–247. IEEE Computer Society Press, 1995.
14. M.A. Palis, J.-C Liou, and D.S.L. Wei. Task clustering and scheduling for distributed memory parallel architectures. *IEEE Transactions on Parallel and Distributed Systems*, 7(1):46–55, January 1996.
15. C.H. Papadimitriou and M. Yannakakis. Towards an architecture-independent analysis of parallel algorithms. *SIAM J. Comput.*, 19:322–328, 1990.
16. D. B. Skillicorn, J. M. D. Hill, and W. F. McColl. Question and answers about BSP. *Scientific Computing*, May 1997.
17. T. Yang and A. Gerasoulis. DSC: Scheduling parallel tasks on an unbounded number of processors. *IEEE Transactions on Parallel and Distributed Systems*, 5(9):951–967, 1994.
18. W. Zimmermann, M. Middendorf, and W. Lowe. On optimal k-linear scheduling of tree-like task graph on LogP-machines. In D. Pritchard and J. Reeve, editors, *The Proceedings of the Fourth International Euro-Par Conference on Parallel Processing (Euro-Par'98)*, LNCS 1470, pages 328–336, Southampton, September 1998. Springer-Verlag.

A Range Minima Parallel Algorithm for Coarse Grained Multicomputers *

H. Mongelli[1] and S. W. Song[2]

[1] Universidade Federal do Mato Grosso do Sul
and Universidade de São Paulo <mongelli@ime.usp.br>
[2] Universidade de São Paulo <song@ime.usp.br>

Abstract. Given an array of n real numbers $A = (a_1, a_2, \ldots, a_n)$, define $MIN(i,j) = \min\{a_i, \ldots, a_j\}$. The range minima problem consists of preprocessing array A such that queries $MIN(i,j)$, for any $1 \leq i \leq j \leq n$, can be answered in constant time. Range minima is a basic problem that appears in many other important problems such as lowest common ancestor, Euler tour, pattern matching with scaling, etc. In this work we present a parallel algorithm under the CGM model (Coarse Grained Multicomputer), that solves the range minima problem in $O(\frac{n}{p})$ time and constant number of communication rounds.

1 Introduction

Given an array of n real numbers $A = (a_1, a_2, \ldots, a_n)$, define $MIN(i,j) = \min\{a_i, \ldots, a_j\}$. The range minima problem consists of preprocessing array A such that queries $MIN(i,j)$, for any $1 \leq i \leq j \leq n$, can be answered in constant time.

The parallel computing model used in this work is the CGM (Coarse Grained Multicomputer). The CGM(n,p) consists of an input of size $O(n)$ and p processors with point-to-point communication. An algorithm under this model consists of supersteps that involve a local computing round and a communication round between processors. In a communication round, each processor can send and/or receive $O(\frac{n}{p})$ data.

This is a more realistic model than the PRAM model for most real parallel computers. The number of communication phases is taken into consideration in the design of algorithms. It has been introduced in [7,8] where algorithms using a constant number of communication rounds were presented for several Computational Geometry problems. Algorithms requiring $O(\log p)$ communication rounds have been presented for the solution of several graph problems[6].

In this work we consider the range minima problem. We present an algorithm under the CGM(n,p) model, that solves the problem requiring a constant number of communication rounds and $O(\frac{n}{p})$ computation time. In the PRAM model, there is an optimal algorithm of $O(\log n)$ time [12]. This algorithm, however, is

---

* Supported by FAPESP (Fundação de Amparo à Pesquisa do Estado de São Paulo) - Proc. No. 98/06138-2, and CNPq - Proc. No. 52 3778/96-1 and CAPES.

not immediately transformable to the CGM model. Berkman et al. [4] describe an optimal algorithm of $O(\log \log n)$ time for a PRAM CRCW. This algorithm motivates the CGM algorithm presented in this paper.

Range minima is a basic problem and appears in other problems. In [12], range minima is used in the design of a PRAM algorithm for the problem of Lowest Common Ancestor (LCA). This algorithm uses twice the Euler-tour problem. The idea of this algorithm can be used in the CGM model, by using the same sequence of steps. In [6], the Euler-tour problem in the CGM model is discussed. Its implementation reduces to the list ranking problem, that uses $O(\log p)$ communication rounds in the CGM model [6]. Applications of the LCA problem, on the other hand, are found in graph problems (see [15]) and in the solution of some other problems. Besides the LCA problem, the range minima problem is used in several other applications [3, 10, 14].

2 The CGM model and some definitions

The CGM model was proposed by Dehne [7, 8]. It is similar to the BSP (Bulk-Synchronous Parallel) model[17]. However, it uses only two parameters: the input size n and the number of processors p, each one with a local memory of size $O(\frac{n}{p})$. The term *coarse grained* means the size of the local memory is much larger than $O(1)$. Each processor is connected by a router that can send messages in a point-to-point fashion. A CGM algorithm consists of an alternate sequence of computing and communication phases separated by a barrier synchronization. A computing phase is equivalent to a superstep in the BSP model. In a computing phase, we usually use the best sequential algorithm in each processor to process its data locally.

In each communication phase or round, each processor exchanges a maximum of $O(\frac{n}{p})$ data with other processors. In the CGM model, the communication cost is modeled by the number of communication phases. Some algorithms for Computational Geometry and graph problems require a constant number or $O(\log p)$ communication rounds [6–9]. Contrary to a PRAM algorithm, that is often designed for $p = O(n^k)$, with $k \in \mathcal{N}$ and each processor receives a small number of input data, in this model we consider the more realistic cases where $n \gg p$. The CGM model is particularly suitable in current parallel machines in which the global computing speed is considerably larger than the global communication speed.

Definition 1 *Consider an array $A = (a_1, a_2, \ldots, a_n)$ of n real numbers. The **prefix minimum** array is the array $P = (\min\{a_1\}, \min\{a_1, a_2\}, \ldots, \min\{a_1, a_2, \ldots, a_n\})$. The **suffix minimum** array is the array $S = (\min\{a_1, a_2, \ldots, a_n\}, \min\{a_2, \ldots, a_n\}, \ldots, \min\{a_{n-1}, a_n\}, \min\{a_n\})$.*

Definition 2 *The **lowest common ancestor** of two vertices u and v of a rooted tree is the vertex w that is an ancestor of u and v and that is farthest from the root. We denote $w = LCA(u, v)$.*

Definition 3 *The **lowest common ancestor problem** consists of preprocessing a rooted tree such that queries of the type $LCA(u,v)$, for any vertices u and v of the tree, can be answered in constant sequential time.*

3 Sequential algorithms

The algorithm for the range minima problem, in the CGM model, uses two sequential algorithms that are executed using local data in each processor. These algorithms are described in the following.

3.1 Algorithm of Gabow et al.

This sequential algorithm was designed by Gabow et al. [10] and uses the **Cartesian tree** data structure, introduced in [18]. An example of Cartesian tree is given in figure 1. In this algorithm we need also an algorithm for the lowest common ancestor problem.

Definition 4 *Given an array $A = (a_1, a_2, \ldots, a_n)$ of n distinct real numbers, the **Cartesian tree** of A is a binary tree whose nodes have as label the values of array A. The root of the tree has as label $a_m = \min\{a_1, a_2, \ldots, a_n\}$. Its left subtree is a Cartesian tree for $A_{1,m-1} = (a_1, a_2, \ldots, a_{m-1})$, and its right subtree is a Cartesian tree for $A_{m+1,n} = (a_{m+1}, \ldots, a_n)$. The Cartesian tree for an empty array is the empty tree.*

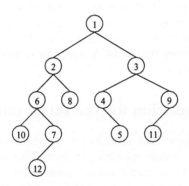

Fig. 1. Cartesian tree corresponding to the array $(10, 6, 12, 7, 2, 8, 1, 4, 5, 3, 11, 9)$.

ALGORITHM 1: Range Minima (Gabow et al.)

1. Build a Cartesian tree for A.
2. Use a linear sequential algorithm for the LCA problem using the Cartesian tree.

The construction of the Cartesian tree takes linear time [10]. Linear sequential algorithms for the LCA problem can be found in [5, 11, 16]. Thus, any query $MIN(i,j)$ is answered as follows. From the recursive definition of the Cartesian tree, the value of $MIN(i,j)$ is the value of the LCA of a_i and a_j. Thus, each range minima query can be answered in constant time through a query of LCA in the Cartesian tree. Therefore, the range minima problem is solved in linear time.

3.2 Sequential algorithm of Alon and Schieber

In this section we describe the algorithm of Alon and Schieber [2] of $O(n \log n)$ time. In spite of its non linear complexity, this algorithm is crucial in the description of the CGM algorithm, as will be seen in section 4. Without loss of generality, we consider n to be a power of 2.

ALGORITHM 2: Range Minima (Alon and Schieber)
1. Construct a complete binary tree T with n leaves.
2. Associate the elements of A to the leaves of T.
3. For each vertex v of T we compute the arrays P_v and S_v, the prefix minimum and the suffix minimum arrays, respectively, of the elements of the leaves of the subtree with root v.

Tree T constructed by algorithm 2 will be called **PS-tree**. Figure 2 illustrates the execution of this algorithm. Queries of type $MIN(i,j)$ are answered as follows. To determine $MIN(i,j)$, $1 \leq i \leq j \leq n$, we find $w = \text{LCA}(a_i, a_j)$ in T. Let v and u be the left and right sons of w, respectively. Then, $MIN(i,j)$ is the minimum between the value of S_v in the position corresponding to a_i and the value of P_u in the position corresponding to a_j. PS-tree is a complete binary tree. In [12, 15] it is shown that a LCA query in complete binary trees can be answered in constant time.

4 The CGM algorithm for the range minima problem

From the sequential algorithms seen in the previous section, we can design a CGM algorithm for the range minima problem. This algorithm is executed in $O(\frac{n}{p})$ time and uses $O(1)$ communication rounds. The major difficulty is how to store the required data structure among the processors so that the queries can be done in constant time, without violating the limit of $O(\frac{n}{p})$ memory, required by the CGM model.

Given an array $A = (a_1, a_2, \ldots, a_n)$, we write $A[i] = a_i$, $1 \leq i \leq n$, and $A[i \ldots j] = \text{subarray}(a_i, \ldots, a_j)$, $1 \leq i \leq j \leq n$.

The idea of the algorithm is based on how queries $MIN(i,j)$ can be answered. Each processor stores $\frac{n}{p}$ contiguous positions of the input array. Thus, given $A = (a_1, a_2, \ldots, a_n)$, we have subarrays $A_i = (a_{i\frac{n}{p}+1}, \ldots, a_{(i+1)\frac{n}{p}})$, for $0 \leq i \leq p-1$. Depending on the location of a_i and a_j in the processors, we have the following cases:

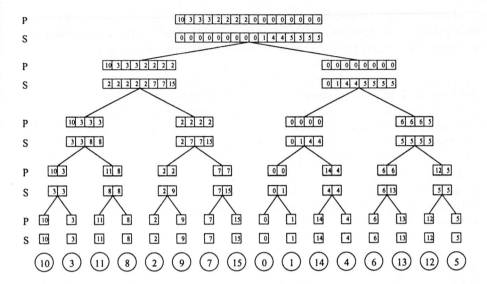

Fig. 2. PS-tree generated by the algorithm of Alon and Schieber for the array $(10, 3, 11, 8, 2, 9, 7, 15, 0, 1, 14, 4, 6, 13, 12, 5)$.

1. if a_i and a_j are in a same processor, the problem domain reduces to the subarray stored in that processor. Thus, we need a data structure to answer this type of queries in constant time. This data structure is obtained in each processor by Algorithm 1 (section 3.1).

2. if a_i and a_j are in distinct processors $p_{\bar{i}}$ e $p_{\bar{j}}$ (without loss of generality, $\bar{i} < \bar{j}$), respectively, we have two subcases:

    (a) if $\bar{i} = \bar{j} - 1$, a_i and a_j are in neighbor processors, $MIN(i,j)$ corresponds to the minimum between the minimum of a_i through the end of array $A_{\bar{i}}$ and the minimum of the beginning of $A_{\bar{j}}$ to a_j. These minima can be determined by Algorithm 1. To determine the minimum of the minima we need one communication round.

    (b) if $\bar{i} < \bar{j} - 1$, $MIN(i,j)$ corresponds to the minimum among the minimum of subarray $A_{\bar{i}}[i - \bar{i}\frac{n}{p} \ldots (i+1)\frac{n}{p}]$, the minimum of subarray $A_{\bar{j}}[\bar{j}\frac{n}{p} + 1 \ldots j - \bar{j}\frac{n}{p}]$ and the minima of subarrays $A_{\bar{i}+1}, \ldots, A_{\bar{j}-1}$. The first two minima are obtained as in the previous subcase. The minima of subarrays $A_{\bar{i}+1}, \ldots, A_{\bar{j}-1}$ are easily obtained by using the Cartesian tree. The minimum among them corresponds to the range minima problem restricted to the array of minima of the data in the processors. Thus we need a data structure to answer these queries in constant time. As the array of minima contains only p values, this data structure can be obtained by Algorithm 2 (section 3.2).

The difficulty of case 2.b is that we cannot construct the PS-tree explicitly in each processor as described in section 3.2, since this would require a memory of size $O(p \log p)$, larger than the memory requirement in the CGM model, which is $O(\frac{n}{p})$, with $\frac{n}{p} \geq p$. To overcome this difficulty we construct arrays P' and S' of $\log p + 1$ positions each, described as follows, to store some partial information of the PS-tree T in each processor. Let us describe this construction in processor i, with $0 \leq i \leq p-1$. Let b_i be the value of the minimum of array A_i. Let v be any vertex of T such that the subtree with root v has b_i as leaf and let d_v be the **depth of v** in T, that is the path length from the root to v, as defined in [1]; and l_v be the **level of v**, which is the height of the tree minus the depth of v, as defined in [1] ($l_v = \log p - d_v$, because tree T has height $\log p$). Array P' (respectively, S') contains in position l_v the value of array P_v (respectively, S_v), of the level l_v of T, in the position corresponding to the leaf b_i. In other words, we have $P'[l_v] = P_v[i \bmod 2^{l_v} + 1]$.

Figure 3 illustrates the correspondence between arrays P' and S' stored in each processor and the PS-tree constructed by the algorithm of Alon and Schieber [2]. In Figure 3.b, the bold face positions in arrays S in each level of the tree correspond to the suffix minimum of $b_0 = 3$ in each level. In this way, we obtain array S' in P_0 (figure 3.c).

ALGORITHM 3: Range Minima (CGM model)
{Each processor receives $\frac{n}{p}$ contiguous positions of array A, partitioned into subarrays $A_0, A_1, \ldots, A_{p-1}$.}
1. Each processor i executes sequentially Algorithm 1 (section 3.1).
2. {Each processor constructs an array $B = (b_i)$ of size p, that contains the minimum of the data stored in each processor.}
   2.1. Each processor i calculates $b_i = \min A_i = \min\{a_{i\frac{n}{p}+1}, \ldots, a_{(i+1)\frac{n}{p}}\}$.
   2.2. Each processor i sends b_i to the other processors.
   2.3. Each processor i puts in b_k the value received from processor k, $k \in \{0, \ldots, p-1\} \setminus \{i\}$.
3. Each processor i executes procedures Construct_P' and Construct_S' (see below) .
   {By observing the description of Algorithm 2 (section 3.2), $P'[k]$ contains position i of the prefix minimum array in level k, $0 \leq k \leq \log p$.}

Given array B, the following procedure constructs array P' of $\log p + 1$ postitions in processor i, for $0 \leq i \leq p-1$. This procedure constructs P' in $O(p)$ ($= O(\frac{n}{p})$) time using only local data. The construction of array S' is done in a symmetric way, considering array B in reverse order.

PROCEDURE 4: Construct_P'.
1. $P'[0] \leftarrow b_i$
2. $pointer \leftarrow i$
3. $inorder \leftarrow 2*i+1$
4. **for** $k \leftarrow 1$ **until** $\log p$ **do**
5. $\quad P'[k] \leftarrow P'[k-1]$

```
6.      if ⌊inorder/2^k⌋ is odd
7.        then for l ← 1 until 2^(k-1) do
8.              pointer ← pointer − l
9.              if P'[k] > B[pointer]
10.                then P'[k] ← B[pointer]
```

To simplify the correctness proof of this procedure, we consider p to be a power of 2 and that, in each processor i, array B contains leaves of a complete binary tree, as in the description of the sequential algorithm of section 3.2. We do not store an entire array in each internal node of this tree, but only the $\log p + 1$ positions of the prefix minimum array in each level corresponding to position i of array B.

The value of the variable *inorder* of line 3 is the value of the in-order number of the leaf containing b_i, obtained in an in-order traversal of the tree nodes. It can easily be seen that these values, from the left to the right, are the odd numbers in the interval $[1, 2p - 1]$.

Theorem 1 *Procedure 4 correctly calculates array P', for each processor i, $0 \le i \le p - 1$.*

Idea of proof. For a processor i, $0 \le i \le p-1$, we prove the following invariant at each iteration of the for loop of line 4: At each iteration k, let T_k be subtree that contains b_i and has root at level k. $P'[k]$ stores the position i of the minimum prefix array of the subarray of B corresponding to the leaves of subtree T_k and the variable *pointer* contains the index, in B, of the leftmost leaf of T_k. □

To determine S' we have a similar theorem with a similar proof.

Lemma 1 *Execution of Procedure 4 in each processor takes $O(\frac{n}{p})$ sequential time.*

Proof. The maximum number of iterations is $\log p$. In each iteration, the value of *pointer* is decremented or remains the same. The worst case is when $i = p-1$. In this case, at each iteration the value of the variable *pointer* is updated. As the number of elements of B is p, the algorithm updates the value of *pointer* at most p times. Therefore, procedure 4 is executed in $O(p) = O(\frac{n}{p})$ sequential time. □

Theorem 2 *Algorithm 3 solves the range minima problem in $O(\frac{n}{p})$ time using $O(1)$ communication rounds.*

Proof. Step 1 is executed in $O(\frac{n}{p})$ sequential time and does not use communication. Step 2 runs in $O(\frac{n}{p})$ sequential time and uses one communication round. By theorem 1, step 3 is executed in $O(\frac{n}{p})$ sequential time and does not require communication. Therefore, algorithm 3 solves the range minima problem in $O(\frac{n}{p})$ time using $O(1)$ communication rounds. □

5 Query processing

In this section, we show how to use the output of algorithm 3 to answer queries $MIN(i,j)$ in constant time. A query $MIN(i,j)$ is handled as follows. Assume that i and j are known by all the processors and the result will be given by processor 0. If a_i and a_j are in a same processor, then $MIN(i,j)$ can be determined by step 1 of algorithm 3. Otherwise, suppose that a_i and a_j are in distinct processors $p_{\bar{i}}$ and $p_{\bar{j}}$, respectively, with $\bar{i} < \bar{j}$. Let $right(\bar{i})$ the index in A of the rightmost element in array $A_{\bar{i}}$, and $left(\bar{j})$ the index in A of the leftmost element in array $A_{\bar{j}}$. Calculate $MIN(i, right(\bar{i}))$ and $MIN(left(\bar{j}), j)$, using step 1. We have two cases:

1. If $\bar{j} = \bar{i} + 1$ then $MIN(i,j) = \min\{MIN(i, right(\bar{i})), MIN(left(\bar{j}), j)\}$.

2. If $\bar{i} + 1 < \bar{j}$, then calculate $MIN(right(\bar{i}) + 1, left(\bar{j}) - 1)$, using step 3 of the algorithm. Notice that $MIN(right(\bar{i}) + 1, left(\bar{j}) - 1)$ corresponds to $\min\{b_{\bar{i}+1}, \ldots, b_{\bar{j}-1}\}$. Thus, $MIN(i,j) = \min\{MIN(i, right(\bar{i})), MIN(right(\bar{i}) + 1, left(\bar{j}) - 1), MIN(left(\bar{j}), j)\}$.

The value of $MIN(right(\bar{i}) + 1, left(\bar{j}) - 1)$ is obtained using step 3, as follows. Each processor calculates $w = \text{LCA}(b_{\bar{i}+1}, b_{\bar{j}-1})$ in constant time. Then determines the level l_w; determines v and u, the left and right sons of w, respectively; determines l, the index of the leftmost leaf in the subtree of u. Processor $\bar{i} + 1$ calculates $S_v(2^{l_w-1} - l + \bar{i})$ and sends this value to processor 0. Processor $\bar{j} - 1$ calculates $P_u(\bar{j} - l)$ and sends this value to processor 0. Processor 0, finally, calculates the minimum of the received minima. In both cases, processor 0 receives the minima of processors $\bar{i}$ and $\bar{j}$.

Notice finally that the correctness of Algorithm 3 results from the observations presented in this section.

6 Concluding remarks

In the CGM(n,p) model, we are interested in designing parallel algorithms that have linear local executing time and that utilize few communication rounds – preferably a constant number. This is because the communication between processors is more expensive than the computing time of the local data.

In this work we present an algorithm, in the CGM(n,p) model, for the range minima problem that runs in $O(\frac{n}{p})$ time using $O(1)$ communication rounds. This algorithm can be used to design a CGM algorithm for the LCA problem, based on the PRAM algorithm of JáJá [12], instead of other PRAM algorithms as in [13, 16] which would give less efficient implementations.

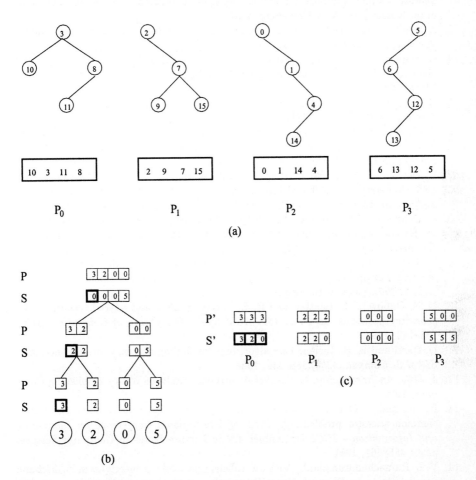

Fig. 3. Execution of algorithm 3 using the array $(10, 3, 11, 8, 2, 9, 7, 15, 0, 1, 14, 4, 6, 13, 12, 5)$. (a) The data distributed in the processors and the corresponding Cartesian trees. (b) PS-trees constructed by the algorithm of Alon and Schieber [2] of section 3.2 for the array $(3, 2, 0, 5)$ of the minima of processors. (c) Arrays P' and S' constructed by step 3 of algorithm 3 corresponding to arrays P and S of T.

References

1. A. V. Aho, J. E. Hopcroft, and J.D. Ullman. *The Design and Analysis of Computer Algorithms*. Addison-Wesley Publishing Company, 1975.
2. N. Alon and B. Schieber. Optimal preprocessing for answering on-line product queries. Technical Report TR 71/87, The Moise and Frida Eskenasy Inst. of Computer Science, Tel Aviv University, 1987.
3. A. Amir, G. M. Landau, and U. Vishkin. Efficient pattern matching with scaling. *Journal of Algorithms*, 13:2–32, 1992.
4. O. Berkman, B. Schieber, and U. Vishkin. Optimal doubly logarithmic parallel algorithms based on finding all nearest smaller values. *Journal of Algorithms*, 14:344–370, 1993.
5. O. Berkman and U. Vishkin. Recursive star-tree parallel data structure. *SIAM J. Comput.*, 22(2):221–242, 1993.
6. E. Cáceres, F. Dehne, A. Ferreira, P. Flocchini, I. Rieping, A. Roncato, N. Santoro, and S. W. Song. Efficient parallel graph algorithms for coarse grained multicomputers and BSP. In *Proc. ICALP'97*, 1997.
7. F. Dehne, A. Fabri, and C. Kenyon. Scalable and architecture independent parallel geometric algorithms with high probability optimal time. In *Proc. 6th IEEE Symposium on Parallel and Distributed Processing*, pages 586–593, 1994.
8. F. Dehne, A. Fabri, and A. Rau-Chaplin. Scalable parallel computational geometry for coarse grained multicomputers. In *Proc. ACM 9th Annual Computational Geometry*, pages 298–307, 1993.
9. F. Dehne and S. W. Song. Randomized parallel list ranking for distributed memory multiprocessors. In *Proc. Second Asian Computing Science Conference (ASIAN'96)*, pages 1–10, 1996.
10. H. N. Gabow, J. L. Bentley, and R. E. Tarjan. Scaling and related techniques for geometry problems. In *Proc. 16th ACM Symp. On Theory of Computing*, pages 135–143, 1984.
11. D. Harel and R. E. Tarjan. Fast algorithms for finding nearest common ancestors. *SIAM J. Comput.*, 13(2):338–335, 1984.
12. J. Jájá. *An Introduction to Parallel Algorithms*. Addison-Wesley Publishing Company, 1992.
13. R. Lin and S. Olariu. A simple optimal parallel algorithm to solve the lowest common ancestor problem. In *Proc. of International Conference on Computing and Information - ICCI'91*, number 497 in Lecture Notes in Computer Science, pages 455–461, 1991.
14. V. L. Ramachandran and U. Vishkin. Efficient parallel triconectivity in logarithmic parallel time. In *Proc. of AWOC'88*, number 319 in Lecture Notes in Computer Science, pages 33–42, 1988.
15. J. H. Reif, editor. *Synthesis of Parallel Algorithms*. Morgan Kaufmann Publishers, 1993.
16. B. Schieber and U. Vishkin. On finding lowest common ancestors: simplification and parallelization. *SIAM J. Comput.*, 17:1253–1262, 1988.
17. L. G. Valiant. A bridging model for parallel computation. *Comm. of the ACM*, 33:103–111, 1990.
18. J. Vuillemin. A unified look at data structures. *Comm. of the ACM*, 23:229–239, 1980.

Deterministic Branch-and-Bound on Distributed Memory Machines*

(Extended Abstract)

Kieran T. Herley[1], Andrea Pietracaprina[2], and Geppino Pucci[2]

[1] Department of Computer Science, University College Cork, Cork, Ireland.
k.herley@cs.ucc.ie
[2] Dipartimento di Elettronica e Informatica, Università di Padova, Padova, Italy.
{andrea,geppo}@artemide.dei.unipd.it

Abstract. The *branch-and-bound* problem involves determining the leaf of minimum cost in a cost-labelled, heap-ordered tree, subject to the constraint that only the root is known initially and that the children of each node are revealed only by visiting their parent. We present the first efficient deterministic algorithm to solve the branch-and-bound problem for a tree T of constant degree on a p-processor Distributed-Memory Machine. Let c^* be the cost of the minimum-cost leaf in T, and let n and h be the number of nodes and the height, respectively, of the subtree $T^* \subseteq T$ of nodes whose cost is at most c^*. When accounting for both computation and communication costs, our algorithm runs in time $O\left(n/p + h(\max\{p, \log n \log p\})^2\right)$ for general values of n, and can be made to run in time $O\left((n/p + h\log^4 p)\log\log p\right)$ for n polynomial in p. For large ranges of the relevant parameters, our algorithm is provably optimal or outperforms the well-known randomized strategy by Karp and Zhang.

1 Introduction

Branch-and-bound is a widely used and effective technique for solving hard optimization problems. It determines the optimum-cost solution of a problem through a selective exploration of a solution tree, whose internal nodes correspond to different relaxations of the problem and whose leaves correspond to feasible solutions. The shape of the tree is generally not known in advance, since the subproblems associated with the nodes are generated dynamically in an irregular and unpredictable fashion. A suitable abstract framework for studying the balancing and communication issues involved in the parallel implementation of branch-and-bound is provided by the *branch-and-bound problem*, introduced in [KZ93], which can be specified as follows. Let T be an arbitrary tree of finite size. Initially, a pointer to the root is available, while pointers to children are revealed only after their parent is *visited*. A node can be visited only if a pointer to it is available, and it is assumed that the visit takes constant time. All nodes of T are labelled with distinct integer-valued *costs*, the cost of each node being strictly less than

---

* This research was supported, in part, by the CNR of Italy under Grant CNR97.03207.CT07 *Load Balancing and Exhausive Search Techniques for Parallel Architectures*.

the cost of its children (heap property). The branch-and-bound problem involves determining the cost c^* of the minimum-cost leaf. Note that any correct algorithm for the branch-and-bound problem must visit all those nodes whose costs are less than or equal to c^*. These nodes form a subtree T^* of T. Throughout the paper, n and h will denote, respectively, the size and the height of T^*.

The efficiency of any parallel branch-and-bound algorithm crucially relies on a balanced on-line redistribution of the computational load (tree-node visits) among the processors. Clearly, the cost of balancing must not be much larger than the cost of the tree-visiting performed. Furthermore, since c^* is not known in advance, one cannot immediately distinguish nodes in T^* (all of which must be visited) from nodes in $T - T^*$ (whose visits represent wasted work). Ensuring that the algorithm visits few superfluous nodes is nontrivial in a parallel setting as it requires considerable coordination between processors.

In this paper, we devise an efficient deterministic parallel algorithm for the branch-and-bound problem on a *Distributed Memory Machine* (DMM) consisting of a collection of processor/memory pairs communicating through a complete network. The model assumes that in one time step, each processor can perform $O(1)$ operations on locally stored data and send/receive one message to/from an arbitrary processor. We consider the weakest DMM variant, also known as *Optical Communication Parallel Computer* (OCPC) in the literature [GJRL93], where concurrent transmissions to the same processor are heavily penalized. Specifically, in the event that two or more processors simultaneously attempt to transmit to the same destination, the processors involved are informed of the collision but no message is received by the destination.

Related Work and New Results A simple sequential algorithm for the branch-and-bound problem is based on the *best-first* strategy, where available (but not yet visited) nodes are stored in a priority queue and visited in increasing order of their cost. The $O(n \log n)$ running time of this simple strategy is dominated by the cost of the $O(n)$ queue operations. In [Fre90], Frederickson devised a clever sequential algorithm to select the k-th smallest item in an infinite heap in $O(k)$ time. The algorithm can be easily adapted to yield an optimal $O(n)$ sequential algorithm for branch-and-bound.

In parallel computation, the branch-and-bound problem has been studied on a variety of machine models. In [KZ93], Karp and Zhang show, by a simple work/diameter argument, that any algorithm for the problem requires at least $\Omega(n/p + h)$ time on any p-processor machine, and devise a general randomized algorithm, running in $O(n/p + h)$ steps, with high probability. Each step of the algorithm entails a constant number of operations on local priority queues per processor, and the routing of a global communication pattern where a processor can be the recipient of $\Theta(\log p / \log \log p)$ messages. A straightforward implementation of this algorithm on a DMM would require $O(\log(n/p) + \log p / \log \log p)$ time per step, with high probability, if both priority queue and contention costs are fully accounted for. The resulting algorithm is nonoptimal for all values of the parameters n, h and p.

Kaklamanis and Persiano [KP95] present a deterministic branch-and-bound algorithm that runs in $O\left(\sqrt{nh} \log n\right)$ time on an n-node mesh. The mesh-specific techniques exploited by the algorithm, coupled with their assumption that the mesh size and problem size are comparable, limits the applicability of their scheme to other parallel architectures.

A deterministic algorithm for the shared-memory EREW PRAM, based on a parallelization of the heap-selection algorithm of [Fre90] appears in [HPP99]. The main result of this paper extends this approach to the distributed-memory DMM with the performance stated in the following theorem.

Theorem 1. *There is a deterministic algorithm running on a p-processor DMM that solves the branch-and-bound problem for any tree T of constant-degree in time*

$$O\left(\frac{n}{p} + h(\max\{p, \log n \log p\})^2\right).$$

When n is polynomial in p, the algorithm can be also implemented to run in time

$$O\left(\left(\frac{n}{p} + h\log^4 p\right)\log\log p\right).$$

Note that when n/p is large with respect to h, a typical scenario in real applications, our algorithm achieves optimal $\Theta(n/p)$ running time. In this case, the implementation of the algorithm is very simple and the big-oh in the running time does not hide any large constant. In contrast, the second implementation is asymptotically better, although more complex, when the values of n/p and h are close, and it is indeed within a mere $O(\log \log p)$ factor of optimal as long as $h = O\left(n/(p\log^4 p)\right)$, which is a weak balance requirement on the solution tree.

The rest of this abstract is organized as follows. Section 2 presents our general branch-and-bound strategy, while Section 3 sketches its DMM implementation.

2 A machine independent algorithm

Consider a branch-and-bound tree T and let s be an integer parameter which will be specified later. For simplicity, we assume that T is binary, although all our results immediately extend to trees of constant degree. We begin by summarizing some terminology introduced in [Fre90]. For a set of tree nodes N, let $Best(N)$ denote the set containing the (at most) s internal or leaf nodes with smallest cost among those in N and their descendants. A set of nodes of the form $Best(N)$ is called a *clan* and nodes themselves are referred to as the clan's *members*. The nodes of T can be organized in a binary *tree of clans* TC as follows. Let r denote the root of T. The root of TC is the clan $R = Best(\{r\})$. Let C be a clan of TC and suppose that $C = Best(N)$ for some set N of nodes of T. Define $Off(C)$ (*offspring*) as the set of tree nodes which are children of members of C but are not themselves members of C, and define $PR(C)$ (*poor relations*) to be the set $N - Best(N)$. Clan C has two child clans C' and C'' in TC, namely $C' = Best(Off(C))$ and $C'' = Best(PR(C))$. Since T is binary, we have $|Off(C)|, |PR(C)| \leq 2s$, for every clan $C \in TC$. If a clan C has exactly s members, its *cost*, denoted by $cost(C)$, is defined as the maximum cost of any of its members; if C has less than s members $cost(C) = \infty$ (note that in this last case, C is a leaf of

$\mathcal{TC}$). Since every node in a clan C costs less than every node in $\mathit{Off}(C) \cup \mathit{PR}(C)$, both $\mathit{cost}(C')$ and $\mathit{cost}(C'')$ are strictly greater than $\mathit{cost}(C)$.

It can be shown that the k-th smallest node of T is a member of one of the $2\lceil k/s \rceil$ clans of minimum cost. Based on such a property, Frederickson [Fre90] develops a sequential algorithm that finds the k-th smallest node in T in linear time by performing a clever exploration of $\mathcal{TC}$ in increasing order of clan cost. Note that once such a node is found, the k nodes of smallest cost in T can be enumerated in linear time as well. By repeatedly applying this strategy for exponentially increasing values of k until the smallest-cost leaf is found, the branch-and-bound problem can be solved for T in time proportional to the size of T^*.

Our branch-and-bound algorithm for the DMM can be seen as an implementation of a general strategy based on a parallel exploration of $\mathcal{TC}$, which was also at the base of the PRAM algorithm presented in [HPP99]. Such a strategy is realized by the generic algorithm BB reported below, which applies to any p-processor machine. In the next section, we will show how to implement algorithm BB efficiently on the DMM.

Let P_i denote the i-th processor of the machine, for $1 \leq i \leq p$. P_i maintains a local variable ℓ_i which is initialized to ∞. Throughout the algorithm, variable ℓ_i stores the cost of the cheapest leaf visited by P_i so far. Also, a global variable ℓ is maintained, which stores the minimum of the ℓ_i's. At the core of the algorithm is a *Parallel Priority Queue* (PPQ), a parallel data structure containing items labelled with an integer-valued key [PP91]. Two main operations are provided by a PPQ: *Insert*, that adds a p-tuple of new items into the queue; and *Deletemin*, that extracts the p items with the smallest keys from the queue. The branch-and-bound algorithm employs a PPQ Q to store clans, using their costs as keys. Together with Q, a global variable q is maintained, denoting the minimum key currently in Q. Initially, Q is empty and a pointer to the root r of T is available.

Algorithm BB:

1. P_1 produces clan $R = \mathit{Best}(\{r\})$ and sets ℓ_1 to the cost of the minimum leaf in R, if any exists. Then, R is inserted into Q, and q and ℓ are set to the cost of R and to ℓ_1, respectively.
2. The following substeps are iterated until $\ell < q$.
   (a) Deletemin is invoked to extract the $k = \min\{p, |Q|\}$ clans $C_1, C_2, \ldots, C_k$ of smallest cost from Q. For $1 \leq i \leq k$, clan C_i is assigned to P_i.
   (b) For $1 \leq i \leq k$, P_i produces the two children of C_i, namely C_i' and C_i'', and updates ℓ_i accordingly.
   (c) Insert is invoked (at most twice) to store the newly produced clans into Q. The values ℓ and q are then updated accordingly.
3. The value ℓ is returned.

Lemma 1. *Algorithm BB is correct. Moreover, the number of iterations of Step 2 required to reach the termination condition is $O\left(n/(ps) + hs\right)$.*

A complete proof of Lemma 1 can be found in [HPP99]. Here, we briefly sketch the main ingredients of the proof. Correctness follows from the observation that at any time during the algorithm, all nodes in T with cost less than or equal to the current value of q

have already been visited and made members of some clan. Therefore, when ℓ becomes smaller than q, the algorithm will have visited at least one leaf (the one with cost ℓ) and all nodes (and, in particular all leaves) with cost less than or equal to $\ell < q$. The upper bound on the number of iterations crucially relies on the fact that clans containing exactly s nodes in T^* (called *good clans*) belong to the first $O(hs)$ levels of the tree of clans $\mathcal{TC}$, and that at each iteration of Step 2 but the last, at least one good clan is extracted from the PPQ. We call an iteration of Step 2 *full*, if exactly p good clans are extracted from the PPQ, and *partial* otherwise. Clearly, there can be no more than $n/(ps)$ full iterations. On the other hand, each partial iteration must complete the visit of all good clans at some level of $\mathcal{TC}$, hence there cannot be more than $O(hs)$ partial iterations.

3 DMM Implementation

In this section, we show how algorithm BB can be efficiently implemented on the DMM. The implementation crucially relies on the availability of fast PPQ operations, which is guaranteed by the following lemma.

Lemma 2. *A PPQ Q storing items of constant size can be implemented on a p-processor DMM so that both Insert and Deletemin operations take $O\left(\log(|Q|/p)\log p\right)$ time.*

Proof (Sketch). Use the p-Bandwidth Heap of [PP91] and store each of the p items held by a heap node in a distinct memory module. Then, the PRAM algorithms for Insert and Deletemin can be run directly by replacing the PRAM merging algorithm with a simple adaptation of bitonic merging for the DMM, yielding the stated time bound.

If we adopted the naive approach of viewing each whole clan (with its $\Theta(s)$ members, offspring and poor relations) as a PPQ item, the complexity of the above operations would increase by a factor of $\Theta(s)$, since each elementary step would entail the actual migration of the clans involved among the processors. To overcome this problem, we store each clan in a distinct *cell* of a *virtual shared memory*, and represent the clan in the PPQ using a constant-size record that includes its cost and its virtual cell's address. Virtual cells are suitably mapped onto the processors' memories, in such a way that the contents of p arbitrary cells can be efficiently retrieved. The mapping must guard against the possibility that the p cells to be accessed be concentrated in a small number of memory modules, which might render the access unacceptably expensive due to memory contention. Based on these ideas, we propose two implementations, distinguished by the choice for s, and whose running times depend on the relative values of n/p and h.

In what follows, we assume that the values of n and h be known in advance to the DMM processors. In the full paper, we will show how to relax such an assumption by means of guessing techniques similar to those employed in [HPP99] for the PRAM implementation.

Implementation 1 Let $s = \max\{p, \log n \log p\}$, and suppose that the $\Theta(s)$ data stored in any virtual cell (i.e., members, offspring and poor relations of some clan) are evenly distributed among the p memory modules, with $O(s/p)$ data per module. Consider an iteration of Step 2 of algorithm BB. Since each clan is represented in the PPQ by a constant-size record, the Deletemin and Insert operations required in Substeps 2.a and 2.c take time $O(\log n \log p)$ time by Lemma 2 (note that Lemma 1 implies that at any time during the algorithm $|Q| = O(nsp)$). The generation of the child clans in Substep 2.b can be performed locally at each processor using Fredrickson's algorithm in $O(s)$ time. Also, all data movements involved in Step 2 can be easily arranged as a set of $O(s)$ fixed permutations that take $O(s)$ time. Thus, each iteration of Step 2 requires time $O(s)$, which yields an $O(n/p + hs^2)$ running time for the entire algorithm. The first part of Theorem 1 follows by plugging in the chosen value for s.

Implementation 2 Note that the above implementation achieves optimal $\Theta(n/p)$ running time when n/p is much larger than h, however it becomes progressively less profitable for unbalanced trees. In the latter scenario, it is convenient to choose a smaller value for s which, however, requires a more sophisticated mechanism to avoid memory contention. In particular, when $s < p$ it becomes necessary to introduce some redundancy in the data representation and to carefully select, for each virtual cell, a subset of processors that store its (replicated) contents, so that any p cells can be retrieved by the processors with low contention at the memory modules. These ideas are explained in greater detail in the rest of the paper.

We assume that n is polynomial in p, that the machine word contains $\Theta(\log p)$ bits, and that each node of the branch-and-bound tree T is represented using a constant number of words. In this fashion, clans, as well as virtual cells, can be regarded as strings of $\Theta(s \log p)$ bits. Each cell is assigned a set of $d = \Theta(\log p)$ distinct memory modules as specified by a suitable *memory map* modelled by means of a bipartite graph $G = (U, V, E)$ with $|U|$ inputs, corresponding to the virtual cells, and $|V| = p$ outputs, corresponding to the DMM modules, and d edges connecting each virtual cell to the d modules assigned to it. The quantity $|U| = p^{O(1)}$ is chosen as an upper bound to the total number of virtual cells ever needed by the algorithm. We call d the *degree* of the memory map.

Consider a set of p newly created clans, $C_1, C_2, \ldots, C_p$ to be stored in the DMM modules, and let processor P_i be in charge of clan C_i. First, a distinct unused cell u_i is chosen for each clan C_i. Then, P_i recodes u_i into a longer string of size $3|u_i|$ and splits it into $k = \epsilon \log p$ *components* ($\epsilon < 1$), each of $3|u_i|/k = \Theta(s)$ bits, by using an information dispersal algorithm [Pre89,Rab89], so that any $k/3$ components suffice to recreate the original contents of u_i. The following lemma, proved in Subsection 3.1, establishes the complexity of these encoding/decoding operations

Lemma 3. *A processor can transform a cell u into k components of $\Theta(s)$ bits each, in $O(s \log k) = O(s \log \log p)$ time, so that u can be recreated from any $k/3$ components within the same time bound.*

After the encoding, P_i replicates each component of u_i into $a = d/k = O(1)$ copies, referred to as *component copies*, and attempts to store the resulting d component copies of u_i into the d modules assigned to the cell by G, in parallel for every $1 \leq i \leq p$. The operation terminates as soon as all processors effectively store at least $2d/3$ component copies each.

Consider now the case when the processors need to fetch the contents of p cells from the DMM modules. Each processor attempts to access the d modules that potentially store the component copies of the cell and stops as soon as *any* $2d/3$ modules are accessed. Although not all accessed modules may effectively contain component copies of the cell, we are guaranteed that at least $d/3$ component copies, hence $k/3$ distinct components, will be retrieved. This is sufficient for each processor to reconstruct the entire cell. The following lemma will be proved in Subsection 3.2.

Lemma 4. *There exists a suitable memory map G such that any set S of p virtual cells can be stored/retrieved in the DMM modules in $O\left(s + \log^2 p \log \log p\right)$ steps.*

Using the above techniques, the extraction of the p clans of minimum cost among those represented in the PPQ Q can be accomplished as follows. First, the addresses of the corresponding cells are extracted from Q in time $O\left(\log(|Q|/p)\log p\right) = O\left(\log^2 p\right)$ and distributed one per processor. Then, by Lemma 4, $k/3$ components for each cell are retrieved in $O\left(s + \log^2 p \log \log p\right)$ time. Finally, the contents of each clan C are reconstructed in $O\left(|C|\log k\right) = O\left(s \log \log p\right)$ time, by Lemma 3. By combining all of the above contributions we obtain a running time of $O\left(s \log \log p + \log^2 p \log \log p\right)$. In a similar fashion one can show that the insertion of p new clans in the queue can be accomplished within the same time bound.

The complexity of algorithm BB is dominated by that of Step 2. Each iteration of this step entails one extraction and at most two insertions of p clans ($O\left(s \log \log p + \log^2 p \log \log p\right)$ time), the generation of at most two new clans per processor ($O(s)$ time), and a number of other simple operations all executable in $O(\log p)$ time. Therefore, each iteration can be completed in time $O\left(s \log \log p + \log^2 p \log \log p\right)$. Since, by Lemma 1, there are $O\left(n/(ps) + hs\right)$ iterations, the second part of Theorem 1 follows by choosing $s = \log^2 p$.

3.1 Proof of Lemma 3

We will only describe the encoding procedure, since the reconstruction procedure embodies similar ideas and exhibits the same running time. Consider a virtual cell u and view it as a rectangular $\Theta(s) \times (k/3)$ bit-array A_u, with every row stored in a separate word. By using an information dispersal algorithm [Pre89,Rab89], each row can be independently recoded into a string of k bits, so that any $k/3$ bits are sufficient to reconstruct the entire row. The resulting $\Theta(s) \times k$ bit-array A'_u is then "repackaged" so that each of its k columns is represented as a sequence of $\Theta(s/k)$ words. Each of these $\Theta(s/k)$-word sequences constitutes a distinct component.

Rather than actually running the information dispersal algorithm, we can use a precomputed look-up table of $2^{k/3} = p^{\epsilon/3}$ entries accessible in constant time. The i-th entry of the table, holds the k-bit encoding of the $(k/3)$-bit binary string corresponding to integer i. Note that this table need only be computed once and made available to each node of the machine.

The repackaging mentioned above can be implemented efficiently by transposing each $k \times k$ block of A'_u (k consecutive rows) in time $O(k \log k)$ through an oblivious sequence of bit manipulations which involve the k words that make up the block. (More details will be provided in the full version of this paper.)

The running time of the encoding procedure is dominated by the repackaging cost, which is $O(s \log k) = O(s \log \log p)$. Note that the encoding is performed locally at each processor and requires no external communication.

3.2 Proof of Lemma 4

Consider a memory map $G = (U, V, E)$ and a set $S \subseteq U$ of p cells, and let $E(S)$ denote the set of edges incident on nodes of S. A *c-bundle* of congestion h for S is a subset $B \subseteq E(S)$ where each $u \in S$ has degree c and any $v \in V$ has degree at most h with respect to the edges in B. The following claim demonstrates that there exists a suitable G which guarantees that a $(2d/3)$-bundle of low congestion exists for every set S of p or fewer cells, and that such a bundle may be determined efficiently.

Claim 1. *For $d = \Theta(\log p)$ sufficiently large, there exists a memory map G of degree d such that, for any given set S of p cells, there is a $(2d/3)$-bundle of congestion $\Theta(d)$. Moreover, one such bundle can be constructed in $O(d \log p) = O(\log^2 p)$ time on the DMM.*

Proof (Sketch). The existence of the bundle is a consequence of the results of [Her96], while its construction can be achieved through the techniques introduced in [UW87].

Let us consider the problem of writing a set S of p cells, $u_1, u_2, \ldots, u_p$ (the problem of reading p cells is similar). Suppose that each processor P_i has encoded the cell u_i it wishes to write into d component copies using the techniques described before, and that the processors have constructed a $(2d/3)$-bundle B of congestion $\Theta(d)$ for S. We assume that P_i knows the edges in B that are incident on cell u_i and on the i-th memory module v_i. It is important to note that the protocol implied by Lemma 1 merely indicates to each processor the locations to which the $2d/3$ component copies of its clan should be written: the actual physical movement of the component copies must be implemented as a separate step. This is a nontrivial task, since each processor must transmit/receive one component copy per edge in B, that is, it must dispatch $2d/3$ component copies and receive $O(d)$ of them. The transmission of copies must be coordinated to avoid "collisions" at receiving processors. Straightforward techniques based on sorting prove too slow for our purpose, so we employ an approach based on the idea that a Δ-edge colouring of B allows one word to be transmitted along each edge $(u, v) \in B$ in $O(\Delta)$ time. Let $\Delta = \Theta(d)$ denote the maximum degree of a node in $U \cup V$ with respect to the edges in B. We have:

Claim 2. *An $O(\Delta)$-edge colouring for B may be computed in $O(\Delta^2 \log \Delta)$ time on the DMM.*

Proof (Sketch). It is easy to modify the construction of Claim 1 so that, together with B, it produces an $O(\Delta^2)$-colouring for B within the same time bound. (We will use the term "colouring" to refer to edge colouring.) We transform this $O(\Delta^2)$-colouring into a (2Δ)-colouring in two stages: first we produce an intermediate $O(\Delta \log \Delta)$-colouring and then refine the intermediate colouring to produce the desired (2Δ)-colouring.

For the first stage we associate a set Colours(x) of $2r - 1 = \Theta(\log \Delta)$ colours from $[1..O(\Delta \log \Delta)]$ with each $x \in [1..\Delta^2]$. This mapping has the property that for any subset D of size Δ chosen from $[1..\Delta^2]$, we may select for each $x \in D$ a *palette* of r colors from Colours(x) such that no two elements of D have the same colour in their palettes. The choice of the mapping is based on the existence of suitable bipartite maps akin to those underlying Claim 1. (Details will be provided in the full version of this paper.) As a consequence, for each edge $(u,v) \in B$ of color $x \in [1..\Delta^2]$ we can select two palettes Left$_{u,v}$, Right$_{u,v} \subset$ Colours(x) of r colours such that Left$_{u,v} \cap$Left$_{u,v'} = \emptyset$ for every $(u,v') \in B$ with $v' \neq v$, and Right$_{u,v} \cap$ Right$_{u',v} = \emptyset$ for every $(u',v) \in B$ with $u' \neq u$. However, there must be at least one colour in Left$_{u,v} \cap$ Right$_{u,v}$. This latter colour from $[1..O(\Delta \log \Delta)]$ is adopted as the intermediate colour for (u,v).

Processor P_i computes sequentially all sets Left$_{u_i,v}$ for each $(u_i,v) \in B$ and all sets Right$_{u,v_i}$ for each $(u,v_i) \in B$. By means of standard techniques [UW87], this task can be accomplished in $O(\Delta \log \Delta)$ local computation time. Then, each set Right$_{u_i,v_j}$ is transmitted from P_j to P_i. Based on the initial $O(\Delta^2)$-colouring of B, this communication is completed in $O(\Delta^2 \log \Delta)$ steps. Overall, the first stage takes $O(\Delta^2 \log \Delta)$ time.

The second stage involves refining the intermediate $O(\Delta \log \Delta)$-colouring to produce a (2Δ)-colouring, using the following standard technique. For each intermediate colour in $[1..O(\Delta \log \Delta)]$ in turn, examine the edges with that colour and recolour each such edge (u,v) with the smallest colour in $[1..2\Delta]$ that does not clash with any new colour of other edges incident on u or v already assigned. It is easy to see that the second stage can be completed in $O(\Delta^2 \log \Delta)$ time. This is also the running time for the entire colouring procedure.

Given a $O(\Delta)$-colouring for B, the component copies corresponding to the bundle can be accessed in $2\Delta = \Theta(\log p)$ phases: during the i-th phase, the component copies corresponding to edges bearing color i are transferred in $O(s/k) = \Theta(s/\log p)$ time. Therefore, the access is completed in overall $O(s)$ time.

Lemma 4 follows by adding up the times needed to determine the bundle B, to compute the $O(\Delta)$-colouring and to access the component copies.

References

[Fre90] G. Frederickson. The Information Theory Bound is Tight for Selection in a Heap. In *Proc. of the 22nd ACM Symp. on Theory of Computing*, pages 26–33, May 1990.

[GJRL93] L.A. Goldberg, M. Jerrum, F.T. Leighton and S. Rao A Doubly Logarithmic Communication Algorithm for the Completely Connected Optical Communication Parallel Computer. In *Proc. of the 5th ACM Symp. on Parallel Algorithms and Architectures*, pages 300-309, June/July 1993.

[Her96] K. Herley. Representing Shared Data on Distributed-Memory Parallel Computers. *Mathematical Systems Theory*, 29:111-156, 1996.

[HPP99] K. Herley, A. Pietracaprina and G. Pucci. Fast Deterministic Parallel Branch-and-Bound. *Parallel Processing Letters*, to appear.

[KP95]   C. Kaklamanis and G. Persiano. Branch-and-Bound and Backtrack Search on Mesh-Connected Arrays of Processors. *Mathematical Systems Theory*, 27:471–489, 1995.

[KZ93]   R.M. Karp and Y. Zhang. Randomized Parallel Algorithms for Backtrack Search and Branch and Bound Computation. *Journal of the ACM*, 40:765–789, 1993.

[PP91]   M.C. Pinotti and G. Pucci. Parallel Priority Queues. *Information Processing Letters*, 40:33–40, 1991.

[Pre89] F.P. Preparata. Holographic Dispersal and Recovery of Information. *IEEE Trans. on Inform. Theory*, IT-35(5):1123–1124, May 1989.

[Rab89] M.O. Rabin. Efficient Dispersal of Information for Security, Load Balancing, and Fault Tolerance. *Journal of the ACM*, 36(2):335–348, April 1989.

[UW87]   E. Upfal and A. Wigderson. How to Share Memory in a Distributed System. *Journal of the ACM*, 34(1):116–127, Jan 1987.

2nd Workshop on Personal Computer Based Networks of Workstations (PC-NOW'99)

Clusters composed of fast personal computers are becoming more and more attractive as cheap and efficient platforms for distributed and parallel applications. The main drawback of a standard NOWs is the poor performance of the standard inter-process communication mechanisms based on RPC, sockets, TCP/IP, Ethernet. Such standard communication mechanisms perform poorly both in terms of throughput as well as message latency. Recently, several prototypes developed around the world have proved that re-visiting the implementation of the communication layer of a standard Operating System kernel, a low cost hardware platform composed of only commodity components can scale up to several tens of processing nodes and deliver communication and computation performance exceeding the one delivered by the conventional high-cost parallel platforms. This workshop provides a forum to discuss issues related to the design of efficient NOWs based on commodity hardware and public domain operating systems as compared to custom hardware devices and/or proprietary operating systems.

Workshop Organizers

G. Chiola (DISI, U. Genoa, I)
G. Conte (CE, U. Parma, I)
L.V. Mancini (DSI, U. Rome, I)

Program Commitee

	M. Baker (CSM, U. Portsmouth, UK)
Program Chair:	G. Chiola (DISI, U. Genoa, I)
	T.-C. Chiueh (CS Dept., SUNY S.B., USA)
	G. Conte (CE, U. Parma, I)
	H.G. Dietz (ECE, Purdue U., USA)
	W. Gentzsch (GENIAS Software GmbH, D)
	G. Iannello (DIS, U. Napoli, I)
	L.V. Mancini (DSI, U. Roma 1, I)
	T.G. Mattson (Intel Microcom. Res. Lab., USA)
	W. Rehm (Informatik, T.U. Chemnitz, D)
	P. Rossi (ENEA HPCN, Bologna, I)
	C. Szyperski (Queensland U. of Tech., AUS)
	T. Sterling (CACR, Caltech, USA)
	Y. Tanaka (PDSP Lab, RWCP, J)
	D. Tavangarian (Informatik, U. Rostock, D)
	B. Tourancheau (LHPC/ENS, U. Lyon, F)

Referees

C. Anglano
M. Baker
M. Bernaschi
G. Chiola
T.-C. Chiueh
G. Ciaccio
G. Conte

F.A. Galvin
W. Gentzsch
G. Iannello
L.V. Mancini
T.G. Mattson
L. Prylli
W. Rehm

P. Rossi
C. Szyperski
Y. Tanaka
D. Tavangarian
B. Tourancheau
R. Westrelin

Accepted Papers

Session 1: Implementation of BSP Clusters

- S.R. Donaldson, J.M.D. Hill, and D.B. Skillicorn "Performance Results for a Reliable Low-latency Cluster Communication Protocol"
- H. Karl "Coscheduling through Synchronized Scheduling Servers — A Prototype and Experiments"

Session 2: Applications and Measurements

- C. Anglano, A. Giordana, and G. Lo Bello "High-Performance Knowledge Extraction from Data on PC-based Networks of Workstations"
- U.K.V. Rajasekaran, M. Chetlur, G.D. Sharma, R. Radhakrishnan, and P.A. Wilsey "Addressing Communication Latency Issues on Clusters for Fine Grained Asynchronous Applications — A case study"
- G. Conte, M. Mazzeo, A. Poggi, P. Rossi, and M. Vignali "Low Cost Databases for NOW"

Session 1: Efficient MPI on Clusters

- T. Takahashi, F. O'Carrol, H. Tezuka, A. Hori, S. Sumimoto, H. Harada, Y. Ishikawa, and P.H. Beckman "Implementation and Evaluation of MPI on an SMP Cluster"

Other Activities

In addition to the presentation of contributed papers two invited talks will be scheduled at the workshop. A concluding session will be devoted to open discussion and to the organization of a new IEEE Task Force on Cluster Computing.

Performance Results for a Reliable Low-Latency Cluster Communication Protocol

Stephen R. Donaldson[1,2], Jonathan M.D. Hill[2], and David B. Skillicorn[3]

[1] Oxford University Computing Laboratory, UK.
[2] Sychron Ltd, 1 Cambridge Terrace, Oxford, UK.
[3] CISC, Queen's University, Kingston, Canada

Abstract. Existing low-latency protocols make unrealistically strong assumptions about reliability. This allows them to achieve impressive performance, but also prevents this performance being exploited by applications, which must then deal with reliability issues in the application code. We present results from a new protocol that provides error recovery, and whose performance is close to that of existing low-latency protocols. We achieve a CPU overhead of $1.5\mu s$ for packet download and $3.6\mu s$ for upload. Our results show that (a) executing a protocol in the kernel is not incompatible with high performance, and (b) complete control over the protocol stack enables (1) simple forms of flow control to be adopted, (2) proper bracketing of the unreliable portions of the interconnect thus minimising buffers held up for possible recovery, and (3) the sharing of buffer pools. The result is a protocol which performs well in the context of parallel computation and the loose coupling of processes in the workstations of a cluster.

1 Introduction

In many respects, recent results from the low-latency communication community are analogous to drag racing. There is a pre-occupation with acceleration and top speed (latency and bandwidth) at the expense of general useability, in the same way that dragsters race on straight stretches of road (point-to-point benchmarks) in which crude braking and steering mechanisms suffice (no error recovery and poor security). Some of the performance claims that clusters of PCs outperform commercial parallel machines are therefore misleading, because the impressive performance results do not necessarily transfer to more general settings.

In this paper we present "dragster style" micro-benchmark results of a low-latency communication protocol used in an implementation of *BSPlib* [9, 15]. However, unlike much other work, the protocol deals fully with reliability issues, and can be used "as is" by application programs. Like other work in this area, we replace existing heavyweight protocol stacks with a lightweight protocol, using a purpose-built device driver to support low-latency communication in a cluster of PCs connected by 100Mbps switched Ethernet. Unlike most other low-latency protocols, we provide the necessary functionality of protocols such as TCP/IP to support SPMD-style parallel computation among a group of processors. The effectiveness of the complete system is tested using the NAS parallel benchmarks. Our results show that for a modest budget of $US12,000, an eight-processor BSP cluster can easily outperform an IBM SP2 (1995), and has performance similar to an SGI Origin 2000 (O2K) (1998) *as observed by application programs.*

One of the issues addressed by this paper is whether error recovery should be handled in user or kernel space. Reliability is not orthogonal to performance, and cannot be designed in later. Protocols that ignore error recovery, such as U-Net [29], GAMMA [7], and BIP [20], are based upon the premise that the link error probability is of the same order as processor failure. This assumption is true when the link is used for point-to-point micro-benchmarks. However, for general patterns of communication between a collection of processors connected by a switch (which is the only sensible *scalable* interconnect) there is the potential for congestion, which leads to dropped packets within the switch. This is indistinguishable from hard link errors. For example, if several processors send a large amount of data to a single processor then, unless some back-pressure notification is made available to the sender, it is always possible to overwhelm a component of the switch.

In our implementation of *BSPlib*, we use a three-tiered approach to providing a general-purpose communication system. At the highest level, we provide the user with the *BSPlib* API, which is implemented on top of a reliable communication middle layer that provides primitives for handshaking and initialisation of the computation on the various processors, mechanisms for orderly and unorderly shutdown, a nonblocking send primitive, a blocking receive primitive and a probe to test for message presence. This middle layer is itself implemented on top of a transport protocol that provides the necessary functionality to support acknowledgement of packets, error recovery and flow control. The lowest layer implements low-latency point-to-point datagram communication by using a kernel module/device driver that services queues that are also manipulated by the user process in user space. By using a network interface card that automatically polls for work when a queue of send descriptors dries up, it is possible to schedule packet transmission without the need for a system call.

We claim that, while the community has concentrated on the "drag-racing" layer, this is neither particularly large (less than 10% of the total code of *BSPlib*), nor difficult to implement. In contrast, the error-recovery component, which has been ignored by many researchers, is considerably more complex and significantly larger. The problems that need to be addressed in the error-recovery layer are: (1) queueing of outgoing packets in case they are dropped; (2) acknowledging packets on send queues; (3) detecting dropped packets at receivers; and (4) initiating retransmission. All of this extra functionality bloats the protocol stack, which affects the performance of micro-benchmarks because packet download consumes more CPU cycles and asynchronous packet upload may incur long running interrupt handlers which upload messages "off the wire." We investigate three different implementations of *BSPlib*: (1) a standard TCP/IP variant in which error recovery is sender-based and performed by TCP within the kernel; (2) one in which error recovery is receiver-based and is performed in user space by *BSPlib*; and (3) one in which error recovery is receiver-based and is performed within the kernel by a BSP device driver. Comparing the three approaches, we demonstrate that a receiver-based protocol is superior to a sender-based one, and that handling recovery in the kernel is better than handling it in user space.

Our solution compares favourably to SP2 and O2K machines with proprietary protocols and to MPI on the cluster. We illustrate this with both micro-kernel benchmarks and the NAS parallel benchmarks.

2 Network messaging layers for *BSPlib*

BSP [23, 25] programs written using the *BSPlib* API [15] consist of a series of *supersteps* which are global operations of the entire machine. Each superstep consists of three sequential phases of: (1) computation, (2) communication, and (3) a barrier synchronisation marked by all processors calling bsp_sync().

BSPlib is implemented on top of a variety of lower-level messaging layers. In this paper we describe three implementations of the messaging layer: (1) the BSPlib/TCP implementation (Section 2.1) that uses BSD stream sockets as an interface to TCP/IP; (2) the BSPlib/UDP implementation (Section 2.2) that uses BSD datagram sockets as an interface to UDP/IP; and (3) the BSPlib/NIC implementation (Section 2.3) that uses the same error recovery and acknowledgement protocol used in the BSPlib/UDP implementation, but replaces the BSD datagram interface with a lightweight packet transmission mechanism that interfaces directly with the network interface card. In all the implementations, any *BSPlib* communications posted during a superstep are delayed until the barrier synchronisation that marks the end of the superstep (as this is a clear performance win [16]). The actual communication in *BSPlib* therefore reduces to the problem of routing a collection of packets between all the processors, within the bsp_sync() procedure that marks the end of the superstep.

2.1 BSPlib/TCP: a messaging layer built upon TCP/IP

The BSD stream socket interface provides a reliable, full-duplex, connection-oriented byte stream between two endpoints using the TCP/IP protocol. As well as providing error recovery to deliver this reliability, TCP/IP uses a sliding window protocol for flow control.

General and reliable transport protocols such as TCP/IP provide a portable implementation path for BSP. However, the functionality of TCP/IP is too rich for BSP-style computation using a dedicated set of processors. For example, the TCP/IP protocol stack cannot take advantage of the nature of the local LAN as the stack also includes support for long-haul traffic, where nothing may be known about the intermediate networks. This makes TCP/IP unsuitable for high-performance computation [10].

BSPlib/TCP uses the send() function to push data into the TCP/IP protocol stack. At the level of the messaging layer, reception of messages is *not* asynchronous. When the higher level requires a packet, the messaging layer waits for a packet to arrive at the process by issuing a select(). When the select()

completes, the received packet is copied into user space by issuing a `recv()`. Of course, message reception is still asynchronous with respect to the lower levels of the TCP/IP protocol stack. However, messages will be buffered within the stack until they have been selected by the messaging layer. Having the actual reception at a lower level makes it particularly difficult for the messaging layer to make any sensible decisions about buffering. For example, if packets are dropped due to insufficient buffer resources, the upper layer is unaware of it, and cannot plan to circumvent the problem by, for example, using global knowledge of the communication pattern. The only way of avoiding an excessive number of packets being dropped by the TCP/IP layer is to associate large buffers with each of the $p-1$ sockets that form the endpoints on each process ($p(p-1)$ in total). This may be an extremely wasteful use of memory if an application uses only skewed communication patterns.

Error recovery in point-to-point protocols over unreliable media is typically accomplished by using timeouts and acknowledgement information that flows back from the receiver to the sender. To improve the usage of the media, TCP/IP attempts to piggy-back such acknowledgements on traffic flowing in the reverse direction on the circuit. In order to do this, acknowledgements may be delayed for a short period in the hope that reverse traffic will be presented for transmission and upon which the acknowledgement can be piggy-backed. Should no reverse traffic be forthcoming within $200ms$, an explicit acknowledgement packet is returned. From the senders point of view, should no acknowledgement be forthcoming for a specified time (usually $1.5s$, with an exponential back-off up to $64.0s$), a timeout will trigger at the sending station which then assumes that either the data was dropped en route or that the acknowledgement (explicit or piggy-backed) was dropped on the return path. In either case, TCP/IP retransmits the data from the point after the last received acknowledgement (a form of go-back-n protocol). These timeouts are quite large as TCP/IP is designed for the long haul traffic hence is not really suitable for intensive bursts of communication between a tightly-coupled group of machines on a fast network.

2.2 BSPlib/UDP: a messaging layer built upon UDP/IP

The BSD datagram socket interface provides an unreliable, full-duplex, connectionless packet delivery mechanism between machines using the UDP/IP protocol. Unlike TCP streams, UDP datagrams have a fixed maximum transmission unit size of 1500 bytes for Ethernet frames. The link between the machines is unreliable as UDP/IP provides no form of message acknowledgement, error recovery, or flow control.

Protocols such as UDP/IP have a shallower protocol stack and therefore have more potential for high performance. However, they do not provide reliable delivery of messages. If user APIs such as *BSPlib* are implemented over UDP/IP, they must handle the explicit acknowledgement of data and error recovery themselves. In the protected kernel/user-space model of UNIX, this must be done in user space.

To implement a reliable send primitive, BSPlib/UDP copies a user data structure referenced in a send call into a buffer that is taken from a free queue. If there are no free buffers available, then the send fails, and it is up to the higher level *BSPlib* layer to recover from this situation[4]. The buffer is placed on a send queue and a communication is attempted using the `sendto()` datagram primitive. Queueing the buffer on the send queue ensures that it does not matter if the packet communicated by `sendto()` fails to reach its destination, as our error-recovery protocol can resend it if necessary. Only when a send has been positively acknowledged by the partner process are buffers reclaimed from the send queue to the free queue.

Message delivery is asynchronous and accomplished by using a signal handler that is dispatched whenever messages arrive at a processor (SIGIO signals). Within the signal handler, messages are copied into user space by using `recvfrom()`. To perform the error recovery as soon as possible, it executes from within the signal handler in *user space*.

Recall that the TCP/IP error recovery protocol is too simplistic for high-performance use in *BSPlib* due to its go-back-n error recovery policy, which may generate a large number of duplicate packets, and its slow-acting timeout mechanisms. These problems are solved in BSPlib/UDP by implementing our own error recovery mechanism that uses a selective retransmission scheme that only resends the packets that are actually dropped, and a sophisticated acknowledgement policy that is based upon available buffer resources and not timeouts [9, 22]. The acknowledgement scheme has both a sender and receiver component to the algorithm. From the sender's perspective, if the number of buffers on a free queue of packets becomes low, then outgoing packets are marked as requiring acknowledgements. This mechanism is triggered by calculating how many packets can be consumed (both incoming and outgoing) in the time taken for a message roundtrip. The sender therefore anticipates when buffer resources will run out, and attempts to force an acknowledgement to be returned just before buffer starvation. This strategy minimises both the number of acknowledgements, and the time a processor stalls trying to send messages. From the receiver's viewpoint, if n is the total number of send and receive buffers, then a receiver will only send a non-piggybacked acknowledgement if it is was requested, and at least $\frac{n}{2p}$ elements have come in over a link since the last acknowledgement (either piggy-backed or explicit).

2.3   BSPlib/NIC: a NIC-based transport layer for *BSPlib*

A Network Interface Card (NIC) provides the hardware interface between a system bus and the physical network medium. BSPlib/NIC uses exactly the same error-recovery and acknowledgement protocol as BSPlib/UDP. The only difference between the implementations is that BSPlib/NIC is built upon a lightweight

---

[4] *BSPlib* recovers by either: (1) receiving any packets that have been queued for *BSPlib*; or (2) asking those links that have a large number of unacknowledged send buffers to return an acknowledgement.

packet transmission mechanism that interfaces directly to the NIC. This is achieved by having a portion of memory used for communication buffers and shared data structures that are mapped into both the kernel and the user's address space.

With the sophistication of modern network interface cards, UDP-like protocols can be implemented to achieve very high bandwidth utilisation of the network. All outgoing packets contain a protocol header which contains piggybacked acknowledgement and error recovery data. As the *BSPlib* layer can queue a potentially large number of packets for transmission by the NIC, the protocol information may become stale as incoming packets update the protocol state. This problem can be alleviated by allowing the NIC to perform a gathered send, whereby the protocol information is only read when the packet is about to be transmitted. Processor usage is also improved by using the features of the NIC to make the communication as asynchronous as possible. For example, the standard technique in device drivers is to use bounded send and receive descriptor rings which the NIC traverses. In our implementation, the NIC traverses an arbitrary sized (linked-list) queue of buffers (the NIC's transmit queue).

In the NIC implementation, the user-space signal handler that was used for asynchronous message delivery in BSPlib/UDP is moved into the kernel as a interrupt handler associated with the interrupt request line used by the NIC. A sequence of receive buffers are serviced by the NIC, such that upon successful packet upload, an interrupt is triggered, and the handler *swaps* the newly-uploaded buffer with a fresh user space mapped buffer from the free queue (this eliminates a memory copy). The handler then inspects the uploaded packet to determine if any error recovery should be triggered *within the kernel*. Finally the packet is placed on a receive queue that can be manipulated in both kernel and user space. As this receive queue is manipulated by the user process, it eliminates the need for a system call, but slightly complicates queueing, as the queue data structure can be concurrently accessed by either the kernel or user process.

3 Benchmarks

We have benchmarked our work at various levels: at the lowest level, microbenchmarks show the raw efficiency of the communication equipment that can be achieved. At a higher, application level we compare our cluster to other popular computers. For these benchmarks we use a *BSPlib* port [17] of the NAS Parallel Benchmarks 2.1 [2]. To compare the efficiency of the cluster and *BSPlib* with other parallel machines, we also include results from the MPI versions of the benchmarks. On the SP2 (66MHz thin-node) and the O2K, results are for proprietary implementations of MPI; on our cluster they are for mpich [14].

3.1 Cluster configuration

The prototype system is a cluster of eight 400MHz Pentium II PC systems each with 128MB of 10ns SDRAM on 100MHz motherboards. Two distinct types of communication are required: slow(er) I/O communication and fast computation data. It simplified coding and debugging to have two separate networks; a control network to deal with I/O, and a network reserved for interprocess application data. The control network uses the standard TCP/IP protocol suite. We concentrate on the data network.

Each of the processors runs the Linux 2.0 Kernel and the driver software is written according to the interfaces documented in [21]. Other than reserving memory for communication, no kernel changes are required. The communication devices used were eight 3COM 3C905B-TX NICs running at 100Mbps [1] and a 100Mbps Cisco 2916XL fast Ethernet switch [8].

3.2 Micro-Benchmarks

The micro-benchmarks are typical and measure the raw bandwidth and latency that can be achieved 'on the wire' (excluding protocol data). We consider three classes: the roundtrip delay between two processors for various message size (Figure 1); link bandwidth between two processors for various message size (Figure 2); and per-packet latency for half-round-trip packets. In this last class, we provide results for short (4 byte) (Figure 3) and large (1400 byte) (Figure 4) packet sizes, and for various number of messages. These benchmarks use the following configurations:

Key	Machine	Library	Transport	Network
PII-BSPlib-NIC-100mbit-wire	PII Cluster	BSPlib	NIC	100BASE-TX, cross-over wire
PII-BSPlib-NIC-100mbit-2916XL	PII Cluster	BSPlib	NIC	100BASE-TX, Cisco 2916XL.
PII-BSPlib-UDP-100mbit-2916XL	PII Cluster	BSPlib	UDP/IP	100BASE-TX, Cisco 2916XL
PII-BSPlib-TCP-100mbit-2916XL	PII Cluster	BSPlib	TCP/IP	100BASE-TX, Cisco 2916XL
PII-MPI-ch_p4-100mbit-2916XL	PII Cluster	mpich	ch_p4	100BASE-TX, Cisco 2916XL
SP2-MPI-IBM-320mbit-vulcan	IBM SP2	IBM's MPI		320 Mbps Vulcan
SP2-MPL-IBM-320mbit-vulcan	IBM SP2	IBM's MPL		320 Mbps Vulcan
O2K-MPI-SGI-700mbit-ccnuma	SGI O2K	SGI's MPI		

Round-trip time: Figure 1 shows the roundtrip time between two processors for increasing message sizes. The experiment performed thousands of samples, and the *minimum* round trip delay time was recorded. This choice of minimum value accounts for the TCP/IP and UDP/IP results being identical since, when no packets are dropped and error recovery is not initiated, the two protocols behave similarly. The difference of approximately $100\mu s$ for small messages between the NIC and UDP/IP results demonstrates the shallowness of the protocol stack. Although this improvement is partly due to the elimination of the system calls, the major benefit is due to the replacement of the user-space signal handler that was used for asynchronous message delivery in BSPlib/UDP, with a kernel-space interrupt handler associated with the interrupt line serviced by the NIC.

As well as benchmarking the NIC-based protocol through a 100Mbps switch, two machines were connected back-to-back using a crossover wire. The purpose of this experiment was to identify the software latencies in the NIC, ignoring the effect of the switch. For 4-byte user messages (which will be 40 bytes including headers, but 60 bytes on the wire due to a 60-byte minimum frame size),

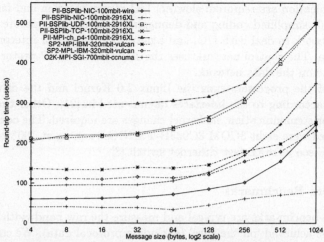

Fig. 1. Round-trip time between two processors as a function of message size

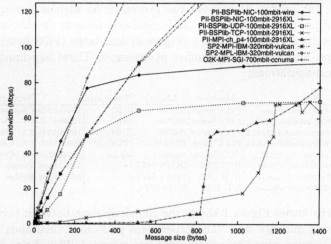

Fig. 2. Link bandwidth between two processors as a function of message size

the experiment PII-BSPlib-NIC-100mbit-wire achieved a roundtrip time of $58\mu s$. This compares to $93\mu s$ for the PII-BSPlib-NIC-100mbit-2916XL experiment, the difference being entirely accounted for by the Cisco 2916XL switch latency [18] of $2 \times 17.2 = 34.4 \mu s \approx 93 - 58 = 35 \mu s$.

From the figure, the latency of the NIC-based protocol running on the cluster is significantly smaller than any other protocol on the cluster, and outperforms the SP2 for messages less than 256 bytes. The loss in latency for larger messages is entirely due to the increasing latency through the switch. These results are consistent with, though slightly better than, the results of U-Net over 100Mbps Ethernet [29], *even with the overhead of queuing outgoing packets in case they are dropped and checking whether recovery should be triggered.*

Bandwidth: Figure 2 shows the sustained bandwidth between two processors for increasing message sizes. The benchmark arranges that one process sends a large number of packets (with their sizes shown on the horizontal axis) to another process, which returns a single small packet after all the packets have been received. The bandwidth on the vertical axis is calculated from the amount of data communicated, and the time between the first packet sent and the small control packet returned. The return latency can be ignored as there are many packets sent by the first processor. However, the traffic includes any reverse communication required to acknowledge the received packets or to recover from errors.

Considering that the peak bandwidths of the O2K and the SP2 are much higher than that of a 100Mbps switch, it is surprising that the cluster can outperform the SP2 for user messages up to 400 bytes, and can match the O2K for messages up to 200 bytes. After these two points, these more-expensive machines perform considerably better.

The BSPlib/UDP and BSPlib/NIC transport layers, which both use the same error recovery and acknowledgement protocol [9], have smooth, predictable bandwidth curves that rise to their asymptotic levels quite early. In contrast, the BSPlib/TCP transport layer and mpich show erratic and late-rising curves. This demonstrates the assertion of the unsuitability of TCP/IP for high-performance computation due to its inappropriate acknowledgement and error-recovery mechanisms. Although this result would suggest that it is worthwhile replacing the naive error recovery used by TCP within the kernel by a more sophisticated scheme running in user space, we will see later that there are some drawbacks to doing this.

Although the peak bandwidth of TCP/IP is greater than UDP/IP in the graphs, BSPlib/UDP can achieve higher bandwidth utilisation (not shown here), although at the expense of a later rising curve.

The observed software latency at the *BSPlib* layer for 1400 bytes of user data is $11.6\mu s$ for packet download. This includes up to $7.8\mu s$ spent in copying the application data structure into a user/kernel space buffer with memcpy(). For packet upload (excluding memory copy into the application), the time for invoking the interrupt handler, doing any necessary error recovery, and uploading the packet is, on average, $3.6\mu s$ per packet. Due to the lightweight nature of this protocol, the observed efficiency on the wire is 93.62% (i.e., 93.62 *Mbps*) for the packet including our protocol headers (i.e., 1436 bytes).

Spray latency: Non-blocking communications can easily fail or suffer from deadlock, if two processes simultaneously push data into their protocol stacks in an attempt to send large amounts of data to each other. The point at which this deadlock occurs depends upon the buffering capacities of both sender and receiver. Quoting from the MPI report: "...*the send start call is local: it returns immediately, irrespective of the status of other processes. If the call causes some system resource to be exhausted, then it will fail and return an error code. Quality implementations of MPI should ensure that this happens only in 'pathological cases'.*" [13]. In MPI, if the error code of the send is ignored and data continues to be sent by a process, deadlock soon occurs. Although well-defined MPI programs are supposed to check the error code, and users deserve everything they get if

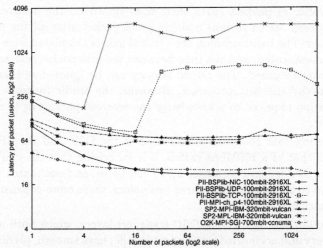

Fig. 3. Latency per packet (half roundtrip) for 4-byte messages as a function of number of packets.

they ignore it, the benchmark described in this section is intended to quantify the effectiveness of the buffering capacity of the various messaging systems by continually sending data in an attempt to exhaust buffers.

Figures 3 and 4 show the gap between successive send calls in the steady state, for 4 byte and 1400 byte messages. The benchmark combines the features of the round-trip and bandwidth benchmarks: a process sends i packets to a receiving process, which reflects them back. However, the application layer on the sender will not receive any of the reflected packets until all packets have been emitted, thus requiring them to be buffered in the protocol stacks or the network. For i packets shown on the horizontal axis, the time shown on the vertical axis is the time between sending the first packet and waiting for the last packet to return, *divided by* i. In the limit, this benchmark measures the latency between two successive send calls. An alternative interpretation, assuming deadlock does not occur, is that it measures the level of pipelining of communication, i.e., at $i = 1$ it is equivalent to the round-trip benchmark; for $i > 1$ it measures the degree of communication overlap due to packets in flight.

Before describing the results of this benchmark, it should be noted that the BSPlib/UDP and BSPlib/NIC messaging layers are used by *BSPlib* in such a way *that this form of deadlock cannot occur*. This is achieved by ensuring that if the protocol stack is unable to accept a send operation, then *BSPlib* will attempt to either: (1) consume a packet if one is available; or (2) request an acknowledgement to be returned on those links that have a large number of unacknowledged send buffers. In BSPlib/TCP, deadlock is avoided by using global flow control via slotting [10] as well as behaving as in point (1) above.

All the figures show that, regardless of packet size, for a large number of messages in flight (i.e., bulk communication as in BSP) BSPlib/NIC outperforms all other configurations, including the SP2 and O2K. All MPI implementations on the Cluster, SP2, and O2K suffer from deadlock when more than 256 packets

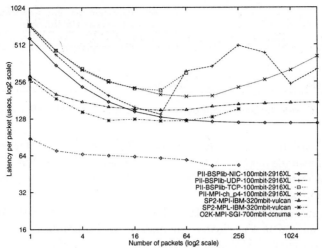

Fig. 4. Latency per packet (half roundtrip) for 1400-byte (approximately full-frame) messages as a function of number of packets

		SP2		Cluster					Origin 2K
	p	MPI	BSPlib	MPI	BSPlib				MPI
		IBM	MPL	mpich	TCP	UDP-1	UDP-2	NIC	SGI
BT	4	31.20	36.44	51.57	50.96	36.52	56.07	56.18	51.10
SP	4	24.89	27.31	33.15	40.86	18.37	41.85	42.36	56.22
MG	8	36.33	36.78	21.02	36.05	31.11	38.25	39.62	36.16
LU	8	38.18	36.50	46.34	55.50	50.15	54.63	63.09	87.34

Table 1. Results in Megaflops/sec *per process* for class-A NAS parallel benchmarks

are injected into the protocol stack. Although the roundtrip time on the cluster is $95\mu s$ for a single 4-byte packet, multiple packets can be injected into the protocol stack every $21\mu s$. While it is possible to do so faster ($8.2\mu s$ is the minimum), the protocol will not settle to a steady state, as the receiver cannot keep up with incoming packets, and overruns start to occur. Therefore $21\mu s$ was achieved by pacing the sender.

Figure 3 shows that the TCP/IP implementations (both BSPlib/TCP and mpich) all have a poor spray latency for large numbers of packets injected into the protocol stack. This is a consequence of the windowing flow control used by TCP where, if communication is run in a state where the window is always exhausted, throughput is considerably reduced. Again, this may lead one to conclude that it is worthwhile replacing the kernel-level TCP error-recovery mechanism, with a more sophisticated scheme in user space.

3.3 NAS Parallel Benchmark Performance

A BSP implementation [17] of version 2.1 of the NAS parallel benchmarks was used to compare the performance of BSPlib/TCP, BSPlib/UDP, and BSPlib/NIC on four and eight processors. Also, a standard MPI implementation of the bench-

	p	computation time	BSPlib/NIC	mpich		BSPlib/TCP		BSPlib/UDP-1		BSPlib/UDP-2	
BT	4	724.6s	24.2s	91.1s	3.8	100.9s	4.2	427.5s	17.7	25.79	1.1
SP	4	458.9s	42.8s	182.2s	4.3	61.2s	1.4	697.8s	16.3	48.96	1.1
MG	8	11.1s	1.1s	4.6s	4.2	2.4s	2.2	4.5s	4.1	1.62	1.5
LU	8	205.9s	30.5s	115.9s	3.8	62.8s	2.1	91.4s	3.0	103.51	3.4

Table 2. Slowdown factors comparing each protocol BSPlib/NIC on the cluster

marks was used to compare the relative performance of the cluster with an 8-processor thin-node 66Mhz SP2, and a 72-processor 195Mhz O2K. Class-A-sized benchmarks were used. Due to memory limitations of the cluster, the FT benchmark was not performed.

Table 1 summarises the results of the NAS benchmarks for the three *BSPlib* protocols running on the cluster, comparing MPI variants of the benchmarks running on the cluster, SP2, and O2K. BSPlib/NIC outperforms mpich and IBM's proprietary implementation of MPI on the SP2 for all the benchmarks. On the O2K, BSPlib/NIC performs better on half of the benchmarks.

As the NAS benchmarks are mostly compute bound, large improvements in communication performance will only be realised as small improvements in the total Mflop/s rating of each of the benchmarks. As we aim to show the improvements of our communication protocol on the cluster compared to MPI, Table 2 gives a breakdown of time spent in computation and communication for each of the protocols running on the cluster. As all the benchmarks on the cluster use the same Fortran compiler (Absoft) and compiler options, and the computational part of the NAS benchmarks were not changed in any of the benchmarks, we assume the computation time spent in all the benchmarks is independent of the protocol (i.e., MPI, BSPlib/NIC, BSPlib/UDP, and BSPlib/TCP) used for communication.

Table 2 shows that BSPlib/NIC produces communication that is approximately 4 times better than mpich on all the NAS parallel benchmarks on the cluster. The improvements of the NIC protocol compared to BSPlib/TCP are not as marked as the mpich results, as some of the improvements are due to global optimisations that are applied in all implementations of *BSPlib* (i.e., packing and scheduling) [16]. As the NAS benchmarks communicate large amounts of data, the latency improvements of BSPlib/NIC over BSPlib/TCP will not be exercised to the fullest, and this accounts only for the approximately two times speedup in communication for these benchmarks.

Although BSPlib/UDP performs better than BSPlib/TCP in the micro benchmarks (see Section 3.2), the results given in the UDP-1 column of Table 1 show extremely poor performance in the computation bound benchmarks SP and BT. The problem with BSPlib/UDP compared to BSPlib/TCP (in which the error recovery is performed within the kernel by the TCP component of the protocol stack, and flow control is used) is that, in order for a message to be received asynchronously, a signal has to be dispatched to the user process and the user process has to be scheduled. To receive the message, the user process then makes

		Data rcvd	Explicit Acks	Data dropped	Duplicate rcvd
BT	UDP-1	893922	7.32%	38964	14913
$p=4$	UDP-2	817471	0.42%	3490	0
	NIC	820233	2.53%	1	0
SP	UDP-1	1561491	6.58%	69076	21882
$p=4$	UDP-2	1440148	0.18%	2534	0
	NIC	1422927	1.00%	14	0
MG	UDP-1	148023	4.21%	2339	1559
$p=8$	UDP-2	141193	0.66%	926	0
	NIC	140113	1.51%	4	0
LU	UDP-1	1968612	2.99%	33681	3819
$p=8$	UDP-2	1955673	2.14%	41878	7655
	NIC	2026933	6.23%	93	1900

Table 3. Packet statistics for the NAS parallel benchmarks

a recvfrom() system call. All this is on top of the height of the UDP/IP protocol stack. A number of effects come into play as a result of this extra path length and the delay in invoking recovery which BSPlib/NIC does not suffer from and which motivates our preference for a kernel level protocol: (1) The unreliable portion of the communiction is not inside the processor. This means that holding on to buffers and performing recovery at a level which includes a reliable path which also takes some time wastes buffer space. As the time between a message arriving at the NIC and its being uploaded into the user application may be large in BSPlib/UDP, more buffers will be required both for the buffering of incomming data and for unacknowledged data in the send queues (a direct result of Little's Law) there is a potential for packet loss due to overruns. (2) Both BSPlib/NIC and BSPlib/UDP reserve the same amount of buffer resources for communication. However, the NIC implementation, which also has a quicker acting interrupt handler, can carefully control their use, whereas BSPlib/UDP can only associate buffers with a point-to-point link by setting an appropriate socket option when the link is created. (3) By processing the protocol is the user space, the number of schedules/dispatches to process the user workload is increased slightly as some of the quanta is consumed processing the protocol. It is even concievable that the extra non-locality in the executable code may slow some processors by introducing TLB and cache stalls (but these effects have not been measured).

Table 3 shows the number of dropped and duplicate packets occurring under the NIC and UDP/IP implementation of *BSPlib* (remember that both protocols use exactly the same error recovery and acknowledgement policy). The marked difference between the error rate of the two protocols stems from the judicious use of buffering resources, which inhibits overruns, in BSPlib/NIC. On close investigation of the BT and SP benchmarks it was found that most of the packets were dropped by only one of the communication calls in each of the programs, yet these calls communicated the largest amount of data, and were performed regularly within each benchmark. Therefore although the error rate for the complete benchmark was only 4.3% for BT and 4.4% for SP, the actual error rate in the problematic section of communication was 99.2% and 81.1% respectively.

On investigation, as the Cisco switch did not drop any packets during any of the benchmarks, it was concluded that the high error rate was caused by overruns occurring due to insufficeint buffer capacity within the IP part of the protocol stack in BSPlib/UDP. The hypothesis was tested by performing a crude form of flow control, whereby the sender was *throttled* so that packets could not be injected into the protocol stack any faster than $120\mu s$ (results are presented in the tables under the heading UDP-2). As can be seen from the UDP-2 data in Table 3, not only does this technique drastically reduce the number of dropped packets, but it reduces the number of redundant messages. This is due to BSPlib/UDP suffering from the problem of stale error recovery information being contained within packet headers due to a large number of packets being queued up for communication by the NIC[5]. Unfortunately, the limitation of the throttling is that it considerably lengthens the round-trip time for packets, although, as can be seen from the UDP-2 column of table 1, it has the desired effect of improving the performance of the NAS benchmarks. From these result we conclude that as slowing down the sender enables the BSPlib/UDP implementation of the NAS benchmarks to produce similar results to the low-latency BSPlib/NIC implementation, the NAS parallel benchmarks are not latency bound.

4 Performance comparison with other NIC protocols

There exist protocols whose hardware requirements are as modest as ours, whose bandwidth is as high as ours, and whose reliability is as great, but there are very few that achieve all three simultaneously. However, all three aspects are critical to building clusters that scale and can be used to execute applications.

The early approaches to low-latency and high-bandwidth communication recognised the redundancy in the protocol stack and hence the necessity of simplifying it, eliminating buffer copies by integration of kernel-space with user-space buffer management, and collapsing a number of network layers. This is the approach used in the protocol of Brustolini [5], based on ATM, which achieves reliable low-latency, high-bandwidth performance close to that of the hardware. However, in that implementation the overhead of sending must be quite high as the sending process blocks until the message is placed on the network.

An approach that bypasses the need for buffer management is the Active Messages of von Eicken *et al.* [28]. In the active message scheme, messages contain the address of a routine in a receiving process which handles the message by sending a response and/or populating the data structures of the computation with the data in the payload. A prototype implementation for a SparcStation over ATM, in which message delivery is reliable, is described by von Eicken *et al.* [27]. Because the message handler routine runs in the user's address space, there has to be some mechanism to make sure that the receiving process is scheduled when a message arrives. In general, this requires some kernel changes. The prototype for the SparcStation solves this scheduling problem by having the receiving process poll for messages. Although active messages are intended to

---

[5] Remember that the BSPlib/NIC alleviates this problem by using a gathered send. This allows low-latency packet download to prepare more than 75 packets for tranmission in the time for a single full frame to be communicated between two machines.

Group	Name	Network	μs	Mbps	Reliable	Ref.
Genova	GAMMA (Active Messages, polling)	100Mbps hub	13	98	×	[7]
Genova	GAMMA (Active Messages, IRQ)	100Mbps hub	17	98	×	[7]
Oxford	PII-BSPlib-NIC-100mbit-wire	100Mbps wire	29	91	√	
Cornell	U-Net/FE (P133, hub)	100Mbps hub	30	97	×	[29]
Cornell	U-Net/FE (P133, switch)	100Mbps switch	40	97	×	[29]
Oxford	PII-BSPlib-NIC-100mbit-2916XL	100Mbps switch	46	91	√	
Oxford	PII-BSPlib-TCP-100mbit-2916XL	100Mbps switch	103	70	√	
Oxford	PII-BSPlib-UDP-100mbit-2916XL	100Mbps switch	105	79	√	
ANL	PII-MPI-ch_p4-100mbit-2916XL	100Mbps switch	147	78	√	
SUNY-SB	Pupa	100Mbps switch	198	62	√	[26]
Utah	Sender Based Protocols	FDDI ring	22	82	√	[24]
	FDDI (theoretical peak)	FDDI ring	510	100	√	
NASA	MPI/davinci cluster FDDI	FDDI ring	818	73	√	[19]
Cornell	Active Messages (write, SS20)	ATM switch	22	44	√	[27]
Cornell	Active Messages (read, SS20)	ATM switch	32	44	√	[27]
Cornell	U-Net/ATM (P133)	ATM switch	42	120	×	[29]
CMU	Simple Protocol Processing	ATM switch	75	48	√	[5]
NASA	MPI/davinci cluster Fore ATM	ATM switch	1005	98	?	[19]
NASA	MPI/davinci cluster SGI ATM	ATM switch	1262	85	?	[19]
Lyon	BIP	Myrinet	4	1008	×	[20]
Princeton	VMMC (Myrinet 2ndG, P166)	Myrinet	10	867	?	[11]
UCB	VIA	Myrinet	25	230	√	[6]
Princeton	VMMC (SHRIMP)	Intel Paragon	5	184	?	[12]
SGI	O2K-MPI-SGI-700mbit-ccnuma	CC-Numa	17	325	√	
Kalsruhe	PULC (port-M PVM)	ParaStation	27	92	√	[3]
IBM	SP2 (theoretical peak)	Vulcan switch	40	320	√	
IBM	SP2-MPL-IBM-320mbit-vulcan	Vulcan switch	55	173	√	
IBM	SP2-MPI-IBM-320mbit-vulcan	Vulcan switch	67	173	√	
	Ethernet (theoretical peak)	10Mbps hub	465	10	√	
NASA	MPI/davinci cluster Ethernet	10Mbps hub	900	8	?	[19]
NASA	MPI/davinci cluster HIPPI	HIPPI	851	265	√	[19]

Table 4. Half roundtrip latency (μs) and Link Bandwidth (Mbps)

be one-sided, this polling activity is a compromise between two-sided and one-sided communication. This is not really an application programming issue as the intention is that the interface be used within a library or in code generated by a parallel-language compiler. BSP communication is also one-sided. Apart from the obvious programming benefits of not having to match send and receive requests, as long as a reasonable number of buffers are deployed, there is no necessity to schedule the distributed processes involved in a computation in synchrony as message disposal is detached from message receipt in BSPlib/NIC. From the results in this paper, there appears to be no detrimental effect from this loose coordination of the processes, as very high percentage bandwidths are being realised. We attribute this to the number of buffers made available, the fact that data can be packed to fill buffers, and the fact that errors are detected early with recovery not relying on process scheduling.

Active Messages have also been used at various other levels in implementing transport mechanisms: GAMMA [7] is an implementation of Active Messages over Fast Ethernet and achieves an impressive 12.7μs one-way latency and a full frame bandwidth of approximately 85% of the medium. This bandwidth rises to an asymptotic bandwidth of 98% of the medium for two processors sending frames back-to-back. However, the assumption is that the medium is reliable and that frames arrive in order.

Table 4 provides a summary of the performances of various research groups' low-latency and high-bandwidth clusters. The U/Net protocol [29] has similar properties to BSPlib/NIC as they both use Fast Ethernet networks, and use similar technique for allocating shared user-kernel memory buffers. As can be seen from the results, the bandwidth and latency of the two are similar. Pupa [26] is also a low-latency communication system over Fast Ethernet. It concentrates on buffer management and provides a reliable transport protocol but does not make assumptions about the network except that packets arrive in order. A missing packet triggers the receiver to initiate recovery. Thus there is an implicit assumption that there is only one route between a pair of nodes. Pupa achieves a one-way latency of $198\mu s$ and achieves a bandwidth of 62Mbps for very large messages (10000 bytes) but achieves less than 60% efficiency for full frame-sized messages.

There are a number of protocols operating over Myrinet [4], for example BIP [20]. BIP (Basic Interface for Parallelism) has similar design objectives to Pupa: to support a message-passing parallel-computing environment (there is an MPI implementation for BIP). No error recovery is supported, but each packet receives a sequence number and contains a checksum, and hence errors can be detected. BIP achieves a good one-way latency of $4.3\mu s$ and an asymptotic bandwidth of 1008Mbps (96% of the available 1056Mbps).

The suitability of some of the network media mentioned in Table 4 for high-performance parallel computing may not be immediately obvious. Certainly, a single wire between two processors is not at all scalable and cannot be taken seriously (we only include such configurations in our our work to illustrate the contribution of the switch to the latency). Also, all of the bus-based schemes are not scalable; this includes the Fast Ethernet implementations that use an Ethernet hub, since the hub merely acts as an an extension of the bus and its bandwidth is divided amongst the communicating processors. Hub implementations imply the use of the CSMA/CD protocol, which makes no promise for reliable delivery, and hence this must be built into some higher protocol. For scalability purposes, the FDDI ring can also be considered a bus because the bandwidth is diluted amongst the communication processes. However, a single FDDI ring can be used in such a way that the higher-level protocols do not have to provide recovery.

Treating error recovery as optional based on the underlying reliability of a point-to-point benchmark is a false argument. Unless some back-pressure notification is made available to the sender, it will always be possible to overwhelm a component of the switch in the presence of unstructured communication patterns. Switches are starting to appear which address this problem, either by augmenting the protocol to send a flow-control packet back along the link, or by holding the cable busy so that the NIC CSMA/CD engine will hold off transmitting the next packet. However, this does not solve edge problems when multiple links are used.

Communication primitives in which the receiver polls on a device, for example in variants of active messages such as GAMMA [7] and U/NET [27], are not one-sided and the measurement of latency may hide a potentially-large loss in

CPU cycles. This loss can arise as a result of the sender running ahead of the receiver. If the send is truly synchronous, then the sender waits for the corresponding receive, else if the sender continues, it may re-invoke a send later and cause an overrun (or an overrun due to another processor sending data to the same destination). If the receiver runs ahead, then time will be wasted polling for the incoming message. Of course depending on the model, such wastage of CPU time may be inevitable, but it does not occur for well-designed BSP computations. Having tightly-coupled communication is difficult in the loosely-coupled environment of clusters; we have adopted a buffer-slackness approach to accommodate this loose coupling.

The switched solutions are the only ones that provide scalability (Myrinet, ATM and switched Fast Ethernet). Of the Fast Ethernet solutions, our results are best in terms of scalability, reliability, latency and bandwidth, and we are competitive with the much more expensive Myrinet solutions.

5 Conclusions

Many low-latency, high-bandwidth protocols for clusters have been developed, and their performance is generally impressive. However, it is not clear that this performance can be exploited in applications, primarily because providing reliability requires adding significant extra capabilities to a protocol.

The important question is: where should this extra functionality be added? The trend in high-performance clusters has been to increase the amount of protocol work that takes place in user space. Our results suggest that this may not be a good thing, because it has a significant effect on computational performance. We show that it is possible to achieve latency and bandwidth comparable to existing protocols while executing in kernel space. The important effect of this is that reliability can be achieved without significant performance degradation.

We document the performance behaviour of this new protocol stack, using low-level benchmarks to explore single-link behaviour, and the NAS parallel benchmarks to demonstrate that the single-link performance scales to applications.

Acknowledgements: Jonathan Hill's work was supported in part by the EPSRC Portable Software Tools for Parallel Architectures Initiative. David Skillicorn is supported in part by the Natural Science and Engineering Research Council of Canada. The authors would like to thank Oxford Supercomputing Centre for providing access to an Origin 2000.

References

1. 3Com. *3C90x Network Interface Cards Technical Reference*, December 1997. Part Number:09-1163000.
2. D. Bailey, T. Harris, W. Saphir, R. van der Wijngaart, A. Woo, and M. Yellow. The NAS parallel benchmarks 2.0. Report NAS-95-020, NASA, December 1995.
3. J. M. Blum, T. M.Warschko, and W. F. Tichy. PULC: Parastation user-level communication. Design and Overview. In *Parallel and Distributed Processing*, volume 1388 of *Lecture Notes in Computer Science*, pages 498–509. Springer, 1998.
4. N. J. Boden, D. Cohen, R. E. Felderman, A. E. Kulawik, C. L. Seitz, J. N. Seizovic, and W.-K. Su. Myrinet — a gigabit-per-second local-area network. http://www.myri.com, November 1994.

5. J. C. Brustolini and B. N. Bershad. Simple protocol processing for high-bandwidth low-latency networking. Technical Report CMU-CS-93-132, School of Computer Science, Carnegie Mellon University, Pittsburgh, PA 15213, 1992.
6. P. Buonadonna, A. Geweke, and D. E. Culler. Implementation and analysis of the Virtual Interface Architecture. In *SuperComputing'98*, 1998.
7. G. Ciacco. Optimal communication performance on Fast Ethernet with GAMMA. In *Parallel and Distributed Processing*, volume 1388 of *Lecture Notes in Computer Science*, pages 534–548. Springer, 1998.
8. Cisco Systems. *Catalyst 2900 series XL installation and configuration guide*, 1997. Part Number: 78-4417-01.
9. S. R. Donaldson, J. M. D. Hill, and D. B. Skillicorn. BSP clusters: high performance, reliable and very low cost. Technical Report PRG-TR-5-98, Programming Research Group, Oxford University Computing Laboratory, September 1998.
10. S. R. Donaldson, J. M. D. Hill, and D. B. Skillicorn. Predictable communication on unpredictable networks: Implementing BSP over TCP/IP. In *EuroPar'98*, LNCS, Southampton, UK, September 1998. Springer-Verlag.
11. C. Dubnicki, A. Bilas, K. Li, and J. Philbin. Design and implementation of virtual memory-mapped communication on myrinet. In *Proceedings of the 11th International Parallel Processing Symposium*, pages 388–396. IEEE, IEEE Press, 1997.
12. C. Dubnicki, L. Iftode, E. W. Felton, and K. li. Software support for virtual memory mapped communication. In *Proceedings of the 10th International Parallel Processing Symposium*. IEEE, IEEE Press, 1996.
13. M. P. I. Forum. *MPI A Message-Passing Interface Standard*, May 1994.
14. W. D. Gropp and E. Lusk. *User's Guide for mpich, a Portable Implementation of MPI*. Mathematics and Computer Science Division, Argonne National Laboratory, 1996. ANL-96/6.
15. J. M. D. Hill, B. McColl, D. C. Stefanescu, M. W. Goudreau, K. Lang, S. B. Rao, T. Suel, T. Tsantilas, and R. Bisseling. BSPlib: The BSP Programming Library. *Parallel Computing*, 24(14):1947–1980, November 1998. see www.bsp-worldwide.org for more details.
16. J. M. D. Hill and D. Skillicorn. Lessons learned from implementing BSP. *Journal of Future Generation Computer Systems*, 13(4–5):327–335, April 1998.
17. A. L. Hyaric. Converting the NAS benchmarks from MPI to BSP. Technical report, Oxford university Computing laboratory, 1997. Available from ftp://ftp.comlab.ox.ac.uk/pub/Packages/BSP/NASfromMPItoBSP.tar
18. Mier Communucations Inc. Product lab testing comparison: 10/100BASET switches, April 1998.
19. National Aeronautics and Space Administration. Summary of recent network performance on davinci cluster. http://science.nas.nasa.gov/Groups/LAN/cluster/latresults/-sumtab.recent.html.
20. L. Prylli and B. Tourancheau. A new protocol designed for high performance networking on Myrinet. In *Parallel and Distributed Processing*, volume 1388 of *Lecture Notes in Computer Science*, pages 472–485. Springer, 1998.
21. A. Rubini. *Linux Device Drivers*. O'Reilly and Associates, 1998.
22. A. Simpson, J. M. D. Hill, and S. R. Donaldson. BSP in CSP: easy as ABC. Technical Report PRG-TR-6-98, Programming Research Group, Oxford University Computing Laboratory, September 1998.
23. D. Skillicorn, J. M. D. Hill, and W. F. McColl. Questions and answers about BSP. *Scientific Programming*, 6(3):249–274, Fall 1997.
24. M. R. Swanson and L. B. Stoller. Low latency workstation cluster communications using sender-based protocols. Technical Report UUCS-96-001, Department of Computer Science, University of Utah, Salt Lake City, UT 84112, USA, 1996.
25. L. G. Valiant. Bulk-synchronous parallel computer. U.S. Patent No. 5083265, 1992.
26. M. Verma and T. cker Chiueh. Pupa: A low-latency communication system for Fast Ethernet, April 1998. Workshop on Personnel Computer Based Network of Workstations held at the 12th International Parallel Processing Symposium and the 9th Symposium on Parallel and Distributed Processing.
27. T. von Eicken, V. Avula, A. Basu, and V. Buch. Low-latency communication over ATM networks using Active Messages. Technical report, Department of Computer Science, Cornell University, Ithaca, NY 14850, 1995.
28. T. von Eicken, D. E. Culler, S. C. Goldstein, and K. E. Schauser. Active Messages: A mechanism for integrated communication and computation. In *The 19th Annual International Symposium on Computer Architecture*, volume 20(2) of *ACM SIGARCH Computer Architecture News*. ACM Press, May 1992.
29. M. Welsh, A. Basu, and T. von Eicken. Low-latency communication over Fast Ethernet. In *EuroPar'96 Parallel Processing: Volume I*, volume 1123 of *Lecture Notes in Computer Science*, pages 187–194. Springer, 1996.

Coscheduling through Synchronized Scheduling Servers— A Prototype and Experiments

Holger Karl[*]

Humboldt-University of Berlin
Institut für Informatik
Unter den Linden 6
10099 Berlin, Germany
`karl@informatik.hu-berlin.de`

Abstract. Predictable network computing still involves a number of open questions. One such question is providing a controlled amount of CPU time to distributed processes. Mechanisms to control the CPU share given to a single process have been proposed before. Directly applying this work to distributed programs leads to unacceptable performance, since the execution of processes on distributed machines is not coordinated in time. This paper discusses how coscheduling can be achieved with share-controlling scheduling servers. The performance impact of scheduling control is evaluated for BSP-style programs. These experiments show that synchronization mechanisms are indispensable and that coscheduling can be achieved for unmodified programs, but also that a performance overhead has to be paid for the control over CPU share.

1 Introduction

In recent years, the potential of interconnected stand-alone workstations for parallel and distributed applications has been widely discussed in research literature (cp. [1] for a classic text). These so-called *networks of workstations* (NOW) exhibit other research issues than dedicated, custom-designed parallel supercomputers. One such issue is predictable runtime of parallel programs.

To achieve predictable runtime, it is necessary to control the allocation of computing resources, e.g. CPU time, of a single machine to a parallel program. Since a workstation might also be used to run interactive applications, parallel programs should not be allowed to interfere with this interactive work arbitrarily, but should leave enough free capacity for interactive users. On the other hand, it might be desirable to dedicate a certain minimum amount of resources to the parallel application. This gives rise to the notion of a contract between (one or

---

[*] This research was sponsored by Deutsche Forschungsgemeinschaft (German Research Council), by the Defense Advanced Research Projects Agency and Rome Laboratory, Air Force Materiel Command, USAF, under agreement number F30602-96-1-0320; and by the National Science Foundation under grant number CCR-94-11590.

several) parallel applications and interactive, local users of a machine, regimenting how resources are allocated to each program and guaranteeing minimum and maximum shares of these resources.

Enforcing such a contract could be done by the operating system, but a middleware approach (not modifying a commodity operating system) is more in accordance with the NOW ideas. Several such middleware systems (often called "scheduling servers") have been described; this paper presents an implementation for the Linux operating system.

However, all other comparable systems are concerned with non-parallel applications running only on a single machine. The case of parallel applications executed in a distributed fashion on a NOW is more challenging. As experiments will show (cp. Section 4.3), a naïve use of distributed scheduling servers results in unacceptable performance. This is commensurate with the well-known fact that distributed scheduling and the coordination of programs can have an immense effect on program efficiency — as evidenced by work on coscheduling. Therefore there is a need to combine controlled resources with coordinated, distributed scheduling. This paper discusses how a set of scheduling servers can enforce a resource contract while at the same time providing coscheduling of parallel programs, combining guaranteed resources with good performance.

Parallel programs conforming to the bulk synchronous parallel-style (BSP, [15]) are used for evaluations. This choice is due to their close relationship with consensus-based methods which have been suggested as a basis for fault-tolerant network computing [11]. Combining these fault tolerance mechanisms with the guaranteed resource allocation for distributed programs described in this paper allows to give (probabilistic) guarantees on the program execution time, even in the presence of faults. This is a step towards predictable network computing on commercial off-the-shelf systems.

The rest of this paper is organized as follows. Section 2 describes related work in more detail, Section 3 considers limitations imposed by commodity operating systems and explains the design of a prototype, Section 4 gives some initial experimental results, Section 5 outlines possibilities for extending this work and Section 6 finally gives some conclusions.

2 Related work

2.1 Controlling CPU share

In real-time operating system, controlling CPU share is relatively straightforward (cp. e.g. [10] for real-time Mach). Other approaches target commodity off-the-shelf operating systems and try not to modify them. [13] describes a *Scheduling Server* for NeXTSTEP. This server is a program that runs with the highest available real-time scheduling priority, and cyclically lowers or raises a controlled program's priority. The controlled program also runs with a high real-time scheduling priority, higher than normal programs, but strictly lower than the scheduling server. The scheduling server itself sleeps practically all the time and only wakes up to schedule the controlled program. The resource share that is given to the program is tunable via the scheduling server. Similarly, the URSched system introduced in [8] uses the fixed-priority scheduling of SUN's Solaris 2.4 operating system to provide smooth video playback. URSched achieves considerably reduced jitter compared to standard time-sharing scheduling strategies.

A comparison between the user-level approaches [8, 13] and the operating system approaches [10] shows that for the price of modifications to the system kernel, more precise accounting of resource usage (in particular, processing time spent in the operating system on behalf of a process) can be achieved. User-level approaches on the other hand allow better flexibility to implement various advanced scheduling schemes on top of existing operating systems [5].

2.2 Coordinated scheduling

The question of coordination of processes belonging to a parallel program has been addressed by work on gang/coscheduling. The original idea of gang scheduling is to schedule distributed threads of a parallel program, running on multiple processors, at the same time [12]. This provides these threads with an environment similar to a dedicated machine and allows them to spin-wait for synchronization and/or communication. This is particularly important for fine-grained programs which would suffer considerably from the context switches entailed by blocking communication. For coarse-grained or highly imbalanced threads, blocking can be beneficial.

Classic gang scheduling research assumes that the parallel machine is dedicated to parallel programs, which basically use it in a time-sharing fashion. This is no longer true for NOWs. In particular, pure spin-waiting is no longer acceptable since this is a waste of resources. [6] and later [2] suggest to use an adaptive two-phase spin-blocking communication, which, in combination with a standard operating system's time-sharing scheduler, results in coordinated scheduling of distributed threads. In this scheme, a communication event (sending or receiving of messages) is considered as an implicit request for coscheduling between sender and receiver. They show that this implicit coscheduling is particularly beneficial for fine-grained programs. The essential property exploited in [6] here is the scheduler's property to give a priority boost to a blocked program after becoming runnable again. [14] argues along similar lines, but uses additional hardware support (a programmable network interface) and system support (a custom-made network driver) to generate additional scheduling events when messages arrive.

[7] gives a performance comparison of gang and coscheduling for data-parallel workloads on a PC cluster. They have implemented gang scheduling via a signal-based distributed user-level scheduler, SCore-D, which is similar to the prototype presented here except that it does not provide controlled resource allocation.

3 Prototype description

The prototype has two main objectives: controlling CPU share with a scheduling server on a single machine, and synchronizing distributed scheduling servers.

3.1 Controlling CPU share

The prototype was developed for the Linux operating system, version 2.0. Newer Linux kernels provide real-time scheduling classes (similar to the schedulers of Solaris or NeXTSTEP) in the form of a system call `set_scheduling_priority()`. Therefore, the basic structure is the same as in [13] or [8]: a high-priority server process acts cyclically upon controlled processes, which run at a medium real-time priority. These real-time scheduling classes have strictly higher priority than time-shared processes, and are therefore not impeded by background load. The server process sleeps almost all the time, and only wakes up periodically to act out its scheduling decisions: activating or deactivating controlled processes. The operating system's time-sharing scheduler remains in control of process that are not under the control of the scheduling server. The amount of time given to each process is an input parameter to the scheduling server. This amount can, e.g., be computed by well-known algorithms like rate-monotonic scheduling, or can be set according to user input or requirements.

There are two basic possibilities for acting out scheduling decisions: manipulating priorities, or explicitly suspending/resuming the controlled process. Manipulating priorities can be done by setting a process' scheduling class to either real-time or standard time-sharing. The natural implementation for the second alternative under Linux is to send SIGSTOP or SIGCONT signals. In the first version, a controlled process may run even if it is not activated by the scheduling server (by competing with normal processes for the CPU via standard time-sharing mechanisms); in the second version, this is not possible.

This prototype uses the signal-based approach since it showed better stability under background load compared to an implementation that manipulates process priorities. Also, it provides cleaner semantics as far as maximum CPU share is concerned. Due to the use of standard signals, this prototype should be easily portable to any operating system that supports fixed-priority scheduling classes.

It should be noted that even though signals are reputed for their long (and hard to predict) delivery latencies, this is generally not true for SIGSTOP or SIGCONT signals. These signals are not actually delivered to the receiving processes but rather are processed by the operating system kernel directly and merely change internal tables of the scheduler.

3.2 Synchronizing distributed schedulers

Work on gang scheduling and coscheduling has conclusively shown the need to coordinate the execution of distributed threads of a parallel application in time to obtain acceptable performance. Since the scheduling server controls the times of execution of these processes, it follows that the distributed servers need to synchronize their scheduling activities.

Ideally, synchronizing the scheduling servers should be done in a decentralized fashion, with only a few and rare control messages. An obvious way to achieve this is to use some of the well-known clock synchronization protocols — one server sending its remaining sleep time to its peers, which adjust to it. But this approach is problematic: it requires sleep times of well under a millisecond to achieve acceptable synchronization. However, Linux on PCs only provides timer interrupts with 10 milliseconds granularity.

This raises the need for generating interrupts that start execution of the operating system scheduler at more or less arbitrary points of time. Message arrivals do generate interrupts. Hence, messages to the scheduling servers can be used to start their execution. In the initial prototype, this fact is exploited by designating one of the scheduling servers as master which sends a short control message to the other slave scheduling servers. If the communication medium is capable of broad- or multicast (e.g. like Ethernet) this control message is no scalability bottleneck. This control message only contains the current slot number in the schedule. The schedule itself has to be communicated between machines only when a parallel application starts or finishes. The complete schedule contains the name of the length of a slot, the number of slots in the schedule, and for each slot, the name of the distributed application that occupies it (if any). The slave servers receive a copy of the master server's schedule, which in turn is can be computed according to various algorithms, e.g. rate-monotonic scheduling or simple round robin. Free slots can be used for local applications, and a number of slots has to be left free to allow the rest of the system to make progress.

Since the slave servers run at highest real-time priority, but are usually blocked in the receive call for this control message, they will be scheduled immediately after message arrival. In this design, only the master server sleeps and wakes up in a time-triggered way; the scheduling activities of the slave servers are driven by message arrival from the master server.

Note that this does not mean that messages between the distributed parts of the application have to pass through the scheduling servers. Indeed, the application itself can completely ignore the fact that it runs under control of such a distributed scheduling server; there is no need to modify it in any way.

Comparing this approach with [14] shows that for the price of modifying the network driver, fast scheduling events can be achieved. Also, no explicit control messages are used. However, no share-control is possible. Indeed, both techniques could be fruitfully combined.

4 Experiments

For an experimental evaluation of the proposed prototype, four Pentium 90-based PCs, running Linux version 2.0.31 (completely unmodified) and connected via standard 10 MBit/s Ethernet, were used. The machines were used exclusively, but some external network traffic during the experiments could not be completely ruled out. Section 4.1 shows results about the stability of a scheduling server-controlled program under background load on a stand-alone machine, Section 4.2 briefly introduces the programming model used for the experiments with distributed programs, Section 4.3 discusses the behavior of a single barrier synchronization and Section 4.4 shows execution time for a parallel application with many synchronization operations. More detailed results can be found in [9].

4.1 Stability

The main purpose of the proposed scheduling server is to allocate a predetermined share of the CPU to a controlled program and also to make sure that this program does not exceed its share. To investigate the stability of this allocation under increasing load on a stand-alone machine, we used a slightly modified version of Linpack as controlled program. This Linpack version computes the actual floating point performance received during its runtime — this number can be interpreted as the relative share of CPU time.

In Figure 1, Linpack's performance under scheduling server control is shown, with 50 runs done for 0, 1, 2, ... simultaneously active background processes each, and CPU share ranging from 10% to 90%. While the behavior is not perfect, the CPU share allocated to Linpack is reasonably stable.

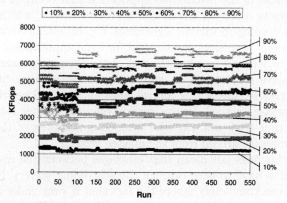

Fig. 1. Linpack under Scheduling Server control with different levels of CPU share and increasing number of background jobs

The reserved CPU share should not exceed about 50% or 60%, since at least the operating system itself needs time for its own operation. Stability degrades considerably beyond this point.

4.2 BSP programming model

A BSP program consists of parallel threads that run independently and synchronize using barrier synchronization. For the following experiments, each machine is assigned one parallel thread, which runs as an individual process. One such thread is characterized by its execution time. This time is drawn from a uniform random distribution $\mathcal{U}(g - (g*v)/2, g + (g*v)/2)$, where g is the granularity (average thread length) of the application, and v represents the load imbalance between parallel threads, expressed in percent of the granularity. Here, granularities of 100 μs, 500 μs, 1 ms, 5 ms, 10 ms and 100 ms are considered, in combination with load imbalances of 0%, 25%, 50%, 75% and 100%. This barrier is repeated n times, the total run time of n cycles forms one measurement, and i of these n iterations are performed for statistical relevance.

To achieve implicit coscheduling, two-phase spin-blocking is typically used: a process spins, constantly checking for message arrivals. If, after a certain spin-time, no message has arrived, the process blocks. This spin-time t is the last parameter for our programming model.[1]

Parallel threads are only allowed to communicate immediately after barrier synchronizations. Here we show results for the following communication patterns:

`barrier`: no communication between threads, only the barrier synchronization,
`complete`: each thread sends a (short, fixed-size) message to every other thread.
   Additionally, after receiving a message, a process has to process this message, which takes 10 ms each.

4.3 Barrier synchronization

Section 4.1 has established acceptable stability for programs running under scheduling server control on a single machine. This section investigates how a BSP-style program performs when run under scheduling server control and distributed over several machines.

As a basis for comparison, Figure 2 gives the average times in microseconds for a single barrier synchronization, i.e. $n = 1$. There is no scheduling server control, blocking communication primitives are used, and the granularities and load imbalances discussed in the previous Section 4.3 serve as parameters.

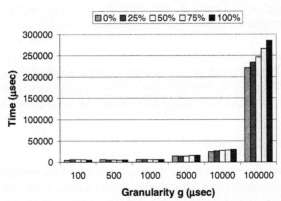

Fig. 2. Performance of a single barrier synchronization without scheduling server, blocking communication

---

[1] This model is similar to the one used in [6], which additionally suggests an adaptive spin-time. Currently, the experiments reported here are limited to fixed spin-times only.

Using unsynchronized scheduling servers on all machines results in the numbers shown in Figure 3. Here as in all following experiments, the scheduling server allocated 2 out of 10 time slices to the controlled program, where one time slice is 20 milliseconds. Evidently, with unsynchronized scheduling servers, the slowdown is much larger than the expected factor of five — an entire scheduling round is necessary to complete a single synchronization. This clearly indicates that unsynchronized scheduling servers are not acceptable for distributed programs.

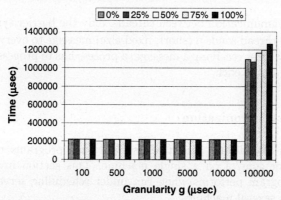

Fig. 3. Performance of a single barrier synchronization with unsynchronized scheduling servers, blocking communication

The effects of the synchronization mechanism among scheduling servers proposed in this paper are shown in Figure 4. Again, these times should be about five times as long as in Figure 2, but they are actually better than this. For coarse-grained programs, this slowdown approaches roughly 4.5. For fine-grained program, it is only about 1.5 to 2; since the program is communication-bound and not computation-bound, the CPU limitation is not as severe. This smaller than expected slowdown can be attributed to both coscheduling effects and to the higher priority used by the processes. Due to their vastly superior performance, only synchronized scheduling servers will be considered henceforth.

4.4 Total execution time

While the previous Section 4.3 has investigated the times needed to perform a *single* barrier synchronization, this section looks at the total execution time of a program that consists of *many* such synchronization operations. The behavior differs somewhat between the `barrier` and the `complete` communication pattern, which are discussed in Section 4.4 and Section 4.4, respectively. Blocking and spin-blocking communication is used, as well as programs running without and with scheduling server control (only synchronized scheduling servers are considered). The scheduling servers were again set to give 20% of a machine's

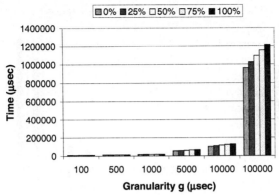

Fig. 4. Performance of a single barrier synchronization with synchronized scheduling server, blocking communication

CPU time to the controlled program. The number of synchronization calls n was again to give reasonable total execution time, but always at least $n = 50$ iterations were performed for one measurement. $i = 50$ measurements were taken to compute average and standard deviation of the execution time.

One of the main purposes of using scheduling servers place is to make program behavior more predictable. A simple measure of predictability is standard deviation of runtime. Since the standard deviation depends on the actual scale and is not unit-less, a slightly more convenient parameter is the variation coefficient, which is defined as the standard deviation divided by the average. The variation coefficient will be discussed for the **complete** communication pattern in Section 4.4.

barrier To abbreviate the exposition, actual runtime is only shown for a base case. For other cases, only the relative numbers are given with respect to this. The base case is blocking communication without scheduling server support; runtimes for this are shown in Figure 5.

Adding scheduling server control to all four distributed program parts should incur a slowdown of at most five. Figure 6 shows the ratio of runtimes between scheduling server support and uncontrolled program. It is notable that for fine-grained programs, the synchronized scheduling servers perform particularly well and stay well below their maximum acceptable ratio of five. With increasing grain size, the ratio approaches this limit.

The comparison between controlled and uncontrolled execution of spin-blocking programs (with spin time 100 microseconds) is shown in Figure 7. Similar to blocking communication, the scheduling server has a smaller slowdown for fine-grained programs and approaches its natural limit for coarse-grained programs. Since there is no communication after a barrier synchronization, using spin-blocking while under scheduling server control does not significantly increase performance.

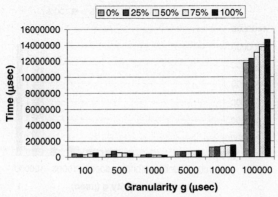

Fig. 5. Runtime of `barrier`, blocking communication, no scheduling server

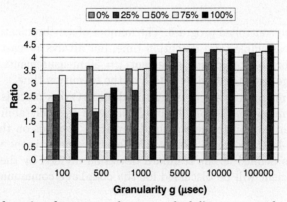

Fig. 6. Ratio of runtime for `barrier` between scheduling server and uncontrolled program, blocking communication

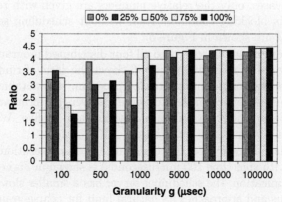

Fig. 7. Ratio of runtime for `barrier` between scheduling server and uncontrolled program, spin-blocking communication (100 ms)

complete This section consider programs with the `complete` communication pattern: after a barrier synchronization has been performed, each process exchanges messages with every other. This puts stronger requirements on coscheduling since messages follow each other immediately. For this pattern, both average runtime and the variation coefficient will be discussed.

Average runtime Figure 8 gives the base case: blocking communication, no scheduling server. It is remarkable that the fine-grained, unbalanced cases are slower than those with larger granularity; this is conjectured to be an artificial byproduct of the experimental environment and needs further verification.

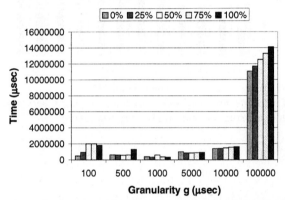

Fig. 8. Runtime of `complete`, blocking communication, no scheduling server

If only blocking communication is used, programs with the `complete` pattern perform less favorably under scheduling server control than programs that only perform barrier synchronizations. For some combinations of g and v, the controlled execution takes up to 6.5 times longer than the uncontrolled one. N.B. that in Figure 6 this was not the case.

This changes, however, if the controlled program uses spin-blocking. As can be seen in Figure 9, now the slowdown is on the same order as what should be expected under server control; indeed, it is slightly better due to the higher priority. This comes as no surprise: a program under server control that uses blocking communication gives up its time slice when blocking on a receive. It does maintain the permission to run, since the scheduling server has not (in general) send the SIGSTOP signal. But it has to be rescheduled by the normal operating system scheduler after a network interrupt occurs. Therefore spinning is clearly beneficial for heavily communicating programs even under scheduling server control. This is substantiated by Figure 10 (compared to Figure 8) — the slowdown is again well under the acceptable maximum of five.

Variation coefficient Since one of the objectives of using scheduling servers is predictable execution of parallel programs, the variation coefficient is an important metric. Naturally, the runtime variation of programs with a complete

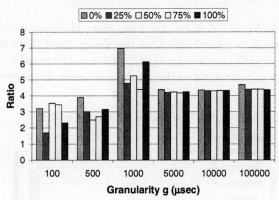

Fig. 9. Ratio of runtime for `complete` between scheduling server and uncontrolled program, spin-blocking communication (100 microseconds)

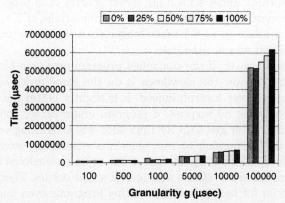

Fig. 10. Runtime of `complete`, spin-blocking communication (100 ms), with scheduling server

communication pattern is subject to a large number of external influences (e.g. collisions on a shared medium).

The discussion on average execution time has shown that spin-blocking communication is recommendable for programs with the complete communication pattern. Hence, Figure 11 shows the variation coefficient for spin-blocking programs with complete communication.

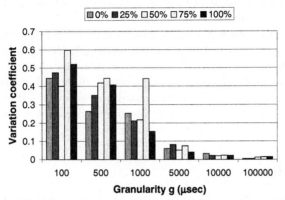

Fig. 11. Variation coefficient of complete's runtime, spin-blocking communication (100 ms), no scheduling server

This should be compared to the variation coefficient of the same programs under scheduling server control: Figure 12 shows these numbers. Under scheduling server control, the variation coefficient is smaller for fine-grained programs and are more or less unchanged for coarse-grained programs. This shows that scheduling servers improve the predictability of program execution.

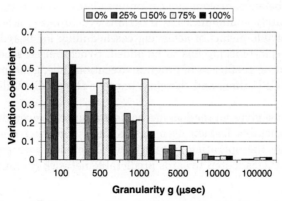

Fig. 12. Variation coefficient of complete's runtime, spin-blocking communication (100 ms), with scheduling server

5 Future work

There are a number of possible extensions. Experiments on a larger cluster to address questions of scalability, multiple distributed programs running under scheduling server control, or with stochastic background load should be performed.

To give guarantees on the execution time, not only the CPU, but other resources like memory (page locking) or network bandwidth should be controlled by the scheduling servers as well.

The coarse timer resolution is a major obstacle. We are currently investigating an extension to the Linux kernel, utime [3], that promises to support micro-second timer resolution. This would enable time-driven synchronization as opposed to the message-based synchronization described in this paper. Other operating systems (e.g. QNX) and/or hardware support (similar to [14]) are other interesting possibilities. Finally, the integration into a resource management scheme like [4] is of practical importance.

6 Conclusions

This paper addressed the need for guaranteed resources for parallel programs on NOWs. It presented a prototypical implementation of a synchronized scheduling server for the Linux operating system. Among several design choices, a signal-based implementation was used due to its stability and portability.

While a straightforward implementation works well for stand-alone programs, the performance impact on parallel programs proved to be disastrous. A synchronization mechanism was suggested that copes with the limited clock resolution of commodity systems and still achieves reasonable performance by coscheduling distributed programs.

Using extensive measurements, the influence of various parameters like granularity, load imbalance, or communication paradigm was investigated. For fine-grained, moderately communicating programs synchronized scheduling servers provide a reasonable means of achieving coscheduling; for heavily communication programs, spin-blocking has to be added. Synchronized scheduling servers reduce the variation coefficient of program execution time, guarantee a predetermined share of CPU time, and therefore improve the predictability of program execution time.

This prototype gives a practical mechanism to execute even distributed programs in a predictable fashion even on standard operating systems, without having to modify the operating system itself. Combined with consensus-based fault tolerance, this is a step towards predictable network computing.

References

1. T. E. Anderson, D. E. Culler, David A. Patterson, et al. A Case for NOW (Networks of Workstations). *IEEE Micro*, 15(1):54–64, February 1995.

2. A. C. Arpaci-Dusseau, D. E. Culler, and A. M. Mainwaring. Scheduling with Implicit Information in Distributed Systems. In *Proc. of SIGMETRICS '98/PERFORMANCE '98 Joint Conf. on Measurement and Modeling of Computer Systems*, pages 233–243. ACM, June 1998.
3. S. Balji, R. Menon, F. Ansari, J. Keinig, and A. Sheth. UTIME — Micro-Second Resolution Timers for Linux. Project homepage at http://hegel.ittc.ukans.edu/projects/utime/index.html, April 1998.
4. A. Baratloo, A. Itzkovitch, Z. Kedem, and Y. Zhao. Just-in-time Transparent Resource Management in Distributed Systems. Technical Report 1998-762, Courante Institute of Mathematical Sciences, New York University, March 1998.
5. H.-H. Chu and K. Nahrstedt. A Soft Real Time Scheduling Server in UNIX Operating System. In R. Steinmetz and L. C. Wolf, editors, *Proc. of Interactive Distributed Multimedia Systems and Telecommunication Services*, pages 153–162, Darmstadt, Germany, September 1997.
6. A. C. Dusseau, R. H. Arpaci, and D. E. Culler. Effective Distributed Scheduling of Parallel Workloads. In *Proc. ACM SIGMETRICS Conference on Measurement and Modeling of Computer Systems*, pages 25–36, Philadelphia, PA, May 1996.
7. A. Hori, F. B. O'Carroll, H. Tezuka, and Y. Ishikawa. Gang Scheduling vs. Coscheduling: A Comparison with Data-Parallel Workloads. In H. R. Arabnia, editor, *Proc. Intl. Conf. on Parallel and Distributed Processing Techniques and Applications*, volume 2, pages 1050–1057, Las Vegas, NV, July 1998. CSREA.
8. J. Kamada, M. Yuharo, and E. Ono. User-level Realtime Scheduler Exploiting Kernel-level Fixed Priority Scheduler. In *International Symposium on Multimedia Systems*, Yokohama, Japan, March 1996.
9. H. Karl. A Prototype for Controlled Gang-Scheduling. Technical Report Informatik Bericht 112, Institut für Informatik, Humboldt-Universität, Berlin, Germany, August 1998.
10. C. Lee, R. Rajkumar, and C. Mercer. Experiences with Processor Reservation and Dynamic QOS in Real-Time Mach. In *Proc. of Multimedia Japan*, April 1996.
11. M. Malek. A Consensus-Based Framework and Model for the Design of Responsive Computing Systems. In G. M. Koob and C.G. Lau, editors, *Foundations of Dependable Computing: Models and Frameworks for Dependable Systems*, chapter 1.1, pages 3–21. Kluwer Academic Publishers, Boston, MA, 1994.
12. J. K. Ousterhout. Scheduling Techniques for Concurrent Systems. In *Proc. 3rd Intl. Conf. Distributed Computing Systems*, pages 22–30, October 1982.
13. A. Polze. How to Partition a Workstation. In *Proc. Eigth IASTED/ISMM Intl. Conf. on Parallel and Distributed Computing and Systems*, pages 184–187, Chicago, IL, October 1996.
14. P. G. Sobalvarro, S. Pakin, W. E. Weihl, and A. A. Chien. Dynamic Coscheduling on Workstation Clusters. Technical Report 1997-017, DEC Systems Research Center, Palo Alto, CA, March 1997.
15. L. G. Valiant. A Bridging Model for Parallel Computation. *Communications of the ACM*, 33(8):103–111, August 1990.

High-Performance Knowledge Extraction from Data on PC-based Networks of Workstations

Cosimo Anglano[1] *, Attilio Giordana[1], Giuseppe Lo Bello[2]

[1] DSTA, Università del Piemonte Orientale, Italy.
[2] Data Mining Laboratory, CSELT, Torino - Italy.

Abstract. The automatic construction of classifiers (programs able to correctly classify data collected from the real world) is one of the major problems in pattern recognition and in a wide area related to Artificial Intelligence, including Data Mining. In this paper we present G-Net, a distributed algorithm able to infer classifiers from pre-collected data, and its implementation on PC-based Networks of Workstations (PC-NOWs). In order to effectively exploit the computing power provided by PC-NOWs, G-Net incorporates a set of dynamic load distribution techniques that allow it to adapt its behavior to variations in the computing power due to resource contention. Moreover, it is provided with a fault tolerance scheme that enables it to continue its computation even if the majority of the machines become unavailable during its execution.

1 Introduction

The induction of classification rules from a database of instances pre-classified by a teacher is a task widely investigated both in Pattern Recognition and in Artificial Intelligence. Image interpretation, medical diagnosis, troubleshooting, text retrieval, and many other applications can be seen as specific instances of this task. The popularity recently gained by the so called *Data Mining* [5, 10] has brought a renovated interest on this topic, and has set up new requirements that started a new trend aimed at developing new algorithms. As a matter of fact, most Data Mining applications are characterized by huge amounts of data that exceed the capabilities of most existing algorithms, such as neural networks, decision trees, and so on. Therefore, a new generation of induction algorithms have been recently proposed in the literature, and the research in this area is currently very hot. Among the emerging approaches, evolutionary algorithms are gaining more and more interest since, on the one hand, they proved to be flexible and easy to apply to a wide variety of problems, obtaining excellent performance, and on the other, they naturally lend themselves to exploit large computational resources such as the ones offered by a network of workstations.

In this paper we will present G-Net [2], a distributed evolutive algorithm for inferring classification rules from data that integrates strategies borrowed from

---

* Corresponding author. Address: Dipartimento di Scienze e Tecnologie Avanzate, Universitá del Piemonte Orientale, Corso Borsalino 54, 15100 Alessandria, Italy. email: mino@di.unito.it

evolutionary computation, genetic algorithms and tabu-search, and we will discuss its implementation on PC-based Networks of Workstations (PC-NOWs). Although the G-Net algorithm is equally suited to various kinds of parallel computing systems having different granularities (ranging from SIMD machines to massively parallel processors), thanks to its architecture that couples a very fine granularity of the computational objects with the absence of global synchronization points, we decided to devise an implementation targeted to PC-NOWs for two main reasons. First, the dramatic advances in low-cost computing technology, together with the availability of new high speed Local Area Networks have enabled the construction of PC-NOWs whose raw processing and communication performance are comparable to those of traditional massively parallel processors. Second, relatively large systems can be built without massive investments in hardware resources, since PCs are relatively inexpensive and, more importantly, already existing machines can be exploited as well. Therefore, by exploiting PC-NOW platforms, G-Net can be used to solve problems of considerable complexity without the need of massive investments in computing resources.

The above advantages, however, do not come for free. In order to exploit the opportunities provided by PC-NOWs, it is necessary to develop applications able to deal with some peculiar features of these platforms, and this makes the development much harder than in the case of traditional parallel systems. In particular, applications must be able to adapt to the occurrence of machine failures during their execution, and to tolerate the effects of resource contention (i.e., variations in the performance delivered by the various PCs) due to the presence of several users sharing the machines. Therefore, the main part of the work required to port G-Net on PC-NOWs platforms was devoted to the development of techniques allowing the application to effectively deal with the above issues.

The paper is organized as follows. Section 2 gives a brief presentation of the G-Net algorithm. In Section 3 we discuss the main issues that we had to face when we designed and implemented G-Net on a PC-NOW. In particular, Section 3.1 discusses the software architecture of our application, Sections 3.2 and 3.3 present the dynamic load balancing techniques devised to deal with heterogeneity and resource contention, while Section 3.4 illustrates the fault-tolerance mechanisms. Section 4 presents some experimental results, and Section 5 concludes the paper and outlines future work.

2 The G-Net Algorithm

In this section we give a brief description of the G-Net algorithm, focusing on the logical behavior of its various computational entities (the interested reader may refer to [2] for a more detailed presentation of its features).

G-Net learns classification rules for a class h of objects starting from a dataset $D = D^+ \cup D^-$, being D^+ and D^- sets of elements belonging and not belonging to h, respectively. In particular, its goal consists in finding a set of classification rules (henceforth referred to as a *classification program* or *classifier*) of the type

$\{\varphi_1 \to h, \varphi_2 \to h, \ldots \varphi_n \to h\}$, where h denotes the class to discriminate and $\varphi \to h$ means "if conditions φ is true for an object x, then $x \in h$".

Symbolic induction algorithms as G-Net essentially combine two types of strategies: Hypothesis (rule) generalization and set-covering. The former strategies concern the way hypotheses are generated, that is, how the hypothesis space is actually searched; different approaches apply different inductive operators for moving through the search space and different evaluation criteria to determine the order in which different alternatives are explored. Set-covering strategies deal instead with the global focus of the search: The coverage of all given positive objects (D^+) and possibly none of the negative ones (D^-). A fundamental issue of learning disjunctive classifiers (i.e., classifier containing more than one classification rule) is the number of different rules (disjuncts) to put into the final classifier. Set-covering strategies try to cover the learning set with the smallest number of classification rules.

In G-Net both hypothesis generalization and set-covering strategies are based on Genetic Algorithms (GAs) [6]; in particular, hypothesis generalization is based on task specific genetic operators, while set-covering is based on cooperative coevolution [12, 11].

Let us now describe in more detail the G-Net algorithm. G-Net has a distributed architecture that encompasses two levels. The lower level is a distributed genetic algorithm that stochastically explores the search space, while the upper level applies a coevolutive strategy that directs the search process of the genetic algorithm towards globally best solutions. The above architecture is schematically depicted in Fig. 1, and encompasses three kind of processes, namely Genetic Nodes (G-Nodes), Evaluator Nodes (E-Nodes), and a Supervisor Node. Each G-

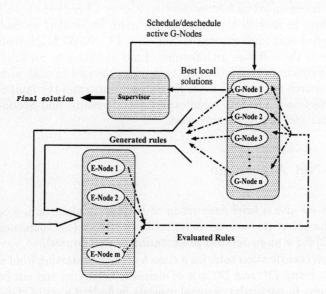

Fig. 1. Distributed Architecture of G-Net

Node corresponds to a particular element $d \in D^+$, and is in charge of finding rules covering d, which are stored into the G-Node's *local memory*, by executing a genetic algorithm schematically working as follows:

1. Select two rules from the local memory with probability proportional to their fitness;
2. Generate two new rules φ_1 and φ_2 by applying suitable genetic operators;
3. Send φ_1 and φ_2 to an E-Node for evaluation in order to obtain their fitness values
4. Receive from the E-Nodes a set of evaluated rules (if available) and stochastically replace some of the rules already in the local memory with them.
5. Go to step 1

As indicated in the above algorithm, E-Nodes perform *rule evaluation*: Each time a G-Node generates a new rule, sends it to an E-Node (since all the E-Nodes use the same dataset D, anyone of them may be used) that matches it against each element of the dataset D, and assigns it a *fitness* value, a numerical quantity expressing the "goodness" of the rule. When the evaluation of a rule is completed, the E-Node broadcasts it (together with its fitness value) to all the G-Nodes, so during Step 4 of the genetic algorithm a G-Node may receive also rules generated by other G-Nodes. G-Nodes and E-Nodes work asynchronously, that is a G-Node can continue generating new rules while already-generated ones are still under evaluation. However, for the sake of convergence, the number of rules that a G-Node can generate without using new rules is limited to a predefined threshold (called the *G-Node credit*).

The Supervisor Node coordinates the execution of the various G-Nodes according to the coevolutive strategy. Its activity basically goes through a cycle (called *macro-cycle*), having period measured in terms of iterations performed by the G-Nodes (*micro-cycles*), in which it receives the best rules found by the G-Nodes, selects (by means of a hill-climbing optimization algorithm) the subset of rules (the *hypothesis pool*) that together give the best classification accuracy for class h, and chooses the subset of G-Nodes that will be *active* (i.e. allowed to execute their genetic algorithm) during the next macro-cycle. As a matter of fact, although in principle all the G-Nodes might be simultaneously active if enough computational resources were available, in real situations the number of processors is limited. Since in a typical Data Mining application the number of G-Nodes ranges from hundred of thousands to several millions, it is evident that even a very large PC-NOW cannot provide enough resources, so each machine will be shared among several (possibly many) G-Nodes. In order to deal with this issue, only a subset of G-Nodes (henceforth referred to as *G-Nodes working pool*) will be activated during each macro-cycle, and in order to ensure a proper coverage of the search space, the composition of the working pool is dynamically varied by the Supervisor (across different macro-cycles) by means of a stochastic criterion that tends to favor those G-Nodes whose best rules belong to hypothesis pool and need further improvement (see [2] for more details). The size of the working pool is specified by the user, and depends on several factors, such as the particular classification task; the duration of a macro-cycle is defined as $G \times \mu$, where G and μ denote the number of active G-Nodes and the number of micro-cycles per G-Node per macro-cycle, respectively.

The execution of the algorithm terminates when a satisfactory global solution is found, that is when the quality of the found classifier does not improve for an assigned number of macro-cycles. Finally, the classifier constructed in this way is validated on a new dataset T such that $D \cap T = \emptyset$.

3 Porting G-Net on PC-based NOWs

Let us now discuss the issues involved in the design of the PC-NOW version of G-Net. Porting G-Net on a traditional parallel computing system is relatively simple, since the distributed architecture of the algorithm can be easily translated into a set of parallel processes. Our implementation, however, was targeted to PC- NOWs, and this required to take into considerations several aspects that can be disregarded in the case of traditional parallel computers.

First of all, we had to choose a software environment allowing to use the ensemble of PCs as a parallel computer. Since one of our goals was portability, we have chosen PVM [13], that is practically available for every platform. This choice, however, is not critical, since PVM is used only for the implementation of a few communication operations and for the heterogeneity support, so replacing it with a different system is straightforward (for instance, an MPI implementation is planned for the near future).

Second, PC-NOWs are characterized by the presence of phenomena not arising in traditional parallel computing systems that make the porting task much more challenging. In particular, since the PCs making up a NOW may deliver different performance from each other (if they are equipped with different processors), and their performance may vary over time (if several users share the machines), it was necessary to adopt a dynamic load distribution policy able to assign to each machine an amount of work proportional to its speed, and to keep the above distribution balanced during the execution of the application.

Finally, the PC-NOW used to execute the application may be relatively unstable, since some of the machines may become unavailable at run-time. Therefore, in order to allow G-Net to continue its execution when one or more machines become unavailable, it was necessary to incorporate into the application a suitable mechanism providing it with some form of fault tolerance.

In the following of this section we present in detail the mechanisms mentioned before. We start with the presentation of the software architecture of our implementation (Sec. 3.1), we continue with the load distribution techniques in Sections 3.2 and 3.3, and we conclude with a brief discussion of the fault tolerance mechanisms (Section 3.4).

3.1 Software Architecture

The software architecture of our implementation of G-Net on PC-NOW platforms, schematically depicted in Fig. 2, encompasses two kinds of processes, namely *Searchers* (that generate rules) and *Matchers* (that perform rule evaluation), besides the Supervisor Process (corresponding to the Supervisor Node and not shown in the figure).

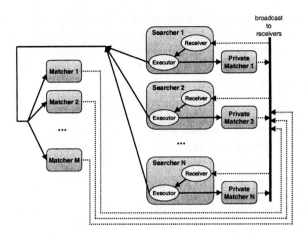

Fig. 2. Software Architecture of G-Net. Ovals correspond to threads, gray boxes to processes, continuous arrows to rules sent for evaluation, and dotted arrows to evaluated rules.

Searcher processes are made up by two threads, called *Executor* and *Receiver*, respectively [2], and split among them the computation load required by the G-Nodes. In particular, each Executor receives a portion of the G-Nodes working pool (called *local working pool*) whose size depends on the computing power allocated to it, multiplexes the processor among its locally active G-Nodes, and alters across different macro-cycles (as the Supervisor Node described in the previous section) the composition of the local working pool. The duration of the macro-cycle of Executor E_i is $wps(E_i) \cdot \mu$, where $wps(E_i)$ is the size of its local working pool. It is worth pointing out that only the G-Nodes working pool is partitioned, and *not* the set of G-Nodes; that is, each Executor chooses the G-Nodes that will be active during a macro-cycle among all the G-Nodes defined in the system.

The Receiver thread is in charge of receiving evaluated rules from the set of Matchers, and acts as a filter for the corresponding Executor. In particular, it does not forward to its Executor all the evaluated rules generated elsewhere (*non-local* rules) since, for the sake of convergence, it is necessary that a) the percentage of non local rules used by an Executor does not exceed a given threshold and b) the difference between the micro-cycle executed by the Executor (on the behalf of its various G-Nodes) and the micro-cycle during which the rule was generated does not exceed a predefined value (i.e., the rule is not too *old*). Finally, Matchers correspond to E-Nodes, and are in charge of evaluating on the dataset D the rules generated by the various Searchers.

---

[2] Since PVM does not provide threads, at the moment Searchers are implemented as a pair of communicating processes executed on the same machine.

The last issue that needs to be discussed is concerned with the choice of the number of Searchers and Matchers. As a general rule, the number of Searchers should be smaller than that of Matchers, since the computationally heavy activity is rule evaluation and not rule generation. Moreover, in order to speed-up the execution as much as possible, the number of Searchers and of Matchers should be chosen in such a way that the rule generation rate and the rule evaluation rate are simultaneously maximized. The (usually) limited number of available machines, however, is often insufficient to simultaneously satisfy the above requirements, so the machine usage must be carefully balanced. As a matter of fact, giving a large number of machines to Searchers increases the rule generation rate, but decreases the amount of computation power assigned to Matchers and, hence, the rule evaluation rate. Conversely, an increase in the number of machines dedicated to Matchers increases the rule evaluation rate, but decreases the generation rate. This choice is further complicated by the fact that the power of the various machines may vary over time as a consequence of resource contention. Therefore, finding the optimal allocation is very hard, if not unfeasible.

In the light of the above considerations, we have followed a different approach: Rather then trying to find an optimal allocation, we require the user to specify (using his/her expertise) the number of Searchers and Matchers (their sum must not exceed the number of available machines), and we use a flexible allocation scheme that allows G-Net to exploit the computing power that may be left unused as a consequence of a poor distribution. In particular, given N Searchers and P hosts $(N < P)$, the Searchers are scheduled on the N fastest machines, while Matchers are executed on the remaining $P - N$ machines. Since the above mapping may yield idle time on the Searchers' machines (this happens when Matchers do not have enough computing power to generate an evaluation rate high enough to avoid that Executors have to wait for evaluated rules to come back), each Executor is provided with a *private* Matcher, which is executed on the same machine with the aim of exploiting the above idle time. The private Matcher can be used only by the corresponding Executor, although the rules it evaluates are broadcast to all the Searchers. Each Executor starts to use the private Matcher instead of a shared one as soon as it determines that the arrival rate of evaluated rules is becoming insufficient to sustain its execution (of course, the Executor has to feed its private Matcher *before* coming to a stopping point). In particular, the Executor monitors the value of its credit, that is the amount of new rules that can be generated without receiving evaluated rules (see Section 2): Each time an Executor passes Step 4 of the genetic algorithm without actually receiving an evaluated rule, its credit is decremented by one, while it is incremented of the same amount each time an evaluated rule is received. When the current credit value drops below 50% of the initial value, the Executor sends the new rules to the private Matcher, otherwise it chooses one of the shared Matchers (the criterion used to choose a shared Matcher is discussed in Section 3.3). Note that the exploitation of the private Matcher increases the rule arrival rate while decreasing the speed of the Executor, so after some time the credit value will exceed the 50% threshold and the Executor will start using again the pool of shared Matchers.

3.2 Dynamically Balancing the Workload among Executors

As mentioned in the previous section, each Executor is associated with a local working pool. The size of this pool must be chosen in such a way that all the Executors complete their macro-cycles approximately at the same rate, otherwise the convergence speed and the quality of the computed classifiers would be hampered. As a matter of fact, if some Executors are excessively slower than the other ones, the rules generated by them will be discarded by the faster Executors because of their age. In order to have all the Executors to proceed approximately at the same pace, the number of G-Nodes assigned to each of them should be proportional to the computing power delivered by the machine on which it is allocated. Moreover, since the above power varies as the load due to other users varies, the local working pool size must vary as well. To achieve this goal, we have developed a dynamic load balancing algorithm based on the *hydrodynamic approach* proposed by Hui and Chanson [8].

The hydrodynamic approach defines a framework for the development of *nearest-neighbor* dynamic load balancing algorithms, that is algorithms that allow the workstations to migrate tasks with their immediate neighbors only. Each workstation W_i is associated with two real functions, the *load* $l_i(t)$ and the *capacity* $c_i(t)$ (reflecting its workload and its speed in processing it at time t, respectively), and is represented as a cylinder having a cross-sectional area given by $c_i(t)$ and containing a quantity of liquid given by $l_i(t)$. The *height* of the liquid in cylinder i at time t is defined as $h_i(t) = l_i(t)/c_i(t)$ (for the sake of readability, in what follows we denote capacity, load, and height of workstation W_i as c_i, l_i and h_i, respectively). The network connections among workstations correspond to infinitely thin liquid channels that allow the liquid to flow from one cylinder to another. The goal of the dynamic load balancing algorithm is to distribute the liquid among the various cylinders (by means of the liquid channels) in such a way that their heights are equal. In [8] Hui and Chanson prove that all the distributed nearest-neighbors algorithms that satisfy certain conditions converge geometrically to the optimal distribution.

The load balancing algorithm developed for G-Net is executed only by Executors. In particular, given a set of Executors $\{E_1, E_2, \ldots, E_n\}$, we associate with each E_i its capacity c_i, measured in terms of micro-cycles executed per second (computed by means of a benchmark executed in absence of contention on the workstation on which it is allocated), and its load l_i, that corresponds to the number of micro-cycles that have to be performed to complete a macro-cycle (computed as $wps(E_i) \cdot \mu$). Consequently, the height h_i corresponds to the amount of seconds needed to E_i to complete a macro-cycle, so the goal of having the same height for all the Executors corresponds, as expected, to keep the Executors proceeding at the same speed.

Initially, $wps(E_i) = c_i \cdot (\sum_{j=1}^{n} l_j / \sum_{j=1}^{n} c_j)$, so all the Executors have the same height $h_i = \mu \cdot \sum_{j=1}^{n} l_j / \sum_{j=1}^{n} c_j$ and, hence, proceed at the same speed. During the execution of each macro-cycle, E_i periodically measures the average speed s_i at which it completes the micro-cycles and broadcasts it to the other Executors (note that s_i may increase or decrease according to the presence of additional jobs). At the end of each macro-cycle, E_i builds a list (the *speed list*) in which all Executors (including itself) are ranked in increasing order according to their speed, and divides it into two parts of equal size. Therefore, the upper part of the list will contain *overloaded* Executors that are slower than the *underloaded* Executors contained into the lower part. If E_i is overloaded, it decreases the size of its local working pool of a suitable amount x of active G-Nodes (see below), i.e. $wps(E_i) = wps(E_i) - x$, and chooses an underloaded Executor to assign the execution of this surplus. If, conversely, E_i is underloaded, it waits for a notification, sent by some overloaded Executor, concerning the amount x of active G-Nodes that must be added to its local working pool (i.e., $wps(E_i) = wps(E_i) + x$). At the end of the balancing step, each Executor broadcasts to all the other Executors the size of its local working pool.

Each overloaded Executor E_i whose position in the list is $pos(E_i)$ ($pos(E_i) < n/2$) balances its workload with the underloaded Executor listed at the position $n - pos(E_i) + 1$. In this way, a given underloaded Executor may be chosen only by a single overloaded Executor (note that the list is the same for all the Executors, since the speeds are globally known), so that excessive fluctuations in the load assigned to underloaded Executors are avoided, and a balanced state should be reached in less balancing steps.

The amount of active G-Nodes sent by the overloaded Executor E_i to an underloaded Executor E_j is computed as $\gamma(c_i * c_j/(c_i + c_j))(h_i - h_j)$, where $\gamma = 0.7$ (the minimum value recommended by [8]), $c_j = 1/s_j$, and h_i is computed by setting $l_j = wps(E_j) \cdot \mu$ (note that s_j and $wps(E_j)$ are globally know to all the Executors).

The above load balancing algorithm satisfies all the conditions required by the hydrodynamic framework [1], so it is guaranteed to converge geometrically to the optimal solution.

3.3 Dynamically Distributing Rule Evaluations among Matchers

As discussed in Section 3.1, when an Executor generates a rule and the private Matcher cannot be used, one of the shared Matchers must be chosen to evaluate it. This choice is performed autonomously by each Executor by means of a *rule distribution policy* whose goal is to reduce, as much as possible, the time required to evaluate the rules in such a way that the amount of time they spend idle is minimized.

Although the problem of Matchers selection is equivalent to that of assigning tasks to servers in distributed systems [7], many of the solutions developed in this context cannot directly be applied to G-Net since they do not deal with heterogeneity and resource contention. As a matter of fact, it is crucial for G-Net that rule distribution is performed is such a way that all the Matchers proceed at the same evaluation rate in spite of possible differences in the speed of the corresponding machines, otherwise it may happen that the work performed by the slower Matchers is completely wasted. To illustrate this phenomenon, consider the simple case were all the Matchers proceed at the same speed, except one which is slower. In this case, if no special care is taken, the rules evaluated by the fast Matchers will allow the Executors' computations to proceed without interruptions. Consequently, when a rule evaluated by the slow Matcher is received by an Executor, it is very likely that it has completed a significant number of micro-cycles since when the rule was generated, so the rule may be too old and, consequently, discarded by the Receiver.

In order to select a suitable rule distribution policy, we have considered three different policies that represent extensions of algorithms developed for task allocation, and we have compared them via simulation. The first policy, henceforth referred to as *work stealing*, was adopted in the first implementation of G-Net [2], and is based on a client/server scheme in which an Executor sends a new rule to a Matcher only if there is a Matcher immediately available for evaluation, otherwise stores it into a local FIFO queue. When a Matcher finishes to evaluate a rule, it signals to the Executor that generated it that it is willing to accept a new rule for evaluation. Each Matcher servers the various Executor in a round robin fashion, in order to avoid that an Executor has exclusive use of the Matcher resources. The rationale behind this algorithm is that faster Matchers will reply more often than slower Matchers and, consequently, will receive more rules to evaluate. Also, if the speed of a Matcher decreases, it will start to reply less often and, consequently, will receive less and less rules to evaluate.

The second policy we consider is the *Join the Shortest Queue* [14] (JSQ) policy. In this case, unlike work stealing, Executors do not store locally a new rule if no free Matchers are available, but rather they send it to a Matcher as soon as it is generated. If the chosen Matcher is busy, the rule waits into a FIFO queue and is evaluated later. Each time an Executor has to send a rule, it chooses the Matcher with the shortest queue, since a shorter queue indicates that the Matcher is able to process its rules faster than Matchers with longer queues. This algorithm is inherently dynamic, since if a Matcher becomes slower, its queue will grow and, as consequence, it will be chosen less frequently. The implementation of the algorithm is straightforward: Each time a Matcher broadcasts an evaluated rule to all the Executors, it piggy-backs the information concerning its queue length on the messages containing the evaluated rule.

Finally, the last policy we analyzed is called *Join the Fastest Queue* (JFQ), and is derived from the Join the Shorted Expected Delay Queue [14]. This policy is similar to the JSQ one, with the difference that now an Executor chooses the Matcher that minimizes the expected time required to evaluate a new rule. In order to allow Executors to perform their choices, each Matcher M_i keeps track of the average time RET_i it needs to evaluate a rule, and computes the expected time that a rule should spend in its queue before it is evaluated as $RET_i \cdot qlen(E_i)$ (where $qlen(E_i)$ denotes the length of its queue). This information is piggy-backed on each evaluated rule and, consequently, broadcast to all the Executors.

In order to evaluate the effectiveness of these policies, we have developed a discrete event simulator, and we have evaluated their performance for two different scenario. As will be shown by our results, the JFQ policy results in the best performance for various system configurations that represent the opposite extremes of the whole spectrum, so it has been chosen for use in G-Net.

Our analysis focused on two performance indices that allow to assess the suitability of each policy to the needs of G-Net. The first one is the average *rule round trip time* (RTT), defined as the time elapsing from when a rule is generated to when the evaluated rule is received by the Executor that generated it, and indicates the speed at which evaluated rules are received by the Executors. The second index is the *evaluation time range*, and is computed as follows. Let ET_{M_i} be the average time elapsing from when a rule is generated to when its evaluation is completed by M_i, and M_{min}, M_{max} the slowest and fastest Matchers, respectively. The evaluation time range is computed as $ET_{M_{max}} - ET_{M_{min}}$, and indicates the maximal difference in the speeds of Matchers and, hence, the probability of discarding rules because of their age. For all the results reported in this Section, the 95% confidence interval is within 3% of the sample mean.

In the first scenario, corresponding to the graphs shown in Figs. 3 and 4, we have kept fixed the number of Executors (8), and we have progressively increased the number of Matchers from 8 to 24. The rule generation speed has been kept equal to a given value S_G for all the Executors in all the experiments. Also, in all the experiments we have used 4 "slow" Matchers, having an evaluation speed 5 times smaller than S_G, while we have progressively increased the number of "fast" Matchers having a speed only 2 times smaller than S_G. As can be

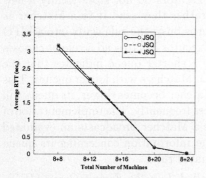 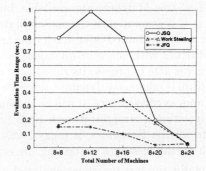

Fig. 3. Average rule RTT **Fig. 4.** Evaluation Time Range

observed from the graph in Fig. 3, all the three policies have practically the same average rule RTT, while the results for the evaluation time range (Fig. 4) are quite different. In particular, we can observe that the JSQ policy has the largest evaluation time range, that corresponds in the highest probability that evaluated rules are discarded because of their age, while the best performance is provided by the JFQ policy.

The second simulation scenario is identical to the first one, the only difference being that the evaluation speed of fast Matchers is 5 times smaller than S_G, while that of slow Matchers is 100 times smaller than SG. From the graphs of

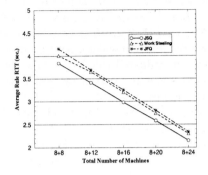

Fig. 5. Average rule RTT

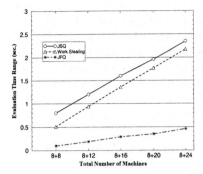

Fig. 6. Evaluation Time Range

Figs. 5 and 6 it is evident that the JFQ policy results in a slightly larger average rule RTT but, as in the first scenario, in a much lower evaluation time range, especially when the number of Matchers gets large.

3.4 Fault Tolerance

As discussed at the beginning of this section, PC-NOWs tend to be relatively unstable, especially if they are made up by machines non-dedicated to the execution of parallel applications (e.g., machines used for teaching purposes that become idle during night hours). In these cases, it may happen indeed that a given machine becomes temporarily or permanently unavailable (for instance, when a user decides to reboot it to run a different operating system). Since the application may have been running for a long time (for realistic data sets the computation may last entire days), it would be an unacceptable waste of resources if, as a consequence of the failure of one or more machines, the application crashed or its execution had to be prematurely terminated.

The basic mechanism employed to allow G-Net to continue its execution, albeit with a possible slowdown or loss of accuracy, when one or more machines fail is nothing else that its stochastic nature and the lack of synchronization points. As a matter of fact, the lack of synchronization points guarantees that the failure of a process will not block other processes, while stochastic search implies that a particular region of the search space is not explored exclusively by a particular process, and consequently the failure of some process does not exclude possibly significant solutions from being found. In order to clarify this point, let us consider how the failure of Searchers and Matchers affect the execution of the entire application.

If the failing machine is running a Matcher, then all the rules enqueued at the Matcher are inevitably lost. This loss, however, is not a major problem because the stochastic nature of the algorithm ensures that losses of new rules do not affect the convergence of the algorithm (since the lost rules have a non null probability of being generated again). Moreover, each Executor does not need to receive back all the rules it generates, since it can continue its execution using also rules generated by somebody else. It is however important that the Executors are informed of the Matcher crash so that they avoid to keep sending it rules. This is achieved by a simple timeout mechanism in which each Executor uses a timer, set to a suitable value each time an evaluated rule is received, for each Matcher: If the timer expires before another rule is received by the above Matcher, the Executor assumes that it is no longer operational.

If, conversely, the failing machine is running a Searcher, then the execution will be certainly slowed down, since the size of the G-Nodes working pool will be reduced by the size of the local working pool of the crashed Executor. However, all the G-Nodes have still the possibility of being selected by the remaining Executors, so the application execution can continue without major problems.

Finally, the last case occurs when the Supervisor process crashes. In this case, the application cannot continue its execution, and has to be prematurely terminated. In order to avoid that the computation has to be restarted from the beginning, G-Net adopts a checkpointing strategy in which periodically its saves the status of its computation. Therefore, when a premature termination occurs, the computation can be restarted from the last checkpoint.

4 Experimental Evaluation

In this section we present some experimental results concerning the performance of G-Net on the *Mutagenesis* problem [9], for which the task consists in learning rules for discriminating substances having cancerogenetic properties on the basis of their chemical structure. This task is a specific instance of the more general problem called *Predictive Toxicology Evaluation*, recently proposed as a challenging Data Mining application [4], and is very hard since the rules are expressed using First Order Logics, and the complexity of matching formulas in First Order Logics limits the exploration capabilities of any induction system.

When assessing the performance of systems for classifiers induction, two different aspects must be considered. The first one is related to the quality of the classifiers that the system is able to find, that is expressed in terms of the number of disjuncts (rules) contained in the classifiers (the smaller the number of disjuncts, the better the classifiers) and of the relative error (misclassification of objects). In [3] we have shown that G-Net is able to find classifiers whose quality is significantly better than those generated by other systems representative of the state of the art.

The second one is instead related to the efficiency of the system, and is measured in terms of the time taken to find the above solution. The graph shown in Fig. 7 reports the speed-up obtained when running G-Net on a PC-NOW made up by 8 PentiumII 400 Mhz PCs, connected by a switched Fast Ethernet. In the above experiments we have kept fixed the number of Searchers

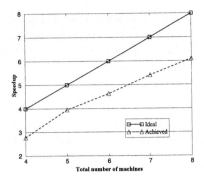

Fig. 7. Speedup achieved by G-Net on the *Mutagenesis* dataset.

(2), while we have progressively increased that of Matchers (from 2 to 6). As can be seen by the graph the speed-up, although sub-optimal (because of the presence of serialization points due to coevolution), can be considered reasonable, meaning that G-Net's performance scaled reasonably well with the number of machines dedicated to the execution of Matchers. This is due to the fact that the evaluation task is significantly demanding in terms of computing power (since, as mentioned before, the Mutagenesis data set is described by means of First Order logics, and the structure of the rules implies that each of them must be evaluated 20^4 times in the worst case). Consequently, the evaluation time dominates over communication time, so an increase in the number of Matchers results in a significant reduction of the execution time.

5 Conclusions and Future Work

In this paper we have presented G-Net, a parallel system for the construction of classifiers, whose implementation has been targeted to PC-based NOWs. G-Net is able to deliver performance higher than other systems of the same type for what concerns both the quality of the found classifiers and the achieved speed-up. Thanks to the dynamic load balancing strategies it incorporates, and to the simple but effective fault tolerance capabilities provided by its stochastic nature, G-Net is able to profitably exploit the computing power provided by non-dedicated PC-based NOWs, with the consequence that it can be effectively used to address real problems characterized by a very high complexity. For instance, at the moment G-Net is being used to solve molecular biology problems that, in order to be solved, required a very high amount of computing power.

Although G-Net can be considered a mature system, several evolution directions, concerning both the algorithm and its PC-NOW implementation, are being considered at the moment. For what concerns the algorithm, we are developing a version of G-Net that extends the numerical capabilities already present, in order to allow it to deal with a broader class of problems. As for the architectural evolutions, for the near future it is planned a G-Net implementation based on MPI rather than on PVM, in order to achieve an even greater portability.

References

1. ANGLANO, C., GIORDANA, A., AND LO BELLO, G. High Performance Knowledge Extraction from Data on PC-based Networks of Workstations. Tech. rep. Available from http://www.di.unito.it/~mino/papers.html.
2. ANGLANO, C., GIORDANA, A., LO BELLO, G., AND SAITTA, L. A Network Genetic Algorithm for Concept Learning. In *Proceedings of the 7th International Conference on Genetic Algorithms* (East Lansing, MI, July 1997), Morgan Kaufman.
3. ANGLANO, C., GIORDANA, A., LO BELLO, G., AND SAITTA, L. An Experimental Evaluation of Coevolutive Concept Learning. In *Proceedings of the 15th International Conference on Machine Learning* (Madison, WI, July 1998), Morgan Kaufman.
4. DEHASPE, L., TOIVONEN, H., AND KING, R. Finding frequent substructures in chemical compounds. In *Proceedings of the 4th International Conference on Knoweldge Discovery and Data Mining* (New York, NY, August 1998), AAAI Press, pp. 30–36.
5. FAYYAD, U. Data Mining and Knowledge Discovery: Making Sense out of Data. *IEEE Expert 11*, 5 (1996), 20–25. October 1996.
6. GOLDBERG, D. *Genetic Algorithms*. Addison-Wesley, Reading, MA, 1989.
7. HARCHOL-BALTER, M., CROVELLA, M., AND MURTA, C. On Choosing a Task Assignment Policy for a Distributed Server System. In *Proceedings of the 10th International Conference on Modelling Techniques and Tools for Computer Performance Evaluation* (1998), Springer-Verlag.
8. HUI, C., AND CHANSON, S. Theoretical Analysis of the Heterogeneous Dynamic Load-Balancing Problem Using a Hydrodynamic Approach. *Journal of Parallel and Distributed Computing 43*, 2 (June 1997).
9. KING, R., SRINIVASAN, A., AND STENBERG, M. Relating chemical activity to structure: an examination of ILP successes. *New Generation Computing 13* (1995).
10. PIATETSKY-SHAPIRO, G., AND FRAWLEY, W., Eds. *Knowledge Discovery in Databases*. AAAI Press / The MIT Press, Menlo Park, CA, 1991.
11. POTTER, M. *The design and analysis of a computational model of cooperative coevolution*. PhD thesis, George Mason University, Fairfax, VA, 1997.
12. POTTER, M., DE JONG, K., AND GREFENSTETTE, J. A coevolutionary approach to learning sequential decision rules. In *6th Int. Conf. on Genetic Algorithms* (Pittsburgh, PA, 1995), Morgan Kaufmann, pp. 366–372.
13. SUNDERAM, V. S., GEIST, G. A., DONGARRA, J., AND MANCHEK, R. The PVM concurrent computing system: evolution, experiences, and trends. *Parallel Computing 20*, 4 (April 1994), 531–45.
14. WEBER, R. On the Optimal Assignment of Customers to Parallel Servers. *Journal of Applied Probability 15* (1978), 406–413.

Addressing Communication Latency Issues on Clusters for Fine Grained Asynchronous Applications - A Case Study*

*Umesh Kumar V. Rajasekaran, Malolan Chetlur,
Girindra D. Sharma, Radharamanan Radhakrishnan,*
and *Philip A. Wilsey*

Computer Architecture Design Laboratory,
Dept. of ECECS, PO Box 210030, Cincinnati, OH 45221–0030
Ph:(513) 556-4779, Fax:(513) 556-7326, phil.wilsey@uc.edu

Abstract. With the advent of cheap and powerful hardware for workstations and networks, a new cluster-based architecture for parallel processing applications has been envisioned. However, fine-grained asynchronous applications that communicate frequently are not the ideal candidates for such architectures because of their high latency communication costs. Hence, designers of fine-grained parallel applications on clusters are faced with the problem of reducing the high communication latency in such architectures. Depending on what kind of resources are available, the communication latency can be improved along the following dimensions: (a) reducing network latency by employing a higher performance network hardware (i.e., Fast Ethernet versus Myrinet); (b) reducing communication software overhead by developing more efficient communication libraries (MPICH versus TCPMPL (our TCP/IP based message passing layer) versus MPI-BIP); (c) rewriting/restructuring the application code for less frequent communication; and (d) exploiting application characteristics by deploying communication optimizations that exploit the application's inherent communication characteristics. This paper discusses our experiences with building a communication subsystem on a cluster of workstations for a fine-grained asynchronous application (a Time Warp synchronized discrete-event simulator). Specifically, our efforts in reducing the communication latency along three of the four aforementioned dimensions are detailed and discussed. In addition, performance results from an in-depth empirical evaluation of the communication subsystem are reported in the paper.

1 Introduction

One of the primary bottlenecks limiting the performance of distributed applications is the communication latency. The cost of communication operations is significantly higher than computation operations because of the overhead involved

---

* Support for this work was provided in part by the Advanced Research Projects Agency under contracts J–FBI–93–116 and DABT63–96–C–0055.

in preparing a message and the electrical delay necessary for signal processing across bandwidth-limited physical network links. The nature of the communication interface in a message passing environment is such that the software overhead (time required for the preparation and authentication of the message) is significantly higher than the hardware overhead (network setup and message propagation time) [5]. Under such circumstances, an important lesson that has been learned is to minimize the frequency of communication, but not necessarily the size of the messages in order to arrive at an efficient implementation. However, a large class of distributed applications tend to have fine-grained asynchronous communication characteristics. This implies that it is not always possible to minimize the frequency of communication and other methods are needed to alleviate the overhead of message latency.

Several strategies for minimizing the communication latencies of distributed applications have been explored. Depending on what kind of cost-performance trade-offs are required, the communication latency can be improved along the following four dimensions: (a) using faster network hardware; (b) using more efficient communication libraries; (c) rewriting or restructuring the application code for less frequent communication; and (d) using communication optimizations that exploit the application's inherent communication characteristics. In this paper, we discuss our experiences with building a communication subsystem on a cluster of three workstations (dual processor Pentium Pros running Linux) for a fine-grained asynchronous application (namely WARPED, a Time Warp synchronized discrete-event simulator). Specifically, our efforts in reducing the communication latency along three of the four aforementioned dimensions are detailed and discussed. As rewriting and restructuring the application code for less frequent communication is relatively a well studied and application-specific method [5], we will not discuss it here in this paper. However, each of the other three dimensions are dealt in detail and supported by empirical analysis. The remainder of this paper is organized as follows. Section 2 overviews parallel discrete event simulation (which is our domain of interest) and describes the experimental environment. Section 3 discusses our experiences with new network hardware (Myrinet) and presents a comparison between different network hardware configurations and their effects on the application. Section 4 illustrates the effect of different communication libraries on the performance of the application and presents a comparison between MPICH, MPI-BIP and TCPMPL (our native TCP/IP based message passing library). Section 5 discusses the effect of two communication optimizations (message aggregation and infrequent polling) that have been developed to further improve the performance of the communication subsystem. Finally, Section 6 presents some concluding remarks.

2 Background

In this section, a brief overview of a class of fine-grained asynchronous applications that is of interest to us is presented. The field of parallel discrete event simulation (PDES) [6] contains several representative examples of fine-grained asynchronous applications. In PDES, the model to be simulated is decomposed into *physical processes* that are implemented as *simulation objects*. Each simulation object is assigned to a *Logical Process* (LP); the simulator is composed of a set of LPs concurrently executing their simulation objects. Simulation objects communicate by exchanging time-stamped messages through the LPs. Thus, each LP (which can be associated with multiple simulation objects) receives messages from other LPs and forwards them to the destination objects. In order to maintain causality, LPs must process messages in strictly non-decreasing time-stamp order [10]. There are two basic synchronization protocols used to ensure that this condition is not violated: (i) *conservative* and (ii) *optimistic*. Conservative protocols [13] avoid causality errors, while optimistic protocols, such as Time Warp [6, 10] allow causality errors to occur, but implement some recovery mechanism.

In a Time Warp simulator such as WARPED [17], each LP operates as a distinct discrete event simulator, maintaining input and output event lists, a state queue, and a local simulation time (called *Local Virtual Time* or LVT). As each LP simulates asynchronously, it is possible for an LP to receive an event from the past (some LPs will be processing faster than others and hence will have local virtual times greater than others) — violating the causality constraints of the events in the simulation. Such messages are called *straggler* messages. On receipt of a straggler message, the LP must rollback to undo some work that has been done. The state and output queue are present to support rollback processing. Rollback involves two steps: (i) restoring the state to a time preceding the time-stamp of the straggler and (ii) canceling any output event messages that were erroneously sent (by sending *anti-messages*). After a rollback, events are re-executed in the proper order.

2.1 The Experimental Environment

The WARPED simulation kernel provides the functionality to develop applications modeled as discrete event simulations [17]. Considerable effort has been made to define a standard programming interface to hide the details of Time Warp from the Application Programming Interface (API). All Time Warp specific activities such as state saving and rollback are performed by the kernel without intervention from the application. Consequently, an implementation of the WARPED interface can be constructed using either conservative [13] or optimistic [6] parallel synchronization techniques. Furthermore, the simulation kernel can operate as a sequential kernel. A more detailed description of the internal structure and organization of the WARPED kernel is available on the www at http://www.ececs.uc.edu/~paw/warped.

The results reported in this paper were obtained by executing four different simulation models on a cluster of three Pentium Pro workstations interconnected by 100Mbps Fast Ethernet and 1.28Gbps Myrinet. To fully test the system, the cluster of workstations chosen for the experiments were not dedicated to the experiments. Five sets of measurements were taken at two different times and the average of these values were then reported. The average granularity[1] (in terms of computation time) of each application is as follows: RAID (20μsecs), SMMP (15μsecs), QUEUE (41μsecs), and PHOLD (20 μsecs). The four models (available in the WARPED distribution) are:

RAID: The RAID application models the RAID Disk Arrays which is a method of providing a vast array of storage [2] with a higher I/O performance than several large expensive disks. This application incorporates a flexible model of a RAID Disk Array and can be configured in various sizes of disk arrays and request generators. Each request is in fact a token that carries information about the number of disks, the number of cylinders, number of tracks, number of sectors, size of each sector and specific information about which stripe to read and parity information.

SMMP: The SMMP application models a shared memory multiprocessor. Each processor is assumed to have a local cache with access to a common global memory (The model is somewhat contrived in that requests to the memory are not serialized — *i.e.,* main memory can have multiple requests pending at any given moment). The model is generated by a C++ program which lets the user adjust the following parameters: (i) the number of processors/caches to simulate, (ii) the number of LPs to generate, (iii) the speed of cache, (iv) the speed of main memory, and (v) the cache hit ratio. The generation program partitions the model to take advantage of the fast intra-LP communication.

QUEUE: The QUEUE application is a reconfigurable queuing library with the essential components needed to model queuing systems. The queuing library consists of *source, queue, server* and *statistics collector* objects. The model and the various parameters of the source, server and queue objects can be set by the modeler through simple configuration files. Different scenarios can be modeled by specifying the desired layout in the configuration file.

PHOLD: The PHOLD model is the parallel hold model first reported by Fujimoto [7]. PHOLD is a synthetic workload model that contains several parameters that are used to test the performance of a parallel simulation. The PHOLD workload model contains the following parameters: number of logical processes, message population, timestamp increment function, movement function, computation grain and an initial configuration.

3 Using Network Hardware to Reduce Network Latency

A cluster of workstations (COW) used for parallel applications consists of low-cost workstations interconnected by a network. Applications running on multiple workstations need a message passing library to communicate information between its peers running on different workstations. The message passing library in turn requires an interconnection network for communicating information across

---
[1] The granularity of the application is the basic unity of computation. It represents the time taken by a simulation process to execute one simulation cycle.

the workstations. This interconnection network can be shared for many purposes or can be used exclusively for parallel applications. Ethernet and Fast Ethernet are one class of interconnection networks that use the ubiquitous Ethernet protocol. The effective utilization of the Ethernet depends on the number of workstations sharing it in shared network. However, the advent of switched Ethernet solutions have helped in improving the utilization of the network.

Alternatively, the Ethernet network can be replaced by a high performance interconnection network hardware to improve the network latency and the effective bandwidth associated with the network. One such high performance interconnection network hardware is Myrinet [1]. Myrinet is a new type of local-area network (LAN) based on the technology used for packet communication and switching on "massively parallel processors" (MPPs). The following are the features of Myrinet which make it a high performance interconnection network suitable for parallel applications: (a) high bandwidth (1.28Gbps), (b) low latency, and (c) host interface that can handle packet traffic. These features of Myrinet help in increasing the communication subsystem performance. In addition, the processor on the Myrinet control board can share (or take control of) the message processing with the main processor. Since Myrinet provides robust communication channels with flow control, packet framing, and several other features, a message passing library using Myrinet for communication directly from the user level would lead to a high performance communication subsystem. There have been several reports comparing the performance of Fast Ethernet clusters with Myrinet clusters [11, 12].

4 Using Efficient Message Passing Libraries

One of the factors that increased the popularity of workstation clusters is the standardization of tools and utilities needed to run distributed applications on workstation clusters. One such standard is the Message Passing Interface (MPI) [8] standard. MPI defines the interface and semantics of a message passing library. Our fine-grained asynchronous parallel simulations use MPICH for communication on workstation clusters. Due to their inherent fine-grained nature, these simulations communicate very frequently during their execution. There are three factors in a message passing software that decide the performance of these applications to a large extent. They are as follows: (a) time consumed by the send operation; (b) one way message latency; and (c) time consumed by the probe operation. The time consumed by the send operation steals the processor cycles away from the application. The one way message latency of the software affects the performance of these applications indirectly. Application characteristics such as rollbacks in Time Warp simulators are affected by the one way message latency. The larger the one way message latency, the more the probability of a rollback. Another factor that affects the performance of these applications is the message probe operation. Since these applications are asynchronous, they probe for messages at regular intervals to check for the arrival of messages. On a successful probe operation, the applications receive the message and resume their work. This section presents the features of MPICH and MPI-BIP. In addition, the following section details how the performance of asynchronous distributed applications can be increased by using efficient message passing libraries.

4.1 MPICH and MPI-BIP

MPICH [9] is a freely available implementation of MPI standard developed by Argonne National Laboratory and Mississippi State University that runs on a wide variety of platforms. The main goals of the architecture of MPICH are: portability and high performance. The design of MPICH was guided by two principles: (a) the maximization of the amount of code that can be shared without compromising performance; and (b) to provide a structure whereby MPICH could be ported to a new platform quickly. Some of the advantages of MPICH are: portability, language bindings for C, Fortran and C++, high performance, heterogeneity and interoperability.

We considered two alternatives in providing an efficient message passing layer to our applications. One way was to have a high cost, high performance interconnection network (such as Myrinet) together with a high-performance message passing layer to take advantage of the faster network hardware. One example of such a message passing layer is the Basic Interface for Parallelism (BIP) [16] protocol. There are other fast messaging layers that provide user-level network access. Some of them are U-Net [19], FM [15], GAMMA [4], and PARMA^2 [12]. While some of these libraries operate over Myrinet, some operate over Fast Ethernet. BIP is a new protocol designed for high performance networking on Myrinet. Prylli et al [16] have designed and implemented BIP such that it exploits the high speed Myrinet network to the fullest, without spending time in system calls or memory copies, giving all the bandwidth (5 μs message latency) to the applications using it. They have also designed and implemented MPI-BIP, a MPI interface on top of BIP to support the portability of parallel applications adhering to the MPI interface. The MPI interface of BIP was one of the primary reasons for us choosing MPI-BIP. This significantly reduced the programming costs involved in using the new message passing library as all our applications are MPI-based. One disadvantage of the BIP library is the exclusive access to the network by only one application running on a workstation [16]. The BIP implementation monopolizes the network interface of one node for each application. Therefore, no other applications can use the network for communication from the same workstation. The other alternative was to have a low cost, reasonable performance interconnection network (such as Fast Ethernet) and to develop a message passing library similar to MPICH but with reduced functionality to increase the performance of the library. This motivated the development of TCPMPL, a TCP/IP based message passing library.

4.2 TCPMPL

The other dimension in which effort (in terms of programming costs) was spent in improving the performance of the communications subsystem, was in the development of a custom message passing library (TCPMPL) over TCP/IP with minimal functionalities. Clearly we did not require the overhead of the extensive library support of MPICH to run our distributed applications. Some of the features of MPICH which we considered as being unnecessary for our purposes are as follows:

- *Heterogeneous Workstations*: MPICH is designed to be run on heterogeneous workstations. Our asynchronous applications run only on homogeneous workstations and therefore, do not require this overhead.
- *Functionalities*: The MPI standard specifies a myriad of functions to encompass all the communication operations performed by parallel applications. Our asynchronous applications use only a limited set of functions in the MPI standard such as send, receive, and probe.
- *Platform Independence*: MPICH has been designed to be portable to multiple hardware platforms. This feature is not necessary if the software is not intended for different hardware platforms.

Therefore, limiting a message passing library to provide only the functionality required by our asynchronous applications has the potential of increasing the communication subsystem's performance. One of the main goals for TCPMPL is portability and high performance. The C++ class interface of TCPMPL is given below:

```
class TCPMPL {
    TCPMPL(void);
    ~TCPMPL();
    void    initTCPMPL(int &argc, char ** &argv);
    int     getId(void) const;
    int     getTCPMPLSize()() const;
    bool    sendData(char *ptr, int nbytes, int id);
    bool    sendDataVector(const struct iovec *vector, int count, int id);
    bool    recvDataFrom(char *ptr, int nbytes, int id);
    bool    probeForData(int *source, int *length, long seconds, long microseconds);
}
```

The following are the minimal requirements of TCPMPL which are required by our asynchronous applications for correct functional execution:

- *SPMD*: TCPMPL must support the Single Program Multiple Data (SPMD) concept. Applications compiled with the library should have the feature of starting the parallel applications on different workstations and synchronize among themselves before executing the application code. This functionality is accomplished by the *initTCPMPL* method.
- *Message Delivery*: The service provided by TCPMPL must be reliable and must ensure ordered delivery. TCPMPL should reliably deliver the messages to the destination process. It should also preserve the ordering (FIFO) of the messages from a particular process.
- *Message Boundary*: TCPMPL should preserve the message boundary, as the applications use it to send messages.

– *Send*, *Probe*, and *Receive*: Application should be able to send a message to any particular process in the group of processes started by TCPMPL. This is handled by the *sendData* and *sendDataVector* communication methods. Since our applications are asynchronous by nature, support for polling for messages that can be received without blocking is required. This is handled by the *probeForData* method. Application processes should be able to receive messages from any other process. This is taken care of by the *recvDataFrom* communication primitive.

One of the foremost design decisions was the choice of the underlying protocol to be used in TCPMPL. There were two choices: (a) TCP - a connection oriented protocol that provides reliability and ordered delivery; and (b) UDP [18] - a connectionless datagram protocol that does not guarantee reliability and ordered delivery. Since our applications require reliable message transmission and ordered delivery, we chose TCP instead of UDP as TCPMPL's underlying protocol. Since the TCP implementation is available as one of the operating system services on most platforms, our library would also be portable to most platforms. Establishment of communication channels for communicating information between the processes involves different design choices. One approach is to establish communication channels between all the processes so that all communication channels are setup before the application starts to execute. A disadvantage of this approach is that not all processes in the group will communicate with each other during their entire execution. The advantage of this approach is that all the communication channels are setup ahead, so there is no overhead during the execution of the application for establishing communication channels. Another design approach would be to establish communication as and when the processes communicate. Resources can be utilized more efficiently by using this approach. The disadvantage is that the control cannot be given to the message passing layer at both the ends for establishment of communication channels since this involves some protocol specific activity which has to be synchronized at both ends. Another approach to minimizing the communication channels would be to have a spanning tree of the processes in which the edges denote the bi-directional communication channels and the nodes represent the processes. This design would decrease the usage of resources but would increase the one way message latency. We wanted TCPMPL to have the minimum one way message latency and lower overhead during the execution of the application. Hence, we chose to have all the communication channels established before the application starts to execute.

In addition, TCPMPL has to preserve message boundaries while delivering a message to the user application. For example, if an user application sends two messages of size s_1 and s_2 bytes, TCPMPL should deliver the first message of size s_1 bytes followed by the message of s_2 bytes to the user. TCPMPL should not deliver the first s_1+s_2 bytes as a single message or in any other combination. To handle message boundaries, TCPMPL sends a message header containing the size of the message that is going to follow the message header. TCPMPL has a primitive error detection mechanism to detect any errors caused by TCP in delivering the messages to the destinations. It sends the sender's id and the destination's id in the message header to detect any errors in the delivery of the messages by TCP. The base version of TCPMPL performed well and certain optimizations were incorporated to further improve the performance of TCPMPL. Some of the optimizations are discussed below:

Reduction in socket I/O: TCPMPL sends a message header for every message that is being sent by the user application. In the base version, TCPMPL sends the message header followed by the user message. Since each write to the socket involves a switch from user-mode to kernel-mode, these calls (one for sending the message header and another call to send the user message) result in a lot of overhead. Hence, both calls were merged to reduce the switching overhead, so that both the writes are performed in one call.

Reduction in copying: When a user application wants to send two buffers together as a single message, the application has to copy the two buffers into a single message and then send it using TCPMPL. To avoid the extra copying, an additional interface to TCPMPL was provided that takes multiple buffers and sends them as a single message. This avoids the extra copying.

Disabling Nagle's algorithm: The size of TCP's header is 20 bytes and IP requires 20 bytes for its header. Therefore, for small message sizes, this 40 byte header wastes the bandwidth of the network. To avoid this, Nagle's algorithm [14] is used in TCP to delay sending of small messages when there is any outstanding segment of data that has not been acknowledged. More data can be accumulated until the acknowledgment for the outstanding packet has arrived. The accumulated data is then sent as a single packet. There is also a delayed acknowledgment timer in TCP which delays the acknowledgment of data being sent so that it can piggy-back with the data in the opposite direction. Nagle's algorithm and delay acknowledgment timer together added a significant delay to the packets. This increases the one way message latency of TCPMPL as the majority of messages transmitted by our fine-grained asynchronous applications were of the order of a few hundred bytes. To avoid this delay, TCPMPL disables Nagle's algorithm throughout the application execution.

Socket Buffer Size: TCP provides both reliability and ordered delivery to the higher layers in the TCP/IP protocol stack. TCP has internal buffers for buffering the user data. Since it assumes the underlying network is not reliable, it buffers the data and retransmits it if the destination process has not received the data. There are buffers at the sending end for this retransmission purposes. There is also a buffer at the receiving end to buffer the data before the application can receive the data. Both sender and receiver buffer sizes were increased to 65536 bytes to increase the performance of TCPMPL. If the buffer size is small, TCP will block until it gets an empty buffer to buffer the data.

Table 1. Round trip times & bandwidth of MPICH, TCPMPL and MPI-BIP

Packet Size bytes	Round Trip Time MPICH μ secs	Round Trip Time TCPMPL μ secs	Round Trip Time MPIBIP μ secs	Bandwidth MPICH Mbps	Bandwidth TCPMPL Mbps	Bandwidth MPIBIP Mbps
128	627	440	38	3.27	4.65	53.89
256	676	560	104	6.06	7.31	39.38
512	868	797	114	9.44	10.28	71.86
1024	1354	1271	124	12.10	12.89	132.13
2048	1977	1893	149	16.57	17.31	219.92
4096	2814	2565	197	23.29	25.55	332.67

4.3 Performance of TCPMPL and MPI-BIP over MPICH

Figure 1 illustrates the performance of four parallel, asynchronous WARPED applications with MPICH, TCPMPL and MPI-BIP. From the figure, it is clear that all applications run with TCPMPL perform much better than the same applications run with MPICH. By tailoring the message passing library to suit our needs, we were able to get on average, a 75% improvement on all the applications tested. These improvements were on a cluster of three workstations (dual processor Pentium Pro) interconnected with Fast Ethernet. Table 1 illustrates the raw performance numbers (round trip times and bandwidth) for all three communication libraries. Specifically, we present the round-trip times and bandwidth of TCPMPL and MPICH on a Fast Ethernet network while round-trip times and bandwidth of MPI-BIP were collected on a Myrinet network.

The experiments on the Myrinet platform were carried out to study the effect of lower message latencies on the fine-grained asynchronous applications. To address this issue, we experimented with a Myrinet network interconnecting the same cluster of workstations. In addition, we chose to use a message passing library that took advantage of the faster underlying network (MPI-BIP). It turns out that MPI-BIP on Myrinet performs about 10 times faster than MPICH on Fast Ethernet. This translates to a further 40% improvement on average over the TCPMPL on Fast Ethernet combination, on all the applications that were tested. Figure 1 shows the same four WARPED applications performing even better with the MPI-BIP on Myrinet combination.

5 Using Communication Optimizations in the Application Layer

So far, we have established that faster hardware and efficient libraries reduce the communication latency. However, the communication latency experienced by distributed applications is inevitable. Cautious exploitation of communication characteristics of applications can further reduce the average communication overhead. For fine-grained asynchronous applications running on clusters, even a small communication latency greatly affects the performance. This fact calls

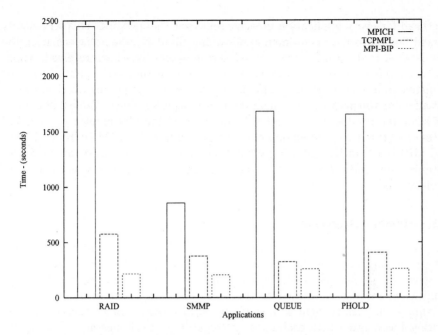

Fig. 1. Effect of message passing libraries and hardware on performance

for optimizations in the application layer that could exploit the communication characteristics to reduce the communication overhead.

Table 2. Communication costs for 128 byte messages

Library	Probe		Receive	Send
	unsuccessful (μ secs)	successful (μ secs)	(μ secs)	(μ secs)
MPICH	22	298	16	101
TCPMPL	21	63	17	44
MPI-BIP	1	12	5	15

In this section, we discuss two optimizations namely *message aggregation* and *infrequent polling* that can be employed in the communication module of fine-grained asynchronous applications. The performance analysis of various message passing libraries showed that communication send operations incur a fixed cost irrespective of the message size. Although timely reception of these messages is desirable in asynchronous applications, there is a local computation that can proceed independent of the reception of these messages. Therefore, by aggregating several application messages into a single physical message, we only incur the communication overhead for the one physical message instead of incurring the overhead for every application message sent over the network. This will clearly benefit fine-grained asynchronous applications. Similarly, the processes must poll the network to determine whether any new messages have been received. Applications poll frequently to receive messages as soon as they arrive, but at the cost of increased overhead. Polling infrequently to reduce the polling overhead without affecting the progress of simulation improves performance. Table 2 illustrates the send, receive and probe overheads (in terms of time) for messages of size 128 bytes of all three communication libraries used in this study (MPICH, TCPMPL and MPI-BIP). In the following subsections, we will discuss in detail these two optimizations and their impact on the communication behavior of asynchronous applications.

5.1 Message Aggregation

Message Aggregation (MA) is an optimization to the Time Warp communication subsystem that matches the communication behavior to the underlying communication subsystem. Using MA, the communication module of each LP collects application messages destined to the same LP, that occur in close temporal proximity, and sends them as a single physical message. Since there is a large overhead associated with each message (regardless of the message size), significant improvement in performance is possible. The decision on when to send the messages is made by the *aggregation policy*.

Clearly, the higher the number of messages aggregated, the greater the reduction in the communication overhead. The longer the messages are delayed, the greater the number of messages aggregated. However, delaying messages excessively harms the performance of the receiving LP. Thus, aggregation policies must balance the potential gain from additional message aggregation, to the potential loss at the receiving LP. It is difficult to determine a static balance between the two factors; the communication behavior of Time Warp simulators is highly dynamic and unpredictable. While static window sizes (in time, or number of messages) for aggregation yield some performance improvement, better overall performance results from dynamic control of the aggregate size.

In specifying the window size, we seek to balance the benefits resulting from aggregating more messages, versus the harm from delaying messages excessively. These two factors are modeled as: (i) Aggregation Optimistic Factor (AOF): AOF is the gain from delaying the messages; it is proportional to the rate of reception of messages to be sent. If AOF is high, a large number of messages are aggregated without excessive delay; and (ii) Aggregation Pessimistic Factor (APF): APF models the harm from delaying the messages. It is proportional to the age of the aggregate. Note that both of these factors vary with the nature of the application, and may change dynamically within the lifetime of the simulation.

Initially, a static policy that aggregates messages for a fixed time, called *Fixed Aggregation Window* (FAW), was developed. The age of the first message received by the aggregation layer is tracked. Once this age reaches a constant value (the size of the window), the aggregate message is sent. The advantage

Table 3. Average Aggregate Size

Application	Msg. Size	# Appl. Messages	# Physical Messages	Avg. Aggregate Size
RAID	132	5133307	3052662	1.68
SMMP	92	977310	870314	1.12
QUEUE	96	1511043	1253779	1.20
PHOLD	92	5413578	320090	1.69

of this policy is its low overhead; only a single check of the current aggregate age (time that the aggregate has been alive) against the constant window size is required. This policy provides a static balance between AOF and APF, making it insensitive to changes in the communication behavior of the application. No matter how high (or low) the message arrival rate is, the fixed window size is used. A fixed window size of 20 (heuristically determined) was used for all experiments with message aggregation. Table 3 illustrates the average aggregate size for all four WARPED examples. Since the chosen window size significantly affects the performance of this policy, dynamic control of the balance between AOF and APF is being investigated [3].

5.2 Infrequent Polling

In asynchronous applications, the process polls the communication layer for messages, receives the messages if any are available, performs some computation and sends some messages. If the message probe operation successfully discovers a message, the message is received, and the network is polled again for messages until no more messages are detected. The optimization is to carry out the message polling periodically instead of polling for messages after every computation. The benefit is that there is a lower chance that the polling is unsuccessful since it is performed infrequently and there is a longer period of time available for messages to arrive between polls.

The analysis of the performance of asynchronous distributed applications that use polling demonstrates that it is possible to enhance the performance by optimizing the message delivery, especially for fine-grained applications. Unfortunately, developing the criteria for optimal polling frequency even under approximate conditions is difficult. Even if it was possible to directly compute the optimal polling frequency, it is likely to vary during the lifetime of the application, and from process to process. Therefore, we suggest a infrequent polling strategy that arrives at the optimal polling frequency heuristically. In this strategy, an initial period for polling is selected at compile time, and enforced on all the processes. The advantage of this policy is that it adds no overhead. For the experiments conducted, each application preferred a unique polling frequency. The following polling frequencies[2] were used for experiments conducted on the Fast Ethernet network: (a) RAID - preferred a polling frequency of 3; (b) SMMP

---

[2] A polling frequency of n implies that a probe is conducted after processing n events.

- preferred a polling frequency of 25; (c) QUEUE - preferred a polling frequency of 10; and (d) PHOLD - preferred a polling frequency of 15. The same polling frequencies were used for TCPMPL and MPICH based experiments as there wasn't any appreciable difference in the unsuccessful polling costs. However, the polling costs for MPI-BIP are much lower than the Fast Ethernet polling costs. So for experiments conducted with MPI-BIP, all applications preferred a polling frequency of 3.

5.3 Performance of Message Aggregation and Infrequent Polling

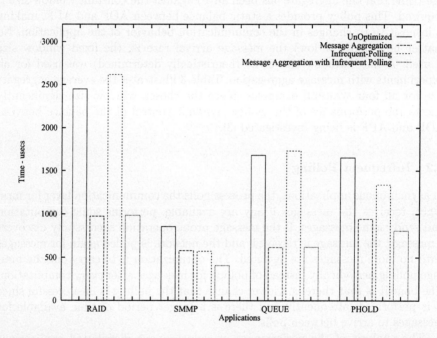

Fig. 2. Effect of optimizations on performance using MPICH on Fast Ethernet

In order to study the effect of message aggregation and infrequent polling on fine-grained asynchronous applications, four WARPED applications were tested with different message passing libraries. Figures 2, 3, and 4 illustrate the effect of the optimizations on the performance of the applications when executed with MPICH, TCPMPL and MPI-BIP respectively.

Figure 2 validates our assertion that aggregation can be beneficial to fine-grained asynchronous applications when there is a significant latency in message transmissions. As seen in the figure, message aggregation (using the MPICH message passing library on a Fast Ethernet network) provides on average 50% performance improvement (over the unoptimized version) on all the applications tested. This improvement can be attributed to the reduction in communication

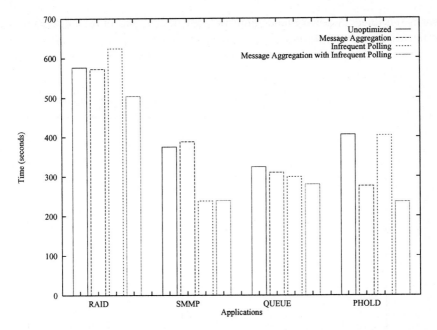

Fig. 3. Effect of optimizations on performance using TCPMPL on Fast Ethernet

overhead as a lower number of physical messages are sent across the network. On the other hand, the infrequent polling optimization does not fare well with the applications tested. The majority of the applications tested (excluding the SMMP model) generate large amounts of messages and infrequently polling for messages actually degrades the performance of the applications. However, when the message aggregation optimization is deployed along with the infrequent polling strategy, there is a significant improvement in performance (57%, *i.e.*, an improvement of 7% over the speedup obtained by having message aggregation alone). This improvement can be attributed to the fact that the message aggregation optimization reduces the number of messages sent across the network and in this scenario, the infrequent polling strategy is beneficial to the application.

Figure 3 illustrates the effect of the optimizations on the performance of the applications when using TCPMPL on a Fast Ethernet network. As TCPMPL significantly reduces the software overhead involved in message transmission, message aggregation does not perform as well as it did with MPICH. As messages are sent with lower latency, deployment of the message aggregation optimization results in only a 8% improvement on average on all the applications tested. The infrequent polling optimization performs much better here than it did with MPICH, mainly because messages are communicated faster and there are occasions where there are no message available in the channel. Barring one application (RAID), infrequent polling provides a small performance improvement for all the other applications tested. So it comes as no surprise that when these two optimizations are combined together, a further performance improvement is obtained. A performance improvement of 26% was obtained when both optimizations were deployed.

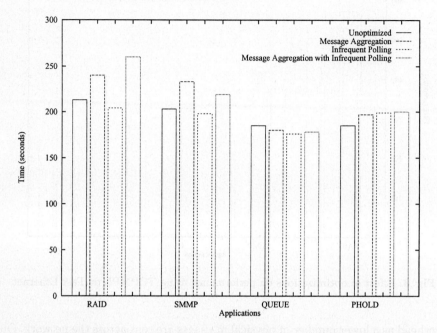

Fig. 4. Effect of optimizations on performance using MPI-BIP on Myrinet

Finally, the effect of the optimizations on the performance of the applications when using the MPI-BIP message passing library on a Myrinet network is illustrated in Figure 4. From the figure, it is clear that on a fast network with a fast message passing library, the optimizations do not improve the performance of the application. Barring a slight performance improvement with infrequent polling, the optimizations (alone and combined) do not significantly improve the performance of the applications. This was expected as the optimizations took advantage of inefficiencies of the Fast Ethernet network and the message passing libraries to improve the performance of the applications. When these inefficiencies were removed (by adding a faster network layer and a fast message passing library), the optimizations were ineffective.

6   Conclusions

The performance of fine-grained asynchronous applications on clusters is directly influenced by the underlying communication subsystem. There is a wide latitude in the choice of hardware and software components for building the communication subsystem. Faster communication networks with efficient communication libraries definitely improve the performance of communication operations. This is clearly demonstrated by the results. We also see that the communication libraries play a major role in the effective utilization of the underlying network bandwidth. This is illustrated by the TCPMPL library which performs better than the MPICH implementation. From this we can conclude that a message passing library with minimal functionality and implementation optimizations such as (i) reduced socket I/O, (ii) reduced copying, and (iii) suitable buffer sizes can significantly improve the performance of fine-grained asynchronous applications. In addition, adhering to the standard interfaces enables the portability of the distributed applications. We also see that, in addition to improving the underlying network and communication libraries, exploiting the application's communication behavior can improve the execution performance. An intelligent combination of these three components in any communication subsystem will significantly improve the performance of a distributed application.

It is clear from the empirical analysis that, even with a relatively inexpensive network (such as Fast Ethernet), improvement in performance of fine-grained applications can be achieved when time (and programming effort) is spent on improving the message passing library and developing application-based communication optimizations. As seen from the figures, TCPMPL along with the two communication optimizations gives performance that is comparable (both in terms of cost and performance) to the MPI-BIP on Myrinet combination. In addition, the time spent in developing a message passing library is justified if one considers the cost of the expensive Myrinet hardware. In this paper, we presented our efforts in finding viable solutions that provide good performance on low-cost hardware technology. In addition, to really appreciate the performance improvement, we compared the Fast Ethernet results with performance results obtained from the same cluster interconnected with the Myrinet network and which uses an optimized message passing library (MPI-BIP).

We also observed that the communication optimizations (message aggregation and infrequent polling) were not able to exploit the application's behavior in high speed networks such as Myrinet. Currently, work is in progress to develop optimizations that can detect application communication characteristics that can be exploited in high speed networks. Clearly there is a wide variety of design choices in building a low cost communication subsystem. The design choices should be made with the cost (money and time) and performance in mind. High speed networks definitely reduce communication latency. However the intelligent combination of the underlying network hardware and efficient message passing libraries coupled with algorithmic optimizations can make COWs a suitable environment for fine-grained asynchronous applications.

References

1. BODEN, N. J., COHEN, D., FELDERMAN, R. E., KULAWIK, A. E., SEITZ, C. L., SEIZOVIC, J. N., AND SU, W.-K. Myrinet – a gigabit-per-second local-area network. *IEEE Micro 15*, 1 (February 1995), 29–36.
2. CHEN, P. M. RAID: High-performance, reliable secondary storage. *ACM Computing Surveys 26*, 2 (June 1994), 145–185.

3. CHETLUR, M., ABU-GHAZALEH, N., RADHAKRISHNAN, R., AND WILSEY, P. A. Optimizing communication in Time-Warp simulators. In *12th Workshop on Parallel and Distributed Simulation* (May 1998), Society for Computer Simulation, pp. 64–71.
4. CIACCIO, G. Optimal communication performance on fast ethernet with gamma. In *12th International Parallel Processing Symposium and 9th Symposium on Parallel and Distributed Processing* (Orlando, Florida, March/April 1998), Springer, pp. 534–548.
5. FELTEN, E. W. Protocol compilation: High-performance communication for parallel programs. Tech. rep., University of Washington — Dept. of Computer Science, 1993.
6. FUJIMOTO, R. Parallel discrete event simulation. *Communications of the ACM 33*, 10 (Oct. 1990), 30–53.
7. FUJIMOTO, R. Performance of time warp under synthetic workloads. *Proceedings of the SCS Multiconference on Distributed Simulation 22*, 1 (Jan. 1990), 23–28.
8. GROPP, W., HUSS-LEDERMAN, S., LUMSDAINE, A., LUSK, E., NITZBERG, B., SAPHIR, W., AND SNIR, M. *MPI: The Complete Reference Volume 2 - The MPI-2 Extension*. MIT Press, 1998.
9. GROPP, W., LUSK, E., DOSS, N., AND SKJELLUM, A. *A High-Performance, Portable Implementation of the MPI Message Passing Interface Standard*, July 1996.
10. JEFFERSON, D. Virtual time. *ACM Transactions on Programming Languages and Systems 7*, 3 (July 1985), 405–425.
11. LAB, S. C. Scl cluster cookbook - technology comparison. (available on the www at http://www.scl.ameslab.gov/Projects/ClusterCookbook/icperf.html).
12. MARENZONI, P., RIMASSA, G., VIGNALI, M., BERTOZZI, M., CONTE, G., AND ROSSI, P. An operating system support to low-overhead communications in NOW clusters. In *Proceedings of Communication and Architectural Support for Network-Based Parallel Computing CANPC97* (San Antonio, Texas, Feb. 1997), vol. 1199, Springer-Verlag, pp. 130–143.
13. MISRA, J. Distributed discrete-event simulation. *Computing Surveys 18*, 1 (Mar. 1986), 39–65.
14. NAGLE, J. Congestion control in TCP/IP internetworks. *Computer Communications Review 14* (Oct 1984), 11–17.
15. PAKIN, S., LAURIA, M., AND CHIEN, A. High performance messaging on workstations: Illinois fast message(FM) for Myrinet. In *Proceedings of Supercomputing '95* (December 1995).
16. PRYLLI, L., AND TOURANCHEAU, B. BIP: A new protocol designed for high performance netowrking on myrinet. In *12th International Parallel Processing Symposium and 9th Symposium on Parallel and Distributed Processing* (Orlando, Florida, March/April 1998), Springer, pp. 472–485.
17. RADHAKRISHNAN, R., MARTIN, D. E., CHETLUR, M., RAO, D. M., AND WILSEY, P. A. An Object-Oriented Time Warp Simulation Kernel. In *Proceedings of the International Symposium on Computing in Object-Oriented Parallel Environments (ISCOPE'98)*, D. Caromel, R. R. Oldehoeft, and M. Tholburn, Eds., vol. LNCS 1505. Springer-Verlag, Dec. 1998, pp. 13–23.
18. STEVENS, W. R. *TCP/IP Illustrated Volume 1: The Protocols*. Addison-Wesley Publishing Company, Reading Massachusetts, March 1996.
19. VON EICKEN, T., BASU, A., BUCH, V., AND VOGELS, W. U-Net: A user-level network interface for parallel and distributed computing. In *Proceedings of the 15th ACM Symposium on Operating Systems Principles* (December 3-6 1995).

Low Cost Databases for NOW

Gianni Conte[1], Michele Mazzeo[2], Agostino Poggi[1],
Pietro Rossi[2], and Michele Vignali[1]

[1] Dipartimento di Ingegneria dell'Informazione
University of Parma, Viale delle Scienze, 43100 Parma, Italy
{conte, poggi, vignali}@ce.unipr.it ,
WWW home page: http://www.ce.unipr.it
[2] ENEA HPCN Project
Via Martiri di Monte Sole, 4, Bologna, Italy
{rossi, mazzeo}@pchpcn1.arcoveggio.enea.it ,
WWW home page: http://www.glue.bologna.enea.it

Abstract. In this paper, we present a system called DONOW (Database on Network of Workstation), that is an example of an implementation and maintenance of a low cost distributed database without performances penalties. The system has a three tier architecture based on a client, a service and a data layer. The client layer allows local and remote users to interact with the system through a Java user friendly interface. The service layer is implemented in C++; it allows the service of user requests and, in particular, the management of queries involving more than one DONOW node. The data layer is based on a well-known freeware relational database management system, that is, MySQL for each DONOW node. DONOW is under experimentation and will be used by Partena, an industry producing very advanced machinery for pharmaceutical products packaging, for the management of the information about their machines, that consists of close to 100000 drawings and related documentation.

1 Introduction

Recent network and computer evolution allows us to obtain low cost high performance by spreading the computation in parallel over a so called Network of Workstations (NOW) [2, 7, 8].

Some NOW architectures have already been presented. However, there is a lack of real application that take advantage of them, mostly because there are not dedicated programming, developing and management tools for NOW architecture.

Databases are one of the most important software products that are used in real applications and that can take advantage of parallelization. Therefore, the availability of database management systems for NOW architectures could be the key for the success of NOW architectures. A database management system for such architectures is a distributed database [16] whose main features should be to take advantage of recent network evolutions to avoid the bottleneck of communication between nodes and to scale out the system partitioning the databases into clusters with different number of nodes.

This paper is a preliminary report of a project, called DIStributed COmputing on network of PC for data base applications (DISCO), whose purpose is the realization of a NOW database management system called Database On NOW (DONOW). The paper, after a brief presentation of related work, describes the software architecture of DONOW and the hardware supporting it and presents the use of the database for the management of the information about machines for pharmaceutical products packaging.

2 Related work

Even if NOW architectures are quite similar to *shared-nothing* distributed architectures for databases [6, 10, 12, 13], the NOW research community is more interested in the analysis of compute-intensive applications than in the analysis of data-intensive applications and the few works centred on data-intensive applications use parallel sort algorithm to benchmark their prototypes [3, 4].

The only interesting work involving the realization of a database management system on a NOW is under development by the Microsoft Scaleable Servers Research Group is realizing a SQL Server for clusters that will scales up to giant hundred gigabyte databases [15]. However, the result is not yet achieved, moreover, it does not offer two fundamental features of distributed databases, that is, transparent partitioning of data and parallel query decomposition.

3 Hardware architecture

The hardware architecture of DONOW currently comprises four nodes and a front-end; in Fig. 1 there is a schematic representation of it. Each node is an Intel Redwood motherboard with dual PentiumII 300Mhz, 128MByte of RAM a Adaptec AIC-7880 Ultra SCSI host adapter and two network interface cards. The SCSI adapter serves a 4.5GByte hard disk from Quantum. Of the two interface cards, one is an Intel EEpro 100 integrated on board, the other is a 3Com905. The EEpro 100 is connected to a 10Base ethernet HUB and through that to the LAN while the 3C905 connects the nodes through a 100BaseTX SWITCH. The ethernet network is primarily used for system administrative tasks and NFS mounting of user directories, while the fast ethernet network, accessible only within the nodes, is used by the high performance applications. The front end, a pentium 200MMX decouples the nodes from internet, acting as a firewall, while the nodes are locally accessible directly from the LAN connected to the HUB and from some computers connected to the SWITCH.

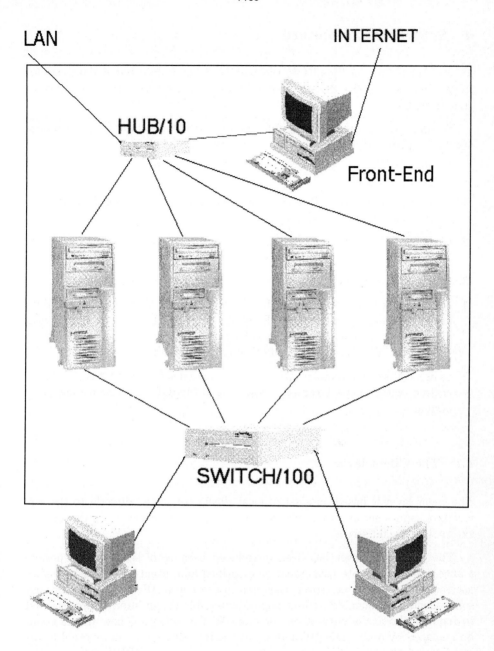

Fig. 1. DONOW cluster architecture

4 Software architecture

As we said before, our goal was the realization of a distributed database for NOW clusters that fits the previous hardware architecture, but also scales for greater clusters.

To guarantee an optimal level of flexibility, usability, scalability, performance and low implementation and maintenance costs, we decided:

1. to realize a multi-layer (three-tier) architecture, distinguishing the client, the service and the data layers;
2. to use a solid pre-existent relational database management system to manage the data in each node of DONOW;
3. to use object-oriented languages for the realization of the software;
4. to use a mix of C++ language, for the critical code of the service and data layers that mainly constraints the performance of the system, and of Java language for the client layer to make accessible the system not only to local clients, but also to remote clients through internet; and
5. to use well-known freeware operating system and software, that is, Linux operating system, gcc++ compiler, jdk 1.1 and MySQL database management system.

4.1 The Client layer

The client layer is based on a set of local clients connected directly to the service layer and a set of remote clients connected to the service layer through an external front-end server.

The front-end server has been introduced because of security and performance reasons. Remote users must be identified and, eventually, refused before accessing the system; moreover, they can interact with DONOW only through some clients implemented as Java applets accessible via an internet browser and so requires the availability of one or more WWW servers. These duties could be managed by each DONOW nodes, but the introduction of an external front-end lightens the DONOW nodes of these duties placing a WWW server on the front-end that manages user authentication and remote clients interaction.

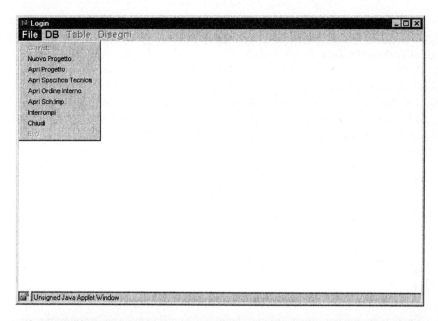

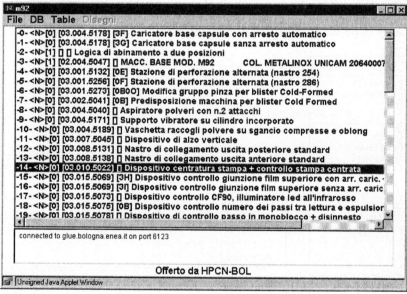

Fig. 2. Two views of DONOW client interface

Clients are Java applications or applets that allow users to interact DONOW database through visual queries without any knowledge about SQL and about the way data are structured in the database. For example, a user of the Partena database can navigate the three of documents related to the design of a machine by simply clicking on the part to expand. Fig. 2 shows a view of the DONOW client interface of the Partena database.

Each client allows a user to interact with the service layer to query and modify stored data; the interaction happens through the exchange of packets, called DoNowPackets.

A DoNowPacket is a serialization of an instance of a class derived from the DoNowPacket abstract class (in its Java meaning). This class has been implemented both in Java and C++, and is able to send/receive itself and guarantees us to have a very flexible asynchronous protocol to exchange control information. Fig. 3 shows the structure of the DoNowPacket abstract class.

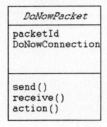

Fig. 3. DoNowPacket class diagram

Each new packet must inherit from DoNowPacket and implement send(), receive() and action() methods. For example, if we have to send an SQL query, it is easy to build a packet, called SimpleQuery, like the one in Fig. 4.

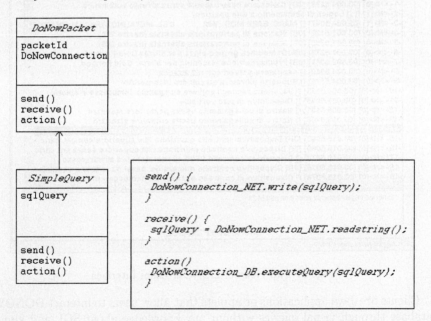

Fig. 4. Example of a simple packet implementation

Each client is connected to the service layer though the internet stream socket protocol. The client does not use primitives of a standard socket libraries, but instances of a class, called DoNowConnection_NET, whose interface masks the use of a specific communication protocol. This implementation simplifies the possible future use of different communication protocols.

4.2 The Service layer

The service layer is composed of a service demon for each DONOW node. Each demon has the main task to handle the action of DoNowPackets. When a DoNow-Packet arrives an action must be taken according to the kind of packet (typically this is a query to be executed or a control message to be sent) and can involve one or more nodes of the data layer.

Clients do not know the number of nodes and their addresses. Clients only know the address of the first DONOW node that is entrusted with balancing the load of the different demons. When it receives a connection request from a client, it returns to the client the address and the listening port of the node with minor load.

Each demon can receive packets from different clients at the same time, therefore, they must be concurrently managed. Due to the importance of performance we choose to handle concurrency by thread. A multi-thread system is more efficient than a multi-process one, because the cost of a context switch is minimal in a threaded architecture. Threads are provided (under Linux) in the form of dynamic linked library. The library used are based on pthread (POSIX thread) version 0.6.

4.3 The data layer

As we said before, the data layer is based on a database management system, that is MySQL [5], for each node of DONOW.

The problem of building a distributed database from scratch is too expensive for our limited human resources and would drive us to a custom solution which will be in few months "out of order software" (OOOS). The choice of a well tested and supported (and documented) database give us a solution with less maintenance problem.

Our solution will be not MySQL dependent, but could be ported to work also with another database. All that the new database must have is a communication layer and a C API; these two characteristics are too important for a database to be not implemented.

MySQL is a multi-thread and sockets based software. That is, it uses threads (and not processes) to handle concurrency and sockets to handle communication between a generic client and the database server.

Because MySQL provides only a C API and we developed critical software layer in C++, we have built a C++ wrapper to its API. This has two main advantages: i) we do not mix C++ with C; and ii) the data layer becomes API independent; in fact, changing API will change only the C++ wrapper and this will have no influence for the rest of the system.

MySQL databases are connected to the service layer through the MySQL protocol. This derives from the fact that is too expensive to rewrite a new high performance MySQL protocol in the first stage of the project. We planned to have a functioning and well modularized prototype of the system at the beginning and then to optimize each component and each protocol. Therefore, at the same way of the interaction between the client and service layers, communication is not based on the use of primitives of MySQL protocol, but on instances of a class, called DoNowConnection_DB, whose interface masks the use of a specific communication protocol and so simplifies the possible future use of different communication protocols.

4.4 Query parallization

A main problem of DONOW system is the management of queries involving data on different nodes. There are two different cases: i) the query can be decomposed in a set of independent subqueries each of them involving data of a single node; and ii) the query cannot be decomposed in a set of independent subqueries each of them involving data of a single node.

In the first case, the execution of the query is simple (Fig. 5 shows a graphical representation of an example of independent subqueries):

1. a client layer node sends a query to a service layer node;
2. the service layer node decomposes the query and sends the subqueries to the data layer nodes involved;
3. the data layer nodes replay the results of the subqueries;
4. the service layer node collects the results and sends the results to the client layer node.

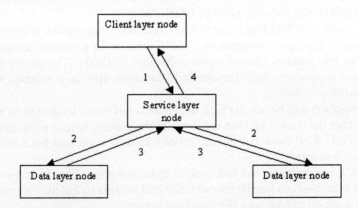

Fig. 5. An example of indenpendent subqueries

In the second case, the execution of the query is quite complex (Fig. 6 shows a graphical representation of an example of non-independent subqueries):

1. a client layer node sends a query to a service layer node;

2. the service layer node decomposes the query into a tree of subqueries: leaf subqueries involve data of a single node, the other subqueries (composition subqueries) work on the results obtained by the queries of their subtree;
3. the service layer node sends the leaf subqueries to the data layer nodes involved;
4. the data layer nodes replay with the size of their results;
5. for each composition subquery:
   (a) the service layer node chooses the data layer node declaring the greatest results size;
   (b) the service layer node moves the other results to the chosen data layer node;
   (c) the service layer node sends the composition subquery to the chosen data layer node;
   (d) the chosen data layer node replay with the size of its results;
6. after the execution of the root composition subquery, the service layer node gets the results of the query and sends them to the client layer node.

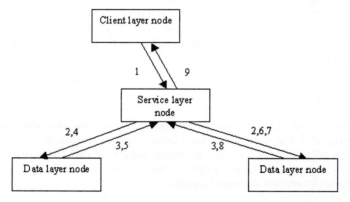

Fig. 6. An example of non-indenpendent subqueries: a join is performed in the data-layer node in the right of figure

5 A case study

The solution proposed arose out of the attempt to satisfy the need of a typical mechanical company, Partena, to manage the information about the design of its products.

Partena produces very advanced machines for pharmaceutical packaging. Such machines are made up of a subset of common parts, more or less standardized, and of a very large number of parts dealing with the specific nature of the needs of their customers. The design of a machine is a top-down process realizing a *tree* of documents each of them corresponding to the design of one of its parts. Parts are of five different types: P02, P03, P13, P40 and P50. P02 are composed of P03, P13, P40 and P50 parts. P03 and p13 are composed of P40 and P50 parts. Finally, P40 and P50 are leaf elements. Therefore, Partena must maintaining an ever growing database given by a graph of documents (some parts and their documentation are reused for realizing different machines) composed of technical designs, specs and ad hoc solutions for the different machines that Partena built. Fig. 7 presents a graphical representation of a part of Partena database.

This, rather quickly growing database, has to be consulted every time a maintenance operation needs to be performed or a new project needs to be undertaken because of the use of common parts in different machines. Presently, the database consists of close to 100000 drawings and is fully kept on paper. Tracking down the complete documentation relative to one single machine is an operation keeping one person busy for about an hour and producing always from 10 to 15 drawings plus several inventory listings, and a handful of technical documents detailing the reasoning behind the adopted solutions.

The database, built so far, is vital to the functioning of the company and is bound to be the bottleneck of their operation if some procedures that scales better with complexity are not adopted. The obvious solution, proposed with this application of DONOW technology, was to develop a way to digitize their database, and provide some form of digital storage and retrieval.

The design goals of such system has taken into account that the database will keep growing at a fast pace, some data mining activity should be made possible to exploit more effectively the historical knowledge associated with the solution of projects realized in the past and the possibility to include, in the near future, several other application with minimal software tunings.

Besides it, other three design goals are: i) to query the database via simple internet connections, ii) to enable remote service units to intervene without asking someone at the headquarters to search materials for them and iii) to produce a cost effective solution both as initial investment and in fixed maintenance costs.

Given the graph structure of Partena database, an object representation and an object-oriented database management system seem be the most suitable solution for their management. This solution is easily feasible if we would manage them on a centralized database management system. However, it is realized with difficulty on a distributed database because of the need of techniques, that are not yet well studied, to partition the graph on different nodes of the database management system guaranteeing a good performance of the system.

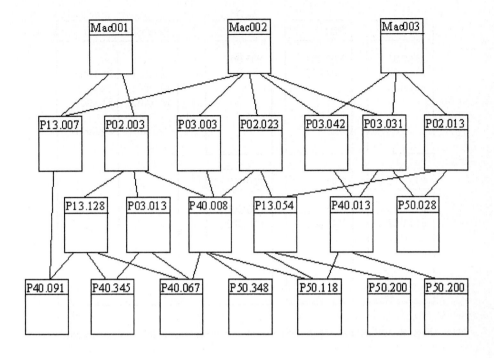

Fig. 7. graphical representation of some Partena machines documents

A relational database is not the most appropriate for graph structures. However, such structures can be managed through the use of join tables [1] obtaining a set of tables that can be easily partitioned for a distributed database management system (see Fig. 8).

We use horizontal partitioning [9] for all the tables, that is, all the nodes manage all the types of table, but each node manages a different subset of records (see Fig. 9). This solution simplifies a lot DONOW software because it does not realized concurrency control and commit protocols to manage distributed transactions. Such protocols are necessary when an operation must modify the data of different nodes of the database management system. Deletion and modification are few used operations and usually acting on a single record, that is, managed by a single node, that is, by a MySQL system. Moreover, the insertion of new data can be decomposed in a sequence of record insertions each of them managed by a MySQL system. However, data insertion is transparent to the user: the DONOW service layer decided on which node inserts the different records.

Machines	P02	P03	PartOf	
Mac001 ..	P02.001 ..	P03.001 ..	...	
Mac002 ..	P02.002 ..	P03.002 ..	Mac001	P13.007
Mac003 ..	...	...	Mac001	P02.003
...			Mac002	P13.007
			...	
			P13.007	P40.091
			...	

P13	P40	P50
P13.001 ..	P40.001 ..	P50.001 ..
P13.002 ..	P40.002 ..	P50.002 ..
...	...	...

Fig. 8. Relational representation of some Partena machines documents

Such a partitioning simplifies a lot query parallelization too. In fact, the duty of a DoNowPacket action is always to send the same query to all the data layer nodes and collects the results.

Currently, Partena is digitizing the design machine documents and we are experimenting the system on a limited set of data. The full data will be available at the end of the year; therefore, complete experimental results will be available at the beginning of the 1999.

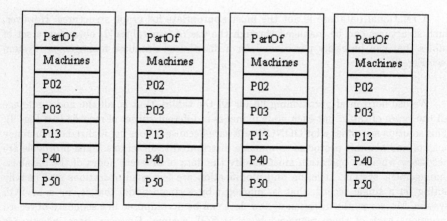

Fig. 9. Partitioning of tables on the DONOW system

6 Conclusions

This paper, presented a system called DONOW (Database on Network of Workstation), that is an example of an implementation and maintenance of a low cost distributed database without performances penalties.

DONOW has a three tier architecture based on a client, a service and a data layer. The client layer allows local and remote users to interact with the system through a Java user friendly interface. The service layer is implemented in C++; it allows the service of user requests and, in particular, the management of queries involving more than one DONOW node. The data layer is based on a well-known freeware relational database management system, that is, MySQL for each DONOW node.

The three layer architecture, the object-oriented implementation through C++ and Java languages, and the use a conventional relational database for each node of the cluster introduces a significant overhead that could in principle reflect itself on a poor performance of the system. For example, the use of a two layer architecture, that eliminates the service layer, where clients would interact directly with MySQL nodes through their JDBC interface, could offer better performance, and certainly would, on modest sized data sets.

However, our choices allowed us to achieve in short time a very important design goal, that is, to have a low cost and flexible prototype that would allow the user to start deploying the new tool, and us to experiment different solutions for the realization of a database management system for NOW architectures.

The exploitation of a relational database management system, that is, MySQL, for each DONOW node, goes a long way towards achieving a short delivery time of a prototype.

The three layer architecture and the use of object-oriented technologies allows us to experiment different solutions reusing most of the software and quickly prototyping different communication protocols and parallelization algorithms without intervening on the user interface. For example, replacing the TCP/IP communication protocol with another protocol causes only the redefinition of the send() and receive() methods.

Finally, the use of Java for the client interface, provides an easy access to the system via internet implementing the client as applet accessible via a WWW server.

Planned future work consists of experimenting different communication protocols between service layers and data partitioning strategies.

As far as protocols are concerned, we are planning to test the performance of PRP [14] and Gamma protocols [11]. Within our framework, that can be simply achieved by specializing the send and receive methods of the DoNowPacket class.

With regard to data partitioning, we are planning to experiment vertical partitioning and some different types of hybrid partitioning not only to increase performance, but also to offer replication of data.

Acknowledgements

We wish to thank the anonymous referees for their valuable comments.

The work has been partially supported by a grant of the European Union (Tender to III/97/31 Lot 5 - Cluster Computing for data Intensive Application) in the framework of the DISCO (DIStributed COmputing on network of PC for data base applications) project and by a grant of the Italian *Ministero dell'Università e della Ricerca Scientifica e Tecnologica (MURST)* in the framework of the MOSAICO (Design Methodologies and Tools of High Performance Systems for Distributed Applications) project.

References

1. Ambler, S.W.: Tips for Mapping Objects to Relational Databases. AmbySoft Inc. (1998) Available from http://www.AmbySoft.com/onlineWritings.html
2. Anderson, T.E., Culler, D.E., Patterson, D.A.: A case for networks of workstations: NOW. *IEEE Micro*, 15(1) (1995) 56–64
3. Arpaci-Dusseau, A.C., Arpaci-Dussuau, R.H., Culler, D.E., Hellerstein, J.M., Patterson D.A.: High-Performance Sorting on Networks of Workstations. In *Proc. SIGMOD '97*. Tucson, AZ (1997) 243–254
4. Arpaci-Dusseau, A.C., Arpaci-Dussuau, R.H., Culler, D.E., Hellerstein, J.M., Patterson D.A.: Searching for the Sorting Record: Experiences in Tuning NOW-Sort. In 2nd SIGMETRICS Symposium on Parallel and Distributed Tools. Welches, OR (1998)
5. Axmark, D.: *MySQL 3.22 Reference Manual.* (1998) Available from http://www.mysql.com
6. Baru, C., Fecteau, G., Goyal, A., Hsiao, H., Jhnigran, A., Padmanabhan, S., Wilson, W.: An Overview of DB2 Parallel Edition. In *Proc. SIGMOD '95*. San Jose, CA (1995) 460-462
7. Basu, A., Buch, V., Vogels, W., von Eicken, T.: A User-Level Network Interface for Parallel and Distributed Computing. In *Proc. 15th ACM Symp. on Operating System Principles* Copper Mountain, CO (1995)
8. Blumrich, M.A., Li, K., Alpert, R., Dubnicki, C., Felten, E.W.: Virtual Memory Mapped Network Interface for the SHRIMP Multicomputer. In *Proc. of the Int. Symp. on Computer Architectures* Chicago, IL (1994) 142-153
9. Bobak, A.R.: *Distributed and Multi-Database Systems.* Artech House (1995)
10. Boral, H., Alexander, W., Clay, L., Copeland, G., et al: Prototyping Bubba, a Highly Parallel database system. *IEEE Trans. on Knowledge Data Engineering.* 2(1) (1990) 44–62
11. Chiola, G., Ciaccio, G.: Implementing a low cost, low latency parallel platform. *Parallel Computing.* 22 (1997) 1703–1717
12. DeWitt, D., Ghandeharizadeh, S., Schneider, D., Bricker, A., et al: The Gamma Database Machine Project. *IEEE Trans. on Knowledge Data Engineering*, 2(1) (1990) 4–24
13. Gerber, R.: Informix Online XPS. In *Proc. SIGMOD '95*. San Jose, CA (1995) 463
14. Marenzoni, P., Rimassa, G., Vignali, M., Bertozzi, M., Conte, G., Rossi, P.: An operating system support to low-overhead communications in NOW clusters. In *Proc. of Communication and Architectural Support for Network-Based Parallel Computing*, San Antonio, TX (1995) 130–143

15. Microsoft: *Clustering Support for Microsoft SQL Server High Availability for Tomorrow's Mission Critical Applications.* (1997) Available from http://research.microsoft.com/ gray/.
16. Ozsu M.T., Valduriez, P.: *Principles of Distributed Database Systems.* Prentice Hall, Englewood Cliff (1991)

Implementation and Evaluation of MPI on an SMP Cluster

Toshiyuki Takahashi, Francis O'Carroll†, Hiroshi Tezuka, Atsushi Hori, Shinji
Sumimoto, Hiroshi Harada, Yutaka Ishikawa, Peter H. Beckman‡

Real World Computing Partnership

{tosiyuki,tezuka,hori,s-sumi,h-harada,ishikawa}@rwcp.or.jp

†MRI Systems, Inc.   ‡Los Alamos National Laboratory

ocarroll@rwcp.or.jp   beckman@lanl.gov

Abstract. An MPI library, called MPICH-PM/CLUMP, has been implemented on a cluster of SMPs. MPICH-PM/CLUMP realizes zero copy message passing between nodes while using one copy message passing within a node to achieve high performance communication. To realize one copy message passing on an SMP, a kernel primitive has been designed which enables a process to read the data of another process. The get protocol using this primitive was added to MPICH. MPICH-PM/CLUMP has been run on an SMP cluster consisting of 64 Pentium II dual processors and Myrinet. It achieves 98 MByte/sec between nodes and 100 MBytes/sec within a node.

1 Introduction

As SMP machines have come into wide use, they are becoming less expensive. Notably, a dual Pentium II machine is now almost the same price as a single Pentium II machine plus the cost of the additional CPU and main memory. This low cost drives construction of cluster systems using dual Pentium II machines. Another advantage is that such a cluster requires less space than a single CPU-based cluster with the same number of CPUs.

In this paper a cluster system whose components are SMPs is called an SMP cluster and a single processor based cluster is called a UP (uniprocessor) cluster. On an SMP cluster, the following programming models are taken into account:

1. Mixture of multi-threaded and message passing models
   A single process runs on each SMP node. A multi-threaded programming model is employed within a process to utilize CPUs within a node, and message passing is used for communication between nodes.
2. Multi-threaded only model
   With the support of a distributed shared memory system on an SMP cluster, a multi-thread programming model is used.
3. Message passing only model
   Running a process on each processor of an SMP node, message passing is employed to communicate with all processes.

In this paper, we consider the third model because it has the advantages of portability and understandability for MPI[9] users. MPI software written for a

UP cluster runs on an SMP cluster without any modification. Programmers do not need to specially consider how to use the SMP node's multiple processors.

We have designed and implemented a high performance MPI library for SMP clusters, called MPICH-PM/CLUMP[1]. MPICH-PM/CLUMP features facilities for zero-copy message transfer between SMP nodes and one-copy message transfer within an SMP node. These facilities reduce message copy overhead, and make high performance possible.

In this paper, first, our cluster system software, called SCore, is briefly introduced in section 2. SCore 2.x, the previous SCore version, supports only UP clusters. In section 3, SCore 2.x is adapted to an SMP cluster, and then the basic performance is evaluated. The extended SCore is called SCore 3.x.

A naive MPI implementation on SMP uses the shared memory OS facility. However, this involves two message copies. A one-copy message transfer in an SMP node is designed in section 4 to better utilize the SMP memory bandwidth. Using this feature, an MPI library for SMP clusters called MPICH-PM/CLUMP is designed and evaluated.

Section 5 evaluates MPICH-PM/CLUMP using the NAS Parallel Benchmarks. Related work is described in section 6.

2 Background

2.1 Hardware Configuration

Table 1. Specification of LBP, an SMP Cluster

Processor	Intel Pentium II
Clock	333 MHz
Chipset	Intel 440LX
Memory/Node	512 MBytes
Number of Processors/Node	2
Number of Nodes	64
Network	Myrinet M2M-PCI32C, M2LM-SW16
Network Bandwidth	full-duplex 1.28+1.28 Gbits/sec/link

Table 1 shows the hardware specification of an SMP cluster named the ACL[2] "Little Blue Penguin" Cluster (LBP) at Los Alamos National Laboratory. It consists of 64 SMP nodes each having two Pentium II processors. The nodes are connected by a Myrinet network[4] which is a high performance network. LBP has eight 16-port switches. Eight ports of each switch are linked to nodes.

---

[1] MPICH using PM on a CLUster of SMPs
[2] Advanced Computing Laboratory

Table 2. Specification of RWC SMPC 0, another SMP Cluster

Processor	Intel Pentium Pro
Clock	200 MHz
Chipset	Intel 450GX
Memory/Node	256 MBytes
Number of Processors/Node	4
Number of Nodes	4
Network	Myrinet M2F-PCI32C, M2F-SW8
Network Bandwidth	full-duplex 1.28+1.28 Gbits/sec/link

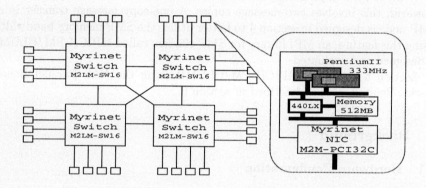

Fig. 1. Network Topology of LBP (32 Nodes)

Seven ports are used to interconnect the switches. The other port is for future expansion. Figure 1 illustrates the network topology of half (32 nodes) of LBP.

Table 2 shows the specification of another SMP cluster named RWC SMP Cluster 0 (SMPC). MPICH-PM/CLUMP supports both LBP and SMPC. As space is limited, only the results on LBP are shown in this paper.

2.2 SCore Cluster System Software

SCore is cluster system software package that realizes a multiple user environment on top of Unix kernels (including SunOS, NetBSD, and Linux). It was implemented by building a device driver and user level programs without any kernel modification. As shown in Figure 2, it consists of a communication driver and handler called PM, a global operating system called SCore, MPI implemented on top of SCore and PM called MPICH-PM, a multi-threaded template library called MTTL, and an extended C++ programming language called MPC++.

SCore assumes the SPMD execution model in which a parallel application is represented by a set of processes. All processes have the same code but may have

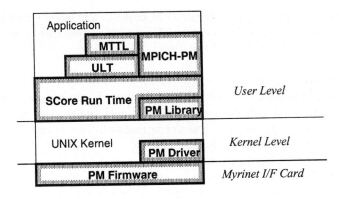

Fig. 2. Architecture of SCore Cluster System Software

different data sets. Each of these processes runs on its own node. We call the set of processes a parallel process, and call each process an element process. SCore has a gang scheduler which runs multiple parallel processes of the multi-user at the same time.

SCore Runtime System The SCore 2.x runtime system manages parallel processes on top of an unmodified Unix operating system kernel. The runtime system provides an interface for communication between element processes. It hides the behavior of the underlying operating system. Table 3 shows some of the primitives in the SCore 2.x runtime system. It also provides primitives for remote memory access. The SCore 2.x runtime system supports UP clusters running Sun OS 4.1.x, NetBSD and Linux operating systems.

Table 3. SCore runtime system primitives

Primitive	Feature
_score_get_send_buf	Get communication buffer for sending
_score_send_message	Send message written in the buffer
_score_peek_network	Test message reception
_score_recv_message	Receive message

PM To extract the performance data of Myrinet, a low level message facility called PM [11, 12, 6] has been implemented. It supports reliable asynchronous message passing on the Myrinet network. PM consists of three parts, code on the Myrinet card, a device driver and a user-level library. By accessing the Myrinet interface directly from a user program, PM achieves low latency and high bandwidth message passing without the overhead of a system call or an interrupt. PM has the following additional features:

Remote Memory Write
Data is written directly to the receiver's memory by the remote memory write mechanism. Since this feature does not require buffering in main memory, so-called zero copy message transfer is realized. This is one of the reasons that PM achieves high bandwidth. [13, 14].

Multiple Communication Channels
PM provides a channel to support a *virtual network*. Each process of a parallel application uses the same channel number to communicate with another. Multiple processes or threads on a SMP node may exclusively use a channel number or may share a channel number.
The number of channels depends on the NIC's hardware resources. In the current implementation, four channels are supported.

Network Context Switch
Since channels are restricted resources, in order to support multiplexing a mechanism of loading and restoring the network context for each channel is provided. This feature is used by the SCore-D gang scheduler which realizes the multi-processes environment.

MPICH-PM MPICH-PM is an MPI library on top of PM, based on MPICH[8]. Using the PM remote memory write feature, MPICH-PM achieves high bandwidth. MPICH-PM has been developed for UP clusters[10]. MPICH-PM uses the *eager* and the *rendezvous* protocols internally, supported by the MPICH implementation.

The *eager* protocol pushes a message into the network as soon as an MPI send primitive is issued. When a message arrives at the receiver, MPICH stores the message into a temporary buffer if an MPI receive primitive has not been issued. When the receive primitive is issued, MPICH copies the message to the receiver's memory area.

The *rendezvous* protocol can realize zero-copy message transfer between nodes. In this protocol, a control message is sent to the receiver when an MPI send primitive is issued. After that, when an MPI receive primitive is issued on the receiver, the control message is accepted by the receive primitive, and then the receiver replies to the sender with a control message requesting transfer. The sender transfers the message using the remote memory write mechanism when it receives the control message. Since there is no copying of the message in main memory, message transfer using this protocol is called zero-copy message transfer.

In general, the *eager* protocol gives better performance when sending a short message while the *rendezvous* protocol is better when sending a long message. The boundary depends on the host machine. MPICH-PM has the ability to change between the two protocols by specifying the boundary at execution time.

3 SCore 3.x

3.1 Design and Implementation

Since the SCore 2.x runtime system assumes a UP cluster, the system assigns only one element process to each node. The new SCore 3.x system is designed to be capable of assigning multiple element processes to each node.

The SCore 3.x interface for message transfer between element processes within a node is the same as the one for message transfer between SMP nodes. An element process is able to communicate with other element processes transparently by specifying an element process number as the target.

The system recognizes whether the target element process is inside the node or not. It transfers the message using shared memory if the target is inside the node, and it transfers the message using PM if the target is outside.

Communication within an SMP node The runtime system allocates a shared memory across element processes in an SMP node. The shared memory is used as a communication buffer between the element processes. When the runtime system transfers a message, a sender process copies the message to the shared memory then a receiver copies the message from the shared memory. To realize mutual access to the buffer, the current implementation uses the i86's test-and-set instruction[3]. Note that it requires two processor-executed copies as a result.

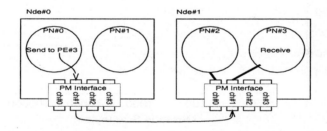

Fig. 3. An example of message transfer within an SMP node

Communication between SMP nodes There are two designs possible to realize communication between an element process and any element process in another SMP node using PM. One is that a single channel is shared by a parallel process. This design has two disadvantages: one is that all processes on a node

---

[3] Buffers for all the destinations should be implemented so that the runtime transfers messages without mutual access.

must mutually access the channel in both the send and the receive operations. Another is that the receiving process must dispatch an arrival message to the destination process. This is because all the senders send messages using the single channel, and thus, the receiver may receive a message sent to another process in the node.

Another design is that element processes on an SMP node have their own receiving channels. When sending a message to an element process on another node, a message is sent using a channel in which the destination process receives the message. This mechanism requires only shared access to a channel for sending messages. The SCore 3.x runtime is implemented based on this design.

Figure 3 illustrates an example of communication between an element process PN#0 on node 0 and an element process PN#3 on node 1. PN#3 is assigned channel 1. When PN#0 sends a message to PN#3, the SCore 3.x runtime on PN#0 uses channel 1 exclusively.

It is also noted that the implementation uses the i86's test-and-set instruction to realize mutual access to the channels.

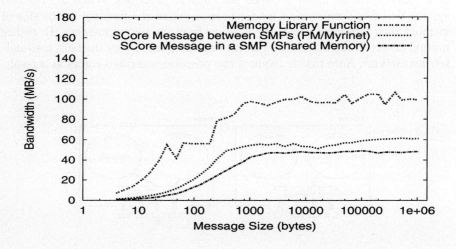

Fig. 4. Basic Performance of SCore 3.x runtime

3.2 Basic Communication Performance

Figure 4 shows the point-to-point bandwidth of the SCore 3.x runtime. For comparison we show bandwidth of the memcpy function provided by a Unix kernel in the graph. The bandwidth of communication in an SMP node is half of the bandwidth of memory copy. This is due to the two processor-executed copies.

4 MPICH-PM/CLUMP

4.1 Design and Implementation

Using SCore 3.x it is not difficult to port MPICH-PM to an SMP cluster. However, we could not achieve the same performance as the memory bandwidth using SCore 3.x when communicating within an SMP node because it involves two memory copies.

MPICH-PM realized zero-copy message transfer between nodes by using the PM remote memory write feature. Inside an SMP node, we could not remove both copies for message transfer, but we could design and implement it using only one copy. We extend MPICH-PM to run on an SMP cluster and support a one-copy message transfer within an SMP node. The extended MPICH-PM is called MPICH-PM/CLUMP.

The _pmVmRead function is introduced in section 4.1 in order to realize one-copy message transfer. The function copies data directly from the virtual memory of another process using a feature implemented inside the PM device driver without any modification of the Linux kernel. Section 4.3 shows the MPICH *get* protocol that implements one copy message transfer using the _pmVmRead function.

Direct Memory Copy between Processes First we show why two copies are required to transfer data between processes in an SMP node. To transfer data between processes, we usually use Unix SYS/V's shared memory or Mach's Memory Object. These require mapping user memory to shared memory. Processes are able to transfer data via the shared memory. We can implement MPI using such a mapping feature. The library has to map the memory used for messages to other processes before the communication so that the sender and the receiver share it. To transfer a message, the sender writes a message to the communication buffer then the receiver reads the message from the buffer. This is the reason why two copies are required using features provided by a Unix kernel.

Since the access cost of main memory is a significant in performance, it is important to reduce the number of copies. To realize one-copy message transfer within a node, we need a feature that makes it possible to copy messages directly from the sender to the receiver without involving the communication buffer. So we design and implement a kernel primitive to read the virtual memory of another process without mapping.

To implement the feature as a kernel primitive, there are two designs:

1. Permit server/client processes to exchange data. First a client negotiates access permission with a server using a well known port or known file name. Then the server permits the client to read or write memory. Only the client process allowed by the server is able to access memory with the access permission. It is dangerous to give write permission to the client, this is the responsibility of the designer of the server process.

2. Permit descendants of a process to exchange data. All Unix processes are ordered by a parent and child relationship. Use this order to inherit permission. Some type of framework is required to specify permission.

Since the reason why we need a memory access facility between processes is the realization of one copy message transfer, we implemented the following simple features based on the latter design.

- Permit descendants of a parent process to mutually read data. The exclusive /dev/pmvm device open by the parent process realizes this. The device implements the direct memory copy operation as an ioctl system call.
- The library function _pmVmRead, realized using the ioctl system call, copies data from the memory of the process specified by arguments to the memory of the current process directly. The following are the declarations of _pmVmRead:

```
int _pmVmRead(src_pid, src_vaddr, dst_vaddr, size)
    int src_pid;
    caddr_t src_vaddr, dst_vaddr;
    size_t size;
```

Implementation of the direct memory copy uses the following three features of the Linux kernel:

1. Looking up the page table for a specified process
2. Paging-in the physical page allocated by other processes
3. Reading the physical page allocated by other processes

In other words, the function can be implemented on all UNIX kernels that support the three features.

4.2 Basic Performance of Direct Memory Copy

Figure 5 shows the bandwidth of _pmVmRead. For comparison, we show the bandwidth of the memcpy function. The result of _pmVmRead shows that the performance is lower than memcpy when sending short messages while the performance is the same as memcpy when sending messages longer than 3 KBytes. The reason is the overhead of the ioctl system call. The _pmVmRead function calls the ioctl to kick the device driver. The call takes 5 usec every time. The overhead is caused by context switches between user level and kernel level.

4.3 The get protocol

We have explained the *eager* and the *rendezvous* protocols used in MPICH in section 2.2. MPICH also defines a *get* protocol. We implemented one-copy message transfer using the *get* protocol with the _pmVmRead function. Figure 6 illustrates one-copy message transfer.

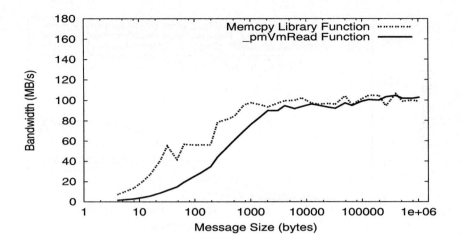

Fig. 5. Bandwidth of _pmVmRead

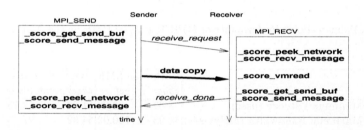

Fig. 6. The get protocol

1. When an MPI send primitive is invoked on the sender, the sender sends a receive_request control message to the receiver.
2. The sender waits for a receive_done control message from the receiver.
3. The receiver receives the receive_request control message and when the MPI receive primitive is invoked, it copies the message from the sender process using the _pmVmRead function.
4. After completing the copy, the receiver sends the receive_done control message to the sender.

5 Evaluation

5.1 Point-to-Point Performance

Table 4 and figure 7 show the point-to-point performance of MPICH-PM/CLUMP measured on LBP. The latency is half of the average round trip time of ping-pong

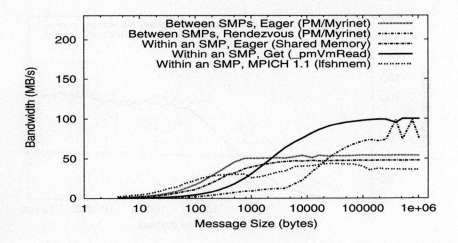

Fig. 7. Point-to-Point Bandwidth of MPICH-PM/CLUMP

Table 4. Point-to-Point Performance of MPICH-PM/CLUMP on the ACL LBP Cluster

	Within an SMP		Between SMPs	
	eager	*get*	*eager*	*rendezvous*
Minimum Latency (usec)	8.62	9.98	11.08	12.75
Maximum Bandwidth (MBytes/s)	48.03	100.02	54.92	97.73
Half Bandwidth Point (KBytes)	0.3	2.5	0.2	21

messages. The bandwidth is the amount of messages transfered. The half bandwidth point is the message length that achieves half of maximum bandwidth. To obtain both results, benchmark programs run for more than one second.

For comparison the graph shows the bandwidth of the lfshmem device that is included in MPICH 1.1 distribution. The lfshmem device has the best performance among the devices included in the distribution. It is noted that the device doesn't have the code to realize communication between SMP nodes.

The crossing point of the *rendezvous* protocol with the *eager* protocol depends on the configuration of the cluster system. On LBP the point is about 30 KBytes. It is natural to attempt to switch protocols depending on the size of messages. However the analysis is not easy, as we should consider the characteristics of each protocol. The memory access load of the *rendezvous* protocol is lighter than that of the *eager* protocol. On the other hand, the *rendezvous* protocol requires additional control messages to synchronize sender and receiver. It is noted that when sending the results of a calculation immediately, the cache system reduces the overhead of memory copies, which will improve the performance of the *eager* protocol.

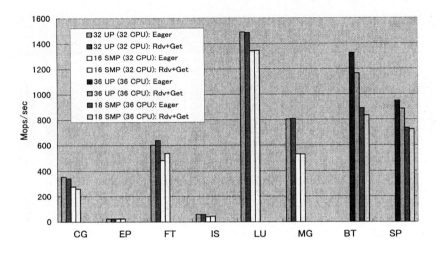

Fig. 8. NPB: Results on 32 CPU or 36 CPU out of LBP

The *get* protocol of MPICH-PM/CLUMP is the best of the protocols within an SMP node when sending messages larger than 4 KBytes. The *get* protocol achieves 100 MBytes/s of bandwidth when sending 128 KBytes while the maximum bandwidth of the *eager* protocol within an SMP node is 48 MBytes/s. The performance is twice that of the lfshmem device. The *get* protocol realizes one-copy transfer of messages. The result shows the protocol achieves its designed performance in practice.

5.2 NAS Parallel Benchmarks

Figure 8 and Figure 9 show the results of NAS Parallel Benchmark 2.3 Class B[5] on LBP. The first graph shows the performance results printed by each NAS benchmark program. The combination of the *rendezvous* protocol and the *get* protocol is slower than the *eager* protocol alone except when running FT and IS. The following reasons are considered.

- Contention of messages reduces the performance of the *rendezvous* protocol drastically.
- When sending calculated results immediately, the cache system reduces the copy overhead of the *eager* protocol.

The second graph compares the results of an SMP cluster and a UP cluster with same number of CPUs. Each bar shows the ratio of the SMP cluster to the UP cluster. The SMP cluster achieves 70% to 100% of the performance of the UP cluster. The combination of the *rendezvous* protocol and the *get* protocol shows good performance compared to that of the *eager* protocol.

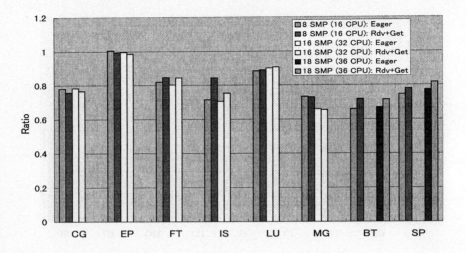

Fig. 9. NPB: Comparison with UP cluster

The following reasons are considered: When using the combination, the *rendezvous* protocol and *get* protocol each handle half of the messages on the SMP cluster. On the other hand, the *rendezvous* protocol handles all the messages on the UP cluster. The SMP and UP clusters each pass the same number of messages to the network. But the number of control messages for the *rendezvous* protocol is half that on UP cluster. Since it is considered that the latency of control messages is significant to the performance when the network is crowded, the number of control messages affects performance.

Intuitively, an SMP cluster is faster than a UP cluster if both clusters have the same number of CPUs because communication within an SMP node is faster than communication across nodes. Using the third programming model, however, some applications that frequently synchronize among processes do not run faster than a UP cluster with the same number of CPUs.

The following reasons are considered.

- Since the synchronization mechanism among processes is realized by message passing, the synchronization cost is dominated by the communication cost. Though communication is faster within a node, the synchronization cost is dominated by communication cost among nodes.
- Total amount of messages of the execution on an SMP cluster is equal to the one on a UP cluster.

Hence, the performance of such applications on the SMP cluster does not exceed the performance of the UP cluster. The NAS Parallel Benchmarks 2.3 are examples of such applications.

6 Related Work

The Active Messages group at the University of California, Berkeley[1], the Fast Messages group at the University of Illinois, Urabana-Champaign[3] and the BIP group at University Claude Bernard Lyon 1[2] are also developing MPI libraries over Myrinet. As far as we know, there is no report describing the results of NAS Parallel benchmarks using these MPI on recent SMP clusters.

The implementation of MPICH 1.1 uses shared memory for communication within an SMP node. It makes two copies to send/receive a message. These copies lower the bandwidth.

From the point of view of SMP implementation of MPI, TOMPI(Threads Only MPI)[7] realized an MPI implementation that copies a message only once when sending the message. TOMPI rewrites an MPI application using a source code translator to run it using multiple threads on an SMP node.

The Linux proc filesystem supports a feature that makes it possible to access the memory of other process. Basically the feature is the same as our kernel primitive _pmVmRead. However the current implementation of the proc filesystem limits the access area to pages resident in main memory. This is one reason why we did not choose it for the MPICH-PM/CLUMP implementation.

From the point of view of the design of a kernel primitive, the Linux pread primitive is able to read any region of a file. If a user process can access another process via a file, the primitive provides the same feature as _pmVmRead.

7 Conclusion

In this paper, we designed and implemented MPICH-PM/CLUMP, an MPI library for SMP clusters based on MPICH 1.0. MPICH-PM/CLUMP supports multiple processes on an SMP node. MPICH-PM/CLUMP realizes zero-copy transfer of messages between SMP nodes, and one-copy transfer of messages within an SMP node. In order to implement MPICH-PM/CLUMP, we first designed and implemented the SCore 3.x runtime. The runtime realizes process management on an SMP cluster. The runtime provides communication primitives to upper level libraries. The primitives realize transparent message transfer between the processes. Secondly, we designed a direct copy facility to realize one-copy message transfer on an SMP node. The facility is implemented inside the PM device driver. Then we used it to implement the *get* protocol in MPICH-PM.

We measured the performance of MPICH-PM/CLUMP using the ACL "Little Blue Penguin" Cluster at Los Alamos National Laboratory. The cluster consists of 64 Dual Pentium II 333MHz machines and a Myrinet network. MPICH-PM/CLUMP achieved 8.62 usec of latency and 100 MBytes/sec of bandwidth within an SMP node. The result is twice the bandwidth of the lfshmem device of the MPICH 1.1 implementation. MPICH-PM/CLUMP achieved 11.08 usec of latency and 98 MBytes/sec of bandwidth between SMP nodes. The resulting bandwidth achieved is close to the node's memory system limit.

We evaluated MPICH-PM/CLUMP using the NAS Parallel Benchmarks 2.3 class B. The ratio of the result on the SMP cluster to the result on a UP cluster that has same number of CPU's is 70% to 100%. As discussed in section 5.2, this is considered to be due to the fact that the network message-based synchronization cost on an SMP cluster is not much lower than that of an equivalent UP cluster, and total message traffic is equal to that of a UP cluster. We need further work to investigate and measure the causes of the disadvantage.

References

1. Message Passing Interface Implementation on Active Messages. http://now.CS.Berkeley.EDU/Fastcomm/MPI/.
2. MPI-BIP An implementation of MPI over Myrinet. http://lhpca.univ-lyon1.fr/mpibip.html.
3. MPI-FM: Message Passing Interface on Fast Messages. http://www-csag.cs.uiuc.edu/projects/comm/mpi-fm.html.
4. Myrinet. http://www.myri.com.
5. NAS Parallel Benchmarks.
6. PM: High-Performance Communication Library. http://www.rwcp.or.jp/lab/pdslab/pm/home.html.
7. Erik D. Demaine. A Threads-Only MPI Implementation for the Development of Parallel Programs. In *11th International Symposium on High Performance Computing Systems (HPCS'97)*, pp. 153–163, July 1997.
8. William Gropp and Ewing Lusk. MPICH Working Note: Creating a new MPICH device using the Channel interface. Technical report, Mathematics and Computer Science Division, Argonne National Laboratory, 1995.
9. Message-Passing Interface Forum. MPI: A message passing interface standard, version 1.1, June 1995.
10. Francis O'Carroll, Hiroshi Tezuka, Atsushi Hori, and Yutaka Ishikawa. The Design and Implementation of Zero Copy MPI Using Commodity Hardware with a High Performance Network. In *ICS'98*, July 1998.
11. Hiroshi Tezuka, Atsushi Hori, and Yutaka Ishikawa. Design and Implementation of PM: a Communication Library for Workstation Cluster. In *JSPP'96 (in Japanese)*. IPSJ, June 1996. (in Japanese).
12. Hiroshi Tezuka, Atsushi Hori, Yutaka Ishikawa, and Mitsuhisa Sato. PM: An Operating System Coordinated High Performance Communication Library. In *High-Performance Computing and Networking '97*, 1997.
13. Hiroshi Tezuka, Atsushi Hori, Francis O'Carroll, Hiroshi Harada, and Yutaka Ishikawa. A User level Zero-copy Communication using Pin-down cache. In *IPSJ SIG Notes*. IPSJ, August 1997. (in Japanese).
14. Hiroshi Tezuka, Francis O'Carroll, Atsushi Hori, and Yutaka Ishikawa. Pin-down Cache: A Virtual Memory Management Technique for Zero-copy Communication. In *IPPS'98*, April 1998.

Fourth International Workshop on Formal Methods for Parallel Programming:
Theory and Applications FMPPTA'99

April 16 1998

in conjunction with
First Merged Symposium IPPS/SPDP 1999 13th International Parallel Processing Symposium & 10th Symposium on Parallel and Distributed Processing
April 12 - April 16, 1998
San Juan, Puerto Rico

Sponsored by
IEEE Technical Computer Society Technical Committee
on Parallel Processing
in cooperation with ACM SIGARCH

Program and Organizing Chair's Message

Seven papers have been accepted for presentation at the Fourth International Workshop on Formal Methods for Parallel Programming: Theory and Applications. This year, the theme of the workshop was Modeling and Proving. Modeling is the process that is carried out to place a problem, both the implementation and specification in a formal framework, while the proving phase verifies the correctness of the implementation with respect to its specification.

Two papers give case studies using the PVS theorem prover for mechanical verification of two tricky concurrent algorithms described by state transition systems. Chernyakhovsky, et. al look at the Time Warp protocol used for distributed discrete event simulation. Havelund describes his experiences verifying the correctness of a classic garbage collection algorithm. In addition to the PVS solution, he also describes an experiment using the Murphi model checker and compares the experience and results of the two approaches.

Additional case studies performed using model checking, this time with a tool called FDR (FDR2) are described in the paper by Simpson, Hill, and Donaldson, and the paper by Creese and Reed. In these cases, the model is described using CSP. Simpson et. al describe a communication protocol developed to be used in the implementation of the Bulk Synchronous Parallel model. Creese and Reed also look at a communication protocol, this time for end-to-end properties of protocols for arbitrary networks. They describe induction techniques that can be used to allow the model checking tool to deal with an unknown number of intermediate nodes.

A theory of compositionality can, in principle, be extremely valuable in both modeling and verification stages, mitigating the complexity of the models and proofs, and delaying the state explosion that limits model checking. Chandy and Charpentier describe their experiences utilizing a theory of compositionality based on existential and universal properties.

In "A Structured Approach to Parallel Programming: Methodology and Models", Massingill describes a way to develop parallel programs from sequential program-like specifications by applying semantics preserving transformations. The transformations have been proved correct, thus the programmer follow a sound process for program development without needing to perform additional proofs in most cases. A similar approach has been used by Attlali, Caromel, and Lippi to allow sequential object-oriented programs to be transformed to parallel programs that are expressed in a parallel extension of the language Eiffel, Eiffel//.

We thank the members of the program committee, and the reviewers for their contribution to ensuring a high quality workshop.

Dominique Méry and Beverly Sanders
January 1999

Table of contents

FROM A SPECIFICATION TO AN EQUIVALENCE PROOF IN OBJECT-ORIENTED PARALLELISM
I. Attali and D. Caromel and S. Lippi

EXAMPLES OF PROGRAM COMPOSITION ILLUSTRATING THE USE OF UNIVERSAL PROPERTIES
M. Chandy and M. Charpentier

A FORMAL FRAMEWORK FOR SPECIFYING AND VERIFYING TIME WARP OPTIMIZATIONS
V. Chernyakhovsky and P. Frey and R. Radhakrihnan and P. Wilsey and P. Alexander and H. Carter

VERIFYING END-TO-END PROTOCOLS USING INDUCTION WITH CSP/FDR
S. Creese and J. Reed

MECHANICAL VERIFICATION OF A GARBAGE COLLECTOR
K. Havelund

A STRUCTURED APPROACH TO PARALLEL PROGRAMMING METHODOLOGY AND MODELS
B. Massingill

BSP IN CSP: EASY AS ABC
A. Simpson and J. Hill and S. Donaldson

From a Specification to an Equivalence Proof in Object-Oriented Parallelism

Isabelle Attali, Denis Caromel, Sylvain Lippi

INRIA Sophia Antipolis, CNRS - I3S - Univ. Nice Sophia Antipolis,
BP 93, 06902 Sophia Antipolis Cedex - France
First.Last@inria.fr

Abstract.
We investigate the use of a TLA specification for modeling and proving parallelization within an object-oriented language.

Our model is based on Eiffel// a parallel extension of Eiffel, where sequential programs can be reused for parallel or concurrent programming with very minor changes. We want to prove that both versions of a given program (sequential and parallel) are equivalent with respect to some properties.

This article presents a description in TLA+ that captures the general Eiffel// execution model, and, as a case-study, specifies a program (a binary search tree) in both its sequential and parallel form. We then prove a property that demonstrates a behavioral equivalence for the two versions.

1 Introduction

In this article, we investigate the use of a specification in TLA [14] and TLA+ [15, 16] for concurrent and parallel object-oriented programming. Our model is based on the Eiffel// language [5, 7], a parallel extension of Eiffel [19].

One interesting aspect of the Eiffel// language is that existing sequential programs can be re-used when designing and programming parallel system: given a sequential system, using program transformations, one can manually generate an equivalent system running in parallel. We then would like to prove that both versions of the system (sequential and parallel) are equivalent.

With a formal definition of the concurrent model, our goal is to describe formally (and prove) parallelization techniques and algorithms for the Eiffel// framework. We want to define formally program transformations, especially parallelizations, from Eiffel to Eiffel// . Thus, we need indeed to prove that transformations preserve the meaning (the semantics, the behavior) of programs.

We already have a formal semantics of Eiffel// [2] based on a transition system whose states represent global configurations of a given system of objects. Objects are either passive or active. The semantics of a program is given by a transition system which represents all possible executions. Using this approach, the major drawback is the tremendous number of configurations, so we are currently studying a method for reducing the number of states, using partial order techniques.

We also investigated the use of π-calculus [20] to model an Eiffel// system [23, 3] (a translation from Eiffel and Eiffel// to π-calculus has been written in Natural Semantics). The π-calculus models the system as a quite low-level, with complexity manifested in a large number of channels and a loss of structure, making it difficult to understand and manipulate the π-terms. In that framework, the next step is to use bisimulation techniques [1] to prove the behavioral equivalence between a sequential and a parallel behavior.

We present here a description in TLA+ that captures in a more natural way the richness of the execution model. From the specification of two versions of a single program (one is sequential, one is parallel), we are able to prove that the two versions are equivalent with respect to some properties. The example that serves our purpose is the management of a binary tree (e.g. for a symbol table) with primitives for insertion and search.

Compared to [18], where a binary tree example is also used, the Eiffel// model allows use of the same code for both sequential and parallel versions; no commit instruction is needed in Eiffel// . This feature is critical for reuse, and is also beneficial for the proof since the actions modeling the program are identical in both cases.

Compared to the work on the $\pi o \beta \lambda$ design language [12], the Eiffel// model features synchronous and asynchronous calls through polymorphism and automatic continuations, which allow to achieve parallelization without changing the source code ($\pi o \beta \lambda$ for instance changes the place of the *return* statement, or needs to add a *delegate* statement). Another major difference seems to be the absence of future-based synchronizations, such as the wait-by-necessity. However, it is worth mentioning that the *island* concept is related to the subsystem notion, but seems different (since objects within an Eiffel// subsystem can perfectly reference the root of another subsystem). Finally, to our knowledge, there is no model and proof of the language within the TLA framework; a comparison would be interesting.

The next section of this article is a brief overview of the concurrent model. Section 3 introduces the main features of TLA and TLA+. In Section 4, we present our description of the Binary Tree example using TLA+. The proof of behavioral equivalence is sketched in Section 5. Finally, Section 6 discusses our contribution and outlines future work.

2 The Eiffel// language

In this section, we give a quick overview of a concurrent language, Eiffel// [7], defined as an extension of Eiffel [19] to support programming of parallel applications. These extensions are not concerned with syntax, but are purely semantic. Both Eiffel and Eiffel// are strongly typed, statically-checked class-based languages. Our purpose here is not to discuss the rationale of Eiffel// (see for instance [2] for a formal specification), and issues related to object-based concurrent languages (the reader can refer to a recent and related design (C++// [8]), and to related research [21, 24, 4, 22, 25]). We present the concurrent features of the model and illustrate it with an example which will serve throughout the paper.

2.1 Features

The Eiffel// model uses the following principles:

- heterogeneous model with both passive and active objects;
- polymorphism between passive and active objects;
- sequential active objects;
- unified syntax between message passing and inter-process communication;
- systematic asynchronous communications towards active objects;
- wait-by-necessity (automatic and transparent futures);
- automatic continuations;
- no shared passive objects (call-by-value between active objects);
- centralized and explicit control by default.

As in Eiffel, the text of an Eiffel// program (or system) is a set of classes, with a distinguished class, the root class. Parallelism is introduced via a particular class named **Process**. Instances of classes inheriting (directly or not) from the **Process** class are active objects. Active objects inherit a default behavior (through the **live** routine, which ensures that requests are treated in a *fifo* order), but this behavior can be redefined by overriding the **live** routine. All other objects are passive. Polymorphism between passive and active objects is possible: an entity which is not declared of an active object type can dynamically refer to an active object heir. In that case, thanks to dynamic binding, a feature-call dynamically becomes an asynchronous communication between active objects (*Inter-Process Communication*).

Cohabitation of active and passive objects leads to an organization in subsystems. Each subsystem contains an active object (root) and the passive objects it references. Within a subsystem, the execution is sequential and communication is synchronous: the target object immediately serves the request and the caller waits for the return of the result. Between subsystems, execution is parallel and communication is asynchronous: the target object (an active object) stores the request in a list of pending requests, and the caller carries on execution.

Synchronization is handled via *wait-by-necessity*, a data-driven mechanism which automatically triggers a wait when an object attempts to use the result of an awaited value (transparent future). The wait-by-necessity, by automatically adding some synchronization, tends to maintain the behavior of a sequential program when doing the parallelization. Finally, the automatic continuation, by automatically taking care of delegation behaviors without blocking the current object, allows reuse of such sequential code in parallel subsystems.

```
class Binary_Tree
  export   insert, search
  feature
  key, info : INTEGER;
  left, right : BINARY_TREE;
  insert (k : INTEGER; i : INTEGER) is
      do -- inserts information i with key k
        if key = 0 then key := k; info := i;
          left.create; right.create;
        elsif k = key then info := i;
        elsif k < key then left.insert(k,i);
        else right.insert(k,i);
        end;
      end;
  search (k : INTEGER) : INTEGER is
      do -- searches for the value of key k
        if key = 0 then result := <notfound>;
        elsif k = key then result := info;
        elsif k < key then result := left.search(k);
        else result := right.search(k);       --(AC)
        end;
      end;
end -- Binary_Tree

class P_Binary_Tree
  repeat Binary_Tree
  inherit Process;
          Binary_Tree redefine left, right;
  feature
    left, right : P_Binary_Tree;
end -- P_Binary_Tree
```

(a) Sequential and Parallel Binary Trees

```
class Client
  feature
    v: INTEGER;
    sequential_behavior is
      local bt: Binary_Tree;
        do
          bt.create;
          bt.insert(3, 6);
          bt.insert(4, 8);
          bt.insert(2, 4);
          bt.insert(6, 12);
          bt.insert(1, 2);
          v := bt.search(4);
          v := v + v
        end; -- sequential_behavior
    parallel_behavior is
      local bt: P_Binary_Tree;
        do
          bt.create;
          bt.insert(3, 6);
          bt.insert(4, 8);
          bt.insert(2, 4);
          bt.insert(6, 12);
          bt.insert(1, 2);
          v := bt.search(4);
          v := v + v       -- (WBN)
        end; -- parallel_behavior
end -- Client
```

(b) Reusing code for a parallel behavior

Fig. 1. An Eiffel// system

2.2 An example

As an illustration, Figure 1 presents an Eiffel// system introduced in [6]. It provides a parallel version of the sequential class Binary_Tree, which describes the management of a sorted binary tree with two routines insert and search: each node of the tree has two children (left and right), an information field (info) and an associated key (key); keys of the left (resp. right) subtree of a node are smaller (resp. greater) than the key of this node. A zero value for the key denotes a leaf of the tree.

To parallelize the binary tree we define the P_Binary_Tree class. It inherits from the Process class and the Binary_Tree class; no other programming is necessary; the full version of the class is actually shown in the Figure 1.a. A client can use the two classes (sequential and parallel binary trees) for both sequential and parallel behaviors, which makes it possible to reuse existing sequential code (using an object bt of class P_Binary_Tree instead of Binary_Tree). In that example, the default *fifo* behavior and wait-by-necessity (labeled in the example with (WBN)) ensure that all insertions are handled in a correct order, and before the search; this allows to preserve the semantics of the sequential system in the parallel execution. Automatic continuation (labeled in the example with (AC)) allows not to block the current node when forwarding the search request to a child, and still ensures that the final result will properly be sent to the objects in which it is needed.

Of course, this equivalence only holds in the case of a single client object; if several clients in different active objects (processes) send insert and search requests, the behavior can be different from the sequential version.

3 The Temporal Logic of Actions

TLA [14] (Temporal Logic of Actions) is a linear-time temporal logic meant to model and reason on concurrent systems. Systems and their properties are described by logical formulas. TLA has been used in a number of varied applications: concurrent algorithms [9] as well as hybrid systems [13].

3.1 TLA

TLA formulas are interpreted as predicates on behaviors where a behavior is an infinite list of states and a state is an assignment of values to variables. Here is the TLA formula defining our "binary tree" system :

$$\Phi \triangleq \mathit{InitEiffel//} \land \mathit{InitClient} \land \Box[\mathbf{updatePendings} \lor \mathbf{send} \lor \mathbf{insert} \lor \mathbf{search}]_w$$

We can distinguish two parts[1]:

---

[1] We could have added a third part called *a fairness condition* which is used to prove liveness properties.

- *InitEiffel// ∧ InitClient*
  This first part specifies the first state of the accepted behavior. For example, on the first state, we have only one object of the Client class;
- □[**updatePendings** ∨ **send** ∨ **insert** ∨ **search**]$_w$
  This second part specifies the different possible steps of the accepted behaviors. A step is made of two states (an initial state and a final state). The predicates **updatePendings**, **send**, **insert**, and **search** are actions on steps. They are built with unprimed and primed variables. Primed variables represent the value of the variable in the final state of the step. For example, the action x'=x+1 accepts all steps where x is incremented by one.
  The symbol □ is the temporal symbol "always" which says that every step is an **updatePendings**, a **send**, an **insert** or a **search** step. The index w is the tuples of all the variables we need. Those variables are called *flexible* because their value can change during the execution of the system. We also use some rigid variables (or constants). For example, **initBinary** specifies the value of an object of class Binary_Tree when it has just been created.

3.2 TLA+

Our specification uses TLA+ [15, 16], a complete language based upon TLA. We are using the module organization of TLA specifications: our module structure is straightforward: each class is modelled as a module (Client, Binary_Tree) and we add an Eiffel// module for the actions and state predicates common to all the specifications of Eiffel// programs.

We also use the "Sequences" module, defined in [17], which defines usual operators on sequences (Head, Tail, concatenation, and length). For example, Seq(S) represents the set of all the sequences of values belonging to S.

Here are some notations used to represent functions and records (as presented in [16]):
- $[A \rightarrow B]$ is the set of all the functions from A into B.
- the notation: $[1 \mapsto "a", 2 \mapsto "b", 3 \mapsto "c"]$ denotes a function from the set {1,2,3} into {"a","b","c"}.
- if f is a function, we note:
    f[i], the application of f to i;
    [f EXCEPT ![i] = j], the function which is equal to f except in i where the returned value is j;
    [f EXCEPT ![i] = @+j] the function which is equal to f except in i where the returned value is $f[i] + j$ (@ denotes $f[i]$).
- records are particular functions which associate a value to a string.
If *rec* is a record, we note *rec.field1* the value of $rec["field1"]$.

4 The Specification

We choose to model the Eiffel (and Eiffel//) programs in three parts using TLA+; some are specific to the binary tree example (definition and use in a sequential or a parallel setting) but one is abstract and generic enough to be used for any example: it captures the features of the Eiffel// programming model. Namely, our specification is expressed in three modules (see Figure 2):

- Eiffel// : state predicates and actions common to all specifications of Eiffel and Eiffel// programs;
- Binary_Tree: actions related to objects of class Binary_Tree;
- Client: actions related to the object of class Client.

All the general and modular aspects of the translation are in the Eiffel// module. It captures the specific semantics of the underlying model, such as asynchronous calls, futures, and wait-by-necessity. All the actions correspond to the management of requests, reply, and automatic continuations.

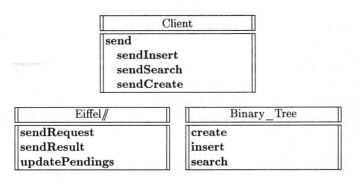

Fig. 2. The modules of the TLA+ specification

The Client and the Binary_Tree modules are specific to this case study. However, each action of the Binary_Tree module corresponds to one instruction of the corresponding class, when viewed from the perspective of a translation from Eiffel// code to TLA. Finally, each action of the Client module represents a call to a feature of Binary_Tree (**sendCreate, sendInsert, sendSearch**).

Please note that the Eiffel// module specifies both the sequential and the parallel cases. For instance, this module specifies synchronous calls in the case of a standard object, or asynchronous one with wait-by-necessity if the target object of a communication is active. As a consequence, the client and Binary_Tree specifications can be the same for both sequential and parallel versions.

We give a brief description of the flexible variables and then explain the state predicates and actions of the three modules.

4.1 Definition of flexible variables

The state function w gives the tuple of all variables in the description. We cannot detail all of them but instead give an intuition of their role in the specification:

$$w \triangleq \langle object, active, requests, curReq, future, pendings, result, synchro, pc \rangle$$

- objects : *object*, *active*
  The *object* variable is a function from object identifiers to objects (records that are object attributes). For instance, a Binary_Tree instance is a record of the form: $[key \mapsto 2, info \mapsto 4, left \mapsto 5, right \mapsto nil]$. The *active* variable is a boolean function that states whenever an object is active or passive.
- list of requests : *requests*, *curReq*
  Each call to an object is modeled by a request; every object has a list of pending requests. The *requests* variable is a function from object identifiers to lists of requests; each request includes the name of the target routine, a list of effective parameters, and a future for the result (see below).
  The *curReq* variable gives the request an object is currently handling.
- sending requests and results: *future*, *pendings*, *result*

  $future \in [futureIDs \rightarrow values \cup \{?\}]$
  Future stores the awaited value corresponding to a given future ID ; "?" means that the result is not yet arrived.

  $pendings \in [objectIDs \rightarrow Seq(futureIDs \times futureIDs)]$
  Pendings is used when the result of a request is itself a future ID, allowing to specify automatic continuations.

  $result \in [objectIDs \rightarrow values \cup futureIDs]$
  Result corresponds to the result Eiffel variable to be returned to the caller at the end of a routine.
- synchronous calls: *synchro*
  The *synchro* variable allows specification of synchronizations arising from synchronous and asynchronous calls. It is a function from object identifiers to two possible sets of values:
  - the value "OK" in case of an asynchronous call;
  - in the case of a synchronous call, a "future" identifier is used to block the calling object (in other words, we introduce an artificial data dependency, specifying an immediate wait on a future that is not yet needed).
- program counter: *pc*
  This is a standard way of ordering instructions in an object (execution is sequential within a subsystem).

4.2 The Binary_Tree module

The Binary_Tree module exports three main actions, each of them corresponds to a routine of the class Binary_Tree (`create`, `insert`, `search`). These actions are a translation of Eiffel// instructions into the TLA specification. The translation seems to be straightforward enough to permit some automatic translation in the future.

The action `create`(objID, attr) creates an instance of the class Binary_Tree, generates a new identifier `id` for this object, updates the value of the attribute attr of the object objID (creation site, this corresponds to the statement `bt.create`) with this new reference. Note that `id` is active whenever objID is.

$$\begin{aligned}
\textbf{create}(objID, attr) \triangleq{} & \exists\, id \in objectIDs \\
& \wedge\ object[id] = \text{FREE} \\
& \wedge\ object' = [object\ \text{EXCEPT}\ ![objID].attr = id, \\
& \hspace{8em} ![id] = initBinary\ ] \\
& \wedge\ active' = [active\ \text{EXCEPT}\ ![id] = active[objID]] \\
& \wedge\ \text{UNCHANGED}\ \langle future, synchro, requests, curReq, pendings, result\rangle
\end{aligned}$$

$$initBinary \triangleq [key \mapsto 0, info \mapsto 0, left \mapsto nil, right \mapsto nil]$$

The action **insert** (resp. **search**) is a disjunction of actions corresponding to the different instructions of the routine `insert` (resp. `search`). In both cases, sequentiality is managed with the *pc* variable.

For instance, action **insert$_4$** corresponds to the statement, `left.create`:

$$\begin{aligned}
\textbf{insert}_4(objID) \triangleq{} & \wedge\ pc[objID] = 4 \\
& \wedge\ pc' = [pc\ \text{EXCEPT}\ ![objID] = 5] \\
& \wedge\ \textbf{create}(objID, left)
\end{aligned}$$

Finally, here is the first action **search$_1$** which corresponds to the test expression `if key=0` :

$$\begin{aligned}
\textbf{search}_1(objID) \triangleq{} & \wedge\ pc[objID] = 1 \\
& \wedge\ \vee\ \wedge\ object[objID].key = 0 \\
& \hspace{2em} \wedge\ pc' = [pc\ \text{EXCEPT}\ ![objID] = 2] \\
& \hspace{1em}\vee\ \wedge\ \neg(object[objID].key = 0) \\
& \hspace{2em} \wedge\ pc' = [pc\ \text{EXCEPT}\ ![objID] = 3] \\
& \wedge\ \text{UNCHANGED}\ \langle future, synchro, requests, curReq, \\
& \hspace{8em} active, object, pendings, result\rangle
\end{aligned}$$

4.3 The Client module

The client object simply sends a "create" request followed by some "insert" requests with finally some "search" requests to the object representing the root of the binary tree. This list of requests which are sent by the client during execution is contained in the *extRequests* rigid variable. The Client module is made of one main action **send** which is a conjunction of actions modeling the possible requests (**sendInsert, sendSearch,** and **sendCreate**). The state predicate *notWaiting* checks whenever an object is blocked by a synchronous call.

$$\begin{aligned}
\textbf{send} \triangleq{} & \wedge\ notWaiting(clientID) \\
& \wedge\ pc[clientID] \in 1..Len(extRequests) \\
& \wedge\ pc' = [pc\ \text{EXCEPT}\ ![clientID] = @ + 1] \\
& \wedge\ \textbf{sendInsert} \vee \textbf{sendSearch} \vee \textbf{sendCreate}
\end{aligned}$$

The Client module imports the two other modules and define the following formula, corresponding to the accepted behavior:

$$\Phi \triangleq InitEiffel // \wedge InitClient \wedge \Box[\textbf{updatePendings} \vee \textbf{send} \vee \textbf{insert} \vee \textbf{search}]_w$$

The parallel behavior is specified with
$$\Phi \wedge active[root] = \text{"true"}$$
while
$$\Phi \wedge active[root] = \text{"false"}$$
specifies a sequential behavior. So every property holding for Φ is true for both behaviors (sequential and parallel).

4.4 The Eiffel// module

The Eiffel// module is the generic part of the specification: it captures all the parallel semantics of the model, and is to be reused when modeling other programs.

Exported actions represent transmission of requests and transmission of results to the caller. These mechanisms, formally described in [2] in Natural Semantics, are expressed in TLA using three main actions: **sendRequest**, **sendResult**, and **updatePendings**.

The action **sendRequest**$(orig, dest, req)$ sends the request req from object $orig$ to object $dest$:

1. the request is first added to the list of requests of object dest;
2. the future value is set to "?" (i.e. unknown, not yet arrived);
3. the synchro value of the caller is set to:
   - OK in case of an asynchronous call,
   - the future itself for a synchronous call, in order to block the caller immediately until the value is returned, as discussed earlier.

This action is expressed in TLA as follows:

sendRequest$(orig, dest, req) \triangleq$
  $\wedge\ requests' = [requests\ \text{EXCEPT}\ ![dest] = @ \circ \langle req \rangle]$
  $\wedge\ future' = [future\ \text{EXCEPT}\ ![req.future] = ?]$
  $\wedge\ \vee\ active[dest] = \text{"true"} \wedge synchro' = [synchro\ \text{EXCEPT}\ ![orig] = \text{OK}]$
     $\vee\ active[dest] = \text{"false"} \wedge synchro' = [synchro\ \text{EXCEPT}\ ![orig] = req.future]$
  $\wedge\ \text{UNCHANGED}\ < active, pendings, curReq >$

The action **sendResult** is performed by an object which sends back the result of the request it is responding to. There are two cases:

– the result is an effective value (not a future): in this case, the function $future$ is updated with the value;
– the result is a future identifier id; in this case, the list $pendings$ is updated with this identifier coupled with the awaited future.

This action is expressed in TLA as follows (with **sendResult$_1$** and **sendResult$_2$** for the two cases):

$\textbf{sendResult}(objID) \triangleq \land pc' = [pc \text{ EXCEPT } ![objID] = 1]$
$\qquad\qquad\qquad\qquad \land curReq' = [curReq \text{ EXCEPT } ![objID] = @ + 1]$
$\qquad\qquad\qquad\qquad \land result' = [result \text{ EXCEPT } ![objID] = \text{OK}]$
$\qquad\qquad\qquad\qquad \land \lor \textbf{sendResult}_1(objID)$
$\qquad\qquad\qquad\qquad\quad\; \lor \textbf{sendResult}_2(objID)$

$\textbf{sendResult}_1(objID) \triangleq \quad \land result[objID] \notin futureIDs$
$\qquad\qquad\qquad\qquad\quad\; \land future' = [future \text{ EXCEPT } !$
$\qquad\qquad\qquad\qquad\qquad [requests[objID][curReq[objID]].future] = result[objID]]$
$\qquad\qquad\qquad\qquad\quad\; \land \text{UNCHANGED } \langle synchro, requests, active, object, pendings \rangle$

$\textbf{sendResult}_2(objID) \triangleq \quad \land result[objID] \in futureIDs$
$\qquad\qquad\qquad\qquad\quad\; \land pendings' = [pendings \text{ EXCEPT } ![objID] =$
$\qquad\qquad\qquad\qquad\qquad @ \circ \langle requests[objID][curReq[objID]].future, result[objID] \rangle]$
$\qquad\qquad\qquad\qquad\quad\; \land \text{UNCHANGED } \langle future, synchro, requests, active, object \rangle$

Finally, the action **updatePendings** is not performed by any object but by something that can be referred to as some runtime action. It specifies the semantics of an automatic continuation, allowing a caller to access the result of the request when several references to futures are chained together, all to receive the same value. For example, when the result of request1 equals the result of request2 which in turns equals the result of request3. The result of **updatePendings** is an update of both *future* and *pendings*:

$\textbf{updatePendings} \triangleq$
$\quad \land \exists\, objID \in objectIDs : \exists\, n \in 1..Len(pendings[objID])$
$\qquad \land future[\, pendings[objID][n][2]\,] \neq ?$
$\qquad \land future' = [future \text{ EXCEPT } ![pendings[objID][n][1]] =$
$\qquad\qquad future[\, pendings[objID][n][2]\,]]$
$\qquad \land pendings' \triangleq [pendings \text{ EXCEPT } ![objID] =$
$\qquad\qquad [i \in 1..Len(@)-1 \mapsto \textbf{if } i < n \textbf{ then } @[i] \textbf{ else } @[i+1]]\,]$
$\quad \land \text{UNCHANGED } \langle synchro, requests, curReq, active, object, pc, result \rangle$

5 The behavioral equivalence proof

Intuitively, we would like to prove that, for a single client, or more generally several clients being in the same subsystem[2] (whether it is the `Binary_Tree` or the `P_Binary_Tree` class), when a result is returned after a search, it is the same in both cases.

Note that this is a partial equivalence. In the general case, it cannot be otherwise due to the non-strict aspect of the Eiffel// model; in particular, the wait-by-necessity implies that a sequential Eiffel system might not terminate when the Eiffel// does. Also note that we do not have a proof of the form *im-*

---

[2] See Section 2.1 for the definition of subsystems.

plementation => *specification* since none of the two versions (sequential and parallel) implies the other.

More specifically, we will prove that, given a list of insert and search requests that is sent by the client object (this list is represented by the *extRequests* rigid variable) and a search request, when an object responds to this request, it gives the same result in the sequential and parallel cases. Figure 3 gives such TLA state predicate that we note E. It expresses the fact that upon a search, when the key is found, the value being returned always corresponds to the value of the last insert request within *extRequests* with the same key.

To simplify the property, we assume that insert and search requests are not "mixed" in the *extRequests* list; we first have all the insert requests and then all the search requests. So, we can easily express that when an object responds to a search(k) request, the result is the info argument of the last insert request with a key argument equal to k.

Figure 3 also gives a graphical overview of the property. For a given search, when the key is found at a node, the information to be returned is equal to the info of the last insert with the same key. Because *extRequests* is a rigid variable, the value returned is always the same, for both the sequential and the parallel cases.

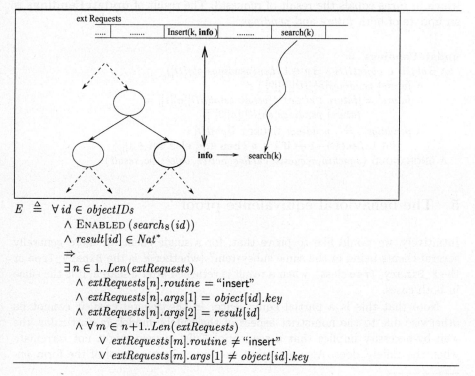

$E \triangleq \forall id \in objectIDs$
$\quad \wedge \text{ENABLED}\ (search_8(id))$
$\quad \wedge \ result[id] \in Nat^*$
$\quad \Rightarrow$
$\quad \exists n \in 1..Len(extRequests)$
$\quad\quad \wedge\ extRequests[n].routine = \text{"insert"}$
$\quad\quad \wedge\ extRequests[n].args[1] = object[id].key$
$\quad\quad \wedge\ extRequests[n].args[2] = result[id]$
$\quad\quad \wedge\ \forall m \in n+1..Len(extRequests)$
$\quad\quad\quad \vee\ extRequests[m].routine \neq \text{"insert"}$
$\quad\quad\quad \vee\ extRequests[m].args[1] \neq object[id].key$

Fig. 3. A behavioral equivalence property

To some extent, the property can be related to bisimulation techniques as used in a π-calculus framework. Please note that we cannot prove direct equivalence between sequential and parallel behaviors, or even that one implies the other.

We do not give the complete proof but rather a scheme. The proof is based on an invariant I, and the principle is the following:

1. find an intermediate invariant I;
2. prove that the initial state verifies I: $InitEiffel// \land InitClient \Rightarrow I$
3. prove that every action preserves I;
4. finally prove that $I \Rightarrow E$.

5.1 The invariant

We first describe the invariant (a state predicate) and then prove that it holds during the execution of the Binary Tree system. This invariant is presented as the conjunction of 20 state predicates. We group those predicates into four sets depending on what kind of property they express (typing, data control and value, structures of data, request flow):

$$I \triangleq \land invTyping \\ \land invDataValue \\ \land invStructure \\ \land invRequest$$

All the sets being defined as:

$$invTyping \triangleq \land searchResultType \\ \land clientType \\ \land insertArgType \\ \land searchArgType \\ \land responderType$$

$$invDataValue \triangleq \land pcValue \\ \land curReqMax \\ \land curReqValue \\ \land leaf \\ \land rightValue \\ \land keyValue \\ \land insertArgValue \\ \land requestsValue \\ \land emptyRequests \\ \land infoValue$$

$$invStructure \triangleq \land descendantOfRoot \\ \land onlyOneFather \\ \land notFatherOfRoot$$

$$invRequest \triangleq \land requestFlow \\ \land rootRequests$$

As an illustration, we mention and detail a few invariants that will be used in the sequel of the paper.

The invariant *descendantOfRoot* states that every **binary** object is a descendant of the root of the tree (*rootID* object) and has one and only one father. More formally, the corresponding TLA predicate is:

$$
\begin{aligned}
descendantOfRoot \triangleq\ & \forall\, id \in objectIDs \\
& \land\ object[id] \in binaryClass \\
& \land\ id \neq rootID \\
& \Rightarrow \\
& \exists\, l \in Seq(objectIDs) \\
& \quad \land\ Head(l) = rootID \\
& \quad \land\ Last(l) = id \\
& \quad \land\ \forall\, n \in 1..Len(l)-1 \\
& \qquad \land\ object[l[n]] \in binaryClass \\
& \qquad \land\ \lor\ object[l[n]].left = l[n+1] \\
& \qquad \lor\ object[l[n]].right = l[n+1]
\end{aligned}
$$

Two invariants capture the tree structure of the list of objects at execution. The invariant *onlyOneFather* denotes the fact that each `binary` object (except the root one) is referenced by a unique object, while *notFatherOfRoot* states that none of them reference the root object.

The invariant *requestFlow* is also central for the proof. It states that each `binary` object has a request list that is a prefix of its father request list.

The informal statement of the invariant *rootRequests* is that the request list at the root object (requests[rootID]) is a sub-list of *extRequests*, starting at the second element (the first one being a special request that create the first node), and finishing at the last element that was sent by the client (pc[clientID]-1).

$$rootRequests \triangleq requests[rootID] = SubSeq(extRequests, 2, pc[clientID] - 1)$$

We prove the invariant I by following the method given in [14]. We have to prove $\Psi \Rightarrow \Box I$ where Ψ is the specification of the Binary Tree system. We use rule $INV1$ (from [14]):

$$\frac{I \land [\mathcal{A}]_w \Rightarrow I'}{I \land \Box[\mathcal{A}]_w \Rightarrow \Box I}$$

$\mathcal{A}$ is the disjunction of all possible actions of the specification. We have to prove $I \land [\mathcal{A}]_w \Rightarrow I'$ by showing that all the formulae of this form: $I \land [\mathcal{A}_i]_w \Rightarrow I'_j$ are valid.

All possible actions are:

- **updatePendings**: 1 case;
- **send**: 3 cases;
- **insert**: 8 cases;
- **search**: 11 cases.

The proof of the invariant is then made of 460 cases, more or less straightforward to prove. Of course, we can not give the proof here.

5.2 Proof of the property

Then, we deduce from this invariant the property E (as shown in Figure 3) and prove that $I \Rightarrow E$.

From the invariant *searchResultType*, we know that when an object replies to a "search" with an element of Nat^* as parameter, it gives the field *info*. So, from *infoValue* and *searchResultType*, when an object replies to a "search(k)" request with an element of Nat^* as parameter, it gives the field *info* of the last request `insert(k, i)` it received:

$infoValue \land searchResultType \Rightarrow searchResult$

where: $searchResult \triangleq \forall id \in objectIDs$
$\land \text{ENABLED}(search_8(id))$
$\land result[id] \in Nat^*$
$\Rightarrow$
$\exists n \in 1..curReq[id]$
$\land requests[id][n].routine = \text{"insert"}$
$\land requests[id][n].args[1] = object[id].key$
$\land requests[id][n].args[2] = result[id]$
$\land \forall m \in n+1..curReq[id]$
$\quad \lor requests[id][m].routine \neq \text{"insert"}$
$\quad \lor requests[id][m].args[1] \neq object[id].key$

From the invariant predicate *rootRequests*, $\forall k \in Nat$, we know that the request list of the object *rootID*, restricted to the requests which have k as first argument, is a prefix of the same restriction of the list *extRequests*.

From invariants *descendantOfRoot* and *requestFlow*, we can show that for every object of class `Binary_Tree`, and every $k \in Nat$, the request list of this object restricted to k is a prefix of the request list of the object *rootID* (also restricted to k). The proof is done by induction on the path length between the object under consideration and the *rootID* object.

This means that for every object of class `Binary_Tree`, and every $k \in Nat$, the request list of this object restricted to k is a prefix of the request list *extRequests* (also restricted to k). Figure 4 specifies the corresponding TLA state predicate, and also gives a graphical overview of the flow of requests within the binary tree.

From *searchResult* and *restrRequests*, when an object replies to a "search(k)" request with an element of Nat^* as parameter, it gives the field *info* of the last request `insert(k, i)` of the request list *extRequests*. Then:

$searchResult \land restrRequests \Rightarrow E$

We have proven the two formulae:

$\Phi \Rightarrow \Box I$
$I \Rightarrow T$

We use the STL4 rule (see page 17 in [14]):

$$\frac{I \Rightarrow T}{\Box I \Rightarrow \Box E} \quad STL4$$

Then, we have:

$$\Phi \Rightarrow \Box E$$

which terminates the proof.

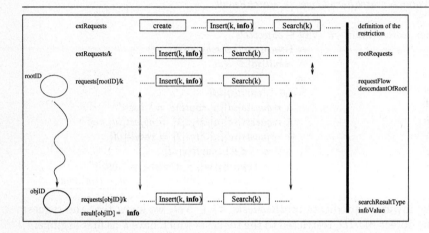

$$rootRequests \land descendantOfRoot \land requestFlow \Rightarrow restrRequests$$

where: $restrRequests \triangleq$
$\forall id \in objectIDs : \forall k \in Nat : object[id] \in btObjects$
$\Rightarrow$
$\forall n \in 1..Len(\ restr(requests[id], k)\ )$
$\quad \land restr(requests[id], k)[n].routine = restr(extRequests, k)[n].routine$
$\quad \land restr(requests[id], k)[n].args = restr(extRequests, k)[n].args$

Fig. 4. Proof scheme of $restrRequests$

6 Conclusion

The specification we present in this paper is rather complex due to low-level primitives needed in TLA+ for expressing the sequence of instructions and the management of continuations. The specification includes a description of synchronous as well as asynchronous message passing and synchronization operations. Object-oriented features of the Eiffel// model (polymorphism between passive and active objects, reuse, inheritance for transforming sequential to parallel behaviors) are captured and used in the TLA specification (typed modules, modeling of dynamic types — reduced in that case to synchronous and asynchronous, reification of requests). The whole specification is made of 3 modules with a total of 30 actions. With this specification, we were able to prove a behavioral equivalence between the sequential version and the parallel version of the `Binary_Tree` example with a standard method using invariants.

The TLA model also provides means to prove other properties such as liveness, freedom from deadlock, or characterizations of sequential (or parallel) executions. We aim at investigating such properties in order to develop more knowledge about object-oriented concurrency and more specifically the Eiffel// model.

Compared a π-calculus description of Eiffel// [3], the TLA specification maintains some structure while π-calculus requires addition of many channels for modeling purposes only; moreover a lot of static typing is lost, due to channel names being sent across channels.

Compared to a specification of Eiffel// using operational semantics [2], the TLA specification appears to be of a lower level (e.g. a program counter has to be explicitly managed in TLA), while the granularity seems comparable (due to continuations). On the other hand, since the specification is usually based on a large number of actions, proofs require a tremendous number of cases for invariants; a mechanization of the proof using TLP [10] should be helpful and would also facilitate further modifications of the specification, within a computer-aided setting (as in [11]).

It still remains to generalize the property that was proved in this paper on a specific example. Intuitively, it seems that we should be able to state a general property that expresses some equivalence between a sequential system and its parallel version when the graph of objects at execution is a tree. The current structuring of the specification (the Eiffel// module contains all the concurrent features of the model itself) should help in that process. Moreover, the fact that the current proof relies so much on invariants that reflect the tree nature of the objects topology should be crucial for the generalization.

Acknowledgment
The authors are grateful to Leslie Lamport and Andrew Wendelborn for helpful comments and discussions.

References

1. R. Amadio, I. Castellani, and D. Sangiorgi. On bisimulations for the asynchronous π-calculus. In *Proc. CONCUR '96*, volume 1119. LNCS, Springer-Verlag, 1996.
2. I. Attali, D. Caromel, S. Ehmety, and S. Lippi. Semantic-based visualization for parallel object-oriented programming. In *Proceedings of OOPSLA'96*, volume 31 of *ACM Sigplan Notices*, San Jose, CA, October 1996. ACM Press.
3. E. Bruneton. Implantation d'un traducteur eiffel// – π-calcul. Rapport de stage deuxième année, Ecole Polytechnique, juillet 1996.
4. *Concurrent Object-Oriented Programming*. Communications of the ACM, 36 (9), 1993. Special issue.

5. D. Caromel. Service, Asynchrony and Wait-by-necessity. *Journal of Object-Oriented Programming*, 2(4):12–22, November 1989.
6. D. Caromel. Concurrency and Reusability: From Sequential to Parallel. *Journal of Object-Oriented Programming*, 3(3):34–42, September 1990.
7. D. Caromel. Towards a Method of Object-Oriented Concurrent Programming. *Communications of the ACM*, 36(9):90–102, September 1993.
8. Denis Caromel, Fabrice Belloncle, and Yves Roudier. *The C++// Language*. MIT Press, 1996. ISBN 0-262-73118-5.
9. D. Doligez. *Conception, réalisation et certification d'un glaneur de cellules concurrent*. PhD thesis, Université Paris VII, 1995.
10. Urban Engberg. The tlp system. http://www.daimi.aau.dk/~urban/tlp/tlp.html.
11. G. Gonthier. Verifying the safety of a practical concurrent garbage collector. In *Proc. of the Eighth International Conference on Computer-Aided Verification*, ACM, New Brunswick, New Jersey, 1996.
12. S. Hodges and C. B. Jones. Non-interferance properties of a concurrent object-based language: Proofs based on an operational semantics. In *Object Orientation with Parallelism and Persistance*. Kluwer Academic Publishers, 1996. ISBN 0-7923-9770-3.
13. L. Lamport. Hybrid systems in tla$^+$. In *Hybrid Systems*, pages 77–102. Springer Verlag LNCS 736, 1993.
14. L. Lamport. The Temporal Logic of Actions. *ACM Transactions on Programming Languages and Systems*, 16(3):872–923, May 1994.
15. L. Lamport. The Module Structure of TLA+. Technical Report 1996-002, DEC SRC, September 1996.
16. L. Lamport. The operators of TLA+. Technical Report 1997-006a, DEC SRC, April 1997.
17. L. Lamport and S. Merz. Specifying and verifying fault-tolerant systems. In H. Langmaack, W. P de Roever, and J. Vytopil, editors, *Formal Techniques in real-time and fault-tolerant systems*, volume 863 of *Lecture Notes in Computer Science*, pages 41–76. Springer-Verlag, 1994.
18. X. Liu and D. Walker. Confluence of processes and systems of objects. In *Proc. of Theory and Practice of Software Development (TAPSOFT'95) 6th International Joint Conference CAAP/FASE*, 1995.
19. B. Meyer. *Object-Oriented Software Construction*. Prentice-Hall, 1988.
20. R. Milner, J. Parrow, and D. Walker. A Calculus of Mobile Processes, i and ii. *Journal of Information and Computation*, 100:1–77, 1992.
21. O. Nierstrasz. The next 700 concurrent object-oriented languages – reflections on the future of object-based concurrency. Object composition, Centre Universitaire d'Informatique, University of Geneva, June 1991.
22. O. Nierstrasz, P. Ciancarini, and A. Yonezawa, editors. *Rule-based Object Coordination*. LNCS 924. Springer-Verlag, 1995.
23. David Sagnol. π-calcul pour les langages à objets parallèles. Rapport de stage de DEA, Université de Nice - Sophia Antipolis, Juin 1995.
24. P. Wegner. Design issues for object-based concurrency. In *Proc. of the ECOOP '91 Workshop on Object-Based Concurrent Computing*, LNCS 612, 1992.
25. G. Wilson and P. Lu, editors. *Parallel Programing Using C++*. MIT Press, 1996. ISBN 0-262-73118-5.

Examples of Program Composition Illustrating the Use of Universal Properties

Michel Charpentier and K. Mani Chandy

California Institute of Technology
Computer Science Department
m/s 156–80, Pasadena, CA 91125
e-mail: {charpov, mani}@cs.caltech.edu

Abstract. This paper uses a theory of composition based on *existential* and *universal* properties. Universal properties are useful to describe components interactions through shared variables. However, some universal properties do not appear directly in components specifications and they must be constructed to prove the composed system. Coming up with such universal properties often requires creativity. The paper shows through two examples how this construction can be achieved. The principle used is first presented with a toy example and then applied to a more substantial problem.

1 Introduction

A goal of compositional systems development is to support the publication of software modules in a repository such as the Web, where each module is published with its specification, and where new modules can be created by composing existing modules. Hardware vendors publish parts lists with specifications, and other vendors compose these parts to create new parts. Personal computers are manufactured in this fashion. We establish properties of a composed system from the specifications of the components; we do not consider how the components are implemented provided they satisfy their specifications.

Systems can be developed in a compositional way whether the development is bottom-up or top-down or some combination of the two. In all cases, a goal is to specify each component so that the component can be used in a wide variety of environments.

We would like a specification of a software module to name only variables used in that module. We prefer not to specify one module using variables named in other modules with which this module may be composed. The reason for this preference is to allow the widest latitude for the environments of a module. Specifying variables in the environment can over-specify the environment.

A property is a predicate on systems. A specification is a desired property of a system. (Usually, a specification is a desired property which is a conjunction of desired properties.) A local property of a system is a property that names only variables of that system. We would like the specification of a system to be a local property of that system.

When we compose components to get larger systems we may find that, luckily for us, all the properties we desire for the composed system can be obtained in a straightforward way from the specifications of the components. In other cases, we may have to be creative in proving system specifications from their component specifications. This paper is an exploration of how we can prove system properties from local component properties.

This paper uses a theory of composition proposed in [5, 6]. This theory is based on *existential* and *universal* property types. A property type is an existential type when it holds in any system in which at least *one* component has the property. A property is a universal type when it holds in any system in which *all* components have the property. Consider a simple example. Imagine that we are putting pieces together in a jigsaw puzzle. An example of a universal property is "the component is entirely dark colored." If we put entirely dark-colored components together we get entirely dark-colored (larger) components. An example of an existential property is: "the component has a light-colored region." A component has a light-colored region if it has a subcomponent with a light-colored region.

Some properties are neither universal nor existential. The earlier work, however, proposes a theory of composition based on universal and existential properties. Conjunctions of these properties are adequate for specifying most concurrent systems. In particular, existential properties seem to play an important role in the specification of distributed systems [3].

Here, we consider the case of a shared memory system. In such systems, a compositional approach must provide means to describe the way components modify shared variables.

Components specifications do not describe their use of shared variables with existential properties. Specifications about one component must make assumptions on the way other components modify shared variables. One way is to use *universal* properties that specify how all components use shared variables.

However, a universal property of one component referring only to local variables and shared variables of that component and not to other components' local variables will generally not be satisfied by all other components, which modify shared variables according to their local variables. This is because each component has a property defined in terms of its local and shared variables, and so these properties are different for different components.

To cope with this difficulty, one approach is to build, from components specifications, a universal property satisfied by all components, which is then a system property. The paper presents an example of that step.

After presenting the programming model used, we first consider a toy example to highlight the difficulties related to universal properties and the way they can be solved. We then apply the same principles to a more important example: a priority mechanism for conflicting processes.

2 Programming Model

The programming model that we use to illustrate the theory is the model used in [6, 3] which is derived from UNITY [4]. A program consists of a set of typed variables, an *initially* predicate which is a predicate on program states, a finite set C of commands, and a subset D of C of commands subjected to a weak fairness constraint: every command in D must be executed infinitely often. The set C contains at least the command *skip* which leaves the state unchanged.

The program composition is defined to be the union of the sets of variables and the sets C and D of the components, and the conjunction of the *initially* predicates. Such a composition is not always possible. Especially, composition must respect variable locality (a variable declared *local* in a component should not be written by another component) and must provide at least one initial state (the conjunction of initial predicates must be logically consistent). We use $F * G$ to denote that programs F and G can be composed. Then, the system resulting from that composition is denoted by $F[\!] G$.

To specify programs and to reason about their correctness, we use the following properties:

$$\begin{aligned}
&\textbf{init } p &&\equiv \textit{initially} \Rightarrow p \\
&\textbf{transient } p &&\equiv \langle \exists c : c \in D : p \Rightarrow wp.c.\neg p \rangle \\
&p \textbf{ next } q &&\equiv \langle \forall c : c \in C : p \Rightarrow wp.c.q \rangle \\
&\textbf{stable } p &&\equiv p \textbf{ next } p \\
&\textbf{invariant } p &&\equiv (\textbf{init } p) \land (\textbf{stable } p)
\end{aligned}$$

We also use the liveness property *leads-to*, denoted by $\mapsto$ and defined by the following rules:

$$\begin{aligned}
&\textit{Transient} &&: [\, \textbf{transient } q \;\Rightarrow\; \textit{true} \mapsto \neg q \,] \\
&\textit{Implication} &&: [p \Rightarrow q] \;\Rightarrow\; [p \mapsto q] \\
&\textit{Disjunction} &&: \textit{For any set of predicates } S: \\
& && \quad [\, \langle \forall p : p \in \mathcal{S} : p \mapsto q \rangle \;\Rightarrow\; \langle \exists p : p \in \mathcal{S} : p \rangle \mapsto q \,] \\
&\textit{Transitivity} &&: [\, p \mapsto q \land q \mapsto r \;\Rightarrow\; p \mapsto r \,] \\
&\textit{PSP} &&: [\, p \mapsto q \land s \textbf{ next } t \;\Rightarrow\; (p \land s) \mapsto (q \land s) \lor (\neg s \land t) \,]
\end{aligned}$$

Note that we use properties with their *inductive* definition and not the definition based on *strongest invariant* [12]. In order to avoid some mix-up, we do not use the *substitution axiom* [4], although we could when dealing with global system properties.

Another element of the theory is the *guarantees* operator, from pairs of properties to properties. Given program properties X and Y, the property X **guarantees** Y is defined by:

$$X \textbf{ guarantees } Y \cdot F \;\triangleq\; \langle \forall G : F * G : (X \cdot F[\!] G) \Rightarrow (Y \cdot F[\!] G) \rangle$$

In this paper, we deal more specifically with universal properties and the *guarantees* operator is not used.

For *existentiality* and *universality*, we use the definition in [3], which is slightly different from the original definition in [6]. For any program property X:

$$\begin{aligned}
&X \text{ is existential} &&\triangleq \langle \forall F, G : F * G : X \cdot F \lor X \cdot G \Rightarrow X \cdot F[\!] G \rangle \\
&X \text{ is universal} &&\triangleq \langle \forall F, G : F * G : X \cdot F \land X \cdot G \Rightarrow X \cdot F[\!] G \rangle
\end{aligned}$$

Properties of type *init*, *transient* and *guarantees* are existential and properties of type *next*, *stable* and *invariant* are universal. The properties *leads-to* are, in general, neither existential nor universal. However, *leads-to* can appear on the right-hand side of a *guarantees* to obtain existential liveness properties other than *transient* [6, 3].

3 The Toy Example

3.1 Informal Description

We consider a set of components sharing a global counter. Each component also uses a local counter. We are interested in the relationship between the local counters and the global counter.

Components increase a global counter C by one each time they perform a certain action a. Therefore, the value of counter C always equals the total number of actions a that have been performed.

In the remainder of the section, we show what difficulties arise when applying a compositional approach to such a problem, and how to solve them. We specify the behavior previously described at the component level, and the correctness of the global system is derived in a compositional way.

3.2 Component Specification

Each component i has a counter c_i of actions a performed. Therefore, it must increase the global counter C each time it increases its counter c_i. The naive specification, corresponding to the case where the system is composed of *one* component i, is:

- **init** $C = c_i \cdot Component_i$
- **stable** $C = c_i \cdot Component_i$

If all components share this specification, we have two problems:

- The initial condition of the global system is $\langle \forall i :: C = c_i \rangle$, from which we cannot deduce the desired property that C equals the sum of the counters c_i;
- If c_i is local to component i and component j has to modify the shared variable C, the property **stable** $C = c_i$ is not satisfied by component j.

To obtain a compositional proof, we have to do a little more work. Initially, C must equal the sum of all the c_i, but expressing this sum is not local to the component. The only way to know the sum, at the component level, is that all c_i are zero (so that the sum does not depend on the number of components[1]). So, the component can have the following local *init* predicate:

$$\textbf{init } c_i = 0 \wedge C = 0 \cdot Component_i$$

If all other components have the same condition, the initial state will satisfy the condition that C equals the sum of the c_i.

Now, we need the property that component i will always increase C and c_i by the same value. Formally:

$$\forall k, N :: c_i = k \wedge C = N \textbf{ next } \langle \exists d : d \geqslant 0 : c_i = k+d \wedge C = N+d \rangle \cdot Component_i$$

---
[1] We could have **init** $C = c_{i_0}$ for the component i_0 and **init** $c_i = 0$ for the others, but this would introduce a dissymmetry.

This is equivalent to:

$$\forall k, N :: C = c_i + N - k \textbf{ next } C = c_i + N - k \cdot Component_i$$

and since k and N are universally quantified, this is equivalent to:

$$\forall k :: \textbf{stable } C = c_i + k \cdot Component_i$$

The last thing we must specify to obtain a compositional specification is what variables are local and what variables are shared. This is achieved through a *local* declaration that allows to (syntactically) check what compositions are valid:

$$\textbf{local } c_i \cdot Component_i$$

Only variables not declared *local* can be written by other components (here, the only non local variable is C). From this *local* declarations, we deduce logical properties in a generic way:

For all variables v, other than c_i and C, $\forall k ::$ **stable** $v = k \cdot Component_i$

Finally, at a logical level, the specification of $Component_i$ becomes:

$$\textbf{init } (c_i = 0 \wedge C = 0) \tag{1}$$
$$\forall k :: \textbf{stable } C = c_i + k \tag{2}$$
$$\textit{For all variables } v, \textit{ other than } c_i \textit{ and } C, \forall k :: \textbf{stable } v = k \tag{3}$$

The set of universal properties are still not shared by other components, but we show, in the next section, how a shared universal property can be deduced from them.

3.3 Correctness Proof

First, from *init* and *local* properties, we observe that:

$$\langle \forall i, j : i \neq j : Component_i * Component_j \rangle$$

Therefore, We can consider a system composed of N components, each component satisfying the previous specification:

$$System = \langle \| i : 0 \leqslant i < N : Component_i \rangle$$

The goal here is to prove global system correctness from the component specifications. This desired property is:

$$\textbf{invariant } C = \sum_{i=0}^{N-1} c_i \cdot System$$

Proof.

$\quad$ {Component specifications, rewriting (2) and (3)}
$\quad\quad$ *For all* i, **init** $(c_i = 0 \land C = 0) \cdot Component_i$
$\quad\quad \land$ *For all* $i, \forall k_1, k_2, \cdots, k_n ::$ **stable** $C = c_i + \sum_{j \neq i} k_j \cdot Component_i$
$\quad\quad \land$ *For all* $i, \forall k_1, k_2, \cdots, k_n ::$ **stable** $\langle \forall j : j \neq i : c_j = k_j \rangle \cdot Component_i$
$\Rightarrow$ {Conjunction of *stable* properties, removing unused dummies}
$\quad\quad$ *For all* i, **init** $(c_i = 0 \land C = 0) \cdot Component_i$
$\quad\quad \land$ *For all* i, **stable** $C = \sum_j c_j \cdot Component_i$
$\Rightarrow$ {*init* properties are existential, *stable* properties are universal}
$\quad\quad$ **init** $\langle \forall i :: (c_i = 0 \land C = 0) \rangle \cdot System$
$\quad\quad \land$ **stable** $C = \sum_j c_j \cdot System$
$\Rightarrow$ {Predicate calculus}
$\quad\quad$ **init** $C = \sum_j c_j \cdot System$
$\quad\quad \land$ **stable** $C = \sum_j c_j \cdot System$
$\Rightarrow$ {Definition of *invariant*}
$\quad\quad$ **invariant** $C = \sum_j c_j \cdot System$

$\hfill \square$

3.4 Lessons from the Toy Example

The proposal of local properties (1) and (2) of $Component_i$ was obtained from an analysis of the kinds of systems in which we expected to embed the component. In this sense, we took a top-down approach to get a local component specification from an anticipated system specification. However, we can now use our local component specification in a variety of systems including those that we have not anticipated.

4 The Priority Mechanism

4.1 Description

We suppose a set of perpetually conflicting components: Each component always wants to perform an action that requires it to have higher priority than all its neighbors. These conflicts are solved by a *priority mechanism*. Such a mechanism should:

- never give priority at the same time to two conflicting components (8);
- give priority to each component in turn (9).

We uses a principle presented in [4]. We give an orientation to the graph of conflicts so that it always remains acyclic, and we use this graph as a priority graph.
$\quad$ Then, a component should:

- wait until it has priority over its neighbors (4);
- yield priority to its neighbors in finite time after receiving priority (5);
- not introduce cycles in the graph (6).

A way not to introduce cycles is that an active node (with a higher priority than its neighbors in the graph), when changing priorities, always gets a lower priority than *all* its neighbors.

4.2 Component Specification

We call $\mathcal{P}$ the (nonoriented) finite graph of neighborhood. Unless explicitly specified, a graph property *prop* is to be understood as $\mathcal{P}.prop$. The graph $\mathcal{P}$ is described by variables $N[i]$:

$$N[i] \triangleq \textit{set of the neighbors of Component}_i$$

We require that $\langle \forall i :: i \notin N[i] \rangle$ (no node is conflicting with itself). We assume $\langle \forall i, j :: i \in N[j] \equiv j \in N[i] \rangle$ is invariant in the system (implementation of variables $N[i]$).

The graph orientation is defined by the arrow $\rightarrow$. The notation $(i \rightarrow j)$ means that component i has priority over component j. This is a boolean value. It can be modified both by i and j and by no other node. Any change must respect the (implementation) invariant: $\langle \forall i, j : j \in N[i] : (i \rightarrow j) \equiv \neg(j \rightarrow i) \rangle$.

The function $Priority.i$ is used to represent the priority of a node i over all its neighbors:

$$Priority.i \triangleq \langle \forall j : j \in N[i] : (i \rightarrow j) \rangle$$

The three properties of component i become:

$\forall b, j :: j \in N[i] \land (i \rightarrow j) = b \land \neg Priority.i \textbf{ next } (i \rightarrow j) = b \cdot Component_i$ (4)

transient $Priority.i \cdot Component_i$ (5)

$Priority.i \textbf{ next } Priority.i \lor \langle \forall j : j \in N[i] : (j \rightarrow i) \rangle \cdot Component_i$ (6)

As previously, we add a locality constraint: A component cannot change edges other than its incoming and outcoming edges:

$\forall b, j, j' :: j \neq i \land j' \neq i \land (j \rightarrow j') = b \textbf{ next } (j \rightarrow j') = b \cdot Component_i$ (7)

4.3 System Specification

Here, we express formally the system specification previously informally stated:

– safety:

invariant $\langle \forall i :: Priority.i \Rightarrow \langle \forall j : j \in N[i] : \neg Priority.j \rangle \rangle \cdot System$ (8)

– liveness:

$$\forall i :: true \mapsto Priority.i \cdot System \quad (9)$$

The proof of safety is trivial. To prove the liveness part, we use the fact that the graph always remains acyclic, and therefore that there is always a node which has the priority. To achieve that, we have to build a global universal property, satisfied by all components, from which we can deduce the graph acyclicity. It corresponds to the step presented in the toy example to obtain the property **invariant** $C = \sum_i c_i$. However, here, the property is more tricky (see property (13)).

4.4 Notations

In order to express this acyclicity, we define the functions[2] $R.i$ and $A.i$:

$$R.i = \{j : j \in N[i] : (i \to j)\}$$
$$A.i = \{j : j \in N[i] : (j \to i)\}$$

and a kind of (nonreflexive) transitive closure $R^*.i$ and $A^*.i$:

$$R^1.i = R.i \qquad \forall n,\ R^{n+1}.i = R^n.i \cup \bigcup_{j \in R^n.i} R.j \qquad R^*.i = \bigcup_{n>0} R^n.i$$

$R^*.i$ is the set of nodes reachable from node i following the graph's edges. $A^*.i$ is defined in the same way and is the set of nodes from which the node i is reachable. We use the following property for all i and j:

$$[i \in R^*.j \equiv j \in A^*.i] \tag{10}$$

Then the graph acyclicity is defined by:

$$Acyclicity \triangleq \langle \forall i :: i \notin R^*.i \rangle$$
$$\equiv \langle \forall i :: i \notin A^*.i \rangle$$

We also use the equivalent definition of $Priority.i$:

$$Priority.i \equiv A^*.i = \emptyset \tag{11}$$

4.5 Construction of a Universal Property

Definition 1. *Let G and G' be two graphs differing only by edge orientation. We say that G' is derived from G through node i_0 if and only if all the edges of i_0 are outcoming in G and incoming in G', all other edges being equal in G and G'.*

$$G \stackrel{i_0}{\leadsto} G' \equiv$$
$$\langle \forall k, k' : k, k' \neq i_0 : G.(k \to k') = G'.(k \to k') \rangle \wedge G.A^*.i_0 = \emptyset \wedge G'.R^*.i_0 = \emptyset$$

Lemma 2. *If a graph G' is derived from a graph G through node i_0, then the reachability of nodes in G' cannot be greater than the union of what they are in G and the singleton $\{i_0\}$.*

$$[G \stackrel{i_0}{\leadsto} G' \Rightarrow \langle \forall i :: G'.R^*.i \subset G.R^*.i \cup \{i_0\} \rangle]$$

Proof. From graph theory. □

Property 12. *The only changes a component i can make in the priority graph are governed by the relation $\stackrel{i}{\leadsto}$.*

$$\forall G :: \mathcal{P} = G \text{ next } \mathcal{P} = G \vee G \stackrel{i}{\leadsto} \mathcal{P} \cdot Component_i \tag{12}$$

---

[2] Defining $R.i$ and $A.i$ as functions instead of relations allows the use of set operators to simplify the writing of some formulas.

Proof. Trivial from the specifications (4), (6) and (7) of component i. □

Property 13 (Universal system property).

$$\forall G :: \mathcal{P} = G \text{ \bf next } \mathcal{P} = G \vee \langle \exists i_0 :: G \overset{i_0}{\leadsto} \mathcal{P} \rangle \cdot System \qquad (13)$$

Proof. From (12), the property is satisfied by every component. Since *next* is universal, this is a system property. □

4.6 Proof of the Liveness Property (9)

Property 14. *A component cannot enter any reachability set before it has priority.*

$$\forall i,j :: A^*.i \neq \emptyset \wedge i \notin R^*.j \text{ \bf next } i \notin R^*.j \cdot System \qquad (14)$$

Proof. From lemma 2 and (13):

$$\forall G, r, i, j ::$$
$$\mathcal{P} = G \wedge R^*.j = r \wedge A^*.i \neq \emptyset \wedge i \notin r$$
$$\textbf{next}$$
$$\mathcal{P} = G \vee \langle \exists i_0 :: G \overset{i_0}{\leadsto} \mathcal{P} \wedge R^*.j \subset r \cup \{i_0\} \wedge i \notin r \rangle \cdot System$$

If $\mathcal{P} = G$, then $R^*.j = r$ and then $i \notin R^*.j$. If not, from $G \overset{i_0}{\leadsto} \mathcal{P}$, we know that $G.A^*.i_0 = \emptyset$. Therefore, $i_0 \neq i$, and since $i \notin r$, we deduce that $i \notin R^*.i$. Using disjunction over G and r, we obtain (14). □

Property 15. *A component with priority will keep its reachability set or its above set empty.*

$$\forall i :: A^*.i = \emptyset \text{ \bf next } A^*.i = \emptyset \vee R^*.i = \emptyset \cdot System \qquad (15)$$

Proof. If a component has priority, its neighbors cannot have priority and, thanks to (4) and (7), cannot change any edge. Therefore, its neighbors cannot set its own priority to *false*. That means that (6) is satisfied by all components. Since it is universal, it is a system property:

$$\forall i :: Priority.i \text{ \bf next } Priority.i \vee \langle \forall j : j \in N[i] :: (j \rightarrow i) \rangle \cdot System$$

Then, just rewriting using $R^*.i$ and $A^*.i$, we obtain exactly (15). □

Property 16. *If it is acyclic initially, the priority graph remains acyclic.*

$$Acyclicity \text{ \bf next } Acyclicity \cdot System \qquad (16)$$

Proof. From (14), choosing $i = j$, we have:

$$\forall i :: A^*.i \neq \emptyset \wedge i \notin R^*.i \text{ \bf next } i \notin R^*.i \cdot System$$

From (15), using $i \in R^*.i \equiv i \in A^*.i$:

$$\forall i :: A^*.i = \emptyset \text{ \bf next } i \notin R^*.i \cdot System$$

From the disjunction of the two above, strengthening the left-hand side of the *next*:

$$\forall i :: i \notin R^*.i \textbf{ next } i \notin R^*.i \cdot System$$

which, from the definition of *Acyclicity*, is exactly (16). □

Lemma 3. *There is at least one maximal node in any nonempty above set of a finite acyclic graph.*

$$[Acyclicity \Rightarrow \langle \forall i : A^*.i \neq \emptyset : \langle \exists j : j \in A^*.i : A^*.j = \emptyset \rangle \rangle]$$

Proof. From graph theory. □

Property 17. *Any nonpriority component has always a priority component above it.*

$$\forall i :: \textbf{invariant } Acyclicity \wedge (A^*.i \neq \emptyset \Rightarrow \langle \exists j : j \in A^*.i : A^*.j = \emptyset \rangle) \cdot System \tag{17}$$

Proof. From lemma 3 and (16). □

Property 18. *Any component with priority eventually escapes every above set.*

$$\forall i, j :: A^*.i = \emptyset \mapsto i \notin A^*.j \cdot System \tag{18}$$

Proof. From the existential characteristics of (5), we have:

$$\forall i :: \textbf{transient } Priority.i \cdot System$$

that rewrites:

$$\forall i :: A^*.i = \emptyset \mapsto A^*.i \neq \emptyset \cdot System$$

From (15) and the above, using PSP:

$$\forall i :: A^*.i = \emptyset \mapsto R^*.i = \emptyset \cdot System$$

Since $i \in A^*.j \equiv j \in R^*.i$:

$$\forall i :: A^*.i = \emptyset \mapsto \langle \forall j :: i \notin A^*.j \rangle \cdot System$$

which is stronger than the required property (18). □

Finally, we prove the liveness correctness (9), which is equivalent to the following property:

Property 19.

$$\forall i :: true \mapsto A^*.i = \emptyset \cdot System \tag{19}$$

Property (14) is equivalent to:

$$\forall i,j :: A^*.i \neq \emptyset \wedge j \notin A^*.i \text{ next } j \notin A^*.i \cdot System$$

which in turn is equivalent to:

$$\forall a,i :: A^*.i = a \neq \emptyset \text{ next } A^*.i \subset a \cdot System$$

In the same way, from (18) we have:

$$\forall a,i,j :: A^*.i = a \wedge j \in A^*.i \wedge A^*.j = \emptyset \mapsto j \notin a \cdot System$$

Applying PSP to the two above, we obtain:

$$\forall a,i,j :: A^*.i = a \wedge j \in A^*.i \wedge A^*.j = \emptyset \mapsto A^*.i \subsetneq a \cdot System$$

Using *leads-to* disjunction over j, it becomes:

$$\forall a,i :: A^*.i = a \wedge \langle \exists j : j \in A^*.i : A^*.j = \emptyset \rangle \mapsto A^*.i \subsetneq a \cdot System$$

From the invariant (17), $A^*.i \neq \emptyset \Rightarrow \langle \exists j : j \in A^*.i : A^*.j = \emptyset \rangle$, and therefore, the previous formula implies:

$$\forall a,i :: A^*.i = a \neq \emptyset \mapsto A^*.i \subsetneq a \cdot System$$

Through induction on the cardinality of $A^*.i$, this gives the liveness correctness (9).

5 Conclusions

This paper explores a methodology for compositional development of systems. The methodology attempts to work with two types of system properties: universal and existential. A goal of the methodology is to specify components using only local properties. In some cases, system properties can be obtained in a straightforward fashion from local component properties. In other cases, creativity is required to derive system properties from local properties.

This case study exposes the use of three kinds of compositional properties:

- an existential property (5);
- a universal property, shared by all components (6);
- a universal property, not shared by other components (4).

The first two types of property are easy to use: they simply hold in the global system when components are gathered. The third one, however, requires creativity and additional work to become useful in the composition step. The case studies help us in exploring compositional steps that appear to be almost mechanical in contrast to steps that require some ingenuity.

The specification of the conflict resolution solution included the property that the graph of the priority relation is an acyclic graph. We could have specified the components in terms of such acyclic graphs, but this would have resulted in component specifications being nonlocal. The specification of one component would include properties about the priority relationship between completely different components. If we specify components using only local properties, then we have to bridge the gap between local properties and the global system property about acyclic graphs. We found no mechanical way of bridging this gap.

The principle used to build a universal shared property is to weaken the component properties so that all components can share the weakened property. This transformation requires some knowledge on how shared variables are modified by other components. This knowledge is provided by other components (universal) specification.

Note that this step leads to *weaken* a property, and is not exactly a *refinement* step in the strict sense of the word [2, 11]. Such transformations, introducing some nondeterminism, seem to appear frequently when dealing with distributed programs [9, 7].

Universal properties seem to be closely related to global safety. In the priority example, the safety correctness is trivial, but we need a strong safety property to prove the liveness part. In [3], a resource allocator example is derived. In that example, all the safety points are local to components and, actually, the example only makes use of existential properties.

Another point worth being noticed is that, in both the toy example and the priority mechanism example, we only make use of *statement* properties (*transient* or *inductive* safety properties). Properties like "always true" are avoided. The theory provides a *guarantees* operator to deal with non *transient* existential properties. However, for universal properties, nothing more than inductiveness is used.

We are currently investigating such questions, both from the theoretical point of view and by applying the theory of composition to a collection of examples. In particular, we are working on developing a theory based on the traditional rely–guarantee approach [8, 13] and relating it to other theories of composition [10, 1].

The vision that drives us is that of modularity at the level used by manufacturers of personal computers, cars and airplanes. Such systems are complex with large numbers of parts. We should be able to compose certain kinds of software modules in the same way.

Just as there have been many generations of airplanes we now are moving towards many generations of user interfaces, and the compositional technologies that the community has learned in building airplanes over many generations are now being used to build user interfaces and other software systems. The trend towards plug-and-play, object systems, and component systems such as Java Beans and Microsoft's DCOM are examples of steps in this direction.

Formal theories that support compositional development of concurrent systems have been proposed. This work is an exploration of a theory based on specifications using only local properties and two types of properties: universal and existential. We believe that this theory is worthy of further investigation because of the extreme simplicity of its foundation and the successful case studies of its use.

References

1. Martffi Abadi and Leslie Lamport. Conjoining specifications. *ACM Transactions on Programming Languages and Systems*, 17(3):507–534, May 1995.
2. R.J.R. Back. Refinement of parallel and reactive programs. Technical report, Marktoberdorf Summer School on Programming Logics, 1992.
3. K. Mani Chandy and Michel Charpentier. An experiment in program composition and proof. Submitted to Formal Methods in System Design, September 1998.
4. K. Mani Chandy and Jayadev Misra. *Parallel Program Design: A Foundation*. Addison-Wesley, 1988.
5. K. Mani Chandy and Beverly A. Sanders. Predicate transformers for reasoning about concurrent computation. *Science of Computer Programming*, 24:129–148, 1995.
6. K. Mani Chandy and Beverly A. Sanders. Reasoning about program composition. Technical Report 96-035, University of Florida, Department of Computer and Information Science and Engineering, 1996.
7. Michel Charpentier. A UNITY mapping operator for distributed programs. In J. Fitzgerald, C.B. Jones, and P. Lucas, editors, *Fourth International Symposium of Formal Methods Europe (FME'97)*, volume 1313 of *Lecture Notes in Computer Science*, pages 665–684. Springer-Verlag, September 1997.
8. Pierre Collette. *Design of Compositional Proof Systems Based on Assumption-Commitment Specifications. Application to* UNITY. Doctoral thesis, Faculté des Sciences Appliquées, Université Catholique de Louvain, June 1994.
9. Mamoun Filali, Philippe Mauran, and Gérard Padiou. Raffiner pour répartir. In *Actes des quatriemes Rencontres francophones du Parallélisme (RenPar'4)*, Villeneuve D'Ascq, France, 1992.
10. S. S. Lam and A. U. Shankar. A theory of interfaces and modules 1: Composition theorem. *IEEE Transactions on Software Engineering*, 20(1):55–71, January 1994.
11. C. Morgan, P. Gardiner, K. Robinson, and T. Vickers. *On the Refinement Calculus*. FACIT. Springer-Verlag, 1994.
12. Beverly A. Sanders. Eliminating the substitution axiom from UNITY logic. *Formal Aspects of Computing*, 3(2):189–205, April–June 1991.
13. Rob T. Udink. *Program Refinement in* UNITY*-like Environments*. PhD thesis, Utrecht University, September 1995.

A Formal Framework for Specifying and Verifying Time Warp Optimizations

Victoria Chernyakhovsky, Peter Frey, Radharamanan Radhakrishnan,
Philip A. Wilsey, Perry Alexander and Harold W. Carter

(vchernya,pfrey,ramanan,paw,alex,hcarter)@ececs.uc.edu
University of Cincinnati, PO Box 210030, Cincinnati, OH 45221-0030

Abstract. Parallel and distributed systems are representative of large and complex systems that require the application of formal methods. These systems are often unreliable because implementors design and develop these systems without a complete understanding of the problem domain; in addition, the nondeterministic nature of certain parallel and distributed systems make system validation difficult if not impossible. To address this issue, the application of formal specification and verification to a class of parallel and distributed software systems is presented in this paper. Specifically, the Prototype Verification System (PVS) is applied to the specification and verification of the Time Warp protocol, a distributed optimistic discrete event simulation algorithm. The paper discusses how the specification of the Time Warp protocol can be mechanized within a general-purpose higher-order theorem proving framework like PVS. In addition, the paper presents the extensibility of the specification to address and verify different aspects and optimizations of the basic Time Warp protocol. As an illustrative example, our experiences in specifying and verifying the infrequent state saving optimization to the basic Time Warp protocol is reported in the paper.

1 Introduction

Discrete-event simulation (DES) is a valuable design tool for several problems in engineering, computer science, economics and military applications. As many of these applications require enormous resources, parallel discrete-event simulation (PDES) is of considerable interest. Several parallel discrete-event simulation algorithms with different attributes have been reported in the literature [7]. This is due to the availability of several simulation paradigms with an assortment of optimizations. Parallel discrete-event simulation paradigms are generally classified as being *conservative* or *optimistic*. Conservative protocols [1] strictly avoid causality errors, while optimistic protocols, such as Time Warp [8] allow causality errors to occur, but implement some recovery mechanism. The Time Warp protocol is based on the *virtual time* paradigm first introduced by Jefferson [8]. Since then, the PDES research community has been actively pursuing the development of algorithmic as well as data structure based optimizations. In addition, as the popularity of the optimistic discrete-event simulation paradigm increases, novel simulation protocols based on the original paradigm are constantly being developed.

However, the complexity and the nondeterministic execution behavior of the Time Warp protocol make it difficult to develop modifications or extensions to the basic protocol. Years of development have shown that even a simple implementation of the basic Time Warp protocol is often error prone due to difficulties in understanding the protocol. Formal specification and verification provides a solution to these problems. Formal specification eliminates ambiguities concerning the algorithm description and clearly identifies the required data structures. This results in better software designs and decreases the introduction of errors in the coding phase. Since software development is often conducted in parallel, a precise definition of the individual modules is required to avoid detection of errors late in the development phase. Formal specification provides these precise interface definitions and allows the designer to state and prove properties of individual program modules.

Despite these advantages, formal specification & verification is rarely applied on large, complex systems and is often used on smaller subsystems (modules) only. The development of module specifications limits their re-usability and a verification of the complete system is not possible. The extensible formal framework for optimistic simulation protocols [4] extendend in this paper, takes a different approach. The specification is modular and encloses the complete Time Warp simulation protocol. Modifications or extensions to the basic protocol can now be included into the framework and only the invariants of the simulation protocol will have to be re-proven. This eliminates error prone assumptions in the design of a subsystem. To illustrate the extensibility of the formal framework, the formal specification and verification of the infrequent state saving [3, 14] optimization to the basic Time Warp protocol was attempted and is reported in this paper. In addition, the framework also provides support for new optimistic simulation algorithm designs. In this case, their correctness can be verified using bisimulation [11]. The complete PVS specification of the Time Warp protocol and the specification of the infrequent state saving optimization to the Time Warp protocol are available off the WWW at http://www.ececs.uc.edu/~pfrey/pvs.

The remainder of the paper is organized as follows. Section 2 describes informally the basic Time Warp protocol and its associated components. A brief overview of the formal specification framework in presented in Section 3. In addition, the mapping of the individual components of the Time Warp protocol to their formal representation is detailed. The changes that were required to formally specify the infrequent state saving optimization is presented in Section 4. Section 5 then briefly describes the verification of the infrequent state saving specification. The conclusions, presented in Section 6, reevaluates the formal representation and the applicability of the formal framework in research and development.

2 Background

In a Time Warp synchronized discrete event simulation, *virtual time* [8] is used to model the passage of the time in the simulation. Changes in the state of the simulation occur as *events* are processed at specific virtual times. In turn, events may schedule other events at future virtual times. The virtual time defines a total order on the events of the system. The simulation state (and time) advances in discrete steps as each event is processed. The simulation is executed via several simulator processes, called simulation objects. Each simulation object is constructed from a physical process (PP), its current state, and three history queues. Figure 1 illustrates the structure of a simulation object. The input and the output queues store incoming and outgoing events respectively. The state queue stores the state history of the simulation object. Each simulation object maintains a clock that records its Local Virtual Time (LVT) and stores the LVT in its current state. Simulation objects interact with each other by exchanging time-stamped event messages. One departure from Jefferson's original definition [8] of Time Warp is that simulation objects are placed into groups called "logical processes" (LPs).

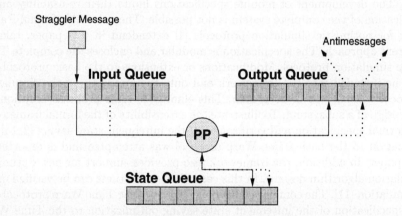

Fig. 1. A Simulation Object in a Time Warp Simulation

In addition, simulation objects in a Time Warp simulation execute their local simulation autonomously, without explicit synchronization. A causality error arises if a simulation object receives a message with a time-stamp earlier than its LVT value (a *straggler* message). In order to allow recovery, the state of the simulation object and the output events generated are saved in history queues as each event is processed. When a straggler message is detected, the erroneous computation must be undone — a *rollback* occurs. The rollback process consists of the following steps: (i) the state of the simulation object needs to be restored to a state prior to the straggler message time-stamp, and (ii) erroneously sent output messages need to be canceled (by sending *anti-messages* to nullify the original message).

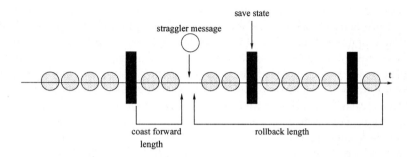

Fig. 2. Rollback and Coasting Forward ($\chi = 4$)

In the original Time Warp algorithm, state is saved after every event execution. While this entails a significant overhead in terms of memory consumed, it is necessary to protect against erroneous optimistic computation. Each simulation object must periodically save its local state such that, in the event of a causality error, a rollback to a correct state is possible. Time Warp objects with large states require considerable memory space as well as CPU cycles for state saving. In general, states are saved after every event execution. However, the check-pointing cost can be reduced by saving the state infrequently. In the simple case, a Time Warp simulator checkpoints every χ events; this scheme is called *Periodic* or *Infrequent Check-pointing* [3, 14]. Figure 2 illustrates a process rolling back upon the receipt of a straggler message (white circle). The shaded circles denote events and the filled boxes denote state saving points. Since check-pointing does not occur after every processed event, a rollback will have to re-execute any intermediate events. This additional processing is called the *coast forward* phase. Thus, an optimal checkpoint interval must balance the cost of check-pointing vs. the cost of the coast forward phase. Excessively large check-pointing periods will have long coast forward phases but low check-pointing cost. The opposite is true for excessively small check-pointing periods. Using periodic state saving, the arrival of a straggler message may require the system to rollback to an earlier state than necessary and coast forward [7]. While coasting forward, no messages are sent out to the other processes in the system (the previously sent messages are correct). The difficulty of periodic state saving is determining an appropriate fixed frequency for check-pointing.

The global progress time of the simulation or Global Virtual Time (GVT), is defined as the lowest of the local virtual time (LVT) values of the simulation objects [7, 10] and the times of all messages in transit. Periodic GVT calculation is necessary to reclaim memory space used by history queues; history items with a time-stamp lower than GVT are no longer needed, and are deleted to make room for new history items.

3 The Formal Specification Framework

The specification is constructed using the Prototype Verification System (PVS). PVS provides a formal specification language, a powerful theorem prover, and a set of pre-defined basic types and axioms in the prelude file. The PVS specification language is based on higher order logic with built-in types such as booleans, integers, reals and uninterpreted (user-defined) types. To represent complex types, PVS supports functions, sets, tuples, records, enumerations, and abstract data types. In order to preserve consistency of the specification, a "conservative extension" style is provided by PVS and was extensively used in this specification. This style combines signature definition and axiomatic description in a single statement preventing the introduction of inconsistencies when new axioms are introduced. The description of the specification uses `Typewriter` font to identify names of axioms and theorems and Sansf Serif for theory names.

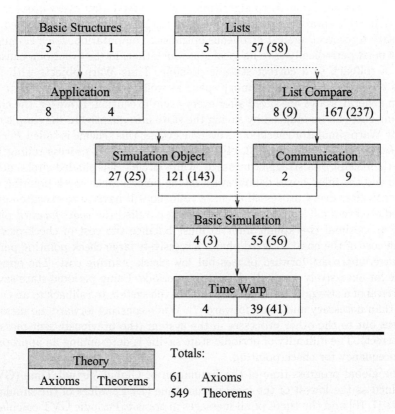

Fig. 3. Hierarchy of Theories

The specification of the infrequent state saving optimization was performed as an extension to an already existing formal specification of the Time Warp protocol [4, 6]. This formal framework follows the original definition of the Time Warp protocol by Jefferson [8] very closely. The specification provides a structured view of the protocol, identifies invariants and other Time Warp requirements (pre and post conditions). The framework is complete as it also provides a complete proof of the protocol. Figure 3 illustrates the hierarchical structure of the theories provided by the formal framework. Some of these theories were modified to include a generic proof of the infrequent state saving algorithm. The figures in the brackets represent the change in the number of the axioms or theorems as a result of the modification to incorporate the infrequent state saving specification. In addition, due to space constraints, only the modifications to the specification are detailed in this paper. For a more in depth description of the framework, please refer to [4] and [6].

Theories Lists and List Compare are parameterized theories and provide the major Time Warp list functionalities. The parameterized theories enable the list specification to be used as state and event lists (input, output) in the Time Warp structure and as message buffers in the simple communication model. The theory Basic Structures provides the generic definition of an Event and State which is annotated by the actual elements of the Application theory. The Application theory provides an abstract definition of simulation time (Time), process state (StateVariables), message content (MsgBody), and process identifiers (ProcessId). Due to this representation, the specification is applicable to any simulation model satisfying the general requirements specified in the Application theory.

The generic list structure and the representation of the application process is combined in the Simulation Object theory. This theory provides the specification of a simulation object (SimObjState) from the elementary Time Warp structure (see Figure 1). At any given virtual time, a state of a simulation object is described by the state of the input queue (inputList), the output queue (outputList), the state queue (stateList), process behavior (process, processOutput), the current state (currentProcessState), and the process identifier (id). In addition, the specification includes a local communication buffer (ether), which is different from the original definition of the Time Warp algorithm, but was included to model the distributed environment. The type definitions of inputList, outputList, stateList, process, processOutput, currentProcessState and id are all defined separately (Basic Structures and Application theories) [6] and are imported into the Simulation Object theory.

The functionality of the simulation object is described by two axioms execute and insertEventsCheck in the Simulation Object theory. These theorems describe the core of the Time Warp algorithm. As Time Warp is based on asynchronous communication, the execution is purely event-driven (*i.e.* based on the arrival of events). This fact is incorporated in the specification as follows: the execution axiom requires events to be processed (pending events). When events are pending for execution, the execute axiom describes the transition of the sim-

```
SimObjState : TYPE
  = [# inputList            : InputList,
       outputList           : OutputList,
       stateList            : StateList,
       process              : Process,
       currentProcessState  : State,
       processOutput        : (ValidProcess?),
       id                   : ProcessId,
       ether                : list[(ValidEvent?)]
  #]
```

ulation object's state during execution. The specification unambiguously identifies the change of state in the outputList, stateList, currentProcessState, and in the ether of the simulation object. Events generated during execution are inserted into the outputList and the ether. The currentProcessState contains the newly generated state and the same state is inserted into the state history (stateList). As each state is inserted into the simulation object history, the frequent state saving paradigm is modeled in the original specification.

The insertEventsCheck axiom defines the insertion of newly arriving events into the simulation object structure. Insertion is defined on a list of arriving events and a snapshot of the simulation object. Newly arriving events are either simply inserted into the input list or in the case of a causality error, a rollback is initiated. A causality error is determined by comparing the smallest incoming event and the LVT of the current process state. The rollback changes the inputList, outputList, stateList, local ether and the currentProcessState as required by the Time Warp protocol. If the Time Warp algorithm uses a frequent state saving protocol, the restored state is the state with a time stamp just smaller than the time stamp of the event responsible for the rollback. As all computed states are stored in the history, forward execution can resume immediately after the rollback is performed. The Basic Simulation and Time Warp theories combine the simple distributed communication model and the simulation object model to complete the basic formal framework. The top most axiom describes the complete Time Warp system and its advancement through wall clock time. As this axiom incorporates the complete Time Warp system, a correctness proof was performed to verify that all proof requirements are satisfied. As the PVS system has a rigorous type checker, all invariants of the algorithm were automatically identified and were proven correct in the original framework [6]. [4] contains a more in depth description of all the proof obligations and the proofs that were attempted. The following section describes how the framework was adapted to incorporate a generic infrequent state saving protocol. Parts of the aforementioned theories are presented in the following section but due to the lack of space, we could not include all the theories discussed. The specification of the protocol is followed by a description of the additional theorems proven during the verification of the protocol.

4 Specification of the Infrequent State Saving Algorithm

The specification of the infrequent state saving algorithm involved modifying both the `execute` and the `insertEventsCheck` axioms. In the frequent state saving strategy, every execution cycle terminates with the storage of the current state in the `stateList`. An infrequent state saving strategy modifies this behavior by storing newly generated states infrequently. This infrequent storage of the simulation object's state is modeled by using the boolean predicate `CheckPointInterval?`. A state is saved only when this predicate evaluates to true. This is an abstract representation of an actual infrequent state saving algorithm as it does not represent any specific interval calculation scheme. This was done purposely because there are several algorithms for the checkpoint interval calculation [3] and a generic specification was required. The interval calculation (static or dynamic) is crucial for the actual performance of the simulator. However, it does not affect the correctness of the infrequent state saving paradigm.

```
% CheckPointInterval? is a predicate which is true when
% infrequent state saving is performed
CheckPointInterval? : bool

% EXECUTION OF THE SIMULATION PROCESS
execute(currentState:(StateAndNewInput?)):(AtLeastAState?) =
  currentState with [
    % no change on the inputList
    ...
    % the state is not always stored in the stateList
    stateList   :=
     IF (CheckPointInterval? ) THEN
       cons( anState with [
               lVT := newlVT(currentState),
               stateVars := process(currentState)(
                             currentInputEvents(currentState),
                             stateVars(currentProcessState(currentState))
                            )
             ],
             stateList(currentState)
           )
     ELSE
       stateList(currentState)
     ENDIF,
    % currentProcessState is advancing
    ...
    % process specification for state change and output stays the same
    % identifier stays the same
    ...
  ]
```

The `CheckPointInterval?` axiom is incorporated in the `execute` axiom. The original axiom was modified to check on the status of the `CheckPointInterval?` predicate to determine whether or not an infrequent state saving strategy is used. This is reflected in the `execute` axiom by the simple "if clause". If the `CheckPoinInterval?` axiom evaluates to true, then it is assumed that the generic checkpoint algorithm has determined the end of a checkpoint interval. In this case, the generated state after the current execution cycle is stored in the state history. Otherwise, the state history remains unchanged. Due to this infrequent checkpointing strategy, the state history may now be incomplete and rollbacks determined by the `InsertEventCheck` axiom have to execute some additional steps to determine the correct state to rollback to. The `inputList`, `outputList`, and `ether` remain unchanged. However, the `stateList` and the `currentProcessState` need to handle the incomplete state history. As shown in the `insertEventsCheck` axiom, this is done with the help of the *coast forward* algorithm.

As introduced in the informal discussion of the infrequent state saving algorithm, coast forwarding is employed to regenerate old states in order to reach the correct state required to process the straggler message in the correct causal order. During coast forwarding, the process advances from the last error free state stored in the history to the state just before the occurrence of the causality error. This behavior is stated in the `coastforward` axiom. By looking at the interface, it can be seen that during coast forwarding, only the state is changed. There are no new events generated as execution resumes from a saved state with the same inputs and stops at the point where the new events have to be processed. The purpose of this whole activity is to regenerate the required correct state.

The progress of the coast forward algorithm is controlled by the measure function which calculates the number of events in the input list between two predefined time stamps. The two time stamps are provided through the events in the `intervalLength` axiom. The use of formal methods automatically required the proof of the termination condition of the coast forward algorithm (since it was stated in a recursive manner). An abstract of this proof is provided in the verification section. The termination condition is reached when no event remains to be unprocessed between the straggler event and the time stamp of the state created during coast forwarding. The `coastforward` axiom uses a restricted type (`legalCoastforward?`) for checking the pre-conditions of the coast forward procedure. This subtype restricts the tuple containing an event and a state to a tuple where it is guaranteed that the event has a time stamp greater than the time stamp of the state. This is true for the state last stored before the straggler event and the actual straggler event. This completes the description of the changes to the original Time Warp specification that were required to specify and prove the infrequent state saving protocol. However, the original Time Warp protocol contained an axiom called `frequentState` which contains the knowledge about the state saving history. Although the frequent state saving property is no longer satisfied and is no longer provable over the

```
insertEventsCheck(stateEvents : (legalInsert?)) : (AtLeastAState?) =
 IF cons?(events(stateEvents)) THEN
  IF lVT(currentProcessState(state(stateEvents)))
     >=
     recvTime(smallest(events(stateEvents)))
  THEN
   state(stateEvents) with [
    inputList    := signInsertList(events(stateEvents),
                    inputList(state(stateEvents))),
    stateList    := expungeAfter(
                       stateBefore(smallest(events(stateEvents)),
                       state(stateEvents)),
                       stateList(state(stateEvents))),
    currentProcessState := coastforward(
    (# event     := smallest(events(stateEvents)),
       currentState := state(stateEvents) WITH [
       inputList   := signInsertList(events(stateEvents),
                                      inputList(state(stateEvents))),
       stateList    := expungeAfter(
                          stateBefore(smallest(events(stateEvents)),
                                      state(stateEvents)),
                          stateList(state(stateEvents))),
       currentProcessState := maximum(expungeAfter(
                                 stateBefore(smallest(events(stateEvents)),
                                             state(stateEvents)),
                                 stateList(state(stateEvents)))) ]
    #)),
    outputList   := remove(
                       getWrongMessages(
                         recvTime(smallest(events(stateEvents))),
                         outputList(state(stateEvents))),
                       outputList(state(stateEvents))
                    ),
    ether        := putAll( ether(state(stateEvents)),
                            getAntimessages(
                              getWrongMessages(
                                recvTime(smallest(events(stateEvents))),
                                outputList(state(stateEvents))
                              )
                            )
                          )
   ]
   ELSE
     state(stateEvents) with [
       inputList := signInsertList(events(stateEvents),
                                    inputList(state(stateEvents)))
     ]
   ENDIF
 ELSE
   state(stateEvents)
 ENDIF
```

simulation's execution, another property of the state history is still accurate and is required for the proof of correctness. The property states the basic property of event driven simulation. Every state in the state history as well as the current state is created as a result of event execution. This information is incorporated in the EventDrivenSimulation axiom. The EventDrivenSimulation axiom establishes two properties: (i) for each state in the stateList there exist at least one event in the inputList with the same timestamp, and (ii) for any currentProcesState of the simulation object during execution there exist an event in the inputList with a corresponding timestamp.

```
coastforward(stateEvent: (legalCoastforward?)): RECURSIVE State =
  IF intervalLength(
       dummyRecv(lVT(currentProcessState(currentState(stateEvent)))),
       event(stateEvent), inputList(currentState(stateEvent))) > 0
  THEN
    coastforward(
    (# event                  := event(stateEvent),
       currentState           := currentState(stateEvent) with [
         currentProcessState :=
           anState with [
             lVT      := newlVT(currentState(stateEvent)),
             stateVars := process(currentState(stateEvent))(
               currentInputEvents(currentState(stateEvent)),
               stateVars(currentProcessState(currentState(stateEvent)))
          )
             ]
]
       #) )
  ELSE
    currentProcessState(currentState(stateEvent))
  ENDIF
MEASURE intervalLength(
       dummyRecv(lVT(currentProcessState(currentState(stateEvent)))),
       event(stateEvent), inputList(currentState(stateEvent)))

EventDrivenSimulation(currentState:(AtLeastAState?)) : bool =
 every( (LAMBDA (y:State): (EXISTS (x:(ValidEvent?)):
     member(x,inputList(currentState)) AND
     lVT(y) = recvTime(x))))(stateList(currentState)) AND
  (EXISTS (event: (ValidEvent?)):
    member(event, inputList(currentState)) AND
    lVT(currentProcessState(currentState)) = recvTime(event)
  )
```

5 Verification of the Infrequent State Saving Algorithm

As seen in Figure 3, several theorems were required to be added to the original specification in order to prove the correctness of the infrequent state saving algorithm in a Time Warp simulation. However, as this proof is built on top of an already existing Time Warp framework, many of the proofs did not require modification. New proofs were only required for the theorems directly affected by the above axioms and newly created type check conditions (TCCs). The coast-forward_TCC4 is just one example of a TCC that was automatically generated by the PVS type-checker. To follow the conservative extension style of specification, a recursive definition of the coast forward procedure was required. This in turn meant that PVS will automatically request a proof of the termination of the recursive definition. The actual proof involved the use of several additional theorems involving properties based on the intervalLength, extensive rewriting and the proof of individual cases. As the intervalLength is defined over the list data structure, most of the proofs concerning intervalLength were simply proven by rewriting and induction. For the sake of brevity, the proof steps are not shown in this paper [1].

Another important type check condition generated by the PVS type checker is the TimeWarpSimulation_TCC2 TCC. It states that the execution of processes results in a valid global system state identifying the ValidSystem? axiom as being another important invariant of the Time Warp protocol. This TCC is generated from the fact that the TimeWarpSimulation axiom needs to transition from one valid system state to another. The proof steps involved in the verification of this invariant are not presented in this paper. It is suffice to note that several sub-proofs needed to be proven. Informally, the proof consists of two major sub-proofs. The first sub-proof verifies that simulation objects always advance forward in virtual time. The second sub-proof verifies that newly generated messages combined with any other unprocessed messages contains timestamps greater than or equal to the earliest unprocessed message timestamp before execution. It should be noted that the original informal specification of the Time Warp protocol does not identify the valid system state property. The formal specification of the complete Time Warp paradigm identified the need for a definition of a valid state, which in turn identified a complete and correct invariant of the algorithm.

6 Conclusion

The formal framework of the Time Warp protocol is intended to (i) improve the comprehension of the protocol, (ii) provide a structured overview with preconditions and postconditions for the elementary building blocks, and (iii) as an initial formal model for research on the protocol. The application of the formal

---

[1] The complete Time Warp specification and the PVS proof files are available at http://www.ececs.uc.edu/~pfrey/pvs.

```
coastforward_TCC4: OBLIGATION
 (FORALL (stateEvent: (legalCoastforward?)):
    intervalLength(
      dummyRecv(lVT(currentProcessState(currentState(stateEvent)))),
      event(stateEvent),
      inputList(currentState(stateEvent)))
      > 0
 IMPLIES
    intervalLength(
      dummyRecv(lVT(currentProcessState(currentState(stateEvent)
        WITH
        [ currentProcessState := anState
            WITH [lVT := newlVT(currentState(stateEvent)),
              stateVars := process(currentState(stateEvent))
                          (currentInputEvents(currentState(stateEvent)),
                           stateVars(currentProcessState(currentState
                           (stateEvent))))
              ]
        ]))),
      event(stateEvent),
      inputList(currentState(stateEvent)
        WITH
        [ currentProcessState := anState
            WITH [lVT := newlVT(currentState(stateEvent)),
              stateVars := process(currentState(stateEvent))
                          (currentInputEvents(currentState(stateEvent)),
                           stateVars(currentProcessState(currentState
                           (stateEvent))))]])
    )
    <
    intervalLength(
      dummyRecv(lVT(currentProcessState(currentState(stateEvent)))),
      event(stateEvent),
      inputList(currentState(stateEvent))))
```

```
TimeWarpSimulation_TCC2: OBLIGATION
 (FORALL (snapshot: (ValidSystem?)):
  ValidSystem?(snapshot WITH [
    commManager :=
      getNewOutput(processes(snapshot)),
    processes :=
      executeAllProcesses((#
        commManager := commManager(snapshot),
        processes :=
          cleanAllEther(processes(snapshot))
    #))]));
```

model in the software development phase has shown some promising results [9, 12]. As many developers today are educated in formal methods, the specification improves the developer's comprehension and the high level overview of the simulation system. In addition, the elimination of implementation issues from the specification leaves the developer the intellectual freedom to choose suitably optimized algorithms satisfying the pre- and postconditions.

The framework is especially conducive for conducting research on the Time Warp algorithm itself. Extensions to the basic Time Warp protocol are in the form of optimizations such as infrequent state saving, rollback optimizations and event list insertion algorithms. The introduction of these optimizations to the basic Time Warp protocol is represented by the addition of the specification of these optimizations to the formal framework. As the formal framework requires several invariants to be proven, any change in the total specification (through extension) will require the invariants to be proven again. The existence of such an extensible framework reduces the specification work to just the optimization specification (as opposed to a complete specification). In addition, as the PVS environment provides methods to view existing proofs and the ability to use proven theorems, modifications to the existing framework can be proven in a shorter time. This claim is supported by the experience gained during several modifications to the basic model (*e.g.*, the effective reuse of the theorems provided by the list specifications and limited set of axioms and theorems that had to be modified to incorporate the specification of the infrequent state saving optimization).

The principal goal of this specification effort was to extend the formal framework to specify and verify in detail the Time Warp protocol and all its optimizations. For this reason, the framework was designed in a modular fashion such that other specifiers could easily add their specifications to the framework. This will allow other specifiers to quickly specify and verify the large set of optimizations to the basic Time Warp protocol in an efficient manner. In this paper, we reported our experiences in specifying and verifying one of the many Time Warp optimizations, namely the infrequent state saving optimization. We intend to continue extending the specification to specify other Time Warp optimizations. Towards this end, efforts are undergoing to specify and verify other Time Warp optimizations such as lazy cancellation [13], several GVT algorithms [2, 10], and event list management optimizations. Also the specification is currently being extended to incorporate the specification and verification of a mixed-mode simulator [4, 5] showing the applicability and extensibility of the specification to other domains.

In short, the formal specification provides a complete verified framework of the original Time Warp protocol and the infrequent state saving optimization. It is extensible in several directions, *i.e.*, to include a more accurate distributed system or be used as specification and verification framework for any Time Warp optimization or extension. In addition, the formal specification provides a description without ambiguities on an abstract level enhancing understanding of the algorithm and the means to prove assumptions of the algorithm with the help of the PVS theorem prover.

References

1. CHANDY, K. M., AND MISRA, J. Asynchronous distributed simulation via a sequence of parallel computations. *Communications of the ACM 24*, 11 (Apr. 1981), 198–206.
2. D'SOUZA, L. M., FAN, X., AND WILSEY, P. A. pGVT: An algorithm for accurate GVT estimation. In *Proc. of the 8th Workshop on Parallel and Distributed Simulation (PADS 94)* (July 1994), Society for Computer Simulation, pp. 102–109.
3. FLEISCHMANN, J., AND WILSEY, P. A. Comparative analysis of periodic state saving techniques in Time Warp simulators. In *Proc. of the 9th Workshop on Parallel and Distributed Simulation (PADS 95)* (June 1995), pp. 50–58.
4. FREY, P. *Protocols for Optimistic Synchronization of Mixed-Mode Simulation.* PhD thesis, University of Cincinnati, August 1998.
5. FREY, P., RADHAKRISHNAN, R., CARTER, H. W., AND WILSEY, P. A. Optimistic synchronization of mixed-mode simulators. In *1998 International Parallel Processing Symposium, IPPS'98* (March 30 – April 3 1998), pp. 694–699.
6. FREY, P., RADHAKRISHNAN, R., WILSEY, P. A., ALEXANDER, P., AND CARTER, H. W. An extensible formal framework for the specification and verification of an optimistic simulation protocol. In *Proceedings of the 32nd Hawaii International Conference on System Sciences (HICSS'99)* (jan 1999), Sony Electronic Publishing Services. (forthcoming).
7. FUJIMOTO, R. Parallel discrete event simulation. *Communications of the ACM 33*, 10 (Oct. 1990), 30–53.
8. JEFFERSON, D. Virtual time. *ACM Transactions on Programming Languages and Systems 7*, 3 (July 1985), 405–425.
9. KANNIKESWARAN, B., RADHAKRISHNAN, R., FREY, P., ALEXANDER, P., AND WILSEY, P. A. Formal specification and verification of the pGVT algorithm. In *FME '96: Industrial Benefit and Advances in Formal Methods* (Mar. 1996), M.-C. Gaudel and J. Woodcock, Eds., vol. 1051 of *Lecture Notes in Computer Science*, Springer-Verlag, pp. 405–424.
10. MATTERN, F. Efficient algorithms for distributed snapshots and global virtual time approximation. *Journal of Parallel and Distributed Computing 18*, 4 (Aug. 1993), 423–434.
11. MILNER, R. *Communication and Concurrency.* International Series in Computer Science. Prentice Hall, New York, NY, 1989.
12. PENIX, J., MARTIN, D., FREY, P., RADHAKRISHNAN, R., ALEXANDER, P., AND WILSEY, P. A. Experiences in verifying parallel simulation algorithms. In *Second Workshop on Formal Methods in Software Practice* (Clearwater Beach, Florida, USA, March 4-5 1998), Co-located with ISSTA98.
13. RAJAN, R., AND WILSEY, P. A. Dynamically switching between lazy and aggressive cancellation in a Time Warp parallel simulator. In *Proc. of the 28th Annual Simulation Symposium* (Apr. 1995), IEEE Computer Society Press, pp. 22–30.
14. RÖNNGREN, R., AND AYANI, R. Adaptive checkpointing in Time Warp. In *Proc. of the 8th Workshop on Parallel and Distributed Simulation (PADS 94)* (July 1994), Society for Computer Simulation, pp. 110–117.

Verifying End-to-End Protocols Using Induction with CSP/FDR

S. J. Creese[1] and Joy Reed[2]*

[1] Oxford University, Oxford UK
[2] Oxford Brookes University, Oxford UK
Sadie.Creese@comlab.ox.ac.uk, jnreed@brookes.ac.uk

Abstract. We investigate a technique, suitable for process algebraic, finite-state machine (model-checking) automated tools, for formally modelling arbitrary network topologies. We model aspects of a protocol for multiservice networks, and demonstrate how the technique can be used to verify end-to-end properties of protocols designed for arbitrary numbers of intermediate nodes. Our models are presented in a version of CSP allowing automatic verification with the FDR software tool. They encompass both inductive and non-inductive behaviours.

1 Introduction

Formal specification and verification of network protocols have focused in the main on link-to-link properties involving the interaction of a fixed number of components, e.g., providing reliable transmission between a sender and receiver using a fixed number of lower layer subcomponents communicating over an unreliable channel. However these techniques are not suitable for verifying end-to-end properties of modern multi-service networks, such as the Internet, which use complex protocols operating with arbitrary numbers of interacting components.

We illustrate a technique encompassing induction for proving correctness properties of network protocols operating with an arbitrary number of components. The technique is suitable for process algebraic, finite-state machine (model-checking) automated tools. Model-checking based techniques do not directly handle unbounded state problems; however this induction technique enables their use for certain classes of such applications, notably end-to-end protocols with arbitrary numbers of intermediate nodes. We illustrate the method using FDR [8, 23], a software package offered by Formal Systems (Europe) Ltd, which allows automatic checking of many properties of finite state systems and the interactive investigation of processes which fail these checks. The tool is based on the mathematical theory of Communicating Sequential Processes(CSP), developed at Oxford University and subsequently applied successfully in a number of industrial applications.

---

* This work was supported by in part by the US Office of Naval Research. Technical staff at Formal Systems (Europe) Ltd provided valuable advice on the use of FDR.

Previous CSP/FDR applications do not address properties of arbitrary numbers of interacting components. The induction technique described here extends the basic method introduced in [22] by employing lazy abstraction and dealing with non-inductive behaviours. It provides a powerful way to establish complex properties of arbitrary network configurations. This technique should prove especially valuable for verifying livelock and deadlock freedom for complex end-to-end network protocols. We illustrate its applicability with an example patterned after the Resource reSerVation Protocol (RSVP) [3, 35], a protocol designed to support resource reservation for high-bandwidth multicasts over IP networks.

We present an overview of formal models of network protocols, CSP and FDR. We describe a reservation protocol, the induction method, and its application to the protocol. We conclude with remarks on applicability. Our FDR source scripts are available upon request.

2 Formal Models of Network Protocols

CSP/FDR belong to the class of formalisms which combine programming languages, and finite state machines. Two similar approaches standardised by ISO for specification and verification of distributed services and protocols are LOTOS [14] (web bibliography [34]) and Estelle [9, 13].

Finite-state techniques are particularly suited for modelling *layered* protocols. Layered protocols are structured as a fixed number of layers, each with fixed service interfaces. Correctness properties for a given layer typically take the form of an assumption of correct service from the immediate lower level in order to guarantee correct service to the immediate higher level. Properties of the entire "protocol stack" can be established by chaining together the service specifications for the fixed number of intermediate layers, ultimately arriving at the service guaranteed by the highest level. In a very natural way, the formal layered model reflects the specification structure of these protocols as adopted by the network and communications community, such as the seven-layer ISO OSI (Open Systems Interconnect) Reference Model. There are numerous examples of formalisations of layered protocols, including Ethernet - CSMA/CD (in non-automated TCSP [7]) (in non-automated algebraic-temporal logic [16]), TCP (in non-automated CSP [10]), DSS1 / ISDN SS7 gateway (in LOTOS [19]), ISDN Layer 3 (in LOTOS [20]), ISDN Link Access Protocol (in Estelle [11]), ATM signalling (in TLT, a temporal logic/UNITY formalism [1]). All of these examples deal with link rather than end system properties.

By way of contrast, we note that an unbounded network topology is modelled with action systems [2] and extended in [29]. Although such deductive-reasoning techniques are not possible for finite-state model checkers such as FDR, the advantage of finite-state methods is that they are fully automatable, indicating when properties are *not* satisfied as well as when they are.

3 CSP, CSP_M and FDR

CSP [12] models a system as a *process* which interacts with its environment by means of atomic *events*. Communication is synchronous; that is, an event takes place precisely when both the process and environment agree on its occurrence. CSP comprises a process-algebraic programming language (see appendix), together with a related series of semantic models capturing different aspects of behaviour. A powerful notion of refinement intuitively captures the idea that one system implements another. Mechanical support for refinement, deadlock and livelock checking is provided by Formal Systems' FDR model checker.

The simplest semantic model identifies a process as the sequences of events, or *traces* it can perform. We refer to such sequences as *behaviours*. More sophisticated models introduce additional information to behaviours which can be used to determine liveness properties of processes.

We say that a process P is a refinement of process S, written $S \sqsubseteq P$, if any possible behaviour of P is also a possible behaviour of S. Intuitively, suppose S (for "specification") is a process for which all behaviours are in some sense acceptable. If P refines S, then the same acceptability must apply to all behaviours of P. The refinement relation is transitive. S can represent an idealised model of a system's behaviour, or an abstract property such as deadlock freedom.

The theory of refinement in CSP allows a wide range of correctness conditions to be encoded as refinement checks between processes. FDR performs a check by invoking a normalisation procedure for the specification process, which represents the specification in a form where the implementation can be checked against it by simple model-checking techniques. When a refinement check fails, FDR provides the user with an illustrative counter example. The definitive source book for CSP/FDR is found in [25].

CSP_M [8, 25, 26] combines the CSP process algebra with an expression language inspired by languages like Miranda-Orwell and Haskell-Gofer and modified to support the idioms of CSP. Hereafter we shall simply refer to CSP.

Unlike most packages of this type, FDR was specifically developed by Formal Systems for industrial applications, initialy used to develop and verify communications hardware (in the Inmos T9000 Transputer and the C104 routing chip). Existing applications include VLSI design, protocol development and implementation, control, signalling, fault-tolerant systems and security. Although the underlying semantic models for FDR do not specifically address time (in contrast to Timed CSP formalism [17, 24, 31]), work has been carried out modelling discrete time with FDR [25, 28], including a class of embedded real-time scheduler implementations [15] and a traffic congestion algorithm [22].

4 A Reservation Protocol

We illustrate the induction technique on a protocol patterned after the RSVP reservation protocol intended for IP based networks. This protocol addresses those requirements associated with a new generation of applications, such as remote video and multimedia conferencing, which are sensitive to the quality of service provided by the network. These applications depend on certain levels of resource (bandwidth, buffer space, etc.) allocation in order to operate acceptably. One strategy for dealing with large demands for bandwidth is to "multicast" from sources to receivers with transmissions along a number of intermediate links shared by "downstream" nodes. The RSVP approach is to create and maintain resource reservations along each link of a previously determined multicast route.

A multicast route consists of multiple sources and receivers, and arbitrary numbers of intermediate nodes forming a path between sources and receivers. Messages carrying reservation requests originate at receivers and are passed upstream towards the sources. Along the way if any node rejects the reservation, a reject message is sent back to the receiver and the reservation message discarded; otherwise the reservation message is propagated as far as the closest point along the way to the source where a reservation level greater than or equal to it has been made. Thus reservations become "merged" as they travel upstream; a node forwards upstream only the maximum request. A binary multicast tree is represented in Figure 3.

5 The Significance of Lazy Abstraction

Our technique reduces unbounded state space to a small finite one by hiding (and thus abstracting away) the unbounded components of our multicast network. We take the perspective of a downstream node in specifying the behaviour it sees on its upstream link.

Abstracting away the network along downstream channels, by hiding the channels using the CSP standard hiding operator \, would have significant consequences for reasoning about behaviour on the visible downstream channel. Consider the process H below; if the event a were to be *eagerly* abstracted, i.e. hidden, then the process H would be in a state where it would be willing to do a b, as if the a had occurred.

$$H = a \to b \to H$$
$$H \backslash a = b \to H$$

By similarly hiding an interface channel, we are not allowing the environment to be silent. Thus any proofs involving such abstraction would carry the "hidden" assumption that the environment is live on all offered events, and any properties thereby established could be significantly devalued.

We wish to establish that the behaviour seen on one downstream channel does not require liveness on other downstream channels. In order to claim that the behaviour over hidden interfaces does not compromise a successful refinement check, we simply show that the refinement holds even in the case of no communications on the hidden channels. By "gluing" $CHAOS$ onto a *downstream* channel, we are effectively abstracting the rest of the downstream network, no matter what its behaviour. The *lazy* abstraction of a from H, $\mathcal{L}_a(H)$, is defined below as the parallel composition of H with $CHAOS$ before hiding a, where $CHAOS$ is the process which may or may not do anything. In the example H above, $\mathcal{L}_a(H)$ is $b \to H \sqcap STOP$, which may fail to offer b (since $CHAOS$ can nondeterministically refuse a, thereby stopping b), in contrast to $H \backslash a$, which always offers b.

$$\mathcal{L}_a(H) = (H \, [| \, a \, |] \, CHAOS(a)) \backslash a$$
$$CHAOS(A) = \sqcap_{a:A} a \to CHAOS(A) \sqcap STOP$$

6 The Method

We model a protocol exhibiting the traffic-reducing aspects of RSVP: the ability of an intermediate node to automatically respond to requests for resources for which previous requests had already established the response. If a reservation is already in place (for a particular source), then any request for the same amount or less can be automatically accepted without any messages forwarded upstream. Similarly, if a request for an amount of resource from a particular source has already been rejected, then any request for the same amount or more can be automatically rejected without forwarding upstream.

FDR is used to perform a *structural induction* to establish end-to-end properties of the protocol through a series of refinement checks. These checks correspond to the base case and the inductive step of proof by induction. In this example, the proposition to be proven for our protocol is:

> An arbitrary number of intermediate traffic-reducing nodes may be placed transparently between sources and end receivers, i.e., a receiver sees the same behaviour from an upstream node connected by an arbitrary number of links to an end source, as it would see from an end source itself.

Let P represent this property. The induction proof obligations take the form:

$$P \sqsubseteq SOURCE \quad (i)$$
$$P \sqsubseteq \mathcal{L}_1^2 \, (P \, [downstream \leftrightarrow upstream] \, NODE) \quad (ii)$$
$$P \sqsubseteq \mathcal{L}_2^1 \, (P \, [downstream \leftrightarrow upstream] \, NODE) \quad (iii)$$

The refinement operator $\sqsubseteq$ refers to refinement in the CSP *failures model*. $\mathcal{L}_i^j(Q)$ is the *lazy abstraction* [25] of the channel $downstream_i$ from the process Q, with $downstream_j$ renamed to $downstream$. By this mechanism, we abstract away unbounded parts of the network further along the $downstream_i$ channel. In the parallel composition $P[downstream \leftrightarrow upstream]NODE$ of (ii) and (iii), the communications between the P and $NODE$ are hidden and happen as soon as they are available according to CSP semantics for the hiding operator. We can safely eagerly (rather than lazily) abstract these communications by hiding the next-hop upstream channel, since this has no effect on our reasoning about communications along a given downstream channel.

Refinement (i) establishes the base case for the induction. Generally this case is simply an end source, although an example in Section 7 requires an intermediate node directly linked to a source.

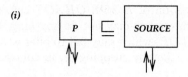

Fig. 1. Base Case: (i) A SOURCE must satisfy property P.

Refinements (ii) and (iii) establish the inductive step. Assuming that P is provided on its *upstream* channel, $NODE$ provides P to each of the *downstream* channels. The linked parallel operator used in the right hand sides of the induction proof obligations given above indicates that $NODE$ communicates on its *upstream* channel in a synchronised fashion with (a node satisfying) P on P's *downstream* channel, with the linked communications hidden. Thus we are able to represent the perspective of the one downstream node, by hiding the other downstream channel and next-hop upstream communications. The refinement checks performed by FDR establish that no matter what communications take place over the *downstream1* channel the property P still holds on channel *downstream2*. Similarly, P holds on *downstream1* regardless of communications on *downstream2*.

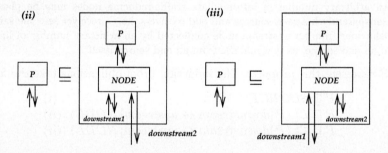

Fig. 2. Inductive Case: The property P connected on its downstream to NODE on NODE's upstream, with all communications hidden apart from the downstream2 interface in (ii), and the downstream1 interface in (iii).

By transitivity of refinement, relations (i) - (iii) allow us to claim that property P is provided along any of the downstream links for an arbitrary binary multicast tree, such as specified below and represented in Figure 3. These relations can be expressed as an `assertion` in FDR, which checks their validity. The techniques can be similarly applied to achieve arbitrary branching degree for sources and receivers.

```
((( SOURCE
    [downstream <-> upstream]
  NODE[[downstream1 <- A, downstream2 <- B]] )
    [A <-> upstream]
  NODE )
    [B <-> upstream]
  NODE[[downstream1 <- C ]])
    [C <-> upstream]
  NODE
```

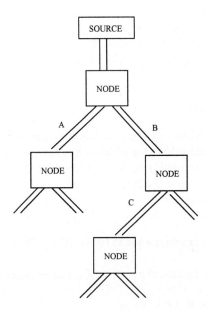

Fig. 3. An example binary multicast. Relations (i) - (iii) ensure that P is provided on each downstream link.

7 The Model

Our protocol operates with a given multicast tree with the source as the root, receivers as leaves and with an arbitrary number of intermediate nodes. A receiver initiates a request for a certain amount of resource by presenting a request to the node immediately upstream. In order to minimise network traffic, reservations requests are merged at intermediate nodes, which automatically respond to any downstream request whenever the appropriate reply is determined by responses previously relayed from above; otherwise the request is relayed upstream and the subsequent response relayed downstream. A receiver is ultimately given an accept (successful) response if all upstream nodes, including the source, have sufficient local resources to satisfy the request, and a reject (failure) response otherwise – due to some node along the path having insufficient local resources.

We build a general model of a network node, and inductively establish appropriate properties. The general communication convention is that an intermediate node has access to bi-directional channels: one upstream towards the source and two downstream towards receivers. We model resources as small integers, and define a single type to distinguish successful reservations from rejections.

```
MAX_R = 3                          // system parameter, set to 3
RESOURCE = {0 . . MAX_R}
datatype RESULT = accept | reject
datatype MESSAGE = request.RESOURCE
                 | reply.RESULT.RESOURCE

channel upstream, downstream, downstream1, downstream2,
        local, local1, local2, A   : MESSAGE
```

A source S with a fixed amount q of local resources is always live on downstream requests, accepting those for which it has adequate resources and rejecting otherwise. Replies are merged, i.e., consecutive identical requests may generate only a single reply. A general SOURCE behaves as a source with an arbitrary amount of local resources q.

```
S(q) = let
   S'(q,{}) = downstream.request?v -> S'(q,{v})
   S'(q,seen) =
       (downstream.request?vv -> S'(q,union(seen,{vv})))
       []
       (|~| v: seen @ if v <= q
          then downstream.reply!accept.v -> S'(q,diff(seen,{v}))
          else downstream.reply!reject.v -> S'(q,diff(seen,{v}))))
within S'(q,{})

SOURCE = |~| q:RESOURCE @ S(q)
```

An intermediate node linking source and receivers has a downstream interface to handle requests from other nodes immediately downstream, which it may or may not forward upstream. The DOWN process is live on downstream requests and upstream replies. It remembers upstream replies in order to automatically issues replies when appropriate. Whenever it cannot already know the response for downstream request it must forward the request upstream. Rather than using a single state variable for recording pending requests, we model a node as a series of "slices", processes handling requests and replies for exactly one resource. This device for avoiding state-space explosion is analogous to using n binary variables, $b_1, \ldots, b_n$ for representing a set containing values drawn from $1 \ldots n$.

In practice this would be implemented by fewer processes sharing state. The slices interleave on downstream requests, downstream replies, and upstream requests. They synchronise on upstream replies, since an accept for an amount ma implies that requests for all $v \leq ma$ should be accepted, and a reject for lr implies that all $v \geq lr$ should be rejected. Inconsistent upstream replies are ignored.

```
Down_slice(v,know,r,repdown,requp) =
   downstream.request.v ->
      (if know
       then Down_slice(v,true,r,true,false)   // merge reply down
       else Down_slice(v,false,r,repdown,true))   // merge req up
[]
   upstream.reply?rr?vv ->
      (if know or (vv>v and rr==reject) or (vv<v and rr==accept)
       then Down_slice(v,know,r,repdown,requp)   //ignore reply
       else Down_slice(v,true,rr,(repdown or requp),false))
[]
   (repdown and know &                          // reply down enabled
       downstream.reply!r!v -> Down_slice(v,true,r,false,false))
[]
   (requp and know==false &                     // request up enabled
       upstream.request!v -> Down_slice(v,false,r,true,false))

DOWNV(v) = Down_slice(v,false,accept,false,false)   // initialised

DOWN = || v: RESOURCE @           // parallel combination of slices
              [{| downstream.request.v, downstream.reply.reject.v,
                  downstream.reply.accept.v, upstream.request.v,
                  upstream.reply, |}]
              DOWNV(v)
```

We model an intermediate node having an upstream channel and two downstream channels as a node with two downstream interfaces and a single local controller behaving as a source. The local controller examines requests relayed from the downstream interfaces, and is prepared to reject automatically or to merge the request upstream. Each downstream interface takes in downstream requests, and is immediately prepared to respond or to merge them to the local controller. Upstream replies are merged downstream. The downstream interfaces interleave downstream requests, downstream replies, and upstream requests, and synchronise on upstream replies.

```
NODE = (((SOURCE [[ downstream.request <- local1.request,
                    downstream.request <- local2.request,
                    downstream.reply.reject <- local.reply.reject,
                    downstream.reply.accept <- upstream.request ]]
                 [|{| local1, local |} |]
          DOWN [[ upstream.request <- local1.request,
                  upstream.reply <- local.reply,
                  downstream <- downstream1,
                  upstream.reply <- upstream.reply ]])
                      \{| local1 |})
                 [|{| local2, local, upstream.reply |} |]
          DOWN [[ upstream.request <- local2.request,
                  upstream.reply <- local.reply,
                  downstream <- downstream2,
                  upstream.reply <- upstream.reply ]])
                      \{| local2, local |}
```

We inductively establish that any collection of intermediate nodes behave as a source on each of its downstream channels. Thus this is an end-to-end property guaranteed by a source to an end receiver. The *base case* is trivial in that we model an end sender as a source, which is certainly refined by itself.

The *inductive case* for each downstream channel, corresponding to 5 (*ii*) and (*iii*) is checked with FDR using the `assert` statement :

```
assert SOURCE [F= (((SOURCE [downstream <-> upstream] NODE)
                  [| {| downstream1 |} |] CHAOS({|downstream1|}))
          \ {|downstream1|})
                  [[downstream2 <- downstream]]

assert SOURCE [F= (((SOURCE [downstream <-> upstream] NODE)
                  [| {| downstream2 |} |] CHAOS({|downstream2|}))
          \ {|downstream2|})
                  [[downstream1 <- downstream]]
```

Also of interest is that the behaviour of an intermediate node along each of its downstream channels (given below as `DownBviour`) can be established inductively, provided that it is ultimately linked to a sender source. The base case is nontrivial in that a simple source does not satisfy the property `DownBviour` since the behaviour of `SOURCE` is less deterministic than `DownBviour`. We require two base cases corresponding to 5(*i*), one for each downstream channel:

```
DownBviour =  (SOURCE [[ downstream <- A ]]
                [| {| A |} |]
                DOWN [[ upstream <- A ]]) \ {| A |}

Basecase = SOURCE [downstream <-> upstream] NODE
```

```
assert DownBviour [F= ((Basecase
                   [|{| downstream2 |}|] CHAOS({|downstream2|}))
                       \ {| downstream2|})
                   [[downstream1 <- downstream]]
assert DownBviour [F= ((Basecase
                   [|{| downstream2 |}|] CHAOS({|downstream2|}))
                       \ {| downstream2|})
                   [[downstream1 <- downstream]]
```

And the inductive case (corresponding to 5 (*ii*) and (*iii*) :

```
assert DownBviour [F= (((DownBviour [downstream <-> upstream] NODE)
                   [|{| downstream1 |} |] CHAOS({|downstream1|}))
                       \ {|downstream1|})
                   [[downstream2 <- downstream]]

assert DownBviour [F= (((DownBviour [downstream <-> upstream] NODE)
                   [|{| downstream2 |} |] CHAOS({|downstream2|}))
                       \ {|downstream2|})
                   [[downstream1 <- downstream]]
```

Our nodes additionally satisfy another property – minimising traffic on the network – by guaranteeing that after some reaction delay after an up reply, they do not issue requests for which they can already know the response. We specify that such a receiver process may repeatedly issue requests, but only for ones not determined by past responses. It remembers the maximum accepted and least rejected amounts, and adjusts these according to new responses:

```
adjust(accept,qq,ma,lr) = if qq < lr then max(qq,ma) else ma
adjust(reject,qq,ma,lr) = if qq > ma then min(qq,lr) else lr
channel tock
```

R specifies that a process must be live on upstream responses, and may or may not issue upstream requests for which it cannot determine the upstream response by past responses. It is allowed some time (tocks) before it must adjust according to new upstream responses.

```
R(ma,lr,t,k,ma',lr') =
  (upstream.reply.accept?vv ->
                RR(ma,lr,0,2,adjust(accept,vv,ma,lr),lr))
  []
  (upstream.reply.reject?vv ->
                RR(ma,lr,0,2,ma,adjust(reject,vv,ma,lr)))
  []
  (((|~| v: RESOURCE @  ma < v and v < lr and t <= k  &
        upstream.request!v -> RR(ma,lr,t,k,ma',lr'))  |~| STOP  )
  [] ( t <= k & tock -> if t = k
        then RR(ma',lr',t+1,k,ma',lr')     // reset
        else RR(ma, lr, t+1,k,ma',lr')  )  // advance clock
```

A receiver REC initialy has maximum accepted ma = 0 and least rejected lr = MAX_R, and must adjust to replies within 2 internal tock cycles:

REC = R(0,MAX_R,0,2,0,MAX_R) \ {| tock |}

An intermediate node satisfies the traffic reducing property on its upstream channel. That is, from the perspective of its upstream node, an intermediate node behaves as a receiver – no matter what the behaviour along the next-hop downstream links. This property is established without using induction simply by lazy abstracting all downstream channels, and asserting that the result refines the REC property:

CHAOSDOWN = CHAOS({|downstream1,downstream2|})

assert REC [F= (NODE[|{| downstream1, downstream2 |}|] CHAOSDOWN)
        \ {|downstream1,downstream2|}

8 Conclusions and Comparison with Other Work

There are a variety of different techniques based on induction for reducing a system of unbounded processes to a finite-state problem. Not surprisingly all of these, as well as ours, rely on proof obligations corresponding to a base case and an inductive case. The methods differ in the particular mechanisms, derived from their underlying process theory, for abstracting away the unbounded parts of the system. Techniques appearing in the literature illustrated on simple protocols include formulations in process algebra applied to a bounded buffer and token ring [33], network grammars and abstraction applied to a token ring algorithm and binary tree parity checker [4], general induction theorem applied to a distributed replication algorithm and dining philosophers [18], PVS model checking applied to a mutual exclusion protocol [27]. Creese and Roscoe [6] are developing a form of induction for CSP/FDR using data independence to handle unbounded collections of processes retaining their separate identities, rather than abstracted away as in the technique presented here. In [5, 22], simpler properties of the RSVP protocol are modelled with CSP/FDR.

Wolper and Lovinfosse [33] establish that network invariants do not always exist, that is, there are true properties for which there are no corresponding network invariants (although they do not give an example of such). The novelty of our work is that it is applied to a complex multiservice network protocol, incorporating both inductive and non-inductive behaviours which very much depend on the data values along visible interfaces. This demonstrates that our technique can be a practical strategy for reducing a large protocol so that it can be handled comfortably by a model checker. The key is to isolate invariant properties amenable to inductive proof, and abstract away non-inductive behaviours. Hence, we isolate interesting properties of each pair of downstream channels alone, and check the induction proof obligations for each pair.

For this protocol there are important properties which are not inductive in the sense defined by refinement relations 5 (*i*) - (*iii*). For example, the traffic reduction property REC does not satisfy these relations. The specified behaviour for this property by necessity allows a fixed amount of reaction delay before new upstream replies must be assimilated for subsequent automatic responses to downstream requests. Such data-dependent behaviour appearing on the right-hand side of the refinement relations increases the reaction delay over that of the left-hand side – in effect, by composing arbitrary nodes satisfying this property this reaction delay becomes unbounded. The most we can show by model checking is that such properties are link-to-link, and hence we rely on a different use of abstraction to establish that

> If a node exhibits desirable behaviour along a link within t for some time t, then an arbitrary composition of nodes exhibits desirable behaviour along the link within t' for some t'.

It is interesting to note that although certain behaviours are inductive in the above sense, it could be significant that they fail to be inductive in a model-checking sense. For some protocols, link reaction delays could accumulate to an unacceptable end-delay.

Our induction technique is naturally suited for verifying certain end-to-end properties of protocols designed to operate with arbitrary numbers of interacting nodes. This is not surprising, since these protocols are inductive by design. The strength of our technique is that it can handle complex protocols exhibiting a mixture of model-checking inductive and non-inductive behaviours.

Finally, we note that one important property which our technique establishes is that our multicast protocol is deadlock free. Our method can be used to model complex end-to-end protocols in order to establish deadlock and divergence freedom, or to reveal unexpected deadlock or livelock for these protocols.

Acknowledgements The authors would like to thank Michael Goldsmith, Bryan Scattergood, Bill Roscoe and Carroll Morgan for valuable advice and discussions. Also appreciation to the reviewers for their careful reading.

References

1. D Barnard and Simon Crosby, The Specification and Verification of an Experimental ATM Signalling Protocol, *Proc. IFIP WG6.1 International Symposium on Protocol Specification, Testing and Verification XV*, Dembrinski and Sredniawa, eds, Warsaw, Poland, June 1995, Chapman Hall.
2. R Butler. A CSP Approach to Action Sysems, DPhil Thesis, Oxford U., 1992.
3. R Braden, L Zhang, S. Berson, S. Herzog and S. Jamin. Resource reSerVation Protocol (RSVP), Ver. 1, Functional Spec. Internet Draft, IETF 1996.
4. E Clarke, O Grumberg and S Jha, Verifying parameterized networks using abstraction and regular languages, *Proc. CONCUR'95*, LNCS 962, Springer 1995.

5. S Creese, An inductive technique for modelling arbitrarily configured networks, MSc Thesis, Oxford U., 1997.
6. SJ Creese and AW Roscoe, Verifying an infinite family of inductions simultaneously using data independence and FDR, (Submitted).
7. J Davies, Specification and Proof in Real-time Systems, D.Phil Thesis, Oxford U., 1991.
8. Formal Systems (Europe) Ltd. Failures Divergence Refinement. *User Manual and Tutorial*, version 2.11.
9. Estelle Specifications, ftp://louie.udel.edu/pub/grope/estelle-specs
10. J Guttman and D Johnson, Three Applications of Formal Methods at MITRE, *Formal Methods Europe*, LNCS873, Naftolin, Denfir, Barcelona '94.
11. R Groz, M Phalippou, M Brossard, Specification of the ISDN Linc Access Protocol for D-channel (LAPD), CCITT Recommendation Q.921, ftp://louie.udel.edu/pub/grope/estelle-specs/lapd.e
12. CAR Hoare. *Communicating Sequential Processes*. Prentice-Hall 1985.
13. ISO Rec. 9074, The Extended State Transition Language (Estelle), 1989.
14. ISO: Information Processing System - Open System Interconnection - LOTOS - A Formal Description Technique based on Temporal Ordering of Observational Behavior, IS8807, 1988.
15. DM Jackson. Experiences in Embedded Scheduling. *Formal Methods Europe*, Oxford, 1996.
16. M Jmail, An Algebraic-temporal Specification of CSMA/CD Protocol, *Proc. IFIP WG6.1 Inter. Sym. on Protocol Spec., Testing and Verification XV*, Dembrinski and Sredniawa, eds, Warsaw Poland, June '95, Chapman Hall.
17. A Kay and JN Reed. A Rely and Guarantee Method for TCSP, A Specification and Design of a Telephone Exchange. *IEEE TSE*. 19,6 1993, pp 625-629.
18. RP Kurshan and M McMillan, A structural induction theorem for processes, *Proc. 8th Symposium on Principles of Distributed Computing*, 1989.
19. G Leon, J Yelmo, C Sanchez, F Carrasco and J Gil, An Industrial Experience on LOTOS-based Prototyping for Switching Systems Design, *Formal Methods Europe*, LNCS 670, Woodcock and Larsen, eds., Odense Denmark, '93.
20. J Navarro and P Martin, Experience in the Development of an ISDN Layer 3 Service in LOTOS, *Proc. Formal Description Techniques III*, J Quemada, JA Manas, E Vazquez, eds, North-Holland, 1990.
21. K Paliwoda and JW Sanders. An Incremental Specification of the Sliding-window Protocol. *Distributed Computing*. May 1991, pp 83-94.
22. J Reed, D Jackson, B Deianov and G Reed, Automated Formal Analysis of Networks: FDR Models of Arbitrary Topologies and Flow-Control Mechanisms, ETAPS-FASE98 Fund. Approaches to Soft. Eng., Lisbon, LNCS 1382 Mar '98.
23. AW Roscoe, PHB Gardiner, MH Goldsmith, JR Hulance, DM Jackson, JB Scattergood. H ierarchical compression for model-checking CSP or How to check 10^{20} dining philosphers for deadlock, Springer LNCS 1019.
24. GM Reed and AW Roscoe, A timed model for communicating sequential processes, Proceedings of ICALP'86, Springer LNCS 226 (1986), 314-323; *Theoretical Computer Science* 58, 249-261.
25. AW Roscoe, **Theory and Practice of Concurrency**, Prentice Hall, 1998.
26. B Scattergood, **Tools for CSP and Timed CSP**, D.Phil Thesis, Oxford U., 1998.
27. N Shankar, Machine-Assisted Verification Usin Automated Theorem Proving and Model Checking, *Math. Prog. Methodology*, ed M Broy.

28. K Sidle, Pi Bus, *Formal Methods Europe*, Barcelona, 1993.
29. J Sinclair, Action Systems, Determinism, and the Development of Secure Systems, PHd Thesis, Open University, 1997.
30. AS Tanenbaum. *Computer Networks*. 3rd edition. Prentice-Hall 1996.
31. J Davies, D Jackson, G Reed, J Reed, A Roscoe, and S Schneider, Timed CSP: Theory and practice. *Proc.REX Workshop, Nijmegen*, LNCS 600, Springer, '92.
32. JS Turner. New Directions in Communications (or Which Way to the Information Age). *IEEE Commun. Magazine*. vol 24, pp 8 -15, Oct 1986.
33. P Wolper and V Lovinfosse, Verifying properties of large sets of processes with network invariants, *Proc. International Workshop on Automatic Verification Methods for Finite-State Machines*, LNCS 407, Springer-Verlaag, 1989.
34. LOTOS Bibliography, http://www.cs.stir.ac.uk/ kjt/research/well/bib.html
35. L Zhang, S Deering, D Estrin, S Shenker and D. Zappala. RSVP: A New Resource ReSerVation Protocol. *IEEE Network*, September 1993.

Appendix A. The CSP Language

The CSP language is a means of describing components of systems, *processes* whose external actions are the communication or refusal of instantaneous atomic *events*. All the participants in an event must synchronise on it. The CSP processes that we use are constructed from the following:

STOP is the simplest CSP process; it never engages in any action, nor terminates.

SKIP similarly never performs any action, but instead terminates successfully, passing control to the next process in sequence (see ; below).

a -> P is the most basic program constructor. It waits to perform the event a and after this has occurred subsequently behaves as process P. The same notation is used for outputs (c!v -> P) and inputs (c?x -> P(x)) of values along named channels.

P |~| Q represents internal choice. It behaves as P or Q *nondeterministically*.

P [] Q represents external or *deterministic* choice. It will offer the initial actions of both P and Q to its environment at first; its subsequent behaviour is like *P* if the initial action chosen was possible only for P, and like Q if the action selected Q. If both P and Q have common initial actions, its subsequent behaviour is nondeterministic (like |~|). STOP [] P behaves as P.

P |[A]| Q represents parallel composition. P and Q evolve concurrently, except that events in A occur only when P and Q agree (i.e. *synchronise*) to perform them.

P ||| Q represents the interleaved parallel composition. P and Q evolve separately, and do not synchronize on their events.

P ; Q is a sequential, rather than parallel, composition. It behaves as P until and unless P terminates successfully: its subsequent behaviour is that of Q.

P \ A is the CSP abstraction or hiding operator. This process behaves as P except that events in set A are hidden from the environment and are solely determined by P; the environment can neither observe nor influence them.

P [[a <- b]] represents the process P with a renamed to b.

P [a <-> b] Q is the linked parallel operator. P and Q synchronise on channels a and b, which have been renamed to the same and hidden.

There are also straighforward generalisations of the choice operators over non-empty sets, written |~| x:X @ P(x) and [] x:X @ P(x).

Mechanical Verification of a Garbage Collector

Klaus Havelund

NASA Ames Research Center
Moffett Field, California, USA
havelund@ptolemy.arc.nasa.gov
http://ic-www.arc.nasa.gov/ic/projects/amphion

Abstract. We describe how the PVS verification system has been used to verify a safety property of a garbage collection algorithm, originally suggested by Ben-Ari. The safety property basically says that "nothing but garbage is ever collected". Although the algorithm is relatively simple, its parallel composition with a "user" program that (nearly) arbitrarily modifies the memory makes the verification quite challenging. The garbage collection algorithm and its composition with the user program is regarded as a concurrent system with two processes working on a shared memory. Such concurrent systems can be encoded in PVS as state transition systems, very similar to the models of, for example, UNITY and TLA. The algorithm is an excellent test-case for formal methods, be they based on theorem proving or model checking. Various hand-written proofs of the algorithm have been developed, some of which are wrong. David Russinoff has verified the algorithm in the Boyer-Moore prover, and our proof is an adaption of this proof to PVS. We also model check a finite state version of the algorithm in the Stanford model checker Murphi, and we compare the result with the PVS verification.

1 Introduction

In [18], Russinoff uses the Boyer-Moore theorem prover to verify a safety property of a garbage collection algorithm, originally suggested by Ben-Ari [1]. We will describe how the same algorithm can be formulated in the PVS verification system [16], and we demonstrate how the safety property can be verified. An earlier related experiment where we verified a communication protocol in PVS is reported in [11].

The garbage collection algorithm, the *collector*, and its composition with a user program, the *mutator*, is regarded as a concurrent system with (these) two processes working on a shared memory. The memory is basically a structure of nodes, each pointing to other nodes. Some of the nodes are defined as *roots*, which are always accessible to the mutator. Any node that can be reached from a root, chasing pointers, is defined as *accessible* to the mutator. The mutator changes pointers nearly arbitrarily, while the collector continuously collects garbage (not accessible) nodes, and puts them into a *free* list. The collector uses a colouring technique for bookkeeping purposes: each node has a *colour* field associated with it, which is either coloured black if the node is accessible or white if not. In order to cope with interference between the two processes, the mutator colours the target node of the redirection black after the redirection. The safety property basically says that *nothing but garbage is ever collected*. Although the collector algorithm is relatively simple, its parallel composition with the mutator makes the verification quite challenging.

An initial version of the algorithm was first proposed by Dijkstra, Lamport, et al. [7] as an exercise in organizing and verifying the cooperation of concurrent processes. They described their experience as follows, citing [18]:

> Our exercise has not only been very instructive, but at times even humiliating, as we have fallen into nearly every logical trap possible ... It was only too easy to design what looked – sometimes even for weeks and to many people – like a perfectly valid solution, until the effort to prove it correct revealed a (sometimes deep) bug.

Their solution involves three colours. Ben-Ari's later solution is based on the same algorithm, but it only uses two colours, and the proof is therefore simpler. Alternative proofs of Ben-Ari's algorithm were then later published by Van de Snepscheut [6] and Pixley [17]. All of these proofs were informal *pencil and paper* proofs. Ben-Ari defends this as follows:

> So as not to obscure the main ideas, the exposition is limited to the critical facets of the proof. A mechanically verifiable proof would need all sorts of trivial invariants ... and elementary transformations of our invariants (... with appropriate adjustments of the indices).

These four pieces of work, however, indeed show the problem with handwritten proofs, as pointed out by Russinoff [18]; the story goes as follows. Dijkstra, Lamport et al. [7] explained how they (as an example of a "logical trap") originally proposed a modification to the algorithm where the mutator instructions were executed in reverse order (colouring before pointer redirection). This claim was, however, wrong, but was discovered by the authors before the proof reached publication. Ben-Ari then later again proposed this modification and argued for its correctness without discovering its flaw. Counter examples were later given in [17] and [6].

Furthermore, although Ben-Ari's algorithm (which is the one we verify in PVS) is correct, his proof of the safety property was flawed. This flaw was essentially repeated in [17] where it yet again survived the review process, and was only discovered 10 years after when Russinoff detected the flaw during his mechanical proof [18]. As if the story was not illustrative enough, Ben-Ari also gave a proof of a liveness property (*every garbage node will eventually be collected*), and again: this was flawed as later observed in [6]. To put this story of flawed proofs into a context, we shall cite [18]:

> Our summary of the story of this problem is not intended as a negative commentary on the capability of those who have contributed to its solution, all of whom are distinguished scientists. Rather, we present this example as an illustration of the inevitability of human error in the analysis of detailed arguments and as an opportunity to demonstrate the viability of mechanical program verification as an alternative to informal proof.

We first informally describe the algorithm. Then we formalize it in PVS as a state transition system similar to the models of, for example, UNITY [5] and TLA [14]. We then outline the PVS proof of the safety property; the paper contains the complete set of invariants and lemmas needed, some of which appear in appendix A. The proof resembles closely the proof in [18] and has the same invariants. We have also verified a finite state version of the garbage collector in the Stanford Murphi model checker [15], and we comment on this extra experiment. The full Murphi code is contained in appendix B.

A main observation is that the PVS proof is surprisingly complex compared to the size of the algorithm proved. It is therefore an excellent case study for the development of techniques that are supposed to automate theorem proving, for example invariant strengthening techniques [4, 3], and abstraction techniques [2, 8]. In [12] we have documented a refinement proof in PVS of the same algorithm. Another observation is that Murphi's execution model forced us to take some concrete design decisions, that could be left undecided and abstract in the PVS specification and proof. Also, we could only verify the algorithm for a particular very small memory with fixed bounds. The fact that the verification of such a small memory causes state explosion also seems a challenge. The advantage of Murphi is of course that it is automatic.

2 Informal Specification

In this section we informally describe the garbage collection algorithm. As illustrated in figure 1, the system consists of two processes, the *mutator* and the *collector*, working on a shared *memory*.

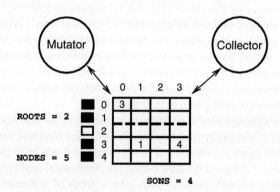

Fig. 1. The Mutator, Collector and Shared Memory

The Memory

The memory is a fixed size array of *nodes*. In the figure there are 5 nodes (rows) numbered 0 – 4. Associated with each node is an array of uniform length of *cells*. In the figure there are 4 cells per node, numbered 0 – 3. A cell is hence identified by a pair of integers (n,i) where n is a node number and where i is called the *index*. Each cell contains a pointer to a node, called the *son*. In the case of a LISP system, there are for example two cells per node. In the figure we assume that all empty cells contain the *NIL* value 0, hence points to node 0. In addition, node 0 points to node 3 (because cell (0,0) does so), which in turn points to nodes 1 and 4. Hence the memory can be thought of as a two-dimensional array, the size of which is determined by the positive integer constants NODES and SONS. To each node is associated a *colour*, black or white, which is used by the collector in identifying garbage nodes.

A pre-determined number of nodes, defined by the positive integer constant ROOTS, is defined as the *roots*, and these are kept in the initial part of the array (they may be thought of as static program variables). In the figure there are two such roots, separated from the rest with a dotted line. A node is *accessible* if it can be reached from a root by following pointers, and a node is *garbage* if it is not accessible. In the figure nodes 0, 1, 3 and 4 are therefore accessible, and 2 is garbage.

There are only three operations by which the memory structure can be modified:

– Redirect a pointer towards an accessible node.
– Change the colour of a node.
– Append a garbage node to the free list.

In the initial state, all pointers are assumed to be 0, and nothing is assumed about the colours.

The Mutator

The mutator corresponds to the user program and performs the main computation. From an abstract point of view, it continuously changes pointers in the memory; the changes being arbitrary except for the fact that a cell can only be set to point to an already accessible node. In changing a pointer the "previously pointed-to" node may become garbage, if it is not accessible from the roots in some alternative way. In the figure, any cell can hence be modified by the mutator to point to anything else than 2. One should think that only accessible cells could be modified, but the algorithm can in fact be proved safe without that restriction. Hence the less restricted context as possible is chosen. The algorithm is as follows:

1. Select a node m, an index i, and an accessible node n, and assign n to cell (m,i).
2. Colour node n black. Return to step 1.

Each of the two steps are regarded as atomic instructions.

The Collector

The collector's purpose is purely to collect garbage nodes, and put them into a *free list*, from which the mutator may then remove them as they are needed during dynamic storage allocation. Associated with each node is a *colour* field, that is used by the collector during it's identification of garbage nodes. Basically it colours accessible nodes *black*, and at a certain point it collects all *white* nodes, which are then garbage, and puts them into the free list. Figure 1 illustrates a situation at such a point: only node 2 is white since only this one is garbage. The collector algorithm is as follows:

1. Colour each root black.
2. Examine each pointer in succession. If the source is black and the target is white, colour the target black.
3. Count the black nodes. If the result exceeds the previous count (or if there was no previous count), return to step 2.
4. Examine each node in succession. If a node is white, append it to the free list; if it is black, colour it white. Then return to step 1.

Steps 1–3 constitutes the *marking* phase and it's purpose is to blacken all accessible nodes. Each iteration within each step is regarded as an atomic instruction. Hence, for example, step two consists of several atomic instructions, each counting (or not) a single node.

The Correctness Criteria

The *safety* property we want to verify is the following: *No accessible node is ever appended to the free list*. In [18], the following *liveness* property is also verified: *Every garbage node is eventually collected*. As in our previous work with a protocol verification in PVS and Murphi [11], we have focused only on safety, since already this requires an effort worth reducing.

3 Formalization in PVS

We have followed the formalization of the algorithm in [18] as much as possible; we have for example used the same names for most of the concepts introduced. This was done in order to create a better basis for comparison, and to avoid introducing errors ourself. We did in fact consider reducing the number of atomic instructions in Rusinoff's formalization, since there seems to be more than in the informal algorithm (some of them are "just" test-and-goto instructions). However, with no changes we feel being on "safe ground".

3.1 The Memory

Basic Memory Operations The memory type is introduced in a theory, parameterized with the memory boundaries, see the figure 2 below. That is, NODES, SONS, and ROOTS define respectively the number of nodes (rows), the number of sons (columns/cells) per node, and the number of nodes that are roots. They must all be positive natural numbers (different from 0). There is also an obvious assumption that ROOTS is not bigger than NODES.

The Memory type is defined as an abstract (non-empty) type upon which a constant and four functions are defined using the AXIOM construct (an alternative would have

```
Memory[NODES:posnat,SONS:posnat,ROOTS:posnat] : THEORY
BEGIN
  ASSUMING
    roots_within : ASSUMPTION ROOTS <= NODES
  ENDASSUMING

  Memory : TYPE+
  NODE   : TYPE = nat
  INDEX  : TYPE = nat
  Node   : TYPE = {n : NODE | n < NODES}
  Index  : TYPE = {i : INDEX | i < SONS}
  Root   : TYPE = {r : NODE  | r < ROOTS}
  Colour : TYPE = bool

  null_array : Memory
  colour     : [NODE -> [Memory -> Colour]]
  set_colour : [NODE,Colour -> [Memory -> Memory]]
  son        : [NODE,INDEX -> [Memory -> NODE]]
  set_son    : [NODE,INDEX,NODE -> [Memory -> Memory]]

  m          : VAR Memory
  n,n1,n2,k  : VAR Node
  i,i1,i2    : VAR Index
  c          : VAR Colour

  mem_ax1 : AXIOM son(n,i)(null_array) = 0

  mem_ax2 : AXIOM colour(n1)(set_colour(n2,c)(m)) =
                  IF n1=n2 THEN c ELSE colour(n1)(m) ENDIF

  mem_ax3 : AXIOM colour(n1)(set_son(n2,i,k)(m)) = colour(n1)(m)

  mem_ax4 : AXIOM son(n1,i1)(set_son(n2,i2,k)(m)) =
                  IF n1=n2 AND i1=i2 THEN k ELSE son(n1,i1)(m) ENDIF

  mem_ax5 : AXIOM son(n1,i)(set_colour(n2,c)(m)) = son(n1,i)(m)
END Memory
```

Fig. 2. The Memory

been to define the memory explicitly as a function from pairs of nodes and indexes to nodes). First, however, some types of nodes, indexes and roots are defined. The types NODE and INDEX are defined just by the natural numbers. Our functions will be applied to arguments of these types.

The types Node and Index are the constrained versions where only natural numbers below respectively NODES and SONS are considered. These latter constrained types are used in axioms where universally quantified variables range over them: we only want our functions to behave correctly within the boundaries of the memory. In addition the type Colour represents *black* with TRUE and *white* with FALSE.

The reason for not using the constrained types in the signatures of functions is that if we did, the PVS type checker would generate TCC's that we could not prove without considering the execution traces that lead to the application of these functions. In fact, some of the invariants that we shall later prove states exactly that these functions are indeed only applied to values that lie within the constrained types. If one really wants to catch such "errors" using type checking, then one needs to define a subtype of well-formed states of the state type that we later will introduce. This, however, is not simple, and may involve strengthening of this well-formedness predicate during the proof of TCC's. We rather prefer to have a PVS specification type checked quickly without too deep proofs.

The memory is read and modified via four functions, and a constant `null_array` represents the initial memory containing 0 in all memory cells (axiom `mem_ax1`). The function `colour` returns the colour of a node. The function `set_colour` assigns a colour to a node. The function `son` returns the pointer contained in a particular cell. That is, the expression `son(n,i)(m)` returns the pointer contained in the cell identified by node n and index i. Finally, the function `set_son` assigns a pointer to a cell. That is, the expression `set_son(n,i,k)(m)` returns the memory m updated in cell (n,i) to contain (a pointer to node) k.

Accessible Nodes In this section we define what it means for a node to be accessible. First, however, we introduce some functions on lists (figure 3).

```
List_Functions[T:TYPE+] : THEORY
BEGIN
  last(l:list[T]|cons?(l)) : RECURSIVE T =
    IF length(l)=1 THEN car(l) ELSE last(cdr(l)) ENDIF
    MEASURE length(l)

  last_index(l:list[T]|cons?(l)) : nat = length(l)-1

  suffix(l:list[T],n:nat |n < length(l)) : RECURSIVE list[T] =
    IF n=0 THEN l ELSE suffix(cdr(l),n-1) ENDIF
    MEASURE length(l)

  last_occurrence(x:T,l:list[T] | member(x,l)):nat =
    epsilon! (idx:nat):
      idx <= last_index(l) AND nth(l,idx) = x AND
      (idx < last_index(l) IMPLIES NOT member(x,suffix(l,idx+1)))
END List_Functions
```

Fig. 3. List Functions

The function `last` returns the last element of a non-empty list, while the function `last_index` returns the index of the last element in a list. So for example if l = cons(5,cons(7,cons(9,null))), then last(l) = 9 and last_index(l) = 2. The other two functions are used only in the proof: one taking the suffix of a list and the other returning the index of the last occurrence of a given element in a list, assuming it exists. The next theory (figure 4) defines the function `accessible`.

The function `points_to` defines what it means for one node, n1, to point to another, n2, in the memory m. The function `pointed` is a predicate on lists of nodes, and is TRUE for a list if for any two successive nodes in the list, the first points to the next in the memory. The function `path` is also a predicate on lists of nodes, and is TRUE for a list if that list represents a non-empty pointed list starting with a root. Finally, a node is `accessible` if it is the last element in some path.

```
Memory_Functions[NODES:posnat,SONS:posnat,ROOTS:posnat] : THEORY
BEGIN
  ASSUMING
    roots_within : ASSUMPTION ROOTS <= NODES
  ENDASSUMING
  IMPORTING List_Functions
  IMPORTING Memory[NODES,SONS,ROOTS]

  m : VAR Memory

  points_to(n1,n2:NODE)(m):bool =
    n1 < NODES AND n2 < NODES AND EXISTS (i:Index): son(n1,i)(m)=n2

  pointed(p:list[Node])(m):bool =
    length(p) >= 2 IMPLIES
      FORALL (i:nat|i<last_index(p)): points_to(nth(p,i),nth(p,i+1))(m)

  path(p:list[Node])(m):bool =
    cons?(p) AND car(p) < ROOTS AND pointed(p)(m)

  accessible(n:NODE)(m):bool =
    EXISTS (p:list[Node]) : path(p)(m) AND last(p) = n
  ...
END Memory_Functions
```

Fig. 4. The Predicate `accessible`

Appending Garbage Nodes In this section we define the operation for appending a garbage node to the list of free nodes, that can be allocated by the mutator. This operation will be defined abstractly, assuming as little as possible about it's behavior. Note that, since the free list is supposed to be part of the memory, we could easily have defined this operation in terms of the functions son and set_son, but this would have required that we took some design decisions as to how the list was represented (for example where the head of the list should be and whether new elements should be added first or last).

The definitions in figure 5 belong to the theory Memory_Functions that we introduced part of in figure 4.

First of all, the predicate closed holds for a memory, if no pointer points outside the memory. The function append_to_free is defined by four axioms, having the following informal explanation:

append_ax1 The appending operation leaves colours unchanged.
append_ax2 The appending operation returns a closed memory when applied to a such.
append_ax3 In appending a garbage node, only that node becomes accessible, and the accessibility of all other nodes stay unchanged.
append_ax4 In appending a garbage node, no pointer from any other garbage node is altered.

3.2 The Mutator and the Collector

The mutator and the collector are introduced in the theory Garbage_Collector in figure 6. First of all, each process has a program counter; the program counter of the

```
m     : VAR Memory
n,f   : VAR Node
i     : VAR Index

closed(m):bool =
  FORALL (n:Node): FORALL (i:Index): son(n,i)(m) < NODES

append_to_free : [NODE -> [Memory -> Memory]]

append_ax1 : AXIOM colour(n)(append_to_free(f)(m)) = colour(n)(m)

append_ax2 : AXIOM closed(m) IMPLIES closed(append_to_free(f)(m))

append_ax3 : AXIOM (NOT accessible(f)(m)) IMPLIES
                    (accessible(n)(append_to_free(f)(m)) IFF
                     (n=f OR accessible(n)(m)))

append_ax4 : AXIOM (NOT accessible(f)(m) AND NOT accessible(n)(m) AND n /= f)
                    IMPLIES
                    son(n,i)(append_to_free(f)(m)) = son(n,i)(m)
```

Fig. 5. The append_to_free Operation

mutator ranges over the type MuPC having two values, while the program counter of the collector ranges over the type CoPC having nine values. The state type is defined as a record type, which contains the program counters, the memory M, and a number of other auxiliary variables (the Q variable is used by the mutator, while BC, OBC, H, I, J, K and L are used by the collector as will be explained below). The initial values of the state variables are defined by the predicate initial.

Now, the mutator and the collector are each defined as a transition relation, being a predicate on pairs of states. Hence, for example if MUTATOR(s1,s2) holds for two states s1 and s2, it means that starting in state s1, the mutator can make a transition into state s2. We shall below show the details of these definitions.

The global transition relation for the whole system, called next, is then defined as the disjunction between the mutator and the collector: in each step, either the mutator makes a move, or the collector does. This corresponds to an interleaving semantics of concurrency.

It is finally possible to define what is a trace of the system: it is a sequence[1] of states where the first state satisfies the initial predicate, and where any two consecutive states are related by the next relation.

The Mutator The mutator has two possible transitions, each defined as a function that when applied to an *old* state yields a *new* state (figure 7). MUTATOR(s1,s2) then holds for two states s1 and s2, if s2 can be obtained from s1, by applying one of the rules.

Each transition function is defined in terms of an IF-THEN-ELSE expression, where the condition represents the guard of the transition (the situation where the tran-

---

[1] A sequence in PVS is modeled as a function from natural numbers to the type of the sequence elements, in this case State. Hence a sequence here represents an infinite enumeration of states. Sequences are defined as part of the PVS prelude.

```
Garbage_Collector[NODES:posnat,SONS:posnat,ROOTS:posnat] : THEORY
BEGIN
  ASSUMING
    roots_within : ASSUMPTION ROOTS <= NODES
  ENDASSUMING
  IMPORTING Memory_Functions[NODES,SONS,ROOTS]

  MuPC : TYPE = {MU0,MU1}
  CoPC : TYPE = {CHI0,CHI1,CHI2,CHI3,CHI4,CHI5,CHI6,CHI7,CHI8}

  State : TYPE =
    [# MU : MuPC, CHI : CoPC, Q : NODE, BC : nat, OBC : nat,
       H  : nat, I : nat, J : nat, K : nat, L : nat, M  : Memory #]

  s,s1,s2 : VAR State

  initial(s):bool =
    MU(s) = MU0 & CHI(s) = CHI0 & Q(s) = 0 & BC(s) = 0 & OBC(s) = 0 &
    H(s) = 0 & I(s) = 0 & J(s) = 0 & K(s) = 0 & L(s)  = 0 & M(s) = null_array

  ...

  MUTATOR(s1,s2):bool = ...

  COLLECTOR(s1,s2):bool = ...

  next(s1,s2):bool = MUTATOR(s1,s2) OR COLLECTOR(s1,s2)

  trace(seq:sequence[State]):bool =
     initial(seq(0)) AND FORALL (n:nat):next(seq(n),seq(n+1))
END Garbage_Collector
```

Fig. 6. The Garbage Collector Components

sition may meaningfully be applied), and where the ELSE part returns the unchanged state, in case the guard is false[2].

The Rule_mutate rule represents the modification of a pointer. It is parameterized with the cell (m, i) to modify, and the node that this cell should hereafter point to, n. These parameters are then existentially quantified over in the definition of MUTATOR, corresponding to a non-deterministic choice[3] of m, i and n. The rule reads as follows: The values m, i and n are arbitrarily selected. If the program counter is MU0, and if the target node n is accessible, then the memory M in the state is updated; Q is set to point to the new target node, and finally the program counter is changed to MU1. The rule Rule_colour_target simply colours the target (now pointed to by Q) of the mutation, and returns control to MU0, enabling another mutation.

---

[2] This allows for *stuttering* where rules are applied without changing the state. If done infinitely often our system would never progress. One way to avoid such behavior is to impose certain fairness constraints on execution traces. We shall, however, not do this since we are only interested in verifying safety properties, where such problems play no role.

[3] The way we model this non-deterministic choice is quite different from the way it is modeled in [18], where the state is extended with an extra component which represents *all the unknown factors that influence the choice*. There are then special transitions to update this component.

```
% MU0 : Redirect arbitrary pointer.

Rule_mutate(m:Node,i:Index,n:Node)(s):State =
  IF MU(s) = MU0 AND accessible(n)(M(s)) THEN
    s WITH [M := set_son(m,i,n)(M(s)), Q := n, MU := MU1]
  ELSE s ENDIF

% MU1 : Colour target of redirection.

Rule_colour_target(s):State =
  IF MU(s) = MU1 THEN
    s WITH [M := set_colour(Q(s),TRUE)(M(s)), MU := MU0]
  ELSE s ENDIF

% ----------------------
% Combining MUTATOR Rules
% ----------------------

MUTATOR(s1,s2):bool =
  (EXISTS (m:Node,i:Index,n:Node): s2 = Rule_mutate(m,i,n)(s1))
  OR s2 = Rule_colour_target(s1)
```

Fig. 7. The Mutator Transitions

The Collector The collector (figures 8, 9, and 10) uses the auxiliary variables BC and OBC for counting black nodes, and H, I, J, K and L for controlling loops. The program counter ranges over the values CHI0 to CHI8.

The Marking Phase (CHI0 ... CHI6)

Root Blackening (CHI0) At CHI0 all the roots from 0 to ROOTS-1 are blackened. The variable K, having the initial value 0, is used to loop through the roots. As soon as the roots have been blackened (K = ROOTS), the propagation phase is started by setting the program counter to CHI1.

Propagation (CHI1, CHI2, CHI3) Here all nodes from 0 to NODES-1 reachable from a root via a pointer are blackened. The variable I, having the initial value 0, is used to loop through the nodes. At CHI2 it is examined whether the current node is black. If not, it is just skipped, and I is increased. If it is black, then at CHI3, all the sons of I are blackened, using the variable J to range over indexes. When I = NODES, all nodes have been processed, and the counting phase is started by setting the program counter to CHI4.

Counting (CHI4, CHI5, CHI6) At CHI4 and CHI5, the black nodes are counted in the variable BC. The variable H, having the initial value 0, is used to loop through the nodes. When the black nodes have been counted, in CHI6, the new count BC is compared to the previous count which is stored in OBC (old black count). If they differ, then the propagation phase is restarted by setting the program counter to CHI1. If they are equal, the appending phase is started at CHI7.

The Appending Phase (CHI7, CHI8) Here all white nodes are appended to the free list, while all black nodes are just coloured white. The variable L, having the initial value 0, is used to loop through the nodes.

```
% -------------
% Blacken Roots
% -------------

% CHI0 : Blacken.

Rule_stop_blacken(s):State =
  IF CHI(s) = CHI0 AND K(s) = ROOTS THEN
    s WITH [I := 0, CHI := CHI1]
  ELSE s ENDIF

Rule_blacken(s):State =
  IF CHI(s) = CHI0 AND K(s) /= ROOTS THEN
    s WITH [M := set_colour(K(s),TRUE)(M(s)), K := K(s) + 1, CHI := CHI0]
  ELSE s ENDIF

% ------------------
% Propagate Colouring
% ------------------

% CHI1 : Decide whether to continue propagating.

Rule_stop_propagate(s):State =
  IF CHI(s) = CHI1 AND I(s) = NODES THEN
    s WITH [BC := 0, H := 0, CHI := CHI4]
  ELSE s ENDIF

Rule_continue_propagate(s):State =
  IF CHI(s) = CHI1 AND I(s) /= NODES THEN
    s WITH [CHI := CHI2]
  ELSE s ENDIF

% CHI2 : (Continue) Check whether node is black.

Rule_white_node(s):State =
  IF CHI(s) = CHI2 AND NOT colour(I(s))(M(s)) THEN
    s WITH [I := I(s) + 1, CHI := CHI1]
  ELSE s ENDIF

Rule_black_node(s):State =
  IF CHI(s) = CHI2 AND colour(I(s))(M(s)) THEN
    s WITH [J := 0, CHI := CHI3]
  ELSE s ENDIF

% CHI3 : (Node is black) Colour each son.

Rule_stop_colouring_sons(s):State =
  IF CHI(s) = CHI3 AND J(s) = SONS THEN
    s WITH [I := I(s) + 1, CHI := CHI1]
  ELSE s ENDIF

Rule_colour_son(s):State =
  IF CHI(s) = CHI3 AND J(s) /= SONS THEN
    s WITH [M := set_colour(son(I(s),J(s))(M(s)),TRUE)(M(s)),
            J := J(s) + 1, CHI := CHI3]
  ELSE s ENDIF

% ----------------
% Count Black Nodes
% ----------------
```

Fig. 8. Collector Transitions (a)

```
% CHI4 : Decide whether to continue counting.

Rule_stop_counting(s):State =
  IF CHI(s) = CHI4 AND H(s) = NODES THEN
    s WITH [CHI := CHI6]
  ELSE s ENDIF

Rule_continue_counting(s):State =
  IF CHI(s) = CHI4 AND H(s) /= NODES THEN
    s WITH [CHI := CHI5]
  ELSE s ENDIF

% CHI5 : (Continue) Count one up if black.

Rule_skip_white(s):State =
  IF CHI(s) = CHI5 AND NOT colour(H(s))(M(s)) THEN
    s WITH [H := H(s) + 1, CHI := CHI4]
  ELSE s ENDIF

Rule_count_black(s):State =
  IF CHI(s) = CHI5 AND colour(H(s))(M(s)) THEN
    s WITH [BC   := BC(s) + 1, H    := H(s) + 1, CHI := CHI4]
  ELSE s ENDIF

% CHI6 : Compare BC and OBC.

Rule_redo_propagation(s):State =
  IF CHI(s) = CHI6 AND BC(s) /= OBC(s) THEN
    s WITH [OBC := BC(s), I := 0, CHI := CHI1]
  ELSE s ENDIF

Rule_quit_propagation(s):State =
  IF CHI(s) = CHI6 AND BC(s) = OBC(s) THEN
    s WITH [L := 0, CHI := CHI7]
  ELSE s ENDIF

% -------------------
% Append to Free List
% -------------------

% CHI7 : Decide whether to continue appending.

Rule_stop_appending(s):State =
  IF CHI(s) = CHI7 AND L(s) = NODES THEN
    s WITH [BC := 0, OBC := 0, K := 0, CHI := CHI0]
  ELSE s ENDIF

Rule_continue_appending(s):State =
  IF CHI(s) = CHI7 AND L(s) /= NODES THEN
    s WITH [CHI := CHI8]
  ELSE s ENDIF

% CHI8 : (Continue) Append if white.

Rule_black_to_white(s):State =
  IF CHI(s) = CHI8 AND colour(L(s))(M(s)) THEN
    s WITH [M := set_colour(L(s),FALSE)(M(s)), L := L(s) + 1,CHI := CHI7]
  ELSE s ENDIF

Rule_append_white(s):State =
  IF CHI(s) = CHI8 AND NOT colour(L(s))(M(s)) THEN
    s WITH [M := append_to_free(L(s))(M(s)), L := L(s) + 1, CHI := CHI7]
  ELSE s ENDIF
```

Fig. 9. Collector Transitions (b)

```
% ------------------------
% Combining COLLECTOR Rules
% ------------------------

COLLECTOR(s1,s2):bool =
      s2 = Rule_stop_blacken(s1)
  OR s2 = Rule_blacken(s1)
  OR s2 = Rule_stop_propagate(s1)
  OR s2 = Rule_continue_propagate(s1)
  OR s2 = Rule_white_node(s1)
  OR s2 = Rule_black_node(s1)
  OR s2 = Rule_stop_colouring_sons(s1)
  OR s2 = Rule_colour_son(s1)
  OR s2 = Rule_stop_counting(s1)
  OR s2 = Rule_continue_counting(s1)
  OR s2 = Rule_skip_white(s1)
  OR s2 = Rule_count_black(s1)
  OR s2 = Rule_redo_propagation(s1)
  OR s2 = Rule_quit_propagation(s1)
  OR s2 = Rule_stop_appending(s1)
  OR s2 = Rule_continue_appending(s1)
  OR s2 = Rule_black_to_white(s1)
  OR s2 = Rule_append_white(s1)
```

Fig. 10. Collector Transitions (c)

4 Theorem Proving in PVS

In this section we outline the proof of correctness for the garbage collector algorithm. First, we formulate in PVS what it means for the collector to be safe. We then outline the technique we have applied to master the relatively big size of the proof. This technique seems general and useful for verifying safety properties since it divides the proof into manageable lemmas. Then we introduce some auxiliary functions (concepts) that are needed during the proof; and finally, we outline the proof itself by listing the needed invariants.

4.1 Formulating the Safety Property

Let us recall the safety property: *no accessible node is ever appended to the free list*. As can be seen from the collector algorithm in figure 9, the `append_to_free` operation is only applied at location CHI8 (in the rule `Rule_append_white`). It is applied to the node `L(s)`, but only if this node is white: `NOT colour(L(s))(M(s))`. Hence, the correctness criteria can be stated as: *Whenever the program counter is* CHI8, *and* L *is accessible, then* L *is black* (and will hence not be appended). This is stated in the theory in figure 11.

The theory defines two predicates and a theorem: the correctness criteria. The predicate `invariant` takes as argument a predicate p on states (`pred[State]` is short for the function space `[State -> bool]`). It returns TRUE if for any execution trace `tr` of the program: the predicate p holds in every position of that trace. The safety property we want to verify is defined by the predicate `safe`. The correctness criteria is then defined by the theorem named `safe`. The dots ... in the theory refers to extra

```
Garbage_Collector_Proof[NODES:posnat,SONS:posnat,ROOTS:posnat] : THEORY
BEGIN
  ASSUMING
    roots_within : ASSUMPTION ROOTS <= NODES
  ENDASSUMING
  IMPORTING Garbage_Collector[NODES,SONS,ROOTS]

  invariant(p:pred[State]):bool =
    FORALL (tr:(trace)): FORALL (n:nat):p(tr(n))
  ...
  safe(s:State):bool =
    CHI(s) = CHI8 AND accessible(L(s))(M(s)) IMPLIES colour(L(s))(M(s))

  safe : THEOREM invariant(safe)
END Garbage_Collector_Proof
```

Fig. 11. The Safety Property

invariants that we needed to add (and prove), in order to prove safe (via what we call *invariant strengthening*).

4.2 The Proof Technique

We now sketch the principle behind the proof technique we applied. All definitions that follow are defined in the theory Garbage_Collector_Proof, which we show part of in figure 11. The predicate we want to prove true in all accessible states is the predicate safe of figure 11. However, this invariant needs to be greatly strengthened (extended) in order to be provable. This extension will be discovered in a stepwise manner during the proof, and not at once.

In principle we could then just go ahead with the proof and extend the invariant whenever we find it necessary. This does, however, in general (for non-trivial examples) lead to a big and unhandy invariant. Also, if we keep extending the invariant as we need, we have to stop the prover for each extension (since we now modify one of the formulae) and then redo what we already had succeeded with. This turns out to be painful and unnecessary. Instead, we split the invariant into lemmas as shall be illustrated below.

The technique allows in addition the (proofs of) sub-invariants to mutually depend on each other in a circular way. For example, suppose that we want to prove the invariant I_1, and that we discover that we need to prove I_2 first, and that further the proof of I_2 depends on the truthness of I_1. Of course this is not a problem if we simply proved that the conjunction $I_1 \wedge I_2$ is an invariant. However, as just stated, we don't want to work with this (potentially big) conjunct, and the proofs of I_1 and I_2 have to be split into lemmas in such a way that this recursion is allowed (note that PVS does not directly allow two lemmas to refer to each other – in their proofs that is). Figure 12 outlines (an illustrative subset of) the definitions and lemmas that we add to the Garbage_Collector_Proof theory in figure 11.

The functions IMPLIES and & are just the corresponding boolean operators lifted to work on state predicates. Next, we define the predicate preserved, with which we can state that a property p is inductive wrt. our program — relative to some invariant I. That is, the expression preserved(I)(p) is true if the predicate p is true in the

```
Garbage_Collector_Proof[NODES:posnat,SONS:posnat,ROOTS:posnat] : THEORY
BEGIN
  ...
  IMPLIES(p1,p2:pred[State]):bool = FORALL (s:State): p1(s) IMPLIES p2(s);

  &(p1,p2:pred[State]):pred[State] = LAMBDA (s:State): p1(s) AND p2(s)

  preserved(I:pred[State])(p:pred[State]):bool =
    (initial IMPLIES p) AND
    FORALL (s1,s2:State): I(s1) AND p(s1) AND next(s1,s2) IMPLIES p(s2)

  s : VAR State

  inv1(s):bool = ...
  inv2(s):bool = ...
  I   : pred[State]
  pi  : [pred[State] -> bool] = preserved(I)

  i_inv1   : LEMMA I IMPLIES inv1
  i_inv2   : LEMMA I IMPLIES inv2
  i_safe   : LEMMA I IMPLIES safe
  p_inv1   : LEMMA pi(inv1)
  p_inv2   : LEMMA pi(inv2)
  p_safe   : LEMMA pi(safe)
  p_I      : LEMMA pi(I)
  correct  : LEMMA invariant(I)
END Garbage_Collector_Proof
```

Fig. 12. The Proof

initial state, and if p is preserved by the next-step relation, under the assumption that the property I holds in the pre-state.

Now we let this I be defined as *unknown*, and let pi be an instantiation of preserved with this I. This I is now supposed to represent the unknown final invariant that we are looking for. For each new invariant we add (like inv1 and inv2 – there are 19 in total), we add three declarations: the definition of the invariant predicate (fx. inv1); a lemma stating that this invariant is implied by I (fx. i_inv1), and finally that the predicate is inductive (fx. p_inv1)[4]. Then during the proof of any pi(...) lemma, we can refer to all the (up to that point introduced) invariants via the I IMPLIES ... lemmas.

Finally, when all invariants have been discovered (no new are needed), we can define I as the conjunction of all the introduced invariants – there are 19 in our case, however inv13, inv16 and safe are logically implied by the rest:

```
I : pred[State] = inv1 & inv2 & inv3 & inv4 & inv5 & inv6 & inv7 & inv8
              & inv9 & inv10 & inv11 & inv12 & inv14 & inv15 & inv17 & inv18 & inv19
```

With this definition all the I IMPLIES ... lemmas can now be proved. Furthermore, we can prove the pi(I) lemma, which directly leads to the proof of the invariant(I) lemma, which again leads to the correctness of the invariant(safe) lemma, and we are done.

The verification of the protocol has been automated as far as possible by defining a set of tactics. One can say that the proof of the 20 invariants is (almost) automated

---

[4] It turns out that some invariants are logical consequences of others, and that for these we can avoid reasoning about the transition relation and just prove the implication.

"up to" lemmas. That is: if some invariant is provable given explicitly as assumptions the other invariants and lemmas it depends on, then it is proved automatically using a single tactic. The obtained level of automatization could be achieved only because of the flexibility provided by the PVS decision procedures. When we say *almost* automated, it means that in some few cases, we needed to assist the prover – always because the PVS (inst?) command did not get instantiations right. The program contains 20 transitions, and with 20 invariants this gives 400 (20*20) proofs, and of these 6 needed manual assistance (two transitions in the proof of inv15 and four in inv17), corresponding to 98.5% automatization. It should though be said that the proofs of lemmas about auxiliary functions were in general not automatic.

```
Memory_Observers[NODES:posnat,SONS:posnat,ROOTS:posnat] : THEORY
BEGIN
  ASSUMING
    roots_within : ASSUMPTION ROOTS <= NODES
  ENDASSUMING
  IMPORTING Memory_Functions[NODES,SONS,ROOTS]

  m : VAR Memory

  <(p1,p2:[NODE,INDEX]):bool =
    LET n1 = PROJ_1(p1), i1 = PROJ_2(p1), n2 = PROJ_1(p2), i2 = PROJ_2(p2) IN
      n1 < n2 OR (n1 = n2 AND i1 < i2);

  <=(p1,p2:[NODE,INDEX]):bool = p1 < p2 OR p1 = p2

  blacks(l,u:NODE)(m) : RECURSIVE nat =
    IF l < u AND l < NODES THEN
      IF colour(l)(m) THEN 1 ELSE 0 ENDIF + blacks(l+1,u)(m)
    ELSE 0 ENDIF
    MEASURE abs(u-l)

  black_roots(u:NODE)(m):bool = FORALL (r:Root| r < u): colour(r)(m)

  bw(n:NODE,i:INDEX)(m):bool =
    n < NODES AND i < SONS AND colour(n)(m) AND NOT colour(son(n,i)(m))(m)

  exists_bw(n1:NODE,i1:INDEX,n2:NODE,i2:INDEX)(m):bool =
    EXISTS (n:Node,i:Index):
      bw(n,i)(m) AND NOT (n,i) < (n1,i1) AND (n,i) < (n2,i2)

  propagated(m):bool = NOT exists_bw(0,0,NODES,0)(m)

  blackened(l:NODE)(m):bool =
    FORALL (n:Node|l <= n): accessible(n)(m) IMPLIES colour(n)(m)
END Memory_Observers
```

Fig. 13. Auxiliary Functions

During the proof, a collection of auxiliary functions[5] are needed, mostly in order to formulate the new invariants, that are introduced to prove the original invariant. These functions are introduced in figure 13.

---

[5] These functions were not spelled out in [18], although their informal descriptions were given. Furthermore, no properties about these functions were presented.

```
inv1(s) :bool = I(s) <= NODES AND ((CHI(s)=CHI2 OR CHI(s)=CHI3)
                    IMPLIES I(s) < NODES)

inv2(s) :bool = J(s) <= SONS

inv3(s) :bool = K(s) <= ROOTS

inv4(s) :bool = H(s) <= NODES AND (CHI(s)=CHI5 IMPLIES H(s) < NODES) AND
                    (CHI(s)=CHI6 IMPLIES H(s) = NODES)

inv5(s) :bool = L(s) <= NODES AND (CHI(s)=CHI8 IMPLIES L(s) < NODES)

inv6(s) :bool = Q(s) < NODES

inv7(s) :bool = closed(M(s))

inv8(s) :bool = (CHI(s)=CHI4 OR CHI(s)=CHI5)
                    IMPLIES BC(s) <= blacks(0,H(s))(M(s))

inv9(s) :bool = CHI(s)=CHI6 IMPLIES BC(s) <= blacks(0,NODES)(M(s))

inv10(s):bool = (CHI(s)=CHI0 OR CHI(s)=CHI1 OR CHI(s)=CHI2 OR CHI(s)=CHI3)
                    IMPLIES OBC(s) <= blacks(0,NODES)(M(s))

inv11(s):bool = (CHI(s)=CHI4 OR CHI(s)=CHI5 OR CHI(s)=CHI6)
                    IMPLIES OBC(s) <= BC(s) + blacks(H(s),NODES)(M(s))

inv12(s):bool = BC(s) <= NODES

inv13(s):bool = CHI(s)=CHI6 IMPLIES OBC(s) <= BC(s)

inv14(s):bool = (CHI(s)=CHI0 OR CHI(s)=CHI1 OR CHI(s)=CHI2 OR CHI(s)=CHI3 OR
                 CHI(s)=CHI4 OR CHI(s)=CHI5 OR CHI(s)=CHI6) IMPLIES
                    black_roots(IF CHI(s)=CHI0 THEN K(s) ELSE ROOTS ENDIF)(M(s))

inv15(s):bool = FORALL (n:Node,i:Index):
                    (((CHI(s)=CHI1 OR CHI(s)=CHI2 OR CHI(s)=CHI3) AND
                     blacks(0,NODES)(M(s)) = OBC(s) AND
                     (n,i) < (I(s),IF CHI(s)=CHI3 THEN J(s) ELSE 0 ENDIF) AND
                     bw(n,i)(M(s)))
                        IMPLIES (MU(s)=MU1 AND son(n,i)(M(s))=Q(s)))

inv16(s):bool = ((CHI(s)=CHI1 OR CHI(s)=CHI2 OR CHI(s)=CHI3) AND
                 blacks(0,NODES)(M(s)) = OBC(s) AND
                 exists_bw(0,0,I(s),IF CHI(s)=CHI3 THEN J(s) ELSE 0 ENDIF)(M(s)))
                    IMPLIES MU(s)=MU1

inv17(s):bool = ((CHI(s)=CHI1 OR CHI(s)=CHI2 OR CHI(s)=CHI3) AND
                 blacks(0,NODES)(M(s)) = OBC(s) AND
                 exists_bw(0,0,I(s),IF CHI(s)=CHI3 THEN J(s) ELSE 0 ENDIF)(M(s)))
                    IMPLIES
                    exists_bw(I(s),IF CHI(s)=CHI3 THEN J(s)
                                        ELSE 0 ENDIF,NODES,0)(M(s))

inv18(s):bool = ((CHI(s)=CHI4 OR CHI(s)=CHI5 OR CHI(s)=CHI6) AND
                 OBC(s) = BC(s) + blacks(H(s),NODES)(M(s)))
                    IMPLIES blackened(0)(M(s))

inv19(s):bool = (CHI(s)=CHI7 OR CHI(s)=CHI8) IMPLIES blackened(L(s))(M(s))
```

Fig. 14. Invariants

The predicates < and <= define lexicographic ordering on node-index pairs, where each such pair identifies a cell in our memory. The projection function PROJ_i (for i ∈ {1,2}) selects the i'th component of a tuple. Hence, for example (2,3) < (3,0).

The rest of the functions are explained as follows. The expression blacks(l,u)(m) returns the number of black nodes between l (included) and u (excluded). In particular, blacks(0,NODES)(m) represents the total number of black nodes in the memory m. The expression black_roots(u)(m) returns true if all the nodes below u are black. In particular, black_roots(ROOTS)(m) if all roots are black. The expression bw(n,i)(m) returns true if node n is black and the son of cell (n,i) is white. The expression exists_bw(n1,i1,n2,i2)(m) returns true if there exists a pointer between (n1,i1) and (n2,i2) from a black node to a white node. The expression propagated(m) returns true if no black node points to a white node. Finally, blackened(l)(m) returns true if all nodes above (and including) l are black if they are accessible.

55 lemmas are needed (and proved) about these functions in order to carry out the proof of the safety property. In addition, 15 lemmas about various general list processing functions are needed. These lemmas are given in appendix A. This should be compared to Russinoff's "over one hundred lemmas characterizing the behavior of relevant functions" in [18]. The invariants defined and proved are given in figure 14. These are the same as in [18]. The proof took 1.5 months of effort.

5 Model Checking in Murphi

In this section, we shortly outline our experience with encoding the garbage collector in the Murphi model checker [15]. The full formal Murphi program is contained in appendix B.

Murphi uses a program model that is similar to those of UNITY [5] and TLA [14], hence the one we have used in our PVS specification. A Murphi program has three components: a declaration of the global variables, a description of the initial state, and a collection of transition rules. Each transition rule is a guarded command that consists of a boolean guard expression over the global variables, and a deterministic statement that changes the global variables. In addition, one can state invariant conditions to be verified.

An execution of a Murphi program is obtained by repeatedly *(1) arbitrarily selecting one of the transition rules where the boolean guard is true in the current state; (2) executing the statement of the chosen transition rule.* The statement is executed atomically: no other transition rules are executed in parallel. Thus state transitions are interleaving and processes communicate via shared variables. The notion of process is not formally supported, but may be thought of as a subset of the transition rules. The Murphi verifier tries to explore all reachable states in order to ensure that all invariants hold. If a violation is detected, Murphi generates a violating trace.

The reader is referred to the appendix for the details of the model. Here we shall focus on the differences between the PVS model and the Murphi model. The two principal advantages of PVS are that (1): in PVS we can verify a parameterized program, whereas in Murphi we are limited to a finite state program; and: (2) in PVS we can be abstract at the algorithmic level, whereas in Murphi we have to make certain design choices. The advantage of Murphi is that verification is automatic, whereas in PVS we have to manually assist the proof.

Infinite Versus Finite State

The first obvious difference concerns the size of the memory. In PVS, the boundaries (NODES, SONS and ROOTS) are unspecified parameters (figure 2), hence the correctness does not depend on their specific values. The garbage collector is therefore verified for any size of memory. In the Murphi program, on the other hand, we have to fix the boundaries to particular natural numbers. In our case, we verified the algorithm with the following values: NODES = 3, SONS = 2 and ROOTS = 1 (figure 15). In this context, Murphi used 803 seconds to verify the invariant, exploring 415633 states. It turned out that Murphi was unable to verify bigger memories within reasonable time (24 hours). An experiment with 4 nodes, two sons and two roots had not terminated after 25 hours and over 5 million states visited.

```
Const
  NODES : 3; SONS   : 2; ROOTS : 1;
Type
  Node : 0..NODES-1; Index : 0..SONS-1; Colour : boolean;
  NodeStruct : Record colour : Colour; cells : Array[Index] Of Node; End;
Var
  M : Array[Node] Of NodeStruct;

Function colour(n:Node):Colour;
  Begin Return M[n].colour; End;

Procedure set_colour(n:Node;c:Colour);
  Begin M[n].colour := c; End;

Function son(n:Node;i:Index):Node;
  Begin Return M[n].cells[i]; End;

Procedure set_son(n:Node;i:Index;k:Node);
  Begin M[n].cells[i] := k; End;
```

Fig. 15. The Memory

Abstractness of Memory

The PVS memory is abstractly specified in terms of a set of axioms (figure 2). Although we think of the memory as an array of two dimensions, this is in fact not required by an implementation. In the Murphi program, on the other hand, we need to choose an implementation of the memory, and we have chosen to model it as a two dimensional array as illustrated in figure 15.

Abstractness of the Append Operation

The PVS append operation is abstractly specified in terms of a set of axioms (figure 5). Hence we have made no decisions for example as to where is the head of the free list, or whether to append elements first or last. In Murphi we are obliged to take such decisions. Figure 16 shows how we have chosen cell (0,0) to be the head of the list, such that new elements are added to the front.

```
Procedure append_to_free(new_free:Node);
Var old_first_free : Node;
Begin
  old_first_free := son(0,0); set_son(0,0,new_free);
  For i:Index Do set_son(new_free,i,old_first_free) EndFor;
End;
```

Fig. 16. The append_to_free Operation

```
Function accessible(n:Node):boolean;
Type Status : Enum{TRY,UNTRIED,TRIED};
Var status : Array[Node] Of Status; s : Node; try_again : boolean;
Begin
  For k:Node Do status[k] := (is_root(k) ? TRY : UNTRIED) EndFor;
  try_again := true;
  While try_again Do
    try_again := false;
    For k:Node Do
      If status[k]=TRY Then
        For j:Index Do
          s := son(k,j);
          If status[s]=UNTRIED Then status[s] := TRY; try_again := true; End;
        EndFor;
        status[k] := TRIED;
      End;
    EndFor;
  End;
  Return status[n]=TRIED
End;
```

Fig. 17. The accessible Predicate

Abstractness of the Accessibility Predicate

The PVS accessible predicate is specified in an abstract manner using existential quantification over paths (figure 4). Such a formulation is not possible in Murphi, where we have to code an algorithm that marks nodes already visited during the examination of accessible nodes. This is to avoid a looping behavior in case of cyclic accessibility in the memory. Figure 17 illustrates the algorithm.

6 Observations

The properties that we formulated and proved in PVS can be divided into two classes: invariants and properties about auxiliary functions; we call the latter just lemmas. There were 20 invariants, the same as in [18], and there were 70 lemmas, whereas [18] has over 100. It's however not clear what the reason for this reduction is, since [18] does not contain the lemmas. 98.5% of our invariant proofs were automatic, once the lemmas and other invariants needed as assumptions were identified. That is, observing that the program has 20 possible transitions and that there were 20 invariants, there were hence 20*20 = 400 transition proofs, whereof 6 needed manual assistance. The assistance always consisted of guiding the instantiation of universally quantified assumptions when the PVS (inst?) command did not succeed in finding the right instantiations. Many of the lemmas needed manual assistance.

The general approach to the proof of an invariant was as follows. The proof would typically fail, the result being a set of unproved sequents. Basically, a generalization of the conjunction of these sequents would form the new invariant to prove, and the process continued. This style of proof was also applied in [11] to a communications protocol. A particular hard problem seems to be the occurrence of loops in this strengthening process, implying possibly infinite strengthening. Here is where generalization is needed in terms of introducing quantifiers into the invariant. The PVS proof took 1.5 months of effort.

Murphi performed the proof automatically in less than 14 minutes, but we had to fix the boundaries of the memory, and we had to make concrete implementation choices at the algorithmic level as well as at the data type level. The size of the memory for which the garbage collector could be model checked was so small that it could represent a boarder case (three nodes, two sons per node, and one root), which hence may give less confidence in what to conclude from the verification result.

We believe that this example provides a good case study for efforts to improve theorem proving techniques as well as model checking techniques. This is mainly due to its small size (lines of code) combined with the difficulty to prove it. The example is in particular a challenge to invariant strengthening techniques such as [4, 3], and abstraction techniques such as [2, 8].

Acknowledgments. Thérèse Hardin provided a stimulating environment at Laboratoire d'Informatique de Paris 6, France, where a major part of the work was carried out. John Rushby, Natarajan Shankar and Sam Owre received me well during my long term visit to SRI, California, USA, and provided me with valuable comments. The Human Capital Mobility grant through which this work was funded came into existence partly due to Patrick Cousot (École Normale Supérieure, Paris). Recently SeungJoon Park (NASA Ames Research Center) has provided comments on using symmetry and abstraction to reduce the state space of the Murphi model.

References

[1] M. Ben-Ari. Algorithms for On-the-Fly Garbage Collection. *ACM Toplas*, 6, July 1984.
[2] S. Bensalem, Y. Lakhnech, and S. Owre. Computing Abstractions of Infinite State Systems Compositionally and Automatically. In *Computer-Aided Verification, CAV'98*, number 1427 in Lecture Notes in Computer Science, pages 319–331. Springer-Verlag, 1998.
[3] S. Bensalem, Y. Lakhnech, and S. Owre. InVeSt: A Tool for the Verification of Invariants. In *Computer-Aided Verification, CAV'98*, number 1427 in Lecture Notes in Computer Science, pages 505–510. Springer-Verlag, 1998.
[4] S. Bensalem, Y. Lakhnech, and H. Saïdi. Powerful Techniques for the Automatic Generation of Invariants. In Rajeev Alur and Thomas A. Henzinger, editors, *Computer-Aided Verification, CAV '96*, number 1102 in Lecture Notes in Computer Science, pages 323–335, New Brunswick, NJ, July/August 1996. Springer-Verlag.
[5] K.M. Chandy and J. Misra. *Parallel Program Design: A Foundation*. Addison Wesley, 1988.
[6] J. L. A. Van de Snepscheut. "Algorithms for On-the-Fly Garbage Collection" Revisited. *Information Processing Letters*, 24, March 1987.
[7] E. W. Dijkstra, L. Lamport, A.J. Martin, C. S. Scholten, and E. F. M. Steffens. On-the-Fly Garbage Collection: An Exercise in Cooperation. *ACM*, 21, November 1978.
[8] S. Graf and H. Saidi. Construction of Abstract State Graphs with PVS. In *Computer-Aided Verification, CAV'97*, Lecture Notes in Computer Science. Springer-Verlag, 1997.
[9] K. Havelund, K. G. Larsen, and A. Skou. Formal Verification of an Audio/Video Power Controller using the Real-Time Model Checker UPPAAL. BRICS, Aalborg University, Denmark. Submitted for publication, October 1998.
[10] K. Havelund, M. Lowry, and J. Penix. Formal Analysis of a Space Craft Controller using SPIN. In *Proceedings of the 4th SPIN workshop, Paris, France*, November 1998.
[11] K. Havelund and N. Shankar. Experiments in Theorem Proving and Model Checking for Protocol Verification. In M-C. Gaudel and J. Woodcock, editors, *FME'96: Industrial Benefit and Advances in Formal Methods*, volume 1051 of *Lecture Notes in Computer Science*, pages 662–681. Springer-Verlag, 1996.
[12] K. Havelund and N. Shankar. A Mechanized Refinement Proof for a Garbage Collector. NASA Ames Research Center. To be published, 1998.
[13] K. Havelund, A. Skou, K. G. Larsen, and K. Lund. Formal Modeling and Analysis of an Audio/Video Protocol: An Industrial Case Study Using UPPAAL. In *Proc. of the 18th IEEE Real-Time Systems Symposium*, pages 2–13, Dec 1997. San Francisco, California, USA.
[14] L. Lamport. The Temporal Logic of Actions. Technical report, Digital Equipment Corporation (DEC) Systems Research Center, Palo Alto, California, USA, April 1994.
[15] R. Melton, D.L. Dill, C. Norris Ip, and U. Stern. Murphi Annotated Reference Manual, Release 3.0. Technical report, Stanford University, Palo Alto, California, USA, July 1996.
[16] S. Owre, S. Rajan, J.M. Rushby, N. Shankar, and M.K. Srivas. PVS: Combining Specification, Proof Checking, and Model Checking. In Rajeev Alur and Thomas A. Henzinger, editors, *Computer-Aided Verification, CAV '96*, number 1102 in Lecture Notes in Computer Science, pages 411–414, New Brunswick, NJ, July/August 1996. Springer-Verlag.
[17] C. Pixley. An Incremental Garbage Collection Algorithm for Multi-mutator Systems. *Distributed Computing*, 3, 1988.
[18] D. M. Russinoff. A Mechanically Verified Incremental Garbage Collector. *Formal Aspects of Computing*, 6:359–390, 1994.

A PVS Lemmas about Auxiliary Functions

```
List_Properties[T:TYPE+] : THEORY
BEGIN
  IMPORTING List_Functions[T]

  e       : VAR T
  l,l1,l2 : VAR list[T]
  p       : VAR pred[T]
  n,k     : VAR nat

  length1 : LEMMA
    cons?(l) IMPLIES length(cdr(l)) = length(l)-1

  length2 : LEMMA
    length(append(l1,l2)) = length(l1) + length(l2)

  member1 : LEMMA
    member(e,l) =
      EXISTS n : (n < length(l) AND nth(l,n)=e)

  member2 : LEMMA
    member(e,l) IMPLIES
      EXISTS (x: nat):
        x <= last_index(l) AND nth(l,x) = e AND
        (x < last_index(l) IMPLIES
          NOT member(e,suffix(l,x+1)))

  car1 : LEMMA
    cons?(l1) IMPLIES car(append(l1,l2)) = car(l1)

  last1 : LEMMA
    length(l)>=2 IMPLIES last(l)=last(cdr(l))

  last2 : LEMMA
    last(cons(e,null)) = e

  last3   : LEMMA
    (length(l) >=2 AND p(car(l)) AND NOT p(last(l)))
    IMPLIES
      EXISTS (i:nat|i<last_index(l)):
        p(nth(l,i)) AND NOT p(nth(l,i+1))

  last4 : LEMMA
    cons?(l2) IMPLIES
      last(append(l1,l2)) = last(l2)

  last5 : LEMMA
    cons?(l) IMPLIES nth(l,last_index(l)) = last(l)

  suffix1 : LEMMA
    (length(l) > 0 AND n <= last_index(l))
    IMPLIES cons?(suffix(l, n))

  suffix2 : LEMMA
    (length(l) > 0 AND n <= last_index(l))
    IMPLIES car(suffix(l,n)) = nth(l,n)

  suffix3 : LEMMA
    (length(l) > 0 AND n <= last_index(l))
    IMPLIES
      last(suffix(l,n)) = last(l)

  suffix4 : LEMMA
    n < length(l) IMPLIES
      length(suffix(l,n)) = length(l) - n

  suffix5 : LEMMA
    n+k < length(l) IMPLIES
      nth(suffix(l,n),k) = nth(l,n+k)
END List_Properties

Memory_Properties[NODES : posnat,
                  SONS  : posnat,
                  ROOTS : posnat] : THEORY
BEGIN
  ASSUMING
    roots_within : ASSUMPTION ROOTS <= NODES
  ENDASSUMING
  IMPORTING List_Properties
  IMPORTING Memory_Functions[NODES,SONS,ROOTS]
  IMPORTING Memory_Observers[NODES,SONS,ROOTS]

  abs(i:int):nat = IF i < 0 THEN -i ELSE i ENDIF

  m              : VAR Memory
  n,n1,n2,k      : VAR Node
  i,i1,i2,j      : VAR Index
  c              : VAR Colour
  x              : VAR nat
  N,N1,N2        : VAR NODE
```

```
I,I1,I2   : VAR INDEX
l,l1,l2   : VAR list[Node]

smaller1 : LEMMA
  NOT (n,i) < (0,0)

smaller2 : LEMMA
  (NOT (n,i) < (k,0) AND (n,i) < (k+1,0))
  IMPLIES n = k

smaller3 : LEMMA
  (n,i) < (k,SONS) IFF (n,i) < (k+1,0)

smaller4 : LEMMA
  (NOT (n,i) < (k,j) AND (n,i) < (k,j+1)) IMPLIES
    (n,i)=(k,j)

closed1 : LEMMA
  closed(null_array)

closed2 : LEMMA
  closed(set_colour(n,c)(m)) = closed(m)

closed3 : LEMMA
  closed(m) IMPLIES closed(set_son(n,i,k)(m))

closed4 : LEMMA
  closed(m) IMPLIES son(n,i)(m) < NODES

blacks1 : LEMMA
  blacks(N1,N2)(set_son(n,i,k)(m)) =
    blacks(N1,N2)(m)

blacks2 : LEMMA
  blacks(N1,N2)(m) <=
    blacks(N1,N2)(set_colour(n,TRUE)(m))

blacks3 : LEMMA
  NOT colour(n2)(m) IMPLIES
    blacks(n1,n2+1)(m) = blacks(n1,n2)(m)

blacks4 : LEMMA
  n1<=n2 AND colour(n2)(m) IMPLIES
    blacks(n1,n2+1)(m) = blacks(n1,n2)(m) + 1

blacks5 : LEMMA
  NOT colour(n1)(m) IMPLIES
    blacks(N1,N2)(m) = blacks(n1+1,N2)(m)

blacks6 : LEMMA
  (n1<N2 AND colour(n1)(m)) IMPLIES
    blacks(N1,N2)(m) = blacks(n1+1,N2)(m) + 1

blacks7 : LEMMA
  N1 <= N2 IMPLIES blacks(N1,N2)(m) <= N2-N1

blacks8 : LEMMA
  (n < N1 OR n >= N2) IMPLIES
    blacks(N1,N2)(set_colour(n,c)(m)) =
    blacks(N1,N2)(m)

blacks9 : LEMMA
  (n >= N1 AND n < N2 AND NOT colour(n)(m))
  IMPLIES
    blacks(N1,N2)(set_colour(n,TRUE)(m)) =
    blacks(N1,N2)(m) + 1

blacks10 : LEMMA
  (blacks(0,NODES)(set_colour(n,TRUE)(m)) =
   blacks(0,NODES)(m))
    IMPLIES colour(n)(m)

blacks11 : LEMMA
  blacks(N,N)(m) = 0

black_roots1 : LEMMA
  black_roots(0)(m)

black_roots2 : LEMMA
  black_roots(N)(set_son(n,i,k)(m)) =
    black_roots(N)(m)

black_roots3 : LEMMA
  black_roots(N)(m) IMPLIES
    black_roots(N)(set_colour(n,TRUE)(m))

black_roots4 : LEMMA
  black_roots(n+1)(set_colour(n,TRUE)(m)) =
    black_roots(n)(m)

bw1 : LEMMA
  closed(m) IMPLIES
    (NOT bw(n1,i1)(m) AND
     bw(n1,i1)(set_son(n2,i2,k)(m)))
    IMPLIES
      (n1,i1)=(n2,i2)
```

```
bw2 : LEMMA
  closed(m) IMPLIES
    (NOT bw(n,i)(m) AND
     bw(n,i)(set_colour(k,TRUE)(m)))
      IMPLIES
    (n=k AND NOT colour(n)(m))
bw3 : LEMMA
  bw(n,i)(m) IMPLIES
    colour(n)(m) AND NOT colour(son(n,i)(m))(m)

exists_bw1 : LEMMA
  exists_bw(N1,I1,N2,I2)(m) IMPLIES
    EXISTS (n:Node,i:Index):
      bw(n,i)(m) AND
      NOT (n,i) < (N1,I1) AND
      (n,i) < (N2,I2)

exists_bw2 : LEMMA
  closed(m) IMPLIES
    (NOT exists_bw(0,0,N2,I2)(m) AND
     exists_bw(0,0,N2,I2)(set_son(n,i,k)(m)))
      IMPLIES
    (NOT colour(k)(m) AND (n,i) < (N2,I2))

exists_bw3 : LEMMA
  (accessible(n)(m) AND
   NOT colour(n)(m) AND
   black_roots(ROOTS)(m))
    IMPLIES
    exists_bw(0,0,NODES,0)(m)

exists_bw4 : LEMMA
  exists_bw(0,0,NODES,0)(m) IMPLIES
    exists_bw(0,0,N,I)(m) OR
    exists_bw(N,I,NODES,0)(m)

exists_bw5 : LEMMA
  closed(m) IMPLIES
    (exists_bw(N,I,NODES,0)(m) AND (n,i) < (N,I))
      IMPLIES
    exists_bw(N,I,NODES,0)(set_son(n,i,k)(m))

exists_bw6 : LEMMA
  closed(m) AND colour(n)(m) IMPLIES
    exists_bw(N1,I1,N2,I2)(set_colour(n,TRUE)(m))
    = exists_bw(N1,I1,N2,I2)(m)

exists_bw7 : LEMMA
  exists_bw(0,0,N+1,0)(m) IMPLIES
    exists_bw(0,0,N,SONS)(m)

exists_bw8 : LEMMA
  exists_bw(N,SONS,NODES,0)(m) IMPLIES
    exists_bw(N+1,0,NODES,0)(m)

exists_bw9 : LEMMA
  (NOT colour(n)(m) AND exists_bw(0,0,n+1,0)(m))
    IMPLIES
    exists_bw(0,0,n,0)(m)

exists_bw10 : LEMMA
  (NOT colour(n)(m) AND exists_bw(n,0,NODES,0)(m))
    IMPLIES
    exists_bw(n+1,0,NODES,0)(m)

exists_bw11 : LEMMA
  (colour(son(n,i)(m))(m) AND
   exists_bw(0,0,n,i+1)(m))
    IMPLIES
    exists_bw(0,0,n,i)(m)

exists_bw12 : LEMMA
  (colour(son(n,i)(m))(m) AND
   exists_bw(n,i,NODES,0)(m))
    IMPLIES
    exists_bw(n,i+1,NODES,0)(m)

exists_bw13 : LEMMA
  NOT exists_bw(N,I,N,I)(m)

points_to1 : LEMMA
  (k /= n2 AND
   points_to(n1,n2)(set_son(n,i,k)(m)))
    IMPLIES
    points_to(n1,n2)(m)

pointed1 : LEMMA
  (NOT member(k,l) AND
   pointed(l)(set_son(n,i,k)(m)))
    IMPLIES
    pointed(l)(m)

pointed2 : LEMMA
  (pointed(l)(m) AND cons?(l) AND
   x <= last_index(l))
    IMPLIES
```

```
    pointed(suffix(l,x))(m)

pointed3 : LEMMA
  pointed(cons(n,l))(m) IMPLIES pointed(l)(m)

pointed4 : LEMMA
  (cons?(l) AND points_to(n,car(l))(m) AND
   pointed(l)(m))
    IMPLIES
    pointed(cons(n,l))(m)

pointed5 : LEMMA
  (cons?(l1) AND cons?(l2) AND
   points_to(last(l1),car(l2))(m) AND
   pointed(l1)(m) AND pointed(l2)(m))
    IMPLIES
    pointed(append(l1,l2))(m)

path1 : LEMMA
  (path(l1)(m) AND
   cons?(l2) AND
   points_to(last(l1),car(l2))(m) AND
   pointed(l2)(m))
    IMPLIES
    path(append(l1,l2))(m)

accessible1 : LEMMA
  (accessible(k)(m) AND
   accessible(n1)(set_son(n,i,k)(m)))
    IMPLIES
    accessible(n1)(m)

propagated1 : LEMMA
  (cons?(l) AND pointed(l)(m) AND
   colour(car(l))(m) AND propagated(m))
    IMPLIES
    colour(last(l))(m)

propagated2 : LEMMA
  propagated(m) = NOT exists_bw(0,0,NODES,0)(m)

blackened1 : LEMMA
  (accessible(k)(m) AND blackened(N)(m))
    IMPLIES
    blackened(N)(set_son(n,i,k)(m))

blackened2 : LEMMA
  blackened(N)(m) IMPLIES
    blackened(N)(set_colour(n,TRUE)(m))

blackened3 : LEMMA
  (black_roots(ROOTS)(m) AND propagated(m))
    IMPLIES
    blackened(0)(m)

blackened4 : LEMMA
  blackened(n)(m) IMPLIES
    blackened(n+1)(set_colour(n,FALSE)(m))

blackened5 : LEMMA
  (NOT accessible(n)(m) AND blackened(n)(m))
    IMPLIES
    blackened(n+1)(append_to_free(n)(m))

blackened6 : LEMMA
  (blackened(n)(m) AND accessible(n)(m)) IMPLIES
    colour(n)(m)
END Memory_Properties
```

B Murphi Formalization

```
Const
  NODES : 3; MAX_NODE : NODES-1;
  SONS  : 2; MAX_SON  : SONS-1 ;
  ROOTS : 1; MAX_ROOT : ROOTS-1;

Type
  NumberOfNodes : 0..NODES;
  Colour : boolean;
  Node   : 0..MAX_NODE;
  Index  : 0..MAX_SON;
  Root   : 0..MAX_ROOT;
  NodeStruct :
    Record
      colour : Colour;
      cells  : Array[Index] Of Node;
    End;

Var
  MU : Enum{MU0,MU1};
  CHI : Enum{CHI0,CHI1,CHI2,CHI3,CHI4,
             CHI5,CHI6,CHI7,CHI8};
  Q : Node;
```

```
BC : NumberOfNodes; OBC : NumberOfNodes;
I,L,H : 0..NODES;
J : 0..SONS; K : 0..ROOTS;

Var M : Array[Node] Of NodeStruct;

Function colour(n:Node):Colour;
Begin Return M[n].colour; End;

Procedure set_colour(n:Node;c:Colour);
Begin M[n].colour := c; End;

Function son(n:Node;i:Index):Node;
Begin Return M[n].cells[i]; End;

Procedure set_son(n:Node;i:Index;k:Node);
Begin M[n].cells[i] := k; End;

Function is_root(n:Node):boolean;
Begin Return n < ROOTS; End;

Function accessible(n:Node):boolean;
Type
  Status : Enum{TRY,UNTRIED,TRIED};
Var
  status : Array[Node] Of Status;
  s : Node; try_again : boolean;
Begin
  For k:Node Do
    status[k] :=
      (is_root(k) ? TRY : UNTRIED)
  EndFor;
  try_again := true;
  While try_again Do
    try_again := false;
    For k:Node Do
      If status[k]=TRY Then
        For j:Index Do
          s := son(k,j);
          If status[s]=UNTRIED Then
            status[s] := TRY;
            try_again := true;
          End;
        EndFor;
        status[k] := TRIED;
      End;
    EndFor;
  End;
  Return status[n]=TRIED
End;

Procedure append_to_free(new_free:Node);
Var
  old_first_free : Node;
Begin
  old_first_free := son(0,0);
  set_son(0,0,new_free);
  For i:Index Do
    set_son(new_free,i,old_first_free)
  EndFor;
End;

Procedure initialise_memory();
Begin
  For n:Node Do
    set_colour(n,false);
    For i:Index Do set_son(n,i,0); EndFor;
  EndFor;
End;

Startstate
Begin
  MU := MU0; CHI := CHI0;
  clear Q; clear BC; OBC := 0;
  clear I; clear J; K := 0;
  clear L; clear H;
  initialise_memory();
End;

-- The Mutator Process --

Ruleset m:Node; i:Index; n: Node Do
  Rule "mutate"
    MU = MU0 & accessible(n) ==>
    set_son(m,i,n); Q := n; MU := MU1;
  End;
End;

Rule "colour_target"
  MU = MU1 ==>
  set_colour(Q,true); MU := MU0;
End;

-- The Collector Process --

Rule "stop_blacken"
  CHI = CHI0 & K = ROOTS ==>
  I := 0; CHI := CHI1;
End;

Rule "blacken"
  CHI = CHI0 & K != ROOTS ==>
  set_colour(K,true);
  K := K+1; CHI := CHI0;
End;

Rule "stop_propagate"
  CHI = CHI1 & I = NODES ==>
  BC := 0; H := 0; CHI := CHI4;
End;

Rule "continue_propagate"
  CHI = CHI1 & I != NODES ==>
  CHI := CHI2;
End;

Rule "white_node"
  CHI = CHI2 & !colour(I) ==>
  I := I+1; CHI := CHI1;
End;

Rule "black_node"
  CHI = CHI2 & colour(I) ==>
  J := 0; CHI := CHI3;
End;

Rule "stop_colouring_sons"
  CHI = CHI3 & J = SONS ==>
  I := I+1; CHI := CHI1;
End;

Rule "colour_son"
  CHI = CHI3 & J != SONS ==>
  set_colour(son(I,J),true);
  J := J+1; CHI := CHI3;
End;

Rule "stop_counting"
  CHI = CHI4 & H = NODES ==>
  CHI := CHI6
End;

Rule "continue_counting"
  CHI = CHI4 & H != NODES ==>
  CHI := CHI5;
End;

Rule "skip_white"
  CHI = CHI5 & !colour(H) ==>
  H := H+1; CHI := CHI4;
End;

Rule "count_black"
  CHI = CHI5 & colour(H) ==>
  BC := BC+1; H := H+1; CHI := CHI4;
End;

Rule "redo_propagation"
  CHI = CHI6 & BC != OBC ==>
  OBC := BC; I := 0; CHI := CHI1;
End;

Rule "quit_propagation"
  CHI = CHI6 & BC = OBC ==>
  L := 0; CHI := CHI7;
End;

Rule "stop_appending"
  CHI = CHI7 & L = NODES ==>
  BC := 0; OBC := 0; K := 0;
  CHI := CHI0;
End;

Rule "continue_appending"
  CHI = CHI7 & L != NODES ==>
  CHI := CHI8
End;

Rule "black_to_white"
  CHI = CHI8 & colour(L) ==>
  set_colour(L,false);
  L := L+1; CHI := CHI7;
End;

Rule "append_white"
  CHI = CHI8 & !colour(L) ==>
  append_to_free(L);
  L := L+1; CHI := CHI7
End;

-- Specification --

Invariant "safe"
  CHI = CHI8 & accessible(L) -> colour(L);
```

A Structured Approach to Parallel Programming: Methodology and Models*

Berna L. Massingill

University of Florida, P.O. Box 116120, Gainesville, FL   32611
blm@cise.ufl.edu

Abstract. Parallel programming continues to be difficult, despite substantial and ongoing research aimed at making it tractable. Especially dismaying is the gulf between theory and the practical programming. We propose a structured approach to developing parallel programs for problems whose specifications are like those of sequential programs, such that much of the work of development, reasoning, and testing and debugging can be done using familiar sequential techniques and tools. The approach takes the form of a simple model of parallel programming, a methodology for transforming programs in this model into programs for parallel machines based on the ideas of semantics-preserving transformations and programming archetypes (patterns), and an underlying operational model providing a unified framework for reasoning about those transformations that are difficult or impossible to reason about using sequential techniques. This combination of a relatively accessible programming methodology and a sound theoretical framework to some extent bridges the gulf between theory and practical programming. This paper sketches our methodology and presents our programming model and its supporting framework in some detail.

1   Introduction

Despite the past and ongoing efforts of many researchers, parallel programming continues to be difficult, with a persistent and dismaying gulf between theory and practical programming. We propose a structured approach to developing parallel programs for the class of problems whose specifications are like those usually given for sequential programs, in which the specification describes initial states for which the program must terminate and the relation between initial and final states. Our approach allows much of the work of development, reasoning, and testing and debugging to be done using familiar sequential techniques and tools; it takes the form of a simple model of parallel programming, a methodology for transforming programs in this model into programs for parallel machines based on the ideas of semantics-preserving transformations and programming archetypes (patterns), and an underlying operational model providing a unified framework for reasoning about those transformations that are difficult or impossible to reason about using sequential techniques.

By combining a relatively accessible programming methodology with a sound theoretical framework, our approach to some extent bridges the gap between theory and

---
* This work was supported by funding from the Air Force Office of Scientific Research (AFOSR) and the National Science Foundation (NSF).

practical programming. The transformations we propose are in many cases formalized versions of what programmers and compilers typically do in practice to "parallelize" sequential code, but we provide a framework for formally proving their correctness (either by standard sequential techniques or by using our operational model). Our operational model is sufficiently general to support proofs of transformations between markedly different programming models (sequential, shared-memory, and distributed-memory with message-passing). It is sufficiently abstract to permit a fair degree of rigor, but simple enough to be relatively accessible, and applicable to a range of programming notations.

This paper describes our programming methodology and presents our programming model and its supporting framework.

2 Our programming model and methodology

Our programming model comprises a primary model and two subsidiary models, and is designed to support a programming methodology based on stepwise refinement and the reuse where possible of the techniques and tools of sequential programming. This section gives an overview of our model and methodology.

The arb model: parallel composition with sequential semantics. Our primary programming model, which we call the *arb model*, is simply the standard sequential model (as defined in [8] or [10], e.g.) extended to include parallel compositions of groups of program elements whose parallel composition is equivalent to their sequential composition. The name (**arb**) is derived from UC (Unity C [3]) and is intended to connote that such groups of program elements may be interleaved in any arbitrary fashion without changing the result. We define a property we call *arb-compatibility*, and we show that if a group of program elements is **arb**-compatible, their parallel composition is semantically equivalent to their sequential composition; we call such compositions *arb compositions*. Since **arb**-model programs can be interpreted as sequential programs, the extensive body of tools and techniques applicable to sequential programs is applicable to them. In particular, their correctness can be demonstrated formally by using sequential methods, they can be refined by sequential semantics-preserving transformations, and they can be executed sequentially for testing and debugging.

Transformations from the arb model to practical parallel languages. Because the **arb** composition of **arb**-compatible elements can also be interpreted as parallel composition, **arb**-model programs can be executed as parallel programs. Such programs may not make effective use of typical parallel architectures, however, so our methodology includes techniques for improving their efficiency while maintaining correctness. We define two subsidiary programming models that abstract key features of two classes of parallel architectures: the *par model* for shared-memory (single-address-space) architectures, and the *subset par model* for distributed-memory (multiple-address-space) architectures. We then develop semantics-preserving transformations to convert **arb**-model programs into programs in one of these subsidiary models. Intermediate stages in this process are usually **arb**-model programs, so the transformations can make use of sequential refinement techniques, and the programs can be executed sequentially. Finally, we indicate how the **par** model can be mapped to practical programming languages for shared-memory architectures and the subset **par** model to practical programming languages for distributed-memory–message-passing architectures. Together, these groups of transformations provide a semantics-preserving path from the original **arb**-model program to a program in a practical programming language.

Supporting framework for proving transformations correct. Some of the transformations mentioned in the preceding section — those within the **arb** model — can be proved correct using the techniques of sequential stepwise refinement (as defined in [10] or [11], e.g.). Others — those between our different programming models, or from one of our models to a practical programming language — require a different approach. We therefore define an operational model based on viewing programs as state-transition systems, give definitions of our programming models in terms of this underlying operational model, and use it to prove the correctness of those transformations for which sequential techniques are inappropriate.

Programming archetypes. An additional important element of our approach, though not one that will be addressed in this paper, is that we envision the transformation process just described as being guided by *parallel programming archetypes*, by which we mean abstractions that capture the commonality of classes of programs, much like the design patterns [9] of the object-oriented world. We envision application developers choosing from a range of archetypes, each representing a class of programs with common features and providing a class-specific parallelization strategy (i.e., a pattern for the shared-memory or distributed-memory program to be ultimately produced) together with a collection of class-specific transformations and a code library of communication or other operations that encapsulate the details of the parallel programs. Archetypes are described in more detail in [15].

Program development using our methodology. We can then employ the following approach to program development, with all steps guided by an archetype-specific parallelization strategy and supported by a collection of archetype-appropriate already-proved-correct transformations.

Development of initial program. The application developer begins by developing a correct program using sequential constructors and parallel composition (||), but ensuring that all groups of elements composed in parallel are **arb**-compatible. We call such a program an ***arb**-model program*, and it can be interpreted as either a sequential program or a parallel program, with identical meaning. Correctness of this program can be established using techniques for establishing correctness of a sequential program.

Sequential-to-sequential refinement. The developer then begins refining the program into one suitable for the target architecture. During the initial stages of this process, the program is viewed as a sequential program and operated on with sequential refinement techniques, which are well-defined and well-understood. A collection of representative transformations (refinement steps) is presented in [15]. In refining a sequential composition whose elements are **arb**-compatible, care is taken to preserve their

arb-compatibility. The result is a program that refines the original program and can also be interpreted as either a sequential or a parallel program, with identical meaning. The end product of this refinement should be a program that can then be transformed into an efficient program in the **par** model (for a shared-memory target) or the subset **par** model (for a distributed-memory target).

Sequential-to-parallel refinement. The developer next transforms (refines) the refined **arb**-model program into a **par**-model or subset-**par**-model program.

Translation for target platform. Finally, the developer translates the **par**-model or subset-**par**-model program into the desired parallel language for execution on the target platform. This step, like the others, is guided by semantics-preserving rules for mapping one of our programming models into the constructs of a particular parallel language.

3 The arb model

As discussed in Section 2, the heart of our approach is identifying groups of program elements that have the useful property that their parallel composition is semantically equivalent to their sequential composition. We call such a group of program elements **arb**-compatible. This section first presents our operational model for parallel programs and then discusses **arb**-compatibility and **arb** composition. In this paper we present this material with a minimum of mathematical notation and no proofs; a more formal treatment of the material, including proofs, appears in [15]. The notation we use for definitions and theorems for our programming models is based on that of Dijkstra's guarded-command language [8].

Program semantics and operational model. We define programs in such a way that a program describes a state-transition system, and show how to define program computations, sequential and parallel composition, and program refinement in terms of this definition. Treating programs as state-transition systems is not a new approach; it has been used by many other researchers (as mentioned in Section 6) to reason about both parallel and sequential programs. The basic notions of a state-transition system — a set of states together with a set of transitions between them, representable as a directed graph with states for vertices and transitions for edges — are perhaps more obviously helpful in reasoning about parallel programs, but they are applicable to sequential programs as well. Our operational model builds on this basic view of program execution, presented in a way specifically aimed at facilitating the stating and proving of our main theorems.

Definition 3.1 (Program). We define a program P as a 6-tuple $(V, L, InitL, A, PV, PA)$, where

- V is a finite set of typed variables. V defines a state space in the state-transition system; that is, a state is given by the values of the variables in V. In our semantics, distinct program variables denote distinct atomic data objects; aliasing is not allowed.
- $L \subseteq V$ represents the *local variables* of P. These variables are distinguished from the other variables of P in two ways: (i) The initial states of P are given in terms of

their values, and (ii) they are invisible outside P — that is, they may not appear in a specification for P, and they may not be accessed by other programs composed with P, either in sequence or in parallel.
- *InitL* is an assignment of values to the variables of L, representing their initial values.
- A is a finite set of *program actions*. A program action describes a relation between states of its input variables (those variables in V that affect its behavior, either in the sense of determining from which states it can be executed or in the sense of determining the effects of its execution) and states of its output variables (those variables whose value can be affected by its execution). Thus, a program action is a triple (I_a, O_a, R_a) in which $I_a \subseteq V$ represents the input variables of A, $O_a \subseteq V$ represents the output variables of A, and R_a is a relation between I_a-tuples and O_a-tuples.
- $PV \subseteq V$ are *protocol variables* that can be modified only by *protocol actions* (elements of PA). (I.e., if v is in PV, and $a = (I_a, O_a, R_a)$ is an action such that $v \in O_a$, a must be in PA.) Such variables and actions are not needed in this section but are useful in defining the synchronization mechanisms of Section 4 and Section 5; the requirement that protocol variables be modified only by protocol actions simplifies the task of defining such mechanisms. Variables in PV can include both local and non-local variables.
- $PA \subseteq A$ are *protocol actions*. Protocol actions may modify both protocol and non-protocol variables.

A program action $a = (I_a, O_a, R_a)$ defines a set of state transitions, each of which we write in the form $s \stackrel{a}{\to} s'$, as follows: $s \stackrel{a}{\to} s'$ if the pair (i, o), where i is a tuple representing the values of the variables in I_a in state s and o is a tuple representing the values of the variables in O_a in state s', is an element of relation R_a. Observe that we can also define a program action based on its set of state transitions, by inferring the required I_a, O_a, and R_a. □

Examples of defining the commands and constructs of a particular programming notation in terms of our model are presented in [15].

Definition 3.2 (Initial states). For program P, s is an *initial state* of P if, in s, the values of the local variables of P have the values given in *InitL*. □

Definition 3.3 (Enabled). For action a and state s of program P, we say that a is *enabled* in s exactly when there exists program state s' such that $s \stackrel{a}{\to} s'$. □

Definition 3.4 (Computation). If $P = (V, L, InitL, A, PV, PA)$, a *computation* of P is a pair $C = (s_0, \langle j : 1 \leq j \leq N : (a_j, s_j) \rangle)$ in which
- s_0 is an initial state of P.
- $\langle j : 1 \leq j \leq N : (a_j, s_j) \rangle$ is a sequence of pairs in which each a_j is a program action of P, and for all j, $s_{j-1} \stackrel{a_j}{\to} s_j$. We call these pairs the state transitions of C, and the sequence of actions a_j the actions of C. N can be a non-negative integer (in which case we say C is finite) or ∞ (in which case we say C is infinite).

- If C is infinite, the sequence $\langle j : 1 \leq j : (a_j, s_j)\rangle$ satisfies the following fairness requirement: If, for some state s_j and program action a, a is enabled in s_j, then eventually either a occurs in C or a ceases to be enabled. □

Definition 3.5 (Terminal state). We say that state s of program P is a *terminal state* of P exactly when there are no actions of P enabled in s. □

Definition 3.6 (Maximal computation). We say that a computation of C of P is a *maximal computation* exactly when either (i) C is infinite or (ii) C is finite and ends in a terminal state. □

Definition 3.7 (Affects). For predicate q and variable $v \in V$, we say that v *affects* q exactly when there exist states s and s', identical except for the value of v, such that $q.s \neq q.s'$. For expression E and variable $v \in V$, we say that v affects E exactly when there exists value k for E such that v affects the predicate $(E = k)$. □

Specifications and program refinement. The usual meaning of "program P is refined by program P'" is that program P' meets any specification met by P. We confine ourselves to specifications that describe a program's behavior in terms of initial and final states, giving (i) a set of initial states s such that the program is guaranteed to terminate if started in s, and (ii) the relation, for terminating computations, between initial and final states. In terms of our model, initial and final states correspond to assignments of values to the program's variables; we make the additional restriction that specifications do not mention a program's local variables. We make this restriction because otherwise program equivalence can depend on internal behavior (as reflected in the values of local variables), which is not the intended meaning of equivalence. We write $P \sqsubseteq P'$ to denote that P is refined by P'; if $P \sqsubseteq P'$ and $P' \sqsubseteq P$, we say that P and P' are equivalent, and we write $P \sim P'$. The following definition and theorem provide a sufficient condition for showing that $P_1 \sqsubseteq P_2$ in our semantics.

Definition 3.8 (Equivalence of computations). For programs P_1 and P_2 and a set of typed variables V such that $V \subseteq V_1$ and $V \subseteq V_2$ and for every v in V, v has the same type in all three sets (V, V_1, and V_2), we say that computations C_1 of P_1 and C_2 of P_2 are *equivalent with respect to* V exactly when: (i) For every v in V, the value of v in the initial state of C_1 is the same as its value in the initial state of C_2; and (ii) either (a) both C_1 and C_2 are infinite, or (b) both are finite, and for every v in V, the value of v in the final state of C_1 is the same as its value in the final state of C_2. □

Theorem 3.9 (Refinement in terms of equivalent computations). For P_1 and P_2 with $(V_1 \setminus L_1) \subseteq (V_2 \setminus L_2)$ (where $\setminus$ denotes set difference), $P_1 \sqsubseteq P_2$ when for every maximal computation C_2 of P_2 there is a maximal computation C_1 of P_1 such that C_1 is equivalent to C_2 with respect to $(V_1 \setminus L_1)$. □

Program composition. Before giving definitions of sequential and parallel composition in terms of our model, we need the following definition formalizing conditions under which it makes sense to compose programs.

Definition 3.10 (Composability of programs). We say that a set of programs $P_1, \ldots, P_N$ *can be composed* exactly when (i) any variable that appears in more than one program has the same type in all the programs in which it appears (and if it is a protocol variable in one program, it is a protocol variable in all programs in which it appears), (ii) any action that appears in more than one program is defined in the same way in all the programs in which it appears, and (iii) different programs do not have local variables in common. □

Sequential composition. The usual meaning of sequential composition is this: A maximal computation of $P_1; P_2$ is a maximal computation C_1 of P_1 followed (if C_1 is finite) by a maximal computation C_2 of P_2, with the obvious generalization to more than two programs. We can give a definition with this meaning in terms of our model by introducing additional local variables $En_1, \ldots, En_N$ that ensure that things happen in the proper sequence, as follows: Actions from program P_j can execute only when En_j is *true*. En_1 is set to *true* at the start of the computation, and then as each P_j terminates it sets En_j to *false* and En_{j+1} to *true*, thus ensuring the desired behavior.

Definition 3.11 (Sequential composition). If programs $P_1, \ldots, P_N$, with $P_j = (V_j, L_j, InitL_j, A_j, PV_j, PA_j)$, can be composed (Definition 3.10), we define their sequential composition $(P_1; \ldots; P_N) = (V, L, InitL, A, PA, PV)$ thus:

- $V = V_1 \cup \ldots \cup V_N \cup L$.
- $L = L_1 \cup \ldots \cup L_N \cup \{En_P, En_1, \ldots, En_N\}$, where $En_P, En_1, \ldots, En_N$ are distinct Boolean variables not otherwise occurring in V.
- $InitL$ is defined thus: The initial value of En_P is *true*. For all j, the initial value of En_j is *false*, and the initial values of variables in L_j are those given by $InitL_j$.
- A consists of the following types of actions:
    - Actions corresponding to actions in A_j, for some j: For $a \in A_j$, we define a' identical to a except that a' is enabled only when $En_j = true$.
    - Actions that accomplish the transitions between components of the composition: (i) Initial action a_{T_0} takes any initial state s, with $En_P = true$, to a state s' identical except that $En_P = false$ and $En_1 = true$. s' is thus an initial state of P_1. (ii) For j with $1 \leq j < N$, action a_{T_j} takes any terminal state s of P_j, with $En_j = true$, to a state s' identical except that $En_j = false$ and $En_{j+1} = true$. s' is thus an initial state of P_{j+1}. (iii) Final action a_{T_N} takes any terminal state s of P_N, with $En_N = true$, to a state s' identical except that $En_N = false$. s' is thus a terminal state of the sequential composition.
- $PV = PV_1 \cup \ldots \cup PV_N$.
- PA contains exactly those actions a' derived (as described above) from the actions a of $PA_1 \cup \ldots \cup PA_N$.

□

Parallel composition. The usual meaning of parallel composition is this: A computation of $P_1 \| P_2$ defines two threads of control, one each for P_1 and P_2. Initiating the composition corresponds to starting both threads; execution of the composition corresponds to an interleaving of actions from both components; and the composition is understood to terminate when both components have terminated. We can give a definition with this meaning in terms of our model by introducing additional local variables that ensure that the composition terminates when all of its components terminate, as follows: As for sequential composition, we introduce additional local variables $En_1, \ldots, En_N$ such that actions from program P_j can execute only when En_j is *true*. For parallel composition, however, all of the En_j's are set to *true* at the start of the computation, so computation is an interleaving of actions from the P_j's. As each P_j terminates, it sets the corresponding En_j to *false*; when all are *false*, the composition has terminated. Observe that the definitions of parallel and sequential composition are almost identical; this greatly facilitates the proof of Theorem 3.15.

Definition 3.12 (Parallel composition). If programs $P_1, \ldots, P_N$, with $P_j = (V_j, L_j, InitL_j, A_j, PV_j, PA_j)$, can be composed (Definition 3.10), we define their parallel composition $(P_1 \| \ldots \| P_N) = (V, L, InitL, A, PV, PA)$, where $V, L, InitL, PV$, and PA are as defined in Definition 3.11, and:

- A consists of the following types of actions:
  - Actions corresponding to actions in A_j, for some j, as defined in Definition 3.11.
  - Actions that correspond to the initiation and termination of the components of the composition: (i) Initial action a_{T_0} takes any initial state s, with $En_P = true$, to a state s' identical except that $En_j = true$ for all j. s' is thus an initial state of P_j, for all j. (ii) For j with $1 \leq j \leq N$, action a_{T_j} takes any terminal state s of P_j, with $En_j = true$, to a state s' identical except that $En_j = false$. A terminating computation of P contains one execution of each a_{T_j}; after execution of a_{T_j} for all j, the resulting state s' is a terminal state of the parallel composition.

□

arb-compatibility. We now turn our attention to defining sufficient conditions for a group of programs $P_1, \ldots, P_N$ to have the property we want, namely that $(P_1 \| \ldots \| P_N) \sim (P_1; \ldots; P_N)$. We first define a key property of pairs of program actions; we can then define the desired condition and show that it guarantees the property of interest.

Definition 3.13 (Commutativity of actions). Actions a and b of program P are said to *commute* exactly when the following two conditions hold: (i) Execution of b does not affect (in the sense of Definition 3.7) whether a is enabled, and vice versa. (ii) It is possible to reach s_2 from s_1 by first executing a and then executing b exactly when it is also possible to reach s_2 from s_1 by first executing b and then executing a. (Observe that if a and b are nondeterministic, there may be more than one such state s_2.) □

*Definition 3.14 (**arb**-compatible).* Programs $P_1, \ldots, P_N$ are **arb-compatible** exactly when they can be composed (Definition 3.10) and any action in one program commutes (Definition 3.13) with any action in another program. □

*Theorem 3.15 (Parallel $\sim$ sequential for **arb**-compatible programs).* If $P_1, \ldots, P_N$ are **arb**-compatible, then $(P_1 || \ldots || P_N) \sim (P_1; \ldots; P_N)$. □

A detailed proof of this theorem is given in [15]. The proof relies on showing that for every maximal computation of the parallel composition there is a computation of the sequential composition that is equivalent with respect to the nonlocal variables, and vice versa.

arb composition. For **arb**-compatible programs $P_1, \ldots, P_N$, then, we know that $(P_1 || \ldots || P_N) \sim (P_1; \ldots; P_N)$. To denote this parallel/sequential composition of **arb**-compatible elements, we write $\mathbf{arb}(P_1, \ldots, P_N)$, where $\mathbf{arb}(P_1, \ldots, P_N) \sim (P_1 || \ldots || P_N)$ (or equivalently $\mathbf{arb}(P_1, \ldots, P_N) \sim (P_1; \ldots; P_N)$). We refer to this notation as "**arb** composition", although it is not a true composition operator since it is properly applied only to groups of elements that are **arb**-compatible. We regard it as a useful form of syntactic sugar that denotes not only the parallel/sequential composition of $P_1, \ldots, P_N$ but also the fact that $P_1, \ldots, P_N$ are **arb**-compatible. Examples of valid and invalid uses of **arb** composition are given in [15].

We also define an additional bit of syntactic sugar useful in improving the readability of nestings of sequential and **arb** composition: $\mathbf{seq}(P_1, \ldots, P_N) \sim (P_1; \ldots; P_N)$.

arb composition has a number of useful properties. It is associative and commutative (proofs given in [15]), and it allows refinement by parts, as the following theorem states. This last property is the justification for our program-development strategy, in which we apply the techniques of sequential stepwise refinement to **arb**-model programs.

*Theorem 3.16 (Refinement by parts of **arb** composition).* We can refine any component of an **arb** composition to obtain a refinement of the whole composition. That is, if $P_1, \ldots, P_N$ are **arb**-compatible, and, for each j, $P_j \sqsubseteq P'_j$, and $P'_1, \ldots, P'_N$ are **arb**-compatible, then $\mathbf{arb}(P_1, \ldots, P_N) \sqsubseteq \mathbf{arb}(P'_1, \ldots, P'_N)$. □

A simpler sufficient condition for arb-compatibility. The definition of **arb**-compatibility given in Definition 3.14 is the most general one that seems to give the desired properties (equivalence of parallel and sequential composition, and associativity and commutativity), but it may be difficult to apply in practice. We therefore give a more-easily-checked sufficient condition for programs $P_1, \ldots, P_N$ to be **arb**-compatible.

Definition 3.17 (Variables read/written by P). For program P and variable v, we say that v is *read by* P if it is an input variable for some action a of P, and we say that v is *written by* P if it is an output variable for some action a of P. □

Theorem 3.18 (**arb**-*compatibility and shared variables*). If programs $P_1, \ldots, P_N$ can be composed (Definition 3.10), and for $j \neq k$, no variable written by P_j is read or written by P_k, then $P_1, \ldots, P_N$ are **arb**-compatible. □

arb composition and programming notations. A key difficulty in applying our methodology for program development is in identifying groups of program elements that are known to be **arb**-compatible. The difficulty is exacerbated by the fact that many programming notations, unlike our operational model's semantics, permit aliasing. Syntactic restrictions sufficient to guarantee **arb**-compatibility do not seem in general feasible. However, we can give semantics-based rules that identify, for program P, supersets of the variables read and written by P, and identifying such supersets is sufficient to permit application of Theorem 3.18. In [15] we discuss such rules for two representative programming notations and present examples of their use.

Execution of arb-model programs. Since for **arb**-compatible program elements, their **arb** composition is semantically equivalent to their parallel composition and also to their sequential composition, programs written using sequential commands and constructors plus (valid) **arb** composition can, as noted earlier, be executed either as sequential or as parallel programs with identical results. **arb**-model programs are easily converted into sequential programs in the underlying sequential notation by replacing **arb** composition with sequential composition; simple syntactic transformations also suffice to convert them to a notation that includes a construct implementing general parallel composition as defined in Definition 3.12. Details and examples are given in [15].

4 The par model and shared-memory programs

As discussed in Section 2, once we have developed a program in our **arb** model, we can transform it into one suitable for execution on a shared-memory architecture via what we call the *par model*, which is based on a structured form of parallel composition with barrier synchronization that we call *par composition*. In this section we sketch how we extend our model of parallel composition to include barrier synchronization, give key transformations for turning **arb**-model programs into **par**-model programs, and briefly discuss executing such programs on shared-memory architectures. Details are given in [15].

A preliminary remark: Behind any synchronization mechanism is the notion of "suspending" a component of a parallel composition until some condition is met — that is, temporarily interrupting the normal flow of control in the component, and then resuming it when the condition is met. We model suspension as busy waiting, since this approach simplifies our definitions and proofs by making it unnecessary to distinguish between computations that terminate normally and computations that terminate in a deadlock situation — if suspension is modeled as a busy wait, deadlocked computations are infinite.

Parallel composition with barrier synchronization. We first expand the definition of parallel composition given previously to include barrier synchronization, a common form of synchronization (see [1], e.g.) based on the notion of a "barrier" that all processes much reach before any can proceed. (A formal specification is given in [15].)

We define barrier synchronization in terms of our model by defining a new command, **barrier**, and extending Definition 3.12. This combined definition (presented in detail in [15]) implements a common approach to barrier synchronization based on keeping a count of processes waiting at the barrier. In the context of our model, we implement this approach using two protocol variables local to the parallel composition: a count Q of suspended components and a flag *Arriving* that indicates whether components are arriving at the barrier or leaving. As components arrive at the barrier, we suspend them and increment Q. When Q equals the number of components, we set *Arriving* to *false* and allow components to leave the barrier. Components leave the barrier by unsuspending and decrementing Q. When Q equals 0, we reset *Arriving* to *true*, ready for the next use of the barrier. This behavior is implemented by the protocol actions of the **barrier** command.

The par model. We now define a structured form of parallel composition with barrier synchronization. Previously we defined a notion of **arb**-compatibility and then defined **arb** composition as the parallel composition of **arb**-compatible components. Analogously, we define a notion of **par**-compatibility and then define **par** composition as the parallel composition of **par**-compatible components.

The idea behind **par**-compatibility is that **par**-compatible elements "match up" with regard to their use of the **barrier** command — that is, they all execute **barrier** the same number of times and hence do not deadlock. (For example, $(a := 1; \textbf{barrier}; b' := a)$ and $(b := 2; \textbf{barrier}; a' := b)$ are **par**-compatible, but they would not be if the **barrier** command were removed from one of them.) A detailed definition of **par**-compatibility, based on our model, is given in [15]. An aspect that should be noted here is that part of the definition involves extending the definition of **arb**-compatibility to require that **arb**-compatible elements not interact with each other via barrier synchronization.

As with **arb**, we write $\textbf{par}(P_1, \ldots, P_N)$ to denote the parallel composition (with barrier synchronization) of **par**-compatible elements $P_1, \ldots, P_N$. Examples of valid and invalid uses of **par** composition are given in [15].

Transforming arb-model programs into par-model programs. The transformation of **arb**-model programs into **par**-model programs involves a relatively small collection of transformations that describe how to "interchange" **arb** composition with the standard constructors of the sequential model, namely sequential composition and the *IF* and *DO* constructs of Dijkstra's guarded-command language [8]. (The ideas are not specific to Dijkstra's guarded-command language; analogous transformations can be defined for any notation that provides equivalent constructs.) Space limitations preclude presenting all the transformations in this paper; we present one of the simpler transformations in this section and note that the full set of transformations is given in [15] along with examples of their use.

*Theorem 4.1 (Interchange of **par** and sequential composition)*. If $Q_1, \ldots, Q_N$ are **arb**-compatible and $R_1, \ldots, R_N$ are **par**-compatible, then

$$\sqsubseteq \begin{array}{l} \mathbf{arb}(Q_1, \ldots, Q_N); \mathbf{par}(R_1, \ldots, R_N) \\ \mathbf{par}(\quad (Q_1; \mathbf{barrier}; R_1), \\ \qquad \ldots, \\ \qquad (Q_N; \mathbf{barrier}; R_N) \qquad\qquad\qquad) \end{array}$$

□

Executing par-model programs. **par** composition as described in this section is implemented by general parallel composition (as described in Section 3) plus a barrier synchronization that meets the specification of [15]. Thus, we can transform a program in the **par** model into an equivalent program in any language that includes constructs that implement composition and barrier synchronization in a way consistent with our definitions (which in turn are consistent with the usual meaning of parallel composition with barrier synchronization). Details and examples are given in [15].

5 The subset par model and distributed-memory programs

As discussed in Section 2, once we have developed a program in our **arb** model, we can transform it into one suitable for execution on a distributed-memory–message-passing architecture via what we call the *subset **par** model*, which is a restricted form of the **par** model discussed in Section 4. In this section we sketch how we extend our model of parallel composition to include message-passing operations, define a restricted subset of the **par** model that corresponds more directly to distributed-memory architectures, discuss transforming programs in the resulting subset **par** model into programs using parallel composition with message-passing, and briefly discuss executing such programs on distributed-memory–message-passing architectures.

Parallel composition with message-passing. We first expand the definition of parallel composition given previously to include message-passing over single-sender–single-receiver channels with infinite slack (defined much as in [1]; a formal specification is given in [15].)

We define message-passing in terms of our model by defining two new commands, **send** and **recv**, and extending Definition 3.12. This combined definition (presented in detail in [15]), like many other implementations of message-passing, represents channels as queues, with message sends and receives modeled as enqueue and dequeue operations. The **send** command includes local protocol variables of type Queue ("outports") and a protocol action that performs an enqueue on one of them; the **recv** command includes local protocol variables of type Queue ("inports") and a protocol action that performs a dequeue on one of them (suspending until the queue is nonempty). (As in

Section 4, we model suspension as busy waiting.) The extended definition of parallel composition includes local protocol variables of type Queue ("channels"), one for each ordered pair of component programs (elements of the composition), and specifies that as part of the composition operation the component programs P_j are modified by replacing inport and output variables with channel variables. That is, the composition "connects" the inports and outports to form channels.

The subset par model. We define the subset **par** model such that a computation of a program in this model may be thought of as consisting of an alternating sequence of (i) blocks of computation in which each component operates independently on its local data, and (ii) blocks of computation in which values are copied between components, separated by barrier synchronization. We refer to the former as *local-computation sections* and to the latter (together with the preceding and succeeding barrier synchronizations) as *data-exchange operations*.

That is, a program in the subset **par** model is a composition $\mathbf{par}(P_1, \ldots, P_N)$, where $P_1, \ldots, P_N$ are subset-**par**-compatible as defined by the following: (i) $P_1, \ldots, P_N$ are **par**-compatible; (ii) the variables V of the composition (excluding the protocol variables representing message channels) are partitioned into disjoint subsets $W_1, \ldots, W_N$ (sets of "local data" as mentioned previously); and (iii) the forms of $P_1, \ldots, P_N$ are such that they "match up" with respect to execution of barriers (as for **par**-compatibility), with the additional restriction that each group of "matching" sections (between barriers) constitutes either a local-computation section or a data-exchange operation. Details and examples are given in [15].

Transforming subset-par-model programs into message-passing programs. We can transform a program in the subset **par** model into a program for a distributed-memory–message-passing architecture by mapping each component P_j onto a process j and making the following additional changes: (i) Map each element W_j of the partition of V to the address space for process j. (ii) Convert each data-exchange operation (consisting of a set of (**barrier**; Q'_j; **barrier**) sequences, one for each component P_j) into a collection of message-passing operations, in which each assignment $x_j := x_k$ is transformed into a pair of message-passing commands. (iii) Optionally, for any pair (P_j, P_k) of processes, concatenate all the messages sent from P_j to P_k as part of a data-exchange operation into a single message.

Such a program refines the original program, as discussed in [15]. An example of applying these transformations is also given in [15].

Executing subset-par-model programs. We can use the transformation of the preceding section to transform programs in the subset **par** model into programs in any language that supports multiple-address-space parallel composition with single-sender–single-receiver message-passing. Details and examples are given in [15].

6 Related work

Other researchers, for example Back [2] and Martin [14], have addressed stepwise refinement for parallel programs. Our work is somewhat simpler than many approaches because we deal only with specifications that can be stated in terms of initial and final states, rather than also addressing ongoing program behavior (e.g., safety and progress properties).

Our operational model is based on defining programs as state-transition systems, as in the work of Chandy and Misra [5], Lamport [12], Manna and Pnueli [13], and others. Our model is designed to be as simple as possible while retaining enough generality to support all aspects of our programming model.

Programming models similar in spirit to ours have been proposed by Valiant [17] and Thornley [16]; our model differs in that we provide a more explicit supporting theoretical framework and in the use we make of archetypes.

Our work is in many respects complementary to efforts to develop parallelizing compilers, for example Fortran D [7]. The focus of such work is on the automatic detection of exploitable parallelism, while our work addresses how to exploit parallelism once it is known to exist. Our theoretical framework could be used to prove not only manually-applied transformations but also those applied by parallelizing compilers.

Our work is also in some respects complementary to work exploring the use of programming skeletons and patterns in parallel computing, for example that of Cole [6] and Brinch Hansen [4]. We also make use of abstractions that capture exploitable commonalities among programs, but we use these abstractions to guide a program development methodology based on program transformations.

7 Conclusions

We believe that our operational model, presented in Section 3, forms a suitable framework for reasoning about program correctness and transformations, particularly transformations between our different programming models. Proofs of the theorems of Section 3, presented in detail in [15], demonstrate that this model can be used as the basis for rigorous and detailed proofs. Our programming model, which is based on identifying groups of program elements whose sequential composition and parallel composition are semantically equivalent, together with the collection of transformations presented in [15] for converting programs in this model to programs for typical parallel architectures, provides a framework for program development that permits much of the work to be done with well-understood and familiar sequential tools and techniques. A discussion of how our approach can simplify the task of producing correct parallel applications is outside the scope of this paper, but [15] presents examples of its use in developing example and real-world applications with good results.

Much more could be done, particularly in exploring providing automated support for the transformations we describe and in identifying additional useful transformations, but the results so far are encouraging, and we believe that the work as a whole constitutes an effective unified theory/practice framework for parallel application development.

Acknowledgments

Thanks go to Mani Chandy for his guidance and support of the work on which this paper is based, and Eric Van de Velde for the book [18] that was an early inspiration for this work.

References

1. G. R. Andrews. *Concurrent Programming: Principles and Practice*. The Benjamin/Cummings Publishing Company, Inc., 1991.
2. R. J. R. Back. Refinement calculus, part II: Parallel and reactive programs. In *Stepwise Refinement of Distributed Systems: Models, Formalisms, Correctness*, volume 430 of *Lecture Notes in Computer Science*, pages 67–93. Springer-Verlag, 1990.
3. R. Bagrodia, K. M. Chandy, and M. Dhagat. UC — a set-based language for data-parallel programming. *Journal of Parallel and Distributed Computing*, 28(2):186–201, 1995.
4. P. Brinch Hansen. Model programs for computational science: A programming methodology for multicomputers. *Concurrency: Practice and Experience*, 5(5):407–423, 1993.
5. K. M. Chandy and J. Misra. *Parallel Program Design: A Foundation*. Addison-Wesley, 1989.
6. M. I. Cole. *Algorithmic Skeletons: Structured Management of Parallel Computation*. MIT Press, 1989.
7. K. D. Cooper, M. W. Hall, R. T. Hood, K. Kennedy, K. S. McKinley, J. M. Mellor-Crummey, L. Torczon, and S. K. Warren. The Parascope parallel programming environment. *Proceedings of the IEEE*, 82(2):244–263, 1993.
8. E. W. Dijkstra and C. S. Scholten. *Predicate Calculus and Program Semantics*. Springer-Verlag, 1990.
9. E. Gamma, R. Helm, R. Johnson, and J. Vlissides. *Design Patterns: Elements of Reusable Object-Oriented Software*. Addison-Wesley, 1995.
10. D. Gries. *The Science of Programming*. Springer-Verlag, 1981.
11. C. A. R. Hoare. An axiomatic basis for computer programming. *Communications of the ACM*, 12(10):576–583, 1969.
12. L. Lamport. A temporal logic of actions. *ACM Transactions on Programming Languages and Systems*, 16(3):872–923, 1994.
13. Z. Manna and A. Pnueli. Completing the temporal picture. *Theoretical Computer Science*, 83(1):97–130, 1991.
14. A. J. Martin. Compiling communicating processes into delay-insensitive VLSI circuits. *Distributed Computing*, 1(4):226–234, 1986.
15. B. L. Massingill. A structured approach to parallel programming. Technical Report CS-TR-98-04, California Institute of Technology, 1998. Ph.D. thesis.
16. J. Thornley. A parallel programming model with sequential semantics. Technical Report CS-TR-96-12, California Institute of Technology, 1996.
17. L. G. Valiant. A bridging model for parallel computation. *Communications of the ACM*, 33(8):103–111, 1990.
18. E. F. Van de Velde. *Concurrent Scientific Computing*. Springer-Verlag, 1994.

BSP in CSP: Easy as ABC

Andrew C. Simpson, Jonathan M. D. Hill, and Stephen R. Donaldson

Oxford University Computing Laboratory, Oxford, United Kingdom

Abstract. In this paper we describe how the language of Communicating Sequential Processes (CSP) has been applied to the analysis of a transport layer protocol used in the implementation of the Bulk Synchronous Parallel model (BSP). The protocol is suited to the bulk transfer of data between a group of processes that communicate over an unreliable medium with fixed buffer capacities on both sender and receiver. This protocol is modelled using CSP, and verified using the refinement checker FDR2. This verification has been used to establish that the protocol is free from the potential for both deadlock and livelock, and also that it is fault-tolerant.

1 Introduction

In this paper we describe how the formal notation Communicating Sequential Processes (CSP) [6, 8] and the associated refinement checker FDR2 (for Failures-Divergences Refinement) [7] have been applied to the analysis of a transport layer protocol used in the implementation of the Bulk Synchronous Parallel model (BSP); the protocol is described in full in [2]. In particular, we describe how FDR2 has been used to establish that the protocol is free from the potential for both deadlock and livelock, and also that it is fault-tolerant.

The structure of this paper is as follows. In Section 2 we provide a brief introduction to BSP, while in Section 3 we describe the protocol with which we are concerned. In Section 4 we provide an introduction to the language of CSP, as well as the concept of refinement, which is fundamental to our formal analysis. Section 5 illustrates how CSP has been used to model the protocol, while in Section 6 we discuss how formal verification has been carried out using FDR2. Finally, Section 7 discusses the key issues raised in the paper, and provides some suitable conclusions.

2 The BSP Programming Model and *BSPlib*

BSP [11, 12, 10] is a model of parallel computation in which the abstract machine is a collection of p processor-memory pairs, an interconnection network or switch through which the processors communicate packets in a point-to-point manner, and a mechanism by which the processors can barrier synchronise.

A BSP program consists of a sequence of parallel *supersteps*, which are made up of three ordered phases. The first phase involves simultaneous local computation in each processor, in which only values stored in the processor's memory are used. The second phase involves communication actions among the processors, which causes the transfer of data. Finally, a barrier synchronisation waits for the completion of all of the communication actions, and then makes any data transferred visible in the local memories of the destination processors.

In the absence of special-purpose BSP programming languages and computers, the BSP programming style can be realised by providing BSP communication primitives as a library which may be called from C or Fortran. Such a library is *BSPlib* [5], which provides a small number of communication primitives for BSP programming in an SPMD (single program, multiple data) manner. Data transfers are initiated by calls to several possible routines, e.g., bsp_put(i,x,y,s), where x is a pointer to a data structure to be communicated, y is a data structure on i where a copy of x will be written, and s is the number of bytes transferred (i.e., the size of x). Barriers are entered by calls to bsp_sync, which is of course blocking: each independent thread invokes it when it has completed its computation, and returns from the invocation when all other threads have also invoked it. Calls to bsp_sync also trigger pending communications, since the implementation chooses to preserve the non-overlapping of computation and communication implied by superstep semantics.

In the next section we give an informal description of the implementation of *BSPlib*, and motivate why the system does not suffer from deadlock. We describe a low-level messaging layer (with fixed buffering capacity) that provides reliable delivery of packets. We aim to prove that this layer *and its use by BSPlib* guarantee freedom from both deadlock and livelock.

3 The messaging layer used by *BSPlib*

BSPlib is implemented on top of a *messaging* layer that provides a certain degree of isolation from the transport layer of any particular machine. The important primitives supplied by this layer are for hand-shaking and initialisation of the computation on the various processors, mechanisms for orderly and unorderly shutdown, a non-blocking send primitive, a blocking receive primitive, and a probe to test for message presence. It is important to realise that this is not a general interface; in particular, the implementation of the messaging layer takes into account how it is used by the *BSPlib* implementation. This restriction in the interaction between the two layers guarantees that a user's BSP programs are deadlock- and livelock-free; the purpose of this paper is to verify this claim.

In modern message passing systems, deadlock can be avoided if non-blocking communications are used in preference to blocking synchronous sends. This absence of deadlock is intuitive, as the use of non-blocking sends removes the circularity of pairing send-receive actions that gives rise to deadlock when communication is synchronous. However, non-blocking communications only guarantee this property if the buffering capacity of the protocol stack exceeds the amount of data that is pushed into the stack. Therefore, non-blocking communication can quite easily fail or suffer from deadlock if, for example, two processes simultaneously push data into their protocol stacks in an attempt to send large amounts of data to each other. The point at which this deadlock occurs depends upon the buffering capacities of both sender and receiver. To quote the MPI report [4]: "...*the send start call is local: it returns immediately, irrespective of the status of other processes. If the call causes some system resource to be exhausted, then it will fail and return an error code. Quality implementations of MPI should ensure that this happens only in 'pathological cases'.*"

The BSP messaging layer implements the semantics of supersteps in the following way:

1. Calls to `bsp_sync` begin the actual transfer of data implied by previous communications in the superstep.
2. A total exchange (i.e., every processor communicates with every other processor) takes place between all the processors, so that each processor can inform the others of the number and size of messages it plans to send to them.
3. Each processor interleaves the sending of outgoing packets with the reception of any incoming packets destined for that processor. The emphasis of the interleaving is placed on reception so that messages need not be queued up at the receiver for a long time, wasting buffer resources.
4. When each processor has received all of the messages it is expecting, it proceeds into the computation phase of the next superstep.

Freedom from deadlock is guaranteed due to the fact that if there is insufficient buffering capacity in the protocol stack when a message is communicated in stages 2 and 3, then the non-blocking send fails and the implementation of *BSPlib* will always receive any packets that arrive. This reclaims the buffers associated with the received packets, which can be re-used for sending data. If this condition were not guaranteed, then deadlock may occur. Given our CSP description of the protocol, this possibility can be automatically detected (by FDR2).

3.1 A lower level transport layer

Despite the fact that *BSPlib* is built using a messaging layer which guarantees the ordered delivery of packets, this messaging layer is implemented upon the UDP/IP transport layer which does not. Therefore, although protocols such as UDP/IP have a shallower protocol stack and have a greater potential for high performance, if user APIs such as *BSPlib* are implemented over UDP/IP, they must handle the explicit acknowledgement of data and error recovery themselves.

To implement a reliable non-blocking send primitive, the implementation of *BSPlib* copies a user data structure—referenced in the send call—into a buffer that is taken from a free queue. If there are no free buffers available, then the send fails and it is up to the higher level *BSPlib* layer to recover from this situation. *BSPlib* recovers either by receiving any packets that have been queued for *BSPlib* or by asking those links that have a large number of unacknowledged send buffers to return an acknowledgement. The buffer is placed on a send queue and a communication is attempted using the `sendto()` datagram primitive. Queueing the buffer on the send queue ensures that it does not matter if the packet communicated by `sendto()` fails to reach its destination, as our error recovery protocol can resend it if necessary. Only when a send has been positively acknowledged by the partner processor are buffers reclaimed from the send queue to the free queue.

Message delivery is asynchronous and accomplished by using a signal handler that is dispatched whenever messages arrive at a processor (SIGIO signals). Within the signal handler, messages are copied into user space by using `recvfrom()`.

We have used our own error recovery mechanism that uses a selective retransmission scheme that only resends the packets that are actually dropped and a sophisticated acknowledgement policy that is based upon available buffer resources and not timeouts [2]. The algorithm associated with the acknowledgement scheme has both a sender and receiver component. From the sender's perspective, if the number of buffers on a free queue of packets becomes low then outgoing packets are marked as requiring acknowledgements. This mechanism is triggered by calculating how many packets can be consumed (both incoming and outgoing) in the time taken for a message round-trip. The sender therefore anticipates when buffer resources will run out, and attempts to force an acknowledgement to be returned just before buffer starvation. This strategy minimises both the number of acknowledgements and the time a processor stalls trying to send messages. From the receiver's viewpoint, if n is the total number of send and receive buffers, then a receiver will only send a non-piggy-backed acknowledgement if it was requested and at least $\frac{n}{2p}$ messages have come in over a link since the last acknowledgement (either piggy-backed or explicitly).

3.2 A protocol for a reliable transport layer

Two types of packets are used in the protocol:

1. Sequenced data packets that contain a fixed-sized header with MAC, sequence number, payload length and type, and piggy-backed recovery information. This is followed by a variable-length payload, the length of which is bounded so that the entire packet is no larger than the physical layer's payload size (1500 bytes for Ethernet).
2. Unsequenced control packets that contain the same fixed-sized header, but no data. Control packets are used for explicit acknowledgements and *prods*, as described below.

Each processor contains a free queue of packets, and send and receive queues associated with each communication channel ($p - 1$ in total). The send queue contains unacknowledged sent data-packets, and the receive queue contains data-packets not yet consumed by the upper *BSPlib* layer.

We describe a protocol which aims to minimise the number of redundant retransmissions of data packets and the number of control packets required to achieve this. As an example, consider two processors—A and B—communicating over a channel that loosely guarantees packet ordering, but may drop packets. Suppose processor A sends the sequence of packets $\langle 0, 1, \ldots, 9 \rangle$ (which will be queued on the send queue that A associates with B until acknowledged) to B, and only the sequence $\langle 0, 1, 4, 5, 6, 9 \rangle$ arrives at B (and are queued onto a receive queue associated with A). If a go-back-n protocol (as used by TCP/IP) is used to recover from the lost packets, then processor A—after learning that the highest in-sequence packet received has sequence number 1—retransmits the packet sequence $\langle 2, 3, 4, 5, 6, 7, 8, 9 \rangle$, which contains four redundant packets. A more sophisticated selective retransmit scheme would only resend the sequence $\langle 2, 3, 7, 8 \rangle$. However, in both cases, it is the responsibility of the recovery protocol to ensure that the sending processor learns of these missing packets. This can either be done by piggy-backing recovery information on packets travelling in the reverse direction, or by sending an explicit control packet that informs the sender of the problem if there is no reverse traffic.

In our selective retransmission protocol, the fixed sized packet header contains two fields for piggy-backed sequence number information:

1. An *acknowledgement* field such that if B sent a message back to A after having received $\langle 0, 1, 4, 5, 6, 9 \rangle$, this field would contain 1 (this allows A to remove all packets up to this value from the send queue).
2. An *end-of-hole* field which specifies the sequence number of the packet after the last in-sequence packet, i.e., 4. When no unaccounted-for packet loss has occurred, the acknowledged sequence number and this field are the same.

If A and B are involved in symmetric communication, then A will learn of the first hole from this piggy-backed data and resend the packets 2 and 3. After these have been received by B, further piggy-backed information will trigger the resending of 7 and 8. To prevent potentially stale in-flight information from being reused to cause repeated resends, a double retransmission of data is delayed for at least a round-trip delay of the network.

If there is no reverse traffic from B to A then if the upper *BSPlib* layer has consumed packets 0 and 1, and requires packet 2, B will *prod* processor A with a control packet. This allows A to learn of the hole and resend packets 2 and 3. In case the control packet goes missing, this prodding is repeated no more frequently than the round trip interval of the network. This prodding can also arise without packet loss when the computation on a processor runs ahead of another and hence enters its communication phase earlier. In this case, the prodded processor responds with an acknowledgement rather than missing data. The prodding processor then realises that it has run ahead and is being impatient, and enters an exponential back-off of prods.

There can be a problem with the protocol described above if the tail of a sequence of packets has been dropped. For example, consider if B only received the packets $\langle 0, 1 \rangle$ and packets $\langle 2, 3, \ldots, 9 \rangle$ have been dropped. Then, after the upper layer consumes the first two packets, it will send a prodding control packet to A with an acknowledgement and end-of-hole number of 1. When A receives this control packet, packets $\langle 0, 1 \rangle$ will be acknowledged and removed from the send queue, but—as there are more packets on the send queue—A will assume that they must have been dropped. Instead of just resending them all (which amounts to a go-back-n protocol), A resends the first and last packets from the send queue (i.e., 3 and 9). If there was no real hole then only two redundant messages are sent. On the other hand, if there really is a hole, then the receiver (eventually) gets an out-of-sequence message and can report the extent of the hole the next time the sender is prodded.

The benefits of this scheme over a more general selective retransmission scheme that could, for example, enumerate all the sequence numbers of packets dropped, is that only one integer field is required in the packet header. In the worst case, as many prodding control packets are generated as holes in the data; in the best, none. The scheme relies upon the observed fact that dropped packets will be clustered into holes. The network used, and the scheme we employ, ensures that packet loss is quite rare (0.05% packet drops are experienced in the NAS benchmarks [3]) and any packets lost are actually dropped by the receivers (for example, when available buffers run low). When buffers run low, only the next expected packets are accepted, but the acknowledgement data on all arriving packets is still inspected so that buffers held in the send queues can be freed and progress is guaranteed. This is proved in the remainder of the paper.

4 Communicating Sequential Processes

4.1 Events and processes

The language of CSP is a notation for describing the behaviour of concurrently-evolving objects, or *processes*, in terms of their interaction with their environment. This interaction is modelled in terms of *events*: abstract, instantaneous, synchronisations that may be shared between several processes.

We use compound events to represent communication. The event name $c.x$ may represent the communication of a value x on a *channel* named c. At the event level, no distinction is made between *input* and *output*. The willingness to engage in a variety of similar events—the readiness to accept input—is modelled at the process level. The same is true of output, which corresponds to an insistence upon a particular event from a range of possibilities.

A *process* describes the pattern of availability of certain events. The simplest process of all is *Stop*, which makes no events available at any time. The prefix process $a \rightarrow P$ is ready to engage in event a; should this event occur, the subsequent behaviour is that of P, which must itself be a process.

An external choice of processes $P \,\square\, Q$ is resolved through interaction with the environment: the first event to occur will determine the subsequent behaviour. If this event was possible for only one of the two alternatives, then the choice will go on to behave as that process. If it was possible for both, then the choice becomes internal.

An internal choice of processes $P \sqcap Q$ will behave either as P or as Q. The result of this choice can be neither influenced nor predicted by the environment.

Both forms of choice exist in an indexed form: for example, $\bigsqcap i : I \bullet P(i)$ is an internal choice between processes $P(i)$, where i ranges over the (finite) indexing set I, while $\Box\, i : I \bullet P(i)$ is the corresponding external choice.

We may denote input in one of two ways. The process $c?x \to P$ is willing initially to accept any value (of the appropriate type) on channel c. On the other hand, if we wish to restrict the set of possible input values to a subset of the type associated with the channel c, then we may write $\Box\, x : X \bullet c.x \to P$. Furthermore, the processes $c?x?y \to P$ and $\Box\, x : X \bullet \Box\, y : Y \bullet c.x.y \to P$ are equivalent (assuming that X and Y the appropriate types).

In the parallel combination $P \parallel Q$, the component processes cooperate upon shared events. To identify which events are shared and which may be performed independently, we associate each component with an interface, or *alphabet*. If $\alpha\, P$ is the alphabet of P, then P will share in every event from this set; conversely, no event from $\alpha\, P$ can take place without P's participation.

We will sometimes wish to take the other extreme and insist that no events are shared, whatever the specified alphabets. We write $P \parallel\!\parallel\!\parallel Q$ to represent the *interleaved* parallel combination of P and Q.

Finally, we may remove events from the interface of a process using the hiding operator. The process $P \setminus A$ behaves exactly as P except that events from the set A no longer require the cooperation of the environment; they are no longer visible to other processes.

4.2 Refinement

The standard notion of refinement for CSP processes, which is defined in [1], is based upon the failures-divergences model of CSP. In this model, each process is associated with a set of behaviours: tuples of sequences and sets that record the occurrence and availability of events.

The *traces* of a process P, given by *traces* $[\![P]\!]$, are finite sequences of events in which that process may participate *in that order*; the *failures* of P, given by *failures* $[\![P]\!]$, are pairs of the form (tr, X), such that tr is a trace of P and X is a set of events which may be refused by P after tr has occurred; and the *divergences* of P, given by *divergences* $[\![P]\!]$, are those traces of P after which an infinite sequence of internal events may be performed.

A process, Q, is said to be a *failures-divergences refinement* of another process, P, written $P \sqsubseteq Q$, if *failures* $[\![Q]\!] \subseteq$ *failures* $[\![P]\!]$ and *divergences* $[\![Q]\!] \subseteq$ *divergences* $[\![P]\!]$.

We may use this theory of refinement to investigate whether a potential design meets its specification. The refinement checker FDR2 [7] does exactly this. A pleasing feature of FDR2 is that if a potential design fails to meet its specification, a counter-example is returned to indicate why this is so.

5 A Formal Description

In this section we illustrate how the protocol described in Section 3 has been specified using the language of CSP.

5.1 Basic types and functions

We assume given types, *Node* and *Message*, which represent the processors associated with the protocol and the set of all messages respectively:

[*Node*, *Message*]

In addition, the type *Index* provides us with a means of associating sequence numbers with messages:

Index == $\mathbb{N}$

We assume the existence of an injective function, *number*, which associates sequence numbers with messages and functions, *sender* and *receiver*, which map messages to their sender and receiver respectively:

$$
\begin{array}{|l}
number : Message \rightarrowtail Index \\
sender : Message \twoheadrightarrow Node \\
receiver : Message \twoheadrightarrow Node
\end{array}
$$

Finally, the free types *Probe_message* and *Probe_result* are concerned with communications associated with 'probes':

Probe_message ::= *prod* | *back_off*
Probe_result ::= *yes* | *no*

The former represent communications between processors checking for the availability of messages, while the latter represent similar communications between the upper and lower layer at given processors.

5.2 Events and channels

Prior to the protocol starting to transfer messages it is necessary for each node to establish exactly which messages it is to send. Thus, a communication of the form $to\_go.n_1.x$ informs node n_1 that x contains the messages which are to be sent (from n_1) and $to\_get.n_1.y$ indicates that y relates to the numbers of the messages which n_1 can expect to receive from other nodes. The event *synch* represents synchronisation between all of the processors after such communication.

The channels *send* and *receive* allow the processors to send and receive messages. As an example, the communication $send.n_1.n_2.m.2.4$ relates to node n_1 sending some message $m \in Message$ to node n_2, with 2 and 4 being the numbers of the last message acknowledged and number of the first message received 'after the window' respectively by n_1 from n_2. The communication $receive.n_2.n_1.m.2.4$ relates to the corresponding receipt.

The channel *send_prompt* models the action of one processor 'prodding' another for data, or replying to an earlier prompt from another processor. If the receiving node (for example, n_2) has no data to send, then its reply might be of the form $send\_prompt.n_1.n_2.back\_off.1.3$. The channel $receive\_prompt$ relates to the receipt of such a prompt.

At a higher level, the channel $send\_to$ initiates the sending of a message from one node to another: it corresponds to the upper layer instructing the lower layer to send a message; the channel can_send is used to indicate whether or not there is buffer space available for such an action. In addition, the channel out passes messages from the lower layer to the upper layer.

Finally, communications on the channel $timeout$ indicate that a timeout (due to non-receipt messages) has occurred at a particular processor. Such timeouts may only occur after an upper layer has sent all of its outgoing messages.

5.3 The upper layer

The upper layer is responsible for keeping track of the messages to be sent to, and to be received, from other nodes. Both kinds of information are represented by partial functions:

$$\begin{array}{|l}
to : Node \nrightarrow \text{seq } Message \\
from : Node \nrightarrow \text{seq } Index \\
\hline
this\_node \notin \text{dom } to \\
this\_node \notin \text{dom } from
\end{array}$$

Initially, each node informs each other of the sequence numbers of the messages which it intends sending. We are not concerned with the explicit modelling of this phase here.

$$Upper(i) = to\_go.i?to \rightarrow to\_get.i?from \rightarrow synch \rightarrow Main(i, to, from)$$

After this initial phase, the behaviour of the upper layer depends upon the state of the system with regard to which messages have been received and sent.

$Main(i, to, from) =$
    if $\forall x : \text{dom } to \bullet to\, x \neq \langle\rangle \land \forall x : \text{dom } from \bullet from\, x \neq \langle\rangle$
    then $Probe(i, to, from)$
    else if $\forall x : \text{dom } to \bullet to\, x \neq \langle\rangle$
        then $All\_Out(i, to, from)$
        else if $\forall x : \text{dom } from \bullet from\, x \neq \langle\rangle$
            then $Receive(i, to, from)$
            else $Upper(i)$

If there are still messages to be sent and to be received, then the upper protocol may try and probe; that is, it looks to see if there are any messages which are ready to be passed from the lower layer:

$Probe(i, to, from) = Probe\_Yes(i, to, from)$
$\square$
$Probe\_No(i, to, from)$

$Probe\_Yes(i, to, from) =$
$\quad probe.i.yes \to out.i?m \to$
$\quad\quad \textbf{if} \quad number\ m = head\ (from\ sender(m))$
$\quad\quad \textbf{then}\ Main(i, to, from \oplus \{sender(m) \mapsto tail(from\ sender(m))\})$
$\quad\quad \textbf{else}\ Main(i, to, from)$

In the above the notation $f \oplus g$ indicates that the function f is *over-ridden* with the function g.

$Probe\_No(i, to, from) =$
$\quad probe.i.no \to$
$\quad\quad can\_send.i.yes \to$
$\quad\quad\quad \square\, n : \{x : \text{dom}\ to \mid to\ x \neq \langle\rangle\} \bullet$
$\quad\quad\quad\quad send\_to.i.n.head(to\ n) \to$
$\quad\quad\quad\quad\quad Main(i, to \oplus \{n \mapsto tail(to\ n)\}, from)$
$\quad\square$
$\quad\quad can\_send.i.no \to Main(i, to, from)$

If there is insufficient space in the buffer for a message to be passed from the upper layer to the lower layer, then the upper layer is forced to re-enter the *Main* process, which will check for incoming messages.

If the next message has arrived at the lower layer, then it is passed to the upper layer. On the other hand, if there is no message waiting to be passed from the lower layer then an outgoing message is passed to the lower layer.

If the node has received all of its messages but some are left to be sent, then the process *All_Out* is invoked:

$All\_Out(i, to, from) =$
$\quad \textbf{if} \quad \exists x : \text{dom}\ to \bullet to\ x \neq \langle\rangle$
$\quad \textbf{then}\ can\_send.i.yes \to$
$\quad\quad\quad \square\, n : \{x : \text{dom}\ to \mid to\ x \neq \langle\rangle\} \bullet$
$\quad\quad\quad\quad send\_to.i.n.head(to\ n) \to$
$\quad\quad\quad\quad\quad All\_Out(i, to \oplus \{n \mapsto tail(to\ n)\}, from)$
$\quad\quad \square$
$\quad\quad\quad can\_send.i.no \to Main(i, to, from)$
$\quad \textbf{else}\ Main(i, to, from)$

Here, the upper layer enters a phase where it continues to pass all messages which are left to be sent to the lower layer.

Finally, if all messages have been sent to the lower layer, but there are still messages to come in, then the upper layer enters a 'blocking receive' phase:

$Receive(i, to, from) =$
　　$probe.i.yes \to out.i?m \to$
　　　　if　　$number\ m = head\ (from\ sender(m))$
　　　　then $Main(i, to, from \oplus \{sender(m) \mapsto tail(from\ sender(m))\})$
　　　　else　$Main(i, to, from)$
　　□
　　$probe.i.no \to Receive(i, to, from)$
　　□
　　$timeout.i \to Receive(i, to, from)$

This process keeps probing for data from the lower layer until messages are available to be output: it is only during this phase that a timeout can occur at the lower layer.

5.4 The protocol

At the lower level, the process *Protocol* is defined as follows.

$Protocol(i, msgs, last, ack, next) =$
　　$Timeout(i, msgs, last, ack, next)$
　　□
　　$Receive(i, msgss, last, ack, next)$
　　□
　　$Send(i, msgs, last, ack, next)$
　　□
　　$Probed(i, msgs, last, ack, next)$
　　□
　　$Can\_Send(i, msgs, last, ack, next)$

In the above, the process *Protocol* is concerned with five parameters. The first, i, identifies the local processor, while *msgs* is the message buffer containing those messages which have been sent to, and received from, other nodes (and have not yet been acknowledged). The functions *last* and *ack* associate with each processor the sequence number of the last message seen (or the first message after a gap, if there is one) and the last message acknowledged respectively. Finally, the function *next* indicates the sequence numbers of the next expected message from each processor.

The process *Timeout* is defined thus:

$Timeout(i, msgs, last, ack, next) =$
　　$timeout.i \to Send\_Prompts(i, msgs, last, ack, next)$

Here, if a timeout occurs then the other processors are prompted for data.

The process *Receive* is described thus:

$Receive(i, msgs, last, ack, next) =$
　　$Receive\_Prod(i, msgs, last, ack, next)$
　　$\Box$
　　$Receive\_Back\_Off(i, msgs, last, ack, next)$
　　$\Box$
　　$Receive\_Message(i, msgs, last, ack, expecting)$

In the above, three communications may be received from another node: a 'prod', a 'back-off' warning, or a 'normal' message. The first of these cases—the receipt of a prod—is described by the following process:

$Receive\_Prod(i, msgs, last, ack, next) =$
　　$receive\_prompt.i?n.prod?a?l \rightarrow$
　　　　if $\quad \neg\, \exists\, m : msgs \bullet receiver(m) = n$
　　　　then $send\_prompt.i.n.back\_off.ack(n).last(n) \rightarrow$
　　　　　　$Protocol(i, msgs, last, ack, next)$
　　　　else $Send\_All(i, msgs, last, ack, next, a, l, n)$

Here, if a prompt is received then either a 'back-off' warning is sent to the prodding node (if there are no missing messages) or all of the missing messages are sent to it; the process *Send_All* (which we do not describe here) takes care of the latter case. In addition to sending any missing messages, this process also updates the local state with respect to those messages which have been acknowledged by the sending processor.

If a 'back-off' message is received from another node then the message numbers of the last message received and the last acknowledgement sent are noted, and any missing messages are sent to that node:

$Receive\_Back\_Off(i, msgs, last, ack, next) =$
　　$receive\_prompt.i?n.back\_off?a?l \rightarrow$
　　　　$Send\_All(i, msgs, last, ack, next, a, l, n)$

If a 'normal' message is received, then the process behaves thus:

$Receive\_Message(i, msgs, last, ack, next) =$
　　$receive.i?n?message?a?l \rightarrow$
　　　　if $\quad \# msgs = capacity(i) - 1$
　　　　　　$\land$
　　　　　　$next(sender(message)) \neq number(message)$
　　　　then $Send\_All(i, msgs, last, ack, next, a, l, n)$
　　　　else $Send\_All(i, msgs \cup \{message\}, last, ack, next, a, l, n)$

If the upper layer wishes to send a message then it is passed to the protocol which, in turn, passes it on:

$Send(i, msgs, last, ack, next) =$
　　$send\_to.i?n?m \rightarrow send.i.n.m.ack(n).last(n) \rightarrow$
　　　　$Protocol(i, msgs \cup \{m\}, last, ack, next)$

The result of a probe by the upper layer is resolved as follows:

$Probed(i, msgs, last, ack, next) =$
if $\exists n : Node \setminus \{i\}; m : msgs \bullet next(n) = number(m)$
then $probe.i.yes \to$
$\quad \Box \, n : \{n : Node \setminus \{i\} \mid \exists m : msgs \bullet next(n) = number(m)\} \bullet$
$\quad out.i.(\mu \, m : msgs \mid number \, m = next(n)) \to$
$\quad Protocol(i, \{m : msgs \mid sender(m) \neq n \lor number(m) \neq next(n)\},$
$\quad\quad\quad last, ack, next \oplus \{n \mapsto next(n) + 1\})$
else $probe.i.no \to Protocol(i, msgs, last, ack, next)$

Finally, the ability for the upper layer to pass a message to the lower layer depends on the current state of the buffer:

$Can\_Send(i, msgs, last, ack, next) =$
if $\#msgs < capacity(i)$
then $can\_send.i.yes \to Protocol(i, msgs, last, ack, next)$
else $can\_send.i.no \to Protocol(i, msgs, last, ack, next)$

5.5 The Protocol

For any processor $n \in Node$, we compose the upper and lower layers thus:

$Node(n) = Protocol(n, \emptyset, last_n, ack_n, next_n) \parallel Upper(n)$

We assume initial values of *last*, *ack* and *next* for each processor. We also assume the existence of a buffer for passing messages between pairs of nodes. For the moment, we assume such a buffer is perfect.

$Buffer(from, to) =$
$\quad send.from.to?m?a?l \to$
$\quad\quad receive.to.from.m.a.l \to Buffer(from, to)$
$\quad \Box$
$\quad send\_prompt.from.to?p?a?l \to$
$\quad\quad receive\_prompt.to.from.p.a.l \to Buffer(from, to)$

The process *Comm* represents the interleaving of all such buffers:

$Comm = \vertiii{} \, i : Node; \, j : Node \mid i \neq j \bullet Buffer(i, j)$

As such, our model of the protocol is given by the process *Comm_System*:

$Comm\_System = (n : Node \bullet \parallel Node(n)) \parallel Comm$

6 Formal Analysis

The refinement checker FDR2 (for Failures Divergences Refinement) [7] has been used in the analysis of our CSP description of the protocol (of which, the essential elements have been described). This tool allows one to verify that potential implementations satisfy their specifications by checking that the former is a refinement of the latter.

We may attempt to establish the deadlock-freedom of *Comm_System* by testing whether it is a refinement of the process DF, where DF is defined by

$$DF = \bigsqcap e : \alpha\, Comm\_System \bullet e \to DF$$

We have established that two safety harnesses help ensure deadlock-freedom: these combine to ensure that there is always sufficient space in the lower layer's message buffer. The first condition ensures that messages can only be passed from the upper layer to the lower layer if there is sufficient space in the buffer (assuming that the upper layer behaves in accordance with messages received from the lower layer). The second condition ensures that if the buffer is reaching its capacity and an out-of-sequence message is received from another processor, then that message is thrown away (but the acknowledgement information is retained). If either of these conditions were to be removed, then the possibility of deadlock arises. (Strictly, of course, removing one (or both) of these conditions will not result in deadlock, but in *livelock*.)

Livelock and deadlock are related phenomena; indeed, the relationship between them is demonstrated by the means in which we apply FDR2 in the analysis of the possibility of livelock. In the context of our specification, we consider the protocol to be livelocked precisely when the lower layer embarks upon an infinite uninterrupted sequence of events, without any communication with the upper layer: from the point of perspective of the upper layer, the system is deadlocked.

If we let $\alpha\, Lower$ denote the set of all communications which the lower layer may observe and $\alpha\, Upper$ denotes the set of all communications which the upper layer may observe, we may test our protocol for livelock by investigating whether $Comm\_System \setminus \alpha\, Lower$ is deadlock-free. We have found this to be the case—assuming perfect communication.

In [9], a method for verifying the fault-tolerance of protocols using CSP and FDR2 is described. This approach involves ensuring that *critical* events are *protected* from potential faults or failures.

By replacing our *Buffer* processes with a process which represents faulty behaviour, we can verify whether our protocol is fault-tolerant. Such faulty buffers may nondeterministically create, drop, or resequence messages at any time. Provided that such behaviour is suitably bounded (i.e., there cannot be an uninterrupted, infinite sequence of dropped packets), we have been able to establish that our protocol is deadlock-free in the context of such faults.

If, however, either of our safety harnesses is removed, then it is no longer the case that our protocol is free from the potential for deadlock. For example, if we fail to ensure that only the next expected message can be received when there is limited space in the buffer, then it is possible for a series of out of sequence messages to arrive and fill up the buffer: this will result in deadlock.

7 Conclusions

We have shown that our protocol is fault-tolerant, and free from the potential for deadlock and livelock. As the protocol uses fixed buffer capacity, there is a potential for deadlock when buffers run low. Deadlock is avoided in this situation due to the manner in which packets are dropped on reception, but after their protocol information has been processed. Further, if buffers become low when sending a message, then the sender guarantees to back-off into the reception of a message. This protection mechanism restricts the ways in which the lower layer is used. It should be noted that this is not overly-restrictive, as *BSPlib* automatically guarantee this restriction. By contrast, an application using BSD stream sockets (i.e., TCP/IP) could easily suffer this fate without due care from the programmer (for example, by mismatching sends and receives, or by simultaneously sending too much data).

References

1. S. D. Brookes and A. W. Roscoe. An improved failures model for communicating processes. In S. D. Brookes, A. W. Roscoe, and G. Winskel, editors, *Proceedings of the NSF-SERC Seminar on Concurrency*, pages 281–305. Springer-Verlag Lecture Notes in Computer Science, volume 197, 1985.
2. S. R. Donaldson, J. M. D. Hill, and D. B. Skillicorn. BSP clusters: high performance, reliable and very low cost. *Parallel Computing: Special Issue on Cluster Computing*, 1998. Submitted for publication.
3. S. R. Donaldson, J. M. D. Hill, and D. B. Skillicorn. Performance results for a reliable low-latency cluster communication protocol. In *PC-NOW '99: International Workshop on Personal Computer based Networks Of Workstations, held in conjunction with IPPS'99 (under submission)*, 1999.
4. Message Passing Interface Forum. *MPI A Message-Passing Interface Standard*, May 1994.
5. J. M. D. Hill, B. McColl, D. C. Stefanescu, M. W. Goudreau, K. Lang, S. B. Rao, T. Suel, T. Tsantilas, and R. Bisseling. BSPlib: The BSP Programming Library. Technical Report PRG-TR-29-97, Oxford University Computing Laboratory, May 1997.
6. C. A. R. Hoare. *Communicating Sequential Processes*. Prentice-Hall International, 1985.
7. A. W. Roscoe. Model checking CSP. In A. W. Roscoe, editor, *A Classical Mind: Essays in honour of C.A.R. Hoare*, pages 353–378. Prentice-Hall International, 1994.
8. A. W. Roscoe. *The Theory and Practice of Concurrency*. Prentice-Hall International, 1997.
9. A. C. Simpson, J. C. P. Woodcock, and J. W. Davies. Safety through security. In *Proceedings of the Ninth International Workshop on Software Specification and Design*, pages 18–24. IEEE Computer Society Press, 1998.
10. D. B. Skillicorn, J. M. D. Hill, and W. F. McColl. Questions and answers about BSP. *Scientific Programming*, 6(3):249–274, Fall 1997.
11. L. G. Valiant. A bridging model for parallel computation. *Journal of the ACM*, 33(8):103–111, August 1990.
12. L. G. Valiant. Bulk-synchronous parallel computer. U.S. Patent No. 5083265, 1992.

4th International Workshop on Embedded HPC Systems and Applications (EHPC'99)

Workshop Co-Chairs

Devesh Bhatt

Honeywell Technology Center, USA
bhatt@htc.honeywell.com

Viktor Prasanna

Univ. of Southern California, USA
prasanna@usc.edu

Preface

The International Workshop on Embedded HPC Systems and Applications (EHPC) is a forum for the presentation and discussion of approaches, research findings, and experiences in the applications of High Performance Computing (HPC) technology for embedded systems. Of interest are both the development of relevant technology (e.g.: hardware, middleware, tools) as well as the embedded HPC applications built using such technology.

In the last three workshops, EHPC'99 has brought together industry, academia, and government researchers/users to explore the special needs and issues in applying HPC technologies to defense and commercial applications.

This years's papers promise to keep that tradition and touch upon each of the topic areas of interest, ranging from the embedded applications to usage of special purpose architectures for obtaining high performance. This year's topics include:

— Algorithms and applications such as radar signal processing, condition based maintenance, automatic vehicle guidance, and heterogeneous sorting
— Benchmarking of embedded HPC architectures and middleware services for communications, QoS, and resource management
— Operating systems and middleware architectures and services, addressing real-time communications monitoring, resource management, and special purpose embedded operating systems
— High-performance architectures such as fiber-optic networks for embedded HPC applications and use of microprocessor features to obtain high performance.

Program Committee

Ashok Agrawala, Univ. of Maryland, USA
Bob Bernecky, NUWC, USA
Hakon O. Bugge, Scali Computer, Norway
Terry Fountain, University College London, UK
Richard Games, MITRE, USA

Farnam Jahanian, Univ. of Michigan, USA
Jeff Koller, USC/Information Sciences Institute, USA
Mark Linderman, USAF Rome Laboratory, USA
Craig Lund, Mercury Computer Systems, Inc., USA
Stephen Rhodes, Adavanced Systems Architectures Ltd., UK
Samuel H. Russ, Mississipi State Univ., USA
Philip Sementilli, Hughes Missile Systems Co., USA
Behrooz Shirazi, University of Texas at Arlington, USA
Anthony Skjellum, Mississipi State Univ., USA
Lothar Thiele, Swiss Federal Institute of Technology, Switzerland
Chip Weems, Univ. of Massachusetts, USA
Sudhakar Yalamanchili, Georgia Tech., USA

Advisory Committee

Keith Bromley, NRaD, USA
Dieter Hammer, Eindhoven Univ. of Technology, The Netherlands
David Martinez, MIT Lincoln Laboratory, USA
Jose Munoz, DARPA/ Information Technology Office, USA
Clayton Stewart, SAIC, USA
Lonnie Welch, Univ. of Texas at Arlington, USA

A Distributed System Reference Architecture for Adaptive QoS and Resource Management

Lonnie R. Welch[1], Michael W. Masters[2], Leslie A. Madden[2], David T. Marlow[2], Philip M. Irey IV[2], Paul V. Werme[2], and Behrooz A. Shirazi[1]

Abstract. This paper deals with large, distributed real-time systems that have execution times and resource utilizations which cannot be characterized *a priori*. (The motivation for our work is provided in part by the characteristics of *combat systems*.) There are several implications of these characteristics: 1) demand space workload characterizations may need to be determined *a posteriori*, and 2) an adaptive approach to resource allocation may be necessary to accommodate dynamic workload changes. Thus, we present a scheme for dynamically managing distributed computing resources by continuously computing and assessing QoS and resource utilization metrics that are determined *a posteriori*. Specifically, the paper presents an adaptive distributed system reference architecture that is suitable for such an approach. This reference architecture provides the capabilities and infrastructure needed to construct multi-component, replicated, distributed object real-time systems that negotiate for a given level of QoS from the underlying distributed computing resources.

Introduction

The majority of real-time computing research has focused on the scheduling and analysis of real-time systems whose timing properties and execution behavior are known *a priori*. This is not without justification, since static approaches to the engineering of real-time systems have utility in many application domains [Sha91]. Furthermore, the pre-deployment guarantee afforded by such approaches is highly desirable. However, there are numerous applications which must operate in highly dynamic environments (such as battle environments), thereby precluding *accurate* characterization of the applications' properties by static models. In such contexts, temporal and execution characteristics can only be known *a posteriori*. (We assume the definition of *a posteriori* contained in [Web]: "from particular instances to a generalization; based on observation or experience; empirical; opposed to *a priori*." For contrast, we also present the definition for *a priori* from [Web]: "from a

---

[1] Computer Science and Engineering Dept., The University of Texas at Arlington, Arlington, TX 76019-0015, {welch|shirazi}@cse.uta.edu. Sponsored in part by DARPA/NCCOSC contract N66001-97-C-8250, and by the NSWC/NCEE contracts NCEE/A303/41E-96 and NCEE/A303/50A-98.

[2] The Naval Surface Warfare Center, Dahlgren, VA, 22448, {MastersMW|MaddenLA|WermePV|MarlowDT|IreyPM}@nswc.navy.mil.

generalization to particular instances; based on theory instead of experience or experiment." We distinguish three classes of analysis: *a priori, a posteriori* and *post mortem*. *A priori* analysis is performed before run-time, *a posteriori* analysis is performed at run-time and *post mortem* analysis is performed after run-time.) Thus, guarantees of real-time performance based on *a priori* characterizations are extraneous. However, the potential benefits of *a posteriori* approaches are significant. These benefits include the ability to function correctly in dynamic environments through adaptability to unforeseen conditions and higher *actual* utilization of computing resources.

This paper deals with large, distributed real-time systems that have execution times and resource utilizations which cannot be characterized *a priori*. (The motivation for our work is provided in part by the characteristics of *combat systems* [Har94, Wel96].) There are several implications of these requirements: 1) demand space workload characterizations may need to be determined *a posteriori* and 2) an adaptive approach to resource allocation may be necessary to accommodate dynamic workload changes. In existing real-time computing models, the execution time of a "job" is often used to characterize workload, and is usually considered to be known *a priori*. Typically, execution time is assumed to be an integer "worst-case" execution time (WCET), as in [Liu73, Wel95, Sha91]. While [Sha91] establishes the utility of WCET-based approaches by listing some of its domains of successful application, others [Leh96, Jah95, Hab90, Kuo97, Sun96, Tia95, Str97, Ste97, Liu91, Abe98, Atl98, Bra98] cite the drawbacks, and in some cases the inapplicability, of the approaches in certain domains. In [Tia95, Leh96, Hab90, Abe98] it is mentioned that characterizing workloads of real-time systems using *a priori* worst-case execution times can lead to poor resource utilization, particularly when the difference between WCET and normal execution time is large. It is stated in [Ste97, Abe98] that accurately measuring WCET is often difficult and sometimes impossible. In response to such difficulties, techniques for detection and handling of deadline violations have been developed [Jah95, Str97, Ste97]. Paradigms which generalize the execution time model have also been developed. Execution time is modeled as a *set* of discrete values in [Kuo97], as an interval in [Sun96], and as a probability distribution in [Leh96, Str97, Tia95, Atl98]. Most models consider execution time to apply to the job atomically; however, some paradigms [Liu91, Str97] view jobs as consisting of mandatory and optional portions; the mandatory portion has an *a priori* known execution time in [Liu91], and the optional portion has an *a priori* known execution time in [Str97]. Most of these approaches assume that the execution characteristics (set, interval or distribution) are known *a priori*. Others have taken a hybrid approach; for example, in [Hab90] *a priori* worst case execution times are used to perform scheduling, and a hardware monitor is used to measure *a posteriori* task execution times for achieving adaptive behavior. In [Bra98, Wel98], resource requirements are observed *a posteriori*, allowing applications which have not been characterized *a priori* to be accommodated. For applications with *a priori* characterizations, the observations are used to refine the *a priori* estimates. These characterizations are then used to drive dynamic resource allocation algorithms.

In this paper we present an approach that is appropriate for systems which experience large variations in workload. The approach dynamically manages a distributed collection of computing resources by continuously computing and assessing QoS and resource utilization metrics that are determined *a posteriori*.

Specifically, the paper presents an adaptive distributed system reference architecture that is suitable for such an approach.

The Distributed System Reference Architecture

Figure 1 depicts a distributed system reference architecture. This reference architecture provides the capabilities and infrastructure needed to construct multi-

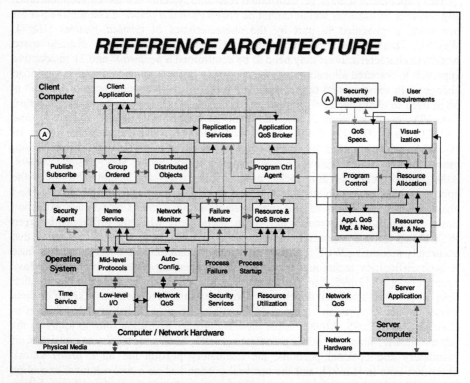

Figure 1. A distributed system reference architecture.

component, replicated, distributed object real-time systems that negotiate for a given level of QoS from the underlying distributed computing resources. The diagram shows the functional architecture structure needed to support distributed application programs for a computer containing a client application. This structure is repeated throughout the computers of the distributed system; in particular, the computer hosting the server application contains a comparable structure. Also executing somewhere in the distributed system's collection of computers is a set of QoS management components that interact with the computers, network components and applications of the distributed system to provide QoS management.

Besides the applications themselves, the components of the reference architecture consist of four primary types: operating systems, network services, high-level communication and state management middleware services, and resource management services. The operating system services are those needed to support real-time applications; operating system services are not discussed further herein. The network services provide low-level network communications and time management capabilities. The middleware components provide the ability to communicate in accord with three high level communication models: publish-subscribe; group programming, with associated ordered multicast and state data synchronization services; and distributed object programming. The QoS management services consist of computing resource and application status monitoring and reporting services, QoS negotiation services, and program control services.

Taken together, these services allow distributed applications to perform allocated processing functions, to communicate with each other, to perform fault detection and recovery activities, to perform load sharing activities, and to negotiate resource utilization and QoS requirements with the underlying distributed system infrastructure. This architecture has been partially implemented and successfully employed for dynamic management of QoS and distributed computing resources within a Navy distributed computing testbed. Features of the testbed include real-time mission critical computing, fault tolerance and scalability.

The components of the architecture are explained in the remainder of this document. Section 3 describes the *network* QoS reference architecture and introduces some additional details not shown in Figure 1. Section 4 explains the *system composition* reference architecture, and in section 4 the *resource and QoS* reference architecture is discussed.

The Network QoS Reference Architecture

The network provides a means of transferring information among the processors that compose a particular system. For the purposes of this Network QoS Reference Architecture (NQRA) there are two types of information passed through the network: (1) lower-level signaling and (2) system information. Lower-level signaling provides the means for network to operate and pass system information correctly. Examples of lower-level signaling are link operations, status and control information, routing information as well as the control information needed to support network QoS negotiation and QoS signaling for other Reference Architecture components. It is assumed that the necessary resources are allocated sufficiently to support the transfer of lower level signaling, as QoS control is exercised over the transfer of system information.

Some network provisioning decisions are performed outside of QoS control; for example, automatic fault recovery may support instantaneous reconfiguration of a host's network connection. QoS control is applied to network fault recovery at a higher level, as the resource manager must consider which processors are using network connections which may have reconfigured, leaving no redundancy and thus no recovery capability for further failures.

Figure 2 depicts the subset of the distributed system reference architecture that is the NQRA. While the NQRA architecture is only shown in the client host, it is assumed that all hosts contain this architecture. Unshaded boxes are utilized by only the NQRA. Shaded boxes are used by the NQRA and the other Reference

Architecture Components. The components of the NQRA within a host computer include: 1) Security Agent: makes network security service requests on behalf of applications; 2) Security Services: provides multi-level network security services such as encryption, registration, and key management; 3) Name Service: provides dynamic logical to physical name translation for hosts and services; 4) Auto Configuration: allows network components to autonomously initialize and dynamically reconfigure; 5) Network Monitor: gathers remote performance metrics and failure notifications; 6) Failure Monitor: collects and acts upon network resource failure reports; 7) Resource & QoS Broker: correlates local/remote network performance metric/failure reports and submits QoS requests to the Network QoS component; gathers and correlates elementary monitoring information for determining system-level QoS and performance metrics; 8) Mid-level protocols: provides basic network communications services; 9) Time Services: provides accurate high-resolution global time; 10) Low-level I/O: underlying I/O and lower-level signaling mechanisms which support Network QoS component functionality (e.g. queue schedulers, packet classifiers, etc.); 11) Network QoS: dynamically allocates network resources to meet QoS requested by other components; 12) Resource Utilization: provides a local view of network resource utilization/performance metrics and failure notifications; also collects elementary host status and load metrics.

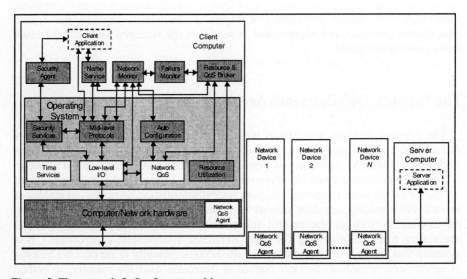

Figure 2. The network QoS reference architecture.

Within a network device (e.g. a switch, router, hub, etc. which is part of the fabric used to interconnect host computers) there is a single NQRA component, the Network QoS Agent which processes and acts upon lower-level signaling and provides status/performance reports back to host nodes which contain all of the NQRA components. The NQRA architecture is applicable to a variety of environments (e.g. LAN, WAN, mobile, etc.). The environment type is defined by the number and type of network devices along the QoS-enabled path that interconnects the communicating hosts.

As an example of a typical interaction of the NQRA components, suppose a communications channel with a given QoS requirement must be established between client and server applications on two different hosts. To communicate with the server, the client first uses the Name Service to find a physical address for it. A security association is established with the server via the security agent which uses the underlying security services component. The QoS requirements for the application are negotiated among the Network QoS Agents in each of the network devices required to establish a QoS-enabled path between the client and server. Finally, with the QoS-enabled path established, communications commences between the client and server applications. If a network device on the QoS enabled path is unable to meet the required QoS, the Network Monitor notifies the Failure Monitor and another QoS-enabled path is established by the Network QoS components and agents. All communications pass through the low-level I/O component as scheduled by the Network QoS component.

The System Composition Reference Architecture

System composition (SC) deals with services that allow for the composition of distributed, real-time, survivable applications. The SC components and their interactions are illustrated in Figure 3. The SC architecture components are: (1) Application: The unique portion of a given distributed process, the application implements specific algorithms of the combat system [Har94, Wel96], and is the driver of a subset of QoS requirements pertaining to real-time, survivability, security, and scalability. (2) Replication Service: Supports application management of replication for the local process, including load sharing, synchronization of replica state data, and replica initialization and termination handling. (3) Distributed Objects: An implementation of the CORBA standard with support for replicated objects, load sharing, and QoS. (4) Group Ordered Communications: One of several available middleware tools that provides a group programming model of interaction, ordered messages within a communications group, atomic message delivery, support for ordered state transfer, and ordered group membership change events; examples of such tools include Ensemble (Cornell), Isis (Stratus), RTCAST (U Mich), and Spread (JHU). (5) Data Distribution Service: Support for potentially high bandwidth, low latency multicast, such as might be provided by a publish/subscribe middleware tool; for example, Salamander (U Mich), RTI NDDS, or TIBCO Rendezvous (potentially with support by an underlying group programming tool). (6) Name Service: A system resource that provides a common, location-independent handle for referencing any other process or application in the distributed system. A hierarchical name reference is highly desirable; such that classes of applications can be referenced (e.g. all track distribution applications) as well as specific instances of an application. Although the name service is a system-level resource, node-level access is necessary to meet real-time service and survivability requirements. (7) Failure Monitor: Detects faults and failures of processes and host resources.

In a typical scenario, an application is initiated by the RM component. During initialization, RM provides initial replication and configuration information to support management of replica state data synchronization. Application security data is communicated via the security service, which in turn updates the system name service with the appropriate security status. A specification of the types of replication supported for different communications channels is provided to the local replication service. This information, combined with the current configuration data, is used by the replication service to initiate state transfer, state data synchronization, and load sharing allocations, as required, via the appropriate middleware services. The initial replica configuration for each of the communications channels (e.g. load sharing, primary/shadow) is reported to the application when all interfaces are initialized and stable. At this point, the application can begin its normal processing.

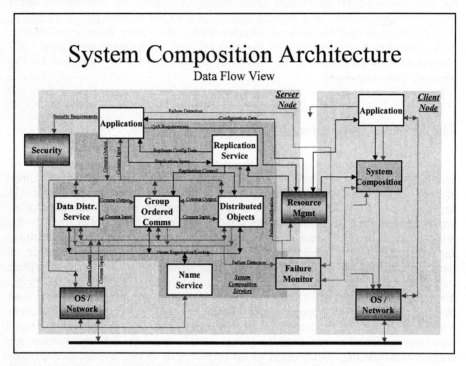

Figure 3. The system composition components and their interactions.

Application-to-application communication occurs via the appropriate middleware interface, either distributed object, group comms, data distribution, or, when necessary, UDP/TCP/IP. It is envisioned that the distributed object and data distribution middleware may eventually utilize group communications services where reliable, ordered, communications support is required. The system name service is used for common, logical, location-independent addressing, and security-level information is provided with each communication to assure that security constraints are not violated. During algorithm execution, the application communicates with the RM component, reporting its ability to meet its QoS requirements, with regard to processing and data latencies.

Application or communication errors detected at the application and middleware levels are report to the failure monitor. Errors occurring elsewhere in the distributed system that directly affect an application's replication or load sharing configuration, are reported by the failure monitor via the local replication service. Load redistribution and replica role changes are made in real-time by the replication service, based on predefined algorithms. Those decisions are reported back to the application. Reconfiguration changes made by RM are also reported to the application via the replication service, so that load redistribution and replica role decisions can be made, as appropriate.

The Resource and QoS Management Reference Architecture

Resource management (RM) and QoS management (QM) negotiate to allocate computing and communication resources in a way that attempts to satisfy all application QoS requirements. The distributed RM and QM reference architecture is shown in Figure 4, and consists of the following components: (1) QoS Specifications: Describe the static features of application software systems, and the distributed hardware and network systems; also describe QoS requirements pertaining to real-time, survivability, security and scalability. (2) Application QoS Broker: Collects elementary time-stamped events from applications and converts them into program-level QoS metrics. Also participates in QoS negotiation. (3) Application QoS Management & Negotiation: Monitors real-time path and application QoS and detects/predicts QoS violations; diagnoses local causes of QoS violations and suggests corrective reallocation actions to the Resource Management & Negotiation component; maintains application resource usage profiles; negotiates QoS with Resource Management & Negotiation component on behalf of managed applications; performs local stability analysis. (4) Resource Management & Negotiation: Oversees the resource management processes; performs global stability analysis; leads QoS negotiation in overloaded conditions. (5) Resource Allocation: Performs system level resource allocation tradeoff analysis via synthesis of (a) host and network utilization metrics and (b) tactical application performance and resource usage metrics. Performs global diagnosis to determine the causes of QoS violations and system faults, analyzes potential corrective actions, and selects specific hardware and software reconfiguration actions. (6) Program Control: Implements application control actions (startup, shutdown, reconfiguration) requested by the Resource Allocation components or by an operator, and automates the control process (e.g., analyzing and handling startup dependencies, configuration of command line arguments, etc...). (7) Program Control Agent: Provides the low-level mechanisms for starting, stopping, and reconfiguring application programs. (8) Visualization: Displays information regarding system configuration and performance; status and QoS performance of the applications, middleware, hosts, and networks; and details of the reallocation decision process.

A general interaction scenario among the architecture components which are directly related to resource and QoS management occurs as follows. The user QoS requirements for the application system are described in terms of the System/QoS Specification language. Specifications are also produced for the software system structure, the software startup and shutdown procedures, and the distributed hardware system. The specifications are used by the Resource Allocation components to initially allocate the applications of the software system to the distributed hardware resources. The initial allocation is enacted by Program Control, which notifies Program Control Agents on each host to start the appropriate applications. Once operational, the distributed applications interact as needed via the Publish/Subscribe, Group Ordered and/or Distributed Objects communication protocols. Application QoS, performance, and statuses are monitored via the Application QoS Broker components, and correlated and analyzed for actual or predicted QoS violations by the Application QoS Management & Negotiation component. Hardware utilization levels and program statuses are monitored by the Resource Utilization and Failure Monitor components. When inadequate QoS, hardware saturation or failure, or abnormal program termination is detected (or predicted), the Resource Allocation and the Resource Management & Negotiation components subsequently select a better allocation of resources and issue requests to the Program Control components to enact the new allocation. In situations where inadequate resources do not exist, QoS negotiation is performed via iteration among Resource Allocation, Resource Management and Negotiation, Application QoS Management & Negotiation, Application QoS Broker and the Client Application.

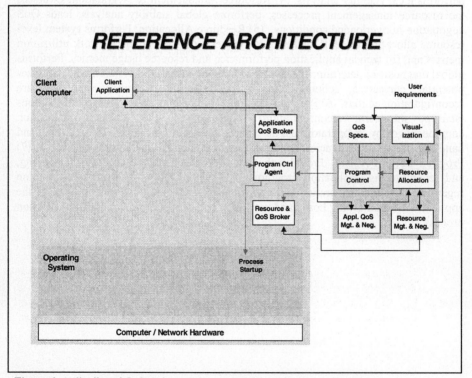

Figure 4. A distributed QoS and resource management reference architecture.

Conclusions and Ongoing Work

This paper presents adaptive QoS and resource management technology for distributed real-time systems with *a-posteriori*-determined workloads. A reference architecture that provides a plan for adaptive QoS and resource management technology within this context was presented. This reference architecture provides the capabilities and infrastructure needed to construct multi-component, replicated, distributed object real-time systems that negotiate for a given level of QoS from the underlying distributed computing resources.

A significant portion of the reference architecture has been realized. As reported in [Wel99], the effectiveness of our approach was demonstrated within an experimental large-scale combat system. The results show that real-time QoS violations were successfully detected, appropriate recovery actions were performed, and required QoS was restored. In addition, we tested the ability to provide survivability services to real-time application systems in a timely manner. Across all tests, the total response time of QoS and resource management services was less than one second, even under heavy tactical and system loads.

Ongoing work includes the realization of additional portions of the architecture, and the experimental assessment and comparison of candidate components in specific portions of the architecture. The experimental assessment efforts include the development of computing and communication benchmarks which exhibit characteristics of dynamic real-time systems.

References

[Abe98] L. Abeni and G. Buttazzo, "Integrating multimedia applications in hard real-time systems," in *Proc. of the 19th IEEE Real-Time Sys. Symposium,* 3-13, IEEE Computer Society Press, 1998.

[Atl98] A. Atlas and A. Bestavros, "Statistical rate monotonic scheduling," in *Proceedings of the 19th IEEE Real-Time Systems Symposium,* 123-132, IEEE Computer Society Press, 1998.

[Bra98] S. Brandt, G. Nutt, T. Berk and J. Mankovich, "A dynamic quality of service middleware agent for mediating application resource usage," in *Proceedings of the 19th IEEE Real-Time Sys. Symposium,* 307-317, IEEE Computer Society Press, 1998.

[Hab90] D. Haban and K.G. Shin, "Applications of real-time monitoring for scheduling tasks with random execution times ," IEEE Trans. on Software Engineering, **16(12)**, December 1990, 1374-1389.

[Har94] Robert D. Harrison Jr., "Combat system prerequisites on supercomputer performance analysis," in *Proceedings of the NATO Advanced Study Institute on Real Time Computing*, NATO ASI Series **F(127)**, 512-513, Springer-Verlag 1994.

[Jah95] F. Jahanian, "Run-time monitoring of real-time systems," in *Advances in Real-time Systems,* Prentice-Hall, 1995, 435-460, edited by S.H. Son.

[Kuo97] T.E. Kuo and A. K. Mok, "Incremental reconfiguration and load adjustment in adaptive real-time systems," *IEEE Transactions on Computers,* **46(12)**, December 1997, 1313-1324.

[Leh96]   J. Lehoczky, "Real-time queueing theory," in *Proceedings of the 17th IEEE Real-Time Systems Symposium,* 186-195, IEEE Computer Society Press, 1996.

[Liu73]   C.L. Liu and J.W. Layland, "Scheduling algorithms for multiprogramming in a hard-real-time environment," *Journal of the ACM,* **20**, 1973, 46-61.

[Liu91]   J.W.S. Liu, K.J. Lin, W.K. Shih, A.C. Yu, J.Y. Chung and W. Zhao, "Algorithms for scheduling imprecise computations," *IEEE Computer,* **24(5)**, May 1991, 129-139.

[Sha91]   L. Sha, M. H. Klein, and J.B. Goodenough, "Rate monotonic analysis for real-time systems," in *Sched. and Res. Mgmt.,* Kluwer, 1991, 129-156, edited by A. M. van Tilborg and G. M. Koob.

[Ste97]   D.B. Stewart and P.K. Khosla, "Mechanisms for detecting and handling timing errors," *Communications of the ACM,* **40(1)**, January 1997,87-93.

[Str97]   H. Streich and M. Gergeleit, "On the design of a dynamic distributed real-time environment," in *Proceedings of the 5th International Workshop on Parallel and Distributed Real-Time Systems,* 251-256, IEEE Computer Society Press, 1997.

[Sun96]   J. Sun and J.W.S. Liu, "Bounding completion times of jobs with arbitrary release times and variable execution times," in *Proceedings of the 17th IEEE Real-Time Systems Symposium,* 2-11, IEEE Computer Society Press, 1996.

[Tia95]   T.S. Tia, Z. Deng, M. Shankar, M. Storch, J. Sun, L.C. Wu and J.W.S. Liu, "Probabilistic performance guarantee for real-time tasks with varying computation times," in *Proceedings of the 1st IEEE Real-Time Technology and Applications Symposium,* 164-173, IEEE Computer Society Press, 1995.

[Web]   *Webster's New World Dictionary of the American Language (College Edition)*, The World Publishing Company.

[Wel95]   L. R. Welch, A. D. Stoyenko, and T. J. Marlowe, "Modeling resource contention among distributed periodic processes specified in CaRT-Spec," *Control Engineering Practice,* **3(5)**, May 1995, 651-664.

[Wel96]   L. R. Welch, Binoy Ravindran, Robert D. Harrison, Leslie Madden, Michael W. Masters and Wayne Mills, "Challenges in Engineering Distributed Shipboard Control Systems," *The IEEE Real-Time Systems Symposium (in Proceedings of the Work-in-Progress Session),* December 1996, 19-22.

[Wel98]   L. R. Welch, B. Ravindran, B. Shirazi and C. Bruggeman, "Specification and analysis of dynamic, distributed real-time systems," in *Proceedings of the 19th IEEE Real-Time Systems Symposium,* 72-81, IEEE Computer Society Press, 1998.

[Wel99]   L. R. Welch, Paul V. Werme , Larry A. Fontenot, Michael W. Masters, Behrooz A. Shirazi, Binoy Ravindran and D. Wayne Mills, "Adaptive QoS and Resource Management Using A Posteriori Workload Characterizations," *Technical Report,* The University of Texas at Arlington, Computer Science and Engineering Department, 1999.

Transparent Real–Time Monitoring in MPI [*]

Samuel H. Russ[1], Rashid Jean–Baptiste[2], Tangirala Shailendra Krishna Kumar[1], and Marion Harmon[2]

[1]Mississippi State University
NSF Engineering Research Center for Computational Field Simulation
Box 9627
Mississippi State, MS 39762 USA
{russ,krishna}@erc.msstate.edu
http://www.erc.msstate.edu

[2]Florida A&M University
Center for Distributive Computing
Florida A&M University
Tallahassee, FL 32307
{rjean,harmon}@cis.famu.edu
http://www.cis.famu.edu

Abstract. MPI has emerged as a popular way to write architecture–independent parallel programs. By modifying an MPI library and associated MPI run–time environment, transparent extraction of timestamped information is possible. The wall–clock time at which specific MPI communication events begin and end can be recorded, collected, and provided to a central scheduler. The infrastructure to create and collect these events has been implemented and tested, and a future architecture that can use this information is described.

1 Introduction and Background

Distributed processing is an increasingly popular architecture for execution. Because it relies on numerous, typically commodity systems, it offers low cost, inherent fault tolerance, and flexibility. A common problem with commodity systems is lack of software support for parallel execution, fault tolerance, and ability to support applications with real–time constraints.

Countless systems exist today to support distributed computing over commodity workstations. They vary in the degree to which individual workstations retain their identity, the mechanism for and degree of support for various features such as task migration and fault tolerance, and support for legacy applications and common programming standards. While a complete review is well outside the scope of this paper, one good survey of cluster computing systems may be found in [1].

One system that supports a variety of desirable features is the Hector Distributed Run–Time Environment under development at Mississippi State University and Florida A&M University [2]. The environment supports task migration, fault tolerance via check-point–and–rollback [3], collection of detailed run–time information [4], and automatic performance optimization. All of these features are provided automatically and transparently to MPI applications. That is, with no source code modifications, MPI applications can run with these features on networked workstations. (A more detailed description of these capabilities is found in the appendix at the end of this article.)

---

[*] This work was funded in part by NSF Grant No. EEC–8907070 and NSF Grant No. EEC–9730381.

An alternate approach to the software–based timestamping described here is to augment the system with special–purpose hardware [5]. The hybrid approach has been shown to be minimally intrusive and offer great accuracy, but at the expense of lower portability. It was decided to maintain Hector's software–only, commodity–only design.

The environment had no support for priority–based scheduling or for applications with real–time constraints, and so it was decided to add such features to the environment. The modifications needed to support the creation and collection of timestamps are detailed below, as is an architecture that can use the information for real–time scheduling and performance optimization.

2 An Infrastructure to Timestamp Events

The next step was to specify the API for sending timestamps and to design a way to collect them. The API was designed to match that of the MPI standard that had already been embraced. Collection was a matter of reusing existing modifications for instrumentation.

2.1 Appearance to the Programmer

It was decided that automatically timestamping every MPI event might overwhelm the collection infrastructure, and so an extension to the MPI standard was made. This extension indicates that the next MPI call that is made is to be timestamped and reported.

The format of the extension is *HCT_RT_Tag(int Marker, char *User_Str)*, where Marker and User_Str are user–defined structures that uniquely specify the tag and are reported via the collection mechanism.

For example, if the beginning and end wall–clock time of an MPI_Send is to be tagged, the code may be modified as follows:

```
...
HCT_RT_Tag(Event++,"About to send a number");
...
MPI_Send(...);
...
```

The value of the "Event" variable, the string "About to send a number", and the wall–clock time of the beginning and ending of the call to MPI_Send are reported. In this example, by incrementing Event every time an MPI event is tagged, every MPI event will have a unique value.

2.2 Modifying MPI Wrapper Functions

Modifications had already been made to Hector's MPI library in order to track CPU time spent inside and outside of MPI. This information is used to estimate time spent communicating and computing. These modifications consisted of "wrapper functions" inserted before and after every MPI call in order to update this instrumentation. The wrappers are invoked by using a special #define structure that redirects the function call from the MPI library to the wrapper library, as shown below in Fig. 1.

Two changes were made to support the real–time instrumentation. First, the modified wrappers checked to see if the HCT_RT_Tag function had been called and, if so, tagged and timestamped the MPI library entry (or exit) and sent the information. Second, it was discovered that some MPI library function calls recursively called other MPI library functions. (For example, MPI_Bcast calls MPI_Send.) The wrappers were modified so

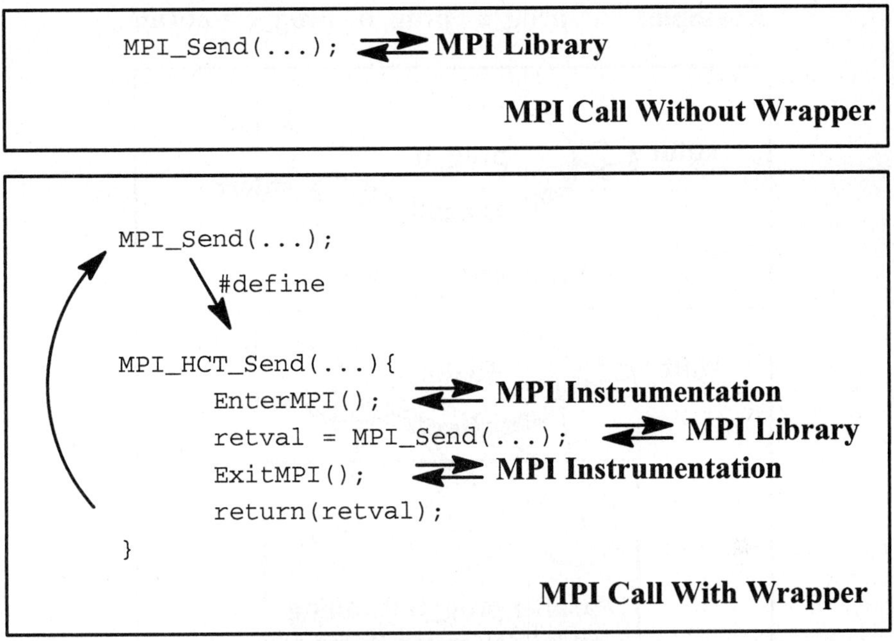

```
MPI_Send(...);   ⇄ MPI Library
```

MPI Call Without Wrapper

```
MPI_Send(...);
    ↘ #define
     ↓
MPI_HCT_Send(...){
    EnterMPI();          ⇄ MPI Instrumentation
    retval = MPI_Send(...);   ⇄ MPI Library
    ExitMPI();           ⇄ MPI Instrumentation
    return(retval);
}
```

MPI Call With Wrapper

Fig. 1. Wrapper Structure

that only the outermost MPI function invocation was timestamped. This also fixed a bug in the CPU–time–gathering infrastructure.

The wrapper itself works as follows. First, when the HCT_RT_Tag() function is called, the user–defined integer and character string are copied into local static variables and a flag is set. Second, when EnterMPI() is called, the flag is checked and, if set, a timestamp is obtained (using gettimeofday()) and the information (timestamp, integer, and string) are sent over the "real–time socket" via a call to write(). Third, when ExitMPI() is called, a process similar to EnterMPI() ensues and then the flag is reset.

3 An Infrastructure to Gather Information

Once the infrastructure was designed to emit timestamped data, the means of collecting the data centrally had to be developed.

3.1 Using Redirection

A system had already been developed to redirect stdout, stderr, and stdin to and from parallel MPI jobs. The redirection system not only collects the stdout correctly, it also supports task migration correctly, and so the stdout stream (and stdin and stderr) is maintained even as tasks migrate. The original purpose of the redirector was to support users who wanted to invoke parallel jobs inside a shell script and who used redirection in legacy shell scripts.

A redirected task works by closing the stdin, stdout, and stderr file descriptors, opening sockets back to the central collection task (called a "redirector"), and using dup() to make sure the file descriptors match. Special command–line arguments notify the task of the hostname and port number of the redirector. By closing and opening the descriptors, all subsequent calls (such as printf and scanf) work correctly. The central collection

Example: prog_a | prog_b | prog_c > outfile

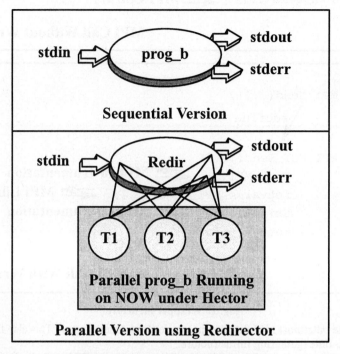

Fig. 2. Example Using the Redirector

task also submits the original job–launch request to Hector, and so acts as a single Unix command–line point of interface. The structure and an example is shown below in Fig. 2.

3.2 Adding the Real–time Stream

The redirector and the tasks were modified to add a fourth stream, nicknamed the "stdtim" stream. Since the opening of sockets was already supported, only incremental changes had to be made to add the real–time socket. Sending is accomplished by converting the timestamp and other information to a single ASCII string and calling a write().

The redirector polls to make sure all sockets have been connected and, once accomplished, polls the sockets and redirects the information streams. The current version of the redirector sends the real–time timestamp information to stdout for ease of debugging. Future versions will send the stream to a real–time scheduler and/or modified optimizer.

3.3 A Complete Example

A simple matrix multiply program was modified (by adding 6 HCT_RT_Tag() calls) to show how real–time information appears to the user. HCT_RT_Tag calls were inserted before several MPI calls, and a distributed version was launched on several workstations under Hector. The complete stdout and stdtim stream was redirected into a file that shows the results below in Fig. 3. The timestamped information is shown in boldface.

Each timestamp has the following format:

Task *Number Enter/Exit*: *String Integer Timestamp*

where *Number* is the task ID of the task sending the timestamp, "*Enter*" or "*Exit*" indicates whether the timestamp was generated entering or exiting the MPI library, *String* is

```
0:  There are 3 other processes
Opening matrix file mm.small
Current working directory = /projects/cps/hector/HCT2.1/taskalloc/mi-
grate/testing
Task 0 Enter:Sending assignment...         0        913757003.257503
Task 0 Exit :Sending assignment...         0        913757003.258122
Task 0 Enter:About to bcast N and K...     1        913757003.259126
0: Got Matrix of 100 X 100 X 100
Task 0 Exit :About to bcast N and K...     1        913757003.259828
Task 0 Enter:Bcast some rows...            2        913757003.426740
Task 0 Exit :Bcast some rows...            2        913757003.427490
Task 0 Enter:About to send some rows...    3        913757003.777638
This is 0.....Multiplying matrices!
Task 0 Exit :About to send some rows...    3        913757003.778076
Task 0 Enter:About to send some rows...    4        913757003.789029
Task 0 Exit :About to send some rows...    4        913757003.789413
This is 0.....Freeing A and B
Task 0 Enter:About to recv some rows...    5        913757004.496749
*--*0:A rows*--*0:A*--*0:B rows*--*0:BThis is 0.....Ready to receive
results from 1
Task 0 Exit :About to recv some rows...    5        913757004.617608
This is 0.....Got results from 1
Task 0 Enter:About to recv some rows...    6        913757004.652495
Task 0 Exit :About to recv some rows...    6        913757004.652667
This is 0.....Ready to receive results from 2
This is 0.....Got results from 2
Start is 2.375451
End is   3.865196
```

Fig. 3. Actual Redirected Output Stream

the user–specified string, *Integer* is the user–specified number, and *Timestamp* is the machine local time in seconds since the epoch. On the Sun workstations used in testing, time was available with microsecond resolution.

4 Architecture of a Complete Real–time Scheduler

The modifications that have been made and tested enable the design of a complete real–time scheduler for distributed computing. Modifications and additions are needed to the optimizer and master allocator.

4.1 Modifying the Optimizer

A real–time scheduler for Hector is under development at Florida A&M University. The scheduler will use user–specified priority information and real–time timestamp information to schedule, optimize, and allocate MPI tasks.

One example of how the timestamp information could be used is to diagnose that the margin by which real–time tasks are meeting specific deadlines is decreasing. This would cause the optimizer to suspend less important tasks so that real–time tasks have more processor power at their disposal. This is very similar to the approach taken in [6], but more general–purpose in the sense that only very minor code modifications are needed and very broad classes of applications are supported.

In the example, Hector would have several ways to respond to a real–time task with a declining performance margin. First, it could suspend less important tasks. It can do this by migrating the less important tasks to slower machines, by suspending them to disk, or by signalling them (which is akin to control–Z). Hector already supports these methods, and so it is simply a matter of giving the real–time monitor the authority to issue the appropriate commands. Second, it could move the real–time tasks to faster machines. This latter strategy will only be successful if the time to migrate is small enough.

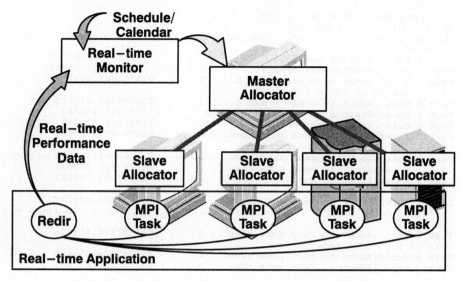

Fig. 4. Hector Running an Application with Real-time Support

4.2 Assembling the Pieces

In Hector, the optimizer is periodically invoked by the central controller (which is called the "master allocator" or "MA"). The optimizer is simply a function that is periodically called and is therefore not persistent. This design was intentional, so that it could easily be tested and replaced. An additional real-time monitor will be needed to collect all of the real-time streams and compare the information to some sort of schedule or calendar, and the modified optimizer can communicate with the real-time monitor. An example of Hector running with complete real-time instrumentation and optimization is shown below in Fig. 4. (More information about Hector's architecture and capabilities may be found in the Appendix at the end of this article.)

The real-time monitor can be implemented either as a standalone process that issues commands to the master allocator or as a separate thread of the MA. The latter is more efficient but requires more extensive modifications. Specifically, the MA is intentionally single-threaded and acts as a point of synchronization for many protocols.

5 Status, Future Work, and Conclusions

As described above, timestamping, sending, and collecting the real-time stream has been implemented and tested. Hector has already demonstrated several sophisticated migration and suspension mechanisms that can be used to remediate performance degradation. The modified optimizer and the creation of a real-time monitor are under development now.

With minimal modifications, a runtime system for MPI programs on networked workstations has been extended to timestamp and collect information about the wall-clock time of MPI events. This infrastructure can enable real-time scheduling of MPI programs with minimal source-code modifications on commodity workstations, and Hector's ability to suspend jobs can be used to free up resources for real-time applications automatically.

References

1. Mark A. Baker, Geoffrey C. Fox, and Hon W. Yau, "Cluster Computing Review", Northeast Parallel Architectures Center, Syracuse University, 16 November 1995. Available via *http://www.npac.syr.edu/techreports/hypertext/sccs-0748/cluster-review.html*.
2. Samuel H. Russ, Jonathan Robinson, Dr. Brian K. Flachs, and Bjorn Heckel, "The Hector Distributed Run–Time Environment", *IEEE Transactions on Parallel and Distributed Systems*, Vol. 9, No. 11, November 1998, pp. 1104–1112.
3. Samuel H. Russ, "An Architecture for Rapid Distributed Fault Tolerance", 3rd International Workshop on Embedded High–Performance Computing, in J. Rolim, Editor, *Parallel and Distributed Processing*, Lecture Notes in Computer Science, vol. 1388, Springer–Verlag, 1998, pp. 925–930.
4. Samuel H. Russ, Brad Meyers, Chun–Heong Tan, and Bjørn Heckel, "User–Transparent Run–Time Performance Optimization", *Proceedings of the 2nd International Workshop on Embedded High Performance Computing*, Associated with the 11th International Parallel Processing Symposium (IPPS 97), Geneva, April 1997.
5. James Harden, Cedell Alexander, Donna Reese, Marlene Evans, and Charles Hudnall, "In Search of a Standards–Based Approach to Hybrid Performance Monitoring,", *IEEE Parallel and Distributed Technology*, Winter, 1995, pp. 61–71.
6. Rakesh Jha, Mustafa Muhammad, Sudhakar Yalamanchili, Karsten Schwan, Daniela Ivan–Rosu, and Chris de Castro, "Adaptive Resource Allocation for Embedded Parallel Applications", *Proceedings of the International Conference on High–Performance Computing*, 1996, pp. 425–431.

Appendix: About Hector

The Hector Distributed Run–Time Environment has been under development at Mississippi State University for about four years. During that time it has demonstrated numerous capabilities relevant to running MPI–based parallel programs on networked resources.

First, it can migrate tasks from one workstation to another. This capability is essential in order to harness idle workstation resources, so that workstations can be released back to their "owner" on demand. It has been tested both for reliability and speed. Time to migrate is dominated by the time to transfer the program state over the network, and is therefore program size divided by network bandwidth.

Second, it can use its migration capability to suspend tasks onto remote disks. This can be used to create "backup copies" of programs that are running in order to provide fault tolerance and to free up resources.

Third, it has an extensive information–gathering infrastructure to remain well–informed about platforms and applications. Each candidate platform runs a small utility task called a "slave allocator" or SA. The SA obtains system–level information from the machine's kernel and application–level information from each task's self–instrumentation and forwards the information periodically to the central master allocator or "MA".

Fourth, it can optimize program allocation by taking into account how busy each machine is, the relative CPU speeds, and other performance–relevant information. One recent addition is an affinity function that tends to map tasks from the same program onto the same machine when possible. The MA performs this optimization, and so is a repository for global performance information.

The MA also acts as a point of synchronization for certain protocols, such as the migration and termination protocols. Its single–threaded, event–driven design make such synchronization straightforward.

Fifth, Hector provides these services transparently (with no source–code modifications) to MPI programs by a specially modified MPI implementation. The implementation permits consistent checkpointing and the ability to switch to shared–memory when two tasks share a machine.

More information about Hector may be found on the Hector Home Page at `http://www.erc.msstate.edu/thrusts/cps/hector/main/index.html`.

DynBench: A Dynamic Benchmark Suite for Distributed Real-Time Systems[1]

Behrooz Shirazi[2], Lonnie Welch[2], Binoy Ravindran[3], Charles Cavanaugh[2], Bharath Yanamula[2], Russ Brucks[2], Eui-nam Huh[2]

[2]University of Texas at Arlington
Department of Computer Science and Engineering
Box 19015
Arlington, TX 76019-0015
{shirazi|welch}@cse.uta.edu
[3]Virginia Tech
The Bradley Dept. of Electrical and Computer Engineering
340 Whittemore Hall (Mail Code 0111)
Blacksburg, VA 24061

Abstract. In this paper we present the architecture and framework for a benchmark suite that has been developed as part of the DeSiDeRaTa project. The proposed benchmark suite is representative of the emerging generation of distributed, mission-critical, real-time control systems that operate in dynamic environments. Systems that operate in such environments may have unknown worst-case scenarios, may have large variances in the sizes of the data and event sets that they process (and thus, have large variances in execution latencies and resource requirements), and may be very difficult to characterize statically, even by time-invariant statistical distributions. The proposed benchmark suite (called DynBench) is useful for evaluation of the Quality of Service (QoS) management and/or Resource Management (RM) services in distributed real-time systems. As such, DynBench includes a set of performance metrics for the evaluation of the QoS and RM technologies in dynamic distributed real-time systems. The paper demonstrates the successful application of DynBench in evaluation of the DeSiDeRaTa QoS management middle-ware.

Introduction and Background

Fig. 1 [Koob98] shows a generic *Quality of Service* (QoS) management architecture for distributed real-time systems. The distributed real-time applications are supported by a set of translucent systems layers that provide communication libraries, group communication facilities, and other support routines. Additionally, the system provides a set of middle-ware components to support QoS negotiation and Resource Management (RM). A typical application requires a set of QoS parameters such as real-time deadlines and fault-tolerance capabilities. On the other hand, the

---

[1] Sponsored in part by DARPA/NCCOSC contract N66001-97-C-8250, and by NSWC/NCEE contracts NCEE/A303/41E-96 and NCEE/A303/50A-98.

system at any given time can provide some level of QoS for the applications, given its available resources. Thus, the QoS negotiation components need to detect the available system resources and reconcile the applications' requirements with the system's resources. The resource management component is then deployed to re-allocate the resources to the applications according to the negotiated QoS.

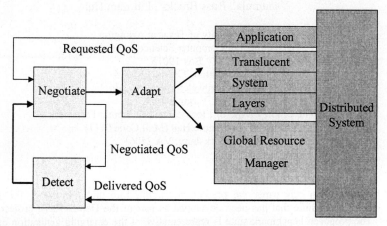

Fig. 1. A generic QoS management architecture for distributed system.

The behavior of a typical distributed real-time system application, shown in Fig. 1, may be dynamic in at least 3 dimensions: (1) the application's input data size and event arrival rates may be unknown; (2) the application may be scalable and fault-tolerant, implying that the number of program components at execution time can vary; and, (3) the application behavior may be influenced by independent events, such as load of a host or network. There are a number of QoS negotiation and RM technologies [H+97, CS+97, DL97, Hon97, RLLS97, RSYJ97, ZBS97] that are deployed in distributed real-time systems. However, there are no standard benchmarks, with the outlined dynamic behaviors, that can be used to evaluate such technologies. The goal of this work is to address this void. In this paper we present (i) the design and development of a benchmark suite and the framework to control the dynamic behavior of the benchmark in a repeatable way, and (ii) the definition and implementation of performance assessment metrics for evaluation of QoS and RM technologies.

To validate and assess QoS negotiation and resource management middle-ware in distributed real-time systems, there needs to be a *generic, repeatable,* and *deterministic* experimental setup. As part of the DeSiDeRaTa (Dynamic, Scalable, Dependable, Real-Time systems) project, we developed *DynBench*, a Dynamic real-time Benchmark suite. DeSiDeRaTa provides the QoS negotiation and RM middle-ware needed to address the needs of dynamic distributed real-time systems [Ravin98a-b, Welch97a, Welch98a-d].

For distributed real-time systems that have strong QoS requirements, a benchmark must have the following properties: (i) scalability, (ii) allowance of varying input rates for unknown environments, (iii) consideration of the unpredictable environment (host or network) behavior, and (iv) provision of facilities for easy collection of metrics from heterogeneous system architectures and networks. DynBench has been designed to have all of the above characteristics.

Traditional computing benchmarks (e.g., Dhrystone [Wei84], Whetstone [Cur76], AIM[Pri89], Linpack [Wei90], Livermore Loops, NAS Kernels, Fhourstones) are inadequate for characterizing dynamic distributed real-time systems. Primarily, they were not designed to exhibit behavior characteristic of control systems, such as periodic, transient and transient periodic activation/deactivation. Also, they focus on component level benchmarking; thus, they do not provide the notion of large grain entities with end-to-end timing constraints. The Rhealstone benchmark [Kar89] also has a small grain focus, but it focuses on system features that are relevant to real-time systems. It measures task switch time, preemption time, interrupt latency time, inter-task message latency, semaphore shuffle time, and deadlock-break time for *components* of real-time systems.

The Hartstone benchmark [Wei89, Kam91, Wei92, Ujv97] builds on the traditional component level synthetic benchmarks by incorporating them into a framework that provides periodic and aperiodic activation of distributed, time constrained communicating tasks. The authors show how to use the benchmark to evaluate system software (real-time O.S. and real-time language run-time system), host hardware, communication system software, node communication hardware, and physical channel characteristics and protocols. The primary use of Hartstone is to determine the workloads under which these various components of the real-time platform "break down"; this is done by incrementally increasing workload (sizes of messages, number of tasks sending/receiving messages, number of messages) until a real-time QoS violation occurs.

Like Hartstone, SWSL [Kis94] and SW [Kis96] provide distributed, communicating periodic and aperiodic tasks that use synthetic workloads. However, the authors provide a more comprehensive framework than in [Wei89, Kam91, Wei92, Ujv97]. Workloads are described using SWSL (synthetic workload specification language) and then compiled by SWG (synthetic workload generator) to produce executables for all processors in a distributed or parallel system. SWSL specifications are compact and flexible, employing data flow graphs at the task level and control constructs and synthetic operations at the operation level. To allow rapid specification of large concurrent systems, SWSL allows instantiation of parameterized objects, and specification of the task-to-processor mapping. Stochastic workloads are generated at run-time via random number generators with specific distributions and parameters. (SWSL provides a library of random number generators with various distributions.) To reduce the time required to perform multiple experiments, SWSL supports a multiple run facility, which uses a set of static parameters to drive a set of test runs.

While there has been significant work performed in benchmarking and synthetic workload generation for real-time systems, DynBench addresses some important issues that have not been dealt with previously. We extend the workload models of Hartstone and SW/SWSL beyond periodics and aperiodics, by including transient periodics (or quasi-continuous paths [Wel98a]). To test dynamic QoS and resource management technology, DynBench applications are dynamically controllable and dynamically relocatable; furthermore, they are dynamically replicable (for survivability and scalability), so the number of replicas can vary arbitrarily at run-time under the control of QoS and resource managers. Previous work did not support dynamic adaptation: in Hartstone, a QoS violation causes the test to terminate; in

SWSL, objects can be replicated, but the number of software components as well as the locations where they execute are fixed statically. The variations in workload can be stochastic in SWSL; in Hartstone, workload can be increased incrementally until a QoS violation occurs. DynBench permits workload to increase and to decrease dynamically, according to online user input, and according to user–defined experiment scenario files; these workload variations can be fully dynamic, i.e., they need not be constrained by time-invariant statistical distributions. Another obvious distinction of DynBench is that it employs real workloads (with features similar to those described in [Dav80, Mol90, Kav92, Har94, Wel96]), instead of synthetic workloads, avoiding the problems of (1) obtaining a realistic synthetic workload and (2) accurately characterizing a realistic workload and its low-level data-dependent behaviors. Also, this has the advantage of allowing a user to "see" that the appropriate QoS is being delivered to the benchmark application system by observing its tactical behavior. (In the C3I Parallel Benchmark [Met96], real workloads are also employed for seven subsystems: terrain masking, hypothesis testing, route optimization, plot-track assignment, synthetic aperture radar imaging, threat analysis, decision support systems and map-image correlation. However, due to their classified nature, the real-time versions of those benchmarks are only available to U.S. government employees and their contractors.) DynBench also provides instrumentation support necessary for resource and QoS management, a feature not found in previous work. Additionally, we have used the "computing cornerstones" presented in [Mast95] as a basis for defining a set of metrics for assessing QoS and resource management technology; and we provide an environment that automatically gathers the metrics.

DynBench Framework

Fig. 2 illustrates the DynBench framework. The DynBench framework controls the dynamic behavior of the benchmark in a repeatable way and collects performance assessment metrics for evaluation of QoS and RM technologies. The lower four components (benchmark, experiment generator, program control, and monitor) comprise the DynBench benchmark suite. The benchmark suite is used to evaluate any QoS/RM technology, represented by the QoS/RM services components.

The *benchmark* component is a distributed real-time control system designed to behave in a dynamic manner. Application instrumentation data collected from the benchmark goes to the *monitor* component, which calculates QoS metrics from the data. The monitor also takes instrumentation data from the QoS/RM services and calculates assessment metrics, storing them in the *assessment metrics file*. The *QoS/RM services* components represent the middle-ware components that perform QoS negotiation and resource management based on the QoS metrics provided by the monitor. The *experiment generator* allows the user to create multiple instances of the benchmark application. This also allows the user to dynamically change the characteristic behavior of the benchmark system at run-time. The *meta spec* file is provided to the experiment generator for creating different instances of the benchmark. The *commands file* allows the user to dynamically change the behavioral characteristics of the benchmark systems and that of the hardware platform in a deterministic and repeatable manner. The *spec file* describes the software QoS requirements as well as the hosts and networks that make up the distributed system.

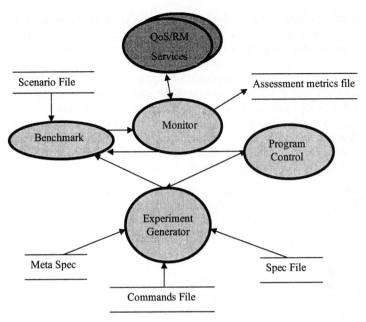

Fig. 2. The DynBench framework.

Benchmark:

The dynamic path paradigm [Wel96, Welch97a] provides a convenient abstraction for capturing the characteristics of dynamic real-time systems. In this paradigm, a path consists of a set of large-grain communicating programs with end-to-end QoS requirements. Fig. 3 depicts a typical real-time control system, such as an anti-air defense, a robotics, or an autonomous vehicle system. Control systems perform actions in response to conditions detected in the environment that they inhabit. They typically consist of sensors that monitor external conditions, and a situation assessment path that filters, correlates and evaluates the sensor data. Events detected by the situation assessment path activate the engagement path that plans and initiates responses via actuators. This in turn activates the monitor and guide path that guides the actions to successful completion. The processing performed by a control system depends on the environment in several ways. The load of the situation assessment path at any point in time is a function of the number of environmental elements detected by the sensors. The loads of the engagement path and the monitor and guide path depend on the rate at which environmental events occur. Thus, the characteristics of a control system's environment are important engineering considerations, making such a system ideal as a dynamic benchmark for distributed real-time systems.

Using the *dynamic path* paradigm, the DynBench benchmark application is modeled after typical distributed real-time military applications such as an air defense subsystem. Fig. 4 shows the three *dynamic paths* from the DynBench benchmark application. The *detect* path (path 1) is a continuous path that performs the role of examining radar sensor data (radar tracks) and detecting potential threats to a defended entity. The sensor data are filtered by software and are passed to two evaluation components, one is software and the other is a human operator. The detection may be performed manually, automatically, or semi-automatically (automatic detection with manual approval of engagement recommendation). When a threat is detected and confirmed, the transient *engage* path (path 2) is activated, resulting in the firing of a missile to engage the threat. After a missile is in flight, the quasi-continuous *guide missile* path (path 3) uses sensor data to track the threat, and issues guidance commands to the missile. The *guide missile* path involves sensor hardware, software for filtering/sensing, software for evaluating and deciding, software for acting, and actuator hardware.

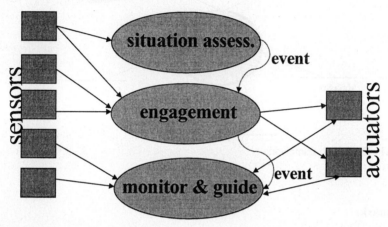

Fig. 3. A typical path-based real-time control system application.

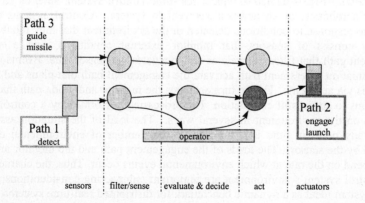

Fig. 4. The DynBench dynamic paths.

The *real-time* characteristics of the three paths are as follows. The continuous *detect* path is a data-driven path, with desired end-to-end cycle latencies for evaluating radar track data and detecting potential threats. The sensor data are provided to the path periodically, but the periodicity can change dynamically. For example, if a potential threat is detected, the frequency of sensing may increase.

The transient *engage* path is a sporadic, high-priority, "one-time" path. It is activated by an event from the *detect* path. Each time the *engage* path is activated, it executes only once. There is an end-to-end QoS timeliness objective on the *engage* path that typically has a higher priority than the timeliness QoS of the *detect* path or the *guide missile* path.

The *guide missile* path is transient periodic (or quasi-continuous) in the sense that it is sporadically activated and deactivated; but when active, it behaves like a continuous path. It is activated by the *missile launch* event. Once activated, the path continuously issues guidance commands to the missile until it detonates. Thus, the guide missile path is an example of a quasi-continuous path. The required completion time for one iteration is dynamically determined based on characteristics of the threat. For example, a fast moving, nearby threat will require issuing more frequent guidance commands than a slower-moving threat that is farther away.

QoS Specification

The DeSiDeRaTa environment includes a specification language called D-Spec [Welch98a] that is used for specification of QoS requirements of distributed systems, such as DynBench, that operate in unknown environments. D-Spec is useful for describing real-time QoS requirements in terms of end-to-end paths through application programs.

Experiment Generator:

In addition to the benchmark application described in the previous section, the DynBench framework includes an experiment scenario generator, which allows construction of multiple instances of the benchmark application using a meta spec file. Further, it allows dynamic changes to the behavioral characteristics of the benchmark systems and that of the hardware platform in a deterministic and repeatable manner using a commands file. The commands file contains a description of a set of events as a function of time. These events include increasing/decreasing the tactical load, killing a program or a host, and increasing/decreasing the load of a host or a network subsystem. The generator interfaces with the program control and resource manager components of the middle-ware, and the benchmark.

Monitor and Program Control:

The Quality of Service (QoS) being received by the DynBench benchmark applications is constantly monitored and passed to the QoS and RM services at their path and subpath levels of abstraction. The path and program latencies are currently the main QoS objectives monitored by the QoS monitor. Similarly, the monitor gets performance data from the QoS/RM services (such as latency of RM operations), computes QoS services assessment metrics, and outputs the metrics to the assessment metrics file. There is a QoS Monitor for each path in the subsystem. The main objective of the Program Control is to communicate a program failure to the RM services or start execution of a program on a host as specified by the Resource Manager.

QoS Assessment Metrics

In order to evaluate different QoS and RM services in distributed real-time systems, there is a need for a set of standard performance metrics. In this section we define the desirable attributes of such performance metrics. Some of these metrics are already implemented in the DeSiDeRaTa middle-ware services using the DynBench application as a testbed. Other metrics are in the process of being implemented. We propose the following QoS/RM services assessment metrics [Mast95]:

- *Performance*
  Performance metrics for QoS/RM services include latencies of different components, whether or not the latencies are bounded and deterministic, complexity of the QoS/RM components, and overhead of the QoS/RM operations.
- *Accuracy*
  Another set of assessment metrics include the degree of accuracy of the QoS/RM services operations in terms of monitoring, detecting, predicting (future trends), and diagnosing application ills.
- *Capacity / Scalability*
  QoS/RM services can be distinguished by maximum capacities for handling faults, tactical load, software components, and hardware components.
- *Quality*
  This is probably one of the most important (as well as one of the most difficult to measure) assessment metrics for evaluation of QoS/RM services. A quality metric may include QoS violation rate, ratio of QoS to resources consumed, stability of the QoS/RM services, and sensitivity of a chosen allocation to: hardware faults, increased tactical load, or increased number of programs.
- *Survivability*
  Is the QoS/RM service itself fault-tolerant and survivable?
- *Openness*
  Does the QoS/RM service have an open architecture, allowing different components to be mixed and matched?
- *Intrusiveness*
  This metric measures the degree of intrusiveness of the QoS/RM service on the design, development and fielding of applications.
- *Testability / Verifiability*
  Is the QoS/RM service easily testable and verifiable?

Experimental Results

We have used the DeSiDeRaTa [Ravin98a-b, Welch97a, Welch98a-d] middleware testbed to conduct different experiments with the DynBench benchmark suite. The experimental system parameters and environment are summarized as follows. The distributed network of hosts consists of two Sun Sparc5 workstations (170MHZ processor, 32 MB RAM, 2GB disk, and Sun Solaris 2.5.1 configurations), named *virginia* and *nujersy*; and two Sun Ultra Sparc I workstations (167MHZ processor, 128MB RAM, 2GB disk, and Sun Solaris 2.5.1 configurations) named *texas* and *desidrta*. The four hosts are connected via a 10/100MB per second Ethernet network. The HCI runs under Windows NT on a Pentium Class machine. The DynBench application consists of the situation assessment path with a latency deadline of 1 second.

This section presents scenarios showing system behavior for the DynBench situation assessment path under 3 different run-time conditions. In two scenarios, the DeSiDeRaTa adaptive resource management middle-ware monitors the real-time QoS of the DynBench application, detects QoS violations, diagnoses cause(s) of poor QoS, and recovers accordingly. In these experiments, the QoS is measured along one dimension, the latency of a path. Experiments that consider survivability QoS are also performed. Throughout these experiments we demonstrate how DynBench can be used to evaluate adaptive resource allocation technology in a dynamic real-time client-server-based system.

In the first test, we examined the scalability capabilities of DynBench and the DeSiDeRaTa middle-ware. The idea is that if we fix all DynBench and system parameters, but increase the tactical load (number of sensed tracks), then the middleware should detect this condition and scale up the DynBench program(s) that is the bottleneck. Fig. 5 depicts the results of this experiment. Here the observed situation assessment path latency is shown as a function of the DynBench cycles. The situation assessment path latency is set to 1 Sec. (1000 milli-sec) and is shown as a straight line. Initially, the Sensor, and Evaluate and Decide (ED) programs run on *nujersy*, Filter Manager (FM) and Evaluate and Decide Manger (EDM) run on *virginia*, and one copy of the Filter (FL) runs on *texas*. The middle-ware components are distributed among the four hosts.

During the first 26 cycles, the tactical load (number of sensed tracks) is set to 1000 and the path is running with latencies below the threshold. At cycle 27, using the experiment generator component of the DynBench, we increased the tactical load by 500. Note that the path latency increases above the threshold because of this dynamic change in the input data set size. The QoS Monitor component of the DeSiDeRaTa middle-ware detects the QoS violation during the [27, 43] cycle interval. In order to avoid threshing and filter out the noise from the observed QoS parameters, the QoS manger is configured to diagnose a QoS violation if there are 15 violations in any consecutive 20-cycle window. Once the QoS violation was diagnosed in cycle 43, another copy of the FL and ED programs were initiated in *desidrta*, causing the observed latency to fall below the threshold in the next cycle. This experiment shows that DynBench is clearly scalable and such a capability can be used to evaluate the scalability of distributed real-time middle-ware.

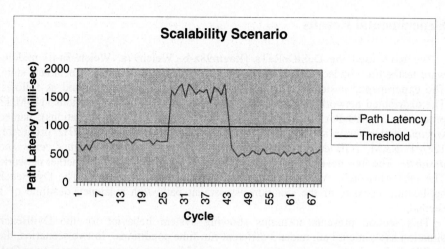

Fig. 5. The situation assessment path latency in the scalability test scenario.

In the second experiment, we examined the adaptability capability of the DeSiDeRaTa middleware. The idea is that if we fix all DynBench and system parameters, but increase the load of a host, then the middle-ware should detect this condition and perhaps move some of the DynBench programs to less loaded computing resources. Fig. 6 depicts the results of this experiment. Initially, FM and EDM programs run on *nujersy*, ED runs on *virginia*, and Sensor and FL run on *desidrta*.

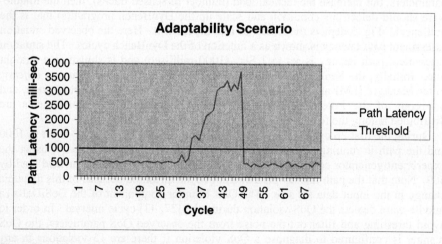

Fig. 6. The situation assessment path latency in the adaptability test scenario.

During the first 33 cycles, the tactical load (number of sensed tracks) is set to 1000 and the path is running with latencies below the threshold. At cycle 34, using the experiment generator component of the DynBench, we increase the load of the *desidrta* host by running 5 copies of a 500x500 matrix multiply application. Note that the path latency increases above the threshold because of this dynamic change in the system resources. The QoS Monitor component of the DeSiDeRaTa middle-ware detects the QoS violation during the [34, 49] cycle interval. Once the QoS violation is diagnosed, the DeSiDeRaTa adaptive RM moves the FL program from *desidrta* to *nujersy* and moves the EDM program from *nujersy* to *virginia*. These moves cause the observed latency to fall below the threshold in the next cycle. This experiment shows that DynBench's experiment generator can inject dynamic events (such as increasing a host's load) during a distributed real-time application's execution; and such a capability can be used to evaluate the adaptability of QoS and resource management services in distributed real-time systems.

In the third experiment, we evaluated the survivability capability of the DeSiDeRaTa middle-ware in terms of detecting faulted application programs and restarting them. The idea is that if we use DynBench's experiment generator to kill an application program, then the middle-ware should detect this condition and restart the application. Fig. 7 presents the results of this experiment. Initially, the Sensor, FM, ED and EDM programs run on *desidrta* and FL runs on *nujersy*.

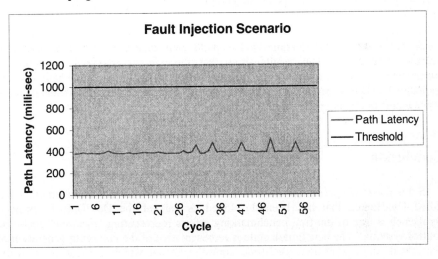

Fig. 7. The situation assessment path latency in the survivability test scenario.

At cycle 27, using the experiment generator component of the DynBench, we killed the ED program on *desidrta*. The QoS Monitor component of the DeSiDeRaTa middle-ware detects this condition in cycle 27 and restarts the ED on *virginia* in the next cycle. DynBench programs are state-less applications; thus, the recovery process is very quick. Once again, this experiment demonstrates the experiment generator's ability to dynamically inject events (such as killing a program) during an application's execution. Such a feature can be used to evaluate the fault detection and resolution capability of QoS and resource management services in distributed real-time systems.

QoS Assessment Metrics Results

Currently DynBench is set up to collect and report the *latencies* of the QoS and resource management services. Table 1 depicts the average latencies of a number of DeSiDeRaTa service components over multiple runs of the middle-ware with

DynBench. It is quite clear that in case of DeSiDeRaTa, all the QoS and resource management services take place in low hundreds of milli-sec., and the QoS violation detection and decision-making process take place in less than 10 milli-sec.

Table 1. DeSiDeRaTa QoS assessment metrics.

Metric	Average Latency (in milli-sec.)
QoS violation detection	1
RM decision-making process	3
RM best host selection for allocation	120
Total RM response time	123
Application start-up time (network dependent)	213

To measure the intrusiveness of the middle-ware, a series of experiments were conducted in which the real-time QoS of paths were measured in isolated and non-isolated conditions. The results show that the average maximal and minimal intrusiveness have been determined to be 3.47% and 0.299% of the path latency, respectively. Due to space limitations we are not able to present the details of these experiments in this extended abstract.

Conclusion

In this paper we presented the architecture and framework for a benchmark suite, called DynBench, that has been developed as part of the DeSiDeRaTa project. DynBench is one of the first benchmarking efforts representing "*dynamic*" real-time control systems. The benchmark suite is representative of the emerging generation of distributed, mission-critical, real-time control systems which operate in dynamic environments. The DynBench experiment generator framework takes into account the dynamic behavior of the application as well as the distributed system on which it runs.

DynBench was shown to be useful for evaluation of the quality of service management and/or resource management services in distributed real-time systems. We presented experimental results that validated DynBench as a credible benchmark for distributed real-time systems. As such, DynBench includes a set of performance metrics for the evaluation of the QoS and RM technologies in dynamic distributed real-time systems. The paper demonstrated the successful application of DynBench in evaluation of DeSiDeRaTa QoS management middle-ware.

References

[CS+97]   S. Chatterjee, J. Sydir, B. Sabata, and T. Lawrence, "Modeling Applications for Adaptive QoS-based Resource Management," 2nd

	IEEE High-Assurance System Engineering Workshop (HASE97), Bethesda, Maryland, August 1997.
[Cur76]	H.J. Curnow and B.A. Wichmann, "A synthetic benchmark," *Computer Journal* 19(1), January 1976, 43-49.
[Dav80]	C.G. Davis, "Ballistic missile defense: a supercomputer challenge," *IEEE Computer,* 30, 1980, 37-46.
[DL97]	Z. Deng and J. W.-S. Liu, "Scheduling real-time applications in an open environment," *Real-Time Systems Symposium*, San Francisco, California, December 1997, pp. 308-319.
[Har94]	Robert D. Harrison Jr., "Combat system prerequisites on supercomputer performance analysis," in *Proceedings of the NATO Advanced Study Institute on Real Time Computing*, NATO ASI Series F(127), 512-513, Springer-Verlag 1994.
[Hon97]	Honeywell Technology, "Adaptive Computing systems Benchmark," http://www.htc.honeywell.com/projects/acsbench/.
[H+97]	J. Huang, etal., "RT-ARM: A real-time adaptive resource management system for distributed mission-critical applications", Workshop on Middleware for *Distributed Real-Time Systems*, RTSS-97.
[Kam91]	N.I. Kamenoff and N.H. Weiderman, "Hartstone distributed benchmark: requirements and definitions," in *Proceedings of the 12th IEEE Real-Time Systems Symposium*, 1991, 199-208.
[Kar89]	Kar, R.P. and Porter, K., "Rhealstone -- A Real Time Benchmarking Proposal," *Dr. Dobbs Journal* 14(2):4-24, February, 1989
[Kav92]	K.M. Kavi and S.M. Yang, "Real-time systems design methodologies: an introduction and a survey," *The Journal of Systems and Software,* April 1992, 85-99, Elsevier Science Publishers.
[Kis94]	D.L.Kiskis and K.G. Shin, "SWSL: A synthetic workload specification language for real-time systems," *IEEE Transactions on Software Engineering,* 20(10), Oct. 1994, 798-811.
[Kis96]	D.L.Kiskis and K.G. Shin, "A synthetic workload for a distributed real-time system," *Journal of Real-Time Systems,* 11(1), July 1996, 5-18.
[Koob98]	G. Koob, "DARPA, Quorum mission statement," http://www.darpa.mil/ito/ research/quorum/index.html.
[Mast95]	Michael W. Masters, "Real-time Computing Cornerstones: A System Engineer's View," Proceedings of the 3rd *Workshop on Parallel and Distributed Real-time Systems*, 1995
[Met96]	R.C. Metzger, B. VanVoorst, L.S. Pires, R. Jha, W. Au, M. Amin, D.A. Castanon and V. Kumar, "C3I parallel benchmark suite: introduction and preliminary results," in *Supercomputing'96*, 1996.
[Mil95]	D.L. Mills, "Improved Algorithms for Synchronizing Computer Network Clocks," IEEE Trans. on Netwroks, 3, June 1995.
[Mol90]	J.J. Molini, S.K. Maimon and P.H. Watson, "Real-time system scenarios," in *Proceedings of The 11th IEEE Real-Time Systems Symposium*, 214-225, IEEE Computer Society Press, 1990.

[Pri89]   W.J. Price, "A benchmark tutorial," *IEEE Micro,* October 1989, 28-43.

[RLLS97]   R. Rajkumar, C. Lee, J. Lehoczky, and D. Siewiorek, "A resource allocation model for qos management," *18th IEEE Real-Time Systems Symposium,* pages 298-307, 1997.

[Ravin98a] B. Ravindran, L.R. Welch, and B. Shirazi, "A Resource Management Middleware for Dynamic, Dependable Real-Time Systems," Tech Report, CSE Dept, UTA, Aug. 1998.

[Ravin98b] B. Ravindran, L.R. Welch, C. Bruggeman, B. Shirazi, C. Cavanaugh, "A Resource Management Model for Dynamic, Scalable, Dependable, Real-Time Systems," Lecture Notes in Computer Science, #1388, Springer Verlag, March 1998, pp. 931-936.

[RSYJ97]   D. Rosu, K. Schwan, S. Yalamanchili, and R. Jha, "On adaptive resource allocation for complex real-time applications," In *Proceedings of the 18th IEEE Real-Time Systems Symposium,* 1997, 320-329.

[Ujv97]   B.G. Ujvary and N.I. Kamenoff, "Implementation of the Hartstone distributed benchmark for real-time distributed systems: results and conclusions," in *Proceedings of the 5th International Workshop on Parallel and Distributed Real-Time Systems,* 98-103, IEEE Computer Society Press, 1997.

[Wei84]   R.P. Weicker, "Dhrystone: a synthetic systems programming benchmark," *Communications of the ACM,* 27(10), 1984, 1013-1030.

[Wei89]   N.H. Weiderman, "Hartstone: synthetic benchmark requirements for hard real-time applications," CMU/SEI-89-TR-23, June 89.

[Wei90]   R.P. Weicker, "An overview of common benchmarks," *IEEE Computer,* 23(12), Dec. 1990, 65-75.

[Wei92]   N.H. Weiderman and N.I. Kamenoff, "Hartstone uniprocessor benchmark: definitions and experiments for real-time systems," *Journal of Real-Time Systems,* 4(4), Dec. 1992, 353-382.

[Wel96]   L. R. Welch, Binoy Ravindran, Robert D. Harrison, Leslie Madden, Michael W. Masters and Wayne Mills, "Challenges in Engineering Distributed Shipboard Control Systems," in *Proceedings of Work-in-Progress Session of The IEEE Real-Time Systems Symposium,* December 1996, 19-22.

[Welch97a] L.R. Welch and B. Shirazi, "DeSiDeRaTa: QoS Management Tools for Dynamic, Scalable, Dependable, Real-Time Systems," *IEEE Workshop on Middleware for Distributed Real-Time Systems and Services,* Dec. 1997, pp. 164-178.

[Welch98a] L.R. Welch, B. Ravindran, B. Shirazi, C. Bruggeman, "Specification and Modeling of Dynamic, Distributed Real-time Systems," Proceedings of the 19th *IEEE Real-Time Systems Symposium* (RTSS'98), Dec. 1998, pp. 72-81.

[Welch98b] L.R. Welch, B. Shirazi, B. Ravindran, C. Bruggeman, "DeSiDeRaTa: QoS Management Technology for Dynamic, Scalable, Dependable, Real-time Systems," 15th *IFAC Workshop - Distributed Computer Control Systems* (DCCS '98), Sept. 1998.

[Welch98c] L.R. Welch, et al., "Adaptive Resource Management for Scalable, Dependable Real-time Systems," in the proceedings of the Work-In-Progress Session of the Fourth IEEE *Real-Time Technology and Applications Symposium* (RTAS'98), June 1998, pp. 3-6.

[Welch98d] L.R. Welch and B. Shirazi, "A Distributed Architecture for QoS Management of Dynamic, Scalable, Dependable, Real-Time Systems", Workshop on *Parallel and Distributed Real-Time Systems*, April 1998, available on 1998 IPPS/SPDP CD-ROM.

[Welch98e] L.R. Welch, B. Shirazi, B. Ravindran, and F. Kamangar, "Instrumentation, Modeling, and Analysis of Dynamic, Distributed Real-time Systems," Tech Rep, CSE Dept, UTA, 1998.

[ZBS97] John A. Zinky, David E. Bakken, and Richard Schantz, "Architectural Support for Quality of Service for CORBA Objects," *Theory and Practice of Object Systems*, 1997.

Reflections on the Creation of a Real-Time Parallel Benchmark Suite

Brian VanVoorst[1], Rakesh Jha[1], Subbu Ponnuswamy[1], Luiz Pires[1],
Chirag Nanavati[1], David Castanon[2]

[1]Honeywell Technology Center
3660 Technology Drive
Minneapolis, MN 55441
{vanvoors, jha, subbu, pires, nanavati}@htc.honeywell.com

[2] ALPHATECH, Inc
50 Mall Road
Burlington, MA 01803
dac@bu.edu

Abstract. Standard benchmark suites are a popular way to measure and compare computers performance. The Honeywell Technology Center has developed two benchmarking suites for parallel computers, the C3I Parallel Benchmark Suite (C3IPBS)[1] and more recently, the Real Time Parallel Benchmark Suite (RTPBS)[2]. The C3IPB suite was similar to other benchmarks but focused on a new domain. The development of the real-time benchmark suite presented a number of new challenges. In this paper we share some of the lessons we have learned about the art of making parallel benchmarks.

1. Introduction

Benchmarks are used to evaluate and compare the performance of systems, using well-defined specifications and metrics. The Real Time Parallel Benchmark (RTPB) suite is a collection of benchmarks designed to help real-time C3I (command, control, communications and intelligence) parallel application designers evaluate parallel computing systems for real-time performance. Real time C3I applications impose unique requirements on a computing platform, and at the time this project was conceived, existing parallel benchmark suites (even those focusing on C3I applications) did not adequately address these needs. The results obtained from this benchmark

---

[1] This work was sponsored by USAF Rome Laboratory, contract number F30602-94-C-0084.

[2] This work was sponsored by DARPA via USAF Rome Laboratory contract number F30602-94-C-0084.

suite should help system architects, integrators and software designers better understand the factors that affect C3I real-time applications on parallel systems. Two goals of this projects were: 1) to help reduce the cost and time of development cycles, and 2) to facilitate discussions between the C3I community and parallel system architects.

Benchmarks suites come in many forms. The RTPB suite is a "pen and paper" benchmark suite like NAS 1.0 parallel benchmarks [15], meaning that the benchmarks themselves are specifications. The specification describes (in great detail sometimes) exactly what the benchmark is: what computations are to be performed, what input should be accepted and what the expected output should be, as well as a several other lower level details. In a sense, the specification states the objective of the benchmark and provides a detailed set of rules. When someone implements a benchmark based on the rules from the specification, then they have created a *benchmark implementation*, which is one realization or instance of the benchmark. Pen and paper benchmarks are used when there are considerable differences in the architectures on which that the benchmark may be implemented. By using the pen and paper approach, we let implementers choose the best possible implementation approach for a target platform. This is important, as we feel it would be improper to mandate one implementation (e.g. MPI) to be run on all machines when it is possible that on a given machine (e.g. a shared memory machine) a different implementation approach could produce better results. A side effect of pen and paper benchmarks is that they require considerable effort, both in developing specifications and in create implementations.

This paper summarizes our work on the benchmark suite and highlights some of the lessons we learned about benchmarking. This paper does not discuss the actual results from the benchmarks, as they have been discussed in other places [2, 4, 6]. The rest of this report is organized as follows: Section 2 describes the goals of the project, Section 3 presents a brief account of our accomplishments, Section 4 describes what we learned, Section 5 presents our conclusions.

2. Goals of the Project

The amount of work to be done by a C3I system is increasing rapidly as the quantity of the available information to be processed increases, and the expected quality of the results increases. Furthermore, the need for C3I systems with ever faster response times drives the need for increasingly shorter real-time deadlines. As an example, next generation sensors will produce tarabytes of data per second that must be processed in real-time as they scan the surface of the planet. To complicate matters, cost constraints force system designers to adopt commodity COTS components as the building blocks for their systems instead of special purpose hardware. The combination of these factors requires the use of commercial parallel processing systems for real-time C3I applications.

A system designer must consider several factors when choosing hardware for an embedded C3I application. Some of the relevant factors that must be considered include: single processor speed, scalability, real-time support and performance, efficiency of the algorithm on the machine, problem mapping, communication and memory requirements, fault tolerance, reliability, size, weight, power and the list goes on. Even if there are just a handful of architectures to consider, the amount of trade-off analysis required is staggering. Because of the amount of work involved, it is usually only the easily available metrics that are used for comparison (size, weight, power, processor speed in Mhz, published communication latency). The most important factor, the interactions between the actual algorithm and the system under study, are rarely considered. The Real Time Parallel Benchmark Suite provides efficient mechanisms to make these comparisons easily.

The benchmarks provide a new metric for comparison that allows systems to be evaluated side-by-side in an "apples-to-apples" manner. This new metric captures the real world interactions between applications and hardware with real input data, and reflects the kind of performance that will be seen when the system is deployed.

Consider the following example of how a system architect might use such a benchmark suite. Suppose a system architect has a C3I application with intensive computational requirements and real-time deadlines. From the system architect's point of view, the best possible situation is that there already exist benchmark results that capture the main computational and real-time requirements of the application. If the results are not available, but a similar benchmark exists, the system architect can port and run the benchmark on the platform of interest. If there is no such benchmark available, the architect can follow the methodology provided to specify a new benchmark and ask that vendors run this benchmark and submit results for comparison. Either way, in short order a fair comparison can be made between the platforms under consideration.

A secondary benefit is that the benchmarks serve to increase awareness and interest in real-time C3I applications on parallel machines. The NAS parallel benchmarks [15], which have become a standard for comparing machines for scientific computing, serve as an example. Because of the popularity of the NAS benchmarks and the emphasis placed on their results, vendors compete to improve their performance for CFD type applications. If the Real Time Parallel Benchmarks can foster the same kind of interest, the DoD will see the benefit in the form of increased attention paid to their applications by researchers and hardware designers.

3. Scope and Accomplishments of the project

We defined ten benchmarks for the RTPB suite - two application level benchmarks (Adaptive beamforming and Multiple hypothesis testing), three kernel level benchmarks, (associative gating, constant false alarm rate, and matched filter) and five low level benchmarks (three clock benchmarks and two communication benchmarks). For each benchmark we created a sample implementation in C using MPI for communication, and collected results on two or more platforms. The kernel and application level benchmarks represent topics of significant interest to the C3I community, namely image processing, image understanding, signal processing, filtering, pattern recognition, and classification problems. The low-level benchmarks are of interest to anyone who is implementing a real-time algorithm on a parallel machine. The details of the benchmarks are discussed in their respective specification documents [1, 3, 5]. A summary of our accomplishments in this project appear below, with cross references to detailed reports.

- A document describing the rationale and criteria for benchmarks selection [8].
- Creation of five low-Level benchmark specifications [1]:
    gclock, rtcomm1, rtcomm2, lclock and lsclock.
- Creation of three-kernel level benchmark specifications [3]:
    CFAR, Associative Gating and Matched Filter.
- Creation of two application-level benchmark specifications [5]:
    Multiple Hypothesis Testing, and Adaptive Beamforming.
- Creation of a benchmark real-time parallel benchmarking methodology [7].
- Creation of a portable and reusable benchmarking harness to facilitate the real-time benchmarking of parallel machines efficiently.
- Creation of data sets, acceptance tests, and results for each of the 10 benchmarks.
- Creation of a portable sample implementation for each of the 10 benchmarks
- Performance results for each of the sample implementations for at least two different machines [2], [4], [6].

4. Lessons we learned about defining, implementing, and running benchmarks

In this section we share some of the lessons we learned from building and running benchmarks. While we primarily focus on the Real Time Parallel Benchmark Suite, many of the same issues were also seen in the creation of the C3I Parallel Benchmark Suite, and perhaps by other benchmark development teams as well. We hope that these lessons will be of benefit to others who create new benchmark suites in the future.

We do not provide results for any particular benchmark or machine, as that is not the purpose of this paper. Detailed observations on the results of each of the benchmarks are described in their respective reports [2,4,6].

Pen and Paper benchmarks are expensive to develop and implement

Pen and paper benchmarks are very flexible as they allow benchmark implementers to choose the best methods to implement the benchmark onto a specific machine. However, this flexibility comes at a cost. Benchmark specifications are expensive to develop, and new implementations are expensive to create. The decision to use Pen and Paper benchmarks or not is a philosophical one. We choose to use Pen and Paper benchmarks because we wanted to be as fair as possible to the variety of architectures used for embedded high performance computing. As standard and portable tools evolve (such as MPI, Open MP, libraries, and other API's) for embedded systems, it may be possible to create "code" benchmarks, where the benchmarker need only run a standard piece of source code on any platform to gather results.

Selection of real-time parallel platforms to benchmark is limited

Most high performance computing centers provide access to platforms that are commercially available, and usually these centers' computers have a workload made up mostly of scientific applications (CFD, computational chemistry, statics and dynamics modeling, etc). These applications are typically run in a batch mode from datasets stored on disk. Few of these machines have real-time operating systems installed (even though vendors may offer real-time environments, the majority of the users have no need for it). Similarly, most of the jobs on these machines are compute-bound, whereas C3I applications with real-time input often are input-bandwidth bound.

Unfortunately, the embedded systems vendors who sell machines with parallel real time environments were often unable or unwilling to allow us free access to their machines for the purposes of benchmarking. We found that these vendors see benchmark suites as an expensive burden and they feel trapped: either they let others benchmark their machines (whom they do not trust to represent their product as well as they could themselves) or they do the benchmarking themselves (at a considerable cost in labor). Therefore most vendors will not report benchmark results unless the vendor is required to when competing for a procurement. The result is that sample implementations get run on publicly accessible machines that may not have a real-time environment. In the end vendors may produce results for their machines, or otherwise independent researchers with access to these platforms may collect results.

It is our recommendation that future benchmarking projects secure funding to pay for time on vendor's machines, or establish partnerships in the program to ensure access to these platforms.

There is a trade-off between portable examples and collecting the best results

Our benchmarks, being pen-and-paper benchmarks, allow the benchmarker considerable freedom in tailoring a solution to the architecture of interest. This can be a very time consuming task. Our approach was to make an example implementation that had reasonable performance and that was very portable. This way others could quickly get the benchmarks to run on a new machine, and have a starting point for optimization.

Because of our interest in portability the RTPB suite examples are provided in C and use MPI for communication. Had OpenMP been available when we were benchmarking we would certainly have done OpenMP implementations, as in some cases a true shared memory implementation may have been more efficient. Unfortunately we did not have the resources to do many custom shared-memory implementations and OpenMP was not available.

Our reusable benchmark harness proved to be a useful tool

At the beginning of our work we created a portable benchmark "harness". The harness provided many of the tools and services we needed for benchmarking platforms that were common to all of our real-time benchmarks. For example, the benchmark harness handled issues of global time, sending periodic input to the benchmark (simulating a real system), collecting results from the benchmark node, timing the computations, checking results, enforcing real-time deadlines, and increasing the input rate. We re-used the harness on each of our application and kernel level benchmarks. Not only did it save us time, it provides each of our sample implementations with a common interface which should be beneficial to those who work with the sample implementation source code. Future benchmark developers may want to consider using our benchmark harness, or building a similar tool to facilitate their development.

C3I applications have different characteristics than traditional scientific applications

As we became more familiar with the C3I (command, control, communications and intelligence) applications we had selected for the Real Time Benchmark Suite, and its predecessor the C3I Parallel Benchmark Suite, we realized that C3I applications were different from traditional scientific. For example, most scientific applications deal with double precision computations (CFD being one example), where most C3I applications do not (such as image processing). Data structures differ too. Many scientific applications work on structured matrices, while many C3I applications use prediction and classification algorithms (such as hypothesis testing and gating) and use data structures like lists and trees. Even the I/O characteristics are different - many scientific applications load in a data set, perform a set of computations, and produce a set of results. Most C3I applications take in data periodically, process it, and deliver the results just in time to receive a new input data.

Since most high-end computers are optimized for scientific applications (the typically workload of a parallel machine) there is a potential problem. The high-end COTS hardware that is readily available for scientific applications is probably not well suited for C3I applications. One can speculate that this may get worse in the future. Many research centers that traditionally have purchased high end parallel machines are investigating the use of PC clusters to run their applications. The side effect of this may be that the low-cost COTS hardware of the future is optimized for desktop applications and gaming.

The lesson to be learned is that not all applications are the same, and there is a price to pay for using commodity parts. This is, after all, why we have different benchmark suites to represent different classes of applications. In this case we have learned that C3I applications are sufficiently different from the application mixes found on a typical parallel machines. With the benchmark suite we can quickly evaluate how different architectures perform on C3I applications.

Most C3I applications require a considerable amount of memory

Many of the kernel and application level benchmarks require nearly 256 Megabytes of memory per node, and in some cases even more. In the Associative Gating benchmark [4] it was noted that 512 Megabytes or more per processor would be required for an efficient distributed memory implementation. We believe that many future C3I applications implemented on parallel machines may be memory constrained on typical machine configurations.

This suggests that an area open for investigation is to determine how sensitive the class of C3I applications is to changes in the memory architecture. Our experience with benchmarking has shown that C3I applications differ from the typical scientific applications used in most parallel benchmarks. Since these more common benchmarks influence procurements and design decisions, it is possible that modern architectures are diverging from the needs of C3I applications. An investigation of the memory needs of C3I applications may influence system design decisions in the future.

Scalability may be limited by factors other than aggregate processing power

Many of our benchmarks increase the rate of input to stress the machine's ability to meet deadlines and establish a breakdown point. When a breakdown point is found for a set of processors, the number of processors is increased and the experiment is performed again. If the benchmark implementation is scalable it is expected that the sustainable input rate would double as the number of processors is doubled. This conventional wisdom does not always hold true as it assumes the reason for a deadline miss is due to insufficient compute resources, and that adding more processors will allow for a shorter deadline. The speed at which input can be processed consistently without missing a deadline can be dependent on many other factors, including bisection bandwidth of the inner-connect and latency for communication. These resources must also scale as the number of processors increases. Furthermore, the number and duration of interruptions and delays introduced by the operating system services must continue to be reduced when using more and more processors to achieve perfect real-time scalability. The opposite is more likely to happen, as the OS must manage communication channels with more and more processors more interruptions are introduced. The result is that a benchmark with a deadline may not scale well even though the computations scale well. This is an accurate reflection of how the system will perform on real applications when deadlines are present.

Acceptance tests for real-time behavior must be carefully constructed

Acceptance tests are used in the RTPB suite to determine if a benchmark is meeting real-time deadlines. Most of our benchmarks receive periodic input and deliver periodic output, with the requirement that the output must be delivered before the next input arrives. Accurately testing this is more challenging than it may seem. The test must take into account measurement errors, granularity of the clock, clock skew and make measurements on the appropriate processor. In some cases the input arrives from a different processor than the output in sent. In other benchmarks the processor to measure is the one that takes the longest time to complete, which may change from iteration to iteration. To address all of these issues we provide a benchmarking methodology document [7].

The importance of putting a benchmark in context to other tasks

When defining a benchmark it is important that its definition be faithful to the real world application it represents. This requires taking careful definition of the input and output requirements. Some questions that need to be considered are, "in the real world is the data read from disk, delivered from another process, or already in memory?", "should we start timing after the data is available on all processors, or when the first processor has the data available?", "do we end timing when the computations are ready and the results are finished, or do we need to include the time necessary to send the results on to the next computation?" "do our timing assumptions change with different architectures, does a shared memory implementation imply something different than a distributed memory implementation?" By addressing these issues a benchmark developer can help ensure that the benchmark is a realistic representation of the real world application.

The challenges in choosing input data

Choosing the input data and the method by which it is delivered to the benchmark is a very important task. We went to great lengths to make sure that the data we were using for the benchmark was representative of the real application. This is not always an easy task. Some computations are data dependent, so in order to exercise a benchmark implementation fairly several different data sets need to be run, which may require a considerable amount of input data and run time. Choosing what these data sets are can be time consuming part of benchmark development.

Another challenge can be the size of the input. Some data sets can be 100's of Megabytes, which makes the distribution of the benchmarks cumbersome. Our solution to the size problem was to create synthetic data set generators; small programs that the user could compile and run at their site to create a specific data set locally.

Deciding how the benchmark will receive the data is perhaps the biggest challenge. Because of the nature of the applications we were representing, most of our benchmarks received input periodically as messages from a processor outside of the computation. This was done to simulate the "real world" where a sensor (a radar for example) would be sending an embedded system data for processing. Managing the transmission of the data was particularly challenging, as the requirements changed from benchmark to benchmark. In some cases the data was broadcast, in others it was sent to a single processor. For some benchmarks the data was being generated by the sending processor, in other cases it was read from disk. Care had to be taken to make sure that the sending processor sent the data at exactly the right time. If the data was delayed the benchmark implementation might miss a deadline through no fault of its own. To make this task more manageable we implemented these tasks in our reusable benchmark harness.

Finally, it is important to remember that different platforms have different representations for data types. Binary input files will fail on some systems if there is an assumption of byte order or of the size of a datatype (e.g. 32 bit vs. 64 bit).

Different clocks on the same system can give dramatically different results

For a benchmark, timing is everything. On a modern machine a programmer has access to many different functions that can be used for timing. Different clocks have different overhead. For example, one MPI implementation tested on a Mercury had a MPI_Wtime() call that gave considerably worse results than the Mercury's hardware clock. Similarly, we found that the SGI special hardware clock reading routines gave much better results than any other SGI operating system level clock calls. It is advisable that benchmarks and programmers familiarize themselves with all the clock options on a platform and select the clock that gives them the best granularity possible with the least overhead.

Machines that run time synchronization protocols can cause bad measurements

We found that machines that run clock synchronization protocols such as NTP can cause bad measurements in benchmarks. These protocols can cause clocks to "jump" to synchronize, which hinders the accurate timing of a benchmark, and leads to unreliable results. Our benchmark harness tests for this type of behavior, and reports it if detected. It is up to the user running the benchmarks to make arrangements to have the NTP service turned off, find a similar platform without NTP, or re-run the benchmark until it completes without clock adjustments during the benchmark execution.

Benchmarks are useful even to those who do not want to do benchmarking

Several other researchers have requested our benchmarks as sample applications to test their systems and subsystems. While this is an unexpected use of the benchmarks, it indicates that the benchmarks are fulfilling a need that was previously unnoticed, namely researchers desire freely accessible C3I applications to demonstrate and test their work. DARPA and other agencies that want to promote research in a particular domain may want to consider the benefit of making representative applications available to researchers. This may encourage more work in the agency's area of interest.

How to guard against benchmark obsolescence

Benchmarks grow old. What takes hours to run today will only take a few minutes to run tomorrow. The good news is that data sets are growing as fast (if not faster) than our computational ability, and that the necessary response time (which often determines real-time requirements) is becoming shorter and shorter. For this reason we have tried to make it possible to scale our benchmarks so that they can represent tomorrows problems as when today's problems are no longer challenging. For the RTPB suite this can be done in two ways: increasing the amount of data to be processed, or creasing deadlines. Furthermore we have created our benchmark methodology to provide guidelines for adding new benchmarks to the RTPB suite. We encourage other benchmark suite developers to consider how they may structure their benchmarks so that they can remain relevant as long as possible.

5. Conclusions

We have developed a real-time parallel benchmark suite for DoD applications, a first of its kind. The Real Time Parallel Benchmark suite fulfills a need and is complementary to work performed at MITRE [11, 14], and the C3I Parallel Benchmark Suite [12, 13] developed at Honeywell. Together these three projects have produced over twenty-five benchmarks for the C3I community. These benchmarks have the potential to assist future system designers in making decisions about the hardware and software they use. They can also be of benefit to other researchers who wish to use them for their own experimentation.

In creating these benchmarks we have identified many important issues that we hope can be of benefit to future benchmark developers.

6. Acknowledgments

We would like to thank the following computing centers for providing their resources: NASA Ames Research Center, Arctic Region Supercomputer Center, Maui High Performance Computing Center, University of Minnesota, Mississippi State University.

We would also like to thank the following individuals for their help and guidance: Ralph Kohler at AFRL, Jose Munoz at DARPA, Robert Bernecky at NRAD, Vipin Kumar at the University of Minnesota, Benoit Champagne at the University of Quebec, Richard Games at MITRE, Craig Lund at Mercury Computer Systems, Glen Ladd and Rob Moore at Hughes Research Labs, and David Bailey at the NASA Ames Research Center.

7. References

1. "Low-Level Benchmark Specifications", Technical Information Report (C006), Rome Laboratory, Contract Number F30602-94-C-0084, August 1998
2. "Low-Level Benchmark Results", Technical Information Report (C007), Rome Laboratory, Contract Number F30602-94-C-0084, August 1998
3. "Kernel Level Benchmark Specifications", Technical Information Report (C008), Rome Laboratory, Contract Number F30602-94-C-0084, August 1998
4. "Kernel-Level Benchmark Results", Technical Information Report (C009), Rome Laboratory, Contract Number F30602-94-C-0084, August 1998
5. "Application Level Benchmark Specifications", Technical Information Report (C0010), Rome Laboratory, Contract Number F30602-94-C-0084, August 1998
6. "Application Level Benchmark Results", Technical Information Report (C0011), Rome Laboratory, Contract Number F30602-94-C-0084, August 1998
7. "Benchmarking Methodology", Technical Information Report (C005), Rome Laboratory, Contract Number F30602-94-C-0084, August 1998
8. "Benchmark Requirements and Selection", Technical Information Report (C004), Rome Laboratory, Contract Number F30602-94-C-0084, August 1998
9. "Benchmark Distribution Plan Report", Technical Information Report (C012), Rome Laboratory, Contract Number F30602-94-C-0084, August 1998
10. "A Real-Time Parallel Benchmark Suite", B. VanVoorst, L. Pires, R. Jha, Proceedings of SIAM's Parallel Processing for Scientific Computing, SIAM, March 1997.
11. "Real-Time Embedded High Performance Computing: Application Benchmarks", C. Brown, M. Flanzbaum, R. Games, J. Ramsdell, MITRE Technical Report, MTR 94B0000145, October 1994.
12. "C3I Parallel Benchmark Suite Final Report", Technical Information Report (A012), Rome Laboratory, Contract Number F30602-94-C-0084, May 1998
13. "The C3I Parallel Benchmark Suite - Introduction and Preliminary Results", R. Metzger, B. VanVoorst, L. Pires, R. Jha, W. Au, M. Amin, D. Castanon, V. Kumar, Proceedings from Supercomputing 96, 1996.
14. "Benchmarking Methodology for Real-Time Embedded Scalable High Performance Computing", R. Games, MITRE Technical Report, MTR96B0000010, March 1996.
15. "NAS Parallel Benchmark (Version 1.0) Results 11-96", Subhash Saini, David Bailey, NASA Ames Technical Report NAS-96-18, November 1996.

Tailor-Made Operating Systems for Embedded Parallel Applications *

Antônio Augusto Fröhlich[1] and Wolfgang Schröder-Preikschat[2]

[1] GMD FIRST
Rudower Chaussee 5
D-12489 Berlin, Germany
guto@first.gmd.de
http://www.first.gmd.de/~guto
[2] University of Magdeburg
Universitätsplatz 2
D-39106 Magdeburg, Germany
wosch@cs.uni-magdeburg.de
http://ivs.cs.uni-magdeburg.de/~wosch

Abstract. This paper presents the PURE/EPOS approach to deal with the high complexity of adaptable operating systems and also to diminish the distance between application and operating system. A system designed according to the proposed methodology may be automatically tailored to satisfy an specific application. In order to enable this, the application must be written referring to the *inflated interfaces* that export the system object repository and then be submitted to an analyzer that will proceed syntactical and data flow analysis to extract a blueprint for the operating system to be generated. This blueprint is then refined by dependency analysis against information about the execution scenario acquired from the user via visual tools. The outcome of this process is a configuration file consisting of *selective realize* keys that will support the compilation of the tailored operating system.

1 Introduction

The boom of embedded systems in the recent years projects a near future replete of complex embedded applications, including navigation, computer vision and particularly automotive systems. Some currently available limousines benefit from over 60 networked processors (μ-controllers) and can be considered parallel systems on wheels. Moreover, many of these new embedded applications will demand performance levels that can only be achieved by parallelization and thus new operating systems and tools are to be conceived.

Our experiences developing run time support systems for ordinary, i.e., non-embedded, parallel applications [9, 1] convinced us that adjectives such as "all

---

* Work partially supported by Federal University of Santa Catarina, by Fundação Coordenação de Aperfeicoamento de Pessoal de Nível Superior and by Deutsche Forschungsgemeinschaft grant no. SCHR 603/1-1.

purpose", "global" and "generic" do not fit together with "high performance" and "deeply embedded", whereas different parallel applications have quite different requirements regarding operating systems. Even apparently flexible designs like μ-kernel based operating systems may imply in waste of resources that, otherwise, could be used by applications. The reality for embedded parallel applications, that usually have to operate under extreme resource constrains, not only in terms of processor cycles, but also of memory consumption, can not be different and therefore we should give each application its own operating system.

This paper presents a proposal to automate the process of generating a tailored operating system for a given (embedded) application. This proposal is presented in the realm of EPOS, an operating system designed to support (high performance) parallel computing on distributed memory architectures. EPOS is actually an extension of PURE [2], that aims to provide a portable, universal runtime executive for resource-sparing (real-time) applications. The following sections describe the EPOS/PURE design approach, a case study to demonstrate the necessity for automated tailored operating systems and a new strategy to achieve them.

2 The Design of PURE

EPOS basic approach to give each application an operating system that closely fulfils its requirements is to analyze the application looking for hints about the nature of the operating system services requested. After collecting information, a custom EPOS will be generated using the building blocks supplied by PURE. In this way, EPOS becomes an extension of PURE, to whom it owes most of its design. Thus, describing EPOS design implies in describing PURE.

The approach followed by PURE is to understand an operating system as a *program family* [8] and to use *object orientation* [10] as the fundamental implementation discipline. The former concept, program families, helps to prevent the design of a monolithic system organization, while object orientation enables the efficient implementation of a highly modular system architecture.

2.1 Incremental System Design

The program family concept does not dictate any particular implementation technique. A so called "minimal subset of system functions" defines a platform of fundamental abstractions serving to implement "minimal system extensions". These extensions are, then, made on the basis of an *incremental system design* [5], with each new level being a new minimal basis, i.e., an *abstract machine*, for additional higher-level system extensions. A true application-oriented system evolves, since extensions are only made on demand, when the implementation of a new system feature that supports a specific application is required. Design decisions are postponed as long as possible. In this process, system construction takes place bottom-up but is controlled in a top-down (application-driven) fashion. In its last consequence, applications become the final system extensions and the traditional boundary between application and operating system disappears. In other words, the operating system extends into the application and vice versa.

Inheritance is the appropriate technique to either introduce new system extensions or replace existing ones by alternate implementations. Either case, the system extensions are customized with respect to specific user demands and will be present at runtime only in coexistence with the corresponding application. Thus, applications are not forced to pay for (operating system) resources that will never be used.

2.2 Revival of Program Families

Applying the family concept in the software design process leads to a highly modular structure. New system features are added to a given subset of system functions. Because of the strong analogy between the notions of "program family" and "object orientation", it is almost natural to construct program families by using an object-oriented framework [6]. Both approaches are, in a certain sense, reciprocal to each other. The minimal subset of system functions in the program family concept has its counterpart in the superclass of the object-oriented approach. Minimal system extensions are thus introduced by means of subclassing. Inheritance and polymorphism are the proper mechanisms to allow different implementations of the same interface to coexist. In the realm of embedded distributed/parallel systems, inheritance must be applicable even in the case of crossing address space, node, and network boundaries [7]. With this design strategy, reusability is significantly enhanced, increasing the commonalties of different family members.

3 The Design of EPOS

Although a tailored EPOS will be comprised of PURE building blocks[1], EPOS extends PURE, towards applications, in order to cope with two growing problems: the system configuration complexity and the distance between application and system software.

3.1 System Object Adapters

The growth in configuration complexity arises from the fact that, as an adaptable operating system, PURE is designed to yield a large number of building blocks, or system objects, which are, in turn, to be put together to compose the tailor-made operating system. To achieve high performance, these system objects must be fine tuned to each of the execution scenarios aimed and therefore a reasonable building block repository will comprise a very large number of elements. This turns system configuration into a nightmare. EPOS approach to the matter is to make system objects adaptable to execution scenarios.

---

[1] New building blocks to support parallel computing are being conceived according to the PURE design approach.

By carefully analyzing the system objects repository, one will promptly realize that those abstractions designed to present the same functionality in different execution scenarios are indeed quite similar. Besides, abstractions conceived to support the same scenario often differ from each other following a pattern. In this way, we propose system abstractions to be implemented as independent from the execution scenario as possible. They would then be put together with the aid of some sort of "glue" specific to each scenario. We named these "glue" *scenario adapters*, since they will adapt an existing system abstraction to a certain execution scenario.

In order to exemplify the use of scenario adapters and also to demonstrate the basis for our assertion that system abstractions implementation follow a pattern along scenarios, we could consider the "semaphore" abstraction for three multi-threaded scenarios: single-task, multi-task-single-processor and multi-task-multi-processor (SMP). For all three execution scenarios, a semaphore abstraction (i.e., the semaphore system object) will involve a counter and a thread queue. Aspects such as crossing the application/operating system boundary in the second and third scenarios or such as synchronizing concurrent invocation of semaphore methods in the third scenario are not intrinsic to the semaphore abstraction, but to the execution scenario. Therefore, such kind of aspects would be better implemented as scenario adapters than as different types of semaphore. Similarly, a "thread" abstraction conceived for the same three scenarios, would show a repetition of code when compared to the semaphore abstraction: crossing system boundary and supporting concurrent execution are intrinsic to the scenarios and not to the abstraction. It is also worth to say that scenario adapters are implemented in a way not to insert overhead on the path from applications to system objects, i. e., when an existing system object fulfils the requirements of a scenario, the scenario adapter simply vanishes.

The approach of writing pieces of software that are independent from certain aspects and later adapting them to a given scenario has been referred to as "Aspect Oriented Programming" [4]. However, there are significant differences in our proposal to make system abstractions adaptable. We are not proposing a language to describe aspects neither tools (*weavers*) to automatically combine aspects and aspect-independent components. The problem we want to tackle is the complexity inherent to adaptable operating systems and we are doing this by reducing the number of system abstractions in detriment to the growth of the number of scenario adapters. We believe this will reduce system complexity because dealing with aspects isolated in scenario adapters is much simpler than dealing with them spread along the system abstraction implementation; and also because new scenario adapters can be derived, automatically or not, from other existing adapters.

3.2 Inflated Class Interfaces

The second goal on EPOS extension of PURE is to diminish the gap between operating system and applications. This gap originates mainly from the fact that the operating system building blocks offered to applications are usually conceived considering the optimization of resource utilization in a bottom-up fashion and not the applications expectances. EPOS adopt *abstract data type declarations* in the form of *inflated class interfaces* as a mechanism to advertise the system abstractions repository to applications and to automate the selection of the proper building blocks when tailoring an operating system.

These inflated class interfaces shall enable the application programmer to express his expectations regarding the operating system simply by writing down well-known system object invocations (system calls in non object oriented systems) while coding the application. By well-know system object invocations we mean that the operating system services will be made available to applications via abstractions commonly accepted by the parallel systems community, such as "threads", "tasks", "address spaces", "channels", "ports", etc. These interfaces are to be defined according to the fundamental law of object orientation that says [3]: look at the real world while looking for objects. The question of where in the real world one can find a "thread" can be easily answered when we realize that threads, just like numbers, are human conventions and that a couple of classical computer science books should comprise most widely accepted conventions. Nevertheless, our users, i.e., embedded parallel application programmers, are welcome to suggest modification or extensions for these interfaces at any time.

With these interfaces in hand, it should no longer be a problem for a skilled parallel application programmer to select how he wants to express process communication, thread synchronization or any other operating system service. It is important to notice that these inflated interfaces will never be implemented as a single class, but as a (possibly huge) set of classes specific to certain scenarios. They are only a mechanism to help programmers to design and implement their applications. An automatic tool shall bind (reduce) the inflated interfaces to specific implementations.

To support system design based on inflated interfaces, we propose two new object-oriented design notations: *partial realize* and *selective realize*. Both relationships take place between an inflated interface and a class that realizes this interface. However, as the name suggest, a class participating in a partial realization implements only a specific subset of the corresponding inflated interface. In this scope, selective realization means that only one of several possible realizations will be connected to the inflated interface at a time. To support selective realization, each class joining the relation is tagged with a key. By changing the value of this key one can select a specific, usually partial, realization for an interface. These two design elements are depicted in figure 1.

As important as design elements, partial and selective realization have their counterparts for system implementation so that tailoring an operating system can be done simply by defining values for selective realize keys. These keys are defined in a single configuration file and dispense the use of both, conditional compilation and "makefile" customization. Furthermore, the implementation of these relationships may be used to bind non object oriented inflated interfaces

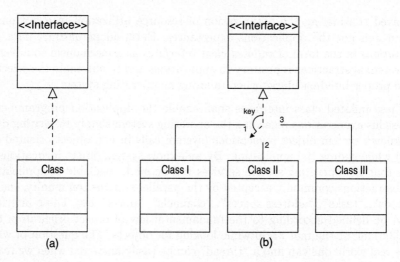

Fig. 1. Partial realize (a) and selective realize (b) relationships.

to object oriented implementations. This is useful, for instance, to bind an application written in Fortran or C to EPOS.

4 Automatically Tailoring an Operating System

With the design and implementation mechanism described so far, we can now consider the automatic generation of an operating system to closely fulfil an embedded parallel application. Our strategy begins top-down at the application, with the programmer specifying the application requirements regarding the operating system by designing/coding the application while referring to the set of inflated interfaces that export the system abstractions repository. An application designed and implemented in this fashion can now be submitted to an analyzer (figure 2) that will conduct syntactical and data flow investigations to determine which system abstractions are really necessary to support the application and how they are invoked. This tool shall generate an operating system blueprint that will, for instance, define the use of multi-tasking instead of single-tasking, of multi-threading instead of single-threading, of insolated protected address spaces instead of a single unprotected address space and so on.

Our primary operating system blueprint is, unfortunately, not complete. We got good hints about how the ideal operating system for a given application should look like, but there are aspects that can not be deduced by analyzing the application. As an example, we could consider the decision of generating an operating system that supports multiple processes with protected address spaces based on a micro-kernel or an operating system for a single, possibly multi-threaded, process to be embedded into (linked to) the application. The fact that the application does not show any evidence that multiple processes may need to run concurrently in a single processor, does not necessarily mean that this

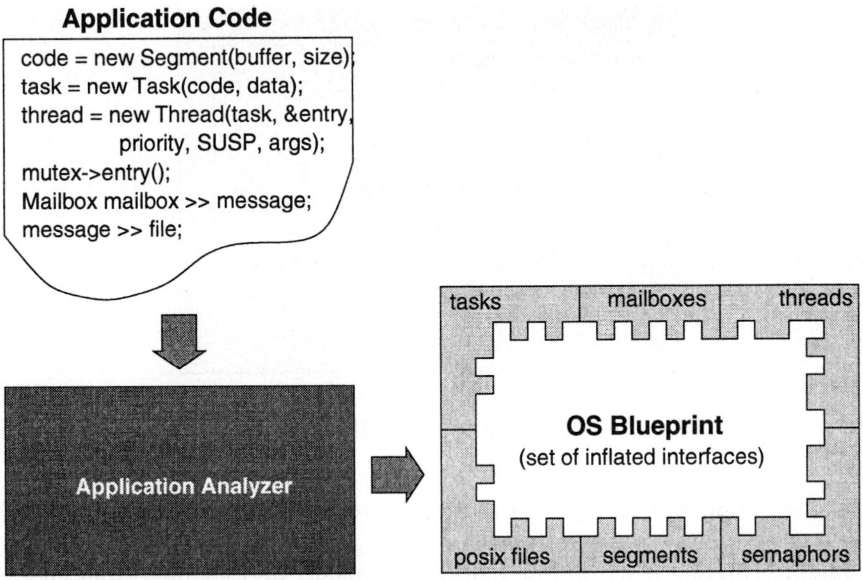

Fig. 2. Extracting an operating system blueprint from the application.

situation will not happen. The multi-task support may be required because the application may need to span more processes than the available number of processors, and this will not be perceivable until the user tell us something about the intended execution scenario. Other factors such as target architecture, number of processors available, network architecture and topology are fundamental to tailor a good operating system, but are usually not expressed inside the application. Therefore, we still need user intervention to describe the application's execution scenario, however, the description of the available resources will be due to the operating system developers and the interaction with the user will be done via visual tools.

Refining the blueprint obtained when analyzing the application with the context information acquired via this visual tool will render a much more precise description of how the ideal operating system for a given application should look like. This refined blueprint is the result of a dependency analysis and is expressed via a configuration file consisting of selective realization key values. With the definition of these keys, the inflated interfaces referred by the application programmer are bound to scenario specific implementations. For example, the inflated thread interface from the first step may have included remote invocation and migration, but reached the final step as a simple single-task, priority scheduled thread for a certain μ-controller. A representation of an application tailored operating system generated according to this model is depicted in the figure 3.

It is important to understand that, at the early stages of the operating system development, very often a required building block will not yet be available. Even

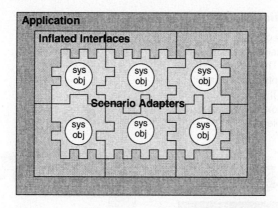

Fig. 3. An operating system tailored to an application.

then, the proposed strategy is of great value, since the operating system developers get a precise description for the missing building blocks. In many cases, a missing building block will be quickly (automatically) adapted from another scenario using the scenario adapters described earlier.

Only if the operating system developers are not able to deliver the requested building blocks in a time considered acceptable by the user, either because a building block with that functionality have not yet been implemented for any scenario or because the requested scenario is radically different from the currently supported scenarios, we will shock the user asking him to select the best option from the available set of system abstractions (scenario adapters) and to adapt his program. In this way, our strategy ends where most tailorable operating systems start. Moreover, after some development effort, the combination of scenario adapters and system abstractions shall satisfy the big majority of parallel embedded applications.

5 A Case Study on Configuration

A good example of how tailoring an operating system to a specific application may improve performance can be observed when selecting the proper member of PURE nucleus family. PURE is made of a nucleus and nucleus extensions. The nucleus implements a *concurrent runtime executive* (CORE) for passive and active objects. By means of minimal *nucleus extensions* (NEXT) features such as application-oriented process and address space models, blocking (thread) synchronization, and problem-oriented (remote) message passing are added to the system. These extensions are present only if demanded by the application. They transform the nucleus into a distributed abstract thread processor. The coarse structure of PURE is depicted in Figure 4.

Depending on the actual application requirements, the nucleus components appear in different configurations. Each of these configurations represent a member of the nucleus family. PURE's current implementation includes six family

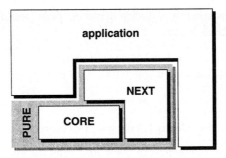

Fig. 4. PURE architecture.

members, each of which implementing a specific operating mode. Each member is built by one or more function blocks and described by a functional hierarchy of these blocks. The function blocks can be reused in various configurations. In order not to restrict the family, design decisions have been postponed as far as possible and encapsulated by higher-level abstractions. As a consequence, PURE can be customized, for example, with respect to the following scenarios:

1. One way of operating the CPU is to let PURE run *interruptedly*. This family member merely supports low-level trap/interrupt handling. The nucleus is free of any thread abstraction. It only provides means for attaching/detaching exception handlers to/from CPU exception vectors (*interruption*).
2. In order to *reconcile* the asynchronously initiated actions of an *interrupt service routine* (ISR) with the synchronous execution of the interrupted program, a minimal extension to 1. was made. The originating family member ensures a synchronous operation of event handlers (*driving*) in an interrupt-transparent manner (*serialization*).
3. The second basic mode of operating the CPU means *exclusive* execution of a single active object. In this situation, the nucleus provides only means for *objectification* of a single thread. The entire system is under application control, whereby the application is assumed to appear as a specialized active object. There is only a single active object run by the system.
4. A minimal extension to 3. leads to *cooperative* thread scheduling. No other design decisions are made except that threads are implemented as active objects and scheduled entirely on behalf of the application (*threading*). There may be many active objects run by the system.
5. Adding support for the serialized execution of thread scheduling functions (*locking*) enables the *non-preemptive* processing of active objects in an interrupt-driven context. Thread scheduling till happens cooperatively, however the nucleus is prepared to schedule threads on behalf of application-level interrupt handlers. Actions of global significance, and enabled by interrupt handlers, are assumed to be synchronized properly (*serialization*).
6. Multiplexing the CPU between threads in an interrupt-driven manner establishes the autonomous, *preemptive* execution of active objects. In this case,

the nucleus is extended by a device driver module (*multiplexing*) taking care of timed thread scheduling.

Referring to figure 4, CORE takes care of interruption, serialization, locking, threading and objectification, while NEXT covers (device) driving and (CPU) multiplexing.

5.1 Functionality vs. Complexity vs. Performance

The PURE system is implemented in C++ and runs (as guest level and in native mode) on i80x86-, i860-, sparc-, and ppc60x-based platforms. A port to C 167-based, "CANned" μ-controllers is in progress. At the time being, the nucleus consists of over 100 classes exporting over 600 methods. Every class implements an abstract data type. Inheritance is employed extensively to build complex abstract data types. For example, the thread control block is made of about 45 classes arranged in a 14-level class hierarchy.

As table 1 shows, the highly modular nucleus structure still results in a small and compact implementation. The numbers were produced using GNU g++ 2.7.2.3 for the i586 running Linux Red Hat 5.0.

Table 1. PURE memory consumption.

family member	size (in bytes)			
	text	data	bss	total
interruptedly	812	64	392	1268
reconcile	1882	8	416	2306
exclusive	434	0	0	434
cooperative	1620	0	28	1648
non-preemptive	1671	0	28	1699
preemptive	3642	8	428	4062

Table 2 shows the number of (i586) CPU clock cycles spent during thread scheduling, i.e., the overhead imposed on applications at run-time due to thread scheduling. The clock frequency in this experiment was 166 MHz and the measurement was made reading the i586 (on-chip) counter register. Again, the above-mentioned C++ compiler was used.

At first sight, these numbers may seem insignificant, but when we realize that some deeply embedded applications (e.g. car engines) require the scheduler to be invoked once every 10 μs, assigning the *preemptive* family member to an application that could run with the *exclusive* family member will imply in a slow down of 15%. We still do not have performance measurements for the family members implemented to support parallel computing, but previous experiments with the PEACE family based operating system [9] showed that applications can improve performance in up to 74% just by being assigned to run with the proper

Table 2. PURE scheduler latency.

family member	CPU cicles
interruptedly	no scheduler
reconcile	no scheduler
exclusive	no scheduler
cooperative	49
non-preemptive	57
preemptive	300

operating system (e.g., a single-task, stand-alone environment supported by a library vs. a multi-task, distributed environment supported by a μ-kernel).

All these performance figures demonstrate the still "featherweight" structure of PURE, although quite a large amount of abstractions (classes, modules, functions) are involved in all the occurrences: "*It is the system design which is hierarchical, not its implementation*" [5]. Besides, these figures demonstrate the necessity to give each application its own operating system.

6 Further Work

The strategy to automatically adapt an operating system to a given (embedded) parallel application proposed by PURE and EPOS can drastically improve application performance, since the application will get only the operating system components it really needs. Besides, these components are fine tuned to the execution scenario aimed. However, our strategy is not able to deliver an *optimal* operating system. Consider, for instance, the decision for a thread scheduling policy: several thread implementations, with different scheduling policies, may fit into the blueprint extracted by our tools, as long as they match the selected interfaces (selectively realize the requested interfaces) and satisfy the dependencies. Nevertheless, it is unnecessary to say that there is an optimal scheduling policy for a given set of threads running in a given scenario.

The decision of which variant of a system abstraction to select when several ones accomplish the application's requirements is, in the current system, arbitrary. Further development of our tools shall include *profiling* primitives to collect run-time statistics for the application. These statistics will then drive operating system reconfigurations towards the optimal. To grant that we can generate an *optimal* system, however, would imply in formal specification and validation of our system objects, what is not in the scope of EPOS, but of the WABE project. WABE aims to implement a workbench for the construction of optimal operating systems. It is now being developed under a DFG project at the Universities of Magdeburg and Potsdam.

7 Conclusion

In this paper we presented the PURE/EPOS approach to deal with the high complexity of adaptable systems, to diminish the distance between application and operating system and to automate the generation of application tailored operating systems. The complexity problem is tackled with the adoption of *scenario adapters*: software structures that support the adaptation of aspect independent system abstractions to specific execution scenarios; and by *inflated class interfaces*: a mechanism to advertise the system abstraction repository with a (very) reduced number of well-known components. Inflated interfaces are also an effective way to diminish the distance between applications and operating systems, since the application programmer is no longer requested to deal with the complex collection of basic system objects directly.

A system designed according to the methodology proposed in this paper can be automatically tailored to satisfy an specific application. In order to enable this, the application must be written referring to the inflated interfaces that export the system abstractions repository and then be submitted to an analyzer. This analyzer will proceed syntactical and data flow analysis to extract a blueprint for the operating system to be generated. The blueprint is then refined by dependency analysis against information about the execution scenario acquired from the user via visual tools. The outcome of this process is a configuration file comprised of *selective realize* keys that will support the compilation of the tailored operating system. These tools are now under development at GDM-FIRST Institute and at the University of Magdeburg.

References

[1] Antônio A. Fröhlich, Rafael Avila, Luciano Piccoli & Helder Savietto. A Concurrent Programming Environment for the i486. In *Proceedings of the 5th International Conference on Information Systems Analysis and Synthesis - InterSymp'96*, Orlando, USA, July, 1996.

[2] Friedrich Schön, Wolfgang Schröder-Preikschat, Olaf Spinczyk & Ute Spinczyk. Design Rationale of the PURE Object-Oriented Embedded Operating System. In *Proceedings of the International IFIP WG 10.3/WG 10.5 Workshop on Distributed and Parallel Embedded Systems - DIPES'98*, Paderborn, Germany, October 1998.

[3] Grady Booch. *Object Oriented Analysis and Design with Applications*. Benjamin/Cummings, Redwood City, CA, USA, 1994.

[4] Gregor Kiczales, John Lamping, Anurag Mendhekar, Chris Maeda, Cristina Lopes, Jean-Marc Loingtier & John Irwin. Aspect-Oriented Programming. In *Proceedings of ECOOP'97, Lecture Notes in Computer Science*, pages 220–242, Springer-Verlag, 1997.

[5] A. N. Habermann, L. Flon, and L. Cooprider. Modularization and Hierarchy in a Family of Operating System s. *Communications of the ACM*, 19(5):266–272, 1976.

[6] J. Cordsen and W. Schröder-Preikschat. Object-Oriented Operating System Design and the Revival of Program Families. In *Proceedings of the Second International Workshop on Object Orientation in Operating Systems*, pages 24–28, Palo Alto, CA, USA, October 1991.

[7] J. Nolte and W. Schröder-Preikschat. Dual Objects — An Object Model for Distributed System Programming. In *Proceedings of the Eigth ACM SIGOPS European Workshop, Support f or Composing Distributed Applications*, 1998.

[8] D. L. Parnas. Designing Software for Ease of Extension and Contraction. *Transaction on Software Engineering*, SE-5(2), 1979.

[9] W. Schröder-Preikschat. *The Logical Design of Parallel Operating Systems*. Prentice-Hall, 1994.

[10] P. Wegner. Classification in Object-Oriented Systems. *SIGPLAN Notices*, 21(10):173–182, 1986.

Fiber-Optic Interconnection Networks for Signal Processing Applications

Magnus Jonsson

Halmstad University, Box 823, S-301 18 Halmstad, Sweden
Magnus.Jonsson@cca.hh.se

Abstract. In future parallel radar signal processing systems, with high bandwidth demands, new interconnection technologies are needed. The same reasoning can be made for other signal processing applications, e.g., those involving multimedia. Fiber-optic networks are a promising alternative and a lot of work have been done. In this paper, a number of fiber-optic interconnection architectures are reviewed, especially from a radar signal processing point-of-view. Two kinds of parallel algorithm mapping are discussed: (i) a chain of pipeline-stages mapped, more or less directly, one stage per computation node and (ii) SPMD (Same Program Multiple Data). Several network configurations, which are simplified due to the nature of the applications, are also proposed.

1 Introduction

The interconnection network is a very important part in a parallel computing system, especially in future radar signal processing systems where high amounts of data should be moved. In this paper, fiber-optic interconnection networks suitable for signal processing applications are reviewed and proposed. Two kinds of parallel algorithm mapping are discussed for the networks as follows. Both cases are taken from the field of radar signal processing, but the discussion holds for, e.g., other signal processing applications with similar mappings too.

The first case considered is a straight forward pipeline chain, shown in Fig. 1, where each box represents a computation node [1]. Each node runs one or several pipeline stages and the arcs only represent the dataflow between modules. The physical topology can, e.g., be a ring or a crossbar switch. For simplicity, the chain is viewed in a form without, e.g, multicasting between some of the nodes. A data cube, that initially comes from the antenna, contains data in three dimensions (channel, pulse, and distance). After the processing of one stage, the new data cube is forwarded to the next node in the chain.

The second case of mapping is to let all processors do the same operation but on different sets of data, i.e., SPMD (Same Program Multiple Data). All the PEs in Fig. 2 then work together on one of the algorithmic pipeline stages at the time. After the processing of one stage, the data cube is redistributed if needed. Incoming data from the antenna is not shown in Fig. 2, but is assumed to be fed into the PEs not effecting the communication pattern shown in the figure. For example, special I/O channels can be used instead of the communication system carrying the traffic indicated by the arcs in the figure.

Although not considered here, inter-module communication during the processing of one stage might occur. The number of times it is needed to redistribute the data cube between the nodes might however be lower than the number of times the data cube is transferred to next module in a pipeline chain. Instead, it might be harder to overlap communication and computation, which implies extra bandwidth requirements [2]. When corner turning the data cube, one half of the data cube has to be transferred across the network bisection, essentially with all-to-all communication [2].

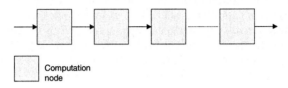

Fig. 1. A pure pipeline chain where one or several stages are mapped on each node

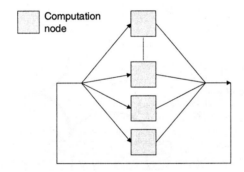

Fig. 2. If the SPMD model is used, all PEs work together on one stage at the time

The rest of the paper is organized as follows. In Section 2 through 6, the following technologies and interconnection architecture classes are reviewed, respectively: (i) fiber-ribbon ring, (ii) WDM star, (iii) WDM ring, (iv) integrated fiber and waveguide solutions, and (v) optical fibers and electronic crossbars. The paper is then concluded in Section 7.

2 Fiber-Ribbon Ring Networks

If fiber-ribbon cables are used to connect the nodes in a point-to-point linked ring network, bit-parallel transfer can be utilized. In such a network one of the fibers in each ribbon is dedicated to carry the clock signal. Therefore, no clock-recovery circuits are needed in the receivers. Other fibers can be dedicated for, e.g., frame synchronization. An example of a fiber-ribbon ring network is the USC POLO Network, which is proposed to be used in COWs (Cluster of Workstations) and similar systems [3]. Integrated circuits have been developed for the network and tests have been performed [4] [5].

As seen in Fig. 3, aggregated throughputs higher than 1 can be obtained in ring networks with support for spatial bandwidth reuse (sometimes called pipeline rings) [6]. This feature can be highly useful in signal-processing applications with a pipelined dataflow, i.e., most of the communication is to the nearest down-stream neighbor. Two fiber-ribbon pipeline ring networks have been reported recently [7]. The first network has support for circuit-switching on 8+1 fibers (data and clock) and packet-switching on an additional fiber [8]. The second network is more flexible, but with a little more complexity, and has support for packet-switching on 8+1 fibers while using a tenth fiber for control packets (see Fig. 4) [9]. The control packets carries MAC (Medium Access Control) information for the collision-less MAC protocol with support for slot-reserving to get guaranteed bandwidth and worst-case latency.

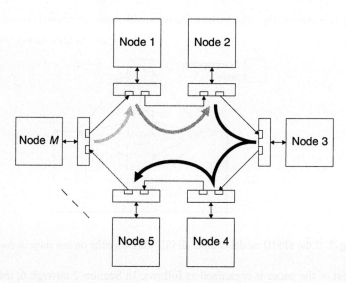

Fig. 3. Example of spatial bandwidth reuse. Node M sends to Node 1 at the same time as Node 1 sends to Node 2, and Node 2 sends a multicast packet to Node 3, 4, and 5

3 WDM Star Networks

A passive fiber-optic star distributes all incoming light, on the input ports, to all output ports. A network with the logical function of a bus is obtained when connecting the transmitting and receiving side of each node to one input and output fiber of the star, respectively. By using WDM (Wavelength Division Multiplexing), multiple wavelength (color of light) channels can carry data simultaneously in the network [10]. To get a flexible network, we need tunable receivers and/or transmitters, i.e., it should be possible to send/listen on an arbitrary channel [11].

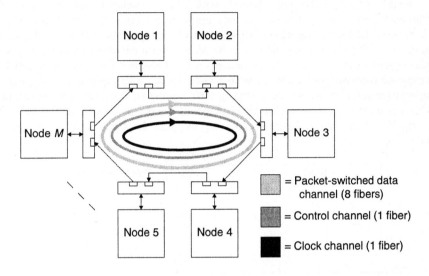

Fig. 4. Control-channel based network built up with fiber-ribbon point-to-point links

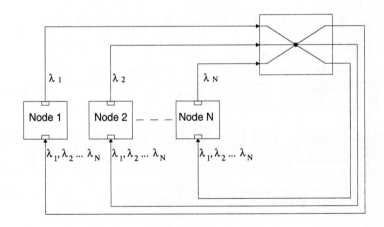

Fig. 5. WDM star network

In Fig. 5, an example of a WDM star network configuration is shown. Each node transmits on a wavelength unique to the node, while the receiver can listen to an arbitrary wavelength. The configuration is used in the TD-TWDMA network (Time-Deterministic Time and Wavelength Division Multiple Access) [1], which has support for guaranteeing real-time services, both in single-star networks [12] and in star-of-stars networks [13]. One can say that this kind of networks implements a distributed crossbar. The flexibility is hence high, and multicast and single-destination traffic can co-exist. The number of wavelengths is practically limited to 16-32 [14], but, as stated above, hierarchical networks with wavelength reuse can be built.

Tunable components (e.g., filters) with tuning latencies in the order of a nanosecond have been reported, but they often have a limited tuning range [15]. At the expense of longer tuning latencies, however, components with a broader tuning range exist [16]. Such components can be used to get a cheaper network for systems where much of the communication patterns remain constant for a longer period, e.g., during the processing of a data cube in a radar system with a pipelined mapping. To support for some more rapidly changing traffic patterns, the nodes can be extended with transmitters and receivers fixed-tuned to a special broadcast channel. This configuration can be compared to having support for both circuit- and packet-switching.

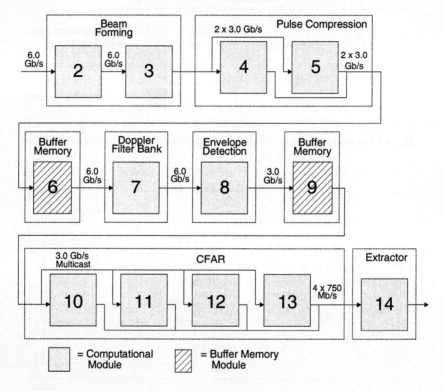

Fig. 6. Data flow between the modules in the sample radar signal processing chain

An example is given below to show that only one input and one output channel (in addition to the broadcast channel) are needed during the processing of a data cube, i.e., the normal minimum time running one pipeline chain (working mode of the radar). In Fig. 6, a signal processing chain similar to those described in [17] and [1] is shown together with its bandwidth demands. Each computational module in the figure contains multiple processors itself. The chain is a good example containing both multicast, one-to-many, and many-to-one communication patterns. The aggregated throughput demand is 45 Gb/s. We leave out all details of the chain, referring to the two other papers, and do only treat the throughput demands here. In the figure, there are 13 nodes. In addition, the antenna is seen as one node (feeds the first node in the chain with data), Node 1, and there is one master node responsible for supervising the whole system and interacting with the user, Node 15. For simplicity we assume an efficient channel bandwidth of 6.0 Gb/s. A feasible allocation scheme of eight channels (in addition to the broadcast channel) is shown in Table 1.

Table 1. A feasible allocation scheme of the wavelength channels in a WDM star network for the sample radar system

Node	Input channel	Output channel
1	-	λ_1
2	λ_1	λ_2
3	λ_2	λ_3
4	λ_3	λ_4
5	λ_3	λ_4
6	λ_4	λ_5
7	λ_5	λ_6
8	λ_6	λ_7
9	λ_7	λ_7
10	λ_7	λ_8
11	λ_7	λ_8
12	λ_7	λ_8
13	λ_7	λ_8
14	λ_8	λ_8
15	λ_8	-

Totally removing the ability of tuning in a WDM star network leads to a multi-hop network [18]. Each node in a multi-hop network transmits and receives on one or a few dedicated wavelengths. If a node does not have the capability of sending on one of the receiver wavelengths of the destination node, the traffic must pass one or several intermediate nodes. The wavelengths can be chosen to get, e.g., a perfect-shuffle network [19], or a network with a pattern similar to the dataflow in a pipelined system. One can also choose to have a network with several topologies embedded, e.g., a ring and a hypercube.

4 WDM Ring Networks

Each node in a WDM ring network has an ADM (Add-Drop Multiplexer) to insert, listen, and remove wavelength channels from the ring. Spatial wavelength reuse can be achieved by removing the transmitted light at the destination node. At high degrees of nearest-downstream-neighbor communication (e.g., in pipeline mapping), throughputs significantly higher than 1 can be achieved with only a few wavelength channels. In the WDMA ring network described in [20], each node is assigned a node-unique wavelength to transmit on. The other nodes can then tune in an arbitrary channel to listen on. This configuration is logically the same as that for the WDM star network with fixed transmitters and tunable receivers.

As discussed for the WDM star network, components with long tuning latencies (ADMs in the case of a ring) can be used for traffic patterns that do not change rapidly. An additional broadcast wavelength dedicated for packet-switching keeps the network flexible. A single 6 Gbit/s channel (in addition to the broadcast channel) is enough for the signal processing chain shown in Fig. 6, if the ADMs in Node 1, 2, 3, 6, 7, 8, and 9 terminates the channel for wavelength reuse.

5 Integrated Fiber and Waveguide Solutions

Fibers or other kinds of waveguides (hereafter commonly denoted as channels) can be integrated to form a more or less compact system of channels. Fibers can be laminated to form a foil of channels, for use as intra-PCB (Printed Circuit Board) or back-plane interconnection systems [21] [22]. Fiber-ribbon connectors are applied to fiber endpoints of the foil. An example is shown in Fig. 7, where four computational nodes are connected in a ring topology. In addition, there is a clock node that distributes clock signals to the four computational nodes via equal-length fibers to keep the clock signals in phase. If one foil is placed on each PCB in a rack, they can be passively connected to each other via fiber-ribbon cables. Using polymer waveguides instead of fibers brings advantages such as the possibility of integrating splitters and combiners into the foil, and the potential for more cost-effective mass-production [21].

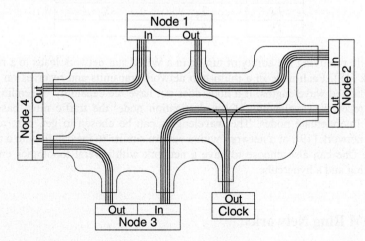

Fig. 7. A foil of fibers connects four computational nodes and a clock node

Integrated systems of channels can be setup and used in a number of configurations, for which some are discussed below. One way is to embed a ring with bit-parallel transmission and the possibility of spatial bandwidth reuse as described in Section 2. Of course, this leads to the same good performance for pipelined data flows as the fiber-ribbon ring network, only changing the medium to a more compact form. Besides pure communication purposes, channels for, e.g., clock distribution (as seen in the example) and flow control can be integrated into the same system.

Another way is to follow the proposed use of an array of passive optical stars to connect processor boards in a multiprocessor system, via fiber-ribbon links, for which experiments with 6 x 700 Mb/s fiber-ribbon links were done (see Fig. 8) [23]. As indicated above, such a configuration can be integrated by the use of polymer waveguides. The power budget can, however, be a limiting factor to the number of nodes and/or the distance. Advantages are simple hardware due to bit-parallel transmission, and the broadcast nature. Many-to-many communication patterns, as used when corner-turning in SPMD mode, map easily on the broadcast architecture as long as the star-array not becomes a bottle-neck.

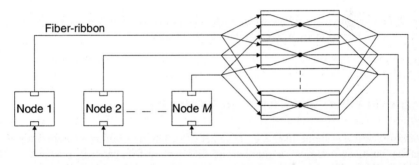

Fig. 8. Array of passive optical stars connects a number of nodes via fiber-ribbons

Electronic crossbars can be distributed on the PCBs and/or placed on a special switch-PCB in a back-plane system, and be connected by integrated parallel channels. A distributed crossbar, instead, can be realized by bit-serial transmissions over a fully connected topology, i.e., there is one channel between each pair of nodes (see Fig. 9). Broadcast is done by driving all the laser-diodes of a node with the same bit-stream. A simple solution is to always drive all the laser-diodes in the transmitting node and couple the photo-diodes together as one incoming channel. This brings an architecture with the support of a single broadcast at the time. By only sending on one fiber when performing single-destination communication will, however, give the opportunity of having multiple transmissions in the system at the same time. The number of nodes, N, in this configuration is limited because the number of fibers grows by N^2. Also, clock-recovery circuits must be involved. After all, the distressed power budget brings an advantage over a system with passive splitters or stars (this holds for, e.g., the point-to-point connected ring also). The flexibility of a crossbar makes the network good for radar systems with the SPMD mode.

Other similar systems include the integration of fibers into a PCB for the purpose of clock distribution [24]. Distribution to up to 128 nodes was demonstrated. The fibers are laminated on one side of the PCB while integrated circuits are placed on the reversed side. The end section of each fiber is bent 90 degrees to lead the light through a via hole to the reversed side of the PCB.

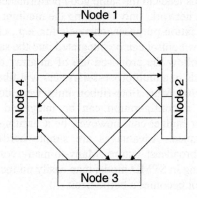

Fig. 9. Fully connected topology where each node has an array of $N-1$ laser diodes and an array of $N-1$ photo diodes, where N is the number of nodes

6 Optical Fibers and Electronic Crossbars

Communication systems like, e.g., Myrinet [25], where arbitrary switched topologies can be built, can support a number of different traffic patterns possible in radar systems. Fiber-ribbons can be used to increase bandwidth while still sending in bit-parallel mode. Bit rates in the order of 1 Gbit/s over each fiber in the ribbon is possible over tens of meters using standard fiber-ribbons. As noted in Section 5, foils of fibers or waveguides (e.g., arranged as ribbons) can be used to interconnect nodes and crossbars on the PCB and/or back-plane level.

An alternative to ribbons is bit-parallel transfer over a single fiber by using WDM. In such configurations each bit in the word, plus the clock signal, is given a dedicated wavelength. Wavelengths can also be dedicated to other purposes like frame synchronization and flow control. Significantly higher bandwidth-distance products can be achieved when using bit-parallel WDM over dispersion-shifted fiber, instead of fiber-ribbons [26]. If, however, only communication over shorter distances exist (e.g., a few meters), the bandwidth-distance product is not necessarily a limiting factor. Transmission experiments with an array of eight pie-shaped VCSELs arranged in a circular area with a diameter of 60 μm, to match the core of a multimode fiber, has been reported [27]. Other works on the integration of components for short-distance (non telecom) WDM links have been reported, e.g., a 4 × 2.5 Gbit/s transceiver with integrated splitter, combiner, filters, etc. [28].

The switch itself can also be modified to increase performance or packing density. A single-chip switch core where fiber-ribbons are coupled directly to optoelectronic devices on the chip is possible [29]. Attaching 32 incoming and 32 outgoing fiber-ribbons with 800 Mb/s per fiber translates to an aggregated bandwidth of 204 Gbit/s through the switch when eight fibers on each link are used for data.

A 16×16 crossbar switch-chip, with integrated optoelectronic I/O, have been implemented for switching of packets transferred using bit-parallel WDM [30]. Each node has two fibers (input and output) coupled to the switch.

Another switch chip with integrated optoelectronic I/O is intended to be attached to each node in a static topology like a multidimensional torus [31]. A special feature of the chips is that potential deadlocks is handled by a global mechanism. The mechanism brings mutual exclusion to let one packet at a time use dedicated hardware to recover from a potential deadlock. Finally, it can be noted that both Myricom (Myrinet) and Mercury (RACEway) are looking at optical technologies for their future products [32].

7 Conclusions

Some of the mentioned networks are summarized in Table 2, where remarks on suitability/performance for the two cases of mapping are made. The increasingly good price/performance ratio for fiber-ribbon links indicates a great success potential for several of the networks discussed. On the other hand, the WDM technique offer flexible multi-channel networks that can be passively implemented. Integrated fiber and waveguide solutions makes the building of compact systems possible, especially for networks like those using fiber-ribbons.

Table 2. Summary of some of the discussed networks

Network	Pipeline	SPMD	Notes
Fiber-ribbon pipeline ring	Good	Moderate	Good for SPMD too if enough bandwidth
WDM star with rapid tuning	Good	Good	Flexible passive network
WDM star with slow tuning and broadcast channel	Good	Poor	WDM star alternative that might be cheaper
Multi-hop WDM star	Moderate	Moderate	Can be optimized for pipelined mapping
WDM ring with rapid tuning	Good	Good	More channels might be needed for SPMD
WDM ring with slow tuning and broadcast channel	Good	Poor	WDM ring alternative that might be cheaper
Fiber ribbons and array of stars	Moderate	Moderate	Power and bandwidth limited
Fully connected topology with broadcast driving	Moderate	Moderate	Bandwidth limited. Grows with N^2
Fully connected topology with flexible driving	Good	Good	Grows with N^2
Optical fibers and electronic crossbars	Good	Good	Optolectronics needed in switch too

8 Acknowledgement

This work is part of the REMAP project, financed by NUTEK, the Swedish National Board for Industrial and Technical Development, and the PARAD project, financed by the KK Foundation in cooperation with Ericsson Microwave Systems AB.

References

1. Jonsson, M., A. Åhlander, M. Taveniku, and B. Svensson, "Time-deterministic WDM star network for massively parallel computing in radar systems," *Proc. Massively Parallel Processing using Optical Interconnections, MPPOI'96*, Lahaina, HI, USA, Oct. 27-29, 1996, pp. 85-93.
2. Teitelbaum, K., "Crossbar tree networks for embedded signal processing applications," *Proc. Massively Parallel Processing using Optical Interconnections (MPPOI'98)*, Las Vegas, NV, USA, June 15-17, 1998, pp. 200-207.
3. Sano, B. J., and A. F. J. Levi, "Networks for the professional campus environment," in *Multimedia Technology for Applications*. B. Sheu and M. Ismail, Eds., McGraw-Hill, Inc., pp. 413-427, 1998, ISBN 0-7803-1174-4.
4. Sano, B., B. Madhavan, and A. F. J. Levi, "8 Gbps CMOS interface for parallel fiber-optic interconnects," *Electronics Letters*, vol. 32, pp. 2262-2263, 1996.
5. *POLO Technical Summary*. University of Southern California, Oct. 1997.
6. Wong, P. C., and T.-S. P. Yum, "Design and analysis of a pipeline ring protocol," *IEEE Transactions on communications*, vol 42, no. 2/3/4, pp. 1153-1161, Feb./Mar./Apr. 1989.
7. Jonsson, M., "Two fiber-ribbon ring networks for parallel and distributed computing systems," *Optical Engineering*, vol. 37, no. 12, pp. 3196-3204, Dec. 1998.
8. Jonsson, M., B. Svensson, M. Taveniku, and A. Åhlander, "Fiber-ribbon pipeline ring network for high-performance distributed computing systems," *Proc. Int. Symp. on Parallel Architectures, Algorithms and Networks (I-SPAN'97)*, Taipei, Taiwan, Dec. 18-20, 1997, pp. 138-143.
9. Jonsson, M., "Control-channel based fiber-ribbon pipeline ring network," *Proc. Massively Parallel Processing using Optical Interconnections (MPPOI'98)*, Las Vegas, NV, USA, June 15-17, 1998, pp. 158-165.
10. Brackett, C. A., "Dense wavelength division multiplexing networks: principles and applications," *IEEE Journal on Selected Areas in Communications*, vol. 8, no. 6, pp. 948-964, Aug. 1990.
11. Mukherjee, B., "WDM-based local lightwave networks part I: single-hop systems," *IEEE Network*, pp. 12-27, May 1992.
12. Jonsson, M., K. Börjesson, and M. Legardt, "Dynamic time-deterministic traffic in a fiber-optic WDM star network," *Proc. 9th Euromicro Workshop on Real Time Systems*, Toledo, Spain, June 11-13, 1997.
13. Jonsson, M., and B. Svensson, "On inter-cluster communication in a time-deterministic WDM star network," *Proc. 2nd Workshop on Optics and Computer Science (WOCS)*, Geneva, Switzerland, Apr. 1, 1997.
14. Brackett, C. A., "Foreword: Is there an emerging consensus on WDM networking?," *Journal of Lightwave Technology*, vol. 14, no. 6, pp. 936-941, June 1996.
15. Kobrinski, H., M. P. Vecchi, E. L. Goldstein, and R. M. Bulley, "Wavelength selection with nanosecond switching times using distributed-feedback laser amplifiers," *Electronics Letters*, vol. 24, no. 15, pp. 969-971, July 21, 1988.

16. Cheung, K.-W., "Acoustooptic tunable filters in narrowband WDM networks: system issues and network applications," *IEEE Journal on Selected Areas in Communications*, vol. 8, no. 6, pp. 1015-1025, Aug. 1990.
17. Taveniku, M., A. Åhlander, M. Jonsson, and B. Svensson, "A multiple SIMD mesh architecture for multi-channel radar processing," *Proc. International Conference on Signal Processing Applications & Technology (ICSPAT'96)*, Boston, MA, USA, Oct. 7-10, 1996, pp. 1421-1427.
18. Mukherjee, B., "WDM-based local lightwave networks part II: multihop systems," *IEEE Network*, pp. 20-32, July 1992.
19. Acampora, A. S., and M. J. Karol, "An overview of lightwave packet networks," *IEEE Network*, pp. 29-41, Jan. 1989.
20. Irshid, M. I., and M. Kavehrad, "A fully transparent fiber-optic ring architecture for WDM networks," *Journal of Lightwave Technology*, vol. 10, no. 1, pp. 101-108, Jan. 1992.
21. Eriksen, P., K. Gustafsson, M. Niburg, G. Palmskog, M. Robertsson, and K. Åkermark, "The Apollo demonstrator – new low-cost technologies for optical interconnects," *Ericsson Review*, vol. 72, no. 2, 1995.
22. Shahid, M. A., and W. R. Holland, "Flexible optical backplane interconnections," *Proc. Massively Parallel Processing using Optical Interconnections (MPPOI'96)*, Maui, HI, USA, Oct. 27-29, 1996, pp. 178-185.
23. Parker, J. W., P. J. Ayliffe, T. V. Clapp, M. C. Geear, P. M. Harrison, and R. G. Peall, "Multifibre bus for rack-to-rack interconnects based on opto-hybrid transmitter/receiver array pair," *Electronics Letters*, vol. 28, no. 8, pp. 801-803, April 9, 1992.
24. Li, Y., J. Popelek, J.-K. Rhee, L. J. Wang, T. Wang, and K. Shum, "Demonstration of fiber-based board-level optical clock distributions," *Proc. 5th International Conference on Massively Parallel Processing using Optical Interconnections (MPPOI'98)*, Las Vegas, NV, USA, June 15-17, 1998, pp. 224-228.
25. Boden, N. J., D. Cohen, R. E. Felderman, A. E. Kulawik, C. L. Seitz, J. N. Seizovic, and W.-K. Su, "Myrinet: a gigabit-per-second local area network," *IEEE Micro*, vol. 15, no. 1, pp. 29-36, Feb. 1995.
26. Bergman, L., J. Morookian, and C. Yeh, "An all-optical long-distance multi-Gbytes/s bit-parallel WDM single-fiber link," *Journal of Lightwave Technology*, vol. 16, no. 9, pp. 1577-1582, Sept. 1998.
27. Coldren, L. A., E. R. Hegblom, Y. A. Akulova, J. Ko, E. M. Strzelecka, and S. Y. Hu, "Vertical-cavity lasers for parallel optical interconnects," *Proc. 5th International Conference on Massively Parallel Processing using Optical Interconnections (MPPOI'98)*, Las Vegas, NV, USA, June 15-17, 1998, pp. 2-10.
28. Aronson, L. B., B. E. Lemoff, L. A. Buckman, and D. W. Dolfi, " Low-cost multimode WDM for local area networks up to 10 Gb/s," *IEEE Photonics Technology Letters*," vol. 10, no. 10, pp. 1489-1491, Oct. 1998.
29. Szymanski, T. H., A. Au, M. Lafrenière-Roula, V. Tyan, B. Supmonchai, J. Wong, B. Zerrouk, and S. T. Obenaus, "Terabit optical local area networks for multiprocessing systems," *Applied Optics*, vol. 37, no. 2, pp. 264-275, Jan. 10, 1998.
30. Krisnamoorthy et al., "The AMOEBA chip: an optoelectronic switch for multiprocessor networking using dense-WDM," Proc. Massively Parallel Processing using Optical Interconnections (MPPOI'96), Maui, HI, USA, Oct. 27-29, 1996, pp. 94-100.
31. Pinkston, T. M., M. Raksapatcharawong, and Y. Choi, "WARP II: an optoelectronic fully adaptive network router chip," *Proc. Optics in Computing (OC'98)*, Brugge, Belgium, June 17-20, 1998, pp. 311-315.
32. Lund, C., "Optics inside future computers," *Proc. Massively Parallel Processing using Optical Interconnections (MPPOI'97)*, Montreal, Canada, June 22-24, 1997, pp. 156-159.

Reconfigurable Parallel Sorting and Load Balancing: HeteroSort

Emmett Davis[1], Dr. Bonnie Holte Bennett[1], Bill Wren [2], Dr. Linda Davis[1],

[1] University of St. Thomas, Graduate Programs in Software, Mail # OSS 301, 2115 Summit Avenue, Saint Paul, MN
EmmettDa@aol.com, bhbennett@stthomas.edu
[2] Honeywell Technology Center, 3660 Technology Drive, Minneapolis, MN 55418
wrenpc@worldnet.att.net

Abstract. HeteroSort load balances and sorts within static or dynamic networks. Upon failure of a node or path, HeteroSort uses a genetic algorithm to minimize the distribution path by optimally mapping the network to physical near neighbor nodes. We include a proof that a final state of HeteroSort (barren best trades of the Wren cycle) is always a detection of termination of sorting. By capturing global system knowledge in overlapping microregions of nodes, HeteroSort is useful in data dependent applications such as data information fusion on distributed processors.

1 Introduction

Dynamic adaptability is a keystone feature for applications implemented on modern networks. Dynamic adaptability is a basis for fault tolerance. A system, which is dynamically adaptive, strives to withstand the assault of hardware glitches, electrical spikes, and component destruction. The research described in this paper set out to develop a high-speed load balancing algorithm, which would balance loads by sorting data across the network of nodes and resulted in developing a reconfigurable system for parallel sorting with dynamic adaptability.

1.1 Dynamic Adaptability

With the increased dependence on distributed and parallel processing to support general as well as safety-critical applications, we must have applications that are fault tolerant. Programs must be able to recognize that current resources are no longer available. Schedulers are employed in the presence of faults to manage resources against program needs using dynamic or fixed priority scheduling for timing correctness of critical application tasks.[1]

We have taken a different approach and refocused on the design of elemental processes such as load balancing and sorting. Instead of depending on schedulers, we design process algorithms where global processes are completed using only local knowledge and recovery resources. This lessens the need for schedulers and eases their workload.

1.2 Information and Data Fusion Applications

Information fusion is the analysis and correlation of incoming data to provide a comprehensive assessment of the arriving reports. Example applications include military intelligence correlation, meteorological data interpretation, and real-time economic news synthesis. Data fusion is the correlation of individual reports from sensor systems. It operates at a lower level than information fusion. Data fusion is often concerned with processing such as common aperture correlation across similar sensors (for example, SAR and IR radar systems).

The first step to information and data fusion applications is often data sorting. Once the data are properly sorted they can be correlated with other data and higher level reports. Data sorting is important to many algorithms, but it is particularly important for parallel and distributed algorithms requiring search because non-local data are either ignored or delay the system's performance while they are retrieved. Furthermore, this is of growing importance now.

For example, in a complex data fusion system individual data from a variety of reports may be input to the system. These inputs may be reports from surveillance in border-patrol for drug enforcement or they may be locations and flight paths for aircraft in an air traffic control scenario. The goal of the processing may be to

---

[1] Examples of these fault tolerant efforts can be found in the work of Jay Strosnider and his colleagues at Department of Electrical and Computing Engineering, Carnegie Mellon University in the Fault- Tolerant Real Time Computing Project. Katcher, Daniel I., Jay K. Strosnider, and Elizabeth A. Hinzelman-Fortino. "Dynamic versus Fixed Priority Scheduling: A Case Study" http://usa.ece.cmu.edu/Jteam/papers/abstracts/tse93.abs.html.
Ramos-Theul, Sandra and Jay K. Strosnider.
"Scheduling Fault Tolerant Operations for Time-Critical Applications." http://usa.ece.cmu.edu/Jteam/papers/abstracts/dcca94.abs.html.
Theul, Sandra. "Enhancing the Fault Tolerance of Real-Time Systems Through Time Redundancy." http://usa.ece.cmu.edu/Jteam/papers/abstracts/thuel.abs.html.

correlate related reports where correlation often depends on spatial or temporal proximity. In either the drug enforcement or the air traffic control scenario the reports are spread across a two- or three-, or four-dimensional spatio-temporal domain. These reports can be more efficiently processed if they are sorted by proximity. Furthermore, if these reports are to be processed by a distributed system, then proximity sorting is essential since different reports may be sent to different processors for correlation. And, the space and time location of the reports is one way to partition the processing across the nodes. Thus, sorting is important in direct and indirect ways to the processing of many algorithms.

1.3 Local Knowledge and Global Processes

An efficient network sort algorithm is highly desirable, but difficult. The problem is that it requires local operations with global knowledge. So, consider a group of data (for example, that of names in a phone directory) was to be distributed across a number of processors (for example, 26). Then an efficient technique would be for each processor to take a portion of the unsorted data and send each datum to the processor upon which it eventually belongs (A's to processor 1, B's to processor 2, ... Z's to processor 26).

A significant practical feature of HeteroSort is that in our experiments it load balances before it finishes sorting. Since HeteroSort detects when the system is sorted, it also detects termination of load balancing. Chengzhong Xu and Francis Lau in Load Balancing in Parallel Computers: Theory and Practice (Boston: Kluwer Academic Publishers, 1997) state:

> From a practical point of view, the detection of the global termination is by no means a trivial problem because there is a lack of consistent knowledge in every processor about the whole workload distribution as load balancing progresses.[2]

Thus the global knowledge that all names beginning with the same letter belong on a prespecified processor facilitates local operations in sending off each datum. The problem, however, is that this does not adequately balance the load on the system because there may be many A's (Adams, Anderson, Andersen, Allen, etc.) and very few Q's or X's. So the optimal loading Aaa-Als on processor 1, Alb-Bix on processor 2, ... Win-Zzz on processor 26) cannot be known until all the data is sorted. So global knowledge (the optimal loading) is unavailable to the local operations (where to send each datum) because it is not determined until all the local operations are finished. HeteroSort combines load balancing within sorting processes.

---

[2] Xu, Chengzhong and Francis C.M. Lau. *Load Balancing in Parallel Computers: Theory and Practice.* (Boston: Kluwer Academic Publishers, 1997), p. 122.

Traditionally, techniques such as hashing have been used to overcome the non-uniform distribution of data. However, parallel hash tables require expensive computational maintenance to upgrade each sort cycle, thus making them less efficient than HeteroSort, which requires no external tables.

1.4 Related Work

Much of the work in this area deals with linear arrays.[3] The general approach is to take linear sort techniques and use either a row major or a snake-like grid overlaid on a regular grid topology of processors.[4] The snake-like grid is used at times with a shear-sort or shuffle sorting program where there is first a row operation and then an alternating column operation. So, either the row or the column connections are ignored in each cycle.

1.5 Objective

The questions we investigated in our research were: "How can heterogeneous network sorting be optimized?" and "What implications and benefits result?"

Our goal was to efficiently load balance and sort data over an arbitrary network. We sought to determine a way to move data around the network quickly. So that, using the telephone directory example, names starting with A and B that were initially assigned to processor 26 would quickly move up to processors 1 or 2 or 3 without having to spend a great deal of time "bubbling up" through the bowels of the network interconnections. Thus, our initial sorting goal was to get data to processors in the neighborhood where they belonged and then to shuffle the local data into the right positions on the right processor all with the correct load balance.

---

[3] Lin, Yen-Chun. "On Balancing Sorting on a Linear Array." *IEEE Transactions on Parallel and Distributed Systems*, vol 4, no 5, pp. 566-571, May 1993.
Thompson, C.D., and H.T. Kung. "Sorting on a Mesh-connected Parallel Computer." *Communication of the ACM*, vol 20, no 40, pp. 263-271, April 1977.

[4] Scherson, Isaac D., and Sandeep Sen. "Parallel Sorting in Two-Dimensional VSLI Models of Computation." *IEEE Transactions of Computers*, vol 38, no 2, pp. 238-249, February 1989.
Gu, Qian Ping, and Jun Gu. "Algorithms and Average Time Bounds of Sorting on a Mesh-Connected Computer." *IEEE Transactions on Parallel and Distributed Systems*, vol 5, no 3, pp. 308-315, March 1994.

2 Approach

HeteroSort is our load balancing and sorting algorithm. Our initial approach was to use four-connectedness (as an example of N-connectedness) for load balancing and sorting. In traditional linear sorts data is either high or low for the processor it is on, and is sent up or down the sort chain accordingly. Our approach differs in that we defined data to be very high, high, low, or very low. In order to do this we first defined a sort sequence across an array of processors as depicted in Figure 1.

1	**2**	3	4
8	**7**	**6**	5
9	**10**	11	12
16	15	14	13

Fig. 1. The sort sequence is overlaid in a snake-like grid across the array of processors. The lowest valued items in the sort will eventually reside on processor 1 and the highest valued items on processor 16. Node 7's four connected trading partners are in bold: 2, 6, 8, and 10. When Node 7 receives its initial state, it sorts and splits the data into four quarters. The lowest quarter goes to Node 2. The next lowest quarter goes to Node 6, the third quarter to Node 8, and the highest quarter goes to node 10. Thus the extremely high and low data are shipped across the coils of the snake network.

Next we defined the four neighbors. This is easily understood by examining Node 7 in the example of sixteen processors shown in Figure 1. The neighbors for Node 7 are 2, 6, 8, and 10. When Node 7 receives its initial data, it sorts it and splits it into four quarters. The lowest quarter goes to Node 2, the next lowest quarter goes to Node 6, the third quarter goes to Node 8, and the highest quarter goes to Node 10. Thus, the extremely high and low data are shipped on "express pathways" across the coils of the snake network.

The trading neighbors Node 2 and Node 10 which are not adjacent on the sort sequence (transcoil neighbors) provide a pathway for very low or very high data to pass across the coils of the snake network into another neighborhood of nodes. This provides an express pathway for extremely ill sorted data to move quickly across the network. The concept of four connectedness is easy to understand with an interior node like Node 7, but other remaining nodes in this example are edge nodes, and their implementation differs slightly.

Table 1. Trading partner list. Determining which data is kept at a node depends on how that node falls among the sort order of its neighbors. For example, node 1 falls below all of its neighbors and thus receives the lowest quarter.

Node	Odd Cycle	Even Cycle	Node	Odd Cycle	Even Cycle
1	**1** 2 4 8 16	**1** 16 4 8 2	9	8 **9** 10 12 16	8 **9** 16 12 10
2	1 **2** 3 7 15	1 **2** 15 7 3	10	7 9 **10** 11 15	9 7 **10** 15 11
3	2 **3** 4 6 14	2 **3** 14 6 4	11	6 10 **11** 12 14	10 6 **11** 14 12
4	1 3 **4** 5 13	3 1 **4** 13 5	12	5 9 11 **12** 13	11 9 5 **12** 13
5	4 **5** 6 8 12	4 **5** 12 8 6	13	4 12 **13** 14 16	12 4 **13** 16 14
6	3 5 **6** 7 11	5 3 **6** 11 7	14	3 11 13 **14** 15	13 11 3 **14** 15
7	2 6 **7** 8 10	6 2 **7** 10 8	15	2 10 14 **15** 16	14 10 2 **15** 16
8	1 5 7 **8** 9	7 5 1 **8** 9	16	1 9 13 15 **16**	15 9 13 1 **16**

Simply put, we use a torus for full connectivity. So nodes along the "north" edge of the array which have no north neighbors are connected (conceptually) to nodes along the "south" edge and vice versa (transedge neighbors). Similarly, a node along the "east" edge are given nodes along the "west" edge as east neighbors and so forth. The odd cycle column of Table 1 summarizes all the nodes of a sixteen node network.

Thus, the use of the torus for four-connectedness provides full connectivity. The result is a modified shear-sort where both row and column connections are used with each round of sorting. Furthermore, ill-sorted data is quickly moved across the network via torus connections.

The "express pathway" is a conceptual map of the sorting network. Ideally, the operating systems supports express pathways, such as in an Intel Paragon system where we first implemented our algorithm. Where this environmental support is missing, the cost of these non-adjacent operations is higher. In those environments where networks have edges, HeteroSort has three strategies. The first is to still implement the conceptual torus at the higher transmission cost. The second is to reconfigure itself to the reality of some nodes having only two or three physical neighbors. A third strategy is particularly useful in heterogeneous environments, where we employ a genetic algorithm to determine the optimal network by minimizing transmission costs.

2.1 Optimization of HeteroSort

HeteroSort's distributed approach can provide an efficient control mechanism for a wide variety of algorithms. It also provides "reconfiguration-on-fault" fault tolerance when a node or network error occurs. HeteroSort automatically reconfigures to account for the failed node(s), and the distributed data is not lost. However, efficient operation requires that major sort axis nodes should reside on near neighbor network physical processors. This minimizes communication costs for efficient operation. And, for a heterogeneous topology, or a homogeneous topology made irregular by failed nodes, automatically achieving this near neighbor configuration for the sort nodes is difficult.

Figure 2 indicates a homogeneous mesh made irregular by two failed nodes. The numbers in the boxes (nodes) indicate the node's position in the sort order.

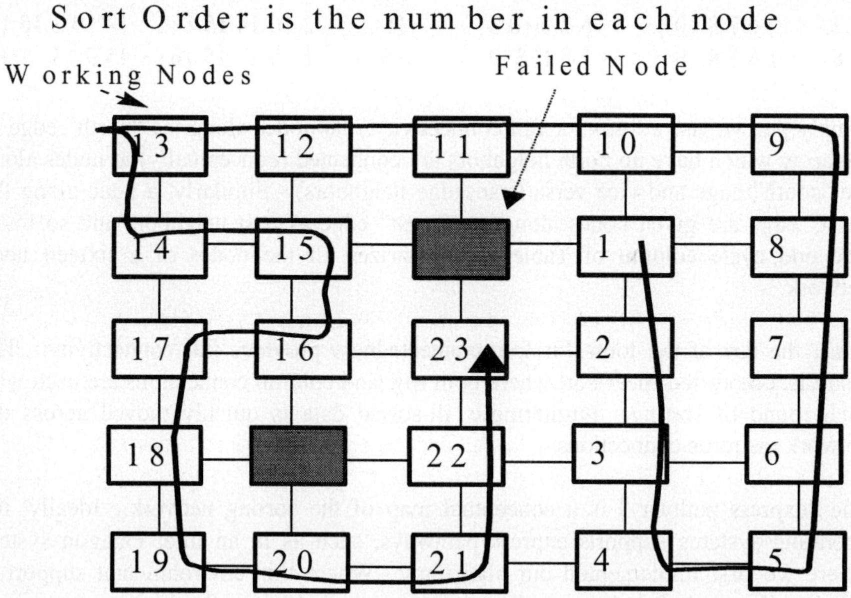

Fig. 2. Sort order is the number of each node. A homogeneous mesh of 25 nodes made irregular by two failed nodes requires a new sort order for efficient performance. The numbers in the boxes (nodes) indicate the node's position in the new sort order. The lowest valued items in the sort will eventually reside on processor 1 and the highest valued items on processor 23. This new order optimizes near neighbor relations.

We assume that a message cannot be sent across a failed node. To provide for online reconfiguration of the node sort order, we have developed an adaptive online sort order optimizer named the Scaleable Adaptive Load-balancing (SAL) Online Optimizer (SOO). Shown in Figure 3 is a result of running SOO. SOO is performed by using a genetic algorithm which minimizes the total path length of the HeteroSort major sort axis, indicated on the figure by the line from one to 23. Note that other possible minimum path sort orders exist. Also note that for some topologies or failure patterns, strict near neighborness is not achievable. For these cases SOO defines the minimum path that includes store-and-forwards or traversals across other nodes. SOO can optimize given any combination of failed nodes and busses.

3 The SOO Genetic Optimization Methodology

In order for the SOO to be most useful in a wide variety of networks, SOO must be able to efficiently adapt the HeteroSort algorithm to irregular topologies. Even if the target network is working perfectly, nodal resources may periodically be available and then become unavailable.

SOO is able to adapt the HeteroSort algorithm architecture to a wide variety of topological changes because SOO is intended to operate online on the target architecture itself. Furthermore, we explicitly designed SOO to be executed in scaleable parallel fashion in order to provide high performance.

The SOO genetic algorithm (GA) is very similar to the GA used for the Traveling Salesperson Problem. In general, GA requires a means for initially creating possible solutions, calculating the fitness of each trial solution, and a means for creating new trial solutions (reproduction). To provide for computing the fitness functions, SOO uses a utility named the Dynamic System Optimizing Simulator (DySOS). DySOS is resident on each node(s) of the network that calculates the new sort order after a network configuration change.

It is assumed that the network HeteroSort is operating on is instrumented so that when the state of the network changes these changes may be input to SOO so that reconfiguration of the HeteroSort may be calculated. The steps employed by SOO in reconfiguring its network are as follows:

1. Topology / Node information is received from the network via instrumentation.

2. DySOS calculates new routing tables from the topology/node information. These tables represent the shortest distance between any two nodes in the network.

3. The new routing tables are loaded into the DySOS fitness function calculator.

4. SOO uses genetic algorithms (GA) to compute the new HeteroSort sort file order.

If the GA is computed in parallel, each node running SOO/DySOS has its own separate population of trial solutions, and computes its own fitness functions independently of the other node(s). After each generation of calculation, each node sends its best solution (most fit) to a command tree which coordinates all SOO nodes. This command node selects the best single solution from all the nodes, and then broadcasts this solution to all the nodes. This best solution is then used to in the next generation of trial solutions. This technique is named Elitism, and enables the many SOO nodes to effectively cooperate in calculating the best solution for the SAL HeteroSort.

A disadvantage of GA is running time. A complex GA solution for a large network can be lengthy, even when performed in parallel. To help obviate this shortcoming, the SOO GA saves its solutions, so that if the changes made to the network are incremental, by using the previous solution space as a start, the next solution can be made more quickly, than if the SOO were started from scratch each optimization period. However, the advantages GA brings to SOO far outweigh the performance shortcomings. These advantages are:

1. Robustness - The SOO and its GA always yields a solution that works for any topology.
2. Arbitrary networks with nonlinear properties can be modeled.
3. The GA is scaleable to high performance.

SOO and DySOS are currently implemented in Visual Basic running on PCs.

4 HeteroSort Operation

Each cycle of HeteroSort has two steps: first - sort, partition and send and second - receive and merge.

The sort-partition-and-send step occurs at system initialization or whenever the second step is complete and new data arrives from a neighbor. In this step the data is sorted in order on each node and partitioned into quarters, copies of which are then sent off to the neighbors in the method described above. The receive-and-merge step occurs when new data arrive. Data are received from neighboring nodes and integrated into the node's current data set. Then the first step begins again.

4.1 Best Trade

The integration process includes the concept of the "best trade." In the "best trade," a copy of a quarter of data to be sent is kept by the sending node. When data arrives from a neighbor it is combined with the copy of the data sent to that neighbor. The new combined list is split in the center and either the top or the bottom half (whichever is appropriate) is kept for the next round of trading. So, for example, node 2 and node 7 may be trading. The appropriate data from node 2 to send to node

8 may be the set (2,5,10). The data from node 8 to node 2 may be (3,4,12). After the data has been sent node 7 combines what it sent with what it received to obtain (2,3,4,5,10,12). Node 7 will keep the top half (5,10,12). Node 2 will have done the same thing keeping (2,3,4).

This is an example of a partial trade in which only some of the data in a given quarter is exchanged. Other alternatives include full trades [as when 2 sends (6,7,8) and 7 sends (1,3,5)] or barren trades [as when 2 sends (1,2,7) and 7 sends (15,16,19)]. A barren best trade occurs when the trading nodes reject all the data their partners have sent.

In all these events the "best trade" policy allows each node to transact the optimal transfer with its trading partner. This process eliminates the unfortunate situation where data is simply swapped in oscillating fashion between neighbors, that is, thrashing

4.2 Transport Optimization

The first locations to complete the sort (i.e., have all the data due them in sorted order) are the extremes or ends of the sort sequence. In our example, nodes 1 and 16 would reach this state first. This occurs because extremely high or extremely low data is passed across "express pathways" (transedge) to these nodes very rapidly.

The end nodes 1 and 16 are receiving extreme data from four other nodes as well as keeping their own extreme data. Thus, they receive the data due them (the extremely high or low data in the system) very quickly. When they have received all their data due them they are finished since data on each node is always maintained in sorted order.

At this point it makes sense to retire these completed nodes from the sort. This has three effects. First, these nodes are now available for other tasks. For example, these nodes now become available for system monitoring tasks like evaluating the performance of the overall system. Second, minimum and maximum data are located early in the process, if this is useful for other processing. Third, the fact that these nodes are retired means that their immediately adjacent neighbors (node 2, the next highest node to 1, and node 15 the next lowest node to 16) now become extremes. These new extremes are now the next nodes to finish, and retire. Thus the system is expected to exhibit a completion pattern from the ends inward. This early completion situation applies to the first and last rows also.

Furthermore, after a number of cycles there is a time when we can switch to a smaller number of trading partners where only the immediately adjacent neighbors (next highest and next lowest) are the trading partners. This is done because most of the data are on their eventual destination nodes or immediate vicinity. Thus, we can maximize linear trading.

4.3 Dual Configuration Cycle Sorting

HeteroSort sorts data by having each node use a combination of directional transfers, both left and right on the linear sort (translinear) and up and down across the coils of the snake network (transcoil). In the torus configuration, the edges of the net itself are bridged (transedge). The use of transcoil and transedge data movement improves the efficiency of the transportation of data when large amounts of the data is highly unsorted. These transcoil and transedge bridges, however, introduce side effects.

HeteroSort uses alternating odd and even cycles to compensate for these side effects. Using inequalities, we can provide insight into the side effects and how the odd and even cycles compensate for the side effects. When HeteroSort using only a single configuration cycle with transcoil and/or transedge data movements and best trading ceases to have productive trades, are the records totally ordered, that is, sorted? In other words, when best trades are no longer productive has the system stalled before full sorting or has full sorting always occurred?

This is important for the implementation of the algorithm, where we depend on local knowledge to govern the system. Barren best trades could indicate that sorting is completed, if we could depend on barren trades indicating only full sorts and not either full or stalled sorts. Experimentally, when the HeteroSort consists of only a single trading cycle (not alternating odd and even cycles) and best trades, the records can stall before they are completely sorted.

4.4 Logical Basis for Dual Cycles

Definition: Let u and v be finite sets of real numbers. The set u is said to be less than or equal to the set v, written $u \leq v$, if every number in u is less than or equal to every number in v. It is easy to see that $\leq$ has the transitive property, i.e. if u, v, w are three finite sets of real numbers satisfying $u \leq v$ and $v \leq w$ then $u \leq w$. To see this let a be any element of u, b be any element of v, and c be any element of w. Since $u \leq v$ and $v \leq w$, we have $a \leq b$ and $b \leq c$. Then, of course, $a \leq c$.

Comment: The symbol $\leq$ is being used in two different ways: to compare sets of numbers and to compare numbers in the usual manner.

Let k be a positive integer. Define Node k to a collection of four finite sets a_k, b_k, c_k, and d_k of real numbers where k is a positive integer. Let N be a positive integer.

Let Node 1, ..., Node N be a collection of N nodes. Refer to the collection of all of the sets a_1, b_1, c_1, d_1, ... a_N, b_N, c_N, d_N as data, and the individual sets as parcels.

Definition: The data on nodes Node 1, ... , Node N is said to be sorted if and only if the following three conditions are satisfied
    1. Each one of the sets a_k, b_k, c_k, d_k, $1 \leq k \leq N$, is in increasing order.
    2. $a_k \leq b_k \leq c_k \leq d_k$ for all k, $1 \leq k \leq N$.
    3. $u \leq v$ whenever
          $u = a_i$, b_i, c_i, or d_i
          $v = a_j$, b_j, c_j, or d_j
          $i < j$, $1 \leq i \leq N$, $1 \leq j \leq N$.

Now assume that there are sixteen nodes Node 1, ..., Node 16 arranged as in Figure 1. Assume further that the parcels will be traded and compared as described in the HeteroSort and Best Trade Sections. A more detailed description of the trading and comparing is given below.

The trading partners' chart for the odd and even cycles is given in Table 1. The construction of the odd cycle column was described in the Approach Section. During the odd cycle, the trading partners are arranged from lowest to highest according to the subscript of the node. The trading partners list for the even cycle is obtained from the odd cycle list by exchanging the order of the highest and lowest partners lower than the node, and exchanging the highest and lowest partners above the node. Three representative examples are given below.

Node 1 order for odd cycle: Node 2, Node 4, Node 8, Node 16. Lower neighbors: none. Higher neighbors: Node 2, Node 4, Node 8, Node 16. To obtain order for the even cycle: exchange Node 2 and Node 16. Order for the even cycle: Node 16, Node 4, Node 8, Node 2.

Node 5 order for odd cycle: Node 4, Node 6, Node 8, Node 12. Lower neighbors: Node 4. Higher neighbors: Node 6, Node 8, Node 12. To obtain order for the even cycle: exchange Node 6 and Node 12. Node 4 is the only lower node. There is no exchange. Order for the even cycle: Node 4, Node 12, Node 8, Node 6.

Node 7 order for odd cycle: Node 2, Node 6, Node 8, Node 10. Lower neighbors: Node 2 and Node 6. Higher neighbors: Node 8 and Node 10. To obtain order for the even cycle: exchange Node 2 and Node 6. Exchange Node 8 and Node 10. Order for the even cycle: Node 6, Node 2, Node 10, Node 8.

The Trading Partners chart (Table 1) can be used to decide which parcels are sent to which neighbors during the odd and even cycles. As an example, consider what happens on Node 5.

During the odd cycle:
>Node 5 sends a_5 to Node 4 and receives c_4 from Node 4.
>Node 5 sends b_5 to Node 6 and receives b_6 from Node 6.
>Node 5 sends c_5 to Node 8 and receives b_8 from Node 8.
>Node 5 sends d_5 to Node 12 and receives a_{12} from Node 12.

During the even cycle:
>Node 5 sends a_5 to Node 4 and receives d_4 from Node 4.
>Node 5 sends b_5 to Node 12 and receives c_{12} from Node 12.
>Node 5 sends c_5 to Node 8 and receives b_8 from Node 8.
>Node 5 sends d_5 to Node 6 and receives a_6 from Node 6.

The best trades on Node k, as described in the section Best Trades, are barren on the odd or even cycle if there are no changes in the parcels a_k, b_k, c_k, and d_k after the best trades have been performed. The question this section seeks to answer is, can barren best trades be used as a criterion to stop the sort? Specifically:
What is the order of the data when the trades are barren during the odd cycle?
What is the order of the data when the trades are barren during the even cycle?

The Barren Best Trades chart (Table 3) is constructed to answer this question. The Barren Best Trades chart lists the relationships which exist among the groups of data on the nodes when the best trades are barren on the odd cycle, and when the best trades are barren on the even cycle. First of all, the data on each node is ordered, i.e. $a_k \le b_k \le c_k \le d_k$. This is true for both cycles.

Barren Best Trade Examples for Odd Cycle, Using the Trading Partners List.

Node 1: Node 1 and Node 2 trade parcels a_1 and a_2. Since the best trade is barren, all of the records in group a_1 must be smaller than those in a_2. This can be expressed with the notation $a_1 \le a_2$. Node 1 and Node 4 trade parcels b_1 and a_4. Since the best trade is barren, $b_1 \le a_4$. Node 1 and Node 8 trade parcels c_1 and a_8. Since the best trade is barren, $c_1 \le a_8$. Node 1 and Node 16 trade parcels d_1 and a_{16}. Since the best trade is barren, $d_1 \le a_{16}$.

Node 2: Node 2 and Node 1 trade parcels a_2 and a_1. Since the best trades are barren, $a_1 \le a_2$. This relationship has already been listed under Node 1 and is not repeated. Redundant relationships are not repeated on the Barren Best Trades Chart. Node 2 and Node 3 trade parcels b_2 and a_3. Since the best trade is barren, $b_2 \le a_3$. Node 2 and Node 7 trade parcels c_2 and a_7. Since the best trade is barren, $c_2 \le a_7$. Node 2 and Node 15 trade parcels d_2 and a_{15}. Since the best trade is barren, $d_2 \le a_{15}$.

Node 5: Node 5 and Node 4 trade parcels a_5 and c_4. This relationship has already been listed under Node 4 and is not repeated. Node 5 and Node 6 trade parcels b_5 and b_6. Since the best trade is barren, $b_5 \le b_6$. Node 5 and Node 8 trade parcels c_5

and b_8. Since the best trade is barren, $c_5 \leq b_8$. Node 5 and Node 12 trade parcels d_5 and a_{12}. Since the best trade is barren, $d_5 \leq a_{12}$.

This process is repeated for the remainder of the sixteen nodes.

Barren Best Trade Examples for Even Cycle, Using the Trading Partners List.
Node 1: Node 1 and Node 16 trade parcels a_1 and d_{16}. Since the best trade is barren, $a_1 \leq d_{16}$. Node 1 and Node 4 trade parcels b_1 and b_4. Since the best trade is barren, $b_1 \leq b_4$. Node 1 and Node 8 trade parcels c_1 and c_8. Since the best trade is barren, $c_1 \leq c_8$. Node 1 and Node 2 trade parcels d_1 and a_2. Since the best trade is barren, $d_1 \leq a_2$.

Node 5: Node 5 and Node 4 trade parcels a_5 and d_4. This relationship has already been listed under Node 4 and is not repeated. Node 5 and Node 12 trade parcels b_5 and c_{12}. Since the best trade is barren, $b_5 \leq c_{12}$. Node 5 and Node 8 trade parcels c_5 and b_8. Since the best trade is barren, $c_5 \leq b_8$. Node 5 and Node 6 trade parcels d_5 and a_6. Since the best trade is barren, $d_5 \leq a_6$.

This process is repeated for the remaining fourteen nodes.

<u>If the best trades are barren on the even cycle, the Wren cycle, the file is sorted.</u>

Proof: If the best trades are all barren on the even cycle, then the Barren Best Trade Chart (Table 3) reveals that the following relationships hold.
$$a_1 \leq b_1 \leq c_1 \leq d_1 \qquad d_1 \leq a_2$$
$$a_2 \leq b_2 \leq c_2 \leq d_2 \qquad d_2 \leq a_3$$
$$a_3 \leq b_3 \leq c_3 \leq d_3 \qquad d_3 \leq a_4 \qquad \ldots$$

$$a_{15} \leq b_{15} \leq c_{15} \leq d_{15} \qquad d_{15} \leq a_{16}$$
$$a_{16} \leq b_{16} \leq c_{16} \leq d_{16}$$

The transitive property of $\leq$ now implies that the data are sorted. Note that not all of the inequalities on the barren Best Trades chart are needed.

Barren best trades on the odd cycle do not guarantee that a sorted order has been achieved. This has been demonstrated experimentally on data sets. Examination of the Barren Best Trades chart for the odd cycle reveals that there are not enough inequalities present to get a comparison of d_1 and a_2 or d_2 and a_3, et cetera through use of transitivity.

The results in this section can be extended to an arbitrary N x N network.

Table 2. Barren Best Trades Chart. This table contains the relations needed to prove that both the odd and even cycles constitute a total order and lead to a full sort. The inequalities immediately following the node numbers hold for both even and odd cycles.

Node Odd Cycle	Even Cycle	Node Odd Cycle	Even Cycle
1 $a_1 \leq b_1 \leq c_1 \leq d_1$		9 $a_9 \leq b_9 \leq c_9 \leq d_9$	
$a_1 \leq a_2$	$a_1 \leq d_{16}$	$b_9 \leq b_{10}$	$b_9 \leq b_{16}$
$b_1 \leq a_4$	$b_1 \leq b_4$	$c_9 \leq b_{12}$	$c_9 \leq b_{12}$
$c_1 \leq a_8$	$c_1 \leq c_8$	$d_9 \leq b_{16}$	$d_9 \leq a_{10}$
$d_1 \leq a_{16}$	$d_1 \leq a_2$		
2 $a_2 \leq b_2 \leq c_2 \leq d_2$		10 $a_{10} \leq b_{10} \leq c_{10} \leq d_{10}$	
$b_2 \leq a_3$	$b_2 \leq c_{15}$	$c_{10} \leq b_{11}$	$c_{10} \leq b_{15}$
$c_2 \leq a_7$	$c_2 \leq b_7$	$d_{10} \leq b_{15}$	$d_{10} \leq a_{11}$
$d_2 \leq a_{15}$	$d_2 \leq a_3$		
3 $a_3 \leq b_3 \leq c_3 \leq d_3$		11 $a_{11} \leq b_{11} \leq c_{11} \leq d_{11}$	
$b_3 \leq a_4$	$b_3 \leq c_{14}$	$c_{11} \leq c_{12}$	$c_{11} \leq b_{14}$
$c_3 \leq a_6$	$c_3 \leq b_6$	$d_{11} \leq b_{14}$	$d_{11} \leq a_{12}$
$d_3 \leq a_{14}$	$d_3 \leq a_4$		
4 $a_4 \leq b_4 \leq c_4 \leq d_4$		12 $a_{12} \leq b_{12} \leq c_{12} \leq d_{12}$	
$c_4 \leq a_5$	$c_4 \leq b_{13}$	$d_{12} \leq b_{13}$	$d_{12} \leq a_{13}$
$d_4 \leq a_{13}$	$d_4 \leq a_5$		
5 $a_5 \leq b_5 \leq c_5 \leq d_5$		13 $a_{13} \leq b_{13} \leq c_{13} \leq d_{13}$	
$b_5 \leq b_6$	$b_5 \leq c_{12}$	$c_{13} \leq c_{14}$	$c_{13} \leq c_{16}$
$c_5 \leq b_8$	$c_5 \leq b_8$	$d_{13} \leq c_{16}$	$d_{13} \leq a_{14}$
$d_5 \leq a_{12}$	$d_5 \leq a_6$		
6 $a_6 \leq b_6 \leq c_6 \leq d_6$		14 $a_{14} \leq b_{14} \leq c_{14} \leq d_{14}$	
$c_6 \leq b_7$	$c_6 \leq b_{11}$	$d_{14} \leq c_{15}$	$d_{14} \leq a_{15}$
$d_6 \leq a_{11}$	$d_6 \leq a_7$		
7 $a_7 \leq b_7 \leq c_7 \leq d_7$		15 $a_{15} \leq b_{15} \leq c_{15} \leq d_{15}$	
$c_7 \leq c_8$	$c_7 \leq b_{10}$	$d_{15} \leq d_{16}$	$d_{15} \leq a_{16}$
$d_7 \leq a_{10}$	$d_7 \leq a_8$		
8 $a_8 \leq b_8 \leq c_8 \leq d_8$		16 $a_{16} \leq b_{16} \leq c_{16} \leq d_{16}$	
$d_8 \leq a_9$	$d_8 \leq a_9$		

6 Results: Graphs

We implemented HeteroSort on two computer environments: a serial simulation and a parallel processor. These implementations worked as predicted. In a serial simulation, HeteroSort sorted 12,288 records on 1,024 nodes in 140 cycles. The data consisted of simulated signal data from various sensors: time, location (two UTM coordinates) and signal frequency. The initial state is where the lowest node has the highest valued records, that is, the data is in the reverse order.

We used a snake pattern of nodes with a torus topology. The sort was simulated on an Intel Pentium Pro, Windows 95 machine using Borland International's Delphi, that is, essentially Pascal with Paradox databases. Figures 4 through 9 show Maple V plots of the results of the initial state, the final 140th cycle, and the 2nd through the Xth cycle's results.

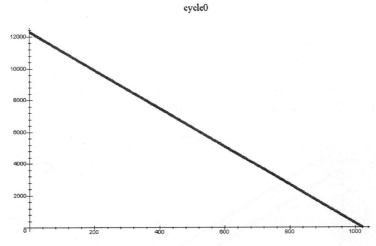

Fig. 3. 12,288 records across 1,024 nodes at the initial 0th cycle of the sort. The x axis is the number of the node and the y axis is the value of data, in this instance, the first UTM location coordinate.

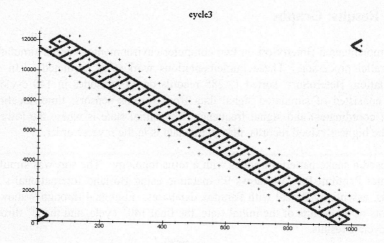

Fig. 4. 12,288 records across 1,024 nodes at the 3^{d} cycle of the sort.

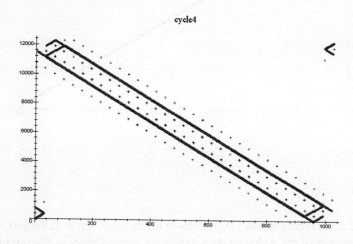

Fig. 5. 12,288 records across 1,024 nodes at the 4th cycle of the sort.

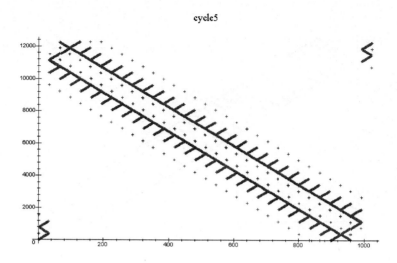

Fig. 6. 12,288 records across 1,024 nodes at the 5th cycle of the sort.

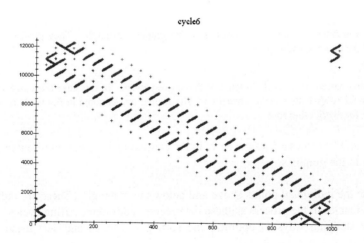

Fig. 7. 12,288 records across 1,024 nodes at the 6th cycle of the sort.

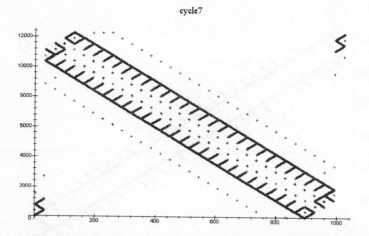

Fig. 8. 12,288 records across 1,024 nodes at the 7th cycle of the sort.

Note the four phase pattern to the borders of the graphs' rectangle. They phases reflect the results of the odd and even cycles.

The hashing pattern reflects the impact of the snake pattern. In this case the 1,024 nodes are ordered in 32 rows with a snake formation. The formation, as noted in Figure 1, reverses the sort order for every other row.

The first and last rows are sorted early in the process and are not represented as other rows are in the graphics.

Also note the individual points above and below the rectangle. These are indications of inefficiencies in the process as these data require additional sorting cycles. We are currently studying ways to optimize the sorting processes and sort these outliers earlier.

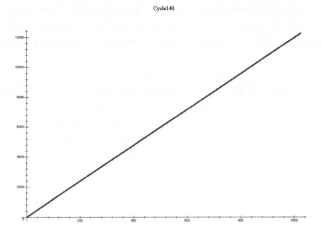

Fig. 9. 12,288 records across 1,024 nodes at the final 140th cycle of the sort.

7 Implementations

The initial serial simulations of nine and sixteen node networks of this algorithm are written in C on a personal computer. These simulations validated our approach and provided insights for scaling up the implementation. They performed as we predicted with the diameter of the network and depth of the connection graphs.

We implemented two parallel implementations in C on an Intel Paragon parallel processor. One implementation has nine nodes, the other sixteen nodes. Again, these implementations validated our predictions. The current simulations are written in Delphi, Borland's Object Pascal. The networks vary from nine to 1,024 nodes.

7.1 Fault Tolerance

The most important aspect of our algorithm is that it does not depend on a regular network topology (as, for example, a traditional shear sort does) because the torus can be superimposed on any physical architecture.

This yields fault tolerance because our system can dynamically reconfigure itself, and easily accommodates "holes" in the connection. All that is required is for HeteroSort to change the partitioning schema in the data, and to stop sending data to a node when it is removed.

Three other aspects of fault tolerance result from this algorithm. First, since only local knowledge is used in the sort, the system is fault tolerant because it does not require global knowledge. Thus, individual nodes continue to operate regardless of the performance (or even existence) of other non-neighbor nodes. Second, since each node keeps a backup copy of the data it sends off to its neighbors, if a node is eliminated during operation of the load balancing and sorting, its neighbors can make up for the loss of data. Third, the natural load balancing of the data during operation of the sorts adds a degree of fault tolerance. With data evenly distributed across nodes, then the loss of a node means the minimal loss of data to the system. The intent is to build minimum weight spanning trees and to use them in improving sort efficiency.

7.2 Future Directions

We currently have the concept of near (adjacent) neighbors and far neighbors (which exist with the implementation of the torus structure). This has implications for implementations on heterogeneous and distributed networks. Specifically, the far neighbors are metaphors for nodes on another processor in a distributed system. So, one component of the sort, partition, and send task could be that the data is partitioned not into equal subsets, but into subsets of a size proportional to the speed of the link to that node.

Furthermore, in heterogeneous architectures, the subset size could also be related to the speed of the corresponding neighbor node. Thus, future enhancements will include an applications kernel that will be resident on each node of the heterogeneous network. Upon startup, each kernel will negotiate with its near neighbor kernels to adjust the size of the exchange list (to be load balanced and sorted). The negotiated value will be a function of each node's own capacity in memory, processing, and its number of neighbors. Upon a fault, the kernels will re-negotiate the exchange files with the surviving near neighbors.

The primary implementations of HeteroSort will be in distributed workstation or parallel processing environments where workstations drop in and out of the shared work pool. In addition, high performance, real-time applications which require clustering, tracking, data fusion and other "assignment intensive" problems with non-uniformly populated data spaces will benefit from our algorithm.

Acknowledgments

This research was partially supported by a grant from the Defense Nuclear Agency 93DNA-3. We gratefully acknowledge this support.

Addressing Real-Time Requirements of Automatic Vehicle Guidance with MMX Technology *

Massimo Bertozzi[1], Alberto Broggi[2], Alessandra Fascioli[1], Stefano Tommesani[1]

[1] Dipartimento di Ingegneria dell'Informazione
Università di Parma, I-43100 Parma, Italy
`bertozzi,fascioli,tommesa@ce.unipr.it`
[2] Dipartimento di Informatica e Sistemistica
Università di Pavia, I-27100 Pavia, Italy
`broggi@dis.unipv.it`

Abstract. This paper describes a study concerning the impact of MMX technology in the field of automatic vehicle guidance. Due to the high speed a vehicle can reach, this application field requires a very precise real-time response.
After a brief description of the ARGO autonomous vehicle, the paper focuses on the requirements of this kind of application: the use of only visual information, the use of low-cost hardware, and the need for real-time processing.
The paper then presents the way these problems have been solved using MMX technology, discusses some optimization techniques that have been successfully employed, and compares the results with the ones of a traditional scalar code.

1 Introduction

During the last years, a large number of research institutes worldwide have been involved in national and international projects about *Automatic Vehicle Guidance*, namely the techniques aimed at automating –entirely or partially– one or more driving tasks. The automation of these tasks carries a large number of benefits, such as: a higher exploitation of the road network, lower fuel and energy consumption, and –of course– improved safety conditions compared to the current scenario.

As a result, a number of prototypes of vehicles embedding systems for Automatic Vehicle Guidance have been designed, implemented, and tested on the road [3]. For being sold on the market, these systems needs to fit three requirements.

1. Since it is expected that an increasing number of vehicles will be equipped with these facilities, the sensors for data acquisition must not interfere with the ones used on other vehicles. This suggests the use of passive sensors, that do not alter the environment; in this case, the *processing of visual information* plays a fundamental role.

---

* The work described in this paper has been carried out under the financial support of the Italian *Ministero dell'Università e della Ricerca Scientifica e Tecnologica* (*MURST*) in the framework of the MOSAICO (Design Methodologies and Tools of High Performance Systems for Distributed Applications) Project and the financial support of the *CNR Progetto Finalizzato Trasporti* under contracts 93.01813.PF74 and 93.04759.ST74.

2. The system must acquire data and produce results in *real-time*, since the vehicle maximum speed is determined by the response time of the system. Therefore a high performance computing engine is needed.
3. On the other hand, the complete system must have a sufficiently *low cost* to allow a widespread use in the car market.

These requirements have been used in developing a prototype system [3] installed on the ARGO autonomous vehicle (figure 1). The ARGO capabilities have been demonstrated to the international scientific community during the first week of June 1998, when ARGO drove itself autonomously for more than 2000 km along the Italian highway network. The percentage of automatic driving was about 95%, up to a maximum speed of about 120 km/h.

The ARGO's system is based on a low-cost computational engine, a standard PC, which processes data coming from a pair of videocameras in real-time.

The needs for both low-cost hardware and real-time image processing as well as the recent enhancements (MMX) of the x86 processors family dedicated to the efficient handling of audio and video data [4, 5] suggest the exploitation of MMX technology and the design of highly optimized algorithm implementations.

This work presents the preliminary results of a study about the use of the MMX technology on intelligent vehicles applications. Particularly, the low level portion of the processing aimed at the detection of obstacles [1] has been implemented exploiting the MMX features. The effectiveness of this implementation makes the whole system very responsive, requiring less than one millisecond to perform the low and middle level portion of obstacle detection.

This paper is organized as follows: the next paragraph introduces a brief description of the Obstacle Detection algorithm used on the ARGO vehicle giving emphasis to the low- and middle-level portions of the processing, since they gain the most benefit from the MMX use, section 3 sketches the MMX programming/optimizing issues, while section 4 presents the MMX implementation of the algorithm. Section 5 ends this work with some concluding remarks.

2 The Obstacle Detection Algorithm

The Obstacle Detection algorithm used on board the ARGO vehicle is based on the *Inverse Perspective Mapping*, IPM [2], geometrical transform widely used in this kind of applications [6–8]: once known all the acquisition intrinsic and extrinsic parameters, the IPM can be used to remap the acquired images in a new domain, called *road domain*, that represents a bird's eye view of the road surface.

The use of the IPM transform on pairs of stereo images (figs. 2.a and 2.b) allows to obtain two patches (*remapped images*) of the road texture (figs. 2.c and 2.d) which can be brought to correspondence exploiting the knowledge of the vision system setup. Any difference in the two remapped images represents a deviation from the starting hypothesis of flat road and thus identifies a potential obstacle, namely anything rising up from the road surface.

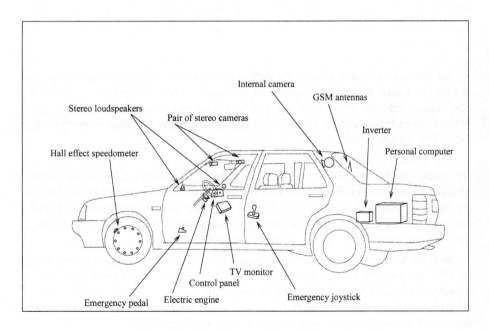

Fig. 1. The ARGO autonomous vehicle and its equipment

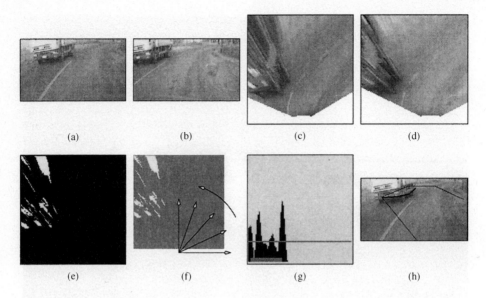

Fig. 2. Obstacle detection: (*a*) left and (*b*) right stereo images, (*c*) and (*d*) the remapped images, (*e*) the difference image, (*f*) the angles of view overlapped with the difference image, (*g*) the polar histogram, and (*h*) the result of obstacle detection using a black marker superimposed on the acquired left image; the thin black line highlights the road region visible from both cameras

An obstacle is detected when the **difference image**, obtained comparing the two remapped images, presents sufficiently large clusters of non-zero pixels having a specific shape. Thus, the low-level portion of the processing consists in the computation of the difference between the remapped images and to the binarization of the resulting image followed by a **noise filtering** to remove isolated pixels.

Due to the different angles of view of the stereo system, the vertical edges of an obstacle generate two triangles in the difference image. Unfortunately, due to their texture, irregular shape, and non-homogeneous brightness, real obstacles produce triangles that are not so clearly defined; nevertheless in the difference image some clusters of pixels having a quasi-triangular shape are anyway recognizable (see fig. 2.e). The obstacle detection process thus is reduced to the localization of pairs of these *triangles*.

A **polar histogram** is used for the detection of triangles: it is computed scanning the difference image with respect to a focus (fig. 2.f) and computing the number of overthreshold pixels for every straight line originating from the focus. The histogram is normalized and low-pass filtered (fig. 2.g). The position of a peak within the histogram determines the angle of view under which the obstacle edge is seen.

A further analysis of the difference image along the directions pointed out by the maxima of the polar histogram allows the determination of obstacle distance [1]. Figure 2.h show the result displayed with a black marker superimposed on a brighter version of the left image.

3 Implementation with MMX Technology

All low-level plus some middle level computations of the discussed algorithm have been implemented exploiting the MMX enhancements of the x86 processors family.

The main characteristics that have been considered in this new implementation are:

- the processing of multiple data according to the SIMD computational model (*data parallelism*), and the use of a specific instruction set dedicated to Image Processing;
- the careful exploitation of the *instruction level parallelism* of the code and the use of the two internal pipelines;
- the reduction of the amount of data in order to fit into the internal processor cache.

The complete obstacle detection process can now be performed is less than 20 ms (more precisely it takes 5 ms), which is the time required to acquire a single image field from a conventional camera. Hence, the whole processing is performed in real-time (at frame rate) during image acquisition, which becomes the system bottleneck.

The next paragraphs introduce the MMX optimizations topic; for a more detailed discussion see also [9].

3.1 MMX Optimization Issues

Major performance gains can be attained with the fastest algorithms and the most efficient data structures, so every code optimization attempt should start at the highest level of abstraction, focusing on more streamlined procedures. Low level code optimization must be used when the algorithms are fully tested, and no more high level optimizations can be applied. Low level optimizations generally produce code that cannot be easily modified, so it is useless to optimize code that could change soon.

We will focus this discussion on a pair of strategies that can easily be implemented and that guarantee the highest performance payoff:

Data alignment: a misaligned access in the data cache or on the bus costs at least three extra clock cycles on the Pentium processor. Given that an aligned memory access might take only 1 cycle to execute (assuming a L1 cache hit), this is a serious performance penalty.
Data alignment is critical when writing memory-bounded MMX code: the execution speed can be boosted by more than 30% by using an efficient memory allocator.

Instruction pairing: the Pentium processor is an advanced superscalar processor: it is built around two general-purpose integer pipelines and a pipelined floating-point unit, allowing the processor to execute two integer instructions simultaneously. The Pentium processor can issue two instructions every clock cycle, one in each pipe. The first logical pipe is referred to as the *U-pipe*, and the second as the *V-pipe*. During decoding of any given instruction, the next two instructions are checked, and, if possible, they are issued such that the first one executes in the U-pipe and the second in the V-pipe. If it is not possible to issue two instructions, then the next instruction is issued to the U-pipe and no instruction is issued to the V-pipe.
Therefore the execution of two MMX instruction in a single clock cycle might double the performance of the code by keeping both pipes busy. Unfortunately,

this is not always possible: i.e. two shift instructions cannot be allocated on the same clock cycle, because the Pentium has only one shifter unit. In addition not all instructions can be routed to both pipes indifferently: table 1 shows the number of functional units and whether they are addressable from the U or the V-pipe.

Operation	Number of Functional Units	Latency	Throughput	Execution Pipes
ALU	2	1	1	U and V
Multiplexer	1	3	1	U or V
Shift/pack/unpack	1	1	1	U or V
Memory access	1	1	1	U only
Integer register access	1	1	1	U only

Table 1. number of functional units and their accessibility from the U and V-pipes

- Two MMX instructions which both use the MMX shifter unit (pack, unpack, and shift instructions) cannot pair since there is only one MMX shifter unit, although they may be issued in either the U-pipe or the V-pipe.
- Two MMX instructions which both use the MMX multiplier unit (pmull, pmulh, pmadd type instructions) cannot pair since there is only one MMX multiplier unit. Multiply operations may be issued in either the U-pipe or the V-pipe but not in both in the same clock cycle.
- MMX instructions which access either memory or the integer register file can be issued in the U-pipe only. Their scheduling in the V-pipe causes a wait and a swap to the U-pipe.

3.2 Tuning MMX code

This section presents the *Loop Interleaving* technique heavily exploited in this project. MMX code usually consists of a short loop iterated many times over an array of data. Often there is no relationship between two iterations of the same loop, so the output of the second iteration does not depend on the output of the first. In fact, they may be executed in any order (even at the same time). The first step is designing the algorithm so that it occupies the lowest number of MMX registers: there are only 8 of them, so they limit the number of iterations that can be executed at the same time. MMX registers that are shared between the iterations (for example, bit masks) should be counted only once. Having calculated the highest number of iterations that can be executed in parallel (N), the loop can be unrolled by N and then instructions can be moved around to maximize the use of both pipelines:
- MMX instructions that access memory are always executed in the U-pipe, so it is better to spread them around and try to exploit the V-pipe with ALU or shift instructions.
- Two shift instructions cannot be issued in the same cycle, but all shifts of the different iterations will end up very near, because they belong to the same stage of the loop. This problem can be solved by introducing a slight delay between

the iterations, so that the group of MMX instructions that are physically near belong to different stages of their respective iterations. This is best suited to long loops, because in short loops the overhead given by the bunch of instructions belonging to the same iteration (which usually do not pair) situated at the top of the loop may offset the benefits of a greater V-pipe exploitation.

4 Implementing the Obstacle Detection Algorithm with MMX

This section describes the MMX implementation of each low and medium level routine for obstacle detection and its performance analysis.

4.1 Computation of the Difference Image

The idea is to compute the absolute difference between the left and the right image, set pixels below a given threshold to zero, then pack each pixel using only one bit per pixel.

The absolute difference can be easily implemented computing two saturated [4] differences and OR'ing them:

```
MOVQ    MM2, MM0 ; make a copy of MM0
PSUBUSB MM0, MM1 ; compute saturated difference one way
PSUBUSB MM1, MM2 ; compute saturated diff. other way
POR     MM0, MM1 ; OR them together
```

The threshold is loaded into register MM7 at startup. Pixels below the threshold are set to zero using a further saturated subtraction, then one bit is extracted from the result, which will be 1 only if its byte is above the threshold.

Since the image is 128×128, the use of one bit per pixel shrink the output data down to 2 kbyte allowing a more efficient use of the Pentium's L1 cache memory. Additional benefits of this choice will be shown in the following. Unfortunately this packing process requires quite a lot of instructions to be implemented, and an additional phase of bit swapping.

To minimize the performance hit, the *packing process* works on 32 bits word on each loop extracting a bit from each saturated byte (namely a single pixel) and then a *bit swapping* process rearranges two 32 bits words into a single double word (64 bits). The Loop Interleaving technique has been applied by running two iterations in parallel, and the resulting dual pipeline exploitation ratio is quite high.

The bit swapping code is simple: it applies a mask to isolate all the bits with index i then copies them into position $7 - i$ (where i is between 0 and 7). The bit swapping can be highly optimized by using only three MMX registers and thanks to a round-robin policy: many flow dependencies can be avoided. Careful instruction scheduling is required due to the limitations on the use of shift instructions: only one shift operation can be issued in each clock cycle.

4.2 Noise Filtering

This routine removes isolated pixels from the difference image. A pixel • is defined as isolated if all o pixels are zero as shown in the following diagram:

	○		○	
○		●		○
	○		○	

The highly packed image format described in the previous paragraph gives more benefits now: this small chunk of data (2 kbyte) fits comfortably into the Pentium MMX's 16 kbyte L1 cache, avoiding slow memory accesses on the system bus. Even more important, the high degree of parallelism allows to process 64 pixels each loop, since each pixel is represented by a bit. This powerful combination, plus some optimized coding, make this process extremely fast.

A1	B1	C1	D1
A2	B2	C2	D2
A3	B3	C3	D3

The target is to evaluate the pixels in the second row (A2 → D2). Applying the rule explained before B2 is isolated where A1, A2, A3 and C1, C2, C3 are all set to zero. A new row is computed by joining all three rows with the OR operator:

A1+A2+A3	B1+B2+B3	C1+C2+C3	D1+D2+D3

this new row is copied twice and shifted once left and once right by one position:

0	A1+A2+A3	B1+B2+B3	C1+C2+C3
A2	B2	C2	D2
B1+B2+B3	C1+C2+C3	D1+D2+D3	0

All that remains to do is joining the first and the third row with the OR operator, saturate the result and then AND'ing the second row with the resulting row. The B2 cell will contain the B2 value only if at least one pixel among A1, A2, A3 and C1, C2, C3 is not zero. This routine can also gain advantage from the Loop Interleaving optimization: the difference image has a width of 128 pixels, which can be packed into two MMX registers. Unrolling the loop by two, a single row in each iteration can be analyzed (actually these two loops are not completely independent, due to the pixels on the edges of the central column that must be exchanged between the loops).

This routine performs very well, with an high instruction pairing ratio and a very good memory access pattern. The difference image's size (2 kbyte) fits perfectly into the L1 cache, therefore most memory accesses do not use the system bus. Given these assumptions, each loop takes 17 CPU cycles to execute, and the whole routine process takes about 2200 cycles. Actually in execution benchmarks it took slightly more than 3000 cycles to execute: this is probably due to L2 cache memory access. According to these numbers, each pixel in the difference image is checked and eventually removed in only 0.19 cycles.

4.3 Computation of the Polar Histogram

This routine builds the polar histogram counting all non-zero pixels on each ray, ranging from 40 to 140 degrees with 1 degree increment and storing the results into an array. It is really hard to design a parallel algorithm that performs this task, and the simplest and fastest way to solve it is using a lookup table. We have not exploited the parallel capabilities of MMX while developing this routine, nonetheless smart coding speeds up the resulting program compared to non-MMX code: the additional 8 MMX registers help reduce register pressure, and MMX shifts are much faster than integer ones. The lookup table is pre-calculated and it is stored with the layout shown in table 2

Number of internal loops (N_1) of first angle	Internal loop block (repeated N_1 times)	...	Number of internal loops (N_n) of last angle	Internal loop block (repeated N_n times)	End of loop (zero internal loops)

Table 2. Layout of the lookup table for polar histogram computation

Address of first memory block	First bit to isolate	Address of second memory block	Second bit to isolate	Address of third memory block	Third bit to isolate

Table 3. Internal loop block for polar histogram computation

	Average loop (clocks)	%	Startup loop (clocks)	%
Difference Image Computation	43177	32.6	83229	36.8
Noise Filtering	3054	2.3	4376	1.9
Polar Histogram Computation	84867	64.2	136645	60.4
Histogram Normalization	463	0.4	970	0.4
Histogram Average Computation	716	0.5	1041	0.5
Total	132277		226261	
Execution time at 200 MHz (ms)	0.661		1.131	

Table 4. execution times of the MMX routines

where the number of internal loops is stored as

Number of internal loops (4 bytes)	Padding (4 bytes set to zero)

and the internal loop block as shown in table 3.

In addition Loop Interleaving technique has been applied again to compute 3 pixels in each loop; an higher level of parallelism may give further performance gains, but we must consider that the number of pixels in each line will be an integer multiple of the unrolling factor, so when using large numbers the overhead due to padding pixels may offset the benefits of longer loops.

Experimental tests showed that both pipes are fully used most of the time. Some optimizations that may not appear obvious are discussed below:

– Memory accesses must be executed in the U-pipe. Each iteration reads 6 quad words: they are three groups of bit mask plus image map accesses. The bit mask quad word defines the block of the image map to read and the position of the desired bit into the block. This data layout wastes some memory, because only 5 bytes are needed to store this piece of information, but for alignment reasons this number has been rounded up to 8 bytes. We have tried to pack all 3 bit positions into a single double word, but the additional computations offset the reduced number of memory accesses. We can reasonably assume that the image map will be stored into the Pentium's L1 cache memory, being small (only 2 kbyte) and recently accessed (it was fully scanned). The position of the bit mask array into the memory hierarchy may vary: on systems with a large L2 cache memory it may still be there, as a remainder of the previous execution, but it could also be in the main memory. The location of the bit mask array is the most important factor when evaluating the performance of

this routine, as the worst case can bring a 70% speed penalty compared to the best case.
- Memory addressing: the address into the image map is extracted from a MMX register and put into an integer register to access memory. This process can cause additional penalties: at least one CPU cycle must pass between the writing of the address into the integer register and the reading from that address. This problem was solved by interleaving another memory access that uses a different integer register: by using two kinds of data (bit masks and image map blocks) we can interleave their memory accesses and avoid the penalty explained above.
- Integer register access: all instructions that access integer registers must be issued to the U-pipe. Two penalties are due to this limitation.
- Loop delay: the three iterations are delayed, so that their shift instructions (which cannot be paired) are located in different zones of the code. There are no penalties due to shift instructions or register dependencies.

In execution benchmarks this routine took slightly more than 100,000 cycles to run when the bit mask array was located into the L2 cache, and slightly less than 170,000 cycles when it was in the main memory. This difference is striking, and shows that memory optimizations with modern CPU's are much more important than they used to be, as they can offer a big payoff, even better than the ones that can be obtained with coding techniques such as instruction scheduling.

Since the performance bottleneck of the previous algorithm is reading the bit mask table from slow memory, a simple way to improve the running speed is shrinking the bit mask table: two adjacent pixels that belong to the same angle can be encoded in a single table entry, using only one shift index. The bit mask table is then split in two parts: the first one is a list of pixel duos, the latter a list of single pixels. In real world benchmarks, this algorithm shrinks the bit mask table to 68808 bytes down from 100466 bytes, and reduces the execution time by 19%, down to 84867 cycles.

Each value in the histogram is the number of non-zero pixels that belong to that angle. However, each angle spans a different area (see figure 2), so if we want the ratio between non-zero pixels and all pixels contained in that area, we will have to scale each value of the resulting histogram. (the scaling factors are computed at startup). In this case the running time is so short (463 cycles on average) that specific optimizations are not worth applying.

Each value contained in the histogram is averaged to smooth spikes. The following formula was implemented (the division by 2 or 4 can be coded as a shift right instruction):

$$\text{Average}(V_i) = \frac{\frac{V_{i-2} + V_{i+2}}{2} + V_{i-1} + V_i + V_{i+1}}{4} \qquad (1)$$

Also this routine's running time is so short (slightly more than 700 cycles) that we cannot see the point in further optimizations.

5 Conclusions

This paper presented a porting of the obstacle detection algorithm used on board of the ARGO autonomous vehicle toward the MMX technology.

The discussed implementation fully satisfies real-time requirements, being able to complete all computations in less than one millisecond.

The major performance hit is given by an empty L2 cache, as table 4 shows: on the startup loop all data lies in the main memory, while on the average loop it is split between L1 and L2 cache memories. The startup condition is more likely in real world execution, because other routines used on the ARGO vehicle (i.e. lane detection and perspective removal) use a lot of memory therefore flushing the data out of cache memories.

No automatic code parallelization or optimization was made exploiting the MMX enhancements: the complete code was written in assembly language by hand and the optimization was done after a careful analysis of the specific instructions sequence. This hard and time-consuming task has been rewarded with an enormous speed-up of the processing: the new MMX-based code runs about 26 times faster than its scalar C version.

References

1. M. Bertozzi and A. Broggi. GOLD: a Parallel Real-Time Stereo Vision System for Generic Obstacle and Lane Detection. *IEEE Trans. on Image Processing*, 7(1):62–81, Jan. 1998.
2. M. Bertozzi, A. Broggi, and A. Fascioli. Stereo Inverse Perspective Mapping: Theory and Applications. *Image and Vision Computing Journal*, 1998(16):585–590, 1998.
3. A. Broggi, M. Bertozzi, A. Fascioli, and G. Conte. *Automatic Vehicle Guidance: the Experience of the ARGO Vehicle*. World Scientific, 1999.
4. Intel Corporation. *Intel Architecture MMX Technology Developers' Manual*. Intel Corporation, 1997. Available at http://www.intel.com.
5. Intel Corporation. *MMX Technology Programmers Reference Manual*. Intel Corporation, 1997. Available at http://www.intel.com.
6. H. A. Mallot, H. H. Bülthoff, J. J. Little, and S. Bohrer. Inverse perspective mapping simplifies optical flow computation and obstacle detection. *Biological Cybernetics*, 64:177–185, 1991.
7. L. Matthies. Stereo vision for planetary rovers: stochastic modeling to near real-time implementation. *Intl. Journal of Computer Vision*, 8:71–91, 1992.
8. M. Ohzora, T. Ozaki, S. Sasaki, M. Yoshida, and Y. Hiratsuka. Video-rate image processing system for an autonomous personal vehicle system. In *Procs. IAPR Workshop on Machine Vision and Applications*, pages 389–392, Tokyo, 1990.
9. S. Tommesani. The Pixel64 Module. Technical report, Dipartimento di Ingegneria dell'Informazione, Università di Parma, May 1998. Available at http://www.ce.unipr.it/~tommesa.

Condition-Based Maintenance: Algorithms and Applications for Embedded High Performance Computing

[1]Bonnie Holte Bennett, [2] Knowledge Partners of Minnesota, Inc., [3] George Hadden, Ph.D.

[1]University of St. Thomas, 2115 Summit Avenue OSS 301, St. Paul, MN 55105,
651-962-5509
bhbennett@stthomas.edu
[2] 9 Salem Lane, Suite 100, St. Paul, MN 55118-4700, 612-720-4960
Bbennett@kpmi.com
[3] Honeywell Technology Center, 3660 Technology Drive, Minneapolis, MN 55418,
612-951-7769
hadden@htc.honeywell.com

Abstract. Condition based maintenance (CBM) seeks to generate a design for a new ship wide CMB system that performs diagnoses and failure prediction on Navy shipboard machinery. Eventually, a variety of shipboard systems will be instrumented with embedded high-performance processors to monitor equipment performance, diagnosis failures, and predict anticipated failures. To illustrate the general principles of our design, we are currently building a research prototype of the CBM system that applies to the shipboard chilled water system.

1 Introduction

System health management is a concerted effort with applicability to airline maintenance operations and support, refineries and other chemical processing facilities, commercial aircraft, and sensor interface modules for the space shuttle operational maintenance system. The current effort is a three-year demonstration of concept scheduled to conclude in late 1999.

A team is accomplishing this work with members from industry, government labs, and academia. Our partners include The Georgia Institute of Technology, and PredictDLI. This paper will concern work on knowledge fusion performed by Knowledge Partners of Minnesota, Inc. (KPMI), rule-based system analysis by Graduate Programs in Software (GPS) at the University of St. Thomas, St. Paul, Minnesota, and system integration by Honeywell technology Center (HTC).

The CBM program is extremely ambitious. Eventual implementations will instrument thousands (perhaps tens of thousands) of replaceable parts on shipboard systems. Fleet-wide, thousands of embedded processors will collect millions of data points per second of data from tens of thousands of locations each. These data will be processed by local software, and the results will be passed on to local "data concentrators" (DCs) that will concentrate reports from a number of different sensors via a variety of specialized analyzers, for example DLI's Expert Alert expert system for vibration analysis. Results from hundreds of DCs per ship will be correlated at a system level in another processor, the Prognostic/Diagnostic Monitoring Engine (PDME). The result is evident: significant data loads, multiple embedded processors, and critical high performance computing needs.

We are currently designing and refining a Machinery Prognostic and Diagnostic System (MPROS) system architecture with open interfaces to provide machinery condition and raw sensor data to other shipboard systems such as ICAS (Integrated Condition Assessment System, a product from IDAX). We have implemented an Object-Oriented Ship Model (OOSM) and knowledge fusion algorithms to integrate the data at PDME level.

The knowledge fusion algorithms currently use class heuristics and Dempster/Schafer probability reasoning to scrutinize incoming reports for correlation with others. We expect to implement a Bayesian Network probability theory when sufficient data exists for a priori dependence calculations. Our current prognostic extrapolation uses worst-case scenarios for failure prediction, but next generation software will use more complex failure analysis using historical data, and learning to refine its estimates over time.

The high-performance computing issues related to this work are challenging, as described above, the data rates are extremely high, data overload is rampant, and we are working to articulate the specific challenges here.

1.1 MPROS

The ONR CBM system, called MPROS (for Machinery Prognostics and Diagnostics System), is a distributed, open, extensible architecture for housing multiple on-line diagnostic and prognostic algorithms. Additionally, our prototype contains four sets of algorithms aimed specifically at centrifugal chillers. These are:

1. DLI's (a company in Bainbridge Island, Washington that has a Navy contract for CBM on shipboard machinery) vibration based expert system adapted to run in a continuous mode.

2. State Based Feature Recognition (SBFR), an HTC-developed embeddable technique that facilitates recognition of time-correlated events in multiple data streams. Originally developed for Space Station Freedom, this technique has been used in a number of NASA related programs.

3. Wavelet-Neural Network (WNN) diagnostics and prognostics developed by Professor George Vachtsevanos at Georgia Tech. This technique is, like DLI's, aimed at vibration data, however, unlike DLI's, their algorithm will excel in drawing conclusions from transitory phenomena rather than steady state data.

4   Fuzzy Logic diagnostics and prognostics also developed by Georgia Tech which draws diagnostic and prognostic conclusions from non-vibrational data.

Since these algorithms (and others we may add later) have overlapping areas of expertise, they may sometimes disagree about what is ailing the machine. They may also reinforce each other by reaching the same conclusions from similar data. In these cases, another subsystem, called *Knowledge Fusion*, is invoked to make some sense of these conclusions. We use a technique called *Dempster-Shafer Rules of Evidence* to combine conclusions reached by the various algorithms. It can be extended to handle any number of inputs.

MPROS is distributed in the following sense: Devices called *Data Concentrators* (DC) are placed near the ship's machinery. Each of these is a computer in its own right and has the major responsibility for diagnostics and prognostics. The algorithms described above run on the DC. Conclusions reached by these algorithms are then sent over the ship's network to a centrally located machine containing the other part of our system – the *Prognostic/Diagnostic/Monitoring Engine* (PDME). KF is located at the PDME. Also on the PDME is the *Object Oriented Ships Model* (OOSM). The OOSM represents parts of the ship (e.g. compressor, chiller, pump, etc.) and a number of relationships among them (part-of, proximity, kind-of, etc.). It also serves as a repository of diagnostic conclusions – both those of the individual algorithms and those reached by KF. Communication among the DC's and the PDME is done using DCOM (Distributed Common Object Modules), a standard developed by Microsoft.

The MPROS program is designed to comprise two phases. The first phase, which we have just finished, will see MPROS installed and running in the lab. The second phase will extend MPROS's capability somewhat and install it on a ship. The current plan is to install it on the Navy Hospital Ship, Mercy, in San Diego.

1.2 Mission

Our central mission in this project is to design a *shipwide* CBM system to predict remaining life of all shipboard mechanical equipment. However, implementation of such a system would be much too ambitious for the current project. In light of this, we have chosen to illustrate the general principles of our design by implementing it in a specific way on the Centrifugal Chilled Water System. The result of this philosophy is that occasionally we will choose a more general way of solving a problem over a "centrifugal chiller-specific" solution.

2 Why Centrifugals?

There were two main reasons for our choice of centrifugal chillers: System complexity and commercial applicability.

These A/C systems combine several rotating machinery equipment types (i.e. induction motors, gear transmissions, pumps, and centrifugal compressors) with a fluid power cycle to form a complex system with several different parameters to monitor. This dictates the requirement for a correspondingly complex and versatile monitoring system. Dynamic vibration signals must be acquired using high sampling rates and complex spectrum and waveform analysis. Slower changing parameters such as temperatures and pressures must also be monitored, but at a lower frequency and can be treated as scalars rather than vectors as with vibration spectra.

All of these monitored parameters and analysis techniques are combined using a versatile diagnostic system. The final product has the inherent capability of diagnosing not just the whole A/C system, but each of its parts as well, making it a potentially very useful tool for monitoring any pump, motor, gearset, or centrifugal compressor in the fleet.

Secondly, the selection of A/C system as the subject will provide a high probability of commercial applicability of the resultant monitoring system. There are a great deal of facilities industrial, military, commercial and institutional that use large centrifugal chiller based A/C systems throughout the US and the world.

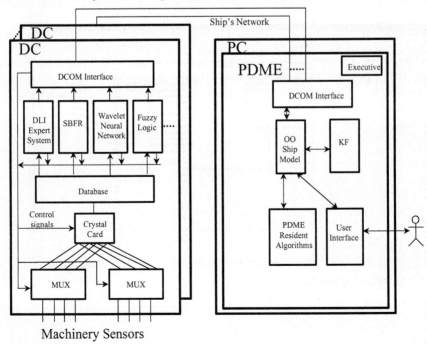

Fig. 1. The MPROS System

3 Software

Figure 1 shows a diagram of the MPROS system. Here we describe the various parts.

3.1 PDME

The Prognostic/Diagnostic Monitoring Engine (PDME) is the logical center of the MPROS system. Diagnostic and prognostic conclusions are collected from DC-resident algorithms as well as PDME-resident algorithms. Fusion of conflicting and reinforcing source conclusions is performed to form a prioritized list for the use of maintenance personnel.

The PDME is implemented on a Windows NT platform as a set of communicating servers built using Microsoft's Component Object Model (COM) libraries and services. Choosing COM as the interface design technique has allowed us to build some components in C++ and others in Visual Basic, with an expected improvement in development productivity as the outcome. Some components were prototyped using Microsoft Excel and we continue to use Excel worksheets and macros to drive some testing of the system. Communications between DC components and PDME components depend on Distributed COM (DCOM) services built into Microsoft's operating systems.

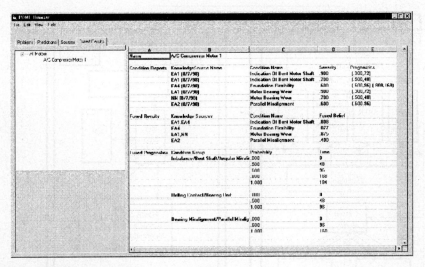

Fig. 2. MPROS User Interface

3.2 User interface

As shown in **Fig. 2.**, an interface to the MPROS conclusions has been built. The sample screen shown indicates that for machine A/C Compressor Motor 1, six condition reports from four different knowledge sources (expert systems) have been received, some conflicting and some reinforcing.

After these reports are processed by the Knowledge Fusion component, the predictions of failure for each machine condition group are shown at the bottom of the screen.

This display is updated as new reports arrive at the PDME and are accumulated in the OOSM.

3.3 Status

The PDME has been implemented and running since August, 1998. This work has included a number of specification, analysis and design tasks. An RF survey was completed on the USNS Comfort, a Navy Hospital ship in San Diego. A failure effects mode analysis (FMEA) was completed and used to select 12 candidate failure modes. This work is being integrated with industry standards such as Machinery Management Open Systems Alliance (MIMOSA), and related work at Penn State and BBN. Significant completed milestones include:

- Demonstrated an object-oriented Level5 Object ™ simulation of Carrier Chiller startup.
- Instrumented a lab with an optical RIMS (Resonant Integrated Micro-Structure – a type of MEMS) for temperature, and are investigating it for accelerometer.
- Installed a state of the art DLI Fulltime ™ system on CVN-72 (Lincoln), the Comfort, and at an HTC lab.
- Completed phase 1 design and implementation.
- Integrated a vibration rule-base, and state based feature recognition for diagnosis.
- Implemented knowlege fusion with prognostic reporting for "time to failure" estimates.
- Designed for integration of Wavelet Neural Net and Fuzzy Logic analysis from Georgia Tech.

3.4 Going Forward

Preparation plans for shipboard deployment include continued testing and monitoring, as the installed system will be disconnected from our labs for months at a time.

4 Object Oriented Ship Model

4.1 Objectives

The Object Oriented Ship Model (OOSM) is a persistent repository for machinery state information used for communication between the various prognostic and diagnostic software modules. In addition to diagnostic and prognostic reports, the OOSM also models the physical, mechanical and energy characteristics of the machinery being monitored. Exposing an integrated programming interface to all of this information eases the development of new knowledge-based algorithms for diagnostics and prognostics.

4.2 Object Model

An object-oriented design was chosen to provide a consistent interface to all the developers using the OOSM to store, retrieve and monitor changes to information of interest to MPROS components.

Entities in the OOSM are modeled as objects with properties and relationships to other entities. Some of the OOSM objects represent physical entities such as sensors, motors, compressors, decks, and ships while other OOSM objects represent more abstract items such as a failure prediction report or a knowledge source. Some common properties include name, manufacturer, energy usage, capacity, and location. Common relationships include "part-of", whole and refers-to.

As part of easing the use of an object-oriented model for developers, the persistence is entirely managed in the background. In fact, save for retrieving the first object in a connected graph of objects, no understanding of the persistence mechanism is necessary.

4.3 Contents

While the preceding section described the form of objects in the OOSM, this section briefly describes the contents. We have modeled a portion of the information about the system under observation in the OOSM. This includes information about the motors, compressors and evaporators in the chillers we are working with. We have also modeled relationships between the failure predictions and the relevant equipment to provide for a future expert system to analyze interactions between equipment subsystems.

As we expand the scope of equipment this system is expected to monitor, the contents of the OOSM will expand to match.

4.4 Applications Programming Interface (API)

As mentioned in preceding sections, a consistent API for developers has been defined. Briefly, it consists of functions to retrieve specific object instances, to view the values of properties, to update their properties and relationships, and to create and delete instances. This API, based upon COM, has been used from C++, Visual Basic, and Java programs to work with the OOSM.

4.5 Events

An event model has been implemented for the OOSM, which allows client programs to be notified of changes to property or relationship values without the need to poll. This facility is provided by OLE Automation events, making it usable from C++, Visual Basic and Java. The Knowledge Fusion component uses this to automatically process failure prediction reports as they are delivered to the OOSM. The PDME browser also uses events to update it's display for users.

4.6 Database Mapping

Persistence of object state in the OOSM is implemented using a relational database. Object types are mapped to tables and properties and relationships are mapped to columns and helper tables. This mapping approach has helped in system debugging and has proven to be very reliable in operation.

4.7 Implementation

The implementation of the OOSM is C++ using Microsoft's Active Template Library (ATL) and Active Data Objects (ADO) library. This approach was chosen because of the control over object lifetime, performance, and reliability.

4.8 Status

The OOSM has been implemented and running since August, 1998.

4.9 Going Forward

As we prepare for shipboard deployment, we plan to exercise scenarios that will likely be encountered. For example, power supply and communications are stable in our labs but may not be the same on board the ships. Simulating the range of problems that may arise will let us improve robustness to the point of long-term unattended operation.

5 Knowledge Fusion

5.1 Objectives

This section describes the objectives for knowledge fusion. It will start by describing the challenges presented by the data received as input for knowledge fusion. Then, it will continue to describe the functions required to deal with these inputs.

Knowledge fusion is the coordination of individual data reports from a variety of sensors. It is higher level than pure "data fusion" which generally seeks to correlate common-platform data. Knowledge fusion, for example, seeks to integrate reports from acoustic, vibration, oil analysis, and other sources, and eventually to incorporate trend data, histories, and other components necessary for true prognostics.

The knowledge fusion components must be able to accommodate inputs which are incomplete, time-disordered, fragmentary, and which have gaps, inconsistencies, and contradictions. In addition, knowledge fusion components will have to be able to collate, compare, integrate, and interpret the data from a variety of sources.

In order to this, it must provide inference control that accommodates a variety of input data, and fusion algorithms with the ability to deal with disparate inputs.

Knowledge fusion follows the general format:

1   New reports arriving to the PDME are posted in the OOSM.

2   New reports posted in the OOSM generate "new data" messages to the knowledge fusion components.

3   The knowledge fusion components access the newly arrived data from the OOSM. They perform knowledge fusion of diagnostic reports and knowledge fusion of prognostic reports.

4   Conclusions from the knowledge fusion components are posted to the OOSM and presented in user displays in the graphical user interface.

5.2 Implementation

To date, two levels of knowledge fusion have been implemented one for diagnostics and one for prognostics.

5.3 Knowledge Fusion for Diagnostics

The current approach for implementing knowledge fusion for diagnostics uses Dempster-Shafer belief maintenance for correlating incoming reports. This is facilitated by use of a heuristic that groups similar failures into logical groups.

Dempster-Shafer theory is a calculus for qualifying beliefs using numerical expressions. Specifically, given two sets of beliefs. For example, given a belief of 40% that A will occur and another belief of 75% that B or C will occur, it will concluded that A is 14% likely, "B or C" is 64% likely and there is 22% of belief assigned to unknown possibilities. This maintenance of the likelihood of unknown possibilities is both a differentiator and a strength of Dempster-Shafer theory. It was chosen over other approaches like Bayes Nets because they require prior estimates of the conditional probability relating two failures. The data is not yet available for the CBM domain.

The system was augmented by a heuristic grouping similar failures into logical groups. This facilitates processing and streamlines operation. The reason for this is that Dempster-Shafer analysis looks at each failure in light of every other failure possible, and this is required to produce the likelihood of unknown possibilities. In the CBM case, this is inadequate because this would assumes mutual exclusivity of failures, in other words, it assumes that any one failure precludes any other failures. However this is not the case in CBM, there can, in fact, be several failures at one time, and two or more of them might be independent of one another. Thus, we developed the concept of logical groups of failures. Failures, which are all part of the same logical groups, are related to each other (for example, one group might be electrical failures, another lubricant failures, etc.).

Moreover, failures within a group might be mistaken for one another, so they are logically related and should share probabilities when they are both under consideration. Note that this does not preclude multiple failures within a group to all be suspect concurrently, it simply ensures that they are tracked correctly and weighted correctly given similar failures which might be related to them.

5.4 Knowledge Fusion for Prognostics

Prognostics are defined in this system as time point, probability pairs, and lists of these pairs. So for example, a prognostic of (3 months, .1) would indicate that the system has a 10% likelihood of failure within 3 months time from now. Subsequently, a prognostic list of ((2 weeks, .1) (1 month, .5) (2 months, .9)) would indicate a likelihood of failure of 10% within 2 weeks, 50% at 1 month and 90% in 2 months. Knowledge fusion for prognostics is the combination of these lists of time and failure likelihoods. Our approach in phase one has been to combine the lists taking the most conservative estimate at any given time period, and interpolating a smooth curve from point to point.

So, for example, if we have a prognostic for a given component that indicates it will perform well for 3 months, then experience some trouble making it as likely to break as not by 4 months, and almost surely break by 5 months: ((3 months, .01) (4 months, .5) (5 months, .99)) and we need to combine this with another report showing that the same component will experience some small trouble at 4 1/2 months ((4.5 months, .12)), then we will ignore the second report, and stick with the first which is more conservative.

If, however, the second report indicates a much higher likelihood of failure ((4.5 months, .95)) then this report would dominate, and the extrapolation of the curve beyond this point would indicate an even earlier demise of the component that the original which would be some time after 5 months.

5.5 Interfaces Provided

The general incoming report format may contain the following data fields (not all reports need use all fields):

- DC ID–Identifyer of the data concentrator source of this report.
- KS ID–Identifyer of the knowledge source generating this report.
- Timestamp–Time that the report was issued.
- Machine ID–Identifyer of the machine for which this report is generated.
- Machine condition–Type of failure (e.g., motor imbalance, motor rotor bar problem, pump bearing housing looseness, etc.).
- Severity–DLI issues a severity rating of slight, moderate, serious, or extreme.
- Belief–Some of DLI's reports have a belief value less than 1.0 (on a scale of 0 to 1).
- Explanation–There may be an explanation about the diagnosis.
- Recommendation–This may indicate what action should be taken.
- Prognostic vector–This vector of time point, probability pairs indicate projected likelihood of failure in the given time period.

5.6 Knowledge Fusion Outputs

Diagnostic knowledge fusion generates a new fused belief whenever a diagnostic report arrives for a suspect component. This updates the belief for that suspect component and for every other failure in the logical group for that component. It also updates the belief of "unknown" failure for that logical group for that component.

Prognostic knowledge fusion generates a new prognostic vector for each suspect component whenever a new prognostic report arrives.

5.7 Resident Algorithms

As the reader can see from Figure 1, the PDME has the capability to host prognostic and diagnostic algorithms. Some reasons for placing the algorithms in the PDME rather than the DC include: the algorithm requires data from widely separate parts of the ship, the algorithm can reason from PDME resident components (a model-based diagnostic and prognostic system, for instance, might use only the OOSM), and so on.

Currently, our Phase 1 system does not place any diagnostic/prognostic algorithms in the PDME – all of them run in the data concentrators.

5.8 Data Concentrator

The data concentrator is a open architecture ODBC compliant relational database designed to store all of the instrumentation configuration information, machinery configuration information, test schedules, resultant measurements, diagnostic results, and condition reports. The database design is a commercially available database.

The DC software is coordinated by an event scheduler. It coordinates standard vibration test and including data acquisition and communication of the results. In similar fashion, the scheduler conducts wavelet and neural network testing and analysis, and state based feature recognition routines to collect and analyze process variables. The data is processed and then sent to an expert system DLL which applies stored rules for each equipment type and derives the diagnoses. The DLL then passes the results back to the DC database. Each of the components extract information from and store data in the DC database which is configured as a database server and can be accessed by client PC's on the network. In this way, the PDME or any other client can command the scheduler to conduct another test and analysis routine.

6 Prognostic/Diagnostic Algorithms

6.1 DLI Expert System

Currently, all standard machinery vibration FFT analysis and associated diagnostics in the Data Concentrator are handled by the DLI expert system. Over the past ten years, this system has been installed in hundreds of industrial and manufacturing facilities in addition to a variety of Navy ships. In one study, it was found that the system exceeds 95% agreement with human expert analysts for machinery aboard the Nimitz class ships. The DLI expert system provides an intermediate level of sensor and knowledge fusion. The frame based rules application method employed allows the spectral vibration features to be analyzed in conjunction with process parameters such as load or bearing temperatures to arrive at a more accurate and knowledgeable machinery diagnosis.

For example since some compressors vibrate more at certain frequencies when unloaded, the DLI expert system rule for bearing looseness can be sensitized to available load indicators (such as pre-rotation vane position) in order to ensure that a false positive bearing looseness call is not made when the compressor enters a low load period of operation. This kind of knowledge and sensor fusion is found again at the PDME that ascertains the relative believability of the various diagnostic systems and derives a reasonable conclusion. It is however advantageous for the vibration expert system to utilize all available known system responses when analyzing the vibration patterns because it minimizes the necessary PDME decisions and improves overall system accuracy.

An elementary level of machinery prognostics has always been provided by the DLI expert system which since its inception, has provided a numerical severity score along with the fault diagnosis. This numerical score is interpreted through empirical methods which map it into four gradient categories.

These categories are Slight, Moderate, Serious and Extreme and correspond to expected lengths of time to failure described loosely as: no foreseeable failure, failure in months, weeks, and days of operation. This approach to prognostics was developed and improved upon during the last 10 years and was refined further for the Data Concentrator and the PDME through the introduction of believability factors for each of the diagnoses. These believability factors are based on DLI's statistical database that demonstrates the individual accuracy of each diagnosis by tracking how often each was reversed or modified by a human analyst prior to report approval.

6.2 Wavelet Neural Networks

The Wavelet Neural Network (WNN) belongs to a new class of neural networks with such unique capabilities as multi-resolution and localization in addressing classification problems. For fault diagnosis, the WNN serves as a classifier so as to classify the occurring faults. CMB uses a fault diagnostic system based on the WNN.

Critical process variables are monitored via appropriate sensors. Features extracted from input data are organized into a feature vector, which is fed into the WNN. Then, the WNN carries out the fault diagnosis task. In most cases, the direct output of the WNN must be decoded in order to produce a feasible format for display or action. Results of the WNN can be used to perform fault diagnosis via classification using information sucha as the peak of the signal amplitude, standard deviation, cepstrum, DCT coefficients, wavelt maps, temperature, humidity, speed, and mass

6.3 State Based Feature Recognition

SBFR is a technique for the hierarchical recognition of temporally correlated features in multi-channel input. It consists of a set of several enhanced finite-state machines operating in parallel. Each state machine can transition based on sensor input, its own state, the state of another state machine, measured elapsed time, or any logical combination of these. This implies that systems based on SBFR can be built with a layered architecture, so that it is possible use them to draw complex conclusions, such as prognostic or diagnostic decisions.

Our implementation of the SBFR system requires very little memory (100 state machines operating in parallel and their interpreter can fit in less than 32K bytes) and can cycle with a period of less than 4 milliseconds. It is thus ideal for embedding into the DC. We have successfully applied SBFR-based diagnostic and prognostic modules to several problems and platforms, including:

- valve degradation and failure prediction in the Space Shuttle's Orbital Maneuvering System,
- imminent seize-up in Electro-Mechanical Actuators through electrical current signature analysis and other parameters, and
- failure prediction in several subsystems (including Control Moment Gyro bearing failure) in the Space Station Freedom's Attitude Determination and Control System.

SBFR embedded in the DC will take as input the raw sensor data and the output of other algorithms (e.g., DLI's vibration analysis and FFTs) and perform trending analysis, feature extraction, and some diagnostics and prognostics on this data. Local knowledge gained by these techniques will be delivered to ICAS and the subsystems contained within ICAS for further analysis and crew alerting. Some level of local alerting (e.g., indicator lights) will also be made available. Under control of the System Executive running in the PDME under ICAS, new finite-state machines may be downloaded into the smart sensor. This will allow the behavior of the sensor to adapt to its data in appropriate ways. It will have, for instance, the capability to take a "closer look" at a problem that has been discovered.

To give a sense for how SBFR works, here is an example. The two state machine system shown in Figure 3 was used to predict a seize-up failure mode in an electro-mechanical actuator (EMA). EMAs are essentially large solenoids meant to replace hydraulic actuators for the steering of rocket engines. Prediction of this fault was done by recognizing stiction in the mechanism. Stiction is recognized by the two state machines in the following way: Machine 0 recognizes spikes in the drive motor current. Machine 1 counts the spikes that are *not* associated with a commanded position change (CPOS). When the count is greater than 4, a stiction condition is flagged, and higher level software (e.g., the PDME) can conclude that a seize-up failure is imminent.

We now consider these machines in more detail. Starting with Machine 1, notice that it has two states labeled **Wait** and **Stiction**. Each state has transitions into and out of it with labels C and A, standing for condition and action, respectively. The condition on one of the transitions out of the **Wait** state is "Status:0 - 0 & CPOS unchanged".

This means that the transition will be taken if both of the following are true: First, the status of Machine 0 (the other one) must be nonzero. This means that a spike has occurred. Second, no command has been issued to change the position of the EMA. If both of these are true, the action associated with this condition is executed. In this transition, the action is 1) to set the status register of Machine 0 back to 0 so that it can continue looking for spikes in parallel with the actions of any other state machines and 2) to increment local variable 1 (Local:1), which represents the number of spikes noticed by Machine 1.

Current SPIKE Machine (Machine 0)

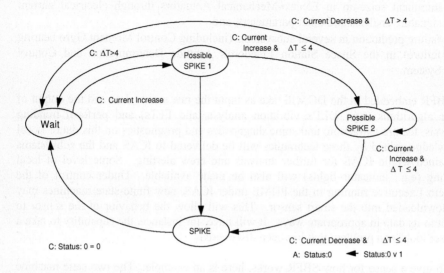

EMA Stiction Machine (Machine 1)

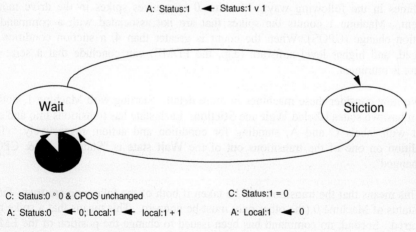

Figure 3. Two state machines

Notice two things: first, a state machine can transition states based on the state of another machine. Second, each machine can have any number of local variables, one of which, the "status", is distinguished by being readable and writeable by any of the state machines.

Another transition out of the **Wait** state occurs if local variable 1 is greater than 4 (i.e., more than four spikes have been noticed). If this transition is taken, the status register of Machine 1 (the current machine) is set to 1. (Actually, only the lowest bit is set to one, since we would like to save the option of using other bits for some other purpose.) Machine 1 then enters the **Stiction** state. At this point some other agent — another state machine or some other software — can recognize that stiction has occurred and take appropriate action (e.g., notify the operator). That agent has the responsibility to then reset Machine 1's status register to 0 allowing the machine itself to set the count back to 0 and start over.

The Spike recognition machine (Machine 0) is somewhat more complex: it has four states and seven transitions. Although it is not necessary to cover this machine in detail, there are a few points to make. The most important is that this spike recognition machine is relatively noise free.

There are time constraints (ΔT in the figure) with which changes in the current must be consistent before a spike can be recognized. Intermediate states (**Possible Spike 1** and **Possible Spike 2**), from which the machine can return to the **Wait** state, exist so that the machine can be sure about whether the fluctuation in the current should be labeled a spike or not. Notice also that some of the transitions can be taken without any explicit actions.

The **Spike** state has one transition out of it. This occurs when Machine 0's status register (the current machine) is reset to 0. Recall that Machine 1 did this after it had recorded the existence of the spike.

The sizes of the current spike machine (Machine 0) and the stiction machine (Machine 1) are respectively 229 and 93 bytes. The interpreter that executes the SBFR system in the DCs is about 2000 bytes long.

7 Communication Protocols

7.1 Failure Prediction Reporting Protocol

One of the goals of the MPROS system is to encourage the incorporation of many diverse expert systems supplying diagnostic and prognostic conclusions based upon similar, overlapping or entirely disjoint sensor readings. At the same time, we recognized that these diverse results must be unified into a meaningful report to the system's users. To this end, a standard protocol has been defined for reporting failure predictions to the PDME for fusion and display.

7.2 Diagnostic Data

1 KnowledgeSourceID: The unique MPROS object ID for the instance of the knowledge source.

2 SensedObjectID: The unique MPROS object ID for the sensed object to which this report applies.

3   MachineConditionID: The unique MPROS object ID for the diagnosed machine condition.

4   Severity: Numeric value in range 0.0 to 1.0 indicating relative severity of machine condition to operation. Maximal severity is 1.0.

5   Belief: Numeric value in range 0.0 to 1.0 indicating belief that this diagnosis is true. Maximal belief is 1.0.

6   Explanation: An optional text string (possibly very long) providing human-readable description of the diagnosis. Where information beyond defined the Machine Condition description is unnecessary or unavailable, this is allowed to be blank.

7   Recommendations: An optional text string (possibly very long) providing human-readable description of the recommended actions to take. Where information beyond defined Machine Condition description is unnecessary or unavailable, this is allowed to be blank.

8   Timestamp: The timestamp for when this report should be considered "effective."

9   Additional Information: An optional text string (possibly very long) providing human-readable additional information.

7.3   Prognostics Vector

Zero to n ordered pairs of the form "(probability, time)." Each pair indicates the probability that the given machine condition will lead to failure of the machine within 'time' seconds from now.

8   Hardware

8.1   Prognostics/Diagnostics/Monitoring Engine

Unlike the Data Concentrator, the PDME is entirely software. It runs on any sufficiently powerful NT machine. Currently, the PDME runs in the office of one of our developers communicating over HTC's data network with the Data Concentrator in the basement.

The DC hardware (figure 4 shows the HTC-installed DC) consists of a PC104 single board Pentium PC (about 6"x6") with flat screen LCD display monitor, a PCMCIA host board, a 4 channel PCMCIA DSP card, two multiplexer (MUX) cards, and a terminal bus for sensor cable connections. The operating system is Win95 and there are connections for keyboard and mouse. Data is stored via DRAM.

The DC is enclosed in a NEMA enclosure with a transparent front door and fans for cooling. Overall dimensions are 10"x12"x4". The system was built entirely with commercial off the shelf components with the exception of the MUX cards which are a DLI hardware subcomponent and the PCMCIA card which was modified from a commercial 2 channel unit to meet the needs of the project.

Data Concentrator (DC)

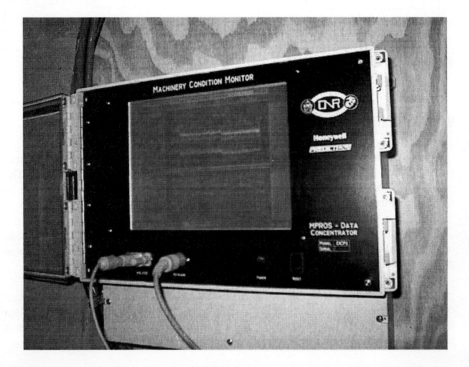

Figure 4. Data Concentrator Installed at HTC

The 4 channel PCMCIA card samples DC and AC dynamic signals. Highest sampling rate exceeds 40,000 Hz and the length of sampled signals is limited only by the PC's storage capacity.

The MUX cards provide power to standard accelerometers and are controlled using the PC's IO port. Each of the 2 MUX cards can switch between 4 sets of 4 channels each yielding up to 32 channels of data. Of those 32 channels, 24 can power standard accelerometers. All channels can be configured to sample DC voltage signals. Additionally, all channels are equipped with an RMS detector which can be configure to provide a digital signal when the RMS of the incoming signal exceeds a programmed value. This allows for real-time and constant alarming for all sensors.

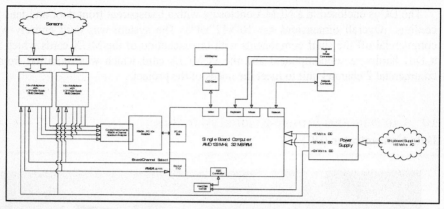

Figure 5. Data Concentrator Hardware

9 Validation

One question we are often asked is "How are you going to prove that your system does what you say it does?" This question, as it turns out, is a quite difficult one. The problem is that we are developing a system that we claim will predict failures in devices, and that in real life, these devices fail relatively rarely.

In fact, for any one failure mode, it is entirely possible that the failure mode (although valid) may never have occurred on any piece of equipment on any ship in the fleet! We have a number of answers to the question:

We are still going to look for the failure modes. We have a number of installed data collectors both on land and on ships. In addition, DLI is collecting time domain data for a number of parameters whenever their vibration-based expert system predicts a failure on shipboard chillers. This will give us data that we can use to test our system.

As Honeywell upgrades its air conditioning systems to be compliant with new non-polluting refrigerant regulations, older chillers become obsolete. We have managed to acquire one of these chillers that HTC replaced. It has been shipped to York, and we are now constructing a test plan to collect data from this chiller through carefully orchestrated destructive testing.

Seeded faults are worth doing. Our partners in the Mechanical Engineering Department of Georgia Tech are seeding faults in bearings and collecting the data. These tests have the drawback that they might not exhibit the same precursors as real-world failures, especially in the case of accelerated tests.

Honeywell, York, DLI, NRL, and WM Engineering have archives of maintenance data that we will take full advantage of in constructing our prognostic and diagnostic models.

Similarly, these partners have human expertise that we are able to tap in building our models.

Although persuasive, these answers are far from conclusive. The authors would welcome any input on how to validate a failure prediction system.

10 Current and Future Related Activities

We have just completed phase 1 of our program. The next phase calls for an installation on a Navy ship. Current plans are to install it on the Mercy, a hospital ship stationed in San Diego. This ship was chosen for a number of reasons:

1. It will be in a climate that is likely to require cooling during the winter shipboard test phase of our centrifugal chiller prognostics and diagnostics system.

2. It is likely to stay stationary and not put out to sea (as, for instance, a carrier would).

3. It contains pieces of equipment that are the target of our prognostics and diagnostics efforts.

4. We have a good working relationship with the crew.

Other activities outside our current contract are:

The destructive centrifugal chiller test. Honeywell has donated a surplus centrifugal chiller for use by the prognostics/diagnostics community. We are in the process of assembling a test plan to take full advantage of this opportunity, both to validate our existing system and to discover new indicators of incipient failure that can be incorporated into future versions of our and others' systems.

1. RSVP (Reduced Shipboard manning through Virtual Presence).

2. Honeywell is teamed with Pennsylvania State University for this CBM follow-on.

3. ACI Integration. Honeywell has volunteered to lead this effort to combine the Advanced Capability Initiatives.

These include CBM, Corrosion, Oil Debris Monitoring, and Human Computer Interaction. Other potential CBM related activities. There are a number of these, both inside and outside our members' companies.

10.1 Future Directions For Knowledge Fusion

Several high-level control extensions are under consideration for future extensions. First, multi-level data is represented the object-oriented ship model. We are not currently exploiting this fully. For example, we could reason about the health of a system based on the health of a constituent part.

Currently, only the parts are tracked. Second, spatial reasoning using the object-oriented ship model could lead us to fuse information about spatially related components. Examples of spatial relations are proximity (for example, a device is vibrating because a component next to it is broken and vibrating wildly) and flow.

Flows are relationships that represent either fluid flow through the system (one component passing fouled fluids on to other components downstream), electrical flow or mechanical flow of physical energy. Third, temporal reasoning components could be implemented to scrutinize failure histories and provide better projections of future faults as they develop.

Two other aspects of knowledge fusion future directions will be the refinement of specific knowledge fusion components for diagnostics and prognostics. For example, Bayes' Nets seem to be a promising approach to diagnostic knowledge fusion when causal relations and a priori relationships can be teased out of historical data for this system.

Prognostic knowledge fusion could be improved with the addition of techniques from the analysis of hazard and survival data. These approaches scrutinize history data to refine the estimates of life-cycle performance for failures. These refined inputs to the prognostic analysis would yield better projections of future failures.

Author Index

Ka-an Agun 697
Hiroyuki Aizu 430
S. Mounir Alaoui 201
Enrique Alba 248
Perry Alexander 1228
Makoto Amamiya 138
Patrick Amestoy 986
Cosimo Anglano 1130
Gabriel Antoniu 496
A. W. Apon 299
F. Arbab 1046
Isabelle Attali 1197

Nader Bagherzadeh 661
S. Bandyopadhyay 938
Peter H. Beckman 1178
Siegfried Benkner 1015
Bonnie Holte Bennett 1386, 1420
Massimo Bertozzi 1409
Jean-Luc Beuchat 700
Ricardo Bianchini 845, 859
Rupak Biswas 968
Brian Blount 1026
Cristina Boeres 1065
Ladislau Boloni 275
Kiran Bondalapati 570
Piotr Borsuk 174
Luc Bouge 466, 496
Fabian Breg 733
Alberto Broggi 1409
Russ Brucks 1335
Robert Brunner 483
Leugim Bustelo 400

J.C. Campelo 384
Denis Caromel 1197
Bryan Carpenter 533, 748
Enrique V. Carrera 845, 859
Harold W. Carter 1228
David Castanon 1350
Charles Cavanaugh 1335

Manuel M. T. Chakravarty 108
Prachya Chalermwat 257
K. Mani Chandy 1215
Morris Chang 697
B. Chappell 389
Michel Charpentier 1215
Saurav Chatterjee 442
Siddhartha Chatterjee 1026
Jian Chen 996
Lei Chen 289
Victoria Chernyakhovsky 1228
Malolan Chetlur 1145
Peng Yin Choo 873
Raymond Clark 353
Mark Clement 781
Gianni Conte 1163
David Coudert 897
S.J. Creese 1243
Jose L. Cruz-Rivera 400
Ireneusz Czarnowski 210

Andreas Dandalis 652
Emmett Davis 1386
Linda Davis 1386
Abram Detofsky 873
Oliver Diessel 579
Peter A. Dinda 309
Pedro Diniz 570
Stephen R. Donaldson 1097, 1299
Lucia M.A. Drummond 18
Phillip Duncan 570
Ahmed El-Amawy 924
Tarek El-Ghazawi 201, 257
Hossam ElGindy 643
C.T.H. Everaars 1046

Luciano G. Fagundes 395
Thomas Fahringer 49
Alessandra Fascioli 1409
Martin Feldman 924
Jose Alberto Fernandez-Zepeda 616

Afonso Ferreira 897
Jose J. Figueroa 400
Eliseu M. Filho 661
Eugene F. Fodor 100
Cyril Fonlupt 239
Javier E. Fonseca-Camacho 400
Geoffrey Fox 748
Ryan Fox 519
Peter Frey 1228
O. Frieder 201
Antonio Augusto Froehlich 1361

Demetris G. Galatopoullos 813
Dennis Gannon 733
Guang R. Gao 138, 968, 1025
Antonio Gentile 400
Apostolos Gerasoulis 1055
P.J. Gil 384
Attilio Giordana 1130
Sergei Gorlatch 79, 123
John Granacki 570
Achim Gratz 706

George Hadden 1420
Mary Hall 570
Mounir Hamdi 911
Ruibing Hao 275
Hiroshi Harada 1178
Wolfram Hardt 703
Marion Harmon 1327
Bernhard Haumacher 718
Klaus Havelund 1258
Gerd Heber 968
Kieran T. Herley 1085
Jonathan M.D. Hill 1097, 1299
Michael Holzrichter 978
Manpyo Hong 887
Atsushi Hori 1178
Jyh-Ming Huang 418
Eui-nam Huh 1335
Pat Hurley 353

Norbert Imlig 679
Minoru Inamori 679
Kentaro Inenaga 138

Germinal Isem 266
P. M. Irey IV 389 , 406, 1316
Yutaka Ishikawa 1178

A. Jaekel 938
Rajeev Jain 570
Cezary Z. Janikow 192
Rashid Jean-Baptiste 1335
Piotr Jedrzejowicz 210
E. Douglas Jensen 353
Rakash Jha 1350
Jan Jonsson 363
Magnus Jonsson 1374
Glenn Judd 781
Kyungkoo Jun 275

Jorg Kaiser 425
Laxmicant Kale 483
Loukas F. Kallivokas 309
Arkady Kanevsky 353
Holger Karl 1115
Meenakshi Kaul 606
David Kearney 579
Peter Keleher 511
Gabriele Keller 108
Mohammed A.S. Khalid 597
Shahadat Khan 289
Kihan Kim 413
Roger L. King 156
Bernd Klainjohann 703
Jaeyong Koh 413
Steffen Kohler 706
Ryusuke Konishi 679
B. Koren 1046
Jacek Koronacki 192
Krishna Kumar 1335
Umesh Kumar 1145
Fadi J. Kurdahi 661
Tei-Wei Kuo 329
Shigeru Kusakabe 138

Kam-yiu Lam 329
Aric B. Lambert 156
Gary C.K. Law 329
Tom Lawrence 353

Jacqueline LeMoigne 257
Ming-hau Lee 661
Claudia Leopold 230
Keqin Li 911
Kin F. Li 289
Weifa Liang 831
Yuan Lin 1036
R. Lin 634
Sylvain Lippi 1197
Mohammad Ali Livani 425
Giuseppe Lo Bello 1130
Peter K. K. Loh 220
Henrik Lonn 363
Ahmed Louri 873
Bruce Lowekamp 309
Guangming Lu 661
Wilfredo E. Lugo-Beauchamp 400

Kevin Maciunas 797
Leslie A. Madden 1316
Mark Maker 748
G. Manimaran 319
Eric G. Manning 289
Elias S. Manolakos 813
Dan C. Marinescu 275
David T. Marlow 389, 406, 1316
Andres Marquez 138
Berna L. Massingill 1284
Michael W. Masters 1316
Akihiro Matsuura 670
John Maurer 353
Michele Mazzeo 1163
Mario Medina 519
Dalila Megherbi 266
Jean-Francois Mehaut 466
Alessandro Mei 652
Rodrigo F. Mello 395
Martin Middendorf 625
Anita Mittal 319
H. Mongelli 1075
Celio E. Moron 395
Z. George Mou 63
Frank Mueller 553
Jan. J. Mulawka 174
Xavier Munos 897

C. Siva Ram Murthy 319

Hidehisa Nagano 670
Akira Nagoya 670, 709
Hiroshi Nakada 679
Tatsuo Nakajima 430
K. Nakano 634
Raymond Namyst 466, 496
Chirag Nanavati 1350
Aline Nascimento 1065
F. Nat-Abdesselam 375
Douglas Niehaus 454
Jarek Nieplocha 533
Joerg Nolte 34

Francis O'Carrol 1178
Luiz S. Ochi 183
K. O'Donoghue 389
David R. O'Hallaron 309
Kiyoshi Oguri 679
S. Olariu 634
Suely Oliveira 978
Ricardo Olivieri 400
Ronald A. Olsson 100
James Oly 519

David Padua 1036
Yi Pan 911
Santosh Pande 4
Mario Pantano 49
Susanna Pelagatti 79, 123
Francois Pellegrini 986
Guy-Rene Perrin 1006
Michael Philippsen 718, 1026
James Phillips 483
Andrea Pietracaprina 1085
M.C. Pinotti 634
Luiz Pires 1350
Agostino Poggi 1163
Subbu Ponnuswamy 1350
Viktor K. Prasanna 652
Geppino Pucci 1085
Jean-Michel Puiatti 688

Chunming Qiao 950

Marlyn Quinones-Cerpa 400

R. Radhakrishnan 1145, 1228
Shankar Radhakrishnan 588
P. Rajagopal 299
V. Rajasekaran 1145
Binoy Ravindran 1335
Vinod E.F. Rebello 1065
Daniel A. Reed 519
Joy Reed 1243
Donna S. Reese 156
Achim Rettberg 703
David L. Rhodes 1055
Gilles Ritter 688
Alexis H. Rivera-Rios 400
Denis Robillard 239
F. Rodriguez 384
Jean Roman 986
Jonathan Rose 597
Pietro Rossi 1163
Olivier Roux 239
Samuel H. Russ 156,1335

Eduardo Sanchez 688, 700
Luis F.G. Sarmenta 763
Sergej Sawitzki 706
W. Schroeder-Preikschat 1361
Martin Schulz 19
J.L. Schwing 634
D. Scott Wills 400
Thomas Seidmann 547
Giuseppe Sena 266
A. Sengupta 938
Franciszek Seredynski 192
J.J. Serrano 384
Tangirala Shailendra 1335
Girindra D. Sharma 1145
Venson Shaw 220
Xiaojun Shen 831
Kang G. Shin 363
Heonshik Shin 413
Behrooz Shirazi 1316, 1335
Huseyin Simitci 519
Andrew C. Simpson 1299
Hertej Singh 661

Aleksander Skakowski 210
David Skillicorn 1097
Quinn Snell 781
W. Song 1075
Rainer G. Spallek 706
Andre Spiegel 93
Vinco Srinivasan 588
Ram Subramanian 4
Shinji Sumimoto 1178
Xian-He Sun 49
Takayuki Suyama 709
Henryk Szreder 210
Tadeusz Szuba 165

Mamadou Tadiou Kone 430
Toshiyuki Takahashi 1178
El-ghazali Talbi 239
Xinan Tang 138
Valerie E. Taylor 996
Hiroshi Tezuka 1178
Kritchalach Thitikamol 511
Alexander Tiountchik 712
Stefano Tommesani 1409
Eduardo Tovar 339
Jerry L. Trahan 616
Elena Trichina 712
Jose M. Troya 248
Wai-Hung Tsang 329

R. Vaidyanathan 616, 924
Brian Van Voorst 1350
Krishnan Varadarajan 483
Iomar Vargas-Gonzales 400
Francisco Vasques 339
Ranga Vemuri 588, 606
Dalessandro Soares Vianna 183
Michelle Viera-Vera 400
Michele Vignali 1163
Frederique Voisin 1006

Paul Wallace 353
Jeff Walrath 588
Piotr Wasiewicz 174
Darren Webb 797
Kyubum Wee 887

Piotr Weglenski 174
Lonnie Welch 1316, 1335
Douglas Wells 353
Andrew Wendelborn 797
Paul V. Werme 1316
Thomas Wheeler 353
Grant Wigley 579
Philip A. Wilsey 1145, 1228
Mathew Wojko 643
Bill Wren 1386

Bharath Yanamula 1335

Ted C. Yang 418
Hongjin Yeh 887
Makoto Yokoo 709
P. Yuste 384

Zhaohua Zhan 49
Xijun Zhang 950
Yun Zhang 353
Heidi Ziegler 570
A. Zomaya 634

Lecture Notes in Computer Science

For information about Vols. 1–1497
please contact your bookseller or Springer-Verlag

Vol. 1498: A.E. Eiben, T. Bäck, M. Schoenauer, H.-P. Schwefel (Eds.), Parallel Problem Solving from Nature – PPSN V. Proceedings, 1998. XXIII, 1041 pages. 1998.

Vol. 1499: S. Kutten (Ed.), Distributed Computing. Proceedings, 1998. XII, 419 pages. 1998.

Vol. 1500: J.-C. Derniame, B.A. Kaba, D. Wastell (Eds.), Software Process: Principles, Methodology, and Technology. XIII, 307 pages. 1999.

Vol. 1501: M.M. Richter, C.H. Smith, R. Wiehagen, T. Zeugmann (Eds.), Algorithmic Learning Theory. Proceedings, 1998. XI, 439 pages. 1998. (Subseries LNAI).

Vol. 1502: G. Antoniou, J. Slaney (Eds.), Advanced Topics in Artificial Intelligence. Proceedings, 1998. XI, 333 pages. 1998. (Subseries LNAI).

Vol. 1503: G. Levi (Ed.), Static Analysis. Proceedings, 1998. IX, 383 pages. 1998.

Vol. 1504: O. Herzog, A. Günter (Eds.), KI-98: Advances in Artificial Intelligence. Proceedings, 1998. XI, 355 pages. 1998. (Subseries LNAI).

Vol. 1505: D. Caromel, R.R. Oldehoeft, M. Tholburn (Eds.), Computing in Object-Oriented Parallel Environments. Proceedings, 1998. XI, 243 pages. 1998.

Vol. 1506: R. Koch, L. Van Gool (Eds.), 3D Structure from Multiple Images of Large-Scale Environments. Proceedings, 1998. VIII, 347 pages. 1998.

Vol. 1507: T.W. Ling, S. Ram, M.L. Lee (Eds.), Conceptual Modeling – ER '98. Proceedings, 1998. XVI, 482 pages. 1998.

Vol. 1508: S. Jajodia, M.T. Özsu, A. Dogac (Eds.), Advances in Multimedia Information Systems. Proceedings, 1998. VIII, 207 pages. 1998.

Vol. 1510: J.M. Zytkow, M. Quafafou (Eds.), Principles of Data Mining and Knowledge Discovery. Proceedings, 1998. XI, 482 pages. 1998. (Subseries LNAI).

Vol. 1511: D. O'Hallaron (Ed.), Languages, Compilers, and Run-Time Systems for Scalable Computers. Proceedings, 1998. IX, 412 pages. 1998.

Vol. 1512: E. Giménez, C. Paulin-Mohring (Eds.), Types for Proofs and Programs. Proceedings, 1996. VIII, 373 pages. 1998.

Vol. 1513: C. Nikolaou, C. Stephanidis (Eds.), Research and Advanced Technology for Digital Libraries. Proceedings, 1998. XV, 912 pages. 1998.

Vol. 1514: K. Ohta, D. Pei (Eds.), Advances in Cryptology – ASIACRYPT'98. Proceedings, 1998. XII, 436 pages. 1998.

Vol. 1515: F. Moreira de Oliveira (Ed.), Advances in Artificial Intelligence. Proceedings, 1998. X, 259 pages. 1998. (Subseries LNAI).

Vol. 1516: W. Ehrenberger (Ed.), Computer Safety, Reliability and Security. Proceedings, 1998. XVI, 392 pages. 1998.

Vol. 1517: J. Hromkovič, O. Sýkora (Eds.), Graph-Theoretic Concepts in Computer Science. Proceedings, 1998. X, 385 pages. 1998.

Vol. 1518: M. Luby, J. Rolim, M. Serna (Eds.), Randomization and Approximation Techniques in Computer Science. Proceedings, 1998. IX, 385 pages. 1998.

1519: T. Ishida (Ed.), Community Computing and Support Systems. VIII, 393 pages. 1998.

Vol. 1520: M. Maher, J.-F. Puget (Eds.), Principles and Practice of Constraint Programming - CP98. Proceedings, 1998. XI, 482 pages. 1998.

Vol. 1521: B. Rovan (Ed.), SOFSEM'98: Theory and Practice of Informatics. Proceedings, 1998. XI, 453 pages. 1998.

Vol. 1522: G. Gopalakrishnan, P. Windley (Eds.), Formal Methods in Computer-Aided Design. Proceedings, 1998. IX, 529 pages. 1998.

Vol. 1524: G.B. Orr, K.-R. Müller (Eds.), Neural Networks: Tricks of the Trade. VI, 432 pages. 1998.

Vol. 1525: D. Aucsmith (Ed.), Information Hiding. Proceedings, 1998. IX, 369 pages. 1998.

Vol. 1526: M. Broy, B. Rumpe (Eds.), Requirements Targeting Software and Systems Engineering. Proceedings, 1997. VIII, 357 pages. 1998.

Vol. 1527: P. Baumgartner, Theory Reasoning in Connection Calculi. IX, 283. 1999. (Subseries LNAI).

Vol. 1528: B. Preneel, V. Rijmen (Eds.), State of the Art in Applied Cryptography. Revised Lectures, 1997. VIII, 395 pages. 1998.

Vol. 1529: D. Farwell, L. Gerber, E. Hovy (Eds.), Machine Translation and the Information Soup. Proceedings, 1998. XIX, 532 pages. 1998. (Subseries LNAI).

Vol. 1530: V. Arvind, R. Ramanujam (Eds.), Foundations of Software Technology and Theoretical Computer Science. XII, 369 pages. 1998.

Vol. 1531: H.-Y. Lee, H. Motoda (Eds.), PRICAI'98: Topics in Artificial Intelligence. XIX, 646 pages. 1998. (Subseries LNAI).

Vol. 1096: T. Schael, Workflow Management Systems for Process Organisations. Second Edition. XII, 229 pages. 1998.

Vol. 1532: S. Arikawa, H. Motoda (Eds.), Discovery Science. Proceedings, 1998. XI, 456 pages. 1998. (Subseries LNAI).

Vol. 1533: K.-Y. Chwa, O.H. Ibarra (Eds.), Algorithms and Computation. Proceedings, 1998. XIII, 478 pages. 1998.

Vol. 1534: J.S. Sichman, R. Conte, N. Gilbert (Eds.), Multi-Agent Systems and Agent-Based Simulation. Proceedings, 1998. VIII, 237 pages. 1998. (Subseries LNAI).

Vol. 1535: S. Ossowski, Co-ordination in Artificial Agent Societies. XV; 221 pages. 1999. (Subseries LNAI).

Vol. 1536: W.-P. de Roever, H. Langmaack, A. Pnueli (Eds.), Compositionality: The Significant Difference. Proceedings, 1997. VIII, 647 pages. 1998.

Vol. 1537: N. Magnenat-Thalmann, D. Thalmann (Eds.), Modelling and Motion Capture Techniques for Virtual Environments. Proceedings, 1998. IX, 273 pages. 1998. (Subseries LNAI).

Vol. 1538: J. Hsiang, A. Ohori (Eds.), Advances in Computing Science – ASIAN'98. Proceedings, 1998. X, 305 pages. 1998.

Vol. 1539: O. Rüthing, Interacting Code Motion Transformations: Their Impact and Their Complexity. XXI,225 pages. 1998.

Vol. 1540: C. Beeri, P. Buneman (Eds.), Database Theory – ICDT'99. Proceedings, 1999. XI, 489 pages. 1999.

Vol. 1541: B. Kågström, J. Dongarra, E. Elmroth, J. Waśniewski (Eds.), Applied Parallel Computing. Proceedings, 1998. XIV, 586 pages. 1998.

Vol. 1542: H.I. Christensen (Ed.), Computer Vision Systems. Proceedings, 1999. XI, 554 pages. 1999.

Vol. 1543: S. Demeyer, J. Bosch (Eds.), Object-Oriented Technology ECOOP'98 Workshop Reader. 1998. XXII, 573 pages. 1998.

Vol. 1544: C. Zhang, D. Lukose (Eds.), Multi-Agent Systems. Proceedings, 1998. VII, 195 pages. 1998. (Subseries LNAI).

Vol. 1545: A. Birk, J. Demiris (Eds.), Learning Robots. Proceedings, 1996. IX, 188 pages. 1998. (Subseries LNAI).

Vol. 1546: B. Möller, J.V. Tucker (Eds.), Prospects for Hardware Foundations. Survey Chapters, 1998. X, 468 pages. 1998.

Vol. 1547: S.H. Whitesides (Ed.), Graph Drawing. Proceedings 1998. XII, 468 pages. 1998.

Vol. 1548: A.M. Haeberer (Ed.), Algebraic Methodology and Software Technology. Proceedings, 1999. XI, 531 pages. 1999.

Vol. 1550: B. Christianson, B. Crispo, W.S. Harbison, M. Roe (Eds.), Security Protocols. Proceedings, 1998. VIII, 241 pages. 1999.

Vol. 1551: G. Gupta (Ed.), Practical Aspects of Declarative Languages. Proceedings, 1999. VIII, 367 pages. 1999.

Vol. 1552: Y. Kambayashi, D.L. Lee, E.-P. Lim, M.K. Mohania, Y. Masunaga (Eds.), Advances in Database Technologies. Proceedings, 1998. XIX, 592 pages. 1999.

Vol. 1553: S.F. Andler, J. Hansson (Eds.), Active, Real-Time, and Temporal Database Systems. Proceedings, 1997. VIII, 245 pages. 1998.

Vol. 1554: S. Nishio, F. Kishino (Eds.), Advanced Multimedia Content Processing. Proceedings, 1998. XIV, 454 pages. 1999.

Vol. 1555: J.P. Müller, M.P. Singh, A.S. Rao (Eds.), Intelligent Agents V. Proceedings, 1998. XXIV, 455 pages. 1999. (Subseries LNAI).

Vol. 1557: P. Zinterhof, M. Vajteršic, A. Uhl (Eds.), Parallel Computation. Proceedings, 1999. XV, 604 pages. 1999.

Vol. 1558: H. J.v.d. Herik, H. Iida (Eds.), Computers and Games. Proceedings, 1998. XVIII, 337 pages. 1999.

Vol. 1559: P. Flener (Ed.), Logic-Based Program Synthesis and Transformation. Proceedings, 1998. X, 331 pages. 1999.

Vol. 1560: K. Imai, Y. Zheng (Eds.), Public Key Cryptography. Proceedings, 1999. IX, 327 pages. 1999.

Vol. 1561: I. Damgård (Ed.), Lectures on Data Security. VII, 250 pages. 1999.

Vol. 1563: Ch. Meinel, S. Tison (Eds.), STACS 99. Proceedings, 1999. XIV, 582 pages. 1999.

Vol. 1567: P. Antsaklis, W. Kohn, M. Lemmon, A. Nerode, S. Sastry (Eds.), Hybrid Systems V. X, 445 pages. 1999.

Vol. 1568: G. Bertrand, M. Couprie, L. Perroton (Eds.), Discrete Geometry for Computer Imagery. Proceedings, 1999. XI, 459 pages. 1999.

Vol. 1569: F.W. Vaandrager, J.H. van Schuppen (Eds.), Hybrid Systems: Computation and Control. Proceedings, 1999. X, 271 pages. 1999.

Vol. 1570: F. Puppe (Ed.), XPS-99: Knowledge-Based Systems. VIII, 227 pages. 1999. (Subseries LNAI).

Vol. 1572: P. Fischer, H.U. Simon (Eds.), Computational Learning Theory. Proceedings, 1999. X, 301 pages. 1999. (Subseries LNAI).

Vol. 1575: S. Jähnichen (Ed.), Compiler Construction. Proceedings, 1999. X, 301 pages. 1999.

Vol. 1576: S.D. Swierstra (Ed.), Programming Languages and Systems. Proceedings, 1999. X, 307 pages. 1999.

Vol. 1577: J.-P. Finance (Ed.), Fundamental Approaches to Software Engineering. Proceedings, 1999. X, 245 pages. 1999.

Vol. 1578: W. Thomas (Ed.), Foundations of Software Science and Computation Structures. Proceedings, 1999. X, 323 pages. 1999.

Vol. 1579: W.R. Cleaveland (Ed.), Tools and Algorithms for the Construction and Analysis of Systems. Proceedings, 1999. XI, 445 pages. 1999.

Vol. 1580: A. Včkovski, K.E. Brassel, H.-J. Schek (Eds.), Interoperating Geographic Information Systems. Proceedings, 1999. XI, 329 pages. 1999.

Vol. 1581: J.-Y. Girard (Ed.), Typed Lambda Calculi and Applications. Proceedings, 1999. VIII, 397 pages. 1999.

Vol. 1582: A. Lecomte, F. Lamarche, G. Perrier (Eds.), Logical Aspects of Computational Linguistics. Proceedings, 1997. XI, 251 pages. 1999. (Subseries LNAI).

Vol. 1586: J. Rolim et al. (Eds.), Parallel and Distributed Processing. Proceedings, 1999. XVII, 1443 pages. 1999.

Vol. 1587: J. Pieprzyk, R. Safavi-Naini, J. Seberry (Eds.), Information Security and Privacy. Proceedings, 1999. XI, 327 pages. 1999.

Vol. 1593: P. Sloot, M. Bubak, A. Hoekstra, B. Hertzberger (Eds.), High-Performance Computing and Networking. Proceedings, 1999. XXIII, 1318 pages. 1999.

Vol. 1594: P. Ciancarini, A.L. Wolf (Eds.), Coordination Languages and Models. Proceedings, 1999. IX, 420 pages. 1999.